Mathematical Physics

in One Dimension

PERSPECTIVES IN PHYSICS

A Series of Reprint Collections

STATISTICAL THEORIES OF SPECTRA: FLUCTUATIONS, Charles E. Porter (Ed.). 1965

QUANTUM THEORY OF ANGULAR MOMENTUM, L. C. Biedenharn and H. van Dam (Eds.). 1965

MAGNETOHYDRODYNAMIC STABILITY AND THERMONUCLEAR CONTAINMENT, A. Jeffrey and T. Taniuti (Eds.). 1966

MATHEMATICAL PHYSICS IN ONE DIMENSION: EXACTLY SOLUBLE MODELS OF INTERACTING PARTICLES, E. Lieb and D. Mattis (Eds.). 1966

In Preparation

COULOMB EXCITATION, K. Alder and A. Winther (Eds.).

Mathematical Physics in One Dimension

Exactly Soluble Models of Interacting Particles

A Collection of Reprints
With Introductory Text by

Elliott H. Lieb
Department of Physics
Northeastern University, Boston, Massachusetts

Daniel C. Mattis
Belfer Graduate School of Science
Yeshiva University, New York, New York

1966

ACADEMIC PRESS New York and London

ACADEMIC PRESS INC.
111 Fifth Avenue, New York, New York 10003

United Kingdom Edition published by
ACADEMIC PRESS INC. (LONDON) LTD.
Berkeley Square House, London W.1

Library of Congress Catalog Card Number: 66-24106

PRINTED IN THE UNITED STATES OF AMERICA

Foreword

The problems of mathematical physics have often had one-dimensional analogs which have turned out to be simpler to solve than those in three dimensions. Specifically, the normal modes of disordered structures and the free energy of an interacting gas are two subjects of investigation which have had an exact solution only in one dimension. Other topics we have considered in the present volume, such as the quantum theory of disordered structures, the quantum theory of interacting fermions or bosons, and the eigenstates of antiferromagnets, are all representative of problems or models which have an exact theory in one dimension but are understood only very approximately in three dimensions. The apparent simplicity of restricting motion to one spatial dimension is well known and has had much appeal, so that there now exists a vast literature on this subject, albeit one which is not indexed under the topic "one dimension" in standard indexing journals and which is therefore hard to research. One of the purposes of the present book is to contribute a bibliography complete to Spring, 1965, which should be sufficient to start the interested researcher or student on his task. We hope this will be of value, although bibliographical aspects were not our prime motivation; for we started this project with two scientific questions in mind, which appear and reappear in all seven chapters of this book.

Why are nontrivial models of physical systems so often soluble in one dimension but not in three dimensions? What do these solutions teach us about the validity of the standard approximations in two or three dimensions? In studying the soluble models, we must know whether we are beating a retreat into pure fantasy (or pure mathematics) or whether this is a possible path to understanding nature. What we did find is that in almost every case the one-dimensional physics is devoid of much structure, and describes a colorless universe much less interesting than our own. A one-dimensional gas was once thought to be incapable even of condensation. Later, with the introduction of infinite range "unphysical" forces, it has been made to condense to a liquid, which is essentially a *tour de force*. Even so, this liquid can never freeze and reveal the third phase of matter—the solid state. In one dimension bosons do not condense, electrons do not superconduct, ferromagnets do not magnetize, and liquids do not freeze. What relation can the exactly soluble models, which we seriously invite the reader to consider in the present book, have with the rich and puzzling problems of the three-dimensional world?

We have found these models to be useful tests of approximate mathematical methods. We have found the solutions of these models to be surprisingly complex and interesting, within the above-mentioned limitations. Physical applications, from chain molecules to field theory, are often and unexpectedly discovered so that the physical arguments against one-dimensional models are no longer very convincing. Their solutions make a contribution to the three-dimensional reality as well: by educating us to the need for rigorous and exact analysis, they lead us away from the easy and simplistic first theories toward a more critical and mathematical approach, and finally to a better definition of reality. For example, for a long time it was thought that the physical explanation of ferromagnetism was as follows: electrons interact with their nearest neighbors by means of an "exchange integral" which tends to line up their spins. This is a problem we considered in one dimension (Chapters 4 and 6) and it was found that these exchange corrections notwithstanding, the ground state of interacting electrons was always *non*magnetic. Therefore the ferromagnetism of real materials (iron, nickel, gadolinium, etc.) must be due to the three-dimensional structure of space, which is responsible for orbital degeneracy, Hund's rules for magnetic atoms, etc. Thus the study of a one-dimensional model pinpointed the probable cause of ferromagnetism in three dimensions. The fact that particles can get around each other is responsible for much of the structure of the ordinary world, and is also responsible for the difficulties which the mathematical physicist encounters in studying it. In one dimension we forfeit some of the structure but we gain the possibility of obtaining an exact solution.

To see why the inability of particles to get around each other is such a crucial feature, consider the following example of interacting spins (or of a binary alloy), in which paradoxically there are no particles at all but spins s_i fixed in space in a linear array. The energies are given by

$$E = \sum_{i=1}^{N-1} s_i s_{i+1}$$

where each independent variable s_i can assume two values: $+1$ or -1 only. Let us define the following variables:

$$t_1 = s_2 s_1, \quad t_2 = s_3 s_2, \quad \ldots, \quad t_{N-1} = s_N s_{N-1}, \quad \text{and} \quad t_N = s_N$$

Each of these variables can also assume the values ± 1 independent of the others, and in terms of these new, more appropriate variables, we find that the energy

$$E = \sum_{i=1}^{N-1} t_i$$

has been expressed as the sum of noninteracting terms. One can see by inspection that E is independent of the last variable, t_N. The free energy is also obtainable almost by inspection. But now try to accomplish this in two- or three-dimensional systems, for example, the two-dimensional Ising net, the energy of which is given by

$$E = \sum_{i,j=1}^{N-1} s_{i,j}s_{i+1,j} + \sum_{i,j=1}^{N-1} s_{i,j}s_{i,j+1}$$

where each of the variables $s_{i,j}$ again assumes two values, ± 1, independent of the others. If we define t's as before, one or the other of the above sums becomes impossibly complicated and the problem can no longer be reduced to a simple set of noninteracting terms. The reader is encouraged to see this for himself. We shall not here discuss Onsager's famous solution of the free energy of this model, which was only possible because of the remarkable simplicity of the $s_{i,j}$'s, nor venture to explain why the three-dimensional extension has never been solved, as these subjects and difficulties are well known. But we shall emphasize the essential feature shown above: the fact that particles can be ordered like integers in the one-dimensional system, enabling even very troublesome, nonlinear, stochastic, and/or many-body problems to be solved, in one dimension only.

Modern physics has many interesting avenues of research and in choosing the main topics for the present survey, we have kept in mind those areas which have been of great interest in the past, as well as those which are rapidly coming into prominence. One of the latter is the study of disordered structures undertaken in Chapter 2, which includes eight important reprinted articles and, in the text, an extensive discussion of the roles of theoretical and numerical analysis. Chapter 1 gives a survey of statistical mechanics and introductory material on phase transitions, although none of it is quite so elementary as the spin system described above. We find that the statistical theory of energy levels—i.e. the one-dimensional charged Coulomb gas—contributes to the understanding of this field, and a discussion is given of this theory. Chapter 3 is a survey of band structure in ordered and disordered materials. Who has not heard of the Kronig-Penney model? We include this as well as the matrix method of Kramers, and build on the results of Chapter 2 in discussing the random alloy. The many-fermion problem is the topic of Chapter 4. There we discover that the eigenstates of interacting fermions can be quite sensible *even when perturbation theory diverges*, and we also find that the energy levels of the interacting system are not too dissimilar from noninteracting model systems which are well understood. Extra material on the related topics of field theory (Thirring model, etc.) will be found in the bibliography of this chapter. As this is not covered in the textual material, the field-theoretic bibliography includes sufficient quotations from the abstracts of the cited papers so that the contents of each will become apparent. Chapter 5 is a theory of the interacting boson gas, and a comparison of exact results with the notions of quasi-particle excitations. Chapter 6 discusses mainly the theory of the *anti*ferromagnetic linear chains, which turn out to have many aspects in common with the soluble boson systems discussed in the previous chapter. The briefest chapter is the last, Chapter 7, and it is devoted to time-dependent phenomena of many-body systems—i.e., classical or quantum-mechanical *dynamics*. Here limitations of space and time conflicted with our desire to do justice to such an important topic, and the reader should thus understand that nothing derogatory is implied by the brevity of the chapter. One of the important questions toward which work in this field has been directed concerns how a large system, started in some arbitrary configuration, achieves thermal equilibrium; unless this occurs in a finite time, the work of Chapter 1 (*equilibrium* statistical mechanics) would be meaningless! Fortunately the reprinted calculations, a blend of theoretical work and computer studies, lay our (rhetorical) fears to rest.

We are very grateful to Dr. Peter J. Price for having made available his selected bibliography on electron tunneling, given in Chapter 3. We thank Dr. Donald W. Jepsen for help in Chapter 7 and Ingeborg Lieb (Mrs. E. Lieb) for her translation of the Takahashi article, Chapter 1. Finally, we are indebted to the authors and journals who, with such friendship and courtesy, have allowed us to reprint the interesting material which it is our privilege to present herein.

E. LIEB AND D. MATTIS

August, 1965

Acknowledgments

We thank the following publishers for permission to reproduce material.

American Institute of Physics and the Physical Review
Cambridge Philosophical Society
Institute of Physics and the Physical Society of London
Philips Research Reports
Royal Society of London
Stichting Physica

This work was partially supported by grants from the U. S. Air Force Office of Scientific Research at Yeshiva University: Nos. AFOSR 713–64 and 508–66 (E.L.); No. AFOSR 107–566 (D.M.).

CONTENTS

• *Chapter 3* •

Electron Energy Bands in Ordered and Disordered Crystals

Reprinted papers

• *Chapter 5* •

The Bose Gas

• *Chapter 6* •

Magnetism

Classical Statistical Mechanics

Chapter I

Classical Statistical Mechanics

Introduction

The basic problem in statistical mechanics is to compute a partition function, Z, from which it is possible to deduce the thermodynamic and other bulk properties of a system such as the magnetization. Other properties of a system, such as correlation functions, can usually be expressed as some derivative of Z with respect to some external potential.

Compared to the quantum-mechanical problem, the classical problem appears to be simplicity itself. Owing to the fact that classically the potential energy and the kinetic energy commute with each other, the latter problem is immediately reduced to quadratures—namely the N-fold configurational integral. In the quantum case, the density matrix, which plays the role of the Boltzmann factor, is not given and must first be calculated before the problem becomes one of computing an N-fold integral. In studying classical systems, therefore, we are really learning something about integrals. Can we, moreover, hope to learn anything about $3N$-dimensional integrals from their N-dimensional counterparts?

At first glance we are inclined to be discouraged. There is the well-known dictum that "one-dimensional systems with short-range forces can never have a phase transition." In point of fact there does not exist a general proof of such a theorem. One can give a heuristic proof along the lines of the Landau-Lifschitz (1958) argument that separate phases in configuration space cannot exist; for if there is a condensed phase consisting of a chain of n atoms, it may be broken at $n - 1$ places with a resultant contribution to the entropy of order log n. But the resultant change in the internal energy is of order unity and hence if the chain is of macroscopic length, the free energy will certainly decrease. Therefore, at most, condensed "clusters" can exist provided these are not macroscopically long. For finite-range pairwise potentials with a *cutoff*, Van Hove (1950) has given a proof that there is no phase transition, but if the potential is not simply a sum of a *finite* number of finite pair, three body, etc., potentials, a phase transition can occur (see the remarks below on the Fisher model). It can also occur if the potential does not have a finite range or if it is singular. See Baur and Nosanow (1962). For the lattice gas with finite-neighbor interaction a proof that there is no phase transition is much simpler, and was given by Rushbrooke and Ursell (1948).

Unfortunately, not all transitions are accompanied by a condensation in configuration space (e.g., superconductivity, superfluidity are not) and it must be admitted that a proper nonexistence proof—one that would show the partition function to be analytic with respect to all the intensive variables of the system—has not been given. One of the major difficulties is trying to formulate a general "system" with short-range "forces" in a precise mathematical way.

When long-range forces are introduced the situation may be radically altered and a phase transition can occur in one dimension. We shall discuss several such models in the sequel. One of them, the Kac-Baker model, displays a phase transition that is not out of keeping with that observed for real gases and thereby offers a primitive beginning to the study of real phase transitions.

Fortunately, however, there is more to statistical mechanics than phase transitions alone. For instance, one can ask about the convergence of the Mayer cluster expansion for a gas. For a long time it was believed that this would converge as long as there was no phase transition. A simple one-dimensional model to be discussed here disproves the conjecture, because, while it has no transition, its Mayer series has only a finite radius of convergence.

Another important quantity is the two-particle correlation function. This is physically interesting because it can be directly measured in x-ray or neutron scattering experiments, but it is also of theoretical interest because most approximation schemes for three-dimensional systems require as an intermediate step, the calculation of this function and/or its related three-, four-, etc., particle counterparts. The apparent fact that one-dimensional fluids have no phase transition would seem to imply that they are in some way simpler than their three dimensional analogs. Consequently, it would seem reasonable indeed to test a general correlation function approximation scheme on a one-dimensional model of a gas for which the exact result is known. If satisfactory agreement is not obtained the approximation scheme can be defended only if some cogent argument is brought forth for the importance of three dimension. Generally speaking, topological considerations are rarely invoked in formulating approximations, and *post hoc* rationalizations of the failure to preduct the properties of simple one-dimensional systems correctly would hardly be convincing.

The Takahashi Nearest-Neighbor Gas

The simplest model is known as the Tonks (1936) gas, after its inventor, and had its generalization given by Takahashi (1942), wherein the total potential is the sum of two body potentials of the form

$$\begin{aligned} v(x) &= \infty && \text{for} \quad |x| < a \quad (\text{"hard core"}) \\ &= \varphi(x - a) && \text{for} \quad 2a > |x| > a \\ &= 0 && \text{for} \quad |x| \geq 2a \end{aligned} \tag{1.1}$$

We have N particles in a "box" of length L with a density $\rho = N/L$, and the configurational partition function Z and free energy F are

$$e^{-\beta F} \equiv Z = 1/N! \int_0^L \cdots \int dx_1 \cdots dx_N \exp\Big(-\beta \sum_{i>j} v(x_i - x_j)\Big) \tag{1.2}$$

where $\beta = (kT)^{-1}$.

We observe that the integral in (1.2) is $N!$ times the integral over the subregion $R_1{:}\, 0 \leq x_1 \leq x_2 \leq \cdots \leq x_N \leq L$. The introduction of the region R_1 is frequently a useful device in the one-dimensional many-body problem and we shall meet with it again elsewhere. *Unfortunately, it has no useful counterpart in higher dimensions.* If we take the original Tonks gas, wherein $\varphi = 0$ and only the hard core remains, the integral over R_1 may be easily accomplished by changing to the variables $y_j = x_j - (j - 1)a$, $l = L - (N - 1)a$, whence

$$Z = \int_0^l dy_N \cdots \int_0^{y_3} dy_2 \int_0^{y_2} dy_1 = l^N/N! \rightarrow \exp N[-\ln \rho + \ln(1 - \rho a) + 1] \tag{1.3}$$

the arrow indicating that we have gone to the limit of a large system.

When $\varphi \neq 0$ we can still go into the region R_1 and we note the important point that because of the third equation in (1.1) in this region the only terms in the potential energy that do not vanish are $v(x_i - x_j)$ where $i = j + 1$. Hence

$$Z = \int_0^l dy_N \cdots \int_0^{y_3} dy_2 \exp[-\beta\varphi(y_3 - y_2)] \int_0^{y_2} dy_1 \exp[-\beta\varphi(y_2 - y_1)] \tag{1.4}$$

The form of (1.4) is an iterated Laplace convolution and therefore, taking the Laplace transform of Z as a function of l, we obtain

$$\bar{Z}(s) = \int_0^\infty Z(l) e^{-sl}\, dl = s^{-2} K(s)^{N-1} \tag{1.5}$$

where

$$K(s) = \int_0^\infty \exp[-sx - \beta\varphi(x)]\, dx. \tag{1.6}$$

To find $Z(l)$ one can invert the Laplace transform (1.5) by using steepest descent which becomes exact for large N. Alternatively (but less rigorously because it involves an assumption of extensivity), one can evaluate (1.5) directly by steepest descent, whence

$$\ln \bar{Z}(s) = -s\bar{l} + \ln Z(\bar{l}) \tag{1.7}$$

where

$$s = \frac{\partial \ln Z(l)}{\partial l}\bigg|_{\bar{l}} = \beta P(\bar{l}) \tag{1.8}$$

P being the pressure of the gas. Hence $K(s)$ is directly related to thermodynamic quantities, and we note that the Laplace transform variable is to be identified with the pressure.

Gursey (1950), unaware of Takahashi's result, independently solved the same problem at a later date. The interested reader will find graphs of typical isotherms in Gursey's paper. Isotherms for the analogous lattice gas, in which one assumes that particles can exist only on a discrete set of points, can be found in Pressman and Keller (1960) together with isotherms for the "spherical model" approximation to the lattice gas.

If we relax the third condition in (1.1) and instead assert that

$$v(x) = 0, \qquad |x| \geq (\nu + 1)a \tag{1.1'}$$

then in the region R_1 any given particle can interact with as many ν other particles on either side of it. The configurational integral, Z, is then not simply an iterated Laplace convolution, but Van Hove (1950) has shown that for large N the problem becomes one of finding the largest eigenvalue of a Fredholm integral equation for a function of $\nu - 1$ variables. He did not attempt to solve this equation for any particular case, but gave a proof that z is necessarily analytic in all intensive variables, and hence that no phase transition can occur in one dimension for potentials that vanish beyond a certain fixed distance.

Thus all these gases are qualitatively similar to the hard-core gas. Returning to (1.1) we see that the activity z is given by

$$z = Z_N/Z_{N+1} = (\rho/1 - \rho a) \exp(\rho a/1 - \rho a) \tag{1.9}$$

As z goes from 0 to ∞, ρ goes from 0 to a^{-1}. In the Mayer theory one seeks a power series for ρ in terms of z. But we note from (1.9) that although ρ is analytic for z on the positive real axis (the physical regime of z), ρ becomes infinite when $za = -e^{-1}$. Owing to this singularity on the negative real axis the power series for ρ has only a finite radius of convergence even though the gas has no phase transition. The distribution of roots of the grand partition function (as a function of activity or fugacity) has been computed by Hiis Hauge and Hemmer (1963).

Another interesting quantity that can be computed exactly for the Takahashi, (nearest-neighbor) gas is the two-particle distribution function defined by

$$g(x_1, x_2) = N(N-1) \frac{\int_0^L \cdots \int dx_3 \cdots dx_N \exp(-\beta \sum_{i<j} v(x_i - x_j))}{\int_0^L \cdots \int dx_1 \cdots dx_N \exp(-\beta \sum_{i<j} v(x_i - x_j))} \tag{1.10}$$

There are also the three-, four-, etc., particle distribution functions which are obvious generalizations of (1.10). These have been computed in detail by Salsburg, Zwanzig, and Kirkwood (1953), and in their Eq. (32) they give g for the simple hard-core Tonks' gas. In the bulk limit g always becomes a function of $x = x_1 - x_2$ only, and in this limit it turns out to be simpler to discuss the Laplace transform of g:

$$\bar{g}(k) = \int_0^\infty g(x)e^{-kx}\,dx \tag{1.11}$$

$$= \rho^2[e^{ka}(k - k\rho a + \rho) - \rho]^{-1} \tag{1.12}$$

Equation (1.12) is only for the hard-core gas and was computed from Eq. (32) of SZK. For the general Takahashi gas, Lebowitz, Percus, and Zucker (1962) have given an interesting formula for $\bar{g}$ in terms of thermodynamic quantities:

$$\bar{g}(k) = \rho\{\exp\beta[\mu(\beta P + k) - \mu(\beta P)] - 1\}^{-1} \tag{1.13}$$

where P is the pressure given by (1.8) and μ is the chemical potential defined by $\beta\mu = \ln z + \frac{1}{2}\ln(\beta h^2/2\pi m)$, where z is given generally by the left side of (1.9), h is Planck's constant, and m is the particle mass. To evaluate (1.13) it is necessary to express μ in terms of P, and when this is done for the hard-core gas one easily obtains the preceding result, (1.12), of SZK.

The historical development of the subject did *not* follow the foregoing logical scheme. Actually the two-particle distribution function for the hard-core gas was first calculated by Zernike and Prins (1927), well before Tonks' paper. They give several graphs of this function as well as graphs of the x-ray scattering function (which is related to an analytic continuation of our $\bar{g}(k)$ above). Sells, Harris, and Guth (1953) showed how one could derive the pair distribution function for the Takahashi gas from the Born-Green integral equation, because for this gas the superposition approximation is exact if (and only if) the particles are ordered; that is, $g^3(x_1, x_2, x_3) = g(x_1, x_2)g(x_2, x_3)$ if $x_1 < x_2 < x_3$. [This is not true of more general potentials, as in (1.1′).] Using this equation they calculated $g(x)$ explicitly for the hard-core case. In their second paper they showed an interesting connection between the Takahashi gas and a Markovian random walk, which in addition to providing a different point of view also simplifies calculations considerably.

There have been many variations on the theme of a one-dimensional gas with (essentially) nearest-neighbor forces. The first work after Tonks' seems to have been that of Nagamiya (1940) (although this predates Takahashi's contribution, it does not contain the same results because Nagamiya considered only very special examples of interparticle forces). He first considers Tonks' hard-core potential quantum mechanically (we shall return to this problem in the Bose gas chapter) and then goes on to consider two other potentials:

$$\begin{aligned} v_2(x) &= \infty &&\text{for} \quad 0 < |x| < a \\ &= \varphi_0 + c(|x| - a) &&\text{for} \quad |x| > a \end{aligned} \tag{1.14}$$

and

$$v_3(x) = \varphi_0 + (K/2)(x - a)^2 \qquad \text{for} \quad -\infty < x < \infty \tag{1.15}$$

The important additional assumption—one that makes these potentials even more unphysical—is that they are presumed to act *only* between nearest neighbors. Thus, in either case the total potential is $V = v(x_2 - x_1) + v(x_3 - x_2) + \cdots + v(x_N - x_{N-1})$. Type II, (1.14), is solved by introducing two fixed particles, one at 0 and the other at L, whereupon the total potential for any configuration is simply $(N + 1)(\varphi_0 - ca) + cL$. This means that we have Tonks' gas plus a van der Waals attraction and it is not surprising that a phase transition occurs. This problem is also solved quantum mechanically for it is only a slight generalization of the first case. For v_3, (1.15), a is the equilibrium separation of hard-core particles joined by springs. If one ignores the nearest-neighbor restriction (which means that one must ensure that $x_{i+1} - x_i > 0$), then the problem is simply one of harmonic oscillators and can therefore be solved even quantum mechanically, as Nagamiya does.

Unfortunately this restriction is indeed overlooked in Nagamiya's paper, so that although the solution is simple it refers to a problem that is neither nearest neighbor nor the usual two-body potential, and its significance is unclear. The correct solution to the problem of rigid rods connected by springs was elegantly given by Koppel (1963). There is no phase transition, although, depending on the choice of parameters, the system can be made to resemble either a gas or a solid. See also Northcote and Potts (1964). Another model of this type that can be made to resemble a solid is given by Mazur and Rubin (1963) who compute Young's modulus and the coefficient of thermal expansion.

Further examples are furnished by Baur and Nosanow (1962), Runnels (1965a,b), Salzberg (1965), Barker (1962), and Kummer (1962). The latter two authors also consider the problem of many parallel one-dimensional systems with crosswise finite range interactions among them. The analysis reduces to an integral equation similar to that found by Van Hove (1950).

The one-dimensional gas with nearest-neighbor forces has also been extended to binary mixtures by Prigogine and Lafleur (1954), Kikuchi (1955), Prigogine (1957), and Longuet-Higgins (1958), and has found use as a model for nonequilibrium statistical mechanics (Klein and Prigogine, 1953). [A minor word of caution is in order regarding this last paper. What the authors refer to as the two-particle distribution function, $\rho(r)$ (Eq. (2.8) in their paper), is not the usual one as we have defined it in (1.10). Our $g(x_1, x_2)$ is the total probability per (unit length)^2 of finding particles at x_1 and x_2 irrespective of the number of particles in between. The $\rho(r)$ of *KP* is only the nearest-neighbor distribution, i.e., the probability of finding two particles at a distance r with none in between. Gursey (1950) also calculated only the nearest-neighbor distribution in his Section 7.]

Finally, we must mention an important model proposed by Fisher (1965). Very long range forces are introduced, which can bring about a phase transition, but the forces are not of the usual two-body type. Rather they are of the "cluster" type, by which is meant that they are similar to the surface forces that one expects to exist among clusters of atoms in a real gas. The total potential energy consists of two parts: the first is an arbitrary, nearest-neighbor pair force with a hard core as in

Eq. (1.1). We know how to handle such a potential exactly. The second part is the cluster potential defined as follows: For any definite configuration of the particles we say that the particles $j, j+1, \cdots, j+k$ belong to a cluster of size $k+1$ if all the distances $x_{j+1} - x_j, \cdots, x_{j+k} - x_{j+k-1}$ are less than a certain amount, c, called the clustering distance. Note that with this definition a three-particle cluster, for example, always has contained within it at least two, and possibly three, two-particle clusters. If a definite particle configuration contains n_k clusters of size k, the total cluster energy is $\Sigma_{k=1}^{N} n_k\omega_k$, where ω_k are certain specified numbers characterizing the system. This clustering energy is a very-many-body interaction and it is of enormously long range. Fisher finds that there is a transition if the sum $\Sigma \exp(-\beta\omega_n)$ converges and that there is no transition if the sum diverges. In other words $kT_c = \lim_{n\to\infty} \omega_n/\log n$. The interesting feature of this model is that different choices of ω as a function of n give rise to a wide variety of types of coexistence curves, some of which are indeed unusual.

The Kac-Baker Model

To summarize the discussion thus far, we have seen that one can solve completely the problem of a finite-range potential and that no phase transition occurs. If we want to have a transition in one dimension we must resort to a potential such as in (1.14) which is completely unphysical because not only is it unbounded, but it is also not a true pair potential. We turn now to a model which does resemble a physical system somewhat and which does have a phase transition. This model has two ingredients: The first is to find a reasonable, integrable potential which has a long range and for which the partition function can be calculated exactly. Normally, such a potential would require solving Van Hove's (1950) integral equation which becomes hopelessly complicated as the range increases. But Kac (1959) showed that for the true pair potential $V = \Sigma_{i<j} v(x_i - x_j)$, where

$$\begin{aligned} v(x) &= \infty && \text{for } |x| < \delta \\ &= -\alpha \exp(-\gamma|x|) && \text{for } |x| > \delta \end{aligned} \tag{1.16}$$

the partition function could be calculated exactly for any finite value of the constants α, δ, and γ in the limit $N \to \infty$. Kac proved that this model has no phase transition. (Because of the infinite range of the potential, Van Hove's proof is not applicable here.) The second ingredient is the observation made by Baker (1961) that if one sets

$$\alpha = \alpha_0\gamma \tag{1.17}$$

(so that the integral of the potential is independent of γ) and then takes the limit $\gamma \to 0$ *after* the limit $N \to \infty$, *then a phase transition of the classic van der Waals type is obtained.* In point of fact, Baker did not treat the gas described by (1.16) but rather the analogous Ising model of ± 1 spins on a lattice. The interaction energy of two spins ν_i and ν_j separated by a distance $x > \delta$ is taken to be $v(x)\nu_i\nu_j$, but the solution of one problem can be generalized to solve the other. Baker's analysis, however, is quite different from Kac's, which we shall now briefly outline.

Clearly the total potential is not of the nearest- (or several) neighbor type so that we cannot obviously use the previous method which essentially involves calculating the effect of adding one particle at the end of the chain (actually, Baker succeeded in analyzing the problem just this way). Instead, Kac observed that the total Boltzmann factor has the simple integral representation

$$\exp\left(-\beta \sum_{i<j} v(t_j - t_i)\right) = e^{-\frac{1}{2}NJ} \int_{-\infty}^{\infty} \cdots \int \prod_{i=1}^{N} dx_i \exp\left(\sqrt{J} \sum_{i=1}^{N} x_i\right) W(x_i) \times \prod_{j=1}^{N-1} P(x_j|x_{j+1};\, t_{j+1} - t_j) \tag{1.18}$$

whenever $t_{j+1} - t_j > \delta$ (for all $j = 1, \cdots, N-1$). Here $J = \beta\alpha$ and

$$W(x) = (2\pi)^{-1/2} \exp(-\tfrac{1}{2}x^2) \tag{1.19}$$

$$P(x|y;\, t) = [2\pi(1 - e^{-2\gamma t})]^{-1/2} \exp[-(y - xe^{-\gamma t})^2/2(1 - e^{-2\gamma t})] \tag{1.20}$$

The right side of (1.18) is a standard Gaussian integral and (1.18) can easily be verified by the well-known formula

$$\int_{-\infty}^{\infty} \cdots \int \prod_{i=1}^{N} dx_i \exp\left(-\sum_{i,j=1}^{N} x_i A_{ij} x_j + \sum_{i=1}^{N} k_i x_i\right) = \pi^{N/2} (\det A)^{-\frac{1}{2}} \exp\left(\tfrac{1}{4} \sum_{i,j=1}^{N} k_i (A^{-1})_{ij} k_j\right) \tag{1.21}$$

in which A is any positive-definite symmetric matrix.

Thus, at the price of introducing N additional variables of integration, the partition function is brought into a form we can handle—namely, a product of nearest-neighbor pair functions, both in the x's and in the t's. As regards the latter variables, the pair function depends on the differences $t_{j+1} - t_j$ only and we can thus accomplish the t integration,

$$\int_{(W-1)\delta}^{L} dt_N \cdots \int_{\delta}^{t_3-\delta} dt_2 \int_0^{t_2-\delta} dt_1$$

by the Laplace transform method as in (1.2) to (1.6). Thus,

$$\bar{Z}(s) = s^{-2} e^{-\frac{1}{2}NJ} \int_{-\infty}^{\infty} \cdots \int \prod_{i=1}^{N} dx_i \exp\left(\sqrt{J} \sum_{i=1}^{N} x_i\right) W(x_i) \prod_{j=1}^{N-1} p_s(x_j|x_{j+1}) \tag{1.22}$$

where

$$p_s(x|y) = \int_{\delta}^{\infty} dt\, e^{-st} P(x|y;\, t) \tag{1.23}$$

Noting that $W(x)P(x|y;\, t)$ is symmetric in x and y, we introduce the symmetric kernel

$$K_s(x, y) = \exp[\tfrac{1}{2}\sqrt{J}(x + y)]W(x)p_s(x|y)[W(x)W(y)]^{-1/2} \tag{1.24}$$

Let the eigenvalues and eigenfunctions of this kernel (which Kac proves is of the Hilbert-Schmidt type) be denoted by λ_j and $\Psi_j(x)$, respectively (they depend on s, of course). In terms of these, (1.22) obviously can be written

$$\bar{Z}(s) = s^{-2}e^{-\frac{1}{2}NJ} \sum_{j=1}^{\infty} \lambda_j^{N-1} C_j \tag{1.25}$$

where

$$C_j = \left\{\int_{-\infty}^{\infty} dx\, [W(x)]^{1/2} \exp(\tfrac{1}{2}\sqrt{J}x)\Psi_j(x)\right\}^2 \tag{1.26}$$

The corresponding Ising problem can be solved in the same way (Kac, 1959). We wish to compute

$$Z = \sum_{\nu_N = \pm 1} \cdots \sum_{\nu_1 = \pm 1} \exp\left(-\beta \sum_{i<j} v_{ij}\nu_i\nu_j\right) \tag{1.27}$$

where

$$v_{ij} = -\alpha \exp(-\gamma|i - j|) \tag{1.16a}$$

Again, the Boltzmann factor in (1.27) may be written as

$$e^{-\frac{1}{2}NJ} \int_{-\infty}^{\infty} \cdots \int \prod_{j=1}^{N} dx_i\, W(x_1) \prod_{j=1}^{N-1} P(x_j|x_{j+1};\, 1) \exp\left(\sqrt{J} \sum_{i=1}^{N} \nu_i x_i\right) \tag{1.18a}$$

but this time there is no need to introduce a Laplace transformation because the $\nu = \pm 1$ summation can be accomplished directly; the last exponential factor in (1.18a) is simply replaced by $2^N \prod_{i=1}^{N} \cosh(\sqrt{J}x_i)$. The analysis proceeds as before, except that the relevant kernel is now

$$K(x, y) = [\cosh(\sqrt{J}x) \cosh(\sqrt{J}y)]^{1/2}W(x)P(x|y;\, 1)[W(x)W(y)]^{-1/2} \tag{1.24a}$$

and the same qualitative conclusions apply.

Since the kernel is positive for $\alpha > 0$, the eigenvalue of largest modulus, λ_1, is nondegenerate and positive, and $\Psi_1(x)$ is nonnegative. Hence C_1 is not zero and in the bulk limit

$$\lim_{N\to\infty} (1/N) \ln \bar{Z}(s) = -\tfrac{1}{2}J + \ln \lambda_1 \tag{1.28}$$

One can also show that λ_1 is an analytic, monotonic function of s and hence no phase transition occurs.

On the other hand, if, after having taken the limit $N \to \infty$ as in (1.28), we now let $\gamma \to 0$ with α_0 fixed as in (1.17), it is found that when $\alpha_0 > 0$ for a certain range of temperature and density (depending on α_0) the largest eigenvalue becomes asymptotically degenerate. This, in itself, does not cause a phase transition but, as one would expect, λ_1 in the degenerate region is not the analytic continuation of

λ_1 in the nondegenerate region. A discontinuous change in derivative occurs on the boundary of the two regions and the isotherms become flat (as for a real gas) when there is degeneracy. This same phenomenon, namely, the onset of degeneracy accompanied by a change in analytic character of a largest eigenvalue, occurs for the phase transition of the *two*-dimensional standard Ising model. This latter problem is in fact soluble just because it can be reduced to a one-dimensional problem.[1]

When $\alpha < 0$ (repulsive case), the kernel (1.24) is symmetric but not Hermitian because $J < 0$. This case has been studied by Newman (1964) who points out that the eigenvalue expansion (1.25), as well as (1.28) may not be justified here. While Newman did not resolve the problem, he did show that in the $\gamma \to 0$ limit the free energy is given correctly by (1.29) below, and hence there is no phase transition. The kernel in Baker's model, Eq. (1.24a), remains Hermitian, so that the parallel analysis should be easier. We suggest it as an exercise for the interested reader.

Despite the complexity of the preceding analysis, the results in the $\gamma \to 0$ limit are remarkably simple and can be deduced directly by the following heuristic argument based, partly, on the work of Penrose and Lebowitz.[2] When γ is very small each particle "feels" many neighbors and, assuming the particles to be more or less uniformly distributed on the line, the exponential part of the potential would be expected to add merely $[N(N-1)/2L]\int v_{\text{exponential}}(x)\,dx = -N\rho\alpha_0$ to the free energy. When we add this to the free energy of a hard-core gas of diameter δ given by (1.3) and divide the total by L, we obtain $\bar{F}$ = configurational part of the free energy per unit volume:

$$\bar{F} = kT\rho[\ln \rho - 1 - \ln(1 - \rho\delta)] - \alpha_0\rho^2 \tag{1.29}$$

For reasons we now discuss, this is in fact the correct answer *provided* α_0, T, ρ, and δ are such that $\ddot{\bar{F}} \equiv \partial^2\bar{F}/\partial\rho^2 > 0$. It is known in general that the *correct* $\ddot{\bar{F}}$ must be nonnegative, to ensure that the pressure, $P = \rho(\partial\bar{F}/\partial\rho) - \bar{F}$, be a nondecreasing function of the density. Hence, whenever the second derivative of (1.29) is negative it means that a phase separation has occurred and (1.29) is to be replaced by the straight tangent line which will thereby make the second derivative vanish. One says that the correct result is the least convex envelope (with respect to ρ) of (1.29), a statement which is equivalent to the Maxwell equal area rule. For sufficiently high temperatures, $\ddot{\bar{F}}$ is always positive, but for low temperatures there are always two (and only two) values of ρ between which $\bar{F}$ must be replaced by a straight line. The critical temperature dividing these two regions is $kT_c = 8\alpha_0/27\delta$ and the critical pressure and density, respectively, are $P_c = \alpha_0/27\delta^2$, $\rho_c = (3\delta)^{-1}$. The equation of state (except in the two-phase region) is

$$P = kT(\rho/1 - \rho\delta) - \alpha_0\rho^2 \tag{1.30}$$

exactly the van der Waals formula.

[1] T. Schultz, D. Mattis, and E. Lieb, *Rev. Modern Phys.* **36,** 856 (1964).

[2] J. Lebowitz and O. Penrose, *J. Math. Phys.* **7,** 98 (1966). For an extension to quantum mechanical systems, see E. Lieb, *J. Math. Phys.* (to be published, 1966).

Let us consider a generalization of (1.16) to an arbitrary number of dimensions $n = 1, 2, 3, \ldots$ and to a rather general two-body potential of the form

$$v(\mathbf{r}) = v_0(\mathbf{r}) - \alpha_0\gamma^n\varphi(\gamma\mathbf{r})$$

where v_0 is any short-range potential with a hard core, for which we suppose $\bar{F}_0$, the free energy per unit volume is known, and $\varphi(\mathbf{r})$ is a function whose integral over all space is unity. Under suitable conditions on v_0 and φ we are led to believe that the total free energy per unit volume should be the least convex envelope of

$$\bar{F} = \bar{F}_0 - \tfrac{1}{2}\alpha_0\rho^2 \tag{1.31}$$

Indeed, this surmise is correct as Lebowitz and Penrose[2] have shown completely rigorously by dividing the total volume $V = L^n$ into boxes of length S such that $b \ll S \ll \gamma^{-1} \ll L$, where b is the range of v_0. By carefully estimating surface effects, (1.31) can be established in the bulk limit $L \to \infty$ followed by $\gamma \to 0$.

Thus, while the exponential potential in one dimension is the only long-range problem that can be analyzed exactly, the phase transition when $\gamma \to 0$ is a completely general feature. A *generalized* van der Waals equation of state is always obtained. The one advantage of the exact solution is that one can show that (1.28) can be differentiated twice to obtain the correct limiting specific heat, whereas Penrose and Lebowitz can rigorously prove only that the first derivative of (1.31) gives the correct pressure and internal energy. But there is no doubt that any finite number of derivatives of (1.31) give the correct bulk limits of the same derivatives of the partition function.

Aside from the intrinsic mathematical beauty and interest of the analysis leading from (1.16) to (1.29), there is some question as to whether the phase transition in the $\gamma \to 0$ limit has any real physical meaning. To be sure, the isotherms have qualitative similarities to those of a real gas, but so do those of van der Waals' semiphenomenological theory. The real problem is to explain why finite-range forces behave like infinitely long range ones. The same problem occurs in trying to explain the success of the molecular field theory of ferromagnetism. As regards the neighborhood of the critical point, however, the situation is different, for even the experimental data is incomplete in this region. Similarly, the Kac-Baker model is not thoroughly understood in the transition region (cf. the papers of Kac, Uhlenbeck, and Hemmer, 1963–1964) and further insight into this model could be extremely valuable.

For generalizations of this potential as well as its applications to the lattice gas, spherical model, and two-dimensional Ising model see Baker (1962), Kac and Helfand (1963), Helfand (1964), and Gersch (1963).

The distribution of roots of the grand partition function for the van der Waals gas [Eq. (1.28)] has been given by Hemmer and Hiis Hauge (1964). For the weak long-range repulsive potential with no hard core [where $\bar{F}$ is given by (1.29) with $\delta = 0$ and $\alpha_0 < 0$] the distribution of roots is given in Hiis Hauge and Hemmer (1963).

The Coulomb Gas and the Statistical Theory of Energy Levels

Every so often in the history of science, a complicated problem is solved simultaneously and independently by two people. This was the case in the invention of calculus by Newton and Leibnitz (ca. 1680), and this, in fact, happened in the interesting but, of course, much less momentous case of the one-dimensional Coulomb gas. The co-discoverers Prager (1962) and Lenard (1961) used essentially the same combinatorial methods to solve this problem, and both announced their results at the 1961 Spring meeting of the American Physical Society.[3] Later, Edwards and Lenard (1962) and Lenard (1963) showed how the problem could be solved much more directly using methods of functional integration, but it is really necessary to understand functional integrals thoroughly if one wants to appreciate the elegance of this second approach. We shall not pursue the matter here and instead refer the reader to Edwards and Lenard's (1962) readable paper.

The potential energy of two charges σ_i and σ_j at positions q_i and q_j, respectively, is taken to be (Lenard's notation) $-2\pi\sigma_i\sigma_j|q_i - q_j|$ and it is supposed that there are altogether $2N$ particles, half of which have charge $+\sigma$, and the others $-\sigma$. We can order the particles as before $0 \leq q_1 \leq \cdots \leq q_{2N} \leq L$, but then we must remember to sum over all possible arrangements of charge $\pm\sigma$. We define a partial sum over charges by

$$\nu_i = \sigma^{-1}\left(\sum_{k=1}^{i} \sigma_k\right) \tag{1.32}$$

whence

$$|\nu_i - \nu_{i-1}| = 1 \qquad \text{and} \qquad \nu_0 = \nu_{2N} = 0 \tag{1.33}$$

the last condition expressing charge neutrality. A little algebra then shows that the total potential energy can be written

$$\Phi = 2\pi\sigma^2 \sum_{i=1}^{2N-1} \nu_i^2(q_{i+1} - q_i) \tag{1.34}$$

and we observe that for a fixed choice of the ν_i's the partition function is of the nearest-neighbor type we have discussed before. It is therefore convenient to take a Laplace transform of $Z(L)$ with respect to L, the length of the container, whence

$$\bar{Z}(\gamma) = (2\pi\sigma^2\beta)^{-2N} \prod_{i=0}^{2N} (\gamma + \nu_i^2)^{-1} \tag{1.35}$$

where $\gamma = s(2\pi\beta\sigma^2)^{-1} = P(2\pi\sigma^2)^{-1}$, P being the pressure and s the transform variable. The γ defined here has nothing to do with the Kac parameter γ of the preceding

[3] *Bull. Am. Phys. Soc.* **6**, 239, 246 (1961).

section, and it is hoped that the unfortunate duplication of notation will not cause confusion. If we define

$$\ln \alpha (\gamma) = - \lim_{N \to \infty} N^{-1} \ln \bar{Z} (\gamma) \tag{1.36}$$

the equation of state given by (1.7) and (1.8) is

$$\beta P/\rho = \tfrac{1}{2}\gamma \, \partial \ln \alpha(\gamma)/\partial\gamma \tag{1.37}$$

Note that $\rho = 2N/L$.

To find α as a function of the dimensionless parameter γ we must sum (1.35) over all distinct choices of the ν's consistent with (1.33). This is a complicated combinatorial problem, handled by a generating function, which is expressed as a continued fraction. This, in turn, can be related to an eigenvalue of Mathieu's equation, viz.,

$$(d^2/dt^2 + 2\alpha^{1/2} \cos t)y(t) = \gamma y(t) \tag{1.38}$$

where $y(t)$ is an even function, periodic in 2π. It is the nodeless solution (designated Ce_0).

Since the combinatorial method of Prager and Lenard is rather involved, we shall present here a simpler derivation of (1.38). Consider the sum over all configurations consistent with (1.33) of the product in (1.35), with the difference that we have exactly j instead of $2N$ variables, and where we relax the requirement that $\nu_j = 0$. Denoting ν_j simply by ν, and denoting the sum by $f_j(\nu)$, we have [ignoring the irrelevant factor of $(2\pi\sigma^2\beta)^{-j}$]

$$(\gamma + \nu^2)f_j(\nu) = f_{j-1}(\nu + 1) + f_{j-1}(\nu - 1).$$

If we now define

$$F_j(t) = \sum_{\nu=-\infty}^{\infty} f_j(\nu)e^{i\nu t}$$

we have

$$[-d^2/dt^2 + \gamma]F_j(t) = 2 \cos t \, F_{j-1}(t)$$

and

$$F_j(t + 2\pi) = F_j(t); \qquad F_0(t) = 1$$

The result we require is $f_{2N}(0) = (2\pi)^{-1} \int_0^{2\pi} F_{2N}(t)\, dt$, and to obtain it we consider the eigenvalue problem for periodic functions on the interval $(0, 2\pi)$:

$$[-d^2/dt^2 + \gamma]Y_n(t) = 2\lambda_n \cos t \, Y_n(t)$$

Expanding the functions $F_j(t)$ in the complete set of eigenfunctions,

$$F_j(t) = \sum_n A_{n,j} Y_n(t)$$

we obtain

$$A_{n,j} = (\lambda_n)^{-j} A_{n,0}$$

Consequently,

$$f_j(0) = (2\pi)^{-1} \sum_n (\lambda_n)^{-j} \Big(\int_0^{2\pi} Y_n(t)\, dt \Big) A_{n,0}$$

Owing to the invariance of $\cos t$ with respect to a translation of π followed by a reflection in the t-axis, it is plain that the negative of every eigenvalue is an eigenvalue while at the same time $A_{n,0}$ and $\int_0^{2\pi} Y_n(t)\, dt$ remain the same. Hence, when j is odd, $f_j(0) = 0$ as it should be. When j is even and very large ($= 2N$) only the smallest eigenvalue and its negative contribute and we obtain (1.38). The above analysis can easily be made rigorous, in the sense of showing that the manipulations with the infinite sums are legitimate.

Expansions for γ in terms of α are known for large γ (high temperature, high pressure, or small charge):

$$P/\rho kT = 1 - (8\gamma)^{-1/2} + (16\gamma)^{-1} - \tfrac{1}{2}(32\gamma)^{-3/2} + \cdots \tag{1.39}$$

and for small γ (low temperature; low pressure, or large charge):

$$P/\rho kT = \tfrac{1}{2} + \tfrac{7}{16}\gamma + \cdots \tag{1.40}$$

Equation (1.39) agrees with the standard Debye-Hückel theory (see Prager, 1962), while (1.40), the strong coupling result, is a bit unusual. It implies that the system behaves like a perfect gas of *half* the total number of particles. Lenard (1961) interprets the result as caused by the formation of neutral bound pairs.

Lenard and Prager also solve the problem with an external electric field.

Baxter (1963) considered the problem of a one-component Coulomb gas in the presence of a neutralizing uniform background with a view to finding out if the equation of state of such a system is similar to that of the more physical two-component system discussed previously. The reason for this is that the assumption(or hope) of similarity is often made for three-dimensional systems because the one-component system is easier to treat. If the N particles each have charge σ and go between $-L$ and L (so that $\rho = N/2L$), and if the background charge density is $(N-1)/2L$, the total potential energy is

$$\phi = -2\pi\sigma^2 \Big(\sum_{i<j} |x_i - x_j| - \frac{N-1}{2L} \sum x_i^2 - \tfrac{1}{6}N(N+1)L \Big) \tag{1.41}$$

Because of the quadratic nature of (1.41) the configurational integral is conveniently performed by shifting variables to eliminate the linear terms. The integral is still not elementary, but it can be shown by an elegant analysis to lead directly to an eigenvalue problem:

$$d/dv\, G(v) = z \exp(-\pi\lambda^2 v^2) G(v+1), \tag{1.42}$$

with boundary conditions

$$G(\infty) = 1; \qquad G(-\infty) = 0 \tag{1.43}$$

Also $G > 0$ and $\lambda^2 = 2\sigma^2\beta\rho^{-1}$. The eigenvalue z gives the equation of state:

$$\beta\rho^{-1}P = 1 - \tfrac{1}{2}\lambda\, d \ln z/d\lambda \tag{1.44}$$

For weak coupling this turns out to be

$$\beta\rho^{-1}P = 1 - (\tfrac{1}{8}\pi)^{1/2}\lambda + O(\lambda^2) \tag{1.45}$$

in agreement with (1.39) and with Debye-Hückel theory. For strong coupling, Baxter obtains

$$\beta\rho^{-1}P = \tfrac{1}{2} + \tfrac{1}{4}\sqrt{2}\lambda \exp(-\tfrac{1}{2}\pi\lambda^2) \tag{1.46}$$

This agrees with (1.40) as far as the first term is concerned, but in this case the pressure approaches its asymptote much faster than for the two-component gas. Nevertheless, on the basis of numerical calculations, Baxter claims that the maximum difference in the pressure for the two systems is only 7% at any temperature. It is interesting to note that the asymptotic value of $\frac{1}{2}$ in (1.46) (which was formerly explained on the basis of bound pairs) is here explained as the pressure to be expected for N harmonic oscillators each having the plasma frequency $\omega_p^2 = 4\pi\sigma^2\rho/m$.

One other feature of these results is noteworthy. For both systems the *strong* coupling expansions, (1.40) and (1.46), can be obtained easily from the exact equations, (1.38) and (1.42). In contrast, for both cases the *weak* coupling expansions (1.39) and (1.45) are difficult to obtain. This odd fact, that an exact solution is difficult to obtain in the regime where a quasi-perturbation technique (such as Debye-Hückel theory or Bogoliubov's theory of the Bose gas) is valid, parallels the result for the exactly soluble Bose gas (Lieb and Liniger, 1963), which is mentioned in the chapter on the Bose gas.

The reader may suspect that there is some underlying connection between the mathematics of the Coulomb gas and the mathematics of the Kac-Baker model. Indeed, Baxter (1964) has developed an elegant method for finding the partition function when the pair potential satisfies a linear differential equation with constant coefficients, of which the above two models are special cases. In a sequel (Baxter, 1965) the underlying similarity is made even clearer.

Another question of great importance is that of the time and spatial (or energy and momentum) variations of the electric field in a plasma. One would like to be able to compare the exact results, which are not available, unfortunately, with those predicted by the linearized Vlasov equation. Eldrige and Feix (1962) have done just this by following the evolution of a one-dimensional plasma on a computer.

Long before the one-dimensional Coulomb gas was solved, Ufford and Wigner (1942) proposed a test of the validity of the Debye-Hückel theory for long-range forces. They investigated a system of N particles on $(0, L)$, as usual, with a two-body potential

$$v(x_1 - x_2) = -\ln|\sin \pi (x_1 - x_2)/L| \tag{1.47}$$

Admittedly, this potential is not very physical, since it depends on the length of the box, but it is certainly long range. Because of the periodicity of (1.47) we can equally well consider the system to be N particles constrained to lie on a circular wire of radius L; since the coordinate system then has no real origin it is unnecessary to introduce a neutralizing background in order to achieve uniformity. Now, notice what happens when $\beta = 2$. The Boltzmann factor is simply

$$\prod_{i<j} \sin^2\pi (x_i - x_j)/L = 2^{-N(N-1)}\Psi_F(x_1, \ldots, x_n)^2 \tag{1.48}$$

where Ψ_F is the usual one-component Fermi function

$$\Psi_F = \det|\exp(ik_ix_j)| \tag{1.49}$$

where

$$k_j = \frac{2\pi}{L}\left(\frac{n+1}{2} - j\right)$$

[To verify (1.48) it is necessary to recognize (1.49) as a Vandermonde determinant.] It is now trivial to do the N-fold integration from 0 to L of the square of a Fermi function, with the result that

$$Z = 2^{-N(N-1)}L^N \tag{1.50}$$

The two-particle correlation is that of a Fermi gas and can be computed easily. Ufford and Wigner display a graphical comparison of this function and the one derived from the Debye-Hückel equation appropriate to this case. The agreement is "fair," and that it is not better is due to the fact that β (which is the coupling constant here) is not small.

Now, with a change of nomenclature we can turn the above system into a physically reasonable one, but it was only very recently that the problem was solved for $\beta \neq 2$. Dyson (1962, part I)—in connection with his study of Wigner's statistical theory of energy levels—pointed out that the potential energy of a pair of two-dimensional charged particles of charge σ, confined to a circular wire of radius R is

$$\begin{aligned} v(\mathbf{r}_i - \mathbf{r}_j) &= -\sigma^2 \ln|\mathbf{r}_i - \mathbf{r}_j| = -\sigma^2 \ln R|\exp(i\theta_i) - \exp(i\theta_j)| \\ &= -\sigma^2 \ln 2|\sin\tfrac{1}{2}(\theta_i - \theta_j)| - \sigma^2 \ln R \end{aligned} \tag{1.51}$$

where $0 \leq \theta_i \leq 2\pi$ is the angle of the ith particle. If we add to the sum on pairs of (1.51) the additional potential energy of a uniform neutralizing background of charge $N\sigma$, we obtain a Boltzmann factor for the system

$$e^{-\beta v} = \exp(-\tfrac{1}{2}\beta' N \ln R)|\Psi_F(\theta_1, \ldots, \theta_N)|^{\beta'} \tag{1.52}$$

where $\beta' = \sigma^2\beta$ and Ψ_F is (1.48) with $L = 2\pi$. It is important that the background have a charge exactly $N\sigma$ and not $(N-1)\sigma$, for example. Otherwise the partition function will not have the usual extensivity properties. To calculate the partition function we must compute

$$\Psi_N(\beta') = (2\pi)^{-N} \int_0^{2\pi} \cdots \int_0^{2\pi} |\Psi_F(\theta_1, \ldots, \theta_N)|^{\beta'} \tag{1.53}$$

We have seen that (1.53) is easy to compute for $\beta' = 2$. Dyson (1962) also calculated it for $\beta' = 1$ and 4 (these three values are what are needed for the statistical theory of energy levels). He conjectured, however, that

$$\Psi_N(S) = \Gamma(1 + \tfrac{1}{2}NS)/[\Gamma(1 + \tfrac{1}{2}S)]^N \tag{1.54}$$

for any integer N and any complex S with $\mathrm{Re}(S) \geq 0$, and was able to convert the conjecture to a simpler algebraic one. This was finally proved by Wilson (1962).

With (1.54) thus established, we can combine it with (1.53) to obtain the partition function

$$\begin{aligned} Z &= (N!)^{-1}(2\pi)^N R^N \Psi_N(\beta') \exp(-\tfrac{1}{2}N\beta' \ln R) \\ &= \exp N\{1 - \ln \rho + \tfrac{1}{2}\beta'[\ln(\pi\rho\beta') - 1] - \Gamma(1 + \tfrac{1}{2}\beta')\} \end{aligned} \tag{1.55}$$

where $\rho = N/2\pi R$. Actually, Dyson considered only the case $R = 1$, but we see that it is easy to consider the general case. As Salzberg and Prager (1963) pointed out, a two-dimensional Coulomb potential (1.51) always has the property that the volume dependence of the partition function factors out in a trivial way, even if the configuration is a one-dimensional circle. Thus, we could have concluded directly from (1.51) that the equation of state is

$$\beta P = \rho[1 - \tfrac{1}{2}\beta\sigma^2] \tag{1.56}$$

but to obtain the dependence of the free energy on temperature is another matter requiring (1.54).

Dyson (1962b) also proposed investigating a model of two-dimensional Coulomb particles confined to an infinitely long straight wire and attracted to the origin by a harmonic potential. The total potential energy in dimensionless units would then be

$$V = -\sum_{i<j} \ln|x_i - x_j| + \tfrac{1}{2}\sum x_j^2 \tag{1.57}$$

Dyson also demonstrated a connection between (1.57) and a Brownian motion problem—a connection which is very similar to the functional integral approach of Edwards and Lenard (1962). As for the partition function of (1.57), all that is available at this time is a conjecture of Mehta and Dyson (1963) which can indeed be verified for $\beta = 1$, 2, or 4:

$$\begin{aligned} Z_N(\beta) &= 1/N!\, (2\pi/\beta)^N \beta^{-\frac{1}{4}N(N-1)\beta}[\Gamma(1 + \tfrac{1}{2}\beta)]^{-N} \\ &\times \left[\prod_{g=1}^{N} \Gamma(1 + \tfrac{1}{2}\beta g)\right] \qquad \text{(conjecture)} \end{aligned} \tag{1.58}$$

Unlike (1.54), Eq. (1.58) has not been proved, although it has been reduced to a purely algebraic problem by Dyson.

It is apparent, from the above two examples, that the statistical theory of energy levels and the conventional statistical mechanical theory of one-dimensional systems have much in common. It would not really be in place to attempt a review here of the former theory (for which see the book by Porter in this series[4]), but it must be remarked that it is the one-dimensional theory *par excellence*! Not only does it have immediate usefulness and validity for real physical systems but, from the mathematical point of view, it has given rise to profound results and makes use of the deepest theorems of analysis. One might almost say that it is a new branch of mathematics in itself.

We cannot close this chapter without pointing out one of the most important and elegant methods in the subject, discovered by Gaudin and Mehta (1960). This method which will, no doubt, appeal to statistical mechanicians because of its combinatorial significance, has made it possible to compute analytically certain correlation functions in the theory.

For complex spectra (such as occur in atomic or nuclear physics) with a great many energy levels, say N, one cannot hope to predict all the levels theoretically, even if the Hamiltonian of the system is known. Nevertheless, one might hope to be able, for example, to find the distribution throughout the spectrum of nearest-neighbor spacings, given that the average level spacing is some constant number D. The idea of the statistical theory of energy levels is that this distribution function ought to be the average, in some suitable sense, of the distribution function obtained for all Hamiltonian that have N energy levels. If a Hamiltonian be represented by a $N \times N$ Hermitian matrix, whose elements are regarded as independent random variables, it is necessary to decide on a probability density function (analogous to the Boltzmann factor of a gas) for any given matrix. One of the most popular assumptions is the Gaussian ensemble of Wigner, i.e., that the relative probability of a matrix H is $C \exp(-\lambda \operatorname{tr} H^2)$, where C is a normalization constant and λ is chosen to make the average level spacing equal to D. Starting with such a weight function one can ask for the probability density function for a specified sequence of energy levels, $x_1, x_2, \ldots, x_N$ which will depend on whether the elements of H are assumed to be real, complex, or quaternion. The quaternions are important for systems with half-odd integer angular momentum, in which case H must really be a $2N \times 2N$ matrix with all eigenvalues doubly degenerate—the so-called Kramers' degeneracy. These three ensembles (real, complex, and quaternion) are termed orthogonal, unitary, and symplectic, respectively, and are designated by a parameter β which is respectively 1, 2, or 4 in the three cases. It can be proved that the probability density of eigenvalues for the three cases is

[4] C. E. Porter, ed., "Statistical Theories of Spectra: Fluctuations." Academic Press, New York, 1965.

$$P(x_1, x_2, \ldots, x_N) = C \prod_{i<j} |x_i - x_j|^\beta \exp\left(-\lambda \sum_1^N x_i^2\right) \tag{1.59}$$

The $\beta = 2$ case is the simplest mathematically, but it is rarely relevant in practice. Usually $\beta = 1$ is required. An alternative hypothesis (Dyson, 1962, part I) is to say that the distribution of the x's (in the small) is the same as the distribution of N angles, $\theta_1, \ldots, \theta_N$ on a circle given by

$$P(\theta_1, \theta_2, \ldots, \theta_N) = C \prod_{i<j} |e^{i\theta_i} - e^{i\theta_j}|^\beta \tag{1.60}$$

To find the relative probability of finding two levels at z_1 and z_2 with none in between, which was our original question, we must set $x_1 = z_1$ and $x_2 = z_2$ and integrate (1.59) over the remaining $N - 3$ variables from $-\infty$ to $+\infty$, exclusive of the region (z_1, z_2). This is obviously difficult to do for $\beta = 1$, but the Mehta-Gaudin method solves the problem. The method can also be applied to the circular orthogonal ensemble, (1.60) [q.v. Dyson (1962, part III)].

We recognize that the polynomial factor in (1.59) is the absolute value of the Vandermonde determinant, and that the remaining exponential factor is a product of single variable functions. The integral of (1.59) over all space is $N!$ times the integral over the region R_1 defined by $x_1 \leq x_2 \leq \ldots \leq x_N$, but in this region the Vandermonde determinant is positive and, therefore, equal to its absolute value. We therefore consider the slightly generalized problem of computing the integral

$$I = \int_{R_1} \cdots \int \prod_{j=1}^N f(x_j) D(x_1, \ldots, x_N)\, dx_1 \ldots dx_N \tag{1.61}$$

where D is the determinant whose elements are

$$D_{ij} = g_i(x_j) \tag{1.62}$$

For the Vandermonde determinant, $g_i(x) = x^{i-1}$.

Now, the trick is to integrate over R_1 in the following way:

$$\int_{-\infty}^{\infty} dx_N \ldots \int_{-\infty}^{x_6} dx_4 \int_{-\infty}^{x_4} dx_2 \int_{x_{N-2}}^{x_N} dx_{N-1} \ldots \int_{x_4}^{x_6} dx_5 \int_{x_2}^{x_4} dx_3 \int_{-\infty}^{x_2} dx_1 \tag{1.63}$$

assuming N is even. Define

$$G_i(x) = \int_{-\infty}^{x} f(x) g_i(x)\, dx \tag{1.64}$$

Integration over x_1 replaces the first column of D by $G_1(x_2)$. Integration over x_3 replaces the third column by $G_i(x_4) - G_i(x_2)$, but since $G_i(x_2)$ already appears in the first column it can be dropped from the third. In like manner the jth column (for odd j) is replaced by $G_i(x_{j+1})$. The resultant determinant is now symmetric in the variables $x_2, x_4, \ldots, x_N$, so that the remaining $\frac{1}{2}N$ integrations can be performed by integrating all variables from $-\infty$ to ∞ and dividing by $(N/2)!$.

Since each variable appears in two columns of the determinant, a little thought will show that the integral is a Pfaffian[5] which, as always, is the square root of the determinant of an antisymmetric matrix. Thus, a little algebra shows that

$$I^2 = \det|M_{ij}| \tag{1.65}$$

where

$$M_{ij} = G_j{}^i - G_i{}^j \tag{1.66}$$

and

$$G_j{}^i = \int_{-\infty}^{\infty} G_j(x) g_i(x) f(x)\, dx \tag{1.67}$$

In order to find the function $P(z_1, z_2)$, the relative probability that there are levels at z_1 and z_2 with none in between, we define the function $R(z_1, z_2)$ to be the relative probability that the interval z_1, z_2 is empty of levels and observe that

$$P(z_1, z_2) = \partial^2/\partial z_1\, \partial z_2\, R(z_1, z_2) \tag{1.68}$$

To compute R, on the other hand, all we need do is replace the weight function $f(x)$ by $f(x)\theta(x)$ where $\theta(x) = 0$ if $z_1 < x < z_2$ and is unity otherwise.

A simple extension of the method can also yield the relative probability that there is a level at z_1 and at z_2, irrespective of how many are in between.

Bibliography

I. Nearest-Neighbor (Takahashi) Gas

J. Barker (1962). "Statistical Mechanics of Almost One-Dimensional Systems," *Australian J. Phys.* **15,** 127.

M. Baur and L. Nosanow (1962). "Phase transitions in One-Dimensional Order-Disorder Systems: Applications to Helix-Random Coil Transition in Polymers," *J. Chem. Phys.* **37,** 153.

M. Fisher (1965). "The Theory of Condensation," *Proc. Centennial Conf. Phase Transformation, Kentucky, 1965.* To be published.

F. Gursey (1950). "Classical Statistical Mechanics of a Rectilinear Assembly," *Proc. Cambridge Phil. Soc.* **46,** 182.

R. Kikuchi (1955). "Theory of One-Dimensional Fluid Binary Mixtures," *J. Chem. Phys.* **23,** 2327.

G. Klein and I. Prigogine (1958). "Sur la Mécanique Statistique des Phénomènes Irréversibles," *Physica* **19,** 74,89.

D. Koppel (1963). "Partition Function for a Generalized Tonks' Gas," *Phys. Fluids* **6,** 609.

M. Kummer (1962). "Über Eindimensionale und Quasieindimensionale Gasmödelle," *Helv. Phys. Acta* **35,** 451.

L. D. Landau and E. M. Lifschitz (1958). "Statistical Physics," p. 482. Pergamon Press, New York.

J. Lebowitz, S. Percus, and I. Zucker (1962). "Radial Distribution Functions in Crystals and Fluids," *Bull. Am. Phys. Soc.* **7,** 415.

[5] The mathematics of Pfaffians is discussed by C. A. Hurst and H. S. Green, *J. Chem. Phys.* **33,** 1059 (1960).

H. Longuet-Higgins (1958). "One-Dimensional Multicomponent Mixtures," *Molecular Phys.* **1,** 83.

J. Mazur and J. Rubin (1963). "Coefficient of Thermal Expansion and Young's Modulus for a One-Dimensional Model of a Solid," *Am. J. Phys.* **31,** 835.

T. Nagamiya (1940). "Statistical Mechanics of One-Dimensional Substances, I," *Proc. Phys.-Math. Soc. Japan* (Nippon Suugaku-Buturigakkwai Kizi, Tokyo) **22,** 705. II, *ibid.* 1034.

R. Northcote and R. Potts (1964). "Isobaric Partition Function of a Generalized Tonks' Gas," *Phys. Fluids* **7,** 475.

W. Pressman and J. Keller (1960). "Equation of State and Phase Transition of the Spherical Lattice Gas," *Phys. Rev* **120,** 22.

I. Prigogine (1957). "The Molecular Theory of Solutions," Chap. VI. North-Holland Publ. Co., Amsterdam, and Wiley (Interscience), New York.

I. Prigogine and S. Lafleur (1954). "On the Statistical Mechanics of Mixtures, I and II." *Bull. Acad. Roy. Belgique Ci. Sci.* **40,** 484, 497.

L. Runnels (1965). "Exact Lattice Model of Liquids: One-Dimensional Gas of Hard Lines." *J. Chem. Phys.* **42,** 217; "One-Dimensional Lattice Gas," *ibid.* 212.

G. Rushbrooke and H. Ursell (1948). "On One-Dimensional Regular Assemblies," *Proc. Cam bridge Phil. Soc.* **44,** 263.

A. Salzberg (1965). "Exact Statistical Mechanics of Gravitational Interactions in One and Two Dimensions," *J. Math. Phys.* **6,** 158.

*Z. Salsburg, R. Zwanzig, and J. Kirkwood (1953). "Molecular Distribution Functions in a One-Dimensional Fluid," *J. Chem. Phys.* **21,** 1098.

R. Sells, C. Harris, and E. Guth (1953). "The Pair Distribution Function for a One-Dimensional Gas," *J. Chem. Phys.* **21,** 1422. Also, C. Harris, R. Sells, and E. Guth "Random Walk Theory of One-Dimensional Gases," *ibid.* 1617.

*H. Takahashi (1942). "Eine Einfache Methode zur Behandlung der Statistischen Mechanik Eindimensionaler Substanzen," *Proc. Phys.-Math. Soc. Japan* (Nippon Suugaku-Buturigakkwai Kizi, Tokyo) **24,** 60.

L. Tonks (1936). "The Complete Equation of State of One, Two, and Three Dimensional Gases of Hard Elastic Spheres," *Phys. Rev.* **50,** 955.

*L. Van Hove (1950). "Sur l'Intégrale de Configuration pour les Systèmes de Particules à Une Dimension," *Physica* **16,** 137.

F. Zernike and J. Prins (1927). "Die Beugung von Röntgenstrahlen in Flüssigkeiten als Effekt der Molekülarordnung" *Z. Physik.* **21,** 184.

II. Kac-Baker Model

G. Baker (1961). "One-Dimensional Order-Disorder Model which Approaches a Phase Transition," *Phys. Rev.* **122,** 1477.

*G. Baker (1962). "Certain General Order-Disorder Models in the Limit of Long-Range Interactions," *Phys. Rev.* **126,** 2071.

R. Baxter (1964). "Many-Body Functions of a One-Dimensional Gas, I," *Phys. Fluids* **7,** 38;

R. Baxter (1965). "One-Dimensional Gases with Hard Core Repulsion," *Phys. Fluids* **8,** 687.

*H. A. Gersch (1963). "One-Dimensional Phase Transition in the Spherical Model of a Gas," *Phys. Fluids* **6,** 599.

E. Helfand (1964). "Approach to a Phase Transition in a One-Dimensional System," *J. Math. Phys.* **5,** 127.

P. Hemmer (1964). "On the van der Waals Theory of the Vapor-Liquid Equilibrium, IV, The Pair Correlation Function and Equation of State for Long-Range Forces," *J. Math. Phys.* **5,** 75.

P. Hemmer and E. Hiis Hauge (1964). "Yang-Lee Distribution of Zeros for a Van der Waals Gas," *Phys. Rev.* **133,** A1010.

P. Hemmer, E. Hiis Hauge and J. Aasen (1966). "Distribution of Zeros of the Grand Partition Function," *J. Math. Phys.* **7,** 35.

E. Hiis Hauge and P. Hemmer (1963). "On the Yang-Lee Distribution of Roots," *Physica* **29,** 1338.

M. Kac (1959). "On the Partition Function of a One-Dimensional Gas," *Phys. Fluids* **2,** 8.

*M. Kac, G. Uhlenbeck, and P. Hemmer (1963). "On the van der Waals Theory of the Vapor-Liquid Equilibrium, I, Discussion of a One-Dimensional Model," *J. Math. Phys.* **4,** 216; (1964). II, "Discussion of the Distribution Functions," *ibid.* **4,** 229. III, Discussion of the Critical Region," *ibid.* **5,** 60.

M. Kac and E. Helfand (1963). "Study of Several Lattice Systems with Long Range Forces," *J. Math. Phys.* **4,** 1078.

D. Newman (1964). "Equation of State for a Gas with Weak Long Range Positive Potential," *J. Math. Phys.* **5,** 1153.

III. The Coulomb Gas and the Statistical Theory of Energy Levels

*R. Baxter (1963). "Statistical Mechanics of a One-Dimensional Coulomb System with a Uniform Charge Background," *Proc. Cambridge Phil. Soc.* **59,** 779.

F. Dyson (1962). "Statistical Theory of the Energy Levels of Complex Systems, I," *J. Math. Phys.* **3,** 140; III, *ibid.* 166; "A Brownian-Motion Model for the Eigenvalues of a Random Matrix," *ibid.* 1191.

S. Edwards and A. Lenard (1962). "Exact Statistical Mechanics of One-Dimensional System with Coulomb Forces, II, the Method of Functional Integration," *J. Math. Phys.* **3,** 778.

*O. Eldridge and M. Feix (1962). "One-Dimensional Plasma Model at Thermodynamic Equilibrium," *Phys. Fluids* **5,** 1076.

*A. Lenard (1961). "Exact Statistical Mechanics of a One-Dimensional System with Coulomb Forces," *J. Math. Phys.* **2,** 682; III, "Statistics of the Electric Field" (1963) *ibid.* **4,** 533.

V. Magalinskii (1965). "Thermodynamics of a One-Dimensional Model System with Coulomb Interactions, *Soviet Phys.—JEPT* **21,** 114 [J. Exptl. Theoret. Phys. (U.S.S.R.) **48,** 167 (1965)].

M. Mehta (1960). "On the Statistical Properties of Level-Spacings in Nuclear Spectra," *Nucl. Phys.* **18,** 395.

M. Mehta and M. Gaudin (1960). "On the Density of Eigenvalues of a Random Matrix," *Nucl. Phys.* **18,** 420.

M. Mehta and F. Dyson (1963). "Statistical Theory of the Energy Levels of Complex Systems, V," *J. Math. Phys.* **4,** 713.

C. Porter, ed. (1965). "Statistical Theory of Spectra, Fluctuations" Academic Press, New York, *Volume I of this Series.*

S. Prager (1962). "The One-Dimensional Plasma," Advances in Chemical Physics IV, Wiley (Interscience), New York, page 201.

A. Salzberg and S. Prager (1963). "Equation of State for a Two-Dimensional Electrolyte," *J. Chem. Phys.* **38,** 2587.

C. Ufford and E. Wigner (1942). "On the Calculation of the Distribution Function," *Phys. Rev.* **61,** 524.

K. Wilson (1962). "Proof of a Conjecture by Dyson," *J. Math. Phys.* **3,** 1040.

A Simple Method for Treating the Statistical Mechanics of One-Dimensional Substances[*,†]

HIDETOSHI TAKAHASHI

Physics Institute, Tokyo Imperial University

(Received 15 November 1941)

Many authors[1,2] have considered the statistical mechanics of one-dimensional substances (linear chains). However, their papers are concerned with special forms of the interaction potential between nearest neighbors, and are either approximate[1] or too complicated. I show here an extremely simple method for treating exactly the statistical mechanics of one-dimensional substances for any form of the interaction. We consider only classical mechanics, and suppose that the interaction occurs only among nearest neighbors.

The configurational integral for the partition function of a one-dimensional substance is given by

$$Z = \int\int_{x_0 < x_1 < \ldots < x_{n-1} < x_n} \cdots\cdots \int \exp \{ - [U(x_1 - x_0) + U(x_2 - x_1) + \ldots + U(x_n - x_{n-1})]/kT\} \, dx_1 \, dx_2 \ldots dx_{n-1},$$

where $x_0, x_1, \ldots x_n$ are the coordinates of the atoms and $U(r)$ represents the interaction potential.

* From Proceedings of the Physico-Mathematical Society of Japan (Nippon Suugaku-Buturigakkwai Kizi Tokyo) **24,** 60 (1942).

† Translated from the German by Ingeborg Lieb.

[1] A. Harasima, *Proc. Phys.-Math. Soc. Japan* **21,** 156 (1939).

[2] T. Nagamiya, *Proc. Phys.-Math. Soc. Japan* **22,** 705 (1940); *ibid.*, 1034 (1940); Y. Muto, *ibid.* **24** (1942).

Instead of using this integral directly, we multiply it by a factor and integrate over x_n, namely

$$\Phi(p) = \int\limits_{x_0 < x_n} Z \exp\left[-p(x_n - x_0)/kT\right] dx_n$$
$$= \iint\limits_{x_0 < x_1 < \ldots < x_n} \cdots\cdots \int \exp\{-[U(x_1 - x_0) + U(x_2 - x_1) + \ldots$$
$$+ U(x_n - x_{n-1}) + p(x_n - x_0)]/kT\}\, dx_1 \ldots dx_n.$$

Setting

$$x_1 - x_0 = r_1, \qquad x_2 - x_1 = r_2, \ldots, \qquad x_n - x_{n-1} = r_n$$

and observing that

$$\partial(x_1, x_2, \ldots, x_n)/\partial(r_1, r_2, \ldots, r_n) = 1,$$

one can simplify the integral in the following way:

$$\Phi(p) = \iint\limits_{r_1, r_2 \ldots, r_n > 0} \cdots\cdots \int \exp\{-[U(r_1) + U(r_2) + \ldots + U(r_n)$$
$$+ p(r_1 + r_2 + \ldots + r_n)]/kT\}\, dr_1\, dr_2 \ldots dr_n$$
$$= \iint\limits_{r_1, r_2 \ldots, r_n > 0} \cdots\cdots \int \exp\{-[U(r_1) + pr_1]/kT\} \exp\{-[U(r_2) + pr_2]/kT\} \ldots$$
$$\exp\{-[U(r_n) + pr_n]/kT\}\, dr_1\, dr_2 \ldots dr_n$$
$$= \left(\int_0^\infty \exp\{-[U(r) + pr]/kT\}\, dr\right)^n.$$

On the other hand

$$\Phi(p) = \int_0^\infty Z \exp\left[-pl/kT\right] dx_n = \int_0^\infty \exp\left[-(F^* + pl)/kT\right] dx_n$$

with

$$l = x_n - x_0,$$

and where F^* represents the configurational part of the free energy. If n now becomes very large, one can replace the integral by the maximum value of the integrand, as is done in statistical mechanics. Then,

$$\Phi(p) \approx \exp\left[-(F^* + pl)/kT\right],$$

whence $\partial/\partial x_n(F^* + pl) = 0$, or $\partial F^*/\partial x_n = -p$.

The last equation is the condition that the pressure of the substance be p. If this condition is fulfilled, then

$$\Phi = e^{-F^*/kT}$$

or

$$\Psi^* = -kT \log \Phi = -nkT \log \int_0^\infty \exp\{-[U(r) + pr]/kT\}\, dr,$$

where Ψ^* is the configurational part of the thermodynamic potential. The entire thermodynamic potential Ψ can be obtained by adding the kinetic part namely from

$$\Psi = \Psi^* - nkT \log [(2\pi mkT)^{\frac{1}{2}}/h].$$

The relation between p and l is obtained directly from

$$l = \frac{\partial\Psi}{\partial p} = \frac{\partial\Psi^*}{\partial p} = \frac{n\int_0^\infty r \exp \{-[U(r) + pr]/kT\}\, dr}{\int_0^\infty \exp \{-[U(r) + pr]/kT\}\, dr}.$$

All other thermodynamic quantities can be derived from Ψ in the known way.

The above expression for l is exactly n times the average atomic distance for a system of only two atoms which attract or repel each other with a potential $U(r)$, and which are also subject to a constant, attractive external force equal to p. *Therefore, the special features of bound systems do not occur in one-dimensional substances.* Examination of one-dimensional substances cannot reveal anything about real three-dimensional substances.

From the above it is clear that Ψ, and also l, must be single-valued functions of p. This means that *the coexistence of two phases is impossible in one-dimensional substances for any choice of the potential.* The same can be concluded from purely qualitative considerations.[3] The result of Harasima,[1] that the coexistence of two phases can occur in the same model, can therefore not be correct. Obviously, the use of bad approximations was responsible for that result.

[3] T. Nagamiya, *ibid.*, p. 1045–1046.

Physica XVI, no 2 Februari 1950

SUR L'INTÉGRALE DE CONFIGURATION POUR LES SYSTÈMES DE PARTICULES À UNE DIMENSION

par L. VAN HOVE

Université de Bruxelles*); actuellement à l'Institute for Advanced Study, Princeton, New Jersey, U.S.A.

Synopsis

The free energy of a one-dimensional system of particles is calculated for the case of non-vanishing incompressibility radius of the particles and a finite range of the forces. It is shown quite generally that no phase transition phenomena can occur under these circumstances. The method used is the reduction of the problem to an eigenvalue problem.

§ 1. *Introduction.* L'intégrale des phases d'un système à une dimension a été étudiée par plusieurs auteurs dans le cas où l'interaction entre les particules s'exerce seulement entre voisins immédiats et a une forme suffisamment simple pour permettre une intégration directe [1]). Aucun phénomène de changement de phase tel que condensation n'a été mis en évidence par ces calculs. Nous avons traité par deux méthodes différentes le cas d'une interaction de forme arbitraire s'exerçant entre voisins immédiats seulement **). A notre insu l'une de ces méthodes avait été utilisée précédemment par T a k a h a s i pour le même problème [2]). De nouveau l'apparition de phénomènes de condensation a été trouvée impossible.

Dans ce qui suit, nous traitons le cas plus général où l'interaction n'est pas limitée aux voisins immédiats. Nous supposons que les particules ont un rayon d'incompressibilité non nul et que le potentiel d'interaction entre deux particules a une portée finie; à part cela nous le supposons simplement continu. Dans ces conditions, le problème peut être remené à la recherche d'une valeur propre d'une équation intégrale. Ceci est un exemple de plus de pro-

*) Département de Physique Chimique.

**) Pour l'aspect mathématique d'une méthode, bien adaptée également au calcul des fonctions de distribution, voir une conférence de l'auteur dans le Bulletin de la Société Mathématique de Belgique (1949), sous presse. L'autre méthode est développée dans le présent travail.

blèmes de mécanique statistique se réduisant à la détermination d'une valeur propre d'une matrice [3]). Dans notre cas cependant, il s'agit d'une équation intégrale; son noyau n'est en général pas symétrique et l'intervalle d'intégration est infini. Par un changement de variables, nous rendons l'intervalle d'intégration fini, et pouvons alors tirer parti de la théorie de Fredholm. D'habitude [4]), l'apparition de changements de phase est liée à la multiplicité de la valeur propre obtenue, et est souvent exclue par suite d'un théorème connu de Frobenius [5]). La situation est ici un peu différente; cependant une extension de ce théorème aux équations intégrales [6]) montre que tout phénomène de changement de phase est encore exclu dans le cas que nous traitons. Il faut noter toutefois que l'existence simultanée d'un rayon d'incompressibilité non nul et d'une portée finie des forces est essentielle pour la validité du raisonnement et de la conclusion.

§ 2. *L'intégrale de configuration.* Soit un système de N particules identiques situées sur un segment rectiligne de longueur L. Leurs positions sont définies par leurs abscisses x_i ($i = 1, \ldots, N$; $0 \leqslant x_i \leqslant L$). L'énergie potentielle du système a l'expression suivante

$$V = \Sigma_{i<j} U(|x_i - x_j|), \tag{2.1}$$

où $U(\xi)$ est une fonction vérifiant

$$\begin{aligned} U(\xi) &= +\infty \quad \text{pour } 0 \leqslant \xi < d_0, \\ U(\xi) &= 0 \qquad \text{pour } \xi \geqslant d_1,\ 0 < d_0 < d_1. \end{aligned} \tag{2.2}$$

Dans l'intervalle $d_0 \leqslant \xi \leqslant d_1$, nous supposons $U(\xi)$ continue et bornée inférieurement. De plus, on ne peut avoir $U(\xi) = +\infty$ que pour $\xi < 2d_0$. La longueur d_0 apparaît comme le diamètre d'incompressibilité des particules; d_1 est la portée maxima des forces d'interaction.

Désignons par m la masse des particules supposées ponctuelles, et par T la température absolue multipliée par la constante de Boltzmann. L'énergie libre de Helmholtz F, par particule, est alors donnée par

$$F/T = -f(N, L) - \log (d_0 h^{-1} \sqrt{2\pi m T}) \tag{2.3}$$

avec

$$\begin{aligned} \exp[N f(N,L)] = Q(N, L) &= (1/N!\, d_0^N) \int_0^L dx_1 \ldots \int_0^L dx_N . \exp(-V/T) \\ &= d_0^{-N} \int_0^L dx_1 \int_{x_1}^L dx_2 \ldots \int_{x_{k-1}}^L dx_k \ldots \int_{x_{N-1}}^L dx_N \exp(-V/T)\ ^*) \end{aligned}$$

*) Le facteur d_0 est introduit uniquement pour que la quantité $Q(N, L)$ soit sans dimension.

Comme on le sait, pour toute suite de longueurs L_N vérifiant $\mathrm{Lim}_{N\to\infty}\ L_N/N = l > d_0$, l'expression $f(N, L_N)$ converge vers une limite finie $f(l)$, fonction croissante de l à dérivée non croissante [7]). Nous allons ramener la détermination de $f(l)$, donc de l'énergie libre F, à la résolution d'une équation intégrale. Remarquons que par suite de (2.2), deux particules voisines ne peuvent jamais se rapprocher à moins de d_0 l'une de l'autre; par conséquent, les particules capables d'interaction avec une particule donnée sont limitées à ses ν voisins de droite et à ses ν voisins de gauche, ν étant le plus grand nombre entier pour lequel $\nu\, d_0 < d_1$. Cette limitation, qui résulte du fait que $d_0 > 0$ et $d_1 < +\infty$, est essentielle pour les développements ci-après.

§ 3. *L'énergie libre à pression constante.* Comme l'a remarqué Takahasi dans le cas où $\nu = 1$ [2]), pour un système à une dimension l'énergie libre à pression constante $G = F + pl$ est plus facile à déterminer que F. Elle est étroitement liée à la transformée de Laplace de $Q(N, L)$:

$$I_N(a) = \textstyle\int_0^\infty Q(N, L)\, e^{-aL}\, \mathrm{d}L. \tag{3.1}$$

En effet, du comportement de $Q(N, L)$ pour $N \to \infty$, il résulte que $\{I_N(a)\}^{1/N}$ converge vers une limite finie

$$\lim_{N\to\infty} \{I_N(a)\}^{1/N} = e^{-g(a)} \tag{3.2}$$

donnée par

$$e^{-g(a)} = \lim_{N\to\infty} \{Q(N, L_N^{(a)})\, \exp\,(-\, a\, L_N^{(a)})\}^{1/N}, \tag{3.3}$$

où $L_N^{(a)}$ est la valeur de L qui réalise le maximum de

$$Q(N, L)\, e^{-aL} = \exp\,[N\, f(N, L) - aL].$$

Comme $f(N, L_N) \to f(l)$ quand $N \to \infty$ et $N^{-1}L_N \to l$, la longueur $N^{-1}L_N^{(a)}$ doit converger vers la valeur de l qui réalise le maximum de $f(l) - al$ et qui est donc définie par

$$df(l)/dl = a. \tag{3.4}$$

L'équation (3.3) donne alors

$$g(a) = al - f(l). \tag{3.5}$$

Tenant compte de (3.4) et (2.3), on voit que la pression p et l'énergie libre à pression constante G sont liées à a et $g(\beta)$ par

$$p/T = a,\quad G/T = g(a) - \log\,((d_0/h)\ \sqrt{2\pi mT}).$$

Au paragraphe usivant nous déterminerons $g(\alpha)$. On en déduira $f(l)$ en inversant la transformation de Legendre contenue dans (3.4) et (3.5):

$$f(l) = \alpha l - g(\alpha), \; l = \mathrm{d}g(\alpha)/\mathrm{d}\alpha. \tag{3.6}$$

Dans le cas particulier où $\nu = 1$, l'énergie d'interaction (2.1) est $V = \Sigma_{i=1}^{N-1} U(\xi_i)$ où les ξ_i désignent les distances entre particules voisines. La transformation de Laplace (3.1) se fait sans peine et (3.2) donne

$$g(\alpha) = -\log\left\{\frac{1}{d_0}\int_0^\infty \exp\left[-\frac{U(\xi)}{T} - \alpha\xi\right] d\xi\right\},$$

d'où

$$f(l) = \alpha l + \log\left\{\frac{1}{d_0}\int_0^\infty \exp\left[-\frac{U(\xi)}{T} - \alpha\xi\right] d\xi\right\},$$

$$l = \frac{\int_0^\infty \xi \exp\left[-U(\xi)/T - \alpha\xi\right] d\xi}{\int_0^\infty \exp\left[-U(\xi)/T - \alpha\xi\right] d\xi}.$$

Ce sont les formules données par Takahasi [2]); nous les avons obtenues également par une autre méthode *). Comme elles expriment l univoquement en fonction de α, il n'y a pas de condensation possible.

§ 4. *Le problème de valeur propre.* Déterminons maintenant $g(\alpha)$ dans le cas général où $\varrho = \nu - 1 > 0$. Dans ce cas, l'énergie d'interaction est

$$V = \Sigma_{i=1}^{N-1} U(\xi_i) + \Sigma_{i=2}^{N-1} U(\xi_{i-1}+\xi_i) + \ldots \Sigma_{i=\nu}^{N-1} U(\xi_{i-\varrho} + \ldots + \xi_{i-1} + \xi_i),$$

les ξ_i désignant de nouveau les distances entre particules voisines. Définissons les fonctions

$$h(\xi_1,\ldots\xi_\varrho) = \exp\left\{-\frac{1}{2T}\Sigma_{i=2}^{\varrho} U(\xi_{i-1}+\xi_i) - \frac{1}{2T}\Sigma_{i=3}^{\varrho} U(\xi_{i-2}+\xi_{i-1}+\xi_i)\ldots \right.$$
$$\left. \ldots - \frac{1}{2T} U(\xi_1 + \ldots \xi_\varrho)\right\},$$

$$k(\xi_1, \ldots \xi_\varrho, \xi_1' \ldots \xi_\varrho') = \exp\left\{-\frac{1}{T}\left[U(\xi_\varrho + \xi_1') + U(\xi_{\varrho-1} + \xi_\varrho + \xi_1') + \right.\right.$$
$$+ U(\xi_\varrho + \xi_1' + \xi_2') + \ldots + U(\xi_1 + \ldots + \xi_\varrho + \xi_1') +$$
$$\left.\left. + U(\xi_2 + \ldots + \xi_\varrho + \xi_1' + \xi_2') + \ldots + U(\xi_\varrho + \xi_1' + \ldots + \xi_\varrho')\right]\right\},$$

$$K(\xi_1, \ldots \xi_\varrho, \xi_1', \ldots \xi_\varrho') = h(\xi_1, \ldots \xi_\varrho)\, k(\xi_1, \ldots \xi_\varrho, \xi_1', \ldots \xi_\varrho')\, h(\xi_1', \ldots \xi_\varrho'),$$

*) Voir la note p. 1.

$$K^{(\alpha)}(\xi_1, \ldots \xi_\varrho, \xi'_1, \ldots \xi'_\varrho) = \exp\left\{-\tfrac{1}{2}\Sigma_{i=1}^{\varrho}\left[\frac{U(\xi_i)}{T} + \alpha\xi_i\right]\right\} \times$$

$$\times\, K(\xi_1, \ldots \xi_\varrho, \xi'_1, \ldots \xi'_\varrho) \times \exp\left\{-\tfrac{1}{2}\Sigma_{i=1}^{\varrho}\left[\frac{U(\xi'_i)}{T} + \alpha\xi'_i\right]\right\}. \quad (4.1)$$

Le noyau itéré $K_n^{(\alpha)}(\xi_1, \ldots \xi_\varrho, \xi'_1, \ldots \xi'_\varrho)$ est défini par la relation de récurrence

$$K_{n+1}^{(\alpha)}(\xi_i, \xi'_1) = \int_0^\infty d\xi''_1 \ldots \int_0^\infty d\xi''_\varrho . K_n^{(\alpha)}(\xi_i, \xi''_i)\, K^{(\alpha)}(\xi''_i, \xi'_i)\,;\; K_1^{(\alpha)} = K^{(\alpha)}.$$

Il permet d'exprimer la fonction $I_N(\alpha)$ donnée en (3.1) à l'aide d'une intégrale dont la multiplicité ne dépend plus de N:

$$I_N(\alpha) = \int_0^\infty d\xi \int_0^\infty d\xi_1 \ldots \int_0^\infty d\xi_\varrho \int_0^\infty d\xi' \int_0^\infty d\xi'_1 \ldots \int_0^\infty d\xi'_\varrho . K_n^{(\alpha)}(\xi_i, \xi'_i)\, h(\xi_i)\, h(\xi'_i).$$

$$.\exp\{-2T^{-1}\Sigma_{i=1}^{\varrho}[U(\xi_i) + U(\xi'_i)]\}.\exp\{-\alpha[\xi + \xi' + \tfrac{1}{2}\Sigma_{i=1}^{\varrho}(\xi_i + \xi'_i)]\},$$

si $N = \nu + n\varrho$. De cette relation et de (3.2), il résulte que

$$\lim_{n\to\infty}\{K_n^{(\alpha)}(\xi_i, \xi'_i)\}^{1/n} = \lim_{n\to\infty}\{I_N(\alpha)\}^{1/n} = e^{-\varrho g(\alpha)}. \quad (4.2)$$

Pour peu que $U(\xi_i) \neq +\infty$, $U(\xi'_i) \neq +\infty$, le premier membre de cette égalité ne dépend plus des variables ξ_i, ξ'_i; on peut s'en assurer à l'aide des considérations qui établissent l'existence de la limite de $\{Q(N, L_N)\}^{1/N}$ quand $N \to \infty$ et $N^{-1}L_N \to l$ [7]).

Comme il arrive fréquemment dans ce genre de problèmes [8]), nous sommes amenés à considérer l'équation intégrale

$$\varphi(\xi_i) = \lambda \int_0^\infty d\xi'_1 \ldots \int_0^\infty d\xi'_\varrho\, K^{(\alpha)}(\xi_i, \xi'_i)\, \varphi(\xi'_i), \quad (4.3)$$

dont $K_n^{(\alpha)}$ est le noyau itéré d'ordre n. On peut y limiter les intégrations à l'intervalle (d_0, ∞). Nous allons la transformer en une équation intégrale à domaine d'intégration borné et à noyau constamment positif. Nous utilisons le changement de variables

$$\eta = q^{(\alpha)}(\xi) = \int_\xi^\infty \exp\{-[T^{-1}U(\xi') + \alpha\xi']\}\, d\xi',$$

dont nous désignons l'inverse par $\xi = r^{(\alpha)}(\eta)$, et le changement de fonction inconnue

$$\varphi(\eta_i) = \exp\{\tfrac{1}{2}\Sigma_1^{\varrho}[T^{-1}U(\xi_i) + \alpha\xi_i]\}\, \varphi(\xi_i), \text{ avec } \eta_i = q^{(\alpha)}(\xi_i).$$

D'après (4.1), l'équation (4.3) devient

$$\varphi(\eta_i) = \lambda \int_0^{\delta(\alpha)} d\eta'_1 \ldots \int_0^{\delta(\alpha)} d\eta'_\varrho\, H^{(\alpha)}(\eta_i, \eta'_i)\, \varphi(\eta'_i), \quad (4.4)$$

avec

$$\delta^{(\alpha)} = \int_{d_0}^\infty \exp\{-[T^{-1}U(\xi') + \alpha\xi']\}\, d\xi',$$

$$H^{(a)}(\eta_i, \eta_i') = K\,[r^{(a)}(\eta_i), r^{(a)}(\eta_1')].\ ^{*})$$

Les noyaux itérée $H_n^{(a)}(\eta_i, \eta_i')$ de (4.4) ne diffèrent des $K_n^{(a)}(\xi_i, \xi_i')$ que par un facteur indépendant de n. La relation (4.2) devient donc

$$e^{-\varrho\, g(a)} = \lim_{n\to\infty} \{H_n^{(a)}(\eta_i, \eta_i')\}^{1/n},$$

où la limite du second membre est indépendante des η_i, η_i'. D'après le théorème de Cauchy-Hadamard, la quantité $e^{\varrho g(a)}$ est donc le rayon de convergence de la série résolvante

$$H^{(a)} + \lambda H_2^{(a)} + \ldots \lambda^{n-1} H_n^{(a)} + \ldots.$$

D'après la théorie de Fredholm, qui s'applique à l'équation (4.4), ce rayon et $|\lambda^{(a)}|$, où $\lambda^{(a)}$ est une valeur propre de valeur absolue minimum de l'équation [8]). Donc $g(a) = \varrho^{-1} \log |\lambda^{(a)}|$.
Remarquons que les noyaux des équations (4.2) et (4.4) ne sont pas symétriques si $\varrho > 1$; on a

$$K^{(a)}(\xi_1, \ldots, \xi_\varrho, \xi_1', \ldots \xi_\varrho') = K^{(a)}(\xi_\varrho', \ldots, \xi_1', \xi_\varrho, \ldots, \xi_1),$$

et une relation pour $H^{(a)}$.

§ 5. *L'impossibilité de changements d'état.* Un changement d'état se traduit par l'existence d'une pression constante sur un intervalle non nul de valeurs de l. Sur un tel intervalle, a doit rester constant quand l varie; d'après (3.6), la dérivée $dg(a)/da$ doit donc présenter une discontinuité. Nous allons voir que cette circonstance est impossible.

D'après la théorie de Fredholm, les valeurs propres de l'équation (4.4) sont les zéros d'une fonction entière $D^{(a)}(\lambda)$. On s'assure facilement que pour Re $(a) > 0$, la fonction $D^{(a)}(\lambda)$ est simultanément holomorphe en a et λ. Comme le noyau de l'équation est positif, il résulte d'un théorème de Jentzsch [6]) que $D^{(a)}(\lambda)$ a une racine de valeur absolue minimum, à la fois unique, simple et réelle. C'est elle qui a été désignée par $\lambda^{(a)}$, et $|\lambda^{(a)}| = \lambda^{(a)} > 0$. Etant donné le caractère analytique de $D^{(a)}(\lambda)$, la racine $\lambda^{(a)}$ est fonction holomorphe de a quand Re $(a) > 0$. Par conséquent la fonction $g(a) = \varrho^{-1} \log \lambda^{(a)}$ est holomorphe. Aucun phénomène de condensation ne peut se produire pour le type de système étudié.

*) Quand $U(\xi)$ est borné pour $d_0 \leqslant \xi \leqslant d_1$, on peut utiliser le changement de variables $\eta = 1/a \exp(-a\xi)$ et le changement de fonction inconnue $\psi(\eta_i) = \exp(a/2\, \Sigma^p\, \xi_i)\, \varphi(\xi_i)$; on obtient aussi un noyau positif. Par contre, quand par exemple $\lim U(\xi) = +\infty$ pour $d_0 < \xi \to d_1$, il faut recourir au procédé utilisé dans le texte.

Reçu 19-12-49.

REFERENCES

1) K. F. Herzfeld and M. Goeppert-Mayer, Journ. chem. Phys. **2**, 38, 1934. L. Tonks, Phys. Rev. **50**, 955, 1936; T. Nagamiya, Proc. phys.-math. Soc. Japan **22**, 705, 1035, 1940.

2) H. Takahasi, Proc. phys.-math. Soc. Japan **24**, 60, 1942.

3) H. A. Kramers and G. H. Wannier, Phys. Rev. **60**, 252, 1941; E. W. Montroll, Journ. chem. Phys. **9**, 706, 1941; E. N. Lassettre and J. P. Howe, Journ. chem. Phys. **9**, 747, 1941; G. S. Rushbrooke and H. D. Ursell, Proc. Cambr. phil. Soc. **44**, 263, 1947.

4) E. W. Montroll, loc. cit. [3]).

5) G. Frobenius, Sitzungsber. Preuss. Akad. Wiss. Berlin, 1908, 471, 1909, 514.

6) R. Jentzsch, Crelles Journ. **141**, 235, 1912.
E. Hopf, Sitzungsber. Preuss. Akad. Wiss. Berlin, 1928, 233

7) L. Van Hove, Physica, sous presse.

8) E. Goursat, Cours d'Analyse Mathématique, vol. III, Paris, 1927, chap. XXXI.

Molecular Distribution Functions in a One-Dimensional Fluid*

ZEVI W. SALSBURG, ROBERT W. ZWANZIG, AND JOHN G. KIRKWOOD
Sterling Chemistry Laboratory, Yale University,† New Haven, Connecticut
(Received February 4, 1953)

Exact expressions are derived for the molecular distribution functions in a one-dimensional fluid whose particles interact with a nearest neighbor pair potential. The pair distribution function for rigid spheres is found to be identical with Frenkel's result. The one-dimensional form of a new set of integral equations for the molecular distribution functions is examined. The superposition principle is found to be exact in a one-dimensional fluid with nearest neighbor interactions.

WE have recently derived a new set of integral equations for the molecular distribution functions.[1] In order to understand the nature of the theoretical and mathematical problems presented by these equations, we felt that it would be profitable to study them for the case of a one-dimensional system. With certain assumptions about the intermolecular potential, an exact solution can be obtained. In this article, we derive exact expressions for the distribution functions in a one-dimensional system whose particles interact only with nearest neighbors and whose potential can be expressed as the sum of pair potentials. We then use these distribution functions to calculate the thermodynamic properties of the fluid. Finally, we discuss the integral equations for a one-dimensional system and use the known distribution function to test these equations for a system of rigid spheres having no attractive forces between them.

As an alternative to the well-known method of calculating the partition function, the thermodynamic functions of a macroscopic system can be obtained from a knowledge of the distribution functions for sets of n molecules and the forces between individual particles. When the potential of intermolecular force is approximated by a sum of pair potentials,

$$V_N=\sum_{i<j}\sum V(R_{ij}),$$

then the pair distribution function plays a central role. It has not yet been possible to calculate the pair distribution function for three-dimensional systems without using an approximation, the validity of which is hard to assess in an *a priori* way. However, in the one-dimensional system with nearest neighbor interactions, the pair and higher-order distribution functions may be evaluated exactly.

The classical statistical mechanical theory of a one-dimensional system has been developed in recent years by several authors.[2-8]

In these investigations, the partition function has been integrated exactly for several special forms of the intermolecular potential which is restricted to nearest neighbor interactions. However, a satisfactory treatment of the distribution function approach is lacking.

I. DERIVATION OF THE DISTRIBUTION FUNCTIONS

We will consider a system of N identical particles, located at the points $R_1, \cdots, R_N$ on a straight line of length L. The probability distribution function in configuration space for such a system in a canonical ensemble is

$$P^{(N)}(R_1, \cdots, R_N)=\exp[\beta(\bar{A}_N-V_N-V_W)],$$
$$\beta=1/kT,$$
$$\exp[-\beta\bar{A}_N]=\int_0^L\cdots\int_0^L \times\exp[-\beta(V_N+V_W)]\prod_{i=1}^N dR_i. \quad (1)$$

T is the absolute temperature and k is the Boltzmann constant. V_N is the potential of intermolecular force, and V_W is the potential of the wall forces, which act on the N particles inside the length L. $\bar{A}_N$ is the configurational contribution to the free energy of the system.

If $\varphi(R_1, \cdots, R_N)$ is a function of the coordinates of the particles, then the expectation value of φ is given by

$$\langle\varphi\rangle_{Av}=\exp[\beta\bar{A}_N]\int_0^L\cdots\int_0^L \varphi(R_1, \cdots, R_N) \times\exp[-\beta(V_N+V_W)]\prod_{i=1}^N dR_i. \quad (2)$$

The most general conveniently applicable method for evaluating multiple integrals of this type is that attributable to Gürsey.[7] This is restricted to potentials that may be expressed as sums of pair potentials be-

* This work was carried out with support from the U. S. Office of Naval Research under Contract Nonr-410(00) with Yale University.
† Contribution No. 1173.

[1] J. G. Kirkwood and Z. W. Salsburg, Trans. Faraday Soc. (to be published).
[2] K. F. Herzfeld and Maria Goeppert-Mayer, J. Chem. Phys. 2, 38 (1934).
[3] L. Tonks, Phys. Rev. 50, 955 (1936).
[4] T. Nagamiya, Proc. Phys.-Math. Soc. Japan 22, 705 (1940); 1034 (1940).
[5] J. Frenkel, *Kinetic Theory of Liquids* (Oxford University Press, New York, 1940), p. 126.
[6] G. Rushbrooke and H. Ursel, Proc. Cambridge Phil. Soc. 44, 263 (1948).
[7] F. Gürsey, Proc. Cambridge Phil. Soc. 46, 182 (1950).
[8] R. Kikuchi, J. Chem. Phys. 19, 1230 (1951).

tween nearest neighbors. If we choose the convention

$$0<R_1<R_2<\cdots<R_N<L,$$

then the potential is

$$V_N=\sum\sum_{i<j} V(R_{ij})=\sum_{j=1}^{N-1} V(R_{j+1}-R_j). \qquad (3)$$

The wall effects are easily handled if one assumes that the walls behave like particles identical with those of the system, fixed at $R=0$ and $R=L$, so that the total potential energy is

$$V_N+V_W=V(R_1)+\sum_{j=1}^{N-1} V(R_{j+1}-R_j)+V(L-R_N). \quad (4)$$

We further assume that the potential $V(R)$ has these properties:

(a) $\lim_{R\to 0} V(R)=\infty$,

(b) $\lim_{R\to\infty} V(R)=0$,

strongly enough that any integrals we may encounter (e.g., Eq. 29) will converge. Condition (a), which is satisfied by most of the commonly used potential functions, insures the maintenance of a given order of particles. For example, particles three and four cannot exchange places under this type of force. Therefore, we are permitted to label the particles so that $R_1<R_2<$, $\cdots$, $<R_N$ and be sure that all available configurations of the system will retain this order.

The method of Gürsey may be used only if we assume that φ is symmetric in the variables $R_1\cdots R_N$. Then

$$\langle\varphi\rangle_{Av}=N!\exp[+\beta\bar{A}_N]$$
$$\times\int_0^L \exp[-\beta V(L-R_N)]dR_N$$
$$\times\int_0^{R_N} \exp[-\beta V(R_N-R_{N-1})]dR_{N-1}$$
$$\times\cdots\int_0^{R_2} \exp[-\beta V(R_1)]\varphi(R_1,\cdots,R_N)dR_1. \quad (5)$$

This identity can be proven by mathematical induction. In a physical sense, it is equivalent to the process of ordering the particles on the line, and prohibiting the exchange of particles. The factor $N!$ corrects for those configurations which are permutations of the order $R_1<R_2<$, $\cdots$, $<R_N$. In the case $\varphi\equiv 1$, we have

$$\exp[-\beta\bar{A}_N]=N!\int_0^L \exp[-\beta V(L-R_N)]dR_N$$
$$\times\int_0^{R_N} \exp[-\beta V_{N-1,N}]dR_{N-1}$$
$$\times\cdots\int_0^{R_2} \exp[-\beta V(R_1)]dR_1. \quad (6)$$

This integral is an $N+1$ fold convolution for the Laplace transform[9] of the function $\exp[-\beta V(R)]$, and one can therefore write

$$\int_0^\infty \exp[-sL-\beta\bar{A}_N]dL=N![\Omega(s)]^{N+1},$$
$$\Omega(s)=\int_0^\infty \exp[-sR-\beta V(R)]dR. \qquad (7)$$

The inversion formula for the Laplace transform[9] yields

$$\exp[-\beta\bar{A}_N]=\frac{N!}{2\pi i}\int_{c-i\infty}^{c+i\infty} e^{Ls}[\Omega(s)]^{N+1}ds, \qquad (8)$$

where c is greater than the real parts of all the singularities of $\Omega(s)$.

We shall now express the molecular distribution functions as expectation values of certain functions over the same ensemble. The probability that the kth molecule will be between R_k and R_k+dR_k is given by

$$P^{(1)}(R_k)dR_k=\underbrace{\int_0^L\cdots\int_0^L}_{(N-1)}$$
$$\times P^{(N)}dR_1,\cdots,dR_{k-1},dR_{k+1},\cdots dR_N\cdot dR_k. \quad (9)$$

By introducing Dirac's δ-function, we can express this probability in the form

$$P^{(1)}(r)=\langle\delta(R_k-r)\rangle_{Av}$$
$$=\int_0^L\cdots\int_0^L \delta(R_k-r)P^{(N)}\prod_{i=1}^N dR_i. \quad (10)$$

The average number density in singlet space (i.e., the probability per unit length that any molecule will be at the point r) is defined by

$$\rho^{(1)}(r)=\sum_{k=1}^N\langle\delta(R_k-r)\rangle_{Av}=\langle\sum_{k=1}^N\delta(R_k-r)\rangle_{Av}. \quad (11)$$

Similarly the average density of ordered pairs is given by

$$\rho^{(2)}(r,r')=\langle\sum_{k=1}^N\sum_{j=1\neq k}^N\delta(R_k-r)\delta(R_j-r')\rangle_{Av}, \quad (12)$$

$\rho^{(2)}$ is the probability per (unit length)^2 that one molecule will be at r and another will be at r'.

Let

$$\varphi=\sum_{k=1}^N\delta(R_k-r),$$

then from Eq. (11) and Eq. (5) we obtain

$$\rho^{(1)}(r)=N!\exp[\beta\bar{A}_N]\sum_{k=1}^N I_k^{(N)}(L,r), \qquad (13)$$

[9] E.g., see G. Doetsch, *Theorie und Anwendung der Laplace Transformation* (Dover Publications, New York, 1943), pp. 157–167.

where

$$I_k^{(N)}(L,r)=\int_0^L \exp[-\beta V(L-R_N)]dR_N \times\cdots\int_0^{R_{k+1}} \delta(R_k-r)\exp[-\beta V_{k,k+1}]dR_k, \times\cdots\int_0^{R_2}\exp[-\beta V(R_1)]dR_1.$$

Again we recognize the convolution form for the Laplace transform and obtain

$$\int_0^\infty e^{-sL}I_k^{(N)}(L,r)dL =[\Omega(s)]^{N-k+1}\int_0^\infty \delta(L-r)h_k(L)e^{-sL}dL, =[\Omega(s)]^{N-k+1}e^{-sr}h_k(r), \quad (14)$$

where

$$h_k(r)=\int_0^r \exp[-\beta V(r-R_{k-1})]dR_{k-1} \times\cdots\int_0^{R_2}\exp[-\beta V(R_1)]dR_1=\frac{\exp[-\beta\bar{A}_{k-1}(r)]}{(k-1)!} =\frac{1}{2\pi i}\int_{c-i\infty}^{c+i\infty} e^{rt}[\Omega(t)]^k dt. \quad (15)$$

The second line follows from the first by the use of Eqs. (6) and (8). After using the inversion formula for the Laplace transform on Eq. (14) and substituting the result together with Eq. (8) into Eq. (13), we obtain

$$\rho^{(1)}(r)=\frac{\sum_{k=1}^{N}\left\{\frac{1}{2\pi i}\int_{c-i\infty}^{c+i\infty} e^{(L-r)s}[\Omega(s)]^{N-k+1}ds\right\}\left\{\frac{1}{2\pi i}\int_{c-i\infty}^{c+i\infty} e^{rt}[\Omega(t)]^k dt\right\}}{\frac{1}{2\pi i}\int_{c-i\infty}^{c+i\infty} e^{Ls}[\Omega(s)]^{N+1}ds}. \quad (16)$$

This function represents the probability per unit length of finding a particle at a distance r away from the wall. Since the wall was replaced by a fixed particle, one may state that $\rho^{(1)}(r)$ is also the probability per unit length that a particle will be found at a distance r from a given particle fixed at r' on a line of length $L+r'$. In this manner, one might anticipate the formal result for $\rho^{(2)}$ in an infinite system.

To get the analogous expression for $\rho^{(2)}$ we let

$$\varphi=\sum_{k=1}^{N}\sum_{j=1\neq k}^{N}\delta(R_k-r)\delta(R_j-r').$$

Then

$$\rho^{(2)}(r,r')=N!\exp[\beta\bar{A}_N]\sum_{k=2}^{N}\sum_{j=1}^{k-1}\times I_{jk}^{(N)}(L,r,r'); \quad r>r'$$
$$=N!\exp[\beta\bar{A}_N]\sum_{k=1}^{N-1}\sum_{j=k+1}^{N}\times I_{kj}^{(N)}(L,r',r); \quad r<r', \quad (17)$$

where

$$I_{jk}^{(N)}(L,r,r')=\int_0^L \exp[-\beta V(L-R_N)]dR_N \times\cdots\int_0^{R_{k+1}}\delta(R_k-r)\exp[-\beta V_{k,k+1}]dR_k, \times\cdots\int_0^{R_{j+1}}\delta(R_j-r')\exp[-\beta V_{j,j+1}]dR_j \times\cdots\int_0^{R_2}\exp[-\beta V(R_1)]dR_1.$$

The Laplace transforms of $I_{jk}^{(N)}$ and $I_{kj}^{(N)}$ can be computed since these functions have the form of a convolution, and one obtains, for example,

$$\int_0^\infty e^{-sL}I_{jk}^{(N)}(L,r,r')dL =[\Omega(s)]^{N-k+1}e^{-rs}I_j^{(k-1)}(r,r'). \quad (18)$$

By using the inversion formula and the known result for $I_j^{(k-1)}(r,r')$ we get

$$I_{jk}^{(N)}(r,r')=\left\{\frac{1}{2\pi i}\int_{c-i\infty}^{c+i\infty} e^{(L-r)s}[\Omega(s)]^{N-k+1}ds\right\} \times\left\{\frac{1}{2\pi i}\int_{c-i\infty}^{c+i\infty} e^{(r-r')t}[\Omega(t)]^{k-j}dt\right\} \times\left\{\frac{1}{2\pi i}\int_{c-i\infty}^{c+i\infty} e^{r'u}[\Omega(u)]^{j}du\right\}$$

and

$$I_{kj}^{(N)}(r',r)=\left\{\frac{1}{2\pi i}\int_{-i\infty} e^{(L-r')s}[\Omega(s)]^{N-j+1}ds\right\} \times\left\{\frac{1}{2\pi i}\int_{c-i\infty}^{c+i\infty} e^{(r'-r)t}[\Omega(t)]^{j-k}dt\right\} \times\left\{\frac{1}{2\pi i}\int_{c-i\infty}^{c+i\infty} e^{ru}[\Omega(u)]^{k}du\right\}. \quad (19)$$

It is shown in Appendix A that asymptotic expressions for $\rho^{(1)}$ and $\rho^{(2)}$ for large N take the following form:

$$\rho^{(1)}(r)=\sum_{k=1}^{N}\frac{\exp[-rc]}{[\Omega(c)]^k}\frac{1}{2\pi i}\int_{c-i\infty}^{c+i\infty}e^{rt}[\Omega(t)]^k dt, \quad (20a)$$

and

$$\rho^{(2)}(r,r')=\sum_{k=2}^{N}\sum_{j=1}^{k-1}\frac{\exp[-rc]}{[\Omega(c)]^k}$$

$$\times\left\{\frac{1}{2\pi i}\int_{c-i\infty}^{c+i\infty}e^{r's}[\Omega(s)]^j ds\right\}$$

$$\times\left\{\frac{1}{2\pi i}\int_{c-i\infty}^{c+i\infty}e^{(r-r')t}[\Omega(t)]^{k-j}dt\right\}, \quad (20b)$$

when $r>r'$.

The constant c is determined by the condition that the function

$$\sigma=\Re\left\{\frac{Ls}{N+1}+\ln\Omega(s)\right\}$$

has a stationary value at the point c on the real axis.

The integrals in Eqs. (20a) and (20b) are most conveniently evaluated by the theory of residues. In order to insure the validity of converting the complex integral into a contour integral around the origin in the complex plane, we assume that

$$\begin{aligned} V(R)&=\infty; & R&<a\\ &=V(R); & a&\le R\le\lambda \\ &=0; & R&>\lambda. \end{aligned} \quad (21)$$

In the case of a hard core model with a finite range of the potential one can, for example, readily show[7] that

$$\frac{1}{2\pi i}\int_{c-i\infty}^{c+i\infty}eL^s[\Omega(s)]^{N+1}ds=\frac{1}{2\pi i}\oint eL^s[\Omega(s)]^{N+1}ds, \quad (22)$$

where $L>(N+1)a$.

Equations (20a) and (20b) include the effects of the wall, which persist for any finite values of r and r'. However, if r and r' are allowed to become infinite in such a manner that $r-r'$ remains fixed, the following expressions, calculated in appendix A, are obtained. The step function $A(R-na)$ occurs since we have explicitly introduced the hard core model, with a diameter (a).

$$\rho^{(1)}=1/la;\quad l=L/[(N+1)a] \quad (23a)$$

$$\rho^{(2)}(R)=\sum_{n=1}^{N-1}\frac{A(R-na)}{la}\frac{1}{2\pi i}$$

$$\times\int_{c-i\infty}^{c+i\infty}e^{R(s-c)}\left[\frac{\Omega(s)}{\Omega(c)}\right]^n ds, \quad (23b)$$

$$\rho^{(2)}(R)=\rho^{(2)}(-R),$$

where

$$R=r-r'$$

$$\begin{aligned} A(x)&=0 & x&<0\\ &=1 & x&\ge 0. \end{aligned}$$

The average density $\rho_N^{(n)}$ of ordered sets n in a canonical ensemble of systems of N particles is given by

$$\rho_N^{(n)}(r_1,\cdots,r_n)$$

$$=\langle\sum_{l_n=1}^{N}\sum_{\substack{l_{n-1}=1\\ \neq l_n}}^{N}\cdots\sum_{\substack{l_1=1\\ \neq l_j}}^{N}\prod_{\substack{i=1\\ (j=2,\cdots,n)}}^{n}\delta(R_{l_i}-r_i)\rangle_{Av}. \quad (24)$$

For a one-dimensional system in which the interactions are restricted to nearest neighbors, a fixed subset of n molecules effectively divides the system into $n+1$ independent subsystems. An extension of the analysis given in this section shows that

$$\rho^{(n)}=\sum_{l_n=n}^{N}\sum_{\substack{l_{n-1}=n-1\\ \neq l_n}}^{l_n-1}\cdots\sum_{\substack{l_1=1\\ \neq l_j}}^{l_2-1}\frac{Q_{N-l_n+1}(L-r_n)}{Q_N(L)}$$

$$\times\prod_{i=1}^{n-1}Q_{l_{i+1}-l_i}(r_{i+1}-r_i)Q_{l_1}(r_1), \quad (25)$$

where $r_n>r_{n-1}>\cdots>r_1$,

$$Q_k(r)=\frac{1}{2\pi i}\int_{-i\infty}^{c+i\infty}e^{rs}[\Omega(s)^k ds.$$

The asymptotic expression for large N takes the form

$$\rho^{(n)}=\sum_{l_n=n}^{N}\sum_{l_{n-1}=n-1}^{l_n-1}\cdots\sum_{l_1=1}^{l_2-1}\frac{\exp[-r_n c]}{[\Omega(c)]^{l_n}}$$

$$\times Q_{l_1}(r_1)\prod_{i=1}^{n-1}Q_{l_{i+1}-l_i}(r_{i+1}-r_i),$$

with $r_n>r_{n-1}>\cdots>r_1$. By substituting the expression for $Q_k(r)$ and rearranging, we obtain

$$\rho^{(n)}=\sum_{i=1}^{n-1}\left\{\sum_{k_i=1}^{N-b_i}K_i(R_{i,i+1},c)\right\}$$

$$\times\sum_{l_1=1}^{N-b_0}\frac{1}{2\pi i}\int_{c-i\infty}^{c+i\infty}e^{r_1(s-c)}\left[\frac{\Omega(s)}{\Omega(c)}\right]^{l_1}ds, \quad (25)$$

where

$$K_i(R_{i,i+1},c)=\frac{1}{2\pi i}\int_{c-i\infty}^{c+i\infty}\exp[R_{i,i+1}(s-c)]\left[\frac{\Omega(s)}{\Omega(c)}\right]^{k_i}ds,$$

$$R_{i,i+1}=r_{i+1}-r_i,$$

$$b_i=\sum_{j=i+1}^{n-1}k_j-i,$$

$$k_i=l_{i+1}-l_i$$

If we now let r_1 approach infinity in such a manner that all $R_{i,i+1}$ are held fixed, then, as the calculation in Appendix A proves, we get

$$\rho^{(n)}=\frac{1}{la}\prod_{i=1}^{n-1}\sum_{k_i=1}^{\infty}A(R_{i,i+1}-k_ia)K_i(R_{i,i+1},c)$$

$$=(la)^{n-2}\prod_{i=1}^{n-1}\rho^{(2)}(R_{i,i+1}) \quad (27)$$

$$l=L/(N+1)a.$$

II. THE EQUATION OF STATE

The pressure p in a thermodynamic system may be calculated from the partition function Q by the relation

$$p=kT\partial \ln Q/\partial L, \quad (28)$$

where L is the length of the system. There are several ways of deriving an expression for the pressure in terms of the distribution functions and the intermolecular potential. For example, we may use the method due to Born and Green.[10]

Since the kinetic contribution to the partition function does not depend on the length L, we can put Eqs. (1) and (28) together with the substitution $y_i=R_i/L$ to obtain

$$p=kT\exp[\beta\bar{A}_N]\frac{\partial\exp[-\beta\bar{A}_N]}{\partial L},$$

$$=kT\exp[\beta\bar{A}_N]\Big\{NL^{N-1}\int_0^1\cdots\int_0^1$$

$$\times\exp[-\beta V_N(Ly_1,\cdots,Ly_n)]\prod_{i=1}^{N}dy_i,$$

$$-\beta L^N\int_0^1\cdots\int_0^1\sum_{k=1}^{N}y_k\frac{\partial V_N}{\partial y_k}\exp[-\beta V_N]\prod_{i=1}^{N}dy_i\Big\}.$$

We have neglected the wall potential in this expression. With the assumption

$$V_N=\sum_{i<j}\sum V(R_{ij}),$$

we find that

$$\frac{pL}{NkT}=1-\frac{\beta}{2N}\int_0^L\int_0^L R\frac{\partial V(R)}{\partial R}\rho^{(2)}(r,r')drdr',$$

where $R=|r-r'|$. As N approaches infinity, at constant N/L, the equation of state takes the form

$$\frac{pL}{NkT}=1-\beta\frac{L}{N}\int_0^\infty R\frac{\partial V(R)}{\partial R}\rho^{(2)}(R)dR, \quad (29)$$

where $\rho^{(2)}(r,r')$ is a function of $R=|r-r'|$ only and is now expressed by $\rho^{(2)}(R)$.

[10] M. Born and H. S. Green, *A General Kinetic Theory of Liquids* (Cambridge University Press, New York, 1949), p. 23.

By using Eqs. (23b) and (29), we obtain

$$\frac{pL}{NkT}=1-\beta\sum_{n=1}^{\infty}\frac{1}{2\pi i}\int_{c-i\infty}^{c+i\infty}\left[\frac{\Omega(s)}{\Omega(c)}\right]^n ds$$

$$\times\int_{na}^{\infty}R\frac{\partial V}{\partial R}e^{R(s-c)}dR. \quad (30)$$

As an illustration of this equation, we consider the one-dimensional system of rigid "spheres" with diameter a, where the intermolecular potential is given by

$$V(R)=\infty;\quad R<a,$$
$$=0;\quad R\geq a.$$

Thus

$$\Omega(s)=\frac{1}{s}\exp[-as].$$

With the use of the theory of residues,[7]

$$\frac{1}{2\pi i}\int_{c-i\infty}^{c+i\infty}e^{Rs}[\Omega(s)]^n ds$$

$$=\frac{1}{2\pi i}\oint e^{Rs}[\Omega(s)]^n ds=\frac{(R-na)^{n-1}}{(n-1)!}. \quad (31)$$

Using Eqs. (23b) and (31) together with the rigid sphere value for c, $c=1/a(l-1)$, we find that for $l>1$,

$$\rho^{(2)}(x)=\frac{1}{la^2}\sum_{k=1}^{\infty}A(x-k)\left(\frac{1}{l-1}\right)^k$$

$$\times\frac{(x-k)^{k-1}}{(k-1)!}\exp\left(-\frac{x-k}{l-1}\right), \quad (32)$$

where $x=R/a$ is a reduced distance. This function can be interpreted as the probability per unit length of finding a particle at a distance $R=xa$ from an arbitrary fixed particle. This function was first derived by Frenkel[5] who used an elementary probability argument.

For this system of rigid "spheres," the equation of state is completely determined by the value of $\rho^{(2)}(x)$ at $x=1$. From Eqs. (29) and (32), we obtain the familiar Tonks'[2] equation of state,

$$pla/kT=l/(l-1). \quad (33)$$

III. INTEGRAL EQUATIONS FOR THE DISTRIBUTION FUNCTIONS

It has not yet been possible to obtain the average densities of ordered sets n for a three-dimensional system by direct integration. However, an alternative approach, which attempts to solve a set of integral equations, has been formulated and is amenable to numerical calculation. In view of the importance of these equations in the three-dimensional problem, we will consider the one-dimensional case, even though we have obtained expressions for the distribution functions by direct integration.

In the theories of Kirkwood,[11] Born-Green,[12] Yvon,[13] and of Mayer,[14] systems of integro-differential equations for the distribution functions have been formulated in several equivalent forms. These systems have the common feature that they involve derivatives of the distribution functions with respect to suitable parameters upon which the potential of intermolecular force is assumed to depend. Integration with respect to these parameters offers difficulties, which have allowed it to be performed only after approximations, such as the superposition assumption[10] in the space of molecular triplets, has been introduced. We shall see that the superposition approximation is only true in the one-dimensional system in which we only consider nearest neighbor interactions. Recently, Kirkwood[1] has derived a new system of integral equations for the distribution functions by a method which does not depend upon the device of differentiation with respect to parameters in the potential of intermolecular force. This method leads directly to a system of integral equations rather than to a system of integro-differential equations.

The set of equations for a one-dimensional system where wall effects are neglected can be written in the form

$$
\begin{aligned}
g^{(n)}(1, 2, \cdots, n) &= \exp[\beta(\mu^e - V_n^{(1)})]g^{(n-1)}(2, \cdots, n) \\
&\quad + \int_0^L K^{(n)}(1, 2, \cdots, n+1) \\
&\qquad \times g^{(n)}(2, \cdots, n+1)dR_{n+1}, \\
K^{(n)}(1, 2, \cdots, n+1) &= \frac{1}{la} f_{1,\,n+1} \exp[\beta(\mu^e - V_n^{(1)})] \\
&\quad \times \left\{1 + \sum_{s=1}^{\nu-1} \frac{1}{(s+1)!} \int_0^L \cdots \int^L \right. \\
&\quad \times \rho^{(n,s)}(2, \cdots, n+1/n+2, \cdots, n+s+1) \\
&\quad \left. \times \prod_{\sigma=n+2}^{n+s+1} f_{1\sigma}dR_\sigma \right\}, \\
\rho^{(n,s)}(2, \cdots, n+1/n+2, \cdots, n+s+1) &= \left(\frac{1}{la}\right)^s \frac{g^{(n+s)}(2, \cdots, n+s+1)}{g^{(n)}(2, \cdots, n+1)}, \\
\exp[-\beta\mu^e] &= 1 + \sum_{s=1}^{\nu} \frac{1}{s!}\left(\frac{1}{la}\right)^s \int \cdots \int \\
&\quad \times g^{(s)}(2, \cdots, s+1) \prod_{\sigma=2}^{s+1} f_{1\sigma}dv_\sigma,
\end{aligned} \tag{34}
$$

and

$$
\begin{aligned}
g^{(n)} &= (la)^n \rho^{(n)}, \\
f_{1\sigma} &= \exp[-\beta V(R_{1\sigma})] - 1, \\
V_n^{(1)} &= \sum_{i=2}^{n} V(R_{1,i}), \\
l &= L/Na.
\end{aligned}
$$

μ^e is the excess chemical potential per molecule. $\nu+1$ is the maximum number of spheres of radius a, which can be packed into a sphere of radius λ [see Eqs. (21)]. $\rho^{(n,s)}$ is the probability per (unit length)^s that s molecules will be at the specified positions $R_{n+2}, R_{n+3}, \cdots, R_{n+s+1}$, if $2, 3, \cdots, n+1$ are fixed, and it is independent of the positions of the remaining $N-n-s-1$ molecules. $\rho^{(n)}$ is, as before, the average density of ordered sets in n tuplet space, and the correlation function $g^{(n)}$ for a subset of n molecules is defined by the last equation in (34). The derivation of these equations is given elsewhere.[1]

In principle one could solve these equations for a system in which one considers more than only nearest neighbor interactions. Since most non-ionic forces have a range of only a few molecular diameters, the number of terms in the sums of Eq. (34) could be taken to be, say, less than six. However, when we restrict our considerations to nearest neighbor interactions, a considerable simplification takes place, for as we have seen in Eq. (27) a fixed molecule effectively divides the system into two independent subsystems. If we fix molecule n of an arbitrary subset of $n+1$ molecules, then the position of molecule $n+1$ is independent of the positions of molecules $1, 2, \cdots, n-1$ and depends only on the position of molecule n. We can express this principle analytically by stating that, for nearest neighbor forces, the conditional probability

$$
\frac{g^{(n+1)}(1, 2, \cdots, n, n+1)}{g^{(n)}(1, 2, \cdots, n)} = \frac{g^{(2)}(n, n+1)}{g^{(1)}(n)}. \tag{35}
$$

This superposition law is exact for a one-dimensional system in which the interactions are restricted to nearest neighbors. That this result is consistent with the system of integral equations can be easily verified by substituting Eq. (35) into Eq. (34).

As an illustration of the nature of these equations, we consider the special case of a system composed of rigid "spheres." Then

$$
\begin{aligned}
f_{1\sigma} &= 0; \quad R_{1\sigma} > a, \\
f_{1\sigma} &= -1; \quad R_{1\sigma} < a,
\end{aligned} \tag{36}
$$

---

[11] J. G. Kirkwood, J. Chem. Phys. 3, 300 (1935).
[12] M. Born and H. S. Green, Proc. Roy. Soc. A188, 10 (1946).
[13] J. Yvon, *Actualités Scientifiques et Industrielles* (Herman et Cie, Paris, 1935), p. 203.
[14] J. E. Mayer, J. Chem. Phys. 15, 187 (1947).

where a is the diameter of a rigid sphere. Although the following discussion could be carried out in terms of an arbitrary intermolecular potential, the simplicity of the expressions for a hard sphere potential aids in the clarity of the presentation. Since we shall only attempt to illustrate the type of integral equation which one encounters, this simple model will be sufficient.

Since $\nu+1$ is the total number of particles that can possibly fit into a closed interval of length $2a$, we see that $\nu=2$. The superposition principle, Eq. (35), terminates the chain of integral equations so that we need only consider the equation for $g^{(2)}$. Equation (35), together with Eq. (34), also yields

$$\rho^{(n,s)}(2, \cdots, n+1/n+2, \cdots, n+s+1) = \left(\frac{1}{la}\right)^{s} \prod_{\sigma=n+1}^{n+s} g^{(2)}(\sigma, \sigma+1). \quad (37)$$

With these simplifications, the equation for $g^{(2)}$ takes the form

$$g^{(2)}(1, 2)=0; \quad R_{12}<a,$$

$$g^{(2)}(1, 2)=\exp[\beta\mu^{e}] + \int_{R_1-a}^{R_1+a} K(1, 2, 3) g^{(2)}(2, 3) dR_3; \quad R_{12}>a,$$

$$K(1, 2, 3)=0; \quad R_{13}>a,$$

$$K(1, 2, 3)=\frac{1}{la}\exp[\beta\mu^{e}] \times\left\{-1+\frac{1}{2la}\int_{R_1-a}^{R_1+a} g^{(2)}(3, 4)dR_4\right\}; \quad R_{13}>a,$$

and

$$\exp[-\beta\mu^{e}]=1-\frac{2}{l} +\frac{1}{2}\left(\frac{1}{la}\right)^2 \int_{R_1-a}^{R_1+a}\int_{R_1-a}^{R_1+a} g^{(2)}(2, 3)dR_3 dR_2. \quad (39)$$

Since $g^{(2)}=g^{(2)}(R_{12})$, that is to say the radial distribution function is a function of only the relative distance between the two molecules, we let

$$R_{12}=xa, \quad R_{23}=sa, \quad R_{34}=ua. \quad (40)$$

In terms of the reduced distances x, s, u the equations take the form

$$g^{(2)}(x)=0; \quad |x|<1,$$

$$g^{(2)}(x)=\exp[\beta\mu^{e}] \times\left\{1+\frac{1}{l}\int_{x-1}^{x+1} K'(x-s)g^{(2)}(s)ds\right\}; \quad |x|>1, \quad (41)$$

where

$$K'(t)=-1+\frac{1}{l}\int_{0}^{t+1} g^{(2)}(u)du; \quad -1<t<1.$$

In addition, we obtain

$$\exp[-\beta\mu^{e}]=1+\frac{1}{l}\int_{-1}^{1} K'(t)dt. \quad (42)$$

A solution of Eq. (41) is given by Eq. (32) and has the explicit form

$$g^{(2)}(x)=l\sum_{k=1}^{\infty} A(x-k)\left(\frac{1}{l-1}\right)^{k}\frac{(x-k)^{k-1}}{(k-1)!} \times\exp\left[-\frac{x-k}{l-1}\right]. \quad (43)$$

It is shown in Appendix B that this is indeed a solution. Straightforward calculation with Eqs. (41)–(43) shows that

$$\begin{aligned} K'(t)&=-\exp[-t/(l-1)]; && 1>t>0,\\ &=-1; && 0>t>-1,\\ &=0; && |t|>1 \end{aligned} \quad (44)$$

and that

$$\exp[-\beta\mu^{e}]=(l-1/l)\exp[1/(1-l)]. \quad (45)$$

This expression for the excess chemical potential μ^{e} can also be obtained by integrating the equation of state, Eq. (33).

The one-dimensional and three-dimensional formulations of the integral equations [see Eq. (34)] differ only by the replacement of the volume per molecule in the latter by the length per molecule. The major simplification that can be made in the one-dimensional problem with nearest neighbor interactions only is the superposition principle given by Eq. (35). The general form of these integral equations can be seen to be a nonhomogeneous linear form of the Fredholm type in which the kernel depends upon the distribution functions. Even under the simplifications introduced by considering a one-dimensional system of rigid spheres this general form is maintained. Therefore, the one-dimensional form of the integral equations may prove to be a useful guide in investigating the solution of the three-dimensional problem. For example, one might decide between the use of two numerical methods on the basis of the investigation of the convergence of both methods in the one-dimensional problem, where the solution is known.

APPENDIX A

In his article,[7] Gürsey investigated the partition function Q_N in the limit of large N by considering the series

$$S(Z)=\sum_{N=0}^{\infty} Q_N Z^N,$$

for then

$$\lim_{N\to\infty} (Q_N)^{1/N+1}=R,$$

where R is the radius of convergence for the series $S(Z)$. However, we will find it more convenient to obtain the asymptotic expansion for large N by a procedure which utilizes the method of steepest decent. The two approaches lead to identical results.

From Eqs. (16) and (19), we see that it is convenient to define a function M_N by

$$M_N=\frac{1}{2\pi i}\int_{c-i\infty}^{c+i\infty} \exp[(N+1)f(a)]\chi(s)ds,$$

$$f(s)=(Ls/N+1)+\ln\Omega(s). \tag{A1}$$

When $\chi(s)=1$, M_N is the partition function

$$\exp[-\beta\bar{A}_N]/N!.$$

In the discussion of the distribution functions the function $\chi(s)$ is taken to be

$$\chi(s)=[\Omega(s)]^k \exp[-rs],$$

as stated in Part I, we first want to obtain

$$\lim_{N\to\infty} M_N.$$

As the path of integration, we take the line $s=c+iy$, which is assumed to be the line of steepest decent, which passes over the saddle point for the function $f(s)$. This saddle point is located at the point c on the positive real axis. For a system of rigid spheres one can show that $c=1/a(l-1)$. For any given potential one can readily obtain c by solving the equation,

$$(\partial f/\partial s)_{s=c}=0.$$

We now expand $f(s)$ in a power series around the point c:

$$\begin{aligned} f(s)&=f(c)+\tfrac{1}{2}(s-c)^2f''(c)+\cdots,\\ &=f(c)-\tfrac{1}{2}y^2f''(c)+\cdots, \end{aligned} \tag{A2}$$

where $iy=(s-c)$. Then

$$M_N \exp[-(N+1)f(c)]$$

$$=\frac{1}{2\pi}\int_{-\infty}^{+\infty} \exp\left[-\frac{N+1}{2}f''(c)y^2\right]\chi(c+iy)dy. \tag{A3}$$

The integral is now in the form considered by G. N. Watson[16] who obtained the asymptotic expansion for such integrals. The first term of this expansion gives

$$M_N \exp[-(N+1)f(c)]\sim\frac{\chi(c)}{(2\pi(N+1)f''(c))^{\frac{1}{2}}}. \tag{A4}$$

By using Eqs. (A1) and (A4) together with Eqs. (16), (17), and (19), we obtain

$$\rho^{(1)}(r)\sim\sum_{k=1}^{N}\frac{\exp[-rc]}{[\Omega(c)]^k}\frac{1}{2\pi i}\int_{c-i\infty}^{c+i\infty} e^{rt}[\Omega(t)]^k dt \tag{A5a}$$

and

$$\rho^{(2)}(r,r')\sim\sum_{k=2}^{N}\sum_{j=1}^{k-1}\frac{\exp[-rc]}{[\Omega(c)]^k}$$

$$\times\left\{\frac{1}{2\pi i}\int_{c-i\infty}^{c+i\infty} e^{r's}[\Omega(s)]^j ds\right\}$$

$$\times\left\{\frac{1}{2\pi i}\int_{c-i\infty}^{c+i\infty} e^{(r-r')t}[\Omega(t)]^{k-j}dt\right\} \quad r>r'.$$

We now wish to consider the limit as r and r' approach infinity in such a manner that $r-r'$ remains constant. From (A5a), we obtain for infinite N

$$\rho^{(1)}(r)=\frac{1}{2\pi}\int_{-\infty}^{\infty} e^{iru}\sum_{k=1}^{\infty}\left[\frac{\Omega(c+iu)}{\Omega(c)}\right]^k du. \tag{A6}$$

Within the radius of convergence of the infinite series, $\rho^{(1)}$ can be represented by

$$\rho^{(1)}(r)=\frac{1}{2\pi}\int_{-\infty}^{\infty}\frac{e^{iur}\Omega(c+iu)}{\Omega(c)-\Omega(c+iu)}du, \tag{A7}$$

and using the principle of analytic continuation, we will take this as the definition of $\rho^{(1)}(r)$ beyond the singular points of the integrand.

From the theory of residues, we obtain

$$\rho^{(1)}(r)=\sum_{\nu}\exp[iu_\nu r]\left\{-\frac{\Omega(c+iu_\nu)}{\Omega'(c+iu_\nu)}\right\}, \tag{A8}$$

where

$$\Omega'(s)=\partial\Omega/\partial s.$$

u_ν is a pole of the integrand of Eq. (A7) which has a positive imaginary part. A particular pole is $u_1=0$. for this is a solution of the equation,

$$\Omega(c)-\Omega(c+iu)=0.$$

Therefore, we obtain

$$\lim_{r\to\infty}\rho^{(1)}(r)=\frac{\Omega(c)}{-\Omega'(c)}$$

$$=\frac{\int_0^\infty \exp[-cR-\beta V(R)]dR}{\int_0^\infty R\exp[-cR-\beta V(R)]dR} \tag{A9}$$

[16] G. N. Watson, Proc. London Math. Soc. (2) 17, 133 (1918).

The second line follows from the definition of $\Omega(c)$, Eq. (7). From the normalization condition on $\rho^{(1)}$,

$$\int_0^L \rho^{(1)}(r)dr = N,$$

we see that

$$\lim_{R\to\infty} \rho^{(1)}(r) = 1/la. \qquad \text{(A10)}$$

After changing the summation in Eq. (A5b), we obtain

$$\rho^{(2)}(r,r')\sim\sum_{n=1}^{N-1} \times\left\{\frac{1}{2\pi i}\int_{c-i\infty}^{c+i\infty} e^{r'(s-c)}\sum_{j=1}^{N-n}\left[\frac{\Omega(s)}{\Omega(c)}\right]^j ds\right\}, \times\left\{\frac{1}{2\pi i}\int_{c-i\infty}^{c+i\infty} e^{(r-r')(s-c)}\left[\frac{\Omega(s)}{\Omega(c)}\right]^n ds\right\}, \qquad \text{(A11)}$$

where $r>r'$ and $n=k-j$. If we adopted the hard core model, Eq. (21), then $n\leq(r-r')/a$, where a is the diameter of the hard core. Then for finite $R=r-r'$ and infinite N we find that

$$\rho^{(2)}(R)=\sum_{n=1}^{\infty}\frac{A(R-na)}{la}\frac{1}{2\pi i} \times\int_{c-i\infty}^{c+i\infty} e^{R(s-c)}\left[\frac{\Omega(s)}{\Omega(c)}\right]^n ds, \qquad \text{(A12)}$$
$$R=r-r'>0.$$

The step function $A(R-na)$ expresses explicitly the restriction imposed by the hard core intermolecular potential.

APPENDIX B

We wish to show here that the correlation function

$$g^{(2)}(x)=l\sum_{k=1}^{\infty}A(x-k)\left(\frac{1}{l-1}\right)^k\frac{(x-k)}{(k-1)!} \times\exp\left[-\frac{x-k}{l-1}\right] \qquad \text{(1B)}$$

satisfies Eq. (41). By using Eqs. (44) and (45), the integral equation can be written in the form

$$g^{(2)}(x)=\frac{l-1}{l}\exp\left[\frac{1}{1-l}\right]\left\{1-\frac{1}{l}\int_{-1}^{0}g^{(2)}(x-u)du, -\frac{1}{l}\int_0^1\exp\left[\frac{u}{1-l}\right]g^{(2)}(x-u)du\right\};$$
$$|x|>1. \qquad \text{(B2)}$$

By using Eq. (B1) we find

$$\int_0^1\exp\left[\frac{u}{1-l}\right]g^{(2)}(x-u)du =l\sum_{k=1}^{\infty}\left(\frac{1}{l-1}\right)^k\frac{\exp\left[-\frac{x-k}{l-1}\right]}{(k-1)!} \times\int_0^1 A(x-k-u)(x-k-u)^{k-1}du \qquad \text{(B3)}$$

$$\int_{-1}^{0}g^{(2)}(x-u)du=l\sum_{k=1}^{\infty}\left(\frac{1}{l-1}\right)^k\frac{1}{(k-1)!} \times\int_{-1}^{0}A(x-k-u)(x-k-u)^{k-1} \times\exp\left[-\frac{x-k-u}{l-1}\right]du.$$

However,

$$\int_0^1 A(x-k-u)(x-k-u)^{k-1}du =\frac{A(x-k-1)}{k}\{(x-k)^k-(x-k-1)^k\} +\frac{A(x-k)A(k+1-x)}{k}(x-k)^k, \qquad \text{(B4)}$$

and

$$\int_{-1}^{0}A(x-k-u)(x-k-u)^{k-1}\exp\left[-\frac{x-k-u}{l-1}\right]du$$
$$=A(k-x)A(x-k+1)(l-1)^k(k-1)!$$
$$+\sum_{\nu=0}^{k-1}A(x-k)\frac{(k-1)!(l-1)^{\nu+1}}{(k-1-\nu)!}(x-k)^{k-1-\nu}$$
$$\times\exp\left[-\frac{x-k}{l-1}\right]-\sum_{\nu=0}^{k-1}\{A(x-k)$$
$$+A(x-k)A(x-k+1)\}\frac{(k-1)!(l-1)^{\nu+1}}{(k-1-\nu)!}$$
$$\times(x-k+1)^{k-1-\nu}\exp\left[-\frac{x-k+1}{l-1}\right].$$

By substituting Eqs. (B4) and (B3) into the right-hand side of (B2) and after interchanging summation signs and rearranging terms, we obtain

$$\begin{aligned}
\text{r.h.s.} &= g^{(2)}(x) - l\sum_{k=1}^{\infty} \\
&\quad \times\{A(x-k)+A(x+1-k)A(k-x)\} \\
&\quad \times\frac{(x-k+1)^{k-1}}{(l-1)^k(k-1)!}\exp\left[-\frac{x-k}{l-1}\right] \\
&\quad +l\sum_{\nu=0}^{\infty}\sum_{k=1}^{\infty}A(x+1-k-\nu) \\
&\quad \times\frac{(x+1-\nu-k)^{k-1}}{(k-1)!(l-1)^k}\exp\left[-\frac{x-k-\nu}{l-1}\right] \\
&\quad -l\sum_{\nu=0}^{\infty}\sum_{k=1}^{\infty}A(x-k-\nu) \\
&\quad \times\frac{(x-k-\nu)^{k-1}}{(k-1)!(l-1)^k}\exp\left[-\frac{x-k-\nu-1}{l-1}\right],
\end{aligned} \tag{B5}$$

$$\begin{aligned}
\text{r.h.s.} &= g^{(2)}(x) - \exp\left[\frac{1}{l-1}\right]g^{(2)}(x+1) \\
&\quad +\exp\left[\frac{1}{l-1}\right]\sum_{\nu=0}^{\infty}g^{(2)}(x+1-\nu) \\
&\quad -\exp\left[\frac{1}{l-1}\right]\sum_{\nu=0}^{\infty}g^{(2)}(x-\nu) \\
&= g^{(2)}(x).
\end{aligned}$$

We have formally proved that the correlation function obtained by direct integration satisfies the integral equation, Eq. (42).

ADDENDUM

An alternative statistical mechanical formulation of a one-dimensional fluid in which the averages are taken for an ensemble whose systems all have a specified temperature, pressure and total number of particles has been given by Takahasi.* The probability density for this ensemble is

$$P^{(N)}(R_1,\cdots,R_N,L)=\exp[\beta(\bar{F}_N-V_N-V_W-pL)],$$

where

$$\begin{aligned}
\exp[-\beta\bar{F}_N] &= \int_0^{\infty}\int_0^{L}\cdots\int_0^{L}\exp[-\beta(V_N+V_W+pL)] \\
&\qquad\qquad \times dR_1,\cdots,dR_N dL \\
&= \int_0^{\infty}e^{-\beta pL}\exp[-\beta\bar{A}_N(L)]dL.
\end{aligned}$$

Since $\exp[-\beta\bar{F}_N]$ is the Laplace transform of $\exp[-\beta\bar{A}_N]$, the utilization of p and T as independent variables in place of L and T eliminates the necessity of considering the inverse of the Laplace transform (see Eqs. (7) and (8)). This method leads directly to

$$\rho^{(1)}(r)=\sum_{k=1}^{\infty}\frac{e^{-\beta pr}}{[\Omega(\beta p)]^k}\frac{1}{2\pi i}\int_{\beta p-i\infty}^{\beta p+i\infty}e^{rt}[\Omega(t)]^k dt.$$

The fact that the constant c introduced by the method of steepest descents is identically equal to βp was used by Gürsey[7] to determine the equation of state. In order to express this result for $\rho^{(1)}(r)$ in terms of the length per molecule, one would have to solve the equation of state given by $L=(\partial\bar{F}_N/\partial p)_{T,N}$ for p [see Eqs. (20)]. Since we wish to stress the integral equations for the distribution functions which are formulated in terms of the canonical ensemble, we have retained this formulation in our discussion.

---

* H. Takahasi, Proc. Phys.-Math. Soc., Japan 24, 60 (1942).

Reprinted from THE PHYSICAL REVIEW, Vol. 126, No. 6, 2071–2078, June 15, 1962
Printed in U. S. A.

Certain General Order-Disorder Models in the Limit of Long-Range Interactions*

GEORGE A. BAKER, JR.†
Department of Physics, University of California, San Diego, La Jolla, California

(Received December 20, 1961)

We investigate in the limit as the range of the interparticle interactions becomes indefinitely great, but is still small compared to the size of the system, the behavior of a large class of order-disorder models. This class includes, for example, the Ising model, the spherical model, and the Gaussian model. We show that when certain general conditions hold and the interparticle interaction is chosen to be the same for all models, but otherwise arbitrary, the energy per particle above the critical temperature has the same limiting value through terms of order $1/R$, where R is a measure of the number of spins in the range of the interaction. We further show why the behavior above the critical point in this limit does not necessarily provide information about the behavior below the critical point. Some examples are worked out which illustrate the above results.

1. INTRODUCTION

THE purpose of this paper is to investigate the behavior of a certain class of order-disorder models as a function of the range of the interaction. We investigate the limiting behavior of these models as the range of interaction becomes indefinitely great, but is still much smaller than the total size of the system. In the second section of this paper we show that above the critical point, when certain general conditions are met, the details of the model (probability distribution of the various states of the system) but not the shape of the interaction are unimportant in the limit of indefinitely long-range forces. The class of models considered is general enough to contain the Ising model, the spherical model, and the Gaussian model. Below the critical point no such general result is obtained. We see that in this region the details of the model affect the energy per spin in leading order.

In the third section we compute the energy per spin for the one-dimensional spherical model (and Gaussian model) with exponential interactions between spins. We verify explicitly the results of the second section for this type of interaction by comparing the results of the third section with the previously known results for the Ising model. We also verify explicitly that the behavior below the critical point is different for the spherical and Ising models.

In the last section of this paper we evaluate the energy per spin of the three-dimensional spherical model as a function of the range of a force which drops off approximately exponentially with distance. Here we may follow the behavior of the known third-order transition with range.

* Supported in part by the U. S. Atomic Energy Commission.
† On leave of absence from Los Alamos Scientific Laboratory, 1961–1962.

We find that the thermal and magnetic properties are the same below the transition point for the one-, two-, and three-dimensional spherical models in the long-range interaction limit.

2. LIMIT OF LONG-RANGE FORCES

We shall illustrate a more general result on the behavior of various order-disorder models by three examples: the Gaussian model, the spherical model, and the Ising model. For these three models (and in fact for a whole class of models) the partition function is the same above the critical point through terms of order $1/R$, where R is the range of the force. Below the transition point there is little or no relationship. We shall use methods introduced previously by Kac[1] and extended by us[2] to show this result. Previously Brout[3] has essentially obtained, among other results, a special case of this result by diagrammatic analysis.

The first step in our program is to rewrite the partition function in a form to permit the sum over all states of the system to be performed easily. Let us suppose that the energy of the system given by a quadratic form (A is assumed symmetric)

$$E/kT=-\tfrac{1}{2}\sum_{i,j}\nu_iA_{ij}\nu_j. \tag{2.1}$$

If we make use of the well-known integration formula,[4]

$$\exp(\tfrac{1}{2}\sum_{i,j}\nu_iA_{ij}\nu_j)=(2\pi)^{-N/2}\int_{-\infty}^{+\infty}\cdots\int\exp(-\tfrac{1}{2}\sum_{i=1}^{N}x_i^2-\sum_{i,j}x_i(A^{\frac{1}{2}})_{ij}\nu_j)\prod_{j=1}^{N}dx_j, \tag{2.2}$$

which holds for any symmetric A_{ij}, then the partition function

$$Z=\sum_{\text{all states}}\exp(-E/kT) \tag{2.3}$$

may be written as

$$Z=(2\pi)^{-N/2}\sum_{\text{all states}}\int_{-\infty}^{+\infty}\cdots\int\exp(-\tfrac{1}{2}\sum_{i=1}^{N}x_i^2-\sum_{i,j}x_i(A^{\frac{1}{2}})_{ij}\nu_j)\prod_{j=1}^{N}dx_j$$

$$=(2\pi)^{-N/2}\int_{-\infty}^{+\infty}\cdots\int\exp(-\tfrac{1}{2}\sum_{i=1}^{N}x_i^2)\mathfrak{M}(x_i)\prod_{j=1}^{N}dx_j, \tag{2.4}$$

where $\mathfrak{M}(x_i)$ is defined by

$$\mathfrak{M}(x_i)\equiv\sum_{\text{all states}}\exp(-\sum_{i,j}x_i(A^{\frac{1}{2}})_{ij}\nu_j). \tag{2.5}$$

Eq. (2.4) is valid whenever (2.3) is. For the examples we are considering, $\mathfrak{M}(x_i)$ is simply evaluated.

In the Ising model the states of the system are given by $\nu_j=\pm1$, for all j. Thus we have at once, normalizing $\mathfrak{M}(0)=1.0$,

$$\mathfrak{M}_I(x_i)=\prod_{j=1}^{N}\{\cosh[\sum_{i=1}^{N}x_i(A^{\frac{1}{2}})_{ij}]\}. \tag{2.6}$$

In the Gaussian model[5] the ν_j are independently and normally distributed with mean zero and unit variance. Thus, again normalizing $\mathfrak{M}(0)=1.0$, we obtain

$$\mathfrak{M}_G(x_i)=\int_{-\infty}^{+\infty}\cdots\int\exp(-\tfrac{1}{2}\sum_{i=1}^{N}\nu_i^2-\sum_{i,j}x_i(A^{\frac{1}{2}})_{ij}\nu_j)\prod_{j=1}^{N}\frac{d\nu_j}{(2\pi)^{\frac{1}{2}}}=\exp(\tfrac{1}{2}\sum_{i,j}x_iA_{ij}x_j). \tag{2.7}$$

In the spherical model[5] all sets of values of the ν_j which satisfy

$$\sum_{j=1}^{N}\nu_j^2=N \tag{2.8}$$

are equally likely. We impose this restriction in the now standard way by use of the Laplace transform formula

$$\delta(x-y)=\frac{1}{2\pi i}\int_{k-i\infty}^{k+i\infty}e^{q(x-y)}dq. \tag{2.9}$$

[1] M. Kac, Phys. Fluids **2**, 8 (1959).
[2] G. A. Baker, Jr., Phys. Rev. **122**, 1477 (1961).
[3] R. Brout, Phys. Rev. **118**, 1009 (1960); and **122**, 469 (1961).
[4] See, for example, M. G. Kendall and A. Stuart, *The Advanced Theory of Statistics* (Hafner Publishing Company, New York, 1958), Vol. 1, Sec. 15.2.
[5] Introduced by T. H. Berlin and M. Kac, Phys. Rev. **86**, 821 (1952).

Thus, approximately normalizing $\mathfrak{M}(0)=1.0$, we have

$$\mathfrak{M}_S(x_i)=\frac{1}{2\pi i}\int_{k-i\infty}^{k+i\infty}\int_{-\infty}^{+\infty}\cdots\int \exp\Big(Nq-\sum_{j=1}^{N} qv_j^2-\sum_{i,j} x_i(A^{\frac{1}{2}})_{ij}v_j\Big)\prod_{i=1}^{N}\frac{dv_i}{(2\pi e)^{\frac{1}{2}}}$$

$$=\frac{1}{2\pi i}\int_{k-i\infty}^{k+i\infty} e^{Nq}\exp\Big(\frac{1}{4q}\sum_{i,j} x_iA_{ij}x_j\Big)(2eq)^{-N/2}dq. \quad (2.10)$$

The value of the q integral is known[6] to be

$$\mathfrak{M}_S(x_i)=\Bigg(\frac{4N}{\sum_{i,j} x_iA_{ij}x_j}\Bigg)^{\frac{1}{2}N-\frac{1}{2}} I_{\frac{1}{2}N-1}([N\sum_{i,j} x_iA_{ij}x_j]^{\frac{1}{2}})(2e)^{-\frac{1}{2}N}. \quad (2.11)$$

For N very large we may use the method of steepest descents[7] to obtain an asymptotic value for (2.10). It is

$$\mathfrak{M}_S(x_i)\sim\{\tfrac{1}{2}[1+(1+4\xi)^{\frac{1}{2}}]\}^{-\frac{1}{2}N}\times\exp\{\tfrac{1}{2}N[(1+4\xi)^{\frac{1}{2}}-1]\}, \quad (2.12)$$

where

$$\xi=(\sum_{i,j} x_iA_{ij}x_j)/N. \quad (2.13)$$

The next step in our program is to examine the behavior of the interaction matrix A_{ij}. We shall now assume that the indices $\mathbf{i}$ and $\mathbf{j}$ are d-dimensional position vectors and that A is a function of $(\mathbf{i}-\mathbf{j})$ only. We may then diagonalize A by introducing the eigenvectors $Z_\mathbf{q}$ and eigenvalues $a(\mathbf{q})$,

$$Z_\mathbf{q}=\frac{1}{N^{\frac{1}{2}}}\sum_\mathbf{j}\exp(2\pi i\mathbf{q}\cdot\mathbf{j})x_\mathbf{j},$$
$$A_{\mathbf{ij}}=\frac{1}{N}\sum_\mathbf{q} a(\mathbf{q})\exp[2\pi i\mathbf{q}\cdot(\mathbf{j}-\mathbf{i})]. \quad (2.14)$$

If $\mathbf{r}$ is the vector between two points and $V(\mathbf{r})$ is the potential, then

$$a(\mathbf{q})=\frac{1}{N}\sum_{\text{lattice}} V(\mathbf{r})\exp[2\pi i\mathbf{r}\cdot\mathbf{q}]. \quad (2.15)$$

In order to maintain the same total strength for $V(\mathbf{r})$ for varying ranges, let

$$V(\mathbf{r})=v(\mathbf{r}/R^{1/d})/R, \quad (2.16)$$

where R is proportional to the number of spins in the range of the interaction. Using (2.16), Eq. (2.15) becomes

$$a(\mathbf{q})=\frac{1}{RN}\sum_{\text{lattice}} v(\boldsymbol{\zeta})\exp(2\pi iR^{1/d}\boldsymbol{\zeta}\cdot\mathbf{q}), \quad (2.17)$$

where $\boldsymbol{\zeta}=\mathbf{r}/R^{1/d}$. We may show by use of the same type of arguments which lead to Riemann's theorem on trigonometric functions[8] that

$$|a(\mathbf{q})|\le M\prod_{j=1}^{d}(1+R^{1/d}q_j)^{-1}, \quad (2.18)$$

where q_j is the jth component of $\mathbf{q}$. We have also assumed that

$$\int |v(\boldsymbol{\zeta})|d\boldsymbol{\zeta}\le M', \quad (2.19)$$

where M and M' are certain finite positive constants.

We are now in a position to show that for $T>T_c$ the partition function for each of the three examples we are considering is

$$Z=(2\pi)^{-\frac{1}{2}N}\int_{-\infty}^{+\infty}\cdots\int \exp\{-\tfrac{1}{2}\sum_\mathbf{q}|Z_\mathbf{q}|^2[1-a(\mathbf{q})]\}\prod dZ_\mathbf{q}+O([\ln R]^{2d}/R^2). \quad (2.20)$$

For the Gaussian model, (2.20) follows at once from (2.4) and (2.7) when we note that

$$\sum_{i,j} x_iA_{ij}x_j=\sum_\mathbf{q} a(\mathbf{q})|Z_\mathbf{q}|^2. \quad (2.21)$$

Eq. (2.20) is exact for the Gaussian model.

For the spherical model we may, assuming ξ to be small, expand (2.12) as

$$\approx\exp(+\tfrac{1}{2}N\xi-\tfrac{1}{4}N\xi^2+\cdots). \quad (2.22)$$

The terms through order ξ in (2.22) together with (2.4) give (2.20). It remains to show that the remainder of the expression is neglectable to the required order. The weighting factor in (2.4) is a positive definite quadratic form. As (2.12) can increase asymptotically as $\xi\to\infty$ at a rate bounded by $\exp[N(\xi)^{\frac{1}{2}}]$ the whole integral converges. If in addition the maximum of $|a(\mathbf{q})|$ is less than 1.0, then the integrand has a single maximum at the origin ($|Z_\mathbf{q}|=0$). Leaving aside factors of the order of 1.0, we may estimate the mean value of ξ^2 by estimating the value of each term of ξ separately and adding them up. Thus, replacing sums with integrals over the

[6] G. A. Campbell and R. M. Foster, *Fourier Integrals for Practical Applications* (D. Van Nostrand Company, Inc., Princeton, New Jersey, 1948), pair 650.0.

[7] See, for instance, H. Jeffreys and B. S. Jeffreys, *Methods of Mathematical Physics* (Cambridge University Press, New York, 1950), Chap. 17.

[8] See, for example, P. Franklin, *A Treatise on Advanced Calculus* (John Wiley & Sons, Inc., New York, 1949), p. 480.

first Brillouin zone, we obtain

$$N\xi^2 \propto N\left[\int \frac{a(\mathbf{q})d\mathbf{q}}{1-a(\mathbf{q})}\right]^2 < \frac{N}{(1-a_{\max})^2}\left[\int a(\mathbf{q})d\mathbf{q}\right]^2. \quad (2.23)$$

By Eq. (2.18), we have

$$\frac{NM^2}{(1-a_{\max})^2}\left[\prod_{j=1}^{d}\int \frac{dq_i}{1+R^{1/d}q_j}\right]^2 = \frac{NM^2}{(1-a_{\max})^2}\left(\frac{\ln R}{R^{1/d}}\right)^{2d}, \quad (2.24)$$

plus terms of lower order, which is the required result. We denote the absolute value of the maximum $a(\mathbf{q})$ by $a_{\max}$. We remark that (leaving factors of order 1.0 aside) one can obtain corresponding bounds for ξ^3, ξ^4, etc. Equation (2.24) may not be the best bound obtainable, but it is sufficient to show that (2.20) is at least asymptotically correct. It is to be noted that this proof fails when $a_{\max}$ reaches 1.0. At this point the integrand of (2.4) no longer possesses a single peak at the origin. We identify

$$a_{\max} = 1.0 \quad (2.25)$$

as the equation for the critical point. For lower temperatures than it, there is little relationship between the models considered here. This point will be made clearer by several examples examined in the following sections of this paper.

We follow the same procedure to establish (2.20) for the Ising model as we did for the spherical model. Expanding the ln cosh terms in the exponent of the partition function in powers of their arguments, we get from (2.6)

$$\mathfrak{M}_I(x_i) \approx \exp\{\tfrac{1}{2}\sum_j [\sum_i x_i(A^{\frac{1}{2}})_{ij}]^2 - \tfrac{1}{12}\sum_j [\sum_i x_i(A^{\frac{1}{2}})_{ij}]^4 + \cdots\}$$

$$= \exp\{\tfrac{1}{2}\sum_{ij} x_iA_{ij}x_j - \tfrac{1}{12}\sum_j [\sum_i x_i(A^{\frac{1}{2}})_{ij}]^4 + \cdots\}. \quad (2.26)$$

We again identify the first term in the exponent as the part of $\mathfrak{M}_I(x_i)$ which we need to complete (2.20). Our problem is again to bound the remainder. Transforming the fourth-order term to the $Z_\mathbf{q}$ basis, we obtain

$$\frac{N}{N^2}\sum_{\mathbf{q}_1,\mathbf{q}_2,\mathbf{q}_3,\mathbf{q}_4}' Z_{\mathbf{q}_1}a^{\frac{1}{2}}(\mathbf{q}_1)Z_{\mathbf{q}_2}a^{\frac{1}{2}}(\mathbf{q}_2)Z_{\mathbf{q}_3}a^{\frac{1}{2}}(\mathbf{q}_3)Z_{\mathbf{q}_4}a^{\frac{1}{2}}(\mathbf{q}_4). \quad (2.27)$$

The prime means the summation extends over only those $\mathbf{q}_i$ for which $\sum_i \mathbf{q}_i$ is an integral multiple of **N**. On the assumption that (2.27) is small, we may compute its approximate magnitude by treating it as a perturbation to the integrand of (2.20). To avoid a spuriously low estimate we compute the root-mean-square value, and include all terms. Thus the error due to the fourth-order term is less than

$$N\left\{\left[\frac{1}{N}\sum_\mathbf{q} E(|Z_\mathbf{q}|^2)a(\mathbf{q})\right]^4\right\}^{\frac{1}{2}}, \quad (2.28)$$

plus terms of lower order in N. However, this is the same expression as we obtained (2.23) for the error in the spherical model. Hence the same bound (2.24) follows under the same restrictions ($a_{\max}<1.0$). Corresponding bounds can be obtained for any order. Thus Eq. (2.20) also holds for the Ising model.

We see from the above special cases that for the asymptotic formula (2.20) to hold the important requirements are: (1) the integrand of (2.4) must have a single peak at the origin; (2) through second order in the x's, $\ln\mathfrak{M}(x)$ must agree with $\ln\mathfrak{M}_G(x)$; and (3) the higher order terms in $\ln\mathfrak{M}(x)$ must becrease more rapidly than $1/R$ in the limit as $R\to\infty$.

Although condition (3) is satisfied for a wide class of functions, it need not always be satisfied as can be seen from the following example. Suppose the states of the system are $\nu_\mathbf{q}=\pm 1$, where $\nu_\mathbf{q}$ are the Fourier transforms of the ν_j. Then

$$\mathfrak{M}(x) = \prod_\mathbf{q}\{\cosh Z_\mathbf{q}[a(\mathbf{q})]^{\frac{1}{2}}\}. \quad (2.29)$$

The partition function (2.4) may be easily evaluated to give

$$Z = \exp\left[\tfrac{1}{2}N\int a(\mathbf{q})d\mathbf{q}\right]. \quad (2.30)$$

Evaluating (2.20), we obtain

$$Z = \exp\left[-\tfrac{1}{2}N\int \ln(1-a(\mathbf{q}))d\mathbf{q}\right]$$

$$\approx \exp\left[\tfrac{1}{2}N\int a(\mathbf{q})d\mathbf{q} - \tfrac{1}{4}N\int a^2(\mathbf{q})d\mathbf{q} + \cdots\right]. \quad (2.31)$$

From (2.18) we see that there is a volume of order R^{-1} near the origin of $\mathbf{q}$ space in which, in general, $a(\mathbf{q})$ is of order 1.0 rather than of order $1/R$. Thus there will be contributions from this region to the higher terms in (2.31) by terms of order $1/R$, which spoils (2.20).

3. EXAMPLE OF THE ONE-DIMENSIONAL SPHERICAL MODEL

The Ising model for one dimension with an exponential interaction between spins has previously been solved exactly and some of its properties analyzed in the long-range limit.[2] If the spin-spin interaction energy is given by

$$(1-r)J\sum_{j=1}^{N-1}\sum_{k=j+1}^{N} r^{k-j-1}\nu_j\nu_k, \quad (3.1)$$

then the energy per particle is found to be asymptotically equal to

$$-\tfrac{1}{2}(1-r)J[-1+(1-2K)^{-\frac{1}{2}}], \tag{3.2}$$

above the critical point, $K_c=\frac{1}{2}$. The constant J is equal to half the maximum possible interaction energy per spin, and $K=J/kT$. Below the critical point the energy is

$$E=-\tanh^2 z, \tag{3.3}$$

where

$$z=2K\tanh z. \tag{3.4}$$

In the Appendix of our previous paper[2] it was stated that the results of the Bragg-Williams approximation differed from these by a factor of two. That was incorrect. They are the same. Equation (A1) of that paper needs a factor of two to be consistent with Eq. (2.8), where K was defined as $\frac{1}{2}$ the maximum possible interaction energy divided by kT.

We shall work out the corresponding results for the spherical model.[5] We will verify explicitly that the results of the previous section are valid in this case of exponential interaction. If we define

$$B_{ij}=r^{|i-j|}+r^{N-|i-j|}, \tag{3.5}$$

then we may write the interaction energy over kT as

$$A_{ij}=-(1-r)K[B_{ij}-(1+r^N)\delta_{ij}]/(2r), \tag{3.6}$$

where δ_{ij} is the Kronecker delta and N is the number of spins in the system. The quantity K is taken to be half the maximum total interaction energy per spin, J, over kT as above. If we introduce the spherical constraint, (2.8), by means of (2.9), we may write the partition function, (2.3), as

$$Z=\frac{1}{2\pi i}\int_{k-i\infty}^{k+i\infty} e^{Nq}dq\frac{1}{(2\pi e)^{N/2}}\int_{-\infty}^{+\infty}\cdots \times\int \exp\Big(-q\sum_{j=1}^{N}\nu_j^2+\sum_{i,j}\nu_i A_{ij}\nu_j\Big)\prod_{j=1}^{N}d\nu_j. \tag{3.7}$$

The integration over the ν's is readily performed, yielding[4]

$$Z=\frac{1}{2\pi i}\int_{k-i\infty}^{k+i\infty} e^{N(q-\frac{1}{2})}dq\,\det|q\delta_{ij}-A_{ij}|^{-\frac{1}{2}}. \tag{3.8}$$

Since the determinant is the product of the eigenvalues and the eigenvalues are, ignoring terms of order r^N,

$$q-(1-r)K\left[\frac{\cos(2\pi j/N)-r}{1+r^2-2r\cos(2\pi j/N)}\right],\qquad j=1,\cdots,N, \tag{3.9}$$

we may, where N is large enough to approximate sums with integrals, write (3.8) as

$$Z\approx\frac{1}{2\pi i}\int_{k-i\infty}^{k+i\infty} dq\exp\left[N(q-\tfrac{1}{2})-\frac{N}{4\pi}\int_0^{2\pi}d\omega \times\ln\left\{q-(1-r)K\left[\frac{\cos\omega-r}{1+r^2-2r\cos\omega}\right]\right\}\right]. \tag{3.10}$$

For $r<1$, we may rewrite the ω integral in (3.10) as

$$\int_0^{2\pi}d\omega\ln\{(1+r^2)q+r(1-r)K-[(1-r)K+2rq]\cos\omega\} -\int_0^{2\pi}d\omega\ln(1+r^2-2r\cos\omega). \tag{3.11}$$

By a standard integration formula[9] we may easily show that the second integral vanishes. Using the same standard formula to do the first integral, we may write for (3.10)

$$Z=\frac{1}{2\pi i}\int_{k-i\infty}^{k+i\infty}dq\exp\left\{N(q-\tfrac{1}{2})-\tfrac{1}{2}N\ln\left[\frac{q(1+r^2)+r(1-r)K+\{[q(1+r^2)+r(1-r)K]^2-[(1-r)K+2rq]^2\}^{\frac{1}{2}}}{2}\right]\right\}.$$

In the limit as $N\to\infty$ we may evaluate $(\ln Z)/N$ exactly by means of the method of steepest descents.[7] The location of the saddle point is given by solving the following equation for q:

$$(2q-1)r=(1-r)\times\left\{-K+\tfrac{1}{2}K\left[q^2-\frac{2r}{1+r}qK-\left(\frac{1-r}{1+r}\right)K^2\right]^{-\frac{1}{2}}\right\}. \tag{3.13}$$

Equation (3.13) was derived by setting the partial derivative of the logarithm of the integrand of (3.12) equal to zero and then simplifying. As (3.13) is a quartic equation it can be solved explicitly for q, but we will not carry out this step. For $r=0$ we recover the known[5] short-range solution,

$$q_s=(K^2+\tfrac{1}{4})^{\frac{1}{2}}. \tag{3.14}$$

For r very near unity, the right-hand side of (3.13) is very small, so long as the radical does not vanish. When this condition is satisfied, we obtain $q_s=\frac{1}{2}$. For r near 1, the radical vanishes for $q\approx 0$ and $q\approx K$. Thus, in the limit as r tends to unity, we obtain

$$\begin{aligned} q_s&=\tfrac{1}{2}\quad (K<\tfrac{1}{2}),\\ q_s&=K\quad (K>\tfrac{1}{2}). \end{aligned} \tag{3.15}$$

The energy per particle implied by the spherical

[9] B. O. Peirce, *A Short Table of Integrals* (Ginn and Company, Boston, 1910), No. 523.

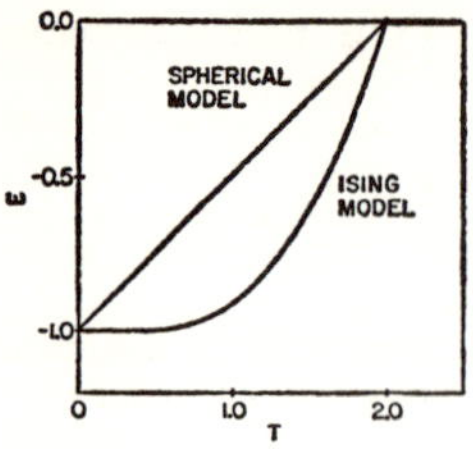

FIG. 1. Limiting energy per spin curves for the one-dimensional spherical and Ising models as the range of the interaction becomes indefinitely great.

model is[5]

$$E=-(q_s-\tfrac{1}{2})kT. \tag{3.16}$$

If we carry the solution of (3.13) to the next higher order in $(1-r)$ above the critical point $(K=\frac{1}{2})$, then, substituting in (3.16) we obtain (3.2) for the energy per particle. Thus we see that the conclusions of Sec. 2 hold for the one-dimensional spherical and Ising models with exponential interactions. When we note that the Gaussian model may be obtained from the spherical model by setting $q=\frac{1}{2}$ instead of integrating over q, we see that the results are also identical for it above the critical point. It should be noted that the next order deviations can be obtained for these three cases are found to be of order $(1/R)^2$ rather than $(\ln R/R)^2$.

In Fig. 1 we illustrate the energies for the spherical and the Ising models below the critical point in the limit as $r\to 1$. One sees that agreement above the critical point does not imply anything about the behavior below the critical point. In fact, although the Gaussian model agrees with the other two models studied above the critical point, it is not even defined below the critical point.

It is important to note that neither the one-dimensional spherical nor the Ising[2] model possesses any critical point for $r<1$, but both possess a discontinuity in the specific heat at $K=\frac{1}{2}$ in the limit as $r\to 1$. Although the first-order term in the $(1-r)$ expansion has a singularity (and, in fact, so do all higher terms), one cannot conclude that the sum of all of them do. As we see from these examples, it need not. Hence caution must be exercised in attempting to draw conclusions about the nature of the transition for $r<1$ from the limiting nature of the transition.

While it is true that if one defines the spontaneous magnetization for $r=1$ as the limit of the spontaneous magnetization as $r\to 1$, one obtains zero below the transition point together with an infinite susceptibility, physically one should consider it to be

$$\mathfrak{M}=\lim_{H\to 0+}\lim_{r\to 1}\lim_{N\to\infty}\frac{\partial(\ln Z)/N}{\partial H}. \tag{3.17}$$

This would correspond to the results of measurements at fixed magnetic field strengths as $r\to 1$. As the introduction of a magnetic field H corresponds to multiplying the integrand of (3.12) by

$$\exp[\tfrac{1}{4}\mu^2H^2N/(q-K)], \tag{3.18}$$

one readily computes that $\mathfrak{M}=\mu[1-T/T_c]^{\frac{1}{2}}$ for $T<T_c$ and 0 for $T\geq T_c$. This result is in agreement with the results for $r\to 1$ in two dimensions and the results for all r in three dimensions.[10]

4. VARIATION WITH THE RANGE OF THE INTERACTION OF A THREE-DIMENSIONAL SPHERICAL MODEL

In the previous section we considered a one-dimensional model in which the interaction energy fell off exponentially with the distance between the spins. We shall generalize that model to three dimensions in the following way. Instead of having the interaction energy fall off exponentially with the distance we shall let it fall off exponentially with $|x|+|y|+|z|$, where x, y, and z are the three components of the separation vector between the two spins. We shall consider the spins to be placed on a simple cubic lattice. The interaction energy over kT may, for this model, be written as

$$A_{ij}=-(1-r)^3K[B\times B\times B-(1+r^N)^3\delta\times\delta\times\delta]/(6r+2r^3), \tag{4.1}$$

where B is defined by (3.5) and $\times$ means the direct product. The quantity K is again taken to be half the maximum total interaction energy per spin, J, over kT. In this model we know from the work of Berlin and Kac[5] that there is a transition for nearest neighbor interactions only $(r=0)$ which corresponds to the onset of spontaneous magnetization. With the model discussed here we can follow the transition as a function of range all the way to the limit of infinite range (though small compared to the size of the system).

The analysis of this model follows very closely that for the one-dimensional problem which was given in the previous section. If we follow steps analogous to (3.7)–(3.11), we obtain for the partition function

$$Z\approx\frac{1}{2\pi i}\int_{k-i\infty}^{k+i\infty}dq\exp\left\{N(q-\tfrac{1}{2})-\frac{N}{16\pi^3}\int\int_0^{2\pi}\int d\omega d\omega' d\omega''\right.$$
$$\left.\times\ln\left[(1+r^2-2r\cos\omega)(1+r^2-2r\cos\omega')(1+r^2-2r\cos\omega'')\left(q+\frac{(1-r)^3K}{6r+3r^3}\right)-\frac{(1-r)^3(1-r^2)^3K}{6r+2r^3}\right]\right\}, \tag{4.2}$$

where reference 9 has again been used and N denotes the total number of spins in the system. In the limit as

[10] The author is happy to acknowledge a fruitful discussion of this point with M. Kac and T. H. Berlin.

$N\to\infty$ we may evaluate $(\ln Z)/N$ exactly by means of the method of steepest descents,[7] if a normal saddle point exists. From the results of Berlin and Kac[5] we know that for $r=0$ a normal saddle point does exist above the critical point, and below the critical point the integrand possesses a cusp instead of a saddle point at the point of the maximum value on a path of steepest descents for the q integration. We shall find that the same situation holds for all $r<1$. If a normal saddle point exists, its location is given by solving the following equation for q:

$$0=1-\frac{1}{2}\left(q+\frac{(1-r)^3K}{6r+2r^3}\right)^{-1}-\frac{1}{2}\frac{(1-r)^3(1-r^2)^3K}{6r+2r^3}\left(q+\frac{(1-r)^3K}{6r+2r^3}\right)^{-1}\frac{1}{\pi^3}\int\int_0^{\pi}\int d\omega d\omega' d\omega''$$
$$\times\left[\left(q+\frac{(1-r)^3K}{6r+2r^3}\right)(1+r^2-2r\cos\omega)(1+r^2-2r\cos\omega')(1+r^2-2r\cos\omega'')-\frac{(1-r)^3(1-r^2)^3K}{6r+2r^3}\right]^{-1}. \quad (4.3)$$

We may perform the integration over ω by means of a standard integration formula,[11] obtaining

$$0=1-\frac{1}{2}\left(q+\frac{(1-r)^3K}{6r+2r^3}\right)^{-1}-\frac{1}{2}\frac{(1-r)^3(1-r^2)^3K}{6r+2r^3}\left(q+\frac{(1-r)^3K}{6r+2r^3}\right)^{-1}\frac{1}{\pi^2}\int_0^{\pi}\int d\omega' d\omega''$$
$$\times\left\{\left[\left(q+\frac{(1-r)^3K}{6r+2r^3}\right)(1+r^2)(1+r^2-2r\cos\omega')(1+r^2-2r\cos\omega'')-\frac{(1-r)^3(1-r^2)^3K}{6r+2r^3}\right]^2\right.$$
$$\left.-4r^2\left(q+\frac{(1-r)^3K}{6r+2r^3}\right)^2(1+r^2-2r\cos\omega')^2(1+r^2-2r\cos\omega'')^2\right\}^{-\frac{1}{2}}. \quad (4.4)$$

We may perform the integration over ω' by letting $w=\cos\omega'$, and factoring the denominator to find its zeros. If we confine our attention to $q\geq K>0$, the ferromagnetic-type interaction case, then we may use formula number 566 of reference 9 to perform the w integration. The corresponding analysis can easily be carried out for $K<0$, the antiferromagnetic case, but we shall not give it here. The result of the w integration is

$$0=1-\frac{3r+r^3}{(6r+2r^3)q+(1-r)^3K}-\frac{(1-r^2)(1-r)^3K}{(6r+2r^3)q+(1-r)^3K}$$
$$\times\frac{1}{\pi^2}\int_0^{\pi}d\omega''\frac{\mathbf{K}\left(\dfrac{16(1-r)^2K[rq+(1-r)^3K/(6+2r^2)](1+r-2r\cos\omega'')}{(1+r)(6+2r^2)\{q(1+r^2-2r\cos\omega'')+[(1-r)^3/(3+r^2)]K(r-\cos\omega'')\}^2}\right)}{|q(1+r^2-2r\cos\omega'')+[(1-r)^3/(3+r^2)]K(r-\cos\omega'')|}, \quad (4.5)$$

where $\mathbf{K}(k^2)=\mathrm{sn}^{-1}(1,k)$ is the complete elliptic integral of the first kind. One should note that (4.5) reduces in an obvious way to the corresponding result of Berlin and Kac[5] when r is set equal to zero. In order for the formula which corresponds to (3.8) to be valid in this derivation, q must be greater than K for the ν integrations to converge. In the one-dimensional case the term corresponding to the integral goes to infinity as $q\to K$ and thus there is a solution for all K and therefore no transition in the one-dimensional case for $r<1$. The same is true in two dimensions as was in one dimension. The equation analogous to (4.5) and (3.13) is, for two dimensions,

$$0=1-\frac{2r}{4rq+(1-r)^2K}-\frac{(1-r)^2K}{[4rq+(1-r)^2K]|q|\pi}\times\mathbf{K}\left(\frac{K[4rq+(1-r)^2K]}{(1+r)^2q^2}\right), \quad (4.6)$$

in which the appropriate term goes to infinity logarithmically as $q\to K$. The limit as $r\to 1$ is $q_s=\frac{1}{2}$ for $K<\frac{1}{2}$ and $q_s=K$ for $K>\frac{1}{2}$, which is exactly the same as for the one-dimensional case. The coefficient of $(1-r)^2$ is, of course, different than that of $(1-r)$ was in the one-dimensional case; hence the dimensionality makes a

[11] No. 300 in reference 9.

difference in the leading order (which vanishes) of the energy per spin above the critical point (critical only in limit as $r \to 1$).

In three dimensions we see from (4.5) that while the integrand is singular at $\omega''=0$, the integral, being essentially $\int \ln\omega'' d\omega''$ near $\omega''=0$, is finite in the limit $q \to K$. Thus, for $r<1$, we obtain a solution, q_s, of (4.5) for K less than K_s as given by

$$K_c=\frac{1}{2}\left(\frac{1-r}{1+r}\right)^3\left\{\frac{1}{2}-\frac{(3+r^2)(1-r^2)}{\pi^2}\int_0^\pi d\omega''\frac{\mathbf{K}\left(\dfrac{4(1-r^2)^2(1+r^2-2r\cos\omega'')}{[(3+r^2)(1+r^2)+r(1-r)^3-(1+r)^3\cos\omega'']^2}\right)}{[3+r^2(1+r^2)+r(1-r)^3-(1+r)^3\cos\omega'']}\right\}. \tag{4.7}$$

It follows for any $r<1$, in exactly the same way as Berlin and Kac[5] demonstrated for $r=0$, that $q_s=K$ for $K>K_c$. K_c varies from[5] $27+18\sqrt{2}-15\sqrt{3}-10.5\sqrt{6} \approx 0.7554396$ to 0.5 as r goes from 0 to 1.0. It should be noted that $(1-r^2)\int$ is of the order of unity and not of order $(1-r)$.

To establish the nature of the transition it will suffice to show that K near K_c is of the form

$$K=K_c+A(q-K_c)^{\frac{1}{2}}+O((q-K_c)). \tag{4.8}$$

By the arguments of Berlin and Kac[5] it then follows at once that the specific heat is continuous and its slope is discontinuous at the critical point for all $r<1$. Thus the transition is of the third order. In the limit as $r \to 1$, we again obtain the same limit as we did in the one- and two-dimensional cases. Again, of course, the coefficient of $(1-r)^3$ differs from that of $(1-r)$ in the one-dimensional case and so the dimensionality makes a difference in the leading order of the energy per spin. Also the spontaneous magnetization persists in the limits as $r \to 1$ in three dimensions. This result follows easily using the methods of Berlin and Kac.[5]

To establish (4.8), we use the expansion[12] for $\mathbf{K}(x)$:

$$\mathbf{K}(x)=\frac{1}{2\pi}\sum_{n=0}^{\infty}\left(\frac{\Gamma(\frac{1}{2}+n)}{n!}\right)^2(1-x)^n \times\left[\ln(1-x)-4\ln 2+4\left(\frac{1}{1}-\frac{1}{2}+\cdots-\frac{1}{2n}\right)\right]. \tag{4.9}$$

It is easy to show that contributions to A can only come from the terms involving $\ln(1-x)$. As these terms are all of the same sign, we can obtain a lower bound on the magnitude of A by considering only the first one. A lower bound is all we need to differentiate A from zero. If we introduce $\delta=(q/K)-1$, substitute the $n=0$ coefficient of $\ln(1-x)$ into (4.5), drop several terms of order (δ) and order (1.0) we reduce the calculation of the lower bound to the magnitude of A to the evaluation of

$$-\frac{(1-r)^4(3+r^2)}{2\pi^2(1+r)^5}\int_0^\pi \frac{d\omega''\ln(\theta-\cos\omega'')}{\gamma-\cos\omega''}, \tag{4.10}$$

where

$$\theta \approx 1+(1-r)^2(3+r^2)(1+r)^{-3}\delta, \qquad \gamma=(3-2r+3r^2)(1+r)^{-2}. \tag{4.11}$$

By the use of some known integrals and series expansions[13] we can compute that

$$K \leq K_c-\left(\frac{1-r}{1+r}\right)^3\frac{(3+r^2)}{4\pi(1+r)}\left[\left(\frac{3+r^2}{1+r}\right)\delta\right]^{\frac{1}{2}}+O(\delta). \tag{4.12}$$

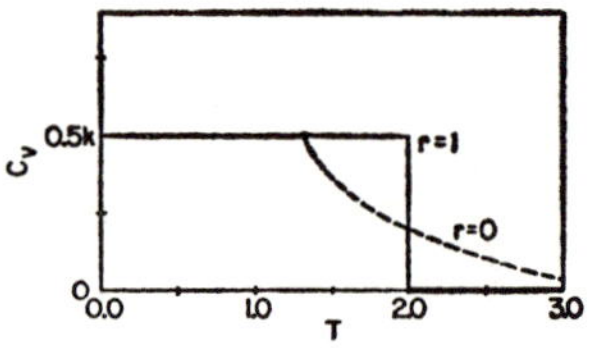

FIG. 2. Specific heat curves for nearest neighbor ($r=0$) and indefinitely long range ($r=1$) interactions for the three-dimensional spherical model. Intermediate values of the range interpolate between the two curves shown and have a discontinuity in slope which becomes progressively sharper as r goes from 0 to 1.

It is to be noted that in the limit as $r \to 0$ we obtain about 6% larger magnitude for A than Berlin and Kac[5] did. This is because in order to demonstrate the proper dependence on $(1-r)$ we had to include a slightly more complicated term than they did to establish the result for $r=0$.

In Fig. 2 we give a sketch of the specific heat curve for the three-dimensional problem for $r=0$ and $r=1$.

[12] E. T. Whittaker and G. N. Watson, *A Course in Modern Analysis* (Cambridge University Press, New York, 1927), Ex. 20, 21, p. 299.

[13] W. Gröbner and N. Hofreiter, *Integraltafel, Zweiter Teil, Bestimmte Integrale* (Springer-Verlag, Berlin, 1958), Nos. 332.24 and 338.13.

Reprinted from *Printed in U.S.A.*
JOURNAL OF MATHEMATICAL PHYSICS VOLUME 4, NUMBER 2 FEBRUARY 1963

On the van der Waals Theory of the Vapor–Liquid Equilibrium. I. Discussion of a One-Dimensional Model

M. Kac, G. E. Uhlenbeck, and P. C. Hemmer
The Rockefeller Institute, New York, New York
(Received 17 September 1962)

For a one-dimensional fluid model where the pair interaction potential between the molecules consists of a hard core and an exponential attraction, Kac has shown that the partition function can be determined exactly in the thermodynamic limit. In Sec. II this calculation is reviewed and further discussed. In Sec. III, we show that in the so-called van der Waals limit when the range of the attractive force goes to infinity while its strength becomes proportionally weaker, a phase transition appears which is described exactly by the van der Waals equation plus the Maxwell equal-area rule. In Sec. IV the approach to the van der Waals limit is discussed by an appropriate perturbation method applied to the basic integral equation. The perturbation parameter is the ratio of the size of the hard core to the range of the attractive force. It is seen that the phase transition persists in any order of the perturbation. The two-phase equilibrium is characterized by the fact that in this range of density, the maximum eigenvalue of the integral equation is doubly degenerate and that the corresponding two eigenfunctions do not overlap. In Sec. V we comment on the relevance of our results for the three-dimensional problem.

I. INTRODUCTION

In this series of papers we intend to present a new discussion of the old theory of van der Waals[1] of the continuity of the gaseous and liquid states of matter. It is well known that the great merit of this theory lies in the fact that it gave the first qualitative kinetic interpretation of condensation phenomena and of the existence of a critical point. On the other hand, it has proved very difficult to make the theory more rigorous and as a result the modern theory[2] of the equation of state of a nonideal gas, has followed more the idea of Kamerlingh Onnes to represent all properties of the gas as power series in the density—the so-called virial expansion. In this way one can take successively into account the interaction of the molecules in pairs, triples, quadruples, etc., and one can derive precise expressions for the successive deviations from the ideal gas laws in terms of the intermolecular potential. Many attempts,[3] thus far unsuccessful, have been made to construct a rigorous theory of condensation phenomena from such expansions. In fact we believe that such a construction is very difficult, if not impossible,[4] and it therefore seems worthwhile to try to reformulate in a more rigorous way, the basic ideas of van der Waals.

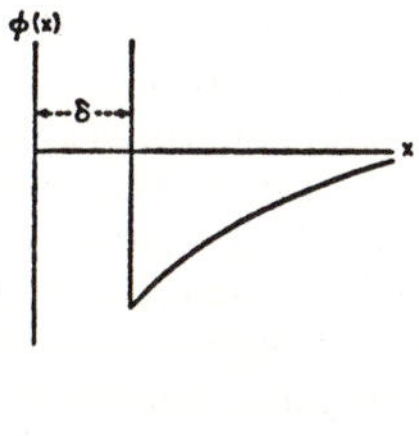

Fig. 1. Interaction potential $\varphi(x)$.

We have attempted to do this starting always from a one-dimensional gas model, first proposed by M. Kac,[5] for which all calculations can be carried out exactly. The model consists of N particles moving on a line of length L and interacting in pairs through a potential $\varphi(x)$ which consists of a hard core of length δ and an exponential attraction (see Fig. 1.) For this model it is possible to give an exact discussion of the partition function in the *thermodynamic limit* $L \to \infty$, $N \to \infty$, $l = L/N$ finite. As shown already by Kac, the problem in this limit can be reduced to the discussion of a linear integral equation with a positive definite, Hilbert–Schmidt kernel of which the maximum eigenvalue determines the

[1] J. D. van der Waals, Dissertation Leiden, 1873. This was expanded in the book: *Die Kontinuität des gasförmigen und flüssigen Zustandes* (Johann A. Barth, Leipzig, Germany, 1899), 2 volumes. Compare also the monograph by J. P. Kuenen, Die Zustandsgleichung (Vieweg, Braunschweig, 1907).

[2] Due mainly to J. E. Mayer. A summary is given in the book by J. E. Mayer and M. G. Mayer, *Statistical Mechanics* (John Wiley & Sons, Inc., New York, 1940), Chaps. 13 and 14. For a recent account see the monograph by G. E. Uhlenbeck and G. W. Ford in *Studies in Statistical Mechanics* (North-Holland Publishing Company, Amsterdam, 1962), Vol. 1, Part B.

[3] These also go back to J. E. Mayer; (see reference 2, Chap. 14). For a more recent discussion, see K. Ikeda, Progr. of Theoret. Phys. (Kyoto) **19**, 653 (1958), and **26**, 173 (1961). Much of the motivation came from the analogy with the Bose–Einstein condensation, which was pointed out by B. Kahn and G. E. Uhlenbeck [Physica **5**, 399 (1938)]. However, we now believe that this analogy is superficial and that there is no real connection between the Bose–Einstein condensation and ordinary condensation phenomena.

[4] The difficulties become especially clear in the formulation of the condensation problem according to C. N. Yang and T. D. Lee, Phys. Rev. **87**, 404 (1952). Compare also the discussion by G. E. Uhlenbeck and G. W. Ford in Chapter 3 of the book *Lectures in Statistical Mechanics* (Proceedings of the Summer Seminar, Boulder, Colorado, 1960, published by the American Mathematical Society, Providence, Rhode Island, 1963.)

[5] M. Kac, Phys. Fluids **2**, 8 (1959).

thermodynamic potential (Gibbs free energy) of the system. These results will be recapitulated in Sec. II.

For finite γ, that is for a finite range of the attractive force, the system does *not* show a phase transition in agreement with all previously known results[6] for one-dimensional systems. However, if one sets $\alpha = \alpha_0\gamma$, and then lets $\gamma \to 0$, (i.e., for a weak but very long-range force) so that the integral

$$\int_0^\infty dx\, \varphi_{\text{attr.}}(x)$$

is finite (we will call this the *van der Waals limit*), a phase transition appears which is described exactly by the van der Waals equation

$$p = kT/(l - \delta) - \alpha_0/l^2, \tag{1}$$

together with the well known Maxwell rule.[7] This will be shown in detail in Sec. III.

In Sec. IV, all the eigenfunctions and eigenvalues of the Kac integral equation are found in this limit by a perturbation procedure with $\gamma\delta$ as the expansion parameter. The phase transition appears as a double degeneracy of the maximum eigenvalue, and the phase transition *persists* when the perturbation calculation is carried further to any finite power of $\gamma\delta$. One should also note that this entire development breaks down near the critical point. For the critical region, a separate discussion is required which will be given in Part III of this series of papers.

In the last section, we will comment on the relevance of our results for the three-dimensional problem, and on the relation to the usual derivation of the van der Waals equation.

II. THE KAC INTEGRAL EQUATION

The partition function for our one-dimensional gas is given by

$$Z(L, T, N) = \frac{1}{\Lambda^N}\cdot\frac{1}{N!}\int_0^L \cdots \int_0^L dt_1 \cdots dt_N$$
$$\times \exp\left[-\frac{1}{kT}\sum_{i<j}\varphi(|t_i - t_j|)\right]$$
$$= \frac{1}{\Lambda^N}\cdot\frac{1}{N!}\int_0^L \cdots \int_0^L dt_1 \cdots dt_N$$
$$\times \exp\left[\nu \sum_{i<j} e^{-\gamma|t_i - t_j|}\right] \prod_{i<j} S(|t_i - t_j|), \tag{2}$$

with $\Lambda^2 = h^2/2\pi mkT$, $\nu = \alpha/kT$, and where the stepfunction $S(x)$ is defined by

$$S(x) = \begin{cases} 0 & \text{for } |x| < \delta \\ 1 & \text{for } |x| > \delta. \end{cases}$$

Since the integrand is symmetric in $t_1, t_2 \cdots t_N$, and because the hard cores impose a linear order of the molecules in L, one can write (2) in the form

$$Z(L, T, N) = \frac{e^{-\frac{1}{2}N\nu}}{\Lambda^N} \int_{0<t_1<t_2\cdots<t_N<L} \cdots \int dt_1 \cdots dt_N$$
$$\times \exp\left[\frac{\nu}{2}\sum_{i=1}^N\sum_{j=1}^N e^{-\gamma|t_i - t_j|}\right]\cdot\prod_{j=1}^{N-1} S(|t_{j+1} - t_j|). \tag{3}$$

Also the attractive part of the integrand can be ordered by making use of the identity

$$\exp\left[\frac{\nu}{2}\sum_{i,j=1}^N \exp(-\gamma|t_i - t_j|)\right]$$
$$= \int_{-\infty}^{+\infty}\cdots\int dx_1 \cdots dx_N$$
$$\times \exp[\nu^{\frac{1}{2}}(x_1 + x_2 + \cdots + x_N)]$$
$$\cdot W(x_1)\prod_{j=1}^N P(x_j \mid x_{j+1}, t_{j+1} - t_j), \tag{4}$$

where

$$W(x) = \frac{1}{(2\pi)^{\frac{1}{2}}}\exp[-\tfrac{1}{2}x^2],$$
$$P(x \mid y, t) = [\{2\pi(1 - e^{-2\gamma t})\}^{\frac{1}{2}}]^{-1} \tag{5}$$
$$\times \exp[-(y - xe^{-\gamma t})^2/2(1 - e^{-2\gamma t})].$$

The motivation and the proof of the identity (4) follows from the observation that $\exp(-\gamma|t_i - t_j|)$ is the covariance of a one-dimensional, Gaussian Markoff process (the so-called *O–U* process), but of course one can also prove (4) directly. By substituting (4) into (3) and by making a Laplace transformation in L, the integrals over the t_i can be separated in pairs since clearly

$$\int_0^\infty e^{-sL} Z(L, T, N)\, dL$$

is of the form

$$\int_0^\infty dLe^{-sL}\int_0^L dt_1 \int_{t_1}^L dt_2$$
$$\times \int_{t_2}^L dt_3 \cdots \int_{t_{N-1}}^L dt_N \prod_{j=1}^{N-1} F_j(t_{j+1} - t_j)$$

[6] That there is no phase transition for a one-dimensional system with only nearest-neighbor interaction, was shown first by F. Gursey, Proc. Cambridge Phil. Soc. **46**, 182 (1950). This was generalized (with the same result) to the case where each molecule interacts with a finite number of neighbors by L. Van Hove, Physica **16**, 137 (1950). Compare also the discussion given by A. Munster in his book *Statistische Thermodynamik* (Springer-Verlag, Berlin, Germany, 1956), Secs. 7.7 and 8.8, where one also finds further references.

[7] Also called the equal-area rule. See Maxwell, *Collected Works*, Dover reprint, Vol. II, p. 425.

$$= \int_0^\infty dt_1 \int_{t_1}^\infty dt_2 \cdots \int_{t_{N-1}}^\infty dt_N$$

$$\times \int_{t_N}^\infty dL e^{-sL} \prod_{j=1}^{N-1} F_j(t_{j+1} - t_j). \tag{6}$$

Setting

$$t_1 = \tau_1,$$

$$t_2 = \tau_1 + \tau_2, \cdots, t_N = \tau_1 + \tau_2 + \cdots + \tau_N,$$

$$L = \tau_1 + \tau_2 + \cdots + \tau_N + \tau_{N+1},$$

one easily shows that (6) becomes

$$\frac{1}{s^2} \prod_{j=1}^{N-1} \int_0^\infty d\tau\, e^{-s\tau} F_j(\tau),$$

so that by putting in the appropriate form for the $F_j(z)$, one obtains (suppressing the temperature T from now on)

$$\int_0^\infty dL e^{-sL} Z(L, N)$$

$$= \frac{1}{\Lambda^N} \frac{e^{-N\nu/2}}{s^2} \int_{-\infty}^{+\infty} \cdots \int dx_1 \cdots dx_N$$

$$\times \exp\,[\nu^{\frac{1}{2}}(x_1 + \cdots + x_N)] W(x_1) \prod_{j=1}^{N-1} p_s(x_j \mid x_{j+1}) \tag{7}$$

with

$$p_s(x \mid y) = \int_s^\infty d\tau\, e^{-s\tau} P(x \mid y, \tau). \tag{8}$$

The ordering of the x_i in successive pairs suggests the introduction of the kernel

$$K_s(x, y) = \frac{W(x) p_s(x \mid y)}{[W(x)W(y)]^{\frac{1}{2}}} \exp\,[(\nu^{\frac{1}{2}}/2)(x + y)] \tag{9}$$

and the corresponding Kac integral equation

$$\int_{-\infty}^{+\infty} dy\, K_s(x, y)\psi(y) = \lambda\psi(x). \tag{10}$$

It is easy to see that $K_s(x, y)$ is symmetric and in addition one can show (for the proof, see Kac[5]):

(a) $K_s(x, y)$ is positive definite, which means that

$$\iint K_s(x, y)\psi(x)\psi(y)\, dx\, dy$$

is always positive, whenever $\psi(x)$ is not identically 0;

(b) $K_s(x, y)$ is a Hilbert–Schmidt kernel, which means that

$$\iint K_s^2(x, y)\, dx\, dy < \infty.$$

From these facts, one can conclude that Eq. (10) has a *discrete* set of *positive* eigenvalues $\lambda_i(s)$ starting from a maximum eigenvalue $\lambda_0(s)$ and converging to zero as $i \to \infty$, that the corresponding eigenfunctions $\psi_i(x, s)$ form a complete orthonormal set, and that the kernel $K_s(x, y)$ can be expanded in the convergent series

$$K_s(x, y) = \sum_i \lambda_i(s)\psi_i(x, s)\psi_i(y, s). \tag{11}$$

Writing Eq. (7) in the form

$$\int_0^\infty dL e^{-sL} Z(L, N)$$

$$= \frac{e^{-\frac{1}{2}N\nu}}{\Lambda^N s^2} \int_{-\infty}^{+\infty} \cdots \int dx_1 \cdots dx_N \exp\left[\frac{\nu^{\frac{1}{2}}}{2}(x_1 + x_N)\right]$$

$$\times [W(x_1)W(x_N)]^{\frac{1}{2}} \prod_{j=1}^{N} K_s(x_j, x_{j+1}),$$

and using Eq. (11), one can integrate over x_2, $x_3 \cdots x_{N-1}$ and obtain

$$\int_0^\infty dL e^{-sL} Z(L, N) = \frac{e^{-\frac{1}{2}N\nu}}{\Lambda^N s^2} \sum_{j=0}^{\infty} \lambda_j^{N-1}(s) A_j^2 \tag{12}$$

with

$$A_j = \int_{-\infty}^{+\infty} dx\, \psi_j(x, s) W^{\frac{1}{2}}(x) e^{\frac{1}{2}\nu^{\frac{1}{2}} x}. \tag{13}$$

If one now forms the grand partition function

$$G(L, z) = \sum_{N=1}^{\infty} Z(L, N)(\Lambda z)^N, \tag{14}$$

it follows from Eq. (12), that

$$\int_0^\infty dL e^{-sL} G(L, z) = \frac{z e^{-\frac{1}{2}\nu}}{s^2} \sum_j \frac{A_j^2}{1 - \lambda_j z e^{-\frac{1}{2}\nu}}, \tag{15}$$

provided $z < e^{\frac{1}{2}\nu}/\lambda_0(s)$. It is clear therefore that the abscissa of convergence of the Laplace transform of $G(L, z)$ is that value of s, for which

$$z = e^{\frac{1}{2}\nu}/\lambda_0(s). \tag{16}$$

Since on the other hand this abscissa is also

$$\lim_{L\to\infty} \frac{1}{L} \ln G(L, z), \tag{17}$$

which has the thermodynamic meaning of p/kT, it follows that the relation between z and the pressure of the gas is given by

$$\lambda_0(p/kT) = e^{\frac{1}{2}\nu}/z. \tag{18}$$

Furthermore, since the thermodynamic meaning of z is the fugacity, which is related to the chemical potential (Gibbs free energy per particle) $\mu(p, T)$ by

$$\mu = kT \ln (\Lambda z), \tag{19}$$

the equation of state follows from

$$l = (\partial\mu/\partial p)_T = -\lambda_0'(s)/\lambda_0(s), \quad s = p/kT, \tag{20}$$

where the prime denotes differentiation after s.

The following statements can be proved concerning the maximum eigenvalue $\lambda_0(s)$ and the eigenvalue spectrum $\lambda_i(s)$ (see Appendix 1):

(a) *All* eigenvalues are monotonically decreasing functions of s and approach zero for $s \to \infty$.

(b) The maximum eigenvalue $\lambda_0(s)$ goes to infinity for $s \to 0$. For no value of s can it be degenerate, and it is an analytic function of s for all real $s > 0$. Finally there is the inequality

$$\lambda_0'(s)^2 - \lambda_0(s)\lambda_0''(s) < 0, \tag{21}$$

which implies that $\lambda_0''(s)$ is always positive.

(c) All other eigenvalues are finite for $s = 0$.

It seems very likely that the curves $\lambda_i(s)$, $i > 0$ will also not cross each other, in which case the spectrum as a function of s may look as shown in Fig. 2.

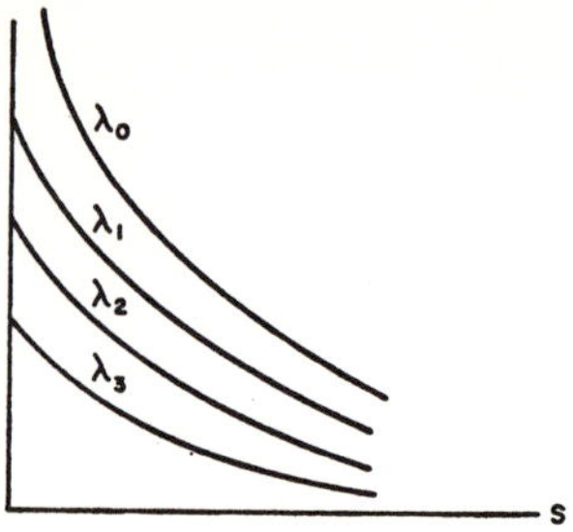

FIG. 2. Eigenvalue spectrum as a function of s.

One can conclude that for any value of z there is an *unique* value s fulfilling Eq. (16), and from Eqs. (20) and (21) it then follows that l is a monotonically decreasing and analytic function of $s = p/kT$ for all positive values of s. There is therefore *no* phase transition. Note that the theory gives directly [see Eqs. (18) and (19)], the Gibbs free energy as a function of the pressure, and that therefore one obtains the specific volume l as a function of the pressure instead of the pressure as a function of l as in the Mayer theory. The connection with the virial expansion of the equation of state is therefore quite complicated. One can show that the maximum eigenvalue $\lambda_0(s)$ can be expanded in the form

$$\lambda_0(s) = e^{\frac{1}{2}\nu}(s^{-1} + a_0 + a_1 s + a_2 s^2 + \cdots). \tag{22}$$

The coefficients a_i can be calculated successively and they are related to the Mayer cluster integrals b_l. One finds

$$a_0 = b_2; \qquad a_1 = b_3 - b_2^2; \tag{23}$$

and so on. For the proof we refer again to Appendix I. From Eqs. (22) and (23) one can then verify that the equation of state (20) is, for small pressures, simply the inversion of the usual virial expansion

$$pl/kT = 1 + B/l + C/l^2 + \cdots,$$

with $B = -b_2$, $C = -2b_3 + 4b_2^2$, etc. We have been unable to find a simple "graphological" characterization of the coefficients a_i.

III. THE VAN DER WAALS LIMIT

We will now study what happens to the maximum eigenvalue $\lambda_0(s)$, if one puts $\alpha = \alpha_0\gamma$ and then goes to the limit $\gamma \to 0$. To do this, we investigate the moments $\sum_i \lambda_i^n(s)$ of the distribution of the eigenvalues $\lambda_i(s)$. One has

$$\sum_i \lambda_i^n = \int_{-\infty}^{+\infty}\!\!\cdots\!\int dx_1 \cdots dx_n$$

$$\times\, K_s(x_1, x_2)K_s(x_2, x_3) \cdots K_s(x_n, x_1)$$

$$= \int_{-\infty}^{+\infty}\!\!\cdots\!\int dx_1 \cdots dx_n$$

$$\times \exp\,[-(\nu_0\gamma)^{\frac{1}{2}}(x_1 + x_2 \cdots + x_n)]$$

$$\times \int_0^{\infty}\!\!\cdots\!\int d\tau_1 \cdots d\tau_n$$

$$\times \exp\,[-s(\tau_1 + \cdots + \tau_n)] \prod_{i=1}^{n} P(x_i \mid x_{i+1}, \tau_i),$$

where $\nu_0 = \alpha_0/kT$ and $x_{n+1} \equiv x_1$. The integrals over the x_i can be carried out and one can then investigate the limit $\gamma \to 0$. Some of the details are given in Appendix II. The result is the following theorem. For $\gamma \to 0$,

$$\lim_{\gamma\to 0} \gamma \sum_i \lambda_i^n = \frac{1}{2\pi}\iint_{-\infty}^{+\infty} d\xi\, d\eta \tag{24}$$

$$\times \left(\exp\,[\eta(2\nu_0)^{\frac{1}{2}}] \int_s^{\infty} d\tau \exp\,[-s\tau - \tfrac{1}{2}(\xi^2 + \eta^2)\tau]\right)^n.$$

One now can reason heuristically as follows. Call

$$f(\xi, \eta) = \exp\,[\eta(2\nu_0)^{\frac{1}{2}}] \int_s^{\infty} d\tau$$

$$\times \exp\,[-s\tau - \tfrac{1}{2}(\xi^2 + \eta^2)\tau]; \tag{25}$$

then since (24) holds for all n, one can expect that for any reasonable function $g(x)$,

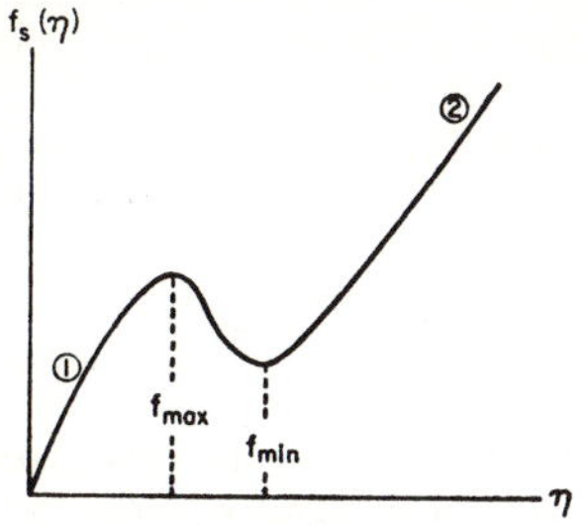

FIG. 3. The function $f_s(\eta)$ in the two-phase region.

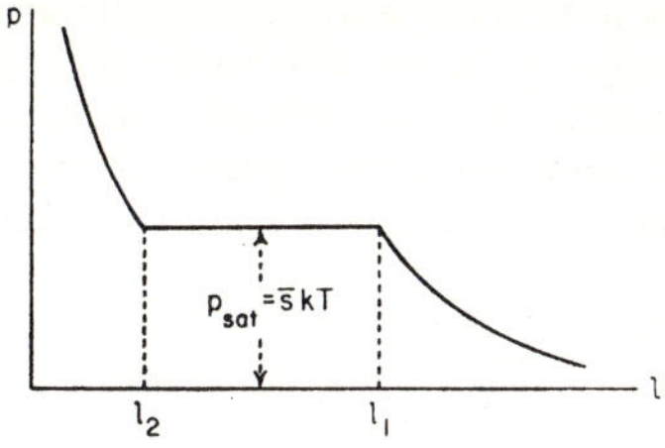

FIG. 5. Isotherm corresponding to Fig. 4.

$$\lim_{\gamma\to 0} \gamma \sum_i g(\lambda_i) = \frac{1}{2\pi} \iint_{-\infty}^{+\infty} d\xi\, d\eta\, g[f(\xi, \eta)]. \quad (26)$$

Take now for $g(x)$ the step function

$$g(x) = \begin{bmatrix} 1 & \text{for} \quad \alpha < x < \beta \\ 0 & \text{otherwise}\ , \end{bmatrix}$$

and let $N_\gamma(\alpha, \beta)$ be the number of eigenvalues between α and β; then one concludes from Eq. (26) that

$$\lim_{\gamma\to 0} \gamma N_\gamma(\alpha, \beta) = \frac{1}{2\pi}$$
$$\times \text{ [Area in } (\xi, \eta) \text{ plane for which } \alpha < f(\xi, \eta) < \beta]. \quad (27)$$

Suppose that $f(\xi, \eta)$ has an *absolute* maximum $\omega(s)$, then it is clear that for $\gamma \to 0$, ω must be a limit point for the series of eigenvalues, because in any interval between $\omega - \epsilon$ and ω there must be in the limit $\gamma \to 0$, an infinite number of eigenvalues because the area $(\omega - \epsilon) < f(\xi, \eta) < \omega$ is finite. It seems therefore reasonable to expect that for $\gamma \to 0$, $\omega(s)$ is the maximum eigenvalue, or in other words that

$$\lim_{\gamma\to 0} \lambda_0(s, \gamma) = \omega(s) = \max_{(\xi,\eta)} f(\xi, \eta) = \max_{(\eta)} F(\eta), \quad (28)$$

with

$$F(\eta) \equiv f(0, \eta)$$
$$= \exp\left[-\delta(s + \tfrac{1}{2}\eta^2) - \ln(s + \tfrac{1}{2}\eta^2) + \eta(2\nu_0)^{\frac{1}{2}}\right],$$

since the maximum of $f(\xi, \eta)$ will always occur for $\xi = 0$.

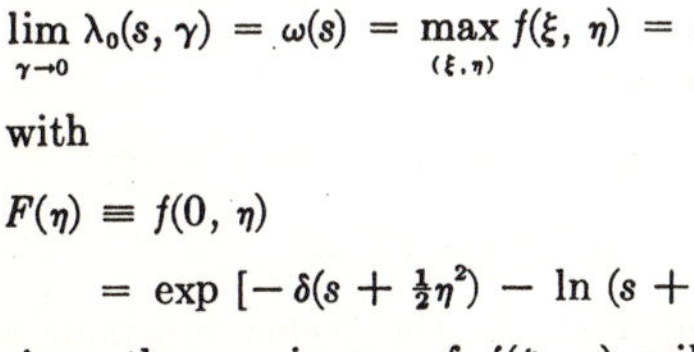

FIG. 4. The maximum eigenvalue, in the van der Waals limit, for a temperature below the critical.

Note that our argument from Eq. (27) does not really prove Eq. (28), since (27) does not exclude that a few discrete eigenvalues remain above $\omega(s)$ for $\gamma \to 0$. However, Eq. (28) is correct and the argument can be made completely rigorous. This is not done, because in the next section, the detailed investigation of the approach to the limit $\gamma \to 0$ implies a rigorous proof of Eq. (28).

To discuss Eq. (28) further, first note that the maximum will occur for that value $\eta(s)$ for which

$$f_s(\eta) \equiv \eta[\delta + 1/(s + \tfrac{1}{2}\eta^2)] = (2\nu_0)^{\frac{1}{2}}. \quad (29)$$

Since $df_s/d\eta = 0$ implies

$$s + \tfrac{1}{2}\eta^2 = (1/2\delta)[1 \pm (1 - 8s\delta)^{\frac{1}{2}}], \quad (30)$$

one must distinguish two cases.

(a) $8s\delta > 1$; $f_s(\eta)$ is then a monotonically increasing function of η, so that (29) has always one solution $\eta(s)$ corresponding to an unique maximum for $F(\eta)$. The equation of state follows from

$$l = -\omega'(s)/\omega(s) = +\delta$$
$$+ [1/(s + \tfrac{1}{2}\eta^2(s))] = +(2\nu_0)^{\frac{1}{2}}/\eta(s), \quad (31)$$

using (29). Putting $s = p/kT$ and $\eta(s) = (2\nu_0)^{\frac{1}{2}}/l$ in Eq. (29), one obtains the van der Waals equation

$$p = kT/(l - \delta) - \alpha_0/l^2.$$

(b) $8s\delta < 1$; $f_s(\eta)$ has then a maximum and a minimum (see Fig. 3), and one easily verifies that for $s = 1/8\delta$, $f_{min} = f_{max} = (\frac{3}{2})(3\delta)^{\frac{1}{2}}$, and that for decreasing s, f_{min} and f_{max} increase monotonically; f_{min} remains finite, while f_{max} goes to infinity for $s \to 0$.

If now $(2\nu_0)^{\frac{1}{2}} < f_{min}$, then again there is only one solution of Eq. (29) corresponding to an unique maximum of $F(\eta)$. This is also the case if $(2\nu_0)^{\frac{1}{2}} > f_{max}$. However if $f_{min} < (2\nu_0)^{\frac{1}{2}} < f_{max}$, then Eq. (29) has *three* roots, the outer two corresponding to local maxima of $F(\eta)$ and the inner one to a local minimum. One must now decide which of the two maxima is the absolute maximum. Clearly if one varies ν_0 at fixed s, or if one varies s (keeping $8s\delta < 1$) at fixed ν_0, there must be a value $\bar{s}(\nu_0)$ for which the two maxima are equal. It is at this value that $\omega(s)$ changes its analytic behavior abruptly, since $\eta(s)$ will suddenly jump from the branch ① to the branch ② of the $f_s(\eta)$ curve (see Fig. 3). In fact it is easy to show that for $s = \bar{s}$, $\omega(s)$ has a discontinuity in slope, and since $l = -\omega'(s)/\omega(s)$ this corresponds to an isotherm with a horizontal piece (see Figs. 4 and 5). Furthermore, since (31) remains valid, *both* the liquid and the vapor part are *still* described by the van der Waals equation. Finally, the equality of the maxima of $F(\eta)$ for $s = \bar{s}$ or of the two values of $\omega(s)$ means the equality of the Gibbs free energy of the two phases for $s = \bar{s}$, which in turn is equivalent to the Maxwell rule.

The critical point is determined by $8s_c\delta = 1$ and $(2\nu_c)^{\frac{1}{2}} = f_{\min}(s_c) = f_{\max}(s_c) = (\frac{3}{2})(3\delta)^{\frac{1}{2}}$, and this leads to the well-known formula

$$v_c = 3\delta; \quad p_c = \alpha_0/27\delta^2; \quad kT_c = 8\alpha_0/27\delta. \tag{32}$$

IV. THE EIGENFUNCTIONS AND EIGENVALUES FOR SMALL γ

Clearly the reason that, for small γ, the eigenvalues $\lambda_n(s, \gamma)$ of the Kac equation crowd together near the value $\omega(s)$, must be related to the fact that for $\gamma \to 0$, $P(x \mid y, t)$ approaches the Dirac δ function $\delta(x - y)$. However if one assumes that for $\gamma \to 0$, the eigenfunctions $\psi_n(x, s, \gamma)$ would remain centered around $x = 0$, as in the usual eigenvalue problems, then it follows from Eq. (10) that for $\gamma \to 0$, the eigenvalues would cluster around $e^{-s\delta}/s$ which is the maximum eigenvalue for the gas of hard rods. This contradiction can only be removed if, for $\gamma \to 0$, the eigenfunctions are centered farther and farther away from the origin. This in fact is the clue for constructing a consistent successive-approximation method for the eigenfunctions and eigenvalues if γ is small.

In the basic integral equation (10), substitute

$$x = x' + \eta(2/\gamma)^{\frac{1}{2}}, \qquad y = y' + \eta(2/\gamma)^{\frac{1}{2}}, \tag{33}$$

where $\eta \equiv \eta(s)$ is the value of η for which the function $F(\eta, s)$ defined by (28) has an absolute maximum. Think *first* of the *one-phase region*, so that $\eta(s)$ is unique. Let

$$h(x') \equiv \psi[x' + \eta(2/\gamma)^{\frac{1}{2}}]; \tag{34}$$

then Eq. (10) can be written in the form

$$\exp\,[\eta(2\nu_0)^{\frac{1}{2}}] \int_{-\infty}^{+\infty} dy' \int_{\delta}^{\infty} d\tau \left[\frac{W(x')}{W(y')}\right]^{\frac{1}{2}} P(x' \mid y', \tau)$$
$$\times \exp\left[-s\tau - \frac{\eta^2}{\gamma}\tanh\left(\frac{\gamma\tau}{2}\right) + (x' + y')\,\frac{(\nu_0\gamma)^{\frac{1}{2}}}{2}\right.$$
$$\left. - \frac{\eta}{2}\left(\frac{2}{\gamma}\right)^{\frac{1}{2}} \tanh\left(\frac{\gamma\tau}{2}\right)(x' + y')\right] h(y') = \lambda h(x'). \tag{35}$$

Clearly now for $\gamma \to 0$, $P(x' \mid y', \tau) \to \delta(x' - y')$ if $h(x')$ is centered around $x' = 0$, the eigenvalues will approach

$$\exp\,[\eta(2\nu_0)^{\frac{1}{2}}] \int_{\delta}^{\infty} d\tau \exp\left[-s\tau - \frac{\eta^2\tau}{2}\right] = \omega(s),$$

as expected. One may say that by the substitution (33) one has "tamed" the integral equation for a perturbation expansion of the form

$$\lambda = \omega(s)\{1 + \mu^{(1)}\gamma + \mu^{(2)}\gamma^2 + \cdots\},$$
$$h(x') = h^{(0)}(x') + \gamma^{\frac{1}{2}}h^{(1)}(x') + \gamma h^{(2)}(x') + \cdots. \tag{36}$$

Since the algebra is involved, we simply indicate how this is done. In the left-hand side of Eq. (35), set

$$y' = x'e^{-\gamma\tau} + \zeta(1 - e^{-2\gamma\tau})^{\frac{1}{2}},$$

and develop the whole integrand in powers of $\gamma^{\frac{1}{2}}$. The integrals over ζ are Gaussian integrals and the integrals over τ can all be expressed in terms of $\omega(s)$ and its derivatives. Up to order $\gamma^{\frac{3}{2}}$, and except for a common factor $\omega(s)$, the left-hand side of (35) becomes

$$h(x') + \gamma l[h''(x') + h(x)(\tfrac{1}{2} - \tfrac{1}{4}x'^2 + \tfrac{1}{2}x'^2\nu_0(l - \delta)^2/l^3)]$$
$$+ [\gamma^{\frac{3}{2}}\nu_0^{\frac{1}{2}}(l - \delta)^2/l][\tfrac{1}{4}x'^3h(x')(1 - 4\nu_0(l - \delta)/l^2)$$
$$- \tfrac{1}{2}x'h(x') - h'(x') - x'h''(x')],$$

where (31) has been used to express $\eta(s)$ in terms of l. If one now introduces the expansions (36) on both sides of Eq. (35) and equates equal powers of $\gamma^{\frac{1}{2}}$, one obtains, in zeroth approximation,

$$\frac{d^2H^{(0)}}{dz^2} + \left[\frac{1}{B}\left(\frac{l}{2} - \mu^{(1)}\right) - \frac{z^2}{4}\right]H^{(0)}(z) = 0, \tag{37}$$

where we have set

$$B^2 = l^2 - 2\nu_0(l - \delta)^2/l, \qquad x' = z(l/B)^{\frac{1}{2}},$$
$$h^{(0)}(x') = H^{(0)}(z). \tag{38}$$

This is the equation for the Weber functions and hence one concludes that

$$\mu_n^{(1)} = \tfrac{1}{2}[l - (2n + 1)B],$$
$$H_n^{(0)}(z) = N_nD_n(z), \tag{39}$$

where N_n is the normalization factor $(B/2\pi l)^{\frac{1}{2}}(n!)^{-\frac{1}{2}}$. Note that the quantity B is related to the compressibility of the gas, since, from the van der Waals equation,

$$B^2 = [-l^2(l - \delta)^2/kT](\partial p/\partial l)_T. \tag{40}$$

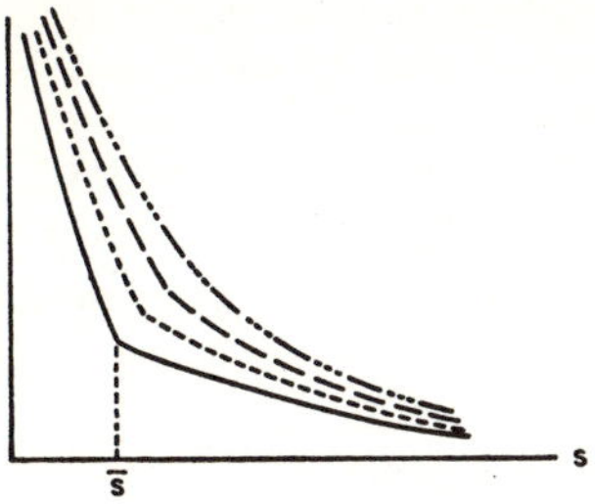

FIG. 6. ——— $\omega(s)$;—···—$\lambda(s,\gamma)$ (exact); -------- $\lambda(s,\gamma)$ to order $\gamma\delta$; — — — $\lambda(s,\gamma)$ to order $(\gamma\delta)^2$.

In the one-phase region $\partial p/\partial l$ is always negative, and therefore B can be taken positive. The maximum eigenvalue corresponds therefore to $n = 0$, and is given up to first order by

$$\lambda_0(s,\gamma) = \omega(s)[1 + \tfrac{1}{2}\gamma(l - B)]. \tag{41}$$

This proves the statement in Sec. III, that the maximum eigenvalue is $\omega(s)$ in the van der Waals limit.

In first approximation one obtains

$$\frac{d^2H_n^{(1)}}{dz^2} + (n + \tfrac{1}{2} - \tfrac{1}{4}z^2)H_n^{(1)} = \frac{\nu_0^{\frac{1}{2}}(l-\delta)^2}{l^{\frac{1}{2}}B^{\frac{1}{2}}}\frac{dH_n^{(0)}}{dz} + \frac{\nu_0^{\frac{1}{2}}(l-\delta)^2}{l^{\frac{1}{2}}B^{\frac{1}{2}}}\left(\frac{l}{2B} - n - \tfrac{1}{2}\right)zH_n^{(0)} - \frac{\nu_0^{\frac{1}{2}}(l-\delta)^3(l-3\delta)}{6l^{5/2}B^{5/2}}z^3H_n^{(0)}. \tag{42}$$

It can be easily verified that the right-hand side is orthogonal to $H_n^{(0)}$, so that (42) has an unique solution. For the explicit form of the solution and also for the second approximation, which will be needed in Part II of this series of papers, we refer to Appendix III.

Let us now consider the *two-phase region*.

It is clear that in this case the argument must be revised, since the $\eta(s)$ in the shift (33) is no longer unique. In fact, since $F(\eta, s)$ has now two equal maxima, there are *two* values of $\eta(s)$ corresponding to the volumes l_1 and l_2 of the saturated vapor and liquid phase (see Fig. 5). For each of these two values one can make the shift (33), and repeat the the argument. Clearly in both cases one will get the *same* zeroth-order eigenvalue $\omega(s)$, but one will get *another* eigenfunction, since the quantity l/B will be different at l_1 and l_2. One must say therefore that for $l_2 < l < l_1$, i.e. in the two-phase region, the maximum eigenvalue in the van der Waals limit is *doubly degenerate*, and in lowest order, the eigenfunction will have the form [since $D_0(z) = \exp(-\frac{1}{4}z^2)$],

$$\psi_0(x,s) = \alpha_1\left(\frac{B_1}{2\pi l_1}\right)^{\frac{1}{4}}\exp\left[-\frac{B_1}{4l_1}\left(x - \eta_1\left[\frac{2}{\gamma}\right]^{\frac{1}{2}}\right)^2\right] + \alpha_2\left(\frac{B_2}{2\pi l_2}\right)^{\frac{1}{4}}\exp\left[-\frac{B_2}{4l_2}\left(x - \eta_2\left[\frac{2}{\gamma}\right]^{\frac{1}{2}}\right)^2\right], \tag{43}$$

where $\eta_1(\bar{s}) = (2\nu_0)^{\frac{1}{2}}/l_1$ and $\eta_2(\bar{s}) = (2\nu_0)^{\frac{1}{2}}/l_2$.

The question now arises as to the appropriate linear combination corresponding to a specific length $l = \xi_1 l_1 + \xi_2 l_2$, (where ξ_1, ξ_2 are the mole fractions of the vapor and liquid phase), so that $\xi_1 + \xi_2 = 1$. We show that the answer is given by

$$\xi_1 = \alpha_1^2 \qquad \xi_2 = \alpha_2^2. \tag{44}$$

To prove this, note first that for small γ, *the two Gaussian functions in* (43) *do not overlap*. Hence from the normalization of $\psi(x,\bar{s})$, one obtains $\alpha_1^2 + \alpha_2^2 = 1$. Furthermore, for any finite γ, it follows from the Kac equation that

$$\lambda_0(s,\gamma) = \iint dx\,dy\,K_s(x,y) \times \psi_0(x,s,\gamma)\psi_0(y,s,\gamma),$$
$$\lambda_0'(s,\gamma) = -\lambda_0(s,\gamma)l = \iint dx\,dy\,\frac{\partial K_s}{\partial s} \times \psi_0(x,s,\gamma)\psi_0(y,s,\gamma) \tag{45}$$

[since the differentiation of the eigenfunctions after s gives zero because the normalization of $\psi_0(x,s,\gamma)$ implies

$$\int \psi_0(x,s,\gamma)\frac{\partial}{\partial s}\psi_0(x,s,\gamma)\,dx = 0].$$

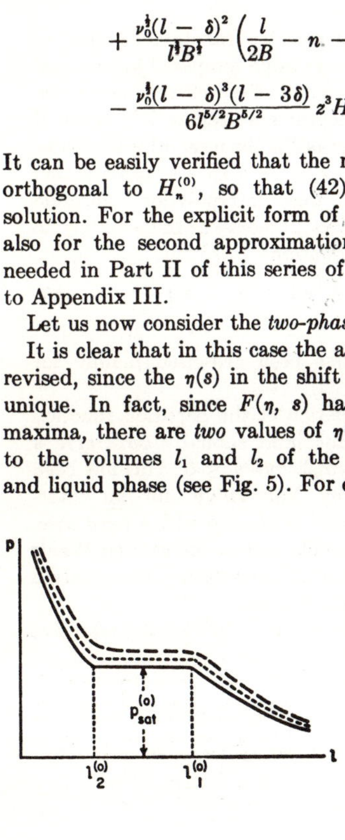

FIG. 7. ——— The van der Waals isotherm ($\gamma = 0$); — — — The exact isotherm; ······· The isotherm to order $\gamma\delta$.

Now take $s = \bar{s}$, and let $\gamma \to 0$. Since for each of the two functions in Eq. (43) one has an equation like (45) with the same s and the same $\lambda_0(s) = \omega(s)$, but with l successively l_1 and l_2, and since one can again neglect the overlap of the two functions, it follows from Eq. (45) that for $\gamma \to 0$,

$$l = \alpha_1^2 l_1 + \alpha_2^2 l_2,$$

which proves Eq. (44).

We conclude this section with the following two remarks.

(a) It is clear that the entire development breaks down near the critical point because at that point, $B \to 0$ and $l_1 \cong l_2$. The overlap of the two eigenfunctions in the two-phase region can therefore no longer be neglected. In the critical region, one has to "tame" the integral equation in a different way, which will be explained in the third part of this series.

(b) It follows from Eq. (41) that up to first order in γ, $\lambda_0(s, \gamma)$ *still* shows a discontinuity in slope, and one can be convinced that the discontinuity *remains* if the perturbation calculation is continued up to *any* finite order in γ. For finite γ, $\lambda_0(s, \gamma)$ is an analytic function of s for $s > 0$. This means that for small γ, the successive-approximation method which we have developed, approximates the analytic function $\lambda_0(s, \gamma)$ by a series of functions which have a discontinuity in slope in the region of s, where the slope of $\lambda_0(s, \gamma)$ varies very quickly (see Fig. 6). This implies that *the phase transition persists* up to any finite order in γ, although the equation of state is then of course different from the van der Waals equation. In fact it is not difficult to show from Eq. (41), that, up to first order in γ, the equation of state in the one-phase region is given by

$$p = \frac{kT}{l - \delta} - \frac{\alpha_0}{l^2} + \frac{\gamma}{2}\left\{kT - \frac{l}{B}\left[kT - \frac{\alpha_0(l^2 - \delta^2)}{l^3}\right]\right\}.$$

The saturated vapor pressure and the densities of the saturated vapor and liquid also change accordingly. One finds

$$p_{\text{sat}} = p_{\text{sat}}^{(0)} + \gamma \frac{kT}{2}\left(\frac{B_1 - B_2}{l_1 - l_2} - 1\right)$$

$$l_i = l_i^{(0)} + \frac{\gamma}{2}(l_i - \delta)^2\left(\frac{l_i}{B}\right)^2 \times \left[\frac{B_1 - B_2}{l_1 - l_2} - \frac{l_i}{B_i}\left(1 - \frac{\alpha_0(l_i^2 - \delta^2)}{kTl_i^3}\right)\right]$$

with $i = 1, 2$.

V. CONCLUDING REMARKS

To judge the relevance of our results for the three-dimensional case, let us first recall the usual derivation[8] of the van der Waals equation from the partition function. One can argue as follows: If the range of the attractive force is very large so that there are many molecules in the action sphere of each molecule, then the potential energy of the attractive forces should for almost all configurations of the molecules, be equal to the average value

$$\Phi_{\text{attr}} = -CN^2/2V, \tag{46}$$

where

$$C = -\int d\mathbf{r}\, \varphi_{\text{attr}}(r).$$

Equation (46) follows because on the average, the potential energy of *one* molecule will be proportional to the number density N/V and is in fact equal to $-CN/V$. The factor two in (46) is of course needed because each molecule interacts with all other molecules. For long-range attractive forces it seems reasonable therefore to approximate the partition function by

$$Z(V, T, N) = \frac{1}{\Lambda^{3N}}\frac{1}{N!}\exp\left[CN^2/2kTV\right] \times \int_V \cdots \int_V d\mathbf{r}_1 \cdots d\mathbf{r}_N \prod_{i<j} S(|\mathbf{r}_i - \mathbf{r}_j|), \tag{47}$$

where $S(x)$ is the same step function as used in Sec. II. Now if V is very large compared to the total proper volume Nv_0, $v_0 = \frac{4}{3}\pi r_0^3$ of the molecules, then clearly the integral approaches V^N, while if V is of the same order as Nv_0 the integral will become strongly zero about as $(V - Nv_0)^N$. Approximation of the integral by the "interpolation" formula $(V - b)^N$ with $b \cong Nv_0$, then leads immediately to the van der Waals equation

$$(p + a/V^2)(V - b) = RT, \tag{48}$$

with $a = \frac{1}{2}CN^2$.

This derivation has been criticized mainly for two reasons:

1. A correct evaluation of the partition function in the thermodynamic limit should *always* give a *stable* isotherm for which $\partial p/\partial v \leq 0$. This has been proved rigorously by Van Hove[9] for an intermolecular force of finite range and with a hard core. Since there is an unstable part in the van der Waals equation below the critical temperature, the equation is in conflict with this theorem.

2. The approximation $(V - b)^N$ for the repulsive

[8] This derivation goes back to L. S. Ornstein, Dissertation Leiden 1908. Van der Waals himself, and then later also Lorentz [*Collected Papers* (Martinus Nyhoff, The Hague, Netherlands, 1935), Vol. 6, p. 40], started from the virial theorem of Clausius. We will come back to this in Part II in this series of papers.

[9] L. Van Hove, Physica 15, 951 (1949). Recently Dr. N. G. van Kampen has pointed out a gap in Van Hove's argument. Apparently the theorem can be proved even under less restrictive conditions by a different method. See a recent paper by D. Ruelle, which will appear in the Helv. Phys. Acta. (to be published).

part of the partition function must be very rough. Even if b is fixed so that the correct second virial coefficient (which makes $b = 4Nv_0$) is obtained, the higher virial coefficients which can be computed exactly, do *not* agree with the van der Waals' values.

Let us now return to our one-dimensional system. Note that by the same reasoning as used above, one would obtain

$$Z(L, T, N) = \frac{1}{\Lambda^N} \frac{1}{N!} \exp\left[\frac{\alpha_0 N^2}{kTL}\right] \times \int_L \cdots \int_L dt_1 \cdots dt_N \prod_{i<j} S(|t_i - t_j|), \quad (49)$$

since C becomes $2\alpha_0$, with the factor two because of the two sides of the attractive force $-\alpha_0\gamma \exp[-\gamma |t|]$. The *great difference* between Eqs. (47) and (49) is that the integral in Eq. (49) can be carried out *exactly* because of the linear order which the hard cores impose. The result is $(L - N\delta)^N$. For a gas of hard rods of length δ, the equation of state as noted by Tonks[10] has precisely the van der Waals form

$$p = kT/(l - \delta). \quad (50)$$

The second objection against the van der Waals equation disappears in one dimension, and it is therefore *no* wonder that for our model in the van der Waals limit, Eq. (48) with $b = N\delta$, $a = \alpha_0 N^2$ is obtained exactly. The only question which remains is the conflict of the form (49) for Z with the Van Hove theorem. From the point of view of the development in Secs. II and III, it is clear that this conflict is due to the fact that in the derivation given above, the thermodynamic and the van der Waals limits have been coupled together, while they should be taken separately and in the proper order. Since we go *first* to the thermodynamic limit for a finite range of the attractive force, and *then* allow the range to go to infinity, we do not conflict with the Van Hove theorem, and we always obtain the stable isotherm[11]). It seems to us therefore, that the first objection against the van der Waals equation is *not* as serious as we have been led to believe.

To summarize this discussion, we conclude with the following statements.

(a) It seems a good idea, following the example of van der Waals, to try to separate the effects of the repulsive and the attractive forces on the equation of state, and to develop all thermodynamic functions in the ratio

$$\rho = \frac{\text{range of repulsive force}}{\text{range of attractive force}}. \quad (51)$$

One expects that for $\rho \ll 1$ one will get a van der Waals-*like* equation of state which would lead to the condensation phenomena and the existence of a critical point. The thermodynamic equilibrium conditions will follow automatically by going first to the thermodynamic limit.

(b) *Only* in *one* dimension does one obtain in the limit $\rho \to 0$, *exactly* the van der Waals equation of state for hard-core repulsion. This should be independent of the precise form of the attractive force. We have been able to confirm this by generalizing the Kac model to the case where the attractive force is of the form

$$\varphi_{\text{attr}}(x) = -\sum_{i=1}^{m} \alpha_i \gamma \exp[-\gamma\sigma_i x]. \quad (52)$$

The calculation of the partition function can be reduced as in Sec. II to an integral equation in m variables and the behavior of the maximum eigenvalue in the limit $\gamma \to 0$ may be considered in the same way as in Sec. III: the *same* equation (28) is obtained, except that $2\nu_0$ is replaced by

$$\frac{1}{kT} \sum_{i=1}^{m} \frac{2\alpha_i}{\sigma_i} = -\frac{2}{kT} \int_0^\infty \varphi_{\text{attr}}(x)\, dx. \quad (53)$$

Thus, again the van der Waals equation is obtained and the van der Waals a is again proportional to the area of the attractive potential.

(c) In *three* dimensions, the equation of state of a gas of hard spheres is certainly *not* $p = kT(v - b)$, although qualitatively the behavior may be similar. Even for very long-range attractive forces one can only hope to get a van der Waals-*like* equation of state. In a sense, the problem of the gas of hard spheres becomes the central problem. Whether such a gas would show a phase transition for densities near close packing (the so-called Kirkwood transition), is a famous open question. We believe that there are strong indications for the existence of such a transition and that this transition might be an idealization (or caricature!) of the fluid-solid transition. Clearly, in one dimension [see Eq. (50)], such a transition does *not* exist, and we believe therefore that even with long-range forces, to enforce co-operation, *one phase transition at most* can occur in one dimension.

Finally, we would like to mention two questions

[10] L. Tonks, Phys. Rev. **50**, 955 (1936).

[11] For our model, the Van Hove theorem is expressed by the inequality (21) for the maximum eigenvalue $\lambda_0(s, \gamma)$, which holds for any value of γ. The isotherm is monotonically decreasing for any γ, and therefore also in the limit $\gamma = 0$, $\partial p/\partial l \leq 0$.

which we have so far been unable to answer in a satisfactory way.

1. What is the relation of our theory with the Yang–Lee theory of condensation? From the existence of a hard core, it follows that the grand partition function $G(L, z)$ must be a polynomial of degree $M = L/\delta$ in z with positive coefficients. Setting

$$G(L, z) = \exp [L\chi(L, z)],$$

one can write

$$\chi(L, z) = \frac{1}{L} \sum_{i=1}^{M} \ln \left(1 - \frac{z}{z_i}\right),$$

where the z_i are the zero's of $G(L, z)$, none of which can be on the positive real axis. One can interpret $\chi(L, z)$ as the complex logarithmic potential of M point charges of strength $1/L$ situated at the points z_i. In the limit $L \to \infty$, the strength of the charges becomes smaller and smaller while the number M increases. Suppose now that in the limit $L \to \infty$, a number of the charges concentrate in a single layer *which crosses the positive real axis*, say at $z = z_0$. The limit function

$$\bar{\chi}(z) = \lim_{L\to\infty} \chi(L, z),$$

which surely exists, will then for positive and real z, consist of two analytic pieces, one for $z < z_0$ and the other for $z > z_0$. At z_0 these two pieces will be continuous but the first derivative will have a discontinuity. Since $\bar{\chi}(z) = p/kT$ one would therefore obtain a curve as in Fig. 4 (with abscissa and ordinates interchanged), and condensation would occur. It is clear that something like this must happen for our model in the van der Waals limit. It must be that for finite γ, the single layer will *not* cross the positive real axis but leave a gap, say, of order γ. For finite γ, $\bar{\chi}(z)$ will then be analytic for all real and positive z, but in the van der Waals limit when the gap closes, condensation occurs. However, we have been unable to substantiate this picture, because it is difficult to discuss the behavior of the eigenvalues of the Kac equation for complex values of s.

2. What is the relation of our theory with the Ursell–Mayer theory? We have already mentioned the problem of the "graphological" interpretation of the coefficients a_i in the expansion of $\lambda_0(s)$ [see Eq. (22)], but there are also other questions. For instance it should be possible to introduce Mayer f functions both for the repulsive and for the attractive parts of the intermolecular potential, which would lead to graphs with two types of lines. The question is then, which simplifications occur if the attractive force is long range, and whether one can learn from the one-dimensional model how to characterize better the van der Waals-like equation in three dimensions.

APPENDIX 1

For the proof of statement (a) see Kac (reference 5, p. 11). There one also finds the proof that

$$\lambda_0(s) \geq e^{\frac{1}{2}\nu} \int_\delta^\infty d\tau \exp [-s\tau + \nu e^{-\gamma\tau}], \qquad (A.1)$$

and since for $s = 0$ the integral diverges, $\lambda_0(s) \to \infty$ for $s \to 0$. Furthermore, since the right hand side of (A.1) can be shown to correspond to the "nearest-neighbor" approximation, one should expect that for small s (i.e. small pressure), $\lambda_0(s)$ becomes equal to it, which implies that for $s \to 0$, $\lambda_0(s) \sim e^{\frac{1}{2}\nu}/s$, and this will be confirmed later when we derive the expansion (22).

The statement that $\lambda_0(s)$ cannot be degenerate for any value of s, follows from the fact that the kernel $K_s(x, y)$ is positive for all x, y and for $s > 0$. Since

$$\lambda_0(s) \geq \frac{\iint dx\, dy\, \varphi(x) K_s(x, y) \varphi(y)}{\int dx\, \varphi^2(x)} \qquad (A.2)$$

for any function $\varphi(x)$, and since the maximum is reached when $\varphi(x) = \psi_0(x)$, $K_s(x, y) > 0$ implies that $\psi_0(x)$ must have the same sign for all x, and we may assume therefore $\psi_0(x) \geq 0$. Furthermore, $\psi_0(x)$ cannot be zero, unless for some particular value of x, say x_0, $K_s(x_0, y) \equiv 0$ in y, which is excluded by $K_s > 0$. Now suppose that for some value of s, $\lambda_0(s)$ was degenerate and corresponded to the two eigenfunctions $\psi_0^{(1)}(x)$ and $\psi_0^{(2)}(x)$. Taking $\psi_0^{(1)}(x) > 0$, one then can construct a linear combination $\varphi(x)$ of $\psi_0^{(1)}$ and $\psi_0^{(2)}$ which is *orthogonal* to $\psi_0^{(1)}$, and which fulfills the equation

$$\lambda_0(s) = \frac{\iint dx\, dy\, \varphi(x) K_s(x, y) \varphi(y)}{\int dx\, \varphi^2(x)}. \qquad (A.3)$$

Since $\psi_0^{(1)}(x)$ cannot be zero, $\varphi(x)$, being orthogonal to $\psi_0^{(1)}$, must be both positive and negative over some intervals, and this yields a contradiction. The absence of a degeneracy of $\lambda_0(s)$ implies that $\lambda_0(s)$ is an analytic function of s for all real $s > 0$. This is because it is known that the eigenvalues $\lambda_i(s)$ are the zero's in λ of the Fredholm determinant

$D(x, \lambda)$, which is an entire function of λ, and is analytic for all real $s > 0$. Therefore the $\lambda_i(s)$ must be analytic functions of s except at those values of s where two or more eigenvalues cross.

Finally the inequality (21) is proved as follows: With normalized eigenfunctions, one finds from

$$\lambda_0(s) = \iint dx\, dy\, \psi_0(x, s) K_s(x, y) \psi_0(y, s),$$

that the quadratic form

$$Q(z) \equiv \lambda_0(s)z^2 + 2\lambda_0'(s)z + \lambda_0''(s)$$
$$= \iint_{-\infty}^{+\infty} dx\, dy\, \psi_0(x, s)\psi_0(y, s) \times \left[z^2 K_s(x, y) + 2z \frac{\partial K_s}{\partial s} + \frac{\partial^2 K_s}{\partial s^2} \right]$$
$$+ \iint_{-\infty}^{+\infty} dx\, dy\, \frac{\partial K_s}{\partial s} \left[\psi_0(x, s) \frac{\partial \psi_0(y, s)}{\partial s} + \psi_0(y, s) \frac{\partial \psi_0(x, s)}{\partial s} \right]. \quad \text{(A.4)}$$

From Eq. (9) for $K_s(x, y)$, it can be seen that the variable s enters only through $p_s(x, y)$, and from Eq. (8) it follows that the first term in Eq. (4) is positive for all z. Also the second term in Eq. (4) is positive, since by using

$$K_s(x, y) = \sum_0^\infty \lambda_n(s)\psi_n(x, s)\psi_n(y, s),$$

and the fact that for any s the $\psi_n(x, s)$ form an orthonormal set, it can be transformed into

$$2\lambda_0(s) \int_{-\infty}^{+\infty} dx \left[\frac{\partial \psi_0(x, s)}{\partial s} \right]^2 - 2 \sum_{n=1}^\infty \lambda_n(s) \left[\int_{-\infty}^{+\infty} dx\, \frac{\partial \psi_0(x, s)}{\partial s}\, \psi_n(x, s) \right]^2,$$

and since $\lambda_n(s) < \lambda_0(s)$ for $n \geq 1$, this is bigger than

$$2\lambda_0(s) \left[\int_{-\infty}^{+\infty} dx \left(\frac{\partial \psi_0}{\partial s} \right)^2 - \sum_0^\infty \left(\int_{-\infty}^{+\infty} dx\, \psi_n \frac{\partial \psi_0}{\partial s} \right)^2 \right],$$

which is zero because of Parseval's relation applied to the function $\partial\psi_0/\partial s$. Hence $Q(z) > 0$ for all z, which implies Eq. (21).

Statement (c), Sec. II follows from the relation

$$\sum_0^\infty \lambda_i = \int_{-\infty}^{+\infty} dx\, K_s(x, x) = \int_\delta^\infty \frac{d\tau}{1 - e^{-\gamma\tau}} \exp\left[-s\tau + \frac{\nu}{2} \frac{1 + e^{-\gamma\tau}}{1 - e^{-\gamma\tau}} \right],$$

which shows that for $s \to 0$, the sum of all the eigenvalues goes to infinity as $e^{\frac{1}{2}\nu}/s$, and since this is also the behavior of $\lambda_0(s)$, all other eigenvalues must be finite for $s = 0$.

To find the expansion (22) for $\lambda_0(s)$, one must make a perturbation calculation around $s = 0$ for the eigenvalue problem

$$\int_{-\infty}^{+\infty} dy\, K_s(x, y)\psi_0(y, s) = \lambda_0(s)\psi_0(x, s). \quad \text{(A.5)}$$

First we expand $K_s(x, y)$ in powers of s. To do this we write $K_s(x, y)$ in the form

$$K_s(x, y) = \exp\left[\tfrac{1}{2}\nu^{\frac{1}{2}}(x + y)\right] \Bigg\{ \frac{e^{-s\delta}}{s} [W(x)W(y)]^{\frac{1}{2}} + \sum_{n=0}^\infty \frac{(-s)^n}{n!} \int_\delta^\infty d\tau\, \tau^n \left(\left[\frac{W(x)}{W(y)} \right]^{\frac{1}{2}} P(x \mid y, \tau) - [W(x)W(y)]^{\frac{1}{2}} \right) \Bigg\}. \quad \text{(A.6)}$$

Now put

$$\lambda_0(s) = (e^{\frac{1}{2}\nu}/s)(a_{-1} + a_0 s + a_1 s^2 + \cdots),$$
$$\psi_0(x, s) = \psi^{(0)}(x) + s\psi^{(1)}(x) + \cdots, \quad \text{(A.7)}$$

and let $\psi^{(0)}(x)$ be normalized to one. Introducing (6) and (7) into (5) and equating the terms proportional to $1/s$, one obtains

$$\int_{-\infty}^{+\infty} dy\, [W(x)W(y)]^{\frac{1}{2}} \times \exp\left[\tfrac{1}{2}\nu^{\frac{1}{2}}(x + y)\right]\psi^{(0)}(y) = a_{-1}\psi^{(0)}(x)e^{\frac{1}{2}\nu}. \quad \text{(A.8)}$$

One easily verifies that this has the solution

$$a_{-1} = 1; \quad \psi^{(0)}(x) = (W(x))^{\frac{1}{2}} \exp\left[-\tfrac{1}{4}\nu + \tfrac{1}{2}\nu^{\frac{1}{2}}x\right]. \quad \text{(A.9)}$$

Notice that if we try to determine in the same way the values of the other eigenvalues $\lambda_n(s)$ at $s = 0$, we would find in this approximation that $a_{-1}^{(n)} = 0$, so that all $\lambda_n(0)$ for $n > 0$ are degenerate, and to find the actual values, we would have to solve an infinite secular determinant. *Only* $\lambda_0(s)$ becomes infinite for $s \to 0$, and only for $\lambda_0(s)$ can we carry the perturbation calculation further.

Equating in Eq. (5) the terms independent of s, one finds

$$\int_{-\infty}^{+\infty} dy\, \exp\left[\tfrac{1}{2}\nu^{\frac{1}{2}}(x + y)\right] \Bigg\{ \Bigg(-\delta[W(x)W(y)]^{\frac{1}{2}} + \int_\delta^\infty d\tau \left[\frac{W(x)}{W(y)} \right]^{\frac{1}{2}} \times [P(x \mid y, \tau) - W(y)] \Bigg) \psi^{(0)}(y) + [W(x)W(y)]^{\frac{1}{2}}\psi^{(1)}(y) \Bigg\}$$
$$= e^{\frac{1}{2}\nu}[\psi^{(1)}(x) + a_0\psi^{(0)}(x)]. \quad \text{(A.10)}$$

Multiply with $\psi^{(0)}(x)$ and integrate over x. Using Eq. (9), the fact that $\psi^{(1)}(x)$ is orthogonal to $\psi^{(0)}(x)$, and the formula

$$\iint_{-\infty}^{+\infty} dx\, dy \exp[\nu^{\frac{1}{2}}(x+y)]W(x)P(x \mid y, \tau) = \exp[\nu(1+e^{-\gamma\tau})],$$

which is a special case of Eq. (4), one obtains

$$a_0 = -\delta + \int_\delta^\infty d\tau\, \{\exp(\nu e^{-\gamma\tau} - 1)\}. \qquad (A.11)$$

The right-hand side can be written as

$$\lim_{L\to\infty} \frac{1}{2L} \iint_L dt_1\, dt_2 \left\{\exp\left[\frac{1}{kT}\varphi(|t_1 - t_2|)\right] - 1\right\},$$

which is the definition of Mayer's b_2. One also easily verifies from Eq. (10), that

$$\psi^{(1)}(x) = -(a_0 + \delta)\psi^{(0)}(x) + e^{-\frac{1}{2}\nu}\int_{-\infty}^{+\infty} dy\, \psi^{(0)}(y) \exp\left[\frac{\nu^{\frac{1}{2}}}{2}(x+y)\right] \times \int_\delta^\infty d\tau \left[\frac{W(x)}{W(y)}\right]^{\frac{1}{2}} [P(x \mid y, \tau) - W(y)].$$

We can go on in this way, but the calculation soon becomes very lengthy. We have verified that in next approximation, one obtains $a_1 = b_3 - b_2^2$, where Mayer's b_3 is defined by

$$b_3 = \lim_{L\to\infty} \frac{1}{6L} \iiint_L dt_1\, dt_2\, dt_3 \times [f_{12}f_{13} + f_{12}f_{23} + f_{13}f_{23} + f_{12}f_{13}f_{23}]$$

with

$$f_{ij} = \exp\left[\frac{1}{kT}\varphi(|t_i - t_j|)\right] - 1.$$

The general expression for the a_i in terms of the Mayer b_l is lacking. To obtain the correct equation of state up to the fourth virial coefficient, a_2 should be equal to $b_4 - 3b_2b_3 + 2b_2^3$, but we have not verified this.

APPENDIX II

To indicate the proof of Eq. (24), it will be sufficient to consider the case $n = 3$. One has

$$\sum_i \lambda_i^3 = \iiint_\delta^\infty d\tau_1\, d\tau_2\, d\tau_3\, e^{-s(\tau_1+\tau_2+\tau_3)} \times \iiint_{-\infty}^{+\infty} dx_1\, dx_2\, dx_3 \exp[-(\nu_0\gamma)^{\frac{1}{2}}(x_1+x_2+x_3)] \times P(x_1 \mid x_2, \tau_1)P(x_2 \mid x_3, \tau_2)P(x_3 \mid x_1, \tau_3),$$

where $P(x \mid y, \tau)$ is given by Eq. (5). Introduce the new variables

$$\xi_1 = (x_2 - x_1 e^{-\gamma\tau_1})(1 - e^{-2\gamma\tau_1})^{-\frac{1}{2}}$$
$$\xi_2 = (x_3 - x_2 e^{-\gamma\tau_2})(1 - e^{-2\gamma\tau_2})^{-\frac{1}{2}}$$
$$\xi_3 = (x_1 - x_3 e^{-\gamma\tau_3})(1 - e^{-2\gamma\tau_3})^{-\frac{1}{2}},$$

with the Jacobian

$$\frac{\partial(\xi_1, \xi_2, \xi_3)}{\partial(x_1, x_2, x_3)} = \{1 - e^{-\gamma(\tau_1+\tau_2+\tau_3)}\} \times \prod_{i=1}^3 (1 - \exp[-2\gamma\tau_i])^{-\frac{1}{2}}.$$

The integral over the x_i then becomes

$$\{(2\pi)^{\frac{3}{2}}[1 - e^{-\gamma(\tau_1+\tau_2+\tau_3)}]\}^{-1} \times \iiint_{-\infty}^{+\infty} d\xi_1\, d\xi_2\, d\xi_3 \exp[-\tfrac{1}{2}(\xi_1^2 + \xi_2^2 + \xi_3^2)] \times \exp\left[\frac{(\nu_0\gamma)^{\frac{1}{2}}}{e^{-\gamma(\tau_1+\tau_2+\tau_3)} - 1} \sum_{i=1}^3 A_i\xi_i\right],$$

with

$$A_i = (1 - e^{-2\gamma\tau_i})^{\frac{1}{2}}[e^{-\gamma(\tau_{i+1}+\tau_{i+2})} + e^{-\gamma\tau_{i+1}} + 1],$$

and $\tau_4 \equiv \tau_1$, $\tau_5 \equiv \tau_2$.

In this form it is convenient to go to the limit $\gamma \to 0$. Since the A_i approach $3(2\gamma\tau_i)^{\frac{1}{2}}$, the integrand of the ξ_i integral becomes independent of γ, and one obtains

$$\lim_{\gamma\to 0} \gamma \sum_i \lambda_i^3 = \iiint_\delta^\infty d\tau_1\, d\tau_2\, d\tau_3\, e^{-s(\tau_1+\tau_2+\tau_3)} \times [(2\pi)^{\frac{3}{2}}(\tau_1+\tau_2+\tau_3)]^{-1} \iiint_{-\infty}^{+\infty} d\xi_1\, d\xi_2\, d\xi_3 \times \prod_{i=1}^3 \exp\left[-\tfrac{1}{2}\xi_i^2 - \frac{3(2\nu_0\tau_i)^{\frac{1}{2}}\xi_i}{\tau_1+\tau_2+\tau_3}\right]. \qquad (a)$$

To uncouple the τ_i integrals, introduce the auxiliary variable

$$\eta = \frac{1}{\tau_1+\tau_2+\tau_3} \sum_{i=1}^3 \xi_i \tau_i^{\frac{1}{2}},$$

and multiply (a) with

$$1 = \int_{-\infty}^{+\infty} d\eta\, \delta\left(\eta - \frac{1}{\tau_1+\tau_2+\tau_3}\sum_{i=1}^3 \xi_i\tau_i^{\frac{1}{2}}\right) = \frac{1}{2\pi}\int_{-\infty}^{+\infty} d\eta \int_{-\infty}^{+\infty} dv \times \exp\left[iv\left(\eta - \frac{1}{\tau_1+\tau_2+\tau_3}\sum_{i=1}^3 \xi_i\tau_i^{\frac{1}{2}}\right)\right],$$

using the Fourier representation of the δ function. Set $v = w(\tau_1 + \tau_2 + \tau_3)$, and integrate first over the ξ_i. We then obtain

$$\lim_{\gamma \to 0} \gamma \sum_i \lambda_i^3 = \frac{1}{2\pi} \int_{-\infty}^{+\infty} d\eta \int_{-\infty}^{+\infty} dw$$
$$\times \left\{ \int_\delta^\infty d\tau \exp\left[-\eta(2\nu_0)^{\frac{1}{2}} + i\tau\eta w - \frac{\tau}{2} w^2 \right] \right\}^3 .$$

Finally, setting $w = \xi + i\eta$, one obtains Eq. (24) for $n = 3$. The proof for arbitrary n follows the same procedure.

APPENDIX III

The solution of Eq. (42) can be written in the form

$$H_n^{(1)}(z) = N_n[P_n D_{n+3}(z) + Q_n D_{n+1}(z) + R_n D_{n-1}(z) + S_n D_{n-3}(z)],$$

where

$$P_n \equiv P = \frac{\nu_0^{\frac{1}{2}}(l - \delta)^3(l - 3\delta)}{18 l^{5/2} B^{5/2}},$$

$$Q_n = 9(n + 1)P + (n + 1)V - W,$$

$$R_n = -9n^2 P - n^2 V + nW,$$

$$S_n = -n(n - 1)(n - 2)P,$$

with

$$V = \nu_0^{\frac{1}{2}}(l - \delta)^2 / l^{\frac{1}{2}} B^{\frac{1}{2}}, \qquad W = \nu_0^{\frac{1}{2}}(l - \delta)^2 / 2 l^{\frac{1}{2}} B^{\frac{1}{2}}.$$

To obtain the equation for the next approximation, the left-hand side of Eq. (35) must be expanded to terms of order γ^2. Using the expansions (36), and equating terms of order γ^2, one obtains

$$\frac{d^2 H_n^{(2)}}{dz^2} + \left(n + \tfrac{1}{2} - \frac{z^2}{4}\right) H_n^{(2)} = \frac{\mu_n^{(2)}}{B} H_n^{(0)}$$
$$+ \frac{\nu_0^{\frac{1}{2}}(l - \delta)^2 l^{\frac{1}{2}}}{4B^{5/2}} \left(\frac{4\nu_0}{3l^2} - 1\right) z^3 H_n^{(1)}$$
$$+ \frac{\nu_0^{\frac{1}{2}}(l - \delta)^2}{l^{\frac{1}{2}} B^{\frac{1}{2}}} z \frac{d^2 H_n^{(1)}}{dz^2} + \frac{\nu_0^{\frac{1}{2}}(l - \delta)^2}{l^{\frac{1}{2}} B^{\frac{1}{2}}}$$
$$\times \left\{ \tfrac{1}{2} z H_n^{(1)} + \frac{B}{l} \frac{dH_n^{(1)}}{dz} \right\}$$
$$+ H_n^{(0)}(z) \left[\frac{\nu_0^2 (l - \delta)^4 (l^2 - 2l\delta - \delta^2)}{8 l^4 B^3} z^4 \right.$$
$$+ \frac{\nu_0 (l + \delta)(l - \delta)^3}{2 l^3 B} \left(n + \tfrac{1}{2} - \frac{l}{2B}\right) z^2$$
$$- \frac{\nu_0}{12 l^2 B} (15 l^3 - 36 l^2 \delta + 30 l \delta^2 - 8\delta^3)$$
$$+ \frac{1}{2B} (2l^2 - 2l\delta + \delta^2)$$
$$\left. \times \left\{ \tfrac{1}{4} + (n + \tfrac{1}{2}) \frac{B}{l} - (n^2 + n + \tfrac{3}{4}) \frac{B^2}{l^2} \right\} \right]$$
$$- \frac{\nu_0 (l - \delta)^3 (l + \delta)}{l^3 B} z \frac{dH_n^{(0)}}{dz},$$

from which $\mu_n^{(2)}$ and then $H_n^{(2)}$ can be determined in the usual way.

Reprinted from *Printed in U.S.A.*
JOURNAL OF MATHEMATICAL PHYSICS VOLUME 5, NUMBER 1 JANUARY 1964

On the van der Waals Theory of the Vapor–Liquid Equilibrium. III. Discussion of the Critical Region

P. C. HEMMER,* M. KAC, AND G. E. UHLENBECK
The Rockefeller Institute, New York, New York
(Received 20 September 1963)

The discussion of the properties of the Kac one-dimensional fluid model presented in Parts I and II of this series of papers breaks down near the critical point. In Sec. II of the present paper we develop a new successive-approximation method for the eigenvalues and eigenfunctions of the Kac integral equation which is valid in the critical region and which connects smoothly with the developments in the one- and two-phase regions given in Part I. The perturbation parameter is $(\gamma\delta)^{1/4}$ where $\gamma\delta$ is the ratio of the ranges of the repulsive and attractive forces. The main physical consequence is that in the critical region the long-range behavior of the two-point distribution function is represented by an infinite series of decreasing exponentials with ranges all of order $1/\gamma(\gamma\delta)^{1/4}$, and with amplitudes of order $(\gamma\delta)^{1/2}$. This leads to deviations from the Ornstein–Zernike theory and to a specific heat anomaly which are discussed in Sec. V. We conclude with some comments on the possible relevance of our results for the three-dimensional problem.

I. INTRODUCTION

As we noted already in the first two parts of this series of papers,[1] the discussion of the approach to the van der Waals limit which was given there breaks down near the critical point. This is evident from the way the quantity B^2, which is proportional to $\partial p/\partial l$, enters into the expansions for the eigenvalues and eigenfunctions of the Kac equation and into the expansions for the short- and long-range behavior of the two-point distribution function. For $B = 0$ all these expansions blow up. It is therefore necessary to develop a new asymptotic treatment of the Kac equation in the critical region and this is presented in this paper.

It turns out that the critical region should be defined by the ranges in temperature and specific volume given by

$$(T - T_c)/T_c \approx (\gamma\delta)^{1/2}, \qquad (l - l_c)/l_c \approx (\gamma\delta)^{1/4}, \tag{1}$$

where T_c and l_c are the van der Waals critical values and where $\gamma\delta$ is again the ratio of the ranges of the repulsive and attractive forces. In Sec. II we will show that in this region one can again "tame" the Kac equation by an appropriate change of variables, and that this leads to a consistent successive-approximation method for the eigenfunctions and eigenvalues. In contrast to the one- and two-phase regions, the expansion parameter is no longer $\gamma\delta$ but $(\gamma\delta)^{1/4}$, and also the zeroth-order eigenfunctions are no longer the harmonic oscillator eigenfunctions ($\equiv$ Weber functions), but correspond to the eigenfunctions of an oscillator for which the potential energy is a fourth-degree polynomial in the deflection. As a result, the eigenfunctions and eigenvalues can no longer be determined explicitly. However, it is possible to discuss qualitatively how the eigenfunctions change when one goes away from the critical region, and it is of special interest to note the difference in behavior when one goes towards the one-phase region or towards the two-phase region. In the first case, the eigenfunctions become again the Weber functions with only slight distortions, but in the second case (two-phase region) they approach the eigenfunctions for a potential which has two equal or almost equal minima. It is well known that for such a two-minimum potential, the lowest eigenvalue is almost doubly degenerate, and this corresponds to the beginning of the degeneracy which is characteristic for the two-phase region.

In Sec. III the consequences are discussed with respect to the form of the isotherms in the critical region. In I we have already pointed out that, to any order in $\gamma\delta$, the successive-approximation method always gives a phase transition and a van der Waals-like equation of state. We now find that in the critical region (i.e., for finite $\gamma\delta$), there is a *qualitative* difference of the isotherm net from the prediction of the simple van der Waals theory, (i.e., for $\gamma\delta = 0$). For finite $\gamma\delta$, the transition from the one-phase to the two-phase regime occurs continuously in the critical region.[2] Strictly speaking,

* Present address: Institute for Theoretical Physics, N.T.H. Trondheim, Norway.

[1] J. Math. Phys. 4, 216, 229 (1963), hereafter referred to as I and II, respectively.

[2] The concept of a critical *region* in contrast to a critical *point* has often been discussed especially in order to explain various anomalous critical phenomena which apparently were in conflict with the van der Waals equation. For a review

for $T < T_c$ the isotherms do *not* have a horizontal portion, and as a result the densities of the two coexistent phases cannot be defined unambiguously. However, although it is difficult to make precise statements because of the uncertainty one always encounters when one tries to join two different asymptotic developments of the same function, it is possible to make a reasonable extrapolation of the locus of the coexistent points from the two-phase region into the critical region. We then find that, for finite $\gamma\delta$, the critical density remains at the van der Waals value, but the critical temperature is lowered by ΔT_c with $\Delta T_c/T_c \cong 0.702(\gamma\delta)^{\frac{1}{2}}$. At the new critical point, the locus of the coexistent points is still parabolic but it is much flatter than the van der Waals theory predicts.[3]

In Sec. IV and V the consequences are discussed with respect to the two-point distribution function. The main result is that in the critical region the long-range behavior of this distribution function is represented by an infinite series of decreasing exponentials. The ranges of these exponentials are all of order $1/\gamma(\gamma\delta)^{\frac{1}{4}}$, and also the amplitudes are all of the same order of magnitude, namely $(\gamma\delta)^{\frac{1}{4}}$. The Ornstein–Zernike theory is therefore *not* valid in the critical region. However, we can show that if one goes away from the critical region towards the one-phase region, our results go over into the expansions of the long-range behavior of the correlation function derived in II, of which the Ornstein–Zernike exponential was the first term. The situation seems to be that, if one approaches the critical point, more and more exponential terms become excited, so to say. *Outside* the critical region these terms are of decreasing order of magnitude, but *in* the critical region they all contribute.

It seems to us that this picture of how the deviations from the Ornstein–Zernike theory develop if one approaches the critical point, is probably also valid in three dimensions, and that it is independent of our assumption of an exponential attractive potential. In fact, we show in Sec. VI that the whole discussion of Sec. IV can be generalized to the case where the attractive potential consists of a sum of m exponentials, and that this does not change the qualitative picture.

In Sec. VII we conclude with a few comments on some recent experimental results on the critical opalescence and on the so-called specific-heat anomaly in order to see whether there are some experimental indications for the existence of a critical region.

II. THE EIGENFUNCTIONS AND EIGENVALUES IN THE CRITICAL REGION

We will develop all quantities around the critical point of the zeroth-order equation of state, that is the van der Waals equation

$$s = 1/(l - \delta) - \nu_0/l^2,$$

in the notation used in I and II. The critical quantities are given by:

$$s_c = 1/8\delta; \quad l_c = 3\delta; \quad \nu_{0c} = \tfrac{27}{8}\delta. \tag{2}$$

Instead of I, Eq. (33), it turns out that around the critical point one can "tame" the basic Kac integral equation [I, Eq. (10)] by the substitution

$$z = (\gamma\delta)^{\frac{1}{4}}[x - \eta_c(2/\gamma)^{\frac{1}{2}}], \tag{3}$$

where $\eta_c \equiv \eta(s_c) = (2\nu_{0c})^{\frac{1}{2}}/l_c$. This is motivated by the fact that in the one-phase region the "tamed" eigenfunctions depend on the variable [see I, Eq. (38)]

$$z = (B/l)^{\frac{1}{2}}[x - \eta(2/\gamma)^{\frac{1}{2}}]. \tag{4}$$

If one now *defines* the critical region in temperature and specific volume by the equations corresponding

of the older literature, see for instance J. P. Kuenen, *Die Zustandsgleichung* (Vieweg, Braunschweig, 1907), Chap. 5, especially for the effects of gravity and of impurities on the critical phenomena. The only attempt we know of to deduce a critical region from the ideas of van der Waals is by G. Bakker [Z. Physik. Chem. 49, 609 (1904)]. He pointed out that there may be a range in temperature where the thickness of the capillary layer is of the same order as $v_1^{\frac{1}{3}} - v_2^{\frac{1}{3}}$. In this region one would not see a meniscus, although the two phases are still present. This is also suggested by Mayer. For a review of his arguments, which are partially formal and partially physical, and are based on the fugacity and virial expansions, see the book by J. E. Mayer and M. G. Mayer [*Statistical Mechanics*, (John Wiley & Sons, Inc., New York, 1940), Chap. 14]. Mayer's conclusion is that there may be a temperature $T_m < T_c$ where the meniscus would disappear, although there are still horizontal parts in the isotherms for temperatures between T_m and T_c. In this temperature range, the isotherms enter the two-phase region with a horizontal tangent. Because of the resulting shape of the coexistence curve near the critical point, this region is often referred to as Mayer's derby hat.

From the experimental point of view the existence of such a critical region is still controversial. For a review of recent experimental results see the book by T. O. Hirschfelder, Ch. F. Curtiss, and R. B. Bird [*Molecular Theory of Gases and Liquids* (John Wiley & Sons, Inc., New York, 1954), Chap. 5, Sec. 2]. See also the work by H. W. Habgood and W. G. Schneider on Xenon [Can. J. Chem. 32, 98, 164 (1954)].

[3] The question of the shape of the coexistence curve near the critical point has been much discussed. Guggenheim [J. Chem. Phys. **13**, 253 (1945)], pointed out that the data are better represented by the law $(v_1 - v_2) \sim (T_c - T)^{\frac{1}{3}}$ than by the van der Waals result $(v_1 - v_2) \sim (T_c - T)^{\frac{1}{2}}$ which follows from the parabolic form of the coexistence curve near the critical point. Compare also M. A. Weinberger and W. G. Schneider, Can. J. Chem. **30**, 422 (1952), and the more recent discussion by B. Widom and O. K. Rice, J. Chem. Phys. **23**, 1250 (1955). The Guggenheim result would imply that at the critical point also $(\partial^3 p/\partial v^3)_T$ would be zero, and B. H. Zimm [J. Chem. Phys. **19**, 1019 (1951)] has given a plausibility argument for the vanishing of *all* derivatives at the critical point. However, also in this respect, there are no results which are generally accepted.

to (1),

$$\nu_0 = \nu_{0c}[1 + \nu_1(\gamma\delta)^{\frac{2}{3}}], \tag{5}$$

$$l = l_c[1 + l_1(\gamma\delta)^{\frac{2}{3}}], \tag{6}$$

then one finds that B/l is of order $(\gamma\delta)^{\frac{1}{3}}$, so that (4) goes over into (3). However, one must of course still show that the substitution (3) works! We proceed in the same way as in I. For the variable y in the Kac equation, one makes the analogous substitution

$$z' = (\gamma\delta)^{\frac{1}{6}}[y - \eta_c(2/\gamma)^{\frac{1}{2}}],$$

and we put

$$z' = ze^{-\gamma\tau} + (\gamma\delta)^{\frac{1}{6}}\zeta(1 - e^{-2\gamma\tau})^{\frac{1}{2}},$$

$$\psi(x) = \psi[(\gamma\delta)^{-1/6}z + \eta_c(2/\gamma)^{1/2}] \equiv (\gamma\delta)^{1/12}H(z),$$

where the factor $(\gamma\delta)^{1/12}$ is introduced in order to keep $H(z)$ normalized to one. Finally we put for s the expansion[4]

$$s = s_c[1 - 3\nu_1(\gamma\delta)^{2/3} + s_1(\gamma\delta) + s_2(\gamma\delta)^{4/3} + \cdots], \tag{7}$$

where the numbers s_1, s_2 $\cdots$ will depend on the given *numbers* ν_1 and l_1 through the equation of state, which is still given by [see I, Eq. (20)]

$$l = -\lambda_0'(s)/\lambda_0(s), \tag{8}$$

in terms of the maximum eigenvalue $\lambda_0(s)$ of the Kac equation.

Making all these substitutions, introducing the expansions (5) and (7), and developing the kernel in powers of $(\gamma\delta)^{1/3}$ up to order $(\gamma\delta)^{5/3}$, one finds, after carrying out the integrations over ζ and τ and using (2),

$$\frac{d^2H}{dz^2}[3(\gamma\delta)^{4/3} - (6)^{1/2}z(\gamma\delta)^{5/3}] - (6)^{1/2}(\gamma\delta)^{5/3}\frac{dH}{dz}$$
$$+ H(z)\Big[3\Lambda + \left(\frac{3}{2}\right)^{1/2} z\Big\{\left(\frac{s_1}{4} - 1\right)(\gamma\delta)^{4/3}$$
$$+ \left(\frac{s_2}{4} - \frac{39}{16}\nu_1^2\right)(\gamma\delta)^{5/3}\Big\} + z^2\Big\{\frac{3\nu_1}{4}(\gamma\delta)^{4/3}$$
$$+ \left(1 - \frac{s_1}{4}\right)(\gamma\delta)^{5/3}\Big\} - \frac{3}{8}\left(\frac{3}{2}\right)^{1/2}\nu_1(\gamma\delta)^{5/3}z^3$$
$$- \frac{1}{16}(\gamma\delta)^{4/3}z^4 + \frac{1}{20}\left(\frac{3}{2}\right)^{1/2}(\gamma\delta)^{5/3}z^5\Big] = 0 \tag{9}$$

where the new eigenvalue parameter Λ is related to the original λ by the equation

$$\frac{\lambda}{\omega(s_c)} = 1 + \frac{9\nu_1}{4}(\gamma\delta)^{2/3} + \frac{3}{2}\left(1 - \frac{s_1}{4}\right)\gamma\delta$$
$$+ \frac{3}{8}\left(-s_2 + \frac{27}{4}\nu_1^2\right)(\gamma\delta)^{4/3} + \cdots - 3\Lambda. \tag{10}$$

From (9) one sees that one obtains a consistent approximation procedure by putting

$$H(z) = H^{(0)}(z) + (\gamma\delta)^{\frac{1}{3}}H^{(1)}(z) + (\gamma\delta)^{\frac{2}{3}}H^{(2)}(z) + \cdots,$$

$$\Lambda = \Theta(\gamma\delta)^{4/3} + \Omega(\gamma\delta)^{5/3} + \Xi(\gamma\delta)^2 + \cdots. \tag{11}$$

In lowest order [i.e., $(\gamma\delta)^{4/3}$] one gets

$$\Big[\frac{d^2}{dz^2} - \frac{z^4}{48} + \frac{\nu_1}{4}z^2 + (6)^{-\frac{1}{2}}\left(\frac{s_1}{4} - 1\right)z + \Theta\Big]H^{(0)}(z) = 0, \tag{12}$$

and in next order [i.e., $(\gamma\delta)^{5/3}$],

$$\Big[\frac{d^2}{dz^2} - \frac{z^4}{48} + \frac{\nu_1}{4}z^2 + (6)^{-\frac{1}{2}}\left(\frac{s_1}{4} - 1\right)z + \Theta\Big]H^{(1)}(z)$$
$$= \Big[\left(\frac{2}{3}\right)^{\frac{1}{2}}\frac{d}{dz}z\frac{d}{dz} - \frac{z^5}{20(6)^{\frac{1}{2}}} + \frac{\nu_1}{8}\left(\frac{3}{2}\right)^{\frac{1}{2}}z^3$$
$$+ \frac{s_1 - 4}{12}z^2 - \left(\frac{2}{3}\right)^{\frac{1}{2}}\left(\frac{s_2}{8} - \frac{39}{32}\nu_1^2\right)z - \Omega\Big]H^{(0)}(z). \tag{13}$$

These equations determine, in principle, the eigenfunctions $H_n^{(0)}(z)$, $H_n^{(1)}(z)$ and the corresponding eigenvalues Θ_n, Ω_n as functions of the constants ν_1, s_1, and s_2.

Turning now to the equation of state, we start from the general expression

$$-\lambda_0'(s) = \iint_{-\infty}^{+\infty} dx\, dy\, \psi_0(x)\psi_0(y)\left[\frac{W(x)}{W(y)}\right]^{\frac{1}{2}}$$
$$\exp\left\{\frac{(\nu_0\gamma)^{\frac{1}{2}}}{2}(x + y)\right\}\int_s^\infty d\tau\, \tau e^{-s\tau}P_\gamma(x \mid y, \tau).$$

By expanding ν_0 and s around the critical values, and by using the same taming substitutions for x and y ,one obtains:

$$-\lambda_0'(s) = \omega(s_c)\Big[3\delta - \delta(6)^{\frac{1}{2}}(\gamma\delta)^{\frac{1}{3}}\int_{-\infty}^{+\infty} dz\, zH_0^{(0)}(z)^2$$
$$+ (\gamma\delta)^{\frac{2}{3}}\Big\{-2\delta(6)^{\frac{1}{2}}\int_{-\infty}^{+\infty} dz\, zH_0^{(0)}(z)H_0^{(1)}(z)$$
$$+ 2\delta\int_{-\infty}^{+\infty} dz\, z^2H_0^{(0)}(z)^2 + \frac{33}{4}\delta\nu_1\Big\} + \cdots\Big]. \tag{14}$$

From (10) and the equation of state (8), one then finds that one must have, by comparing the result for l with the defining equation (6),

[4] It follows from the van der Waals equation and Eqs. (5) and (6), that in (7) the coefficient of the term of order $(\gamma\delta)^{\frac{2}{3}}$ must be $(-3\nu_1)$. It is also not difficult to show that only in this way one obtains a consistent approximation procedure for the Kac equation.

$$\int_{-\infty}^{+\infty} dz\, zH_0^{(0)}(z)^2 = -l_1(\tfrac{3}{2})^{\frac{1}{2}} \tag{15}$$

$$-2(\tfrac{2}{3})^{\frac{1}{2}} \int_{-\infty}^{+\infty} dz\, zH_0^{(0)}(z)H_0^{(1)}(z) + \frac{2}{3}\int_{-\infty}^{+\infty} dz\, z^2 H_0^{(0)}(z)^2 + \tfrac{1}{2}\nu_1 = 0, \tag{16}$$

and these equations determine, in principle, the dependence of s_1 and s_2 on ν_1 and l_1.

III. ISOTHERM NET IN THE CRITICAL REGION

Since the equation of state is expressed in such an implicit fashion in terms of the lowest eigenvalue and eigenfunction of Eq. (12) which can not be determined in closed form, it is convenient for the discussion to make use of the following physical analogy. Writing Eq. (12) in the form

$$d^2H^{(0)}/dz^2 + [\Theta - V(z) + Fz]H^{(0)} = 0, \tag{17}$$

with

$$V(z) = \tfrac{1}{48} z^4 - (\tfrac{1}{4}\nu_1)z^2, \quad F = (s_1 - 4)/4(6)^{\frac{1}{2}}, \tag{18}$$

one sees that Eq. (12) can be looked upon as a Schrödinger equation for a particle moving in the potential $V(z)$ and under the influence of a quasi-"electric" field of strength F. Furthermore it is easy to show that from Eq. (15) it follows that

$$l_1 = (\tfrac{2}{3})^{\frac{1}{2}}\, \partial\Theta_0/\partial F. \tag{19}$$

Hence l_1 is proportional to the "electric moment" produced by F, and the discussion of the isotherm becomes equivalent to the discussion of the Stark effect for the lowest state of the motion of the particle in the potential $V(z)$.

Consider first the case when F is small. Since $V(z) = V(-z)$, the lowest eigenfunction $H_0^{(0)}(z)$ is also symmetric in z when $F = 0$. Hence one has a quadratic Stark effect, so that $l_1 = 0$ for $F = 0$, which follows also immediately from (15). By the usual perturbation theory one finds

$$\Theta_0(F) = \Theta_0 - F^2 \sum_{n=1}^{\infty} \frac{1}{\Theta_n - \Theta_0} \times \left[\int_{-\infty}^{+\infty} dz\, zH_0^{(0)}H_n^{(0)}\right]^2 + O(F^4), \tag{20}$$

where the Θ_n and $H_n^{(0)}$ are the eigenvalues and eigenfunctions of the unperturbed equation. Therefore for $F = 0$, $\partial^2\Theta_0/\partial F^2 < 0$, and not only $\partial\Theta_0/\partial F = 0$, but also $\partial^3\Theta_0/\partial F^3 = 0$. This implies that at the critical density, that is for $l_1 = 0$, all isotherms have an inflection point with a negative slope. In fact one obtains from (19) and (20)

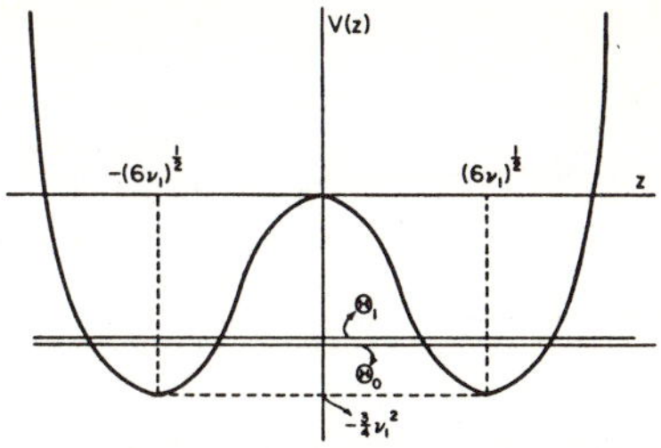

FIG. 1. The potential $V(z)$ of Eq. (17) for a temperature much below the critical temperature.

$$\left(\frac{\partial s}{\partial l}\right)_{l_1=0} = -\frac{(\gamma\delta)^{\frac{1}{2}}}{4\delta^2} \Big/ \sum_{n=1}^{\infty} \frac{1}{\Theta_n - \Theta_0} \times \left[\int_{-\infty}^{+\infty} dz\, zH_0^{(0)}H_n^{(0)}\right]^2, \tag{21}$$

which is always negative. The isotherms have therefore *no* horizontal part in the critical region. However the behavior of the slope (21) as function of the temperature is quite different for positive or negative ν_1, that is, for $T < T_c$ or $T > T_c$. Clearly, for negative ν_1 ($T > T_c$), the potential $V(z)$ has only a minimum at $z = 0$, and for $\nu_1 \ll -1$, $V(z)$ becomes a parabola and the eigenfunctions become the Weber functions. For $\nu_1 \ll -1$, it is simple to develop a perturbation method for the Θ_n and $H_n^{(0)}(z)$ and hence for the slope (21). One finds for $\nu_1 \ll -1$

$$\left(\frac{\partial s}{\partial l}\right)_{l_1=0} = \frac{(\gamma\delta)^{\frac{1}{2}}}{4\delta^2}\nu_1\left[1 + \frac{1}{2(-\nu_1)^{\frac{3}{2}}} + O(-\nu_1)^{-3}\right]. \tag{22}$$

It is also easy to see that this is in agreement with the equation of state in the one-phase region,

$$s = \frac{1}{l-\delta} - \frac{\nu_0}{l^2} + \frac{\gamma}{2}\left[1 - \frac{l}{B}\left\{1 - \frac{\nu_0(l^2-\delta^2)}{l^3}\right\}\right], \tag{23}$$

with

$$B^2/l^2 = 1 - 2\nu_0(l-\delta)^2/l^3,$$

which was given in I, Sec. IV. Using (5) and (6) one obtains from (23) the expansion (7) for s with

$$s_1 = 6\nu_1 l_1 - \tfrac{3}{2}l_1^3 + 4 - 6l_1(3l_1^2 - 4\nu_1)^{-\frac{1}{2}}, \tag{24}$$

from which (22) follows.

The discussion is quite different for the case $\nu_1 > 0$, that is $T < T_c$. The potential $V(z)$ now has *two* minima (see Fig. 1), and for $\nu_1 \gg 1$, this leads to the well-known near degeneracy of the lowest energy level corresponding to the symmetric

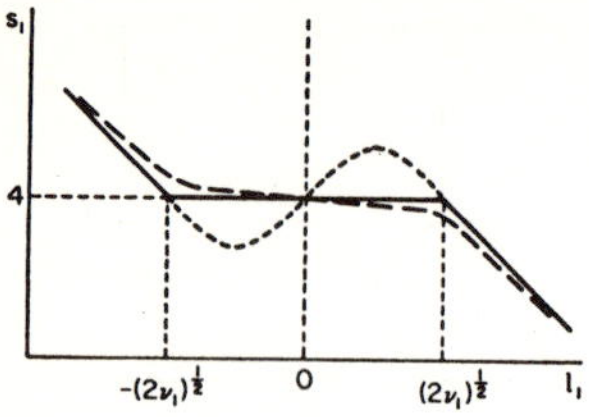

FIG. 2. Shape of the isotherm below the critical point; ——— Eq. (30); - - - - van der Waals; ‒ ‒ ‒ ‒ Eq. (30a).

and antisymmetric combination of the two lowest harmonic-oscillator eigenfunctions around the two minima. Using the WKB method[5] one finds

$$\Theta_1 - \Theta_0 \cong [(2\nu_1)^{\frac{1}{2}}/\pi] \exp[-(2\nu_1)^{\frac{3}{2}}],$$

$$\int_{-\infty}^{+\infty} dz\, zH_0^{(0)}H_1^{(0)} \cong (6\nu_1)^{\frac{1}{2}}.$$

Only the first term in the sum in (21) contributes, so that for $\nu_1 \gg 1$,

$$\left(\frac{\partial s}{\partial l}\right)_{l_1=0} \cong -\frac{(\gamma\delta)^{\frac{1}{2}}}{12\pi\delta^2(2\nu_1)^{\frac{1}{2}}} \exp[-(2\nu_1)^{\frac{3}{2}}]. \qquad (25)$$

The slope of the isotherm drops therefore very fast to zero if ν_1 increases.

Consider next the case when F and therefore l_1 are not small. Again, to discuss $s_1(l_1, \nu_1)$, one must distinguish between the cases $\nu_1 \ll -1$ and $\nu_1 \gg 1$. For temperatures *above* $T_c(\nu_1 \ll -1)$, the "potential" function $V(z) - Fz$ has still only one minimum at $z = z_0$, determined by

$$\tfrac{1}{12}z_0^3 - (\tfrac{1}{2}\nu_1)z_0 - F = 0. \qquad (26)$$

The "energy" Θ_0 is then given approximately by

$$\Theta_0 = V(z_0) - Fz_0 + 2^{-\frac{3}{2}}(z_0^2 - 2\nu_1)^{\frac{1}{2}}, \qquad (27)$$

where the last term represents the "zero-point energy" $\frac{1}{2}\hbar\omega$ of the oscillation around z_0 [with $\hbar = 1$, $m = \frac{1}{2}$, potential energy $\frac{1}{4}\omega^2(z - z_0)^2 = \frac{1}{2}(d^2V/dz^2)_0(z - z_0)^2$, so that $\omega = 2^{\frac{1}{2}}(d^2V/dz^2)_0^{\frac{1}{2}} = 2^{-\frac{1}{2}}(z_0^2 - 2\nu_1)^{\frac{1}{2}}$]. From (19) and (27) one gets

$$(\tfrac{3}{2})^{\frac{1}{2}}l_1 = -z_0 + 2^{-\frac{3}{2}}z_0(z_0^2 - 2\nu_1)^{-\frac{1}{2}}\,\partial z_0/\partial F$$

$$= -z_0 + 2^{\frac{1}{2}}z_0/(z_0^2 - 2\nu_1)^{\frac{3}{2}},$$

using (26). In *zeroth* approximation, the minimum $z_0 = -(\frac{3}{2})^{\frac{1}{2}}l_1$, and substituting this in (26), and expressing F in s_1 by (18), one obtains the first three terms of the equation of state (24) which also follow from the van der Waals equation. One can say that these van der Waals terms correspond to the "classical" approximation of the Stark effect. In *first* approximation, z_0 becomes

$$z_0 = -(\tfrac{3}{2})^{\frac{1}{2}}l_1[1 - 2(6)^{\frac{1}{2}}l_1/(3l_1^2 - 4\nu_1)^{\frac{3}{2}}]$$

and this leads to the last term in the equation of state (24), which represents therefore, so to say, the first "quantum" correction to the van der Waals equation. Note that the corresponding result for the energy levels Θ_n is given by

$$\Theta_n = -\tfrac{9}{64}l_1^4 + \tfrac{3}{8}l_1\nu_1 - \nu_1/(3l_1^2 - 4\nu_1)^{\frac{1}{2}}$$

$$+ \tfrac{1}{2}n(3l_1^2 - 4\nu_1)^{\frac{1}{2}}, \qquad (28)$$

and that in this approximation the eigenfunctions are the Weber functions $D_n(y)$, where

$$y = 2^{-\frac{1}{2}}(3l_1^2 - 4\nu_1)^{\frac{1}{4}}\{z + (3/2)^{\frac{1}{2}}l_1\}. \qquad (29)$$

It is also not difficult to find the higher approximations. The results are collected in Appendix I, since they are needed in Sec. IV.

For temperatures *below* $T_c(\nu_1 \gg 1)$, the potential function $V(z) - Fz$ can have *two* minima which will be of unequal depth and the sign of F [or of $(s_1 - 4)$] determines which minimum is deeper. If in zeroth approximation we assume again that the energy Θ_0 is given by the value of the potential function at the deeper minimum, then one finds (in exactly the same way as above) for the equation of state

$$l_1 = -2\nu_1^{1/2}\,\mathrm{sign}\,(s_1 - 4)$$

$$- (s_1 - 4)/12\nu_1 + O(\nu_1^{-5/2}), \qquad (30)$$

and this follows also from the van der Waals equation [first three terms in (24)] plus the Maxwell rule (see Fig. 2). Again one can say that the van der Waals equation is the "classical" approximation of the corresponding Stark-effect problem. Especially the horizontal portion of the isotherm at $s_1 = 4$ is due to the fact that the deeper minimum switches from the right to the left when F changes sign. Adding to the minimum energy, the zero-point energy does *not* change the qualitative picture; the horizontal portion remains. The finite slope of the isotherm, which we discussed before [see Eq. (25)] is due to a second "quantum mechanical" effect, namely the penetration of the barrier between the two minima if they become of almost equal depth. By using the WKB method one can show that this penetration effect changes Eq. (30) to

$$l_1 = -2\nu_1^{\frac{1}{2}}(s_1 - 4)/[(s_1 - 4)^2 + \epsilon^2]^{\frac{1}{2}}$$

$$- (s_1 - 4)/12\nu_1, \qquad (30a)$$

where

$$\epsilon = (2^{\frac{3}{2}}/\pi) \exp[-(2\nu_1)^{\frac{3}{2}}]$$

and this agrees with Eq. (25).

[5] Compare D. M. Dennison and G. E. Uhlenbeck, Phys. Rev. 41, 313 (1932).

So far we have discussed the isotherm net mainly for $\nu_i \ll -1$ and $\nu_1 \gg 1$. Although it is qualitatively clear how the isotherms change going through the critical region, it is difficult to make precise statements if ν_1 is of order one, and it is clearly *not* possible to answer unambigously the question of what the shape of the coexistence curve is in the critical region. However, it seems reasonable to say that, because of the symmetry of the function $V(z)$, the critical density is not changed if $\gamma\delta$ is finite, but that the critical temperature is lowered because for $\nu_1 > 0$, $V(z)$ will have two minima, which is somehow an indication for the separation of the two phases. One can then define the new critical temperature as that value of ν_1 for which the lowest energy level is just at zero, which is the value of $V(z)$ at the maximum. To determine this value one can use the variant of the WKB method which was developed by Kramers and Ittmann.[6] With this method one finds that the relation between ν_1 and Θ_0, for $l_1 = 0$ and Θ_0 close to zero, is given by

$$\nu_1 = \left(\frac{3\pi}{16}\right)^{\frac{2}{3}} + \frac{|\lambda|}{6}\left(\frac{2}{3\pi}\right)^{\frac{1}{3}}\left[\ln (36\pi) + C + \frac{\pi}{2}\right], \qquad (31)$$

where $\lambda = 2\Theta_0\nu_1^{-\frac{1}{2}}$ and C is the Euler constant. Hence $\nu_1 = (\frac{3}{16}\pi)^{\frac{2}{3}} = 0.702$ if $\Theta_0 = 0$, so that the critical temperature changes by ΔT_c with $\Delta T_c/T_c = 0.702(\gamma\delta)^{\frac{1}{2}}$. For still lower temperatures, that is, for $\nu_1 > 0.702$, the lowest eigenfunction will have two maxima. It is tempting to identify the positions of these maxima with the value of $l_1(\frac{3}{2})^{\frac{1}{2}}$ for the incipient "gas" and "liquid" phase. Doing this one obtains

$$l_1^2 = 4\nu_1^{-\frac{1}{2}}|\lambda|. \qquad (32)$$

Eliminating $|\lambda|$ between (31) and (32) then gives for the extrapolated coexistence curve[7] the equation

$$\nu_1 = \left(\frac{3\pi}{16}\right)^{\frac{2}{3}} + \frac{1}{48}\left[\ln (36\pi) + C + \frac{\pi}{2}\right]l_1^2$$

$$= 0.702 + 0.143l_1^2. \qquad (33)$$

This is still a parabola, but it is appreciably flatter than the van der Waals result $\nu_1 = \frac{1}{4}l_1^2$ (see Fig. 3).

IV. THE TWO-POINT DISTRIBUTION FUNCTION IN THE CRITICAL REGION

To determine the long-range behavior of the two-point distribution function in the critical region we will follow the same method as used in II, Sec. V for the one-phase region. However, since in the latter region $\bar{n}_2(x; l)$ decays to $1/l^2$ according to the Ornstein–Zernike exponential exp $(-B\gamma x/l)$ [II, Eq. (54)], and since in the critical region B/l is of order $(\gamma\delta)^{\frac{1}{4}}$, one should expect that in the critical region the range in x is of order $1/\gamma(\gamma\delta)^{\frac{1}{4}}$. We therefore replace, in the basic formula [II, Eq. (9a)] for the Laplace transform of $\bar{n}_2(x; l)$, the variable σ by $\sigma\gamma(\gamma\delta)^{\frac{1}{4}}$ and we start from

$$l\int_0^\infty dx\, \bar{n}_2(x; l) \exp [-\sigma\gamma(\gamma\delta)^{\frac{1}{4}}x]$$

$$= -1 + \sum_{n=0}^{\infty} \frac{\lambda_0(s)}{\lambda_0(s) - \lambda_n(s + \sigma\gamma(\gamma\delta)^{\frac{1}{4}})} b_n^2, \qquad (34)$$

with

$$b_n = \int_{-\infty}^{+\infty} dx\, \psi_0(x; s)\psi_n(x; s + \sigma\gamma(\gamma\delta)^{\frac{1}{4}}).$$

We first compute b_n^2 up to order $(\gamma\delta)^{\frac{1}{2}}$. Introduce the variable z of Eq. (3) and the expansion (11) for the eigenfunctions. Notice that $s + \sigma\gamma(\gamma\delta)^{\frac{1}{4}}$ is obtained from the expansion (7) for s by adding 8σ to s_2, so that the equation (12) for $H^{(0)}$ is *not* affected by the change of s to $s + \sigma\gamma(\gamma\delta)^{\frac{1}{4}}$. Using the orthogonality and normalization relations, one obtains

$$b_n^2 = \delta_{n0} + (\gamma\delta)^{\frac{1}{2}}\left\{\int_{-\infty}^{+\infty} dzH_0^{(0)}(z, s) \times [H_n^{(1)}(z, s + \sigma\gamma(\gamma\delta)^{\frac{1}{4}}) - H_n^{(1)}(z, s)]\right\}^2$$

$$+ 2(\gamma\delta)^{\frac{1}{2}}\delta_{n0}\int_{-\infty}^{+\infty} dz\{H_0^{(0)}(z, s) \times [H_0^{(2)}(z, s + \sigma\gamma(\gamma\delta)^{\frac{1}{4}}) - H_0^{(2)}(z, s)]$$

$$+ H_0^{(1)}(z, s)[H_0^{(1)}(z, s + \sigma\gamma(\gamma\delta)^{\frac{1}{4}}) - H_0^{(1)}(z, s)]\}. \qquad (35)$$

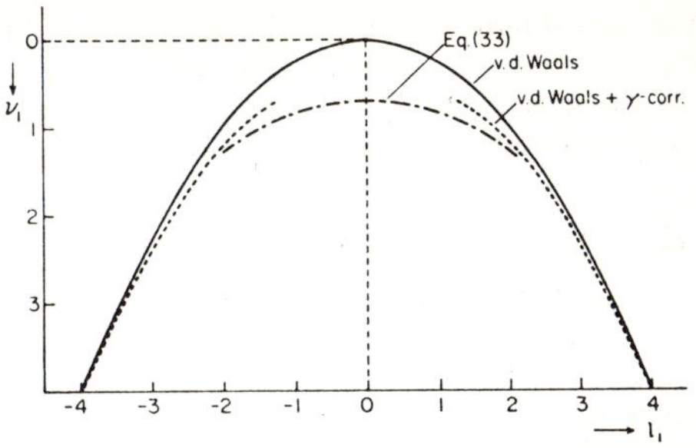

FIG. 3. Shape of the coexistence curve near the critica point; ——— van der Waals; – – – – van der Waals + γ corrections [Eq. (23)]; ·—·— Eq. (33).

[6] H. A. Kramers and G. P. Ittmann, Z. Physik 58, 217 (1929). See especially Sec. 6.

[7] Note that the equation for the coexistence curve as derived from the van der Waals equation plus the corrections of $O(\gamma)$ (see I, Sec. 4) diverges at the critical point, so that a direct extrapolation is not possible.

The second term can be calculated as follows. From Eq. (13) for $H^{(1)}$ and the fact that $H_n^{(0)}$ and Θ_n are not affected by the change $s \to s + \sigma\gamma(\gamma\delta)^{\frac{1}{3}}$, one shows that

$$\left[\frac{d^2}{dz^2} - \frac{z^4}{48} + \frac{\nu_1 z^2}{4} + \frac{s_1 - 4}{4(6)^{\frac{1}{2}}} z + \Theta_n\right]$$
$$\times [H_n^{(1)}(z, s + \sigma\gamma(\gamma\delta)^{\frac{1}{3}}) - H_n^{(1)}(z, s)]$$
$$= [-(\tfrac{2}{3})^{\frac{1}{2}}\sigma z - \Omega_n(\sigma) + \Omega_n(0)]H_n^{(0)}(z, s), \tag{36}$$

where $\Omega_n(\sigma) \equiv \Omega_n(s + \sigma\gamma(\gamma\delta)^{\frac{1}{3}})$. Multiplying (36) with $H_0^{(0)}(z, s)$ and integrating gives for $n \neq 0$

$$\int_{-\infty}^{+\infty} dz H_0^{(0)}(z, s)\{H_n^{(1)}(z, s + \sigma\gamma(\gamma\delta)^{\frac{1}{3}}) - H_n^{(1)}(z, s)\}$$
$$= -(\tfrac{2}{3})^{\frac{1}{2}} \frac{\sigma}{\Theta_n - \Theta_0} \int_{-\infty}^{+\infty} dz\, z H_0^{(0)}(z, s) H_n^{(0)}(z, s). \tag{37}$$

To calculate next the terms in (34) containing the eigenvalues, it turns out that we need to extend the expansion (10) up to order $(\gamma\delta)^2$. This leads to

$$\lambda_0(s) - \lambda_n(s + \sigma\gamma(\gamma\delta)^{\frac{1}{3}}) = \omega(s_c)$$
$$\times \Big\{3(\Theta_n - \Theta_0 + \sigma)(\gamma\delta)^{4/3} + 3[\Omega_n(\sigma) - \Omega_0(0)](\gamma\delta)^{5/3}$$
$$+ 3\Big[\Xi_n(\sigma) - \Xi_0(0) + \frac{11}{4}\nu_1\sigma\Big](\gamma\delta)^2 + \cdots\Big\},$$

which gives

$$3(\gamma\delta)^{4/3}\sigma \frac{\lambda_0(s)}{\lambda_0(s) - \lambda_n(s + \sigma\gamma(\gamma\delta)^{\frac{1}{3}})}$$
$$= \frac{\sigma}{\sigma + \Theta_n - \Theta_0} - \frac{(\gamma\delta)^{\frac{1}{3}}\sigma}{(\sigma + \Theta_n - \Theta_0)^2}$$
$$\times [\Omega_n(\sigma) - \Omega_0(0)] + \frac{(\gamma\delta)^{\frac{2}{3}}\sigma^2}{(\sigma + \Theta_n - \Theta_0)^2}$$
$$\times \Big\{+\frac{11\nu_1}{4} + \frac{9\nu_1}{4}\Big(1 + \frac{\Theta_n - \Theta_0}{\sigma}\Big)$$
$$+ \frac{[\Omega_n(\sigma) - \Omega(0)]^2}{\sigma[\sigma + \Theta_n - \Theta_0]} - \frac{\Xi_n(\sigma) - \Xi(0)}{\sigma}\Big\}. \tag{38}$$

Notice also that from (36) for $n = 0$ one gets, by multiplying with $H_0^{(0)}(z, s)$ and integrating,

$$\Omega_0(\sigma) - \Omega_0(0)$$
$$= -(\tfrac{2}{3})^{\frac{1}{2}}\sigma \int_{-\infty}^{+\infty} dz\, z H_0^{(0)}(z, s)^2 = \sigma l_1, \tag{39}$$

where we have used the equation of state (15).

Substituting all these partial results in (34) one finds after some simple rearrangements

$$l_c^2\sigma\gamma(\gamma\delta)^{\frac{1}{3}} \int_0^{\infty} dx \exp[-\sigma\gamma(\gamma\delta)^{\frac{1}{3}}x]\Big[\bar{n}_2(x; l) - \frac{1}{l^2}\Big]$$
$$= (\gamma\delta)^{\frac{2}{3}} \frac{2}{3} \sum_{n=1}^{\infty} \frac{\sigma}{\sigma + \Theta_n - \Theta_0}$$
$$\times \Big[\int_{-\infty}^{+\infty} dz\, z H_0^{(0)} H_n^{(0)}\Big]^2 + (\gamma\delta)^{\frac{2}{3}} R. \tag{40}$$

Here R is a collection of terms for which we refer to Appendix II, where it will be proved that in fact $R = 0$. Of course Eq. (40) is valid only in the critical region, so that l is always given by Eq. (6), and the eigenvalues Θ_n and eigenfunctions $H_n^{(0)}$ must be determined from Eq. (12). With $R = 0$ one can invert the Laplace transform, and one finds

$$\bar{n}_2(x, l) = \frac{1}{l^2} + \frac{2}{3}\frac{(\gamma\delta)^{\frac{1}{3}}}{l_c^2} \sum_{n=1}^{\infty} \Big[\int_{-\infty}^{+\infty} dz\, z H_0^{(0)} H_n^{(0)}\Big]^2$$
$$\times \exp[-(\Theta_n - \Theta_0)\gamma x(\gamma\delta)^{\frac{1}{3}}]. \tag{41}$$

This is the final result. It shows that in the critical region the decay of $\bar{n}_2(x, l)$ to the uncorrelated value $1/l^2$ is represented by an infinite series of decreasing exponentials which are all of order $(\gamma\delta)^{\frac{1}{3}}$ and have ranges of order $1/\gamma(\gamma\delta)^{\frac{1}{3}}$.

Before starting a detailed discussion of Eq. (41) we conclude with some remarks.

(a) The fluctuation theorem [see II, Eq. (34a)] can be checked as follows. From (41) one gets:

$$\int_0^{\infty} dx\Big[\bar{n}_2(x, l) - \frac{1}{l^2}\Big] = \frac{2}{3}\frac{\delta}{l_c^2}(\gamma\delta)^{-\frac{2}{3}}$$
$$\times \sum_{n=1}^{\infty} \frac{1}{\Theta_n - \Theta_0}\Big[\int_{-\infty}^{+\infty} dz\, z H_0^{(0)} H_n^{(0)}\Big]^2,$$

and from (21) one sees that to order $(\gamma\delta)^{-\frac{2}{3}}$, the right-hand side is equal to $(-\frac{1}{2}l^3)/(\partial s/\partial l)$ and this is also the leading term of the fluctuation integral.

(b) In the range $x \sim 1/\gamma$, Eq. (41) gives, to lowest order,

$$\bar{n}_2(x, l) = \frac{1}{l^2} + \frac{2}{3l_c^2}(\gamma\delta)^{\frac{1}{3}}$$
$$\times \sum_{n=0}^{\infty}\Big\{\int_{-\infty}^{+\infty} dz[z + (\tfrac{3}{2})^{\frac{1}{2}}l_1]H_0^{(0)}H_n^{(0)}\Big\}^2$$
$$= \frac{1}{l^2} + \frac{2}{3l_c^2}(\gamma\delta)^{\frac{1}{3}} \int_{-\infty}^{+\infty} dz[z + (\tfrac{3}{2})^{\frac{1}{2}}l_1]^2 H_0^{(0)}(z)^2, \tag{42}$$

where the equation of state (15) has been used to extend the summation from zero to infinity which then allows the application of Parseval's theorem. One can derive (42) also directly from the basic formula [II, Eq. (9a)] replacing σ by $\sigma\gamma$ and expanding to order $(\gamma\delta)^{\frac{1}{3}}$, but we will leave this to the reader. Eq. (42) clearly shows that in the critical region $1/\gamma$ is *not* the ultimate range, since in this range $\bar{n}_2(x, l)$ still differs from $1/l^2$ by a constant of order $(\gamma\delta)^{1/2}$.

(c) To check the virial theorem [see II, Eq. (40a)]

$$s = \frac{1}{l} + \delta\bar{n}_2(\delta^+, l) - \nu_0\gamma^2 \int_\delta^\infty dx\, xe^{-\gamma x}\bar{n}_2(x; l)$$

in the critical region, one still needs the correction of order $(\gamma\delta)^{1/2}$ in the short-range behavior of $\bar{n}_2(x; l)$. In Appendix III we prove that

$$\bar{n}_2(\delta^+, l) = \frac{1}{l(l-\delta)} + \frac{9}{4l_c^2}(\gamma\delta)^{1/2} \times \int_{-\infty}^{+\infty} dz[z + (\tfrac{3}{2})^{1/2}l_1]^2 H_0^{(0)}(z)^2. \tag{43}$$

Using (42) and the expansions (5), (6), and (7) for ν_0, l, and s, it is then easy to check the virial theorem up to order $(\gamma\delta)^{1/2}$.

V. DISCUSSION OF EQ. (41)

We will show that Eq. (41) goes over into the results obtained in II for the long-range behavior of $\bar{n}_2(x; l)$ in the one- and two-phase region.

Consider first the two-phase region. We found (see II, Sec. VI) that there, in the range $x \sim 1/\gamma$ and up to order γ,

$$\bar{n}_2(x, l) = \frac{\xi_l l_l}{l}\left[\frac{1}{l_l^2} + \frac{\nu_0(l_l-\delta)^4\gamma}{l_l^5 B_l}\exp\left(-\frac{B_l}{l_l}\gamma x\right)\right] + \frac{\xi_g l_g}{l}\left[\frac{1}{l_g^2} + \frac{\nu_0(l_g-\delta)^4\gamma}{l_g^5 B_g}\exp\left(-\frac{B_g}{l_g}\gamma x\right)\right], \tag{44}$$

where ξ_l, ξ_g are the mole fractions of the liquid and vapor phase which have the specific volumes l_l and l_g, so that

$$\xi_l + \xi_g = 1, \qquad \xi_l l_l + \xi_g l_g = l.$$

Using the expansion (5) for ν_0 and the expansion (6) both for l_l and l_g, one finds

$$B_i/l_i = (\gamma\delta)^{1/4}(\tfrac{3}{4}l_{1i}^2 - \nu_1)^{1/2}, \quad \text{with } i = l \text{ or } g, \tag{45}$$

$$\frac{1}{l}\left(\frac{\xi_l}{l_l} + \frac{\xi_g}{l_g}\right) = \frac{1}{l^2} + \frac{\xi_l\xi_g(l_{1g} - l_{1l})^2}{l_c^2}(\gamma\delta)^{1/2} + O(\gamma\delta).$$

Furthermore, near the critical point, $l_{1l} = -2\nu_1^{1/2}$ and $l_{1g} = +2\nu_1^{1/2}$, since the van der Waals coexistence curve is given by $l_1^2 = 4\nu_1$. Hence $B/l = (\gamma\delta)^{1/4}(2\nu_1)^{1/2}$ both for the liquid and the vapor, so that the two exponentials in (44) become identical. Finally one has

$$l_1 = \xi_l l_{1l} + \xi_g l_{1g} = 2\nu_1^{1/2}(\xi_g - \xi_l),$$

which gives

$$\left.\begin{matrix}\xi_l\\ \xi_g\end{matrix}\right\} = \frac{1}{2}\left[1 \mp \frac{l_1}{2\nu_1^{1/2}}\right].$$

Substituting all this in (44) one obtains

$$\bar{n}_2(x, l) = \frac{1}{l^2} + \frac{2}{3}\frac{(\gamma\delta)^{1/2}}{l_c^2}\left\{6\nu_1 - \tfrac{3}{2}l_1^2 + \frac{1}{(2\nu_1)^{1/2}} \times \exp[-(2\nu_1)^{1/2}\gamma(\gamma\delta)^{1/4}x]\right\}. \tag{46}$$

This follows also from Eq. (41) when $\nu_1 \gg 1$. Let us verify it only in the symmetric case ($l_1 = 0$). We saw in Sec. III that then the energy levels Θ_n are a series of narrow doublets corresponding to the symmetric and antisymmetric combinations of the oscillator eigenfunctions around the two minima. In the series (41) only the odd values of n contribute, and in first approximation only $n = 1$ and $n = 3$ need to be considered. One finds

$$\Theta_1 - \Theta_0 \cong 0, \qquad \Theta_3 - \Theta_0 \cong (2\nu_1)^{1/2};$$

$$\int dz\, zH_0^{(0)}H_1^{(0)} \cong (6\nu_1)^{1/2}, \qquad \int dz\, zH_0^{(0)}H_3^{(0)} \cong (2\nu_1)^{-1/4};$$

so that Eq. (41) becomes equal to (46) with $l_1 = 0$.

Consider next the transition to the one-phase region, which is of greater interest since more detailed results are available [see II, Sec. V, Eqs. (54) and (55)]. For $\nu_1 \ll -1$, or better, for $(3l_1^2 - 4\nu_1) \gg 1$, we found in Sec. III that in first approximation [see Eq. (28)]

$$\Theta_n - \Theta_0 = \tfrac{1}{2}n(3l_1^2 - 4\nu_1)^{1/2},$$

and that the eigenfunctions are just the Weber functions $D_n(y)$, where y is given by Eq. (29). It is then easy to verify that the first term of the sum in Eq. (41) becomes

$$\frac{4}{3l_c^2}\frac{(\gamma\delta)^{1/2}}{(3l_1^2 - 4\nu_1)^{1/2}}\exp[-\tfrac{1}{2}(3l_1^2 - 4\nu_1)^{1/2}\gamma x(\gamma\delta)^{1/4}],$$

and that this is also precisely what the Ornstein–Zernike term [see II, Eq. (54)]

$$[\gamma\nu_0(l-\delta)^4/l^5 B]\exp[-(B/l)\gamma x]$$

becomes in the critical region up to order $(\gamma\delta)^{1/2}$. One only has to use (45) for B/l, and replace ν_0 and l by their critical values.

In the next approximation, using the results derived in Appendix I, one finds

$$\Theta_n - \Theta_0 = \tfrac{1}{2}n(3l_1^2 - 4\nu_1)^{1/2} + \frac{1}{2(3l_1^2 - 4\nu_1)}\left[n^2 + n - \frac{2l_1^2}{3l_1^2 - 4\nu_1}(5n^2 - n)\right],$$

and up to order ϵ^2,

$$\int_{-\infty}^{+\infty} dz\, zH_0^{(0)}H_n^{(0)} = (\tfrac{3}{4}l_1^2 - 4\nu_1)^{-1/4} \times \left\{\delta_{n1}\left[1 + \frac{\epsilon^2}{3}\frac{12\nu_1 - 5l_1^2}{3l_1^2 - 4\nu_1}\right] - 2(\tfrac{2}{3})^{1/2}\frac{\epsilon l_1}{(3l_1^2 - 4\nu_1)}\delta_{n2}\right\},$$

omitting terms which do not contribute after squaring. In this order, therefore, the first and second term of the sum in Eq. (41) contribute. One obtains

$$\bar{n}_2(x;l) - \frac{1}{l^2} = \frac{(\gamma\delta)^{1/3}}{l_c^2}\left[\frac{4}{3}\frac{1}{(3l_1^2 - 4\nu_1)^{1/2}}\right.$$
$$+ \frac{4}{3}\frac{l_1^2 + 4\nu_1}{(3l_1^2 - 4\nu_1)^{5/2}}\gamma x(\gamma\delta)^{1/3} - \left.\frac{8}{9}\frac{5l_1^2 - 12\nu_1}{(3l_1^2 - 4\nu_1)^3}\right]$$
$$\times \exp\left[-\tfrac{1}{2}(3l_1^2 - 4\nu_1)^{1/2}\gamma x(\gamma\delta)^{1/3}\right]$$
$$+ \frac{(\gamma\delta)^{2/3}}{l_c^2}\frac{32}{9}\frac{l_1^2}{(3l_1^2 - 4\nu_1)^3}$$
$$\times \exp\left[-(3l_1^2 - 4\nu_1)^{1/2}\gamma x(\gamma\delta)^{1/3}\right] + \cdots, \quad (47)$$

and one can verify that the last three terms in (47) correspond precisely to the three corrections of order γ^2 to the Ornstein–Zernike result in the one-phase region [see II, Eq. (55)] if these corrections are computed in the critical region up to order $(\gamma\delta)^{2/3}$. This confirms the general picture mentioned in the introduction, and it shows again that the Ornstein–Zernike exponential is the leading term only *outside* the critical region.

Since in the experimental investigation of the critical opalescence one observes the Fourier transform of the two-point distribution function, it is of interest to discuss also for our model the function

$$\tilde{g}(k) = \int_{-\infty}^{+\infty} dx\, e^{ikx} g(x)$$
$$= l\int_{-\infty}^{+\infty} dx\, e^{ikx}\left[\bar{n}_2(x,l) - \frac{1}{l^2}\right].$$

For the Ornstein–Zernike exponential (in any number of dimensions), $\tilde{g}(k) \sim 1/(k^2 + b^2)$, where b^2 is proportional to the compressibility, and it is therefore customary to plot $1/\tilde{g}(k) \equiv f(k)$ vs k^2, since this should be a straight line intersecting the $f(k)$ axis at a point which goes to zero at the critical point if the Ornstein–Zernike theory is correct. Since Eq. (41) is of the form

$$g(x) = \sum_{n=0}^{\infty} a_n e^{-b_n x},$$

where the a_n and b_n are positive, we obtain for our model, in the critical region,

$$\frac{1}{f(k)} = \tilde{g}(k) = \sum_0^{\infty}\frac{2a_n b_n}{b_n^2 + k^2}.$$

For k^2 large compared to the b_n^2, $f(k)$ is almost linear in k^2. In fact,

$$f(k) = \frac{k^2}{2\sum a_n b_n} + \frac{\sum a_n b_n^3}{2(\sum a_n b_n)^2} + O\left(\frac{1}{k^2}\right). \quad (48)$$

To study the curvature for smaller k^2, note that one can write

$$\frac{f''}{2f^3} = \left[\sum_n \frac{a_n b_n}{(b_n^2 + k^2)^2}\right]^2$$
$$- \left(\sum_n \frac{a_n b_n}{b_n^2 + k^2}\right)\left[\sum_n \frac{a_n b_n}{(b_n^2 + k^2)^3}\right],$$

where the primes denote differentiations after k^2. Hence, by Schwartz's inequality, one concludes that $f'' \leq 0$, so that the curve is convex towards the k^2 axis.

For the discussion of the temperature dependence of the Ornstein–Zernike plot, we will restrict ourselves to the case that $l_1 = 0$ (critical density) and that $\nu_1 \ll -1$. One then can use Eq. (47); since the last term vanishes, one can write $g(x)$ as one exponential by putting the second term back in the exponent, and one obtains

$$g(x) = a_1 e^{-b_1 x},$$

with

$$a_1 = \frac{2}{3l_c}\frac{(\gamma\delta)^{1/3}}{(-\nu_1)^{1/2}}\left[1 - \frac{1}{4(-\nu_1)^{3/2}} + \cdots\right],$$
$$b_1 = (\gamma\delta)^{4/3}\frac{(-\nu_1)^{1/2}}{\delta}\left[1 + \frac{1}{4(-\nu_1)^{3/2}} + \cdots\right].$$

Since the slope of the Ornstein–Zernike plot is $\frac{1}{2}a_1 b_1$, one sees that the temperature dependence drops out up to order $(-\nu_1)^{3/2}$, so that the straight parts of the plot are very nearly parallel. From (48) one further sees that the intersection with the $f(k)$ axis is $b_1/2a_1$, or

$$f(0) = \tfrac{9}{4}(\gamma\delta)^{2/3}(-\nu_1)[1 + 1/2(-\nu_1)^{3/2} + \cdots].$$

Since $-(\gamma\delta)^{2/3}\nu_1 = (T - T_c)/T_c$ to this order in $(\gamma\delta)$ one sees that, in first approximation, $f(0) \sim (T - T_c)$ which is the Ornstein–Zernike result. However, closer to the critical point, $f(0)$ bends upwards and one can show [using the exact Eq. (41)] that $f(0)$ is finite for $T = T_c$. The deviations from the Ornstein–Zernike theory which follow from our model, look therefore about as shown in Fig. 4.

Finally we will discuss briefly the so-called *specific heat anomaly*.[8] It is well known that according to the van der Waals theory the specific heat at constant volume is constant ($= \frac{3}{2}k$) in the one-phase region and increases suddenly if one crosses at a fixed density into the two-phase region. Especially at the critical density one finds

[8] We are indebted to Dr. J. V. Sengers for pointing out to us the interest of discussing this anomaly for our model.

$$\Delta c_v = \begin{cases} 0 \quad \text{for} \quad T > T_c, \\ \frac{9}{2}k[1 - \frac{28}{25}(T_c - T)/T_c + \cdots] \\ \qquad\qquad \text{for} \quad T < T_c. \end{cases} \tag{49}$$

Since the experimental results are quite different (this is the anomaly), it is of interest to see what the implications are of Eq. (41) for the specific heat. We start from the equation

$$\epsilon = \frac{kT}{2} - \frac{\alpha_0}{l_c} - \frac{2}{3}\frac{\alpha_0}{l_c}(\gamma\delta)^{\frac{1}{2}} \times \int_{-\infty}^{+\infty} dz[z + l_1(\tfrac{3}{2})^{\frac{1}{2}}]^2 H_0^{(0)}(z)^2, \tag{50}$$

which follows from the general result [II, Eq. (41)] by introducing the expression (42) for $n_2(x; l)$. Since in the critical region

$$\frac{\partial}{\partial T} = -\frac{\alpha_0}{kT_c^2}\frac{1}{\nu_{0c}(\gamma\delta)^{\frac{1}{2}}}\frac{\partial}{\partial\nu_1},$$

one obtains

$$\Delta c_v \equiv \left(\frac{\partial\epsilon}{\partial T}\right)_l - \frac{k}{2} = \frac{3}{4}k\frac{\partial}{\partial\nu_1}\int_{-\infty}^{+\infty} dz\, z^2 H_0^{(0)2}(z, \nu_1, l_1)$$
$$= -3k\left(\frac{\partial^2\Theta_0}{\partial\nu_1^2}\right)_{l_1} + \frac{3k}{8}l_1\left(\frac{\partial^2 s_1}{\partial\nu_1^2}\right)_{l_1}, \tag{51}$$

by differentiating Eq. (12) for $H_0^{(0)}(z)$ after ν_1 at constant l_1, and by using the equation of state (15). This shows that our model gives corrections to the specific heat which are of order k in the critical region, so that in this sense our model gives an anomalous behavior of c_v. To discuss Eq. (51) further, consider first the case $l_1 = 0$. It is easy to show that then

$$\Delta c_v = -3k\left(\frac{\partial^2\Theta_0}{\partial\nu_1^2}\right)_{l_1}$$
$$= \frac{3k}{8}\sum_{n=1}^{\infty}\frac{1}{\Theta_n - \Theta_0}\left[\int_{-\infty}^{+\infty} dz\, z^2 H_0^{(0)}H_n^{(0)}\right]^2,$$

and by using the results of Sec. III one finds for $\nu_1 \ll -1$

$$\Delta c_v = \tfrac{3}{8}k[1/(-\nu_1)^{\frac{1}{2}}] + \cdots,$$

and for $\nu_1 \gg 1$[9]

$$\Delta c_v = \tfrac{9}{2}k - \tfrac{3}{4}k[1/(2\nu_1)^{\frac{1}{2}}] + \cdots.$$

At the critical density our model therefore smooths out the discontinuity (49) in c_v which follows from the van der Waals equation. Next, consider the

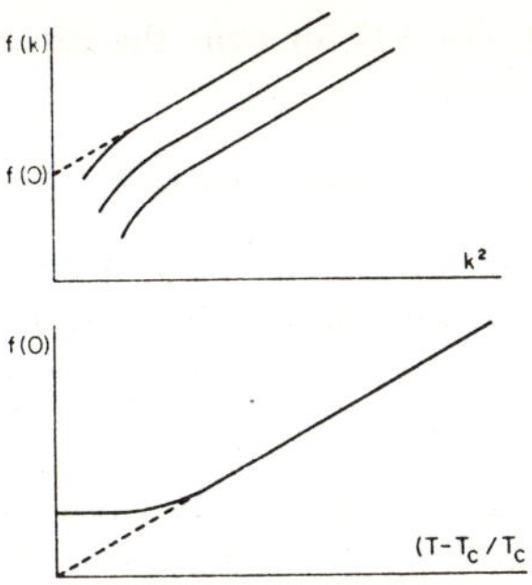

FIG. 4. Ornstein-Zernike plot according to Eq. (47).

dependence of Δc_v on l_1 at fixed ν_1. From (51) follows

$$[\partial(\Delta c_v)/\partial l_1]_{\nu_1} = \tfrac{3}{8}k(\partial^2 s_1/\partial\nu_1^2)_{l_1}, \tag{52}$$

using

$$8(\partial\Theta_0/\partial l_1)_{\nu_1} = l_1(\partial s_1/\partial l_1)_{\nu_1},$$

which is a consequence of (19). Eq. (52) is just a transcription of the thermodynamic formula

$$(\partial c_v/\partial v)_T = T(\partial^2 p/\partial T^2)_v,$$

and it shows that, at the critical density, Δc_v has always an extremum, because $(s_1 - 4)$ is an odd function of l_1 so that $\partial^2 s_1/\partial\nu_1^2$ is zero for $l_1 = 0$. Furthermore, for small l_1,

$$\Delta c_v = \Delta c_v(l_1 = 0) + l_1^2\tfrac{3}{16}k(\partial^3 s_1/\partial\nu_1^2\,\partial l_1)_{l_1=0} + \cdots,$$

which gives for $\nu_1 \ll -1$

$$\Delta c_v = \Delta c_v(l_1 = 0) - \tfrac{27}{64}k[l_1^2/(-\nu_1)^{5/2}] + \cdots,$$

and for $\nu_1 \gg 1$

$$\Delta c_v = \Delta c_v(l_1 = 0) - (27k/8\pi)(2\nu_1)^{\frac{1}{2}} \times \exp[-(2\nu_1)^{\frac{1}{2}}]l_1^2 + \cdots,$$

so that the extremum is probably always a maximum. The maximum becomes sharper when one approaches the critical point from the one-phase region, and then becomes rapidly very flat in the two-phase region. Presumably, the width of the maximum is smallest for ν_1 of order one.

VI. GENERALIZATION OF EQ. (41)

In order to show that qualitatively the behavior of the two-point distribution function in the critical region does not depend sensitively on the form of the attractive potential (provided it has a long range), we will generalize the discussion of Sec. IV to the case where the attractive potential consists of a sum of m exponentials. Since the formulas are very long, we will only present an outline of the

[9] To derive this result it is simpler not to use (51), but to go back to the general result [See II, Eq. (41)] and then use Eq. (46) for $\bar{n}_2(x; l)$.

calculations. As in II, Sec. VII, we write the attractive potential in the form

$$\varphi_{\text{attr}}/kT = -\gamma \sum_{i=1}^{m} \nu_i \exp(-\gamma\sigma_i x),$$

where $\nu_i = \alpha_i/kT$, and we start to tame the Kac integral equation in the same way as in II, except that we replace η by $\eta_c \equiv \eta(s_c)$. One expands the van der Waals $\nu_0 = \sum_i (\nu_i/\sigma_i)$ [see II, Eq. (79)] and s as in (5) and (7), and one then introduces new variables y_m by an orthogonal transformation which diagonalizes, just as in II, Eq. (75), the matrix

$$M_{ij} = \sigma_i^2\delta_{ij} - \sigma_i\sigma_j(c_ic_j)^{\frac{1}{2}},$$

except that now,

$$c_i = \frac{2\nu_{ic}(l_c - \delta)^2}{\sigma_i l_c^3} = \frac{8\nu_{ic}}{27\sigma_i\delta} = \frac{\nu_{ic}}{\sigma_i\nu_{0c}} = \frac{\nu_i}{\sigma_i\nu_0},$$

where $\nu_{ic} = \alpha_i/kT_c$. It follows that $\sum c_i = 1$, in agreement with II, Eq. (80), since $B = 0$, and from the discussion given in II, this implies that the smallest eigenvalue $A_m = 0$, so that

$$\sum_{i,j} M_{ij}z_iz_j = \sum_{k=1}^{m-1} A_k y_k^2.$$

The variable y_m plays a special role; *only* in this variable one must make the second, or critical-region "taming",

$$y_m = z/(\gamma\delta)^{\frac{1}{3}},$$

and one then gets a consistent successive-approximation scheme with the parameter $(\gamma\delta)^{1/3}$. Up to order $(\gamma\delta)^{4/3}$ one obtains[10]

$$\gamma\delta \sum_{k=1}^{m-1}\left(\frac{\partial^2 h}{\partial y_k^2} - \frac{A_k}{4}y_k^2 h\right) + (\gamma\delta)^{4/3}\left[\frac{\partial^2 h}{\partial z^2} - \frac{(\mu z)^4}{48}h + \frac{\nu_1(\mu z)^2}{4}h + \frac{\mu z}{6^{\frac{1}{2}}}\left(\frac{s_1}{4} - \sum_{i=1}^{m}\sigma_i\right)h\right] + (\gamma\delta)^{4/3}\frac{\mu z h}{2(6)^{\frac{1}{2}}}\sum_{k=1}^{m-1} A_k y_k^2 + \Lambda h = 0, \tag{53}$$

which replaces Eq. (9) up to this order. In (53), h is a function of $y_1, y_2, \cdots, y_{m-1}, z$, and the new eigenvalue Λ is defined by

$$3\Lambda = 1 + \frac{9\nu_1}{4}(\gamma\delta)^{2/3} + \tfrac{3}{2}(\gamma\delta)\left(\sum_{i=1}^{m}\sigma_i - \frac{s_1}{4}\right) + (\gamma\delta)^{4/3}\left(-\frac{3}{8}s_2 + \frac{81}{32}\nu_1^2\right) - \frac{\lambda}{\omega(s_c)}, \tag{54}$$

[10] In this equation, ν_1 should not be confused with $\nu_1 = \alpha_1/kT$. From now on, only the critical quantities ν_{ic} will appear and ν_1 will always mean the number defined by Eq. (5).

which replaces Eq. (10). Finally the quantity μ is defined by

$$\sum_{i=1}^{m}\sigma_i^2 a_{im}^2 = \left(\sum_{i=1}^{m}\sigma_i c_i^{\frac{1}{2}} a_{im}\right)^2 \equiv \mu^2,$$

where a_{ik} is the orthogonal matrix, which diagonalizes M_{ij}. By introducing instead of the y_k the new variables w_k by

$$y_k = w_k\{1 + [\mu z/2(6)^{\frac{1}{2}}](\gamma\delta)^{\frac{1}{3}}\},$$

Eq. (53) can be written (always up to order $(\gamma\delta)^{4/3}$) in the form

$$\gamma\delta\left[1 - \frac{\mu z}{6^{\frac{1}{2}}}(\gamma\delta)^{\frac{1}{3}}\right]\sum_{k=1}^{m-1}\left[\frac{\partial^2 h}{\partial w_k^2} - \frac{A_k}{4}w_k^2 h\right] + (\gamma\delta)^{4/3}\left[\frac{\partial^2 h}{\partial z^2} - \frac{(\mu z)^4}{48}h + \frac{\nu_1(\mu z)^2}{4}h + \frac{\mu z}{6^{\frac{1}{2}}}\left(\frac{s_1}{4} - \sum_i \sigma_i\right)h\right] + \Lambda h = 0, \tag{55}$$

and hence it becomes separable. It is easy to verify that (55) is fulfilled by putting

$$\Lambda = \gamma\delta\sum_{k=1}^{m-1} A_k^{\frac{1}{2}}(n_k + \tfrac{1}{2}) + \Theta(\gamma\delta)^{4/3},$$
$$h(w_1\cdots w_{m-1}, z) = \prod_{k=1}^{m-1} N_{n_k}D_{n_k}(A_k^{\frac{1}{4}}w_k)H^{(0)}(z), \tag{56}$$

if $H^{(0)}(z)$ and Θ satisfy the equation

$$\left\{\frac{d^2}{dz^2} - \frac{(\mu z)^4}{48} + \frac{\nu_1(\mu z)^2}{4} + \frac{\mu z}{6^{\frac{1}{2}}}\left[\frac{s_1}{4} - \sum_{i=1}^{m}\sigma_i + \sum_{k=1}^{m-1}A_k^{\frac{1}{2}}(n_k + \tfrac{1}{2})\right] + \Theta\right\}H^{(0)} = 0, \tag{57}$$

which is of the same form as (12). In fact by introducing the new variables

$$\hat{z} = z\mu^{2/3}, \qquad \hat{\nu}_1 = \nu_1\mu^{-2/3}, \qquad \hat{\Theta} = \Theta\mu^{-4/3},$$
$$\hat{s}_1 = 4 + \frac{1}{\mu}\left(s_1 - 4\sum_{i=1}^{m}\sigma_i + 2\sum_{k=1}^{m-1}A_k^{\frac{1}{2}}\right), \tag{58}$$

and by putting $H^{(0)}(z) \equiv \mu^{\frac{1}{3}}\hat{H}(\hat{z})$, Eq. (57) becomes

$$\left[\frac{d^2}{d\hat{z}^2} - \frac{\hat{z}^4}{48} + \frac{\hat{\nu}_1\hat{z}^2}{4} + \frac{\hat{z}}{6^{\frac{1}{2}}}\left(\frac{\hat{s}_1}{4} - 1 + \frac{1}{\mu}\sum_{k=1}^{m-1}A_k^{\frac{1}{2}}n_k\right) + \hat{\Theta}_{\mathbf{n}}\right]\hat{H}_{\mathbf{n}}(\hat{z}) = 0, \tag{59}$$

which is quite similar to (12). The only difference is that, because of the extra term containing the n_k, the eigenvalues $\hat{\Theta}_{\mathbf{n}}$ and eigenfunctions $\hat{H}_{\mathbf{n}}$ will depend on the whole set of "quantum" numbers $\mathbf{n} \equiv n_1, n_2 \cdots n_m$. Note still, that with the definitions (58) and with

$$\hat{\gamma} = \mu\gamma, \tag{60}$$

Eq. (5) for the critical temperature region can be written as

$$\nu_0 = \nu_{0c}[1 + \hat{\nu}_1(\hat{\gamma}\delta)^{\frac{1}{2}}],$$

while Eq. (7) for the pressure region becomes

$$s = s_c\left[1 - 3\hat{\nu}_1(\hat{\gamma}\delta)^{\frac{1}{2}} + \left(\hat{s}_1 - 4 + \frac{4}{\mu}\sum_{i=1}^{m}\sigma_i - \frac{2}{\mu}\sum_{k=1}^{m-1}A_k^{\frac{1}{2}}\right)(\hat{\gamma}\delta) + \hat{s}_2(\hat{\gamma}\delta)^{4/3} + \cdots\right],$$

with $\hat{s}_2 = s_2\mu^{-4/3}$. Finally, Eqs. (54) and (56) for the eigenvalue λ_n can be combined to

$$\frac{\lambda_n(s)}{\omega(s_c)} = 1 + \frac{9}{4}\hat{\nu}_1(\hat{\gamma}\delta)^{\frac{1}{2}} + \tfrac{3}{2}\hat{\gamma}\delta \times\left[1 - \frac{\hat{s}_1}{4} - \frac{1}{\mu}\sum_{k=1}^{m-1}A_k^{\frac{1}{2}}(2n_k + \tfrac{1}{2})\right] + (\hat{\gamma}\delta)^{4/3}\left(-3\hat{\Theta}_n - \frac{3}{8}\hat{s}_2 + \frac{81}{32}\hat{\nu}_1^2\right) + \cdots$$

Clearly it follows from (59) that the *maximum* eigenvalue and corresponding eigenfunction, for which the whole set $\mathbf{n} = 0$, obey in the capped variables exactly the same equation as before. Therefore the equation of state is still determined by

$$\int_{-\infty}^{+\infty} d\hat{z}\,\hat{z}\hat{H}_{0\cdots0}^2(\hat{z}) = -(\tfrac{3}{2})^{\frac{1}{2}}\hat{l}_1, \tag{61}$$

where $\hat{l}_1$ is defined by:

$$l = l_c[1 + \hat{l}_1(\hat{\gamma}\delta)^{\frac{1}{2}}],$$

similar to Eq. (6).

Turning now to the behavior of the two-point distribution function in the range of order $1/\hat{\gamma}(\hat{\gamma}\delta)^{\frac{1}{2}}$ up to order $(\hat{\gamma}\delta)^{\frac{1}{2}}$, it is not difficult to prove that also in this case the calculation goes in the same way as before. The starting point is the obvious generalization of Eq. (34). One then shows that *only* the terms with $n_1 = n_2 = \cdots = n_{m-1} = 0$ contribute to order $(\hat{\gamma}\delta)^{\frac{1}{2}}$, and since in this case the equation for $\hat{H}_{0,0\cdots0,\,n_m}(\hat{z})$ is the same as Eq. (12) for $H_n(z)$, one obtains, in complete analogy with Eq. (41), that, in the general case,

$$\tilde{n}_2(x;l) = \frac{1}{l^2} + \frac{2}{3}\frac{(\hat{\gamma}\delta)^{\frac{1}{2}}}{l_c^2}\sum_{n=1}^{\infty}\left[\int_{-\infty}^{+\infty} d\hat{z}\,\hat{z}\hat{H}_0\hat{H}_n\right]^2 \times \exp\left[-(\hat{\Theta}_n - \hat{\Theta}_0)x\hat{\gamma}(\hat{\gamma}\delta)^{\frac{1}{2}}\right], \tag{62}$$

where $\hat{H}_n$ is the *same* function of $\hat{z}$ as $H_n^{(0)}$ is of z in (41), and $\hat{\Theta}_n$ is equal to the Θ_n occurring in (41).[11] The only essential change is therefore the replacement of γ by $\hat{\gamma} = \mu\gamma$. Now we saw in II [see the equation following (85)] that, at the critical point,

$$a_{im} = (c_i^{\frac{1}{2}}/\sigma_i)/[\sum_i (c_i/\sigma_i^2)]^{\frac{1}{2}}.$$

Hence,

$$\mu^2 = \sum_{i=1}^{m}\sigma_i^2 a_{im}^2 = \frac{\sum_i c_i}{\sum_i (c_i/\sigma_i^2)} = \frac{\sum_i (\alpha_i/\sigma_i)}{\sum_i (\alpha_i/\sigma_i^3)}$$

since $c_i = (\alpha_i/\sigma_i)(1/\nu_0 kT)$. Therefore we can write

$$\frac{1}{\hat{\gamma}^2} = \frac{1}{\mu^2\gamma^2} = \frac{\frac{1}{2}\int_{-\infty}^{+\infty} dx\,x^2\varphi_{\text{attr}}(|x|)}{\int_{-\infty}^{+\infty} dx\,\varphi_{\text{attr}}(|x|)}, \tag{63}$$

so that $1/\hat{\gamma}$ is just the range R of the Ornstein–Zernike theory [see II, Eq. (72)]. The argument given in Sec. V, showing that the first term of the series (41) becomes the Ornstein–Zernike exponential in the one-phase region, can therefore be repeated in exactly the same way in the general case. We have not investigated in detail what happens in the higher approximations, but it seems very likely that qualitatively at least everything will remain the same.

VII. CONCLUDING REMARKS

It is well known that the Fourier transform of the Ornstein–Zernike exponential [the function $\tilde{g}(k)$ of Sec. V] is independent of the number of dimensions, and one may hope therefore that the deviations from the Ornstein–Zernike theory which follow from our model with regard to $\tilde{g}(k)$ will at least qualitatively be the same in three dimensions. This is confirmed by the results of P. C. Hemmer, who has succeeded in generalizing to three dimensions the discussion of the long-range behavior of the correlation function reported in II.[12] It is therefore of interest to inquire whether there are any experimental indications of the deviations from the Ornstein–Zernike theory discussed in Sec. V [see especially Fig. 4].[13] Unfortunately, at present the experiments are still not decisive. The recent work

[11] These Θ_n should not be confused with the Θ occurring in Eq. (57), which is related to $\hat{\Theta}$ by Eq. (58).

[12] This work will be reported in Part IV of this series.

[13] The question of the validity of the Ornstein–Zernike theory close to the critical point has been discussed especially by M. S. Green [J. Chem. Phys. **33**, 1403 (1960)]. Although we do not agree with his theoretical arguments, we are very much indebted to Dr. Green for telling us about his work, and about the experimental results indicating deviations from the O–Z theory. Compare also the review by O. K. Rice in *Thermodynamics and Physics of Matter* (Princeton University Press, Princeton, New Jersey, 1955), Sec. E, where one finds a very complete bibliography.

by Thomas and Schmidt[14] on the critical opalescence in Argon does not show any deviation from the Ornstein–Zernike theory. On the other hand, deviations have been reported for various binary mixtures and these appear to point in the same direction as those shown in Figure 4.[15] It seems to us of great interest to see whether more refined experiments will show deviations from the Ornstein–Zernike theory, since such deviations *near* the critical point are perhaps the clearest indication of the existence of a critical *region.*

With regard to the specific-heat anomaly, the experiments[16] clearly show that c_v is *not* a constant in the one-phase region, and that it has a pronounced maximum as function of the density at constant temperature near the critical point. The maximum occurs close to the critical density, and this aspect of the anomaly agrees therefore qualitatively with the results for our model. However, the temperature dependence is quite different. Instead of smoothing out of the van der Waals discontinuity predicted by our model, recent experiments for Argon[17] show that at the critical density, c_v has as function of the temperature an asymmetric logarithmic singularity at the critical temperature. It seems therefore that in three dimensions, one has both a critical region *and* a critical point! Whether this is an universal phenomenon, more or less independent of the intermolecular forces, remains to be seen.

APPENDIX I. HIGHER APPROXIMATIONS OF THE EIGENFUNCTIONS NEAR THE ONE-PHASE REGION

For a systematic application of the perturbation theory to the basic equation (12) when $\nu_1 \ll -1$, it is best to replace z by the variable y of Eq. (29). Eq. (12) then becomes

$$\left[\frac{d^2}{dy^2} - (3l_1^2 - 4\nu_1)^{-3/2}\frac{y^4}{6} + (3l_1^2 - 4\nu_1)^{-5/4} \times \frac{l_1 y^3}{3^{1/2}} - \frac{y^2}{4} + \alpha y + \vartheta\right]K(y) = 0, \qquad \text{(A1)}$$

where

$$K(y) = (\tfrac{3}{4}l_1^2 - \nu_1)^{-1/8}H^{(0)}(z),$$

$$\alpha = \frac{2}{3^{1/2}}(3l_1^2 - 4\nu_1)^{-3/4}\left(\frac{s_1}{4} - 1 + \tfrac{3}{8}l_1^3 - \tfrac{3}{2}\nu_1 l_1\right),$$
$$\vartheta = (3l_1^2 - 4\nu_1)^{-1/2} \times \left(2\Theta + l_1 - \frac{3}{32}l_1^4 + \frac{3}{4}l_1^2\nu_1 - \frac{s_1}{4}l_1\right). \qquad \text{(A2)}$$

For the small perturbation parameter, choose

$$\epsilon = (3l_1^2 - 4\nu_1)^{-3/4}$$

Note that $l_1/(3l_1^2 - 4\nu_1)^{1/2}$ is always of order one. Write (A1) in the form

$$\left[\frac{d^2}{dy^2} - \frac{\epsilon^2 y^4}{6} - \left(1 - \frac{4\nu_1}{3l_1^2}\right)^{-1/2}\frac{\epsilon y^3}{3} - \frac{y^2}{4} + \alpha(\epsilon)y + \vartheta(\epsilon)\right]K(y) = 0,$$

and expand

$$K(y) = K^{(0)}(y) + \epsilon K^{(1)}(y) + \epsilon^2 K^{(2)}(y) + \cdots,$$
$$\alpha(\epsilon) = \alpha_0 + \epsilon\alpha_1 + \epsilon^2\alpha_2 + \cdots,$$
$$\vartheta(\epsilon) = \vartheta^{(0)} + \epsilon\vartheta^{(1)} + \epsilon^2\vartheta^{(2)} + \cdots.$$

In zeroth order, one gets

$$[d^2/dy^2 - \tfrac{1}{4}y^2 + \alpha_0 y + \vartheta^{(0)}]K^{(0)} = 0. \qquad \text{(A3)}$$

Since the equation of state (15) implies that for the ground state

$$\int_{-\infty}^{+\infty} dy\, yK_0(y)^2 = 0, \qquad \text{(A4)}$$

it follows that α_0 must be zero, and hence

$$K_n^{(0)}(y) = N_n D_n(y),$$
$$\vartheta_n^{(0)} = n + \tfrac{1}{2}. \qquad \text{(A5)}$$

In first order one then gets

$$\left[\frac{d^2}{dy^2} - \frac{y^2}{4} + n + \frac{1}{2}\right]K_n^{(1)}(y) = \left[-\left(1 - \frac{4\nu_1}{3l_1^2}\right)^{-1/2}\frac{y^3}{3} - \alpha_1 y - \vartheta_n^{(1)}\right]K_n^{(0)}. \qquad \text{(A6)}$$

Since $K_n^{(0)}(y)^2$ is symmetric in y, multiplication of (A6) with $K_n^{(0)}$ and integration over y gives

$$\vartheta_n^{(1)} = 0. \qquad \text{(A7)}$$

For $n = 0$, one gets from (A6)

$$K_0^{(1)}(y) = \frac{1}{9}\left(1 - \frac{4\nu_1}{3l_1^2}\right)^{-1/2} D_3(y) + \left[\left(1 - \frac{4\nu_1}{3l_1^2}\right)^{-1/2} + \alpha_1\right]D_1(y).$$

---

[14] J. E. Thomas and P. W. Schmidt, J. Chem. Phys. (to be published).

[15] See especially D. McIntyre, A. Wims, and M. S. Green, J. Chem. Phys. **37**, 3019 (1962).

[16] The most complete data are for Argon, but other substances show the same behavior. See J. M. H. Levelt, Dissertation, University of Amsterdam, 1958.

[17] M. I. Bagatskii, A. V. Voronel, and V. G. Gusak, Zh. Eksperim. i Teor. Fiz. **43**, 728 (1962). We are indebted to Dr. M. Fisher for pointing out this reference to us.

Since it follows from (A4) that

$$\int_{-\infty}^{+\infty} dy\, yK_0^{(0)}K_0^{(1)} = 0,$$

one concludes that

$$\alpha_1 = -[1 - \tfrac{4}{3}(\nu_1/l_1^2)]^{-\frac{1}{2}}. \tag{A8}$$

From (A2) together with (A5), (A7) and (A8) one obtains the results (24) and (28) for the equation of state and the energy Θ_n. For the first-order correction to the eigenfunction, one then gets from (A6)

$$K_n^{(1)}(y) = N_n\left(1 - \frac{4\nu_1}{3l_1^2}\right)^{-\frac{1}{2}}\left[\frac{1}{9}D_{n+3} + nD_{n+1} - n(n-1)D_{n-1} - \frac{n(n-1)(n-2)}{9}D_{n-3}\right].$$

In second order one obtains

$$\left[\frac{d^2}{dy^2} - \frac{y^2}{4} + n + \frac{1}{2}\right]K_n^{(2)}(y) = -\left(1 - \frac{4\nu_1}{3l_1^2}\right)^{-\frac{1}{2}} \times\left(y - \frac{y^3}{3}\right)K_n^{(1)}(y) + \left(\frac{y^2}{6} - \alpha_2 y - \vartheta_n^{(2)}\right)K_n^{(0)}(y), \tag{A9}$$

from which one finds

$$\vartheta_n^{(2)} = n^2 + n + \tfrac{1}{2} - [3l_1^2/(3l_1^2 - 4\nu_1)] \times\left(\frac{10}{3}n^2 - \frac{2}{3}n + \frac{2}{9}\right), \tag{A10}$$

and from (A4) one concludes that

$$\int_{-\infty}^{+\infty} dy\, y(K_0^{(1)2} + 2K_0^{(0)}K_0^{(2)}) = 0,$$

which leads to $\alpha_2 = 0$. One then can determine $K_n^{(0)}(y)$ in the usual way; and so on!

APPENDIX II. COMPLETION OF THE PROOF OF EQ. (41)

The quantity R left undefined in Eq. (40) is given by

$$R = \frac{2}{3}\sum_{n=1}^{\infty}\frac{\sigma(\sigma - \Theta_n + \Theta_0)}{(\Theta_n - \Theta_0)^2}\left[\int_{-\infty}^{+\infty} dz\, zH_0^{(0)}H_n^{(0)}\right]^2 + 2\int_{-\infty}^{+\infty} dz\, H_0^{(0)}\left\{H_0^{(2)}(z, s + \sigma\gamma(\gamma\delta)^{\frac{1}{2}}) - H_0^{(2)}\right\} + 2\int_{-\infty}^{+\infty} dz\, H_0^{(1)}\left\{H_0^{(1)}(z, s + \sigma\gamma(\gamma\delta)^{\frac{1}{2}}) - H_0^{(1)}\right\} - \frac{\Xi_0(\sigma) - \Xi_0(0)}{\sigma} - \frac{\nu_1}{2}, \tag{A11}$$

where the arguments of the eigenfunctions which are not indicated are always meant to be z, s. This can be simplified by using the fact that

$$\sum_{n=0}^{\infty} b_n^2 = 1.$$

Substituting (35) and using (37) shows that the second and third terms in (A11) are equal to

$$-\frac{2}{3}\sum_{n=1}^{\infty}\frac{\sigma^2}{(\Theta_n - \Theta_0^2)}\left[\int_{-\infty}^{+\infty} dz\, zH_0^{(0)}H_n^{(0)}\right]^2,$$

so that one gets

$$R = -\tfrac{2}{3}\sigma\sum_{n=1}^{\infty}\frac{1}{\Theta_n - \Theta_0}\left[\int_{-\infty}^{+\infty} dz\, zH_0^{(0)}H_n^{(0)}\right]^2 - \frac{\Xi_0(\sigma) - \Xi_0(0)}{\sigma} - \frac{\nu_1}{2}. \tag{A12}$$

One now has to calculate $\Xi_0(\sigma) - \Xi_0(0)$, for which the equation for $H_n^{(2)}(z)$ is needed, which requires the extension of the expansion in Eq. (9) to order $(\gamma\delta)^2$. We will not give the explicit expression, but only note that the general form of the equations for the successive order eigenfunctions $H_0^{(2)}(z)$ is

$$L_1H_0^{(0)} = 0, \tag{A13a}$$

$$L_1H_0^{(1)} = L_2H_0^{(0)}, \tag{A13b}$$

$$L_1H_0^{(2)} = L_2H_0^{(1)} + L_3H_0^{(0)}, \tag{A13c}$$

where the operators L_1 and L_2 can be read of Eqs. (12) and (13). The replacement of s by $s + \sigma\gamma(\gamma\delta)^{\frac{1}{2}}$ does not affect L_1 but changes L_2 and L_3, and we indicate this by writing $L_2(\sigma)$, $L_3(\sigma)$ for the new operators. The eigenvalue Ξ_0 occurs in L_3, and to evaluate $\Xi_0(\sigma) - \Xi_0(0)$ one needs $L_3(\sigma) - L_3$. From the extension of the expansion (9) to order $(\gamma\delta)^2$ one finds

$$[L_3(\sigma) - L_3]H_0^{(0)}(z) = \frac{2\sigma}{3}z^2H_0^{(0)}(z) - [\Xi_0(\sigma) - \Xi_0(0)]H_0^{(0)}(z),$$

so that

$$\Xi_0(\sigma) - \Xi_0(0) = \frac{2\sigma}{3}\int_{-\infty}^{+\infty} dz\, z^2H_0^{(0)2} - \int_{-\infty}^{+\infty} dz\, H_0^{(0)}[L_3(\sigma) - L_3]H_0^{(0)}.$$

The last integral can be transformed with the help of (A13c), and then one gets

$$\Xi_0(\sigma) - \Xi_0(0) = \frac{2\sigma}{3}\int_{-\infty}^{+\infty} dz\, z^2H_0^{(0)2} + \int_{-\infty}^{+\infty} dz\, H_0^{(0)}[L_2(\sigma)H_0^{(1)}(z, s + \sigma\gamma(\gamma\delta)^{\frac{1}{2}}) - L_2H_0^{(1)}]$$

$$= -\frac{\nu_1\sigma}{2} + \sigma(\tfrac{2}{3})^{\frac{1}{2}} \int_{-\infty}^{+\infty} dz\, zH_0^{(0)}H_0^{(1)} - \sigma(\tfrac{2}{3})^{\frac{1}{2}}$$

$$\times \int_{-\infty}^{+\infty} dz\, zH_0^{(0)}[H_0^{(1)}(z, s + \sigma\gamma(\gamma\delta)^{\frac{1}{2}}) - H_0^{(1)}]$$

$$+ \int_{-\infty}^{+\infty} dz\, H_0^{(0)}L_2[H_0^{(1)}(z, s + \sigma\gamma(\gamma\delta)^{\frac{1}{2}}) - H_0^{(1)}], \tag{A14}$$

where we have used the equation of state (16), the relation

$$L_2(\sigma) = L_2 - \sigma(\tfrac{2}{3})^{\frac{1}{2}}z + \Omega_0(0) - \Omega_0(\sigma), \tag{A15}$$

and the fact that

$$\int_{-\infty}^{+\infty} dz\, H_0^{(1)}(z, s')H_0^{(0)}(z, s) = 0$$

for $s' = s$ and for $s' = s + \sigma\gamma(\gamma\delta)^{\frac{1}{2}}$.

Substituting (A14) in (A12) one obtains

$$R = -\frac{2\sigma}{3}\sum_{n=1}^{\infty}\frac{1}{\Theta_n - \Theta_0}\left[\int_{-\infty}^{+\infty} dz\, zH_0^{(0)}H_n^{(0)}\right]^2$$

$$-\left(\frac{2}{3}\right)^{\frac{1}{2}}\int_{-\infty}^{+\infty} dz\, zH_0^{(0)}H_0^{(1)}$$

$$+\left(\frac{2}{3}\right)^{\frac{1}{2}}\int_{-\infty}^{+\infty} dz\, zH_0^{(0)}[H_0^{(1)}(z, s$$

$$+ \sigma\gamma(\gamma\delta)^{\frac{1}{2}}) - H_0^{(1)}]$$

$$-\frac{1}{\sigma}\int_{-\infty}^{+\infty} dz\, H_0^{(0)}L_2[H_0^{(1)}(z, s$$

$$+ \sigma\gamma(\gamma\delta)^{\frac{1}{2}}) - H_0^{(1)}]. \tag{A16}$$

Now express $H_0^{(1)}(z, s + \sigma\gamma(\gamma\delta)^{\frac{1}{2}})$ in terms of the complete set of functions $H_n^{(0)}$. From (A13b) written with s replaced by $s + \sigma\gamma(\gamma\delta)^{\frac{1}{2}}$, one obtains

$$H_0^{(1)}(z, s + \sigma\gamma(\gamma\delta)^{\frac{1}{2}})$$

$$= -\sum_{n=1}^{\infty}\frac{H_n^{(0)}}{\Theta_n - \Theta_0}\int_{-\infty}^{+\infty} dz\, H_n^{(0)}L_2(\sigma)H_0^{(0)}.$$

Eliminate in this way all the functions $H^{(1)}$ from the expression (A16) for R. Using again (A15) and the fact that the operator L_2 is Hermitian, one finds the desired result

$$R = 0.$$

APPENDIX III. PROOF OF EQ. (43)

We start from the equation

$$\bar{n}_2(\delta^+; l) = \frac{e^{-s\delta}}{l\lambda_0(s)}\iint_{-\infty}^{+\infty} dx\, dy\, \psi_0(x)\psi_0(y)\left[\frac{W(x)}{W(y)}\right]^{\frac{1}{2}}$$

$$\times P_\gamma(x \mid y, \delta) \exp\{\tfrac{1}{2}(\nu_0\gamma)^{\frac{1}{2}}(x + y)\}, \tag{A17}$$

which is an exact consequence of the general formula [see II, Eq. (28a)] for the Laplace transform of $\bar{n}_2(x; l)$. To show this, multiply Eq. (28a) of II by $\sigma \exp(\sigma\delta)$ and go the limit $\sigma \to \infty$. The left-hand side becomes $l\bar{n}_2(\delta^+; l)$. Using in the right-hand side the series [see II, Eq. (29)] for the resolvent $R_{s+\sigma}$, and the expression [See I, Eq. (9)] for the kernel $K_{s+\sigma}(x, y)$, it is easy to see that the first term of the series leads to (A17) since

$$\lim_{\sigma\to\infty} \sigma e^{\sigma\delta}p_s(x, y) = e^{-s\delta}P_\gamma(x \mid y, \delta),$$

and that all the further terms in the series vanish in the limit $\sigma \to \infty$.

One now "tames" the integral in (A17) by the same substitutions as used in Sec. II, replacing the variables x, y by z and ζ. Expanding the integrand in powers of $(\gamma\delta)^{\frac{1}{2}}$, using the expansions (5), (6), (7) and (11) for ν_0, l, s, and the eigenfunction $H_0(z)$, one finds up to order $(\gamma\delta)^{\frac{1}{2}}$

$$\bar{n}_2(\delta^+; l) = \frac{1}{2\delta l}\left[1 - \tfrac{3}{2}l_1(\gamma\delta)^{\frac{1}{2}}\right.$$

$$\left. + \tfrac{3}{2}(\gamma\delta)^{\frac{1}{2}}\int_{-\infty}^{+\infty} dz\, z^2H_0^{(0)}(z)^2\right], \tag{A18}$$

where also Eqs. (15) and (16) have been used. It is easy to verify that, to order $(\gamma\delta)^{\frac{1}{2}}$, (A18) is equivalent to Eq. (43).

Reprinted from
THE PHYSICS OF FLUIDS
Volume 6, Number 5, May 1963

One-Dimensional Phase Transition in the Spherical Model of a Gas

H. A. GERSCH
School of Physics, Georgia Institute of Technology, Atlanta, Georgia
(Received 19 December 1962)

Thermodynamic properties and molecular pair distribution functions are calculated for the one-dimensional spherical model of a gas with an attractive interaction outside of a hard core, in the limit as the range of attractive forces becomes infinite. In this limit, a phase transition appears, and the equation of state consists of analytically different parts. The calculations are made using the canonical ensemble which permits a description of the two-phase region. Properties in this region are found to have a simple description in terms of the coexistence of gas and liquid phases.

I. INTRODUCTION

EXACT calculations of partition functions for systems exhibiting phase transitions are very few in number. Examples are the Bose–Einstein condensation of an ideal gas,[1] the two-dimensional Ising model,[2] and the spherical model.[3] The Ising and the spherical models for magnetic systems have their counterparts as models of a lattice gas.[4] The partition function and phase transition for the spherical model of a three-dimensional lattice gas with nearest neighbor interactions has been treated recently.[5]

Here we consider the one-dimensional system with an interaction characterized by a hard core and a weak long-range attraction. It is well known that a one-dimensiona system with forces having finite range will show no phase transition.[6] However, as the range of interaction becomes infinite, a phase transition is possible. The existence of such transitions has already been demonstrated for the one-dimensional Ising and spherical models for the ferromagnet,[7] as well as a one-dimensional gas model.[8]

[1] F. London, Phys. Rev. 54, 947 (1938).
[2] L. Onsager, Phys. Rev. 65, 117 (1944).
[3] T. H. Berlin and M. Kac, Phys. Rev. **86**, 821 (1952).
[4] C. N. Yang and T. D. Lee, Phys. Rev. **87**, 410 (1952).
[5] W. Pressman and J. B. Keller, Phys. Rev. **120**, 22 (1960). H. A. Gersch and T. H. Berlin, Phys. Rev. **127**, 2276 (1962).
[6] L. Van Hove, Physica **15**, 951 (1949).
[7] G. A. Baker, Phys. Rev. **126**, 2071 (1962).
[8] G. E. Uhlenbeck, P. C. Hemmer, and M. Kac (to be published).

We evaluate the canonical partition function and molecular pair distribution function for the one-dimensional gas in the spherical model in the limit as the range of interaction becomes infinite. The calculation is carried out for finite range; finally the range is allowed to become infinite. Discontinuities in thermodynamic properties appear only after the final limiting process. If the limit of infinite range is carried out too early in the calculation, the equation of state so obtained has unstable loops of the van der Waals type. The proper limiting process yields a first-order phase transition. Thermodynamic properties and the pair distribution function in the transition region are simply describable in terms of the coexistence of gas and liquid phases.

In Sec. II, the canonical ensemble for the spherical lattice gas is introduced. Section III contains the evaluation of the canonical partition function as the range of the attractive interaction becomes infinite. In Sec. IV, the thermodynamic properties and the correlation function (equivalent to the molecular pair distribution function) are evaluated. Some concluding remarks are contained in Sec. V.

II. CANONICAL ENSEMBLE FOR SPHERICAL LATTICE GAS

The lattice gas is a cell model of an imperfect gas in which the volume V containing N particles is subdivided into K cells, each of fixed volume δ, $\delta = V/K$. The volume δ has to be chosen sufficiently small to reproduce the essential features of the interparticle potential energy. The simplest choice is to take δ small enough so that configurations having more than one particle per cell contribute negligibly to the partition function. This amounts to assuming the interaction between particles in the same cell to be infinite. One then gets the configurational partition function for the lattice gas to be

$$\frac{Z_N}{N!} = \delta^N \sum_{\{n_j\}} \exp\left(\frac{-\beta}{2} \sum_{\substack{j,k=1 \\ j\neq k}}^{K}\sum U_{jk} n_j n_k\right), \qquad (2.1)$$

where $\beta = 1/kT$, the sum is over all sets of integers n_j, $n_j = 0, 1$ subject to the restriction $\sum_{j=1}^{K} n_j = N$. U_{jk} is the interaction between particles in the jth and kth cells.

It proves to be convenient to introduce new variables x_j defined by

$$x_j = 2n_j - 1. \qquad (2.2)$$

In terms of these variables, the configuration partition function for the lattice becomes

$$\frac{Z_N}{N!} = \delta^N \exp\left(\frac{-N\beta}{2} \sum_{\substack{j=1 \\ j\neq k}}^{K} U_{jk}\right)$$

$$\cdot \sum_{\{x_j\}} \exp\left[\frac{\beta}{16} \sum_{\substack{j,k=1 \\ j\neq k}}^{K}\sum U_{jk}(x_j - x_k)^2\right], \qquad (2.3)$$

where the summation over the $x_j = \pm 1$ is made subject to the restriction $\sum_{j=1}^{K} x_j = 2N - K$.

The spherical model of the lattice gas is obtained by treating the discrete variables x_j as if they were continuous, and replacing the sum over coordinates x_j by an integration throughout the volume Ω of the $K - 1$ dimensional figure described by $-1 \leq x_j \leq +1$ and $\sum_{j=1}^{K} x_j = 2N - K$. Thus, the configurational partition function for the spherical model is

$$\frac{Z_N}{N!} = A\delta^N \exp\left(\frac{-N\beta}{2} \sum_{\substack{j=1 \\ j\neq k}}^{K} U_{jk}\right) \int_{\Omega} \cdots \int dx_1 \cdots dx_K$$

$$\cdot \exp\left[\frac{\beta}{16} \sum_{\substack{j,k=1 \\ j\neq k}}^{K}\sum U_{jk}(x_j - x_k)^2\right], \qquad (2.4)$$

where A is a normalization constant, to be chosen later.

Finally, the region of integration over Ω is replaced by the intersection of the hyperplane $\sum_{j=1}^{K} x_j = 2N - K$ with the K dimensional sphere $\sum_{j=1}^{K} x_j^2 = K$. This now gives the configurational partition function for the spherical model as

$$\frac{Z_N}{N!} = A\delta^N \exp\left(\frac{-N\beta}{2} \sum_{\substack{j=1 \\ j\neq k}}^{K} U_{jk}\right) \int_{\substack{\sum x_j = 2N-K \\ \sum x_j^2 = K}} \cdots \int dx_1 \cdots dx_K$$

$$\cdot \exp\left[\frac{\beta}{16} \sum_{\substack{j,k=1 \\ j\neq k}}^{K}\sum U_{jk}(x_j - x_k)^2\right]. \qquad (2.5)$$

An asymptotically exact calculation of the configurational partition function defined in (2.5) is possible because we can make a rotation to normal coordinates $\{y_j\}$ such that

$$\sum_{j=1}^{K} x_j^2 = \sum_{j=1}^{K} y_j^2 \qquad (2.6)$$

and

$$\sum_{\substack{j,k=1 \\ j\neq k}}^{K}\sum U_{jk}(x_j - x_k)^2 = \sum_{j=1}^{K} (\nu_1 - \nu_j) y_j^2. \qquad (2.7)$$

The transformation matrix t can be taken as

$$t_{jk} = K^{-\frac{1}{2}}\left[\cos\frac{2\pi}{K}(j-1)(k-1) + \sin\frac{2\pi}{K}(j-1)(k-1)\right], \qquad (2.8)$$

where the characteristic vectors are normalized to unity, and the ν_k are given by

$$\nu_k = 2 \sum_{p=2}^{\frac{1}{2}(K+1)} U_{1p} \cos\frac{2\pi}{K}(k-1)(p-1). \qquad (2.9)$$

The transformation brings (2.5) to the form

$$\frac{Z_N}{N!} = \delta^N A$$

$$\cdot \exp\left(\frac{-N\beta\nu_1}{2}\right) \int_{\sum_{j=2}^{K} y_j^2 = 4N(1-N/K)} \cdots \int dy_2 \cdots dy_K$$

$$\cdot \exp\left[\frac{\beta}{8} \sum_{j=2}^{K} (\nu_1 - \nu_j) y_j^2\right]. \qquad (2.10)$$

The following physical meaning may be attributed to the normal coordinates. We note that the expression for the expectation value of the total potential energy is

$$\langle U\rangle = \frac{N}{2} \sum_{\substack{j=1 \\ j\neq k}}^{K} U_{jk} - \frac{1}{8} \sum_{j=2}^{K} (\nu_1 - \nu_j)\langle y_j^2\rangle. \qquad (2.11)$$

The first term on the right-hand side of (2.11) represents the lowest potential energy for attractive forces, coming from the configuration of one particle in each cell. The second term in (2.11) represents the increase in the energy necessary to break the system up into clusters. For a given specific volume $v = K\delta/N$ and temperature T, there will be a certain distribution of mean square amplitudes $\langle y_i^2\rangle$ among the normal modes. A large value of $\langle y_i^2\rangle$ for a particular mode implies a large expectation value for the corresponding configuration of the x_i. For the modes with small index j, which correspond to long wavelengths, the energy for a fixed amplitude y_i is small. In these modes, there are large groups of cells for which each cell of the group has the same sign for the variable x_i. These variables are continuous; we do not have $x_i = \pm 1$ only. If we agree to identify $x_i > 0$ with $x_i = 1$ in the lattice gas, and $x_i < 0$ with $x_i = -1$, then for these configurations there are large groups of cells containing particles and large groups of empty cells. Clearly then, long wavelength modes correspond to clusters of large numbers of particles. As the index j increases, the energy for fixed amplitude y_i increases, and the regions for which the x_i are all of the same sign become smaller, so that the number of particles in a cluster is smaller.

The restriction on the region of integration of the variables in (2.10) may be relaxed by means of the delta function, using the representation

$$\delta[R - (\sum y_i^2)^{\frac{1}{2}}] = \frac{2R}{2\pi i}\int_{s_0 - i\infty}^{s_0 + i\infty} ds \exp [s(R^2 - \sum y_i^2)]. \qquad (2.12)$$

Because $R = [4N(1 - N/K)]^{\frac{1}{2}}$ is positive we may write

$$\frac{Z_N}{N!} = \frac{B}{2\pi i}\int_{s_0 - i\infty}^{s_0 + i\infty} ds \exp (sR^2) \cdot \prod_{j=2}^{K}\left[\pi^{\frac{1}{2}}\left(s - \frac{\beta}{8}(\nu_1 - \nu_j)\right)^{-\frac{1}{2}}\right], \qquad (2.13)$$

where $B = A\delta^N \exp(-N\beta\nu_1)$ and s_0 is chosen real, positive, and large enough to make $s - \frac{1}{8}\beta(\nu_1 - \nu_j) > 0$ for all j.

We are interested in the limit $N \to \infty$, $K \to \infty$ and the limiting form of the integrand in (2.13) is required. The product may be written

$$\prod_{j=2}^{K}\left[s - \frac{\beta}{8}(\nu_1 - \nu_j)\right]^{-\frac{1}{2}} = \exp\left\{-\frac{1}{2}\sum_{j=2}^{K}\ln\left[s - \frac{\beta}{8}(\nu_1 - \nu_j)\right]\right\}. \qquad (2.14)$$

Let

$$G_n(s) = \lim_{K\to\infty} K^{-1}\sum_{j=2}^{K}\ln\left[s - \frac{\beta}{8}(\nu_1 - \nu_j)\right] = \frac{1}{(2\pi)^n}\int_0^{2\pi}\cdots\int d\omega_1\cdots d\omega_n \cdot \ln\left[s - \frac{\beta}{8}[\nu(0) - \nu(\omega)]\right], \qquad (2.15)$$

where n is the number of dimensions and

$$\nu(\omega) = 2\sum_{j=2} U_{1j} \cos \omega(j - 1). \qquad (2.16)$$

Let $\bar{s}$ denote the algebraically smallest value of $\frac{1}{8}\beta[\nu(0) - \nu(\omega)]$. If the s plane is cut from $s = -\infty$ to $s = -\bar{s}$ along the real axis, then the integrand in (2.15) is analytic in the cut plane. The behavior of the integral in the neighborhood of $s = -\bar{s}$ is characterized by the density of eigenvalues $[\nu(0) - \nu(\omega)]$ in the neighborhood of the algebraically smallest value. This behavior incompletely describes the behavior of (2.14) in the neighborhood of $s = -\bar{s}$ because the singular nature of the integral in (2.15) has in general nothing to do with the singular nature of a single term in the sum in (2.14). We then more accurately replace (2.15) by

$$G_n(s) \to (2/K)\ln(s + \bar{s}) + G_n(s), \qquad (2.17)$$

where the 2 arises because the algebraically smallest eigenvalue is doubly degenerate.

The configurational partition function may be written as

$$\frac{Z_N}{N!} = \frac{B}{2\pi i}\pi^{\frac{1}{2}(K-1)}\int\frac{ds}{s + \bar{s}}\exp[Ng_n(s)], \qquad (2.18)$$

where

$$g_n(s) = 4(1 - \delta/v)s - (v/2\delta)G_n(s) \qquad (2.19)$$

with $\delta/v = N/K$. Applying the method of steepest descents to the integral we find that

$$Z_N = \frac{B\pi^{\frac{1}{2}(K-1)}\exp[Ng_n(s_s)]}{(s_s + \bar{s})[\partial^2 g_n/\partial s^2]_{s_s}^{\frac{1}{2}}} \qquad (2.20)$$

if a saddle point s_s can be found such that s_s is real, positive, to the right of the singularities of the integrand, and with

$$[\partial g_n/\partial s]_{s_s} = 0, \qquad [\partial^2 g_n/\partial s^2]_{s_s} > 0. \qquad (2.21)$$

The constant A shall be determined by normalizing to the lattice gas when $U_{jk} = 0$ and $N = \frac{1}{2}K(v = 2\delta)$. Then, as shown in Appendix A,

$$\ln A = (Nv/2\delta)\ln 2/\pi e. \qquad (2.22)$$

This yields, for the limiting free energy per particle

$$-\beta\psi_n = (v/2\delta)\ln 2/e + \ln\delta - \tfrac{1}{2}\beta\nu(0) + g_n(s_n) + \tfrac{1}{2}n\ln(2\pi m/\beta h^2). \qquad (2.23)$$

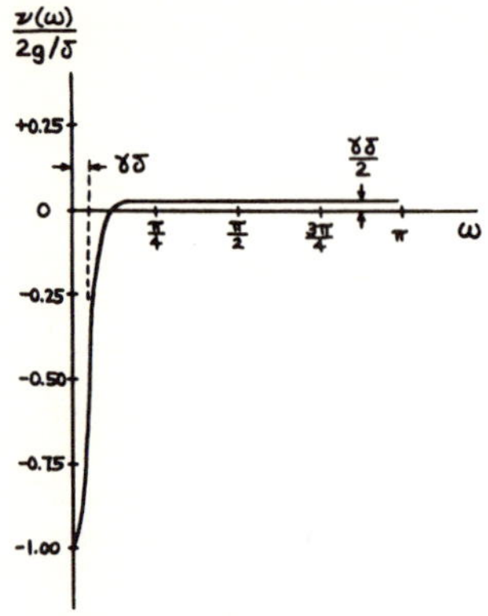

FIG. 1. Behavior of $\nu(\omega)$, the Fourier transform of the interaction energy, as a function of ω for $\gamma\delta \ll 1$

III. INFINITE RANGE POTENTIAL IN ONE DIMENSION

We choose the interaction potential between particles in the jth and kth cells to be

$$\begin{aligned} U_{jk} &= -g\gamma \exp\left[-\gamma\delta\,|k-j|\right] \quad \text{if} \quad k-j \neq 0, \\ U_{jk} &= \infty \quad \text{if} \quad k-j = 0. \end{aligned} \tag{3.1}$$

The repulsive core is consistent with our choice of cell volume equal to δ. As γ approaches zero, the range of the attractive interaction becomes infinite, but its strength goes to zero in such a way that the average potential energy per particle remains finite. We may see this by calculating $\nu(0)$ for this potential,

$$\begin{aligned} \nu(0) &= 2 \sum_j U_{1j} \\ &= -2g\gamma(e^{\gamma\delta} - 1)^{-1}. \end{aligned} \tag{3.2}$$

As $\gamma \to 0$, $\nu(0) \to -2g/\delta$, a finite interaction strength. Note that by disregarding the finite upper limit in the sum we are essentially assuming the interaction range, although infinite, is still negligible compared with the spatial extent of the system. For $\nu(\omega)$, we have the result

$$\begin{aligned} \nu(\omega) &= 2 \sum_{j=2} U_{1j} \cos\omega(j-1) \\ &= -2g\gamma \frac{e^{\gamma\delta}\cos\omega - 1}{e^{2\gamma\delta} - 2e^{\gamma\delta}\cos\omega + 1}. \end{aligned} \tag{3.3}$$

As γ approaches zero, $\nu(\omega)$ has the shape shown in Fig. 1. Starting at the value $-2g/\delta$ at $\omega = 0$, $\nu(\omega)$ falls to zero for values of ω of the order of $\gamma\delta$, and then remains essentially constant at the value $-g\gamma$. This behavior is understandable in terms of our previous remarks concerning the physical meaning of the normal modes. The variable ω can be written as $2\pi\delta/\lambda$ where λ is the wavelength of the normal mode configuration, which is the same as the length of the clusters of particles. When $\omega \ll \gamma\delta$ so that $\lambda \gg 1/\gamma$, the configuration of atoms correspond to clusters of length much greater than the interaction range, so that their energy is essentially the same as the ground state energy, and $\nu(0) - \nu(\omega) = 0$. As ω becomes of order $\gamma\delta$, the wavelength of the clusters corresponding to the normal modes is comparable to the interaction range, and consequently their energy is significantly larger than the lowest energy state. Finally when $\omega \gg \gamma\delta$, the particles are in clusters whose length is much smaller than the interaction range, so that their energy becomes essentially independent of ω.

The equation which determines the saddle point s_s follows from (2.21),

$$\begin{aligned} 8\frac{\delta}{v}\left(1 - \frac{\delta}{v}\right) &= I(s_s) \\ &= \frac{1}{2\pi}\int_0^{2\pi} \frac{d\omega}{s_s - \frac{1}{8}\beta[\nu(0) - \nu(\omega)]}. \end{aligned} \tag{3.4}$$

This equation determines the distribution of mean square amplitudes among the normal modes. It is clear from (2.10) and (2.12) that

$$\langle y_j^2 \rangle \to \langle y^2(\omega) \rangle = (2\{s_s - \tfrac{1}{8}\beta[\nu(0) - \nu(\omega)]\})^{-1}. \tag{3.5}$$

The algebraically smallest value for $\frac{1}{8}\beta[\nu(0) - \nu(\omega)]$, which we have previously denoted by $\bar{s}$, is clearly equal to zero.

We require the solution to the saddle point equation in the limit as γ approaches zero. The calculation of $I(s_s)$ in (3.4) is straightforward, and is carried out in Appendix B. The essential qualitative features are the following. As γ approaches zero, the integrand of $I(s_s)$ in (3.4) depends on ω as shown in Fig. 2. It starts at the value $1/s_s$ for $\omega = 0$, and falls to the value $[s_s + (\beta g/4\delta)(1 + \frac{1}{2}\gamma\delta)]^{-1}$

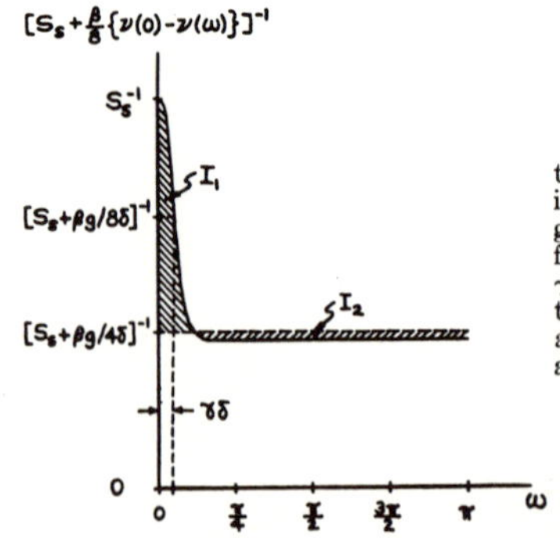

FIG. 2. Qualitative behavior of the integrand of $I(s_s)$ given in (3.4) as a function of ω for $\gamma\delta \ll 1$. Contributions depicted as I_1 and I_2 are evaluated in Appendix B.

for ω of the order of $\gamma\delta$, remaining essentially constant at this value for all greater ω values. The integral $I(s_s)$ can be expressed in terms of the areas I_1 and I_2 depicted in Fig. 2 as

$$I(s_s) = (s_s + \beta g/4\delta)^{-1} + I_1 - I_2. \tag{3.6}$$

The calculation given in Appendix B evaluates I_1 and I_2, with the result that (3.6) becomes, for small γ,

$$8\frac{\delta}{v}\left(1 - \frac{\delta}{v}\right) = \left(s_s + \frac{\beta g}{4\delta}\right)^{-1} \cdot\left\{1 + \frac{\beta g\gamma}{8}\left[\left\{s_s\left(s_s + \frac{\beta g}{4\delta}\right)\right\}^{-\frac{1}{2}} - \left(s_s + \frac{\beta g}{4\delta}\right)^{-1}\right]\right\}. \tag{3.7}$$

The second term on the right-hand side of (3.7) becomes infinite as $s_s \to \bar{s} = 0$ if $\gamma \neq 0$. Therefore, if $\gamma \neq 0$, the saddle point never coincides with the branch point of the function $g_1(s)$, there is no "sticking" of the saddle point s_s, and no phase transition. It is only in the $\gamma \to 0$ limit that the phase transition occurs. We will find essentially different limiting behaviors for s_s for different temperature and density regions. To delineate these regions, let us first put $\gamma = 0$ in (3.7), resulting in the solution for s_s

$$s_s = \frac{1}{8(\delta/v)(1 - \delta/v)} - \frac{\beta g}{4\delta}. \tag{3.8}$$

In order for a normal saddle point to exist, and also for (3.8) to be valid in the $\gamma \to 0$ limit, s_s must be greater than zero. This implies the inequality

$$8\frac{\delta}{v}\left(1 - \frac{\delta}{v}\right) < \frac{4\delta}{\beta g}. \tag{3.9}$$

As δ/v ranges from zero to one, the left-hand side of (3.9) has its maximum value, equal to 2, when $v = 2\delta$. Thus a normal saddle point will exist for all specific volumes v if $(4\delta/\beta g) > 2$. If $(4\delta/\beta g) < 2$, a normal saddle point exists only for those specific volumes lying outside the region whose boundary is given by the two volumes v_l and v_g which are solutions to

$$8\frac{\delta}{v}\left(1 - \frac{\delta}{v}\right) = \frac{4\delta}{\beta g}. \tag{3.10}$$

These two solutions v_l and v_g obey the law of constant diameter characteristic of the lattice gas, $\delta/v_l + \delta/v_g = 1$.[4] It is interesting to note that the 3-dimensional spherical model of a gas with nearest neighbor interactions exhibits a phase transition where the relation between v_l, v_g and β is identical with (3.10) if one replaces the interaction strength $2g/\delta$ of the present case by $\epsilon/(0.50546)$, where ϵ is the nearest neighbor interaction energy.[5]

We now define the critical temperature T_c and critical volume v_c by the relations

$$\frac{4\delta}{\beta_c g} = \frac{4k\delta T_c}{g} = 2, \qquad v_c = 2\delta. \tag{3.11}$$

For $T < T_c$ and $v_l \le v \le v_g$ we must evaluate s_s for finite γ and then finally let $\gamma \to 0$. Returning for this purpose to (3.7) we rewrite this equation in terms of reduced temperature and volume variables,

$$T^* = \frac{T}{T_c} = \frac{2k\delta}{g}T, \qquad v^* = \frac{v}{v_c} = \frac{v}{2\delta}. \tag{3.12}$$

Equation (3.7) becomes

$$\frac{4}{v^*}\left(1 - \frac{1}{2v^*}\right) = \frac{2T^*}{1 + 2T^*s_s} \cdot\left[1 + \frac{\gamma\delta}{2}\left\{\frac{1}{[2T^*s_s(1 + 2T^*s_s)]^{\frac{1}{2}}} - \frac{1}{1 + 2T^*s_s}\right\}\right]. \tag{3.13}$$

In obtaining the behavior of s_s as $\gamma \to 0$, the terms on the right-hand side of (3.13) proportional to γ become important only for $s_s \to 0$. Therefore in these terms we can make the approximation $2T^*s_s \ll 1$, obtaining

$$\frac{1}{[2T^*s_s(1 + 2T^*s_s)]^{\frac{1}{2}}} - \frac{1}{1 + 2T^*s_s} \cong [2T^*s_s]^{-\frac{1}{2}}. \tag{3.14}$$

This approximation yields the saddle point equation

$$\frac{4}{v^*}\left(1 - \frac{1}{2v^*}\right) = \frac{2T^*}{1 + 2T^*s_s}\left[1 + \frac{\gamma\delta}{2}(2T^*s_s)^{-\frac{1}{2}}\right]. \tag{3.15}$$

Now make the replacement

$$z = (2T^*s_s)^{\frac{1}{2}}. \tag{3.16}$$

Then we have

$$z^3 + (1 - x)z - \tfrac{1}{2}\gamma\delta x = 0, \tag{3.17}$$

where

$$x = \frac{T^*}{(2/v^*)(1 - 1/2v^*)}. \tag{3.18}$$

Reference to (3.10), which, written in terms of T^*, v^* is simply $x - 1 = 0$, shows the significance of the parameter x. Namely, $x > 1$ refers to the region of specific volumes and temperatures for which a normal saddle point exists for $\gamma = 0$, $x = 1$ gives the values v_g and v_l which form the boundaries of this region, and $x < 1$ denotes the region inside, where a normal saddle point does not exist for $\gamma = 0$.

The solution to (3.17) which yields the real, positive s_s is

$$z = A_+ + A_-,$$

$$A_\pm = \left\{\frac{\gamma\delta x}{4} \pm \left[\left(\frac{\gamma\delta x}{4}\right)^2 + \frac{(1-x)^3}{27}\right]^{\frac{1}{2}}\right\}^{\frac{1}{3}}. \quad (3.19)$$

As $\gamma \to 0$, the relative size of the two terms under the square root sign in (3.19) determine the limiting behavior of z, and hence of s_s. There are the following cases. First consider the normal region in which $x > 1$. If

$$\frac{(x-1)^3}{27} \gg \left(\frac{\gamma\delta x}{4}\right)^2, \quad (3.20)$$

we obtain

$$s_s = \frac{1}{2T^*}\left[x - 1 + \frac{\gamma\delta}{2}\frac{1}{(x-1)^{\frac{1}{2}}}\right]. \quad (3.21)$$

As $\gamma \to 0$, this reduces to (3.8) for the saddle point in the normal region. However, if (still for $x > 1$) we have

$$\frac{(x-1)^3}{27} \ll \left(\frac{\gamma\delta x}{4}\right)^2, \quad (3.22)$$

then

$$s_s = \frac{1}{2T^*}\left(\frac{\gamma\delta x}{2}\right)^{\frac{2}{3}}. \quad (3.23)$$

Equation (3.23) holds as well for $x = 1$, i.e., on the boundary of the transition region. The second case is for $x < 1$, inside the transition region. If

$$\frac{(1-x)^3}{27} \ll \left(\frac{\gamma\delta x}{4}\right)^2, \quad (3.24)$$

then

$$s_s = \frac{1}{2T^*}\left(\frac{\gamma\delta x}{2}\right)^{\frac{2}{3}} \quad (3.25)$$

which is identical with (3.23). However, if the reverse inequality holds,

$$\frac{(1-x)^3}{27} \gg \left(\frac{\gamma\delta x}{4}\right)^2, \quad (3.26)$$

then

$$s_s = \frac{1}{2T^*}\left[\frac{\gamma\delta x}{2(1-x)}\right]^2. \quad (3.27)$$

These different behaviors of the saddle point in the different regions of specific volume and temperature have as a consequence different behaviors of thermodynamic properties and pair distribution function, as we show in Sec. IV.

IV. THERMODYNAMIC PROPERTIES AND PAIR DISTRIBUTION FUNCTION

Equation (2.23) gives ψ, the Helmholtz free energy per particle for the one-dimensional system as

$$-\beta\psi = \frac{v}{2\delta}\ln\frac{2}{e} + \ln\delta - \frac{\beta}{2}\nu(0) + 4\left(1 - \frac{\delta}{v}\right)s_s$$
$$- \frac{v}{2\delta}G(s_s) + \tfrac{1}{2}\ln\left(\frac{2\pi m}{\beta h^2}\right), \quad (4.1)$$

where

$$G(s_s) = \frac{1}{2\pi}\int_0^{2\pi} d\omega \ln\{s_s - \tfrac{1}{8}\beta[\nu(0) - \nu(\omega)]\}. \quad (4.2)$$

From ψ, we get the pressure, utilizing the relation $P = (-\partial\psi/\partial v)_T$ and the saddle point equation (3.4),

$$\delta\beta P = \tfrac{1}{2}\ln(2/e) + 4(\delta/v)^2 s_s - \tfrac{1}{2}G(s_s). \quad (4.3)$$

An expression for $G(s_s)$ sufficiently accurate for the $\gamma \to 0$ limit can be obtained from (3.4) and (3.15) for $I(s_s)$, since

$$dG(s_s)/ds_s = I(s_s).$$

By straightforward integration we obtain

$$G(s_s) = \ln[s_s + (1/2T^*)] + \gamma\delta\{(2T^*s_s)^{\frac{1}{2}}$$
$$- (2T^*s_s)/2 - (2T^*s_s)^{\frac{3}{2}}/2\}. \quad (4.4)$$

We may now calculate the pressure in the $\gamma \to 0$ limit for the different volume, temperature regions described in Sec. III. In the normal region, where $x > 1$, and where inequality (3.20) is obeyed, we have s_s given by (3.21), which yields the following for the pressure:

$$\frac{\delta P}{kT} = -\tfrac{1}{2} + \tfrac{1}{2}\ln\left[\frac{8}{v^*}\left(1 - \frac{1}{2v^*}\right)\right]$$
$$+ \frac{1}{4(v^* - \frac{1}{2})} - \frac{1}{2T^*v^{*2}} + O(\gamma\delta), \quad (4.5)$$

where $O(\gamma\delta)$ denotes correction terms which vanish in the limit $\gamma \to 0$ like $\gamma\delta$. In (4.5), the first three terms on the right represent the contribution to the pressure from a lattice gas with no interaction between particles in different cells; the last term contains the sole contribution to the pressure from the attractive interactions. Figure 3 shows the pressure for several values of T^*—Eq. (4.5) gives P outside the transition region in the figure.

For the region close to the transition curve $x = 1$, described by (3.22) and (3.24), i.e.,

$$\frac{|1-x|^3}{27} \ll \left(\frac{\gamma\delta x}{4}\right)^2, \quad (4.6)$$

we find for the pressure

$$\delta P/kT = -\tfrac{1}{2} + \tfrac{1}{2}\ln 4T^* + O(\gamma\delta)^{2/3}, \qquad (4.7)$$

while well inside the transition region, according to (3.27)

$$\delta P/kT = -\tfrac{1}{2} + \tfrac{1}{2}\ln 4T^* + O(\gamma\delta)^2. \qquad (4.8)$$

Equations (4.7) and (4.8) enable us to complete the isotherms in the transition region in Fig. 3. As these equations show, the horizontal portion of P vs v^* is reached as $\gamma \to 0$ more quickly [deviations vanish like $(\gamma\delta)^2$] than is the discontinuity in $(dP/dv)_T$ at the transition volume [deviations vanish like $(\gamma\delta)^{2/3}$].

Had we gone to the $\gamma \to 0$ limit too early in the calculation, then we would have (improperly) used throughout the entire volume–temperature plane the saddle point equation (3.8) and the corresponding expression for the pressure, (4.5). This would yield, inside the transition region, the van der Waals loops indicating instability. These are shown dotted in Fig. 3.

The energy per particle for the system follows from the relation $E = T^2\, d(-\psi/T)/dT$. Using (4.1) for ψ, we obtain

$$E = \frac{kT}{2} + \frac{\nu(0)}{2} - \frac{v}{16\delta}\nu(0)I(s_s)$$
$$+ \frac{v}{16\delta}\frac{1}{2\pi}\int_0^{2\pi}\frac{\nu(\omega)\,d\omega}{s_s - \frac{1}{8}\beta[\nu(0) - \nu(\omega)]}. \qquad (4.9)$$

The last integral on the right-hand side is evaluated in Appendix B. Using the result, and replacing $I(s_s)$

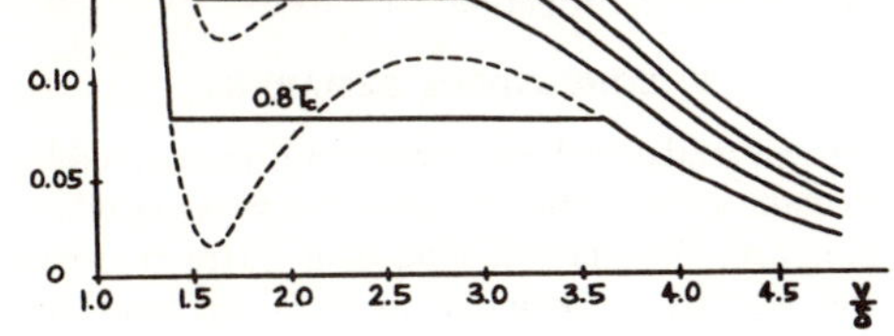

Fig. 3. Plot of the pressure P from the one-dimensional gas vs the specific volume v for various temperatures near the critical temperature T_c, obtained from (4.5) and (4.7). The dotted portions result from improper use of the saddle point equation (3.8) inside the transition region.

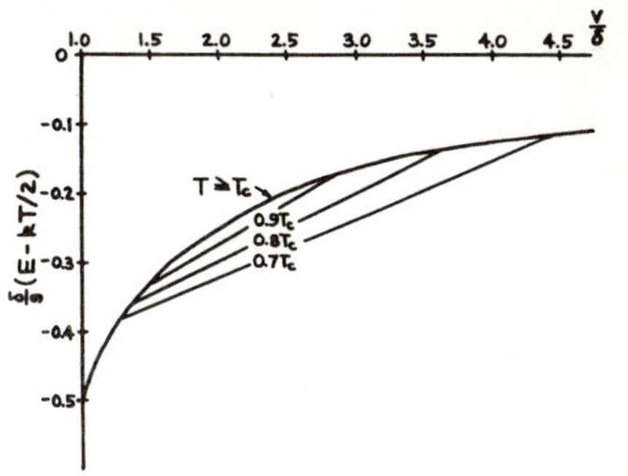

Fig. 4. The potential energy per particle, $E - kT/2$, as a function of specific volume v for temperatures close to the critical temperature.

from the saddle point equation (3.4) yields

$$E - \frac{kT}{2} = \frac{\delta\nu(0)}{2v} - \frac{vg\gamma}{16\delta[s_s(s_s + \beta g/4\delta)]^{\frac{1}{2}}}. \qquad (4.10)$$

The last term on the right side of (4.10) vanishes as $\gamma \to 0$ outside the transition region, so that there we have

$$E - \frac{kT}{2} = \frac{-g}{v} = \frac{-g}{2\delta v^*}. \qquad (4.11)$$

Inside the transition region, where the saddle point is given by (3.27), we have the result for $\gamma \to 0$,

$$E - \frac{kT}{2} = \frac{-g}{\delta} + \frac{gv^*T^*}{2\delta}. \qquad (4.12)$$

The energy E as a function of v^* for various T^* is shown in Fig. 4. The energy decreases linearly with v^* inside the transition region. Equation (4.12) can be written in a manner which emphasizes the coexistence of two phases in this region,

$$E - \frac{kT}{2} = \frac{-g}{v_g}\frac{N_g}{N} - \frac{g}{v_l}\frac{N_l}{N}, \qquad (4.13)$$

with the fractions N_g/N and N_l/N of gas and liquid given by

$$\frac{N_g}{N} = \frac{v - v_l}{v_g - v_l}, \qquad \frac{N_l}{N} = \frac{v_g - v}{v_g - v_l}. \qquad (4.14)$$

Finally we calculate the correlation function between two cells, $\langle n_j n_k\rangle$. For this cell model, $\langle n_j n_k\rangle$ plays the role of the molecular pair distribution function $n_2(\mathbf{r}_1, \mathbf{r}_2) = N(N-1)P_2(\mathbf{r}_1, \mathbf{r}_2)$. Here $P_2(\mathbf{r}_1, \mathbf{r}_2)$ is the probability density for a specified particle to be in the volume element $d\mathbf{r}_1$ and another specified particle to be in the volume element $d\mathbf{r}_2$ surrounding $\mathbf{r}_2$. From their definitions, it follows that

$$\langle n_j n_k \rangle = \iint_\delta d\mathbf{r}_{1_j}\, d\mathbf{r}_{2_k}\, n_2(\mathbf{r}_{1_j}, \mathbf{r}_{2_k}), \qquad (4.15)$$

where the position vector $\mathbf{r}_{1_j}$ for one particle is integrated over cell j, that for the other particle over cell k.

In terms of the normal coordinates y_l, the correlation function is given by

$$\langle n_j n_k \rangle = (\delta\rho)^2 + \frac{1}{4} \sum_{l=1}^{K-1} t_{jl} t_{kl} \langle y_l^2 \rangle. \qquad (4.16)$$

Making use of (2.8) for the components t_{jl} of the characteristic vectors and (3.5) for the mean square amplitude $\langle y_l^2 \rangle$ this becomes

$$\langle n_j n_k \rangle = (\delta\rho)^2 + H_{jk}/4 \qquad (4.17)$$

where

$$H_{jk} = (4\pi)^{-1} \int_0^{2\pi} \frac{e^{i\omega(k-j)}\, d\omega}{s_s - \frac{1}{8}\beta[\nu(0) - \nu(\omega)]}. \qquad (4.18)$$

The integral is evaluated in Appendix B, with the result

$$H_{jk} = \frac{\delta_{jk}}{2[s_s + (\beta g/4\delta)]} + \frac{\beta g \gamma}{16 s_s^{\frac{1}{2}}[s_s + (\beta g/4\delta)]^{\frac{1}{2}}} \cdot \exp\left\{\frac{-\gamma \delta s_s^{\frac{1}{2}}\, |k - j|}{[s_s + (\beta g/4\delta)]^{\frac{1}{2}}}\right\} \qquad (4.19)$$

or, in terms of reduced units

$$H_{jk} = \frac{\delta_{jk}}{2[s_s + (1/2T^*)]} + \frac{\gamma\delta}{8T^* s_s^{\frac{1}{2}}[s_s + (1/2T^*)]^{\frac{1}{2}}} \cdot \exp\left\{\frac{-\gamma \delta s_s^{\frac{1}{2}}\, |k - j|}{[s_s + (1/2T^*)]^{\frac{1}{2}}}\right\}. \qquad (4.20)$$

In the normal region, where s_s is given by (3.21), H_{jk} is

$$H_{jk} = \frac{T^* \delta_{jk}}{x} + \frac{\gamma \delta T^*}{2x^{\frac{1}{2}}(x - 1)^{\frac{1}{2}}} \cdot \exp\left[-\gamma\delta\left(\frac{x-1}{x}\right)^{\frac{1}{2}} |k - j|\right]. \qquad (4.21)$$

In (4.21), as $\gamma \to 0$, the second term vanishes, and there is no correlation of particles in different cells.

Close to the normal region, where we have s_s given by (3.23) and (3.25),

$$H_{jk} = T^* \delta_{jk} + \left[\left(\frac{\gamma\delta}{2}\right)^2 \frac{1}{x}\right]^{\frac{1}{3}} T^* \cdot \exp\left[\frac{-(\gamma\delta)^{4/3}}{2^{\frac{1}{3}}} |k - j|\right]. \qquad (4.22)$$

In this region, as $\gamma \to 0$, again the correlation between particles in different cells vanishes, although more slowly than in the normal region.

Finally, inside the transition region, with s_s given by (3.27) we have

$$H_{jk} = T^* \delta_{jk} + \frac{T^*(1 - x)}{x} \cdot \exp\left[\frac{-(\gamma\delta)^2 x}{2(1 - x)} |k - j|\right]. \qquad (4.23)$$

Here we see that as $\gamma \to 0$, the second term on the right side of (4.23) does not vanish, so that there is a constant correlation between particles in different cells

$$\langle n_j n_k \rangle = (\delta\rho)^2 + T^*(1 - x)/4x, \qquad j \neq k. \qquad (4.24)$$

This relation can be rewritten to again emphasize the coexistence of gas and liquid phases:

$$\langle n_j n_k \rangle = \frac{\delta}{v}\left[\frac{\delta}{v_g}\frac{N_g}{N} + \frac{\delta}{v_l}\frac{N_l}{N}\right] \qquad (4.25)$$

where the fractions of gas and liquid, N_g/N and N_l/N are as defined in (4.14). The value of $\langle n_j n_k \rangle$ between different cells as given in (4.25) is just what one would expect asymptotically for a two phase region. Through the general relation between the energy per particle and the correlation function,

$$E - \frac{kT}{2} = \frac{1}{2N} \sum_{k \neq j} \sum U_{jk} \langle n_j n_k \rangle = \frac{v}{2\delta} \sum_k U_{jk} \langle n_j n_k \rangle. \qquad (4.26)$$

We find for the normal region, where $\langle n_j n_k \rangle = (\delta/v)^2$ as $\gamma \to 0$,

$$E - kT/2 = -g/v \qquad (4.27)$$

which is, of course, identical with our previous result for the energy in this region, (4.11). In the transition region, use of (4.26) for $\langle n_j n_k \rangle$ yields an expression for the energy identical with (4.13). In the same way, by using the general relation between pressure and correlation function, our previous results for the behavior of the pressure may be obtained again from the above expressions for the correlation function.

V. CONCLUDING REMARKS

The cell model we have employed here has enabled us to make a complete separation between the effects of the hard core and the long-range attractive part of the interaction potential as (4.5) indicates. We get in this way an equation of state which in the normal region is of the form of van der Waals equation. Such clean-cut separation between repulsive and attractive parts of the interaction is most likely not possible for a more realistic model.[8]

It is clear that in order to get correct results, the limit $\gamma \rightarrow 0$ must be taken at the end of the calculation for the partition function. The saddle point equation obtained by putting $\gamma = 0$ at an earlier stage (3.8), is correct only outside the transition region. Its unwarranted use throughout the entire temperature volume plane gives rise to the characteristic van der Waals loops indicating improper evaluation of the partition function. The limit $\gamma \rightarrow 0$ inside the transition region can be described in terms of a modified path of steepest descent in the complex s plane. Such a treatment has been carried out for a different cell model than the one discussed here.[9]

The simplicity of the description of the transition region in terms of the coexistence of gas and liquid phases, as given for example by (4.13) and (4.25) is rather interesting. No such simple description was possible for the transition found in the three-dimensional model with nearest neighbor interactions.[5] Of course, no actual geometrical separation of the two phases is possible in our model where surface effects and gravitational fields are neglected.

ACKNOWLEDGMENTS

The author is priviledged to acknowledge his gratitude to the late Professor T. H. Berlin for informing him of the recent work with long-range attractive interactions, and for preliminary discussions which led to the treatment presented here.

APPENDIX A

In this Appendix, we evaluate the normalization constant A, which first appears in (2.10), showing that A is given by (2.22),

$$\ln A = (Nv/2\delta) \ln (2/\pi e).$$

We choose A so that we get the same value for $Z_N/N!$ for the lattice gas as for the spherical model when $U_{ik} = 0$ and $N = K/2$ or $v = 2\delta$. The configurational partition function for the lattice gas is

$$\frac{Z_N}{N!} = \delta^N \frac{K!}{N!(K-N)!}, \tag{A1}$$

while the spherical model yields

$$\frac{Z_N}{N!} = A\delta^N \pi^{K/2-1} \frac{[4N(1-N/K)]^{K/2}}{\Gamma(K/2+1)}. \tag{A2}$$

Equating these two for $N = K/2$, we get

$$A = \pi(2/\pi e)^{K/2}[2(1+2/K)]^{\frac{1}{2}}. \tag{A3}$$

In the limit $N \rightarrow \infty$, $K \rightarrow \infty$, $K/N = v/\delta$, this becomes

$$\ln A = (Nv/2\delta) \ln (2/\pi e) \tag{A4}$$

which is the desired result.

APPENDIX B

In this Appendix, we evaluate the integral for $2H_{jk}$ given in (4.18):

$$2H_{jk} = (2\pi)^{-1} \int_0^{2\pi} \frac{e^{i\omega(k-j)}\, d\omega}{s_s - \frac{1}{8}\beta[\nu(0) - \nu(\omega)]} \tag{B1}$$

for $\gamma\delta \ll 1$, and show that the result is (4.19). This result will be applicable as well to the evaluation of the integral $I(s_s)$ in (3.4),

$$I(s_s) = \frac{1}{2\pi} \int_0^{2\pi} \frac{d\omega}{s_s - \frac{1}{8}\beta[\nu(0) - \nu(\omega)]}, \tag{B2}$$

since $I(s_s) = 2H_{jj}$. In this way we will also show that the result given in (3.7) for $I(s_s)$ is correct.

First rewrite (B1) in the form

$$2H_{jk} = \frac{1}{s_s - \frac{1}{8}\beta\nu(0)} \left\{ \frac{1}{2\pi} \int_{-\pi}^{\pi} e^{i\omega(k-j)}\, d\omega - \frac{1}{2\pi} \int_{-\pi}^{\pi} \frac{\frac{1}{8}\beta\nu(\omega) e^{i\omega(k-j)}}{s_s - \frac{1}{8}\beta[\nu(0) - \nu(\omega)]}\, d\omega \right\} \tag{B3}$$

or

$$2H_{jk} = [s_s - \tfrac{1}{8}\beta\nu(0)]^{-1}[\delta_{jk} - J_{jk}(s_s)] \tag{B4}$$

where

$$J_{jk} = \frac{1}{2\pi} \int_{-\pi}^{\pi} \frac{\frac{1}{8}\beta\nu(\omega) e^{i\omega(k-j)}\, d\omega}{s_s - \frac{1}{8}\beta[\nu(0) - \nu(\omega)]}. \tag{B5}$$

Into this expression for $J_{jk}(s_s)$ we insert the value for $\nu(\omega)$ given in (3.3),

$$\nu(\omega) = -2g\gamma[e^{\gamma\delta} \cos\omega - 1] \cdot [e^{2\gamma\delta} - 2e^{\gamma\delta} \cos\omega + 1]^{-1}, \tag{B6}$$

resulting in the following integral for $J_{jk}(s_s)$:

$$J_{jk} = \frac{-\beta g\gamma}{16\pi} \int_{-\pi}^{\pi} \frac{(e^{\gamma\delta}\cos\omega - 1)e^{i\omega(k-j)}}{Z - Q\cos\omega}\, d\omega \tag{B7}$$

where

$$Z = P(1 + e^{2\gamma\delta}) + \beta g\gamma/8,$$
$$Q = e^{\gamma\delta}[2P + \beta g\gamma/8], \tag{B8}$$
$$P = s_s + (\beta g\gamma/8)(e^{\gamma\delta} - 1)^{-1}.$$

The integral $J_{jk}(s_s)$ is evaluated by contour integration over the path indicated in Fig. 5. The

[9] J. L. Strecker, Ph.D. thesis, The Johns Hopkins University, 1961 (unpublished).

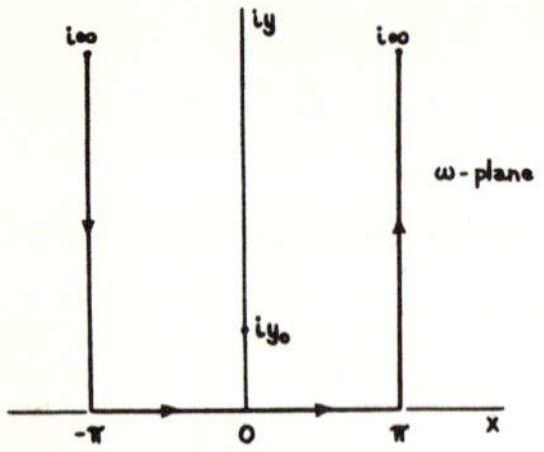

FIG. 5. Graph of the contour in the complex ω plane used for the evaluation of the integral in Eq. (B7). The pole of the integrand along the imaginary axis is denoted by y_0.

integrals over the vertical paths in the figure cancel, while the integral from $\pi + iy$ to $-\pi + iy$ goes to zero as y approaches infinity, if $k \neq j$. The poles of the integrand of (B7) are located by

$$\cos \omega_0 = Z/Q. \tag{B9}$$

Since Z/Q is real and greater than one, ω_0 is purely imaginary, $\omega_0 = iy_0$, and the single pole of the integrand inside our contour is given by

$$\cosh y_0 = Z/Q. \tag{B10}$$

The result for $J_{ik}(s_s)$, valid for $j \neq k$, is

$$J_{ik}(s_s) = \frac{-\beta g\gamma}{8Q} \frac{e^{\gamma\delta} \cosh y_0 - 1}{\sinh y_0} e^{-y_0|k-j|}. \tag{B11}$$

We require $J_{ik}(s_s)$ for $\gamma\delta \ll 1$, when Z/Q can be approximated as

$$\frac{Z}{Q} \cong 1 + \frac{(\gamma\delta)^2 s_s}{2(s_s + \beta g/4\delta)}. \tag{B12}$$

With this approximation, J_{ik} becomes, for $j \neq k$,

$$J_{ik}(s_s) \cong \frac{-\beta g\gamma}{8[s_s(s_s + \beta g/4\delta)]^{\frac{1}{2}}} \cdot \exp\left[\frac{\gamma\delta s_s^{\frac{1}{2}} |k - j|}{(s_s + \beta g/4\delta)^{\frac{1}{2}}}\right]. \tag{B13}$$

When $j = k$, the integral from $\pi + iy$ to $-\pi + iy$ does not approach zero as y approaches infinity, and we must add to (B13) the contribution from this part of the contour. As $y \to \infty$ we have $\nu(\omega + iy) \to g\gamma$, so that integral (B5) along this part of the contour is

$$\frac{1}{2\pi}\int_{\pi+i\infty}^{-\pi+i\infty} \frac{(\beta g\gamma/8)\, d\omega}{s_s + (\beta g/4\delta)[1 + \gamma\delta/2]} \cong \frac{-\beta g\gamma}{8(s_s + \beta g/4\delta)}. \tag{B14}$$

Combining (B13) and (B14), we have the result

$$J_{ik}(s_s) = \frac{\beta g\gamma}{8}\left\{\frac{\delta_{jk}}{s_s + (\beta g/4\delta)} - \frac{1}{[s_s(s_s + \beta g/4\delta)]^{\frac{1}{2}}} \exp\left[\frac{(\gamma\delta)s_s^{\frac{1}{2}} |k - j|}{(s_s + \beta g/4\delta)^{\frac{1}{2}}}\right]\right\} \tag{B15}$$

and

$$H_{ik} = \frac{1}{2(s_s + \beta g/4\delta)}\left\{\left(1 - \frac{\beta g\gamma}{8(s_s + \beta g/4\delta)}\right)\delta_{ik} + \frac{\beta g\gamma}{8[s_s(s_s + \beta g/4\delta)]^{\frac{1}{2}}} \exp\left[\frac{(\gamma\delta)s_s^{\frac{1}{2}} |k - j|}{(s_s + \beta g/4\delta)^{\frac{1}{2}}}\right]\right\}. \tag{B16}$$

Neglecting the second term in the bracketed expression of (B16), which is negligible compared to one for small γ, this is identical with (4.19). For $I(s_s) = 2H_{ii}$ we have

$$I(s_s) = \frac{1}{s_s + \beta g/4\delta}\left\{1 + \frac{\beta g\gamma}{8}\left[\frac{1}{s_s^{\frac{1}{2}}(s_s + \beta g/4\delta)^{\frac{1}{2}}} - \frac{1}{s_s + \beta g/4\delta}\right]\right\} \tag{B17}$$

which is identical with (3.7).

Reprinted from the Journal of Mathematical Physics, Vol. 2, No. 5, 682–693, September–October, 1961
Printed in U. S. A.

Exact Statistical Mechanics of a One-Dimensional System with Coulomb Forces

A. Lenard
Plasma Physics Laboratory, Princeton University, Princeton, New Jersey
(Received March 27, 1961)

A system consisting of an equal number of positively and negatively charged "sheets" is considered in thermal equilibrium, with motion restricted to one dimension. The configurational part of the partition function can be represented as a sum of terms, each a simple algebraic expression. The summation is performed with the technique of generating functions. The asymptotic form in the limit of an infinite system is obtained from the pole of the generating function closest to the origin. This pole is the solution of a certain transcendental equation for which an explicit analytic representation in terms of an infinite continued fraction is available. It is shown that this equation is identical with the characteristic equation associated with the even Mathieu functions of even order.

In the limit, when the ratio of interparticle force to pressure is small, the system behaves as an ideal gas, the deviations from this state being expandable in powers of the square root of this ratio. In the opposite limit of large ratio, the particles associate in pairs of opposite charge, thus behaving like an ideal gas of neutral "molecules" which have an internal vibrational degree of freedom.

The analysis may be generalized to include the effect of a constant external electric field. For a given pressure there is a critical field which can never be surpassed without disrupting equilibrium.

1. INTRODUCTION

The calculation of thermodynamic properties of systems from their atomic properties is generally frustrated by two difficulties. The first one is analytical: The computation or, at least, approximation of certain integrals ("cluster" integrals, contributions of "graphs," etc.) in terms of which the partition function is expressed algebraically. The other difficulty is algebraic, and is frequently associated with very complicated combinatorial problems, i.e., the systematic classification and eventual summing up of these contributions to the partition function. Usually these problems can be overcome only in the limit when some physical parameter is very small or large, and it is only in a few examples when an exact solution, encompassing the whole range of macroscopic parameters, is possible. This situation lends some interest to the study of simplified models which, while not realistic, still show some features of physical problems and, at the same time, are susceptible to exact analysis. It was in this spirit that one-dimensional systems have been investigated in the past.

The subject of the present paper is the statistical mechanical study of a one-dimensional system with a particularly simple long-range interaction law. It turns out that a complete discussion of the macroscopic behavior of this system is possible in spite of the fact that the statistical mechanics—at least in its combinatorial aspects—is not entirely trivial. We shall see that the absence of analytic problems follows from the elementary nature of all integrals into which the partition function can be decomposed. The combinatorics, in turn, can be mastered by special methods whose exposition is the principal purpose of this article.

There is added interest in this model on account of the long range of the interparticle force. The force law of the model to be discussed can be "physically" realized. It is the law of force between parallel, infinite, electrically charged plane condenser plates. Our system is then an assembly of a great number of such plates or sheets (referred to in the following as "particles") all restricted in their motion to the direction perpendicular to their planes, freely able to cross each other, and carrying fixed electrical surface charges (Sec. 2). It turns out that for the discussion of the statistical mechanics of this model the most useful ensemble is not the canonical ensemble of Gibbs, but rather a modified ensemble in which the pressure and temperature are given fixed values and the volume is derived as a phase average. While the general theory of such an ensemble has been given before,[1] it is probably not so familiar that a short summary should be superfluous (Sec. 3).

It will be shown how the partition function can be broken up into additive contributions of simple algebraic expressions each of which represents the phase integral over a certain region in phase space, a "configuration" (Sec. 4). The combinatorial problem is then the problem of summing these expressions over all configurations. With the aid of a simple pictorial representation of configurations we shall show that the natural technique for handling these sums is the method of generating functions (Sec. 5). After investigating the analytic nature of these generating functions we are led directly to the asymptotic form of the partition function for the limit of a very large number of particles. The limiting thermodynamic potential per particle appears to be simply expressible in terms of the solution of a certain transcendental equation whose left-hand side can be represented as an infinite continued fraction (Sec. 6). We shall then establish the correspondence between this equation and a classical boundary value problem of the Sturm-Liouville type which has been investigated in great detail (Sec. 7).

Making use of this mathematical correspondence, we shall then investigate the thermodynamics in two limiting cases (Secs. 8 and 9). The physical significance of certain "irreducible" configurations will be discussed

[1] W. B. Brown, Mol. Phys. 1, 68 (1958).

(Sec. 10). Finally, the generalization will be indicated to the case when an external electric field is present. This can be treated without new methods and with little extra complication (Sec. 11).

After this paper had been completed, the author learned that Professor S. Prager of the University of Minnesota had independently and simultaneously treated the present problem. This work is to be published in the 1961 volume of *Advances in Chemical Physics* edited by I. Prigogine.

2. DEFINITION OF THE MODEL

Consider a system of plane sheets (referred to in the following as "particles") freely movable in the x direction. Suppose that they all carry a surface mass density of magnitude unity, and that the ith one carries an electric surface charge density σ_i. The Newtonian equations of motion for this dynamical system are

$$d^2q_i/dt^2 = 2\pi\sigma_i \sum_j \sigma_j \,\mathrm{sg}(q_i - q_j), \quad (1)$$

where q_i denotes the position of the ith particle, and

$$\mathrm{sg}(x) = \begin{cases} 1 & x>0 \\ 0 & x=0 \\ -1 & x<0. \end{cases}$$

The terms in the sum represent the electric force exerted by the particles on a selected one. The analytic simplicity of the model is a consequence of this force law; the force on a particle is constant during an interval in which no other particle passes over it. While this fact seems not to be of much help in comprehending the gross features of a truly dynamical problem, it is indicative of great simplifications in the treatment of thermal equilibrium.

Our model is related to that investigated most recently by Dawson.[2] The difference is that we deal with particles of both signs of charge, whereas Dawson considers particles of one sign of charge embedded in an oppositely charged, uniform, neutralizing background. While for many purposes this difference is relatively insignificant, it is quite crucial for the possibility of calculating the partition function. This has its roots in the fact that in our model the potential energy is linear in displacements, while in Dawson's model it is quadratic.

For the purposes of statistical mechanics it is necessary to have the Hamiltonian function which leads to Eq. (1). One may verify directly that this is

$$H(q_1,q_2,\cdots;p_1,p_2,\cdots) = \tfrac{1}{2}\sum_i p_i^2 - 2\pi \sum_{i<j}\sum \sigma_i\sigma_j |q_i - q_j|. \quad (2)$$

To complete the specification of the model we stipulate that $2N$ particles are given, N of which are charged positively and N negatively. All charges have the same magnitude $|\sigma_i| = \sigma$. The sums in Eqs. (1) and (2) run from 1 to $2N$.

[2] J. M. Dawson, Phys. Rev. **113**, 383 (1959).

3. THE CONSTANT PRESSURE ENSEMBLE

The fundamental probability density in the phase space of the system will be taken as follows

$$\rho(q_1,\cdots,p_{2N}) = \exp\{\theta^{-1}[G - Pq_{\max} - H(q_1,\cdots,p_{2N})]\}, \quad (3)$$

when $q_i \geq 0 (1 \leq i \leq 2N)$; $\rho = 0$ otherwise. $q_{\max}$ is the largest among the q_i; it is to be regarded as a phase function. θ, P, and G are numerical parameters which are identified with the temperature, the pressure, and the Gibbs potential, respectively. If $f = f(q_1,\cdots,p_{2N})$ is any phase function, its average is

$$\langle f\rangle = (N!)^{-2}\int_{-\infty}^{\infty} d^{2N}p \int_0^{\infty} d^{2N}q f\rho. \quad (4)$$

In particular $\langle 1\rangle = 1$, and this normalization relation defines G as a function of P and θ. Further important quantities are the average volume[3] $V = \langle q_{\max}\rangle$, the average energy $U = \langle H\rangle$ and the entropy $S = \langle \ln\rho\rangle$. Between these quantities the following relations hold

$$S = -(\partial G/\partial\theta)_P, \quad (5)$$

$$V = (\partial G/\partial P)_\theta, \quad (6)$$

and

$$U = G - PV + \theta S. \quad (7)$$

The reader will notice that we do not limit the system to a finite region (the "box") in the space of the coordinates as it is done in the standard canonical ensemble of Gibbs. Our probability distribution corresponds to a physical situation where the system is bounded on one side by a rigid wall and on the other by a freely movable "piston," that is to say, a device which exerts a constant pressure on the particle with largest coordinate. Thus the "volume" $q_{\max}$ is not fixed but fluctuates, and only its average has thermodynamic significance. The relation with respect to volume of the constant pressure and the constant volume ensembles is closely analogous to the relation with respect to energy of the canonical and microcanonical ensembles.

The aim of the following calculations is to obtain G as a function of θ and P, so that Eqs. (5)–(7) can be used to determine the other thermodynamic functions. An important simplification can be achieved immediately. *G depends in a nontrivial way only on a single variable.* The reason for this is the fact that the potential energy is a homogeneous function of first degree in the coordinates. Let $\epsilon_i = \mathrm{sg}(\sigma_i)$ be dimensionless charge

[3] Extensive thermodynamic quantities refer to unit surface area of the sheets. Thus V has the dimension of length. θ has the dimension of energy per unit area.

numbers, and let $\gamma=P/2\pi\sigma^2$. Also introduce dimensionless coordinates $x_i=2\pi\sigma^2 q_i/\theta$. In terms of these we obtain

$$G=-\theta\ln\{[(2\pi\theta)^{\frac{1}{2}}]^{2N}(\theta/2\pi\sigma^2)^{2N}Q(\gamma)\}, \quad (8)$$

where

$$Q(\gamma)=(N!)^{-2}\int_0^\infty d^{2N}x \times\exp\{-\gamma x_{\max}-\Phi(x_1,\cdots,x_{2N})\} \quad (9)$$

and

$$\Phi(x_1,\cdots,x_{2N})=-\sum_{1\leq i<j\leq 2N}\sum \epsilon_i\epsilon_j|x_i-x_j|. \quad (10)$$

The dependence on the pressure is entirely contained in the function $Q(\gamma)$. The dimensionless ratio γ plays a very important role, as we shall see. The temperature dependence is given explicitly by Eq. (8). It is important to remember that we are only interested in the asymptotic form of the thermodynamic functions in the limit of a very large number of particles. In this limit, the quantity G as well as all other extensive quantities become proportional to N. Thus it is expedient to introduce the Gibbs potential per particle

$$\mu=\lim_{N\to\infty}(G/2N). \quad (11)$$

Ignoring irrelevant constants, we have then

$$\mu=\mu(\theta,P)=\tfrac{1}{2}\theta\ln[\alpha(\gamma)/\theta^3], \quad (12)$$

where the function $\alpha(\gamma)$ is defined by

$$\ln\alpha(\gamma)=-\lim_{N\to\infty}\frac{\ln Q}{N}. \quad (13)$$

The existence of this limit will be demonstrated later (Sec. 6). In terms of the function $\alpha(\gamma)$, the equation of state is then

$$v=\lim_{N\to\infty}(V/2N)=\frac{\theta}{P}\left(\frac{\gamma}{2}\frac{d\ln\alpha}{d\gamma}\right). \quad (14)$$

The internal energy per particle is

$$u=\lim_{N\to\infty}(U/2N)=\theta\left(\frac{3}{2}-\frac{\gamma}{2}\frac{d\ln\alpha}{d\gamma}\right). \quad (15)$$

These follow directly from Eqs. (5)–(7) and (12). We also have the exact relation

$$u+Pv=\tfrac{3}{2}\theta. \quad (16)$$

4. INTEGRATION OVER CONFIGURATIONS

Consider the potential energy function Φ defined by Eq. (10). In any region of coordinate space where the relative order according to magnitudes of the coordinates $x_1, x_2, \cdots, x_{2N}$ is fixed, Φ is a homogeneous linear function of these coordinates. The coefficients of this linear function depend on the order of the coordinates. A simple consideration shows that the integral (9) is elementary. This is so because the class of functions consisting of exponentials of linear functions multiplied by polynomials is closed under the operation of indefinite integration. That is, however, not the same thing to say that the task of evaluating Q is trivial, since it is necessary to develop a systematic algorithm to deal with the very great number of integrations involved. Fortunately this can be done in a way which we now proceed to explain.

Imagine coordinate space broken up into $(2N)!$ regions, each corresponding to a definite linear succession of the $2N$ particles. These regions can be united into groups, each group containing $(N!)^2$ regions differing from each other only in permutation of the positively charged particles among themselves and the negatively charged particles among themselves. Such a group of regions will be called a "configuration." There are

$$\frac{(2N)!}{(N!)^2}=(-4)^N\binom{-\frac{1}{2}}{N} \quad (17)$$

configurations. Each is characterized by a definite linear succession of plus and minus signs. The potential energy remains unchanged when positively or negatively charged particles are permuted among themselves. Hence, if we conventionally relabel the particles in each region so that

$$0\leq x_1\leq x_2\leq\cdots\leq x_{2N}, \quad (18)$$

the potential energy becomes a linear function of the x_i with coefficients that depend *on the configuration only*. We may then rewrite the formula (9) in the following way

$$Q=\sum_C\int_0^\infty dx_{2N}\int_0^{x_{2N}}dx_{2N-1}\cdots\int_0^{x_2}dx_1 \times\exp\{-\gamma x_{2N}-\Phi_C(x_1,x_2,\cdots,x_{2N})\}. \quad (19)$$

Here, the summation goes over all configurations. Φ_C is the potential energy function for a definite configuration. We shall now show how an extraordinarily simple form for Φ_C may be obtained.

A configuration is uniquely characterized by the sequence $\epsilon_1, \epsilon_2, \cdots, \epsilon_{2N}$, where N of the ϵ_i are $+1$, N are -1. An alternative sequence of integers which can be used equally well to characterize a configuration is defined by

$$\begin{cases}\nu_0=0\\ \nu_i=\sum_{k=1}^{i}\epsilon_k. \quad (1\leq i\leq 2N),\end{cases} \quad (20)$$

ν_i is the total charge, in units of σ, up to and including the ith particle. We may think of the ν_i as a sequence of integers subject to the sole conditions

$$\begin{cases} \nu_0=\nu_{2N}=0, \\ |\nu_i-\nu_{i-1}|=1, \quad (1\leq i\leq 2N). \end{cases} \tag{21}$$

The utility of the ν_i lies in the result of the following simple transformation

$$\begin{aligned} \Phi_C &= -\sum\sum_{1\leq i<j\leq 2N} \epsilon_i\epsilon_j|x_i-x_j| \\ &= -\sum_{1\leq i\leq 2N} x_i\epsilon_i\Big(\sum_{1\leq j<i}\epsilon_j-\sum_{i<j\leq 2N}\epsilon_j\Big) \\ &= -\sum_{1\leq i\leq 2N} x_i(\nu_i-\nu_{i-1})(\nu_{i-1}+\nu_i) \\ &= \sum_{1\leq i\leq 2N} x_i(\nu_{i-1}^2-\nu_i^2). \end{aligned} \tag{22}$$

The potential energy appears as a sum of terms each depending *on the coordinate of one particle only*. That this must be so is indeed obvious physically; within a configuration a particle moves quite independently of the others, the force on it depending only on how many positive or negative particles are on either side of it.

It is convenient to write Φ_C with a slightly different arrangement of terms

$$\Phi_C=\sum_{1\leq i\leq 2N} \nu_{i-1}^2(x_i-x_{i-1}), \tag{23}$$

where $x_0=0$. This way of writing exhibits the non-negative nature of Φ, a fact which is not immediately visible from the definition (10). Now substitute (23) into (19); the integral then becomes the Laplace transform of the convolution of the functions $\exp(-\nu_i^2x)$ $(i=0, 1, 2, \cdots, 2N-1)$, and this is equal[4] to the product of the Laplace transforms of these functions

$$\prod_{0\leq i\leq 2N-1}\int_0^\infty dx \exp[-(\gamma+\nu_i^2)x] = \prod_{0\leq i\leq 2N-1}\frac{1}{\gamma+\nu_i^2}. \tag{24}$$

Therefore we have the explicit algebraic formula

$$Q=\sum_C \frac{1}{(\gamma+\nu_0^2)(\gamma+\nu_1^2)\cdots(\gamma+\nu_{2N-1}^2)}. \tag{25}$$

The sum over all configurations C is a sum over all sequences ν_i satisfying conditions (21).

Before considering the problem of this summation, we shall only remark that the motivation for the use of the constant-pressure ensemble was, of course, precisely the possibility of doing the integrals in this simple manner. The appropriateness of the method of Laplace transformation and its interpretation in terms of the constant-pressure ensemble for one-dimensional systems has been noted before.[5]

[4] D. V. Widder, *Advanced Calculus* (Prentice-Hall, Inc., Englewood Cliffs, New Jersey, 1947), Chap. XIII.

5. THE SUM OVER CONFIGURATIONS

In this section the dependence of Q on N, hitherto treated as a constant, becomes important. Therefore, in the following we shall indicate this by a subscript.

In order to investigate the nature of the sum (25), it is convenient to make use of a simple pictorial representation. Each configuration can be associated uniquely with a certain graph.

A graph is simply the plot of ν_i versus i. For the sake of easier visualization we draw this as a continuous function, but only the "vertices," i.e., points with integer coordinates, have a significance. A graph is thus a broken line consisting of sections of slope ± 1. Each factor in a product $\prod_i(\gamma+\nu_i^2)^{-1}$ is assigned to a definite vertex and vice versa. An example of a graph is shown on Fig. 1. This graph represents a product in the sum (25) for $N=9$. The first and the last vertices are fixed at $\nu=0$. Note that there is no factor associated with the the last vertex.

FIG. 1. The graph of the configuration +++−+−−−−−++++−+−−.

The fundamental idea for grasping the nature of this summation is the decomposition of graphs into simpler subgraphs with a corresponding decomposition of the associated products.

We start with the concept of an "irreducible" configuration. A configuration is called irreducible if $\nu_i\neq 0$ for all $i\neq 0$, $2N$, i.e., if its graph has only its endpoints in common with the i axis. Let $\bar{Q}_n$ stand for *the sum total of products belonging to all irreducible configurations with 2n vertices*. Consider now an arbitrary, in general reducible, configuration. Its graph can be regarded as a succession of irreducible graphs, each starting at one of the vertices with $\nu=0$. The expression associated with the whole graph is the product of the expressions associated with its irreducible components. We now carry out the summation over all graphs of a given length in three steps. First we hold the k (≥ 1) vertices with $\nu=0$ fixed and let the others vary in all possible ways without falling on the i axis. This gives a product

$$\bar{Q}_{n_1}\bar{Q}_{n_2}\cdots\bar{Q}_{n_k},$$

where $2n_1$ is the number of vertices in the first irre-

[5] Z. W. Salsburg, R. W. Zwanzig, and J. G. Kirkwood, J. Chem. Phys. 21, 1098 (1953), especially the end of this paper.

ducible component, $2n_2$ in the second, etc. Next, still holding k fixed, we sum over all possible *positive* integers $n_1, n_2, \cdots, n_k$, subject to the condition

$$n_1+n_2+\cdots+n_k=N. \tag{26}$$

Finally, we sum over k from 1 to N. It is clear that all configurations have been counted, and counted only once. Thus we get the following important combinatorial formula

$$Q_N=\sum_{k=1}^{N}\sum_{(n_1+n_2+\cdots+n_k=N)}\bar{Q}_{n_1}\bar{Q}_{n_2}\cdots\bar{Q}_{n_k}. \tag{27}$$

The physical significance of the quantities $\bar{Q}_n$ will be explored later (Sec. 8). The first few of them are

$$\begin{aligned}\bar{Q}_1&=\frac{2}{\gamma(\gamma+1)}\\ \bar{Q}_2&=\frac{2}{\gamma(\gamma+1)^2(\gamma+4)}\\ \bar{Q}_3&=\frac{2}{\gamma(\gamma+1)^2(\gamma+4)^2(\gamma+9)}+\frac{2}{\gamma(\gamma+1)^3(\gamma+4)^2}.\end{aligned} \tag{28}$$

We shall now show how the relationship (27) between the sequences Q_n and $\bar{Q}_n$ may be given a very simple form.

Let z be a complex variable, and let us introduce the *generating functions*

$$Q(z)\equiv 1+\sum_{n=1}^{\infty}Q_nz^n \tag{29}$$

and

$$\bar{Q}(z)\equiv\sum_{n=1}^{\infty}\bar{Q}_nz^n. \tag{30}$$

It is easy to see that, for sufficiently small $|z|$, these converge. We have, indeed, the following inequalities

$$\bar{Q}_n<Q_n<\frac{1}{(n!)^2}\int_0^\infty d^{2n}x e^{-\gamma x_{\max}} =\frac{(2n)!}{(n!)^2}\frac{1}{\gamma^{2n}}=\left(-\frac{4}{\gamma^2}\right)^n\binom{-\frac{1}{2}}{n}. \tag{31}$$

The integral is just Eq. (9) with Φ deleted, and the inequality follows from $\Phi\geq 0$. Therefore, the series for

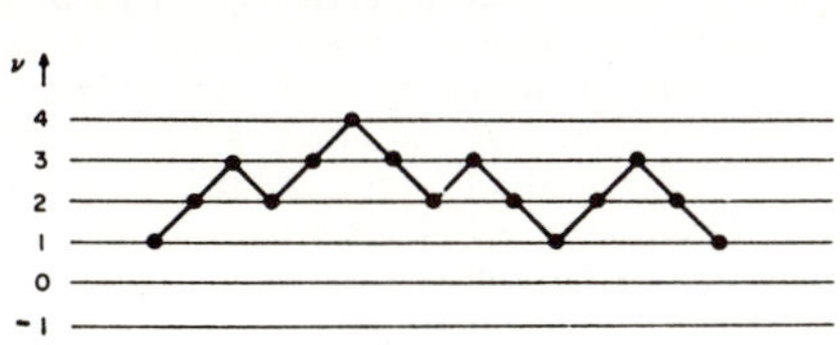

FIG. 2. A graph belonging to $G_8^{(1)}$.

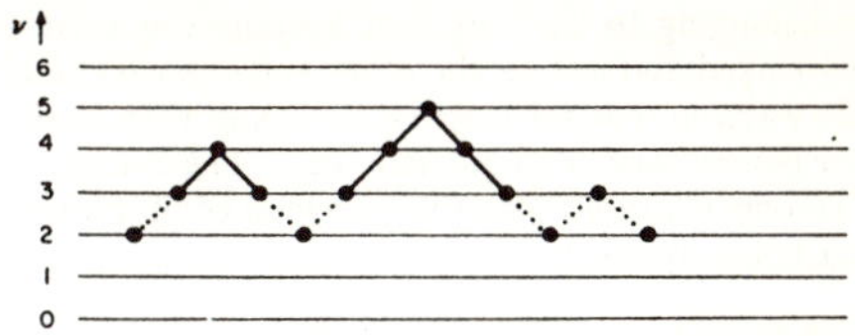

FIG. 3. A term in $G_7^{(2)}$ is equal to a term in the product $G_2^{(3)}G_2^{(3)}G_1^{(3)}/(\gamma+4)^4$.

$\bar{Q}(z)$ is majorized by that of $Q(z)$, and the latter is majorized by

$$\sum_{n=0}^{\infty}\left(-\frac{4z}{\gamma^2}\right)^n\binom{-\frac{1}{2}}{n}=\frac{1}{(1-4z/\gamma^2)^{\frac{1}{2}}}, \tag{32}$$

which converges for $|z|<\gamma^2/4$. Now multiply Eq. (27) by z^N and sum over N. A rearrangement of the order of summations is justified by the convergence of the series involved, and one obtains

$$Q(z)=[1-\bar{Q}(z)]^{-1}. \tag{33}$$

This simple analytic relation replaces the combinatorial formula (27).

It is convenient at this point to introduce the function $G^{(1)}(z)$ by the relation

$$\bar{Q}(z)=2\gamma^{-1}G^{(1)}(z) \tag{34}$$

and the corresponding sequence $G_n^{(1)}$ ($n=1, 2, \cdots$) generated by it

$$G^{(1)}(z)=\sum_{n=1}^{\infty}G_n^{(1)}z^n. \tag{35}$$

In terms of graphs, $G_n^{(1)}$ is the sum total of the products from all graphs with $2n-1$ vertices such that *the minimal ν value $\nu=1$ is assumed at the first as well as at the last vertex* (*not excluding the possibility of $\nu=1$ in between*). An example of such a graph is shown in Fig. 2. The factor γ^{-1} in Eq. (34) is just the single omitted vertex $\nu=0$ which would make such a graph represent an irreducible configuration. The factor 2 arises from the fact that such configurations occur in pairs related to each other by reflection of their graphs on the $\nu=0$ axis (or interchange of positive and negative charges).

In a similar way we may define generating functions $G^{(2)}(z)$, $G^{(3)}(z)$, $\cdots$ in terms of sequences $G_n^{(2)}$, $G_n^{(3)}$, $\cdots$ ($n=1, 2, \cdots$). We define $G_n^{(r)}(r\geq 1)$ as the sum total of products from all graphs with $2n-1$ vertices such that *the minimal ν value $\nu=r$ is assumed at the first as well as at the last vertex* (*not excluding the possibility $\nu=r$ in between*). We shall now show how the successive functions $G^{(1)}(z)$, $G^{(2)}(z)$, $\cdots$ are related to each other. If $n=1$, we have $G_n^{(r)}=1/(\gamma+r^2)$. Let then $n\geq 2$, and consider an arbitrary graph belonging to $G_n^{(r)}$ (see Fig. 3). Such a graph can be regarded as an alternation of vertices $\nu=r$ with smaller graphs that belong to some $G_m^{(r+1)}$ $(1\leq m\leq n-1)$. We may now sum over all graphs belonging to $G_n^{(r)}$ by first keeping the vertices with $\nu=r$ fixed and letting the other vertices vary in all possible ways so that for *them* $\nu\geq r+1$, and then letting the $\nu=r$ points vary in all possible ways. The reader will easily convince himself that the following combinatorial formula holds for $n\geq 2$,

$$G_n^{(r)}=\sum_{k=1}^{n-1}\frac{1}{(\gamma+r^2)^{k+1}}\times\sum_{(n_1+n_2+\cdots+n_k=n-1)}G_{n_1}^{(r+1)}G_{n_2}^{(r+1)}\cdots G_{n_k}^{(r+1)}. \tag{36}$$

Now, just as we have reduced Eq. (27) to the simple analytic relation (33) between the generating functions, so we may show by precisely the same sort of calculation that Eq. (36) implies

$$G^{(r)}(z)=\frac{z}{\gamma+r^2-G^{(r+1)}(z)}. \tag{37}$$

The interchange in orders of summation which is necessary for the derivation of (37) is justified by the convergence of the series involved. This follows from the obvious inequality

$$G_n^{(r+1)}<G_n^{(r)}, \tag{38}$$

which allows the series $G^{(r)}(z)$ to majorize $G^{(r+1)}(z)$. There is, therefore, a circle at least as large as $|z|=\gamma^2/4$, within which all series $G^{(r)}(z)$ converge.

The equation (37) which is of the nature of a recursion formula may be used to develop $G^{(1)}(z)$ into an infinite continued fraction. We have, in fact, proved the following

Theorem 1. *The configurational integral Q_N is the coefficient of z^N in the Maclaurin series of the function*

$$Q(z)=[1-\bar{Q}(z)]^{-1},$$

where $\bar{Q}(z)$ is the infinite continued fraction

$$\bar{Q}(z)=\frac{2}{\gamma}\cdot\cfrac{z}{\gamma+1-\cfrac{z}{\gamma+4-\cfrac{z}{\gamma+9-\cdots}}} \tag{39}$$

6. THE ANALYTIC NATURE OF $Q(z)$ THE ASYMPTOTIC LIMIT $N\to\infty$

It is now necessary to penetrate more deeply into the analytic nature of $Q(z)$. Cauchy's integral formula gives

$$Q_N=\frac{1}{2\pi i}\oint\frac{dz}{z^{N+1}}Q(z), \tag{40}$$

where the path of integration is a small closed curve surrounding the origin in a positive sense. We wish to study the asymptotic nature of Q_N as $N\to\infty$, and this requires some information on the singularities of $Q(z)$. It is well to point out here that the infinite continued fraction (39) is to be regarded as a formal expression. Its Maclaurin series defines an analytic function only wherever the series converges. Now, in the theory of continued fractions[6] such an expression is regarded as the limit of its finite approximating fractions which, in general, exists beyond the radius of convergence of the Maclaurin series and defines the analytic continuation of the latter. We shall, however, not need this theory, since all the desired results follow already from the more elementary treatment below. This consists of a series of simple propositions culminating in Theorem 2. Their proofs depend only on generally familiar facts of the theory of analytic functions.

Lemma 1. *Let a_1, a_2, $\cdots$ be a sequence of positive numbers, $a_i\geq a_1$, and let*

$$f(z)=\cfrac{z}{a_1-\cfrac{z}{a_2-\cfrac{z}{a_3-\cdots}}},$$

then the Maclaurin series of $f(z)$ converges at least inside the circle $|z|=\frac{1}{4}a_1^2$. For the proof let us compare the Maclaurin series of $f(z)$ with the series which is obtained when all a_i are replaced by a_1. From $a_i\geq a_1$ it follows that this comparison series has positive coefficients not smaller than those of $f(z)$. It is the Maclaurin series of

$$g(z)=\cfrac{z}{a_1-\cfrac{z}{a_1-\cdots}}$$

which is easily seen to be

$$g(z)=\tfrac{1}{2}[a_1-(a_1^2-4z)^{\frac{1}{2}}].$$

This series converges inside $|z|=\frac{1}{4}a_1^2$, hence so does $f(z)$.

Lemma 2. *Let $F(z,w)$ be a rational function of two variables, and let $f(z)$ be an analytic function, regular in a closed region R and not vanishing identically. Then the function $F(z,f(z))$ is an analytic function of z, regular in R with the possible exception of a finite number of poles.* $F(z,f(z))$ can have a singularity in R only where its denominator vanishes. But the latter is an analytic function in R, regular at all points, and not vanishing identically. Thus it can only have a finite number of zeros. These are the poles of $F[z,f(z)]$.

[6] H. S. Wall, *Analytic Theory of Continued Fractions* (D. Van Nostrand Company, Inc., Princeton, New Jersey, 1948).

Lemma 3. *Let $f(z)$ be an analytic function vanishing at the origin and having a Maclaurin expansion with real positive coefficients. Furthermore let $f(z)$ have a pole on the circle of convergence of its Maclaurin series. Then the equation*

$$f(z)=1$$

has a real positive solution $z=z_0$ inside the circle of convergence, and any other solution of this equation has an absolute value greater than z_0. Let $z=a$ be the pole whose existence is assumed, and let $z\to a$ from the inside of the circle of convergence. Then $|f(z)|\to\infty$. But

$$f(|z|)\geq|f(z)|$$

because the coefficients of the series are positive. This shows that $f\to\infty$ as well when z moves to the radius of convergence on the positive real axis. Since $f(0)=0$, there must be a positive real value z_0 such that $f(z_0)=1$, and indeed only one, since f is monotone increasing on the positive real axis. Let $z=z_1$ be any other solution of the equation, which may be assumed to lie inside the circle of convergence, since otherwise the last statement of the lemma follows. But then z_1 cannot be real and positive, and we have the strict inequality

$$f(|z_1|)>|f(z_1)|=1=f(z_0),$$

showing that $|z_1|>z_0$. This proves Lemma 3.

Theorem 2. *$Q(z)$ is a meromorphic function. It has a simple pole $z=\alpha$ on the positive real axis such that all of its other poles have greater absolute value. Proof.* From Lemma 1 it follows that the function $G^{(r)}(z)$ has a Maclaurin expansion converging at least inside the circle $|z|=\frac{1}{4}(\gamma+r^2)^2$. $Q(z)$ may be represented as a rational function of z and $G^{(r)}(z)$ by means of Eqs. (33), (34), and the first $r-1$ of the Eqs. (37). Lemma 2 shows that $Q(z)$ is regular in the circle $|z|<\frac{1}{4}(\gamma+r^2)^2$ with the possible exception of a finite number of poles. But this circle can be chosen as large as desired by choosing r sufficiently large. Thus $Q(z)$ is regular in the whole open complex plane with the exception of poles, i.e., it is meromorphic. Evidently the poles of $Q(z)$ are the solutions of the equation

$$\bar{Q}(z)=1. \qquad (41)$$

But $\bar{Q}(z)$ is also a meromorphic function [and so are all the functions $G^{(r)}(z)$; for the same reason as $Q(z)$], so that it has a pole on the circle of convergence of its Maclaurin series. It satisfies the other conditions of Lemma 3 as well. The existence and properties of the number α of the theorem then follow from the conclusion of the lemma. That the pole of $Q(z)$ at $z=\alpha$ is simple follows from $\bar{Q}'(\alpha)>0$, which is quite evident as this quantity is a convergent series of positive terms. This completes the proof of Theorem 2.

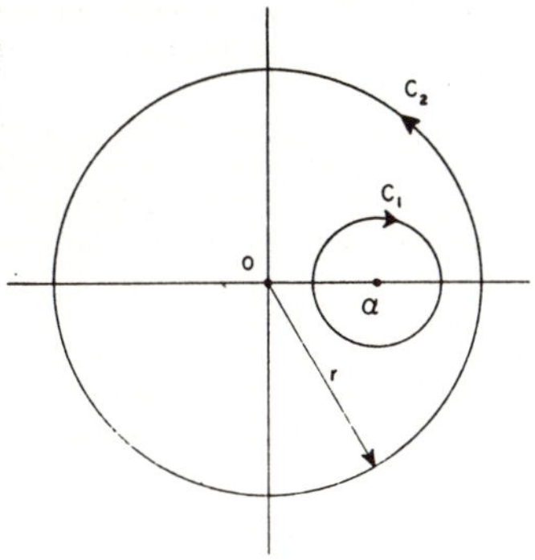

FIG. 4. Path of integration for Q_N.

After this preparation it is a simple matter to find the asymptotic behavior of Q_N as $N\to\infty$. In view of Theorem 2, the path of integration in Eq. (40) can be deformed into closed curves C_1 and C_2 with the following properties (Fig. 4). C_1 is any closed curve, traversed in the negative directions, and surrounding only one pole $z=\alpha$ of $Q(z)$. C_2 is a circle around the origin with a radius $r>\alpha$, such that it too surrounds only the pole $z=\alpha$. The integral over C_1 is just the residue at the pole, while the integral over C_2 is bounded in absolute value by $2\pi r$ multiplied by the maximum absolute value of the integrand. This leads immediately to the following asymptotic result[7]

$$Q_N\alpha^N=\frac{1}{\alpha\bar{Q}'(\alpha)}+O\{(\alpha/r)^N\}, \quad (N\to\infty). \qquad (42)$$

We actually need only a somewhat weaker statement, namely one obtained by taking the logarithm

$$\ln Q_N=-N\ln\alpha+O(1), \quad (N\to\infty). \qquad (43)$$

This establishes the limit (13), and thus completes the computation of the thermodynamic potential (Sec. 3). We summarize the result in:

Theorem 3. *The thermodynamic properties of the system are given by the Gibbs potential per particle (12) where the quantity α is the smallest real positive root of Eq. (41) whose left-hand member is the infinite continued fraction (39).*

7. THE CHARACTERISTIC VALUE PROBLEM ASSOCIATED WITH EQ. (41)

The further study of the model then depends on investigating the properties of the function $\alpha=\alpha(\gamma)$. This function is monotone increasing, because the left-hand side of Eq. (41) $\bar{Q}=\bar{Q}(z,\gamma)$ is an increasing function of z on the positive real axis within the radius of convergence of the series (30), and a decreasing function of γ. It is easy to see by glancing at (39) that in the

[7] We use the symbol O in the mathematically precise sense of indicating a quantity less in absolute value than the argument multiplied by some number independent of the variable which goes to the limit.

limit $\gamma\to 0$ we must have $\alpha(\gamma)\sim\gamma/2$. The behavior in the opposite limit $\gamma\to\infty$ is a more sophisticated question. One is tempted to say, in a heuristic manner, that the denominators $\gamma+1$, $\gamma+4$, $\cdots$ should all be replaced by γ in the limit $\gamma\to\infty$. This leads to replacing $\bar{Q}$ by the periodic infinite fraction

$$\frac{2}{\gamma}\,\frac{z}{\gamma-\cfrac{z}{\gamma-\cdots}}=1-(1-4z/\gamma^2)^{\frac{1}{2}}, \tag{44}$$

so that solving Eq. (41) one gets $\alpha(\gamma)\sim\frac{1}{4}\gamma^2$ $(\gamma\to\infty)$. However, it is clear that for *some* denominators this replacement is not justified, no matter how large (but finite) γ is, so that this simple argument is certainly incorrect.[8] It happens that the result just "derived" is correct, but a more refined asymptotic analysis is needed to establish it rigorously. There is, of course, the further problem of obtaining not just the dominant term but correction terms in this limiting case.

In this respect, it is fortunate that Eq. (41) plays an important role in a classical characteristic value problem of the Sturm-Liouville type which has been investigated very thoroughly in the mathematical literature. Thus it turns out that—from our point of view—it is only necessary to establish this connection, and then the known properties of the equation and its solution may be taken over and used to find the thermodynamic implications. For reasons that will become apparent in a moment it is not $\alpha=\alpha(\gamma)$ but the inverse function $\gamma=\gamma(\alpha)$ that is usually studied. Not only are the asymptotic series of this function known for the limits $\alpha\to 0$ and $\alpha\to\infty$, but it has even been tabulated numerically.

Let, temporarily, $\sqrt{\alpha}$ and γ be two independent positive numbers, and consider the problem of fixing them so that the differential equation

$$\left\{\frac{d^2}{dt^2}+2\alpha^{\frac{1}{2}}\cos t-\gamma\right\}y(t)=0 \tag{45}$$

shall have an even solution of period 2π not vanishing identically. We shall show that the condition for this is precisely

$$\bar{Q}(\alpha,\gamma)=1. \tag{41}$$

Indeed, let us expand $y(t)$ in a Fourier series

$$y(t)=\tfrac{1}{2}A_0+\sum_{r=1}^{\infty}A_r\cos rt, \tag{46}$$

and substitute into the differential equation. A simple calculation shows that the coefficients must be related by

$$\begin{cases}\tfrac{1}{2}\gamma A_0=\alpha^{\frac{1}{2}}A_1\\ (\gamma+r^2)A_r=\alpha^{\frac{1}{2}}(A_{r-1}+A_{r+1}), \quad (r\geq 1).\end{cases} \tag{47}$$

On account of the homogeneous nature of the problem, only the ratios of the coefficients need be determined; the successive ratios are then related by

$$\frac{\alpha^{\frac{1}{2}}A_r}{A_{r-1}}=\frac{\alpha}{\gamma+r^2-\alpha^{\frac{1}{2}}A_{r+1}/A_r}, \quad (r\geq 1). \tag{48}$$

We now note that this sequence of equations is identical with (37) provided there we set $z=\alpha$; hence we get

$$\alpha^{\frac{1}{2}}A_r/A_{r-1}=G^{(r)}(\alpha),$$

$$A_r=A_0\frac{G^{(1)}(\alpha)G^{(2)}(\alpha)\cdots G^{(r)}(\alpha)}{\alpha^{\frac{1}{2}r}}. \tag{49}$$

If we now substitute A_1 thus obtained into the first of Eq. (47) we have precisely Eq. (41), as was to be shown. To justify the validity of these formal manipulations, it is necessary to show that the Fourier series (46) converges, together with at least its first two derivatives. For this purpose we may use the inequality

$$G^{(r)}(z)<\tfrac{1}{2}\{\gamma+r^2-[(\gamma+r^2)^2-4z]^{\frac{1}{2}}\}, \tag{50}$$

derived in the way indicated by the proof of Lemma 1. It is valid for all z for which the right-hand side has a convergent Maclaurin expansion; in particular, for $z=\alpha$ if r is sufficiently large. Furthermore, $1-[1-x]^{\frac{1}{2}}<x$ for all positive $x<1$; so that

$$G^{(r)}(\alpha)<\frac{2\alpha}{\gamma+r^2}<\frac{2\alpha}{r^2}, \quad (r \text{ sufficiently large}).$$

This gives the asymptotic upper bound

$$A_r=0\{(2\sqrt{\alpha})^r/(r!)^2\}, \quad (r\to\infty), \tag{51}$$

which shows that the Fourier series (46) converges fast enough so that it can be differentiated term by term any number of times.

The differential equation (45) is known as the Mathieu equation.[9] Its even periodic solutions are the functions ce_0, ce_2, ce_4, $\cdots$ in the notation of reference 9. We are interested in that solution for which the two parameters in the equation γ and $2\sqrt{\alpha}$ tend simultaneously to zero, i.e., the function $y(t)$ degenerates into a constant. This property identifies uniquely the function ce_0. The situation is summarized as follows:

Theorem 4. *The relation between the quantities α and γ (see Theorem 3) is the same as the relation between*

[8] An indication of trouble is that (44) has a branch point and not a pole on its circle of convergence. Furthermore, the solution of Eq. (41) appears *on* the circle of convergence and not inside of it, as it must for any γ no matter how large.

[9] E. T. Whittaker and G. N. Watson, *A Course of Modern Analysis* (Cambridge University Press, New York, 1927), Chap. XIX.

the two parameters of the Mathieu equation (45) when the solution is the even Mathieu function of zero order.[10]

8. THE LIMIT $\gamma\to\infty$

We shall first consider the limit in which the effect of the interparticle force is small. This is the limit in which the mean kinetic energy is large compared to the mean potential energy. Another way of expressing the same state of affairs is to say that the pressure P is large compared to the "pressure" $2\pi\sigma^2$ exerted by one particle on another. This is the limit $\gamma=P/2\pi\sigma^2\to\infty$.

We have already remarked (cf. Sec. 7 above) that this limit presents peculiar difficulties when looked at purely from the point of view of solving Eq. (41). The correspondence with the Mathieu equation (Theorem 4) throws some further light on why this is so. The reader will observe that since we expect $\gamma\sim2\sqrt{\alpha}$ as $\alpha\to\infty$, the problem of Eq. (45) becomes one of finding the solution of a differential equation whose highest derivative term is multiplied by a small parameter. What is involved is an asymptotic analysis of the way in which the nature of the equation and its solution changes in the limit.[11]

This has first been done by Ince[12]; and we shall not attempt to reproduce his somewhat complicated analysis, but merely quote his result[10]:

$$\gamma=2\alpha^{1/2}-\alpha^{1/4}+\frac{1}{16}+\frac{1}{256\alpha^{1/4}}+\cdots,\quad(\gamma,\alpha\to\infty).\tag{52}$$

This formula can be inverted

$$\alpha=\frac{\gamma^2}{4}\left\{1+(8\gamma)^{-1/2}+(32\gamma)^{-1}-\frac{1}{256\sqrt{2}\gamma^{3/2}}+\cdots\right\}^4.\tag{53}$$

Now substitute into (12), this gives

$$\mu=\frac{\theta}{2}\left\{-3\ln\theta+2\ln\gamma+\left(\frac{2}{\gamma}\right)^{1/2}-\frac{1}{8\gamma}+\frac{1}{192\sqrt{2}\gamma^{3/2}}+\cdots\right\}.\tag{54}$$

We shall recognize the meaning of the dominant term by calculating the equation of state (14)

$$\frac{Pv}{\theta}=1-\frac{1}{(8\gamma)^{1/2}}+\frac{1}{16\gamma}-\frac{1}{256\sqrt{2}\gamma^{3/2}}+\cdots,\tag{55}$$

which shows that the system with increasing γ tends more and more to an ideal gas state. It is characteristic that corrections are expandable in inverse powers of $\sqrt{\gamma}$. The internal energy per particle is given by Eq. (16)

$$u=\frac{\theta}{2}\left\{1+\frac{1}{(2\gamma)^{1/2}}-\frac{1}{8\gamma}+\frac{1}{128\sqrt{2}\gamma^{3/2}}+\cdots\right\}.\tag{56}$$

The first term is just the mean kinetic energy, the rest the mean potential energy. If we write the latter as $\sigma\phi$ where ϕ is the mean electric potential at the location of a particle, we get for the dominant term

$$\phi=\tfrac{1}{2}\theta(\pi/P)^{1/2}+\cdots.\tag{57}$$

9. THE LIMIT $\gamma\to0$

In some ways the investigation of the opposite limit $\gamma\to0$ is simpler; $\alpha=\alpha(\gamma)$ can be expanded in an asymptotic series in powers of γ. Making use of the theory of Mathieu functions we could, of course, quote the pertinent results; but in view of the simplicity, it is worth while to give here the derivation of at least the first two terms in this series.

Consider two functions

$$X(z)=\frac{2}{\gamma}\,\frac{z}{\gamma+1-[z/(\gamma+4)]}$$

and

$$Y(z)=\frac{2}{\gamma}\,\cfrac{z}{\gamma+1-\cfrac{z}{\gamma+4-\cfrac{z}{\gamma+1-\cfrac{z}{\gamma+4-\cdots}}}}$$

$$=\frac{1}{\gamma}\{\gamma+4-[(\gamma+4)^2-4z/(\gamma+1)]^{1/2}\}.$$

For real positive z, small enough that the Maclaurin series of $Y(z)$ converges, we have the inequalities [cf. Eq. (39)]

$$X(z)<\bar{Q}(z)<Y(z).\tag{58}$$

Thus α, which is the root of the equation $\bar{Q}(z)=1$, is bounded below by the root of $Y(z)=1$, and above by the root of $X(z)=1$. The latter two equations can be solved explicitly, and one obtains the inequalities

$$\frac{\gamma(\gamma+1)(\gamma+8)}{4(\gamma+4)}<\alpha(\gamma)<\frac{\gamma(\gamma+1)(\gamma+4)}{3\gamma+8}.\tag{59}$$

These hold for all γ. Now, the Maclaurin series in powers of γ of the two extreme members of (59) have the same first two terms; hence, by letting $\gamma\to0$, we see that these are also the first two terms in the development of α

$$\alpha(\gamma)=\tfrac{1}{2}\gamma+\tfrac{7}{16}\gamma^2+0(\gamma^3).\tag{60}$$

The thermodynamic functions are then given by the

[10] The reader should be warned that there is no standard normalization for the Mathieu functions. Therefore, we have refrained from writing out the relation $\gamma=\gamma(\alpha)$ explicitly. This is easily done by comparing our Eq. (45) with the standardized form adopted by any particular author.

[11] See also remarks of footnote 8.

[12] E. L. Ince, Proc. Roy. Soc. Edinburgh **46**, 316 (1926).

formulas of Sec. 3: The Gibbs potential per particle

$$\mu=\tfrac{1}{2}\theta\{-3\ln\theta+\ln\tfrac{1}{2}\gamma+\tfrac{7}{8}\gamma+0(\gamma^2)\}, \tag{61}$$

the equation of state

$$Pv/\theta=\tfrac{1}{2}+\tfrac{7}{16}\gamma+\cdots, \tag{62}$$

and the internal energy per particle

$$\mu=\theta\{1-\tfrac{7}{16}\gamma+\cdots\}. \tag{63}$$

In the first approximation, the system appears as an ideal gas of N (and not $2N$) particles. This at the first sight surprising result can be explained as follows. For small values of the parameter γ, those parts of coordinate space have a relatively high probability for which the potential energy is close to its minimum value. Formula (23) shows that this is the case, in the first instance, for configurations for which the numbers ν_i^2 assume their minimum value. The restrictions (21) imply that these are the configurations for which

$$\nu_i=\begin{cases}0 & (i \text{ even})\\ \pm 1 & (i \text{ odd}).\end{cases} \tag{64}$$

An example is shown on Fig. 5. These are the configurations in which *particles are arranged in successive pairs, each pair having zero total charge.* We may say that the

FIG. 5. A configuration of minimal potential energy.

system is an assembly of neutral "molecules" made up of two oppositely charged ions. In the configurations of the type (64), these molecules do not overlap, so that there is no force between them. Mathematically, restriction to these configurations means omitting all terms of the sum (27) save those with $n_1=n_2=\cdots=1$.

This interpretation in terms of neutral molecules is helpful even beyond the dominant terms of the limit $\gamma\rightarrow 0$. For instance, if we take the next term into account we have the effect of "binary interactions." The reader will have no difficulty in checking that, in this approximation, irreducible configurations contributing to $\bar{Q}_1$ and $\bar{Q}_2$ must be taken into account, the latter being the contribution from overlap-configurations of two molecules. In still higher approximations, triple interactions etc., come into play. This train of thought leads to a better understanding of the opposite limit $\gamma\rightarrow\infty$ too (Sec. 8). There *the number of particles that are in simultaneous interaction is large.* We shall come back to this point further below (Sec. 10).

We conclude with a remark on the internal energy. The dominant term $2N\theta$ shows that, on the average, we have the same amount $2N(\theta/2)$ of kinetic energy as potential energy. However, only half of this, $N\theta/2$, is the kinetic energy of the translational degrees of freedom of the molecules. The fact that, for the internal motion, the mean potential energy is twice the mean kinetic energy is the consequence of the proportionality of the potential energy to the absolute value of the separation.

10. SIGNIFICANCE OF IRREDUCIBLE CONFIGURATIONS

We have already seen that the concept of an irreducible configuration plays an important role both in the mathematics (Sec. 5) and the physical interpretation (Sec. 9). Here we shall consider some further points that illuminate the statistical structure of the system.

The reason for the importance of the irreducible configurations is that they are the *minimal electrically neutral parts* of a general configuration. They cannot be further subdivided into successive neutral parts. They are the noninteracting building blocks of the system. In contrast to other systems that can somehow be decomposed into noninteracting entities, one must remember that this decomposition for our system is not determined locally. It is unique only if the *whole* system is so decomposed. An extra positive particle on one end and an extra negative one on the other end will change totally the decomposition throughout the system. The physical reason for this is, of course, that the interparticle force is independent of distance. Nevertheless, the irreducible configurations should be thought of as having definite physical significance. Their average length is a measure of the efficacy of electrostatic shielding.

Let us introduce a random variable n_1 equal to half the number of particles in the irreducible configuration furthest to the left, say. n_1 is a definite phase function, constant over configurations (cf. Sec. 4), and indeed over all those configurations for which the first irreducible component has the same number of particles. It has thus a definite probability distribution which we shall now consider. *The probability that $n_1=n$ ($n=1, 2, \cdots, N$) is that fraction of the sum (27) for which $n_1=n$.* In other words

$$\text{Prob}\{n_1=n\}=\bar{Q}_nQ_{N-n}/Q_N, \quad (1\leq n\leq N), \tag{65}$$

where $Q_0=1$ by convention. The reader will easily verify that the sum of these probabilities is 1 in virtue of Eq. (33). We are actually interested in the asymptotic form of this distribution as $N\rightarrow\infty$. This is quite simply obtained from the fundamental result (42)

$$P_n\equiv\lim_{N\rightarrow\infty}\text{Prob}\{n_1=n\}=\bar{Q}_n\alpha^n. \tag{66}$$

The normalization is now just the characteristic

equation (41). We can now form various moments

$$\langle n_1\rangle = \sum_{n=1}^{\infty} nP_n = \alpha\bar{Q}'(\alpha), \qquad (67)$$

etc. All moments of this probability distribution are finite, because α is inside the radius of convergence of the series (30) and so term by term differentiation is justified any number of times. In fact, using an analysis quite analogous to that of Sec. 6, it is easy to show that P_n decreases exponentially with n in the limit $n\to\infty$. P_n is, of course, a function of γ, and when we want to call attention to this we will write $P_n(\gamma)$.

It is of interest now to ask about the limits of P_n as $\gamma\to\infty$ and $\gamma\to 0$. $P_n(\infty)$ can be derived from (66) and (53) together with the definition of $\bar{Q}_n$ as a sum over irreducible configurations (Sec. 5).

$$P_n(\infty) = \sum_{\substack{C \text{ irreducible} \\ \text{with } 2n \text{ vertices}}} \lim_{\gamma\to\infty} \left(\frac{\gamma^2}{4}\right)^n \times \frac{1}{(\gamma+\nu_0^2)(\gamma+\nu_1^2)\cdots(\gamma+\nu_{2n-1}^2)}$$

$$= \left(\frac{1}{4}\right)^n \times \left(\begin{matrix}\text{number of irreducible configurations} \\ \text{with } 2n \text{ vertices}\end{matrix}\right).$$

The computation of the number in the parentheses is accomplished by the remark that its generating function is related to the generating function of the number (17) of all configurations with $2N$ vertices precisely as $\bar{Q}(z)$ is related to $Q(z)$, Eq. (33). These generating functions can be summed because they are binomial series, and one obtains after a simple calculation

$$\begin{aligned} P_n(\infty) &= (-1)^{n-1}\binom{\frac{1}{2}}{n} \\ &= \frac{1}{2n-1}\,\frac{1\cdot 3\cdot 5\cdots(2n-1)}{2\cdot 4\cdot 6\cdots(2n)}. \end{aligned} \qquad (68)$$

This probability distribution can be interpreted very simply. $\gamma\to\infty$ is the limit in which the effect of the interparticle force vanishes (Sec. 8). Thus, all configurations become equally likely. In the limit, $N\to\infty$ when our system is infinitely large; this amounts to playing "heads and tails" with the successive signs of charge indefinitely. The question about the probability $P_n(\infty)$ of the first irreducible configuration having $2n$ vertices is then equivalent to asking: In a game of "heads and tails," what is the probability that the accumulated net gain of one side falls to zero for the first time after $2n$ tosses? This problem has been well discussed in elementary probability theory.[13]

[13] W. Feller, *An Introduction to Probability Theory and Its Applications* (John Wiley & Sons, Inc., New York, 1957), Chap. III. Equation (4.7) of this chapter is identical with our (68).

This limiting probability distribution has a property that is important for a deeper understanding of the limit $\gamma\to\infty$ for our model. *All moments of (68) diverge.* The reason is that

$$P_n(\infty) \sim [2n(\pi n)^{\frac{1}{2}}]^{-1}, \quad (n\to\infty), \qquad (69)$$

so that the decrease with n is not fast enough for the series expressing the moments to converge. But we have just seen that all moments of $P_n(\gamma)$ are finite for any finite γ. Thus we conclude

$$\left\{\begin{matrix} \langle n_1\rangle, \quad \langle n_1^2\rangle, \cdots \to \infty \\ \text{as} \quad \gamma\to\infty. \end{matrix}\right. \qquad (70)$$

Now, the first moment $\langle n_1\rangle$ may be regarded as a measure of electrostatic shielding in the sense that it is the average number of particles within a distance over which electrostatic forces are neutralized (cf. discussion above). Hence, *in the limit $\gamma\to\infty$, the number of particles within the shielding distance tends to infinity.* This is one customary definition of this so-called "plasma limit."

The fact that the reach of the electrostatic forces becomes larger in the same limit as their effect becomes smaller is only seemingly paradoxical. It is related to the fact that in the coin tossing game the probability of long leads is relatively high.[14] Our model with γ finite corresponds to a game in which the probability of winning on a toss of coin is larger for the party that has accumulated a net loss. In such a game the persistence of long leads[14] would not occur. The electrostatic forces, by pulling in neutralizing charges, prevent their propagation through large distances.

Let us briefly look at the limit $\gamma\to 0$. For all n, the quantity $\gamma\bar{Q}_n(\gamma)$ remains finite in the limit. Since $\alpha\sim\gamma/2$, we therefore get

$$\begin{aligned} P_1(\gamma) &= 1-0(\gamma), \\ P_n(\gamma) &= 0(\gamma^{n-1}) \quad (n\geq 2), \end{aligned} \qquad (71)$$

as $\gamma\to 0$. This result confirms the picture of the system in this limit, as discussed above (Sec. 9). It indicates that an approximation including terms of order γ^{n-1} corresponds to neglecting irreducible configuration of more than $2n$ particles.

11. EXTERNAL ELECTRIC FIELD

In closing we shall consider the generalization of the foregoing to the situation when the system is immersed in a constant, homogeneous electric field $\mathcal{E}$. The only change in the basic equation is that we add a term

$$-\mathcal{E}\textstyle\sum_i \sigma_i q_i \qquad (72)$$

to the Hamiltonian (2) representing the potential energy of the charged particles in the external field. The reader will verify without difficulty that the

[14] Reference 13, Sec. III.5.

important formula (23) is replaced by

$$\Phi_C = \sum_{1\leq i\leq 2N} (\nu_{i-1}^2+2\eta\nu_{i-1})(x_i-x_{i-1}), \tag{73}$$

where $\eta=\mathcal{E}/(4\pi\sigma)$. It is clear that the whole of Secs. 4, 5, and 6 can be taken over with minor changes only. The factor belonging to a vertex is now

$$1/(\gamma+2\eta\nu+\nu^2). \tag{74}$$

Since this is not invariant under a change of sign of ν, we must treat the graphs below the $\nu=0$ axis separately from the graphs above the axis. It is not difficult to see that the formula which replaces (34) is now

$$\bar{Q}(z;\gamma,\eta)=\gamma^{-1}\{G^{(1)}(z;\gamma,\eta)+G^{(1)}(z;\gamma,-\eta)\}, \tag{75}$$

where

$$G^{(1)}(z;\gamma,\eta)=\cfrac{z}{\gamma+2\eta+1-\cfrac{z}{\gamma+4\eta+4-\cfrac{z}{\gamma+6\eta+9-\cdots}}} \tag{76}$$

As before, the Gibbs potential per particle is given by Eq. (12), where now $\alpha=\alpha(\gamma,\eta)$ is the smallest positive solution of $\bar{Q}(\alpha;\gamma,\eta)=1$. The thermodynamic variable conjugate to the electric field is the electric dipole moment $M=\langle\sum_i \sigma_i q_i\rangle$. Its value per particle is

$$\lim_{N\to\infty}\frac{M}{2N}=-\frac{\theta}{8\pi\sigma}\left(\frac{\partial \ln\alpha}{\partial\eta}\right)_\gamma. \tag{77}$$

There is an important restriction on the allowable

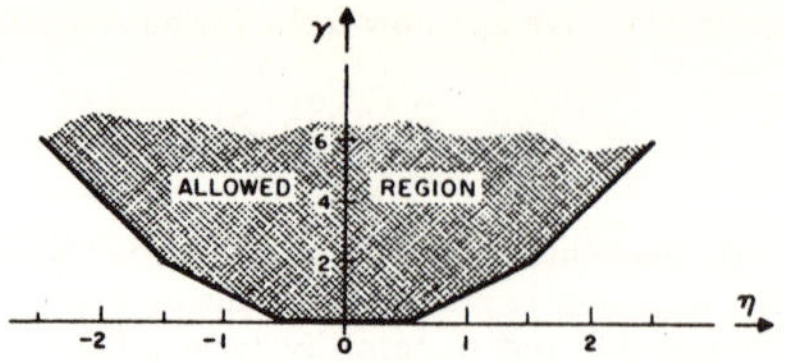

FIG. 6. Allowed region in the (γ,η) plane.

values of γ and η. The convergence of all integrals in (19) requires that all factors (74) shall be positive. Thus γ and η must be such that *the quadratic form*

$$f(x)=x^2+2\eta x+\gamma \tag{78}$$

shall be positive for all integer values of the variable x. An elementary consideration shows that the necessary and sufficient condition for this is

$$\gamma>\eta_0(2\eta-\eta_0), \tag{79}$$

where $\eta_0=\eta_0(\eta)$ is the nearest integer to η. The region allowed in the (γ,η) plane can be described in simple geometrical terms: It is bounded below by those tangents of the parabola $\gamma=\eta^2$ which touch at points with integer η values (Fig. 6). The physical explanation for this restriction is that for too large an electric field the external pressure cannot overcome the force of the field on a suitable charge fluctuation near the "piston."

In closing we may mention that, here too, it is possible to convert the equation $\bar{Q}(\alpha)=1$ into a characteristic value problem using a slight generalization of the technique of Sec. 7. A study of this problem would result in information on the thermodynamics of the system in the presence of an electric field.

One-Dimensional Plasma Model at Thermodynamic Equilibrium

O. C. ELDRIDGE AND M. FEIX*
John Jay Hopkins Laboratory for Pure and Applied Science,
General Atomic Division of General Dynamics Corporation,
San Diego, California
(Received March 2, 1962)

The one-dimensional motion of a number of plane charged sheets is followed with an electronic computer. The energy in the electric field is computed for cases where the number of particles in a Debye length vary from 1/4 to 10. These results agree with an exact theory given by Lenard. The Fourier components of the field energy and of the square of the current density are measured by averaging over a long time interval. These results are compared with values predicted by the linearized Vlasov equation.

I. INTRODUCTION

THE nonrelativistic electrodynamics of charged particles in one dimension is sufficiently simple so that the motion of a large number of particles can be followed in detail by an electronic computer. It is useful to think of such a calculation as a mathematical experiment. This approach is not new, but it has recently been used effectively for problems in plasma physics by Dawson[1] and Buneman.[2] Our models are very similar to theirs.[3] In these models, particles are represented by thin, charged sheets which move along an axis of finite length and which interact only through the electric field. In Sec. II we describe the details of the models. In Sec. III we give some results for systems at thermal equilibrium.

II. DESCRIPTION OF THE MODEL

We consider a system of N particles which may be visualized as charged sheets constrained to move along an interval L of the x axis and which extend to infinity in the y, z plane. The motion of a particle with charge σ_i and mass m_i per unit area is given by the equations

$$\frac{d^2x_i(t)}{dt^2} = \frac{\sigma_i}{m_i} E[x_i(t), t], \tag{1}$$

and

$$\frac{\partial E(x, t)}{\partial x} = 4\pi\rho(x, t) = 4\pi \sum_{i=1}^{N} \sigma_i\, \delta[x - x_i(t)]. \tag{2}$$

The field is then

$$E(x, t) = 2\pi \sum_{i=1}^{N} \sigma_i \epsilon[x - x_i(t)] + F(t), \tag{3}$$

where

$$\epsilon(x) = \begin{Bmatrix} 1 \\ 0 \\ -1 \end{Bmatrix} \quad \text{for} \quad \begin{Bmatrix} x > 0 \\ x = 0 \\ x < 0 \end{Bmatrix},$$

and $F(t)$ is an arbitrary function of time to be determined by the boundary conditions.

Model 1

In this model the electric field at the position of any particle is a constant so long as the trajectories of two particles do not cross. We impose periodic boundary conditions on the particles; when an electron or an ion moves across one boundary, it reappears at the other boundary. The boundary then represents only a point from which displacements are measured.

In Fig. 1(a) we show the electric field for a particular configuration. In Fig. 1(b) we show the field a short time later, after the trajectories of two particles have crossed and one particle has crossed a boundary. The field at the boundary changes as particles cross it, but is a constant for other times. More explicitly $F(t)$ does not change except when a sheet crosses a boundary, when it jumps by $\pm 4\pi\sigma$ depending on the sign of the particle and whether it leaves through the left or right boundary.

The second of Maxwell's equations is a consequence of Eqs. (1) and (3), or can be derived from charge conservation and Eq. (3).

$$\frac{\partial E(x, t)}{\partial t} = -4\pi J(x, t) = -4\pi \sum_{i=1}^{N} \sigma_i v_i(t)\, \delta[x - x_i(t)]. \tag{4}$$

* Permanent address: Association Euratom, CEA, Fontenay aux Roses, Seine.
[1] J. Dawson, Phys. Fluids 5, 445 (1962).
[2] O. Buneman, Phys. Rev. 115, 503 (1959); J. Nuclear Energy, Part C 2, 119 (1961).
[3] The field is different if a background of neutralizing charge is present.

The conservation of the total energy and momentum of the system also follow:

$$\frac{dU}{dt} = \frac{d}{dt}\left[\sum_{i=1}^{N} \tfrac{1}{2} m_i v_i^2(t) + \frac{1}{8\pi}\int_0^L dx\, E(x,t)^2\right] = 0, \tag{5}$$

and

$$\frac{dP}{dt} = \frac{d}{dt}\left[\sum_{i=1}^{N} m_i v_i(t)\right] = 0. \tag{6}$$

This model seems appropriate for the description of the central region of a large homogeneous plasma. The finite length of the system corresponds to a maximum finite wavelength of the electric field. There is an infinite wavelength component of the electric field associated with the potential difference across the interval L:

$$E_0(t) = +\sum_i 2\pi\sigma_i \frac{x_i(t)}{L} + F(t) = -\frac{\Delta\phi(t)}{L}. \tag{7}$$

A macroscopic electric field cannot be maintained in this system. If a field is imposed at initial time, the field and associated current oscillate until damped to the thermal level.

Our computer was the IBM 7090. Given the initial positions, velocities, and accelerations of the particles, the machine computes the times at which neighboring pairs of particles will cross. The accelerations are constant until this time. The machine then chooses the shortest of these crossing times, recomputes the coordinates of the particles affected by the collision, and proceeds to find the next pair which cross.

This method is fairly fast and very accurate. For all cases the total energy and momentum of the system have been constant to five significant figures.

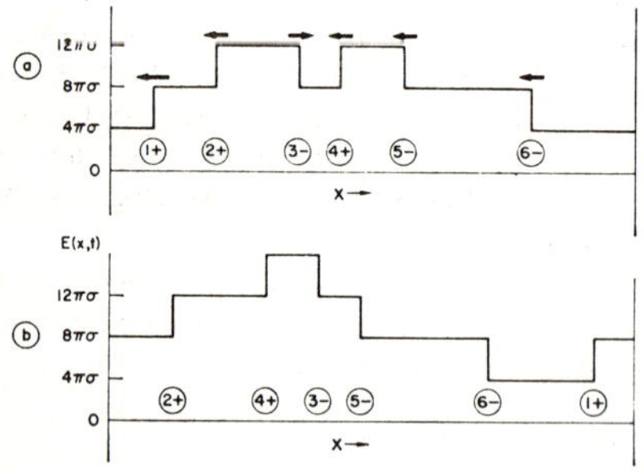

FIG. 1. (a) A simple configuration of the electric field. (b) The same system a short time later, after particles 3 and 4 have crossed and particle 1 has crossed the boundary.

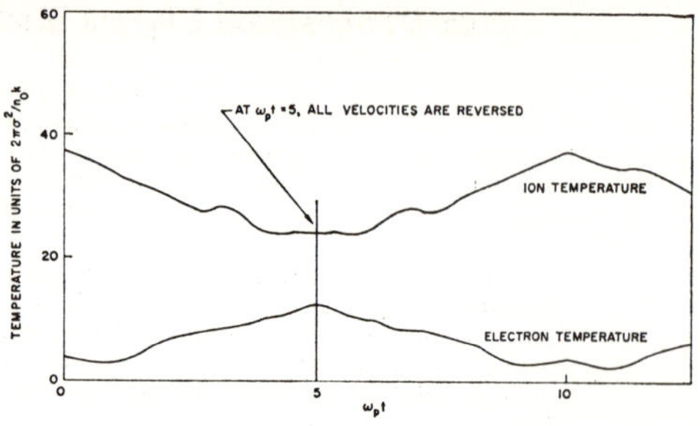

FIG. 2. A check on the accuracy of the machine calculations. The temperatures of a system of 100 electrons and ions are shown. At time $\omega_p t = 5$, all velocities are reversed.

A further check on the accuracy of the method is shown in Fig. 2. The electrons and ions of a system of 100 particles have initially been given Maxwellian velocity distributions at different temperatures. Until the time $t = 5/\omega_p$, the temperatures approach each other. At this time we stopped the program and reversed all the velocities. At time $t = 10/\omega_p$ the temperatures and all coordinates have their initial values to four significant figures.

Model 2

This model is the same as the first except that the particles are reflected by the boundaries. It applies to a finite system; one may think of the boundaries as the sharp edge of a strong magnetic field. All the same equations apply away from the boundaries, but the total momentum is not conserved.

III. THEORY OF A ONE-DIMENSIONAL PLASMA AT THERMAL EQUILIBRIUM

We know that at thermal equilibrium the velocity distributions of both electrons and ions are Maxwellian with the same temperature. More interesting are the properties of the electric field. We can compute three quantities. The simplest is the energy density of the electric field, averaged over time and space:

$$W = \frac{1}{T}\int_0^T dt \frac{1}{L}\int_0^L \frac{E^2(x,t)}{8\pi}\, dx. \tag{8}$$

Next we can analyze the partition of energy among different wavelengths, averaged over the time:

$$W(k) = \frac{\langle E(k)E(k)^*\rangle}{8\pi} = \frac{1}{8\pi T}\int_0^T dt \frac{1}{L}\left|\int_0^L dx \exp(-ikx)E(x,t)\right|^2. \tag{9}$$

The most precise description involves the partition of energy according to both wavelength and frequency:

$$W(k,\omega) = \frac{1}{8\pi}\langle E(k,\omega)E(k,\omega)^*\rangle = \frac{1}{8\pi}\frac{1}{LT} \cdot \left|\int_0^L dx \int_0^T dt \exp(i\omega t - ikx)E(x,t)\right|^2. \qquad (10)$$

The last description is a complete one for an ergodic field.[4]

Linearized Theory

In the case of a one-dimensional plasma an exact theory has been provided by Lenard[5] which gives W, the average field energy, for all values of density and temperature. For $W(k)$ and $W(k, \omega)$ only a linearized theory is available. Moreover this theory is only valid in the fluid limit, when the number of particles in a Debye length is much greater than one. The linearized theory has been given by Bohm and Pines[6] for $W(k)$ and by many others[7–9] for $W(k, \omega)$.

The simplest way to calculate $W(k, \omega)$ at thermal equilibrium is to apply the Nyquist theorem.[8] If $\langle|J(k,\omega)|^2\rangle$ represents the fluctuations of the current,

$$\langle|J(k,\omega)|^2\rangle = (\theta/\pi)\sigma(k,\omega), \qquad (11)$$

where $\sigma(k, \omega)$ is the real part of the transconductivity $Y(k, \omega)$ defined as

$$Y(k,\omega) = \frac{\omega}{4\pi i}\frac{1-H(k,\omega)}{H(k,\omega)} = \sigma(k,\omega) + i\Phi(k,\omega), \qquad (12)$$

with

$$H(k,\omega) = 1 + \frac{\omega_p^2}{k} P\int \frac{dF/dv}{\omega - kv}\,dv + i\pi\frac{\omega_p^2}{k^2}\frac{dF}{dv}\bigg|_{v=\omega/k}.$$

Only the real part of the transconductivity is needed since $Y(k, \omega)$ satisfies the Kramers–Kronig relations.[10] Then $\Phi(k, \omega)$ is a redundant quantity which can be deduced from $\sigma(k, \omega)$.

We shall need two integrals:

$$I_1 = \int_{-\infty}^{\infty} d\omega\, \sigma(k,\omega) = \frac{\omega_p^2}{4}, \qquad (13)$$

and

$$I_2(k) = \int_{-\infty}^{\infty} d\omega \frac{\sigma(k,\omega)}{\omega^2} = \frac{1}{4}\frac{1}{1+(kD)^2}, \qquad (14)$$

with

$$\frac{1}{D^2} = -\omega_p^2\int_{-\infty}^{\infty}\frac{dv}{v}\frac{dF}{dv}.$$

The only assumption needed to obtain Eqs. (12)–(14) is that $F(v)$ is an even function of v, but the connection between fluctuations and the conductivity implies thermal equilibrium.[11] Rostoker[7] has extended the theory to the nonequilibrium case.

The energy in the electric field is deduced from $|\rho(k,\omega)|^2$ or $|J(k,\omega)|^2$ by using Maxwell's equations:

$$|E(k,\omega)|^2 = [(4\pi)^2/k^2]\,|\rho(k,\omega)|^2 = [(4\pi)^2/\omega^2]\,|J(k,\omega)|^2. \qquad (15)$$

For the Maxwell distribution

$$F(v) = (m/2\pi\theta)^{\frac{1}{2}} \exp(-\tfrac{1}{2}mv^2/\theta), \qquad (16)$$

we find

$$W(k,\omega) = \frac{2\pi n\sigma^2}{k^3}\frac{F(\omega/k)}{|H(k,\omega)|^2}, \qquad (17)$$

$$W(k) = \tfrac{1}{2}\theta[1+(kD)^2]^{-1}, \qquad (18)$$

and

$$W = n\pi\sigma^2 D = \tfrac{1}{2}\theta(1/2D), \qquad (19)$$

with the Debye length D given by $D^2 = (\theta/4\pi n\sigma^2)$.

We can also calculate the function

$$\langle|J(k)|^2\rangle = \frac{1}{T}\int_0^T dt \cdot \left|\frac{1}{L}\int_0^L dx \exp(-ikx)J(x,t)\right|^2 = \frac{\theta\omega_p^2}{4\pi}. \qquad (20)$$

In all these results n is the total number of particles, electrons and ions, per unit length, and $\omega_p^2 = 2\pi\sigma^2(n_i + n_e)(1/m_e + 1/m_i)$.

In Figs. 3–5 we give the results of two machine calculations for Model 1 with periodic boundary conditions. These problems are characterized by

nD	N	Averaging time $\omega_p T$
7.590	300	5–73
2.456	200	5–50

Figures 3 and 4 show, respectively, $W(k)$ and $\langle|J(k)|^2\rangle$ for the first case. Figure 5 gives $W(k)$ for the second case. The agreement with theory is

4 C. Shannon and W. Weaver, *The Mathematical Theory of Communication* (University of Illinois Press, Urbana, Illinois, 1949).
5 A. Lenard, J. Math. Phys. **2**, 682 (1961).
6 D. Bohm and D. Pines, Phys. Rev. **85**, 338 (1952).
7 N. Rostoker, Nuclear Fusion **1**, 101 (1961).
8 T. Hagfors, J. Geophys. Research **66**, 1699 (1961).
9 J. Dougherty and D. Farley, Proc. Roy. Soc. (London) **A259**, 79 (1960).
10 C. Kittel, *Elementary Statistical Physics* (John Wiley & Sons, Inc., New York, 1958).
11 R. Callen and T. Welton, Phys. Rev. **80**, 250 (1950).

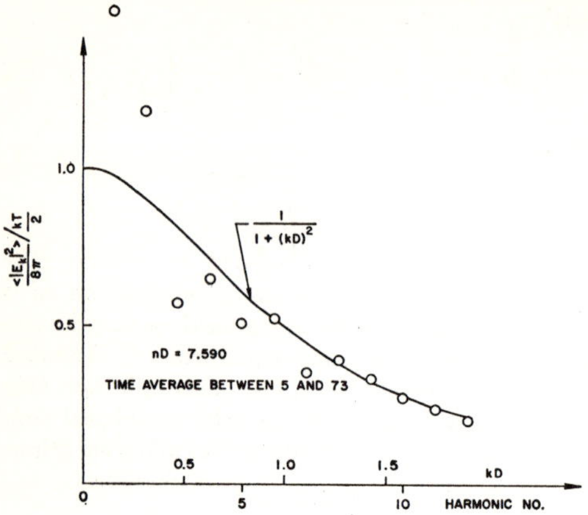

FIG. 3. The energy in each Fourier component of the electric field averaged between times $\omega_p t = 5$–73. The circles are experimental points. $nD = 7.590$.

better for the short wavelength modes than for the modes with wavelength greater than the Debye length. The characteristic behavior in time of the long wavelength modes is illustrated in Fig. 6. The fluctuations become much more important for the shorter wavelengths. The linearized theory tells us that, for long wavelengths, the memory of the system is very long, so that these modes are probably not in equilibrium. We consider here the Landau damping as a measure of the time necessary to forget the initial conditions, but it is probable that nonlinear

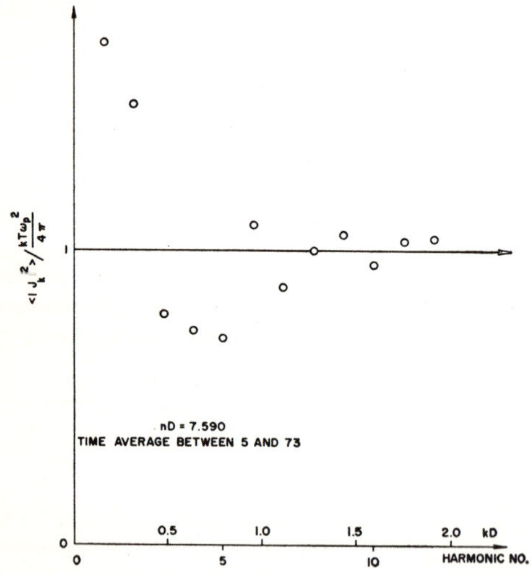

FIG. 4. $\langle |J_k|^2 \rangle$ for $nD = 7.590$. An average was taken between times $\omega_p t = 5$–73. The circles are experimental points.

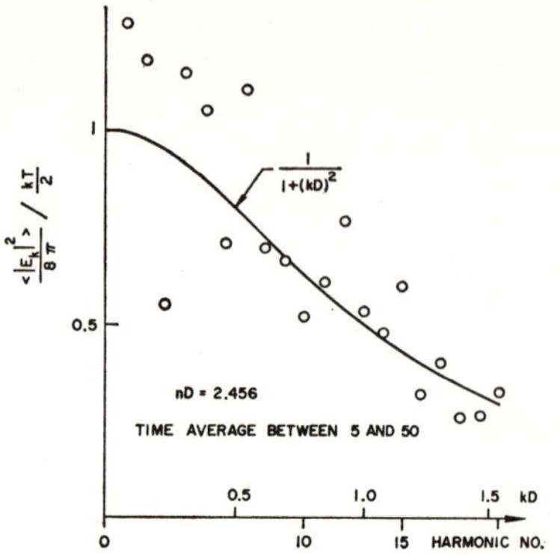

FIG. 5. W_k for $nD = 2.456$. The time average was taken between $\omega_p t = 5$–50.

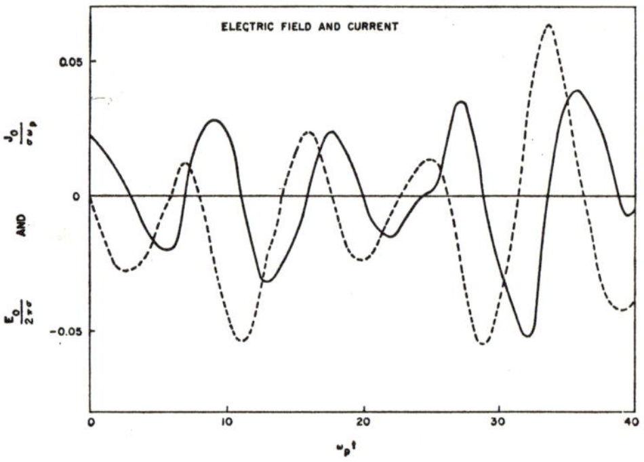

FIG. 6. An illustration of the time dependence of the long wavelength components of the electric field and current. The dotted line shows the $k = 0$ mode of the field; the solid line shows the same mode of current.

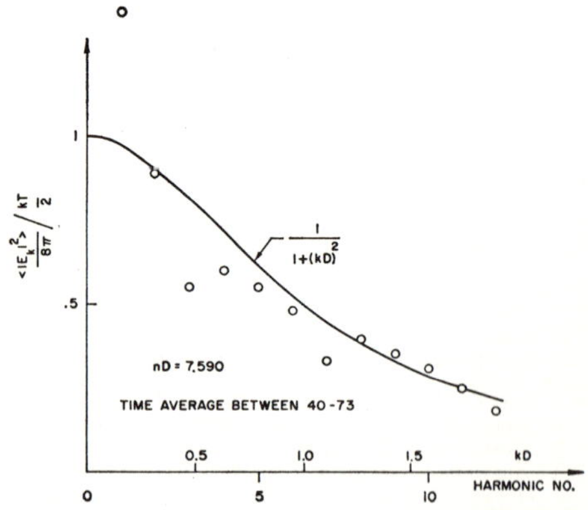

FIG. 7. W_k for $nD = 7.590$. The time average was taken between $\omega_p t = 40$–73.

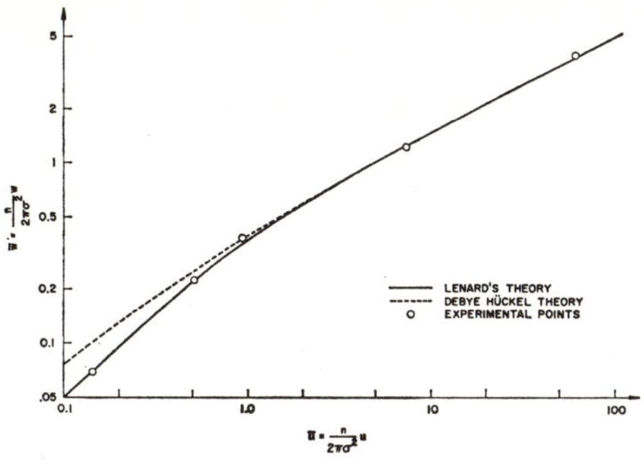

FIG. 8. The field energy per particle as a function of the total energy per particle. The exact theory, the Debye-Hückel theory, and five experimental points are shown.

effects play an important role. To test whether initial conditions are important, we have also varied the time at which the averaging is started. In Fig. 7 $W(k)$ is given for case 1, averaged over the time interval $\omega_p t = 40$–73. The behavior of the long-wavelength components is closer to theory, but large deviations persist.

Exact Theory for W

We have two advantages in discussing the total energy in the field. First, an exact theory has been given by Lenard[5] for any number of particles in the Debye length. Secondly, the fluctuations are smaller, as in all integrated quantities.

Instead of the number of particles in the Debye length, Lenard introduces the parameter $\gamma = P/2\pi\sigma^2$, where P is the pressure, and gives the temperature θ and the internal energy per particle u as functions of γ. These functions involve the characteristic values of the Mathieu equation. Unhappily this equation is not normalized, so we have adopted the normalization of the most complete tables available.[12] We introduce a new parameter s and define the function $be_0(s) = b(s)$.[13] We find then, following Lenard,

$$\bar{u} = (n/2\pi\sigma^2)u = \tfrac{3}{2}(\gamma/\xi) - \gamma, \tag{21}$$

and

$$\bar{\theta} = (n/2\pi\sigma^2)\theta = \gamma/\xi, \tag{22}$$

where

$$\gamma = \frac{s}{8}\,[1 - 2b(s)/s] \tag{23}$$

and

$$\xi = \frac{1 - 2b(s)/s}{1 - 2\,db(s)/ds}. \tag{24}$$

The field energy per particle w is then

$$\bar{w} = (n/2\pi\sigma^2)w = \bar{u} - \tfrac{1}{2}\bar{\theta} = (\gamma/\xi) - \gamma. \tag{25}$$

We can now find $\bar{w}$ as a function of $\bar{u}$. This curve is shown in Fig. 8. On the same figure we have plotted the linearized theory, the Debye–Hückel[14] theory, which may be written in these dimensionless units as

$$\bar{u} = \bar{w} + 4\bar{w}^2, \tag{26}$$

and five "experimental points." The first two points are the ones for which we have shown the Fourier expansion $W(k)$. The numerical data are given in Table I, along with the values calculated from both the linearized and the exact theory.

TABLE I. The comparison with theory of the measured values of the energy in the electric field.

Case	$\bar{u}$	$\bar{w}$ (linearized theory)	$\bar{w}$ (exact theory)	$\bar{w}$ (experimental)
1	61.9	3.81	3.81	3.85
2	7.43	1.24	1.23	1.19
3	0.928	0.372	0.359	0.384
4	0.506	0.252	0.222	0.228
5	0.143	0.102	0.0700	0.0703

We notice that the Debye–Hückel theory agrees with the exact theory to the point $\bar{u} \simeq 1$, where about 40% of the energy is in the field. This is a little surprising for a theory in which the basic assumption is $\bar{w} \ll \bar{u}$. This fact has already been noticed by Balazs.[15] The first three points cannot distinguish between the two theories, but the last two agree very well with Lenard's theory and confirm the equipartition of energy between particles and field when the parameter γ approaches zero.

ACKNOWLEDGMENTS

The authors wish to acknowledge helpful discussions with Marshall Rosenbluth and Andrew Lenard and the assistance of Mary Jane Oring and Carl Bell who performed the computer calculations.

This work was carried out under a joint General Atomic–Texas Atomic Energy Research Foundation program on controlled thermonuclear reactions.

[12] National Bureau of Standards, *Tables Relating to Mathieu Functions* (Columbia University Press, New York, 1951).

[13] The function $be_0(s)$ is defined in reference 12, p. 2.

[14] P. Debye and G. Hückel, Physik. Z. **24**, 305 (1923).

[15] N. Balazs, Phys. Fluids **4**, 1259 (1961).

Proc. Camb. Phil. Soc. (1963), **59**, 779
With 2 text-figures
Printed in Great Britain

Statistical mechanics of a one-dimensional Coulomb system with a uniform charge background

BY R. J. BAXTER

Research School of Physical Sciences, The Australian National University, Canberra

Communicated by K. J. LE COUTEUR

(*Received* 24 *September* 1962)

Abstract. This paper obtains an exact description of the statistical mechanics of a one-dimensional system of charged particles moving in a uniform neutralizing background of positive charge. These results are compared with the behaviour of a one-dimensional system of equal numbers of positive and negative charged particles. Although the thermodynamics of the two systems do differ, the discrepancy is small enough to indicate that the assumption of a uniform background of positive charge, common in statistical mechanical treatments of a plasma in equilibrium, may provide a good approximation to a real system of discrete charges.

1. *Introduction.* The calculation of the classical partition function of a plasma is frustrated from the beginning by the divergence due to the infinite negative energy of an ion pair in contact. Two devices are often used to remove this divergence, one being to replace the Coulomb potential by one which remains finite and differs from it only within a distance of the order of the radius of the smallest Bohr orbit (1, 2), the other being to replace the discrete positive ions by a uniform background of positive charge (3, 4). Although the latter alternative must be further from reality, it has the advantage that the partition function depends in a non-trivial way only on a single variable.

Owing to the inherent analytic and algebraic difficulties, the partition function has not been calculated exactly using either of these assumptions, but each is found to yield the Debye–Huckel theory in the high-temperature limit. It would be of interest to estimate the error introduced by regarding the positive charges as a uniform background by some model in which the partition function could be calculated both with this approximation and without it.

Lenard (5) has investigated the case of a one-dimensional system of equal numbers of positive and negative charges. As the Coulomb potential in one dimension is proportional to the distance between the interacting particles, it has no singularities and Lenard was able to obtain an exact description of the thermodynamics of the system.

The purpose of this paper is to investigate the statistical mechanics of a one-dimensional system of negative charges moving in the field of a neutralizing positive charge background. Although, as Lenard points out, this introduces quadratic terms into the potential energy, the algebraic difficulties may be overcome and an exact description of the thermodynamics obtained. This description has in common with Lenard's the fact that it is based on an eigenvalue relation.

One is then in a position to compare the results of the two theories. It is found that they agree in the high-temperature limit (as in the three-dimensional case) and the difference between them is moderately small for all temperatures. If one assumes that these results are analogous to those for a three-dimensional plasma, it would appear that the assumption of a uniform positive charge background provides a good approximation to the equilibrium behaviour, provided the temperature is high enough for quantum-mechanical effects to be negligible.

2. *Summary of Lenard's results.* Lenard considers a one-dimensional system of $2N$ particles, each with a mass of magnitude unity, such that the ith particle carries a charge σ_i. N of the particles are charged positively and N negatively, and all charges have the same magnitude $|\sigma_i| = \sigma$.

Then if x_i, p_i denote the position and momentum of the ith particle, the interaction potential between the ith and jth particles is

$$-2\pi\sigma_i\sigma_j|x_i - x_j|$$

and it follows immediately that the total Hamiltonian of the system is

$$\tfrac{1}{2}\sum_i p_i^2 - 2\pi \sum_{i<j}\sum \sigma_i \sigma_j |x_i - x_j|,$$

where the summations run from 1 to $2N$.

Lenard uses the constant pressure ensemble and obtains the density ρ and the internal energy per particle U in terms of the pressure P by the relations

$$\frac{P}{\rho kT} = \frac{\gamma}{2}\frac{d}{d\gamma}\log\alpha(\gamma), \tag{1}$$

$$\frac{U}{kT} = \frac{3}{2} - \frac{\gamma}{2}\frac{d}{d\gamma}\log\alpha(\gamma), \tag{2}$$

where k is Boltzmann's constant, T the temperature,

$$\gamma = P/2\pi\sigma^2 \tag{3}$$

and $\alpha(\gamma)$ is given by the condition that the equation

$$\left\{\frac{d^2}{dt^2} + 2\alpha^{\frac{1}{2}}\cos t - \gamma\right\} y(t) = 0$$

has the even Mathieu functions of zero order as a solution.

In order to facilitate comparison of these results with those of this paper, which uses the canonical ensemble of Gibbs, the parameter λ is defined by

$$\lambda = (2\sigma^2/\rho kT)^{\frac{1}{2}}. \tag{4}$$

Then from (3)

$$P/\rho kT = \pi\gamma\lambda^2 \tag{5}$$

and (1) becomes

$$\lambda^2 = \frac{1}{2\pi}\frac{d}{d\gamma}\log\alpha(\gamma). \tag{6}$$

In the high-temperature limit ($\lambda \ll 1, \gamma \gg 1$)

$$\alpha(\gamma) = \tfrac{1}{4}\gamma^2 + (\tfrac{1}{2}\gamma)^{\frac{2}{3}} + O(\gamma). \tag{7}$$

Substituting into (6) and using (5), one obtains

$$P/\rho kT = 1 - (\tfrac{1}{8}\pi)^{\frac{1}{2}}\lambda + O(\lambda^2), \tag{8}$$

which shows that at high temperatures the system behaves as a perfect gas.

In the low-temperature limit ($\lambda \gg 1, \gamma \ll 1$)

$$\alpha(\gamma) = \tfrac{1}{2}\gamma + \tfrac{7}{16}\gamma^2 + O(\gamma^3) \tag{9}$$

and proceeding as before, one obtains

$$P/\rho kT = \frac{1}{2} + \frac{7}{32\pi\lambda^2} + O\left(\frac{1}{\lambda^4}\right). \tag{10}$$

Equation (10) shows that at low temperatures the system behaves as a perfect gas of N (and not $2N$) particles. Lenard is able to explain this by supposing that in this limit all the ions are tightly bound in pairs, each pair having zero total charge. The system may then be regarded as one of N neutral molecules.

It will be seen later that the behaviour of the system in either the high-temperature or the low-temperature limit is not affected by replacing the discrete positive ions by a uniform charge background.

3. *Definition of the model.* Consider a system of N particles, each of mass m and carrying a charge $-\sigma$, freely moveable along a line of length $2L$, together with a uniform distribution of positive charge of magnitude $(N-1)\sigma/2L$ along this line. Alternatively, this may be regarded as a system of parallel charged condenser plates, movable along their common perpendicular and able to cross one another, together with a uniform background of charge of opposite sign.

Note that the system is not exactly neutral, but has a total charge $-\sigma$. This ensures that each charge moves in an external field of total charge zero. As the only quantities of interest relate to the limit when N, L are infinitely large but the density

$$\rho = (N-1)/2L$$

is held fixed, this non-neutrality should be of no consequence.

Taking the line as x-axis, its mid-point as origin, then if the ith particle has momentum p_i and position x_i, the total potential energy of the system is $2\pi\sigma^2\Phi$, where

$$\Phi = -\sum_{i<j}|x_i - x_j| + \rho\sum_i x_i^2 + \text{terms independent of the } x_i. \tag{11}$$

(Unless otherwise indicated, all summations in this paper run from 1 to N.)

It is a feature of the statistical mechanics of a system with two-body interactions that an arbitrary constant may not in general be added to the interaction potential, as this produces a factor of type $\exp(-N^2)$ in the partition function, causing the intensive thermodynamic variables to cease to have a well-defined limit for large N. Thus care must always be taken to choose the constant terms in the potential so that physically meaningful results are obtained. In this case it transpires that the third term on the right side of (1) must be

$$(N-1)\,N(N+1)/12\rho.$$

The total Hamiltonian of the system is

$$H = \frac{1}{2m}\sum_i p_i^2 + 2\pi\sigma^2\Phi \tag{12}$$

and the canonical partition function is

$$Z_N = \frac{1}{h^N}\int_{-\infty}^{\infty}\cdots\int dp_1\ldots dp_N\int_{-L}^{L}\cdots\int dx_1\ldots dx_N\exp(-H/kT). \tag{13}$$

The pressure P and the internal energy per particle U are obtained from Z_N by the relations

$$P = -\frac{kT\rho^2}{N}\frac{\partial}{\partial\rho}\log Z_N, \tag{14}$$

$$U = \frac{kT^2}{N}\frac{\partial}{\partial T}\log Z_N. \tag{15}$$

To obtain Z_N from equation (13), the momentum integrations are first performed, and as Φ is a symmetric function of the variables $x_1, \ldots, x_N$, the integral over N-dimensional x-space may be replaced by $N!$ times the integral over that portion of space in which $x_1 > x_2 > \ldots > x_N$. Hence one obtains that

$$Z_N = (2\pi mkT/h^2)^{\frac{1}{2}N}\,N!\,S_N/\rho^N, \tag{16}$$

where
$$S_N = \rho^N\int_{-L}^{L}dx_1\int_{-L}^{x_1}dx_2\ldots\int_{-L}^{x_{N-1}}dx_N\exp(-2\pi\sigma^2\Phi/kT). \tag{17}$$

In this region of space,

$$\Phi = \sum_i[(2i-N-1)\,x_i + \rho x_i^2] + (N-1)\,N(N+1)/12\rho, \tag{18}$$

so that, changing to normal coordinates

$$v_i = \frac{2i-N-1}{2} + \rho x_i,$$

$$\Phi = \sum_i v_i^2/\rho.$$

From the definition of λ given in (4), it follows that

$$S_N = \int_{-(N-1)}^{0}dv_1\int_{-(N-2)}^{v_1+1}dv_2\ldots\int_{0}^{v_{N-1}+1}dv_N\exp(-\pi\lambda^2\sum_i v_i^2). \tag{19}$$

In order that the intensive thermodynamic variables should be uniquely defined for large N, one expects the limit as N tends to infinity of S_{N-1}/S_N to exist and be strictly positive.

If
$$\lim S_{N-1}/S_N = z_0 \tag{20}$$

it is apparent that z_0 is a function only of λ. Substitution into equations (14), (15), (16) yields

$$\frac{P}{\rho kT} = 1-\frac{\lambda}{2}\frac{d}{d\lambda}\log z_0, \tag{21}$$

$$\frac{U}{kT} = \frac{1}{2}+\frac{\lambda}{2}\frac{d}{d\lambda}\log z_0. \tag{22}$$

4. *An eigenvalue relation for* z_0. It is convenient to define the functions $S_N(v)$ of v by

$$S_0(v) = 1,$$

$$S_N(v) = \int_{-(N-1)}^{v} dt \exp(-\pi\lambda^2 t^2)\, S_{N-1}(t+1). \tag{23}$$

Then $S_N(0)$ is equal to S_N as defined by (17) and hereafter the single expression S_N without an argument will be taken to mean $S_N(0)$.

Clearly

$$\frac{d}{dv}S_N(v) = \exp(-\pi\lambda^2 v^2)\, S_{N-1}(v+1), \tag{24}$$

so that if $y(v)$ is defined by

$$y(v) = \sum_{N=0}^{\infty} z^N S_N(v), \tag{25}$$

then

$$\frac{d}{dv}y(v) = z\exp(-\pi\lambda^2 v^2)\, y(v+1). \tag{26}$$

Equation (26) is a difference-differential equation whose solution tends to a finite limit as v tends to either plus or minus infinity. If the limit at plus infinity is specified, then the solution is uniquely determined. In particular, it is convenient to define the solution of (26) which tends to one as v tends to plus infinity to be $G(v)$. The general solution is then $G(v)$ multiplied by some constant factor.

As v tends to minus infinity, each $S_N(v)$ tends to a finite limit which satisfies the inequality

$$\left|\lim_{v\to-\infty} S_N(v)\right| < \frac{1}{N!}(2\lambda)^{-N}, \tag{27}$$

so that as $y(v)$ is known to tend to a limit as v tends to minus infinity, this limit is given by (25), (27) to be the sum of an absolutely convergent series and therefore finite. Hence $y(v)$ is given by the equation

$$y(v) = G(v)\, y(-\infty)/G(-\infty). \tag{28}$$

From (20), the series (25) for $y(0)$ has radius of convergence z_0. That is, if z lies between 0 and z_0, $y(0)$ is finite, but $y(0)$ becomes infinite when z equals z_0. As $G(0)$ is always finite, it follows from (28) that $G(-\infty)$ is non-zero for z between 0 and z_0, but equals zero when z equals z_0.

In Fig. 1 $G(v)$ is plotted for $\lambda = 0{\cdot}25$ and various values of z. One sees that $G(-\infty)$ does exhibit the required behaviour and the least value of z for which $G(-\infty)$ is zero corresponds to a function $G(v)$ which is positive for all finite v.

Thus the function $z_0(\lambda)$ required for substitution in equations (21), (22) is the eigenvalue $z = z_0$ of the equation

$$\frac{d}{dv}G(v) = z\exp(-\pi\lambda^2 v^2)\, G(v+1), \tag{29}$$

such that $G(v)$ satisfies the boundary conditions

$$\lim_{v\to\infty} G(v) = 1, \tag{30}$$

$$\lim_{v\to-\infty} G(v) = 0, \tag{31}$$

and $G(v)$ is positive for all v.

5. *The limit* $\lambda \to 0$. Putting

$$x = \lambda v,$$

$$G = \exp\left(-\frac{1}{\lambda}\int_x^\infty u(t)\,dt\right). \tag{32}$$

Equation (29) becomes

$$u(x)\exp(-u(x)) = z\exp(-\pi x^2)\exp I, \tag{33}$$

where

$$I = \frac{1}{\lambda}\int_x^{x+\lambda}(u(t)-u(x))\,dt. \tag{34}$$

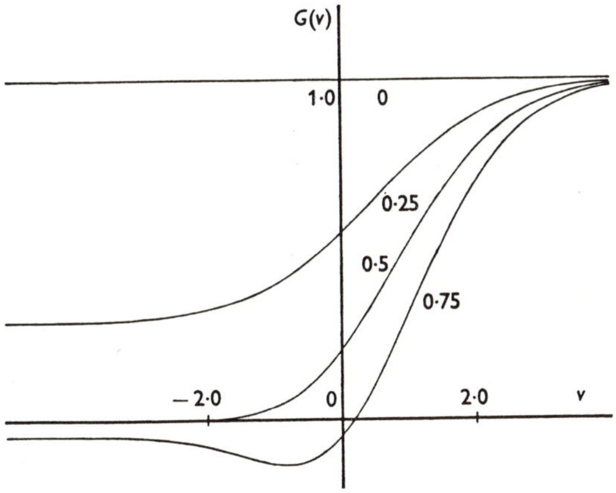

Fig. 1. The function $G(v)$ for $\lambda = 0{\cdot}25$ and $z = 0, 0{\cdot}25, 0{\cdot}5, 0{\cdot}75$.

If $u(x)$ is a differentiable function, I is of order λ. One would therefore expect that in the limit of λ small, I may be put equal to zero. Thus if z is held fixed and λ allowed to tend to zero,

$$\lim_{\lambda\to 0} u(x) = u_0(x),$$

where $u_0(x)$ is given by

$$u_0(x)\exp(-u_0(x)) = z\exp(-\pi x^2), \tag{35}$$

provided that

$$\lim_{\lambda\to 0}\frac{1}{\lambda}\int_x^{x+\lambda}(u_0(t)-u_0(x))\,dt = 0. \tag{36}$$

When $z\exp(-\pi x^2) > e^{-1}$, (35) gives only complex solutions for u_0. As (29) is linear, $G(x)$ may then be understood to be the real part of the right side of (32). As $u_0(x)$ is a continuous even function of x which tends to zero as x tends to infinity, it can only assume those values which lie on the contour shown in Fig. 2 (or on the contour obtained by inverting the imaginary axis). With $u_0(x)$ so defined, (36) is satisfied for all points x, including those corresponding to $u_0 = 1$, where $u_0(x)$ is not differentiable.

If $z < e^{-1}$, $u_0(x)$ is real and has derivatives of all orders at every point, so that $u(x)$ may be expanded (using (33)) as a power series in λ with coefficients which are real non-singular functions of x. Then (32) shows that $G(x)$ is always positive.

If $z > e^{-1}$, $u_0(x)$ is complex over a range independent of λ and hence when λ is small $G(x)$ is in this region a rapidly oscillating function with a large number of zeros.

It follows that the eigenvalue z_0 has the limiting behaviour

$$\lim_{\lambda \to 0} z_0 = e^{-1}.$$

When $\lim_{\lambda \to 0} z = e^{-1}$, (35) gives

$$u_0(x) = 1 - (2\pi)^{\frac{1}{2}} |x|, \tag{37}$$

when

$$|x| \ll 1.$$

For values of x removed from zero $u(x)$ may be developed as a power series in λ with real coefficients. It follows that for such values of x, G can have no zeros. However, as $u_0(x)$ is not differentiable at the origin, it is necessary to consider in more

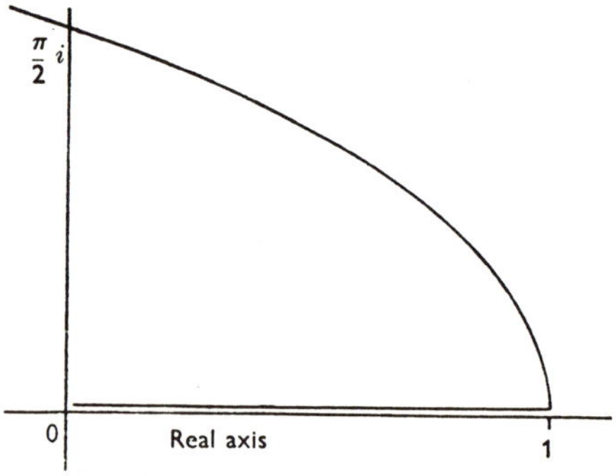

Fig. 2. The contour of possible complex values of $u_0(x)$.

detail the behaviour of G when x^2 is of the same order as I, that is when x is of the order of $\lambda^{\frac{1}{2}}$. To do this it is convenient to apply another change of variable, namely

$$s = (2\pi)^{\frac{1}{2}} \lambda^{-\frac{1}{2}} x \tag{38}$$

and to put

$$G = w(s) \exp(x/\lambda). \tag{39}$$

In order to investigate the limiting behaviour of z_0 in more detail one may put

$$z = e^{-1}(1 + (\tfrac{1}{2}\pi)^{\frac{1}{2}} c\lambda), \tag{40}$$

and allow λ to tend to zero, c being kept constant.

With these substitutions, (29) becomes

$$w(s) + (2\pi)^{\frac{1}{2}} \lambda^{\frac{1}{2}} \frac{d}{ds} w(s) = (1 + (\tfrac{1}{2}\pi)^{\frac{1}{2}} c\lambda) \exp(-(\tfrac{1}{2}\pi)^{\frac{1}{2}} \lambda s^2)\, w(s + (2\pi)^{\frac{1}{2}} \lambda^{\frac{1}{2}}). \tag{41}$$

Expanding the right side as a power series in $\lambda^{\frac{1}{2}}$ and neglecting terms of order $\lambda^{\frac{3}{2}}$ or less, (41) becomes

$$\left(\frac{d^2}{ds^2} + c - s^2\right) w(s) = 0. \tag{42}$$

When s is large it may be expected that the value of $u(x)$ given by (32), (38), (39) and (42) will be equal to the value of $u_0(x)$ given by (37). This implies that

$$w(s) \sim \exp(-\tfrac{1}{2}s^2), \quad s \text{ large and positive,} \tag{43}$$

$$w(s) \sim \exp(\tfrac{1}{2}s^2), \quad s \text{ large and negative.} \tag{44}$$

The solutions of (42), subject to the boundary condition (43), are the well-known parabolic cylinder functions. The boundary condition (44) is always satisfied except when c is a positive odd integer. Milne (6) has investigated the zeros of these functions and finds that if n is the smallest non-negative integer such that

$$c < 2n+1$$

then $w(s)$ has n zeros. Hence as G must have a continuous dependence on c, the eigenvalue z_0 is given by $c = 1$, so that

$$z_0 = e^{-1}(1+(\tfrac{1}{2}\pi)^{\frac{1}{2}}\lambda + O(\lambda^2)). \tag{45}$$

Substituting this expression in (21),

$$P/\rho kT = 1-(\tfrac{1}{8}\pi)^{\frac{1}{2}}\lambda + O(\lambda^2). \tag{46}$$

One sees that in the high-temperature limit (λ small), the system behaves as a perfect gas and (from (8)) has to first order in λ the same behaviour as a system of discrete positive and negative charges.

6. *The limit* $\lambda \to \infty$. In this case the limiting behaviour is readily investigated by solving by recursion the integral equation derived from (29) and the first boundary condition

$$G(v) = 1 - z\int_v^\infty \exp(-\pi\lambda^2 t^2)\, G(t+1)\, dt \tag{47}$$

and then putting $G(-\infty) = 0$.

Calculating $G(v)$ by this method to second order in z, and putting minus infinity for v, one obtains

$$G(-\infty) = 1 - \frac{z}{\lambda} + \frac{z^2}{\lambda}\int_{1/\sqrt{2}}^\infty dv \exp(-\pi\lambda^2 v^2) = 0. \tag{48}$$

Neglecting terms of type $\exp(-\pi\lambda^2)$ or less, equations (48) give

$$z_0 = \lambda + \lambda^2 \int_{1/\sqrt{2}}^\infty dv \exp(-\pi\lambda^2 v^2), \tag{49}$$

so that (21) yields

$$P/\rho kT = \frac{1}{2} + \frac{\lambda}{2\sqrt{2}} \exp(-\tfrac{1}{2}\pi\lambda^2). \tag{50}$$

Equations (10) and (50) show that in the low-temperature limit a system of negative charges moving in a uniform background of positive charge behaves in the same way as a system of discrete positive and negative charges. This behaviour is the same as that of a system of N simple harmonic oscillators, each with the 'plasma frequency' ω_p given by

$$\omega_p^2 = 4\pi\sigma^2\rho/m. \tag{51}$$

7. *Comparison of the two theories.* The function $G(v)$ defined by equations (29), (30) has been obtained for various values of λ by numerical integration using an IBM 1620 digital computer, the value of z being chosen so that $G(v)$ is positive and tends to zero as v tends to minus infinity. The eigenvalue z_0 obtained in this manner is tabulated as a function of λ in Table 1. These values were then used to obtain $P/\rho kT$ as a function of λ from (23).

Lenard's function $\alpha(\gamma)$ was obtained from tables of eigenvalues of the Mathieu equation (7), and these values were then used to obtain $P/\rho kT$ as a function of λ from equations (1), (3) and (4).

The functions $P/\rho kT$ of λ obtained in these two ways are tabulated in Table 1. It is apparent that the error introduced into $P/\rho kT$ by regarding the positive charges as a uniform background is not greater than 7 % for any value of λ. It would therefore appear that this approximation does not radically affect the thermodynamics of the system and solutions obtained by it for three-dimensional plasmas may well provide a good representation of a plasma in equilibrium, provided that the temperature of the system is high enough for quantum-mechanical effects to be ignorable.

Table 1. *The function z_0 of λ and the equation of state of a one-dimensional plasma*

		$P/\rho kT$	
λ	z_0	Discrete charges	Uniform background
0·0	0·36788	1·00000	1·00000
0·1	0·41609	0·93735	0·93951
0·2	0·46859	0·87481	0·88334
0·3	0·52544	0·81271	0·83144
0·4	0·58668	0·75311	0·78377
0·5	0·65229	0·70023	0·74026
0·6	0·72219	0·65713	0·70087
0·7	0·79626	0·62395	0·66551
0·8	0·87431	0·59903	0·63413
0·9	0·95611	0·58034	0·60664
1·0	1·04136	0·56620	0·58295
1·1	1·12970	0·55535	0·56297
1·2	1·22075	0·54688	0·54652
1·3	1·31409	0·54018	0·53339
1·4	1·40928	0·53480	0·52323
1·5	1·50593	0·53041	0·51566

The values of $P/\rho kT$ given in columns 3 and 4 were obtained from the results of Lenard and of this paper respectively.

REFERENCES

(1) Green, H. S. *J. Plasma Physics and Thermonuclear Fusion*, **1** (1961), 69.
(2) Morita, T. *Progr. Theoret. Phys.* **22** (1959), 757.
(3) Salpeter, E. E. *Ann. Physics*, **5** (1958), 183.
(4) Abe, R. *Progr. Theoret. Phys.* **22** (1959), 213.
(5) Lenard, A. *J. Mathematical Phys.* **2** (1961), 682.
(6) Milne, A. *Proc. Edinburgh Math. Soc.* **33** (1915), 48.
(7) National Bureau of Standards. *Tables relating to Mathieu functions* (Columbia: New York, 1951).

The Disordered Chain of Harmonic Oscillators

Chapter 2

The Disordered Chain of Harmonic Oscillators

Introduction

In the past decade a great deal of work—both analytic and computational—has been devoted to the problem of a disordered lattice of harmonic oscillators. The greater portion of this work, by far, has been on the one-dimensional chain, because until this relatively simple case is thoroughly understood the much more complicated two- and three-dimensional calculations will be open to question.

The relevance of this problem to physics stems primarily from two sources: one is a desire to understand substances, such as glass, having an atomic arrangement with a large degree of disorder, bordering on randomness. The second concerns an attempt to understand the effect of a relatively small number of defects (impurities, vacancies, etc.) on the vibrational spectrum of an otherwise regular lattice. In this introductory chapter we shall be primarily concerned with the former problem, for that is the one requiring the more novel mathematical approach and is also the problem more nearly akin to the general many-body problem discussed in the remainder of this book.

The standard lattice problem may be formulated as follows: Consider a chain of N atoms (where N is very large) interacting through harmonic, nearest-neighbor forces. Let the mass of the jth atom be m_j, and the spring constant coupling the jth atom to the $(j + 1)$th atom to be K_j. These quantities are to be regarded as independent random variable described by probability density function $f(m)$ and $h(K)$. Any given lattice, with a definite assignment of masses and spring constants, will have (in the limit of large N) a normal mode spectrum described by an integrated distribution $N(\omega^2) = N^{-1} \times$ (the number of modes with frequency less than ω). One can also define the density function $D(\omega^2) = (\partial/\partial\omega^2)N(\omega^2)$. The question to be answered is this: What is the average, over all chains, of the function $N(\omega^2)$, in the limit of $N \to \infty$?

The problems in the above category that have received the most attention are: (a) all the spring constants have a common value, K, while the masses alone are random; (b) the masses have the same value, m, and the spring constants are variable. The mathematical problem is quite similar in the two cases, and we shall therefore concentrate on (a). Furthermore, the particular example of (a) that has received most attention is where only two masses, m and M (with $m < M$), occur with probabilities p and $1 - p$, respectively.

While the above problems are difficult enough, it should be noted that many realistic models are still more complicated. On the one hand, more-than-nearest neighbor interactions are usually involved, and on the other hand, the probability of finding a mass m at the jth site is generally not uncorrelated with (or independent of) the probability of finding a mass m' at a neighboring site. In addition, in higher dimensions than one, another kind of disorder appears, namely, the topological configuration of the atoms may change. Loosely speaking, vacancies and interstitials can occur and produce important effects on the spectrum of glass. It is difficult, if not impossible, to produce a similar effect in one dimension, for removing or inserting an atom in a chain is merely equivalent to changing a mass or a spring constant.

In the following we shall concentrate primarily on exact analytic results, for that is the main theme of this book. For a broader overall picture of the field, especially with regard to what is known for the three-dimensional problem, we refer the reader to the earlier review articles by Dean (1963) and by Maradudin and Weiss (1958). Since the theme of this book is the many-body problem, we also omit the problems of one or two impurities in an otherwise pure chain which are treated in these reviews and which we also include in our bibliography. Our typical chain has n_j atoms of each species j, and we shall be interested in the rather more difficult situation of n_j, and $N \equiv \Sigma\, n_j$, all tending to infinity with fixed ratios $p_j \equiv n_j/N$.

Dyson's Method

The well-known eigenvalue equation for ω^2 is

$$-M_j\omega^2 x_j = K_j(x_{j+1} - x_j) + K_{j-1}(x_{j-1} - x_j) \tag{2.1}$$

where $x_0 = x_{N+1} = 0$ if we wish to have fixed end boundary conditions. The N values of ω^2 can be obtained by finding the determinant of an $N \times N$ tridiagonal matrix, but a direct assault on the problem by this route is unprofitable. Dyson

(1953), who was the first to bring the problem into a tractable form, observed that it was easier to try to calculate a one-parameter *additive* (or linear function) of the eigenvalues, namely,

$$\Omega(x) = \lim_{N\to\infty} (2N-1)^{-1} \sum_{j=1}^{N} \ln(1 + x\omega_j^2) \tag{2.2}$$

$$= \int_0^\infty \ln(1 + x\mu)\, D(\mu)\, d\mu \tag{2.3}$$

It is of the greatest importance to note that $\Omega(x)$ is a linear sum of functions of ω^2. The exact function is unimportant, but linearity is important, for the average of Ω over all chains is given by Eq. (2.3) wherein the ensemble average value of D is to be used. Occasionally, it is claimed that the average value of a nonadditive function, such as $\exp[\Omega(x)]$, is equal to the same function of the average value, $\exp[\overline{\Omega}(x)]$. Such claims are without foundation, but there is a temptation to believe them since it is often relatively easy to compute the averages of such nonlinear functions. [Faulkner and Korringa (1961) and Mahanty (1961), for example, have exploited the average of $\exp[\Omega(x)]$, but numerical results based on this formalism by Flinn, Maradudin, and Weiss (1961) disagree both with very accurate machine computations and with more rigorous theories.]

Returning to (2.3), it is an elementary exercise in complex variables to deduce that

$$D(1/z) = -z^2 \operatorname{Re}[(i\pi)^{-1} \lim_{\epsilon\to 0} \Omega'(-z + i\epsilon)] \tag{2.4}$$

where $\Omega'(x) = d\Omega(x)/dx$. In other words, $\Omega'(x)$ is essentially the Stieltjes transform of $D(\mu)$ which means that we must find the jump across the cut on the negative real axis of the *analytic continuation* of $\Omega'(x)$. It is this point that makes the Dyson method awkward for numerical computation. For a uniform chain with all $m_j = m$ and all $k_j = k$ one finds

$$\Omega(x) = 2 \ln\{\tfrac{1}{2}[(1 + 4Kxm^{-1})^{1/2} + 1]\} \qquad \text{(uniform chain)} \tag{2.5}$$

To find $\Omega(x)$, Dyson proves that

$$\Omega(x) = \lim_{N\to\infty} N^{-1} \sum_{p=1}^{2N-1} \ln(1 + \xi_j) \tag{2.6}$$

where the ξ_j satisfy the recursion relation

$$\xi_j = x\lambda_j/(1 + \xi_{j+1}) \tag{2.7}$$

and

$$\lambda_{2j-1} = K_j/m_j; \qquad \lambda_{2j} = K_j/m_{j+1} \tag{2.8}$$

Dyson's derivation of (2.6) to (2.8) is complicated, but simpler versions are to be found in Bellman (1956), des Cloizeaux (1957), and Englman (1958).

Now we come to the main point. From (2.7) we observe that the value of ξ at any point in the chain is determined entirely by the properties of the chain to the right of the point in question. We suppose (unproved) that if we are far from the end of the chain, the distribution of ξ's that will occur along the chain is independent of the details at the beginning of the chain. Let us call this density function $F(\xi)$. Dyson now distinguishes two cases.

Type I. Each of the $2N$ parameters λ_j, defined by (2.8), is an independent random variable with the probability density function $G(\lambda)$. We then obtain the following integral equation for F:

$$F(\xi) = \int_0^\infty F(\xi')\, G\left(\frac{\xi(1+\xi')}{x}\right)\frac{1+\xi'}{x}\, d\xi' \tag{2.9}$$

together with the normalization condition

$$\int_0^\infty F(\xi)\, d\xi = 1 \tag{2.10}$$

From (2.6)

$$\Omega(x) = 2\int_0^\infty F(\xi)\ln(1+\xi)\, d\xi \tag{2.11}$$

Type II. Each mass m_j is an independent random variable with probability density $G(m)$, the spring constants being fixed and equal. In this case we must deal with the variable $\eta_j = 1/\xi_{2j}$, whence

$$F(\eta) = xK\int_0^\infty G\{xK[\eta - \eta'(1+\eta')^{-1}]\}\, F(\eta')\, d\eta' \tag{2.12}$$

and

$$\int_0^\infty F(\eta)\, d\eta = 1 \tag{2.13}$$

Here,

$$\Omega(x) = \int_0^\infty F(\eta)\, d\eta \int_0^\infty G(m)\, dm \ln[1 + \eta^{-1} + xKm^{-1}] \tag{2.14}$$

It is to be noted that in either case, F depends parametrically on x, so that the x dependence of Ω as given by (2.11) or (2.14) is quite complicated. When it is recalled that an analytic continuation of Ω, and hence of F, is required, it will be clear that the solution of (2.9) or (2.12) is only part of the problem.

The Type I problem is really quite artificial because it requires that the N masses and N spring constants be distributed in a complicated manner. Nevertheless, *it is only for this problem that a complete analytic solution has ever been obtained.* Dyson is able to solve the problem completely for the class of density functions, the generalized Poisson distributions,

$$G_n(\lambda) = \frac{n^n}{(n-1)!}\lambda^{n-1}\exp(-n\lambda) \tag{2.15}$$

where n is any positive integer, and he finds that the distribution function $N(\omega^2)$ is a very smooth, smeared out function with none of the characteristic kinks associated with the diatomic chain.

For a Type II problem, let us assume that n different masses $m_1, \ldots, m_n$ can occur, with probabilities p_i, respectively. The integral equation (2.12) becomes

$$F(\eta) = \sum_{i=1}^{n} p_i[1 - \eta + (m_i/xK)]^{-2}F\{[1 - \eta + (m_i/xK)]^{-1} - 1\} \qquad (2.16)$$

On the right side of (2.16), F is to be regarded as zero if its argument is negative. Equation (2.16) looks simple, especially for the two-mass case, *but it is probably one of the most complicated equations of a function of a single variable in mathematical physics.*

The two-mass version of (2.16) can in fact be solved for sufficiently small x (corresponding to high frequencies) but this solution is useless because to obtain the analytic continuation of Ω onto any one point of the negative x-axis we need the solution for *all* positive x. Nevertheless, the solution is amusing and does not appear in any of the published literature; therefore we outline it below.

Let the two masses be m and M, with probability p and $1 - p$, respectively, and with $m < M$. Let $a = m/xK$ and $A = M/xK$, whence $a < A$. It is convenient to define the integral of $F(\eta)$, denoted $H(\eta)$,

$$H(\eta) = \int_0^\eta F(\eta')\, d\eta' \qquad (2.17)$$

whence

$$H(\eta) = pH\left(\frac{\eta - a}{1 - \eta + a}\right) + (1 - p)H\left(\frac{\eta - A}{1 - \eta + A}\right) \qquad (2.18)$$

where H is monotonically increasing with $H(0) = 0$; $H(\infty) = 1$, and where $H[(\eta - a)/(1 - \eta + a)]$ vanishes for $\eta < a$ and is 1 for $\eta \geq 1 + a$. The right side of (2.18) is zero for $\eta < a$ and consequently $H(\eta) = 0$ for $\eta < a$. But this means the right side, as well as $H(\eta)$, is zero for $\eta < a + a/1 + a$. Continuing in this way we see that $H(\eta) = 0$ for $\eta < \eta_1$, where

$$(\eta_1 - a)/(1 - \eta_1 + a) = \eta_1 \qquad \text{or} \qquad \eta_1 = \tfrac{1}{2}[a + (a^2 + 4a)^{1/2}]$$

Conversely, $H(\eta) = 1$ for $\eta \geq 1 + A$ and hence, by repeating the above argument, $H(\eta) = 1$ for $\eta > \gamma_1$, where $\gamma_1 = \frac{1}{2}[A + (A^2 + 4A)^{1/2}]$. From these results we see that the first term on the right side of (2.18) is 1 for $\eta > \eta_2 = a + \gamma_1/(1 + \gamma_1)$, while the second term is zero for $\eta < \gamma_2 = A + \eta_1/(1 + \eta_1)$. Therefore, if x is such that $\gamma_1 - \eta_1 < 2(A - a)$, then $\gamma_2 > \eta_2$ and we have the result

$$H(\eta) = p \qquad \text{for} \qquad \eta_2 < \eta < \gamma_2 \qquad (2.19)$$

Thus we know H everywhere except for the regions $\eta_1 < \eta < \eta_2$ and $\gamma_2 < \eta < \gamma_1$. These regions may be "filled in" by using (2.19) again and again. Using (2.19) on the right side of (2.18) we see that

$$H(\eta) = p^2 \quad \text{for} \quad a + \eta_2/(1 + \eta_2) < \eta < a + \gamma_2/(1 + \gamma_2)$$

and

$$H(\eta) = p(1 - p) \quad \text{for} \quad A + \eta_2/(1 + \eta_2) < \eta < A + \gamma_2/(1 + \gamma_2)$$

And so on, and so on, ... A plot of $H(\eta)$ thus consists of an infinite number of straight lines between any two of which there are again an infinite number of straight lines. There is, however, no discontinuity in H, for every gap eventually fills in. It is hard to imagine a more pathologic function! Since F is the derivative of H, it consists of an infinite number of delta functions with points of accumulation of delta functions. In turn there are an infinite number of points of accumulation having an infinite number of accumulations—and so on *ad infinitum.* The characteristic function $\Omega(x)$ could therefore be approximated to any desired accuracy by "playing the above game" long enough, but the analytic continuation of Ω is another matter.

Other Rigorous Solutions

We turn now to a simpler method due to Schmidt (1957) (see also Hori and Asahi, 1957). From (2.1) we see that if we know x_1, and if $x_0 = 0$, we can find all the succeeding x's recursively. Defining

$$z_{n+1} = x_{n-1}/x_n \tag{2.20}$$

we have, from (2.1),

$$z_{n+1} = K_n(K_n + K_{n-1} - M_n\omega^2 - K_{n-1}z_n)^{-1} \tag{2.21}$$

with the boundary conditions

$$z_1 = 0; \qquad z_{N+1} = \infty \tag{2.22}$$

Equation (2.22) is the condition for the allowed frequencies. Schmidt now makes use of an old theorem that is true in one dimension for the solutions to a single particle Schrödinger equation, as well as for solutions to (2.1). Namely, *the eigenvector corresponding to the* n*th eigenvalue (measured in ascending order) has precisely* $n - 1$ *nodes.* A node is a change of sign between some x_j and its neighbor x_{j+1}, with the convention that zero is to be considered a negative number. Alternately, the set $z_1, \ldots, z_N$ contains exactly n negative or zero members (with infinity being regarded as positive) for the nth frequency. There are various proofs of this theorem; one of them is given by Schmidt (1957). If we imagine (2.21) to be iterated N times, starting with $z_1 = 0$, then z_{N+1} is a function of ω^2. As ω^2 increases, z_{N+1} will pass through zero each time an eigenvalue occurs, and for ω between the nth and $(n + 1)$th eigenvalues there will be n negative or zero z's in the chain. It is as though a node comes in through z_{N+1} and then remains in the chain. Therefore, for *any* value of ω the number of negative or zero z's in the set is equal to $N(\omega^2)$. If we assume that several different masses can occur, as in the type II problem, and if we assume as before that the distribution of z's eventually settles down to a distribution $w(z)$ independent of the details at the beginning of the chain, we have Schmidt's equation:

$$w(z) = z^{-2} \sum_{j=1}^{n} p_j w(2 - K^{-1} m_j \omega^2 - z^{-1}) \tag{2.23}$$

$$\int_{-\infty}^{\infty} w(z)\, dz = 1, \qquad w(z) > 0$$

with

$$N(\omega^2) = \int_{-\infty}^{0} w(z)\, dz \tag{2.24}$$

Schmidt actually proves that the integrated function, $W(z) = \int_0^z w(z')\, dz'$, does finally become independent both of N and of the details at the beginning of the chain, and that the solution to the analog of (2.23) for W is unique *provided* that no p_j is identically unity. The advantage of Schmidt's equation over Dyson's is obvious: the function $w(z)$ depends directly on ω^2 and not on some intermediate variable x. Also, no analytic continuation is required, the answer being given by (2.24) directly in terms of w. There is, however, a formal connection between Dyson's equation (2.16) and Eq. (2.23). If we set $F(z^{-1} - 1) \equiv z^2 R(z)$, then substitution into (2.16) yields Eq. (2.23) with $R(z)$ replacing $w(z)$ and with ω^2 replaced by $-\omega^2$. The function $w(z)$ thus seems to be the analytic continuation into the negative ω^2 plane of Dyson's F, but the relationship between the two functions, which have different meanings, has not yet been investigated in the literature.

While the solution to Schmidt's equation for the two-mass case might appear as mysterious as the solution to Dyson's equation, Agacy (1964) has now obtained it numerically; see his reprinted paper for the results.

A slightly different approach to the problem is provided by the 2×2 transfer matrix method which has been discussed in great detail by Hori and Asahi (1957), Matsuda (1962), and in the succeeding papers of these authors. There are many possibilities, and one is to introduce the N variables $y_j = K_j(x_{j+1} - x_j)$ in addition to the N x's. From these we form a set of N (two-dimensional) vectors $V^j = (x_j, y_j)$ which satisfy the equations

$$V^j = T^j V^{j-1} \tag{2.25}$$

where

$$T^j = \begin{pmatrix} 1 & (K_{j-1})^{-1} \\ -m_j\omega^2 & 1 - m_j\omega^2 (K_{j-1})^{-1} \end{pmatrix} \tag{2.26}$$

Thus

$$V^N = H V^0 \tag{2.27}$$

with

$$H = T^N T^{N-1} \cdots T^1 \tag{2.28}$$

Here, it is convenient to consider cyclic chains, whence $x_0 = x_N$, $x_{N+1} = x_1$, and $K_0 = K_N$ in (2.1). This implies that $V^0 = V^N$ and, therefore, H is required to have an eigenvalue $+ 1$. Since H is unimodular (det $= 1$), an eigenvalue $= 1$ is equivalent to

$$\text{trace } H = 2 \tag{2.29}$$

as the eigenvalue equation. [In the present language, the Faulkner-Korringa method mentioned before is equivalent to taking trace $\bar{H} = 2$ as the "average eigenvalue equation," where $\bar{H}$ is the average of H over all chains of the ensemble.]

Special Theorems

The reader is referred to the paper by Hori and Matsuda (1964) and succeeding papers for a penetrating discussion of the properties of these T matrices in the complex plane. With their use, Hori (1964a) is able to prove a Saxon-Hutner type theorem (see Chapter 3)—namely, if H is composed of p isotopically different types of T matrices (in any order) such that for each type a chain composed purely of that type above would have a certain spectral gap, then any frequency which is mutually forbidden to *every* pure chain is forbidden to H. The word *isotopic* is important: it means that the masses can be different, but not the spring constants. The T^j matrices can refer, however, to matrices compounded from two different atoms. Thus, a chain composed of two different types of diatomic molecules AB or CD will satisfy the theorem if the four atoms differ only in their masses.

We turn now to a theorem of Matsuda (1964). Consider the two-mass, type II, random chain with $M > m$. Each eigenfrequency, ω_j, of such a chain will be a monotonic decreasing function of M. Therefore, the maximum frequency must always be less than that for a pure m chain—namely, $2(K/m)$. The general eigenfrequency can always be written as

$$\omega^2 = (4K/m) \cos^2 \tfrac{1}{2} q \tag{2.30}$$

where $0 \leq q \leq \pi$. For a pure m chain the allowed values of q would be $n\pi/(N + 1)$, where $n = 1, 2, \ldots, N$. Matsuda's theorem is directed toward the question: For a given value of q and M, can we ever construct a chain, of any length, such that (2.30) is *exactly* an eigenvalue of that chain? The theorem, which is proved using the properties of the above transfer matrices, is as follows:

Matsuda's Theorem. If q is π times any rational number L/J such that $L < J$ and L and J are taken to be relatively prime, then if M is greater than a certain critical value given by

$$M_c = m[1 + \cot(\pi/2J) \tan(\pi L/2J)] \tag{2.31}$$

no chain, except for a set of measures zero as $N \to \infty$, can have a frequency given by (2.30).

Based on this theorem, Matsuda has proposed an unproved conjecture (which, unfortunately, is referred to as Matsuda's theorem).

Matsuda's Conjecture. If the heavy mass is greater than the critical value given by (2.31) for $q = \pi(L/J)$, then the eigenfrequency density function, $D(\omega^2)$, is zero when ω^2 is given by (2.30).

This conjecture is completely supported by machine calculations, but the machine calculations seem to show that there may be even more points of zero density than are accounted for by the conjecture.

A curious feature of Matsuda's conjecture is that when the heavy mass becomes infinite *all* rational q's become "special." The infinite mass ratio case was solved exactly by Domb, Maradudin, Montroll, and Weiss (1959, part II), and it turns out that eigenvalues exist *only* for rational q's. In this case the chain is divided into "islands" of light masses which vibrate independently. If C is the concentration of light masses, the probability of having an island of n light masses is $C^n(1 - C)^2$, and this island can have n frequencies with q values equal to $\pi[S/(n + 1)]$ where $S = 1, 2, \ldots, n$. Any fixed q equal to $\pi(L/J)$ can occur either in an island with $n + 1 = J$, or $n + 1 = 2J$, etc. The total probability of this frequency is therefore $(1 - C)^2C^{J-1}(1 - C^J)^{-1}$. The density of states function, $D(\omega^2)$, is thus a series of delta functions at the rational q's with the amplitudes of these delta functions anything but a continuous function of ω. This example is a salutary reminder that the average spectrum is not a smooth function of frequency and that highly sophisticated approximation methods are required to analyze a disordered system of harmonic oscillators. Indeed, in the paper mentioned above, Domb *et al.* (1959) were primarily interested in developing an approximation method called the moment-trace method. Partly on the basis of the infinite mass case, adduced by themselves, and partly on the basis of actual machine computation for finite mass random chains, it has become clear that such an approximation is not able to predict the many sharp peaks in the true spectrum, in one dimension. Their idea is that the trace of the nth power of the dynamical matrix given by (2.1) is equal to the nth moment of the distribution, i.e. $\Sigma_1^N (\omega^2)^n = \int_0^\infty D(u)u^n\,du$. The average value of the nth moment is then equal to the average value of the nth trace, which is tedious (but in principle simple) to calculate. Knowing all the moments, we could then calculate $D(\omega^2)$ by standard inversion methods. The difficulty, however, is that since $D(\omega^2)$ is far from being a smooth function, a very large number of moments are needed for an accurate calculation.

An important corollary of Matsuda's theorem is the following theorem, due partly to Borland (1964).

Borland's Theorem. If ω and q are given as before, and if the heavy mass exceeds the critical value (2.31), then the distribution function $n(\omega^2)$ is equal to that of the infinite mass case at the same value of ω.

Actually, Borland (1964) proved the theorem only for the special case $L = 1$. His ingenious method of proof is quite independent of Matsuda's theorem and can be simply generalized to arbitrary L and J, Lieb (unpublished). Subsequently, Jepsen (private communication) suggested a simpler proof which makes it clear that Borland's generalized theorem is equivalent to Matsuda's theorem—that each is a corollary of the other. The proof is as follows:

Consider the jth eigenfrequency, ω_j, as a function of the heavy mass M. It is trivial to prove that ω_j is a continuous monotonically nonincreasing function of M. But Matsuda's theorem states that if $M > M_c$, no eigenfrequency is exactly equal to the right side of Eq. (2.30). Therefore, as M increases beyond M_c, the number of eigenfrequencies less than the right side of (2.30) must remain constant. Conversely, if the number of eigenfrequencies $\leq$ the right side of (2.30) is independent of M for $M > M_c$ (Borland's theorem), then no eigenfrequency ω_j can be exactly this value unless it "sticks," which is to say, unless ω_j is independent of M. But ω_j cannot in fact be independent of M unless the corresponding eigenvector is zero at *every* heavy mass site. The only chains with this property are those satisfying two conditions: (1) a pure light mass chain of that length could have an eigenfrequency given by (2.30); (2) the heavy masses are located only at those points of the chain which would be nodal points when only light masses are present. Such chains clearly constitute a set of measure zero as $N \rightarrow \infty$.

To complete the catalog of exact, analytic results we mention two results of Dean (1964). These are the counterparts for the harmonic oscillator problem of results obtained previously by Borland for the problem of electrons in a disordered potential (see Chapter 3). The first is a theorem on the diatomic lattice $mMmMmM$ . . . with random spring constants given by a probability density function $h(K)$ such that $h = 0$ unless $K_1 < K < K_2$, where K_1 and K_2 are two fixed numbers. If all the K's in a chain were equal to each other, we know there would be a gap in the spectral density $D(\omega^2)$ between the acoustic and optic branches. The question is how much disorder in the spring constants can be tolerated so that a well-defined (but narrower) gap can exist? A sufficient condition is given by Dean's theorem.

Dean's Theorem. If $mK_2 < MK_1$, then no eigenfrequencies exist for $2K_2/M < \omega^2 < 2K_1/m$.

The second result of Dean (1964) has, like Borland's, not been proved rigorously, for it relies on an "ergodic hypothesis." It is claimed that unless the chain is ordered, or unless $\omega = 0$, *all* vibrational modes are localized. Dean does, however, present numerical evidence supporting his conjecture.

In this introduction we have restricted our discussion to analytic results on the random chain and the first section of the bibliography, "General Theoretical," lists only papers devoted primarily to this question. But the interested reader should also turn to papers in the second section, "Numerical Calculations," in order to learn what is, in fact, the answer to the problem, and to gain an understanding of the complexities of the spectrum which analysis must try to elucidate. It has been found necessary to treat chains having lengths upwards of several thousand atoms in order to obtain reliable numerical data at the high frequency end of the spectrum. The third section, "Approximation Methods," contains papers primarily devoted to general approximation schemes which are tested in one dimension. The fourth section, "Defects," is concerned with the problem of a few impurities in an otherwise regular one-dimensional chain. For obvious reasons it is frequently possible to obtain exact, analytic answers to problems in this field.

The remaining sections are mainly, but not entirely, concerned with special aspects of regular systems. Section V, "End Effects and Finite Chains," contains some papers devoted to the study of long molecules with interactions that are not limited to the nearest-neighbor type. Among the papers in section VI, "Equilibrium Thermodynamics," are those by Maradudin *et al.* (1958) on the spectrum of a regular diatonic lattice with more than two atoms per unit cell. Also included is the paper by Pirenne and Renson (1962) in which are tabulated detailed, exact, analytic and numerical results of the well-known integral expressions for the thermodynamic properties of a simple monatomic regular chain.

Closely allied to the subject matter of the last section are the subjects of non-equilibrium thermodynamics and time-dependent phenomena. Vibrating systems have frequently served as model systems for discussing important fundamental questions on ergodicity and irreversibility. Papers devoted to these questions have, however, been listed in the bibliography of Chapter 7.

The reader should be aware of two things when referring to the bibliography of this chapter. First, a paper listed in one section may often have parts related to another section; the grouping is admittedly arbitrary in many cases. Second, the number of papers touching on the subject of harmonic oscillators in one dimension is so vast that there must be many omissions. These have been accidental rather than intentional, and we apologize in advance both to the authors of such papers, and to our readers.

Bibliography

I. General Theoretical

T. Asahi and J. Hori (1963). "Some Applications of the Method of Transfer-Matrix to the Vibration of Lattices with Free Boundaries," Chapter D20 *in* "Lattice Dynamics," *Proc. Intern. Conf., Copenhagen* (R. Wallis, ed.). Pergamon Press, Oxford.

R. Bellman (1956). "Dynamics of a Disordered Linear Chain," *Phys. Rev.* **101,** 19.

*R. E. Borland (1964). "Some Exact Results for the Vibrational Spectrum of a Disordered Chain," *Proc. Phys. Soc. (London)* **83,** 1027.

J. des Cloizeaux (1957). "Spectre de Fréquences d'une Chaine Linéaire Désordonée," *J. Phys. Radium* **18,** 131.

P. Dean (1956). "The Spectral Distribution of a Jacobian Matrix," *Proc. Cambridge Phil. Soc.* **52,** 752.

P. Dean (1959). "On Disordered One-Dimensional Crystals," *Proc. Phys. Soc. (London)* **73,** 413.

P. Dean (1963). "Dynamics of Disordered Lattices," Chapter D19 *in* "Lattice Dynamics," *Proc. Intern. Conf., Copenhagen* (R. Wallis, ed.). Pergamon Press, Oxford.

*P. Dean (1964). "Vibrations of Glass-Like Disordered Chains," *Proc. Phys. Soc.* **84,** 727.

*F. J. Dyson (1953). "The Dynamics of a Disordered Linear Chain," *Phys. Rev.* **92,** 1331.

R. Englman (1958). "The Eigenvalues of a Randomly Distributed Matrix," *Nuovo Cimento* **10,** 615.

J. Hori (1957). "On the Vibration of Disordered Linear Lattice, II," *Progr. Theoret. Phys. (Kyoto)* **18,** 367.

J. Hori (1961). "A New Picture of the Vibration of Linear Lattices," *J. Phys. Soc. Japan* **16,** 23.

J. Hori (1964a). "Structure of the Spectra of Disordered Systems," *Progr. Theoret. Phys. (Kyoto)* **31,** 522.

J. Hori (1964b). "Remarks on Matsuda's Theorem," *Progr. Theoret. Phys.* (*Kyoto*) **31,** 940.

J. Hori (1964c). "Structure of the Spectra of Disordered Systems, II" and "Structure of the Spectra of Disordered Systems, III," *Progr. Theoret. Phys.* (*Kyoto*) **32,** 371 and 471.

J. Hori and T. Asahi (1957). "On the Vibration of Disordered Linear Lattice," *Progr. Theoret. Phys.* (*Kyoto*) **17,** 523.

J. Hori and M. Fukushima (1964). "Why Have the Vibrational Frequency Spectra of Disordered Chains Many Distinct Peaks?" *J. Phys. Soc. Japan* **19,** 296.

*J. Hori and H. Matsuda (1964). "Structure of the Spectra of Disordered System, I, Fundamental Theorems," *Progr. Theoret. Phys.* (*Kyoto*) **32,** 183.

J. Mahanty (1961). "On the Vibration Spectrum of a Disordered Linear Lattice, I," *Nuovo Cimento* **19,** 46.

A. Maradudin and G. H. Weiss (1958). "The Disordered Lattice Problem: A Review," *J. Soc. Ind. Appl. Math.* **6,** 302.

H. Matsuda (1962). "The Transfer Matrix Method in the Theory of Normal Vibrations of Chain Molecules," *Progr. Theoret. Phys.* (*Kyoto*) **23,** 22.

*H. Matsuda (1964). "Special Frequencies in the Vibrational Spectra of Disordered Chains," *Progr. Theoret. Phys.* (*Kyoto*) **31,** 161.

*H. Schmidt (1957). "Disordered One-Dimensional Crystals," *Phys. Rev.* **105,** 425.

II. Numerical Calculations

*R. L. Agacy (1964). "The Vibrational Spectrum of a Disordered Linear System," *Proc. Phys. Soc.* **83,** 591.

P. Dean (1960). "Vibrational Spectra of Diatomic Chains," *Proc. Royal Soc.* **A254,** 507.

P. Dean (1961). "Vibrational Spectra of Diatomic Chains, II," *Proc. Royal Soc.* **A260,** 263.

C. Domb, A. A. Maradudin, E. W. Montroll, and G. H. Weiss (1959). "The Vibration Spectra of Disordered Lattices," *J. Phys. Chem. Solids* **8,** 419.

P. Flinn, A. Maradudin, and G. Weiss (1961). Westinghouse Res. Labs. Sci. Paper 029-G000-P10.

J. L. Martin (1961). "The Vibrational Spectra of Disordered Chains," *Proc. Roy. Soc.* **A260,** 139.

III. Approximation Methods

R. W. Davies and J. S. Langer (1963). "Self-Consistent Field Approximation for the Frequency Spectrum of a Disordered Chain," *Phys. Rev.* **131,** 163.

C. Domb, A. A. Maradudin, E. W. Montroll, and G. H. Weiss (1959). "Vibration Frequency Spectra of Disordered Lattices, I, Moments of the Spectra for Disordered Linear Chains," *Phys. Rev.* **115,** 18; "Vibration Frequency Spectra of Disordered Lattices, II, Spectra of Disordered One-Dimensional Lattices," *ibid.*, 24.

C. Domb (1963). "On One-Dimensional Vibrating Systems," *Proc. Roy. Soc.* **A276,** 418.

J. S. Faulkner and J. Korringa (1961). "Electron Energy Bands of One-Dimensional Random Alloys," *Phys. Rev.* **122,** 390.

J. S. Langer (1961). "Frequency Spectrum of a Disordered One-Dimensional Lattice," *J. Math. Phys.* **2,** 584.

W. von Schlup (1963a). "Die Regularisierung der interpolierten Clusterentwicklung für die Eigenfrequenzverteilung einer ungeordneten binären Kette," *Helv. Phys. Acta* **36,** 886.

W. von Schlup (1963b). "Eine störungstheoretische Methode zur Lösung der Dyson-Schmidt-Gleichung," *Helv. Phys. Acta* **36,** 570.

E. J. Woll, Jr. (1965). "Evaluation of Lattice Thermal Conductivity for a Disordered One-Dimensional Harmonic Crystal," *Phys. Rev.* **137,** A95.

IV. Defects

M. D. Bacon, P. Dean, and J. L. Martin (1962). "Defect Modes in Two-Component Chains," *Proc. Phys. Soc.* (*London*) **80,** 174.

R. L. Bjork (1957). "Impurity-Induced Localized Modes of Lattice Vibration in a Diatomic Chain," *Phys. Rev.* **105,** 456.

J. Mahanty, A. A. Maradudin, and G. H. Weiss (1958). "Vibrational Thermodynamic Properties of Lattices with Defects, I," *Progr. Theoret. Phys.* (*Kyoto*) **20,** 369.

E. W. Montroll and R. B. Potts (1955). "Effect of Defects on Lattice Vibrations," *Phys. Rev.* **100,** 525.

E. W. Montroll and R. B. Potts (1956). "Effect of Defects on Lattice Vibrations: Interaction of Defects and an Analogy with Meson Pair Theory," *Phys. Rev.* **102,** 72.

S. Takeno and J. Hori (1962). "Continuum Approximation for the Motion of a Heavy Particle in One- and Three-Dimensional Lattices," *Progr. Theoret. Phys.* (*Kyoto*) (Suppl.) **23,** 177.

R. F. Wallis and A. A. Maradudin (1960). "Impurity Induced Infrared Lattice Vibration Absorption," *Progr. Theoret. Phys.* (*Kyoto*) **24,** 1055.

V. End Effects and Finite Chains

D. C. Gazis and R. F. Wallis (1962). "Surface Vibrational Modes in Crystal Lattices with Complex Interatomic Interactions," *J. Math. Phys.* **3,** 190.

*H. Matsuda, K. Okada, T. Takase, and T. Yamamoto (1964). "Theory of Normal Vibrations of Chain Molecules with Finite Length," *J. Chem. Phys.* **41,** 1527.

H. B. Rosenstock (1955). "On the Optical Properties of Solids," *J. Chem. Phys.* **23,** 2415.

H. B. Rosenstock (1958). "Vibrations of a One-Dimensional Ionic Lattice," *Phys. Rev.* **111,** 755.

T. Simanouti and S. Mizushima (1949). "The Constant Frequency Raman Lines of N-Paraffins," *J. Chem. Phys.* **17,** 1102.

R. F. Wallis (1957). "Effect of Free Ends on the Vibration Frequencies of One-Dimensional Lattices," *Phys. Rev.* **105,** 540.

M. C. Tobin (1955). "Selection Rules for Normal Modes of Chain Molecules," *J. Chem. Phys.* **23,** 891.

VI. Equilibrium Thermodynamics

C. Domb and C. Isenberg (1962). "The Calculation of Thermodynamic Functions in Lattice Dynamics," *Proc. Phys. Soc.* (*London*) **79,** 659.

S. Kashiwamura (1964). "A Stationary State in One-Dimensional Lattice with a Few Impurities," *Progr. Theoret. Phys.* (*Kyoto*) **31,** 162.

A. A. Maradudin, P. Mazur, E. W. Montroll, and G. H. Weiss (1958). "Remarks on the Vibrations of Diatomic Lattices," *Rev. Modern Phys.* **30,** 175.

A. Maradudin and G. H. Weiss (1958). "On the Vibrations of a Generalized Diatomic Lattice," *J. Chem. Phys.* **29,** 631.

J. Pirenne and P. Renson (1962). "Thermodynamic Properties of the Mono-Atomic Linear Chain," *Physica* **28,** 233.

G. Weiss and A. Maradudin (1958). "Thermodynamic Properties of a Disordered Lattice," *J. Phys. Chem. Solids* **7,** 327.

Reprinted from THE PHYSICAL REVIEW, Vol. 92, No. 6, 1331–1338, December 15, 1953
Printed in U. S. A.

The Dynamics of a Disordered Linear Chain*

FREEMAN J. DYSON
Institute for Advanced Study, Princeton, New Jersey
(Received August 17, 1953)

By a disordered chain we mean a chain of one-dimensional harmonic oscillators, each coupled to its nearest neighbors by harmonic forces, the inertia of each oscillator and the strength of each coupling being a random variable with a known statistical distribution law. A method is presented for calculating exactly the distribution-function of the frequencies of normal modes of vibration of such a chain, in the limit when the chain becomes infinitely long. For some special examples, in which the distribution law of the oscillator parameters is assumed to be of exponential form, the frequency spectra are calculated analytically. The theory applies equally well to a chain of masses connected by elastic springs and making mechanical vibrations, or to an electrical transmission line composed of alternating inductances and capacitances with random characteristics.

I. INTRODUCTION

CONSIDER a chain of N masses, each coupled to its nearest neighbors by elastic springs obeying Hooke's law. We shall study the longitudinal vibrations of the chain, all motions being supposed to take place in one dimension so that each mass is described by a single coordinate. Since the coupling forces are linear in the displacements, the most general vibration is a superposition of $(N-1)$ normal modes, each having a characteristic frequency. The object of this paper is to present a method for determining accurately the spectrum or distribution function of the characteristic frequencies of the chain, in the limit as the number of masses N becomes very large. As is well known, the knowledge of this distribution function enables all the thermodynamical properties of the chain to be deduced immediately.

In the case when the masses and the strengths of the springs are all equal, the calculation of the frequency spectrum is elementary. In Sec. IV of this paper we give explicit formulas for the frequency spectrum in the most general case when the masses and spring-constants are arbitrary. The case of equal masses and springs here serves as a check.

In Secs. V–VI we consider the physically interesting case in which the masses and springs are unequal but are distributed along the chain in a random way. This means that we know the probability that a given mass or spring-constant has a particular value, and that this probability is the same at every point in the chain. There are several different ways of defining precisely how the randomization of the masses and springs is to be understood; for example, the masses may be independent random variables with a known probability distribution, while the springs are all equal; or each mass may be correlated with the strengths of the two neighboring springs, and so on. In every case, given the probability distribution for masses and springs, our method leads to an exact determination of the spectrum of normal frequencies. To illustrate the method, one particular class of probability distributions is worked out in detail, and the corresponding frequency spectra are obtained explicitly.

These calculations were begun in response to a question of C. Kittel, who was concerned with the thermal properties of glass. Glass may be considered roughly to be a disordered array of coupled harmonic oscillators in 3 dimensions. The systems considered in this paper are models of a "one-dimensional glass," a disordered array of atoms in one dimension. It is not, of course, to be expected that the results of this paper have any immediate application to the 3-dimensional problem. But it seems to us remarkable that the 1-dimensional problem can be solved exactly, and we publish this analysis in the hope that the methods will be useful in

* Supported in part by the U. S. Office of Naval Research.

discussing other disordered systems of a less idealized character.

II. DEFINITIONS

Let particle number j in the chain have mass m_j, and let its displacement from its equilibrium position be x_j. Let the elastic modulus of the spring between particles j and $(j+1)$ be K_j. Then the equations of motion of the system are

$$m_j\ddot{x}_j = K_j(x_{j+1}-x_j)+K_{j-1}(x_{j-1}-x_j). \tag{1}$$

It is convenient to introduce new variables

$$y_j = m_j^{\frac{1}{2}}x_j \tag{2}$$

and new constants $\lambda_1, \lambda_2, \cdots, \lambda_{2N-2}$ given by

$$\lambda_{2j-1}=K_j/m_j, \quad \lambda_{2j}=K_j/m_{j+1}. \tag{3}$$

Then the equations of motion take the form

$$\ddot{y}_j = (\lambda_{2j-1}\lambda_{2j})^{\frac{1}{2}}y_{j+1}+(\lambda_{2j-3}\lambda_{2j-2})^{\frac{1}{2}}y_{j-1} - (\lambda_{2j-1}+\lambda_{2j-2})y_j, \tag{4}$$

and the coefficient-matrix is now symmetric. Next we define variables $z_1, z_2, \cdots, z_{N-1}$, by

$$\dot{z}_j = \lambda_{2j}^{\frac{1}{2}}y_{j+1}-\lambda_{2j-1}^{\frac{1}{2}}y_j, \tag{5}$$

so that (4) becomes

$$\dot{y}_j = \lambda_{2j-1}^{\frac{1}{2}}z_j-\lambda_{2j-2}^{\frac{1}{2}}z_{j-1}. \tag{6}$$

Let variables $u_1, u_2, \cdots, u_{2N-1}$ be defined by

$$u_{2j-1}=y_j, \quad u_{2j}=z_j. \tag{7}$$

Then Eqs. (5) and (6) together may be written

$$\dot{u}_j = \lambda_j^{\frac{1}{2}}u_{j+1}-\lambda_{j-1}^{\frac{1}{2}}u_{j-1}. \tag{8}$$

The characteristic frequencies ω_j of the chain are, therefore, the characteristic roots of the $(2N-1)\times(2N-1)$ matrix Λ whose elements are given by

$$\Lambda_{j+1,j} = -\Lambda_{j,j+1} = i\lambda_j^{\frac{1}{2}}, \tag{9}$$

all other elements being zero. There is one zero root corresponding to the degenerate mode in which all the x_j are equal; the remaining roots occur in $(N-1)$ pairs, the members of a pair being $+\omega_j$ and $-\omega_j$.

The spectrum of characteristic frequencies is given by the function $M(\mu)$ which is defined as the proportion of the roots ω_j for which $\omega_j^2 \le \mu$. As $N\to\infty$ we expect that $M(\mu)$ will become a smooth differentiable function, and then a density of characteristic frequencies can be defined by

$$D(\mu) = (dM/d\mu). \tag{10}$$

Our first task is to determine the $M(\mu)$ and $D(\mu)$ corresponding to given λ_j.

We study the function

$$\Omega(x) = \lim_{N\to\infty}(2N-1)^{-1}\textstyle\sum_j \log(1+x\omega_j^2) = \displaystyle\int_0^\infty \log(1+x\mu)D(\mu)d\mu, \tag{11}$$

as a function of the complex variable x. The logarithm is defined as the branch of the function which is real for real positive x. Then the integral (11) is convergent and defines an analytic function of x over the whole x plane, the negative real axis from 0 to $(-\infty)$ being excluded. As x tends from above onto a point $(-z)$ on the negative real axis, the imaginary part of $\log(1+x\mu)$ tends to zero if $z\mu<1$ and to $i\pi$ if $z\mu>1$. Therefore, (11) gives

$$\mathrm{Re}[(i\pi)^{-1}\lim_{\epsilon\to0}\Omega(-z+i\epsilon)] = \int_{1/z}^\infty D(\mu)d\mu = 1-M(1/z), \tag{12}$$

$$D(1/z) = -z^2\,\mathrm{Re}[(i\pi)^{-1}\lim_{\epsilon\to0}\Omega'(-z+i\epsilon)], \tag{13}$$

$$\Omega'(x) = (d\Omega/dx) = \int_0^\infty \mu(1+x\mu)^{-1}D(\mu)d\mu. \tag{14}$$

According to (12) or (13), the spectrum of characteristic frequencies is determined by the limiting values of $\Omega(x)$ on the negative real axis. We call $\Omega(x)$ the characteristic function of the chain.

III. CALCULATION OF THE CHARACTERISTIC FUNCTION[1]

Consider a chain for which all the ω_j^2 are less than a fixed bound B, and let $B|x|<1$; these restrictions will be removed later. Then the logarithm in (11) may be expanded in powers of x, giving

$$\Omega(x) = \lim_{N\to\infty}(2N-1)^{-1}S(x), \tag{15}$$

$$S(x) = \sum_{n=1}^\infty (-1)^{n-1}(x^n/n)\sum_j \omega_j^{2n} = \sum_{n=1}^\infty (-1)^{n-1}(x^n/n)\,\mathrm{Spur}(\Lambda^{2n}). \tag{16}$$

According to (9), the spur of Λ^{2n} is a sum of terms $Q(\sigma)$, one corresponding to each cycle σ, a cycle consisting of $2n$ integers

$$j_0, j_1, j_2, \cdots, j_{2n-1}, j_{2n}=j_0 \tag{17}$$

lying between 1 and $2N-1$ and satisfying for each m

$$j_{m+1}=j_m\pm1. \tag{18}$$

[1] The analysis of this section might have been shortened by using known results in the theory of Jacobi matrices. See A. Wintner, *Spektraltheorie der Unendlichen Matrizen* (Leipzig, 1929), pp. 69–73. For this remark the author is indebted to Professor Kac.

The term $Q(\sigma)$ is equal to the product

$$\prod_{j=1}^{2N-1} (\lambda_j)^{q(j)}, \tag{19}$$

where $q(j)$ is the number of times that the step $(j\to j+1)$ occurs in the cycle σ; the step $(j+1\to j)$ must also occur $q(j)$ times in σ. Hence, we may write

$$\mathrm{Spur}(\Lambda^{2n})=\sum_q R(q)\prod_j(\lambda_j)^{q(j)}, \tag{20}$$

where the summation is over all possible sets of integers $q(j)$ whose sum is n, and $R(q)$ is the number of cycles σ which exist corresponding to a given set of $q(j)$. The evaluation of $R(q)$ is carried out in Appendix I. The result is as follows. Let $q(a)$ be the first nonvanishing $q(j)$, and let $q(b)$ be the last. Then

$$R(q)=2n[q(a)]^{-1}\left[\prod_{j=a+1}^{b} L(j)\right], \tag{21}$$

$$L(j)=(q(j)+q(j-1)-1)!/[(q(j))!(q(j-1)-1)!]. \tag{22}$$

If we substitute from (20) and (21) into (16), we find

$$S(x)=\sum_{a,b} S(a,b), \tag{23}$$

$$S(a,b)=-2\sum_q[q(a)]^{-1}(-x\lambda_a)^{q(a)}\times\left[\prod_{j=a+1}^{b} L(j)(-x\lambda_j)^{q(j)}\right]. \tag{24}$$

In Eq. (23) the summation is over all integers a, b satisfying $1\leq a\leq b\leq 2N-1$. In Eq. (24) the summation is over all sets of integers $q(j)$ which are nonzero for $a\leq j\leq b$. That is to say, each $q(j)$ for $a\leq j\leq b$ is summed over all positive integral values.

The summation over $q(b)$ in Eq. (24) can immediately be performed, because by Eq. (22) the sum is an elementary binomial expansion with exponent $(-q(b-1))$. Therefore,

$$S(a,b)=-2\sum_q{}'(q(a))^{-1}(-x\lambda_a)^{q(a)}\times\left[\prod_{j=a+1}^{b-1} L(j)(-x\lambda_j)^{q(j)}\right][(1+x\lambda_b)^{-q(b-1)}-1], \tag{25}$$

where the summation is now over all sets of integers $q(j)$ which are nonzero on $a\leq j\leq b-1$. That is to say,

$$S(a,b)+S(a,b-1)=S_1(a,b-1), \tag{26}$$

where $S_1(a,b-1)$ is the sum $S(a,b-1)$ with the variable $x\lambda_{b-1}$ replaced by $[x\lambda_{b-1}/(1+x\lambda_b)]$. If we apply the same reduction to $S_1(a,b-1)$, we obtain

$$S(a,b)+S(a,b-1)+S(a,b-2)=S_2(a,b-2), \tag{27}$$

where $S_2(a,b-2)$ is $S(a,b-2)$ with $x\lambda_{b-2}$ replaced by $[x\lambda_{b-2}/(1+x\lambda_{b-1}/(1+x\lambda_b))]$. Continuing in this way, we find

$$\sum_{j=a}^{b} S(a,j)=S_{b-a}(a,a), \tag{28}$$

where $S_{b-a}(a,a)$ is $S(a,a)$ with $x\lambda_a$ replaced by the continued fraction

$$\xi(a,b)=x\lambda_a/(1+x\lambda_{a+1}/[\cdots(1+x\lambda_{b-1}/(1+x\lambda_b))\cdots]). \tag{29}$$

Now from (24),

$$S(a,a)=-2\sum_{q(a)}[q(a)]^{-1}(-x\lambda_a)^{q(a)}=2\log(1+x\lambda_a). \tag{30}$$

Therefore,

$$\sum_{j=a}^{b} S(a,j)=2\log(1+\xi(a,b)). \tag{31}$$

We may now write $b=2N-1$ and make $N\to\infty$ in (29) and (31). This gives

$$\sum_{j=a}^{\infty} S(a,j)=2\log(1+\xi(a)), \tag{32}$$

where $\xi(a)$ is the infinite continued fraction

$$\xi(a)=x\lambda_a/(1+x\lambda_{a+1}/(1+x\lambda_{a+2}/(\cdots. \tag{33}$$

If we sum over all a and use (15) and (23),

$$\Omega(x)=\lim_{N\to\infty} N^{-1}\sum_{a=1}^{2N-1}\log[1+\xi(a)]. \tag{34}$$

Equations (33) and (34) give the explicit representation of $\Omega(x)$ for an arbitrary chain with given coefficients λ_j. The result was proved only under the assumptions that all the ω_j^2 were less than a fixed bound B and that $B|x|<1$. However, the integral (11) defining $\Omega(x)$, and the continued fractions (33), are convergent for all real positive x and for any set of positive coefficients λ_j. Since these expressions are analytic functions of x, it is easy to prove by an analytic continuation argument that Eq. (34) holds for all real positive x and without restriction on the ω_j.

IV. THE FREQUENCY SPECTRUM FOR AN ARBITRARY CHAIN

To determine the frequency spectrum for a chain with given λ_j, it is only necessary to calculate $\Omega(x)$ from Eqs. (33) and (34), and then to use (13) or (14) to find the spectrum. The second step, however, requires the analytic continuation of $\Omega(x)$ from positive to negative real values. If $\Omega(x)$ can be obtained in closed analytic form for positive x, the continuation process is usually easy. For example, consider the case of an infinite chain of equal masses m linked by springs of equal modulus K. In this case all λ_j are equal to $\lambda=(K/m)$, and all $\xi(a)$ are equal to the infinite continued fraction

$$\xi=x\lambda/(1+x\lambda/(\cdots. \tag{35}$$

Since ξ satisfies

$$\xi = x\lambda/(1+\xi) \tag{36}$$

and is small for small x, we find

$$\xi = \tfrac{1}{2}[(1+4x\lambda)^{\frac{1}{2}}-1], \tag{37}$$

and by (34),

$$\Omega(x) = 2\log\{\tfrac{1}{2}[(1+4x\lambda)^{\frac{1}{2}}+1]\}. \tag{38}$$

If we differentiate (38) we obtain

$$\Omega'(x) = x^{-1}[1-(1+4x\lambda)^{-\frac{1}{2}}]. \tag{39}$$

This function is analytic and real for $x > -(4\lambda)^{-1}$. It can be continued analytically through the upper half plane to values $x < -(4\lambda)^{-1}$, where it becomes

$$\Omega'(x) = x^{-1}[1+i(-1-4x\lambda)^{-\frac{1}{2}}]. \tag{40}$$

Hence, Eq. (13) gives the spectrum of characteristic frequencies

$$\begin{aligned} D(\mu) &= (1/\pi)(4\lambda\mu-\mu^2)^{-\frac{1}{2}}, \quad &\mu<4\lambda, \\ D(\mu) &= 0, &\mu>4\lambda, \end{aligned} \tag{41}$$

a result which in this case can easily be checked by an elementary calculation.

In general we will not be able to calculate $\Omega'(x)$ in closed analytic form, and so we require an explicit formula for $D(\mu)$ in terms of the values of $\Omega'(x)$ for positive x. This formula will enable $D(\mu)$ to be calculated if $\Omega'(x)$ is only given numerically or approximately, so that a direct use of analytic continuation is impossible. The formula is

$$D(\mu) = (2\pi^2\mu)^{-1}\int_{-\infty}^{\infty}(\cosh\pi\alpha)d\alpha \times \int_0^{\infty}(x\mu)^{-\frac{1}{2}}\cos[\alpha\log(x\mu)]\Omega'(x)dx. \tag{42}$$

Its derivation is given in Appendix II. The x integration is to be carried out first and the α integration second. Taken in this order, the double integration will always be convergent for values of μ at which $D(\mu)$ exists and is continuous. With Eqs. (33), (34), and (42), we have in principle an exact determination of the frequency spectrum of an arbitrary chain.

V. DISORDERED CHAINS

We now apply the preceding theory to the case of a disordered chain, i.e., an infinite chain whose elements are distributed in a random way according to some known probability law. We consider two types of disordered chain, differing in the way in which the randomization of the elements is defined. Type I is mathematically the simpler, whereas Type II provides the closer approximation to a real chain of randomly arranged atoms.

Type I. Each of the parameters λ_j defined by Eq. (3) is an independent random variable with the probability distribution function $G(\lambda)$.

In this case each of the quantities $\xi(a)$ defined by Eq. (33) will have a probability distribution $F(\xi)$, the same for all a. Now

$$\xi(a) = x\lambda_a/(1+\xi(a+1)) \tag{43}$$

and the variables λ_a and $\xi(a+1)$ are uncorrelated, since $\xi(a+1)$ depends only on $\lambda_{a+1}, \lambda_{a+2}, \cdots$. If we equate the probability distributions of the left and right sides of Eq. (43), we find an integral equation for $F(\xi)$,

$$F(\xi) = \int_0^{\infty} F(\xi')G[\xi(1+\xi')/x]((1+\xi')/x)d\xi'. \tag{44}$$

The solution of (44) for given $G(\lambda)$ can in some cases be obtained in closed form (see Sec. VI). In all cases the solution can be obtained numerically by inserting an arbitrary trial function $F(\xi')$ on the right of (44) and iterating the equation repeatedly. The series of successive iterates will converge rapidly to the true $F(\xi)$, in consequence of the good convergence of the continued fractions (33). When we have found the solution $F(\xi)$ of (44), and have normalized it by

$$\int_0^{\infty} F(\xi)d\xi = 1, \tag{45}$$

the characteristic function of the chain is given by Eq. (34) and is

$$\Omega(x) = 2\int_0^{\infty} F(\xi)\log(1+\xi)d\xi. \tag{46}$$

From this the frequency spectrum may be found as described in Sec. IV.

The frequency spectrum determined in this way is strictly an average over a statistical ensemble of chains, each chain in the ensemble having definite values of the λ_j. But by an argument familiar in the statistical mechanics of systems containing many particles, the same frequency spectrum will be found for an arbitrary chain chosen out of the ensemble, except for an exceptional class of chains whose total probability tends to zero as the number of atoms N tends to infinity. Therefore, we may say that this frequency spectrum is the correct spectrum for a single disordered chain of infinite length.

Type II. Each mass m_j is an independent random variable with distribution function $G(m)$, the spring-constants K_j being fixed and equal.

In this case the λ_j are given by

$$\lambda_{2j-1} = \lambda_{2j-2} = Km_j^{-1}. \tag{47}$$

We introduce the variables

$$\eta_j = [\xi(2j)]^{-1}. \tag{48}$$

Then, from Eq. (33) we derive the recurrence formula

$$\eta_j = (m_{j+1}/xK) + [\eta_{j+1}/(1+\eta_{j+1})]. \tag{49}$$

The variables η_{j+1}, m_{j+1} now are uncorrelated. Hence, Eq. (49) leads to an integral equation

$$F(\eta)=xK\int_0^\infty G[xK(\eta-(\eta'/(1+\eta')))]F(\eta')d\eta', \quad (50)$$

defining the distribution function $F(\eta)$ of each of the η_j. The characteristic function (34) is given now by

$$\Omega(x)=\lim N^{-1}\sum_{j=1}^{N}\log[(1+\xi(2j))(1+\xi(2j-1))], \quad (51)$$

since $\xi(2j)$ and $\xi(2j-1)$ will have different distribution functions. From Eq. (33) we find

$$(1+\xi(2j))(1+\xi(2j-1))=1+\xi(2j)+(xK/m_j), \quad (52)$$

and here the variables $\xi(2j)$ and m_j are uncorrelated. If we insert Eq. (52) into (51), the characteristic function becomes

$$\Omega(x)=\int_0^\infty F(\eta)d\eta\int_0^\infty G(m)dm\log[1+\eta^{-1}+xKm^{-1}], \quad (53)$$

where $F(\eta)$ is defined as the solution of (50).

An interesting special case of a Type II chain is a chain composed of two kinds of atoms with masses m, M, distributed at random in the proportion p: $(1-p)$. This corresponds to the distribution function

$$G(m')=p\delta(m'-m)+(1-p)\delta(m'-M). \quad (54)$$

The equation (50) for $F(\eta)$ then reduces to a difference equation,

$$\begin{aligned}F(\eta)=p[1-\eta+(m/xK)]^{-2}F[(1-\eta+(m/xK))^{-1}-1]\\+(1-p)[1-\eta+(M/xK)]^{-2}\\\times F[(1-\eta+(M/xK))^{-1}-1], \quad (55)\end{aligned}$$

which can be solved very rapidly by iteration since no integration is involved. The characteristic function (53) becomes

$$\begin{aligned}\Omega(x)=\int_0^\infty F(\eta)d\eta[p\log(1+\eta^{-1}+xKm^{-1})\\+(1-p)\log(1+\eta^{-1}+xKM^{-1})]. \quad (56)\end{aligned}$$

Only a moderate amount of numerical work would be needed to calculate the frequency spectrum for any given values of m, M and p.

A chain in which the masses are all equal while the spring strengths are independent random variables can be treated by exactly the same method as a Type II chain. It is only necessary to interchange the roles of $\xi(2j)$ and $\xi(2j-1)$. If we go beyond Types I and II, we could consider a more general type of chain, in which the masses m_j are independent random variables, while the spring strength K_j is a known function of the two adjoining masses m_j and m_{j+1}. This would be a model for a chain composed of different kinds of atoms arranged at random, the strength of the bond between two atoms depending on the chemical nature of the two atoms. The frequency spectrum of such a chain can also be calculated by the methods of this paper, only the formulas become rather more complicated.

VI. A SPECIAL FAMILY OF DISORDERED CHAINS

In this section we calculate explicitly the spectrum of normal frequencies for a special family of disordered chains. This will serve as an illustration, to show quantitatively the effect which a given degree of disorder has upon the spectrum. We consider a chain C_n of Type I, in which each λ_j is an independent random variable with the probability distribution

$$G_n(\lambda)=[n^n/(n-1)!]\lambda^{n-1}e^{-n\lambda} \quad (57)$$

The integer n takes the values 1, 2, 3, $\cdots$. The distribution G_n has mean value 1 and standard deviation $n^{-\frac{1}{2}}$. Thus, C_1 is a highly disordered chain, C_2 is less disordered, and in the limit as $n\to\infty$, C_n becomes the uniform chain with all $\lambda_j=1$. For large n Eq. (57) takes asymptotically the Gaussian form

$$G_n(\lambda)\sim(n/2\pi)^{\frac{1}{2}}\exp[-\tfrac{1}{2}n(\lambda-1)^2]. \quad (58)$$

We choose these C_n for the illustration for reasons of mathematical convenience only. They happen to have frequency spectra which can be calculated to the end analytically. And although they do not correspond closely to any known physical situation, they illustrate clearly enough the behavior of disordered chains in general.

If we substitute Eq. (57) into Eq. (44), we find an integral equation for $F(\xi)$ which has the exact solution

$$F_n(\xi)=K_n^{-1}\xi^{n-1}(1+\xi)^{-n}e^{-n\xi/x}, \quad (59)$$

where K_n is a function of x determined by the normalization condition (45). Hence, Eq. (46) gives

$$\Omega_n(x)=2L_n(x)/K_n(x), \quad (60)$$

$$L_n(x)=\int_0^\infty \xi^{n-1}(1+\xi)^{-n}\log(1+\xi)e^{-n\xi/x}d\xi, \quad (61)$$

$$K_n(x)=\int_0^\infty \xi^{n-1}(1+\xi)^{-n}e^{-n\xi/x}d\xi. \quad (62)$$

We next have to carry out the analytic continuation of L_n and K_n to negative x. This is done in Appendix III. Hence, if we use Eq. (12), we find the following analytic expression for the frequency spectrum of C_n,

$$M_n(z)=\frac{G_1^2-F_0G_2e^{-nz}+[F_0F_2-F_1^2+((\pi^2/6)-t_{n-1})F_0^2]e^{-2nz}}{[G_1+(F_1-(\log nz+s_{n-1}+\gamma)F_0)e^{-nz}]^2+\pi^2F_0^2e^{-2nz}} \quad (63)$$

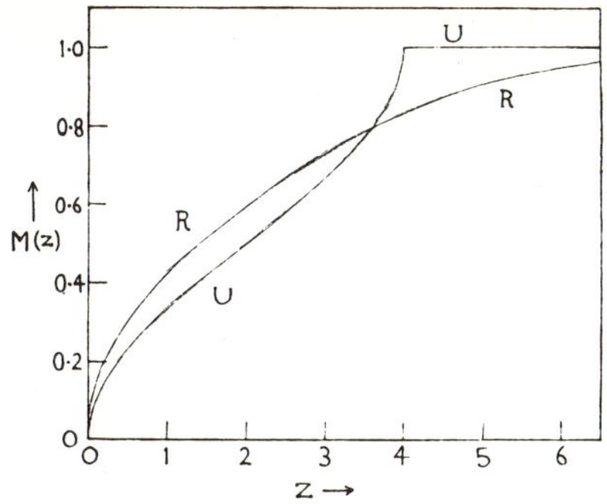

FIG. 1. Integral spectrum of characteristic frequencies for a uniform chain (curve UU) and for a random chain in which each spring-parameter λ has the distribution-function $\exp(-\lambda)$ (curve RR). $M(z)$ gives the proportion of frequencies ω for which $\omega^2 \leq z$.

Here, F_0, F_1, and F_2 are finite polynomials defined by

$$F_0 = \sum_{j=0}^{n-1} \binom{n-1}{j} (-nz)^j / j!, \tag{64}$$

$$F_1 = \sum_{j=0}^{n-1} \binom{n-1}{j} [(-nz)^j / j!] s_j, \tag{65}$$

$$F_2 = \sum_{j=0}^{n-1} \binom{n-1}{j} [(-nz)^j / j!] (s_j^2 + t_j), \tag{66}$$

with

$$s_j = \sum_{l=1}^{j} l^{-1}, \quad t_j = \sum_{l=1}^{j} l^{-2}. \tag{67}$$

The G_1 and G_2 are integral functions of z with power-series expansions

$$G_1 = \sum_{j=0}^{\infty} \binom{n-1+j}{j} [(-nz)^j / j!] s_j, \tag{68}$$

$$G_2 = \sum_{j=0}^{\infty} \binom{n-1+j}{j} [(-nz)^j / j!] (s_j^2 + t_j), \tag{69}$$

and γ is Euler's constant. Asymptotically for large z, G_1 and G_2 have semiconvergent series expansions

$$G_1 = [(-1)^n / (n-1)!] \sum_{j=n}^{\infty} (nz)^{-j} [(j-1)!]^2 / (j-n)!, \tag{70}$$

$$G_2 = 2[(-1)^n / (n-1)!] \sum_{j=n}^{\infty} (nz)^{-j} \times [(j-1)!]^2 s_{j-1} / (j-n)!. \tag{71}$$

From Eq. (63) we can find at once the limiting behavior of M_n for small z,

$$M_n(z) \sim [(\pi^2/6) - t_{n-1}][\pi^2 + (\log nz + s_{n-1} + \gamma)^2]^{-1}. \tag{72}$$

This rises very rapidly as z increases from zero, showing that a disordered chain has a much greater proportion of very low characteristic frequencies than a uniform chain. The limiting behavior of M_n for large z is found from Eqs. (70) and (71) to be

$$M_n(z) \sim 1 - 2(\log nz - s_{n-1} + \gamma) e^{-nz} \times (nz)^{2n-1} [(n-1)!]^{-2}. \tag{73}$$

In Fig. 1 the numerical value of $M_1(z)$ is plotted together with the corresponding function for the uniform chain which is by Eq. (41)

$$M_\infty(z) = \pi^{-1} \arccos[1 - \tfrac{1}{2}z], \quad z < 4, \qquad = 1, \quad z > 4. \tag{74}$$

From Eq. (63) we can also calculate the form of the frequency spectrum for large n, i.e., for a chain composed of only slightly fluctuating elements. In this case we represent the spectrum by a complex contour integral which is evaluated by the method of steepest descent; details are given in Appendix IV. The results are the following. For any fixed z in the range $0 < z < 4$, we have for sufficiently large n,

$$M_n(z) \sim \pi^{-1} \arccos[1 - \tfrac{1}{2}z] + (2\pi n)^{-1} [(4/z) - 1]^{-\frac{1}{2}}. \tag{75}$$

For fixed $z > 4$ and large n,

$$M_n(z) \sim 1 - \pi^{-1} \alpha \exp[-\alpha - 2n(\sinh\alpha - \alpha)], \tag{76}$$

$$\alpha = \operatorname{arg\,cosh}[\tfrac{1}{2}z - 1]. \tag{77}$$

At the critical point $z = 4$ and for large n,

$$M_n(z) \sim 1 - [\Gamma(\tfrac{1}{3})]^{-2} [12/n]^{\frac{1}{3}} = 1 - (0.32\cdots) n^{-\frac{1}{3}}. \tag{78}$$

The errors in Eqs. (75), (76), (78) are in each case of higher order in $(1/n)$ than the last term given. From these results it is seen how the spectrum approaches the limiting form (74) as $n \to \infty$.

VII. APPLICATION TO TRANSMISSION LINES

Consider an ideal loss-free transmission line composed of a series of inductances $L_1, L_2, \cdots$ with a capacitance C_j between each pair L_j and L_{j+1}. The equation of motion for the current I_j in the inductance L_j is

$$L_j d^2 I_j / dt^2 = C_j^{-1} (I_{j+1} - I_j) + C_{j-1}^{-1} (I_{j-1} - I_j). \tag{79}$$

This is identical with Eq. (1), only with L_j replacing m_j and C_j^{-1} replacing K_j. Therefore, the whole of the theory of this paper applies without alteration to a transmission line whose elements L_j and C_j are statistically random variables. The function $M(z)$ gives the integrated frequency distribution of the normal modes of propagation of current in the line.

In conclusion, the author wishes to thank Professor Kittel for suggesting this problem to him, Professor Luttinger for some useful discussions, and the University of California for its hospitality during the summer of 1953 when the work was done.

APPENDIX I. EVALUATION OF $R(q)$

Define $R(q, c)$ to be the number of cycles σ corresponding to a given set of $q(j)$, and beginning and ending with a given $j_0=j_{2n}=c$. Suppose first $c<b$. From each of these σ we can derive a new cycle σ' by simply omitting from σ the steps $(b\rightarrow b+1)$ and $(b+1\rightarrow b)$, since by hypothesis the step $(b+1\rightarrow b+2)$ does not occur in σ. These σ' will all belong to the set of integers q', where $q'(b)=0$ and $q'(j)=q(j)$ for $j<b$. The number of the σ' is $R(q', c)$. From a given σ' we derive the parent σ by inserting $q(b)$ pairs of steps $(b\rightarrow b+1)$, $(b+1\rightarrow b)$ distributed in any way among the $q(b-1)$ places where the integer b occurs in σ'. So the number of σ corresponding to a given σ' is the number of ways of distributing $q(b)$ identical objects among $q(b-1)$ boxes, which is $L(b)$ defined by Eq. (22). Hence,

$$R(q, c)=L(b)R(q', c). \tag{A.1}$$

Proceeding repeatedly in the same way, we shall reduce q to the set of integers q'' defined by $q''(j)=0$ for $j>c$, $q''(j)=q(j)$ for $j\leq c$. We find

$$R(q, c)=\left[\prod_{j=c+1}^{b} L(j)\right]R(q'', c). \tag{A.2}$$

We now start reducing q'' from the other end by omitting steps $(a\rightarrow a+1)$ and $(a+1\rightarrow a)$. In this way we eventually reach q''' defined by $q'''(j)=0$ for $j>c$ or $j<c-1$, $q'''(j)=q(j)$ for $j=c$ or $j=c-1$. We find

$$R(q'', c)=\left[\prod_{j=a+1}^{c-1} L'(j)\right]R(q''', c), \tag{A.3}$$

$$L'(j)=(q(j)+q(j-1)-1)!/ [(q(j)-1)!(q(j-1))!]. \tag{A.4}$$

But $R(q''', c)$ is just the number of ways of arranging $q(c)$ pairs of steps $(c\rightarrow c+1)(c+1\rightarrow c)$ and $q(c-1)$ pairs $(c\rightarrow c-1)(c-1\rightarrow c)$, beginning and ending at c. This number is

$$R(q''', c)=(q(c)+q(c-1))!/[(q(c))!(q(c-1))!]. \tag{A.5}$$

If we put together (A.2), (A.3), and (A.5), we obtain

$$R(q, c)=(q(c)+q(c-1))(q(a))^{-1}\left[\prod_{j=a+1}^{b} L(j)\right]. \tag{A.6}$$

We now sum (A.6) over all values of c from a to $b+1$. Then $\sum q(c)=\sum q(c-1)=n$, and so Eq. (21) of the text is verified.

APPENDIX II. DERIVATION OF EQUATION (42)

Consider the Fourier transform of $x^{\frac{1}{2}}\Omega'(x)$ considered as a function of the variable $\log x$, namely

$$r(\alpha)=\int_{-\infty}^{\infty} x^{\frac{1}{2}}\Omega'(x)\exp[-i\alpha\log x]d(\log x). \tag{A.7}$$

If we use Eq. (14), this becomes

$$\begin{aligned} r(\alpha)&=\int_0^{\infty} D(\mu)d\mu\int_{-\infty}^{\infty} dy2\mu e^{y-2i\alpha y}[1+\mu e^{2y}]^{-1}\\ &=\int_0^{\infty}\mu^{\frac{1}{2}}D(\mu)d\mu e^{i\alpha\log\mu}\int_{-\infty}^{\infty} e^{-2i\alpha y}\operatorname{sech}y dy\\ &=(\pi\operatorname{sech}\pi\alpha)\int_{-\infty}^{\infty}\mu^{\frac{1}{2}}D(\mu)e^{i\alpha\log\mu}d(\log\mu). \end{aligned} \tag{A.8}$$

Thus, $[r(\alpha)\cosh\pi\alpha]$ is also the Fourier transform of $\mu^{\frac{1}{2}}D(\mu)$ as a function of $(\log\mu)$. If we invert the transform (A.8), we obtain

$$\mu^{\frac{1}{2}}D(\mu)=(2\pi^2)^{-1}\int_{-\infty}^{\infty}\exp(-i\alpha\log\mu)r(\alpha)\cosh\pi\alpha d\alpha, \tag{A.9}$$

and substitution for $r(\alpha)$ from (A.7) gives Eq. (42).

APPENDIX III. ANALYTIC CONTINUATION OF K_n AND L_n

We write

$$I(q)=I(q, n, y)=\int_0^{\infty}\xi^{n-1}(1+\xi)^{q-n}e^{-\xi y}d\xi, \tag{A.10}$$

where $y=n/x$. For positive y, this is an analytic function of q for every q, and in particular

$$K_n(x)=I(0),\quad L_n(x)=(dI/dq)_{q=0}. \tag{A.11}$$

We evaluate $I(q)$ assuming $q>n-1$, in which case

$$I(q)=J(q)+R(q), \tag{A.12}$$

where

$$\begin{aligned} J(q)&=\int_{-1}^{\infty}\xi^{n-1}(1+\xi)^{q-n}e^{-\xi y}d\xi\\ &=e^y\sum_{j=0}^{n-1}(-1)^j\binom{n-1}{j}\int_0^{\infty}z^{q-1-j}e^{-zy}dz\\ &=e^y y^{-q}\sum_{j=0}^{n-1}(-1)^j\binom{n-1}{j}\Gamma(q-j)y^j, \end{aligned} \tag{A.13}$$

$$\begin{aligned} R(q)&=-\int_0^1(-\xi)^{n-1}(1-\xi)^{q-n}e^{\xi y}d\xi\\ &=(-1)^n\sum_{j=0}^{\infty}(y^j/j!)\\ &\quad\times[\Gamma(n+j)\Gamma(q-n+1)/\Gamma(q+j+1)]. \end{aligned} \tag{A.14}$$

The series in (A.13) and (A.14) are convergent and represent functions of y which are analytic over the whole y plane except for the simple branch point at $y=0$ produced by the factor y^{-q}. As functions of q, $J(q)$ and $R(q)$ are analytic, but each has a simple pole at $q=0$ which only cancels in the sum $I(q)$. Hence, we may evaluate K_n and L_n using (A.13) and (A.14), expanding these expressions in powers of q and retaining terms of order -1, 0 and 1. The terms in q^{-1} cancel as they should. The terms in q^0 and q^1 give, respectively,

$$K_n(x)=e^y\sum_{j=0}^{n-1}\binom{n-1}{j}(y^j/j!)(s_j-\gamma-\log y)$$

$$+\sum_{j=0}^{\infty}\binom{n+j-1}{j}(y^j/j!)(s_j-s_{n-1}), \quad \text{(A.15)}$$

$$2L_n(x)=e^y\sum_{j=0}^{n-1}\binom{n-1}{j}(y^j/j!)[(s_j-\gamma-\log y)^2$$

$$+t_j+\pi^2/6]-\sum_{j=0}^{\infty}\binom{n+j-1}{j}(y^j/j!)$$

$$\times[(s_j-s_{n-1})^2+t_j+t_{n-1}], \quad \text{(A.16)}$$

in which γ is Euler's constant, and s_j, t_j are defined by Eq. (67). These formulas show K_n and L_n explicitly as analytic functions of $y=n/x$, with a logarithmic branch point at $y=0$. In order to obtain the analytic continuation of K_n and L_n to a point $(-x)$ on the negative real axis, going through the upper half of the x plane, we have only to replace, in Eqs. (A.15) and (A.16), y by $(-y)$ and $\log y$ by $(\log y - i\pi)$. This leads at once, by Eq. (12), to the result stated in Eq. (63).

APPENDIX IV. METHOD OF STEEPEST DESCENT

We consider the analytic continuation of the function $K_n(x)$ given by Eq. (62) through the upper half-plane to a point $x=-z^{-1}$ on the negative real axis. This continuation may be written as a contour integral

$$K_n(-z^{-1})=\int_0^{-\infty}[F(\xi)]^n\xi^{-1}d\xi, \quad \text{(A.17)}$$

$$F(\xi)=\xi(1+\xi)^{-1}e^{z\xi}, \quad \text{(A.18)}$$

and the path of integration passes from 0 to $(-\infty)$ above the singularity at $\xi=-1$. Similarly, by Eq. (61)

$$L_n(-z^{-1})=\int_0^{-\infty}[F(\xi)]^n\log(1+\xi)\xi^{-1}d\xi. \quad \text{(A.19)}$$

We choose the path of integration to pass over the lowest saddle point of $F(\xi)$ between the minima at 0 and $(-\infty)$. The saddle points η are given by the quadratic equation

$$\eta^2+\eta+z^{-1}=0. \quad \text{(A.20)}$$

Suppose first $0<z<4$. Then, there is one saddle point

$$\eta=\tfrac{1}{2}[-1+i((4/z)-1)^{\frac{1}{2}}] \quad \text{(A.21)}$$

in the upper half-plane. The path of integration crosses this saddle at an angle of 135° to the positive real axis. For large n we calculate K_n and L_n by considering contributions to the integrals from the neighborhood of the saddle only. If we use Eq. (12), this gives the result (75). Next suppose $z>4$. Then there are two saddle points on the real axis

$$\eta_+=\tfrac{1}{2}[-1+(1-(4/z))^{\frac{1}{2}}],$$
$$\eta_-=\tfrac{1}{2}[-1-(1-(4/z))^{\frac{1}{2}}]. \quad \text{(A.22)}$$

The path of integration goes along the real axis over η_+ from 0 to η_-, then breaks off at 90° to the real axis and goes from η_- to $(-\infty)$ above $\xi=-1$. For large n the real parts of K_n and L_n come from the integral near η_+, but the imaginary parts come from contributions in the neighborhood of η_-. In this way we find the result (76), which tends to 1 for large n because $|F(\eta_-)|<|F(\eta_+)|$. Finally, consider the case $z=4$. Then there is one saddle point at $\eta=-\tfrac{1}{2}$, at which the first two derivatives of $F(\xi)$ vanish. The path of integration goes from 0 to η along the real axis, then breaks off at an angle of 120° and goes to $(-\infty)$. The integral in the neighborhood of η now gives Eq. (78).

Reprinted from THE PHYSICAL REVIEW, Vol. 105, No. 2, 425–441, January 15, 1957
Printed in U. S. A.

Disordered One-Dimensional Crystals

HELMUT SCHMIDT*
Department of Physics, University of California, Berkeley, California, and Department of Physics, University of Michigan, Ann Arbor, Michigan

(Received June 11, 1956)

A mathematical method is developed which gives fairly generally the density of eigenstates for one-dimensional disordered systems. The method is applied first to a disordered linear chain of elastically coupled masses. The results for the energy spectrum are closely related to those obtained by Dyson.

Then we consider the electronic energy-states in a one-dimensional disordered crystal, represented by a series of δ-function potentials of different strengths, randomly distributed.

We solve the resulting functional equation explicitly in that case which corresponds to a uniform crystal with a small amount of impurities; that is, we find the shape of the impurity bands.

I. INTRODUCTION

THE energy levels of an electron in a pure crystal can be computed in principle if the crystal potential is known. In this paper we will study the case in which the crystal consists of different atoms, randomly distributed.

We restrict our considerations to a one-dimensional crystal model and hope that this gives in some respects a qualitatively correct description of real three-dimensional crystals.

We shall use a general mathematical method which also applies to similar problems. One of these is the determination of the eigenfrequencies of a linear chain which consists of elastically coupled atoms with randomly varying masses. This problem is interesting in itself and most suitable for demonstrating the mathematical method involved. Therefore, we will consider it first.

II. FREQUENCY SPECTRUM OF A DISORDERED CHAIN

(a) Description of the Model

Suppose that we have an alloy containing i different kinds of atoms $A^1, \cdots, A^i$ with the masses $m^1, \cdots, m^i$. Let p^i be the fraction of atoms A^i, $\sum p^i = 1$. As a one-dimensional model of this crystal we choose a chain of elastically coupled masses $m_1, m_2, m_3, \cdots$, where the nth mass m_n can assume the values $m^1, \cdots, m^i$ with the probabilities $p^1, \cdots, p^i$. Suppose for the present that these probabilities are independent of the nature of the atoms which occupy the neighboring places $(n-1)$ and $(n+1)$. (The case where a correlation between neighboring atoms exists will be discussed in Appendix II.) Furthermore we will assume that the forces between all neighboring atoms can be described by the same elastic constant k.

Now consider such a chain of many, say N, atoms and assume as the boundary conditions that the chain is fixed at the points which correspond to the place numbers $n=0$ and $n=N+1$. Then the equation of motion for this chain may be written as

$$m_n\ddot{u}_n = k(u_{n+1}+u_{n-1}-2u_n) \text{ for } n=1, 2, \cdots N, \quad \text{(II, 1a)}$$

$$u_0 = u_{N+1} = 0. \quad \text{(II, 1b)}$$

Here $u_1 \cdots u_N$ are the displacements of the masses $m_1 \cdots m_N$ from their equilibrium position and $u_0 = u_{N+1} = 0$ are merely introduced to establish the validity of (II, 1a) also for $n=1$ and $n=N$.

With $M_n \equiv m_n/k$, $M^i = m^i/k$ and $\ddot{u}_n = -\omega^2 u_n$, Eq. (II, 1) gives

$$u_n(2-M_n\omega^2) = u_{n+1}+u_{n-1} \quad \text{for } n=1, 2, \cdots N, \quad \text{(II, 2)}$$
$$u_0 = u_{N+1} = 0.$$

The secular equation for the eigenvalues ω_ν^2 of ω^2 can now be written as

$$\begin{vmatrix} 2-M_1\omega^2 & -1 & & \\ -1 & 2-M_2\omega^2 & -1 & \\ & -1 & 2-M_3\omega^2 & -1 \\ & & -1 & 2-M_N\omega^2 \end{vmatrix} = 0.$$

We could try to determine the possible eigenvalues ω_ν^2 from this equation and then by averaging over all possible chains find the mean distribution of the eigenvalues. This way, however, is mathematically difficult. We have found a different method which seems considerably simpler.

(b) New Formulation of the Eigenvalue Problem

We write Eq. (II, 2) in matrix form:

$$\begin{pmatrix} u_n \\ u_{n+1} \end{pmatrix} = \begin{pmatrix} 0 & 1 \\ -1 & 2-M_n\omega^2 \end{pmatrix} \begin{pmatrix} u_{n-1} \\ u_n \end{pmatrix} \quad \text{for } n=1, 2, \cdots N, \quad \text{(II, 3)}$$
$$u_0 = u_{N+1} = 0.$$

From this it follows that

$$\begin{pmatrix} u_N \\ u_{N+1} \end{pmatrix} = \begin{pmatrix} 0 & 1 \\ -1 & 2-M_N\omega^2 \end{pmatrix} \cdots \times \begin{pmatrix} 0 & 1 \\ -1 & 2-M_1\omega^2 \end{pmatrix} \begin{pmatrix} u_0 \\ u_1 \end{pmatrix}. \quad \text{(II, 4)}$$

* Present address: Department of Physics, University of California, Berkeley, California.

Now instead of considering $u_0=0$ and $u_{N+1}=0$ as fixed numbers in Eq. (II, 4) for ω^2, we fix only $u_0=0$. Then with $u_0=0$ and $u_1=\cos\omega t$, Eq. (II, 4) gives $u_{N+1}=u_{N+1}(\omega^2)\cos\omega t$ as a function of ω^2 and t, and whenever $u_{N+1}(\omega^2)=0$, then both the boundary conditions are satisfied and $\omega^2=\omega_\nu^2$ is an eigenvalue of our problem.

To study the ω^2 dependence of $u_{N+1}(\omega^2)$, we look at the two-dimensional vectors

$$\mathbf{v}_n(\omega^2)\equiv\begin{pmatrix}u_{n-1}\\ u_n\end{pmatrix} \qquad \text{(II, 5)}$$

in an x-y plane; $\mathbf{v}=\begin{pmatrix}x\\ y\end{pmatrix}$. These vectors have a remarkable property: With increasing ω^2 the vector $\mathbf{v}_n(\omega^2)$ rotates uniformly in the positive direction around the zero point in the x-y plane, i.e., if $\varphi_n/2$ is the angle between the positive x-direction and the vector $\mathbf{v}_n$, then $\varphi_n(\omega^2)$ increases monotonically with ω^2.

To prove this we define

$$z_n(\omega^2)=u_{n-1}/u_n=\tan(\varphi_n/2). \qquad \text{(II, 6)}$$

Then (3) implies

$$z_{n+1}=1/(2-M_n\omega^2-z_n) \quad \text{for} \quad n=1, 2, \cdots N. \qquad \text{(II, 7)}$$

with the boundary condition

$$z_1=0, \quad z_{N+1}=\infty. \qquad \text{(II, 8)}$$

Equations (II, 6) and (II, 7) establish a continuous connection between φ_n and φ_{n+1}. We shall make this connection unique by requiring in accordance with (II, 6) and (II, 7) that

$$\varphi_n=2\pi h+\pi \quad \text{implies} \quad \varphi_{n+1}=2\pi h+2\pi. \qquad \text{(II, 9)}$$

(Here and in the following, h is any integer.)

If we furthermore put $\varphi_1=0$ (in accordance with the boundary condition $u_0=0$) then $\varphi_n=\varphi_n(\omega^2)$ is a well-defined function of ω^2. The second boundary condition $u_{N+1}=0$ is satisfied whenever

$$\varphi_{N+1}(\omega^2)=\pi+2\pi h \quad \text{(boundary condition).} \qquad \text{(II, 10)}$$

Now from (II, 6) and (II, 7) it follows that

$$\left(\frac{\partial\varphi_{n+1}}{\partial\varphi_n}\right)_{\text{for constant }\omega^2}=z_{n+1}^2\left(\frac{1+z_n^2}{1+z_{n+1}^2}\right)\geq 0,$$

$$\left(\frac{\partial\varphi_{n+1}}{\partial\omega^2}\right)_{\text{for constant }\varphi_n}=2M_n\left(\frac{z_{n+1}^2}{1+z_{n+1}^2}\right)\geq 0.$$

Therefore with $\varphi_1=0$ the functions $\varphi_2(\omega^2)$, $\varphi_3(\omega^2)$, $\cdots$ are monotonic nondecreasing functions of ω^2.

Let now ω^2 increase from ω_a^2 to ω_b^2. Then $\varphi_{N+1}(\omega^2)$ increases monotonically from $\varphi(\omega_a^2)$ to $\varphi(\omega_b^2)$ and whenever in this region $\varphi_{N+1}(\omega^2)=\pi+2\pi h$, then both the boundary conditions are satisfied and $\omega^2=\omega_\nu^2$ is an eigenvalue. Therefore the number of eigenvalues in the interval $\omega_a^2<\omega^2<\omega_b^2$ is (with an error smaller than unity) given by

$$\frac{1}{2\pi}[\varphi_{N+1}(\omega_b^2)-\varphi_{N+1}(\omega_a^2)] \equiv N[M(\omega_b^2)-M(\omega_a^2)], \qquad \text{(II, 11)}$$

where $M(\omega^2)$ is defined as

$$M(\omega^2)\equiv(1/N)\times\{\text{number of eigenvalues } \omega_\nu^2 \text{ with } \omega_\nu^2<\omega^2\}. \qquad \text{(II, 12)}$$

Then[1]

$$M(\omega^2)=(1/2\pi N)\varphi_{N+1}(\omega^2)+\text{const.} \qquad \text{(II, 13)}$$

(c) Chains with Randomly Distributed Atoms

So far we have discussed only one chain with a specified distribution of the masses $M^1, \cdots M^i$ over the places (1), $\cdots$ (N). Now we consider a large number of such chains with different random distributions of the masses. For each of these chains the vector

$$\mathbf{v}_n(\omega^2)=\begin{pmatrix}0 & 1\\ -1 & 2-M_{n-1}\omega^2\end{pmatrix}\cdots\begin{pmatrix}0 & 1\\ -1 & 2-M_1\omega^2\end{pmatrix}\begin{pmatrix}0\\ 1\end{pmatrix}$$

is a different function of ω^2.

Let us describe the probability distribution of the vector $\mathbf{v}_n(\omega^2)=\begin{pmatrix}u_{n-1}\\ u_n\end{pmatrix}$ over the different directions in in the x-y plane by means of a distribution function for $z_n=u_{n-1}/u_n$. We define:

$$w_n[z]dz \equiv \frac{\text{No. of chains for which } z<z_n(\omega^2)<z+dz}{\text{total number of chains}}. \qquad \text{(II, 14)}$$

Then we can easily obtain a relationship between the distribution functions $w_n[z]$ and $w_{n+1}[z]$ for z_n and z_{n+1}:

If in all chains the nth place were occupied by the same mass $M_n=M^i$, then z_{n+1} would be a definite function of z_n:

$$z_{n+1}=1/(2-M^i\omega^2-z_n); \quad z_n=2-M^i\omega^2-(1/z_{n+1}).$$

---

[1] If all the masses are equal, $M_n=M$, then

$$\mathbf{v}_{N+1}=\begin{pmatrix}0 & 1\\ -1 & 2\alpha\end{pmatrix}^N\begin{pmatrix}0\\ 1\end{pmatrix}, \quad \text{with} \quad \alpha=1-M\omega^2/2.$$

The eigenvalues of the matrix $\begin{pmatrix}0 & 1\\ -1 & 2\alpha\end{pmatrix}$ are $\lambda_{1,2}=\alpha\pm(\alpha^2-1)^{\frac{1}{2}}$. If these eigenvalues are real, i.e., $|\alpha|>1$ or $M\omega^2>4$, then for increasing n, $\mathbf{v}_n$ converges to an eigenvector of the matrix. This eigenvector does not satisfy the second boundary condition and therefore no eigenvibrations can exist in this region.

If $|\alpha|<1$, however, then $\lambda_{1,2}=e^{\pm i\beta/2}$ with real β, and

$$\begin{pmatrix}0 & 1\\ -1 & 2\alpha\end{pmatrix}^N\begin{pmatrix}0\\ 1\end{pmatrix}=\mathbf{a}\cos(N\beta/2)+\mathbf{b}\sin(N\beta/2),$$

where $\mathbf{a}$ and $\mathbf{b}$ can be expressed by the eigenvectors of the matrix and depend slowly (compared with $N\cdot\beta$) on β. Therefore $\varphi_{N+1}(\omega^2)\approx N\cdot\beta$ and

$$M(\omega^2)=\frac{\beta}{2\pi}+\text{const}=-\frac{1}{\pi}\arcsin\left\{1-\frac{M\omega^2}{2}\right\}+\tfrac{1}{2}.$$

Therefore, in this special case, the probability of finding z_{n+1} in an interval dz_{n+1} must be equal to the probability of finding z_n in the corresponding interval, i.e.,

$$w_{n+1}[z_{n+1}]dz_{n+1}=w_n[z_n]dz_n,$$

or

$$w_{n+1}[z]=\frac{1}{z^2}w_n[2-M^i\omega^2-\tfrac{1}{z}].$$

Since actually the nth place can be occupied by all masses $M^1\cdots M^i$ with the probabilities $p^1\cdots p^i$, the actual relationship between $w_{n+1}[z]$ and $w_n[z]$ is

$$w_{n+1}[z]=\sum_i p^i\cdot\frac{1}{z^2}w_n\left[2-M^i\omega^2-\frac{1}{z}\right]. \quad \text{(II, 15)}$$

According to the boundary condition $z_1=0$ for all chains, we have

$$w_1[z]=\delta(z). \quad \text{(II, 15a)}$$

With this the functions $w_n[z]$ are completely determined.

If for large n the functions $w_n[z]$ approach a function $w[z]$, then $w[z]$ must satisfy

$$\begin{aligned}
&\text{(a)}\quad w[z]=\textstyle\sum_j p^j(1/z^2)w[2-M^j\omega^2-(1/z)],\\
&\text{(b)}\quad w[z]\geq 0, \qquad\qquad \text{(II, 16)}\\
&\text{(c)}\quad \int_{-\infty}^{\infty} w[z]dz=1.
\end{aligned}$$

(The relations (b) and (c) follow from the definition of $w_n[z]$ as a probability density.)

We shall see, however, in Appendix I that $w[z]$ may become an extremely singular function. (We shall give there an example where $w[z]$ is not continuous in any interval of the real axis.)

Therefore it is convenient to introduce

$$W_n[z]=\int_0^z w_n[z']dz'. \quad \text{(II, 17)}$$

Here we consider $W_n[z]$ as a many-valued function of z. (This convention is necessary in order to obtain the simple Eq. (II, 18) for $W_n[z]$.) The main branch $\tilde{W}_n[z]$ of $W_n[z]$ is obtained by integrating $\int_0^z w[z']dz'$ on the direct way without touching infinity (suppose $z\neq\infty$). The other branches are obtained by adding or subtracting $\int_0^\infty w_n[z]dz+\int_{-\infty}^0 w_n[z]dz=1$ several times. We see that $\tilde{W}_n[0]=0$ and that for all branches $W_n[0]$ in an integer.

Now Eq. (II, 15) can be written as

$$W_{n+1}[z]=\textstyle\sum_j p^j W_n[2-M^j\omega^2-(1/z)] -\tilde{W}_n[-\infty]. \quad \text{(II, 18)}$$

Here the same branch of W_n has to be taken for all values of j. Then $W_{n+1}[0]=W_n[-\infty]-\tilde{W}_n[-\infty]$ is an integer and therefore one branch of W_{n+1}, namely $\tilde{W}_{n+1}$ vanishes at $z=0$ in agreement with $W_{n+1}[z] =\int_0^z w_{n+1}[z']dz'$. Instead of (II, 15a), we can write

$$W_1[z]=\begin{cases}\tfrac{1}{2} & \text{for } z>0\\ 0 & \text{for } z=0\\ -\tfrac{1}{2} & z<0.\end{cases} \quad \text{(II, 18a)}$$

Then the functions $W_n[z]$ are completely determined by (II, 18) and (II, 18a). Now it can be shown (see Appendix I) that

$$\lim_{n\to\infty}\frac{W_1[z]+W_2[x]+\cdots+W_n[z]}{n}=W[z] \quad \text{(II, 19)}$$

exists (II, 19) and that $W[z]$ is a continuous function. This function satisfies the relations:

$$\begin{aligned}
&\text{(a)}\quad W[z]=\textstyle\sum_j p^j W[2-M^j\omega^2-(1/z)]-\tilde{W}[-\infty],\\
&\text{(b)}\quad W[z] \text{ is a monotonic nondecreasing function of } z, \qquad \text{(II, 20)}\\
&\text{(c)}\quad \tilde{W}[\infty]-\tilde{W}[-\infty]=1;\quad \tilde{W}[0]=0.
\end{aligned}$$

[The relations (b) and (c) are satisfied already by each $W_n[z]$ and (a) follows from (II, 18) and (II, 19).] It can be shown (see Appendix I) that $W[z]$ is uniquely determined by (II, 20).

Differentiating (II, 20), we obtain with $w[z]=(d/dz) \times W[z]$ Eqs. (II, 16) for $w[z]$, provided that $W[z]$ is a differentiable function. In cases where $W[z]$ may not be differentiable, we shall sometimes say "$w[z]$ is given by Eq. (II, 16)," as synonymous to "$W[z] =\int_0^z w[z']dz'$ is given by Eq. (II, 20)."

By means of (II, 18), (II, 18a), and (II, 19) the function $\tilde{W}[z]$ can be calculated by iteration.[2] This iteration method can be assumed to converge quite well except in some limiting cases. In Sec. IV we shall give an approximate explicit expression for $W[z]$ in such a limiting case.

As a last step we have to express $M(\omega^2)$ of Eq. (12) by means of the distribution functions $w[z]=w[z,\omega^2]$, or $W[z]=W[z,\omega^2]$ (it is convenient to indicate the ω^2 dependence explicitly).

Let us consider for a specified chain the numbers

$$\varphi_{N+1}(\omega^2)>\varphi_N(\omega^2)>\cdots>\varphi_2(\omega^2)>\varphi_1(\omega^2)=0. \quad \text{(II, 21)}$$

Write $\varphi_{N+1}(\omega^2)=2\pi H+\epsilon$, where $0\leq\epsilon<2\pi$ and H is a large integer (we assume that N is very large). Now compare with (II, 21) the numbers

$$\varphi_{N+1}(\omega^2)\geq 2\pi H>2\pi(H-1)>\cdots >4\pi>2\pi>\varphi_1(\omega^2)=0. \quad \text{(II, 22)}$$

Since $2\pi>\varphi_{n+1}-\varphi_n>0$ [this is seen from (II, 6), (II, 7), (II, 9)], it must occur for H different values of n that φ_n and φ_{n+1} include an integral multiple of 2π, i.e., that $\varphi_{n+1}\geq 2\pi h>\varphi_n$. From (II, 6), (II, 7), and (II, 9)

[2] In Sec. IV(b) a different iteration method for $w[z]$ is developed.

it is seen that the latter condition is satisfied if and only if $-\pi \leq \varphi_n - 2\pi h < 0$ or if $-\infty \leq z_n < 0$. Therefore H is equal to the number of z_n's with negative sign in the series $z_N, z_{N-1}, \cdots z_1 = 0$. On the other hand—averaging over many chains—this number is given by

$$N\int_{-\infty}^{0} w[z]dz = -N\bar{W}[-\infty].$$

Therefore, for large N,

$$(1/2\pi N)\bar{\varphi}_{N+1} = \bar{H}/N = -\bar{W}[-\infty]$$

and with (13) we obtain the final result for $M(\omega^2)$ as defined in (II, 12):

$$\bar{M}(\omega^2) = -\bar{W}[-\infty, \omega^2]. \tag{II, 23}$$

[The additive constant in (II, 13) is seen to be zero by checking (II, 23) for $\omega^2=0$. Then $z_{n+1}=1/(2-z_n)$ and $z_n \to 1$ for $n \to \infty$. Therefore $\int_{-\infty}^{0} w_n[z_n]dz_n = -W_n[-\infty, \omega^2] \to 0$. On the other hand, (II, 12) gives $M(\omega^2)=0$ for $\omega^2=0$.]

(d) Summary

We define

$$F[\alpha^1, \cdots \alpha^i; p^1, \cdots p^i] = \int_{-\infty}^{0} w[z]dx, \tag{II, 24}$$

where $w[z]$ is uniquely determined by

$$w[z] = \sum_i p^i \frac{1}{z^2} w[2\alpha^i - (1/z)],$$

$$w[z] \geq 0, \tag{II, 25}$$

$$\int_{-\infty}^{\infty} w[z]dz = 1.$$

Now the relative number of eigenvalues with $\omega_r^2 < \omega^2$ for a long random chain is

$$\bar{M}(\omega^2) = F[\alpha^1, \cdots \alpha^i; p^1, \cdots p^i], \tag{II, 26}$$

with $\alpha^i = 1 - M^i\omega^2/2$. Here p^i is the fraction of atoms with mass M^i.

Similar results were obtained for the first time by Dyson.[3] Dyson's results, though equivalent, are slightly more complicated mathematically. In particular the relationship (II, 26) is much simpler than Dyson's expression for the distribution function.

III. ELECTRONIC ENERGY LEVELS IN ONE-DIMENSIONAL DISORDERED POTENTIALS

In this section, we consider the electronic energy levels in one-dimensional models of a disordered crystal consisting of different atoms $A^1, \cdots A^i$. We restrict ourselves to models which satisfy the following conditions:

1. The atoms $A_1, A_2, \cdots A_N$ in the one-dimensional model shall be equidistant; let the nth atom A_n lie at $x=nl$.

2. The atom A_n can be any one of the $A^1, \cdots A^i$. Considering many systems, we assume that the probability of A_n being an A^i is p^i, independent of the neighboring atoms. (The more general case where correlations between neighboring atoms exist is discussed in Appendix II.)

3. The potential $V(x)$ in the interval $(n-\frac{1}{2})l < x < (n+\frac{1}{2})l$ shall be determined completely by the nth atom, $V(x) = U^i(x-n\cdot l)$ in $(n-\frac{1}{2})l < x < (n+\frac{1}{2})l$, if $A_n = A^i$.

The general method for calculating the density of electronic energy states will be described in (c). As a preliminary we shall consider in (a) and (b) two special potential types for which the results can be stated more explicitly. Here the crystal potential is represented by a succession of equidistant δ functions of different strengths:

$$V(x) = \sum_n (\hbar^2/2m) V_n \cdot \delta(x - n\cdot l).$$

Here V_n characterizes the potential of the nth atom and can assume the values $V^1, \cdots V^i$ corresponding to the different atoms $A^1, \cdots A^i$. Now we consider separately the cases

(a) $V_n > 0$ for all n,

(b) $V_n < 0$ for all n.

We shall prove that the density of eigenstates in these cases can be expressed by the function $F[\alpha^1 \cdots \alpha^i; p^1 \cdots p^i]$ of Eq. (II, 24).

(a) $V(x) = \Sigma_n (\hbar^2/2m) V_n \delta(x-nl)$; $V_n > 0$

We write the wave function in the interval $(n-1)l < x < nl$ as $\psi = A_n e^{ikx_n} + B_n e^{-ikx_n}$, with $x_n = x - l(n-\frac{1}{2})$ where $k = [(2m/\hbar^2)E]^{\frac{1}{2}}$ and E is the energy. Then $\begin{pmatrix} A_{n+1} \\ B_{n+1} \end{pmatrix}$ and $\begin{pmatrix} A_n \\ B_n \end{pmatrix}$ are connected by the relationship

$$\begin{pmatrix} A_{n+1} \\ B_{n+1} \end{pmatrix} = \begin{bmatrix} \left(1+\frac{V_n}{2ik}\right)e^{ikl} & \frac{V_n}{2ik} \\ -\frac{V_n}{2ik} & \left(1-\frac{V_n}{2ik}\right)e^{-ikl} \end{bmatrix} \begin{pmatrix} A_n \\ B_n \end{pmatrix}$$

$$\equiv \mathfrak{M}_n \begin{pmatrix} A_n \\ B_n \end{pmatrix}. \tag{III, 1a}$$ [4]

Assuming as boundary conditions the vanishing of $\psi(x)$ at $x=l/2$ and $x=(N+\frac{1}{2})l$ these boundary condi-

[3] F. J. Dyson, Phys. Rev. 92, 1331 (1953).

[4] If all the potentials are equal: $V_n = V_0, \mathfrak{M}_n = \mathfrak{M}_0$, then the regions in which eigenstates exist are given by $|\lambda_{1,2}|=1$, where $\lambda_{1,2}$ are the eigenvalues of $\mathfrak{M}_0$: $|\mathfrak{M}_0 - \lambda_{1,2}| = \lambda_{1,2}^2 - 2\lambda_{1,2}[\cos kl + (V_0/2k)\sin kl] + 1 = 0$. $|\lambda_{1,2}|=1$ is equivalent to $|\cos kl + (V_0/2k) \times \sin kl| < 1$.

tions can be written as

$$A_1+B_1=0,\quad A_{N+1}+B_{N+1}=0. \qquad \text{(III, 1b)}$$

Now we introduce

$$\tau_n=-iA_n/B_n. \qquad \text{(III, 2)}$$

Then (III, 1) can be written as

$$\tau_{n+1}=(c_n\tau_n+1)/(\tau_n+c_n^*), \quad \text{where} \quad c_n=i[1+(2ik/V_n)]e^{ikl} \qquad \text{(III, 3a)}$$

$$\tau_1=\tau_{N+1}=i \quad \text{(boundary conditions)}. \qquad \text{(III, 3b)}$$

To solve this boundary problem we define with $\tau_1=i$ and Eq. (III, 3a) the functions of k: $\tau_1=i$, $\tau_2(k)$, $\cdots\tau_{N+1}(k)$. Then the k values of the electronic eigenstates are given by $\tau_{N+1}(k)=i$. From (III, 3a) with $\tau_1=i$ it is seen by induction that $|\tau_n|=1$ for all n values. Therefore we can introduce angles φ_n by

$$\tau_n=e^{i\varphi_n}. \qquad \text{(III, 4)}$$

The continuous relation between φ_n and φ_{n+1} given by (III, 3a) and (III, 4) can be made unique by the auxiliary requirement that

$$\varphi_n=\tfrac{1}{2}\pi-kl+2\pi h_0 \quad \text{shall imply} \quad \varphi_{n+1}=\tfrac{1}{2}\pi+kl+2\pi h_0, \qquad \text{(III, 5)}$$

with the same integer h_0 in both expressions. If furthermore we put $\varphi_1=\pi/2$ (in accordance with the boundary condition $\tau_1=i$), then $\varphi_n=\varphi_n(k)$ is a well-defined function of k (for given V_n's); and whenever $\varphi_{N+1}=\frac{1}{2}\pi+2\pi h$, then $\psi(x)$ vanishes at $x=(N+\frac{1}{2})l$ and both the boundary conditions are satisfied. From (III, 3), (III, 4), and (III, 5) with $V_n>0$ it follows by a short calculation that $\varphi_n(k)$ is a monotonically nondecreasing function of k. Therefore in an interval $k_1<k<k_2$ the boundary condition $\varphi_{N+1}(k)=\frac{1}{2}\pi+2\pi h$ is satisfied (with an error smaller than 1) $[\varphi_{N+1}(k_2)-\varphi_{N+1}(k_1)]/2\pi$ times, i.e., the number of eigenstates in the interval $k_1<k<k_2$ is

$$N[M(k_2)-M(k_1)]=[\varphi_{N+1}(k_2)-\varphi_{N+1}(k_1)]/2\pi \qquad \text{(III, 6)}$$

with

$$M(k')\equiv(1/N)\times\{\text{No. of eigenstates with } k<k'\} =[\varphi_{N+1}(k')/2\pi]+\text{const.} \qquad \text{(III, 7)}$$

Suppose now that we have many random chains of δ potentials, where V_n can assume the values $V^1, \cdots V^i$ with the probabilities $p^1, \cdots p^i$. We look at φ_n (for a specified large n) in many such chains. We define $w(\varphi)d\varphi$ as the probability that φ_n lies in *any one* of the intervals

$$2\pi h+\varphi<\varphi_n<2\pi h+\varphi+d\varphi,$$

with $h=0$ or ±1 or $\pm2\cdots$. For this function $w(\varphi)$ we can easily derive a functional equation. In order to write this equation most conveniently, we must express φ_n as a function of φ_{n+1}. We shall write

$$\varphi_n=\varphi_{n+1}-D^j(\varphi_{n+1}), \quad \text{provided that } A_n=A^j. \qquad \text{(III, 8)}$$

From (III, 3a), (III, 4), and (III, 5) we could easily derive an explicit expression for the so defined functions $D^j(\varphi)$. We shall need, however, in the following only two properties of the functions $D^j(\varphi)$ which are easily verified:

$$D^j(\varphi+2\pi)=D^j(\varphi),\quad D^j(\tfrac{1}{2}\pi+kl)=2kl. \qquad \text{(III, 9)}$$

Assuming now that $w(\varphi)$ is the distribution function for φ_n as well as for φ_{n+1} (for large enough n), we obtain by the same reasoning as in Sec. II for $w(\varphi)$:

$$w(\varphi)d\varphi=\textstyle\sum_j p^j w(\varphi-D^j(\varphi))d(\varphi-D^j(\varphi)). \qquad \text{(III, 10)}$$

It is convenient to introduce

$$W(\varphi)=\int_{\frac{1}{2}\pi+kl}^{\varphi} w(\varphi')d\varphi', \qquad \text{(III, 11)}$$

where $w(\varphi)$ is considered as a periodic function, $w(\varphi+2\pi)=w(\varphi)$ [this follows from the above definition of $w(\varphi)$]. Then $W(\varphi)$ is a single-valued function defined for all real values of φ and satisfies the relations:

(a) $W(\varphi)=\sum_j p^j W(\varphi-D^j(\varphi))-W(\frac{1}{2}\pi-kl)$,

(b) $W(\varphi+2\pi)=W(\varphi)+1$, (III, 12)

(c) $W(\frac{1}{2}\pi+kl)=0$,

(d) $W(\varphi)$ is a monotonically nondecreasing function of φ.

[(III, 12)(a) is obtained from (III, 10) by integration; note that $\varphi=\frac{1}{2}\pi+kl$ implies that $\varphi-D^j(\varphi)=\frac{1}{2}\pi-kl$ and that $W(\frac{1}{2}\pi+kl)=0$. (III, 12)(b) and (d) follow from the definition of $w(\varphi)$ as a probability density and (III, 12)(c) follows from (III, 11).]

In order to express $M(k)$ of Eq. (III, 7) by means of $W(\varphi)$, we shall use a method different from that applied in II but more generally applicable. From (III, 7) we have

$$\begin{aligned}2\pi NM(k)+\text{const}&=\varphi_{N+1}(k)\\&=[\varphi_{N+1}(k)-\varphi_N(k)]+[\varphi_N(k)-\varphi_{N-1}(k)]\\&\quad+\cdots+[\varphi_2(k)-\varphi_1(k)].\end{aligned} \qquad \text{(III, 13)}$$

We see that $2\pi M(k)+\text{const}$ equals the average value over different chains of $\varphi_{n+1}(k)-\varphi_n(k)$ (for large enough n). This average value can be written as

$$\langle\varphi_{n+1}-\varphi_n\rangle_{\text{Av}} =\sum_j p^j\int(\varphi_{n+1}-\varphi_n)_j w(\varphi_n)d\varphi_n. \qquad \text{(III, 14)}$$

Here the subscript j indicates that φ_n and φ_{n+1} are related by $\varphi_n=\varphi_{n+1}-D^j(\varphi_{n+1})$ as in Eq. (III, 8), i.e., $(\varphi_{n+1}-\varphi_n)_j$ equals $(\varphi_{n+1}-\varphi_n)$ provided that $V_n=V^j$.

Integration by parts gives from (III, 14)

$$\langle \varphi_{n+1}-\varphi_n\rangle_{Av}=\sum_j p^j(\varphi_{n+1}-\varphi_n)_j W(\varphi_n)\Big|_{\frac{1}{2}\pi-kl}^{\frac{1}{2}\pi-kl+2\pi}$$

$$-\sum_j p^j\int_{\frac{1}{2}\pi-kl}^{\frac{1}{2}\pi-kl+2\pi}\left(\frac{d\varphi_{n+1}}{d\varphi_n}-1\right)_j W(\varphi_n)d\varphi_n$$

$$=2kl+\int_{\frac{1}{2}\pi-kl}^{\frac{1}{2}\pi-kl+2\pi} W(\varphi_n)d\varphi_n$$

$$-\int_{\frac{1}{2}\pi+kl}^{\frac{1}{2}\pi+kl+2\pi}\sum_j p^j W(\varphi_{n+1}-D^j(\varphi_{n+1}))d\varphi_{n+1}.$$

[We could have integrated in (III, 14) over any interval of length 2π. The present choice of the limits of integration is convenient because at these limits $(\varphi_{n+1}-\varphi_n)_j=2kl$, independent of j.]

By means of the functional Eq. (III, 12a) for $W(\varphi)$, the last integral can be written as $-\int\{W(\varphi)+W(\frac{1}{2}\pi-kl)\}d\varphi$, and we finally obtain

$$(1/2\pi)\langle\varphi_{n+1}-\varphi_n\rangle_{Av}=-W(\tfrac{1}{2}\pi-kl)=M(k). \quad \text{(III, 15)}$$

[Note that the additive constant in (III, 13) vanishes because for $k\to 0$ $W(\frac{1}{2}\pi-kl)\to W(\frac{1}{2}\pi+kl)=0$.] With this result our problem is reduced to finding $W(\varphi)$ from (III, 12). We could calculate $W(\varphi)$ directly from (III, 12) by an iteration method. We shall see now, however, that $W(\varphi)$ is closely related to $W[z]$ in (II, 20) and that consequently $M(k)$ can be expressed by means of $F[\alpha^1\cdots\alpha^i;p^1\cdots p^i]$ of (II, 24). We define

$$z_n=\frac{1-ie^{i(-\varphi_n+kl)}}{1-ie^{i(-\varphi_n-kl)}}=\frac{\cos kl-\sin\varphi_n}{1-\sin(kl+\varphi_n)}. \quad \text{(III, 16)}$$

Then (III, 3) and (III, 4) imply

(a) $z_{n+1}=1/(2\alpha_n-z_n)$ with

$$\alpha_n=\cos kl+(V_n/2k)\sin kl. \quad \text{(III, 17)}$$

(b) $z_1=z_{N+1}=-1$ (boundary conditions).

For a better understanding of the transformation (III, 16) note that $\varphi'=\frac{1}{2}\pi-kl$ and $\varphi''=\frac{1}{2}\pi+kl$ are transformed into $z'=\infty$ and $z''=0$. Therefore $z_n=\infty$ implies that $z_{n+1}=0$ because $\varphi_n=\varphi'$ implies that $\varphi_{n+1}=\varphi''$. It follows that the relation between z_{n+1} and z_n must have the form $z_{n+1}=c/(b-z_n)$ and furthermore c can be made unity. From (III, 17) we can draw a simple interesting conclusion: Consider a region where none of the pure A^j crystals $(j=1,2,\cdots i)$ has an energy level, i.e., a region where $|\alpha^j|=|\cos kl+(V^j/2k)\sin kl|>1$ for all j.[4] Now for a mixed crystal we obtain from (III, 17a) with $z_1=-1$ successively $|z_2|<1$, $|z_3|<1$, $\cdots|z_{N+1}|<1$, because $|\alpha_n|>1$. Therefore the second boundary condition $z_{N+1}=-1$ cannot be satisfied. That means that in this region the mixed crystal has no electronic states either.

This result was conjectured first by Saxon and Hutner and proved subsequently by Luttinger.[5]

Let us now define $w[z]$ by $w[z]|dz|=|w(\varphi)d\varphi|$ where z and φ are connected by Eq. (III, 16). (Note that $d\varphi/dz>0$ if $0<kl<\pi$ and $d\varphi/dz<0$ for $\pi<kl<2\pi$). Then Eq. (III, 10) gives for $w[z]$:

$$w[z]=\sum_j p^j w\left[2\alpha^j-\frac{1}{z}\right]\cdot\frac{1}{z^2}, \quad \text{with } \alpha^j=\cos kl+\frac{V^j}{2k}\sin kl;$$

furthermore

$$\int_{-\infty}^{\infty} w[z]dx=1 \quad \text{and} \quad w[z]\geq 0.$$

The same function $w[z]$ which is uniquely determined by these relations, occurred in (II, 25), only with a different meaning of the α^i's. To express $M(k)$ in terms of $w[z]$, we write $kl=2\pi h+\delta$ with $0\leq\delta<2\pi$. Then Eq. (III, 15) with $W(kl)=0$ gives

$$M(k)=W(\tfrac{1}{2}\pi+2\pi h+\delta)-W(\tfrac{1}{2}\pi-2\pi h-\delta)$$

$$=2h+W(\tfrac{1}{2}\pi+\delta)-W(\tfrac{1}{2}\pi-\delta)$$

$$=2h+\int_{\frac{1}{2}\pi-\delta}^{\frac{1}{2}\pi+\delta} w(\varphi)d\varphi$$

$$=\begin{cases} 2h+\int_{-\infty}^{0} w[z]dz & \text{for } 0\leq\delta\leq\pi \\ 2h+1+\int_{0}^{\infty} w[z]dz & \text{for } \pi\leq\delta<2\pi. \end{cases}$$

Therefore we have the final result: If $kl=2\pi h+\delta$ (i. we are in the region of the $(2h+1)$th and $(2h+2)$th band), then

$$M(k)=\begin{cases} 2h+F[\alpha^1\cdots\alpha^i;p^1\cdots p^i] & \text{for } 0\leq\delta\leq\pi \\ 2h+2-F[\alpha^1\cdots\alpha^i;p^1\cdots p^i] & \text{for } \pi\leq\delta<2\pi. \end{cases} \quad \text{(III, 18)}$$

Here $F[\;]$ is defined as in (II, 24) and $\alpha^j=\cos kl+(V^j/2k)\sin kl$.

(b) $V(x)=-\sum_n(\hbar^2/2m)V_n\delta(x-nl)$; $V_n>0$

A single one-dimensional potential hole $V(x)=-(\hbar^2/2m)V_0\delta(x)$ with $V_0>0$ has exactly one bound electron state with the binding energy $(\hbar^2/2m)(V_0^2/4)$. Therefore we expect that for the crystal potential considered there exist states with negative total energy. We shall consider here these states only. Then the wave function in the interval $(n-1)l<x<nl$ can be written as

$$\psi(x)=A_ne^{ikx_n}+B_ne^{-ikx_n}, \quad \text{with } x_n=x-l(n-\tfrac{1}{2}).$$

[5] J. M. Luttinger, Philips Research Repts. 6, 303 (1951).

Here k is real and $E=-(\hbar^2/2m)k^2$ is the total energy. The relation between $\begin{pmatrix}A_n\\B_n\end{pmatrix}$ and $\begin{pmatrix}A_{n+1}\\B_{n+1}\end{pmatrix}$ is easily found:

$$\begin{pmatrix}A_{n+1}\\B_{n+1}\end{pmatrix}=\begin{bmatrix}\left(1-\dfrac{V_n}{2k}\right)e^{kl} & -\dfrac{V_n}{2k}\\ \dfrac{V_n}{2k} & \left(1+\dfrac{V_n}{2k}\right)e^{-kl}\end{bmatrix}\begin{pmatrix}A_n\\B_n\end{pmatrix}$$

$$\equiv\mathfrak{M}_n\begin{pmatrix}A_n\\B_n\end{pmatrix}. \quad \text{(III, 19a)}$$

The boundary conditions $\psi=0$ at $x=l/2$ and $x=(N+\frac{1}{2})l$ can be written as

$$A_1+B_1=0, \qquad A_{N+1}+B_{N+1}=0. \quad \text{(III, 19b)}$$

If all potentials are equal, $V_n=V_0$, $\mathfrak{M}_n=\mathfrak{M}_0$, then the regions in which eigenstates exist are given by $|\lambda_{1,2}|=1$ or $|\alpha_0(k)|=|-\cosh(kl)+(V_0/2k)\sinh(kl)|<1$, where $\lambda_{1,2}$ are the eigenvalues of $\mathfrak{M}_0$. In the case where $|\alpha_0(0)|>1$ or $V_0l>4$, no eigenstates with $k\approx0$ are possible; i.e., the energy of all states is negative, and we have a band with N bound states which correspond for very large values of V_0l to the bound states of N single potential holes. For $V_0l<4$ however, the energy band will be broadened so as to include some states with positive energy. Therefore the number of states with negative energy, which we are considering here, will be smaller than N.

In the following, we shall assume that $lV_n<6$ for all n. This is a very weak assumption [e.g., for $l=10^{-8}$ cm it means that the binding energy in a single potential hole, $|E_b|=(\hbar^2/2m)(V_0^2/4)$ is smaller than $(\hbar^2/8ml^2)\times36=28$ ev]. However, it simplifies the following considerations since it implies that $d\alpha_n(k)/dk<0$ (for all n and k values), where $\alpha_n=-\cosh(kl)+(V_n/2k)\sinh(kl)$. Now we apply to (III, 19) the transformation

$$z_n=-(A_ne^{-kl}+B_n)/(A_n+B_ne^{-kl}). \quad \text{(III, 20)}$$

This gives

$$z_{n+1}=1/(2\alpha_n-z_n), \text{ with } \alpha_n=-\cosh(kl)+(V_n/2k)\sinh(kl), \quad \text{(III, 21a)}$$

$$z_1=z_{N+1}=1 \quad \text{(boundary conditions)}. \quad \text{(III, 21b)}$$

If we put $z_1=1$, then by (III, 21a) $z_2(k), z_3(k), \cdots z_{N+1}(k)$ are functions of k and $z_{N+1}(k)=1$ is the condition for an eigenstate. Furthermore the $z_n(k)$ are monotonically nondecreasing functions of k for $V_nl<6$ (because then $d\alpha_n/dk<0$). It follows then as in (II) that the number of eigenvalues in an interval $k_1<k<k_2$ is given by $N[M(k_2)-M(k_1)]$, where

$$M(k)=F[\alpha^1\cdots\alpha^i;\, p^1\cdots p^i]+\text{const}, \text{ with } \alpha^i=-\cosh(kl)+(V^i/2k)\sinh(kl). \quad \text{(III, 22)}$$

(c) General Method

Now we develop the general method for calculating the density of electronic energy levels in a potential which satisfies the conditions 1–3.[6] In the interval $(n-\frac{1}{2})l<x<(n+\frac{1}{2})l$ the wave function $\psi(x)$ is determined by the values of $\psi(x)$ and $\psi'(x)$ at one point. The values of $\psi(x), \psi'(x)$ at the ends of the interval are related by

$$\begin{pmatrix}\psi(x)\\ \psi'(x)\end{pmatrix}_{x=(n+\frac{1}{2})l}=\mathfrak{M}_n(k)\begin{pmatrix}\psi(x)\\ \psi'(x)\end{pmatrix}_{x=(n-\frac{1}{2})l}. \quad \text{(III, 23)}$$

Here the matrix $\mathfrak{M}_n(k)$ is determined by the potential in the nth interval. Since we assumed this potential to depend on the nature of the nth atom only, we can write

$$\mathfrak{M}_n(k)=\mathfrak{M}^i(k) \quad \text{provided that} \quad A_n=A^i.$$

The matrices $\mathfrak{M}^1(k), \cdots\mathfrak{M}^i(k)$ shall be assumed known. They depend on the electron energy. k shall be any parameter which characterizes this energy. If we proceed from one cell to the next, then $\psi(x), \psi'(x)$ change continuously with x. Therefore

$$\begin{pmatrix}\psi(x)\\ \psi'(x)\end{pmatrix}_{x=(N+\frac{1}{2})l}=\mathfrak{M}_N(k)\mathfrak{M}_{N-1}(k)\cdots\mathfrak{M}_1(k)\begin{pmatrix}\psi(x)\\ \psi'(x)\end{pmatrix}_{x=l/2} \quad \text{(III, 24)}$$

We can assume that $\psi(x)$ is real and therefore define φ_n by

$$\tan(\tfrac{1}{2}\varphi_n)=\left.\frac{\psi(x)}{\psi'(x)}\right|_{x=(n-\frac{1}{2})l} \quad \text{(III, 25)}$$

Then (III, 23) and (III, 25) imply a relationship between φ_n and φ_{n+1}. This relationship can be made unique by an auxiliary requirement [see Eq. (III, 5)]. We write the so-defined unique connection between φ_n and φ_{n+1} as

$$\varphi_n=\varphi_{n+1}-\Delta^i(\varphi_{n+1}) \text{ provided that } A_n=A^i. \quad \text{(III, 26)}$$

The functions $\Delta^i(\varphi)$ can be computed from the matrices $\mathfrak{M}^i$ and are periodic: $\Delta^i(\varphi+2\pi)=\Delta^i(\varphi)$. As our boundary conditions we shall assume $\psi(x)=0$ at $x=l/2$ and $x=(N+\frac{1}{2})l$ or, in terms of φ,

$$\varphi_1=0, \quad \varphi_{N+1}=2\pi h.$$

If we define by (III, 26) with $\varphi_1=0$ functions $\varphi_1=0$, $\varphi_2(k), \cdots\varphi_{N+1}(k)$, then the eigenstates of the systems are given by $\varphi_{N+1}(k)=2\pi h$. Considering now many random systems, we describe the probability distribution of φ_n by a density function $w(\varphi)$. This function

[6] In this section we shall omit some proofs which are obtained by the same reasoning as the corresponding proofs in (II) or (IIIa).

satisfies [see Eq. (III, 10)]

$$\begin{aligned} &w(\varphi)d\varphi=\textstyle\sum_j p^j w(\varphi-\Delta^j(\varphi))d(\varphi-\Delta^j(\varphi)),\\ &w(\varphi)\geq 0,\\ &\int_0^{2\pi} w(\varphi)d\varphi=1,\\ &w(\varphi+2\pi)=(\varphi). \end{aligned} \tag{III, 27}$$

For $W(\varphi)=\int_d^\varphi w(\varphi')d\varphi'$, where d is an arbitrary constant, we obtain

$$\begin{aligned} &W(\varphi)=\textstyle\sum_j p^j W(\varphi-\Delta^j(\varphi))+C(k),\\ &W(\varphi)\quad\text{is monotonic in }\varphi,\\ &W(\varphi+2\pi)=W(\varphi)+1,\\ &W(d)=0. \end{aligned} \tag{III, 28}$$

The relations (III, 28) determine the function $W(\varphi)$ and the constant $C(k)$ (for fixed k) uniquely. Furthermore $W(\varphi)$ and $C(k)$ can be calculated from (III, 28) by an iteration method. We note that $C(k)$ is independent of the choice of d and that $C(k)$ can be written as

$$C(k)=\sum_j p^j \int_{-\Delta^j(0)}^0 w(\varphi)d\varphi. \tag{III, 29}$$

In order to find the mean value of $\varphi_{N+1}(k)$, we write

$$\begin{aligned}\bar{\varphi}_{N+1}(k)&=N\langle\varphi_{n+1}-\varphi_n\rangle_{\text{Av}}\\ &=N\sum_j\int p^j(\varphi_{n+1}-\varphi_n)_j w(\varphi_n)d\varphi_n \quad\text{(for large } N \text{ and } n).\end{aligned}$$

The subscript j indicates that φ_{n+1} and φ_n are related by $\varphi_n=\varphi_{n+1}-\Delta^j(\varphi_{n+1})$. Integration by parts gives, with (III, 28), the result

$$\frac{1}{2\pi N}\bar{\varphi}_{N+1}(k)=C(k). \tag{III, 30}$$

Let us assume first that for all possible chains $\varphi_{N+1}(k)$ increases with k monotonically. Then for k increasing from k_1 to k_2, $\varphi_{N+1}(k)$ increases from $\varphi_{N+1}(k_1)$ to $\varphi_{N+1}(k_2)$ monotonically. Hence in this interval the boundary condition $\varphi_{N+1}(k)=2\pi h$ is satisfied $(1/2\pi)\times\{\varphi_{N+1}(k_2)-\varphi_{N+1}(k_1)\}$ times and therefore the average number of eigenvalues in $k_1<k<k_2$ is

$$\frac{1}{2\pi}\{\bar{\varphi}_{N+1}(k_2)-\bar{\varphi}_{N+1}(k_1)\}=N\{C(k_2)-C(k_1)\}.$$

This condition that $\varphi_{N+1}(k)$ is monotonic in k for all possible chains was satisfied in the special cases considered in II, III(a), and III(b), and there we used this property of $\varphi_{N+1}(k)$ for obtaining the results. Now we shall extend our theory to cases where this condition is not satisfied, i.e., where the functions $\varphi_{N+1}(k)$ for the individual chains need not be monotonic. In these cases we can determine $(1/2\pi N)\bar{\varphi}_{N+1}(k)=C(k)$ as before, but now $C(k)$ also need not be monotonic. Assuming however that $C(k)$ is a smooth function of k, we can divide the k-values into intervals in each of which $C(k)$ is monotonic.

Let us consider such an interval $k_1<k<k_2$ where the function $C(k)$ is monotonic, assume for example that $C(k)$ increases with k in (k_1,k_2). Now there may be chains for which $\varphi_{N+1}(k)$ does not increase with k in (k_1,k_2). We shall show however that the fraction of these chains goes to zero when N and the number of chains compared approach infinity.

Let N, m, s be large numbers such that $ms=N+1$ and write $\varphi_{N+1}(k)$ as a sum of s terms:

$$\begin{aligned}\varphi_{N+1}(k)&=(\varphi_m-\varphi_1)+(\varphi_{2m}-\varphi_m)+\cdots\\ &\quad+(\varphi_{(\nu+1)m}-\varphi_{\nu m})+\cdots+(\varphi_{sm}-\varphi_{(s-1)m})\\ &=t_1+t_2+\cdots+t_{\nu+1}+\cdots+t_s.\end{aligned}$$

Averaging over all chains, we obtain

$$\bar{t}_{\nu+1}=\langle\varphi_{(\nu+1)m}-\varphi_{\nu m}\rangle_{\text{Av}}=mC(k),\quad\text{independent of }\nu.$$

Consider now for a moment those of the chains only for which $\varphi_{\nu m}$ (for one specified ν) lies in a certain small interval. Then for these chains the numbers $\varphi_{\nu m}$, $\varphi_{\nu m+1}$, $\varphi_{\nu m+2}$, $\cdots$ will soon approach the probability distribution given by $w(\varphi)$, independent of the value of $\varphi_{\nu m}$; i.e., almost all of the $\varphi_{\nu m}$, $\varphi_{\nu m+1}$, $\varphi_{\nu m+2}$, $\cdots\varphi_{\nu m+m}$ are (for large n) distributed according to $w(\varphi)$. Therefore the average of $(\varphi_{(\nu+1)m}-\varphi_{\nu m})$ taken over the restricted class of chains with specified $\varphi_{\nu m}$ will also be $mC(k)$, independent of the value of $\varphi_{\nu m}$. Now in the sum

$$\begin{aligned}\frac{d}{dk}\varphi_{N+1}(k)&=\frac{d}{dk}(\varphi_m-\varphi_1)+\cdots+\frac{d}{dk}(\varphi_{sm}-\varphi_{(s-1)m})\\ &=t_1'+\cdots+t_s',\end{aligned}$$

averaging over all chains again, each term has a positive average value and the average value $\bar{t}_{\nu+1}'$ is independent of t_ν'. Therefore, for large s, the fraction of chains for which the sum in negative approaches zero, i.e., for almost all chains $\varphi_{N+1}(k)$ increases with k and we have the final result:

In a region $k_1<k<k_2$ where the mean value $\bar{\varphi}_{N+1}(k)=2\pi NC(k)$ is a monotonic function of k, there the number of eigenvalues is given by

$$N|C(k_2)-C(k_1)|=(1/2\pi)|\bar{\varphi}_{N+1}(k_2)-\bar{\varphi}_{N+1}(k_1)|.$$

There may be chains for which $\varphi_{N+1}(k)$ has not the monotonic character of $\bar{\varphi}_{N+1}(k)$ but the fraction of these chains goes to zero and hence they can be neglected.

IV. IMPURITY BANDS

Introduction

Suppose that we have a crystal of A atoms with a small amount of B-atoms. Then in an energy region where the pure A crystal has no allowed electronic states, there may occur an "impurity band" of electron states due to the B atoms. Impurity bands can occur also in the frequency spectrum of an elastic chain: Suppose the chain consists of atoms A with mass M and a small number of atoms B with a lighter mass m. Then above the frequency limit of the pure M chain, there appears a band of frequencies due to the masses m.

If the B atoms in the crystal are far apart from each other, then the impurity band will be very narrow and for increasing concentration of the B atoms the width of the impurity band will increase.

By means of the general methods developed in the previous sections, it is possible in principle to calculate the density of eigenstates everywhere and in particular in the impurity band. But the iteration method used in II and III, though always convergent, converges sufficiently quickly only so long as the fraction q of B atoms is not too small.

In this section, however, we shall consider the case where the concentration of B atoms is very small. Therefore we have to apply here a different method. We shall approach the problem first, in (a), by a simple approximate method. Later, in (b), we shall develop a more accurate theory based on the general results of Secs. II and III.

It will be sufficient to consider for the present the impurity bands of the elastic chain only. The generalization to electron impurity bands is then straightforward.

(a) *Approximate Method*

Here we calculate the impurity bands in an over-simplified model of the elastic chain. We cannot expect this calculation to give quantitatively good results. The reasons for considering this model however are: (1) Since we obtain explicit results here, we can check the validity of an approximation method which becomes essential in (b). (2) Our over-simplified model can also be applied to three-dimensional problems. It is therefore interesting to check the usefulness of this model by comparison with the more realistic model of (b).

Assume that we have a chain of elastically coupled masses M (fraction p) with a small number of lighter masses m (fraction q). Then each m-mass in the chain will generally be embedded in many M-masses and will therefore in first approximation behave like a mass m when embedded in an infinite chain of M masses. In a second approximation which will be studied here, we take into account not only one mass m but also the next nearest m mass, neglecting all other m-masses in the chain.

Suppose that the two m masses considered are separated by s M masses. Then for calculating the frequencies due to these two m masses we consider them in our simplified model as embedded in an infinite chain of M masses, and again separated by s M masses.

Now we shall calculate the eigenfrequencies due to the two m masses for such a system. Then, by averaging over these systems with the correct statistical weight for all possible s values, we shall obtain an approximate expression for the energy spectrum of a disordered chain.

Looking at one m mass in a random chain, the probability that the next m mass in one direction of the chain is separated by s M masses is clearly $P_1(s)=qp^s$ $(s=0, 1, 2, \cdots)$. Likewise the probability that the next m mass in the other direction is separated by s M masses is $P_2(s)=qp^s$. Therefore the probability that the next m mass in one or the other direction is separated by s m masses becomes $P(s)=P_1(s)P_2(>s)+P_2(s)P_1(>s)+P_1(s)P_2(s)$, or

$$P(s)=2qp^{2s+1}+q^2p^{2s} \quad (\approx 2qp^{2s} \text{ for } q\ll p). \tag{IV, 1}$$

Now we consider a chain consisting successively of: K_0 masses M, 1 mass m, s masses M, 1 mass m, K masses M, where K is very large. As in II, let u_n be the displacement of the nth mass and define

$$\mathbf{v}_n=\begin{pmatrix} u_{n-1} \\ u_n \end{pmatrix}; \quad T_\alpha=\begin{pmatrix} 0 & 1 \\ -1 & 2\alpha \end{pmatrix}; \quad T_\beta=\begin{pmatrix} 0 & 1 \\ -1 & 2\beta \end{pmatrix},$$

$$\text{with } 2\alpha=2-M\omega^2, \quad 2\beta=2-m\omega^2. \tag{IV, 2}$$

Now the displacement vectors $\mathbf{v}_1$ and $\mathbf{v}_{2K+s+3}=\mathbf{v}_{\text{end}}$ at the beginning and the end of the chain are related by

$$\mathbf{v}_{\text{end}}=T_\alpha{}^K T_\beta T_\alpha{}^s T_\beta T_\alpha{}^K \mathbf{v}_1. \tag{IV, 3}$$

We want to calculate the eigenfrequencies due to the two m masses above the frequency limit of the pure M chain. In this region T_α has real eigenvalues:

$$T_\alpha\mathbf{v}'=\lambda'\mathbf{v}', \quad T_\alpha\mathbf{v}''=\lambda''\mathbf{v}'', \tag{IV, 4}$$

with

$$\lambda'=\alpha+(\alpha^2-1)^{\frac{1}{2}}, \quad \lambda''=\alpha-(\alpha^2-1)^{\frac{1}{2}}, \quad \lambda''<-1<\lambda'<0,$$

$$\mathbf{v}'=\begin{pmatrix} 1 \\ \lambda' \end{pmatrix}, \quad \mathbf{v}''=\begin{pmatrix} 1 \\ \lambda'' \end{pmatrix}.$$

If now the m masses vibrate with a frequency above the frequency limit of the M chain, then the amplitude of the M masses must decrease exponentially towards the ends of the chain, i.e.,

$$\left.\begin{matrix} T_\alpha{}^n\mathbf{v}_{K+s+3} \\ T_\alpha{}^{-n}\mathbf{v}_{K+1} \end{matrix}\right\}\to 0 \text{ for large } n.$$

This is possible only if

$$\mathbf{v}_{K+s+3}=\mathbf{v}'\times\text{const},$$
$$\mathbf{v}_{K+1}=\mathbf{v}''\times\text{const}.$$

Therefore the eigenfrequencies are given by the relation

$$\mathbf{v}'=T_\beta T_\alpha{}^s T_\beta\mathbf{v}''\times\text{const}, \tag{IV, 5}$$

with

$$S=\begin{pmatrix}1 & -\lambda'\\ 1 & -\lambda''\end{pmatrix};\; Sv'=\begin{pmatrix}1\\0\end{pmatrix}\times\text{const};\; Sv''=\begin{pmatrix}0\\1\end{pmatrix}\times\text{const};$$

$$ST_\alpha S^{-1}=\begin{pmatrix}\lambda'^2 & 0\\ 0 & 1\end{pmatrix}\times\text{const}.$$

We obtain from (IV, 5)

$$\begin{pmatrix}1\\0\end{pmatrix}=ST_\beta S^{-1}\begin{pmatrix}\lambda'^{2s} & 0\\ 0 & 1\end{pmatrix}ST_\beta S^{-1}\begin{pmatrix}0\\1\end{pmatrix}\times\text{const},$$

and inserting

$$ST_\beta S^{-1}=\begin{pmatrix}(B+1)\lambda'^2 & -\lambda'^2\\ 1 & B-1\end{pmatrix}\times\text{const},$$

$$\text{with}\quad B=\frac{(\alpha^2-1)^{\frac{1}{2}}}{\beta-\alpha},\tag{IV, 6}$$

we finally obtain as conditions for the eigenfrequencies:

$$(1-B)^2=\lambda'^{2(s+1)}$$

or

$$B=\frac{(\alpha^2-1)^{\frac{1}{2}}}{\beta-\alpha}=1\pm|\lambda'|^{s+1}.\ \text{(Note that } |\lambda'|<1.\text{)}\tag{IV, 7}$$

It is easily seen that this equation [where α, β, and λ' are given by (IV, 2) and (IV, 4)] determines exactly two eigenfrequencies ω_s^+ and ω_s^- above the limit frequency of the M chain, provided that $m<\frac{1}{2}M$. For $M>m>\frac{1}{2}M$ two such solutions exist only if s is large enough. (Consider as an illustration the case $s=0$ where the two m masses are neighbors. If $m<\frac{1}{2}M$, then two vibrations exist where the masses m swing against each other or in phase. In the latter case the effective swinging mass is $2m$ and therefore this type of localized vibration in the chain is possible only for $2m<M$.)

When we average now over all possible values of s, the density of eigenfrequencies in the impurity band can be written as

$$\mu(\omega^2)=\tfrac{1}{2}\textstyle\sum_s P(s)\{\delta(\omega^2-\omega_s^{+2})+\delta(\omega^2-\omega_s^{-2})\},\tag{IV, 8}$$

where

$$\mu(\omega^2)d\omega^2=\frac{\text{No. of frequencies in } d\omega^2}{\text{No. of } m \text{ masses in the chain}}.$$

Here we have assumed that either $m<M/2$ and hence for each s two frequencies $\omega_s^\pm$ (above the frequency limit of the pure M-chain) exist, or that we consider only large s values. If $m<M/2$ is satisfied, then $\mu(\omega^2)=0$ beneath the frequency limit of the M chain and $\int_0^\infty\mu(\omega^2)d\omega^2=1$ [since $\sum P(s)=1$] i.e., the number of frequencies in the impurity band equals the number of m masses. For large values of s, i.e., if the two m masses are far apart, ω_s^+ and ω_s^- approach, according to (IV, 7), a frequency $\bar\omega$, given by $B=1$ or

$$\bar\omega^2=\frac{4}{M}\frac{1}{\vartheta(2-\vartheta)},\quad\text{with}\quad \vartheta=\frac{m}{M}.\tag{IV, 9}$$

This frequency $\bar\omega$ is therefore the resonance frequency of a single mass m in a long chain of masses M. [The frequency limit of the pure M chain is given by $\omega^2=(4/M)<\bar\omega^2$.]

If the concentration q of the m masses decreases then the probability of finding two m masses close to each other (e.g., $s<10$) decrease with q^2. Now only two m masses close to each other can give a frequency much different from $\bar\omega$. Therefore, for low impurity concentration q, almost all frequencies lie near to $\bar\omega$.

Let us consider this case, where frequencies much different from $\bar\omega$ can be neglected, in more detail. The relation (IV, 6) can be written near $\omega=\bar\omega$ as

$$B-1=\left(\frac{\vartheta(1-\vartheta/2)}{(1-\vartheta)^2}\right)\left(\frac{\omega^2-\bar\omega^2}{\bar\omega^2}\right),\tag{IV, 10}$$

with $\bar\omega^2$ and ϑ from (9). Using $c=B-1$ as our new coordinate, the eigenfrequencies are determined [see Eq. (IV, 7)] by

$$c=B-1=\pm|\lambda'|^{s+1}.\tag{IV, 11}$$

Therefore the density of frequencies $\mu[c]=\mu(\omega^2)d\omega^2/dc$ is given by

$$\begin{aligned}\mu[c]&=\tfrac{1}{2}\sum_{s=0}^{\infty}P(s)\{\delta(c-|\lambda'|^{s+1})+\delta(c+|\lambda'|^{s+1})\}\\ &=\sum_{s=1}^{\infty}qp^{2s}\{\delta(c-|\lambda'|^{s})+\delta(c+|\lambda'|^{s})\}.\end{aligned}\tag{IV, 12}$$

This function $\mu[c]$ is highly singular at $c\approx 0$ (note that $|\lambda'|<1$ and that therefore the δ functions accumulate around $c=0$). In a real chain these δ functions will be smeared out a little by the interactions between more than two masses which we have neglected here. Therefore it is reasonable to replace $\mu[c]$ for $c\approx 0$ by a smoothed function $\tilde\mu[c]$. In order to determine $\mu[c]$ we form $\int\mu[c]dc$, replace this by a smoother function, and differentiate again: If $c=|\lambda'|^{s_0}$, where s_0 is a positive integer, then

$$\int_0^{c+0}\mu[c']dc'=\sum_{s=s_0}^{\infty}qp^{2s}=\tfrac{1}{2}p^{2s_0}=\tfrac{1}{2}|c|^{4\epsilon},$$

with

$$\epsilon=\ln p/2\ln|\lambda'|=q\Big/2\ln\left|\frac{2-\vartheta}{\vartheta}\right|.\tag{IV, 13}$$

[Here $\lambda'=\alpha+(\alpha^2-1)^{\frac{1}{2}}$ has been replaced by its value at $\omega=\bar\omega$: $\lambda'=-\vartheta/(2-\vartheta)$.] Therefore the most natural way of smoothing $\int_0^c \mu(c')dc'$ is to replace it by

$$\int_0^c \tilde{\mu}[c']dc' = \tfrac{1}{2}|c|^{4\epsilon} \quad \text{for} \quad c>0.$$

This gives

$$\tilde{\mu}[c] = 2\epsilon|c|^{4\epsilon-1}.$$

The same result is obtained by replacing the s summation in (IV, 12) by an integration. This is justified if besides $p\approx 1$ also $|\lambda'|\approx 1$ is satisfied. A simple calculation gives

$$\int_{s=1}^{\infty} ds\, qp^{2s}\{\delta(c-|\lambda'|^s)+\delta(c+|\lambda'|^s)\} = 2\epsilon|c|^{4\epsilon-1} \qquad (\text{if } |c|<|\lambda'|),$$

with ϵ as in (IV, 13). This replacement of the sum by an integral can be described quasi-physically as follows: Actually the numbers s of M atoms between two nearest m atoms is an integer, occurring with the probability $P(s) = 2qp^{2s}$ (for $p\approx 1$). We calculate, however, *as if* s could assume any positive value with the probability density $2qp^{2s}$. This procedure is justified mathematically only if $|\lambda'|\approx 1$. In our example it gives good results, however, also if this condition is not satisfied since even then the result agrees with the preceding smoothing method. The procedure of replacing s by a continuous parameter will be used again in (b). We shall assume there also that the condition $|\lambda'|\approx 1$ is not so essential for the result.

Concluding our discussion of the simplified model—in which we take into account only the interaction between two nearest masses m at a time—we see that the density of eigenfrequencies in the impurity band can be described by

$$\mu[c]\approx\tilde{\mu}[c] = 2\epsilon|c|^{4\epsilon-1}, \qquad \text{(IV, 14)}$$

with

$$c = \frac{\vartheta(1-\vartheta/2)}{(1-\vartheta)^2}\,\frac{\omega^2-\bar{\omega}^2}{\bar{\omega}^2}, \quad \epsilon = \frac{q}{2\ln|(2-\vartheta)/\vartheta|}, \quad \vartheta = \frac{m}{M},$$

provided that the following conditions are satisfied: (1) q must be small enough such that almost all eigenfrequencies lie near $\bar{\omega}$; (2) ω must lie near to $\bar{\omega}$ such that the approximation (IV, 10) is valid and that $\mu(c)$ can be replaced by $\tilde{\mu}[c]$.

The previous theory could be improved by considering the vibrations of 3, 4, 5 ··· masses m in a chain of masses M. This way of approach to the problem is troublesome, however. Therefore we shall use in the next part (b) a different method based on the results of Secs. II and III. We shall obtain there a better expression for $\mu[c]$, which is different from (IV, 14).

(b) *More Accurate Method*

In order to calculate the impurity bands of an M chain containing some lighter masses m (fraction $q\ll 1$) we begin here—according to the general theory of Secs. II and III—by calculating the distribution function $w[z]$. Equations (II, 16a) and (II, 20a) for $w[z]$ and $W[z]$ read now

$$\text{(a)} \quad w[z] = \frac{p}{z^2}w\left[2\alpha-\frac{1}{z}\right]+\frac{q}{z^2}w\left[2\beta-\frac{1}{z}\right],$$

$$\text{(b)} \quad W[z] = pW\left[2\alpha-\frac{1}{z}\right]+qW\left[2\beta-\frac{1}{z}\right]+\text{const}, \qquad \text{(IV, 15)}$$

where p and q are the fractions of M- and m-masses in the chain and α, β are given by (IV, 2). We are considering a region above the frequency limit of the M chain, i.e., $\alpha<-1$. Furthermore we can assume $-1<\beta<+1$, since for $\alpha<-1$ an $\beta<-1$ no eigenfrequencies exist.

It is convenient to introduce

$$x = (z-\lambda')/(z-\lambda''),$$

$$w(x)dx = w[z]dz; \quad W(x) = W[z], \qquad \text{(IV, 16)}$$

where λ', λ'' are defined as in (IV, 4). Then $w(x)$ is determined by [see (IV, 15a) and (II, 16)]:

$$\text{(a)} \quad w(x) = p\lambda''^2 w(\lambda''^2 x)+q\frac{\partial T_\beta^{-1}(x)}{\partial x}w[T_\beta^{-1}(x)],$$

$$\text{(b)} \quad w(x)\geq 0, \qquad \text{(IV, 17)}$$

$$\text{(c)} \quad \int_{-\infty}^{\infty} w(x)dx = 1.$$

with

$$T_\beta^{-1}(x) = \frac{x(B-1)+\lambda'^2}{-x+(B+1)\lambda'^2}; \quad B = \frac{(\alpha^2-1)^{\frac{1}{2}}}{\beta-\alpha}. \qquad \text{(IV, 18)}$$

Before we go into details it is important to obtain a qualitative idea of the behavior of $w(x)$: As before, let u_n be the displacement of the nth mass and define the numbers

$$x_n = (z_n-\lambda')/(z_n-\lambda''),$$

where

$$z_n = u_{n-1}/u_n.$$

Then the relationship between x_{n+1} and x_n is

$$x_{n+1} = T_\alpha(x_n) = \lambda'^2 x_n,$$

or

$$x_{n+1} = T_\beta(x_n) = [(B+1)\lambda'^2 x_n-\lambda'^2]/[x_n+(B-1)],$$

depending on whether the nth mass is M or m. Now $w(x)dx$ is the probability that x_n (for large n) lies in the interval $(x, x+dx)$, when we compare either many chains at the same n or many different n-values at one long random chain. In the case of a pure M chain ($q=0$), the relation between x_{n+1} and x_n is always $x_{n+1}=\lambda'^2 x_n$; and since $\lambda'^2<1$, x_n will go to zero for $n\to\infty$. Therefore we have $w(x)=\delta(x)$ for $q=0$.

For very small values of q one might assume $w(x)$ to differ only slightly from $\delta(x)$ because in the series $x_1, x_2, x_3, \cdots$ the relation between successive x values is mostly $x_{n+1}=\lambda'^2 x_n$, which tend to push x to zero. Only the transformation $x_{n+1}=T_\beta(x_n)$, which occurs with the small probability q, may push x out of the neighborhood of zero again. But this effect of a transformation T_β will generally be canceled by the many following transformations T_α.

This argument fails, however, if $T_\beta(0)\approx\infty$. Here again many of the x_n's will be near to zero, but a transformation T_β now pushes these x_n's near to infinity. An x_n near enough to infinity, however, cannot be transformed into the neighborhood of zero, even by as many as $1/q$ successive transformations T_α.

Hence in this case the x_n's accumulate around $x=\infty$ as well as around $x=0$ and $w(x)$ differs from zero appreciably at $x\approx 0$ *and* $x\approx\infty$. We see that even a small number of lighter masses m in a chain of masses M may change $w(x)$ considerably if the "resonance condition" $T_\beta(0)=\infty$ is satisfied. From (IV, 18) it follows that $T_\beta(0)=\infty$ or $T_\beta^{-1}(\infty)=0$ is equivalent to $B=1$. Therefore $T_\beta(0)=\infty$ means [see part (a)] that ω equals the resonance frequency of a single m mass in an M chain. [The same result can be obtained easily without referring to part (a).]

We return to Eqs. (IV, 15). In deriving these equations in Sec. II we obtained an iteration method for $W[z]$. Now we derive Eq. (IV, 15) once more in order to obtain a new and here more useful iteration method.

Considering a chain of M and m masses, let us define the numbers z_n as before. Then for a particular chain

$$z_{n+1}=1/(2\alpha-z_n)\equiv T_\alpha[z_n],\quad 2\alpha=2-M\omega^2$$

or

$$z_{n+1}=1/(2\beta-z_n)\equiv T_\beta[z_n],\quad 2\beta=2-m\omega^2$$

according to whether the nth mass M_n is M or m. Assume that $M_n=m$ for $n=s_1, s_2, \cdots (s_{n+1}>s_n)$ and that all other places are occupied by masses M. Now consider among the z_n's only $z_{s1}, z_{s2}, \cdots$ and write for convenience $z_{s_n}\equiv z_{(n)}$. Then

$$z_{(n+1)}=T_\alpha^{(s_{n+1}-s_n-1)}T_\beta[z_{(n)}].$$

Let $w_{(n)}[z]$ be the distribution function for $z_{(n)}$ if we average over many chains.

Assuming for a moment that in all chains there would be the same number $s=(s_{n+1}-s_n-1)$ of M masses between the nth m mass and the $(n+1)$th m mass, we would obtain

$$z_{(n+1)}=T_\alpha^s T_\beta[z_{(n)}],$$

and consequently, for

$$W_{(n)}[z]=\int_0^z w_{(n)}[z']dz':$$

we obtain

$$W_{(n+1)}[z]=W_{(n)}[T_\beta^{-1}T_\alpha^{-s}[z]]+\text{const.}$$

Since it occurs, however, with the probability $P'(s)=qp^s$ that the nth and $(n+1)$th m masses are separated by s M masses, we obtain

$$W_{(n+1)}[z]=\sum_{s=0}^{\infty} qp^s W_{(n)}[T_\beta^{-1}T_\alpha^{-s}[z]]+\text{const.}\quad \text{(IV, 19)}$$

This gives a new iteration method for $W[z]$, since again

$$\lim_{n\to\infty}(W_{(1)}[z]+W_{(2)}[z]+\cdots+W_{(n)}[z])\times\frac{1}{n}=W[z].$$

[This can be proved in the same way as (II,19).] It is easily verified that the equation

$$W[z]=\sum_{s=0}^{\infty} qp^s W[T_\beta^{-1}T_\alpha^{-s}[z]]+\text{const}$$

is equivalent to

$$W[z]=pW[T_\alpha^{-1}[z]]+qW[T_\beta^{-1}[z]]+\text{const},$$

which is Eq. (15b). Introducing x and $W(x)$ as in (IV, 16), we can write (IV, 19) as

$$W_{(n+1)}(x)=\sum_{s=0}^{\infty} qp^s W_{(n)}(T_\beta^{-1}(\lambda''^{2s}x))+\text{const.}\quad \text{(IV, 20)}$$

So far our equations are quite general. Now we assume that q is very small. Then the number s of M masses between successive m masses is mostly very large, since the probability for a certain value s is $P'(s)=qp^s$. Here we make the following approximation: We calculate *as if* s could assume any positive value with the probability density qp^s. This is equivalent to replacing the s-summation in (IV, 20) by an integration. [This is mathematically justified if, besides $p\approx 1$, also $|\lambda''|\approx 1$. We may assume, however, as explained in (a), that the requirement $|\lambda''|\approx 1$ is not very important for the result.] Then we obtain for $W(x)$, from (IV, 20),

$$W(x)=\int_0^{\infty} qp^s W(T_\beta^{-1}(\lambda''^{2s}x))ds+\text{const};\quad \text{(IV, 21)}$$

and, differentiating, we get

$$w(x)=\int_0^{\infty} qp^s w(T_\beta^{-1}(\lambda''^{2s}x))\frac{dT_\beta^{-1}(\lambda''^{2s}x)}{dx}ds,$$

or, with

$$\tau=\lambda''^{2s}x,\quad \epsilon=q/\ln\lambda''^2,$$

$$w(x)=\int_x^{(+\infty)x}\epsilon\frac{1}{\tau}\left|\frac{x}{\tau}\right|^{\epsilon-1} w(T_\beta^{-1}(\tau))\frac{dT_\beta^{-1}(\tau)}{d\tau}d\tau.$$

The last equation can be written as

$$w(x)=\int_{-\infty}^{\infty}L(\tau,x)\frac{dT_\beta^{-1}(\tau)}{d\tau}w(T_\beta^{-1}(\tau))d\tau \qquad \text{(IV, 22)}$$

$$=\int_{-\infty}^{\infty}L(T_\beta(\tau'),x)w(\tau')d\tau',$$

with

$$L(\tau,x)=\epsilon|x|^{\epsilon-1}/|\tau|^{\epsilon} \quad \text{for}$$
$$0<x<\tau \quad \text{and} \quad \tau<x<0 \qquad \text{(IV, 22')}$$
$$=0 \text{ otherwise.}$$

We want to solve Eq. (IV, 22) for $w(x)$ under the assumption that q is very small and that the frequency considered lies near the resonance frequency of a single m mass in an M chain (i.e., $B\approx1$). In order to do this we note:

1. For any function $g(\tau)$ which vanishes in a neighborhood of $\tau=0$, we obtain from (IV, 22')

$$\int_{-\infty}^{\infty}L(\tau,x)g(\tau)d\tau=\left\{\begin{matrix}c_1|x|^{\epsilon-1} & \text{for} & x>0\\ c_2|x|^{\epsilon-1} & \text{for} & x<0\end{matrix}\right\} \text{ near } x=0,$$

where c_1 and c_2 are constants.

2. We have seen that $w(x)$ is very small except near $x=0$ and $x=\infty$. Therefore, in (22), $[dT_\beta^{-1}(\tau)/d\tau]\times w(T_\beta^{-1}(\tau))$ is very small near $\tau=0$. [From (IV, 18), with $B\approx1$, it is seen that $T_\beta^{-1}(0)$ is neither near to 0 nor near to ∞.]

Looking now at $w(x)$ on the left of Eq. (IV, 22), we see that $w(x)$ near $x=0$ must be almost equal to a function $\Delta(x)$ defined as

$$\begin{aligned}\Delta(x)&=aq|x|^{\epsilon-1} && \text{for } 0<x<1\\ &=bq|x|^{\epsilon-1} && \text{for } -1<x<0 \qquad \text{(IV, 23)}\\ &=0 && \text{otherwise,}\end{aligned}$$

where a and b are constants. Inserting into both sides of (IV, 22) such a function $w(x)$ with

$$\begin{aligned}w(x)&=\Delta(x) && \text{for } -1<x<1,\\ &=0 && \text{except near } x=0 \text{ and } x=\infty,\end{aligned}$$

where we note that $\Delta(x)$ is very small for $x\rightarrow\pm1$, we obtain from (IV, 22) near $x=0$:

$$\Delta(x)=\int_{-1}^{1}L(T_\beta(x'),x)\Delta(x')dx' + \int_{x'\approx\infty}L(T_\beta(x'),x)w(x')dx',$$

or

$$\Delta(x)=\int_{-1}^{1}L(T_\beta(x'),x)\Delta(x')dx'+dL(T_\beta(\infty),x), \qquad \text{(IV, 24)}$$

with

$$d=\int_{x'\approx\infty}w(x')dx'.$$

[Here we used that $L(x_0,x)$ depends very slowly on x_0 near $x_0=T_\beta(\infty)$.]

Now the constants a, b, and d can be determined as follows: from $\int_{-\infty}^{\infty}w(x)dx=1=\int_{-1}^{1}\Delta(x)dx+d$, it follows that

$$(a+b)\ln\lambda''^2+d=1. \qquad \text{(IV, 25)}$$

Two further relations between a, b, and d are obtained from (IV, 24) by comparing both sides for $x>0$ and $x<0$. These three equations for a, b, and d give us, after a short calculation[7]:

$$\frac{1}{a}=2\ln\lambda''^2+\mathcal{g}_3-\mathcal{g}_2-\frac{\mathcal{g}_1\mathcal{g}_3}{\ln\lambda''^2} \quad \text{if } B<1, \qquad \text{(IV, 26a)}$$

$$\frac{1}{a}=2\ln\lambda''^2+\mathcal{g}_2\frac{\ln\lambda''^2-\mathcal{g}_3}{\ln\lambda''^2-\mathcal{g}_2} \quad \text{if } B>1, \qquad \text{(IV, 26b)}$$

$$\mathcal{g}_1=q\int_0^1|(x+|c|)|^{\epsilon}|x|^{\epsilon-1}dx\approx\tfrac{1}{2}\ln\lambda''^2(1+|c|^{2\epsilon}),$$

$$\mathcal{g}_2=q\int_{-|c|}^{0}|(x+|c|)|^{\epsilon}|x|^{\epsilon-1}dx\approx\ln\lambda''^2|c|^{2\epsilon},$$

$$\mathcal{g}_3=q\int_{-1}^{-|c|}|(x+|c|)|^{\epsilon}|x|^{\epsilon-1}dx\approx\tfrac{1}{2}\ln\lambda''^2(1-|c|^{2\epsilon}),$$

where $|c|=|B-1|$. (The approximate expressions for the integrals can be used since $q\ll1$.)

As the last step in our calculation, we show that $M(\omega^2)$ of Eq. (II, 12) can be expressed by the constant a only:

$$M(\omega^2)=q(1-a\ln\lambda''^2)+\text{const.} \qquad \text{(IV, 27)}$$

---

[7] Let us consider, for example, the case $B<1$. In the region $x>0$, Eq. (IV, 24) gives

$$aq|x|^{\epsilon-1}=\epsilon|x|^{\epsilon-1}\left\{d|T_\beta(\infty)|^{-\epsilon}+b\int_{-1}^{0}q|x'|^{\epsilon-1}|T_\beta(x')|^{-\epsilon}dx' + a\int_0^{1-B}q|x'|^{\epsilon-1}|T_\beta(x')|^{-\epsilon}dx'\right\}.$$

Since

$$T_\beta(x')=[(B+1)\lambda'^2x'-\lambda'^2]/[x'+(B-1)]\approx-\lambda'^2/[x'+(B-1)]$$

for $x'\approx0$ and $B\approx1$, and since furthermore $|T_\beta(\infty)|^{\epsilon}\approx1$ and $|\lambda'^2|^{\epsilon}\approx1$ for $q\ll1$, our equation can be written as

$$a=(\ln\lambda''^2)^{-1}\left\{d+bq\int_0^1|x''|^{\epsilon-1}|x''+|B-1||^{\epsilon}dx'' + aq\int_{-|B-1|}^{0}|x''|^{\epsilon-1}|x''+|B-1||^{\epsilon}dx''\right\}.$$

Similarly, considering (IV, 24) for $x<0$, we obtain

$$b=(\ln\lambda''^2)^{-1}aq\int_{-1}^{-|B-1|}|x|^{\epsilon-1}|x+|B-1||^{\epsilon}dx.$$

Now, with (IV, 25), the constant a can be calculated. This gives us the result (IV, 26a).

This result is easily obtained from the relation

$$M(\omega^2)=\int_{-\infty}^{0} w[z]dz+\text{const}$$

$$=-\int_{\lambda'^2}^{1} w(x)dx+\text{const [see Eq. (IV, 16)].}$$

With $w(x)=aq|x|^{\epsilon-1}$ in the region $\lambda'^2<x<1$, we obtain

$$M(\omega^2)=\text{const}-aq\epsilon^{-1}(1-\lambda'^{2\epsilon})$$
$$\approx\text{const}-aq\ln\lambda''^2 \quad \text{for} \quad \epsilon\approx 0.$$

This derivation of (IV, 27) is questionable, however. We know that $w(x)$ equals $aq|x|^{\epsilon-1}$ for small values of $x>0$ and we know that $w(x)$ is small like $aq|x|^{\epsilon-1}$ near $x=1$, but we have not proved that $aq|x|^{\epsilon-1}$ is a good approximation for $w(x)$ near $x=1$.

To prove (IV, 27) properly, we use here the same ideas which led us in Sec. II to the relation $M(\omega^2)=\int_{-\infty}^{0}w[z]dz+\text{const}$. We define ψ_n by $x_n=\tan(\frac{1}{2}\psi_n)$ and make the relation between ψ_{n+1} and ψ_n ($\psi_{n+1}=T_\alpha\{\psi_n\}$ or $\psi_{n+1}=T_\beta\{\psi_n\}$) unique by postulating that $T_\alpha\{0\}=0$; $2\pi<T_\beta\{\frac{1}{2}\pi\}<2\pi+\frac{1}{2}\pi$. Now $M(\omega^2)+\text{const}$ is equal to the prnbability that ψ_{n+1} and ψ_n include a multiple of 2π. For $\psi_{n+1}=T_\alpha\{\psi_n\}$ this is impossible. For $\psi_{n+1}=T_\beta\{\psi_n\}$ it occurs whenever x_n lies outside the interval $0<x_n<T_\beta^{-1}(0)=1/(B+1)$. Since the probability of finding x_n outside this interval is

$$1-\int_{0}^{(B+1)^{-1}} aq|x|^{\epsilon-1}dx\approx 1-a\ln\lambda''^2,$$

and since furthermore the relation $\psi_{n+1}=T_\beta\{\psi_n\}$ occurs with the probability q, the relationship (IV, 27) follows.

The final result can now be stated explicitly:

$$M(\omega^2)=1-\frac{q}{(\frac{3}{2}-\frac{1}{2}|c|^{2\epsilon})^2} \quad \text{for} \quad \omega^2<\bar{\omega}^2 \qquad \text{(IV, 28)}$$

$$=1-q\frac{1-|c|^{2\epsilon}}{(\frac{3}{2}-\frac{1}{2}|c|^{2\epsilon})^2} \quad \text{for} \quad \omega^2>\bar{\omega}^2,$$

with

$$\epsilon=\frac{q}{\ln\lambda''^2}=\frac{q}{2\ln|(2-\vartheta)/\vartheta|}, \quad \vartheta=\frac{m}{M}<1,$$

$$c=B-1$$

$$=\left(\frac{\vartheta(1-\vartheta/2)}{(1-\vartheta)^2}\right)\left(\frac{\omega^2-\bar{\omega}^2}{\bar{\omega}^2}\right), \quad \bar{\omega}^2=\frac{2}{M}\left(\frac{1}{\vartheta(1-\vartheta/2)}\right).$$

Comparing the result (IV, 28) with Eq. (IV, 14), we note:

1. Both equations were derived under the conditions $q\ll 1$ and $\omega\approx\bar{\omega}$.

2. Under these conditions the Eq. (IV, 14) was a rough approximation; Eq. (IV, 28), however, can be assumed to give very good results.

3. The density function $\mu[c]$ of (IV, 14) and the corresponding $m[c]\equiv(1/q)(d/dc)M(\omega^2)$ of (IV, 28) are both very sharply increasing functions near $c=0$ with weak long wings in the region of large $|c|$ values.

4. If q is very small, then the fraction of eigenvalues of the impurity band which lie in these wings [where (IV, 14) and (IV, 28) do not hold] is also very small.

5. Equation (IV, 14) differs essentially from (IV, 28). Note in particular that near $c=0$ $\mu[c]$ is proportional to $|c|^{4\epsilon-1}$ while $m[c]$ is there proportional to $|c|^{2\epsilon-1}$.

In cases where q is not quite so small that (IV, 28) can be applied, a better solution $w(x)$ of (IV, 22) might be useful. Such a solution could furthermore be improved by the iteration method (IV, 20).

The generalization of our results to electronic impurity bands presents no difficulties. Let us consider the model of Sec. III(b). Assume that the crystal consists of two types of atoms A and B, occurring with the probabilities p and q ($q\ll 1$), and let the potentials of these atoms be determined by V^p and V^q. If now $V^q>V^p$, then below the lowest energy state of the pure A crystal there are energy states due to the B atoms. As shown in (III, b) the number of eigenvalues k_ν with $k_\nu<k$ is

$$NM(k)=N\int_{-\infty}^{0} w[z]dz+\text{const},$$

where $w[z]$ satisfies Eq. (IV, 15) with

$$\alpha=-\cosh(kl)+(V^p/2k)\sinh(kl),$$
$$\beta=-\cosh(kl)+(V^q/2k)\sinh(kl). \qquad \text{(IV, 29)}$$

Starting from Eq. (IV, 15), we can now proceed as before with (IV, 29) and obtain the result analogous to (IV, 28):

$$M(k)=\text{const}+\begin{cases} -\dfrac{q}{(\frac{3}{2}-\frac{1}{2}|c|^{2\epsilon})^2} & \text{for} \quad B<1 \\ -q\dfrac{1-|c|^{2\epsilon}}{(\frac{3}{2}-\frac{1}{2}|c|^{2\epsilon})^2} & \text{for} \quad B>1, \end{cases}$$

where now with α, β from (IV, 29),

$$c=B-1, \qquad B=(\alpha^2-1)^{\frac{1}{2}}/(\beta-\alpha),$$
$$\epsilon=q/\ln\lambda''^2, \quad \lambda''=\alpha-(\alpha^2-1)^{\frac{1}{2}}.$$

In the special case $V^p=0$, we obtain

$$c=(k-\bar{k})/\bar{k},$$
$$\epsilon=q/\bar{k}l,$$

where $\bar{k}=\frac{1}{2}V^q$ is the k value at the resonance.

In conclusion I would like to thank Professor Luttinger, who suggested this problem to me, for many valuable discussions and help in completing the manuscript. Furthermore I am indebted to the American Academy of Science and to the Foreign Operations Administration for making my temporary stay in the United States possible.

APPENDIX I

We add some remarks concerning the solution $w[z]$ of Eqs. (II, 16). We write these equations as

$$w[z]=\sum_j p^j z^{-2} w[T_j^{-1}[z]],$$
$$w[z]\geq 0,\tag{A, 1}$$
$$\int_{-\infty}^{\infty} w[z]dz=1,$$

with

$$T_j^{-1}[z]=2\alpha^j-z^{-1},\quad 2\alpha^j=2-M^j\omega^2.\tag{A, 1a}$$

It is sometimes more convenient to consider $w(\varphi)$, defined by

$$w(\varphi)d\varphi=w[z]dz,\tag{A, 2}$$

with

$$z=\tan(\tfrac{1}{2}\varphi).\tag{A, 3}$$

Then we define the operators T_j^{-1} acting on φ as

$$T_j^{-1}(\varphi)\equiv 2\ \mathrm{arc\ tan}\left\{2\alpha^j-\frac{1}{\tan(\varphi/2)}\right\},\tag{A, 4a}$$

and make the continuous relationship $\varphi'=T_j^{-1}(\varphi)$ unique by requiring

$$T_j^{-1}(0)=\pi.\tag{A, 4b}$$

Now (A, 1) can be written in terms of $w(\varphi)$ as

$$w(\varphi)=\sum_j p^j w(T_j^{-1}(\varphi))\ dT_j^{-1}(\varphi)/d\varphi,$$
$$w(\varphi)\geq 0,$$
$$\int_0^{2\pi} w(\varphi)d\varphi=1,\tag{A, 5}$$
$$w(\varphi+2\pi)=w(\varphi).$$

For $W(\varphi)=\int_0^\varphi w(\varphi')d\varphi'$, this gives

(a) $W(\varphi)=\sum_j p^j W[T_j^{-1}(\varphi)]-W(-\pi)$,

(b) $W(\varphi)$ is monotonically non-decreasing in φ, (A, 6)

(c) $W(\varphi+2\pi)=W(\varphi)+1$,

(d) $W(0)=0$.

We show first that, provided a solution $W(\varphi)$ of (A, 6) exists, this solution is continuous (α) and unique (β).

(α) Assume that $W(\varphi)$ has a jump of height d at $\varphi=\varphi_0$ and that no higher jump does occur. Then $W(\varphi)$ has a jump d also at all points $\varphi_0+2\pi h$. From (A, 6a) it follows that $W(\varphi)$ must also have a jump d at all points $T_j^{-1}(\varphi_0)+2\pi h$ (note that $\sum p^j=1$). In the same way it is seen that $W(\varphi)$ must have a jump d at all points $\varphi_0+2\pi h$, $T_j^{-1}(\varphi_0)+2\pi h$, $T_k^{-1}(T_j^{-1}(\varphi_0))+2\pi h$, $\cdots$. Provided that we have at least two different α^j-values, there can be found an infinite number of such points in the interval $0\leq\varphi<2\pi$ at which $W(\varphi)$ should jump by the amount d. This is impossible since $W(\varphi)$ is monotonic and $W(2\pi)-W(0)=1$. Hence $W(\varphi)$ must be continuous.

(β) Assume that there are two different functions satisfying (A, 6). Then the difference $U(\varphi)$ satisfies:

$$\begin{aligned}&U(\varphi)=\sum_j p^j U(T_j^{-1}(\varphi))-U(-\pi),\\ &U(\varphi+2\pi)=U(\varphi),\\ &U(0)=0;\quad |U(\varphi)|<1,\\ &U(\varphi)\ \text{is continuous.}\end{aligned}\tag{A, 7}$$

Let $U(\varphi)$ assume its maximum at φ_1 and its minimum at φ_2. Then, from (A, 7),

$$-U(-\pi)=U(\varphi_1)-\sum p^j U(T_j^{-1}(\varphi_1))\geq 0,$$
$$-U(-\pi)=U(\varphi_2)-\sum p^j U(T_j^{-1}(\varphi_2))\leq 0,$$

hence $U(-\pi)=0$. It follows further that $U(\varphi)$ must reach its maximum on a whole set S_1 of points:

$$S_1=\{\varphi_1+2\pi h,\ T_j^{-1}(\varphi_1)+2\pi h,\ T_k^{-1}(T_j^{-1}(\varphi_1))+2\pi h,\ \cdots\},$$

and its minimum value at a set S_2:

$$S_2=\{\varphi_2+2\pi h,\ T_j^{-1}(\varphi_2)+2\pi h,\ T_k^{-1}(T_j^{-1}(\varphi_2))+2\pi h,\ \cdots\}$$

We can assume that $|\alpha^1|<1$ and $\alpha^1\neq\alpha^2$ (the case where $|\alpha^j|>1$ for all j has no physical interest). Then it is easily seen that for the sets S_1 and S_2 the point $\varphi=\pi$ is a common point of accumulation. [If there exists an integer n that $T_1^{-n}[z]$ is the unit transformation, then $T_1^{-(n-1)}=T_1$. From (A, 1a) we see that $T_2^{-1}T_1[z]=z+2(\alpha^2-\alpha^1)$. Therefore repeated application of the operator $T_2^{-1}T_1^{-(n-1)}$ transforms any z to infinity or any φ to π as near as we want. Therefore $\varphi=\pi$ is a common point of accumulation of S_1 and S_2. If $T_1^{-n}[z]$ is not the unit transformation for any integer n, then the points $T_1^{-n}[\varphi]+2\pi h$ lie dense everywhere (for any φ_0) and the points of S_1 and S_2 lie dense everywhere.]

Since $U(\varphi)$ is continuous, it follows that

$$U(\pi)=U(\varphi_1)=U(\varphi_2),\quad \text{i.e.,}\quad U(\varphi)=U(0)=0.$$

Now we can construct the solution $W(\varphi)$ of Eq. (A, 6). Let $W_0(\varphi)$ be any function which satisfies

$$W_0(\varphi+2\pi)=W_0(\varphi)+1,$$
$$W_0(0)=0, \qquad (A, 8)$$
$W_0(\varphi)$ is monotonically nondecreasing in φ.

Then define, with $W_0(\varphi)$ from (A, 8), the functions $W_n(\varphi)$, $W^n(\varphi)$ by

$$W_{n+1}(\varphi)=\sum_j p^j W_n(T_j^{-1}(\varphi))-W_n(-\pi), \qquad (A, 9)$$

$$W^n(\varphi)=[W_0(\varphi)+W_1(\varphi)+\cdots+W_n(\varphi)]/(n+1). \qquad (A, 10)$$

The functions $W_n(\varphi)$ and $W^n(\varphi)$ also satisfy $(A, 8)$. Furthermore, we see from (A, 9) and (A, 10) that

$$|W^n(\varphi)-\sum_j p^j W^n(T_j^{-1}(\varphi))+W^n(-\pi)| = \left|\frac{W_{n+1}(\varphi)-W_0(\varphi)}{n+1}\right| \leq \frac{1}{n+1}. \qquad (A, 11)$$

Since the functions $W^n(\varphi)$ are monotonic and bounded in $0\leq\varphi\leq 2\pi$, there exists a convergent subsequence $W^{c_1}(\varphi)$, $W^{c_2}(\varphi)$, $\cdots$ converging to a function $W(\varphi)$. From (A, 11), we see that $W(\varphi)$ satisfies (A, 6). Since furthermore no other solution of (A, 6) can exist, the series $W^1(\varphi)$, $W^2(\varphi)$, $\cdots$ must also converge to $W(\varphi)$. Having proved that $W[z]$ exists and is a continuous function, we shall now demonstrate as an example that $w[z]$ can be highly singular.

Assume that $|\alpha^1|<1$; $|\alpha^2|>1$. Then $z=T_2^{-1}[z]$ has two real solutions z', z'' with $|z'|<1$, $|z''|>1$. Assume now further that $p^{(2)}|z'|^{-2}>1$ and finally that T_1^{-n} is not the unit transformation for any integer n. Now $w[z]$ provided that it exists, satisfies

$$w[z]=p^{(1)}z^{-2}w[T_1^{-1}[z]]+p^{(2)}z^{-2}w[T_2^{-1}[z]]+\cdots,$$
$$w[z]\geq 0, \quad \int_{-\infty}^{\infty} w[z]dz=1. \qquad (A, 12)$$

Since all terms in (A, 12) are non-negative, this equation with $z=z'=T_2^{-1}[z']$ and $p^{(2)}|z'|^{-2}>1$ implies that

$$w[z']=0 \quad \text{or} \quad w[z']=\infty. \qquad (A, 13)$$

If $w[z']=0$, then from (A, 12) it follows that $w=0$ at $T_1^{-n}[z']$ for all integers $n>0$. These points lie dense everywhere. Hence $w[z]$ vanishes on some points of every interval. But $w[z]$ cannot vanish throughout any interval (this would imply $w[z]\equiv 0$ but $\int w[z]dz=1$). Therefore $w[z]$ cannot be continuous in any interval. The same result is obtained if $w[z']=\infty$. Here it is seen that $w[z]$ cannot be finite throughout any interval.

APPENDIX II

So far we have assumed that the atoms in our one-dimensional crystal are distributed at random. Now we consider the case where a correlation between neighboring atoms exists.

Suppose we have a two-component crystal containing A and B atoms occurring with the probabilities p and q. Comparing now many one-dimensional crystals, we call those crystals which happen to have an A atom at the nth place A crystals (n=any fixed number) and the crystals with a B atom at the nth place B crystals. If there is a correlation between neighboring atoms, then among the A crystals the probabilities of finding an A or B atom on the $(n-1)$th place may be different from p and q. We call these probabilities P_{AA} and P_{AB}. In the same way we define among the B crystals the probabilities of finding on the $(n-1)$th place an A or B atom as P_{BA} and P_{BB}. It is seen easily that

$$P_{AA}+P_{AB}=1,$$
$$P_{BA}+P_{BB}=1, \qquad (A, 14)$$
$$qP_{BA}=pP_{AB}.$$

We want to determine the electronic energy levels for the general crystal model discussed in (III, c). [This general formalism applies also to the vibrations of a linear chain of atoms]. The numbers φ_n [see (III, c)] for the different crystals depend only on the nature of the atoms preceding the nth atom [i.e., on the $(n-1)$th, $(n-2)$th, $\cdots$1st atom].

Since now there is an interaction between the nth atom and the $(n-1)$th atom, the distribution functions for φ_n may be different among the A crystals and among the B crystals. Let $w_A(\varphi_n)$ $[w_B(\varphi_n)]$ be the distribution functions for φ_n in the A $[B]$ crystals, where the nth place is occupied by an A $[B]$ atom. Instead of one distribution function $w(\varphi)=w_A(\varphi)=w_B(\varphi)$, we have now two such functions. We can easily find a set of difference equations for $w_A(\varphi)$ and $w_B(\varphi)$: Let us consider the A crystals first. Here the distribution of φ_n is given by $w_A(\varphi_n)$. The distribution of φ_{n-1} is $w_A(\varphi)$ or $w_B(\varphi)$ according to whether the $(n-1)$th place is occupied by an A or B atom. These two possibilities occur with the probability P_{AA} and P_{AB}. Therefore, as in (II, 15), we obtain

$$w_A(\varphi)d\varphi=P_{AA}w_A(\varphi-\Delta^A(\varphi))d(\varphi-\Delta^A(\varphi)) + P_{AB}w_B(\varphi-\Delta^B(\varphi))d(\varphi-\Delta^B(\varphi)),$$

or with

$$W_A(\varphi)=\int_0^\varphi w_A(\varphi')d\varphi'; \quad W_B(\varphi)=\int_0^\varphi w_B(\varphi')d\varphi',$$

$$W_A(\varphi)=P_{AA}W_A(\varphi-\Delta^A(\varphi)) + P_{AB}W_B(\varphi-\Delta^B(\varphi))+C_A(k). \qquad (A, 15a)$$

Considering the B crystals we obtain in the same way

$$W_B(\varphi)=P_{BA}W_A(\varphi-\Delta^A(\varphi)) + P_{BB}W_B(\varphi-\Delta^B(\varphi))+C_B(k). \qquad (A, 15b)$$

Equations (A, 15), together with the relations

$$W_A(\varphi+2\pi)=W_A(\varphi)+1,$$
$$W_A(0)=0, \qquad \text{(A, 16)}$$
$$W_A(\varphi)=\text{monotonically nondecreasing in } \varphi,$$

and the same relations for W_B, determine again the functions W_A and W_B. These functions, as well as the constants $C_A(k)$ and $C_B(k)$, can be calculated from (A, 15) by iteration. Finally, for $M(k)+\text{const}=(1/2\pi)\times\langle\varphi_{n+1}-\varphi_n\rangle_{\text{Av}}$, which gives us the relative number of eigenstates below k, we obtain

$$(1/2\pi)\langle\varphi_{n+1}-\varphi_n\rangle_{\text{Av}}=p\int(\varphi_{n+1}-\varphi_n)_A w_A(\varphi_n)d\varphi_n$$
$$+q\int(\varphi_{n+1}-\varphi_n)_B w_B(\varphi_n)d\varphi_n.$$

A short calculation using (A, 15) gives the result

$$M(k)+\text{const}=pC_A(k)+qC_B(k).$$

Special Frequencies in the Vibrational Spectra of Disordered Chains

Hirotsugu Matsuda

Department of Chemistry
Kyoto University, Kyoto

October 21, 1963

Consider a chain of N atoms where the displacement u_j of the jth atom satisfies the equation of motion

$$m_j\ddot{u}_j=k_j(u_{j+1}-u_j)+k_{j-1}(u_{j-1}-u_j),\qquad (j=1,2,\cdots,N). \tag{1}$$

Here, m_j equals either M or M' $(M<M')$, and $k_j=0$ (for $j=0$, N), K (otherwise). Let $Ng(\omega)d\omega$ be the number of normal frequencies in the interval $(\omega, \omega+d\omega)$ in the limit as $d\omega\to 0$

Recently, Dean[1] has shown by the machine calculation for various values of N up to maximum of 32000 that $g(\omega)$ has a very complex structure when two kinds of masses M and M' are randomly distributed, so that the moment method[2] is inadequate for obtaining $g(\omega)$. Although he showed *a posteriori* that a number of peaks in $g(\omega)$ can be associated with various types of defects in otherwise ordered chains, no *a priori* theory seems to have been proposed which elucidates such peculiar features of $g(\omega)$. In this letter we point out that such complexity of the spectra is rather to be expected, since there must be a number of *special frequencies* at which $g(\omega)=0$ when the mass ratio $Q=M'/M$ is sufficiently large.

Let ω_0 be the maximum frequency of the monatomic infinite chain where every atom has mass M, that is, $\omega_0=2\sqrt{K/M}$. In this infinite chain let us replace a non-zero fraction of atoms (that is, an infinite number of atoms) by those of mass M' in any way whatever. In such an infinite chain we can prove that if $\omega/\omega_0=\sin\left(p\frac{\pi}{2}\right)$ where p is a rational number satisfying $0<p<1$, then corresponding to any p there exists a finite $Q(p)$ such that $g(\omega)=0$ when $Q>Q(p)$ independent of the concentration ratio and the arrangement of atoms. Further we can show that $Q(1/2)=2$.

This property can be proved as follows: According to the transfer matrix method[3] the normal frequency is determined by

$$(\boldsymbol{T}_N\boldsymbol{T}_{N-1}\cdots\boldsymbol{T}_2\boldsymbol{T}_1)_{11}=0, \tag{2}$$

where

$$\boldsymbol{T}_i=\begin{pmatrix} 2-\omega^2m_i/K & -1\\ 1 & 0\end{pmatrix}\quad \text{for } i\neq 1, N, \tag{3}$$

$$\boldsymbol{T}_j=\begin{pmatrix} 1-\omega^2m_j/K & -1\\ 1 & 0\end{pmatrix}\quad \text{for } j=1 \text{ or } N, \tag{4}$$

and $(\dots)_{11}$ denotes the (1, 1)-element of the matrix. We may write $\boldsymbol{T}_i=\boldsymbol{T}$ for $m_i=M$, $\boldsymbol{T}'$ for $m_i=M'$ in (3). Referring to Ledermann's theorem,[4] without affecting $g(\omega)$ we can assume that the matrix $\boldsymbol{T}_N\boldsymbol{T}_{N-1}\cdots\boldsymbol{T}_1$ in (2) essentially consists of the infinite product of matrices of the form $\boldsymbol{T}'^{n'}\boldsymbol{T}^n$, where $\pm\boldsymbol{T}^n$ is not a unit matrix. We obtain

$$\boldsymbol{T}'^{n'}\boldsymbol{T}^n=\frac{(-1)^{n'}}{\operatorname{sh}\psi\sin\lambda}\begin{pmatrix} c(n'+1)\sin\{(n+1)\lambda-\varphi(n'+1)\} & -c(n'+1)\sin\{n\lambda-\varphi(n'+1)\}\\ -c(n')\sin\{(n+1)\lambda-\varphi(n')\} & c(n')\sin\{n\lambda-\varphi(n')\}\end{pmatrix}, \tag{5}$$

where λ and ψ are determined by

$$\omega=2\sqrt{K/M}\,\sin(\lambda/2)=2\sqrt{K/M'}\,\operatorname{ch}(\psi/2), \tag{6}$$

and

$$c(n)=\sqrt{\operatorname{sh}^2 n\psi+\operatorname{sh}^2(n-1)\psi+2\operatorname{sh}n\psi\operatorname{sh}(n-1)\psi\cos\lambda}\ , \tag{7}$$

$$\sin\varphi(n)=\sin\lambda\,\mathrm{sh}(n-1)\psi/c(n). \qquad (8)$$

On the other hand it can be proved, for instance by the formulation developed in reference 5), that

$$\lim_{N\to\infty}|(\boldsymbol{A}_N\boldsymbol{A}_{N-1}\cdots\boldsymbol{A}_2\boldsymbol{A}_1)_{11}|=\infty, \qquad (9)$$

when

$$\boldsymbol{A}_i=\begin{pmatrix} K_i/H_i & 2/H_i \\ (K_i\overline{K}_i-H_i^2)/2H_i & \overline{K}_i/H_i \end{pmatrix} \qquad (10)$$

and $|H_i^2/(K_i+\overline{K}_{i-1})(K_{i+1}+\overline{K}_i)|<\frac{1}{4}$

$$(i=2, 3, \ldots, N-1), \qquad (11)$$

where H_i, K_i and $\overline{K}_i$ are real finite numbers. Taking $\boldsymbol{T}'^{n'}\boldsymbol{T}^n$ given in (5) for $\boldsymbol{A}_i$ we can show that if $\psi\gg1$ and

$$\mathop{\mathrm{Min}}_{s}|\sin n\lambda|\gg\sqrt{2}\,e^{-\psi/2}$$

$$(n\lambda\neq s\pi,\ s\text{: integer}), \qquad (12)$$

then the condition (11) is satisfied, so that Eq. (2) does not hold. For given $\lambda=p\pi$ from (6) by making Q sufficiently large the condition (12) can be made satisfied. When $\lambda=\pi/2$, we can show that if ψ is real then the condition (11) is satisfied, so that from (6) we have $Q(1/2)=2$. Q.E.D.

The condition (12) indicates that rational numbers p with smaller denominator and values closer to 1 are favourable for giving *special frequencies*. In view of this, Dean's results seem to suggest that when $Q=2$, in addition to $p=(1/2)$ with $(\omega/\omega_0)^2=(1/2)$, rational numbers $p=(2/3)$, $(3/4)$, $(4/5)$, $(5/6)$ and $(6/7)$ may correspond to *special frequencies* with $8(\omega/\omega_0)^2\simeq6$, 6.82, 7.24, 7.46 and 7.61, respectively.

1) P. Dean, Proc. Roy. Soc. **A 254** (1960), 507.
2) E. W. Montroll, J. Chem. Phys. **10** (1942), 218.
C. Domb, A. A. Maradudin, P. Mazur, E. W. Montroll and G. H. Weiss, J. Phys. Chem. Solids **8** (1959), 419.
3) H. Matsuda, Prog. Theor. Phys. Suppl. No. 23 (1962), 22.
4) W. Ledermann, Proc. Roy. Soc. **A 182** (1944), 362.
5) H. Matsuda, Prog. Theor. Phys. **27** (1962), 811.

PROC. PHYS. SOC., 1964, VOL. 83

Some exact results for the vibrational spectrum of a disordered chain*

R. E. BORLAND

Mathematics Division, National Physical Laboratory,
Teddington, Middlesex

MS. received 2nd March 1964

Abstract. It is shown that at certain frequencies, which are infinite in number, the integrated vibrational spectrum can be found exactly for an infinite random two-component chain of elastically coupled masses.

1. Introduction

The problem of finding the averaged vibrational spectrum of an ensemble of disordered two-component infinite chains of elastically coupled masses has attracted the attention of many workers. In particular, Dean (1960) found by a machine calculation on a long sample chain that the spectrum is very irregular. This work has recently been confirmed by Agacy (1964) who solved numerically a functional equation due to Schmidt (1957) which gives directly the averaged spectrum.

This problem has resisted any attempt at a complete analytical solution. However, we have found certain frequencies at which the integrated spectrum can be found exactly in the case where the mass ratio is not less than 2. There is an infinite number of these frequencies but unfortunately they are not closely spaced except at the high frequency end of the spectrum so that these exact results do not show analytically that the spectrum is peaky. On the other hand these results show that the numerical results of Dean and Agacy are accurate to better than 0·2% at these frequencies.

2. The pure chain

We are concerned with chains of masses of two types elastically coupled by springs all of the same elastic constant k. The equation for the nth mass M_n is

$$(2-M_n\omega^2/k)U_n = U_{n+1}+U_{n-1} \tag{1}$$

where U_n is the displacement from equilibrium of the nth mass and ω is the angular frequency of the mode of vibration.

We write

$$Z_n = U_{n-1}/U_n$$

so that

$$Z_{n+1} = \frac{1}{(2-M_n\omega^2/k-Z_n)}. \tag{2}$$

We define the integrated spectrum $N(\omega_0{}^2)$ to be the number of eigenvalues less than $\omega_0{}^2$ per mass. Schmidt (1957) has shown that $N(\omega^2)$ is given by the fraction of negative

Z's obtained by successively applying equation (2); this ignores an end effect which tends to zero as the number of masses tends to infinity. This property forms the basis of the method of this paper.

Before we consider the disordered chain we need to derive some results for a pure chain where $M_n = m$ for all n. Although some of these results have been derived many times before, we rederive them here in order to draw attention to certain properties.

Equation (1) in matrix form for $M_n = m$ is

$$\begin{bmatrix} U_n \\ U_{n+1} \end{bmatrix} = \begin{bmatrix} 0 & 1 \\ -1 & 2-m\omega^2/k \end{bmatrix} \begin{bmatrix} U_{n-1} \\ U_n \end{bmatrix} = \mathbf{A} \begin{bmatrix} U_{n-1} \\ U_n \end{bmatrix}.$$

$\mathbf{A}$ can be diagonalized by the similarity transformation

$$\mathbf{Q}^{-1}\mathbf{A}\mathbf{Q} = \begin{bmatrix} e^{i\theta} & 0 \\ 0 & e^{-i\theta} \end{bmatrix}$$

where $\cos\theta = 1 - m\omega^2/2k$ and

$$\mathbf{Q} = \begin{bmatrix} 1 & 1 \\ e^{i\theta} & e^{-i\theta} \end{bmatrix}.$$

Thus

$$\mathbf{A}^s = \mathbf{Q} \begin{bmatrix} e^{is\theta} & 0 \\ 0 & e^{-is\theta} \end{bmatrix} \mathbf{Q}^{-1}.$$

On multiplying the matrices we find that

$$\begin{bmatrix} U_{n+s-1} \\ U_{n+s} \end{bmatrix} = \frac{1}{\sin\theta} \begin{bmatrix} -\sin(s-1)\theta, & \sin s\theta \\ -\sin s\theta, & \sin(s+1)\theta \end{bmatrix} \begin{bmatrix} U_{n-1} \\ U_n \end{bmatrix}. \tag{3}$$

We shall now consider a particular frequency ω_J for which

$$\theta = (J-1)\pi/J \qquad J \text{ is an integer} \geqslant 3$$
$$m\omega_J{}^2/k = 2+2\cos(\pi/J).$$

Inserting θ in equation (3) we find that

$$\begin{bmatrix} U_{n+s-1} \\ U_{n+s} \end{bmatrix} = \frac{(-)^{s+1}}{\sin(\pi/J)} \begin{bmatrix} \sin\{(s-1)\pi/J\}, & \sin(s\pi/J) \\ -\sin(s\pi/J), & -\sin\{(s+1)\pi/J\} \end{bmatrix} \begin{bmatrix} U_{n-1} \\ U_n \end{bmatrix}.$$

It follows that

$$Z_{n+s} = \frac{Z_n \sin\{(s-1)\pi/J\} + \sin(s\pi/J)}{-Z_n \sin(s\pi/J) - \sin\{(s+1)\pi/J\}}. \tag{4}$$

We shall use the notation that

$$Z_{n+s} = T^s(Z_n).$$

From equation (4) we see that

$$T^J(Z_n) = Z_n \qquad \text{for all } Z_n$$
$$T^s(0) < T^{s-1}(0) \qquad 1 \leqslant s \leqslant J-2$$
$$\lim_{\varepsilon \to -0} T^{J-1}(\epsilon) = +\infty.$$

This latter limit we shall choose as the value for $T^{J-1}(0)$.

Now consider the effect of T on a point α where $T(0) < \alpha < 0$. We have

$$T^s(0) < T^{s-1}(\alpha) < T^{s-1}(0) \qquad 1 \leqslant s \leqslant J-2$$
$$-\infty < T^{J-2}(\alpha) < T^{J-2}(0)$$
$$0 < T^{J-1}(\alpha) < +\infty.$$

The inequalities result from the property that $T(Z_n)$ is a monotonically increasing function of Z_n except at the point $Z_n = 2 - m\omega_J^2/k$ where $T(Z_n)$ jumps from $+\infty$ to $-\infty$. Let $\{X_s\}$ be the set of points lying in the open interval $(T^s(\alpha),\ T^s(0))$; $\{X_{s+J}\} = \{X_s\}$. We then see that

$$Z_n \in \{X_s\}$$

implies that

$$T(Z_n) \in \{X_{s+1}\}.$$

Thus for the infinite chain of masses m, $N(\omega_J{}^2) = (J-1)/J$ because Z_n is negative for exactly $J-1$ steps out of every J, if Z_1 lies in one of the sets $\{X_s\}$. We can always choose Z_1 to lie in one of the sets $\{X_s\}$, since the end effect is not important in an infinite chain.

Henceforward we shall take α to have a particular value, namely

$$\alpha = -\frac{1}{1+2\cos(\pi/J)} \tag{5}$$

which has the desired property that

$$T(0) < \alpha < 0.$$

This value of α is chosen because, as will be seen, the corresponding sets $\{X_s\}$ have very useful properties in the mixed random chain. Using (4) we find that

$$T^{J-2}(\alpha) = -\{1+2\cos(\pi/J)\}. \tag{6}$$

3. Mixed chain

Let us now replace some of the masses m by heavier masses M, where $M = 2m$. We shall denote the corresponding bilinear transformation (2) for $M_n = M$, by T_{H} so that

$$Z_{n+1} = T_{\mathrm{H}}(Z_n) = \frac{-1}{2+4\cos(\pi/J)+Z_n} \tag{7}$$

at the frequency ω_J that we are considering. What we have hitherto called T we shall now denote by T_{L} to indicate that it is the transformation due to the lighter of the two masses.

From equations (6) and (7) we can readily verify that

$$T_{\mathrm{H}}\, T_{\mathrm{L}}{}^{J-2}(\alpha) = \alpha \qquad T_{\mathrm{H}}(+\infty) = 0.$$

We note that the singularity in $T_{\mathrm{H}}(Z_n)$ occurs at $Z_n = -2-4\cos(\pi/J)$ which is less than $T_{\mathrm{L}}{}^{J-2}(\alpha)$. It therefore follows from the monotonicity of $T_{\mathrm{H}}(Z_n)$ that if

$$Z_n \in \text{the open interval } (T_{\mathrm{L}}{}^{J-2}(\alpha), +\infty)$$

then

$$T_{\mathrm{H}}(Z_n) \in \text{the open interval } (\alpha, 0).$$

All the sets $\{X_s\}$ are contained in the open interval $(T_L^{J-2}(\alpha), +\infty)$ and the open interval $(\alpha, 0)$ is simply the set $\{X_0\}$ so we have that if

$$Z_n \in \{X_s\}$$

then

$$T_H(Z_n) \in \{X_0\}.$$

Furthermore it is clear from equation (2) that this property is true not only for $M = 2m$ but for all $M \geqslant 2m$.

This is the essential property that enables us to solve the problem at these discrete frequencies ω_J. We know that if $Z_n \in \{X_s\}$ then $Z_{n+1} \in \{X_{s+1}\}$ or $\{X_0\}$ according to whether $M_n = m$ or M respectively. We also know that Z_n is always negative unless $Z_n \in \{X_{J-1}\}$, so that $N(\omega_J{}^2)$ for the mixed system is equal to unity minus the fraction of Z's that lie in the set $\{X_{J-1}\}$. We can use this result to determine $N(\omega_J{}^2)$ for any arrangement of the two masses. In the following section we shall treat the particular case of the completely random chain.

4. Completely random chain

Consider the completely random chain where c is the concentration of light masses. Let f_s be the probability that $Z_n \in \{X_s\}$. We note that

$$f_s = f_{s+J} \text{ and } \sum_{0}^{J-1} f_s = 1.$$

We have

$$\begin{aligned} f_s &= cf_{s-1} \qquad s \neq 0 \text{ (or integral multiple of } J) \\ f_0 &= (1-c)+cf_{J-1}. \end{aligned}$$

Hence

$$\begin{aligned} f_0 &= (1-c)+c^J f_0 \\ f_0 &= \frac{1-c}{1-c^J}. \end{aligned}$$

Therefore

$$f_{J-1} = \frac{f_0-1+c}{c} = \frac{c^{J-1}(1-c)}{1-c^J}$$

and

$$N(\omega_J{}^2) = 1-f_{J-1} = \frac{1-c^{J-1}}{1-c^J}.$$

We earlier restricted J to integral values not less than 3 in order to avoid complicating the proof by having to insert special conditions to include $J = 2$. However, the modifications to the proof are only trivial and the result holds. The relevant sets for $J = 2$ are

$$\begin{aligned} \{X_1\} &= (1, +\infty) \\ \{X_0\} &= (-1,0). \end{aligned}$$

Our results for the random chain with $M \geqslant 2m$ show that the spectrum is not bounded by a maximum frequency which is less than the maximum frequency for a pure chain of light masses. This conflicts with the results found by Flinn, Maradudin and Weiss (1961, Westinghouse Res. Lab. Sci. Paper 029–G000–P10); it would therefore appear that the assumption made in their paper cannot be justified. It is of interest to compare the exact results with those obtained by Dean and by Agacy (private communications). The results of Dean were obtained using a sample chain of 32 000 masses and Agacy's results were obtained using a grid of 40 points to solve Schmidt's functional equation numerically.

$$M = 2m \qquad c = \tfrac{1}{2}$$

J	$m\omega_J^2/k$	$N(\omega_J^2)$ Exact formula	$N(\omega_J^2)$ Dean	$N(\omega_J^2)$ Agacy
2	2·000	0·6667	0·6663	0·6667
3	3·000	0·8571	0·8569	0·8571
4	3·414	0·9333	0·9332	0·9333
5	3·618	0·9677	0·9682	0·9684
6	3·732	0·9841	0·9840	0·9843
7	3·802	0·9921	0·9920	0·9906

It can be seen that for these frequencies the error is never greater than 0·2% and is in general much smaller. This does not of course imply that the accuracy of the results of Dean and Agacy is as high for other frequencies, because the frequencies ω_J are by their nature very special.

5. Other types of disorder

The results of §§ 2 and 3 can be used to derive the values of $N(\omega_J^2)$ for arrangements of the light and heavy masses other than the purely random arrangement. We shall quote without proof the results for a chain having short-range order and a chain having long-range order. The proofs merely involve an application of elementary probability theory and some tedious algebra. Both these chains have equal numbers of light and heavy masses. The short-range order case is obtained by specifying that q is the probability of an unlike linkage, and the long-range order case is obtained by specifying that p is the probability that a heavy mass occupies an even lattice site.

For short-range order

$$N(\omega_J^2) = 1 - \frac{1}{2}\frac{q(1-q)^{J-2}}{1-(1-q)^J}.$$

For long-range order

$$N(\omega_J^2) = \frac{1-\frac{1}{2}\{p(1-p)\}^{\frac{1}{2}J-1}}{1-\{p(1-p)\}^{J/2}} \qquad J \text{ even}$$

$$N(\omega_J^2) = \frac{1-\frac{1}{2}\{p(1-p)\}^{(J-1)/2}-\frac{1}{2}\{p(1-p)\}^{J-1}}{1-\{p(1-p)\}^J} \qquad J \text{ odd.}$$

6. Summary

We can sum up our results for the purely random case by stating the following theorem.

Let ω_J^2 be the eigenvalue parameter for which the integrated spectrum of an infinite chain of masses m is $(J-1)/J$, where J is an integer. Then the integrated spectrum at

ω_J^2 for the infinite random chain containing a concentration c of masses m and a concentration $1-c$ of masses M is $(1-c^{J-1})/(1-c^J)$, provided that $M \geqslant 2m$, and the elastic coupling between the masses is the same for both systems. The result also holds for a multi-component chain in which c is the concentration of light masses and all other masses are not less than twice the light mass. This generalization is obvious from the proof we have given.

Finally, we have indicated that it is relatively simple to find exact results for other types of disorder at these special frequencies ω_J.

Acknowledgments

The work described above has been carried out as part of the research programme of the National Physical Laboratory and this paper is published by permission of the Director.

References

Agacy, R. L., 1964, *Proc. Phys. Soc.*, **83**, 591.
Dean, P., 1960, *Proc. Roy. Soc.* A, **254**, 507.
Schmidt, H., 1957, *Phys. Rev.*, **105**, 425.

Progress of Theoretical Physics, Vol. 32, No. 2, August 1964

Structure of the Spectra of Disordered Systems. I

——Fundamental Theorems——

Jun-ichi HORI and Hirotsugu MATSUDA*

Department of Physics, Hokkaido University, Sapporo
**Department of Chemistry, Kyoto University, Kyoto*

(Received April 20, 1964)

Two theorems are presented, which play a fundamental role in discussing the spectra of disordered systems. The first theorem is valid for any mixed lattice which is described by a set of a number of 2×2 non-singular transfer-matrices with real traces, arranged regularly or randomly. It states that, if a value of frequency or energy lies in one of the spectral gaps for every constituent regular lattice, and if the fixed points of the transformations induced by these matrices are so arranged that there exists a trapping region, the density of frequency- or energy-spectrum of the mixed lattice vanishes at this frequency or energy-value.

For some special forms of the transfer-matrices, which often appear in physical problems, the latter condition may be stated in terms of a simple notion "phase", so that we get the second theorem. Various theorems of Saxon-Hutner-type are to be deduced from these theorems, according to the nature, or the manner of description, of individual systems.

§1. Introduction

In 1949, Saxon and Hutner[1] investigated the electronic energy spectrum of binary Kronig-Penney model, and showed that the energy-value which is forbidden in either of the regular lattices composed of one of the constituent potentials remains to be forbidden in the mixed lattice as well. Later in 1951, Luttinger[2] showed that this statement is true if two potentials are arranged in a random manner. Since then, the statement has often been referred to as "Saxon-Hutner's Theorem", and its validity for systems other than the binary Kronig-Penney model has often been argued. Yet there seems to be no well-established proof or dis-proof of the theorem for general systems, and its limit of validity still remains to be ambiguous.

One of the purposes of the present series of papers is to make it clear, from a general point of view, for what kind of systems the Saxon-Hutner's theorem is valid. It will be shown that several theorems of Saxon-Hutner-type may be valid, according to the nature and to the manner of description of the system under consideration.*) At first in the present paper, two general theorems

*) Main contents of the series have already been reported in two letters to the editor of Prog. Theor. Phys.[3),4)]

are presented, one of which is valid for every ordered or disordered system described by 2×2 non-singular transfer-matrices*) with real traces, and the other is valid for the systems described by real 2×2 matrices or by complex matrices of the form

$$\boldsymbol{Q}=\begin{pmatrix} A & B \\ B^* & A^* \end{pmatrix}. \tag{1·1}$$

It will be shown in the following papers how the general theorems reduce to various theorems of Saxon-Hutner-type for individual systems such as the disordered Kronig-Penney models with randomly varying strength of potentials and/or interval between them, and the isotopically or non-isotopically disordered polyatomic chains with nearest-neighbor harmonic interactions.

§ 2. The first theorem

Consider the system which is described by 2×2 transfer-matrix

$$\boldsymbol{Q}=\begin{pmatrix} a & b \\ c & d \end{pmatrix}, \tag{2·1}$$

where the matrix elements contain the value of frequency or energy as a parameter, and are in general complex. Let a state vector $\psi=(\psi_1, \psi_2)$ be transformed to another vector $\psi'=(\psi_1', \psi_2')$ by the matrix $\boldsymbol{Q}$. If one introduces the quantities z and z' by

$$z=\psi_1/\psi_2, \quad z'=\psi_1'/\psi_2', \tag{2·2}$$

we have

$$z'=(az+b)/(cz+d). \tag{2·3}$$

For non-singular matrix, i.e. if $ad-bc\neq0$, this is the transformation known as Möbius transformation or linear transformation, in the theory of functions of complex variable. It induces a one-to-one mapping on the complex plane, which is conformal and transforms any circle to another circle. If in particular all the elements are real, every real z is mapped into a real z'.

If a point z is transformed by (2·3) into itself, i.e. if $z'=z$, this is called a "fixed point." It is obvious that there are in general two fixed points z_- and z_+, and they are the ratios of components of the eigenvectors ψ_- and ψ_+, respectively, of the matrix $\boldsymbol{Q}$. If $a+d$ and $\det Q$ are real and $|a+d|^2>4(\det \boldsymbol{Q})$, the eigenvalues $\theta_\pm$, of $\boldsymbol{Q}$, are real and distinct, so that any vector except the eigenvector corresponding to the eigenvalue with smaller modulus is transformed by $\boldsymbol{Q}$ to a vector which is nearer to the eigenvector associated with the eigenvalue

*) For instance, electronic or vibrational states of a linear chain can conveniently be described by a set of matrices called transfer-matrices.[5),6)]

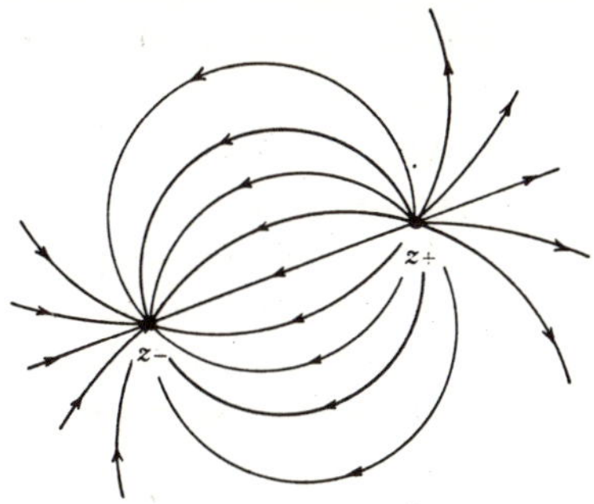

Fig. 1. Flow lines in the hyperbolic transformation.

with larger modulus than the original vector. Let us attach the suffix $-$ to the quantities associated with the eigenvalue of larger modulus. Thus every vector ψ except ψ_+ is "attracted" to ψ_-, and every vector except ψ_- is "repulsed" from ψ_+, under the operation of $\boldsymbol{Q}$. Correspondingly, every complex number except z_+ approaches z_-, while every number except z_- goes away from z_+.

When the eigenvalues $\theta_\pm$ are real and distinct, the Möbius transformation is called hyperbolic. In hyperbolic cases, the flow lines on the complex plane along which the points move under the transformation are as shown in Fig. 1. These are circles which pass through two fixed points. The points z_- and z_+ play the role of sink and source, respectively, of this vector field. The attractive or repulsive property of the fixed points is thus clearly illustrated. Circles which are orthogonal to the flow lines represent a series of successively mapped circles.

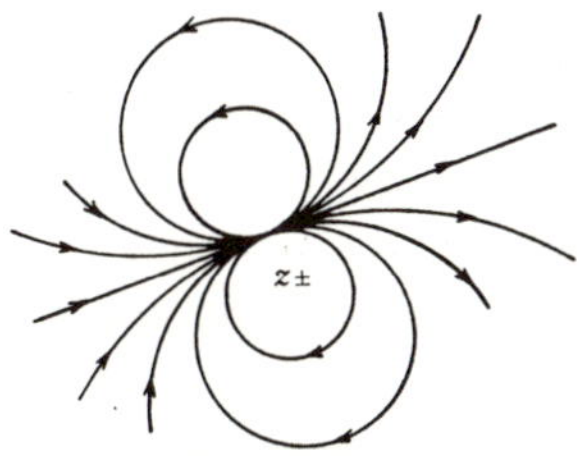

Fig. 2. Flow lines in the parabolic tranformation.

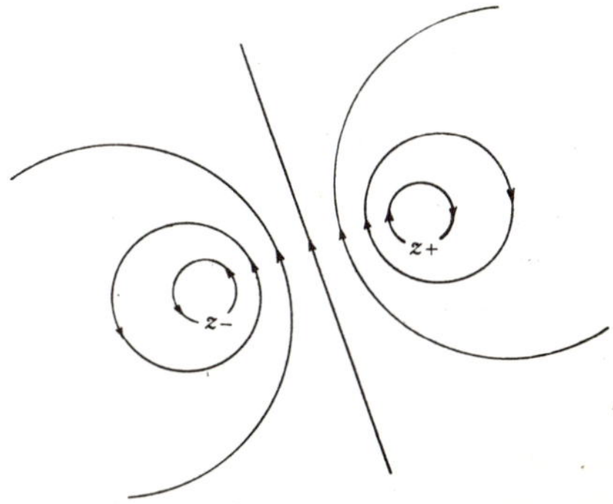

Fig. 3. Flow lines in the elliptic transformation.

We shall call the points z_- and z_+ "sink" and "source" points, respectively. The vectors ψ_- and ψ_+ correspond respectively to the right limit and left limit vectors previously introduced by Hori and Fukushima.[7] But here we shall call them "sink" and "source" vectors, respectively, following the terminology used above.

It is clear that if the transformation is hyperbolic for a given value of the frequency or energy, the spectral density of the regular lattice which is described by the matrix under consideration vanishes at this frequency- or energy-value.

In other words, the frequency or energy for which the transformation induced by the matrix $\boldsymbol{Q}$ is hyperbolic lies in one of the spectral gaps of the regular lattice described by $\boldsymbol{Q}$.*)

If $a+d$ and $\det Q$ are real and $|a+d|^2 \leqq 4(\det \boldsymbol{Q})$, the transformation is called elliptic or parabolic, according as $|a+d|^2 < 4(\det \boldsymbol{Q})$ or $|a+d|^2 = 4(\det \boldsymbol{Q})$. In the latter case, there is only one fixed point and the flow lines are as shown in Fig. 2. One sees that any point is finally attracted to the fixed point, so that it cannot return to the original position. Hence the spectral density should be zero at the frequency or energy for which the transformation is parabolic. On the contrary, in the elliptic case every point can return to the original position, since each flow line is a closed circle, as shown in Fig. 3, so that the spectral density at the frequency or energy at which the transformation is elliptic does not vanish in general. Thus we see that the necessary and sufficient condition for a given frequency or energy to lie in one of the spectral gaps is that the transformation is hyperbolic or parabolic at this frequency or energy, as far as $a+d$ is real.

Next let us turn to consider a mixed (ordered or disordered) lattice, which is described by a number of transfer-matrices

$$\boldsymbol{Q}^{(i)} = \begin{pmatrix} a^{(i)} & b^{(i)} \\ c^{(i)} & d^{(i)} \end{pmatrix}, \qquad i=0,\ 1,\ 2,\ \cdots,\ M-1, \tag{2·4}$$

where the index (i) specifies a set of possible parameter-values, M being the number of different set. Let us specify the quantities pertaining to $\boldsymbol{Q}^{(i)}$ by the same index (i), for example as $\psi_\pm^{(i)}$, $\theta_\pm^{(i)}$, $z_\pm^{(i)}$, etc. Further let us call the regular lattice which is described by one of the $\boldsymbol{Q}^{(i)}$'s, say $\boldsymbol{Q}^{(j)}$, the j-th constituent (regular) lattice of the given mixed lattice. From what was said above, one sees that if for a given value of frequency or energy the transformations are hyperbolic or parabolic for all i, that is, $(a^{(i)}+d^{(i)})$'s are real and $|a^{(i)}+d^{(i)}|^2 \geqq 4(\det \boldsymbol{Q}^{(i)})$ for all i, this frequency or energy lies in one of the spectral gaps for every constituent regular lattice.

The state vector is transformed successively by a series of regularly or randomly arranged transfer-matrices $\boldsymbol{Q}^{(i)}$'s. If the transformations are hyperbolic or parabolic for every i, and if the arrangement of fixed points is such that there is a region from which no flow line runs out, an arbitrary point comes eventually into this region during the successive transformations, and after it has once come into, it can never escape from it. We shall call such a region "trapping region" of the set of transfer-matrices $\{\boldsymbol{Q}^{(i)}\}$, $i=0, 1, 2, \cdots, M-1$. An example of the trapping region is shown in Fig. 4.

It is clear that, if for a given frequency- or energy-value such a trapping region exists, the spectral density of the mixed lattice described by the set

*) Actually, in this case there can be no wave which satisfies periodic boundary conditions.

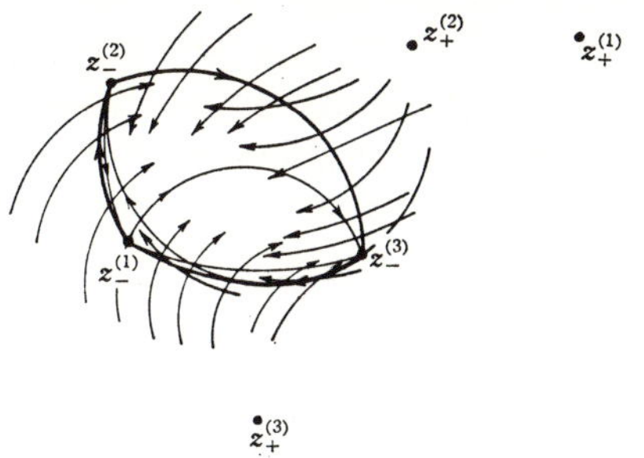

Fig. 4. An example of the trapping region.

$\{Q^{(i)}\}$, $i=0, 1, \cdots, M-1$, must be zero at this value of frequency or energy. Because the existence of the trapping region indicates that the resultant transformation can not be elliptic. Thus we reach

Theorem I. If for a given frequency- or energy-value the transformations induced by the matrices $Q^{(i)}$'s are hyperbolic or parabolic for all i, that is, if this frequency or energy lies in one of the spectral gaps for every constituent regular lattice, and if a trapping region exists for the set of matrices $\{Q^{(i)}\}$, $i=0, 1, 2, \cdots, M-1$, the spectral density of the mixed lattice vanishes at the given frequency or energy.

§ 3. The second theorem

In most of the physical problems, the transfer-matrices are either real or have the form (1·1). In these cases, the condition of the existence of trapping region can be stated in more concrete and practical terms.

Consider first the latter case. The condition for the transformation induced by Q to be hyperbolic or parabolic reduces in this case to

$$(\mathrm{Re}A)^2 \geqq \det Q. \tag{3·1}$$

(In what follows we shall omit the index (i) as far as the particular value of i is not concerned.) Fixed points are given by

$$z_\pm = (\theta_\pm - A^*)/B^* = B/(\theta_\pm - A). \tag{3·2}$$

It is seen that if the transformation is hyperbolic or parabolic, $z_\pm$ lie on the unit circle. Hence we may write $z_\pm = \exp(i\delta_\pm)$, $\delta_\pm$ being real. The condition for a trapping region to exist is that the intervals which are spanned by two sets of phases $\{\delta_+^{(i)}\}$ and $\{\delta_-^{(i)}\}$, $i=0, 1, \cdots, M-1$, do not overlap with each other.

In the case of real matrices, the condition under which the transformation is to be hyperbolic or parabolic is given by

$$|\mathrm{Trace}\, Q|^2 \geqq 4(\det Q). \tag{3·3}$$

In this case the fixed points $z_\pm$ lie on the real axis, so that the condition for the existence of the trapping region becomes equivalent to the condition that the intervals spanned by two sets of fixed points, $\{z_+^{(i)}\}$ and $\{z_-^{(i)}\}$ $(i=0, 1, \cdots, M-1)$ do not overlap with each other. If we define the "phase" by

$$\delta_\pm = 2\,\mathrm{Tan}^{-1} z_\pm, \tag{3·4}$$

this condition becomes formally the same as in the above case.

Let us call the phase interval spanned by the set $\{\delta_-^{(i)}\}$ and $\{\delta_+^{(i)}\}$ "sink-phase" and "source-phase" intervals, respectively. If these do not overlap with each other, the former plays the role of the "trapping interval." Thus we reach

Theorem II. If (A) a given value of frequency or energy lies in one of the spectral gaps for every constituent regular lattice, and (B) the sink-phase and source-phase intervals do not overlap with each other, the spectral density of the mixed lattice described by the set of matrices $\{\boldsymbol{Q}^{(i)}\}$ vanishes at this frequency or energy.

§ 4. Concluding remarks

General theorems obtained above play a fundamental role in discussing the spectra of disordered systems. It will be shown in the following papers that for fairly wide variety of systems, such as mentioned in the final paragraph of § 1, the condition (B) in the second theorem is in fact redundant, that is, it is a necessary consequence of the condition (A). In other words, the theorem without the condition (B) is valid for such systems.

The theorem without the condition (B) might seem, at first sight, to be equivalent to the Saxon-Hutner's theorem in an ordinary sense. It is however merely of Saxon-Hutner-type, and its content is in fact fairly wider than that of the ordinary Saxon-Hutner's theorem. The reason is that the choice of the set of matrices $\{\boldsymbol{Q}^{(i)}\}$ describing the mixed lattice is in a large extent arbitrary, so that it may occur for example that some of the constituent lattices do not correspond to any regular lattice composed of one of the component atom. In this respect it is to be remarked that the theorems obtained above are valid irrespective of the value of the determinants of $\boldsymbol{Q}^{(i)}$'s. In many physical problems the matrices are unimodular, so that $\det \boldsymbol{Q}=1$. There may occur, however, cases in which some of the matrices are non-unimodular. Constituent lattices corresponding to non-unimodular matrices have no real physical significance; in particular they have in general no true eigenfrequencies or eigen-energies. It is merely for convenience that we regard the frequency- or energy-regions in which the transfer-matrix has complex eigenvalues and consequently the induced transformation is elliptic as if they were bands of eigenfrequencies or energies, and call the remaining regions "spectral gaps."

Thanks to such flexibility, various useful theorems may be deduced from our fundamental ones, according to the nature of the system under consideration, on the one hand, and to the manner of description, on the other. It is especially striking that the theorem of Matsuda on the special frequencies in the vibrational spectrum of disordered chain is deduced as one of the special cases of the second theorem.[4),8)] Full discussions on these aspects of our theorems will be given in the following papers.

References

1) D. S. Saxon and R. A. Hutner, Philips Res. Rep. **4** (1949), 81.
2) J. M. Luttinger, Philips Res. Rep. **6** (1951), 303.
3) J. Hori, Prog. Theor. Phys. **31** (1964), 522.
4) J. Hori, Prog. Theor. Phys. **31** (1964), in press.
5) E. H. Kerner, Phys. Rev. **95** (1954), 687; Proc. Phys. Soc. (London) **69A** (1956), 234.
H. Schmidt, Phys. Rev. **105** (1957), 425.
6) J. Hori and T. Asahi, Prog. Theor. Phys. **17** (1957), 523.
J. Hori, Prog. Theor. Phys. **18** (1957), 367.
H. Matsuda, Prog. Theor. Phys. Suppl. No. **23** (1962), 22.
7) J. Hori and M. Fukushima, J. Phys. Soc. Japan **19** (1964), 296.
8) H. Matsuda, Prog. Theor. Phys. **31** (1964) 161.

PROC. PHYS. SOC., 1964, VOL. 83

The vibrational spectrum of a disordered linear system*

R. L. AGACY

Mathematics Division, National Physical Laboratory, Teddington, Middlesex

Communicated by P. Dean; *MS. received 23rd December* 1963

Abstract. Schmidt's functional equation determining the vibrational spectrum of a disordered, linear system is solved numerically for some chosen parameters. The resulting spectrum is found to agree very well with the corresponding spectrum obtained by Dean in 1960 but to differ appreciably from the results of the moment-trace method.

1. Introduction

The problem of finding the vibrational spectrum for an ensemble of disordered, linear, two-component systems has been considered by Domb, Maradudin, Montroll and Weiss (1959 a, b). Their moment-trace technique essentially involves polynomial approximations to the spectra and the results obtained appear to show that the frequency distributions are fairly smooth curves.

However, Dean (1960) using a Sturm-sequence method on finite chains constructed by a pseudo-random number generating programme has obtained, in marked contrast to the moment-trace method, spectra which are very 'peaky', exhibiting a quite irregular structure.

In the present paper a straightforward numerical technique is employed to obtain the solution of a functional equation derived by Schmidt (1957) for the ensemble of infinite disordered diatomic chains of elastically coupled atoms with nearest-neighbour interaction; the elastic constant is taken to be the same for any pair of atoms. From the solution of this equation the distribution of eigenvalues is easily obtained. The parameters chosen correspond to a case considered by Domb *et al.* and by Dean. The spectrum obtained bears a very close resemblance to that of Dean and is not in agreement with the results of the moment-trace method.

2. Functional equation

We consider a two-component chain of N atoms in which M_H and M_L are the masses of the heavy and light atoms respectively. Nearest-neighbour forces only are taken into account and are described by the same elastic constant k. The equation of motion for the nth atom is given by

$$m_n \ddot{u}_n = k(u_{n+1}+u_{n-1}-2u_n)$$

where $m_n = M_H$ or M_L, and u_n is the displacement of the nth atom from its equilibrium position. We impose the rigid-wall boundary conditions

$$u_0 = u_{N+1} = 0.$$

Using the relation $\ddot{u}_n = -\omega^2 u_n$, the secular equations determining the eigenvalues of ω^2 are

$$u_n(2-m_n\omega^2k^{-1}) = u_{n+1}+u_{n-1} \qquad (n = 1, \ldots, N). \tag{1}$$

We are concerned with the limit as $N \to \infty$ of the distribution of eigenvalues averaged over all chains of N atoms in which the probability of occurrence of a heavy atom at a lattice point is c. This limiting distribution (integrated density of eigenvalues) is the number of eigenvalues per atom less than ω^2 and is denoted by $M(\omega^2)$.

Schmidt (1957) has shown that $M(\omega^2)$ may be determined from the solution of the functional equation

$$W(\phi) = cW[T_{\mathrm{H}}^{-1}(\phi)]+(1-c)W[T_{\mathrm{L}}^{-1}(\phi)]-W(-\pi) \tag{2a}$$

subject to

$$W(\phi+2\pi) = W(\phi)+1 \tag{2b}$$

and

$$W(\phi) \text{ is a monotonic increasing function of } \phi. \tag{2c}$$

The function $T_i^{-1}(\phi)$ is defined by

$$T_i^{-1}(\phi) = 2 \arctan [2-M_i\omega^2k^{-1}-\cot \tfrac{1}{2}\phi], \qquad i = H, L \tag{3a}$$

and is to be a continuous function of ϕ, the relationship being made unique by specifying that†

$$T_i^{-1}(0) = -\pi. \tag{3b}$$

The function $W(\phi)$ also depends on the vibrational frequency ω, but for convenience this is not explicitly indicated here, so that (2) is valid for a fixed value of ω^2.

If $\frac{1}{2}\phi_n$ is the principal value of $\arctan(u_{n-1}/u_n)$ and $\phi_1 = 0$, then $W(\phi+d\phi)-W(\phi)$ is the limit as $n \to \infty$ of the probability that ϕ_n lies in the interval $(\phi, \phi+d\phi)$.

Schmidt has proved that $W(\phi)$ is a unique, continuous function of ϕ except possibly when

$$|1-\tfrac{1}{2}M_i\omega^2k^{-1}| > 1 \text{ for } i = H \textit{ and } L.$$

However, as there are no eigenvalues when the above inequality is satisfied we shall not be interested in parameters satisfying this condition.

The integrated density of eigenvalues $M(\omega^2)$ for the ensemble of infinite chains is given by $-W(-\pi)$ corresponding to the particular frequency ω; we indicate the ω^2 dependence here and write this as $-W(-\pi, \omega^2)$. The density of eigenvalues $\rho(\omega^2)$ is then given by $dM(\omega^2)/d(\omega^2)$.

3. Numerical procedure

It is sufficient to calculate $W(\phi)$ for the range $0 \leqslant \phi \leqslant 2\pi$, since from (2b) it can easily be obtained for any other values of ϕ. From (3) it follows that for this range of ϕ

$$-\pi \leqslant T_i^{-1}(\phi) \leqslant \pi.$$

Let

$$\phi_i' = \begin{cases} T_i^{-1}(\phi) & 0 \leqslant T_i^{-1}(\phi) < \pi \quad (4a) \\ T_i^{-1}(\phi)+2\pi & -\pi \leqslant T_i^{-1}(\phi) < 0 \quad (4b) \end{cases}$$

so that $0 \leqslant \phi_i' < 2\pi$.

† Schmidt's uniqueness condition (A, 4*b*) is not consistent with (A, 6) (*a*) in his paper.

We divide the range 0 to 2π into an even number N of equal intervals of length $h\,(=2\pi/N)$. We now approximate $W(\phi)$ by a continuous function linear in these intervals, with values W_r at the points rh where r is an integer $1 \leqslant r \leqslant N-1$ and for the ends of the range we put $W_0 = 0$, $W_N = 1$ as required by (2) and (3). Then for the point rh we can, from (2), (3) and (4), obtain a linear equation for the values W_s ($s = 1, \ldots, N-1$). This equation will, for example, contain terms such as $-c$ or $c-1$ if (4*b*) is satisfied for $i = H$ or L by virtue of (2*b*). Letting r run from 1 to $N-1$ we obtain a set of $N-1$ linear equations in the $N-1$ unknowns W_s.

A programme was written for the ACE computer of the National Physical Laboratory which set up and solved the $N-1$ equations, obtaining the solution W_s for $s = 1, \ldots, N-1$. The parameters $M_H/k = 2$, $M_L/k = 1$, $c = 0{\cdot}5$ are used throughout the paper.

Figure 1, as an example, illustrates the convergence of the numerical solutions for $W(\phi)$ upon increasing the number of intervals. Here $\omega^2 = 2{\cdot}56$ and $N = 20$, 40 and 80.

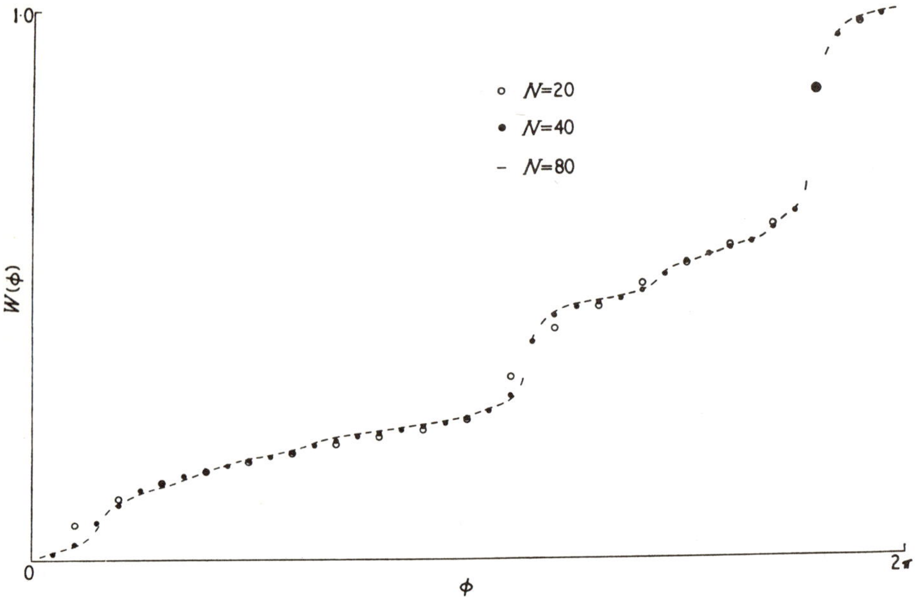

Figure 1. $W(\phi)$ as approximated to by the numerical process for (i) $N = 20$ intervals, open circles; (ii) $N = 40$ intervals, full circles; (iii) $N = 80$ intervals, broken curve; $\omega^2 = 2{\cdot}56$.

The object of the calculation of $W(\phi)$ is to obtain $M(\omega^2)$ which is simply given by

$$M(\omega^2) = -W(-\pi, \omega^2) = 1 - W(\pi, \omega^2)$$

that is $1 - W_{N/2}(\omega^2)$ by the numerical approximation. Examination of $M(\omega^2)$ over a range of ω^2 for $N = 20$, 40 and 80 showed that the average error in $M(\omega^2)$ is approximately proportional to $N^{-3/2}$ with a mean error of about $0{\cdot}001$ for $N = 40$.

Figure 2 shows $M(\omega^2)$ plotted for $N = 40$ for steps $\delta\omega^2 = 0{\cdot}04$ in the range $\omega^2 = 0$ to 4 (cf. model 1, figure 1, Dean 1960).

Owing to the irregular nature of $M(\omega^2)$, numerical differentiation is unsuitable for the purposes of calculating the density of eigenvalues and so $\rho(\omega^2)$ is presented in the form of a histogram determined by

$$\rho(\omega^2 + \tfrac{1}{2}\delta\omega^2) = \{M(\omega^2 + \delta\omega^2) - M(\omega^2)\}/\delta\omega^2$$

for finite increments of $\delta\omega^2$.

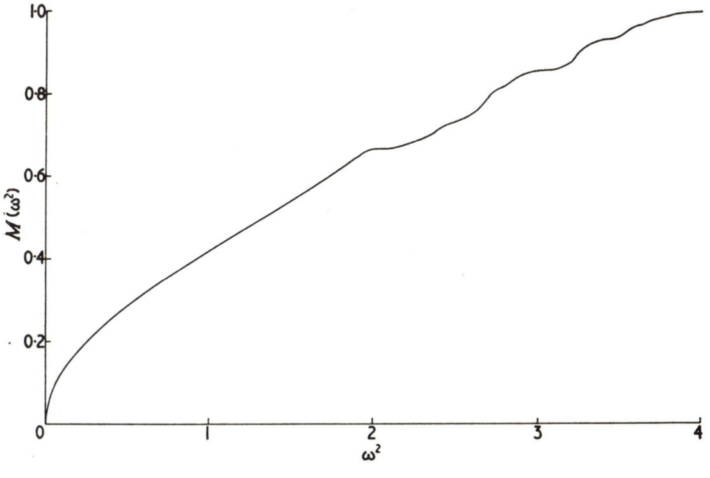

Figure 2. Integrated density of eigenvalues $M(\omega^2)$.

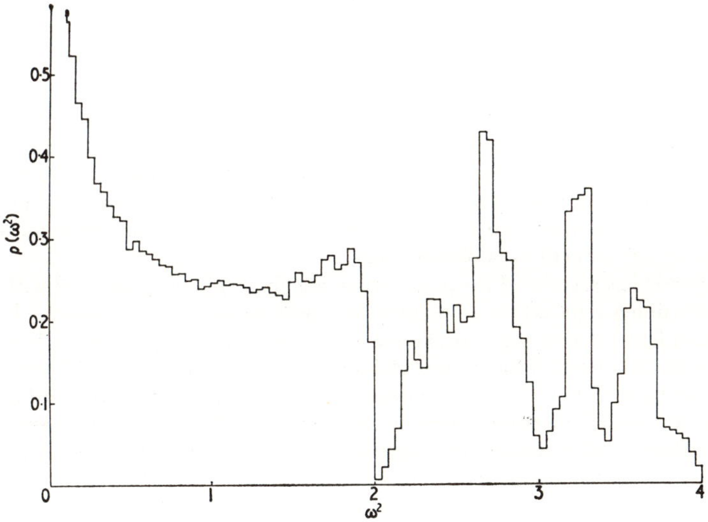

Figure 3. Histogram of density of eigenvalues $\rho(\omega^2)$ for $N = 20$.

Figure 3 shows $\rho(\omega^2)$ plotted for $N = 20$ and $\delta\omega^2 = 0{\cdot}04$. Figure 4 shows the histogram for $N = 40$, with a broken line histogram for $N = 80$ superimposed in the range $\omega^2 = 2{\cdot}44$ to $3{\cdot}00$.

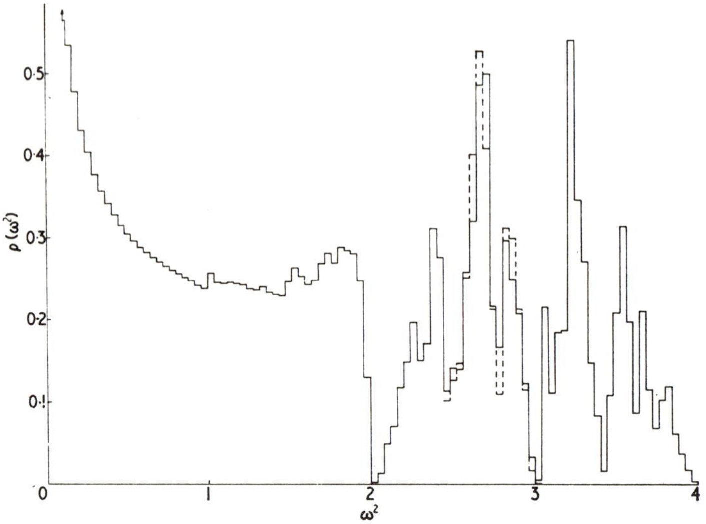

Figure 4. Histogram of density of eigenvalues $\rho(\omega^2)$ for $N = 40$ shown by the full line. Superimposed in the range $\omega^2 = 2{\cdot}44$ to $3{\cdot}00$ is the histogram for $N = 80$ shown by the broken line.

4. Discussion

From the results we have obtained it can be seen that by increasing the number of intervals, and hence the accuracy of the results, the spectrum becomes more and more 'peaky'. Our spectrum closely resembles that of Dean (1960), which is shown as the full curve in figure 5. Dean's results were obtained using the Sturm-sequence method applied to a sample chain of 32 000 atoms. The broken curve of figure 5 shows the frequency distribution based on the moment-trace method of Domb *et al.* (1959 a) when twenty moments are used to approximate the spectrum. Thus we may regard as justified Dean's representation of the ensemble of *infinite* chains by his *finite* chains in order to calculate the spectrum. The moment-trace method suffers from an inadequate approximation of the spectrum by a 'low' order polynomial and is hence incapable of producing such fine detail as is obtained here.

Dean (1961) has associated the peaks occurring above the range $\omega^2 = 2$ with various clusters containing light atoms embedded in a host lattice of heavy atoms. It will be noticed from figure 4 that slight peaks are also present in the range $\omega^2 = 1$ to 2. The association of these peaks with various clusters is not quite so obvious. However, it can be seen from figure 5 that this frequency range is forbidden for the ordered HL lattice. Thus it seems natural that peaks arise because in this range LH or HL pairs assist in causing localized modes in much the same way as the H atoms surrounding clusters are mainly responsible for causing localized modes above $\omega^2 = 2$.

The main value of this work is that it provides a fairly reliable calculation of the average spectrum for the infinite one-dimensional disordered system, since the only error which is involved is that which is associated with the numerical solution of the functional equation (2). Better or more quickly convergent numerical techniques, for example quadratic rather than linear approximation in the intervals, may be used for determining the spectrum more accurately. However, we have here been concerned

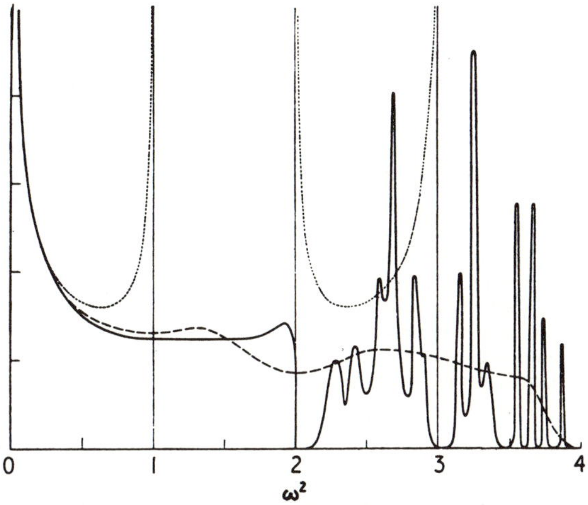

Figure 5. Density of eigenvalues: (i) Sturm-sequence method of Dean, full curve; (ii) Twenty moment approximation of Domb *et al.*, broken curve; (iii) Ordered diatomic chain, dotted curve. (From Dean 1960.)

with demonstrating that Schmidt's functional equation, which gives the correctly averaged spectrum for the ensemble of infinite chains, is amenable to very simple numerical treatment and is a satisfactory method for obtaining the spectrum.

Acknowledgments

I am grateful to Mr. R. E. Borland for bringing the problem to my attention and for his kind assistance all along. The work described above has been carried out as part of the research programme of the National Physical Laboratory and this paper is published by permission of the Director of the Laboratory.

References

Dean, P., 1960, *Proc. Roy. Soc.* A., **254,** 507.
—— 1961, *Proc. Roy. Soc.* A., **260,** 263.
Domb, C., Maradudin, A. A., Montroll, E. W., and Weiss, G. H., 1959 a, *J. Phys. Chem. Solids*, **8,** 419.
—— 1959 b, *Phys. Rev.*, **115,** 18, 24.
Schmidt, H., 1957, *Phys. Rev.*, **105,** 425.

PROC. PHYS. SOC., 1964, VOL. 84

Vibrations of glass-like disordered chains*

P. DEAN

Mathematics Division, National Physical Laboratory, Teddington, Middlesex

MS. received 23rd June 1964

Abstract. Monatomic and alternating diatomic chains, with nearest neighbour harmonic interactions, are considered; in each case the force constants are governed by a probability distribution function which is continuous and bounded in a well-defined region and zero elsewhere. It is shown that the lattice modes are, in general, spatially localized. In the case of the diatomic chain the condition for a band gap is derived and the positions of the spectral branches are found. Spectra are calculated explicitly for monatomic and diatomic chains of a certain class, and typical displacement eigenvectors for the monatomic chain are depicted.

1. Introduction

Although a great deal of information is available on the infra-red and Raman spectra (Simon 1960) and the optical and thermodynamic constants of various glasses, there has been no explicit theoretical investigation of the problem of atomic vibrations in glass-like models. The nearest approach has been the example treated in § VI of the paper by Dyson (1953), but Dyson's system is entirely unrealistic in view of certain strong correlations which exist between the atomic masses and force constants of his linear chain.

The main difficulty, in two and three dimensions, stems from the lack of a periodic base lattice. The information available, on the structure of glasses, is statistical in character, and there is no obvious way in which this can be incorporated into a conventional lattice dynamical theory. The problem thus differs quite radically from the case of alloy-type disorder, in which considerable progress has been made in the past few years.

In one dimension there is obviously no difficulty associated with the absence of a well-defined base lattice, and the problem can be formulated in the same way as for the disordered alloy. The difference, here, between the glass and the alloy is that the former is characterized by a continuous probability distribution function for interatomic distances (or, what is essentially the same, for interatomic forces), whereas the latter is characterized by discrete probabilities relating to the existence of the various atomic species at the lattice sites. Of course, one cannot simulate a real glass by a one-dimensional model; however, such models do show some significant results which appear to be a consequence of the imposed disorder rather than of the dimensionality. In this paper we consider two models: a monatomic chain and a diatomic (alternating) chain. The interatomic forces are taken to be harmonic and to act between nearest neighbours only. The force constants are governed by continuous bounded probability distribution functions with sharp cut-off points.

In many respects the problem of the lattice vibrations of disordered systems is analogous to the problem of the behaviour of electrons in a disordered array of potential wells. We make use of this analogy in a proof of the localization of lattice modes. Thus

727

the work of § 2.2 follows a proof, by Borland (1963), of the spatial localization of the electronic wave functions in a one-dimensional system of potential wells separated by distances which are governed by a continuous distrib··tion function.

2. The monatomic chain

2.1. *Basic equations*

We consider the longitudinal vibrations of a chain of N similar atoms of mass m. Let x_i be the displacement from equilibrium of the ith atom, and γ_i the force constant relating to the atoms i and $i+1$. For vibrations of angular frequency ω, the relation connecting successive atomic displacements is

$$-m\omega^2 x_i = \gamma_i(x_{i+1}-x_i)+\gamma_{i-1}(x_{i-1}-x_i). \tag{1}$$

In order to be specific we take the boundary conditions to be those of free ends; equation (1) then applies to all atoms of the chain ($i = 1, 2, ..., N$) provided we take $\gamma_0 = \gamma_N = 0$. The $N-1$ positive values of ω^2 for which the equations (1) have (non-trivial) solutions define the normal modes of vibration of the chain. Of particular interest is the ω^2 distribution function $D(\omega^2)$ in the limit of large N. $D(\omega^2)$ is defined by the relation $D(\omega^2)d(\omega^2) = dM(\omega^2)$, where $M(\omega^2)$ is the proportion of normal mode squared frequencies less than ω^2.

In this paper we are interested in the ensemble of chains for which the γ_i are independent random quantities distributed according to a continuous bounded probability density function $P(\gamma)$, having sharp cut-off points. Thus, for all i, $P(\gamma_i)d\gamma_i$ is the probability, over the ensemble, that the ith force constant lies in the range γ_i to $\gamma_i+d\gamma_i$, and we take $P(\gamma_i)$ to be identically zero except in a region $\gamma_a \leqslant \gamma_i \leqslant \gamma_b$, with $\gamma_a \geqslant 0$. As $N \to \infty$, the distribution function for force constants of any individual chain of the ensemble tends to $P(\gamma)$, except for a set of chains of measure zero.

We shall find it convenient to work in terms of new variables $y_i = x_{i+1}-x_i$. The equations (1), for an arbitrary chain, can then be recast in the form

$$\left.\begin{gathered} -\gamma_{i-1}y_{i-1}+(2\gamma_i-m\omega^2)y_i-\gamma_{i+1}y_{i+1} = 0 \\ i = 1,2,\ldots,N-1 \\ (\gamma_0 = \gamma_N = 0) \end{gathered}\right\}. \tag{2}$$

The normal modes of the chain correspond to those values of $m\omega^2$ which are eigenvalues of the $N-1 \times N-1$ tridiagonal matrix (a_{ij}) defined by

$$\begin{aligned} a_{i,i-1} &= -\gamma_{i-1} \\ a_{ii} &= 2\gamma_i \\ a_{i,i+1} &= -\gamma_{i+1}. \end{aligned}$$

The leading principal minors of the matrix $a_{ij}-m\omega^2\delta_{ij}$ form a Sturm sequence (Burnside and Panton 1918). Let $g_i(\omega^2)$ be the leading principal minor of order i; then, expanding this determinant in terms of the elements of the last row and column, we obtain

$$\left.\begin{gathered} g_i = (2\gamma_i-m\omega^2)g_{i-1}-\gamma_{i-1}\gamma_i g_{i-2} \\ i = 2,3,\ldots,N-1 \\ g_0 \equiv 1; \quad g_1 = 2\gamma_1-m\omega^2 \end{gathered}\right\}. \tag{3}$$

Sturm's theorem states that the number of changes in sign between successive members of the set $g_i(\omega^2)$ ($i = 0, 1, 2, ..., N-1$) is equal to the number of normal mode frequencies less than ω. Rearranging the relations (3), and using the fact that the γ_i are all positive, we obtain the result that the number of normal mode frequencies smaller than ω is equal to the number of negative s_i, where

$$s_i = g_i/\gamma_i g_{i-1}. \tag{4}$$

(If g_{i-1} is exactly zero we replace it by an arbitrary small positive number in the definition of s_i.)

The s_i are connected by the recurrence formula

$$\left.\begin{aligned} s_i &= 2-\frac{m\omega^2}{\gamma_i}-\frac{1}{s_{i-1}}, \quad i = 2, 3, \ldots (N-1) \\ s_1 &= 2-\frac{m\omega^2}{\gamma_1} \end{aligned}\right\}. \tag{5}$$

This formula is particularly useful as s_{i-1} and γ_i are uncorrelated, s_{i-1} depending only on $\gamma_{i-1}, \gamma_{i-2}, \ldots, \gamma_{N-1}$. We may therefore use the product rule to relate the probability distribution function of s_i to the probability distributions of γ_i and s_{i-1}. It is reasonable to assume that, in the limit of large N, the probability distributions of s_{i-1} and s_i are equal (cf. § 2.2). Let $A(s)$ be this distribution; then, equating probabilities of s_i in both sides of (5), we obtain

$$A(s_i) = m\omega^2 \int_{-\infty}^{+\infty} A(s) P\left(\frac{m\omega^2}{2-s_i-1/s}\right)\left(2-s_i-\frac{1}{s}\right)^{-2} ds. \tag{6}$$

The integrand is identically zero except in the region

$$\gamma_a \leqslant \frac{m\omega^2}{2-s_i-1/s} \leqslant \gamma_b.$$

Consequently the ranges of s which contribute towards the integral are

$$\left.\begin{aligned} &\left(2-s_i-\frac{m\omega^2}{\gamma_b}\right)^{-1} \leqslant s \leqslant \left(2-s_i-\frac{m\omega^2}{\gamma_a}\right)^{-1}, \quad \text{for } \frac{m\omega^2}{2-s_i} < \gamma_a, \quad \text{or} \quad \frac{m\omega^2}{2-s_i} > \gamma_b \\ &-\infty < s \leqslant \left(2-s_i-\frac{m\omega^2}{\gamma_a}\right)^{-1} \text{ and } \left(2-s_i-\frac{m\omega^2}{\gamma_b}\right)^{-1} \leqslant s < +\infty \quad \text{for } \gamma_a < \frac{m\omega^2}{2-s_i} < \gamma_b \end{aligned}\right\} \tag{7}$$

In the limit $N \to \infty$, the proportion of negative s_i is equal to $M(\omega^2)$. We have, therefore,

$$M(\omega^2) = \int_{-\infty}^{0} A(s)\, ds \tag{8}$$

provided $A(s)$ is normalized to unity, i.e. $\int_{-\infty}^{+\infty} A(s)\, ds = 1$.

The ratios of successive atomic displacements are simply related to the s variables. On comparing the equations (2) and (5), we note that s_i can be written as

$$s_i = \frac{(x_{i+2}-x_{i+1})\gamma_{i+1}}{(x_{i+1}-x_i)\gamma_i}. \tag{9}$$

Substituting for $\gamma_{i+1}(x_{i+2}-x_{i+1})$ from the $i+1$th equation (1) then gives, after a little manipulation,

$$r_i = \frac{x_{i+1}}{x_i} = 1-\frac{m\omega^2 s_{i-1}}{\gamma_i(s_{i-1}-1)}. \tag{10}$$

If $Q(r_i)$ is the probability distribution of r_i it follows, from (10), that

$$Q(r_i) = m\omega^2 \int_{\gamma_a}^{\gamma_b} P(\gamma) A\left(\frac{1-r_i}{1-r_i-m\omega^2/\gamma}\right)\gamma^{-1}\left(1-r_i-\frac{m\omega^2}{\gamma}\right)^{-2} d\gamma. \tag{11}$$

2.2. *Localization of lattice modes*

In this section we follow, almost exactly, the argument of Borland (1963) in his proof of the spatial localization of electronic wave functions in a disordered system of potential wells. In fact, it is only necessary here to establish an equivalence between the parameters and equations in the two problems; Borland's arguments can then be taken over for the precise establishment of the localization result.

We define a quantity θ_i, for each of the chains of the ensemble, by the relation

$$\left.\begin{aligned} \tan\theta_i &= s_i \\ -\tfrac{1}{2}\pi \leqslant \theta_i &< \tfrac{1}{2}\pi \ (i = 1, 2, \dots, N-1) \end{aligned}\right\}. \tag{12}$$

We refer to θ_i as the *phase* at the ith atom; it is analogous to the reduced phase ϵ_i of Borland. The relation between successive phases is, from (5),

$$\left.\begin{aligned} \tan\theta_i &= 2-\frac{m\omega^2}{\gamma_i}+\tan(\theta_{i-1}+\tfrac{1}{2}\pi) \\ i &= 2, 3, \dots, N-1 \end{aligned}\right\}. \tag{13}$$

Now, from (5), (10) and (12)

$$r_i = \frac{s_i-1}{1-1/s_{i-1}} = \frac{\tan\theta_i - 1}{\tan(\theta_{i-1}+\tfrac{1}{2}\pi)+1};$$

therefore

$$\begin{aligned} r_i^2 &= \frac{\sin^2\theta_{i-1}}{\cos^2\theta_i}\frac{(1-\sin 2\theta_i)}{(1-\sin 2\theta_{i-1})} \\ &= \frac{d\theta_{i-1}}{d\theta_i}\frac{(1-\sin 2\theta_i)}{(1-\sin 2\theta_{i-1})}. \end{aligned} \tag{14}$$

Consequently

$$\frac{x_{n+1}^2}{x_i^2} = \prod_{j=i}^{n} r_j^2 = \frac{d\theta_{i-1}}{d\theta_n}\frac{(1-\sin 2\theta_n)}{(1-\sin 2\theta_{i-1})}. \tag{15}$$

This relation, connecting the displacements at the $n+1$th and ith sites, is equivalent to Borland's equation (8). From this result we note that if, as n increases, θ_n becomes increasingly insensitive to changes in θ_{i-1} then the function $x_{n+1}^2/(1-\sin 2\theta_n)$ must increase as we proceed along the chain in the direction of increasing n. We now follow the argument of § 3 of Borland's paper and show that, subject to an ergodic assumption, the phase does possess this property.

We consider, now, the ensemble of chains in the case of large N. We specify a boundary condition at only one end of the chains, at $i = 1$. The distribution of θ_1 (over the ensemble) then has a specific form $B_1(\theta_1)$, say. From (13) it follows that the distribution $B_2(\theta_2)$ of θ_2 is given by

$$B_2(\theta_2) = \int_{-\pi/2}^{+\pi/2} B_1(\theta_1) P\left\{\frac{m\omega^2}{2+\tan(\theta_1+\frac{1}{2}\pi)-\tan\theta_2}\right\} \frac{m\omega^2 \sec^2\theta_2\, d\theta_1}{\{2+\tan(\theta_1+\frac{1}{2}\pi)-\tan\theta_2\}^2}$$

$$= \int_{-\pi/2}^{+\pi/2} K(\theta_1, \theta_2) B_1(\theta_1)\, d\theta_1 \tag{16}$$

where

$$K(\theta_1, \theta_2) = P\left\{\frac{m\omega^2}{2+\tan(\theta_1+\frac{1}{2}\pi)-\tan\theta_2}\right\} \frac{m\omega^2 \sec^2\theta_2}{\{2+\tan(\theta_1+\frac{1}{2}\pi)-\tan\theta_2\}^2}. \tag{17}$$

The kernel $K(\theta_1, \theta_2)$ is the probability distribution function of θ_2, given that the phase at the first atom is θ_1. This kernel is a non-negative bounded function in the domain $(-\frac{1}{2}\pi, +\frac{1}{2}\pi)$, and $\int_{-\pi/2}^{+\pi/2} K(\theta_1, \theta_2) d\theta_2 = 1$.

Given $B_2(\theta_2)$ we can, from (13), derive $B_3(\theta_3)$, $B_4(\theta_4)$, ... In general

$$B_j(\theta_j) = \int_{-\pi/2}^{+\pi/2} K(\theta_{j-1}, \theta_j) B_{j-1}(\theta_{j-1})\, d\theta_{j-1} \qquad (j = 2, 3, \dots, N) \tag{18}$$

and therefore

$$B_{n+1}(\theta_{n+1}) = \int_{-\pi/2}^{+\pi/2} K_{n+1,i}(\theta_i, \theta_{n+1}) B_i(\theta_i)\, d\theta_i \tag{19}$$

where

$$K_{n+1,i}(\theta_i, \theta_{n+1}) = \int_{-\pi/2}^{+\pi/2} \dots \int K(\theta_n, \theta_{n+1}) K(\theta_{n-1}, \theta_n) \dots K(\theta_i, \theta_{i+1})\, d\theta_{i+1}\, d\theta_{i+2} \dots d\theta_n. \tag{20}$$

$K_{ji}(\theta_i, \theta_j)$ is defined recursively by

$$K_{ji}(\theta_i, \theta_j) = \int_{-\pi/2}^{+\pi/2} K_{j-1,i}(\theta_i, \theta_{j-1}) K(\theta_{j-1}, \theta_j)\, d\theta_{j-1} \qquad (j = i+2, i+3, \dots, N) \tag{21}$$

with $K_{i+1,i}(\theta_i, \theta_{i+1}) = K(\theta_i, \theta_{i+1})$.

Provided the kernel K satisfies certain conditions (Fréchet 1934) then, as n is increased, the kernel $K_{n+1,i}(\theta_i, \theta)$ converges uniformly to a limiting function $B(\theta)$ independent of θ_i; moreover, $B(\theta)$ is the unique solution of the equations

$$\left.\begin{aligned} B(\theta) &= \int_{-\pi/2}^{+\pi/2} K(\theta', \theta) B(\theta')\, d\theta' \\ &\int_{-\pi/2}^{+\pi/2} B(\theta)\, d\theta = 1 \end{aligned}\right\} \tag{22}$$

The Fréchet conditions are, in fact, satisfied for the kernel K of this paper—this is shown in the appendix—and the property $K_{n+1,i}(\theta_i, \theta) \to B(\theta)$, as n increases, is therefore valid. Thus, as we proceed along the chains of the ensemble in the direction of increasing n, the phase distribution function settles down to the function $B(\theta)$ independently of any initially specified phase distribution.

This result is formally identical with the equivalent result for the phase ϵ of Borland (1963). In fact, the differences throughout the whole of both proofs of localization manifest themselves only as points of detail. For example, our relation (15) is basically similar to Borland's equation (8) although not identical. Again, while the explicit form of our kernel K differs from Borland's kernel K—the phases θ and ϵ are, of course, quite different quantities—the formalism relating to the phase distributions is very similar in both problems. In fact, it transpires that there is an exact correspondence of all the stages in the two proofs of localization. There is little point in repeating all the detailed arguments of Borland here, and in the following we mention only the main steps of the proof for the vibrational case and discuss the details only when they differ significantly from the electronic case.

The stationary phase density function $B(\theta)$ has been derived for solutions satisfying a boundary condition at the end $i = 1$ of the chains. There is, similarly, a stationary phase density function $C(\theta)$ for solutions satisfying a single boundary condition at the other end of the chains. In the following paragraph we prove that, in general, $C(\theta) \neq B(\theta)$.

The relation (13) links successive phases, but we must now regard θ_{i-1} as a function of θ_i and γ_i (noting that then θ_i depends on $\gamma_{i+1}, \gamma_{i+2}, \ldots$ and is independent of γ_i). Following the derivation outlined in equations (16)–(22), we find

$$\left.\begin{aligned} C(\theta) &= \int_{-\pi/2}^{+\pi/2} K'(\theta', \theta) C(\theta')\, d\theta' \\ \int_{-\pi/2}^{+\pi/2} C(\theta)\, d\theta &= 1 \end{aligned}\right\} \tag{23}$$

where

$$K'(\theta', \theta) = P\left\{\frac{m\omega^2}{2+\tan(\theta+\frac{1}{2}\pi)-\tan\theta'}\right\} \frac{m\omega^2 \operatorname{cosec}^2\theta}{\{2+\tan(\theta+\frac{1}{2}\pi)-\tan\theta'\}^2}. \tag{24}$$

Changing to new variables α and α', defined by

$$\left.\begin{aligned} \tan(\alpha+\tfrac{1}{2}\pi) &= -\tan\theta \\ \tan(\alpha'+\tfrac{1}{2}\pi) &= -\tan\theta' \end{aligned} \qquad -\tfrac{1}{2}\pi \leqslant \alpha, \alpha' < +\tfrac{1}{2}\pi\right\} \tag{25}$$

we find that (23) transforms to

$$\left.\begin{aligned} D(\alpha) &= \int_{-\pi/2}^{+\pi/2} K(\alpha', \alpha) D(\alpha')\, d\alpha' \\ \int_{-\pi/2}^{+\pi/2} D(\alpha)\, d\alpha &= 1 \end{aligned}\right\} \tag{26}$$

where $D(\alpha)$ is the distribution of α, and K is the kernel (17). It follows that $D(\alpha) = B(\alpha)$. Since $D(\alpha) = C(\theta)\, |d\theta/d\alpha|$, the distributions B and C, if equal, would be defined by the equation

$$B(\alpha) = B(\theta)\left|\frac{d\theta}{d\alpha}\right| \tag{27}$$

and the upper relation (25), *independently* of $P(\gamma)$. Clearly, then, $B(\theta) \neq C(\theta)$ if disorder is present.†

Now, following Borland, we introduce the hypothesis that the distribution function for phases over the sites of a typical chain of the ensemble approaches, in the limit $N \to \infty$, the appropriate ensemble phase distribution function for almost all values of ω^2 and initially specified phase. This leads (cf. Borland) to a $\theta_n - \theta_{i-1}$ relationship

† A normalized solution of (27) is $B(\theta) = (4-\nu^2)^{1/2}/\pi(2+\nu\sin 2\theta)$. This is the θ distribution for the periodic chain (no disorder present), corresponding to the squared frequency $(4\gamma/m)(\nu+2)$.

similar to that between ϵ'_{N-1} and ϵ_0 in Borland's figure 2. This, in turn, implies (from (15)) diverging solutions from the boundaries for almost all values of ω^2 and initial phase.

We can use similar matching arguments to those in § 6 of Borland's paper in order to show how eigenfunctions occur from solutions of the kind we have found. We now work in terms of an *unreduced* phase ϕ_i, defined similarly to θ_i in (12) and (13) except that if $m\pi \leqslant \phi_i < (m+\frac{1}{2})\pi$, then $(m+\frac{1}{2})\pi \leqslant \phi_{i+1} < (m+\frac{3}{2})\pi$, and if $(m-\frac{1}{2})\pi \leqslant \phi_i < m\pi$, then $(m-\frac{1}{2})\pi < \phi_{i+1} < (m+\frac{1}{2})\pi$, with $-\frac{1}{2}\pi \leqslant \phi_1 < +\frac{1}{2}\pi$ $(m = 0, 1, 2, \ldots)$. From the definition of ϕ_i we have

$$\begin{aligned}\sec^2\phi_i\frac{\partial\phi_i}{\partial\omega^2} &= -\frac{m}{\gamma_i}+\operatorname{cosec}^2\phi_{i-1}\frac{\partial\phi_{i-1}}{\partial\omega^2}\\ &= -\frac{m}{\gamma_i}-\frac{m}{\gamma_{i-1}}\cot^2\phi_{i-1}-\frac{m}{\gamma_{i-2}}\cot^2\phi_{i-1}\cot^2\phi_{i-2}-\ldots-\frac{m}{\gamma_{j+1}}\cot^2\phi_{i-1}\ldots\\ &\quad\ldots\cot^2\phi_{j+1}+\cot^2\phi_{i-1}\cot^2\phi_{i-2}\ldots\cot^2\phi_{j+1}\operatorname{cosec}^2\phi_j\frac{\partial\phi_j}{\partial\omega^2}\end{aligned} \tag{28}$$

and also

$$\begin{aligned}\operatorname{cosec}^2\phi_i\frac{\partial\phi_i}{\partial\omega^2} &= \frac{m}{\gamma_{i+1}}+\sec^2\phi_{i+1}\frac{\partial\phi_{i+1}}{\partial\omega^2}\\ &= \frac{m}{\gamma_{i+1}}+\frac{m}{\gamma_{i+2}}\tan^2\phi_{i+1}+\frac{m}{\gamma_{i+3}}\tan^2\phi_{i+1}\tan^2\phi_{i+2}+\ldots+\frac{m}{\gamma_k}\tan^2\phi_{i+1}\ldots\\ &\quad\ldots\tan^2\phi_{k-1}+\tan^2\phi_{i+1}\ldots\tan^2\phi_{k-1}\sec^2\phi_k\frac{\partial\phi_k}{\partial\omega^2}.\end{aligned} \tag{29}$$

If follows, from these relations, that if the unreduced phase at some point j is a constant, independent of ω^2, then the unreduced phase at a point i is a monotonic decreasing function of ω^2 if $i > j$, and a monotonic increasing function of ω^2 if $i < j$. This property of ϕ is similar, though in the opposite sense, to the monotonic property of Borland's absolute phase (cf. Borland 1963, p. 539). Having established this property, the proof of the matching of solutions satisfying boundary conditions at opposite ends of a long random chain is almost identical with Borland's proof (the opposite sense of the monotonic property of ϕ producing a trivial difference); we shall therefore not repeat the discussions of § 6 of Borland's paper here.

Note that the proof of the localization of the eigenfunctions of the glass-like chain does not apply in the limiting cases of complete order, or $\omega^2 = 0$, as then the kernel K (equation (17)) becomes unbounded.

2.3. *A particular class of disordered chains*

The class of monatomic disordered chains for which

$$\left.\begin{aligned}P(\gamma) &= \text{constant} = \frac{1}{\gamma_b-\gamma_a}\text{ for }\gamma_a < \gamma < \gamma_b\\ &= 0,\text{ elsewhere}\end{aligned}\right\} \tag{30}$$

was investigated numerically. Frequency spectra were obtained by computing the s_i (from (5)) for particular long disordered chains containing 7900 atoms†, and noting the

† 7900 atoms, not 8000 atoms, due to certain machine storage limitations.

proportion of negative s_i at various values of ω^2. The γ_i values, subject to the distribution (30), were generated by a computer programme incorporating a random number routine.

Typical spectra are shown as the histograms of figure 1. It can be seen that these spectra are all fairly featureless, the high frequency singularity of the periodic lattice

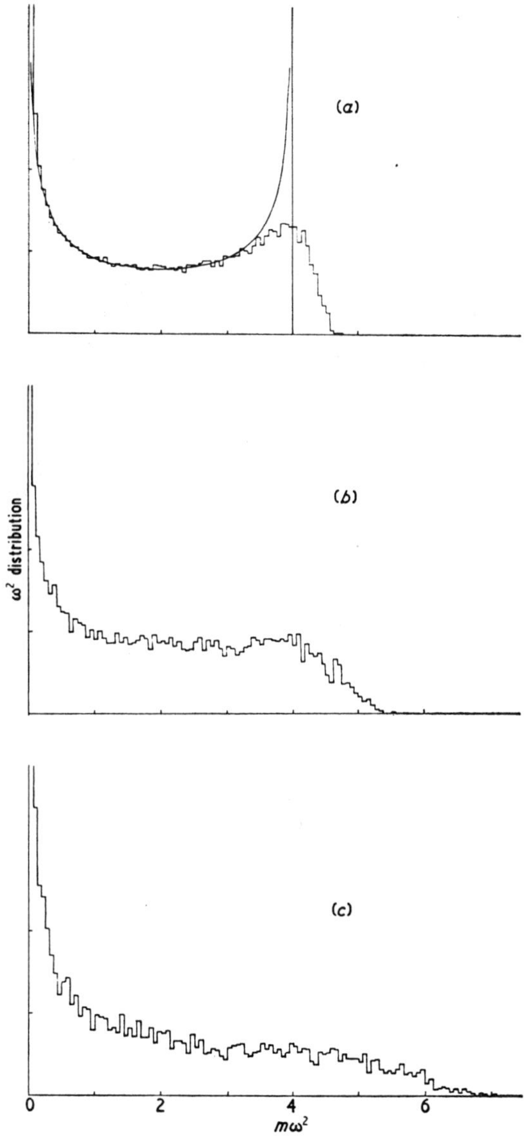

Figure 1. Spectra of squared frequencies for monatomic chains in which the distribution of force constants is of the form (30). (*a*) $\gamma_a = \frac{3}{4}$, $\gamma_b = 1\frac{1}{4}$; (*b*) $\gamma_a = \frac{1}{2}$, $\gamma_b = 1\frac{1}{2}$; (*c*) $\gamma_a = 0$, $\gamma_b = 2$. The smooth curve in (*a*) is the spectrum for the ordered chain ($\gamma = 1$).

becoming rounded and progressively less obvious as the disorder increases. The small local fluctuations in the histograms are due to the finiteness of the chains used; this we have established by comparisons with histograms for other chains having the same statistical properties. The upper limit of values of ω^2 for the chains (30) is given by $m\omega^2 = 4\gamma_b$. This corresponds to the maximum possible frequency of sections of chain in which the force constants are all close to their maximum values γ_b.

It is of interest to note that the other computational possibility—to solve the homogeneous integral equation (6), and then to use the relation (8)—leads to an ensemble averaged spectrum, whereas the method we have used here gives the spectra of particular long disordered chains. In the case of the analogous problem for the disordered diatomic

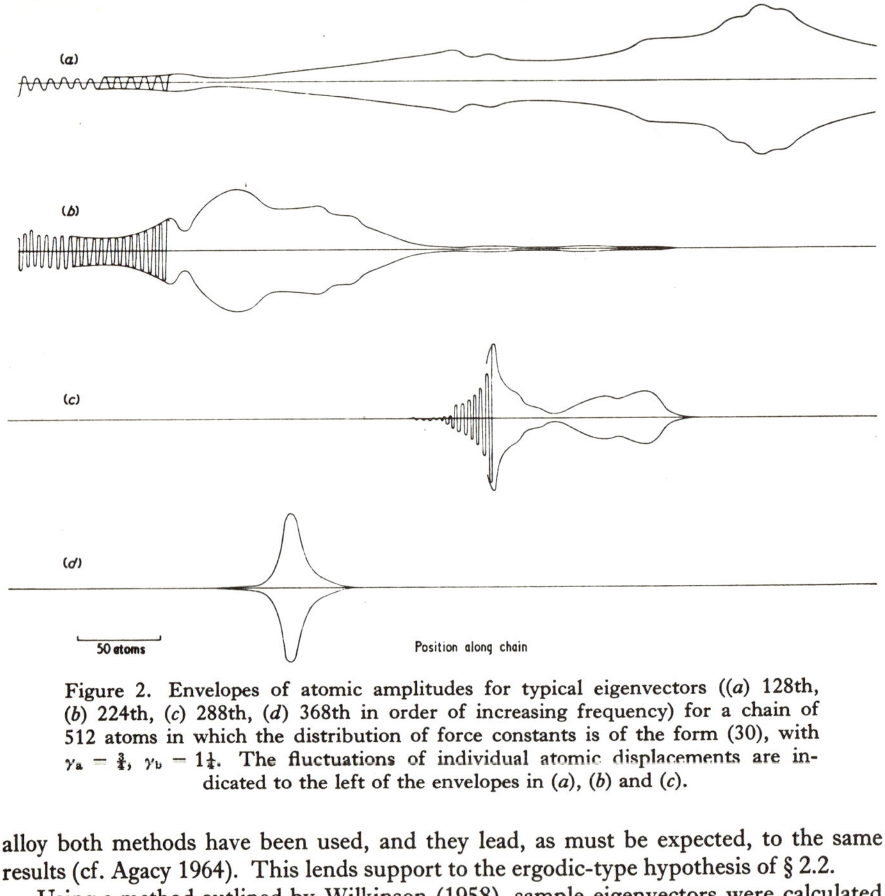

Figure 2. Envelopes of atomic amplitudes for typical eigenvectors ((*a*) 128th, (*b*) 224th, (*c*) 288th, (*d*) 368th in order of increasing frequency) for a chain of 512 atoms in which the distribution of force constants is of the form (30), with $\gamma_a = \frac{3}{4}$, $\gamma_b = 1\frac{1}{4}$. The fluctuations of individual atomic displacements are indicated to the left of the envelopes in (*a*), (*b*) and (*c*).

alloy both methods have been used, and they lead, as must be expected, to the same results (cf. Agacy 1964). This lends support to the ergodic-type hypothesis of § 2.2.

Using a method outlined by Wilkinson (1958), sample eigenvectors were calculated for the class of disordered chains described by (30). It is found that, in general, the localization of the vectors increases with frequency, and with the degree of disorder of the chain. In figure 2 we have drawn the approximate envelopes of four typical vectors for a disordered chain ($\gamma_a = \frac{3}{4}$; $\gamma_b = 1\frac{1}{4}$) of 512 atoms. These are the vectors of the 128th, 224th, 288th and 368th modes (if the modes are enumerated in order of increasing frequency) as we go from (*a*) to (*d*) in the diagram. In (*a*), (*b*) and (*c*) the fluctuations of atomic displacements are indicated diagrammatically at the left of the envelopes.

The figure shows clearly how intense the localization becomes at high frequencies, even when the disorder is sufficiently small for the spectrum to be very similar (figure 1(*a*)) to the spectrum for the periodic lattice. Away from the regions of localization in figures 2(*b*), (*c*) and (*d*) the amplitudes of the atomic displacements are not, of course, exactly zero, although they are extremely small.

3. The diatomic chain

3.1. *Basic equations*

We now study the vibrations of a diatomic chain of the form ...ABABAB... in which the force constants are governed by a continuous bounded probability distribution $P(\gamma)$, as in § 2.1. Again, $P(\gamma)$ is identically zero except in the region $\gamma_a \leqslant \gamma \leqslant \gamma_b$, and $\gamma_a \geqslant 0$.

We take the atomic masses to be m at the even numbered sites of the chain, and M at the odd numbered sites. The equations of motion for all even i are then identical with (1) whereas, in the equations of odd i, M must be substituted for m on the left-hand side. If, now, we consider the i and $i+1$th equations for the case of odd i, and subtract the former from the latter after division by m or M, we find

$$-\frac{\gamma_{i-1}}{M}y_{i-1}+\left(\frac{\gamma_i}{m}+\frac{\gamma_i}{M}-\omega^2\right)y_i-\frac{\gamma_{i+1}}{m}y_{i+1}=0 \qquad (i\text{ odd}) \qquad (31a)$$

where $y_i = x_{i+1}-x_i$. The analogous equation for $i+1$ (which is even), instead of i, is

$$-\frac{\gamma_i y_i}{m}+\left(\frac{\gamma_{i+1}}{m}+\frac{\gamma_{i+1}}{M}-\omega^2\right)y_{i+1}-\frac{\gamma_{i+2}}{M}y_{i+2}=0 \qquad (i+1\text{ even}). \quad (31b)$$

The normal mode squared frequencies of a chain with free end boundary conditions therefore correspond to the eigenvalues of the $N-1\times N-1$ tridiagonal matrix (b_{ij}) with elements

$$b_{ii} = \gamma_i\left(\frac{1}{m}+\frac{1}{M}\right), \qquad (\text{all } i)$$

$$b_{i,i+1} = -\frac{\gamma_{i+1}}{m}, \qquad (i\text{ odd})$$

$$= -\frac{\gamma_{i+1}}{M}, \qquad (i\text{ even})$$

$$b_{i,i-1} = -\frac{\gamma_{i-1}}{M}, \qquad (i\text{ odd})$$

$$= -\frac{\gamma_{i-1}}{m}, \qquad (i\text{ even}).$$

Following the method of § 2.1, it is easy to show that the number of normal mode frequencies smaller than ω is equal to the number of negative s_i, where

$$\left.\begin{aligned} s_i &= \frac{1}{m}+\frac{1}{M}-\frac{\omega^2}{\gamma_i}-\frac{1}{M^2 s_{i-1}}, \quad (i\text{ odd})\\ s_i &= \frac{1}{m}+\frac{1}{M}-\frac{\omega^2}{\gamma_i}-\frac{1}{m^2 s_{i-1}}, \quad (i\text{ even})\\ s_1 &= \frac{1}{m}+\frac{1}{M}-\frac{\omega^2}{\gamma_1} \end{aligned}\right\}. \qquad (32)$$

As $M \to m$, we note that (32) becomes

$$s_i = \frac{2}{m} - \frac{\omega^2}{\gamma_i} - \frac{1}{m^2 s_{i-1}}$$

$$s_1 = \frac{2}{m} - \frac{\omega^2}{\gamma_1}$$

showing that the s_i for the diatomic chain corresponds to the parameter s_i/m of § 2.

As s_{i-1} and γ_i are uncorrelated, we can easily equate the probabilities of both sides of (32). Let $B_0(s_i)$ be the probability distribution function of s_i (over an ensemble of very long chains) for odd i, and $B_e(s_i)$ the corresponding distribution for even i; then, from (32), we have

$$\left.\begin{aligned} B_0(s_i) &= \omega^2 \int B_e(s) P\left(\frac{\omega^2}{1/m + 1/M - s_i - 1/M^2 s}\right)\left(\frac{1}{m} + \frac{1}{M} - s_i - \frac{1}{M^2 s}\right)^{-2} ds \\ B_e(s_{i-1}) &= \omega^2 \int B_0(s) P\left(\frac{\omega^2}{1/m + 1/M - s_{i-1} - 1/m^2 s}\right)\left(\frac{1}{m} + \frac{1}{M} - s_{i-1} - \frac{1}{m^2 s}\right)^{-2} ds \end{aligned}\right\} \tag{33}$$

where the range of integration is effectively limited by the cut-off behaviour of $P(\gamma)$. B_0 and B_e may be found from these equations (or, after an explicit integration, an equivalent single-integral equation) by fairly standard numerical methods, given $P(\gamma)$. The integrated spectrum $M(\omega^2)$ is then given by

$$M(\omega^2) = \tfrac{1}{2} \int_{-\infty}^{0} \{B_0(s) + B_e(s)\}\, ds \tag{34}$$

provided

$$\int_{-\infty}^{+\infty} B_0(s)\, ds = \int_{-\infty}^{+\infty} B_e(s)\, ds = 1.$$

3.2. *Condition for a band gap*

In this section we derive a condition for the existence of a gap separating the spectrum of the disordered diatomic chain into bands corresponding to the acoustic and optical branches of the ordered chain. We find the limits of the two bands when a gap does exist.

In the case of the ordered chain a value of ω^2 in the band gap leads to a complete 'locking' of the s_i at two points, one positive, one negative, such that successive s_i alternate between the two values. When disorder is introduced into the system we expect these two particular values to extend into locking regions of finite extent, necessarily located on opposite sides of the origin. Let us therefore consider a frequency in the band gap for a disordered chain, and let $s_i, s_{i\pm 2}, \ldots$ and $s_{i\pm 1}, s_{i\pm 3}, \ldots$ be confined to positive and negative regions, respectively, such that

$$\left.\begin{aligned} &\sigma_2 < s_{i\pm 2p} < \tau_2 \\ &\sigma_1 < s_{i-1\pm 2p} < \tau_1 \\ &\tau_1 < 0 < \sigma_2 \end{aligned} \quad (p = 0, 1, 2, \ldots) \right\}. \tag{35}$$

We take i to be odd in the following; this implies (as one can show on *a posteriori* grounds), from (32), that $M > m$.

8

Now, since $s_{i-1} > \sigma_1$, and s_i is a monotonic increasing function of s_{i-1} ($s_{i-1} \neq 0$, here), the smallest possible value for s_i is

$$\sigma_2 = \alpha_{\min} - \frac{1}{M^2 \sigma_1} \tag{36a}$$

where $\alpha_{\min}$ is the minimum value of $\alpha = 1/m + 1/M - \omega^2/\gamma$. Similarly, as $s_i > \sigma_2$, the minimum value of s_{i+1} is

$$\sigma_1 = \alpha_{\min} - \frac{1}{m^2 \sigma_2}. \tag{36b}$$

A similar argument relating the maximum values of s_{i-1}, s_i and s_{i+1} leads to

$$\left.\begin{aligned} \tau_2 &= \alpha_{\max} - \frac{1}{M^2 \tau_1} \\ \tau_1 &= \alpha_{\max} - \frac{1}{M^2 \tau_2} \end{aligned}\right\}. \tag{37}$$

The equations (36) lead to the following quadratic for σ_2:

$$\alpha_{\min}\sigma_2^2 + \left(\frac{1}{M^2} - \frac{1}{m^2} - \alpha_{\min}^2\right)\sigma_2 + \frac{\alpha_{\min}}{m^2} = 0. \tag{38}$$

The conditions for real roots are

(i) $$\alpha_{\min}^2 \geqslant \left(\frac{1}{m} + \frac{1}{M}\right)^2, \quad \text{i.e. } \omega^2 \leqslant 0 \text{ or } \omega^2 \geqslant 2\gamma_a\left(\frac{1}{m} + \frac{1}{M}\right)$$

or

(ii) $$\alpha_{\min}^2 \leqslant \left(\frac{1}{m} - \frac{1}{M}\right)^2, \quad \text{i.e. } \frac{2\gamma_a}{M} \leqslant \omega^2 \leqslant \frac{2\gamma_a}{m}.$$

Similarly, the quadratic equation for τ_2, from (37), leads to further necessary conditions on ω^2, viz.

(iii) $$\omega^2 \leqslant 0 \text{ (as before) or } \omega^2 \geqslant 2\gamma_b\left(\frac{1}{m} + \frac{1}{M}\right)$$

or

(iv) $$\frac{2\gamma_b}{M} \leqslant \omega^2 \leqslant \frac{2\gamma_b}{m}.$$

Further analysis shows that the two regions (σ_1, τ_1) and (σ_2, τ_2) lie on opposite sides of the origin only if (ii) and (iv) are valid. (i) and (iii) correspond to cases in which the regions are both either positive ($\omega^2 \leqslant 0$) or negative ($\omega^2 \geqslant 2\gamma_b(1/m + 1/M)$); in fact, they define the regions outside the spectrum. The conditions, then, for ω^2 to lie in the band gap are

$$\frac{2\gamma_a}{M} < \omega^2 < \frac{2\gamma_a}{m} \quad \text{and} \quad \frac{2\gamma_b}{M} < \omega^2 < \frac{2\gamma_b}{m}$$

or, more simply,

$$\frac{2\gamma_b}{M} < \omega^2 < \frac{2\gamma_a}{m} \qquad (M > m). \tag{39}$$

Clearly, a band gap analogous to the gap between the acoustic and optical branches of the spectrum of the ordered system exists only if

$$m\gamma_b < M\gamma_a \qquad (M > m). \tag{40}$$

The properties (39) and (40) are independent of the particular form of $P(\gamma)$, provided only that it is zero outside the range $\gamma_a \leqslant \gamma \leqslant \gamma_b$. Note, from an earlier remark, that the

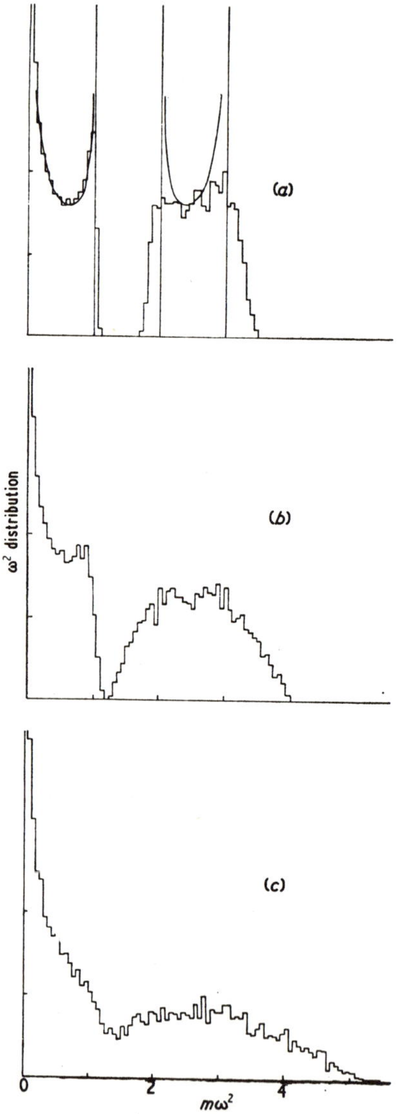

Figure 3. Spectra of squared frequencies for diatomic alternating chains in which the distribution of force constants is of the form (30). The atomic masses are m and $2m$. (*a*) $\gamma_a = \frac{3}{4}$, $\gamma_b = 1\frac{1}{4}$; (*b*) $\gamma_a = \frac{1}{2}$, $\gamma_b = 1\frac{1}{2}$; (*c*) $\gamma_a = 0$, $\gamma_b = 2$. In (*a*) the smooth curves depict the spectrum of the ordered chain ($\gamma = 1$).

maximum frequency of the entire spectrum is given by

$$\omega_{\mathrm{max}}^2 = 2\gamma_{\mathrm{b}}\left(\frac{1}{m}+\frac{1}{M}\right). \tag{41}$$

This frequency corresponds to the maximum possible frequency of a section of chain in which all the force constants take on their maximum values γ_{b}.

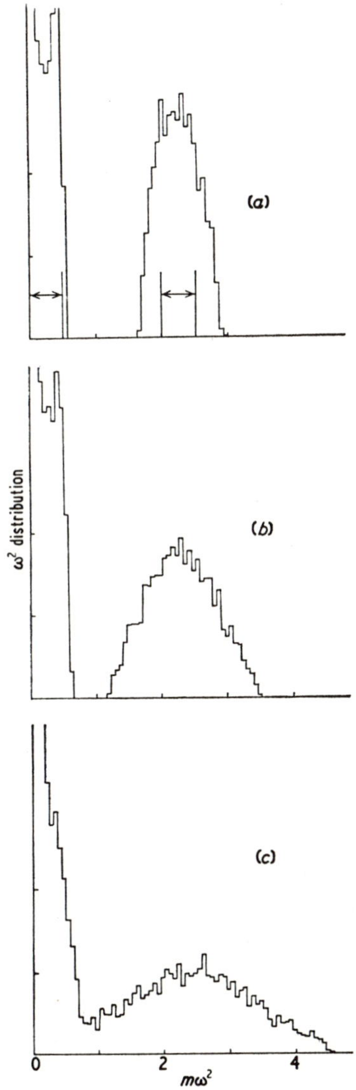

Figure 4. Spectra of squared frequencies for diatomic alternating chains in which the distribution of force constants is of the form (30) and the atomic masses are m and $4m$. (*a*) $\gamma_{\mathrm{a}} = \frac{3}{4}$, $\gamma_{\mathrm{b}} = 1\frac{1}{4}$; (*b*) $\gamma_{\mathrm{a}} = \frac{1}{2}$, $\gamma_{\mathrm{b}} = 1\frac{1}{2}$; (*c*) $\gamma_{\mathrm{a}} = 0$, $\gamma_{\mathrm{b}} = 2$. In (*a*) the regions of the acoustic and optical bands of the spectrum of the ordered chain ($\gamma = 1$) are indicated by horizontal arrows.

3.3. *Localization of lattice modes for the diatomic chain*

One can use a similar method to that outlined in § 2.2 to prove that the lattice modes for the diatomic disordered chain are, in general, localized. The formulae are a little more complicated and, in fact, one must be careful to treat the variables and the distribution functions relating to even and odd lattice sites differently. For example, two basic kernels $L(\theta_i, \theta_{i+2})$ and $M(\theta_{i+1}, \theta_{i+3})$ (i is odd) must be used in place of the single kernel (17) for the monatomic chain. One can then prove that the θ distributions for odd and even sites tend, as in § 2.2, to limiting functions and, again using an ergodic-type hypothesis, show that the ratios of squares of odd-site or even-site vibrational amplitudes become small, or large, with increasing distance, depending upon the direction of the single boundary condition. The other steps of the proof follow similar though, again, rather more complicated lines than in § 2.2.

3.4. *A particular class of diatomic disordered chains*

In this section we consider diatomic alternating chains with the distribution (30) for force constants. Spectra for such systems were computed, using the method outlined in § 2.3, over a range of mass ratios and degrees of order. Typical results, for the cases of mass ratio 2:1 and 4:1, are shown in figures 3 and 4. These spectra are, as in the monatomic case (figure 1) fairly smooth in form, all the local fluctuations of the histograms being the consequence of the finiteness (7900 atoms) of the chains used. The smooth curve in figure 3(*a*) is the density of squared frequencies for the ordered diatomic chain of mass ratio 2:1. In figure 4(*a*) the positions of the acoustic and optical bands for the ordered chain of mass ratio 4:1 are indicated by the horizontal arrows.

The criteria of § 3.2 for the existence and positions of band gaps show gaps to exist in the spectra for the infinite chains corresponding to the cases of figures 3(*a*), 4(*a*) and 4(*b*). The limits of the gaps in these cases are $\omega^2 = \frac{5}{4}$ and $\frac{3}{2}$ (figure 3(*a*)), $\omega^2 = \frac{5}{8}$ and $\frac{3}{2}$ (figure 4(*a*)), and $\omega^2 = \frac{3}{4}$ and 1 (figure 4(*b*)). In the diagrams the gaps appear to be a little wider than indicated theoretically, but this is consistent with the ω^2 distributions falling very rapidly to zero near the band gaps. A similar situation exists near the high frequency end of each spectrum; the histograms of figures 3 and 4 (and also of figure 1) seem to end a little short of the theoretical maximum squared frequencies (41).

The spectrum for the chain of infinitely large mass ratio and γ distribution (30) can be calculated exactly. In this case the heavy masses become rigid walls (of zero vibrational frequency) which separate the light masses into $\frac{1}{2}N$ independent single-atom systems. The frequency of vibration for a typical light atom (i) is given by

$$\omega^2 = \frac{\gamma_i + \gamma_{i-1}}{m}. \tag{42}$$

It follows that, as $N \to \infty$, the ω^2 distribution for the light atoms is

$$D_{\mathrm{L}}(\omega^2) = m \int P(\gamma) P(m\omega^2 - \gamma)\, d\gamma \tag{43}$$

and, if $P(\gamma)$ is given by (30), then

$$\left.\begin{aligned} D_{\mathrm{L}}(\omega^2) &= \frac{m(m\omega^2 - 2\gamma_{\mathrm{a}})}{(\gamma_{\mathrm{b}} - \gamma_{\mathrm{a}})^2}, && 2\gamma_{\mathrm{a}} \leqslant m\omega^2 \leqslant \gamma_{\mathrm{a}} + \gamma_{\mathrm{b}} \\ &= \frac{m(2\gamma_{\mathrm{b}} - m\omega^2)}{(\gamma_{\mathrm{b}} - \gamma_{\mathrm{a}})^2}, && \gamma_{\mathrm{a}} + \gamma_{\mathrm{b}} \leqslant m\omega^2 \leqslant 2\gamma_{\mathrm{b}} \\ &= 0 && \text{elsewhere} \end{aligned}\right\}. \tag{44}$$

Consequently, the total spectrum is

$$D(\omega^2) = \delta(\omega^2)+\tfrac{1}{2}D_{\mathrm{L}}(\omega^2) \tag{45}$$

where the δ function at the origin contains half the frequencies of the system.

There is clear evidence of the approach towards the triangular form (44) in figure 4; and here the mass ratio is only 4:1. The triangular shape of the upper band is very apparent indeed in spectra computed for the case of mass ratio 8:1.

4. Discussion

The lack of feature in the spectra computed in §§ 2.3 and 3.4 seems to be a consequence not so much of the explicit form of $P(\gamma)$ but, rather, of the continuity of this distribution in the range γ_{a} to γ_{b}. The smoothness of these spectra is in marked contrast to the highly detailed spectra of the disordered alloy (Dean 1960, 1961, Agacy 1964, Hori and Fukushima 1964). However, in the case of the alloy various local configurations of atoms occur again and again throughout the structure; associated with these specific local groupings are highly localized modes which contribute towards certain peaks in the spectrum. In the glass-like chains considered in this paper there are no such specific local structures which occur relatively frequently; consequently, there are no associated modes which contribute frequencies to special small intervals. Obviously, then, the diffuseness of the spectrum for a glass-like chain is a consequence of the smoothness of $P(\gamma)$ rather than of its specific form. Although the spectra of the systems considered in this paper depend upon $P(\gamma)$, the overall features will not vary greatly provided that γ_{a} and γ_{b} remains constant and $P(\gamma)$ is a moderately smooth function in this region. We have already shown that, for the diatomic chain, the condition for the existence of a band gap, and the positions of the two branches of the spectrum when a gap does occur, depend only on the values of γ_{a} and γ_{b} and not on the explicit form of $P(\gamma)$.

The work of this paper on the localization of lattice modes, and also the evidence from work on the disordered alloy, suggests that localization is a necessary consequence of the presence of disorder. It seems that when disorder (in any form) exists in a system the lattice modes are localized. Obviously, the extent of localization depends upon the degree of disorder in the system; for example, in the limit $\omega^2 \to 0$ the localization becomes small and the modes become lattice waves.

The qualitative results of this paper (for example, the localization of lattice modes, smooth spectra, and distinct 'acoustic' and 'optical' bands in the presence of appreciable disorder) probable carry over into two and three dimensions. In fact, there is some experimental evidence for the distinguishability of two spectra branches in diatomic glasses (Simon 1960). However, real glasses are far more complicated structures than any direct three-dimensional analogue of the models of this paper—thus, many common glasses contain strongly coupled molecular groups; obviously, quite elaborate models are needed to simulate such systems.

Acknowledgments

I wish to acknowledge helpful discussions with Mr. R. E. Borland. Mr. N. F. Bird wrote the programme for the computation of eigenvectors. The work described in this paper has been carried out as part of the research programme of the National Physical Laboratory and this paper is published by permission of the Director.

Appendix

The proof of the Fréchet conditions is almost identical with the proof set out in the appendix of Borland (1963), and the reader should refer to that paper for the main outline. However, the kernel K of the present paper is a non-negative function of t and x (in the notation of Borland) rather than a strictly positive function. An additional possibility thereby arises in the proof that y = constant is the only solution of Borland's equation (37) for an eigenvalue of modulus unity. This possibility occurs in the case when arg$y$ is not constant. The modified proof is as follows.

If arg$y$ is not constant, then

$$|\gamma|\,|y| = |\int Ky\,dt| \leqslant \int K|y|dt \leqslant |y|_{\max}.$$

As this is true for all x it follows that

$$|\gamma|\,|y|_{\max} \leqslant |y|_{\max}$$

and therefore $|\gamma| \leqslant 1$. Thus it is possible that $|\gamma| = 1$. In this eventuality we have

$$|y| = |\int Ky\,dt| \leqslant \int K|y|dt \leqslant |y|_{\max}$$

and from this relation we see that, at those values of x for which $|y(x)| = |y|_{\max}$, we must have

$$|\int Ky\,dt| = \int K|y|dt$$

(implying that arg$y$ is a constant where $K \neq 0$), and

$$\int K|y|dt = |y|_{\max}$$

(implying that $|y| = |y|_{\max}$ where $K \neq 0$). In the next paragraph we use this fact that if $|\gamma| = 1$, and x is such that $|y(x)| = |y|_{\max}$, then $|y| = |y|_{\max}$ and arg$y$ = constant at those values of t for which $K \neq 0$.

Now, if $|y(x)| = |y|_{\max}$ for $x = x_1$, then from the result just proved, $|y(x)| = |y|_{\max}$ and arg$y$ = constant for all x in the region R_1 for which $K(x_1, x) \neq 0$. From the explicit form of K in (17) we see that R_1 is the region defined by

$$2-\frac{m\omega^2}{\gamma_a}+\tan\left(x_1+\frac{\pi}{2}\right) < \tan x < 2-\frac{m\omega^2}{\gamma_b}+\tan\left(x_1+\frac{\pi}{2}\right).$$

Since we have now shown that $|y(x)| = |y|_{\max}$ for all x in R_1 it follows, by a second application of the result of the last paragraph, that $|y(x)| = |y|_{\max}$ and arg$y$ = constant for all x in a region R_2 which is the range of x for which $K(x', x) \neq 0$ when x' lies in R_1. From (17) we find R_2 to be

$f_{2a} < \tan x < f_{2b}$, if R_2 contains no discontinuity in $\tan x$

$f_{2a} < \tan x < \frac{1}{2}\pi$ and $\frac{1}{2}\pi \leqslant \tan x < f_{2b}$ } otherwise

where

$$f_{2a} = 2-\frac{m\omega^2}{\gamma_a}-\frac{1}{2-m\omega^2/\gamma_a+\tan(x_1+\frac{1}{2}\pi)}$$

and

$$f_{2b} = 2 - \frac{m\omega^2}{\gamma_b} - \frac{1}{2 - m\omega^2/\gamma_b + \tan(x_1 + \frac{1}{2}\pi)}.$$

Proceeding in this way we find successive regions R_2, R_3, ... in which $|y(x)| = |y|_{max}$ and $\arg y$ = constant. R_i is defined in terms of R_{i-1} as the region of x for which $K(x', x) \neq 0$ when x' lies in R_{i-1}. From the explicit form of K we can deduce that R_i is the region

$$f_{ia} < \tan x < f_{ib}, \quad \text{if } R_i \text{ contains no discontinuity in } \tan x$$

$$\left.\begin{array}{l} f_{ia} < \tan x < \frac{1}{2}\pi \\ \text{and} \quad -\frac{1}{2}\pi \leqslant \tan x < f_{ib} \end{array}\right\}, \quad \text{if } R_i \text{ contains one discontinuity in } \tan x$$

$$-\tfrac{1}{2}\pi \leqslant x < \tfrac{1}{2}\pi, \quad \text{if } R_i \text{ contains a repeated discontinuity in } \tan x$$

where

$$f_{ia} = 2 - \frac{m\omega^2}{\gamma_a} - \frac{1}{f_{i-1,a}}$$

and

$$f_{ib} = 2 - \frac{m\omega^2}{\gamma_b} - \frac{1}{f_{i-1,b}} \qquad i > 2,$$

with f_{2a} and f_{2b} as defined previously.

As i increases the regions R_i soon encompass the whole domain $(-\frac{1}{2}\pi, +\frac{1}{2}\pi)$ of x provided $0 < m\omega^2/\gamma_b < 4$, i.e. provided ω^2 lies within the spectral band. Thus for an ω^2 in the band, the conditions $|y| = 1$ and $\arg y \neq$ constant lead to a contradiction and cannot therefore represent a solution.

In other respects (but with the exception of the remarks of the third paragraph of Borland's appendix) the proof of the Fréchet conditions is as given by Borland.

References

Agacy, R. L., 1964, *Proc. Phys. Soc.*, **83**, 591.
Borland, R. E., 1963, *Proc. Roy. Soc.* A, **274**, 529.
Burnside, W. S., and Panton, A. W., 1918, *Theory of Equations* (London: Longmans, Green), p. 198.
Dean, P., 1960, *Proc. Roy. Soc.* A, **254**, 507.
—— 1961, *Proc. Roy. Soc.* A, **260**, 263.
Dyson, F. J., 1953, *Phys. Rev.*, **92**, 1331.
Fréchet, M., 1934, *Quart. J. Math.*, **5**, 106.
Hori, J., and Fukushima, M., 1964, *J. Phys. Soc. Japan*, **19**, 296.
Simon, I., 1960, *Modern Aspects of the Vitreous State*—1 (London: Butterworth's), pp. 120–151.
Wilkinson, J. H., 1958, *Computer J.*, **1**, 90.

Reprinted from THE JOURNAL OF CHEMICAL PHYSICS, Vol. 41, No. 6, 1527–1541, 15 September 1964
Printed in U. S. A.

Theory of Normal Vibrations of Chain Molecules with Finite Length

H. MATSUDA, K. OKADA, T. TAKASE, AND T. YAMAMOTO
Department of Chemistry, Kyoto University, Kyoto, Japan
(Received 2 April 1964)

This paper is concerned in the end effects on the normal vibrations of a chain molecule with finite length. In Part I we study the mathematical problem of how to solve the eigenvalue problem of the *GF* matrix of the chain molecule with finite length. Making use of the fact that the force range can be regarded finite we reduce the problem to solving a determinantal equation usually of much lower order than that of the *GF* matrix through the introduction of auxiliary variables and transfer matrices. We obtain the simple formula for giving an admissible phase difference in which the end effect is taken into account. In Part II, using this formula we analyze the infrared data of Snyder on crystalline normal paraffins C_3H_8 through $C_{30}H_{62}$. Our analyses indicate that the formula is generally useful for making extrapolation, interpolation, assignment, and determination of frequency-phase relations from the experimental data of homologous series of chain molecules.

INTRODUCTION

CONSIDER an infinite chain molecule built up by linking together identical units in such a way that each unit is transformed geometrically into the next by the operation $H(a, \theta)=$ translation through a distance a along a certain axis plus rotation through an angle θ about the same axis. The set of all operations H^n ($n=$ any integer) constitutes an infinite group which is isomorphic with the infinite cyclic group C_∞. Every normal mode of vibration of such a molecule can be classified by the phase difference φ between the motions of adjacent units. There exists a set of relations called frequency-phase relations which connect each normal frequency with φ. There are selection rules which give a set of values for φ the normal modes corresponding to which are infrared or Raman active. On the basis of the model of such an infinite molecule a number of quantitative comparisons have been made between the measured and calculated infrared frequencies of high polymers such as planar zigzag polyethylene or helical polypeptides, etc.[1–3]

Recently, Snyder *et al.* made excellent systematic infrared measurements of normal paraffins C_3H_8 through $C_{30}H_{62}$.[4] For such a finite chain molecule there inevitably is a deviation from the symmetry of the cyclic group, so that the above selection rules fail to apply, and end effects must be considered. Whitcomb, Nielsen, and Thomas have pointed out[5] that in such a finite chain molecule the normal modes fall into two classes; end vibrations in which the amplitude falls off exponentially from the ends towards the middle, and chain vibrations in which it varies harmonically as in a standing wave.

Snyder *et al.* assumed[4] on the grounds of a simple coupled oscillator theory that in n-paraffins each chain vibration has a phase difference φ given by

$$\varphi=k\pi/(N-1) \tag{1}$$

(k is an integer, N is the number of carbon atoms). By suitably assigning the k value for each band, they found that for larger chains the observed frequency can be regarded as a function of the single parameter φ, though some deviations were observed for shorter chains.

Tasumi, Shimanouchi, and Miyazawa derived[3] the frequency-phase curves for the infinite polymethylene chain from a modified Urey–Bradley force field. They compared these curves with the observed vibrational frequencies of long-chain n-paraffins basing on the

[1] T. Simanouchi, J. Chem. Phys. **17**, 734 (1949); C. Y. Liang, S. Krimm, and G. B. B. M. Sutherland, *ibid.*, **25**, 543 (1956); S. Krimm, C. Y. Liang, and G. B. B. M. Sutherland, *ibid.*, p. 549 (1956); J. R. Nielsen & A. H. Woollett, *ibid.* **26**, 1391 (1957).
[2] P. Higgs, Proc. Roy. Soc. (London) **A220**, 472 (1953); T. Miyazawa, J. Chem. Phys. **35**, 693 (1961).
[3] M. Tasumi, T. Shimanouchi, and T. Miyazawa, J. Mol. Spectry. **9**, 261 (1962); **11**, 422 (1963).
[4] R. G. Snyder, J. Mol. Spectry. **4**, 411 (1960); **7**, 116 (1961); R. G. Snyder and J. H. Schachtschneider, Spectrochim. Acta **19**, 85 (1963).
[5] S. E. Whitcomb, H. H. Nielsen, and L. H. Thomas, J. Chem. Phys. **8**, 143 (1940).

assumption (1). On the other hand, Schachtschneider and Snyder have tried[6] to determine the force field from the data for shorter chains by straightforwardly solving the eigenvalue problem by the *GF*-matrix method.[7] In this calculation they did not take direct recourse to the notion of the phase difference.

Here, the following questions may naturally arise. When the length of a chain molecule is finite, as is always the case in actual molecules, what significance can we attach to the phase difference φ? To what extent is the assumption (1) correct? If the chain molecule is long enough, an infinite chain model may usually be good. But what quantitative criterion do we use to judge whether it is "long enough"?[8]

We have started studying end effects on the normal frequency of a chain molecule in order to answer these questions. Theimer has already discussed end effects obtaining some interesting results for *n*-paraffins.[9] He used a perturbation method, but we do not consider that it is most suitable for these problems. Szigeti has discussed torsional vibrations of long-chain hydrocarbons of finite length without using a perturbation method.[10] Their treatments, however, were restricted to special models.

On the other hand, Schmidt[11] and Hori and Asahi[12] have introduced the transfer matrix method to treat the problem of the vibration of a disordered linear chain. This method is very effective when the most parts of the chain consist of identical units, because the repetition of identical units corresponds to the repeated multiplication of identical matrices.

We have tried to generalize this method in such a way that it is applicable to chain molecules in general, which are more complex than the simple linear chains previously considered. Some of our early results were reported at the 1962 International Symposium on Molecular Structure and Spectroscopy in Tokyo.[13] One of the present authors has reviewed the transfer matrix method which includes our generalization (hereafter referred to as I).[14]

[6] J. H. Schachtschneider and R. G. Snyder, Spectrochim. Acta **19**, 117 (1963).

[7] E. B. Wilson, Jr., J. Chem. Phys. **7**, 1047 (1939); **9**, 76 (1941); E. B. Wilson, Jr., J. C. Decius, and P. C. Cross, *Molecular Vibrations* (McGraw-Hill Book Company, Inc., New York, 1955).

[8] The criterion strongly depends on what physical quantity we are interested in. Rosenstock has pointed out that a one-dimensional diatomic chain with free ends has "absorption intensity per unit frequency range" significantly different from that of the infinite chain even if the chain length tends to infinity. Therefore, we may say that $N=\infty$ is not large enough for the above quantity. See H. B. Rosenstock, J. Chem. Phys. **23**, 2415 (1955); **27**, 1194 (1957).

[9] O. Theimer, J. Chem. Phys. **27**, 408, 1041 (1957).

[10] B. Szigeti, Proc. Phys. Soc. (London) **B65**, 19 (1952); Trans. Faraday Soc. **48**, 400 (1952); **49**, 132 (1952); Proc. Roy. Soc. (London) **A264**, 198, 212 (1961).

[11] H. Schmidt, Phys. Rev. **105**, 425 (1957).

[12] J. Hori and T. Asahi, Progr. Theor. Phys. **17**, 523 (1957).

[13] T. Yamamoto, H. Matsuda, and K. Okada, International Symposium on Molecular Structure and Spectroscopy (Tokyo) 1962, Preprints C 202.

[14] H. Matsuda, Progr. Theoret. Phys. (Kyoto) Suppl. **23**, 22 (1962).

In these expositions the equations of motion were referred to the spatially fixed orthogonal coordinate axes. In obtaining normal frequencies of a chain molecule, however, the well-known Wilson's *GF*-matrix method[7] which allows to use the internal coordinates is more convenient. Therefore, in this paper we formulate our theory using the *GF*-matrix method from the outset. This is by no means a trivial extension of the formulation given in I, because we can neither assume that G is a diagonal matrix nor that GF is a symmetric matrix.

Part I is devoted to the mathematical problem how to solve the *GF* matrix of a chain molecule and the problem as to what is the characteristic properties of the solution. Some of the results are similar to those given in I, but we emphasize that their implications are different since they are obtained here under broader assumptions. At the end of Part I we recapitulate the formula which is important for analyzing the experimental data.

In Part II, using the above formula we analyze Snyder's data[4] in detail. We find that our formula makes a considerable correction to (1) even for $N\simeq 30$. The results of these analyses amply indicate that our formula is very convenient for making assignment, extrapolation, and interpolation of spectra of such homologous series as investigated by Snyder.

PART I

1. General Formulation

Let us represent the displacements of atoms of a chain molecule from their equilibrium positions by a column vector $\mathbf{u}^0$. The components of $\mathbf{u}^0$ may be internal coordinates or may be those referred to spatially fixed orthogonal coordinate axes. The secular equation for normal vibrations may be set up by the method of Wilson[7] in terms of the matrices F and G, such that the potential and kinetic energies are, respectively,

$$\mathbf{V}=\tfrac{1}{2}\mathbf{u}^{0t}\mathbf{F}\mathbf{u}^0 \tag{2}$$

and

$$\mathbf{K}=\tfrac{1}{2}\dot{\mathbf{u}}^{0t}\mathbf{G}^{-1}\dot{\mathbf{u}}^0. \tag{3}$$

Here and in the following $\mathbf{A}^t$ denotes the transpose of matrix or vector $\mathbf{A}$. The equation to be solved for the angular frequency ω is

$$|\mathbf{GF}-\omega^2\mathbf{1}|=0, \tag{4}$$

where $\mathbf{1}$ is the unit matrix.

We divide the whole vector space spanned by $\mathbf{u}^0$ into N subspaces and write

$$\mathbf{u}^0=\begin{pmatrix}\mathbf{u}_1^0\\ \mathbf{u}_2^0\\ \cdot\\ \cdot\\ \cdot\\ \mathbf{u}_N^0\end{pmatrix}, \tag{5}$$

where $\mathbf{u}_j^0$, $(j=1, 2, \cdots, N)$, is a p_j-dimensional vector which is the projection of $\mathbf{u}^0$ onto the jth subspace.

We define the submatrix corresponding to the subspaces j and k such that

$$\mathbf{U}_{j,k}^0 \equiv (\mathbf{GF}-\omega^2\mathbf{1})_{j,j+k}. \quad (6)$$

We assume that there exists an integer q such that

$$\mathbf{U}_{j,k}^0=\mathbf{G}_{j,j+k}=\mathbf{F}_{j,j+k}=0$$
$$\text{for } |k|>q, \quad (j=1, 2, \cdots, N). \quad (7)$$

We call q force range.

Then, solving Eq. (4) is equivalent to obtaining ω's which give nontrivial solutions for the simultaneous linear equations

$$\sum_{k=-q}^{q} \mathbf{U}_{j,k}^0 \mathbf{u}_{j+k}^0=0, \quad (j=1, 2, \cdots, N). \quad (8)$$

Here $\mathbf{U}_{j,k}^0$ is not necessarily a square matrix, so that Eq. (8) for a particular value of j can not necessarily be solved for $\mathbf{u}_{j+k}^0$. In order to make a solution possible, we proceed as follows. Adding $(p-p_j)$-dimensional virtual coordinates $\mathbf{u}_j'$ to $\mathbf{u}_j^0$, we define p-dimensional vectors

$$\mathbf{u}_j \equiv \begin{pmatrix} \mathbf{u}_j^0 \\ \mathbf{u}_j' \end{pmatrix}, \quad (j=1, 2, \cdots, N), \quad (9)$$

where $p=\max p_j$. Further we introduce p-dimensional auxiliary vectors $\mathbf{u}_{N+k}$, $(k=1, 2, \cdots, q)$. We refer to $\mathbf{u}_j^0$ as actual coordinates in contrast with virtual and auxiliary coordinates.

Corresponding to the above extension of vector spaces, we define p-dimensional square matrices $\mathbf{U}_{j,k}$ such that

$$\mathbf{U}_{j,k} \equiv \begin{pmatrix} \mathbf{U}_{j,k}^0 & 0 \\ 0 & 0 \end{pmatrix}, \quad \text{for } k\neq 0, \pm q, \quad (10)$$

$$\mathbf{U}_{j,0} \equiv \begin{pmatrix} \mathbf{U}_{j,0}^0 & 0 \\ 0 & (2\kappa-\omega^2)\mathbf{1}_{p-p_j} \end{pmatrix}, \quad (11)$$

where $\mathbf{1}_p$ denotes the p-dimensional unit matrix, and κ is a nonzero constant.

When $j=1, 2, \cdots, N-q$,

$$\mathbf{U}_{j,\pm q} \equiv \begin{pmatrix} \mathbf{U}_{j,\pm q}^0 & 0 \\ 0 & -\kappa\mathbf{1}_{p-p_{j\pm q}} \end{pmatrix} \quad \text{for } p_j=p_{j\pm q}, \quad (12)$$

$$\mathbf{U}_{j,\pm q} \equiv \begin{pmatrix} \mathbf{U}_{j,\pm q}^0 & -\kappa\mathbf{1}_{p_j-p_{j\pm q}} & 0 \\ 0 & 0 & -\kappa\mathbf{1}_{p-p_j} \end{pmatrix} \quad \text{for } p_j>p_{j\pm q}, \quad (12')$$

$$\mathbf{U}_{j,\pm q} \equiv \begin{pmatrix} \mathbf{U}_{j,\pm q}^0 & 0 \\ -\kappa\mathbf{1}_{p_{j\pm q}-p_j} & 0 \\ 0 & -\kappa\mathbf{1}_{p-p_{j\pm q}} \end{pmatrix} \quad \text{for } p_j<p_{j\pm q}. \quad (12'')$$

When $j=N-q+1, N-q+2, \cdots, N$, we define $\mathbf{U}_{j,-q}$ just as in (12), (12′), and (12″), whereas

$$\mathbf{U}_{j,q} \equiv K'\mathbf{1}_p, \quad (13)$$

where K' is a nonzero constant.

From the above definitions it can be seen that the equations

$$\sum_{k=-q}^{q} \mathbf{U}_{j,k}\mathbf{u}_{j+k}=0, \quad (j=1, 2, \cdots, N), \quad (14)$$

give the same relationships between actual coordinates $\mathbf{u}_j^0$'s as (8) in the limit as $\kappa\to 0$ if and only if

$$\mathbf{u}_{N+k}=0, \quad (k=1, 2, \cdots, q). \quad (15)$$

Thus, our problem is reduced to that of how to obtain the nontrivial solutions satisfying (15) from among the solutions which satisfy (14).

If

$$\det \mathbf{U}_{j,q}\neq 0, \quad (16)$$

then Eq. (14) for given j can be solved for $\mathbf{u}_{j+q}$. When (16) does not hold, we slightly modify $\mathbf{U}_{j,q}^0$ in order to have (16) and after obtaining final results let $\mathbf{U}_{j,q}^0$ resume its original values. This procedure is legitimate, since ω must be a continuous function of the set of elements of $\mathbf{U}_{j,k}^0$. In the following we always assume that (16) is satisfied, so that $\mathbf{U}_{j,q}^{-1}$ exists.

Defining

$$\mathbf{v}_j \equiv \begin{pmatrix} \mathbf{u}_j \\ \mathbf{u}_{j-1} \\ \cdot \\ \cdot \\ \cdot \\ \mathbf{u}_{j-2q+1} \end{pmatrix}, \quad (j=q, q+1, \cdots, N+q), \quad (17)$$

we have

$$\mathbf{v}_{j+q}=\mathbf{T}_j\mathbf{v}_{j+q-1}, \quad (18)$$

where

$$\mathbf{T}_j \equiv \begin{pmatrix} -\mathbf{U}_{j,q}^{-1}\mathbf{U}_{j,q-1} & -\mathbf{U}_{j,q}^{-1}\mathbf{U}_{j,q-2} & \cdots & -\mathbf{U}_{j,q}^{-1}\mathbf{U}_{j,-q} \\ \mathbf{1}_p & 0 & \cdots & 0 \\ 0 & \mathbf{1}_p & \cdots & 0 \\ \cdot & \cdot & \cdots & \cdot \\ \cdot & \cdot & \cdots & \cdot \\ \cdot & \cdot & \cdots & \cdot \\ 0 & 0 & \cdots \mathbf{1}_p & 0 \end{pmatrix} \quad (19)$$

Thus, we obtain

$$\mathbf{v}_{N+q}=\mathbf{T}_N\mathbf{T}_{N-1}\cdots\mathbf{T}_2\mathbf{T}_1\mathbf{P}\mathbf{w}_q, \quad (20)$$

where

$$\mathbf{P}\equiv\begin{pmatrix}\mathbf{1}_{pq}\\ 0\end{pmatrix},\qquad \mathbf{w}_j\equiv\begin{pmatrix}\mathbf{u}_j\\ \mathbf{u}_{j-1}\\ \cdot\\ \cdot\\ \cdot\\ \mathbf{u}_{j-q+1}\end{pmatrix} \tag{21}$$

The conditions (15) reduce to

$$\mathbf{A}(\omega)\mathbf{w}_q=0, \tag{22}$$

where

$$\mathbf{A}(\omega)\equiv\mathbf{P}^t\mathbf{T}_N\mathbf{T}_{N-1}\cdots\mathbf{T}_2\mathbf{T}_1\mathbf{P}, \tag{23}$$

$$\mathbf{P}^t=(\mathbf{1}_{pq}\quad 0). \tag{24}$$

The eigenfrequency ω is thus determined by

$$\det\mathbf{A}(\omega)=0. \tag{25}$$

2. Basic Transfer Matrix

In actual models of chain molecules, by suitably choosing N subspaces we may usually assume that for a majority of parts, Matrices $\mathbf{G}_{j,j+k}$ and $\mathbf{F}_{j,j+k}$ do not depend on j:

$$\mathbf{G}_{j,j+k}=\mathbf{G}_k,\qquad \mathbf{F}_{j,j+k}=\mathbf{F}_k, \tag{26}$$

and

$$\mathbf{U}_{j,k}{}^0=\mathbf{U}_k{}^0,\qquad (k=-q,\cdots,q), \tag{26'}$$

where $\mathbf{G}_k$, $\mathbf{F}_k$, and $\mathbf{U}_k{}^0$ are $\bar{p}(=p_j\leqq p)$-dimensional square matrices. We call the jth subspace satisfying (26) and (26′) the basic part, and $\mathbf{T}_j=\mathbf{T}$ the basic transfer matrix.

From (10), (11), (12), (18), (19), and (26), when j is the basic part, the operation of $\mathbf{T}_j$ on $\mathbf{v}_{j+q-1}$ leaves both actual and virtual coordinate spaces invariant. Actually, by separating both spaces such that

$$\mathbf{v}_i=\begin{pmatrix}\mathbf{v}_i{}^0\\ \mathbf{v}_i{}'\end{pmatrix}, \tag{27}$$

where

$$\mathbf{v}_i{}^0\equiv\begin{pmatrix}\mathbf{u}_i{}^0\\ \mathbf{u}_{i-1}{}^0\\ \cdot\\ \cdot\\ \cdot\\ \mathbf{u}_{i-2q+1}{}^0\end{pmatrix},\qquad \mathbf{v}_i{}'\equiv\begin{pmatrix}\mathbf{u}_i{}'\\ \mathbf{u}_{i-1}{}'\\ \cdot\\ \cdot\\ \cdot\\ \mathbf{u}_{i-2q+1}{}'\end{pmatrix}, \tag{27'}$$

we may write from (18)

$$\mathbf{v}_{j+q}{}^0=\mathbf{T}^0\mathbf{v}_{j+q-1}{}^0, \tag{28}$$

$$\mathbf{v}_{j+q}{}'=\mathbf{T}'\mathbf{v}_{j+q-1}{}', \tag{29}$$

where

$$\mathbf{T}^0=\begin{pmatrix} -\mathbf{U}_q{}^{0-1}\mathbf{U}_{q-1}{}^0 & -\mathbf{U}_q{}^{0-1}\mathbf{U}_{q-2}{}^0 & \cdots & -\mathbf{U}_q{}^{0-1}\mathbf{U}_{-q}{}^0\\ \mathbf{1}_{\bar{p}} & 0 & \cdots & 0\\ 0 & \mathbf{1}_{\bar{p}} & \cdots & 0\\ \cdot & \cdot & \cdots & \cdot\\ \cdot & \cdot & \cdots & \cdot\\ \cdot & \cdot & \cdots & \cdot\\ 0 & 0 & \cdot\cdot\mathbf{1}_{\bar{p}} & 0 \end{pmatrix} \tag{30}$$

and

$$\mathbf{T}'=\begin{pmatrix} \mathbf{T}_{q-1}{}' & \mathbf{T}_{q-2}{}' & \cdots & \mathbf{T}_0{}' & \cdots & \mathbf{T}_{-q}{}'\\ \mathbf{1}_{p-\bar{p}} & 0 & \cdots & 0 & \cdots & 0\\ 0 & \mathbf{1}_{p-\bar{p}} & \cdots & 0 & \cdots & 0\\ \cdot & \cdot & \cdots & \cdot & \cdots & \cdot\\ \cdot & \cdot & \cdots & \cdot & \cdots & \cdot\\ \cdot & \cdot & \cdots & \cdot & \cdots & \cdot\\ 0 & 0 & \cdots & 0 & \cdot\cdot\mathbf{1}_{p-\bar{p}} & 0 \end{pmatrix}, \tag{31}$$

$$\mathbf{T}_k{}'=\begin{cases} 0 & (k\neq 0,\ -q),\\ (2-\omega^2/\kappa)\mathbf{1}_{p-\bar{p}} & (k=0),\\ -\mathbf{1}_{p-\bar{p}} & (k=-q). \end{cases} \tag{32}$$

We are concerned in the eigenvalues and eigenvectors of $\mathbf{T}$ such that

$$\mathbf{T}\mathbf{v}^{(\alpha)}=\exp(i\lambda_\alpha)\mathbf{v}^{(\alpha)}. \tag{33}$$

The above consideration indicates that (33) is reducible to the eigenvalue problems of $2\bar{l}$-dimensional ($\bar{l}=\bar{p}q$) actual coordinate spaces,

$$\mathbf{T}^0\mathbf{v}_0{}^{(\alpha)}=\exp(i\lambda_\alpha)\mathbf{v}_0{}^{(\alpha)},\qquad (\alpha=-\bar{l},\ -\bar{l}+1,\ \cdots,\ -1,\ 1,\ \cdots,\ \bar{l}), \tag{34}$$

and to those of $2(l-\bar{l})$-dimensional virtual coordinate spaces,

$$\mathbf{T}'\mathbf{v}'^{(\beta)}=\exp(i\lambda_\beta)\mathbf{v}'^{(\beta)}\qquad (\beta=-l,\ -l+1,\ \cdots,\ -\bar{l}-1,\ \bar{l}+1,\ \cdots,\ l). \tag{35}$$

Here, we note that the definition of basic parts together with the corresponding values of $\bar{p}$ and q naturally depends on the original choice of subspaces. We call the basic part with the smallest possible value of $\bar{p}$ primitive. If there are enough number of primitive basic parts, by uniting consecutive q primitive basic parts we obtain a basic part with the force range $q=1$. We call such basic part a unit basic part. As we find.

the eigenvalue problem (34) in the actual space is most efficiently solved when the basic part is primitive, whereas the eigenvalue problem (35) in the virtual space can easily be solved when the basic part is the unit part. We first discuss the eigenvalue problem of $\mathbf{T}^0$ of the primitive basic part, next the eigenvalue problem of $\mathbf{T}'$ of the unit basic part, and then discuss the properties of the basic transfer matrix $\mathbf{T}$ in general.

Consider Eq. (34). From (27′) and (28) we have

$$\mathbf{T}^0\begin{pmatrix}\mathbf{u}_{j+q-1}{}^{0(\alpha)}\\ \mathbf{u}_{j+q-2}{}^{0(\alpha)}\\ \cdot\\ \cdot\\ \cdot\\ \mathbf{u}_{j-q}{}^{0(\alpha)}\end{pmatrix}=\begin{pmatrix}\mathbf{u}_{j+q}{}^{0(\alpha)}\\ \mathbf{u}_{j+q-1}{}^{0(\alpha)}\\ \cdot\\ \cdot\\ \cdot\\ \mathbf{u}_{j-q+1}{}^{0(\alpha)}\end{pmatrix}. \tag{36}$$

Comparing (34) with (36) and noting (30), we obtain

$$\mathbf{v}_0{}^{(\alpha)}=\mathbf{u}_0{}^{(\alpha)}\times\begin{pmatrix}1\\ \exp(-i\lambda_\alpha)\\ \cdot\\ \cdot\\ \cdot\\ \exp[-i\lambda_\alpha(2q-1)]\end{pmatrix}\equiv\begin{pmatrix}\mathbf{u}_0{}^{(\alpha)}\\ \exp(-i\lambda_\alpha)\mathbf{u}_0{}^{(\alpha)}\\ \cdot\\ \cdot\\ \cdot\\ \exp[-i\lambda_\alpha(2q-1)]\mathbf{u}_0{}^{(\alpha)}\end{pmatrix}, \tag{37}$$

where $\mathbf{u}_0{}^{(\alpha)}$ is the $\bar{p}$-dimensional vector satisfying

$$\mathbf{R}(\omega,\lambda_\alpha)\mathbf{u}_0{}^{(\alpha)}=\mathbf{0}, \tag{38}$$

$$\mathbf{R}(\omega,\lambda_\alpha)\equiv\sum_{k=-q}^{q}\mathbf{U}_k{}^0\exp(ik\lambda_\alpha). \tag{39}$$

From (6), (26), and (26′), we have

$$\mathbf{U}_k{}^0=\sum_{k'}\mathbf{G}_{k'}\mathbf{F}_{k-k'}-\omega^2\delta_{0,k}\mathbf{1}_{\bar{p}}, \tag{40}$$

where $\delta_{j,k}$ is the Kronecker symbol.

Substituting (40) into (39), we obtain

$$\mathbf{R}(\omega,\lambda_\alpha)=\mathbf{G}(\lambda_\alpha)\mathbf{F}(\lambda_\alpha)-\omega^2\mathbf{1}_{\bar{p}}, \tag{41}$$

where

$$\mathbf{G}(\lambda)\equiv\sum_k\mathbf{G}_k\exp(ik\lambda),$$

$$\mathbf{F}(\lambda)\equiv\sum_k\mathbf{F}_k\exp(ik\lambda). \tag{42}$$

Thus λ_α is determined as a function of ω^2 by

$$|\mathbf{G}(\lambda_\alpha)\mathbf{F}(\lambda_\alpha)-\omega^2\mathbf{1}_{\bar{p}}|=0. \tag{43}$$

This is the $\bar{p}$-dimensional determinantal equation, and is of the same form as that obtained by Higgs for infinite helical molecules.[2] Calling λ_α the αth phase we refer to the relation between ω and λ_α determined by (43) as a frequency-phase relation. Higgs considered only real phases between 0 and π and tried to obtain ω as a function of a single phase. Here we rather regard λ_α as a function of ω and are interested in not only real phases but also complex phases.[15] This is important in discussing infrared absorption of finite chain molecules, because there is no simple selection rule for the phase in contrast with the infinite chain molecule considered by Higgs.

Since $\mathbf{G}_k$ and $\mathbf{F}_k$ are real matrices, we have

$$\mathbf{G}(\lambda)^*=\mathbf{G}(-\lambda^*)\quad\text{and}\quad\mathbf{F}(\lambda)^*=\mathbf{F}(-\lambda^*), \tag{44}$$

where $\mathbf{A}^*$ denotes the complex conjugate of $\mathbf{A}$. Since $\mathbf{G}$ and $\mathbf{F}$ are symmetric matrices, from (26) we have

$$\mathbf{G}_k{}^t=\mathbf{G}_{-k}\quad\text{and}\quad\mathbf{F}_k{}^t=\mathbf{F}_{-k}. \tag{45}$$

Hence

$$\mathbf{G}(\lambda)^t=\mathbf{G}(-\lambda)\quad\text{and}\quad\mathbf{F}(\lambda)^t=\mathbf{F}(-\lambda). \tag{45'}$$

Therefore, if λ_α satisfies (43), so do $-\lambda_\alpha$, $\pm\lambda_\alpha{}^*$.

Now, we may label $2\bar{l}$ eigenvalues of $\mathbf{T}^0$ as follows:

$$\lambda_\alpha\equiv\varphi_\alpha-i\psi_\alpha,\qquad(\alpha=-\bar{l},-\bar{l}+1,\cdots,-1,1,\cdots,\bar{l}). \tag{46}$$

Here φ_α and ψ_α are real and satisfy

$$\lambda_\alpha=-\lambda_{-\alpha}, \tag{47}$$

$$\psi_\alpha\geqq\psi_\beta\quad\text{for}\quad\alpha>\beta, \tag{47'}$$

$$\psi_\alpha=0\quad\text{for and only for}\quad s\geqq\alpha\geqq-s, \tag{47''}$$

$$\pi\geqq\varphi_\alpha\geqq0\quad\text{for}\quad s\geqq\alpha>0. \tag{47'''}$$

From (47), (47′), and (47″) we get

$$\psi_\alpha>0\quad\text{for}\quad\alpha>s,$$

$$\psi_\alpha<0\quad\text{for}\quad\alpha<-s. \tag{48}$$

There is a one-to-one correspondence between the eigenvalues of $\mathbf{T}^0$ and $\mathbf{T}^{0t}$. Therefore, there must be the row vector $\tilde{\mathbf{v}}_0{}^{(\alpha)}$ such that

$$\tilde{\mathbf{v}}_0{}^{(\alpha)}\mathbf{T}^0=\exp(i\lambda_\alpha)\tilde{\mathbf{v}}_0{}^{(\alpha)}$$
$$(\alpha=-\bar{l},-\bar{l}+1,\cdots,-1,1,\cdots,\bar{l}). \tag{49}$$

If there is no degeneracy[16] in the eigenvalues of $\mathbf{T}^0$, by putting suitable normalization factors we obtain the relation

$$\tilde{\mathbf{v}}_0{}^{(\alpha)}\mathbf{v}_0{}^{(\alpha')}=\delta_{\alpha,\alpha'}\qquad(\alpha,\alpha'=-\bar{l},\cdots,-1,1,\cdots,\bar{l}). \tag{50}$$

Writing

$$\tilde{\mathbf{v}}_0{}^{(\alpha)}\equiv(\tilde{\mathbf{v}}_{1}{}_0{}^{(\alpha)},\tilde{\mathbf{v}}_{2}{}_0{}^{(\alpha)},\cdots,\tilde{\mathbf{v}}_{2q}{}_0{}^{(\alpha)}), \tag{51}$$

where $\tilde{\mathbf{v}}_{j}{}_0{}^{(\alpha)}$ is a $\bar{p}$-dimensional row vector, we obtain

[15] This viewpoint is the same as that of Whitcomb *et al.* See Ref. 5.

[16] In the following we consider only this case. We hope it will make no serious restriction on the applicability of our theory.

from (30) and (49):

$$\exp(i\lambda_\alpha)\tilde{\mathbf{v}}_{2q}^{0(\alpha)} = -\tilde{\mathbf{v}}_1^{0(\alpha)}\mathbf{U}_q^{0-1}\mathbf{U}_{-q}^0,$$

$$\exp(i\lambda_\alpha)\tilde{\mathbf{v}}_{2q-1}^{0(\alpha)} = \tilde{\mathbf{v}}_{2q}^{0(\alpha)} - \tilde{\mathbf{v}}_1^{0(\alpha)}\mathbf{U}_q^{0-1}\mathbf{U}_{-q+1}^0,$$

$$\cdot \quad \cdot \quad \cdot$$

$$\exp(i\lambda_\alpha)\tilde{\mathbf{v}}_1^{0(\alpha)} = \tilde{\mathbf{v}}_2^{0(\alpha)} - \tilde{\mathbf{v}}_1^{0(\alpha)}\mathbf{U}_q^{0-1}\mathbf{U}_{q-1}^0. \qquad (52)$$

Hence noting (39) and (41) we get

$$\tilde{\mathbf{v}}_1^{0(\alpha)}\mathbf{U}_q^{0-1}[\mathbf{G}(\lambda_\alpha)\mathbf{F}(\lambda_\alpha) - \omega^2\mathbf{1}_{\bar{p}}] = \mathbf{0}. \qquad (53)$$

From (45′) and (53) we have

$$[\mathbf{F}(-\lambda_\alpha) - \omega^2\mathbf{G}^{-1}(-\lambda_\alpha)]\mathbf{G}(-\lambda_\alpha)[\tilde{\mathbf{v}}_1^{0(\alpha)}\mathbf{U}_q^{0-1}]^t = \mathbf{0}. \quad (54)$$

On the other hand, we have from (38), (41), and (47)

$$[\mathbf{F}(-\lambda_\alpha) - \omega^2\mathbf{G}^{-1}(-\lambda_\alpha)]\mathbf{u}^{0(-\alpha)} = \mathbf{0}. \qquad (55)$$

Comparing (54) and (55) we may put

$$\tilde{\mathbf{v}}_1^{0(\alpha)} = c_\alpha\mathbf{u}^{0(-\alpha)t}\mathbf{G}(\lambda_\alpha)^{-1}\mathbf{U}_q^0 = c_\alpha'\mathbf{u}^{0(-\alpha)t}\mathbf{F}(\lambda_\alpha)\mathbf{U}_q^0, \quad (56)$$

where c_α and c_α' are constants. Substituting (56) into (52) we get

$$\tilde{\mathbf{v}}_j^{0(\alpha)} = c_\alpha\mathbf{u}^{0(-\alpha)t}\mathbf{G}(\lambda_\alpha)^{-1}\sum_{k=1}^{j}\exp[i\lambda_\alpha(j-k)]\mathbf{U}_{q-k+1}^0 \quad (57)$$

$$= -c_\alpha\mathbf{u}^{0(-\alpha)t}\mathbf{G}(\lambda_\alpha)^{-1}$$

$$\times\sum_{k=0}^{2q-j}\exp[-i\lambda_\alpha(2q-j+1-k)]\mathbf{U}_{-q+k}^0. \quad (57')$$

Combining (57) and (57′) we may write $\tilde{\mathbf{v}}^{0(\alpha)}$ as follows:

$$\tilde{\mathbf{v}}^{0(\alpha)} = c_\alpha\mathbf{u}^{0(-\alpha)t}\mathbf{G}(\lambda_\alpha)^{-1}$$

$$\times(1 \quad \exp(i\lambda_\alpha)\cdots\exp[i(q-1)\lambda_\alpha])$$

$$\times(\overline{\mathbf{U}}_1 - \exp(-iq\lambda_\alpha)\overline{\mathbf{U}}_{-1}), \quad (58)$$

where

$$\overline{\mathbf{U}}_1 \equiv \begin{pmatrix} \mathbf{U}_q^0 & \mathbf{U}_{q-1}^0 & \cdots & \mathbf{U}_1^0 \\ & \mathbf{U}_q^0 & \cdots & \mathbf{U}_2^0 \\ & & \cdot & \cdot \\ & \mathbf{0} & \cdot & \cdot \\ & & & \cdot \\ & & & \mathbf{U}_q^0 \end{pmatrix},$$

$$\overline{\mathbf{U}}_{-1} \equiv \begin{pmatrix} \mathbf{U}_{-q}^0 & & & \\ & & & \mathbf{0} \\ \mathbf{U}_{-q+1}^0 & \mathbf{U}_{-q}^0 & & \\ \cdot & \cdot & \cdot & \\ \cdot & \cdot & \cdot & \\ \cdot & \cdot & \cdot & \\ \mathbf{U}_{-1}^0 & \mathbf{U}_{-2}^0 & \cdots & \mathbf{U}_{-q}^0 \end{pmatrix}. \quad (58')$$

The constant c_α is to be determined by the normalization (50). From (37), (58), and (58′) we have

$$\tilde{\mathbf{v}}^{0(\alpha)}\mathbf{v}^{0(\alpha)}$$

$$= c_\alpha\mathbf{u}^{0(-\alpha)t}\mathbf{G}(\lambda_\alpha)^{-1}[q\mathbf{U}_q^0 + (q-1)\exp(-i\lambda_\alpha)\mathbf{U}_{q-1}^0$$

$$+\cdots+(-q)\exp(-2qi\lambda_\alpha)\mathbf{U}_{-q}^0]\mathbf{u}^{0(\alpha)}. \quad (59)$$

Noting (40), (42), and (55) we can write

$$\tilde{\mathbf{v}}^{0(\alpha)}\mathbf{v}^{0(\alpha)} = c_\alpha\exp(-iq\lambda_\alpha)I(\omega,\lambda_\alpha), \qquad (60)$$

where

$$I(\omega,\lambda_\alpha) \equiv i^{-1}\mathbf{u}^{0(-\alpha)t}[\mathbf{F}'(\lambda_\alpha)$$

$$+\omega^2\mathbf{G}(\lambda_\alpha)^{-1}\mathbf{G}'(\lambda_\alpha)\mathbf{G}^{-1}(\lambda_\alpha)]\mathbf{u}^{0(\alpha)}$$

$$= i^{-1}\mathbf{u}^{0(-\alpha)t}[\mathbf{F}'(\lambda_\alpha)$$

$$+\omega^{-2}\mathbf{F}(\lambda_\alpha)\mathbf{G}'(\lambda_\alpha)\mathbf{F}(\lambda_\alpha)]\mathbf{u}^{0(\alpha)},$$

$$\mathbf{F}'(\lambda) \equiv d\mathbf{F}(\lambda)/d\lambda, \qquad \mathbf{G}'(\lambda) \equiv d\mathbf{G}(\lambda)/d\lambda. \qquad (60')$$

From (50) and (60) we obtain

$$c_\alpha = \exp(iq\lambda_\alpha)/I(\omega,\lambda_\alpha). \qquad (61)$$

Using (45′) we can easily show that

$$I(\omega,\lambda_\alpha) = -I(\omega,-\lambda_\alpha). \qquad (62)$$

When λ_α is real, from (38) we may put

$$\mathbf{u}^{0(\alpha)*} = \mathbf{u}^{0(-\alpha)}. \qquad (63)$$

Hence we have in this case

$$I^*(\omega,\lambda_\alpha) = I(\omega,-\lambda_\alpha) = -I(\omega,\lambda_\alpha). \qquad (64)$$

Therefore, when λ_α is real $I(\omega,\lambda_\alpha)$ is purely imaginary.

Since the unit basic transfer matrix in the actual space is $\mathbf{T}^{0q}$, its eigenvectors are the same as those of the primitive basic transfer matrix $\mathbf{T}^0$ in the actual space and its eigenvalues are given by $\exp(iq\lambda_\alpha)$, $(\alpha = -\bar{l}, -\bar{l}+1, \cdots, -1, 1, \cdots, \bar{l})$.

Let us turn to the virtual coordinate space. Consider Eq. (35). When we are considering the unit basic part, since then $q=1$, we have from (31) and (32) in the limit as $\kappa\to 0$

$$\mathbf{T}' \to -\frac{\omega^2}{\kappa}\begin{pmatrix} \mathbf{1}_{l-\bar{l}} & 0 \\ 0 & 0 \end{pmatrix}. \qquad (65)$$

Therefore, in this limit writing

$$\mathbf{v}'^{(\beta)} \equiv \begin{pmatrix} \mathbf{v}_1'^{(\beta)} \\ \mathbf{v}_2'^{(\beta)} \\ \cdot \\ \cdot \\ \cdot \\ \mathbf{v}_{2(l-\bar{l})}'^{(\beta)} \end{pmatrix}, \qquad (66)$$

we may put

$$|\exp(i\lambda_\beta)| = \infty, \qquad \mathbf{v}_k'^{(\beta)} = \delta_{k,l-\beta+1}$$

$$\text{for } \beta = l, l-1, \cdots, \bar{l}+1, \quad (67)$$

and

$$|\exp(i\lambda_\beta)| = 0, \qquad \mathbf{v}_k'^{(\beta)} = \delta_{k,l-\beta-2\bar{l}}$$

$$\text{for } \beta = -l, -l-1, \cdots, -\bar{l}-1. \quad (67')$$

The corresponding left-side eigenvectors $\tilde{\mathbf{v}}'^{(\beta)}$ of $\mathbf{T}'$ in this limit is just the transpose of $\mathbf{v}'^{(\beta)}$:

$$\tilde{\mathbf{v}}'^{(\beta)}=\mathbf{v}'^{(\beta)t}. \tag{68}$$

From the above consideration we can state that if $\mathbf{v}^{(\alpha)}$ and $\tilde{\mathbf{v}}^{(\alpha)}$ are the right-side and left-side eigenvectors of the transfer matrix T of the unit basic part, they are given as follows:

When $l\geqq|\alpha|\geqq\bar{l}+1$,

$$\mathbf{P}_a\mathbf{v}^{(\alpha)}=0,\qquad \mathbf{P}_v\mathbf{v}^{(\alpha)}=\mathbf{v}'^{(\alpha)};$$
$$\mathbf{P}_a\tilde{\mathbf{v}}^{(\alpha)}=0,\qquad \mathbf{P}_v\tilde{\mathbf{v}}^{(\alpha)}=\tilde{\mathbf{v}}'^{(\alpha)}, \tag{69}$$

where $\mathbf{P}_a$ and $\mathbf{P}_v$ are the projection operators onto the actual and virtual spaces, respectively. The corresponding eigenvalues satisfy in the limit as $\kappa\rightarrow0$

$$|\exp(i\lambda_\alpha)|=\infty\quad\text{for}\quad l\geqq\alpha\geqq\bar{l}+1,$$
$$|\exp(i\lambda_\alpha)|=0\quad\text{for}\quad -l\leqq\alpha\leqq-\bar{l}-1. \tag{69'}$$

When $|\alpha|\leqq\bar{l}$,

$$\mathbf{P}_a\mathbf{v}^{(\alpha)}=\mathbf{v}_0^{(\alpha)},\qquad \mathbf{P}_v\mathbf{v}^{(\alpha)}=0;$$
$$\mathbf{P}_a\tilde{\mathbf{v}}^{(\alpha)}=\tilde{\mathbf{v}}_0^{(\alpha)},\qquad \mathbf{P}_v\tilde{\mathbf{v}}^{(\alpha)}=0, \tag{70}$$

and the corresponding eigenvalues are given by $\exp(iq\lambda_\alpha)$, which satisfy

$$|\exp(iq\lambda_\alpha)|>1\quad\text{for}\quad \bar{l}\geqq\alpha\geqq s+1,$$
$$|\exp(iq\lambda_\alpha)|=1\quad\text{for}\quad s\geqq\alpha\geqq-s,$$
$$|\exp(iq\lambda_\alpha)|<1\quad\text{for}\quad -s-1\geqq\alpha\geqq-\bar{l}. \tag{70'}$$

We note that the set of $\mathbf{v}^{(\beta)\prime}$'s with $l\geqq\beta\geqq\bar{l}+1$ is invariant in the limit as $\kappa\rightarrow0$ under the operation of the primitive basic transfer matrix $\mathbf{T}$, because from (31), (32), (66), and (67) we get in this limit

$$\mathbf{T}'\mathbf{v}'^{(\beta)}=\mathbf{v}'^{(\beta-p)}\qquad\text{for}\quad l\geqq\beta\geqq\bar{l}+p+1,$$
$$\mathbf{T}'\mathbf{v}'^{(\beta)}=-(\omega^2/\kappa)\mathbf{v}'^{(\beta-p+l-\bar{l})}\qquad\text{for}\quad \bar{l}+p\geqq\beta\geqq\bar{l}+1; \tag{71}$$

$$\tilde{\mathbf{v}}'^{(\beta)}\mathbf{T}'=\tilde{\mathbf{v}}'^{(\beta+p)}\qquad\text{for}\quad l-p\geqq\beta\geqq\bar{l}+1,$$
$$\tilde{\mathbf{v}}'^{(\beta)}\mathbf{T}'=-(\omega^2/\kappa)\tilde{\mathbf{v}}'^{(\beta+p+\bar{l}-l)}\qquad\text{for}\quad l\geqq\beta\geqq l-p+1. \tag{71'}$$

We obtain from (50), (66), (67), and (68) that

$$\tilde{\mathbf{v}}^{(\alpha)}\mathbf{v}^{(\alpha')}=\delta_{\alpha,\alpha'}. \tag{72}$$

Therefore, if we define

$$\mathbf{S}\equiv(\mathbf{v}^{(l)},\mathbf{v}^{(l-1)},\cdots,\mathbf{v}^{(-l)}), \tag{73}$$

then

$$\mathbf{S}^{-1}=\begin{pmatrix}\tilde{\mathbf{v}}^{(l)}\\ \tilde{\mathbf{v}}^{(l-1)}\\ \cdot\\ \cdot\\ \cdot\\ \tilde{\mathbf{v}}^{(-l)}\end{pmatrix}, \tag{74}$$

and the unit basic transfer matrix $\mathbf{T}$ can be diagonalized as

$$\mathbf{S}^{-1}\mathbf{T}\mathbf{S}=\mathbf{\Lambda}\equiv\begin{pmatrix}\exp(i\lambda_l) & & & & 0\\ & \exp(i\lambda_{l-1}) & & & \\ & & \cdot & & \\ 0 & & & \cdot & \\ & & & & \cdot \\ & & & & \exp(-i\lambda_l)\end{pmatrix} \tag{75}$$

Finally, we note that the value of s, which is half a number of eigenvalues with absolute value 1 of the unit basic transfer matrix, is just the number of real phases between 0 and π corresponding to a given value of ω in the frequency-phase relation defined by (43) with respect to the eigenvalue problem of $\mathbf{T}^0$ of the primitive basic part.

3. End Effects

In this section we consider the homologous series of chain molecules consisting of N subspaces with the transfer matrices such that

$$\mathbf{T}_j=\mathbf{T},\qquad (j=N_1+1,N_1+2,\cdots,N-N_2),$$
$$\mathbf{R}_1\equiv\mathbf{T}_{N_1}\mathbf{T}_{N_1-1}\cdots\mathbf{T}_2\mathbf{T}_1,$$
$$\mathbf{R}_2\equiv\mathbf{T}_N\mathbf{T}_{N-1}\cdots\mathbf{T}_{N-N_2+1}, \tag{76}$$

where N_1, N_2, $\mathbf{R}_1$, $\mathbf{R}_2$ and $\mathbf{T}$ are independent of N, and $\mathbf{T}$ is the primitive basic transfer matrix.

According to (23) we may write

$$\mathbf{A}(\omega)=\mathbf{P}^t\mathbf{R}_2\mathbf{T}^{\bar{N}}\mathbf{R}_1\mathbf{P}$$
$$=\mathbf{P}^t\mathbf{R}_2\mathbf{T}^{\bar{N}-q[\bar{N}/q]}(\mathbf{T}^q)^{[\bar{N}/q]}\mathbf{R}_1\mathbf{P}, \tag{77}$$

where $\bar{N}=(N-N_1)-N_2$, and $[\bar{N}/q]$ is the integer part of $\bar{N}/q$. Since the unit basic transfer matrix $\mathbf{T}^q$ is diagonalized as shown in (75), we obtain

$$\mathbf{A}(\omega)=\sum_{\alpha=-l}^{l}\exp(i\lambda_\alpha\bar{N})\mathbf{x}^{(\alpha)}\tilde{\mathbf{y}}^{(\alpha)}, \tag{78}$$

where

$$\mathbf{x}^{(\alpha)}\equiv\mathbf{P}^t\mathbf{R}_2\mathbf{T}^{\bar{N}-q[\bar{N}/q]}\mathbf{v}^{(\alpha)}\exp\{-i\lambda_\alpha(\bar{N}-q[\bar{N}/q])\}, \tag{79}$$
$$\tilde{\mathbf{y}}^{(\alpha)}\equiv\tilde{\mathbf{v}}^{(\alpha)}\mathbf{R}_1\mathbf{P}. \tag{79'}$$

From (78) we get

$$\det\mathbf{A}(\omega)=\sum_{\alpha_1>\alpha_2>\ldots>\alpha_l}\exp[i\bar{N}\sum_{n=1}^{l}\lambda_{\alpha_n}]f(\alpha_1,\alpha_2,\cdots,\alpha_l), \tag{80}$$

where

$$f(\alpha_1,\alpha_2,\cdots,\alpha_l)\equiv\det(\mathbf{x}^{(\alpha_1)}\mathbf{x}^{(\alpha_2)}\cdots\mathbf{x}^{(\alpha_l)})\det\begin{pmatrix}\tilde{\mathbf{y}}^{(\alpha_1)}\\ \tilde{\mathbf{y}}^{(\alpha_2)}\\ \cdot\\ \cdot\\ \cdot\\ \tilde{\mathbf{y}}^{(\alpha_l)}\end{pmatrix}, \tag{81}$$

and the summation extends over all the possible sets of

$$\alpha_1, \alpha_2, \cdots, \alpha_k, \cdots, \alpha_l, \qquad (\alpha_k = -l, -l+1, \cdots, l).$$

In the limit as $\kappa \to 0$, however, in view of (69′) the eigenvalue equation (25) is equivalent to

$$\sum_{\bar{l} \geq \alpha_1 > \alpha_2 > \cdots > \alpha_{\bar{l}} \geq -\bar{l}} \exp\left[i\bar{N} \sum_{n=1}^{\bar{l}} \lambda_{\alpha_n}\right] \times f(l, l-1, \cdots, \bar{l}+1, \alpha_1, \cdots, \alpha_{\bar{l}}) = 0. \quad (82)$$

Noting the note made just above (71), we may redefine $\mathbf{x}^{(\alpha)}$ and $\tilde{\mathbf{y}}^{(\alpha)}$ in Eq. (82) such that

$$\mathbf{x}^{(\alpha)} \equiv \mathbf{P}'\mathbf{R}_2\mathbf{v}^{(\alpha)}, \quad (83)$$

$$\tilde{\mathbf{y}}^{(\alpha)} \equiv \tilde{\mathbf{v}}^{(\alpha)}\mathbf{R}_1\mathbf{P}. \quad (83')$$

This indicates that we may forget about unit basic transfer matrix if we remember that $\mathbf{v}^{(\alpha)}$ and $\tilde{\mathbf{v}}^{(\alpha)}$ are given by (69) and (70).

When $\bar{N} \gg 1$, we can further reduce the eigenvalue equation (82) to

$$D_0(\omega) + D_1(\omega) = 0, \quad (84)$$

$$D_0(\omega) = \sum_{s \geq \alpha_1 > \alpha_2 > \cdots > \alpha_s \geq -s} g(\alpha_1, \alpha_2, \cdots, \alpha_s) \times \cos\left[\bar{N} \sum_{n=1}^{s} \varphi_{\alpha_n} + \delta(\alpha_1, \cdots, \alpha_s)\right], \quad (85)$$

$$D_1(\omega) = O[\exp(-N\psi_{s+1})], \quad (85')$$

where

$$g(\alpha_1, \alpha_2, \cdots, \alpha_s) \equiv |f(l, l-1, \cdots, s+1, \alpha_1, \alpha_2, \cdots, \alpha_s)| \quad (86)$$

and

$$\delta(\alpha_1, \alpha_2, \cdots, \alpha_s) \equiv \arg f(l, l-1, \cdots, s+1, \alpha_1, \alpha_2, \cdots, \alpha_s). \quad (86')$$

Here, we have used the relation

$$f^*(l, l-1, \cdots, s+1, \alpha_1, \alpha_2, \cdots, \alpha_s) = f(l, l-1, \cdots, s+1, -\alpha_1, -\alpha_2, \cdots, -\alpha_s). \quad (87)$$

Therefore, to the approximation of neglecting $\exp(-N\psi_{s+1})$ compared with 1, we have the approximate eigenvalue equation

$$D_0(\omega) = 0. \quad (88)$$

When $s=0$, we have

$$D_0(\omega) = f(l, l-1, \cdots, 2, 1). \quad (89)$$

If the frequency ω satisfying (88) and (89) is in the range with $s=0$, it does not depend on $\bar{N}$ in the present approximation. It corresponds to the end vibration. In this frequency range there can be no chain vibration.

When $s=1$, from (85), (86), (86′), and (87) we get

$$D_0(\omega) = 2g(1) \cos[\bar{N}\varphi_1 + \delta(1)]. \quad (90)$$

Therefore, the approximate eigenvalue equation (88) gives

$$g(1) = 0 \quad (91)$$

or

$$\varphi_1 = k\pi/\bar{N} + [\tfrac{1}{2}\pi - \delta(1)]/\bar{N}, \qquad (k\text{: integer}). \quad (91')$$

Equation (91) corresponds to the end vibration and Eq. (91′) to the chain vibration.[5] Here φ_1 is the function of ω determined by the basic transfer matrix. It is the phase difference of the normal vibration with frequency ω between the adjacent primitive basic parts. On the other hand, $\delta(1)$ is the function of ω determined not only by the transfer matrix of the basic part but also by those of the end parts. It is to be noted, however, that both φ_1 and $\delta(1)$ are independent of N.

Equation (91′) which gives the admissible phase differences of chain vibration is more accurate than (1). The replacement of the denominator $N-1$ by $\bar{N}$ is not significant, since it merely corresponds to the change in the functional dependence of $\delta(1)$ on ω and the change in $\delta(1)$ is $O(N^{-1})$. The quantity $[\frac{1}{2}\pi - \delta(1)]/\bar{N}$ represents the end effect on the phase difference. We call this quantity a phase shift and $[\frac{1}{2}\pi - \delta(1)]$ a primitive phase shift. In Part II we analyze Snyder's data on *n*-paraffins using Eq. (91′).

In view of (81) and (86′) we note that

$$\delta(1) = \arg\det(\mathbf{x}^{(l)}\mathbf{x}^{(l-1)}\cdots\mathbf{x}^{(1)}) + \arg\det\begin{pmatrix} \tilde{\mathbf{y}}^{(l)} \\ \tilde{\mathbf{y}}^{(l-1)} \\ \cdot \\ \cdot \\ \cdot \\ \tilde{\mathbf{y}}^{(1)} \end{pmatrix}. \quad (92)$$

This shows that the primitive phase shift can be regarded as the sum of two independent contributions from both ends of the chain molecule.

When $s=2$, we obtain

$$D_0(\omega) = g_+ \cos(\bar{N}\varphi_+ + \delta_+) + g_- \cos(\bar{N}\varphi_- + \delta_-) + g_0, \quad (93)$$

where

$$g_+ \equiv 2g(2, 1), \qquad \delta_+ \equiv \delta(2, 1), \qquad \varphi_+ \equiv \varphi_2 + \varphi_1,$$

$$g_- \equiv 2g(2, -1), \qquad \delta_- \equiv \delta(2, -1), \qquad \varphi_- \equiv \varphi_2 - \varphi_1,$$

$$g_0 \equiv g(1, -1) + g(2, -2). \quad (94)$$

Equation (93) is not so simple as (90). But in some cases it is reducible to the form of (90). Indeed, when $g_+ = g_- \neq 0$ and $g_0 = 0$, (88) and (93) become

$$\cos[\bar{N}\varphi_2 + \tfrac{1}{2}(\delta_+ + \delta_-)] \cos[\bar{N}\varphi_1 + \tfrac{1}{2}(\delta_+ - \delta_-)] = 0. \quad (95)$$

Hence we obtain

$$\varphi_1 = k_1\pi/\bar{N} + \{\pi - (\delta_+ - \delta_-)\}/(2\bar{N}), \quad (96)$$

and

$$\varphi_2 = k_2\pi/\bar{N} + \{\pi - (\delta_+ + \delta_-)\}/(2\bar{N}), \quad (96')$$

where k_1 and k_2 are integers.

Without going into more complex cases where $s=3$, we conclude this section noting the following facts: If all the real phases φ_α's$(1\leqq\alpha\leqq s\leqq 2)$ corresponding to a given ω are close either to zero or to π and no complex phase $\lambda_\beta(\bar{l}\geqq|\beta|\geqq s+1)$ is close to any φ_α, then every case is reducible to $s=1$ and the phase shift vanishes, that is,

$$\varphi_\alpha \cong k_\alpha\pi/\bar{N}, \tag{97}$$

where k_α is an integer. Because in this case from (37), (38) and (39), for $|\alpha|\leqq\bar{l}$, $\mathbf{v}_0^{(\alpha)}$ can be taken almost real and from (56), (64), and (62), $\tilde{\mathbf{v}}_0^{(\alpha)}$ is almost purely imaginary. Hence $\mathbf{x}^{(\alpha)}$ is almost real and $\tilde{\mathbf{y}}^{(\alpha)}$ is almost purely imaginary. Then we may put for $|\alpha|\leqq s$

$$\mathbf{x}^{(-\alpha)}=\mathbf{x}^{(\alpha)*}\cong\mathbf{x}^{(\alpha)},\qquad \tilde{\mathbf{y}}^{(-\alpha)}=\tilde{\mathbf{y}}^{(\alpha)*}\cong-\tilde{\mathbf{y}}^{(\alpha)}, \tag{98}$$

$$\arg(\mathbf{x}^{(\alpha)})\cong 0,\qquad \arg(\tilde{\mathbf{y}}^{(\alpha)})\cong\tfrac{1}{2}\pi\ \mathrm{sign}(\alpha). \tag{99}$$

Substituting (98) and (99) into (86) and (86′) and noting (87) and that $\mathbf{x}^{(\beta)}$ and $\tilde{\mathbf{y}}^{(\beta)}$ are real for $|\beta|\geqq\bar{l}+1$, we obtain

$$g(1,-1)\cong 0,$$

$$g(2,-2)\cong 0,$$

$$g(2,1)\cong g(2,-1), \tag{100}$$

and

$$\delta(1)\cong\tfrac{1}{2}\pi,\qquad \delta(2,1)\pm\delta(2,-1)=\pi. \tag{101}$$

Therefore, from (88), (90), (93), (100), and (101) we get

$$D_0(\omega)\propto\prod_{\alpha=1}^{s}\sin(\bar{N}\varphi_\alpha)=0, \tag{102}$$

which proves our assertion.[17]

4. Summary of Part I

Consider an infinite helical chain molecule just discussed at the beginning of the Introduction. We cut off both sides of the infinite molecule retaining N identical units and modify the end parts. The structure of the modified parts are assumed to be independent of N.

For such finite chain molecules the frequency-phase relation of the original infinite molecule is still meaningful, but only a limited number of phases are admissible for normal vibrations. In the frequency-phase diagram of the infinite chain in which the phase varies from 0 to

[17] If some λ_β is close to φ_α, then $x^{(\beta)}$ and $\tilde{y}^{(\beta)}$ will be close to $x^{(\alpha)}$ and $\tilde{y}^{(\alpha)}$, respectively. Then it may happen that the higher order terms neglected in the approximation (98) must be considered in evaluating $g(\alpha_1,\alpha_2)$ and $\delta(\alpha_1,\alpha_2)$. In such cases (97) does not necessarily hold. We shall discuss these cases in a separate paper.

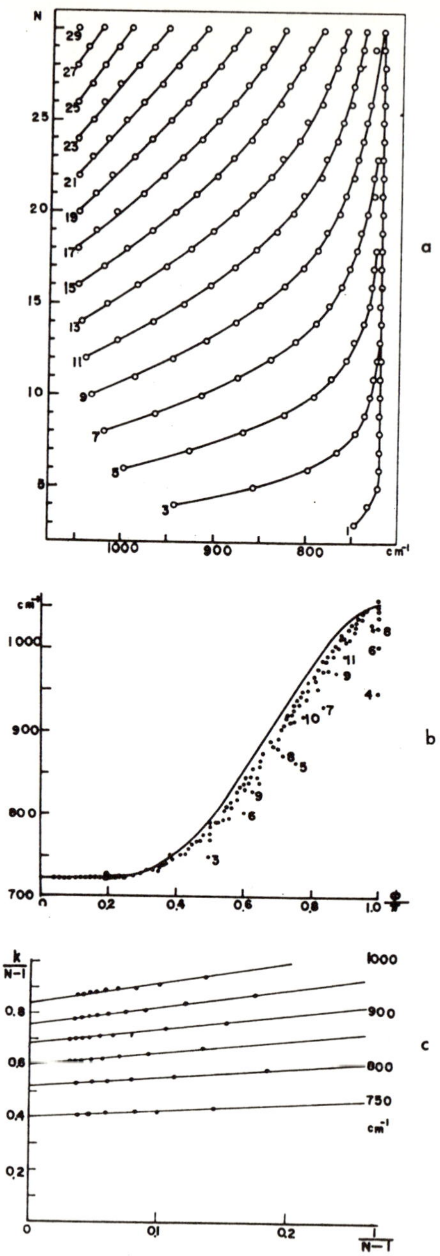

FIG. 1. (a) Methylene rocking–twisting mode array for n-paraffins. C_3H_8 through n-$C_{30}H_{62}$ (k values at left). (b) Frequency-phase curve for methylene rocking–twisting modes. Solid curve is obtained by our analysis of experimental data. Solid circles are due to Snyder *et al.* (Data points for selected n-paraffins indicated: $3=C_3H_8$, etc.) (c) Phase-shift analysis for methylene rocking–twisting modes.

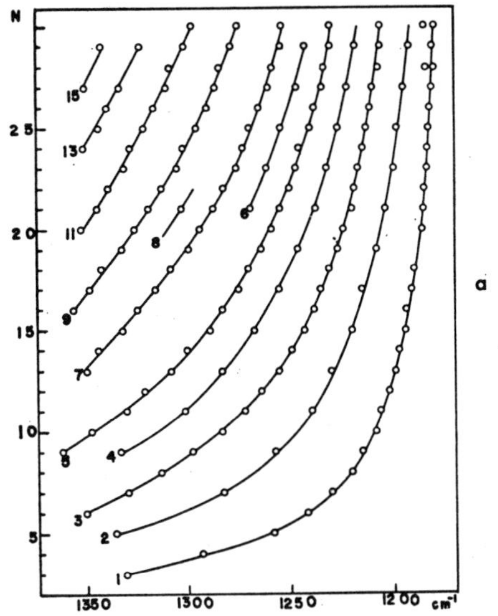

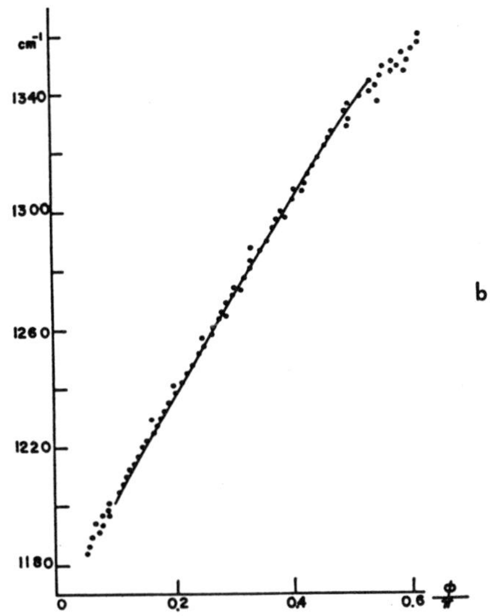

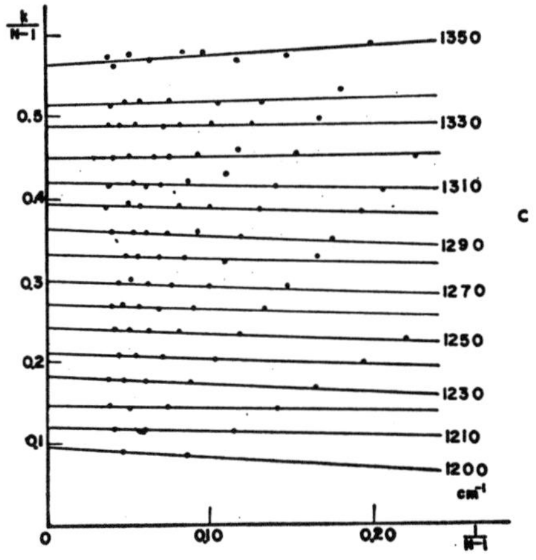

FIG. 2. (a) Methylene wagging mode array for *n*-paraffins. C_3H_8 through *n*-$C_{30}H_{62}$ (*k* values at left). (b) Frequency-phase curve for methylene wagging modes. Solid curve is obtained by our analysis of experimental data. Solid circles are due to Snyder *et al.* (c) Phase-shift analysis for methylene wagging modes.

π, the number of phases s corresponding to a given frequency ω determines the character of the normal vibration with that frequency. In the approximation $\exp(-N\psi_{s+1}) \ll 1$ we may state as follows: When $s=0$ there exists no chain vibration. When $s=1$ the admissible phase φ for the chain vibration must satisfy the relation

$$\varphi = (\pi/N)k + (\bar{\delta}/N), \tag{103}$$

where k is an integer, and φ and $\bar{\delta}$ are functions of ω. The primitive phase shift $\bar{\delta}$ depends on the structure of the end parts, whereas φ depends only on the structure of the original infinite molecule. Both quantities are independent of N.

When $s \geqq 2$ the phase must satisfy more complicated equations such as (93), but there is a reducible case in which the simple formula (103) is still approximately valid.

PART II

As mentioned in the Introduction, Snyder *et al.* have given the infrared spectra of homologous series of fully extended *n*-paraffin molecules in crystal.[4] On the basis of their analysis and the result of the theory of Tasumi *et al.*,[3] we may roughly divide the whole

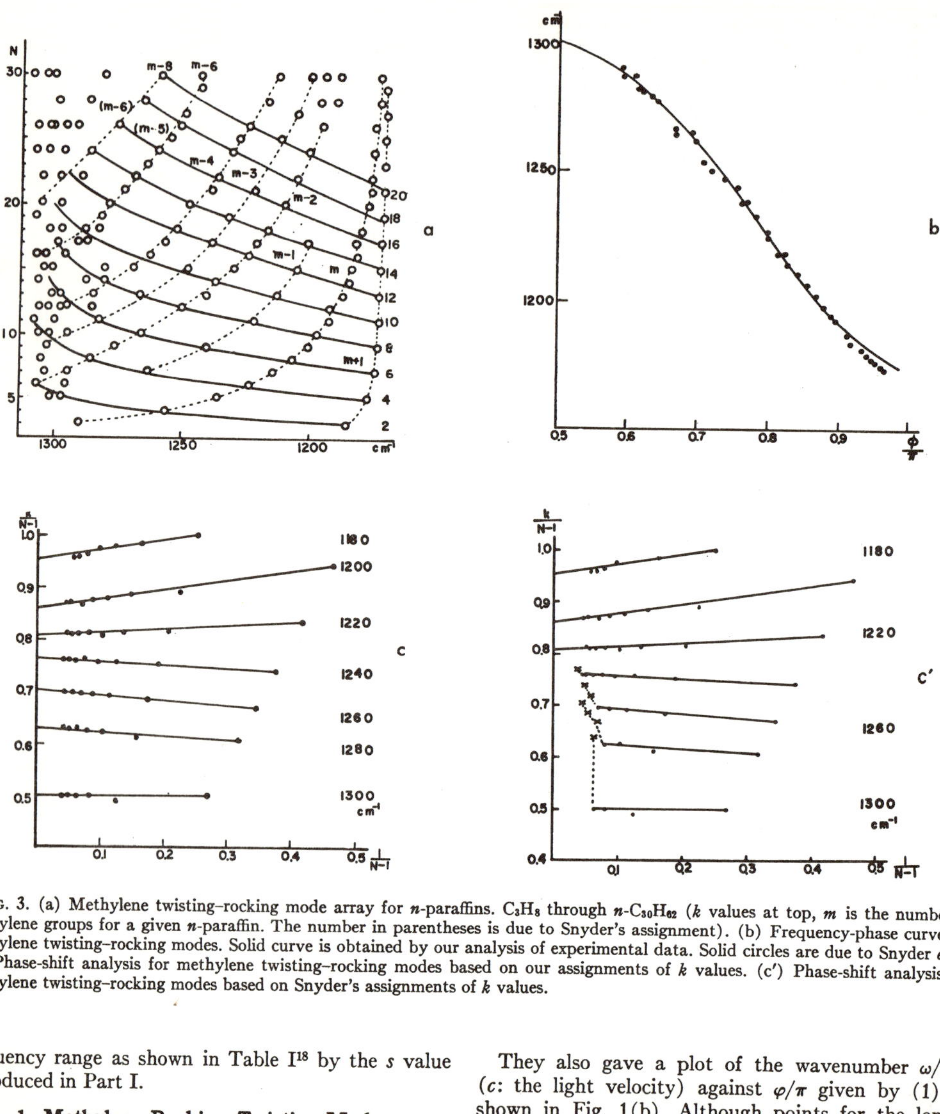

FIG. 3. (a) Methylene twisting–rocking mode array for *n*-paraffins. C_3H_8 through *n*-$C_{30}H_{62}$ (*k* values at top, *m* is the number of methylene groups for a given *n*-paraffin. The number in parentheses is due to Snyder's assignment). (b) Frequency-phase curve for methylene twisting–rocking modes. Solid curve is obtained by our analysis of experimental data. Solid circles are due to Snyder *et al.* (c) Phase-shift analysis for methylene twisting–rocking modes based on our assignments of *k* values. (c′) Phase-shift analysis for methylene twisting–rocking modes based on Snyder's assignments of *k* values.

frequency range as shown in Table I[18] by the *s* value introduced in Part I.

1. Methylene Rocking–Twisting Modes

Snyder *et al.* gave a complete array of methylene rocking modes for C_3H_8 through *n*-$C_{30}H_{62}$ as shown in Fig. 1(a). The assignment of *k* integers introduced in (1) to bands in the spectrum amounts merely to counting bands in the order of increasing frequency, remembering that only bands corresponding to an odd *k* are infrared active for reasons of symmetry.

[18] The number in the bracket is chosen with reference only to the calculation of Tasumi *et al.* and is not experimentally verified.

They also gave a plot of the wavenumber $\omega/2\pi c$ (*c*: the light velocity) against φ/π given by (1) as shown in Fig. 1(b). Although points for the longer molecules tend to converge to a common curve, points for the shorter molecules tend to be displaced to high values of φ/π. These displacements are considered to result from the interaction between methylene rocking and methyl out-of-plane rocking that is the end effect.

As shown in Table I, $s=1$ in the frequency range between 700 and 950 cm^{-1}, where the rocking modes exist. Therefore, according to our theory the phase φ must be given by (103) in this frequency range. From Fig. 1(a) we can assign the value of N (not

TABLE I. s values of a polymethylene chain.

Wavenumber (cm^{-1})	Mode of chain vibration	s value
0–200	Skeletal torsion Skeletal deformation	4
200–500	Skeletal deformation	2
500–700		0
700–950	CH_2 rocking	1
950–1150	CH_2 rocking Skeletal stretching	1, 3, (4)[a]
1150–1300	CH_2 wagging CH_2 twisting	2, (3)
1300–1400	CH_2 wagging	1
1400–1500	CH_2 bending	0, 1
1500–2850		0
2850–2950	CH_2 symmetric and anti-symmetric stretching	0, 1, 2
2950–		0

[a] See Ref. 18.

necessarily an integer) for each set of the frequency and k value. Using the value of N thus obtained, we plot in Fig. 1(c) the value of $k/(N-1)$ against $1/(N-1)$ at several frequencies.[19]

Every point for the fixed frequency is almost in a straight line. Since Eq. (103) indicates that for a fixed frequency k is a linear function of N, this shows that Eq. (103) is in good agreement with experiments. The gradient of the straight line is $(\varphi-\bar{\delta})/\pi$ by (103). This gradient accounts for the deviation of the phase for shorter chains from that given by (1). The small deviation from the straight line might be attributed to the intermolecular interaction in the crystal.

For the frequency range above 950 cm^{-1}, $s=3$, since there exists a skeletal stretching mode the frequency-phase curve corresponding to which has a minimum. The good fit of the formula (91′) even for 1000, 1037.5 and 1050 cm^{-1} in Fig. 1(c) may be due to the fact that the coupling with this mode is weak.

Since $k_\pi/(N-1)$ for $N\to\infty$ equals the phase φ by (103), the extrapolation by the straight line to $N\to\infty$ in Fig. 1(c) gives experimental frequency-phase diagram shown by the solid line in Fig. 1(b), which is more reliable than that obtained by merely connecting points for longer molecules using (1).

Thus, we find that the scattered points even for such shorter molecules as $N\cong5$ can be quantitatively correlated by the gradient of the straight line in Fig. 1(c), or the phase-shift. On the other hand, we find that the molecules with $N\cong30$ are not necessarily long enough from which we can deduce the frequency-phase diagram by using (1). The discrepancy amounts to about 13 cm^{-1} for $0.7<\varphi/\pi<0.85$ in Fig. 1(b).

We call the analysis based on (103), which uses the straight line as in Fig. 1(c), phase-shift analysis.

[19] As mentioned in Sec. 3 of Part I we may as well plot k/N against $1/N$, obtaining also straight lines. Then the gradient would be $[\delta(1)-\frac{1}{2}\pi]$. In this paper we have used $(N-1)$ in the denominator simply because it is convenient to compare our analyses with those by Snyder based on (1).

2. Methylene Wagging Modes

In Fig. 2(a) is given the band progression for wagging modes due to Snyder *et al.* We have done the phase-shift analysis in Fig. 2(c) just as for the rocking modes. The gradient of the straight line is almost zero. This corresponds to the fact that all the points plotted in Fig. 2(b) using (1) are close to the frequency-phase curve shown by the solid line which is obtained by the extrapolation in Fig. 2(c).

3. Methylene Twisting–Rocking Modes

Figure 3(a) shows the band progression for twisting modes. The dotted lines represent the series with the k value indicated there which were assigned by Snyder *et al.* Using the k values given by them we have constructed Fig. 3(c′) for the phase analysis as before. We find then some abrupt deviation from the straight lines for higher frequencies.

This deviation might be presumed to be due to the fact that in the corresponding frequency range, $s\neq1$, as shown in Table I. But in view of the success of the phase-shift analysis based on (103) for methylene rocking and wagging modes even when $s\neq1$ and of the linearity for lower frequencies in the twisting modes shown in Fig. 3(c′), we rather have looked for the cause of discrepancies in the wrong assignments of the

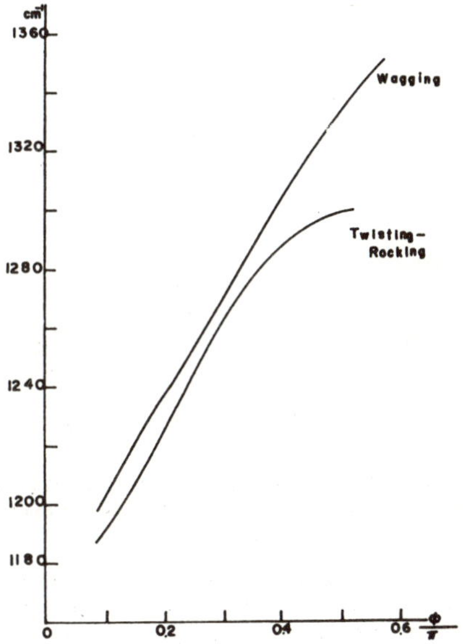

FIG. 4. Frequency-phase curves for methylene wagging and twisting–rocking modes.

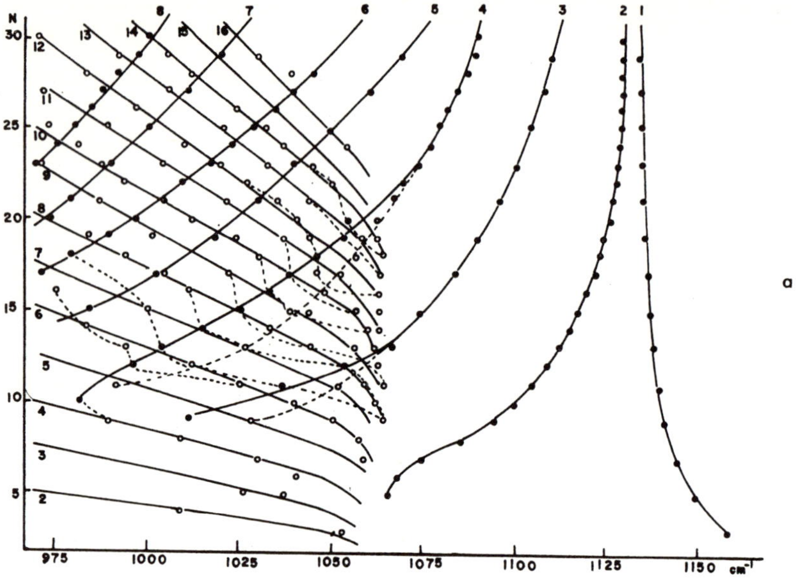

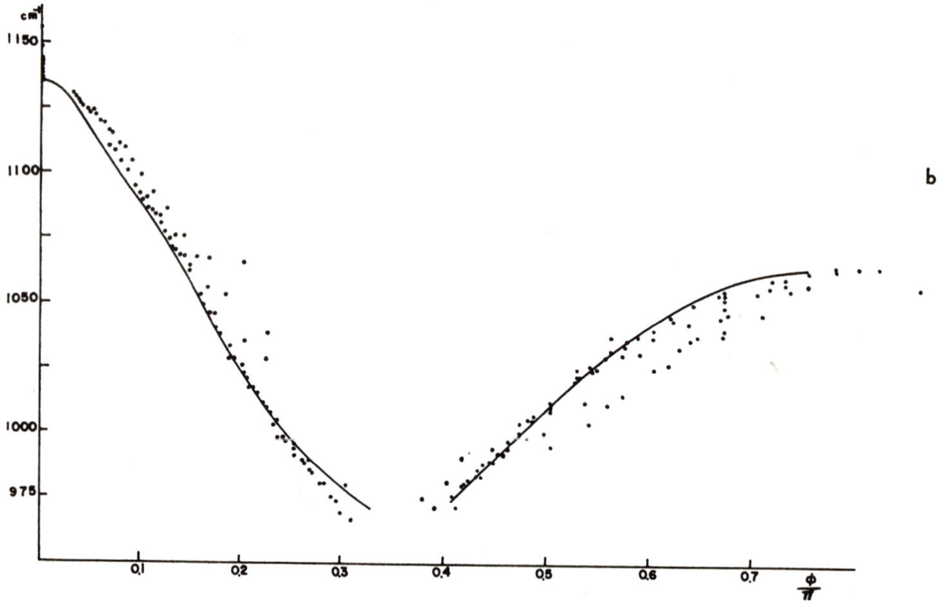

FIG. 5. (a) C–C stretching mode array for *n*-paraffins. C_3H_8 through *n*-$C_{30}H_{62}$ (*k* values at top). (b) Frequency-phase curves for C–C stretching modes. Solid curves are obtained by our analysis of experimental data. Solid circles are due to Snyder *et al.*

k values. Actually if we change $k=m-5$ to $m-6$ and $k=m-6$ to $m-8$ as indicated by the solid lines with the common *k* values in Fig. 3(a), then every point of a given frequency is again on a straight line as shown in Fig. 3(c). Since in Fig. 3(a) $k=m, m-2, m-4, m-6$ bands are observed for almost every molecule, while molecules with even *N* have no bands with $k=m+1$, $m-1$, $m-3$, our new assignments of *k* values will be reasonable from the consideration of intensity, too. Because it may be more natural to assume that bands

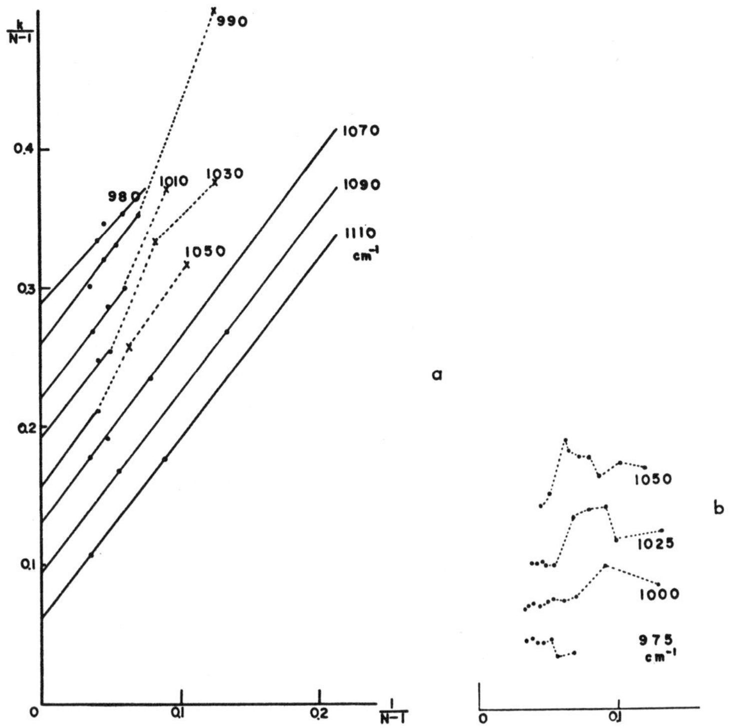

FIG. 6. Phase-shift analysis for C–C stretching modes based on Snyder's assignments of k values. (a) Region A. (b) Region B.

with $k=m-5$, $m-7$ are too weak to be clearly observed than to assume as in Snyder's assignments that bands with $k=m-5$ are observed for every molecule.

Although our new assignments do not change the essential feature of the frequency-phase relations shown in Fig. 3(b), we find that the phase-shift analysis not only gives reliable frequency-phase relations but also it may be useful as a critérion for assigning the k values.

In Fig. 4 we have given the frequency-phase relations for wagging and twisting modes for comparison's sake, where we have used the phase $\pi-\varphi$ for the latter. We have not, however, succeeded in interpreting the scattered points near 1300 cm^{-1} in Fig. 3(a).

4. Skeletal Stretching Modes

Considering that the frequency-phase curve of the C–C stretching modes has a minimum from the theory, Snyder *et al.* divided the phase region at the minimum into two regions, A and B. Region A covers the frequency range from 950–1150 cm^{-1} while Region B covers only the range from 950–1065 cm^{-1}. Assuming that each region has its own series of k values, they have given the band progression as shown in Fig. 5(a).

Using the k values given by them we have constructed the diagrams for phase-shift analysis in Figs. 6(a) and 6(b). Since the obtained diagrams are not straight lines, we again suspect the wrong assignments of k values. As shown by the solid lines in Fig. 5(a) it is possible to correct the assignments of k integers without breaking the guiding rule obeyed by Snyder that in Region B, only bands with even k are active in even-numbered molecules. Of course, for this purpose it is necessary to exchange points between the two regions in making the reassignments.

If we use our corrected assignments, the linearity of the diagrams is again surprisingly good both in Regions A and B as shown in Figs. 7(a) and 7(b). By the extrapolation to $N\rightarrow\infty$ in these figures we have obtained the frequency-phase curve shown in Fig. 5(b).

CONCLUDING REMARKS

We have developed the general method for obtaining normal vibrations of chain molecules. The method is based on the use of virtual and auxiliary coordinates and transfer matrices. The frequency range is classified by the s-value. When $s=1$, the end effect on the chain vibrations can be simply represented by the phase shift. This fact is confirmed by analyzing all the modes of n-paraffins for which Snyder *et al.* assigned k values. We have found that the phase-shift analysis is very

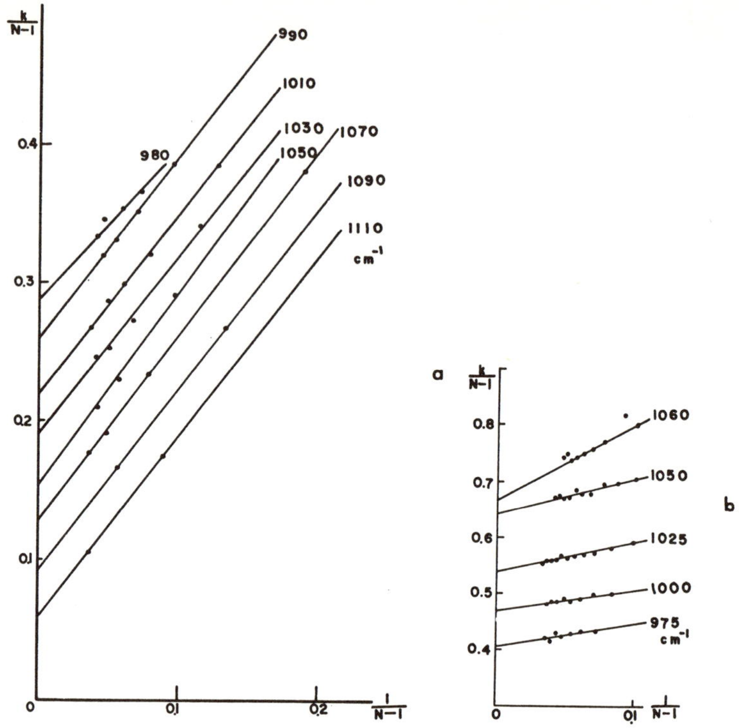

FIG. 7. Phase-shift analysis for C–C stretching modes based on our assignments of k values. (a) Region A. (b) Region B.

useful for making extrapolation, interpolation, assignment, and determination of frequency-phase relations from the experimental data even when $s\neq 1$.

The quantities φ_1 and $\delta(1)$ are the functions of frequency. The former quantity is to be determined by (43) and (46), the latter one by (92) if the force field of the system is given. Tasumi *et al.*[3] have calculated φ_1 from the modified Urey–Bradley force field. We have obtained both φ_1 and $\delta(1)$ from the experimental data, but have not attempted to calculate them from the assumed force field. The success of the phase-shift analysis indicates, however, that we can safely assume that (i) the intermolecular interaction can be neglected, (ii) the end group of the molecule does not significantly affect the force constants of the internal parts, and (iii) the force constants are almost independent of the length of the molecule in crystalline n-paraffins.

If these assumptions are valid for other polymethylene chains with different end groups, these chain molecules will have the same φ_1 but in general different $\delta(1)$. Recently, Machida[20] has made the phase-shift analysis on his infrared data of $CH_3(CH_2)_nCONH_2$ and $CH_3(CH_2)_nN(CH_3)_3Br$, and Takenaka[21] has analyzed his data on $CH_3(CH_2)_nCOOM$(M: Na, K, Li, $\cdots$). These analyses have given a common φ_1 and different $\delta(1)$, so that they seem to support the above assumptions in these molecules.

In this paper we are interested mainly in the end effects on the frequencies of chain vibrations. The applicability of the transfer matrix method introduced in Part I, however, is restricted neither to the end effects nor to the normal frequencies. Some discussions on the end effects on the infrared absorption intensities were already made in I.

Finally, we note, since the transfer matrix method is a convenient mathematical technique for solving the eigenvalue problem of matrices, it may also be applied to other problems such as electronic states of chain molecules than those of normal vibrations.

ACKNOWLEDGMENTS

We express our sincere gratitude to Dr. R. G. Snyder for informing us of his experimental results prior to the publication. We are indebted to Dr. K. Machida for helpful discussions.

[20] K. Machida (private communication).

[21] T. Takenaka (private communication).

Electron Energy Bands in Ordered and Disordered Crystals

Chapter 3

Electron Energy Bands in Ordered and Disordered Crystals

General Considerations about Periodic Structures

In this chapter we study the solutions of the one-particle Schrödinger equation:

$$\left\{-\frac{\hbar^2}{2m}\frac{\partial^2}{\partial x^2}+\sum_n V_n(x-x_n)\right\}\Psi(x) = E\Psi(x) \tag{3.1}$$

for various choices of the potential energy function $V(x)$. This study is central in the independent-particle approximation to the structure of matter, and it is possible to carry it out much further in one dimension than in three, although the general conclusions are the same. The general theory of the properties of such wave equations is given by Eckart (1948) in the first of the reprinted works in this chapter.

The most familiar idealization is the "perfect" crystal structure, when each of the potentials $V_n(x - x_n)$ is centered about regularly spaced points

$$x_n = na \tag{3.2a}$$

and all have identical shape:

$$V_n(x - x_n) = V(x - na) \tag{3.2b}$$

In this case the wave functions which are the solutions of Eq. (3.1) have the familiar *Bloch form*,

$$\Psi(x) = e^{ikx}u_{t,k}(x) \tag{3.3a}$$

with $u(x)$ a periodic function,

$$u_{t,k}(x) = u_{t,k}(x + a) = u_{t,k}(x + 2a) = \text{etc.} \tag{3.3b}$$

Plane waves are special cases of the Bloch functions (when the periodic part $u(x)$ is a constant). The distribution of the allowed energies E will be one of the principal objects of our study. They are found to lie within dense "allowed" bands separated by "energy gaps" or "forbidden bands." The allowed bands shall be labeled by the band index $t = 1, 2, \ldots$ according to increasing energy; the states within each allowed band are further labeled by k, the "crystal momentum" quantum number, which has values in the range

$$-\pi/a < k \leq +\pi/a \tag{3.4}$$

These rather general and well-known facts are often illustrated by means of a simple and exactly soluble model due to Kronig and Penney (1931), and it is appropriate that we reprint in this chapter the interesting and original work by these authors—oft quoted but rarely read. In this paper it is noted that the reflection coefficient for an electron normally incident upon a periodic crystal must be unity (perfect reflection) when the energy is in the forbidden range, and less than unity in the allowed bands. A somewhat more detailed treatment of tunneling through a finite slab with periodic structure has recently been given by Pschenichnov (1962).

The Bloch function approach is suited only to periodic structures, and must be somewhat generalized before aperiodic structures can be studied. The necessary first step was taken by Kramers (1935), and generalized by James (1949a), who remarked that for any given value of the energy E there were two independent solutions of Eq. (3.1) in each cell (the nth cell is the neighborhood of the point x_n). Thus the construction of a continuous solution to Eq. (3.1) over the entire space could be reduced to a matching problem at the boundaries separating neighboring cells, which immediately brought Kramers to the study of 2×2 "transfer matrices." Let $y_1(x)$ and $y_2(x)$ be the two linearly independent solutions to the Schrödinger equation with a given energy and a given potential $V_n(x)$. Choose them so that $y_1(0) = 1$, and $y_1'(0) = 0$ and $y_2(0) = 0$, $y_2'(0) = 1$. Inside the nth cell the solution to the Schrödinger equation must be $S_1y_1(x - x_n) + S_2y_2(x - x_n)$ where S_1 and S_2 are two numbers which may be thought of as two components of a vector $A = (S_1, S_2)$, and where x_n is regarded as the beginning (left-hand side) of the nth cell. If the nth cell has length a, then $\Psi = S_1y_1(x - x_n + a) + S_2y_2(x - x_n + a)$ at the right-hand edge of the cell. But at this edge, Ψ and its derivatives must be continuous, so that we obtain the equation

$$A_{n+1} = \alpha_n A_n \tag{3.5}$$

where α is a 2×2 transfer matrix defined by

$$\alpha_n = \begin{bmatrix} y_1(x - x_n + a) & y_2(x - x_n + a) \\ y_1'(x - x_n + a) & y_2'(x - x_n + a) \end{bmatrix} \equiv \begin{bmatrix} \alpha_{11} & \alpha_{12} \\ \alpha_{21} & \alpha_{22} \end{bmatrix} \tag{3.6}$$

and A_n is the A vector in the nth cell. The Wronskian of the two solutions is

$$W(x) = y_1(x)y_2'(x) - y_1'(x)y_2(x) \tag{3.7}$$

and, since this quantity is independent of x, we deduce that

$$\det\|\alpha\| = \alpha_{11}\alpha_{22} - \alpha_{12}\alpha_{21} = 1 \tag{3.8}$$

That is, the *transfer matrix* α is a complex 2×2 unimodular matrix. In a periodic structure its two eigenvalues must be simply $e^{\pm ika}$, as the eigenfunctions can undergo a phase change only in going from cell to cell. Therefore we must ask about the existence of such exponential solutions to the determinantal equation:

$$\det\begin{Vmatrix} \alpha_{11} - e^{ika} & \alpha_{12} \\ \alpha_{21} & \alpha_{22} - e^{ika} \end{Vmatrix} = 0 \tag{3.9}$$

Equations (3.8) and (3.9) taken together yield

$$\tfrac{1}{2}(\alpha_{11} + \alpha_{22}) = \cos ka \tag{3.10}$$

which allows a solution for a real k provided the matrix elements are in the range

$$-2 \leq \alpha_{11} + \alpha_{22} \leq 2 \tag{3.11}$$

For values larger than $+2$ or smaller than -2, k must be complex. Finally, if periodic boundary conditions are imposed such that the wave function in the $(N + 1)$st cell is required to be identical to that in the 1st cell, then we have as a further constraint:

$$(e^{ika})^N = 1 \qquad \text{which implies} \qquad k = (2\pi/Na) \times \text{integer} \tag{3.12}$$

[For the case of nonperiodic potential, we demand (as a condition for the allowed energies) that the matrix from the 1st to the Nth cell (namely $\alpha = \alpha_N\alpha_{N-1} \cdots \alpha_2\alpha_1$) have an eigenvalue equal to unity. Since α is unimodular, this condition is equivalent to trace $\alpha = 2$.] When the number N of cells is allowed to become infinite, the k's become a continuous variable constrained to the range (3.4), and by definition the allowed bands are those for which the inequality (3.11) is obeyed. In addition to calculating individual energy levels, we shall be interested in the density of energy levels in the allowed bands, the so-called "density of states."

The study of k as a function of the energy E (taken as a complex variable) was an important next step forward taken by James (1949a) as it allowed an understanding to develop about the nature of wave functions in the forbidden bands, and so was essential in understanding the effects of impurities, applied force fields, etc., on the exact eigenfunctions of Schrödinger's equation. This study was taken up once more, and developed quite far in the next of the reprinted papers by Kohn (1959).

Kohn's (1959) work is concerned with the analytic properties of the eigenfunctions $\Psi_{t,k}(x)$ and their eigenvalues in the complex k-plane. He readily proves that $\alpha_{11} + \alpha_{22}$ is an entire function of E, and by analyzing this function he is able to show that the energy has a finite strip of analyticity in the complex k-plane, with simple square-root-type branch cuts on the edge of this strip. Denoting the width of the strip for the tth band by $\bar{h}_t$, he was now in a position to prove an important result concerning the *Wannier functions*, the famous linear combinations of Bloch functions of such particular importance in solid state physics. It is no digression to briefly discuss the Wannier functions $a_t(x)$ at this point: they are restricted Fourier transforms of the Bloch functions,

$$a_t(x - na) \equiv (a/2\pi)^{1/2} \int_{-\pi/a}^{+\pi/a} dk e^{-inak} \Psi_{t,k}(x) \tag{3.13}$$

and because two such functions of different band index t or cell n are orthogonal, the set (3.13) provides a convenient orthonormal basis for expanding electronic functions in the lattice. The Wannier functions are often conceived as the solid state equivalent to *atomic orbitals*, the electron being constrained to the neighborhood of the nth cell rather than to the nth atom. If this suggestive picture is to be correct, then it is required that far from their central atom, these functions decay exponentially with distance. This was generally believed to be the case, but it was Kohn (1959) who first proved the important result:

$$a_t(x) \underset{|x| \to \infty}{\sim} \exp(-\bar{h}_t |x|) \tag{3.14}$$

and thus related the asymptotic properties of the Wannier functions to the analytic properties of the energy in the complex plane.

Special Periodic Structures

The band structure for special choices of the potential $V(x)$ has been studied in so many mathematical and physical publications that there is hardly any need to treat the subject here. For example, the choice

$$V(x) = \cos 2\pi\, x/a \tag{3.15}$$

turns Eq. (3.1) into the well-known Mathieu equation, which is treated in many books on differential equations. But we should mention a thorough treatment of three-dimensional band structure by Slater (1952), based on the separable potential $U(\mathbf{r})$:

$$U(\mathbf{r}) = V(x) + V(y) + V(z) \tag{3.16}$$

Because the kinetic energy is also separable, it is clear that the wave functions $\Psi(\mathbf{r})$ can be factored as follows:

$$\Psi = \phi(x)\chi(y)\xi(z) \tag{3.17}$$

and therefore Slater (1952) was able to reduce the three-dimensional problem to known properties of Mathieu functions.

In view of the physical importance of the Coulomb potential and of the hard-core potential in three dimensions, both of which are singular at the origin, it is interesting to note the one-dimensional singular potential $V(x) = -\csc^2(\pi x/a)$

studied by Scarf (1958), who was able to solve Schrödinger's equation for this special case. Even this pathologic potential gives results which, aside from rather subtle differences, are in agreement with the well-known properties of the Kronig-Penney (1931) model and of the Mathieu equation. All in all, the periodic problem may be said to be qualitatively, and quantitatively, well understood from many different points of view. A general overview is given in Eckart's (1948) analysis of wave theory, but especially in Brillouin's (1946) famous little book.

Slight Deviations from Periodicity

As we have seen, the properties of periodic Hamiltonians are well understood in all their various aspects. However, as soon as the slightest deviations from perfect periodicity are considered, one finds that this poses very intricate mathematical problems which are soluble only in particular instances.

One such soluble problem is concerned with the existence of a surface. In one dimension, the surface is described by some nonperiodic end boundary condition first considered by Tamm (1932), and later by Shockley (1939) and many others. As a consequence of special end boundary conditions one may find that a state in each band becomes localized in the neighborhood of the end cell, much as a Wannier function; the remaining wave functions in each band remain similar to the Bloch functions, with, perhaps, reduced amplitude near the end. This general subject of end effects is becoming once more a very fashionable topic of study for the following reasons: in three dimensions, the surface states spread into surface *bands* because of translational invariance parallel to the surface. It is well known that the electronic band structure is one of the determining factors in establishing the physical crystallographic structure of materials, and if the band structure near the surfaces is sufficiently different from that in the bulk one should expect to find surfaces with crystallographic properties quite different from those of the bulk. Recent experiments[1] show this to be so, and have raised many new and interesting problems.

Another related subject which has been studied (James, 1949b) is the "perturbed" periodic system, such as an impurity cell in an otherwise periodic array. If the impurity potential is slowly varying, it cannot mix wave functions belonging to different bands; but within each band, the eigenfunctions can no longer change by the simple exponential phase factor e^{ika} in going from cell to cell, and must undergo a change in amplitude as well as a phase change depending on the distance from the impurity. The single impurity problem has been re-examined by Kuliasko (1964a), who has also studied interactions between two impurities (Kuliasko, 1964b).

If the impurity potential is rapidly varying, i.e. if the change in potential from cell to cell is comparable to the forbidden energy gaps, then wave functions from the different bands will mix to a significant extent and we have the phenomenon known as *tunneling*. Zener (1934) first considered the effects of an applied electric field on an otherwise periodic chain, and the current which results when an applied field

[1] L. H. Germer, "The Structure of Physical Surfaces." *Sci. Am.* **212,** No. 3, 32 (1965).

pulls electrons out of the valence band and into the conduction band of an insulator is now known as "Zener tunneling," or "Zener breakdown."

The peculiar non-ohmic dependence of the Zener current on the applied field F can be understood on the basis of simple arguments: an electric field F results in a potential energy eFx, and therefore a wave packet in the conduction band which is separated by an energy gap E_g from a wave packet in the valence band at the same point in space will be degenerate with a wave packet in the valence band at a point further down-field by a distance x, where

$$x = E_g/eF \tag{3.18}$$

The current is (approximately) proportional to the overlap of the two degenerate wave packets in question, hence we use Kohn's formula for the exponential decay of wave functions in the forbidden energy region, Eq. (14), to obtain the tunneling current J:

$$J \propto \exp\left(-\bar{h}E_g/eF\right) \tag{3.19}$$

[The reader may wish to compare this with Zener's (1934) original derivation which is reprinted in the present chapter.] In the bibliography we also supply a selected list of papers on related topics: phonon-assisted tunneling, Esaki diodes, and barrier penetration in the normal or superconducting states.

Large Deviations from Periodicity

The pioneering study of several mathematical models of alloys undertaken by Saxon and Hutner (1949) led to their famous conjecture on the band structure of binary alloys: Energy levels that are forbidden in pure metal A and also in pure metal B *are forbidden in all alloys of A with B.*

An elegant proof of this conjecture was given by Luttinger (1951), but the proof is applicable only to Kronig-Penney type chains in which the A and B atoms are characterized by equally spaced delta-function potentials of strengths V_A and V_B. Both James and Ginzbarg (1953) and Kerner (1954) discuss other types of binary chains, and find that the conjecture of Saxon and Hutner is *not* always obeyed. This is a very important point, for an insulator is characterized by a filled valence band separated by a finite energy gap from the empty conduction band; if it were exact, therefore, the S-H conjecture would have a direct bearing on which alloys of insulating materials are themselves insulators, and which will become metallic.

The same questions arise in considering monatomic chains with arbitrary, or random, interatomic spacings—such as we might study in a model of a liquid. Can we, by melting an insulating solid, obtain a metallic liquid, or vice versa? Aspects of this problem have been studied by various authors. A plausible one-dimensional analog of a liquid, in which interatomic distances were randomly distributed about a mean value, was considered by Eisenschitz and Dean (1957). This work was based on the tight-binding approximation, i.e., Wannier functions in different bands were not supposed to mix, and matrix elements among Wannier functions in the same band were supposed to vanish, except for nearest-neighbor overlap. Formally, the Hamiltonian (3.1) is assumed to possess the following matrix structure in the representation of Wannier functions (3.13):

$$\int a_t^*(x - x_n)Ha_{t'}(x - x_n)\,dx = A_{t,n}\,\delta_{t,t'}$$

$$\int a_{t'}^*(x - x_{n+1})Ha_{t'}(x - x_n)\,dx$$

$$= \int a_{t'}^*(x - x_n)Ha_t(x - x_{n+1})\,dx = B_{t,n}\,\delta_{t,t'} \tag{3.20}$$

In a perfectly periodic structure the parameters A_t and B_t are constant, and the energy eigenvalues are

$$E_t(k) = A_t + 2B_t \cos ka \tag{3.21}$$

But in a liquid $A_{t,n}$ and $B_{t,n}$ are random variables and the matrix (3.20) cannot be diagonalized so easily; Eisenschitz and Dean (1957) adapt Dyson's (1953) method for solving the disordered chain, a method we have already discussed in Chapter 2. The principal object of their calculation is the density of states function $D(E)$ and we shall explain here why this function[2] can be readily calculated in one dimension but not in three.

It is a mathematically well-known property of Schrödinger's equation (3.1) that the solutions E can be enumerated in ascending order according to the number of nodes (zeroes of the wave function), so that if some $\Psi_{t,k}(x)$ vanishes precisely n times on the interval 0, L, then there are precisely $n(E_{t,k}) = n$ eigenvalues lower than $E_{t,k}$. As a consequence let us define the "integrated density of states function" $M(E)$,

$$M(E) \equiv \lim_{N\to\infty} (n(E)/N), \qquad N = \text{number of atoms} \tag{3.22}$$

of which the derivative is evidently the density of states,

$$D(E) = dM(E)/dE \tag{3.23}$$

Calculations of $M(E)$ and hence of $D(E)$ were first performed on relatively short ($N < 100$) finite-length disordered chains by James and Ginzbarg (1953), Landauer and Helland (1954), Lax and Phillips (1958), and perhaps others. Such calculations could be considered complementary to the rigorous theories about infinite-length disordered chains elaborated by Eisenschitz and Dean (1957), Schmidt (1957), Frisch and Lloyd (1960), and Borland (1961). Among the most significant results are the following by Borland (1961): defining p in forms of the quantum number k by

$$E_k \equiv \hbar^2p^2/2m \tag{3.24a}$$

[2] Also often denoted $\rho(E)$ or $N(E)$ in the literature; the notation here and below also differs from Chapter 2, but the careful reader will not be misled.

he found energy gaps corresponding to gaps in the allowed values of p in the range given by the inequalities

$$n\pi + 2q \leq pa \leq n\pi - pb \qquad \text{(attractive case)} \qquad (3.24b)$$

$$n\pi \leq pa \leq n\pi - pb + 2q \qquad \text{(repulsive case)} \qquad (3.24c)$$

where a and $a + b$ are, respectively, the minimum and maximum interatomic separation between nearest-neighboring identical delta-function "atoms" of strength K_0, with

$$q \equiv \tan^{-1} K_0/2k \qquad (3.24d)$$

For $K_0 < 0$, Eq. (3.24b) applies, while for $K_0 > 0$, Eq. (3.24c) applies. The integer n would be the band index in the case of the ordered lattice.

One important feature is that the existence of these gaps was established regardless of the degree of order or disorder; and provided the distances of nearest and furthest approach, a and $a + b$, were restricted, the low-lying energy band structure was shown to remain qualitatively unchanged in proceeding from the ordered to the disordered state; the second, or higher, energy gaps might, however, not lie within the inequalities (3.24) and might therefore disappear easier in the process of disordering. The transition: insulator $\leftrightarrow$ metal in the process of melting or freezing can be seen, according to the above, to depend, among other factors, on the number of filled bands and thus, ultimately, on the valency of the constituent atoms. These results of Borland and the earlier work of Luttinger (1951) are considerably extended by Dworin (1965).

Questions about the existence of energy gaps and of "localized states" were also considered in papers of Sah and Eisenschitz (1959), Frisch and Lloyd (1960), Borland (1960, 1961), Mott and Twose (1961), Roberts and Makinson (1962), and Morrison (1962). The work of Frisch and Lloyd (1960), which we reprint, assumes a Poisson distribution of distances between nearest-neighbor atoms, so that Borland's inequalities do not apply. Indeed, these authors do not find any energy gaps but a rather smooth density of states, reminiscent of Dyson's (1953) results in the vibrating chain problem (in the particular case when the mechanical elements were also distributed according to the Poisson distribution). They analyze the band structure indirectly, first showing that the advance of the phase of the wave function in proceeding from cell to cell defines a stationary Markov process which may then be determined as the solution to a differential-difference equation: this also is reminiscent of Dyson's (1953) and Schmidt's (1957) equations. Subsequently, Morrison (1962) gave further mathematical refinements to this work. One general conclusion which all these investigations seem to have established is that the existence of energy gaps depends to a great degree on short-range order, and is generally independent of the degree of long-range order or lack thereof. Roberts and Makinson (1962) and later Borland (1963) investigated the nature of the eigenfunctions of the disordered chain and found such solutions to be strongly localized, i.e., they found a substantial part of the wave function lay in a fixed spatial neighborhood. This implies that the normalization is independent of the length of the chain, in contradistinction to ordinary wave-like solutions (which must be normalized by a factor $N^{-1/2}$ and which are sensitive to the choice of the end boundary condition); see Agacy and Borland (1964) which is reprinted herein.

The great interest in the use of Green's functions in the quantum-statistical mechanics of the theory of solids has led to many applications in the study of disordered structures. In a recent study, Halperin (1965) calculates the spectral density function $A(k, E)$ defined by

$$A(k, E) = L^{-1}\left\langle\sum_i \left|\int_0^L dx\, e^{ikx}\Psi_i(x)\right|^2 \delta(E - E_i)\right\rangle \tag{3.25}$$

where E_i and Ψ_i are the ith eigenvalue and ith normalized eigenfunction of H, and $\langle\ \rangle$ indicates an average over all random variables.

The function $A(k, E)$, which is the imaginary part of the one-particle Green function, spreads out according to the degree of disorder; but in the limit of a perfectly ordered chain A is proportional to a sum of delta functions, one for each band. Using the familiar representation of the delta function

$$\delta(x - x') = (2\pi)^{-1} \int_{-\infty}^{+\infty} dk \exp ik(x - x') \tag{3.26}$$

one readily obtains the density function,

$$N(E) = (2\pi)^{-1} \int_{-\infty}^{+\infty} dk\, A(k, E) \tag{3.27}$$

Halperin calculates $A(k, E)$ by methods closely related to the earlier work of Frisch and Lloyd (1960) and Borland (1961), as the solution of an equivalent Markov process.

With this we terminate the present study of electrons in disordered chains, but it is clear that a definitive chapter will not be written before many years, and before a host of complex and interesting questions have been laid to rest (many of them already posed in connection with Chapter 2).

Bibliography

I. Ordered Periodic Structures

G. Allen (1953). "Band Structures of One-Dimensional Crystals with Square-Well Potentials," *Phys. Rev.* **91,** 531.

L. Brillouin (1946). "Wave Propagation in Periodic Structures," McGraw-Hill, New York.

*C. Eckart (1948). "Approximate Solution of One-Dimensional Wave Equations," *Rev. Modern Phys.* **20,** 399.

T. A. Hoffman (1951). "Linear Chain: Binary Systems," *Acta Phys. Acad. Sci. Hung.* **1,** 175.

*H. M. James (1949). "Energy Bands and Wave Functions in Periodic Potentials," *Phys. Rev.* **76,** 1602.

E. H. Kerner (1954). "Periodic Impurities in a Periodic Lattice," *Phys. Rev.* **95,** 687.

*W. Kohn (1959). "Analytic Properties of Bloch Waves and Wannier Functions," *Phys. Rev.* **115,** 809.

H. A. Kramers (1935). "Das Eigenwertproblem im Eindimsionalen Periodischen Kraftfelde," *Physica* **2,** 483.

*R. deL. Kronig and W. G. Penney (1931). "Quantum Mechanics of Electrons in Crystal Lattices," *Proc. Roy. Soc.* (*London*) **A130,** 499.

E. Pshenichnov (1962). "Tunnel Effect through a System of Identical Potential Barriers," *Soviet Phys.–Solid State* **4,** 819.

*F. L. Scarf (1958). "New Soluble Energy Band Problem," *Phys. Rev.* **112,** 1137.

J. C. Slater (1952). "Soluble Problem in Energy Bands," *Phys. Rev.* **87,** 807.

II. Slight Deviations from Perfect Order

H. M. James (1949). "Electronic States in Perturbed Periodic Systems," *Phys. Rev.* **76,** 1611.

G. F. Koster and J. C. Slater (1954). "Wave Functions for Impurity Levels," *Phys. Rev.* **95,** 1167.

F. Kuliasko (1964a). "Influence of an Impurity on the Energy Levels in an Infinite One-Dimensional Periodic Lattice," *Physica* **30,** 2185.

F. Kuliasko (1964b). "Coupling of Impurities in the One-Dimensional Rectangular Well," *Physica* **30,** 2180.

P. Sah and R. Eisenschitz (1960). "Electrons in Weak Interaction with a Chain of Atoms," *Proc. Phys. Soc.* (*London*) **75,** 700.

W. Shockley (1939). "Surface States Associated with a Periodic Potential," *Phys. Rev.* **56,** 317.

III. Tunneling Phenomena (condensed bibliography)

A. *Zener Process*

*C. Zener (1934). "A Theory of Electrical Breakdown of Solid Dielectrics," *Proc. Roy. Soc.* (*London*) **145,** 523.

P. Argyres (1962). *Phys. Rev.* **126,** 1386.

W. Franz (1959). *Z. Natürforsch.* **14a,** 415.

W. Houston (1940). *Phys. Rev.* **57,** 184.

E. Kane (1959). *J. Phys. Chem. Solids* **12,** 181.

L. Keldysh (1958). *Soviet Phys.—JETP* **6,** 763; **7,** 665.

J. Krieger (1966). *Ann. Phys.* (N.Y.) **36,** 1.

B. *Tunneling in P-N Junctions* (*Esaki process*)

P. Butcher, D. Hum, and E. Pike (1964). *Proc. Roy. Soc.* (*London*) **A280,** 185.

D. Fredkin and G. Wannier (1962). *Phys. Rev.* **128,** 2054.

R. Haering and E. Adams (1961). *J. Phys. Chem. Solids* **19,** 8.

E. Kane (1961). *J. Appl. Phys.* **32,** 83.

P. Price and J. Radcliffe (1959). *IBM J. Res. Develop.* **3,** 364.

R. Shuey (1965). *Phys. Rev.* **137,** A1268.

C. *Barrier Transmission in Normal or Superconducting Materials*

J. Bardeen (1961). *Phys. Rev. Letters* **6,** 57.

M. Cohen, L. Falicov, and J. Phillips (1962). *Phys. Rev. Letters* **8,** 316.

W. Harrison (1961). *Phys. Rev.* **123,** 85.

V. Heine (1965). *Phys. Rev.* **138,** A1689.

R. Prange (1963). *Phys. Rev.* **131,** 1083.

P. Price (1962). *In* "Physics of Semiconductors" (Exeter Conf. Proc.), p. 99. Institute of Physics and Physical Society, London.

IV. Chains with Arbitrary Disorder

*R. Agacy and R. Borland (1964). "Electronic Structure of One-Dimensional Random Alloy," *Proc. Phys. Soc. (London)* **84,** 1017.

R. Borland (1961a). "One-Dimensional Chains with Random Spacing between Atoms," *Proc. Phys. Soc. (London)* **77,** 705.

*R. Borland (1961b). "Existence of Energy Gaps in One-Dimensional Liquids," *Proc. Phys. Soc. (London)* **78,** 926.

R. Borland (1963). "Nature of the Electronic States in Disordered One-Dimensional Systems," *Proc. Roy. Soc. (London)* **A274,** 529.

R. Borland and N. Bird (1964). "Calculation of Density of Electron States and Degree of Localization in One-Dimensional Liquid Model," *Proc. Phys. Soc. (London)* **83,** 23.

J. des Cloizeaux (1955). "Bandes Interdites et Bandes Permises dans les Semi-Conducteurs Impurs et les Alliages Désordonnés," *J. Phys. Radium* **16,** 320.

*L. Dworin (1965). "Existence of Energy Gaps in the Spectrum of a One-Dimensional Atomic Chain," *Phys. Rev.* **138,** A1121.

S. F. Edwards (1961). "Electronic Structure of Disordered Systems," *Phil. Mag.* (8) **6,** 617.

R. Eisenschitz and P. Dean (1957). "Electronic Levels of a Model of Liquid Potassium," *Proc. Phys. Soc. (London)* **70,** 713.

J. Ford (1959). "Density of Electronic Energy Levels for a One-Dimensional Liquid," *J. Chem. Phys.* **30,** 1546.

*H. L. Frisch and S. P. Lloyd (1960). "Electron Levels in a One-Dimensional Random Lattice," *Phys. Rev.* **120,** 1175.

B. Halperin (1965). "Green's Functions for a Particle in a One-Dimensional Random Potential," *Phys. Rev.* **139,** A104.

H. James and A. Ginzbarg (1953). "Band Structure in Disordered Alloys and Impurity Semiconductors," *J. Phys. Chem.* **57,** 840.

R. Landauer and J. Helland (1954). "Electronic Structure of Disordered One-Dimensional Chain," *J. Chem. Phys.* **22,** 1655.

M. Lax and J. Phillips (1958). "One-Dimensional Impurity Bands," *Phys. Rev.* **110,** 41.

*J. Luttinger (1951). "Wave Propagation in One-Dimensional Structures," *Philips Res. Repts.* **6,** 303.

R. Makinson and A. Roberts (1960). "Zone (i.e., Band) Theory of Liquids," *Australian J. Phys.* **13,** 437.

R. Makinson and A. Roberts (1961). "Localized Eigenstates in One-Dimensional Liquids," *Proc. Phys. Soc. (London)* **79,** 222 (Letter).

J. Morrison (1962). "On the Number of Electron Levels in a One-Dimensional Random Lattice," *J. Math. Phys.* **3,** 1023.

A. Roberts and R. Makinson (1962). "Electron Eigenstates in a One-Dimensional Liquid," *Proc. Phys. Soc. (London)* **79,** 630.

D. Saxon and R. Hutner (1949). "Some Electronic Properties of a One-Dimensional Crystal Model," *Philips. Res. Repts.* **4,** 81.

*H. Schmidt (1957). "Disordered One-Dimensional Crystals," *Phys. Rev.* **105,** 425.

REVIEWS OF MODERN PHYSICS VOLUME 20, NUMBER 2 APRIL, 1948

The Approximate Solution of One-Dimensional Wave Equations*

CARL ECKART

University of California, Marine Physical Laboratory, San Diego, California

CONTENTS

INTRODUCTION

THE rigorous and general solution of linear partial differential equations with constant coefficients can be obtained without difficulty, in the form of a Fourier integral.[1,2] However, the very generality of the Fourier integral makes it difficult to interpret in any detail, unless it can be evaluated in terms of elementary functions. The need for a simple and systematic method of approximate evaluation must have been felt very early. The Fresnel method of zones[3] is such an approximation, although the geometric derivation usually associated with it obscures its generality. The abstract reformulation necessary for its wider application was accomplished by Lord Kelvin[4] and has become known as the method of stationary phase. In this form, the close mathematical relation between Fresnel's ideas and those of Debye's method of steepest descent[5] becomes apparent.

The latter has been extensively considered by mathematicians. As a general method for approximating Fourier integrals, it has only one defect: the spectrum function must be susceptible of analytic extension into the complex plane. Formally, this is a very weak restriction; practically, it is a forbidding one, but it may be essential to consider the matter more fully than has been done here.

The Kelvin-Fresnel methods do not operate in the complex plane, and hence are more readily applied to many problems. An excellent summary of the work in this field prior to 1914 has been given by T. H. Havelock.[6] Since that time, wave theory has been very actively studied, under the impetus provided by atomic theory on the one hand, and by radio and sound on the other. Recently, there also has been a marked interest in geophysical wave problems.

One important relationship has become clear

* This work was supported by Contracts N6-ori-111 and NObs-2074 with the Office of Naval Research and the Bureau of Ships, Navy Department.

[1] A. G. Webster, *Partial Differential Equations of Mathematical Physics* (Julius Springer Verlag, Berlin, 1933), p. 44.

[2] H. Bateman, *Partial Differential Equations* (Dover Publications, New York, 1944), Chapter III.

[3] A. Fresnel, *Oeuvres* (Paris, 1866), tome I.

[4] Lord Kelvin, Proc. Roy. Soc. 42, 80 (1887).

[5] P. Debye, Math. Ann. 67, 535 (1908).

[6] T. H. Havelock, *The Propagation of Waves in Dispersive Media* (Cambridge University Press, Teddington, England, 1914).

in the years since Havelock's summary was prepared: the method of stationary phase is also the mathematical foundation for the Hamilton-Jacobi ray theory. The exposition of this relationship will therefore be the principal subject of the following pages.

This review is based largely on notes accumulated over a period of years. Some attempt has been made to give the original sources, but no systematic search of the literature has been made. After the major part of the manuscript had been prepared, an unexpected problem was encountered in applying the method of stationary phase to a dissipative medium. Sections 11, 12, and 13 are devoted to this problem, without, however, reaching definite conclusions.

1. EXAMPLES OF PHYSICAL WAVE EQUATIONS

The various kinds of physical phenomena known as waves are all described by partial differential equations. The simplest of these are the linear equations with constant coefficients, and it is only these that will be discussed here. A further simplification results from the exclusion of equations having more than two independent variables—the time, t, and one space coordinate, x. This restriction is not basic, however.

Waves on a stretched string, sound waves in a tube of uniform cross section, water waves in a shallow uniform canal, and many other phenomena, are governed by an equation of the form

$$(\partial^2\psi/\partial x^2)-(1/c_0^2)(\partial^2\psi/\partial t^2)=0, \qquad (1.1)$$

where c_0 is a constant, while ψ is the dependent variable.

The propagation of electric currents along a uniform wire is governed (1) by the more general equation

$$\frac{\partial^2\psi}{\partial x^2}-\frac{1}{c_0^2}\frac{\partial^2\psi}{\partial t^2}-\frac{2a}{c_0}\frac{\partial\psi}{\partial t}+(k_0^2-a^2)\psi=0, \qquad (1.2)$$

where c_0, a, and k_0 are constants defined as follows:

$$c_0^2=1/LC,$$
$$2a/c_0=RC+LS,$$
$$a^2-k_0^2=RS,$$

C = capacity per unit length, L = inductance per unit length, R = resistance per unit length, S = leakance per unit length. This equation, or special cases of it, also appears in other connections. Diffusion is not usually considered as a wave phenomenon for reasons that will appear below, but is governed by

$$D(\partial^2\psi/\partial x^2)=\partial\psi/\partial t, \qquad (1.3)$$

which is a particular case of Eq. (1.2).

The de Broglie equation for electron waves[7] is

$$\frac{\partial^2\psi}{\partial x^2}-\frac{1}{c_0^2}\frac{\partial^2\psi}{\partial t^2}-\frac{c_0^2}{\alpha^2}\psi=0,$$
$$\alpha=\hbar/m. \qquad (1.4)$$

This equation also governs the motion of long gravitational waves[8] on a rotating disk; c_0^2/α^2 is the Coriolis parameter, and $c_0^2=gh$, where h is the water depth, and g the acceleration of gravity.

Schrödinger's equation is

$$\partial\psi/\partial t=\tfrac{1}{2}i\alpha(\partial^2\psi/\partial x^2), \qquad (1.5)$$

and has a strong formal relationship[9] to that for flexural waves on a rod:

$$\partial^2\psi/\partial t^2=-\tfrac{1}{4}\alpha^2(\partial^4\psi/\partial x^4), \qquad (1.6)$$

although α has a different meaning in the two cases.

All of these equations have special solutions of the form

$$\psi(x,t)=\exp i(kx-nt), \qquad (1.7)$$

for, on substituting this function into any one of them, the result reduces to

$$\psi W(k,n)=0,$$

where W is a different function in each case; and this equation will be satisfied for all x and t if the (real or complex) constants k and n are roots[10] of the characteristic equation.

$$W(k,n)=0. \qquad (1.8)$$

---

[7] P. A. M. Dirac, *Quantum Mechanics* (The Clarendon Press, Oxford, England, 1935), p. 253.

[8] H. U. Sverdrup, "Dynamics of tides on the North Siberian shelf," Geofys. Publikas. 4 (1927).

[9] E. U. Condon and P. M. Morse, *Quantum Mechanics* (McGraw-Hill Book Company, Inc., New York, 1929), p. 26.

[10] H. Bateman, *Partial Differential Equations of Mathematical Physics* (Dover Publications, New York, 1944), p. 101.

Symbolically, the partial differential equations may all be written

$$W\left(\frac{1}{i}\frac{\partial}{\partial x}, \frac{1}{i}\frac{\partial}{\partial t}\right)\psi = 0. \qquad (1.9)$$

In what follows, no particular assumption will be made concerning the function W, although the above examples will be used for illustration.

In these examples W is always a polynomial. In some generalized forms of wave motion, such as surface waves in moderately deep water, W may be a transcendental function of its arguments:

$$W = n^2 - gk \tanh kh, \qquad (1.10)$$

where g is the gravitational acceleration, and h the water depth.[11] Ripples of short wave-length are also influenced by the surface tension, T, and, approximately,

$$W = n^2 - |k|(g + Tk^2/\rho), \qquad (1.11)$$

where ρ is the density of the fluid.[11]

2. THE HAMILTONIAN FUNCTIONS AND THE LOGARITHMIC DECREMENTS

Equation (1.7) may be solved for n as a function of k. Thus, Eq. (1.1) leads to the characteristic equation

$$-k^2 + n^2/c_0^2 = 0,$$

whence $n = \pm c_0 k$. Equation (1.2) leads to

$$-k^2 + n^2/c_0^2 + 2ian/c_0 - a^2 + k_0^2 = 0,$$

or

$$n = c_0(-ia \pm [k^2 - k_0^2]^{\frac{1}{2}}).$$

Equation (1.3) yields $n = iDk^2$. Equation (1.4) leads to $n = \pm[k^2 + k_0^2]^{\frac{1}{2}}$ where $k_0 = c_0/\alpha$, while Eq. (1.5) results in the single-valued function $n = \frac{1}{2}\alpha k^2$ and Eq. (1.6) results in $n = \pm\frac{1}{2}\alpha k^2$.

These examples serve to illustrate two facts: for a given real value of k, there will, in general, be several values of n, and these generally will be complex.

For this reason, it is convenient to replace n by $n - ia$ in all the above formulae, and to write the characteristic equation in the form

$$W(k, n - ia) = 0, \qquad (2.1)$$

where k, n, and a are all real numbers. Then, in general, there will be several single-valued solutions of Eq. (2.1).

$$\begin{aligned} n &= H_1(k), \quad a = D_1(k); \\ n &= H_2(k), \quad a = D_2(k); \\ &\cdots\cdots, \quad \cdots\cdots. \end{aligned} \qquad (2.2)$$

The real functions H_1, H_2, etc., are called the Hamiltonian functions of the wave equation, while the functions D_1, D_2, etc., are called its logarithmic decrements.

The Eq. (1.6) then becomes

$$\psi = \exp\{i[kx - H(k)t] - D(k)t\}, \qquad (2.3)$$

where H and K are any pair of associated functions.

It will be noted that the diffusion equation is characterized by a Hamiltonian that vanishes identically. This is perhaps the first indication of the fundamental role of the Hamiltonian in determining the character of the solutions of partial differential equations. Figure 1 shows

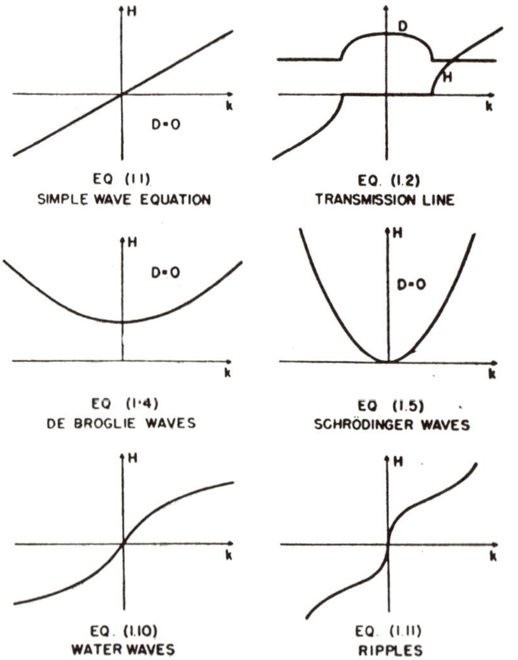

FIG. 1. Graphs of the Hamiltonian function and the related logarithmic decrement for some kinds of wave motion. In general, there will be several pairs of such functions, corresponding to different roots of the equation $W(k, n - ia) = 0$.

---

[11] L. M. Milne-Thomson, *Theoretical Hydrodynamics* (MacMillan and Company, Ltd., London, 1938), p. 376.

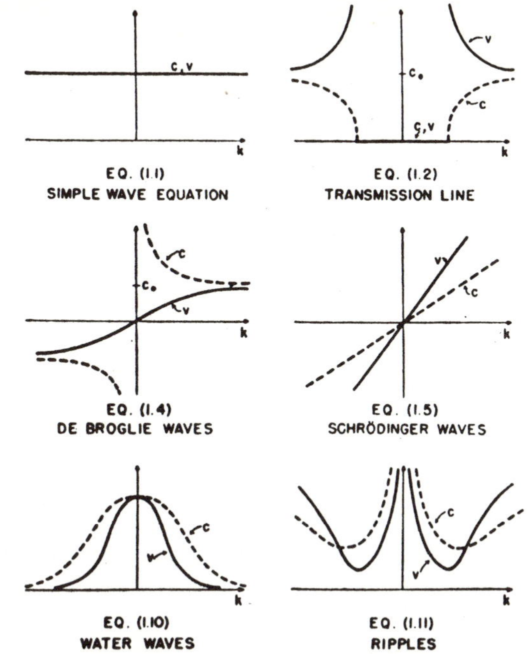

Fig. 2. Graphs of the phase and group velocities for some kinds of wave motion. In general, there will be one pair of these graphs for each Hamiltonian.

one set of H and D functions for some of the wave equations given above.

The real part of the expression on the right of Eq. (2.3) is

$$\exp\{-D(k)t\}\cos\{kx-H(k)t\}.$$

The roots of this quantity are given by

$$kx-H(k)t=(m+\tfrac{1}{2})\pi,\quad m=0,\pm 1,\cdots,$$

or

$$x-c(k)t=\text{constant},$$

where

$$c(k)=H(k)/k. \tag{2.4}$$

The roots, therefore, move with the velocity, $c(k)$, which is called the phase velocity of the waves. Figure 2 shows graphs of $c(k)$ for some of the wave equations given above.

3. THE FOURIER SOLUTION OF THE INITIAL VALUE PROBLEM

Consider the function

$$\psi=\frac{1}{2\pi}\sum\int_{-\infty}^{\infty}F(k)\times\exp\{i[kx-H(k)t]-D(k)t\}dk, \tag{3.1}$$

where the sum is to be extended over all of the Hamiltonian functions. The integrands are clearly solutions of the wave equation. If, moreover, the functions F_1, F_2, etc., which appear in the integrands are such that it is permissible to differentiate under the integral sign a sufficient number of times, ψ itself will satisfy the wave Eq. (1.9).

In the following sections, the multiplicity of Hamiltonians will be ignored as a non-essential complication. For the same reason, the logarithmic decrement will be set equal to zero; both of these complicating features of the Eq. (3.1) can readily be reintroduced at any stage. The further development of the theory will therefore be based on the single integral

$$\psi(x,t)=\frac{1}{2\pi}\int_{-\infty}^{\infty}F(k)\times\exp\{i[kx-H(k)t]\}dk. \tag{3.2}$$

If the time, t, is set equal to zero, Eq. (3.2) becomes

$$\psi(x,0)\equiv\psi_0(x)=\frac{1}{2\pi}\int_{-\infty}^{\infty}F(k)\exp(ikx)dk. \tag{3.3}$$

An application of Fourier's theorem to this equation results in

$$F(k)=\int_{-\infty}^{\infty}\psi_0(x)\exp(-ikx)dx. \tag{3.4}$$

Thus, F is determined by the initial value of ψ, and conversely.**

In general, $F(k)$ will be a complex number, and may be written

$$F(k)=\phi(k)\exp[i\theta(k)], \tag{3.5}$$

where ϕ and θ are real. The function ϕ is called the (amplitude) spectrum of the wave, and θ its eikonal.[12,13] The further abbreviation

$$\Theta(k,x,t)=\theta(k)+kx-H(k)t \tag{3.6}$$

** In deriving Eq. (3.4), it has been explicitely assumed that there is only one Hamiltonian. If there are several, it becomes necessary to complicate the derivation by considering a certain number of the quantities $\partial\psi/\partial t$, $\partial^2\psi/\partial t^2$, $\cdots$; the functions F_1, $F_2\cdots$ are then determined by the initial values of these time derivatives in addition to the initial value of ψ itself.

[12] M. Born, *Optik* (Julius Springer Verlag, Berlin, 1933), Chapter II.

[13] C. Manneback, *Travaux de l'Assemblée Generale de l'Union Radio Scientifique Internationale* (London, 1934, published Brussels, 1935).

is convenient, and reduces Eq. (3.2) to the form

$$\psi(x, t) = \frac{1}{2\pi} \int_{-\infty}^{\infty} \phi(k) \exp[i\Theta(k, x, t)]dk. \quad (3.7)$$

It will be noted that if $\psi_0(x) = \psi_0^*(x)$ (where the asterisk indicates the complex conjugate) Eq. (3.4) shows that F is real, and therefore that the eikonal vanishes identically. Solutions characterized by an eikonal which is constant, or at most a linear function of k, are of special importance in many applications; they are called centered waves. Waves whose initial form is characterized by the above type of symmetry are common in practical problems. It should be noted that centered waves do not necessarily originate from an instantaneous point source, although it will be shown that they have some properties in common with this fictional type.

4. THE METHOD OF STATIONARY PHASE, AND THE EIKONAL

The function Θ is called the phase of the integrand of Eq. (3.7). It should be noted that x and t are constant during the integration, while k is the variable. As k (and Θ) varies, the exponential factor will oscillate, with the period 2π in Θ. If, in any interval on the k axis, Θ increases by many times 2π, while ϕ remains sensibly constant, the contribution of this interval to the value of ψ will be disproportionately small compared to its length. In fact, it can be shown that under certain conditions, the value of the integral is determined almost entirely by the values of the integrand near the stationary points of the phase Θ. (See Section 15.)

These are the values of k that satisfy the

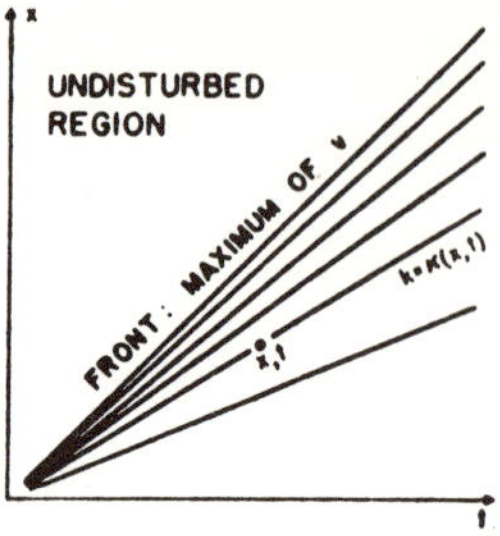

FIG. 4. The $x-t$ diagram for a centered wave, showing the front associated with a maximum of the group velocity.

equation

$$(\partial/\partial k)\Theta(k, x, t) = 0. \quad (4.1)$$

In general, there will be several such values of k:

$$k = \kappa_1(x, t), \quad k = \kappa_2(x, t), \quad \cdots, \quad (4.2)$$

but for the present, it will be supposed that there is only one: $k = \kappa(x, t)$. Near this point, the approximate value of Θ is

$$\Theta(k, x, t) = S(x, t) - \tfrac{1}{2}R(x, t)(k-\kappa)^2, \quad (4.3)$$

where

$$S(x, t) = \Theta[\kappa(x, t), x, t],$$
$$R(x, t) = -\partial^2\Theta(\kappa, x, t)/\partial\kappa^2. \quad (4.4)$$

To a certain approximation, it is permissible to set

$$\phi(k) = \phi(\kappa).$$

Hence, Eq. (3.7) becomes, approximately

$$\psi(x, t) = \phi(\kappa) \exp[iS(x, t)]$$
$$\cdot \frac{1}{2\pi} \int_{-\infty}^{\infty} \exp[-\tfrac{1}{2}iR(k-\kappa)^2]dk. \quad (4.5)$$

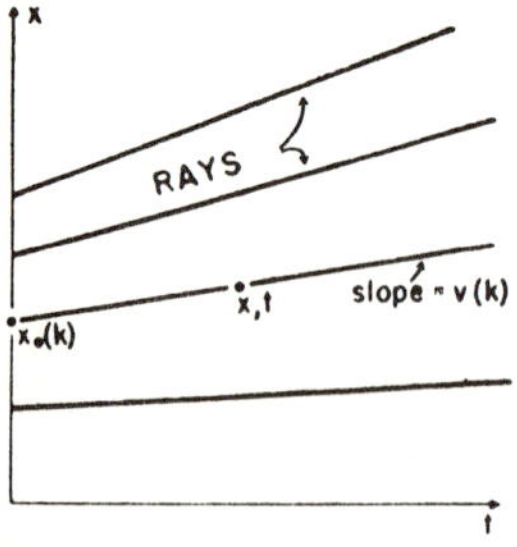

FIG. 3. The rays in the $x-t$ diagram. This diagram is the basis for the graphical method of finding the function $\kappa(x, t)$, described in the text.

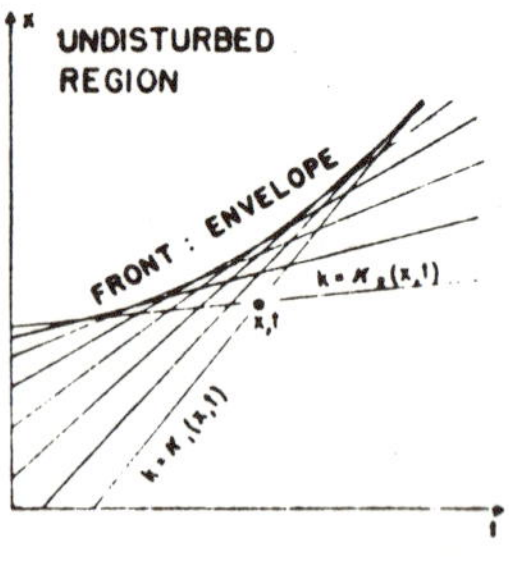

FIG. 5. The $x-t$ diagram for a general wave, showing the front and the two roots, κ_1 and κ_2, of the equation $\partial\Theta/\partial\kappa = 0$, which are associated with an envelope of the family of rays.

The integral in Eq. (4.5) is the complete[14] Fresnel integral:

$$\frac{1}{2\pi}\int_{-\infty}^{\infty}\exp[-\tfrac{1}{2}iRu^2]du=(2\pi iR)^{-\frac{1}{2}}. \quad (4.6)$$

Hence, the approximate value of ψ is

$$\psi(x,t)=[2\pi iR]^{-\frac{1}{2}}\phi(\kappa)\exp[iS]. \quad (4.7)$$

The calculations necessary to estimate the error in this approximation are outlined in Section 15.

If there is more than one root of Eq. (4.1)—i.e., more than one point of stationary phase—the single term on the right of Eq. (4.7) must be replaced by a sum of terms of the same structure. Other methods of treating the case of two stationary points are discussed in Section 7.

If the logarithmic decrement is not zero, Eq. (4.7) is only slightly more complicated:

$$\psi(x,t)=[2\pi iR]^{-\frac{1}{2}}\phi(\kappa)\exp[-D(\kappa)t+iS]. \quad (4.8)$$

5. GROUP VELOCITY AND THE HAMILTON-JACOBI FUNCTION

Using Eq. (3.6), one finds that the condition for stationary phase (Eq. (4.1)) becomes

$$x=x_0(k)+v(k)t, \quad (5.1)$$

where

$$x_0=-d\theta/dk \quad (5.2)$$

and

$$v=dH/dk. \quad (5.3)$$

Equation (5.1) suggests that something starts at $x_0(k)$ at time $t=0$, and travels to x with the velocity $v(k)$. Obviously, this "something" is characterized by the wave number k; it is called, for reasons that will not become clearer below, the "group of waves" having that wave number. The initial coordinate, x_0, is called the origin of the group, and v the group velocity. Equations (5.2) and (5.3) relate these two quantities to the eikonal and the Hamiltonian, respectively. Figure 2 shows graphs of phase and group velocity for the various examples.

The Eq. (5.1) can be made the basis of a graphic solution for $\kappa(x,t)$, as is indicated in Fig. 3. It represents a family of straight lines—the rays—in the x,t plane. Each of these rays is characterized by a particular value of k, and intersects the x axis at $x_0(k)$. Its slope is $v(k)$. To determine $\kappa(x,t)$, one has only to note the value of k for the ray that passes through the point x,t of the diagram.

In the case of a centered wave, x_0 is independent of k—say equal to zero—and the rays all emanate from the origin of the x,t diagram, as shown in Fig. 4. In this case, κ is obtained by solving the equation

$$v(\kappa)=x/t$$

and κ is thus a function of the ratio x/t only.

In those cases where v has a maximum and the wave is centered, there will be a region into which rays penetrate. The ray of greatest speed is called the advancing front. In the same way, the case of a minimum of group velocity results in a receding front.

It is also possible that $v(k)$ has several extrema, in which case even centered waves will have a rather complicated ray diagram and there will be several fronts, some advancing and others receding. An interesting case of this sort has been discussed in detail by C. L. Pekeris.[15] In the following, the discussion will be much simplified by assuming that v has only one extremum and that there is only one front.

Figure 5 also shows a region of the x,t plane that is not reached by any rays. Since there are no real roots of Eq. (5.1) for values of x,t in this region, there will be no points of stationary phase, and the integral of Eq. (3.7) is zero to the present approximation. (See Section 15.) This approximate result is not valid too near the boundary of the undisturbed region, as will be shown in Sections 7 and 8.

It may also happen that even if v has no extremum, several rays pass through a single point of the diagram, as shown in Fig. 5. In this case, Eq. (5.1) will have several roots, $k=\kappa_1(x,t),\cdots$, and Eq. (4.7) will have several terms, as already noted. As is evident from an inspection of Fig. 5, multiple roots of Eq. (5.1) are accompanied by an envelope of the family of rays, and an undisturbed region beyond the envelope. Such fronts

[14] E. Jahnke and F. Emde, *Funktionentafeln mit Formeln und Kurven* (B. G. Teubner, Leipzig and Berlin, 1938; Dover Publications, New York, 1943), p. 35.

[15] C. L. Pekeris, "Theory of propagation of explosive sound in shallow water," Report of the Columbia University Division of War Research, No. 6.1-sr1131-1891, January 1945. (PB 31063).

may be either advancing or receding, but do not have a constant velocity; they will also be discussed further in Section 7.

Turning now to the function

$$S(x, t) = \Theta(\kappa, x, t) \\ = \theta(\kappa) + \kappa x - H(\kappa)t, \qquad (5.4)$$

which is known as the Hamilton-Jacobi function: for, on differentiating Eq. (5.4), one obtains

$$\partial S/\partial x = \kappa + [x - x_0(\kappa) - v(\kappa)t]\partial\kappa/\partial x,$$
$$\partial S/\partial t = -H(\kappa) + [x - x_0(\kappa) - v(\kappa)t]\partial\kappa/\partial t.$$

Because κ is a root of Eq. (5.1), it follows that

$$\partial S/\partial x = \kappa, \quad \partial S/\partial t = -H(\kappa), \qquad (5.5)$$

whence elimination of κ results in the Hamilton-Jacobi[16] equation

$$(\partial S/\partial t) + H(\partial S/\partial x) = 0. \qquad (5.6)$$

This is also variously known as the geometric wave equation, and as the "equation of the characteristics"[10] (not to be confused with the characteristic equation, Eq. (1.8)).

The Hamilton-Jacobi function enters into Eq. (4.7) in the factor

$$\exp(iS).$$

Equation (5.5) thus shows that, near x, t, the disturbance is approximately a sinusoidal wave of frequency $\nu = H(\kappa)$, and wave number κ. The individual wave crest thus moves with the phase velocity

$$(\partial x/\partial t)_S = \nu/\kappa = H(\kappa)/\kappa = c(\kappa) \qquad (5.7)$$

as it passes through the point x, t. The crests are not constantly associated with the same wave number; rather, the wave number is associated with the ray as has already been seen.

Since c is thus expressed as a function of x and t, the space-time locus of an individual wave crest can be obtained by solving the differential equation:

$$dx/dt = c[\kappa(x, t)]. \qquad (5.8)$$

It follows from the above that the solution of this equation will be a curve in the x, t plane, whose equation is $S(x, t) = \text{const}$. In general, this curve cuts across many rays. This result has a simple physical interpretation: the individual crest does not move with a constant velocity, nor does the separation between successive crests remain constant during their motion. This separation between successive crests is appropriately called the wave-length, λ, and varies from place to place and from time to time. For any given pair of crests, it is given approximately (but only approximately) by the equation

$$\lambda = 2\pi/\kappa(x, t), \qquad (5.9)$$

and thus varies as they move across the rays. This approximation is valid only when the value of κ does not change appreciably from one crest to the next.

The amplitude of the wave crests is (at least partially) determined by the spectrum amplitude $\phi(\kappa)$. If the attention be focused not on the motion of the individual wave crests, but on the motion of the point where the amplitude ϕ, or what is the same thing, the wave number κ, has a given value, another velocity will be obtained. This is

$$(\partial x/\partial t)_\phi = -(\partial\phi/\partial t)/(\partial\phi/\partial x), \\ = -(\partial\kappa/\partial t)/(\partial\kappa/\partial x).$$

But

$$\frac{\partial\kappa}{\partial t} = \frac{\partial^2 S}{\partial x\partial t} = -\frac{\partial}{\partial x}H(\kappa) = -v(\kappa)\frac{\partial\kappa}{\partial x};$$

hence,

$$(\partial x/\partial t)_\phi = v(\kappa). \qquad (5.10)$$

The solution of this equation is a straight line in the x, t diagram, as is obvious from Figs. 2 to 4: κ is constant along any ray, and therefore also ϕ. Moreover, the slope of the ray passing through x, t is $v(\kappa)$.

These considerations lead to the following description of the disturbance: the individual crests move with a variable velocity and amplitude. Both are determined by the ray which the crest is momentarily crossing; the wave-length and one factor, ϕ, in the expression for the amplitude are the same at all points of a given ray. The motion of the individual crests is along the curves $S = \text{const}$. The slope of any member of this family, at the point where it crosses the ray characterized by the wave number κ, is $c(\kappa)$. There are thus two families of curves in the x, t diagram: the rays, which are straight lines, and the crest paths, all of which intersect a given ray at the same angle.

[16] N. Born, *Atommechanik* (Julius Springer Verlag, Leipzig, 1925).

If several rays pass through a given point of the x, t diagram, there will be several trains of waves passing through it simultaneously, each moving according to its own appropriate laws. The interference of these various trains may result in complicated patterns. This will be considered in more detail in Section 7.

Another consequence of Eq. (5.5) is that

$$\partial\kappa/\partial t = \partial^2 S/\partial x\partial t = -(\partial/\partial x)H(\kappa) = -\partial\nu/\partial x$$

or

$$(\partial\kappa/\partial t) + (\partial\nu/\partial x) = 0, \qquad (5.11)$$

which is an equation derived by Rossby[17] and Munk[18] from elementary considerations concerning the motion of wave crests.

Finally, it should be noted that $S(x, t)$ satisfies the Hamilton-Jacobi partial differential equation, no matter what form the eikonal $\theta(k)$ may be given, provided only that κ is a root of Eq. (5.1). Thus, a by-product of these considerations is a graphical method of solving partial differential equations of the first-order and constant coefficients.

6. THE RESOLUTION OF THE WAVES INTO A SPECTRUM

Equation (4.7) shows that the amplitude of the waves at x and t is determined directly by the spectrum ϕ and by the function R, but only indirectly by the initial value of the wave function ψ. Obviously, this cannot be true for all values of the time. At instants immediately following $t=0$, the function $\psi(x, t)$ must differ only slightly from its initial value $\psi_0(x)$. This immediately suggests that the approximate value of ψ yielded by the method of stationary phase will be accurate only for large values of t.

Physically, the reason for this is to be found in the different velocity $v(k)$ with which the various Fourier components are propagated. When it is necessary to consider many components as being present at a given time and place, the extremum of the phase function Θ will be very shallow and the above approximation will not be good. Only when the differing velocity of the Fourier components has dispersed the disturbance into a well-resolved spectrum will the above approximation be an accurate one.

The mathematics of this physical argument is best approached by a study of the function R. Using Eqs. (3.6), (4.4), (5.2), and (5.3), it follows that

$$R = -\partial^2\Theta/\partial\kappa^2,$$

$$= (dx_0/d\kappa) + t(dv/d\kappa), \qquad (6.1)$$

$$= (\partial x/\partial\kappa)_t, \qquad (6.2)$$

$$= 1/(\partial\kappa/\partial x)_t. \qquad (6.3)$$

Of these three alternative forms, Eq. (6.2) is the most enlightening. Referring to Fig. 4, it is easily seen that $\partial x/\partial\kappa$ will be large where the rays are widely spaced (for given increments of k or κ) and small where they are poorly resolved. Since the amplitude is inversely proportional to $R^{\frac{1}{2}}$, it follows that the wave disturbance is "spread thin" and has a small amplitude at those places where the space occupied by a given spectral interval is large. The function R may be appropriately called the resolution of the wave spectrum.

Equation (6.1) may also be written

$$R = R_0 + \alpha t, \qquad (6.4)$$

where

$$\alpha = dv(\kappa)/d\kappa$$

$$= d^2H(\kappa)/d\kappa^2 \qquad (6.5)$$

will be recognized as the coefficient of dispersion of the medium. Since α and R_0 are functions of κ only, they will both be constant along a given ray. Thus, the resolution of a given part of the spectrum increases linearly with time from an initial value, at the rate α. For sufficiently large values of t, therefore, $|R|$ must always be great; for small values of t, its magnitude will depend on R_0. These considerations are also obvious from a consideration of the rays in the x, t diagram.

7. WAVE FRONTS

Since α and R_0 may have opposite signs, there may come a time when $R=0$ for a given ray. Reference to Fig. 5 and the interpretation $R=\partial x/\partial\kappa$, shows that this will occur at the envelope of the family of rays, if there is such an envelope. Equation (4.7) is clearly not a good

[17] C. G. Rossby, *On the Propagation of Frequencies and Energy in Certain Types of Oceanic and Atmospheric Waves*, J. Meteorol. **2**, No. 4, 187 (1945).

[18] W. H. Munk, *Increase in the Period of Waves Traveling Over Large Distances, with Applications to Tsunamis, Swell and Seismic Surface Waves*, Trans. Am. Geophys. Union, April 1947.

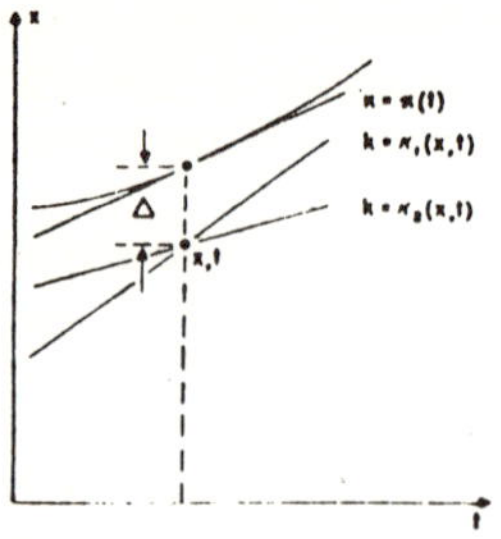

FIG. 6. Notation for the extrema and the point of inflection of Θ, which are associated with an envelope (cf. Figs. 5 and 7).

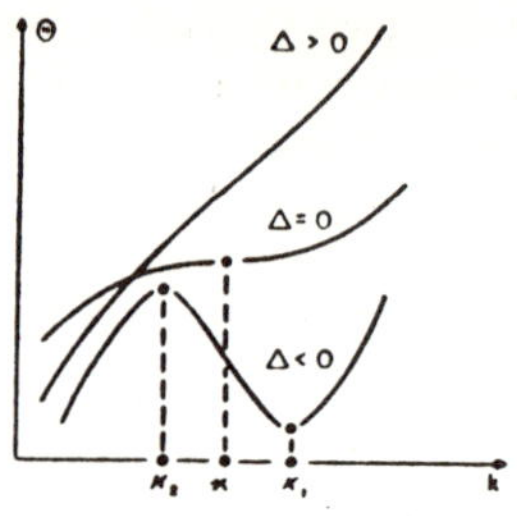

FIG. 7. Schematic graphs of Θ as a function of k for various values of Δ.

approximation to ψ at the wave front, since it yields the absurd value $\psi=\infty$ when $R=0$. It thus becomes necessary to develop a more suitable approximation for this region.

The front is a curve in the x, t plane which separates it into two regions: one is the disturbed region, in which at least two rays pass through each point, the other is the undisturbed region, in which there are no rays. This is the typical description, but more elaborate cases are possible. Only this typical case will be treated here; for the moment it will also be supposed that, at the wave front, κ is not infinite.

The details of the situation are illustrated in Figs. 6 and 7. Let Δ be the distance from the momentary position of the front to the point x, t, counted positive when x is in the undisturbed region. Figure 7 shows schematic graphs of the phase function Θ for $\Delta>0$, $\Delta=0$, and $\Delta<0$. When $\Delta>0$, no ray passes through x, t, and correspondingly, the graph of Θ has no stationary point. When $\Delta<0$, two rays pass through x, t, and Θ, correspondingly, has one maximum and one minimum at $k=\kappa_1$ and $k=\kappa_2$. For $\Delta=0$, Θ has a minimax for $k=K$.

In the previous sections, Θ has been approximated by a quadratic in k, in the neighborhood of its stationary point. Figure 7 makes it obvious that for small values of Δ it will be better to approximate it as a cubic in k.

Let K be the value of k for which Θ has a point of inflection, so that K is a root of

$$-(\partial^2\Theta/\partial k^2)=R_0(k)+\alpha(k)t=0. \qquad (7.1)$$

From this equation, or from Fig. 6, it is clear that K is a function of t only. The cubic approximation to Θ is then

$$\begin{aligned}\Theta(k,x,t)&=\Theta(K,x,t)+(k-K)(\partial\Theta/\partial K)\\&\quad+\tfrac{1}{3}!(k-K)^3(\partial^3\Theta/\partial K^3),\\&=\theta(K)+Kx-H(K)t\\&\quad+(k-K)[x-x_0(K)-v(K)t]\\&\quad-\tfrac{1}{3}!(k-K)[R_0'(K)+\alpha'(K)t]\\&=S_0(x,t)+(k-K)\Delta(x,t)\\&\quad+\tfrac{1}{3}(k-K)^3\delta^3(t),\end{aligned} \qquad (7.2)$$

where S_0, Δ, and δ are obvious abbreviations. It will be noted that $\Delta(x,t)$ is the distance of x from the momentary position of the wave front as already defined. S_0 is formally similar to the Hamilton-Jacobi function, but does not satisfy the same partial differential equation. The real quantity

$$\delta(t)=-\{\tfrac{1}{2}[R_0'(K)+\alpha'(K)t]\}^{\frac{1}{3}} \qquad (7.3)$$

is a scale factor which, it will be seen, determines the steepness of the front.

Arguments similar to those which lead to Eq. (4.5) now lead to the approximate expression

$$\psi(x,t)=\phi(K)\exp[iS_0]$$
$$\cdot\frac{1}{2\pi}\int_{-\infty}^{\infty}\exp\{i[(k-K)\Delta+\tfrac{1}{3}(k-K)^3\delta^3]\}dk. \qquad (7.4)$$

The integral which appears in this equation is essentially Airy's integral:

$$\mathrm{Ai}(x)=\frac{1}{2\pi}\int_{-\infty}^{\infty}\exp[i(xu+\tfrac{1}{3}u^3)]du, \qquad (7.5)$$

so that Eq. (7.4) becomes

$$\psi(x,t)=(\phi/\delta)\mathrm{Ai}(\Delta/\delta)\exp(iS_0). \qquad (7.6)$$

The Airy function can be evaluated[19] in terms of Bessel functions of order $\frac{1}{3}$. Another definition of the function has been discussed by Kramers[20] in connection with the "WKB method" without apparently noting the identity of the function with Airy's. For sufficiently large positive values of ξ, the approximations

$$\mathrm{Ai}(\xi)=\tfrac{1}{2}\pi^{-\frac{1}{2}}\frac{\exp(-\frac{2}{3}\xi^{\frac{3}{2}})}{\xi^{\frac{1}{4}}}, \tag{7.7}$$

$$\mathrm{Ai}(-\xi)=\tfrac{1}{2}\pi^{-\frac{1}{2}}\frac{\cos(\frac{2}{3}\xi^{\frac{3}{2}}-\frac{1}{4}\pi)}{\xi^{\frac{1}{4}}}, \tag{7.8}$$

are valid. Figure 8 has been prepared using these equations to supplement a recently published table[21] of this function for small absolute values of the argument; ordinates for values of $x<-2.4$ may be considerably in error because of the approximate nature of Eq. (7.8). Eight-place tables have been prepared by the BAAS and are being published as this is written.[22]

8. WAVE FRONTS OF CONSTANT VELOCITY, FIRST KIND

It is clear from Fig. 5 that the envelope of the family of rays will be, in general, a curve, so that the velocity of the wave front will not be constant. There is one limiting case, however, in which the envelope becomes a straight line and the wave front moves with constant velocity. This arises when the group velocity has an extremum for some finite value of k—say k_0—and the wave is centered. In the case of water waves governed by Eq. (1.10), the velocity is a maximum for $k_0=0$. Ripples governed by Eq. (1.11) have a minimum group velocity for k_0 given by Eq. (8.9) below.

For a centered wave

$$\Theta=kx-H(k)t, \tag{8.1}$$

and, therefore,

$$R=\alpha(k)t, \tag{8.2}$$

so that $R=0$ only at $t=0$, or at some value of k for which the dispersion $\alpha=dv/dk=0$. The former case is to be considered in Section 10, so that only the latter will be treated here. Letting $c_0=c(k_0)$, $v_0=v(k_0)$, etc., it is seen that

$$S_0=k_0(x-c_0t), \tag{8.3}$$

$$\Delta=x-v_0t, \tag{8.4}$$

$$\delta=-\{\tfrac{1}{2}\alpha_0't\}^{\frac{1}{3}}. \tag{8.5}$$

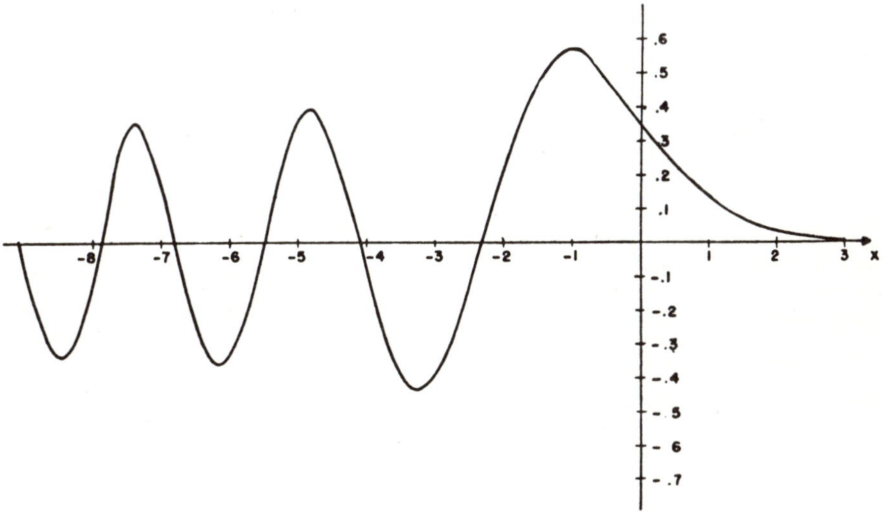

FIG. 8. Graph of the Airy function, Ai(x), defined by Eq. (7.5).

[19] G. N. Watson, *A Treatise on the Theory of Bessel Functions* (Cambridge University Press, New York, 1944), pp. 188, 201, 659.
[20] H. A. Kramers, Zeits. f. Physik **39**, 828 (1926).
[21] P. M. Woodward, A. M. Woodward, R. Hensman, H. Davies, and N. Gamble, Phil. Mag. **37**, 236 (1946).
[22] *Mathematical Tables*, Volume B (to be published for the British Association for the Advancement of Science at the University Press, Cambridge).

Hence the disturbance in the neighborhood of the wave front will be given by

$$\psi=(\phi_0/\delta)\mathrm{Ai}[(x-v_0t)/\delta]\cdot\exp[ik_0(x-c_0t)]. \quad (8.6)$$

The significant point to be noted is that the scale factor δ increases as the cube root of the time.

This is of special significance in case $k_0=0$: for then the "wave-length" of the disturbance will be determined by the oscillations of the Airy integral, and will be determined empirically by measuring the interval between its maxima. The first two of these intervals are

$$L_1=3.8\delta,$$
$$L_2=2.6\delta.$$

Thus the "wave-length" near the front will increase as the cube root of t. In the case of long surface waves, Eq. (1.9) yields

$$\alpha_0'=-g^{\frac{1}{2}}h^{5/2}=-v_0h^2,$$

so that

$$\delta=\{\tfrac{1}{2}h^2v_0t\}^{\frac{1}{3}}>0, \quad \text{for } t>0. \quad (8.7)$$

The situation near this front is very peculiar. It may be described as a carrier wave which is amplitude-modulated. However, contrary to the customary case, the carrier has a longer (infinitely longer!) wave-length than the modulation. Consequently, the empirically determined "wave-length" bears no relation whatever to the spectrum of the disturbance. Such wave fronts are of considerable importance as they constitute the seismic sea waves, commonly called "tidal waves." Observations of these are in qualitative,[18] and possibly quantitative, agreement with the present calculations.

The case of ripples governed by Eq. (1.11) is somewhat more normal. For these, the minimum of the group velocity is at

$$k_0=0.393(g\rho/T)^{\frac{1}{2}}. \quad (8.8)$$

The corresponding velocities are

$$v_0=1.086(gT/\rho)^{\frac{1}{4}}, \quad (8.9)$$

$$c_0=1.713(gT/\rho)^{\frac{1}{4}}, \quad (8.10)$$

while $\delta<0$ (receding front), where

$$-k_0\delta=0.768(gt/v_0)^{\frac{1}{3}}. \quad (8.11)$$

When t is much greater than v_0/g, the wave-length L_1 will be greater than $2\pi/k_0$, and the modulation will have a greater wave-length than the carrier. However, for small values of t the situation is similar to that of the gravitational waves. The question of the extent to which these formulae are applicable for small t will be considered in Section 10.

9. WAVE FRONTS OF CONSTANT VELOCITY, SECOND KIND

While the ray diagram of Fig. 4 for centered waves applies to the Eqs. (1.2) and (1.4), the previous analytical procedures do not. This is because the limiting value of the group velocity is reached for $k=\infty$, and it is not possible to use Taylor's series in this neighborhood. Instead of a cubic approximation to Θ, the expansion

$$H(k)=c_0k+b/k+\cdots \quad (9.1)$$

is valid in this neighborhood. In the case of a centered wave, $\theta=0$, so that approximately

$$\psi(x,t)=\frac{1}{2\pi}\int_{-\infty}^{\infty}\phi(k)\times\exp[ik(x-c_0t)-ibt/k]dk \quad (9.2)$$

when $|x-c_0t|$ is sufficiently small.

The value of this integral depends strongly on the behavior of ϕ for large values of k. Only the special case

$$\phi(k)\to(ik)^{-n-1} \quad (9.3)$$

will be considered. Then the approximation

$$\psi(x,t)=\frac{1}{2\pi}\int(ik)^{-n-1}\times\exp[ik(x-c_0t)-ibt/k]dk \quad (9.4)$$

will be valid. However, because $(ik)^{-n-1}$ becomes infinite for $k=0$, the path of integration must be deformed into the complex k plane, so as to pass above the origin.

The integral of Eq. (9.4) can then be evaluated in terms of Bessel functions. The substitution

$$u=k/bt,$$
$$\xi=bt(x-c_0t)$$

reduces Eq. (9.4) to

$$\psi=(bt)^{-n}f(\xi), \quad (9.5)$$

where

$$f(\xi)=\frac{1}{2\pi}\int(iu)^{-n-1}\exp[i(u\xi-1/u)]du. \quad (9.6)$$

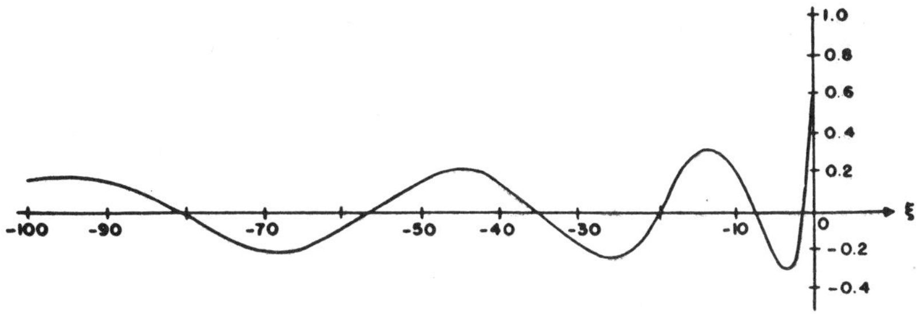

FIG. 9. Graph of the function $J_0(2\xi^{\frac{1}{2}})$.

The rigorous formula

$$f(\xi)=0, \quad \text{where } \xi>0 \tag{9.7}$$

can be proven readily. For negative values of the argument it can be shown that

$$f(-\xi)=\xi^{n/2}J_n[2(\xi)^{\frac{1}{2}}], \tag{9.8}$$

where J_n is Bessel's function of order n. Graphs of these functions for $n=0$ and 1 are shown in Figs. 9 and 10.

10. UNRESOLVED WAVES

There are circumstances under which none of the methods described above will lead to a suitable approximation. The precise delineation of these conditions will be omitted, but by now their general nature should be fairly clear. In each of the three approximate methods, it has been assumed that $\phi(k)$ may be treated as a constant, while $\Theta(k, x, t)$ is approximated by a polynomial in k. In other words, it has been assumed that the principal variations in the integrand are caused by Θ, those caused by ϕ being either non-existent or else canceling out in the end result, so that they may be ignored from the beginning.

In the present section, it will be assumed that $\phi(k)$ has a sharp maximum at $k=k_0$. In this case, it is fairly certain that its variation will dominate over that of Θ.

It is convenient to return to Eq. (3.2),

$$\psi(x, t)=\tfrac{1}{2}\pi\int_{-\infty}^{\infty} F(k)\exp[ikx-iH(k)t]dk,$$

or the following calculation: since $\phi=|F|$ has a maximum at $k=k_0$, the Hamiltonian may profitably be expanded by Taylor's theorem near this point:

$$\begin{aligned}H(k)&=H(k_0)+H'(k_0)(k-k_0)\\&\quad+\tfrac{1}{2}H''(k_0)(k-k_0)^2+\cdots,\\&=c_0k_0+v_0(k-k_0)+\tfrac{1}{2}\alpha_0(k-k_0)^2+\cdots,\end{aligned} \tag{10.1}$$

where c_0, v_0, and α_0 have meanings that is obvious from the previous discussions. For the first, it is convenient to neglect the quadratic term; doing this and substituting in Eq. (3.2),

$$\psi(x, t)=\exp[ik_0t(v_0-c_0)] \cdot\tfrac{1}{2}\pi\int_{-\infty}^{\infty} F(k)\exp[ik(x-v_0t)]dk. \tag{10.2}$$

Comparing this with Eq. (3.3),

$$\psi_0(x)=\tfrac{1}{2}\pi\int_{-\infty}^{\infty} F(k)\exp(ikx)dk,$$

it is seen that

$$\psi(x, t)=\psi_0(x-v_0t)\exp[ik_0t(v_0-c_0)]. \tag{10.3}$$

This remarkably simple result can be put in a more familiar form, if we write

$$\psi_0(x)=\Psi(x)\exp ik_0x, \tag{10.4}$$

where, by definition, Ψ is the modulation, and $\exp ik_0x$ is the carrier wave. It can be shown that (when the maximum of ϕ is sufficiently sharp) Ψ

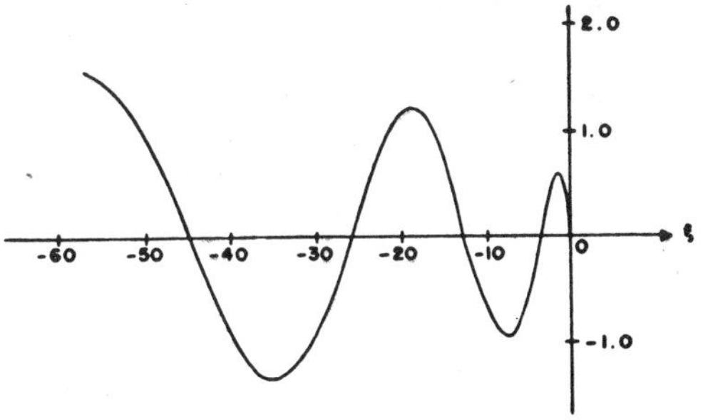

FIG. 10. Graph of the function $\xi^{\frac{1}{2}}J_1(2\xi^{\frac{1}{2}})$.

is a slowly varying function of x in comparison to $\exp ik_0x$. Then Eq. (10.3) becomes

$$\psi(x, t) = \Psi(x - v_0t) \exp[ik_0(x - c_0t)]. \quad (10.5)$$

Thus, under these conditions, we recover the well-known result that the carrier is propagated with the phase velocity, while the modulation is propagated with the group velocity and without distortion.

Equation (10.5) obviously results because the dispersion, α_0, has been neglected in this calculation. If the quadratic terms are included in Eq. (10.1), Eq. (3.2) becomes ($\xi = x - v_0t$):

$$\psi(x, t) = \exp[ik_0(x - c_0t)] \cdot \tfrac{1}{2}\pi \int_{-\infty}^{\infty} F(k) \times \exp\{i(k - k_0)\xi - (i\alpha_0/2)t(k - k_0)^2\} dk$$

$$= \Psi(\xi, t) \exp[ik_0(x - c_0t)]. \quad (10.6)$$

The function $\Psi(\xi, t)$ remains to be studied more carefully; clearly

$$\Psi(\xi, 0) = \psi_0(\xi), \quad (10.7)$$

but its changes with time must be investigated.

A change of variable reduces Ψ to the form

$$\Psi(\xi, t) = \tfrac{1}{2}\pi \int_{-\infty}^{\infty} F(k + k_0) \times \exp[ik\xi - (i\alpha_0/2)tk^2] dk, \quad (10.8)$$

and the general principles of Section 3 then show that it satisfies a wave equation in ξ and t, with the Hamiltonian $\frac{1}{2}\alpha_0k^2$. This uniquely determines the differential equation

$$\partial\Psi/\partial t = (i\alpha_0/2)(\partial^2\Psi/\partial\xi^2), \quad (10.9)$$

which is, except for the replacement of x by ξ and α by α_0, identical with Schrödinger's equation (Eq. (1.5)).

Equations (10.7) and (10.9) serve to determine Ψ completely, but the present assumptions concerning F make it necessary to find a new method of calculating it approximately. In case Ψ is an analytic function of t, the Taylor expansion,

$$\Psi(\xi, t) = \Psi(\xi, 0) + t(\partial\Psi/\partial t)_{t=0} + \tfrac{1}{2}t^2(\partial^2\Psi/\partial t^2)_{t=0} + \cdots,$$

may be used. But, because of Eq. (10.9),

$$(\partial^n\Psi/\partial t^n)_{t=0} = (i\alpha_0/2)^n(\partial^{2n}\psi_0/\partial\xi^{2n}).$$

Hence, the expansion

$$\Psi(\xi, t) = \psi_0(\xi) + (i\alpha_0t/2)(d^2\psi_0/d\xi^2) + \tfrac{1}{2}(i\alpha_0t/2)^2(d^4\psi_0/d\xi^4) + \cdots \quad (10.10)$$

is obtained. For sufficiently small values of α_0t, it often converges very rapidly; for large values of α_0t, the approximations on which it is based fail, and it is useless to attempt to transform it into a more rapidly convergent series. Fortunately, it is clear that for very large values of α_0t the variation of Θ will again dominate over that of ϕ, and one of the previous approximations can be used.

Consequently, Eq. (10.10) should be considered merely as indicating how the process of distortion begins. The different phase velocities of the Fourier components do not have appreciable effect until the ratio

$$\alpha_0t(d^2\psi_0/d\xi^2)/\psi_0$$

becomes appreciable. When it is comparable to unity, the original modulation will have been distorted, probably beyond recognition. When this ratio is much greater than unity, the divergence of the rays in the x, t diagram will have produced a complete resolution of the wave into its periodic components.

The considerations of this section have neglected the possible effects of dissipation. These can be included approximately by multiplying Eq. (10.6) by the factor $\exp[-D(k_0)t]$.

11. THE BOUNDARY VALUE PROBLEM

In the general wave equation

$$W\left(\frac{1}{i}\frac{\partial}{\partial x}, -\frac{1}{i}\frac{\partial}{\partial t}\right)\psi = 0, \quad (1.8)$$

the variables x and $-t$ enter in a symmetric manner. This symmetry is lost in Fourier's solution of the initial value problem:

$$\psi(x, t) = \tfrac{1}{2}\pi \int_{-\infty}^{\infty} F(k) \exp[i(kx - H(k)t)] dk, \quad (3.2)$$

$$F(k) = \int_{-\infty}^{\infty} \psi_0(x) \exp(-ikx) dx. \quad (3.3)$$

This problem can be formulated as follows:

Initial Value Problem: Given that at the initial instant $t = 0$, $\psi = \psi_0(x)$, and that it satisfies Eq. (1.8) for all later times, find $\psi(x, t)$ for all $t > 0$.

Because of the symmetry of Eq. (1.8), there is a second problem, which is of fundamental importance in transmission line theory. It reads:

Boundary Value Problem: Given that at the boundary point $x=0$, $\psi=\psi_1(t)$, and that it satisfies Eq. (1.8) for all points to the right of the boundary, find $\psi(x, t)$ for all $x>0$.

Having solved the initial value problem, the boundary value problem presents no new mathematical difficulties. Its approximate solutions can be obtained from the foregoing pages by replacing x by $-t$ and conversely.

However, there are certain consequences of this procedure that do not appear to have been noticed,*** so that it is profitable to examine the first stages of the boundary value problem in some detail. The symmetry of x and $-t$ has not yet been lost in Eq. (1.6): the special solutions

$$\psi=\exp(ikx-int), \tag{1.6}$$

$$W(k, n)=0, \tag{1.7}$$

are equally useful in the boundary value problem.

The next step in the initial value problem was the solution of this last equation for n as a function of k. In the boundary value problem, this is reversed, and k is expressed as a function of n. If there are no dissipative terms, it follows that

$$k=G(n) \tag{11.1}$$

and that G will have the same central importance as the Hamiltonian function had.

When there is dissipation, this step would be preceded by replacing

$$W(k, n-ia)=0 \tag{2.1}$$

by the equation

$$W(k+ib, n)=0, \tag{11.2}$$

and solving for the real quantities k and b as functions of the real variable n:

$$k=G(n), \quad b=A(n).$$

The function $A(n)$ is the attenuation and is analogous to the logarithmic decrement $D(k)$ of the initial value problem.

The case $A(n)=0$, $D(k)=0$ is of such fundamental importance that physicists have studied it almost to the exclusion of the more general one.

*** They were called to the writer's attention by Dr. R. W. Raitt, after the preceding sections had been written.

This case may be illustrated by

$$W(k, n)=-k^2+n^2/c_0^2$$

or

$$H(k)=c_0k, \quad G(n)=n/c_0.$$

Then G is the inverse function to H, and G and H have the same graph in the k, n plane. In this (and in more general cases of the same sort)

$$c=H(k)/k=n/G(n),$$

and

$$v=dH/dk=1/(dG/dn),$$

always provided that $W(k, n)=0$.

Physically, these last two results mean that the same phase and group velocities govern the propagation of the waves, regardless of whether one solves the initial or the boundary value problems. This numerical equality of two quantities with mathematically different definitions is the justification for ascribing a physical reality to them and calling them "it" rather than "they." Moreover, it follows that the family of rays is the same for the two problems. In the one, the rays are specified by their intercept on the x axis, in the other, by their intercept on the t axis, but the ray passing through a given point has the same slope.

All of this is altered when there are dissipative terms. In the k, n diagram, the equations $n=H(k)$ and $k=G(n)$ have different graphs; in the x, t diagram there are two families of rays. This casts serious doubt on the validity of the previous results in this case, for a given solution of the wave equation may be uniquely specified either by its initial or by its boundary values. The method of stationary phase will yield different approximations in the two cases; in general, at least one of these will be poor.

12. FORMAL THEORY OF MULTIPLE HAMILTONIANS†

The difficulty just discussed is partly inherent in the differential equations, but results mainly from the use of the approximate method of stationary phase, as will be shown in the next section. The formal aspects of the problem are related to the multiplicity of Hamiltonians encountered in solving a general partial differ-

† This section contains unpublished material supplied by Professor Marcel Riesz of the University of Lund; it is a pleasure to acknowledge his kind permission to use it in this review.

ential equation. This multiplicity was noted explicitly by Dirac,[23] and led to the prediction of the positron. In terms of the ray diagram, it leads to several families of rays in the x, t diagram, each given by Eqs. (5.1) to (5.3) with their appropriate Hamiltonians. Physically, it leads to the appearance of several superposed trains of waves, although the converse is not true. A single Hamiltonian can also give rise to several trains if several rays pass through a single point (cf. Fig. 5).

The mathematical theory of these matters is radically different in the cases of conservative and dissipative media. The former will be considered first. Then the characteristic equation, Eq. (1.8), may be factored into

$$W(k, n) = P \prod_{\alpha=1}^{N} [H_\alpha(k) - n], \qquad (12.1)$$

where the Hamiltonians H_α are distinct real functions of k. The quantity P may be a function of k and/or n, but, by definition, has no roots other than trivial ones. Often W will be a polynomial, of degree N in n and degree K in k; then P will be a constant. Only this case will be discussed here.

There will then be N sets of equations similar to Eqs. (5.1) to (5.3), and any one of them may be written

$$\frac{dx}{dt} = \frac{\partial H_\beta(k)}{\partial k}, \quad \frac{dk}{dt} = \frac{\partial H_\beta(k)}{\partial x} = 0. \qquad (12.2)$$

Each of these sets gives rise to a family of rays, so that there are N of these.

Now consider the canonical equations

$$\begin{aligned} dx/d\tau &= \partial W/\partial k, & dk/d\tau &= -\partial W/\partial x = 0, \\ dn/d\tau &= \partial W/\partial t = 0, & dt/d\tau &= -\partial W/\partial n, \end{aligned} \qquad (12.3)$$

whose interpretation will shortly appear. The function W will be called the comprehensive Hamiltonian. They have the integral $W(k, n) = \text{constant}$, and this constant may have any real value whatever; comparing this with Eq. (12.1) it is seen that only those solutions for which it is zero (the null lines of the comprehensive Hamiltonian) will be of interest. It will be shown that the family of null lines consists of all N families of rays, and no others.

The first and last of the Eqs. (12.3) may be written

$$\begin{aligned} \frac{dx}{d\tau} &= \sum_{\alpha=1}^{N} P_\alpha \frac{\partial H_\alpha}{\partial k}, \\ \frac{dt}{d\tau} &= \sum_{\alpha=1}^{N} P_\alpha, \end{aligned} \qquad (12.4)$$

where

$$P_\alpha(kn) = W(k, n)/[H_\alpha(k, n) - n]. \qquad (12.5)$$

When Eq. (12.1) is satisfied, one and only one of the factors $H_\alpha - n$ will vanish; if this is the one for $\alpha = \beta$, all of the P_α except P_β will vanish, and Eqs. (12.4) reduce to

$$\begin{aligned} dx/d\tau &= P_\beta(\partial H_\beta/\partial k), \\ dt/d\tau &= P_\beta; \end{aligned} \qquad (12.6)$$

eliminating the parameter τ, these reduce exactly to Eq. (12.2).

Thus far, the discussion has been directed toward the ray theory appropriate to the initial value problem. Turning now to the boundary value problem in a conservative medium, Eq. (11.1) implies that Eq. (1.8) can also be factored into

$$W(k, n) = Q \sum_{\alpha=1}^{K} [G_\alpha(n) - k], \qquad (12.7)$$

where K is not necessarily equal to N, and Q will again be constant. Proceeding as before, the quantities

$$Q_\beta = W(k, n)/[G_\beta(n) - k] \qquad (12.8)$$

are defined. Because of Eq. (12.1), all but one of them vanish, and hence Eqs. (12.3) reduce to

$$\begin{aligned} dx/d\tau &= Q_\beta(k, n), \\ dt/d\tau &= Q_\beta(k, n)(\partial G_\beta/\partial n), \end{aligned} \qquad (12.9)$$

or, finally to

$$dt/dx = \partial G_\beta/\partial n, \qquad (12.10)$$

in keeping with the alternative formula for the group velocity.

It has thus been shown that the multiplicity of ray families introduced in connection with the Fourier solutions are all determined by a single set of Hamiltonian equations, Eqs. (12.1) and (12.3). But this is the case only for a conservative

[23] P. A. M. Dirac, *Quantum Mechanics* (Clarendon Press, Oxford, 1935), second edition, p. 252.

medium; for a dissipative medium, $W(k, n)$ assumes complex values for real values of k, n, and cannot be resolved into real factors.†† Formally, Eqs. (12.1), (12.3), and (12.10) may still be set up, but their solutions will lead to complex values of x and t. It is not clear that their solutions have any relevance to the Fourier solution of the wave equation.

Rather, this solution leads to the separation of the complex Hamiltonian into H_α and D_α or into G_α and A_α as discussed in Sections 2 and 11. While this is decidedly relevant to the Fourier solution, it makes Eqs. (12.2), (12.3), and (12.10) entirely distinct and unrelated. The curves (in the $x-t$ diagram) determined by each set of equations are different, and those determined by Eq. (12.3) have complex values of the coordinates.

13. ARBITRARINESS OF THE METHOD OF STATIONARY PHASE

Apart from the formal difficulties of treating dissipative media in the same manner as conservative, there are certain difficulties common to the two cases, which are inherent in the method of stationary phase. The mathematical discussion can best be based on the integral

$$I=\int_a^b \exp[w(z)]dz, \tag{13.1}$$

where w is a function of the complex variable $z=x+iy$, analytic at all points of a region, Γ,

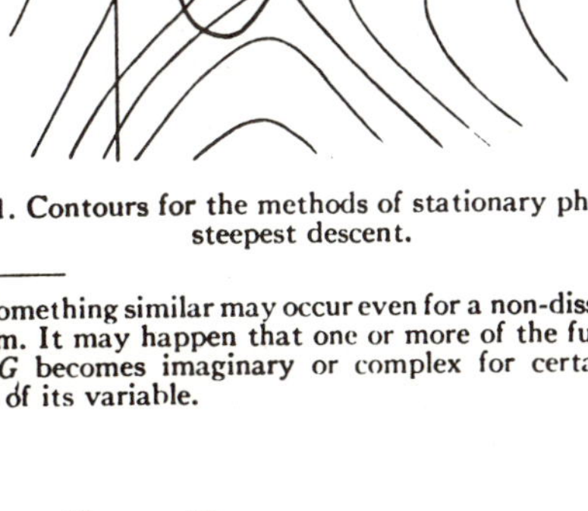

FIG. 11. Contours for the methods of stationary phase and steepest descent.

†† Something similar may occur even for a non-dissipative medium. It may happen that one or more of the functions H or G becomes imaginary or complex for certain real values of its variable.

that includes the segment ab of the x axis. Setting $w(z)=u(x, y)+iv(x, y)$ where u and v are real, the function w may be graphically represented by plotting the contours $v=$const. as shown by the light curves on Fig. 11. The contours $u=$const. intersect the plotted curves orthogonally.

In Eq. (13.1), the path of integration may be along the real axis, but the value of I is the same for any path that runs from a to b and remains entirely within the region Γ. The method of stationary phase can be applied to any one of these paths, and will ordinarily result in a different *approximate* expression for I in each case, despite the fact that the *exact* value of I is independent of the path.

The reason why the difficulty did not obtrude itself in the case of waves in a non-dissipative medium now becomes clear: then the initial value and boundary value problems lead to integrals that can be transformed into each other by a real change of variable. In the dissipative case, the necessary change of variable becomes complex, and must be accompanied by a deformation of the path of integration.

It is of interest to investigate these various approximate values of I more closely. Let a general path of integration be specified by giving x and y as functions of s, the distance along the curve. Then the angle, $\alpha(s)$, between the tangent to the curve and the x axis is given by

$$dx/ds=\cos\alpha, \quad dy/ds=\sin\alpha, \quad dz/ds=\exp(i\alpha), \tag{13.2}$$

and hence

$$I=\int_{s_a}^{s_b} \exp[u+i(v+\alpha)]ds. \tag{13.3}$$

The phase of the integrand is $v+\alpha$, and its stationary point (if any) is determined by the equation

$$\partial v/\partial x \cos\alpha+\partial v/\partial y \sin\alpha+d\alpha/ds=0. \tag{13.4}$$

The geometric interpretation of this equation is simplest if, at the stationary point, the path is a straight line: then it states that the path is tangent to that v-contour which passes through the stationary point. This is shown by the path 1 in Fig. 11; the result could also have been obtained by inspection.

Let $x=\xi$, $y=\eta$, $s=s_0$ be the stationary point in this path; then the expression of the phase near this point will be

$$v=v(\xi,\eta)+\alpha_0-(s-s_0)^2/2r^2+\cdots, \quad (13.5)$$

where, as a consequence of the Cauchy-Riemann equations,

$$r(\xi,\eta)=[(\partial^2u/\partial\xi\partial\eta)\cos 2\alpha_0 - (\partial^2v/\partial\xi\partial\eta)\sin 2\alpha_0]^{-\frac{1}{2}}. \quad (13.6)$$

Hence, the method of stationary phase yields the approximation

$$I=(2\pi)^{\frac{1}{2}}r(\xi,\eta) \times\exp[w(\xi+i\eta)+i(\alpha_0-\pi/4)]. \quad (13.7)$$

The disconcerting fact is that the point ξ, η may be chosen arbitrarily in Γ, and the path of integration then deformed so that ξ, η becomes the point of stationary phase. Thus, one is left with a double infinity of approximate values of I, each of which usually differs from every other. Under such conditions an investigation is indicated to determine which of the many approximate values is most accurate. Such an investigation is beyond the scope of this review, but even if it were carried out, it would still leave the physicist with an awkward but interesting problem.

It has been shown above that the Hamilton-Jacobi theory of waves, inclusive of the concept of group velocity, is based on a single one of these approximations. It is unlikely that this one will always be the best. Can the Hamilton-Jacobi theory be extended so as to yield a better approximation to the solutions of the wave equation? It appears likely that this can be done, but if so, it will modify the concept of group velocity in a peculiar manner. It has been customary to consider it to be a property of the medium, and the same for all disturbances in that medium. In such an extended Hamilton-Jacobi theory, the group velocity will be strongly dependent on the nature of the disturbance.

14. THE METHOD OF STEEPEST DESCENT

There is yet another approximate value of I, which is obtained by choosing the path of integration so that it passes through the saddle point S, Fig. 11, if such a saddle point exists. The coordinates of this point are the roots of

$$\partial v/\partial x=0, \quad \partial v/\partial y=0. \quad (14.1)$$

If the path is straight at S, then Eq. (13.4) will be satisfied for all values of α. Moreover, because of the Cauchy-Riemann equations, one will also have

$$\partial u/\partial x=0, \quad \partial u/\partial y=0 \quad (14.2)$$

at S. Hence, in addition to Eqs. (13.5) and (13.6), one will also have

$$u=u(\xi,\eta)-(s-s_0)^2/2r_1^2+\cdots, \quad (14.3)$$

where

$$r_1=[-(\partial^2v/\partial\xi\partial\eta)\cos 2\alpha_0 - (\partial^2u/\partial\xi\partial\eta)\sin 2\alpha_0]^{-\frac{1}{2}}. \quad (14.4)$$

The value of α_0 can be chosen so as to simplify the remaining calculations. If it is chosen so that $1/r=0$, the value of r_1, becomes

$$r_1=r_0=\{[\partial^2v/\partial\xi\partial\eta]^2+[\partial^2u/\partial\xi\partial\eta]^2\}^{\frac{1}{4}}. \quad (14.5)$$

This is equivalent to requiring that α_0 be chosen so as to make $(1/r_1)$ a maximum. Hence this procedure is usually called the method of steepest descent[5] since it makes the maximum of u as sharp as possible.

The corresponding approximate value of I is

$$I=(2\pi)^{\frac{1}{2}}r_0(\xi\eta)\exp[w(\xi+i\eta)+i\alpha_0]. \quad (14.6)$$

The value of α_0 may also be chosen so that $1/r_1=0$; this value differs from the previous by $\pi/4$, and makes $r=r_0$, so that Eq. (12.7) becomes identical with Eq. (13.6). The method of steepest descent thus appears as a singular case of the method of stationary phase. It is likely that this singular case will often yield a better approximation than any of the others, but, it cannot be used as the basis for the Hamilton-Jacobi theory unless the saddle point happens to be on the real axis.

15. THE REMAINDER IN THE METHOD OF STATIONARY PHASE

The earliest work on this problem is again Fresnel's justification of his method of zones. Unfortunately, the widely current accounts of his argument leave much to be desired. They almost invariably involve the same reasoning by which it was "proven" that the hare cannot overtake the tortoise. There are more recent studies of the problem[24–26] which are more satisfactory.

In order to reduce the problem to its simplest mathematical terms, consider the integral

$$I(p)=\int_a^b \phi(k)\exp[ip\theta(k)]dk, \qquad (15.1)$$

where ϕ and θ are real functions of k, and p is a real parameter. In the foregoing applications the phase of the integrand was a linear function of the two parameters x and t, and it was seen that the approximations resulting from the method of stationary phase were valid only when these parameters were such as to make the phase a rapidly varying function of k. This last phrase is not very precise, nor is it easy to reword it precisely. Therefore, the parameter p has been introduced to simplify the problem: it is required to find an expansion for I in descending powers of p, and to investigate the remainder of this series after n terms, assuming that ϕ and θ are finite, continuous and adequately differentiable in the interval a, b.

The simplest case arises when θ has no extrema in a, b. Then a change of variable reduces the integral to

$$I(p)=\int_\alpha^\beta f(\theta)\exp(ip\theta)d\theta, \qquad (15.2)$$

where $\alpha=\theta(a)$, $\beta=\theta(b)$, and

$$f(\theta)=\phi(k)/\theta'(k). \qquad (15.3)$$

Because of our assumption, $f(\theta)$ will also be sufficiently regular for the following procedures, whereas it would become infinite if $\theta'(k)$ were to vanish somewhere in the interval of integration.

An integration by parts reduces Eq. (14.2) to

$$I(p)=(1/ip)[f(\beta)e^{i\beta}-f(\alpha)e^{i\alpha}]+i\mathfrak{R}_1/p, \qquad (15.4)$$

where

$$\mathfrak{R}_1=\int_\alpha^\beta f'(\theta)\exp[ip\theta]d\theta. \qquad (15.5)$$

An iteration of this process leads to

$$\mathfrak{R}_1=(1/ip)[f'(\beta)e^{i\beta}-f'(\alpha)e^{i\alpha}]+i\mathfrak{R}_2/p,$$

where

$$\mathfrak{R}_2=\int_\alpha^\beta f''(\theta)\exp[ip\theta]d\theta.$$

Thus, the problem of expansion is solved, except for the trouble involved in differentiating Eq. (15.3) with respect to θ. In particular, it is seen that the series begins with a term proportional to $1/p$.

In order to discuss the remainders, it will be sufficient for the present purposes to deal only with $\mathfrak{R}_1$, since it is quite typical. The definition of an integral results in the inequality

$$|\mathfrak{R}_1|\leqslant\int_\alpha^\beta |f'(\theta)|d\theta, \qquad (15.6)$$

which shows that the extrema of the function f are again the central elements of the problem. If $f(\theta)$ has no extrema in the interval of integration, the value of the integral is simply $|f(\beta)-f(\alpha)|$. However, if f has one extremum, at $\theta=\theta_1$, the value of the integral increases to $|f(\beta)-f(\theta_1)|+|f(\theta_1)-f(\alpha)|$, and if it has m extrema, there will be $m+1$ such terms.

In fact, the integral in Eq. (14.6) is essentially a measure of the extent to which the function f varies or oscillates in the interval α to β. This concept is adumbrated by such phrases as "f is a slowly varying function," etc. In order to be able to use it conveniently, the notation

$$\int_\alpha^\beta |f'(\theta)|d\theta=Os[f(\theta),\alpha,\beta] \qquad (15.7)$$

will be introduced. Hence the remainder after n terms satisfies the inequality

$$|\mathfrak{R}_n|/p^n\leqslant(1/p^n)Os[f^{(n-1)}(\theta),\alpha,\beta]. \qquad (15.8)$$

The situation is materially altered if the phase has a stationary point in the interval of integration, for then the oscillation of $f(\theta)$ becomes infinite. Let $k=c$, $\theta(c)=\gamma$ be an extremum of $\theta(k)$ and the only stationary point in the interval

[24] L. Brillouin, Ann. Ecole Normale 2, 33 (1916).
[25] Van der Corput, Compositio Mathematica 1, 15 (1934); 3, 328 (1936).
[26] J. Bijl, Dissertation, Groningen, 1937.

a, b; then Eq. (15.1) can be written

$$I(p)=\phi(c)\int_a^b \exp[ip\theta(k)]dk$$
$$+\int_\gamma^\beta f(\theta)\exp[ip\theta]d\theta$$
$$+\int_\alpha^\gamma f(\theta)\exp[ip\theta]d\theta, \quad (15.9)$$

where it is to be remembered that k is a two-valued function of θ, and where now

$$f(\theta)=[\phi(k)-\phi(c)]/\theta'(k). \quad (15.10)$$

The integrands of the second and third integrals are then sufficiently regular so that the preceding results can be applied to them at once: each has an expansion whose first term is proportional to $1/p$, and the remainder after n terms is subject to Eq. (15.8). It is, therefore, only necessary to discuss the integral

$$I_0(p)=\int_a^b \exp[ip\theta(k)]dk. \quad (15.11)$$

A new variable ξ, which is a single-valued function of k, is defined by the equation

$$\theta(k)=\theta(c)+\tfrac{1}{2}\theta''(c)\xi^2. \quad (15.12)$$

(If c were a minimax, $\theta''(c)$ would be zero, and this and the following equations would have to be altered in a manner analogous to the considerations of Section 7.) It is easily seen that, for $k=c$, $\xi=0$ and $d\xi/dk=1$, so that

$$I_0(p)=\exp[ip\theta(c)]$$
$$\times\int_{\xi_a}^{\xi_b}\exp[(ip/2)\theta''(c)\xi^2]d\xi$$
$$+\exp[ip\theta(c)]\int_{\xi_a}^{\xi_b}((d\xi/dk)-1)$$
$$\times\exp[(ip/2)\theta''(c)\xi^2]d\xi. \quad (15.13)$$

The first integral is the incomplete Fresnel integral,[14] and requires no further discussion here. For large values of ξ_b and $-\xi_a$ it approaches the complete Fresnel integral in a known manner:

$$\int_{\xi_a}^{\xi_b}\exp[(ip/2)\theta''(c)\xi^2]d\xi$$
$$\rightarrow[2\pi i/p\theta''(c)]^{\frac{1}{2}}. \quad (15.14)$$

Hence, the expansion of this integral will have its first term proportional to $1/p^{\frac{1}{2}}$ rather than to $1/p$, as has been the case with the others. The second integral is conveniently treated by changing to the variable

$$\eta=\tfrac{1}{2}\theta''(c)\xi^2, \quad (15.15)$$

which reduces it to two integrals of the same type as the last two in Eq. (15.9). Its expansion therefore begins with a term proportional to $1/p$.

In principle, the results of this section should make it possible to find the answers to the mathematical questions raised in Section 13. However, the analytical complications are so great that it is to be hoped that some more elegant treatment can be found.

Quantum Mechanics of Electrons in Crystal Lattices.

By R. de L. Kronig and W. G. Penney, University of Groningen.

(Communicated by R. H. Fowler, F.R.S.—Received November 13, 1930.)

Introduction.—Through the work of Bloch our understanding of the behaviour of electrons in crystal lattices has been much advanced. The principal idea of Bloch's theory is the assumption that the interaction of a given electron with the other particles of the lattice may be replaced in first approximation by a periodic field of potential. With this model an interpretation of the specific heat,* the electrical and thermal conductivity,† the magnetic susceptibility,‡ the Hall effect,§ and the optical properties|| of metals could be obtained. The advantages and limitations inherent in the assumption of Bloch will be much the same as those encountered when replacing the interaction of the electrons in an atom by a suitable central shielding of the nuclear field, as in the work of Thomas and Hartree.

* Bloch, 'Z. Physik,' vol. 52, p. 555 (1928).

† Bloch, 'Z. Physik,' vol. 52, p. 555 (1928), vol. 53, p. 216 (1929), and vol. 59, p. 208 (1930); Peierls, 'Ann. Physik,' vol. 4, p. 121 (1930), and vol. 5, p. 244 (1930).

‡ Bloch, 'Z. Physik,' vol. 53, p. 216 (1929).

§ Peierls, 'Z. Physik,' vol. 53, p. 255 (1929).

|| Kronig, 'Proc. Roy. Soc.,' A. vol. 124, p. 409 (1929).

In the papers quoted a number of general results were given regarding the behaviour of electrons in any periodic field of potential. To obtain a clearer idea of the details of this behaviour with a view to the application in special problems, however, it appeared worth while to investigate the mechanics of electrons in periodic fields of potential somewhat similar to those met with in practice and of such nature that the energy values W and eigenfunctions ψ of the wave-equation can actually be computed. It is the purpose of this article to discuss a case where the integration is possible. In Section 1 the energy values and in Section 2 the wave-functions in their dependence on the binding introduced by the potential field are discussed for the one dimensional problem. In Section 3 the matrix elements of the linear momentum, which furnish the electric current associated with the various stationary states as well as the probability of radiative transitions between these states, are evaluated. In Section 4 the results are extended to the three dimensional case and those features considered which one may expect to find in the case of more general periodic fields of potential. Section 5 deals with some applications to physical problems.

1. *The Energy Values.*—The potential field which we shall consider is essentially that shown in fig. 1. Later on we shall pass to the limit $b = 0$

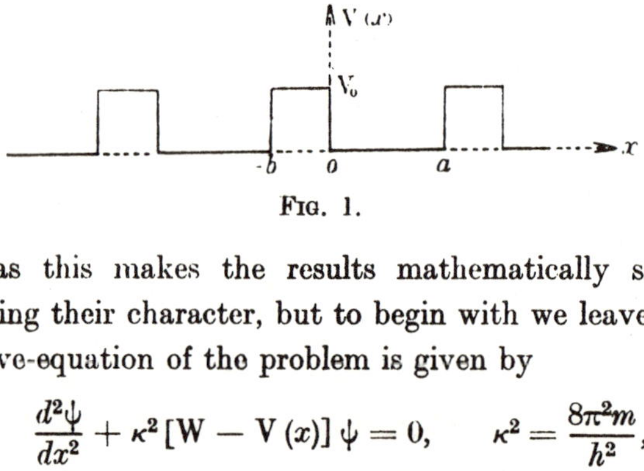

FIG. 1.

and $V_0 = \infty$ as this makes the results mathematically simpler without essentially altering their character, but to begin with we leave both b and V_0 finite. The wave-equation of the problem is given by

$$\frac{d^2\psi}{dx^2} + \kappa^2 [W - V(x)] \psi = 0, \qquad \kappa^2 = \frac{8\pi^2 m}{h^2}, \tag{1}$$

and following Bloch* we enquire after the solutions ψ which are periodic over a distance $L = G(a + b)$, where G is a large integer. As he has shown, these solutions must be of the form

$$\psi(x) = u(x) e^{i\alpha x}, \qquad \alpha = 2\pi k/L, \tag{2}$$

where k is any integer and $u(x)$ a function periodic in x with the period $(a + b)$. To find $u(x)$ we can hence confine ourselves to one period of the lattice, say

* Bloch, 'Z. Physik,' vol. 52, p. 555 (1928).

from $x = -b$ to $x = a$, and moreover, since we later shall make $V_0 = \infty$, we may assume $0 < W < V_0$. Substituting the expression (2) in equation (1) imposes on u in the range $-b \leqq x \leqq 0$ the condition

$$\frac{d^2u}{dx^2} + 2i\alpha \frac{du}{dx} - (\alpha^2 + \gamma^2)\, u = 0$$

with the solution

$$u = Ae^{(-i\alpha+\gamma)x} + Be^{(-i\alpha-\gamma)x}, \tag{3}$$

while in the region $0 \leqq x \leqq a$ we have

$$\frac{d^2u}{dx^2} + 2i\alpha \frac{du}{dx} - (\alpha^2 - \beta^2)\, u = 0$$

with the solution

$$u = Ce^{i(-\alpha+\beta)x} + De^{i(-\alpha-\beta)x} \tag{4}$$

Here β and γ are given by

$$\beta = \kappa\sqrt{W}, \qquad \gamma = \kappa\sqrt{V_0 - W}. \tag{5}$$

In particular β and γ are real quantities and may be assumed positive without loss of generality.

The constants A, B, C, D are to be so chosen that the solutions in the two regions have the same value and the same first derivative for $x = 0$, while from the periodic nature of u it follows that the solution (3) and its first derivative for $x = -b$ shall be equal respectively to the solution (4) and its first derivative for $x = a$. The four linear homogeneous equations resulting for the constants from these requirements are

$$\begin{aligned}
A+B &= C+D,\\
(-i\alpha+\gamma)\,A + (-i\alpha-\gamma)\,B &= i(-\alpha+\beta)C + i(-\alpha-\beta)D,\\
Ae^{(i\alpha-\gamma)b} + Be^{(i\alpha+\gamma)b} &= Ce^{i(-\alpha+\beta)a} + De^{i(-\alpha-\beta)a}\\
(-i\alpha+\gamma)Ae^{(i\alpha-\gamma)b} + (-i\alpha-\gamma)\,Be^{(i\alpha+\gamma)b} &= i(-\alpha+\beta)\,Ce^{i(-\alpha+\beta)a}\\
&\quad + i(-\alpha-\beta)\,De^{i(-\alpha-\beta)a}
\end{aligned}$$

and can be satisfied only if the quantities β and γ, which according to equation (5) are directly related to the energy value W, satisfy the relation

$$\frac{\gamma^2 - \beta^2}{2\beta\gamma} \sinh \gamma b \sin \beta a + \cosh \gamma b \cos \beta a = \cos \alpha\,(a + b).$$

Passing now to the limit where $b = 0$ and $V_0 = \infty$ in such a way that $\gamma^2 b$ stays finite and calling

$$\lim_{\substack{b \to 0 \\ \gamma \to \infty}} \frac{\gamma^2 ab}{2} = P,$$

this relation becomes

$$P \sin \beta a / \beta a + \cos \beta a = \cos \alpha a, \tag{6}$$

a transcendental equation for βa.

To discuss the roots of this equation we have plotted its left-hand side as a function of βa in fig. 2, assuming for P the value $3\pi/2$. The values of βa

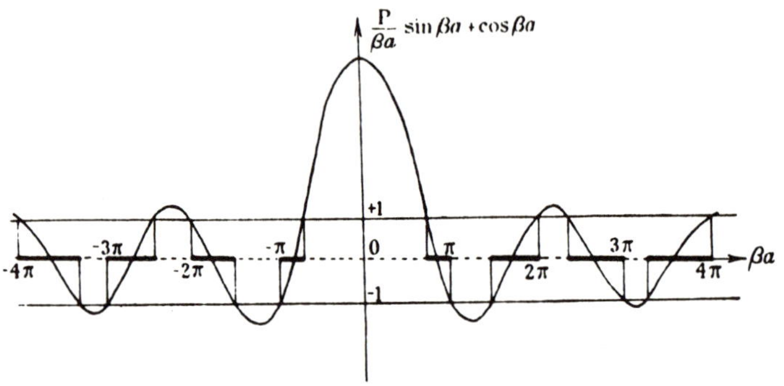

FIG. 2.

satisfying equation (6) are obtained as the projections on the βa-axis of the intersections of the curve with a straight line drawn at the distance $\cos \alpha a$ parallel to this axis. Since $\cos \alpha a$ lies between -1 and $+1$, and since the ordinates of the maxima of the curve have an absolute magnitude greater than 1, then, upon varying α by giving k different values, the curve is found to consist of the portions between the parallels ± 1 on which there lie intersections and the remaining portions outside these parallels on which there lie no intersections. By projection the βa-axis too is divided into portions containing permissible values of βa (drawn heavily in fig. 2) and portions not containing such values. Letting L approach infinity allows us according to equation (2) to vary α continuously, and the permissible values of βa will then fill the heavily drawn portions of the βa-axis continuously. *The energy values which an electron moving through the lattice may have, hence form a spectrum consisting of continuous pieces separated by finite intervals.*

It is of interest to study the influence of P upon this spectrum. If P vanishes, the curve of fig. 2 goes over into $\cos \beta a$, the forbidden intervals of the βa-axis disappear, and we have one continuous spectrum of all energy values from 0 to ∞. This is the limiting case of *free electrons*. Increasing P we obtain the forbidden intervals, the ratio of the length of these intervals to that of the adjoining allowed portions decreasing as we pass to large values of βa. We may say that fast electrons have less trouble to pass the potential barriers than slow electrons and can be considered more nearly as free. Letting P approach infinity reduces the allowed portions of the βa-axis to the points $n\pi$ ($n = \pm 1, \pm 2, \ldots$). The energy spectrum now becomes discrete, the energy values

$$W = n^2h^2/8ma^2, \qquad (n = 1, 2, \ldots)$$

being those of an electron confined to move between two impenetrable potential barriers at a distance a apart. *The electrons* are caught between the potential walls, they *have become bound.*

Fig. 3 shows the change in the energy spectrum during the transition from the case of free to that of bound electrons. In the region $0 \leq P \leq 4\pi$, $P/4\pi$ has been chosen as abscissa, while for $4\pi \leq P \leq \infty$ $4\pi/P$ is used. The shaded area represents the allowed values of $(\beta a/\pi)^2$.

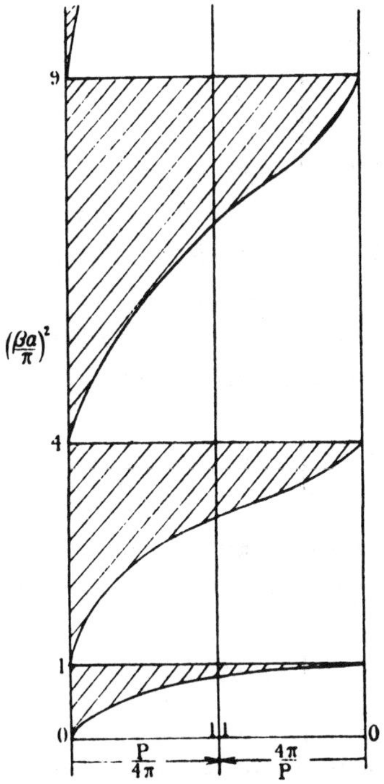

FIG. 3.

2. *The Wave-functions.*—We come now to consider the *wave-functions* of our problem. Since b in the limit is reduced to zero, we are only concerned with the solution (4). Solving the linear equations for the constants A, B, C, D in this limiting case gives us

$$D = -C\frac{1 - e^{-i(\alpha-\beta)a}}{1 - e^{-i(\alpha+\beta)a}}. \qquad (7)$$

u is given by equation (4) in the cell extending from 0 to a, while in the cell extending from ra to $(r + 1)a$ it will be given by

$$u = Ce^{i(-\alpha+\beta)(x-ra)} + De^{i(-\alpha-\beta)(x-ra)},$$

being periodic with the period a. According to equation (2) ψ in this cell will then be given by

$$\psi = Ce^{i\beta x + ir(\alpha-\beta)a} + De^{-i\beta x + ir(\alpha+\beta)a}, \qquad (8)$$

with C and D related by equation (7). If we require that ψ be normalised over a distance equal to its period L, we must have

$$CC^* = \frac{1}{2L}\frac{1 - \cos(\alpha + \beta)a}{1 - \cos\alpha a\cos\beta a + \frac{1}{P}(\cos\alpha a - \cos\beta a)^2}, \qquad (9)$$

the asterisk denoting the conjugate.

From equation (6) it appears that if with a given permissible value of βa a certain value of αa satisfies this equation, then also $\alpha a + 2n\pi$ and $-\alpha a + 2n\pi$ (n an integer) will satisfy it. We see, however, from equations (7) and (8) that by substituting $\alpha a + 2n\pi$ for αa we do not obtain any new wave-functions, while substituting $-\alpha a + 2n\pi$ for αa we get the same wave-function as that belonging to $-\beta a$, $-\alpha a$. We may hence, without loss of generality, accept the convention that if $n\pi \leqq \beta a \leqq (n+1)\pi$, we shall associate with it that value of αa for which $n\pi \leqq \alpha a \leqq (n+1)\pi$. In this way there is associated with every permissible value of βa one and only one value of αa, and we may use α to distinguish the stationary states.

We shall next investigate the influence of the value of P on the wave-functions. If we make P vanishingly small, then according to the convention just introduced βa will equal αa, and according to equation (7) D will vanish excepting for $\alpha a = n\pi$, in which case $D = \pm C$. The wave-functions according to equation (8) will go over into $e^{\pm i\alpha x}$ excepting for the special case just mentioned, in which they become equal to sin αx and cos αx so that we obtain the *wave-functions of free electrons.* At the same time the significance of α becomes apparent. For P approaching infinity, β takes the values $n\pi$, $D = -C$ according to equation (7), and we get from equation (8) the *wave-functions of electrons confined in their motion between impenetrable potential walls* at a distance a apart.

If we let L increase, the energy values, as mentioned before, come closer together, and for very large values of L we may enquire after the *density distribution of the energy values* in the allowed regions of βa. According to equation (2) this is evidently given by

$$\rho(\beta a) = \frac{L}{2\pi}\frac{d\alpha}{d\beta} = \frac{L}{2\pi \sin \alpha a}\left(\frac{P}{(\beta a)^2}\sin \beta a - \frac{P}{\beta a}\cos \beta a + \sin \beta a\right)$$

with $\rho(\beta a)$ so normalised that

$$\int_{(a\beta)\,\mathrm{min.}}^{(\beta a)\,\mathrm{max.}} \rho(\beta a)\, d\beta a = \frac{L}{2},$$

the integration extending over any one of the allowed regions of βa. Fig. 4 shows $\rho(\beta a)/L$ in the first of these for a value $P = 3\pi/2$, while for free electrons ($P = 0$) one obtains instead a horizontal line with an ordinate $1/2\pi$ for all values of βa. The binding has thus the effect of concentrating the stationary states at the limits of an allowed region.

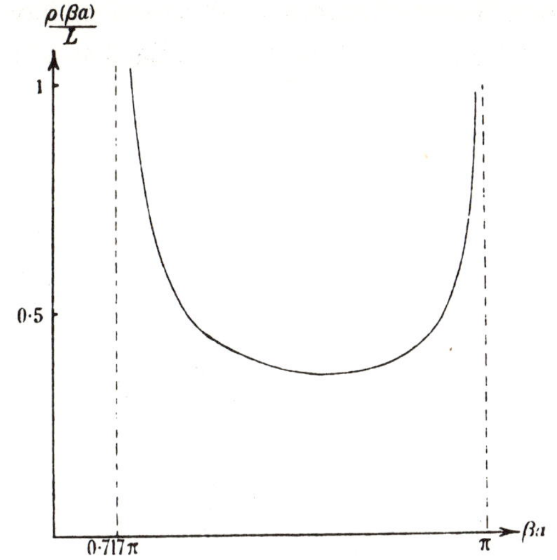

FIG. 4.

3. *The Linear Momentum.*—According to quantum mechanics the matrix elements of the linear momentum of an electron belonging to two stationary states α and α' are given by

$$p\,(\alpha, \alpha') = \frac{h}{2\pi i}\int \psi_\alpha{}^* \frac{\partial \psi_{\alpha'}}{\partial x}\,dx,$$

the integration extending over one period L of the functions ψ_α and $\psi_{\alpha'}$. Introducing for ψ_α and $\psi_{\alpha'}$ their expressions given by equations (7), (8), (9) and performing the integration it is found that $p\,(\alpha, \alpha')$ vanishes unless αa and $\alpha' a$ differ by an integral multiple of 2π. Thus if the state α lies in the first allowed region of the positive βa-axis, then $p\,(\alpha, \alpha')$ is different from zero for only one state α' in the first, one in the third, one in the fifth positive allowed region, etc., and one in the second, one in the fourth, one in the sixth negative allowed region, etc., viz., those states for which $\alpha' a = \alpha a + 2n\pi$. A simple calculation gives us for the square of the absolute value of these non-vanishing matrix elements :

$$p\,(\alpha, \alpha')\,p\,(\alpha', \alpha) = \left(\frac{h}{\pi a}\,\frac{\beta\beta'}{\beta'^2 - \beta^2}\right)^2$$

$$\times \frac{\sin^2 \alpha a\,(\cos \beta a - \cos \beta' a)^2}{\left[1 - \cos \alpha a \cos \beta a + \frac{1}{P}(\cos \alpha a - \cos \beta a)^2\right]\left[1 - \cos \alpha a \cos \beta' a + \frac{1}{P}(\cos \alpha a - \cos \beta' a)^2\right]} \quad (10)$$

To make the significance of equation (10) clearer we have computed from it in the first place $\pi a p\,(\alpha, \alpha)/h$ and plotted it in fig. 5 as a function of βa for the

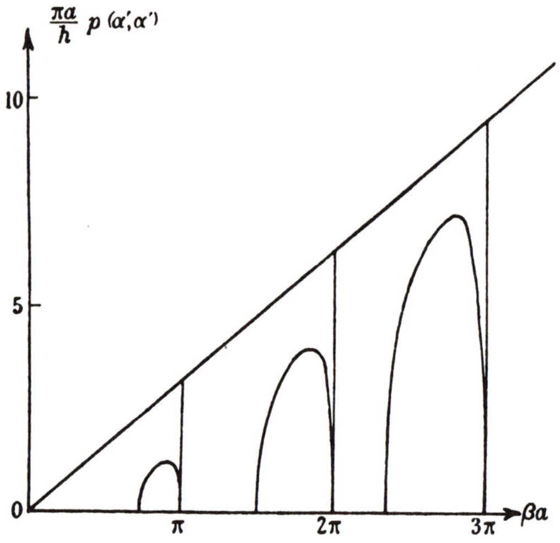

FIG. 5.

values of βa lying in the first, second and third allowed regions of the positive βa-axis, taking again P equal to $3\pi/2$. $p\,(\alpha, \alpha)$ represents the *time average of the linear momentum* of an electron in the stationary state α. As one easily sees

$$p\,(-\alpha, -\alpha) = -\,p\,(\alpha, \alpha)\,.$$

Since the energy of the stationary states α and $-\alpha$ is the same, one may enquire if to a linear combination of the wave-functions ψ_α and $\psi_{-\alpha}$ there might correspond a linear momentum of greater absolute value, but from a simple calculation one finds that p lies then between $p\,(-\alpha, -\alpha)$ and $p\,(\alpha, \alpha)$. The straight line going through the origin in fig. 5 gives us $\pi a p\,(\alpha, \alpha)/h$ for the case of free electrons, showing that the potential barriers have the effect of reducing the linear momentum for a given value of the energy. Indeed they make it vanish when P approaches infinity. $p\,(\alpha, \alpha)$ also furnishes us the *electric current* associated with the stationary state α, whose time average is obtained by multiplying $p\,(\alpha, \alpha)$ by e/m, the ratio of the charge of the electron to its mass.

Furthermore we have plotted in fig. 6 the values of $\pi^2 a^2 p\,(\alpha, \alpha')\,p\,(\alpha', \alpha)/h^2$ as a function of βa when α is a state in the first allowed region while α' is that state in the second negative, third positive and fourth negative allowed region (curves 1, 2 and 3 respectively) for which the above quantity is different from zero. Again P has been taken equal to $3\pi/2$. For vanishing P all the curves are reduced to zero. For infinite P the results agree with those for an electron confined to move between impenetrable walls a distance a apart. The quantities

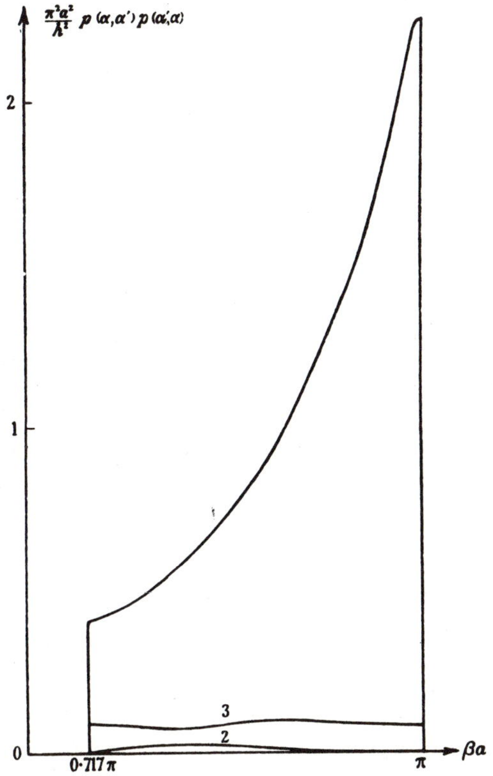

Fig. 6.

$p(\alpha, \alpha')$ determine the probability of radiative transitions between the stationary states, and one sees from fig. 6 that the transitions to the second negative region far outweigh all others.

4. *Extension of the Theory.*—The results just given can be generalised directly so as to apply to the case of a three-dimensional lattice in which the potential is the sum of three terms $V(x)$, $V(y)$, $V(z)$, each of which depends upon its co-ordinate in the same way as $V(x)$ on x in the one dimensional case. This means that the space is divided by infinitely thin potential barriers into cubical cells with an edge a. The wave-equation can then be separated

into three equations of the form (1), each involving only a single co-ordinate, viz.,

$$\left.\begin{aligned} \frac{d^2\psi_1}{dx^2} + \kappa^2[W_1 - V(x)]\psi_1 = 0 \\ \frac{d^2\psi_2}{dy^2} + \kappa^2[W_2 - V(y)]\psi_2 = 0 \\ \frac{d^2\psi_3}{dz^2} + \kappa^2[W_3 - V(z)]\psi_3 = 0 \end{aligned}\right\} \tag{11}$$

the total energy of the system and its wave-function being given by

$$W = W_1 + W_2 + W_3, \tag{12}$$

$$\psi = \psi_1 \psi_2 \psi_3 \tag{13}$$

respectively. Each of these equations can be treated in exactly the same way as equation (1). The stationary states will now be characterised by three quantities β_1, β_2, β_3, related to W_1, W_2, W_3 through equations analogous to equation (5), or by the corresponding quantities α_1, α_2, α_3.

In the one-dimensional case we have seen that for $P = 0$ the energy W is a continuous function of αa, while for $P \neq 0$ W has discontinuities for $\alpha a = n\pi$ ($n = \pm 1, \pm 2, \ldots$). Similarly we shall have here that for $P = 0$, W is a continuous function of $\alpha_1 a$, $\alpha_2 a$, $\alpha_3 a$, while for $P \neq 0$ the function becomes discontinuous on the surfaces $\alpha_1 a = n_1\pi$, $\alpha_2 a = n_2\pi$, $\alpha_3 a = n_3\pi$ ($n_1, n_2, n_3 = \pm 1, \pm 2, \ldots$). The detailed behaviour of the energy values as well as of the wave-functions when P is varied follows directly from equations (12) and (13) and the results of the previous sections. The same remark applies to the matrix elements of the linear momentum. It is interesting to note in this connection that $p_x(\alpha_1, \alpha_2, \alpha_3\,;\ \alpha_1', \alpha_2', \alpha_3')$ is different from zero only provided $\alpha_1' a = \alpha_1 a + 2n\pi$, $\alpha_2' a = \alpha_2 a$, $\alpha_3' a = \alpha_3 a$ (n integer) with similar results holding for p_y and p_z.

A number of general qualitative features encountered in the special problem discussed may be expected to occur also for other periodic fields of potential. The falling apart of the energy spectrum into continuous regions separated by finite intervals has been met with previously in the case of a potential given by $V(x) = A \cos 2ax$, discussed by Strutt,* who also mentions that a similar

* M. J. O. Strutt, 'Ann. Physik,' vol. 86, p. 319 (1928). Attention may be called here to the fact that the energy regions considered by him as allowed are just the ones excluded for translatory motion through the lattice, and *vice versa*. The energy spectra of the equations corresponding to the potential $A \cos 2ax$ and to a potential as shown in fig. 1 for the particular case $a = b$ have also been discussed in connection with problems

effect arises for any one dimensional periodic field of potential according to theorems proved by Haupt.† The stationary states, just as in our problem, can be characterised by a value α, the coefficient of the exponent in the function $e^{i\alpha x}$ which results from the wave-function when the potential field is reduced to zero. The energy W of the stationary states again has discontinuities at the points $\alpha a = n\pi$ ($n = \pm 1, \pm 2, \ldots$). In an analogous manner the surfaces $\alpha_1 a = n_1\pi$, $\alpha_2 a = n_2\pi$, $\alpha_3 a = n_3\pi$ (n_1, n_2, $n_3 = \pm 1, \pm 2, \ldots$) will be surfaces of discontinuity for the function $W(\alpha_1 a, \alpha_2 a, \alpha_3 a)$ in the general three-dimensional case.

Another result, valid for any periodic field of potential, is concerned with the matrix elements of the linear momentum $p(\alpha_1, \alpha_2, \alpha_3\,;\ \alpha_1', \alpha_2', \alpha_3')$‡. It may be shown that these can be different from zero only if $\alpha_1'a = \alpha_1 a + 2n_1\pi$, $\alpha_2'a = \alpha_2 a + 2n_2\pi$, $\alpha_3'a = \alpha_3 a + 2n_3\pi$ (n_1, n_2, n_3 integers). For the wave-functions in the two states α_1, α_2, α_3 and α_1', α_2', α_3' according to Bloch (*loc. cit.*) are given by

$$\psi_{\alpha_1\alpha_2\alpha_3} = u_{\alpha_1\alpha_2\alpha_3}\, e^{i(\alpha_1 x + \alpha_2 y + \alpha_3 z)},$$

$$\psi_{\alpha_1'\alpha_2'\alpha_3'} = u_{\alpha_1'\alpha_2'\alpha_3'}\, e^{i(\alpha_1' x + \alpha_2' y + \alpha_3' z)},$$

where the u's are periodic in x, y, z with the period a. Introducing these functions in the expression for the matrix element of p_x, say, gives us

$$p_x(\alpha_1, \alpha_2, \alpha_3\,;\ \alpha_1', \alpha_2', \alpha_3') = \frac{h}{2\pi i}\int \psi^*_{\alpha_1\alpha_2\alpha_3}\frac{\partial \psi_{\alpha_1'\alpha_2'\alpha_3'}}{\partial x}\,dx\,dy\,dz$$

$$= \frac{h}{2\pi i}\int u^*_{\alpha_1\alpha_2\alpha_3}\left[\frac{\partial u_{\alpha_1'\alpha_2'\alpha_3'}}{\partial x} + i u_{\alpha_1'\alpha_2'\alpha_3'}\right] e^{i[(\alpha_1'-\alpha_1)x + (\alpha_2'-\alpha_2)y + (\alpha_3'-\alpha_3)z]}\,dx\,dy\,dz$$

$$= \frac{h}{2\pi i}\int u^*_{\alpha_1\alpha_2\alpha_3}\frac{\partial u_{\alpha_1'\alpha_2'\alpha_3'}}{\partial x} e^{i[(\alpha_1'-\alpha_1)x + (\alpha_2'-\alpha_2)y + (\alpha_3'-\alpha_3)z]}\,dx\,dy\,dz,$$

the second term in the bracket not contributing anything on account of the orthogonality of the two wave-functions. From the periodicity of the u's it follows that we may write

$$u^*_{\alpha_1\alpha_2\alpha_3}\frac{\partial u_{\alpha_1'\alpha_2'\alpha_3'}}{\partial x} = \sum_{n_1 n_2 n_3 = -\infty}^{\infty} A_{n_1 n_2 n_3}\, e^{\frac{2\pi i}{a}(n_1 x + n_2 y + n_3 z)}$$

of classical physics by van der Pol and Strutt, "Phil. Mag.," vol. 5, p. 18 (1928). The advantage of the potential field considered in our article as compared with the field A cos $2ax$ lies in the fact that only elementary functions occur, making the evaluation of the various matrix elements very easy.

† O. Haupt, 'Math. Ann.,' vol. 79, p. 281 (1919).

‡ See also Kronig, *loc. cit.*

and introducing this in p_x leads immediately to the conclusion stated above. For two stationary states not separated by a surface of discontinuity of W the matrix element p hence always vanishes.

5. *Applications.*—A problem which we shall investigate here with the help of the one-dimensional model is the reflection of electrons of a given velocity falling from vacuum on to the lattice. The potential will then be as shown in fig. 7, for if our model is to represent an actual lattice, the potential of the

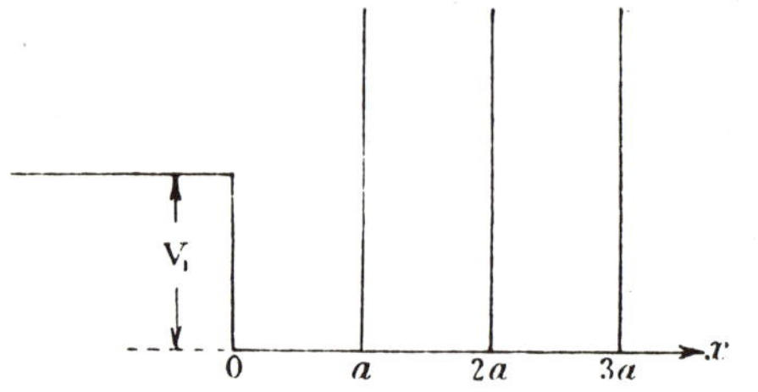

FIG. 7.

bottom of the lattice must be assumed to lie below that outside by a certain amount V_1. We enquire after a solution representing an incident and a reflected beam of electrons for negative values of x and a transmitted beam for positive values of x. For $x < 0$ we shall have thus:

$$\psi = e^{i\alpha_0 x} + Ae^{-i\alpha_0 x}, \tag{14}$$

while for $x > 0$ the solution is given by equation (8). α_0 and β are related by

$$\alpha_0^2 + \kappa^2 V_1 = \beta^2,$$

since the right and left-hand sides of this equation represent respectively the energies of the electron on the left and right of the point $x = 0$, which must be equal. The requirement that the function and its first derivative shall be continuous for $x = 0$ gives us the values of A in equation (14) and of C in equation (8).

AA* measures the intensity of the reflected beam, that of the incident beam being equal to 1, and is hence equal to the coefficient of reflection R. We find that

$$R = \frac{(\cos \alpha a - \cos \beta a)^2 + \left(\sin \alpha a - \frac{\alpha_0}{\beta} \sin \beta a\right)^2}{(\cos \alpha a - \cos \beta a)^2 + \left(\sin \alpha a + \frac{\alpha_0}{\beta} \sin \beta a\right)^2} \tag{15}$$

in the allowed regions of βa, while $R = 1$ for the forbidden regions of βa. The quantity α in equation (15) is the one associated with β in the manner described in the earlier sections. In fig. 8 R is represented as a function of the velocity

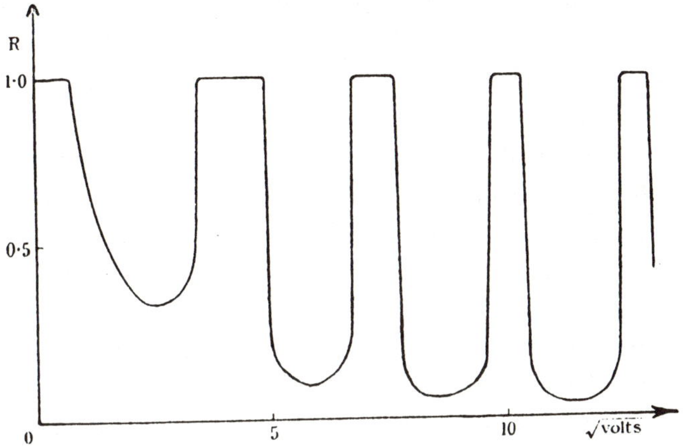

FIG. 8.

of the incident beam measured in $\sqrt{\text{volts}}$, which is proportional to α_0. For the binding constant P we have again taken $3\pi/2$, while for V_1 and a we have assumed the values $V_1 = 15$ volts, $a = 2 \cdot 2 \,.\, 10^{-8}$ cm. (suggested by the work of Rupp quoted below). The most essential feature of the diagram is, that regions of partial reflection alternate with regions of total reflection, and that these latter have a finite breadth, decreasing as the velocity of the particles gets greater. For very large values of the velocity total reflection takes place only when the lattice constant a is very nearly an integral multiple of the de Broglie wave-length of the incident particles, the regions of total reflection then becoming quite narrow.*

Measurements of the reflection and transmission of electrons in crystal lattices have been performed by various observers.† A quantitative comparison of their results with those obtained here is made difficult by the following circumstances: (1) If we had made our calculation with the three-dimensional

* Morse, 'Phys. Rev.,' vol. 35, p. 1310 (1930) has investigated the reflection and scattering of electrons by a crystal in which the potential depends upon the three co-ordinates through terms of the type A cos $2ax$. He obtains for the reflection coefficient at perpendicular incidence a curve essentially of the same type as that shown in fig. 8.

† Davisson and Germer, 'Phys. Rev.,' vol. 30, p. 705 (1927); 'Proc. Nat. Acad.,' vol. 14, pp. 317, 619 (1928); Rupp, 'Ann. Physik,' vol. 85, p. 981 (1928), vol. 1, p. 801 (1929), vol. 3, p. 497 (1929), and vol. 5, p. 453 (1930); 'Z. Physik,' vol. 61, p. 587 (1930); Thomson, 'Proc. Roy. Soc.,' A, vol. 117, p. 600 (1928), vol. 119, p. 651 (1928), and vol. 125, p. 352 (1929).

lattice, we would have obtained, besides the reflected and transmitted beams, diffracted beams which reduce the intensities of the former; (2) the periodic field of potential considered here differs from that actually present in a metal; (3) in the experiments it is found that some of the incident electrons suffer energy losses due to inelastic collisions with the electrons in the lattice, a phenomenon not provided for in our model. This effect also reduces the intensities of the reflected and transmitted beams. Nevertheless the principal features which appeared in our investigation, viz., the finite breadth of the reflection maxima, the decrease of this breadth with increasing velocity of the incident electrons and the decrease in the ordinates of the reflection minima, can be recognised in the experimental data (see in particular the work of Rupp).

An explanation can be given here also of a phenomenon recently observed by Rupp.* He finds that when the velocity of the incident electrons is gradually increased, new radiations appear in the soft X-ray spectrum of the substance bombarded at about the same velocities at which the reflection coefficient has a maximum. This can be understood if it be remembered that when the velocity begins to exceed that corresponding to the upper limit of a forbidden interval in the energy spectrum of the crystal, the impinging electron can enter into a new region of allowed energy values, giving it new possibilities for radiative transitions. According to this view the excitation threshold should differ from the reflection maximum by an amount equal to half the top-breadth of the maximum (lying toward higher velocities), but since in Rupp's measurements the maxima are already rather sharp the difference is probably obliterated by effects such as the inhomogeneity in the velocities, the imperfections of the crystal, etc. Perhaps, also, the constant energy losses of electrons impinging upon incandescent metals as observed by Rudberg† may be interpreted as corresponding to the transfer of the conduction electrons to the higher allowed regions of energy.

It is hoped to investigate later other physical properties of the model discussed as they appear in the phenomena mentioned at the beginning of this article.

Summary.

1. It is shown that the wave-equation representing the motion of an electron in a periodic field of potential can be integrated in terms of elementary functions when the potential takes the form of a series of equidistant rectangular barriers.

* 'Naturwiss.,' vol. 18, p. 880 (1930).

† Rudberg, 'Proc. Roy. Soc.,' A, vol. 127, p. 111 (1930).

When the breadth b of these barriers is made infinitely small and their height V_0 infinitely large, the results become particularly simple, the influence of the barriers depending then only on the product bV_0.

2. In the one dimensional problem for this limiting case the spectrum of permissible energy values is found to consist of continuous regions separated by finite intervals. By varying the quantity bV_0 from zero to infinity we pass from the case of free to that of bound electrons, and can thus study the changes in the allowed and forbidden ranges of the energy and in the wave-functions during this transition.

3. An investigation of the matrix elements of the linear momentum shows that the electrons can pass through the lattice and that there exists for them the possibility of transition to other stationary states under emission or absorption of radiation, the electrons thus having at the same time the characteristic properties of free and of bound electrons.

4. An investigation of the reflection of electrons by a crystal represented by the field of potential considered leads to results in qualitative agreement with the experimental facts. An explanation of a phenomenon recently observed by Rupp is given, and the possible connection of the theory with measurements of Rudberg is pointed out.

Reprinted from THE PHYSICAL REVIEW, Vol. 112, No. 4, 1137–1140, November 15, 1958
Printed in U. S. A.

New Soluble Energy Band Problem*

FREDERICK L. SCARF
Department of Physics, University of Washington, Seattle, Washington
(Received July 24. 1958)

The Schrödinger equation for $V(x)=-V_0\csc^2(\pi x/a)$ is solved. $E(k)$ and the effective mass are examined along with the wave functions for various values of the parameter (V_0a^2). It is found that the wave functions have intermediate forms even when $m^*\cong m$, $m^*\cong\infty$, and that the weak- and tight-binding approximations are not valid in these regions.

I. INTRODUCTION

THIS note contains a discussion of the periodic solutions of Schrödinger's equation with $V(x)=-V_0\csc^2(\pi x/a)$. Although $V(x)$ is singular,[1] even and odd elementary cell solutions are chosen to make the potential equivalent to one which has repulsive core cutoffs at the lattice sites. The problem is then completely soluble and the effective mass, allowed energies, band edge wave functions, and many other quantities can be expressed as simple functions of V_0, a and the crystal momentum $\hbar k$.

The motivation for the investigation is threefold. The study of any new, analytically soluble crystal model has pedagogical interest. Furthermore, the complete orthonormal set of periodic wave functions can be used as a basis for perturbation calculations. Finally, since $V(x)$ becomes infinite at $x=na$, the potential exaggerates the actual Coulomb singularity which occurs at each site in a real crystal; therefore, it is of interest to compare the use of customary approximation techniques with the exact treatment.

When $V_0a^2\to 0$, $m^*\to m$ (all k within the bands) and as V_0a^2 tends to a (finite) critical value, $m^*\to\infty$ (all k). This suggests that the weak- and tight-binding approximations, respectively, should be valid in these regions, but it is actually found that they are not appropriate. The result illustrates Slater's conjecture[2] that the wave function may be quite complicated even when m^* and $E(k)$ have values which indicate that the particle is almost free or that it is tightly bound.

II. GENERAL SOLUTION

The potential $V=-V_0\csc^2(\pi x/a)$ is shown in Fig. 1. In terms of the quantities $\lambda^2=2mEa^2/\pi^2\hbar^2$ and $(\frac{1}{4}-s^2)=2mV_0a^2/\pi^2\hbar^2$, the one-dimensional Schrödinger equation is

$$\pi^2\psi''+a^2[\lambda^2+(\tfrac{1}{4}-s^2)\csc^2(\pi x/a)]\psi=0. \tag{1}$$

For $0<x\leq a/2$, two independent solutions of Eq. (1) are expressible in terms of hypergeometric functions,

$$\begin{aligned} u(x)&=[\sin(\pi x/a)]^{\frac{1}{2}+s}\times\\ &\quad {}_2F_1(\tfrac{1}{4}+\tfrac{1}{2}s+\tfrac{1}{2}\lambda,\ \tfrac{1}{4}+\tfrac{1}{2}s-\tfrac{1}{2}\lambda;\ 1+s;\ \sin^2(\pi x/a)),\\ g(x)&=[\sin(\pi x/a)]^{\frac{1}{2}-s}\times\\ &\quad {}_2F_1(\tfrac{1}{4}-\tfrac{1}{2}s+\tfrac{1}{2}\lambda,\ \tfrac{1}{4}-\tfrac{1}{2}s-\tfrac{1}{2}\lambda;\ 1-s;\ \sin^2(\pi x/a)). \end{aligned} \tag{2}$$

In order that the problem be specified completely, $u(-x)$, $g(-x)$ also must be defined. Recently, it has been shown that the choice $u(-x)=Cu(x)$, $g(-x)=Dg(x)$, $CD=-1$, corresponds to a specific limiting definition, $V(0)=\lim V(x,x_0)$, $|x|<x_0$, as $x_0\to 0$.[3] In particular, for $0\leq s<\frac{1}{2}$, the values $C=-1$, $D=1$ imply $V(x,x_0)\simeq Ax_0^{-2}$, $|x|<x_0$. Although many other specifications are possible, we shall confine the discussion to this choice of symmetries. Thus, u is odd and g is even, in accord with the behavior as $V_0\to 0$ [for $s=\frac{1}{2}$, $g=\cos(\pi\lambda x/a)$, $\lambda u=\sin(\pi\lambda x/a)$].

The most general periodic wave function can be written in the self-matching form suggested by James,[4]

$$\psi(x)=\frac{(r)^N}{1+\rho^{\frac{1}{2}}}\left[\frac{g(x-Na)}{g_0}+\rho^{\frac{1}{2}}\frac{u(x-Na)}{u_0}\right],$$
$$(N-\tfrac{1}{2})a<x\leq(N+\tfrac{1}{2})a, \tag{3}$$

where $r=(1+\rho^{\frac{1}{2}})/(1-\rho^{\frac{1}{2}})$, $\rho=(g_0'u_0/g_0u_0')$ and g_0, u_0, g_0', u_0' represent $g(x)$, $u(x)$ and their first derivatives, all evaluated at $x=a/2$. Equation (3) is normalized so

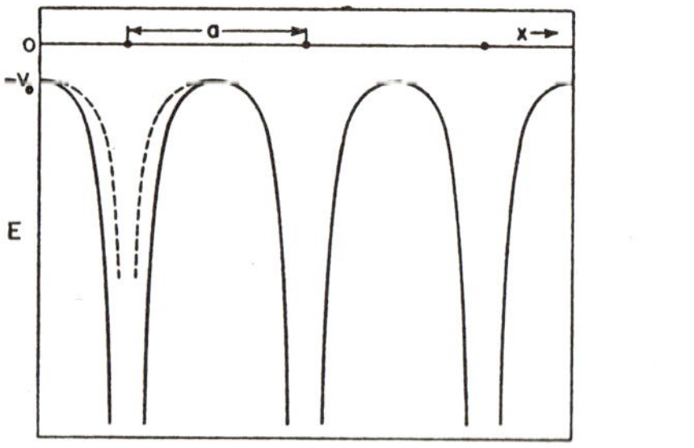

FIG. 1. The potential $V(x)=-V_0\csc^2(\pi x/a)$. For comparison, the dashed line shows a short-range Coulomb-like interaction $[V=-V_0|\csc(\pi x/a)|]$.

* Partially supported by the U. S. Atomic Energy Commission.
[1] K. M. Case, Phys. Rev. **80**, 797 (1950).
[2] J. C. Slater, Phys. Rev. **87**, 807 (1952).
[3] F. L. Scarf, Phys. Rev. **109**, 2170 (1958).
[4] H. M. James, Phys. Rev. **76**, 1602 (1949).

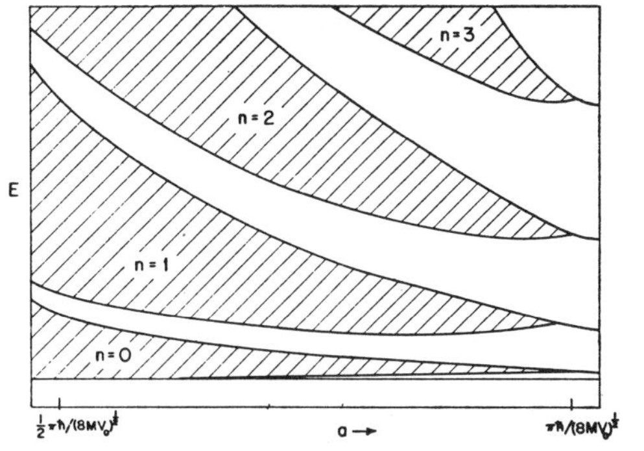

FIG. 2. The first four allowed bands (shaded area) as a function of lattice spacing, a, for fixed V_0. When $V_0a^2>\pi^2\hbar^2/8m$, the potential is genuinely singular and the positions of the (nearly degenerate) bands are sensitive functions of the cutoff; the curves shown in this region correspond to repulsive-core cutoffs.

that $\psi(a/2)=1$. Defining

$$\rho^{\frac{1}{2}}=i\tan(ka/2), \tag{4}$$

the wave function may be written in the Bloch form, $\psi_k(x)=f_k(x)\exp(ikx)$, with

$$f_k(x)=\left[(\cos\tfrac{1}{2}ka)\frac{g(x-Na)}{g_0}+i(\sin\tfrac{1}{2}ka)\frac{u(x-Na)}{u_0}\right]\times\exp[-ik(x-Na+\tfrac{1}{2}a)]. \tag{5}$$

Allowed energies occur for real k and the band edges are obtained from $\rho_s(\lambda)$. Using the well known expressions for ${}_2F_1\ (a,b;c;1)$ in g_0, u_0, etc., one arrives at

$$\rho_s(\lambda)=\tan[\tfrac{1}{2}\pi(\tfrac{1}{2}-s-\lambda)]\tan[\tfrac{1}{2}\pi(\tfrac{1}{2}-s+\lambda)]. \tag{6}$$

Equations (4) and (6) finally yield

$$E_n(k)=\frac{\hbar^2}{2ma^2}[\cos^{-1}(\sin\pi s\cos ka)]^2, \tag{7}$$

where the nth principal value of the inverse cosine is to be taken. The edges of the nth band have $k=0$, π/a (reduced wave number) and $E_n^{(\pm)}=\pi^2\hbar^2(n+\frac{1}{2}\pm s)^2/2ma^2$. For $s=\frac{1}{2}$, Eq. (7) gives $E=\hbar^2k^2/2m$, and when $s=0$ the bands become degenerate. Figure 2 shows the variation of $E_n(k)$ with a for fixed V_0.

Equation (7) also leads to expressions for the group velocity of a wave packet,

$$\langle v\rangle_{nk}=\frac{1}{\hbar}\frac{\partial E}{\partial k}=(2E/m)^{\frac{1}{2}}(\csc^2\pi s-\cos^2 ka)^{-\frac{1}{2}}\sin ka, \tag{8}$$

and the effective mass of a particle,

$$m/m^*=(m/\hbar^2)\partial^2E/\partial k^2=(\sin^2\pi s\csc^2\pi\lambda-\mathrm{ctn}^2\pi\lambda+\pi\lambda\cos^2\pi s\csc^2\pi\lambda\cos\pi\lambda), \tag{9}$$

As $s\to\frac{1}{2}$, $m^*\to m$ and $\langle v\rangle\to(2E/m)^{\frac{1}{2}}$ while for $s=0$, $m^*=\infty$ and $\langle v\rangle=0$.

Many properties of the hypergeometric series are known and can be employed to discuss the wave functions for this problem. In particular, for $k=0$, $\pi/2a$, and π/a, ψ_{nk} may be expressed as a simple polynomial in the quantity $y=\sin^2[\pi(x-Na)/a]$, $(N-\frac{1}{2})a<x\leq(N+\frac{1}{2})a$. All of these k values lead to Jacobi polynomials [see Eqs. (2), (7)] and, in terms of the generating function,[5] the non-normalized wave functions are

$$\psi_{n,0}=(y-y^2)^{\frac{1}{4}+\frac{1}{2}s}d^n[(y-y^2)^{n-s}]/dy^n, \tag{10}$$

$$\psi_{n,\pi/a}=(y-y^2)^{\frac{1}{4}-\frac{1}{2}s}d^n[(y-y^2)^{n+s}]/dy^n\times\exp(\pm\pi iN), \tag{11}$$

and

$$\psi_{n,\pi/2a}=[\phi_s\pm i\phi_{-s}]\exp[\mp i(N-\tfrac{1}{2})\pi/2],$$

$$\phi_s=\frac{2^{-s}y^{\frac{1}{4}+\frac{1}{2}s}(1-y)^{\frac{1}{4}-\frac{1}{2}s}}{\Gamma(-s-n)\Gamma(\frac{1}{2}-\frac{1}{2}s-\frac{1}{2}n)\Gamma(1-\frac{1}{2}s+\frac{1}{2}n)}\frac{d^n}{dy^n}\times[(1-y)^{n+s}y^{n-s}]. \tag{12}$$

These typical solutions illustrate how ψ varies with V_0, a, x and with the quantum numbers. The Fourier transforms of Eqs. (10), (11) may also be evaluated. In general, $\psi(p)$ has a simple pole on the real axis.

There are some other noteworthy features of the results. In any one-dimensional problem, the band edges are defined by the succession of values $\rho=(0,\infty)$, $(\infty,0)$, etc., as the energy increases. In familiar cases (for instance, the Kronig-Penney model[6] and the Slater model[2]), the quantities (g_0',g_0), (u_0',u_0) vanish in turn as E increases, and bands are of the Wigner-Seitz[7] form; pure g (or pure u) at top *and* bottom.

In our case, as E increases these quantities vanish in the order (g_0',u_0'), (g_0,u_0), (g_0',u_0'), etc. Thus, the bands are all pure g at the bottom $(k=0)$ and pure u at the top $(k=\pi/a)$. This type of behavior was anticipated in the general discussion presented by James[4]; it is not a direct consequence of the singular nature of $V(x)$ and in terms of a unit cell centered at $x=a/2$, the wave functions have conventional symmetry.

III. ORIGIN OF THE DEGENERACY

As $s\to 0$, all energy bands become degenerate, $\langle v\rangle_{nk}$ vanishes, and m^* tends to infinity. Since these effects occur when an electron is tightly bound near an ion, it might be inferred that such a situation arises here for a finite value of the lattice spacing.

However, as $s\to 0$, $u/u_0\to g/g_0$ if $Na<x\leq(N+\frac{1}{2})a$ and $u/u_0\to -g/g_0$ when $(N-\frac{1}{2})a<x\leq Na$. Thus, for

[5] P. M. Morse and H. Feshbach, *Methods of Theoretical Physics* (McGraw-Hill Book Company, Inc., New York, 1955), pp. 780–781.
[6] R. de L. Kronig and W. G. Penney, Proc. Roy. Soc. (London) A130, 499 (1931).
[7] E. P. Wigner and F. Seitz, Phys. Rev. 43, 804 (1933).

$s=0$,

$$\psi_{nk}(x)=g_0^{-1}g(x-Na)\exp(iNka),\quad Na<x\leq(N+1)a. \tag{13}$$

This form resembles the tight-binding wave function of Bloch[8]

$$\Psi_{nk}(x)=g_0^{-1}g(x-Na)\exp(iNka),\quad (N-\tfrac{1}{2})a<x\leq(N+\tfrac{1}{2})a, \tag{14}$$

but there is an important difference. Bloch's function is a sum of "atomic" functions centered about the lattice points, $x=0$, $\pm a$, $\pm 2a$, etc., while Eq. (13) represents a sum of functions centered midway between the ions at $x=\pm\frac{1}{2}a$, $\pm\frac{3}{2}a$, etc. Thus, the limit as $s\to 0$ does *not* correspond to tight binding about the ions.

In order to understand this it is useful to refer to Fig. 3(a) where conventional tightly bound levels are shown for Slater's model. When $E\ll V_{\max}$, ψ and $(d\psi/dx)$ are both small within the barrier so that there is little energy difference between the $k=0$ and $k=\pi/a$ states.

Moreover, if one takes seriously the repulsive cores which were implied in the choice of C and D, Fig. 3(b) illustrates that ordinary tight binding occurs again in the new model, with phase changes at the maxima of $V(x)$. However, it must be recognized that V is a maximum *at* the lattice point, not at $x=a/2$.

Even without considering the repulsive cores, it can be seen that narrow bands, small $\langle v\rangle$ and large m^* may occur whenever $\psi\simeq 0$ at some point $(x=x_1)$ in each unit cell. This can arise because $E\ll V(x_1)$, as in Fig. 3. However, $\psi(x_1)$ may also be very small because the potential at x_1 is strongly attractive. Thus, the existence of nearly degenerate bands and large effective mass does not uniquely imply an atomic situation. [For $V_0a^2\geqslant\pi^2\hbar^2/8m$, with a cutoff at some finite x_0, the bands have finite width. As V_0a^2 increases in this singular region, the levels become negative and continue to fall until $E\ll V(\frac{1}{2}a)$. At this point, one has conventional tight binding with wave functions localized near $x=Na$.]

The "transition" of Fig. 2 occurs for a finite value of a because V_0 is held fixed. If V_0a^2 were constant, there would clearly be no transition and the problem would correspond to a set of unpolarizable x^{-2} potentials. However, if $V_0=Ba^{-\epsilon}$, $\epsilon<2$, then $V(x)$ represents a set of polarizable cores. In the neighborhood of the ion $(x\ll\frac{1}{2}\pi a)$ the core potential seen by the electron is

$$V_{\text{core}}\simeq -V_0a^2/\pi^2x^2=-V_{\text{eff}}/x^2, \tag{15}$$

with $V_{\text{eff}}(a)=Ba^{2-\epsilon}/\pi^2$. Thus, V_{eff} decreases with a and the bands suddenly broaden as the singular boundary is passed.

Although infinite m^* for finite a never occurs in a real crystal, it is possible to find narrow bands without having "atomic" wave functions. The extreme behavior

[8] F. Bloch, Z. Physik 52, 555 (1928).

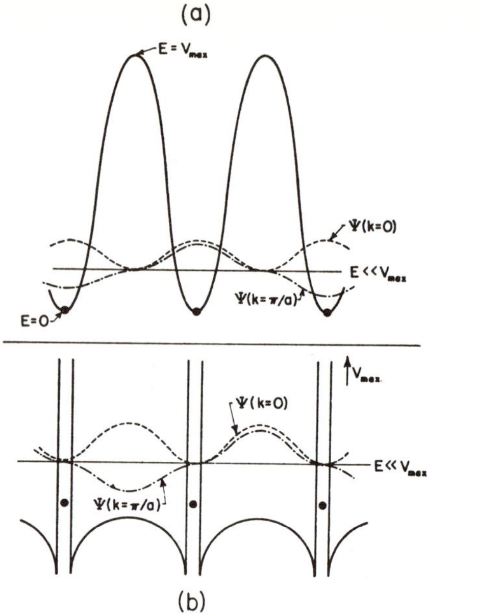

FIG. 3. (a) The wave functions (dashed curves) of tightly bound levels with $V=V_0\cos(\pi x/a)$ (solid curve). (b) The wave functions (dashed curves) of tightly bound levels with $V=-V_0\csc^2(\pi x/a)$ for $|x|>x_0$, $V=$repulsive core for $|x|<x_0$ (solid curve).

discussed above is, of course, a consequence of the singular nature of $V(x)$ and of the fact that the problem is one-dimensional.

IV. WEAK BINDING

When $V_0=0$ $(s=\frac{1}{2})$, all quantities discussed in Sec. II have the free-particle form: $E=\hbar^2k^2/2m$ and $\psi=\exp[i\pi\lambda(x-\frac{1}{2}a)/a]$. For small values of the potential strength, $E_n(k)$, $\langle v\rangle_{nk}$ and m^* may be expanded in convergent power series in the parameter $(\frac{1}{2}-s)$. For instance, the first two terms in the Taylor expansion of $E_n(k)$ are

$$E_n(k)=\frac{\hbar^2k^2}{2m}\left[1+\frac{\pi^2(\frac{1}{2}-s)^2}{ka\tan ka}+\cdots\right]. \tag{16}$$

Weak binding, or second-order perturbation theory, gives[9]

$$E_n(k)=\frac{\hbar^2k^2}{2m}\left[1+(V_0a^2)^2\times\sum_{\nu\neq 0}\frac{|\alpha_\nu|^2}{2\pi\nu k^2a^2(ka-2\pi\nu)}+\cdots\right], \tag{17}$$

with

$$\alpha_\nu=(V_0a)^{-1}\int_0^a dx\,V(x)\exp(2\pi i\nu x/a). \tag{18}$$

[9] C. Kittel, *Introduction to Solid State Physics* (John Wiley and Sons, Inc., New York, 1956), second edition, pp. 583–585.

Equations (16) and (17) have several similarities. They each possess a second-order pole at $k=0$, first-order poles at $ka=2\pi\nu$, $\nu\neq 0$, and a quadratic dependence on the strength, V_0a^2. Furthermore, both treatments predict that for s near $\frac{1}{2}$, the effective mass at a band edge ($E=E_a$) is given by

$$m/m^*=(1+4E_a/\Delta E), \tag{19}$$

where ΔE is the width of the gap.

These properties suggest that the weak-binding approximation is valid when $V_0a^2 \ll \pi^2\hbar^2/8m$. That it is actually not appropriate can immediately be seen by inspection of Eq. (18); if $V(x)=-V_0\csc^2(\pi x/a)$, all of the integrals diverge.

This anomaly is associated with the factor $[\sin(\pi x/a)]^{\frac{1}{2}-s}$ in the wave function. The expansion $\psi(x)=\Sigma(V_0)^\nu\phi^{(\nu)}(x)$ is convergent only for $\sin(\pi x/a) > \exp[1/(s-\frac{1}{2})]$. In fact,

$$\lim_{V_0\to 0}\psi(V_0, x=0)=0, \quad \text{but} \quad \lim_{x\to 0}|\psi(V_0=0, x)|=1.$$

Thus, for any finite V_0, $\psi(0)=0$ and the wave function does not resemble a plane wave.

The wave functions of highly excited levels still reflect the singularity at the origin. For $\lambda\gg 1$ (any value of s), the asymptotic expansion is[10]

$$\psi_{nk}(x)=\left[\frac{\cos(ka/2)\cos(\lambda\pi[x-Na]/a-\frac{1}{4}\pi+\frac{1}{2}\pi s)}{\cos(\frac{1}{2}\lambda\pi-\frac{1}{4}\pi+\frac{1}{2}\pi s)} + i\frac{\sin(ka/2)\cos(\lambda\pi[x-Na]/a-\frac{1}{4}\pi-\frac{1}{2}\pi s)}{\cos^d(\frac{1}{2}\lambda\pi-\frac{1}{4}\pi-\frac{1}{2}\pi s)}\right] \times \exp i(N-\tfrac{1}{2})ka+O(\lambda^{-1}), \tag{20}$$

if the normalization, $\psi(a/2)=1$, is retained. Equation (20) represents a plane wave when $V_0=0$, but for *any* finite value of V_0 and any λ, it does not. It may be noted that $\psi(0)$, as given by Eq. (20), is generally not zero. However, the first term in Eq. (20) refers to $g(x)$ (even) while the second describes $u(x)$ (odd); thus the asymptotic form of u is discontinuous, and g has a discontinuity in slope. This merely illustrates that $\psi(0)$ must always be defined to be zero, and that the expansion is not valid at the origin.

These properties of the solutions show that the weak-binding approximation cannot be used for a singular potential, although it can reproduce certain features of $E(k)$.

V. CONCLUSION

When the $E(k)$ curves for an electron in a periodic potential are examined, the inference is frequently made that the wave functions resemble plane waves, or that they can be approximated by Bloch-type atomic functions. Slater showed that the Mathieu-function model gives rise to bands with clear-cut weak- or tight-binding $E(k)$ characteristics, but that the wave functions have a complex intermediate form for wide ranges of the parameters. It has been conjectured that this is a general property of crystalline wave functions. A possible criticism of such a generalization is that Slater's sinusoidal potential is much smoother than that in a real crystal, and that it has nothing like a Coulomb singularity near the ion.

The present discussion involves a potential which is unphysical in the opposite direction; the extreme singularity at the origin and the one-dimensional nature of the problem exaggerate the actual situation and introduce new effects. However, once again it is shown that the wave functions can be quite complex even when $E(k)$ seems to represent a nearly free, or a tightly bound particle.

[10] A. Erdélyi *et al.*, *Higher Transcendental Functions* (McGraw-Hill Book Company, Inc., New York, 1955), Vol. I, p. 77.

Reprinted from THE PHYSICAL REVIEW, Vol. 76, No. 11, 1602–1610, December 1, 1949
Printed in U. S. A.

Energy Bands and Wave Functions in Periodic Potentials*

HUBERT M. JAMES
Purdue University, Lafayette, Indiana
(Received July 26, 1949)

In preparation for treatment of the one-dimensional wave equation with perturbed periodic potentials, this paper discusses the case of perfectly periodic potentials. Solutions for the entire range of x are constructed by joining together "cell solutions"—solutions defined for a single period of the potential. Several types of cell solution are discussed. The concept of "self-matching cell solutions" leads to a simple demonstration of the band structure of permitted energy levels. The effective momentum of the particle is defined for energies in the forbidden bands, and a parameter $\sigma(E)$ of interest in the treatment of perturbed periodic systems is introduced. Expansions are given for solutions of the wave equation near the band edges, in both the permitted and forbidden bands.

I. INTRODUCTION

THE wave mechanics of particles in perfectly periodic potentials can be regarded as completely understood, at least in principle. Many years before the advent of wave mechanics, Hamel[1] and Haupt[2] discussed, from the mathematician's point of view, an equation essentially identical with the one-dimensional wave equation for a particle of mass m in a periodic potential $W(x)$:

$$-\frac{\hbar^2}{2m}\frac{d^2\psi(x)}{dx^2}+W(x)\psi(x)=E\psi(x). \qquad (1.1)$$

They were particularly concerned with the "oscillation theorems" for this equation, relating to the dependence on E of the existence or non-existence of bounded solutions $\psi(x)$. They showed that all solutions are bounded for certain ranges of E (the "permitted energy bands" for electrons in crystals), separated, in general, by ranges of energy for which no solutions are bounded ("forbidden bands"). This wave equation, and the corresponding three-dimensional one, have since been studied by a number of methods suited to various special purposes; it will suffice to mention here the work of Bloch,[3] Brillouin,[4] and Kramers.[5]

The present paper deals only with periodic potentials, but is intended to lay the foundation for a study of the one-dimensional wave equation with perturbed periodic potentials. These perturbations may consists of slowly varying additions to the potential, such as are considered in a succeeding paper, or of radical but localized changes in the potential, which are to be treated in a paper now in preparation. In either case it is useful and illuminating to construct solutions for the whole range of the variable x by fitting together solutions valid for single periods of the unperturbed potential. As applied to periodic potentials this method is well known.[6] It is here discussed from a very general point of view, with the following purposes:

(1) To give a particularly clear and elementary demonstration of the band structure of permitted energy levels.

(2) To discuss the properties of all solutions of the wave equation, including those which do not satisfy the boundary conditions of wave mechanics. Consideration of solutions in the forbidden ranges of energy is required in the treatment of perturbed periodic potentials. Particular attention will be paid to the forms of solutions with energies near the band edges.

(3) To introduce a new parameter, $\sigma(E)$, which, like the effective momentum $p(E)$, depends on and partially characterizes the periodic potential. This parameter plays an important role in the theory of perturbed periodic potentials.

The discussion will also exhibit some of the advantages to be gained from consideration of several alternative sets of solutions of the wave equation for single periods of the potential (cell solutions).

II. SPECIAL TYPES OF CELL SOLUTION

Let $W(x)$ be a periodic function with period a:

$$W(x)=W(x+na), \quad n=\pm1, \pm2, \cdots. \qquad (2.1)$$

The range of x will be divided into periods or *cells* of length a, such that in the nth cell

$$na\leq x\leq(n+1)a. \qquad (2.2)$$

The potential has then the same form in each cell. It is often convenient to choose the origin of coordinates so that the potential in each cell is symmetrical about the center; this is not always possible, nor is it necessary for application of the methods to be described.

We now fix attention on the zeroth cell, $0\leq x\leq a$, and on a particular value of E. The wave equation is then a linear second-order differential equation with two linearly independent solutions. This section deals with the properties of three special pairs of independent cell solutions.

---

* This work was supported in part by Signal Corps Contract W36-039-SC-32020 with the Department of Physics at Purdue University. It has been circulated as a Signal Corps report, but appears here in slightly enlarged form, with some changes of notation and sign conventions.

[1] G. Hamel, Math. Ann. **73**, 371 (1912).
[2] O. Haupt, Math. Ann. **76**, 67 (1914); **79**, 278 (1919).
[3] F. Bloch, Zeits. f. Physik **52**, 555 (1928).
[4] L. Brillouin, Comptes Rendus **191**, 198, 292 (1930); J. de physique VII, **1**, 377 (1930).
[5] H. A. Kramers, Physica **2**, 483 (1935).
[6] R. deL. Kronig and W. G. Penney, Proc. Roy. Soc. **130**, 499 (1931).

A. The y Solutions

Let $y_1(E;x)$ and $y_2(E;x)$ be solutions of Eq. (1.1) such that

$$\left.\begin{aligned} y_1(E;0)=1,\quad y_1'(E;0)=0,\\ y_2(E;0)=0,\quad y_2'(E;0)=1.\end{aligned}\right\} \qquad (2.3)$$

These solutions are always real and can be obtained by numerical integration, if necessary.

We define

$$\left.\begin{aligned} y_1(E;a)=c_{11}(E),\quad y_2(E;a)=c_{12}(E),\\ y_1'(E;a)=c_{21}(E),\quad y_2'(E;a)=c_{22}(E),\end{aligned}\right\} \qquad (2.4)$$

where the prime denotes the derivative with respect to x. The constancy in x of the Wronskian of y_1 and y_2,

$$W(y_1,y_2)=y_1(E;x)y_2'(E;x)-y_1'(E;x)y_2(E;x) \qquad (2.5)$$

(or of any pair of solutions of Eq. (1.1) with the same E) requires that

$$c_{11}(E)c_{22}(E)-c_{12}(E)c_{21}(E)=1. \qquad (2.6)$$

Henceforth, the dependence of the y's and c's on E will be understood, but made explicit only on occasion.

One can easily construct from the functions $y_1(x)$ and $y_2(x)$, defined only for the range $0\leq x\leq a$, solutions of the periodic problem for all x. Since the potential has the same form in every cell, $y_1(x-na)$ and $y_2(x-na)$ will be independent solutions in the nth cell. Within each cell an arbitrary solution ψ of the wave equation can be expressed as an appropriate linear combination of the corresponding cell solutions:

$$\psi(x)=a_n y_1(x-na)+b_n y_2(x-na),$$
$$\left\{\begin{aligned} na\leq x\leq (n+1)a\\ n=0,\pm1,\pm2,\cdots.\end{aligned}\right. \qquad (2.7)$$

Continuity of ψ and its first derivative at the boundary between the nth and $(n+1)$st cells demands that

$$a_{n+1}=c_{11}a_n+c_{12}a_n,\quad b_{n+1}=c_{21}a_n+c_{22}b_n. \qquad (2.8)$$

In matrix form this becomes

$$\begin{pmatrix} a_{n+1}\\ b_{n+1}\end{pmatrix}=(c_{ij})\begin{pmatrix} a_n\\ b_n\end{pmatrix} \qquad (2.9)$$

By Eq. (2.6), the matrix (c_{ij}) always has an inverse $(c_{ij})^{-1}$. It follows that

$$\begin{pmatrix} a_n\\ b_n\end{pmatrix}=(c_{ij})^n\begin{pmatrix} a_0\\ b_0\end{pmatrix} \qquad (2.10)$$

for all n, positive or negative; the constants a_n and b_n characteristic of any given solution are thus easily determined as soon as a_0 and b_0 are given.

It is a matter of elementary matrix algebra to discuss the conditions under which a_n, b_n, and ψ remain bounded as $n, x\rightarrow\pm\infty$, and ψ is therefore physically acceptable. An equivalent but more picturesque discussion of the band structure of energy levels will be given later, using another set of basic cell solutions.

Use of the y cell solutions is particularly advantageous when the problem to be treated is not strictly periodic. If the range of x is divided into cells (which need not even have the same extent), one can define for the mth cell the solutions y_{1m} and y_{2m} satisfying Eq. (2.3) at the left edge of the cell, and can determine the corresponding constants $c_{ij}^{(m)}$. The conditions of continuity at the boundary between the mth and $(m+1)$st cell involve only the constants $c_{ij}^{(m)}$, and the generalization of Eq. (2.10) is (for positive n)

$$\begin{pmatrix} a_n\\ b_n\end{pmatrix}=(c_{ij}^{(n-1)})\cdots(c_{ij}^{(1)})(c_{ij}^{(0)})\begin{pmatrix} a_0\\ b_0\end{pmatrix} \qquad (2.11)$$

B. The Solutions g and u

When the cell division can be made in such a way that $W(x)$ is symmetrical about the center of each cell, it is convenient to consider, for each E, another set of real solutions, $g(E;x)$ and $u(E;x)$, such that

$$\left.\begin{aligned} g(E;a/2)=1,\quad g'(E;a/2)=0,\\ u(E;a/2)=0,\quad u'(E;a/2)=1.\end{aligned}\right\} \qquad (2.12)$$

These solutions need be defined only for the zeroth cell, about the center of which they are symmetric and antisymmetric, respectively.[7] We shall write

$$\left.\begin{aligned} g(E;0)=g(E;a)=g_0(E);\\ g'(E;0)=-g'(E;a)=g_0'(E);\\ u(E;0)=-u(E;a)=u_0(E);\\ u'(E;0)=u'(E;a)=u_0'(E).\end{aligned}\right\} \qquad (2.13)$$

The constancy in x of $W(g;u)$ assures that

$$g_0(E)u_0'(E)-g_0'(E)u_0(E)=1. \qquad (2.14)$$

The dependence of these quantities on E will hereafter be indicated only as required.

One can, of course, express the solutions g and u in terms of y_1 and y_2, and conversely. From consideration of magnitudes and slopes at $x=0$, it is obvious that

$$\left.\begin{aligned} g(x)=g_0y_1(x)+g_0'y_2(x),\\ u(x)=u_0y_1(x)+u_0'y_2(x).\end{aligned}\right\} \qquad (2.15)$$

Solution of these equations, with use of Eq. (2.14), yields

$$\left.\begin{aligned} y_1(x)=u_0'g(x)-g_0'u(x),\\ y_2(x)=-u_0g(x)+g_0u(x).\end{aligned}\right\} \qquad (2.16)$$

On setting $x=a$ and using Eq. (2.13), one finds

$$\left.\begin{aligned} c_{11}=u_0'g_0+g_0'u_0=c_{22},\\ c_{12}=-2u_0g_0,\\ c_{21}=-2u_0'g_0'.\end{aligned}\right\} \qquad (2.17)$$

---

[7] These solutions have been employed in the discussion of surface states of crystals by W. Shockley, Phys. Rev. **56**, 317 (1939).

It is thus proved that for symmetric cell potentials $y_1(a)=y_2'(a)$.

The values of the c's are determined by the values of g_0, g_0', u_0, u_0', but the converse is not true; the c's remain unchanged when one replaces the latter quantities by κg_0, $\kappa g_0'$, $\kappa^{-1}u_0$, $\kappa^{-1}u_0'$, respectively. It does, however, follow from Eqs. (2.14) and (2.17) that

$$\frac{g_0'}{g_0}=\frac{1-c_{11}}{c_{12}}=-\frac{c_{21}}{1+c_{11}},\quad \frac{u_0'}{u_0}=\frac{c_{21}}{1-c_{11}}=-\frac{1+c_{11}}{c_{12}}. \tag{2.18}$$

As with the y's, an arbitrary solution of the wave equation with periodic potential can be expressed in terms of the cell solutions $g(x)$ and $u(x)$:

$$\psi(x)=\alpha_n g(x-na)+\beta_n u(x-na),\quad \begin{cases} na\leq x\leq (n+1)a, \\ n=0,\pm1,\pm2,\cdots. \end{cases} \tag{2.19}$$

Continuity in the magnitude and slope of ψ at $x=na$ requires

$$\left.\begin{aligned} \alpha_n g_0-\beta_n u_0&=\alpha_{n+1}g_0+\beta_{n+1}y_0, \\ -\alpha_n g_0'+\beta_n u_0'&=\alpha_{n+1}g_0'+\beta_{n+1}u_0'. \end{aligned}\right\} \tag{2.20}$$

Solving these equations for α_{n+1} and β_{n+1}, with use of Eq. (2.14), one finds

$$\begin{pmatrix}\alpha_{n+1}\\ \beta_{n+1}\end{pmatrix}=\begin{pmatrix} g_0u_0'+g_0'u_0 & -2u_0u_0' \\ -2g_0g_0' & g_0u_0'+g_0'u_0\end{pmatrix}\begin{pmatrix}\alpha_n\\ \beta_n\end{pmatrix} \tag{2.21}$$

Like Eq. (2.9), this can (but will not here) be made the basis of a discussion of the band structure of the energy levels.

In the treatment of non-periodic problems g and u are less conveniently used than are y_1 and y_2, since the matrix that transforms the coefficients α_n and β_n into α_{n+1} and β_{n+1} depends on the potential in both cells n and $n+1$.

C. "Self-Matching" Cell Solutions, $f_\pm(x)$

The problem of constructing wave functions for an infinite periodic potential by connecting cell solutions can be made particularly simple by the use of solutions $f(E;x)$, defined for $0\leq x\leq a$, which have the same slope-to-magnitude ratio at the two ends of the cell:

$$\frac{f'(E;0)}{f(E;0)}=\frac{f'(E;a)}{f(E;a)}=\sigma(E). \tag{2.22}$$

To normalize these functions, it will be required that

$$f(E;0)=1. \tag{2.23}$$

In passing from the left end of the cell to the right, the solution will increase in magnitude by the factor

$$\frac{f(E;a)}{f(E;0)}=r(E). \tag{2.24}$$

Such solutions will here be called "self-matching" or, for brevity, f solutions. They derive their importance from the following property: given a *single* self-matching solution in the zeroth cell, one can construct from it a continuous solution of the wave equation through the whole range of x. If $f(x)$ is a cell solution in the zeroth cell, $c_1f(x-a)$ will be a cell solution in the first cell, $c_2f(x-2a)$ a cell solution in the second cell, and so on. All these cell solutions will have the same slope-to-magnitude ratio at the ends of their respective cells. Therefore, if c_1, c_2, $\cdots$ are so chosen that the cell solutions have the same magnitudes at their common cell boundaries, they will have there also the same slopes: they will fit together smoothly to form a continuous solution through the whole range of x. Since each solution changes by a factor r in passing from the left end of the cell to the right end, the magnitudes can be made continuous by choosing $c_n=r^n$, for all n, positive or negative. Thus one obtains from the given cell solution a solution of the wave equation for all x:

$$\psi(E;x)=r^nf(E;x-na)\begin{cases} na<x<(n+1)a \\ n=0,\pm1,\pm2,\cdots. \end{cases} \tag{2.25}$$

This solution has the same form in every cell; it changes from cell to cell only by a (possibly complex) multiplying constant.

It is now necessary to investigate the existence of self-matching cell solutions. If one exists, it must be expressible as a linear combination of the special solutions y_1 and y_2 already discussed, which always exist:

$$f(x)=ay_1(x)+by_2(x). \tag{2.26}$$

From Eqs. (2.3) and (2.23) it follows that $a=1$; Eq. (2.22) then requires

$$\sigma=b=(c_{21}+bc_{22})/(c_{11}+bc_{12}). \tag{2.27}$$

Solving for b, with use of Eq. (2.6), one finds two values:

$$\sigma_\pm=\frac{c_{22}-c_{11}\mp\{(c_{11}-c_{22})^2+4c_{12}c_{21}\}^{\frac12}}{2c_{12}} =\frac{c_{22}-c_{11}\mp\{(c_{11}+c_{22})^2-4\}^{\frac12}}{2c_{12}}. \tag{2.28}$$

Thus, the only self-matching solutions are

$$f_\pm(E;x)=y_1(E;x)+\sigma_\pm y_2(E;x). \tag{2.29}$$

If σ_+ and σ_- are distinct, there are exactly two self-matching solutions. If $\sigma_+=\sigma_-$, there is only one such solution; this will occur when $c_{11}+c_{22}=\pm2$.

The values of r for $f_\pm(E;x)$,

$$r_\pm(E)=\frac{c_{11}+c_{22}\mp\{(c_{11}+c_{22})^2-4\}^{\frac{1}{2}}}{2}, \tag{2.30}$$

can be written as

$$r_\pm(E)=(1\mp\rho^{\frac{1}{2}})/(1\pm\rho^{\frac{1}{2}}), \tag{2.31}$$

where

$$\rho=(c_{11}+c_{22}-2)/(c_{11}+c_{22}+2). \tag{2.32}$$

The definition of $f_\pm(E;x)$ will be made complete by the convention that

$$\left.\begin{aligned}\rho^{\frac{1}{2}}&=|\rho^{\frac{1}{2}}| \quad \text{if } \rho\geq 0,\\ \rho^{\frac{1}{2}}&=-i|\rho^{\frac{1}{2}}| \quad \text{if } \rho<0,\end{aligned}\right\} \tag{2.33}$$

with a similar definition of $\{(c_{11}+c_{22})^2-4\}^{\frac{1}{2}}$ in Eq. (2.28).

It is important to note that

$$r_+r_-=1. \tag{2.34}$$

If the cell potential is symmetrical, one can also express any self-matching solution in terms of $g(x)$ and $u(x)$:

$$f(x)=\alpha g(x)+\beta u(x). \tag{2.35}$$

For symmetrical cell potentials one has, by Eqs. (2.17) and (2.18),

$$\rho(E)=(c_{11}-1)/(c_{11}+1)=g_0'u_0/g_0u_0'. \tag{2.36}$$

Equation (2.22) demands

$$(\beta/\alpha)^2=(g_0g_0')/(u_0u_0')=(g_0/u_0)^2\rho. \tag{2.37}$$

The self-matching solutions then become

$$f_\pm(E;x)=\frac{1}{1\pm\rho^{\frac{1}{2}}}\left\{\frac{g(E;x)}{g_0}\pm\rho^{\frac{1}{2}}\frac{u(E;x)}{u_0}\right\}, \tag{2.38}$$

with

$$\sigma_\pm(E)=\pm\left(\frac{g_0'u_0'}{g_0u_0}\right)^{\frac{1}{2}}=\pm\frac{u_0'}{u_0}\rho^{\frac{1}{2}}. \tag{2.39}$$

The relation

$$\sigma_-=-\sigma_+ \tag{2.40}$$

is valid in general only for symmetrical cell potentials.

III. GENERAL SOLUTIONS OF THE PERIODIC WAVE EQUATION: BAND STRUCTURE OF THE PERMITTED ENERGY LEVELS

The self-matching cell solutions provide the basis for a particularly simple discussion of the properties of the general solution of the periodic wave equation. In Section II it has been shown that from any self-matching solution one can construct a corresponding ψ which changes only by a multiplying factor r from cell to cell. For each E we have then the solutions:

$$\psi_\pm(x)=r_\pm{}^n f_\pm(x-na), \quad \begin{cases} na<x<(n+1)a,\\ n=0,\pm1,\pm2,\cdots.\end{cases} \tag{3.1}$$

(The dependence of $f_\pm$ and $\psi_\pm$ on the parameter E is to be understood.) If f_+ and f_- are not identical, ψ_+ and ψ_- are independent solutions of a second-order differential equation: the general solution is accordingly expressible as

$$\psi(x)=C_+\psi_+(x)+C_-\psi_-(x). \tag{3.2}$$

The character of the general solution, and in particular its boundedness or non-boundedness, depends in any case on the value of ρ, which in turn depends on E. We shall here consider the characteristics of the ψ's as they depend on ρ, deferring for the moment a discussion of the occurrence of the various values of ρ.

Case 1: $\rho>0$

In this case $\rho^{\frac{1}{2}}$, $r_\pm$, β/α, $f_\pm$, and $\psi_\pm$ are all real. If $0<\rho<1$, $r_->1$. Then ψ_- increases in magnitude from cell to cell with increasing n; it may oscillate within each cell, but with an amplitude that increases exponentially as n, $x\to+\infty$. As $x\to-\infty$, $\psi_-\to 0$ exponentially. On the other hand, since $0<r_+<1$, $\psi_+\to 0$ exponentially as $x\to+\infty$, but increases exponentially as n, $x\to-\infty$. The general solution of the wave equation is thus unbounded as $x\to\infty$ unless $C_-=0$, and is unbounded as $x\to-\infty$ unless $C_+=0$; there is no physically acceptable solution of the wave equation.

If $\rho>1$, $r_-<-1$. The functions ψ_+ and ψ_- change sign from cell to cell, but it remains true that ψ_+ and ψ_- show exponential increases in amplitude as $x\to-\infty$ or $x\to+\infty$, respectively. Again there is no physically acceptable solution of the wave equation.

Case 2: $\rho<0$

In this case $\rho^{\frac{1}{2}}$ is imaginary, β/α is imaginary, and $f_\pm$ and $\psi_\pm$ are complex. Since $1+\rho^{\frac{1}{2}}$ and $1-\rho^{\frac{1}{2}}$ are complex conjugates, their ratio has modulus 1:

$$r_\pm=e^{\pm i\theta}. \tag{3.3}$$

Thus, ψ_+ and ψ_- change phase from cell to cell, but do not increase in magnitude: both are bounded as $x\to\pm\infty$. The same is true of the general solution ψ, for all values of C_+ and C_-; every solution of the wave equation is physically acceptable.

Case 3: $\rho=0$

If $\rho=0$, $c_{11}+c_{22}=2$, then $\sigma_+=\sigma_-$ [by Eq. (2.28)] and there is only one self-matching solution, with $r=1$. The corresponding ψ repeats itself exactly from cell to cell; it is periodic with period a, and is physically acceptable. (A normalization different from that of Eq. (2.23) may be necessary if σ becomes infinite as $\rho\to 0$; see Section V.)

An independent solution can be constructed as follows. As $\rho\to 0$, ψ_+ and ψ_- approach the common periodic form mentioned in the preceding paragraph. An independent solution for every $\rho\neq 0$ is given by $\psi_+(x)$

$-\psi_-(x)$. By Eqs. (3.1) and (2.29) we have

$$\psi_+(x)-\psi_-(x)=(r_+^n-r_-^n)y_1(x-na)\cdot +(\sigma_+r_+^n-\sigma_-r_-^n)y_2(x-na), \quad [na<x<(n+1)a]. \tag{3.4}$$

As $\rho\to 0$, this sequence of solution approaches zero everywhere, proportionally to $\sigma_+-\sigma_-$, as is evident from inspection of its form in the zeroth cell. Thus there is a solution of the form $(\psi_+-\psi_-)/(\sigma_+-\sigma_-)$ for each ρ, and this sequence approaches a non-zero limit, itself a solution of the wave equation, as $\rho\to 0$. A simple calculation with Eqs. (2.28), (2.31), (2.32) shows that

$$\begin{aligned}\psi_0(x)&=\lim_{\rho\to 0}\left\{\frac{\psi_+(x)-\psi_-(x)}{\sigma_+-\sigma_-}\right\}\\ &=nc_{12}y_1(x-na)+\left[1+\frac{n}{2}(c_{22}-c_{11})\right]y_2(x-na),\\ &\qquad [na<x<(n+1)a],\end{aligned} \tag{3.5}$$

is the desired independent solution for $\rho=0$. This increases in magnitude linearly as n, $x\to\pm\infty$.

The general solution of the wave equation in this case is expressible as

$$\psi(x)=C_0\psi_0(x)+C_+\psi_+(x). \tag{3.6}$$

It will be bounded only if $C_0=0$. Thus when $\rho=0$ there is (aside from an arbitrary multiplying constant) only one physically acceptable solution of the wave equation, approached by both ψ_+ and ψ_-.

Case 4: $\rho=\infty$

If $\rho=\infty$, $c_{11}+c_{22}=-2$, $\sigma_+=\sigma_-$, and there is only one self-matching solution, with $r=-1$. The corresponding ψ, approached by both ψ_+ and ψ_- as $\rho\to\infty$, changes from cell to cell only by a change in sign; it is periodic with period $2a$, and is physically acceptable.

An independent solution can be constructed as in Case 3. One finds here

$$\begin{aligned}\psi_0(x)&=\lim_{\rho\to 0}\left\{\frac{\psi_+(x)-\psi_-(x)}{\sigma_+-\sigma_-}\right\}\\ &=(-1)^{n+1}nc_{12}y_1(x-na)\\ &\quad+(-1)^n\left[1-\frac{n}{2}(c_{22}-c_{11})\right]y_2(x-na),\\ &\qquad [na\leq x\leq (n+1)a].\end{aligned} \tag{3.7}$$

This special solution also increases linearly in amplitude as x, $n\to\pm\infty$; so does the general solution (Eq. 3.6)) of the wave equation, unless $C_0=0$. Thus for $\rho=\infty$ there is only one independent physically acceptable solution of the wave equation, approached by both ψ_+ and ψ_-.

In summary, all solutions of the wave equation are physically acceptable for E such that $\rho<0$; no solutions are acceptable for E such that $\rho>0$; and there is just one independent physically acceptable solution for E such that $\rho=0$, $\pm\infty$.

To complete the proof of the band structure of the permitted energies in periodic potentials, it is only necessary to prove that ρ is positive for some ranges of E, negative for others. A discussion of this problem for the general case can be based on the work of Kramers.[8] Here it will suffice to consider the simpler case of symmetrical cell potentials, for which

$$\rho(E)=g_0'u_0/g_0u_0'.$$

The behavior of g_0, g_0', u_0, u_0' as functions of E follows from the most familiar properties of the solutions of the one-dimensional wave equation. The solutions $g(x)$ and $u(x)$ are convex to the x axis where $W(x)>E$, and concave (oscillatory) where $W(x)<E$. For very low E (certainly for $E<W_{\min}$) the regions of convexity are dominant, and one must have g_0, $u_0'>0$, g_0', $u_0<0$. As E increases, $g(x)$ and $u(x)$ become oscillatory, and then oscillate with decreasing wave-length; with increasing E, maxima and zeros appear alternately at $x=0$ and move monotonely to increasing x. It follows that g_0, g_0', u_0, u_0' oscillate between positive and negative values. It is easily seen that, as E increases, g_0' is the first of these quantities to vanish, and that its zeros thereafter alternate with zeros of g_0; similarly, u_0' vanishes before u_0, and these zeros alternate thereafter.

From the relation $g_0u_0'-g_0'u_0\equiv 1$ it follows that con-

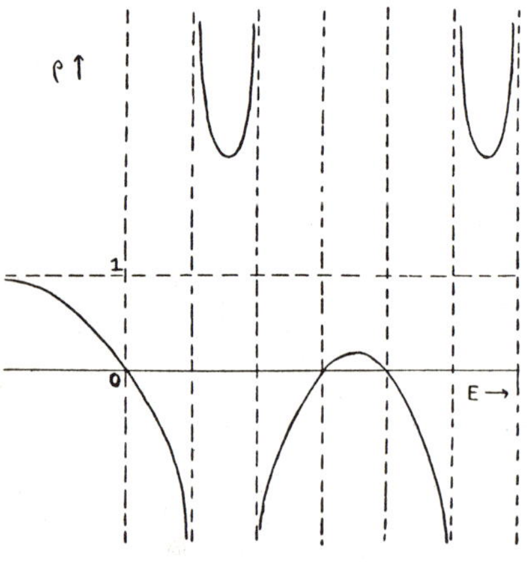

FIG. 1. Qualitative character of $\rho(E)$ for any periodic problem.

[8] See reference 6. If one chooses Kramers' ϕ_1, and ϕ_2 as the y_1 and y_2 of this paper, his α_{11} and α_{22} become identical with the c_{11} and c_{22} defined here. Kramers shows that $\alpha_{11}+\alpha_{22}$, or $c_{11}+c_{22}$, is an oscillatory function of E, crossing the range from -2 to $+2$ (corresponding to $\rho>0$) and from time to time assuming values greater than 2 or less than $-2(\rho<0)$ It follows easily that ρ varies as illustrated in Fig. 1.

secutive zeros of g_0u_0' cannot be separated by a single zero of $g_0'u_0$, or conversely; if this were to occur, g_0u_0' and $-g_0'u_0$ would necessarily both be negative in some range of E, which is impossible. From this, and the facts mentioned in the preceding paragraph, it follows that the quantities under consideration vanish in the following order:

$$g_0',\ (g_0,\ u_0'),\ (g_0',\ u_0),\ (g_0,\ u_0'),\ (g_0',\ u_0)\cdots.$$

The order (within a parenthesis) of the zeros of g_0 and u_0' or of g_0' and u_0 is not fixed, but varies with the character of the potential considered; indeed these zeros may coincide.

The nature of the function $\rho(E)$ is determined sufficiently by consideration of the sequence of its zeros and infinities, together with the fact that it cannot assume the value 1, since $g_0'/g_0=u_0'/u_0$ would imply $g(x)\equiv u(x)$. The preceding paragraphs show that, for low E, $\rho>0$. As E increases, ρ goes through zero with g_0', and thereafter has alternate infinities and zeros in pairs. Its behavior must therefore be that shown in Fig. 1. Positive values of ρ, forbidden bands of E, occur below the first zero of g_0', and thereafter between consecutive zeros of g_0 and u_0', and of g_0' and u_0. If these zeros should happen to coincide, the forbidden band in question will disappear. The constant potential is a very special periodic potential for which all forbidden bands except the lowest disappear in this way.

IV. SPECIAL SOLUTIONS OF THE PERIODIC WAVE EQUATION; EFFECTIVE MOMENTUM

We now define the quantity

$$p(E)=(h/ia)\ \ln r_+(E), \tag{4.1}$$

to be identified as the effective momentum associated with the energy E. In a permitted band, for which $r_+=e^{i\theta}$ [Eq. (3.3)],

$$p=(h/a)\theta \tag{4.2}$$

is a real quantity. In a forbidden band, where r_+ is real, $p(E)$ must be imaginary or complex. In any case,

$$r_\pm{}^n=\exp\left(\pm\frac{i}{h}pna\right) \tag{4.3}$$

The solutions $\psi_\pm$ of the wave equation, as defined by Eq. (3.1), can now be expressed as

$$\psi_\pm(x)=\exp[\pm(i/h)px]\times\{\exp[\mp(i/h)p(x-na)]f_\pm(x-na)\},\quad \begin{cases} na<x<(n+1)a,\\ n=0,\ \pm1,\ \pm2\cdots.\end{cases} \tag{4.4}$$

Defining

$$P_\pm(E;x)=\exp[\mp(i/h)p(E)(x-na)]f_\pm(E,x-na),\quad \begin{cases} na<x<(n+1)a,\\ n=0,\ \pm1,\ \pm2\cdots,\end{cases} \tag{4.5}$$

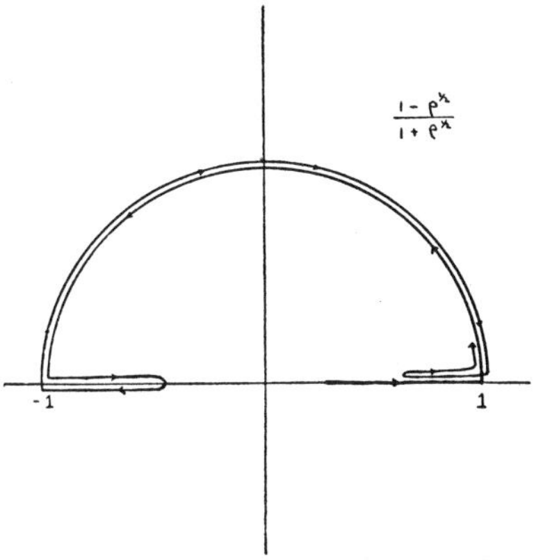

FIG. 2. Path in the complex plane of the point $(1-\rho^{\frac{1}{2}})/(1+\rho^{\frac{1}{2}})$ as E increases.

we can write these solutions as

$$\psi_\pm(E;x)=\exp\left[\frac{i}{h}p(E)x\right]P_\pm(E;x). \tag{4.6}$$

It is obvious that P_+ and P_- are periodic in x with period a: these solutions of the periodic wave equation are thus of the familiar form discovered by Floquet. The role of $p(E)$ in these solutions is sufficient to identify this quantity as the effective momentum of crystal theory, here extended by the assignment of complex values for energies in the forbidden bands.

The logarithm in Eq. (4.1) is multiple-valued, with values differing by multiples of $2\pi i$; thus $p(E)$ is defined only to multiples of $2\pi h/a=h/a$. Use of the convention expressed in Eq. (2.33), and of the principal value of the logarithm, limits p in the permitted bands to the positive values assumed by the "reduced momentum." Figure 2 sketches the path traversed in the complex plane by the point $(1-\rho^{\frac{1}{2}})/(1+\rho^{\frac{1}{2}})$, as E increases; Fig. 3 shows the corresponding values of the real part p_r and imaginary part p_i of p. If one takes $\rho^{\frac{1}{2}}=\pm i|\rho^{\frac{1}{2}}|$, with alternating $-$ and $+$ signs in the consecutive permitted bands, one assures for p the behavior sketched in Fig. 4. Here p takes on the values of the total effective momentum in the permitted bands, and reduces to the true momentum when the periodic potential becomes constant.

With these conventions, ψ_+ and ψ_- have in the permitted band the character of complex waves progressing to the right and left, respectively, multiplied by appropriate periodic functions. In the forbidden bands, on the other hand, they are real exponentials, attenuating to the right or left, respectively, multiplied by functions with period a or $2a$, depending on the value of p_r.

Let p denote the quantity defined in Fig. 3. We now

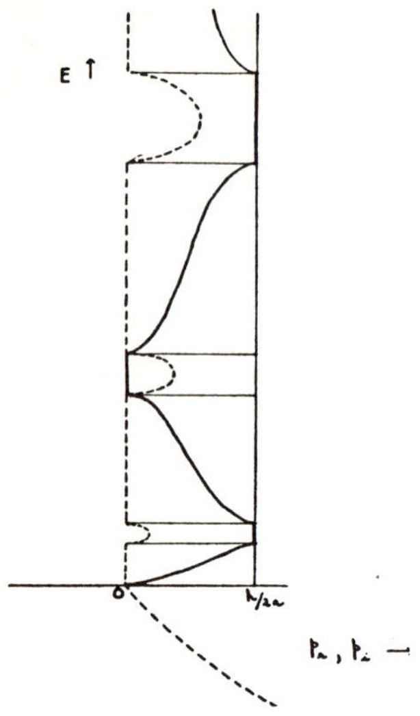

FIG. 3. Dependence on E of the real part p_r (solid line) and imaginary part p_i (dashed line) of the reduced effective momentum.

define

$$\left.\begin{aligned} p_d &= p, \text{ if } 0 \leq p \leq h/4a, \\ p_d &= p-(h/2a), \text{ if } (h/4a) < p \leq h/2a. \end{aligned}\right\} \quad (4.7)$$

This may be expressed also as

$$p_d = p - p_e, \quad (4.8)$$

where p is defined as in Fig. 3 *or* Fig. 4, and p_e is its value at the nearest band edge. The behavior of p_d is indicated in Fig. 5; it is pure real in the permitted bands, and pure imaginary in the forbidden bands.

In terms of p_d, one can write

$$\psi_\pm(E;x) = \exp\left(\pm\frac{i}{h}p_d x\right)P_{d\pm}(E;x), \quad (4.9)$$

where

$$P_{d\pm}(E;x) = \exp\left(\pm\frac{i}{h}p_e x\right)P_\pm(E;x) \quad (4.10)$$

is periodic with period a or $2a$, according as p_e is 0 or $h/2a$. At every band edge P_{d+} and P_{d-} approach a common form, that of the physically acceptable solution at that band edge.

V. SOLUTIONS FOR ENERGIES NEAR A BAND EDGE

The following discussion of the problem for energies near a band edge will deal mainly with symmetric cell potentials.

A band edge occurs whenever ρ vanishes or becomes infinite. In the first case one has (Eqs. (4.1); (2.31))

$$p(E) \cong (2hi/a)\rho^{\frac{1}{2}}, \quad (\rho \text{ small}), \quad (5.1)$$

and in the second

$$p(E) \cong (h/2a) + (2hi/a)\rho^{-\frac{1}{2}} \quad (\rho \text{ very large}). \quad (5.2)$$

Then

$$\left.\begin{aligned} p_d^2 &\cong -(4h^2/a^2)\rho, \quad (\rho \text{ small}), \\ p_d^2 &\cong -(4h^2/a^2)1/\rho, \quad (\rho \text{ large}). \end{aligned}\right\} \quad (5.3)$$

The quantity p_d^2 is positive in a permitted band, but negative in a forbidden band.

A band edge occurs when any one of the quantities g_0, g_0', u_0, u_0' vanishes. Sufficiently near this band edge one can treat the vanishing quantity as linear in E, and can neglect the dependence on E of the three non-vanishing quantities, replacing them by corresponding values at the band edge: g_e, g_e', u_e, u_e'. That is, one can treat ρ and ρ^{-1}, and hence p_d^2, as linear functions of E in this region, writing

$$E - E_e = (p_d^2/2m^*). \quad (5.4)$$

The constant of proportionality m^* will vary from one band edge to the other. It is, of course, the effective mass of the particle at the band edge, and is positive for the lower edge of a permitted band, negative for the upper edge.

Solutions of the wave equation near the band edges can be approximated as follows.

The solutions $g(E;x)$ and $u(E;x)$ (like $y_1(x)$ and $y_2(x)$) can be expanded in powers of $E-E_0$ about any value of E_0. We shall here carry out such an expansion about the band-edge energy E_e. The periodic wave equation can be written as

$$\frac{d^2\psi(x)}{dx^2} + \frac{2m}{h^2}[E_e - W(x)]\psi(x) = -\frac{2m}{h^2}(E-E_e)\psi(x). \quad (5.5)$$

As independent solutions of the homogeneous equation

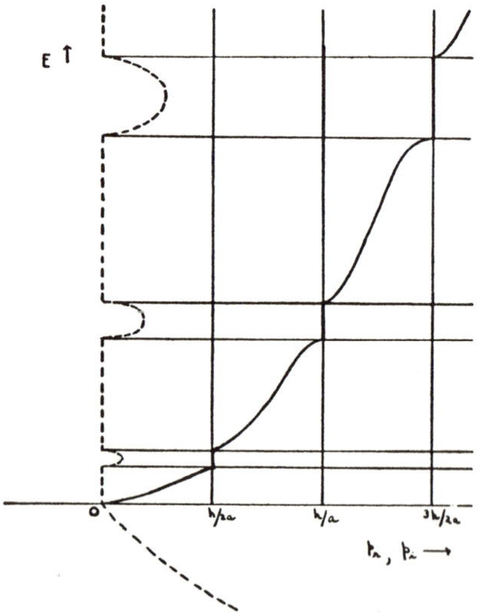

FIG. 4. Dependence on E of the real part p_r (solid line) and imaginary part p_i (dashed line) of the total effective momentum.

we have $g(E_e;x)=g_e(x)$ and $u(E_e;x)=u_e(x)$, such that

$$g_e(x)u_e'(x)-g_e'(x)u_e(x)\equiv 1. \tag{5.6}$$

If one replaces the quantity on the right by an arbitrary function $\phi(x)$, the inhomogeneous equation has as its general solution.

$$\psi=Ag_e(x)+Bu_e(x)-g_e(x)\int_{a/2}^{x}\phi(x)u_e(x)dx+u_e(x)\int_{a/2}^{x}\phi(x)g_e(x)dx. \tag{5.7}$$

Let

$$\xi=(2m/h^2)(E-E_e). \tag{5.8}$$

Replacing $\phi(x)$ by $-\xi g(x)$ and choosing the constants A and B so that ψ satisfies the defining conditions for $g(x)$, one finds

$$g(x)=g_e(x)+\xi\left\{g_e(x)\left\{\int_{a/2}^{x}g(x)u_e(x)dx-u_e(x)\int_{a/2}^{x}g(x)g_e(x)dx\right\}\right. . \tag{5.9}$$

Similarly,

$$u(x)=u_e(x)+\xi\left\{g_e(x)\int_{a/2}^{x}u(x)u_e(x)dx-u_e(x)\int_{a/2}^{x}u(x)g_e(x)dx\right\}. \tag{5.10}$$

Let

$$\left.\begin{aligned}g(x)&=g_e(x)+\sum_{\tau=1}\xi^\tau g_\tau(x),\\ u(x)&=u_e(x)+\sum_{\tau=1}\xi^\tau u_\tau(x).\end{aligned}\right\} \tag{5.11}$$

Substituting these expansions into Eqs. (5.9) and (5.10), and equating coefficients of equal powers of ξ, one finds

$$g_{\tau+1}(x)=g_e(x)\int_{a/2}^{x}g_\tau(x)u_e(x)dx-u_e(x)\int_{a/2}^{x}g_\tau(x)g_e(x)dx, \tag{5.12}$$

$$u_{\tau+1}(x)=g_e(x)\int_{a/2}^{x}u_\tau(x)u_e(x)dx-u_e(x)\int_{a/2}^{x}u_\tau(x)g_e(x)dx. \tag{5.13}$$

By repeated quadratures one can thus carry the series expansions as far as is desired.[9]

[9] The same general method has been used by L. Brillouin, Quart. App. Math. 6, 167 (1948).

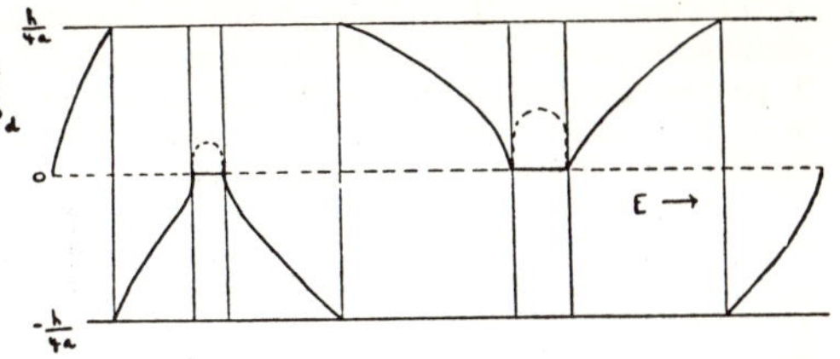

FIG. 5. Dependence on E of the real part (solid line) and imaginary part (dashed line) of the effective momentum p_d.

Let

$$I_{gg}=\int_0^{a/2}g_e(x)g_e(x)dx, \tag{5.14}$$

and let I_{gu} and I_{uu} be similarly defined. Then one finds easily

$$\left.\begin{aligned}g_0&=g_e+\xi[-I_{gu}g_e+I_{gg}u_e]+\cdots,\\ g_0'&=g_e'+\xi[-I_{gu}g_e'+I_{gg}u_e']+\cdots,\\ u_0&=u_e+\xi[-I_{uu}g_e+I_{gu}u_e]+\cdots,\\ u_0'&=u_e'+\xi[-I_{uu}g_e'+I_{gu}u_e']+\cdots.\end{aligned}\right\} \tag{5.15}$$

Use of these relations with Eqs. (5.3), (5.4), and (5.6) leads to expressions for m^* that depend on which of the four quantities vanishes at the band edge:

$$\begin{aligned}\frac{m^*}{m}&=\pm(2/a)^2u_eu_e'I_{gg}, &&\begin{cases}+ \text{ if } g_e=0\\ - \text{ if } g_e'=0\end{cases}\\ \frac{m^*}{m}&=\pm(2/a)^2g_eg_e'I_{uu}. &&\begin{cases}+ \text{ if } u_e=0\\ - \text{ if } u_e'=0\end{cases}\end{aligned} \tag{5.16}$$

In the same way one can show that

$$\left.\begin{aligned}y_1(x)&=y_{1e}(x)+\sum_{\tau=1}\xi^\tau y_{1\tau}(x),\\ y_2(x)&=y_{2e}(x)+\sum_{\tau=1}\xi^\tau y_{2\tau}(x),\end{aligned}\right\} \tag{5.17}$$

where

$$\begin{aligned}y_{1,\tau+1}(x)&=y_{1e}(x)\int_0^x y_{1\tau}(x)y_{2e}(x)dx-y_{2e}(x)\int_0^x y_{1\tau}(x)y_{1e}(x)dx,\\ y_{1,\tau+1}(x)&=y_{1e}(x)\int_0^x y_{2\tau}(x)y_{2e}(x)dx-y_{2e}(x)\int_0^x y_{2\tau}(x)y_{1e}(x)dx.\end{aligned} \tag{5.18}$$

The functions $f_\pm(E;x)$ and $P_\pm(E;x)$ cannot be expanded in powers of $E-E_e$, but they can be expanded in powers, sometimes including the first negative one, of p_d. For example, when the cell potential is symmetrical, the functions $y_1(x)$ and $y_2(x)$ in

$$f_\pm(E;x)=y_1(x)\pm\sigma_+(E)y_2(x) \tag{5.19}$$

are expansible in even powers of p_d, but σ_+ is expansible in odd powers. The required expansions depend on whether it is g_0, g_0', u_0, or u_0' that vanishes at the edge in question. We note here only the leading terms in the expansions for σ_+:

$$\left.\begin{aligned}
g_e=0\quad &\sigma_+=+\frac{2ih}{a}\frac{u_e'}{u_e}p_d^{-1}+O(p_d),\\
g_e'=0\quad &\sigma_+=-\frac{ia}{2h}\frac{u_e'}{u_e}p_d+O(p_d^3),\\
u_e=0\quad &\sigma_+=\frac{2ih}{a}\frac{g_e'}{g_e}p_d^{-1}+O(p_d),\\
u_e'=0\quad &\sigma_+=-\frac{ia}{2h}\frac{g_e'}{g_e}p_d+O(p_d^3).
\end{aligned}\right\} \tag{5.20}$$

Expansions for $f_\pm$ and $P_\pm$ will be given only for the case of a band edge with $g_e'=0$, where these functions remain finite:

$$f_\pm(x)=\frac{g_e(x)}{g_e}\pm\frac{iap_d}{2h}\left\{\frac{g_e(x)}{g_e}-\frac{u_e(x)}{u_e}\right\}+\cdots, \tag{5.21}$$

$$P_\pm(x)=\frac{g_e(x)}{g_e}\pm\frac{ip_d}{h}\left\{\left(\frac{a}{2}-x\right)\frac{g_e(x)}{g_e}-\frac{a}{2}\frac{u_e(x)}{u_e}\right\}+\cdots. \tag{5.22}$$

In both expansions the third term is independent of the subscript sign.

In treating the perturbed-periodic wave equation one needs to know the behavior of $\{P_{d+}(E;x)+P_{d-}(E;x)\}/2$ and $\{P_{d+}(E;x)-P_{d-}(E;x)\}/2\sigma(E)$ near the band edges. By Eq. (4.5), (4.10) and (5.19), one has

$$P_{d\pm}(E;x)=\exp\left(\mp\frac{i}{h}p_dx\right)[y_1(x)\pm\sigma_+y_2(x)],\qquad (0<x<a). \tag{5.23}$$

Then

$$\left.\begin{aligned}
\frac{P_{d+}(E;x)+P_{d-}(E;x)}{2}&=\cos\left\{\frac{p_dx}{h}\right\}y_1(x)-i\sigma_+\sin\left\{\frac{p_dx}{h}\right\}y_2(x),\\
\frac{P_{d+}(E;x)-P_{d-}(E;x)}{2\sigma_+}&=-\frac{i}{\sigma_+}\sin\left\{\frac{p_dx}{h}\right\}y_1(x)+\cos\left\{\frac{p_dx}{h}\right\}y_2(x).
\end{aligned}\right\} \tag{5.24}$$

Expanding the quantities on the right in powers of p_d, with use of Eq. (5.20), one obtains

$$\left.\begin{aligned}
\frac{P_{d+}(E;x)+P_{d-}(E;x)}{2}&=\left[y_{1e}(x)-\frac{i}{h}(p_d\sigma_+)_e xy_{2e}(x)\right]+O(p_d^2),\\
&\qquad (0<x<a)\\
\frac{P_{d+}(E;x)-P_{d-}(E;x)}{2\sigma_+}&=\left[-\frac{i}{h}\left(\frac{p_d}{\sigma_+}\right)_e xy_{1e}(x)+y_{2e}(x)\right]+O(p_d^2),
\end{aligned}\right\} \tag{5.25}$$

where

$$\left.\begin{aligned}
(p_d\sigma_+)_e&=\frac{2ih}{a}\frac{u_e'}{u_e} &&\text{if } g_e=0,\\
&=0 &&\text{if } g_e'=0,\\
&=\frac{2ih}{a}\frac{g_e'}{g_e} &&\text{if } u_e=0,\\
&=0 &&\text{if } u_e'=0,
\end{aligned}\right\} \tag{5.26}$$

and

$$\left.\begin{aligned}
\left(\frac{p_d}{\sigma_+}\right)_e&=0 &&\text{if } g_e=0,\\
&=\frac{2ih}{a}\frac{u_e}{u_e'} &&\text{if } g_e'=0,\\
&=0 &&\text{if } u_e=0,\\
&=\frac{2ih}{a}\frac{g_e}{g_e'} &&\text{if } u_e'=0.
\end{aligned}\right\} \tag{5.27}$$

Similar expansions exist for $\{\psi_+(E;x)+\psi_-(E;x)\}/2$ and $\{\psi_+(E;x)-\psi_-(E;x)\}/2\sigma$; the leading terms are identical with those given in Eq. (5.25).

R 172 Philips Res. Rep. 6, 303-310, 1951

WAVE PROPAGATION IN ONE-DIMENSIONAL STRUCTURES

by J. M. LUTTINGER 539.11:534.2

Summary

A conjecture of Saxon and Hutner on the forbidden energy levels of any substitutional alloy of a certain one-dimensional crystal model is proven. Analogous results are derived for wave transmission down a line loaded with two-terminals.

Résumé

On démontre la verité d'une hypothèse de Saxon et Hutner sur les niveaux d'énergie défendus pour tout alliage de remplacement d'un certain modèle de cristal undimensionnel. On tire des conclusions analogues pour la transmission d'une onde dans un ligne de dipôles.

Zusammenfassung

Es wird die Gültigkeit einer Vermutung von Saxon und Hutner bezüglich der unzulässigen Energiepegel bei irgendeiner Ersatzlegierung eines bestimmten eindimensionalen Kristallmodells bestätigt. Ähnliche Ergebnisse werden für Wellenübertragung in einer mit Zweipolen belasteten Leitung abgeleitet.

Introduction

In a recent interesting paper, Saxon and Hutner [1]) have been led to a conjecture concerning the energy levels in binary solid solutions. The result they found in several special cases was the following. Suppose one has a pure crystal of A-type atoms and a pure crystal of B-type atoms (both being represented by δ-function potentials according to the model of Kronig and Penney). Then the forbidden energy levels that are common to pure A and B crystals will also be present in any arbitrary substitutional alloy of A and B atoms. In section 1, a proof of this statement is given. We set the problem up in a slightly more general manner, using a different technique, than Saxon and Hutner. In section 2, it is shown that the propagation of waves on a transmission line loaded with two-terminals leads to a very similar mathematical formalism, and consequently to a similar theorem.

1. Proof of the Saxon-Hutner conjecture

Let us consider a crystal made up of atoms of type A whose potential does not extend out of the cell in which the atom is contained. This assumption will make it possible to discuss in a simple way alloys of atoms of type A with those of type B, which are also assumed to have this latter property.

It proves convenient to introduce the y-solutions of James [2]). These are the solutions $y_1(E,x)$, $y_2(E,x)$ of the Schrödinger equation in the first cell that assume the initial values

$$y_1(E,0) = 1, \qquad y_1'(E,0) = 0,$$
$$y_2(E,0) = 0, \qquad y_2'(E,0) = 1,$$

where a prime denotes differentiation with respect to x. With James, we define

$$A_{11} = y_1(E,a), \qquad A_{12} = y_2(E,a),$$
$$A_{21} = y_1'(E,a), \qquad A_{22} = y_2'(E,a),$$

in which a is the cell length. The constancy of the Wronskian of y_1 and y_2 implies that $A_{11}A_{22} - A_{12}A_{21} = 1$, which is the same as saying that the determinant of the matrix $A = (A_{ik})$ is unity.

Now let us say that we have a solution

$$\Psi_1(x) = a_1 y_1(x) + b_1 y_2(x)$$

in the first cell. Then, since y_1 and y_2 are linearly independent, and the potential is periodic, the solution in the second cell may be written

$$\Psi_2(x) = a_2 y_1(x-a) + b_2 y_2(x-a) .$$

However, the continuity of Ψ and its first derivative implies that $a_2 = A_{11}a_1 + A_{12}b_1$, $b_2 = A_{21}a_1 + A_{22}b_1$, or, in matrix notation,

$$\begin{pmatrix} a_2 \\ b_2 \end{pmatrix} = \begin{pmatrix} A_{11} & A_{12} \\ A_{21} & A_{22} \end{pmatrix} \begin{pmatrix} a_1 \\ b_1 \end{pmatrix} .$$

We may continue this process indefinitely until we come to the $(n+1)$th cell, giving

$$\begin{pmatrix} a_{n+1} \\ b_{n+1} \end{pmatrix} = A^n \begin{pmatrix} a_1 \\ b_1 \end{pmatrix} .$$

Let us consider a ring of n atoms. Then the solution in the $(n + 1)$th cell must be the same as in the first cell. That is, we must have for the allowed energies

$$A^n \begin{pmatrix} a_1 \\ b_1 \end{pmatrix} = \begin{pmatrix} a_1 \\ b_1 \end{pmatrix} . \tag{1}$$

This means that the allowed energy values correspond to the eigenvalue 1 of the matrix A^n, which in turn shows that they must correspond to the nth roots of unity for the matrix A. That is, the allowed energy levels are given by setting the eigenvalues of A equal to $\exp(\pm ik\pi/n)$, where k is an integer satisfying $0 \leqslant k \leqslant n-1$.

When n is very large the different values of k give us, for $k\pi/n$, practically the continuum from 0 to π. Now this is what we usually mean by an "allowed band" in a finite crystal, so that the allowed band is determined by those values of the energy which make the eigenvalues of A complex numbers whose absolute value is unity. Since the determinant of A is unity, the secular equation for the eigenvalues λ of A reads $\lambda^2 - (\mathrm{tr}A)\lambda + 1 = 0$, or

$$\lambda_\pm = \tfrac{1}{2}[\mathrm{tr}A \pm \sqrt{(\mathrm{tr}A)^2 - 4}]\,.$$

Now, if $|\mathrm{tr}A| \leqslant 2$, then $\lambda_+ = \lambda_-^*$ (the asterisk denoting the conjugate) and $|\lambda_+|^2 = |\lambda_-|^2 = 1$, so that we have allowed values. On the other hand, if $|\mathrm{tr}A| > 2$, one of the eigenvalues is of absolute value greater than unity and the other of absolute values less than unity. That means that the condition for a forbidden level is simply

$$|\mathrm{tr}A| > 2\,.$$

For simplicity, let us now assume that the potential of an A atom is symmetric about the position of the atom, and let us further assume that each atom is located at the centre of its corresponding cell. In this case it is easy to see (James, loc. cit.) that $A_{11} = A_{22}$, so that the matrix A takes the form

$$A = \begin{pmatrix} A_{11} & A_{12} \\ A_{21} & A_{11} \end{pmatrix}, \tag{2}$$

with $A_{12}A_{21} = A_{11}{}^2 - 1$. The condition for forbidden levels then becomes

$$|A_{11}| > 1\,.$$

We are interested in the form A takes for forbidden energies, and therefore we consider the latter case. If $|A_{11}| > 1$, then we may certainly write $A_{11} = \pm \cosh a$, where a is positive. Let us define a matrix $\tilde{A}$ by the relationship

$$\tilde{A} = \pm\,(A \mp U \cosh a)/\sinh a\,, \quad A = \pm\,(U \cosh a + \tilde{A} \sinh a)\,,$$

where U is the unit matrix. Now, by direct computation we see that $\mathrm{tr}\,\tilde{A} = 0$, $\det \tilde{A} = -1$. Since every matrix satisfies its own secular equation we have further

$$\tilde{A}^2 - (\mathrm{tr}\,\tilde{A})\,\tilde{A} + (\det \tilde{A})\,U = O\,,$$

or $\tilde{A}^2 = U$. From this it follows immediately that

$$A = \pm \exp(a\tilde{A})\,. \tag{3}$$

The matrix $\tilde{A}$ may be put in a somewhat more elegant form. We have

$$\tilde{A} = \pm \{A_{11}{}^2 - 1\}^{-\frac{1}{2}} \begin{pmatrix} 0 & A_{12} \\ A_{21} & 0 \end{pmatrix},$$

$\pm$ depending on whether $A_{11} > 1$ or $A_{11} < -1$. Now, $A_{12}A_{21} = A_{11}{}^2 - 1 = (A_{11} + 1)(A_{11} - 1)$. Therefore we may put $A_{12} = (A_{11} + 1)\,\lambda_A$, $A_{21} = (A_{11} - 1)/\lambda_A$, so that finally

$$\tilde{A} = \begin{pmatrix} 0 & \sqrt{\dfrac{A_{11}+1}{A_{11}-1}}\,\lambda_A \\ \sqrt{\dfrac{A_{11}-1}{A_{11}+1}}\,\lambda_A^{-1} & 0 \end{pmatrix} \tag{4}$$

A very simple expression for λ_A may be obtained by introducing even and odd (with respect to the centre of the cell) solutions as follows [2]. Let, in the first cell, $g(E,x)$ be an even solution of the Schrödinger equation such that $g(E,a/2) = 1$, and let $u(E,x)$ be an odd solution such that $u'(E,a/2) = 1$. (It is only possible to introduce these functions when the potential in each unit cell is symmetric about the centre of the cell.) Further, define

$$g = g(E) = g(E,0), \qquad u = u(E) = u(E,0),$$
$$g' = g'(E) = g'(E,0), \qquad u' = u'(E) = u'(E,0).$$

Then it is easy to see (James, loc.cit.) that

$$A_{11} = A_{22} = u'g + ug', \quad A_{12} = -2\,ug, \quad A_{21} = -2\,u'g', \quad u'g - ug' = 1,$$

and consequently

$$A_{12} = -2\,ug = (A_{11} + 1)\lambda_A = (u'g + ug' + 1)\lambda_A = 2\,u'g\lambda_A,$$

or

$$\lambda_A = -\,u/u'\,. \tag{5}$$

Exactly the same process may be carried out for the case of a lattice of pure B atoms. In the regions forbidden for pure B we may write the corresponding matrix $\boldsymbol{B}$ as

$$\boldsymbol{B} = \pm \exp(\beta\tilde{\boldsymbol{B}})\,, \tag{6}$$

$\pm$ being taken when tr $\boldsymbol{B} > 2$ or < -2 respectively; β is a positive real number, and the matrix $\tilde{\boldsymbol{B}}$ is of the form

$$\tilde{\boldsymbol{B}} = \begin{pmatrix} 0 & \sqrt{\dfrac{B_{11}+1}{B_{11}-1}}\,\lambda_B \\ \sqrt{\dfrac{B_{11}-1}{B_{11}+1}}\,\lambda_B^{-1} & 0 \end{pmatrix}, \tag{7}$$

$\lambda_B = -u/u'$, u being evaluated for the potential of atoms of type B.

Now let us consider an arbitrary alloy of A-type and B-type atoms. The most general alloy may be represented by a ring which consists first of

r_1 A atoms, then s_1 B atoms, r_2 A atoms, s_2 B atoms, etc., until r_k A atoms, s_k B atoms. Here the total number of atoms present, n, is given by

$$\sum_{\nu=1}^{k} (r_\nu + s_\nu) = n .$$

Because of our assumption that the potential vanishes outside the cell in which the atom is found, one sees easily that the conditions (1) on the allowed energy levels is replaced by

$$\boldsymbol{B}^{s_k} \boldsymbol{A}^{r_k} \dots \boldsymbol{B}^{s_1} \boldsymbol{A}^{r_1} \begin{pmatrix} a_1 \\ b_1 \end{pmatrix} = \begin{pmatrix} a_1 \\ b_1 \end{pmatrix}.$$

This means as before that the matrix $\boldsymbol{B}^{s_k} \boldsymbol{A}^{r_k} \dots \boldsymbol{B}^{s_1} \boldsymbol{A}^{r_1}$ must have the eigenvalue unity. However, its determinant is also unity, and thus both eigenvalues must be unity. Consequently, the condition for an allowed energy under periodic boundary conditions is

$$\operatorname{tr} (\boldsymbol{B}^{s_k} \boldsymbol{A}^{r_k} \dots \boldsymbol{B}^{s_1} \boldsymbol{A}^{r_1}) = 2 . \tag{8}$$

Using now (3) and (6), which are the forms $\boldsymbol{A}$ and $\boldsymbol{B}$ taken in their forbidden zones, we shall show that (8) cannot be satisfied, under the additional hypothesis that we are dealing with δ-function potentials of the Kronig-Penney model.

For convenience let L denote the absolute value of the left-hand side of equation (8). Then, substituting (3) and (6), we obtain

$$L = \left|\operatorname{tr} \left\{ e^{s_k \beta \tilde{\boldsymbol{B}}} \dots e^{r_1 a \tilde{\boldsymbol{A}}} \right\}\right| =$$
$$= \left|\operatorname{tr} \left\{ (\boldsymbol{U} \cosh s_k\beta + \tilde{\boldsymbol{B}} \sinh s_k\beta) \dots (\boldsymbol{U} \cosh r_1 a + \tilde{\boldsymbol{A}} \sinh r_1 a) \right\}\right| ,$$

where we have used, e.g.,

$$e^{r_1 a \tilde{\boldsymbol{A}}} = \boldsymbol{U} \cosh r_1 a + \tilde{\boldsymbol{A}} \sinh r_1 a .$$

Let us think of this equation as multiplied out. Since $\operatorname{tr} \tilde{\boldsymbol{A}} = \operatorname{tr} \tilde{\boldsymbol{B}} = 0$ and $\tilde{\boldsymbol{A}}^2 = \tilde{\boldsymbol{B}}^2 = \boldsymbol{U}$, all the terms containing an odd number of matrices will give zero. Thus we get a series of the form

$$L = \left|\operatorname{tr} \left[\sum_{\nu\ 0}^{k} (\tilde{\boldsymbol{A}} \tilde{\boldsymbol{B}})^\nu q_\nu \right]\right| \tag{9}$$

where all the coefficients q_ν are positive, since they are sums and products of hyperbolic functions of positive arguments.

In order to investigate (9) more closely it is necessary to investigate $\operatorname{tr} (\tilde{\boldsymbol{A}} \tilde{\boldsymbol{B}})^\nu$. Now the product $\tilde{\boldsymbol{A}} \tilde{\boldsymbol{B}}$ is diagonal, and one easily sees that

$$\tilde{A}\tilde{B} = \begin{pmatrix} \varkappa & 0 \\ 0 & 1/\varkappa \end{pmatrix}, \qquad \varkappa = \sqrt{\left|\frac{(A_{11}+1)(B_{11}-1)}{(A_{11}-1)(B_{11}+1)}\right| \frac{\lambda_A}{\lambda_B}}.$$

Therefore,

$$(\tilde{A}\tilde{B})^\nu = \begin{pmatrix} \varkappa^\nu & 0 \\ 0 & \varkappa^{-\nu} \end{pmatrix}, \qquad \mathrm{tr}\,(\tilde{A}\tilde{B})^\nu = \varkappa^\nu + \varkappa^{-\nu}$$

If $\varkappa$ is positive the function $\varkappa^\nu + \varkappa^{-\nu}$ has a minimum of 2, and we then find

$$\mathrm{tr}\,(\tilde{A}\tilde{B})^\nu \geqslant 2\,. \tag{10}$$

Let us temporarily assume that $\varkappa$ *is* positive, and see what the consequences are. Using (10) in (9) we find $L \geqslant 2 \sum_{\nu=0}^{k} q_\nu \geqslant 2q_0$. However, as one easily sees q_0 is of the form $\cosh r_1\alpha \cosh s_1\beta \ldots \cosh r_k\alpha \cosh s_k\beta$ plus other non-negative terms which arise from cross terms in the multiplication of the original expressions for A^{r_1}, B^{s_1}, etc. Since all the hyperbolic cosines are greater than unity, clearly $q_0 > 1$ and we have $L > 2$, which contradicts (8). Therefore we have shown (for $\varkappa > 0$) that the energy levels forbidden for A and B are forbidden for any alloy of A and B. It only remains to verify the assumption that $\varkappa$ is always positive in common forbidden bands for A and B. Here we must for the first time make use of the fact that we are dealing with the Kronig-Penney model. If the potentials of A and B are just δ-functions of different strengths, then the odd solution u will be completely uneffected *) by the presence of this potential. Therefore the solution u is the same for A and B, and thus $\lambda_A = \lambda_B$, which implies $\varkappa \geqslant 0$ and completes the proof of the theorem.

2. Transmission line loaded with two-terminals

Suppose we have a section of a transmission line of length l, which has a two-terminal network of impedance $Z/2$ placed in the centre of each of its circuits (see fig. 1). Clearly there will be a linear relationship between the voltage and current at the output (V_2, I_2) and the voltage and current at the input (V_1, I_1). We may write

$$\begin{pmatrix} V_2 \\ I_2 \end{pmatrix} = \begin{pmatrix} A_{11} & A_{12} \\ A_{21} & A_{22} \end{pmatrix} \begin{pmatrix} V_1 \\ I_1 \end{pmatrix}.$$

The matrix A may be thought of as a product of three parts: The matrix that gives (V, I) immediately to the left of the impedance $Z/2$, the matrix that transforms from immediately to the left to immediately to the right,

*) See ref. (1), eq. (25). Since $u(a/2) = 0$ and from this equation u' is continuous at the origin, it follows that the solution u is exactly the same as for the case of no δ-function potential at the origin.

and finally the matrix that transforms from immediately to the right to the end of the line. These transformation matrices are all very well known [3]), and we obtain

$$A = \begin{pmatrix} \cos(\frac{1}{2}\beta l) & -ik\sin(\frac{1}{2}\beta l) \\ -(i/k)\sin(\frac{1}{2}\beta l) & \cos(\frac{1}{2}\beta l) \end{pmatrix} \begin{pmatrix} 1 & -Z \\ 0 & 1 \end{pmatrix} \begin{pmatrix} \cos(\frac{1}{2}\beta l) & -ik\sin(\frac{1}{2}\beta l) \\ -(i/k)\sin(\frac{1}{2}\beta l) & \cos(\frac{1}{2}\beta l) \end{pmatrix},$$

where $\beta = \omega\sqrt{LC}$, $k = \sqrt{L/C}$, L = inductance per unit length of line, C = capacitance per unit length of line, and $\omega/2\pi$ = frequency.

Multiplying these matrices out we obtain*), if we take the impedance Z to be a pure reactance ($Z = i\varkappa$),

$$A = \begin{pmatrix} \cos\beta l - \dfrac{\varkappa}{2k}\sin\beta l & -ik\left\{\sin\beta l + \dfrac{\varkappa}{2k}(1+\cos\beta l)\right\} \\ -\dfrac{i}{k}\left\{\sin\beta l - \dfrac{\varkappa}{2k}(1-\cos\beta l)\right\} & \cos\beta l - \dfrac{\varkappa}{2k}\sin\beta l \end{pmatrix}; \quad (11)$$

det $A = 1$, as it must.

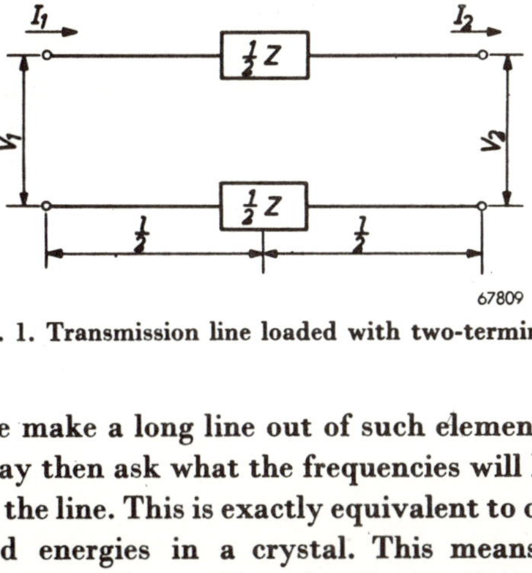

Fig. 1. Transmission line loaded with two-terminals.

Now suppose we make a long line out of such elements by joining them end-to-end. We may then ask what the frequencies will be that allow wave propagation down the line. This is exactly equivalent to our former question as to the allowed energies in a crystal. This means that the allowed frequencies are given by the condition $|\mathrm{tr}A| \leqslant 2$. Just as in section 1, we may write the matrix A in the form $A = \pm\exp(a\tilde{A})$ where all the definitions are just as before. However, it turns out to be more convenient to write $\tilde{A}$ a little differently. We now put

$$A_{12} = (A_{11} - 1)\,\mu_A, \qquad A_{21} = (A_{11} + 1)/\mu_A.$$

*) This matrix is equivalent to the matrix A for the Kronig-Penney model if we make the following correspondence: $\beta \to \sqrt{(8\pi^2 mE/h^2)}$, $l \to a$ (lattice constant), $\varkappa \to -aU_0/C\omega$ ($ah^2U_0/\pi^2 8m$ = strength of δ-function potential), $Z \to aU_0/i\omega C$. This means that the Kronig-Penney model is equivalent to a transmission line loaded with condensers of size aU_0/C.

Then we may $\tilde{A}$ write as follows:

$$\tilde{A} = \begin{pmatrix} 0 & \sqrt{\dfrac{A_{11}-1}{A_{11}+1}}\,\mu_A \\ \sqrt{\dfrac{A_{11}+1}{A_{11}-1}}\,\mu_A^{-1} & 0 \end{pmatrix}.$$

Substituting the values of A_{12} and A_{11} from equation (11), we find

$$\mu_A = ik \cot(\tfrac{1}{2}\beta l).$$

This is independent of the nature of the reactance $\varkappa$, and depends only on the nature of the transmission line used. Proceeding exactly as in section 1, we are led to the following theorem: *The stop bands common to a line loaded with equally spaced reactances $\varkappa$ and a line of identically spaced reactances $\varkappa'$, are also present in a line made up of any mixture of the reactances $\varkappa$ and $\varkappa'$ spaced in the same way.*

This result may be considered as a generalization of that of Saxon and Hutner.

In conclusion, I would like to thank Dr R. Jost for reading and suggesting improvements in the original manuscript.

Department of Physics,
University of Wisconsin,
Madison, Wisconsin (U.S.A.),

March, 1951

REFERENCES

1) D. S. Saxon and R. A. Hutner, Philips Res. Rep. **4**, 81-122, 1949.
2) H. James, Phys. Rev. (2) **76**, 1603-1610, 1949.
3) L. Brillouin, "Wave propagation in periodic structures", McGraw-Hill, New York 1946), p. 218 ff.

PHYSICAL REVIEW VOLUME 115, NUMBER 4 AUGUST 15, 1959

Analytic Properties of Bloch Waves and Wannier Functions*

W. KOHN
Department of Physics, Carnegie Institute of Technology, Pittsburgh, Pennsylvania
(Received March 16, 1959)

The one-dimensional Schrödinger equation with a periodic and symmetric potential is considered, under the assumption that the energy bands do not intersect. The Bloch waves, $\varphi_{n,k}$, and energy bands, $E_{n,k}$, are studied as functions of the *complex* variable, k. In the complex plane, they are branches of multivalued analytic and periodic functions, φ_k, and E_k, with branch points, k', off the real axis. A simple procedure is described for locating the branch points. Application is made to the power series and Fourier series developments of these functions. The analyticity and periodicity of $\varphi_{n,k}$ has some consequences for the form of the Wannier functions. In particular, it is shown that for each band there exists one and only one Wannier function which is real, symmetric or antisymmetric under an appropriate reflection, and falling off exponentially with distance. The rate of falloff is determined by the distance of the branch points k' from the real axis.

INTRODUCTION

1

THE motion of an electron in a crystal lattice is governed by a Schrödinger equation with a periodic potential. In one dimension we may write it as

$$[-d^2/dx^2+V(x)]\psi(x)=E\psi(x), \tag{1.1}$$

where

$$V(x+b)=V(x). \tag{1.2}$$

In the present paper, we assume further that $V(x)$ is symmetric,

$$V(-x)=V(x). \tag{1.3}$$

The nature of the solutions of (1.1) for real values of E has been described in a definitive paper by Kramers.[1] For certain so-called allowed ranges of E, (1.1) has solutions of the form

$$\varphi_{n,k}(x)=u_{n,k}(x)e^{ikx}, \quad n=0, 1, 2, \cdots \tag{1.4}$$

where $u_{n,k}(x)$ has the same periodicity as $V(x)$ and k is real. These are the well-known Bloch waves. The corresponding eigenvalues of (1.1) are the so-called energy bands,

$$E=E_{n,k}, \quad n=0, 1, 2, \cdots. \tag{1.5}$$

The allowed ranges of E are separated by forbidden ranges in which the solutions of (1.1) have the form (1.4), but with k purely imaginary. These solutions are seen therefore to have an exponential behavior. The present paper is restricted to cases in which all allowed bands are separated by forbidden bands of *finite* width.[2]

In some connections, the question of the behavior of $\varphi_{n,k}$ and $E_{n,k}$, regarded as functions of the *complex* variable

$$k=g+ih, \tag{1.6}$$

arises. Examples of such cases are the expansion of $E_{n,k}$ and $\varphi_{n,k}$ in powers of k,[3] and the theory of the motion of Bloch electrons in a magnetic field.[4]

The properties of $E_{n,k}$ and $\varphi_{n,k}$, regarded as analytic functions of k, are discussed in Secs. 3 and 4 of the present paper. One finds that the usual energy bands, $E_{n,g}$, are the traces on the $g-\mathrm{Re}(E)$ plane of a multivalued analytic function, E_k. The latter can be represented on a Riemann surface with an infinite sequence of sheets, S_n. Each sheet, S_n, is connected to S_{n-1} and S_{n+1} at a series of branch points of order 1[5] which lie off the real axis. Thus by going into the complex k-plane, one can pass continuously from one energy band to another. (See Fig. 4.) A similar, but slightly more complex, situation is found in the case of $\varphi_{n,k}$. One useful result is that the phase of $\varphi_{n,g}$ can be so chosen that it is a periodic function of g, with period $2\pi/b$ and, if continued into the complex k-plane, is analytic in a strip of finite width $2\bar{h}_n$, enclosing the real axis.

The analytic properties of $\varphi_{n,k}$ have some interesting consequencies concerning the properties of the Wannier function

$$a_n(x)\equiv\left(\frac{b}{2\pi}\right)^{\frac{1}{2}}\int_{-\pi/b}^{\pi/b}\varphi_{n,g}(x)dg. \tag{1.7}$$

These are discussed in detail in Secs. 5–9. Here we mention only the following results. For each energy band, n, there exists one and only one Wannier function with all of the following properties: 1. It is real. 2. It is symmetric or antisymmetric about either $x=0$ or $x=b/2$. 3. It falls off exponentially as

$$a_n(x)\sim\exp(-\bar{h}_n x), \tag{1.8}$$

where $\bar{h}_n$ is the half-width of the strip of analyticity of $\varphi_{n,k}$. The magnitude of $\bar{h}_n$ is determined by a simple procedure. No other choice of phase leads to an exponential decrease more rapid than (1.8). On the other

* A preliminary note was published by W. Kohn and S. Michaelson, Proc. Phys. Soc. (London) 72, 301 (1958).
[1] H. A. Kramers, Physica 2, 483 (1935).
[2] The limiting case of completely free electrons is thereby excluded.
[3] J. Bardeen, J. Chem. Phys. 6, 1367 (1938).
[4] W. Kohn, Proc. Phys. Soc. (London) 72, 1147 (1958).
[5] The order of a branch point is defined as $\nu-1$, where ν is the number of different values the function can assume in its vicinity. Thus near a branch point of order 1 the function is double-valued.

hand, many choices of phase lead to a less rapid decrease.

While the results of this paper are obtained under fairly restrictive conditions, such as the symmetry of $V(x)$ and no intersection of energy bands, it is clear that some of them will continue to hold when one or the other of these conditions is relaxed. Many of our results may also be expected to carry over to the case of three dimensions.

I. BLOCH WAVES AND ENERGY BANDS

2. Preliminaries

We consider the Schrödinger equation

$$-d^2\psi/dx^2+V(x)\psi=E\psi, \tag{2.1}$$

where $V(x)$ is periodic with period b, symmetric about $x=0$, and sectionally continuous. We denote by $\psi_1(x,E)$ and $\psi_2(x,E)$ the two independent solutions of (2.1) which are defined by the initial conditions

$$\psi_1(0,E)=1,\quad \psi_1'(0,E)=0, \tag{2.2}$$

$$\psi_2(0,E)=0,\quad \psi_2'(0,E)=1. \tag{2.3}$$

It is then well known[6] that $\psi_1(x,E)$ and $\psi_2(x,E)$ are entire functions of the complex variable E, for all x.

We look for solutions of (2.1) which are multiplied by a constant factor λ when x is replaced by $x+b$, $\psi(x+b)=\lambda\psi(x)$. This is equivalent to requiring that

$$\psi(b)=\lambda\psi(0), \tag{2.5}$$

$$\psi'(b)=\lambda\psi'(0). \tag{2.6}$$

We can write

$$\psi=\alpha\psi_1+\beta\psi_2, \tag{2.7}$$

substitute into (2.5) and (2.6), and eliminate α and β. This gives the well-known result[1]

$$\lambda^2-2\mu(E)\lambda+1=0, \tag{2.8}$$

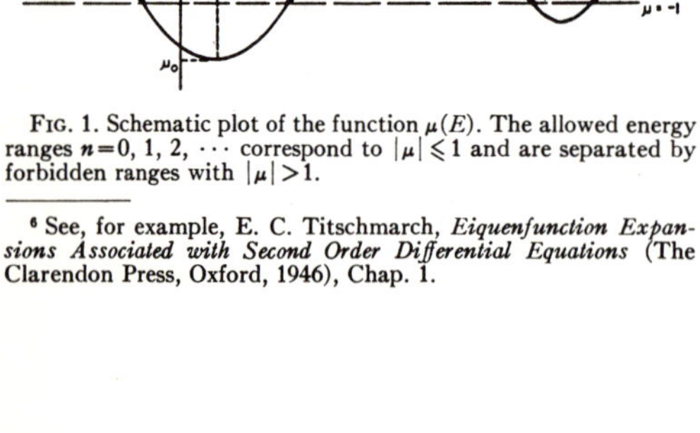

FIG. 1. Schematic plot of the function $\mu(E)$. The allowed energy ranges $n=0, 1, 2, \cdots$ correspond to $|\mu|\leqslant 1$ and are separated by forbidden ranges with $|\mu|>1$.

[6] See, for example, E. C. Titschmarch, *Eiquenfunction Expansions Associated with Second Order Differential Equations* (The Clarendon Press, Oxford, 1946), Chap. 1.

where

$$\mu(E)\equiv\tfrac{1}{2}[\psi_1(b,E)+\psi_2'(b,E)]. \tag{2.9}$$

In our case of a symmetric potential we can also, instead of (2.7), take $\psi_1(x)$ and $\psi_1(b-x)$ as fundamental system, which gives again a result of the form (2.8) with

$$\mu(E)=\psi_1(b,E).^7 \tag{2.10}$$

This expression is a little simpler to use than (2.9). We see that *$\mu(E)$ is an entire function of E.*

We can write λ in the form

$$\lambda=e^{ikb}, \tag{2.11}$$

where, by (2.8), k is determined as a function of E by the equation

$$\cos kb=\mu(E). \tag{2.12}$$

If k is real, it is called the wave number, and the corresponding function ψ is a Bloch wave of the form (1.4).

The properties of $\mu(E)$, considered as a function of the real variable E, have been established by Kramers.[1] For the case where no two allowed energy bands intersect, a schematic plot is shown in Fig. 1. The following properties are relevant for our purposes

$$\lim_{E\to-\infty}\mu(E)=+\infty\,; \tag{2.13}$$

$$\lim_{E\to+\infty}[\mu(E)-\cos(E^{\frac{1}{2}}b)]=0; \tag{2.14}$$

$$d\mu/dE=0 \quad\text{at}\quad E=E_n,\quad n=0,1,2,\cdots \tag{2.15}$$

where, if no two allowed energy regions touch—the case considered here—

$$\mu_{2m}\equiv\mu(E_{2m})<-1, \tag{2.16}$$

$$\mu_{2m+1}\equiv\mu(E_{2m+1})>+1, \tag{2.17}$$

and

$$\lim_{n\to\infty}\left[E_n-\left(\frac{(n+1)\pi}{b}\right)^2\right]=0. \tag{2.18}$$

The zeros E_n of $d\mu/dE$ are simple.

We shall now show that the real zeros E_n of the entire function $(d\mu/dE)$ are its only zeros. We begin by writing

$$E=\kappa^2, \tag{2.19}$$

and

$$\kappa=\sigma+i\tau, \tag{2.20}$$

and look for the zeros of

$$d\mu/d\kappa=2\kappa(d\mu/dE). \tag{2.21}$$

Now as E or $\kappa\to\infty$, it can be shown[6] that

$$\psi_1(b,E)=\cos\kappa b+O(e^{|\tau|b}/|\kappa|), \tag{2.22}$$

[7] I am indebted to Mr. L. Glasser for bringing this result to my attention. See also, E. Kamke, *Differentialgleichungen* (Akademische Verlagsgesellschaft, Leipzig, 1943), Vol. I, Art. 2.22.

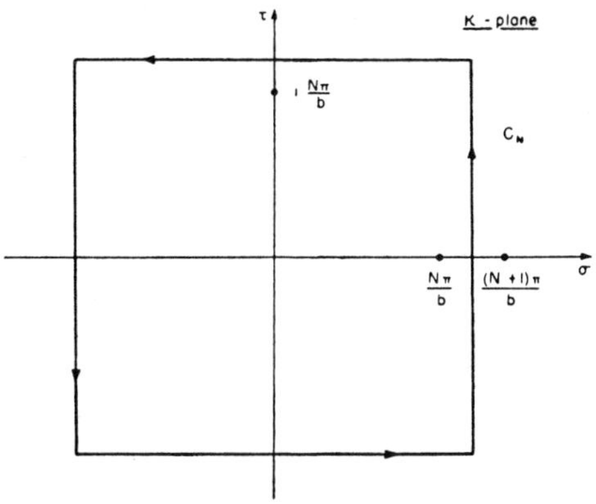

FIG. 2. Contour used to establish the reality of the roots of $d\mu/dE$.

and similarly

$$\partial\psi_1(b,E)/\partial\kappa = -b\sin\kappa b + O(e^{|\tau|b}/|\kappa|). \qquad (2.23)$$

These equations are simply a quantitative expression of the fact that for large E (real or complex), $\psi_1(x,E) \sim \cos(E^{\frac{1}{2}}x)$. Because of Eq. (2.10), Eq. (2.23) implies that for large κ

$$d\mu/d\kappa = -b\sin\kappa b + O(e^{|\tau|b}/|\kappa|). \qquad (2.24)$$

For perfectly free electrons we have, of course,

$$d\mu^{(0)}/d\kappa = -b\sin\kappa b. \qquad (2.25)$$

We now define the function

$$F(\kappa) \equiv \frac{d\mu/d\kappa}{d\mu^{(0)}/d\kappa} = \frac{d\mu/d\kappa}{-b\sin\kappa b}. \qquad (2.26)$$

This is a single-valued analytic function of κ with zeros at the zeros of $d\mu/d\kappa$ and poles at the zeros of $\sin\kappa b$, i.e., at

$$\kappa = j\pi/b, \quad j = 0, \pm 1, \cdots. \qquad (2.27)$$

Now consider $F(\kappa)$ on the square contour C_N of side $(N+\frac{1}{2})\pi/b$ shown in Fig. 2. As $N \to \infty$, κ on C_N approaches ∞. Furthermore, the contours C_N avoid the zeros (2.27) of the denominator of $F(\kappa)$. Hence, in view of (2.24) one sees easily that as $N \to \infty$, $F(\kappa) \to 1$ for all points on C_N. It follows that for sufficiently large N,

$$\oint_{C_N} \left[\frac{d}{d\kappa}\ln F(\kappa)\right] d\kappa = 0, \quad N \geqslant N_0. \qquad (2.28)$$

Now by a well-known theorem of analysis

$$\frac{1}{2\pi i}\oint_{C_N} \left[\frac{d}{d\kappa}\ln F(\kappa)\right] d\kappa = Z_N - P_N, \qquad (2.29)$$

where Z_N and P_N are the number of zeros and poles, respectively, of $F(\kappa)$ inside C_N. Further we know from (2.26) and (2.27) that $P_N = 2N+1$. Hence

$$Z_N = 2N+1, \quad N \geqslant N_0. \qquad (2.30)$$

But in view of (2.18) and (2.21), the real zeros of $d\mu/dE$ and the zero contributed by the factor κ in (2.21) total exactly $2N+1$. Hence, $d\mu/dE$ can have no other zeros.

3. The Energy Function

The connection between E and k (real or complex) is given by Eq. (2.12),

$$\cos kb = \mu(E), \qquad (3.1)$$

where the function $\mu(E)$ is defined by Eq. (2.9) or (2.10) and is a single valued, entire function of E. Equation (3.1) shows at once that

$$E_k = E_{-k}, \qquad (3.2)$$

and

$$E_{k+2\pi/b} = E_k. \qquad (3.3)$$

Further, since $\mu(E)$ is real for real E, we have

$$\mu(E^*) = [\mu(E)]^*. \qquad (3.4)$$

Hence, on taking the complex conjugate of (3.1), one finds

$$E_{k^*} = (E_k)^*. \qquad (3.5)$$

To obtain further insight into the analytic structure of E_k, let us for a moment consider the function $E(\mu)$, the inverse of the entire function $\mu(E)$. $E(\mu)$ is an analytic function of μ except where $d\mu/dE = 0$. In the vicinity of such a point, we have

$$\mu = \mu_n + \alpha_n(E-E_n)^2 + \cdots, \qquad (3.6)$$

where α_n does not vanish. Hence

$$E = E_n + \frac{1}{\alpha_n^{\frac{1}{2}}}(\mu-\mu_n)^{\frac{1}{2}}. \qquad (3.7)$$

Thus each extremum μ_n of the $\mu(E)$-plot in Fig. 1 represents a branch point of order 1[5] of the function $E(\mu)$ (see Fig. 3). Since $d\mu/dE$ has no other zeros, these are the only singularities of $E(\mu)$ in the finite μ-plane. If we start on the real μ axis at $\mu = +\infty$, with a real value of E, proceed to the left to μ_0, around μ_0 to μ_1, around μ_1 to μ_2, etc., as indicated on Fig. 3, we trace

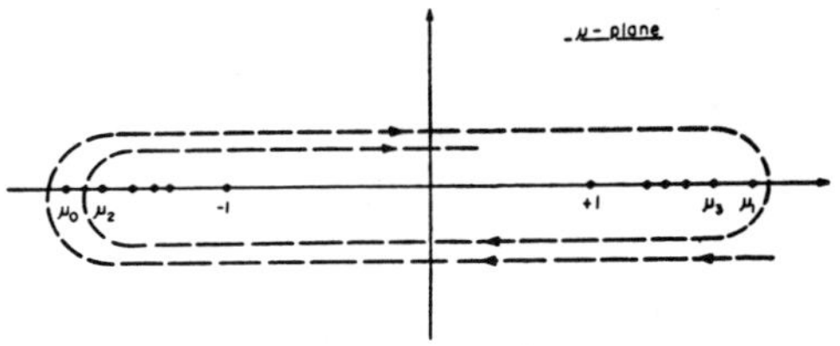

FIG. 3. The singularities of the function $E(\mu)$.

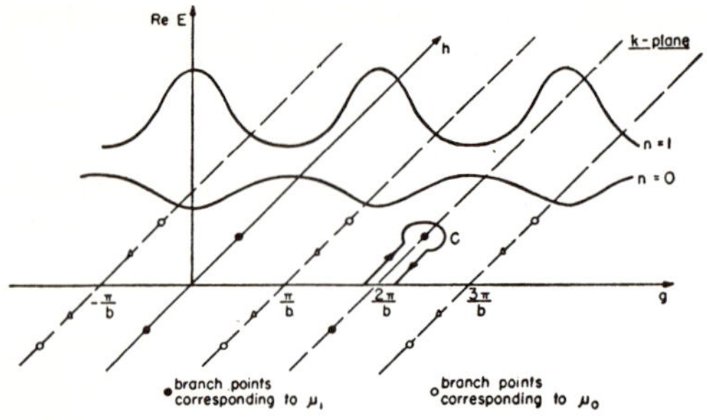

FIG. 4. Schematic representation of the k-plane ($k=g+ih$), the energy bands E_{ng}, and the branch points of E_k.

out the curve $\mu(E)$ of Fig. 1. It is clear that $E(\mu)$ can be represented on a Riemann surface with sheets $\bar{S}_n$ corresponding to the different bands n. $\bar{S}_0$ is connected to $\bar{S}_1$ by the branch point at μ_0, $\bar{S}_1$ to $\bar{S}_2$ by the branch point at μ_1, etc. The values of $E(\mu)$ on $\bar{S}_n$ for real μ between -1 and $+1$ correspond to the allowed energy band n of Fig. 1.

Now let us return to the properties of E as a function of k, as defined by (3.1). Since

$$dE/dk=(dE/d\mu)(-b\sin kb), \tag{3.8}$$

we see that E is an analytic function of k except at those points for which $(d\mu/dE)$ vanishes, i.e., where

$$\cos k_n b=\mu_n. \tag{3.9}$$

Of course, for a given μ_n, (3.9) has many solutions. Since for even n, $\mu_n<-1$, and for odd n, $\mu_n>+1$, the k_n which solve (3.9) have the form

$$k_{2m}=\pm[(2j+1)\pi/b+ih_{2m}], \quad j=0,\pm1,\cdots \tag{3.10}$$

$$k_{2m+1}=\pm(2j\pi/b+ih_{2m+1}), \quad j=0,\pm1,\cdots \tag{3.11}$$

where

$$h_n=|\cosh^{-1}|\mu_n||. \tag{3.12}$$

From (3.7) and (3.1), we see that near one of the k_n

$$E=E_n+\beta_n(k-k_n)^{\frac{1}{2}}, \tag{3.13}$$

where β_n does not vanish. Hence, the k_n are also branch points of order 1. We see from (3.9) and (3.12) that the position of these branch points can be directly read off the "Kramers plot" in Fig. 1. In particular, since $\lim_{n\to\infty}|\mu_n|=1$, we see from (3.12) that

$$\lim_{n\to\infty} h_n=0, \tag{3.14}$$

so that for very high-lying bands the branch points lie very close to the real k-axis. Figure 4 shows a schematic diagram of the k-plane, the energy bands $E_{n,g}$ and the branch points of E_k.

Clearly, the function E_k may be represented on a Riemann surface with an infinite sequence of sheets S_n, in such a way that the energy bands, $E_{n,g}$, are the values of E_k for real k on the sheet S_n. Each sheet S_n is connected to S_{n+1} by the infinite sequence of the branch points k_n given by (3.10) and (3.11). Thus, if one starts at a real value of k on the energy band n, passes around one of the branch points k_n, and returns to the real axis, one is then on the band $n+1$ (see path C in Fig. 4, which would take one from $n=1$ to $n=2$). Similarly, by passing around one of the points k_{n-1} one arrives on band $n-1$. By drawing branch lines from each k_n to $\pm i\infty$ (away from the real axis; see Fig. 4), the Riemann surface is divided into elements on each of which E_k is periodic with period $2\pi/b$ and single valued.

It is useful to represent E as a function of yet another variable, namely, λ, Eq. (2.11). By (2.8) we have

$$\mu=\frac{1}{2}\left(\lambda+\frac{1}{\lambda}\right). \tag{3.15}$$

We write

$$E=E(\mu)=E(\tfrac{1}{2}(\lambda+1/\lambda))\equiv\mathcal{E}(\lambda), \tag{3.16}$$

defining the function $\mathcal{E}(\lambda)$. As

$$\frac{d\mathcal{E}(\lambda)}{d\lambda}=\frac{dE(\mu)}{d\mu}\frac{1}{2}\left(1-\frac{1}{\lambda^2}\right), \tag{3.17}$$

we see that $\mathcal{E}(\lambda)$ is an analytic function of λ except at $\lambda=0$ and ∞, and at those points $\lambda_n^{\pm}$ where $dE/d\mu$ is infinite. In view of (2.8), they are given by

$$\lambda_n^{\pm}=\mu_n[1\pm(1-\mu_n^{-2})^{\frac{1}{2}}], \quad n=0,1,2,\cdots \tag{3.18}$$

where we take the square root to be positive. Figure 5 shows the positions of the singularities in the λ-plane. (The unit circle corresponds to real values of k.) The points $\lambda_n^{\pm}$ are again branch points of order 1,[5] and $\mathcal{E}(\lambda)$ can be represented on a Riemann surface with one sheet corresponding to each band. This surface may be divided into elements on which $\mathcal{E}(\lambda)$ is single-valued by drawing on the nth sheet branch lines joining λ_{n-1}^- and λ_n^- to the origin, and λ_{n-1}^+ and λ_n^+ to $\pm\infty$ away from the origin. The function $\mathcal{E}_n(\lambda)$, corresponding to the nth sheet has a Laurent expansion

$$\mathcal{E}_n(\lambda)=\sum_{l=-\infty}^{\infty}a_{n;l}\lambda^l, \tag{3.19}$$

whose coefficients are clearly real. Its region of convergence is

$$1/\bar{\lambda}_n<|\lambda|<\bar{\lambda}_n, \tag{3.20}$$

where

$$\begin{aligned}\bar{\lambda}_n&\equiv\min[|\lambda_n^+|,|\lambda_{n-1}^+|]\\&=\min[|\mu_n|+(\mu_n^2-1)^{\frac{1}{2}},\ |\mu_{n-1}|+(\mu_{n-1}^2-1)^{\frac{1}{2}}].\end{aligned} \tag{3.21}$$

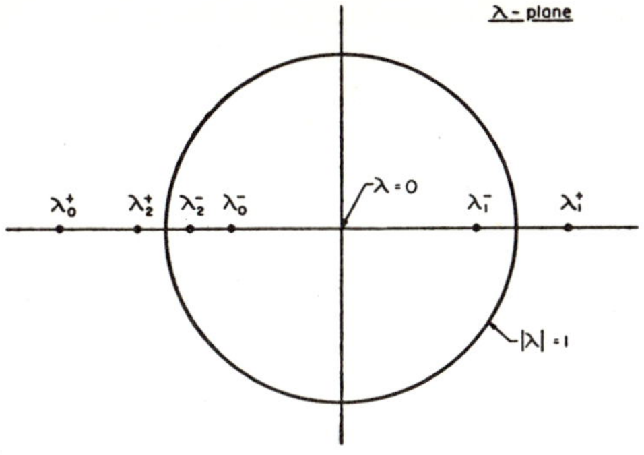

FIG. 5. Singularities of $\mathcal{E}(\lambda)$. The unit circle corresponds to real values of k.

In terms of the quantity

$$\bar{h}_n \equiv \min(h_n, h_{n-1}) \qquad (3.22)$$

[see Eqs. (3.10) and (3.11)], we can also write

$$\exp(-b\bar{h}_n) < |\lambda| < \exp(-b\bar{h}_n). \qquad (3.20')$$

Finally, we see from Eq. (3.16) that $\mathcal{E}_n(\lambda) = \mathcal{E}_n(\lambda^{-1})$ so that (3.19) simplifies to

$$\mathcal{E}_n(\lambda) = a_{n;l} + \sum_{l=1}^{\infty} a_{n;l}\left(\lambda^l + \frac{1}{\lambda^l}\right). \qquad (3.23)$$

Let us now mention some applications of our results.

Power Series Expansion of $E_{n,g}$

From what has been said about the function E_k, it is clear that the expansion of $E_{n,g}$ will converge for

$$|g| < r_n, \qquad (3.24)$$

where

$$r_0 = [(\pi/b)^2 + h_0^2]^{\frac{1}{2}},$$
$$r_{2m} = \min([(\pi/b)^2 + h_{2m}^2]^{\frac{1}{2}}, h_{2m-1}), \quad m = 1, 2, \cdots \qquad (3.25)$$
$$r_{2m+1} = \min(h_{2m+1}, [(\pi/b)^2 + h_{2m}]^{\frac{1}{2}}), \quad m = 0, 1, 2, \cdots.$$

The h_n may be obtained from the "Kramers plot" of Fig. 1, by means of Eq. (3.12). In particular, for $n=0$, the expansion converges over the entire first Brillouin zone, $|g| \leqslant \pi/b$.

For nearly free electrons ($V \to 0$), $h_n \to 0$ and, hence, the radius of convergence $r_n \to 0$ for all bands except the lowest. On the other hand, in the tight-binding limit ($b \to \infty$), it can be shown (see Appendix) that

$$\lim_{b\to\infty} h_n = (-\epsilon_n)^{\frac{1}{2}}, \qquad (3.26)$$

where ϵ_n is the eigenvalue of the atomic wave function corresponding to band n. Hence

$$\lim_{b\to\infty} \frac{h_n}{(\pi/b)} = \infty, \qquad (3.27)$$

so that for large enough b, the power series of $E_{n,g}$ for arbitrary n certainly converge over the entire first Brillouin zone.

Fourier Expansion of $E_{n,g}$

We have seen that the energy can be expressed as a Laurent expansion in λ [Eq. (3.23)]. Writing $\lambda = \exp(igb)$ gives the Fourier series

$$E_{n,g} = a_{n;l} + 2\sum_{l=1}^{\infty} a_{n;l} \cos(lgb). \qquad (3.28)$$

Now from the region of convergence (3.22) of the Laurent expansion (3.23) we conclude that

$$\lim_{l\to\infty} a_{n;l}\lambda^l = \begin{cases} 0 & \text{if } |\lambda| < \bar{\lambda}_n \\ \infty & \text{if } |\lambda| > \bar{\lambda}_n, \end{cases} \qquad (3.29)$$

where $\bar{\lambda}_n$, Eq. (3.21), can be directly obtained from the "Kramers plot," Fig. 1. Again, one finds good convergence in the tight-binding limit, poor convergence for nearly free electrons.

4. Bloch Functions

The term Bloch functions is usually applied to solutions of the Schrödinger equation (1.1) (with real E), which have the form (1.4) (with real $k=g$), and are normalized in the sense

$$(2\pi/b)\int_0^b |\varphi_{n,g}(x)|^2 dx = 1. \qquad (4.1)$$

In the present context, we shall apply the term to the more general class of functions ψ, which have the following properties: They are solutions of (1.1) with complex E; they are quasi-periodic, in the sense that

$$\psi(b) = \lambda\psi(0); \quad \psi'(b) = \lambda\psi'(0); \qquad (4.2)$$

and they are normalized in a manner which, for $|\lambda| = 1$ reduces to

$$(2\pi/b)\int_0^b |\psi(x)|^2 dx = 1, \quad |\lambda| = 1. \qquad (4.3)$$

For $|\lambda| = 1$ (i.e., $\lambda = e^{igb}$), these functions are just the ordinary Bloch functions.

Now let E be given and let us look for a solution of (1.1) which satisfies the quasi-periodic condition (4.2). Such a solution can be written as

$$\psi(x) = \alpha\psi_1(x) + \beta\psi_2(x). \qquad (4.4)$$

Substitution into (4.2) gives the quadratic equation (2.8) for λ, and the ratio

$$\frac{\alpha}{\beta}=\frac{\psi_2(b)}{\lambda-\psi_1(b)}=\frac{\psi_2(b)}{\frac{1}{2}(\lambda-1/\lambda)}. \tag{4.5}$$

Hence, we may write

$$\psi(x)=\frac{1}{N^{\frac{1}{2}}}\left[\psi_2(b)\psi_1(x)+\frac{1}{2}\left(\lambda-\frac{1}{\lambda}\right)\psi_2(x)\right]$$

$$=\frac{1}{N^{\frac{1}{2}}}\chi(x), \tag{4.6}$$

thus defining the function $\chi(x)$. The normalization $N^{-\frac{1}{2}}$ has to be chosen in accordance with the requirement (4.3). One's first inclination would be to set N equal to

$$\left(\frac{2\pi}{b}\right)\int_0^b\left|\psi_2(b)\psi_1(x)+\frac{1}{2}\left(\lambda-\frac{1}{\lambda}\right)\psi_2(x)\right|^2 dx. \tag{4.7}$$

However, this is not an analytic function of E. Instead, we choose

$$N=\frac{2\pi}{b}\int_0^b\left[\psi_2(b)\psi_1(x)+\frac{1}{2}\left(\lambda-\frac{1}{\lambda}\right)\psi_2(x)\right]$$
$$\times\left[\psi_2(b)\psi_1(x)+\frac{1}{2}\left(\frac{1}{\lambda}-\lambda\right)\psi_2(x)\right]dx, \tag{4.8}$$

which reduces to (4.7) for $|\lambda|=1$ and hence satisfies (4.3), and furthermore is an analytic function of E.

This expression for N can be considerably simplified. For this purpose we define

$$\chi(x,E)\equiv\psi_2(b,E)\psi_1(x,E)+\frac{1}{2}\left(\lambda(E)-\frac{1}{\lambda(E)}\right)\psi_2(x,E), \tag{4.9}$$

$$\tilde{\chi}(x,E)\equiv\psi_2(b,E)\psi_1(x,E)+\frac{1}{2}\left(\frac{1}{\lambda(E)}-\lambda(E)\right)\psi_2(x,E). \tag{4.10}$$

Here $\lambda(E)$ is one of the roots of Eq. (2.8). Clearly,

$$N=(2\pi/b)\int_0^b\chi(x,E)\tilde{\chi}(x,E)dx. \tag{4.11}$$

We now take the equations

$$\begin{aligned}&[-d^2/dx^2+V-E]\chi(x,E)=0,\\&[-d^2/dx^2+V-E-\delta E]\tilde{\chi}(x,E+\delta E)=0,\end{aligned} \tag{4.13}$$

multiply the first by $\tilde{\chi}(x, E+\delta E)$, the second by $\chi(x,E)$, subtract and integrate from 0 to b. This gives

$$\delta E\int_0^b dx\,\chi(x,E)\tilde{\chi}(x, E+\delta E)$$
$$=\left[\tilde{\chi}(x, E+\delta E)\frac{\partial}{\partial x}\chi(x,E)-\chi(x,E)\frac{\partial}{\partial x}\tilde{\chi}(x, E+\delta E)\right]_0^b$$
$$=\psi_2(b, E+\delta E)\frac{1}{2}\left(\lambda(E)-\frac{1}{\lambda(E)}\right)\left(\frac{\lambda(E)}{\lambda(E+\delta E)}-1\right)$$
$$-\psi_2(b,E)\frac{1}{2}\left(\frac{1}{\lambda(E+\delta E)}-\lambda(E+\delta E)\right)$$
$$\times\left(\frac{\lambda(E)}{\lambda(E+\delta E)}-1\right), \tag{4.14}$$

where in the last step we have used the "periodicity" properties of χ and $\tilde{\chi}$. We now divide by δE and let $\delta E\rightarrow 0$. This gives

$$N=\frac{2\pi}{b}\psi_2(b,E)\left(\frac{1}{\lambda(E)^2}-1\right)\frac{d\lambda(E)}{dE}. \tag{4.15}$$

Finally, noting that

$$\frac{d\mu}{dE}=\frac{d}{dE}\left[\frac{1}{2}\left(\lambda+\frac{1}{\lambda}\right)\right]=\frac{1}{2}\left(1-\frac{1}{\lambda^2}\right)\frac{d\lambda}{dE}, \tag{4.16}$$

we obtain

$$N=-\frac{4\pi}{b}\psi_2(b,E)\frac{d\mu}{dE}, \tag{4.17}$$

and hence (4.6) becomes

$$\psi(x)=\chi(x)\Big/\left[-\frac{4\pi}{b}\psi_2(b)\frac{d\mu}{dE}\right]^{\frac{1}{2}}, \tag{4.18}$$

where $\chi(x)$ is defined by Eq. (4.9). For $\lambda=e^{iqb}$, this function reduces to a familiar normalized Bloch wave whose phase is such that

$$\text{Im}\psi(0)=0, \quad \lambda=e^{iqb}. \tag{4.19}$$

Bloch Waves Regarded as Analytic Functions of λ

We now wish to study the analytic behavior of $\psi(x)$ as a function of λ, from which the functional dependence on k is then easily deduced. Since E is a function of λ, we may regard ψ_1, ψ_2, and ψ as functions of λ.

$$\begin{aligned}&\psi_1(x)\rightarrow\psi_1(x,\lambda);\quad \psi_2(x)\rightarrow\psi_2(x,\lambda);\\&\psi(x)\rightarrow\psi(x,\lambda).\end{aligned} \tag{4.20}$$

The possible singularities of $\psi(x,\lambda)$ are the points $\lambda_n^{\pm}$, where $d\mu/dE=0$ and $\mathcal{E}(\lambda)$ is singular, as well as the points where

$$\psi_2(b,\lambda)=0, \tag{4.21}$$

which makes N vanish.

We begin by showing that at the latter points $\psi(x,\lambda)$ is, in fact, regular. First we note that (4.21) can occur only for real E. For

$$E=\frac{\int_0^b \psi_2^*(x,\lambda)\left(-\frac{d^2}{dx^2}\right)\psi_2(x,\lambda)dx+\int_0^b V(x)|\psi_2(x,\lambda)|^2dx}{\int_0^b |\psi_2(x,\lambda)|^2dx}$$

$$=\frac{\int_0^b \left|\frac{\partial\psi(x,\lambda)}{\partial x}\right|^2 dx+\int_0^b V(x)|\psi_2(x,\lambda)|^2dx}{\int_0^b |\psi_2(x,\lambda)|^2dx}, \tag{4.23}$$

where (4.21) was used in the last integration by parts. Now since E is real, so is $\psi_2(x,\lambda)$. Further, in view of the symmetry of the potential [and of Eq. (4.21)], we must have

$$\psi_2(b-x,\ \lambda)=\pm\psi_2(x,\lambda). \tag{4.24}$$

Now differentiate this equation with respect to x and set $x=0$. This gives

$$-\psi_2'(b,\lambda)=\pm\psi_2'(0,\lambda)=\pm 1. \tag{4.25}$$

Since also

$$-\psi_2(b,\lambda)=\pm\psi_2(0,\lambda)=0,$$

we see that $\psi_2(x,\lambda)$ is a periodic solution with $\lambda=\pm 1$. Hence we have the preliminary conclusion that (4.21) can occur only at $\lambda=\pm 1$.

Let us take $\lambda=+1$, and study $\psi_2(b,\lambda)$ in the vicinity of $\lambda=1$. We write

$$\frac{d\psi_2(b,\lambda)}{d\lambda}=\frac{\partial\psi_2(b)}{\partial E}\frac{dE}{d\mu}\frac{d\mu}{d\lambda}. \tag{4.26}$$

Now by differentiating the Schrödinger equation satisfied by $\psi_2(x)$ with respect to E, it is easily shown that at $\lambda=1$,

$$\partial\psi_2(b)/\partial E=\int_0^b \psi_2^2 dx=N_0, \tag{4.27}$$

which is nonvanishing. Also $\partial E/\partial\mu$ is finite and nonvanishing. Finally

$$\frac{d\mu}{d\lambda}=\frac{\partial}{\partial\lambda}\left[\frac{1}{2}\left(\lambda+\frac{1}{\lambda}\right)\right]=\frac{1}{2}\left(1-\frac{1}{\lambda^2}\right). \tag{4.28}$$

Hence

$$\psi_2(b,\lambda)=\tfrac{1}{2}N_0\frac{dE}{d\mu}(\lambda-1)^2+\cdots, \tag{4.29}$$

and therefore

$$N=-\frac{2\pi}{b}N_0(\lambda-1)^2+\cdots. \tag{4.30}$$

Now we examine the function $\chi(x)$ of (4.18). At $\lambda=1$, $\psi_2(b)=0$ and $\psi_1(b)=\mu=1$, so that the numerator

$$\chi(x,\lambda)\equiv\psi_2(b)\psi_1(x)+\frac{1}{2}\left(\lambda-\frac{1}{\lambda}\right)\psi_2(x)=0,\quad \lambda=1. \tag{4.31}$$

To determine its behavior near $\lambda=1$, we note that by (4.29)

$$\partial\psi_2(b)/\partial\lambda=0. \tag{4.32}$$

Hence, near $\lambda=1$

$$\chi(x,\lambda)=(\lambda-1)\psi_2(x). \tag{4.33}$$

Combining with (4.30), we see that $\psi(x,\lambda)$ is regular near $\lambda=1$ and

$$\psi(x,1)=\pm\left(\frac{b}{2\pi N_0}\right)^{\frac{1}{2}} i\psi_2(x,1), \tag{4.34}$$

provided that $\psi_2(b,1)=0$. A completely analogous situation holds if $\psi_2(b,\lambda)$ vanishes at $\lambda=-1$.

The only singularities of $\psi(x,\lambda)$ occur then at the points $\lambda_n^\pm$ given by Eq. (3.18). Let us examine $\psi(x,\lambda)$ in the vicinity of such a point,

$$\lambda_n^\pm\equiv\lambda'. \tag{4.35}$$

For a given x, the function $\chi(x,\lambda)$ in (4.18) behaves like

$$\chi(x,\lambda)=A_1(x)+B_1(x)(\lambda-\lambda')^{\frac{1}{2}}+\cdots, \tag{4.36}$$

as ψ_1 and ψ_2 are entire functions of E, but the function $E=\mathcal{E}(\lambda)$ has a branch point of order 1 at λ'. In the denominator the factor ψ_2 behaves as

$$\psi_2(b)=A_2+B_2(\lambda-\lambda')^{\frac{1}{2}}+\cdots, \tag{4.37}$$

while

$$d\mu/dE=2\alpha_n(E-E_n)=A_3(\lambda-\lambda')^{\frac{1}{2}}. \tag{4.38}$$

Hence $\psi(x,\lambda)$ behaves as

$$\psi(x,\lambda)=\frac{C(x)}{(\lambda-\lambda')^{\frac{1}{4}}}[1+D(x)(\lambda-\lambda')^{\frac{1}{2}}+\cdots]. \tag{4.39}$$

Thus we see that the points $\lambda_n^\pm$ are branch points of order 1[5] of the un-normalized function $\chi(x,\lambda)$, but branch points of order 3[5] of the normalized function $\psi(x,\lambda)$.

Now let us go back to Fig. 5 of the λ-plane. The function $\chi(x,\lambda)$ has branch points of order 1 at the same locations $\lambda_n^\pm$ as $\mathcal{E}(\lambda)$. It may be represented on the same Riemann surface as $\mathcal{E}(\lambda)$, with one sheet corresponding to each energy band n. If, for instance, we start at $\lambda=1$ with $\chi(x,1)$ corresponding to band $n=1$, go around λ_1^+ and return to $\lambda=1$, we then have the value of $\chi(x,1)$ for $n=2$.

On the other hand, the function $\psi(x,\lambda)$ requires a Riemann surface with twice as many sheets as $\mathcal{E}(\lambda)$. For each sheet on which ψ has the values $\psi_n(x,\lambda)$, there is a second sheet on which it has the values $-\psi_n(x,\lambda)$. Each branch point λ' connects four sheets, $\sigma_n^\pm$ and $\sigma_{n+1}^\pm$. Suppose we start at $\lambda=1$ with $\psi(x,1)$ corresponding to band $n=1$, on sheet σ_1^+. Passing once around λ_1^+ and returning to $\lambda=1$ gives $\psi(x,1)$ for band $n=2$, on σ_2^+; two encirclements give $-\psi(x,1)$ for band $n=1$, on σ_1^-; three give $-\psi(x,1)$ for band $n=2$, on σ_2^-; and four give again $\psi(x,1)$ for band $n=1$, on σ_1^+.

Behavior of $\psi(x,\lambda)$ in the Vicinity of the Unit Circle $\lambda=1$

As the unit circle corresponds to real values of k, it is of special interest.

Suppose we start on the unit circle with a particular value of $\mathcal{E}(\lambda)$ and a corresponding $\psi(x,\lambda)$. If we remain in the vicinity of the unit circle, $\mathcal{E}(\lambda)$ remains single valued, hence so does $\psi_1(x,\mathcal{E}(\lambda))$ and $\psi_2(x,\mathcal{E}(\lambda))$ and therefore also the un-normalized function $\chi(x,\lambda)$, Eq. (4.6).

We now turn to the normalizing factor $N^{-\frac{1}{2}}$. Since for $|\lambda|=1$, E is real and λ of the form $\exp(igb)$ we see from the original definition, Eq. (4.8), that N is real and $\geqslant 0$. Furthermore, by Eq. (4.17), since $d\mu/dE\neq 0$ on $|\lambda|=1$, we see that N can vanish only where $\psi_2(b)$ vanishes. As we have seen, this can only occur at $\lambda=\pm 1$, and where it occurs $\psi_2(b,\lambda)$ has a double root in λ.

This leads one to distinguish two cases: (A) $\psi_2(b)$ vanishes at neither $\lambda=\pm 1$, or at both; (B) $\psi_2(b)$ vanishes at either $\lambda=\pm 1$, but not at both. In case (A), $N^{-\frac{1}{2}}(\lambda)$ and hence $\psi(x,\lambda)$ returns to its original value on going around the unit circle. Hence, we can write $\psi(x,\lambda)$ as a Laurent expansion in λ

$$\psi(x,\lambda)=\left(\frac{b}{2\pi}\right)^{\frac{1}{2}}\sum_{l=-\infty}^{\infty}\alpha_n^{(l)}(x)\lambda^l, \tag{4.40}$$

which converges in the ring (3.20′). In case (B), $N^{-\frac{1}{2}}(\lambda)$ and hence $\psi(x,\lambda)$ change sign on going around the unit circle. Therefore, we can write $\psi(x,\lambda)$ as a Laurent series of the following form in $\lambda^{\frac{1}{2}}$,

$$\psi(x,\lambda)=\left(\frac{b}{2\pi}\right)^{\frac{1}{2}}\sum_{l=-\infty}^{\infty}\beta_n^{(l)}(x)\lambda^{(2l+1)/2} \tag{4.41}$$

converging again in the ring (3.20′).

Behavior of $\varphi_{n,k}$ as a Function of k

This is easily obtained from the functional dependence of $\psi(x,\lambda)$ on λ by the substitution $\lambda=\exp(ibk)$.

We find that $\varphi_{n,k}$ is a branch of a many-valued function φ_k, whose branch points have the same locations as those of E_k (see Sec. 3 and Fig. 4). In the vicinity of such a branch point, k',

$$\varphi_k\approx C'/(k-k')^{\frac{1}{2}}. \tag{4.42}$$

Thus four Riemann sheets are connected at each branch point. Two of these correspond to a band n and two to the band $n+1$. On the two sheets corresponding to a given n, φ_k differs only by sign.

We now have to distinguish several cases, corresponding to different possible behavior of $\varphi_{n,k}$ at $k=0$ and π/b (or $\lambda=\pm 1$).

Case A. $\varphi_{n,0}(0)$ and $\varphi_{n,\pi/b}(0)$ both not zero or both zero. From (4.40) we have the Fourier expansion

$$\varphi_{n,k}(x)=\left(\frac{b}{2\pi}\right)^{\frac{1}{2}}\sum_{l=-\infty}^{\infty}\alpha_n^{(l)}(x)e^{ilbk}, \tag{4.43}$$

which represents an analytic function of k in the strip

$$|\mathrm{Im}k|<\bar{h}_n, \tag{4.44}$$

where $\bar{h}_n$ is defined in Eq. (3.22). Clearly, $\varphi_{n,g}(x)$ is *periodic* in g with period $2\pi/b$:

$$\varphi_{n,g+2\pi/b}(x)=\varphi_{n,g}(x). \tag{4.45}$$

Now let us further distinguish two subcases.

(A1) $\qquad \varphi_{n,0}(0)\neq 0,\quad \varphi_{n,\pi/b}(0)\neq 0.$

Then from (4.6) and the fact that $N^{-\frac{1}{2}}(\lambda)=N^{-\frac{1}{2}}(\lambda^{-1})$, we find directly the following symmetry properties

$$\varphi_{n,g}^*(x)=\varphi_{n,g}(-x), \tag{4.46}$$

$$\varphi_{n,-g}(x)=\varphi_{n,g}(-x). \tag{4.47}$$

(A2) $\qquad \varphi_{n,0}(0)=0,\quad \varphi_{n,\pi/b}(0)=0.$

Here, since $N^{-\frac{1}{2}}(\lambda)=-N^{-\frac{1}{2}}(\lambda^{-1})$, we find

$$\varphi_{n,g}^*(x)=\varphi_{n,g}(-x), \tag{4.48}$$

$$\varphi_{n,-g}(x)=-\varphi_{n,g}(-x). \tag{4.49}$$

Case B. $\varphi_{n,0}(0)$ or $\varphi_{n,\pi/b}(0)$ is zero, but not both. From (4.41) we have the Fourier expansion

$$\varphi_{n,k}(x)=\left(\frac{b}{2\pi}\right)^{\frac{1}{2}}\sum_{l=-\infty}^{\infty}\beta_n^{(l)}(x)\exp\left(i\frac{2l+1}{2}bk\right), \tag{4.50}$$

which represents an analytic function of k in the strip (4.44). In this case $\varphi_{n,g}(x)$ is *antiperiodic* in g with period $2\pi/b$:

$$\varphi_{n,g+2\pi/b}(x)=-\varphi_{n,g}(x). \tag{4.51}$$

We may again distinguish two subcases.

(B1) $\qquad \varphi_{n,0}(0)\neq 0,\quad \varphi_{n,\pi/b}(0)=0.$

In this case one finds that $\varphi_{n,g}(x)$ has the symmetry relations (4.46), (4.47).

(B2) $\qquad \varphi_{n,0}(0)=0,\quad \varphi_{n,\pi/b}(0)\neq 0.$

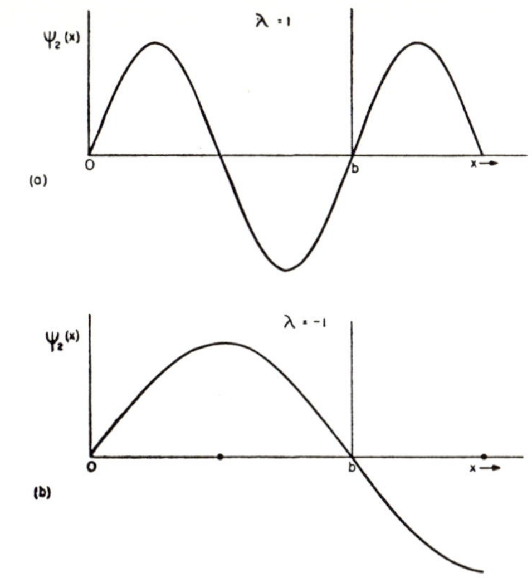

FIG. 6. Schematic plots of $\psi_2(x)$ for $\lambda=\pm1$ and $\psi_2(b)=0$.

In this case one finds that $\varphi_{n,g}(x)$ has the symmetry relations (4.48) and (4.49).

Power Series Expansion of $\varphi_{n,g}$

Since the singularities of $\varphi_{n,k}$ have the same locations as those of $E_{n,k}$, our remarks about the power series expansion of $E_{n,g}$ can be taken over without change.

Coordinate Shift Through Half Period

Let us consider a new pair of basis functions, $\bar{\psi}_1(x)$ and $\bar{\psi}_2(x)$, referring to the other center of symmetry of $V(x)$, namely $b/2$:

$$\bar{\psi}_1(b/2)=1, \quad \bar{\psi}_1'(b/2)=0, \tag{4.52}$$

$$\bar{\psi}_2(b/2)=0, \quad \psi_2'(b/2)=1. \tag{4.53}$$

Elementary considerations using the symmetry of the potential give the following results: At $\lambda=+1$, $\psi_2(b)$ and $\bar{\psi}_2(\frac{1}{2}b+b)$ are either both zero or both not zero [see Fig. 6(a)]; while at $\lambda=-1$, one and only one of $\psi_2(b)$ and $\bar{\psi}_2(\frac{1}{2}b+b)$ [see Fig. 6(b) and use the fact that bands do not intersect]. Hence, if the normalized Bloch waves $\bar{\varphi}_{n,g}$ are defined by the requirement that $\bar{\varphi}_{n,g}(b/2)$ be real and that $\bar{\varphi}_{n,k}(x)$ be analytic for $k=g$, then depending on whether

$$\varphi_{n,g+2\pi/b}(x)=\pm\varphi_{n,g}(x), \tag{4.54}$$

$\bar{\varphi}_{ng}$ has the property

$$\bar{\varphi}_{n,g+2\pi/b}(x)=\mp\bar{\varphi}_{n,g}(x). \tag{4.55}$$

Thus in all cases one of the functions, $\varphi_{n,g}$ and $\bar{\varphi}_{n,g}$, is periodic with period $2\pi/b$, while the other is antiperiodic.

Remark on Other Choices of Phase

We have discussed two particular assignments of phase suggested by the symmetries of $V(x)$ about $x=0$ and $b/2$. Other choices are of course possible. For complex k we can write the most general Bloch function as

$$\Phi_{n,k}(x)=e^{i\theta(k)}\varphi_{n,k}(x), \tag{4.56}$$

where, to preserve the normalization for real k, we must have

$$\mathrm{Im}\theta(g)=0. \tag{4.57}$$

However, it is obvious that all possible $\Phi_{n,k}$ have singularities at the singularities of $E_{n,k}$. For suppose the contrary, i.e., that $\Phi_{n,k}$ is regular at such a point. Then so would be the function

$$\frac{(-d^2/dx^2+V)\Phi_{n,k}}{\Phi_{n,k}}=E_{n,k}, \tag{4.58}$$

which is a self-contradiction. On the other hand, $\theta(k)$ can of course introduce singularities of its own.

II. WANIER FUNCTIONS

5. Preliminaries

Wannier functions,[8] $a_n(x)$, are localized linear combinations of all the Bloch waves of a given band. In the following paragraphs, we shall be speaking only about one band at a time and, therefore, generally suppress the band index, n. Let then $\Phi_g(x)$ be a normalized Bloch wave of as yet arbitrary phase. Then the corresponding Wannier function is defined

$$a(x)\equiv\left(\frac{b}{2\pi}\right)^{\frac{1}{2}}\int_{\pi/b}^{\pi/b}\Phi_g(x)dg. \tag{5.1}$$

Its shape depends on the variation with g of the phase of Φ_g. From the form (1.4) of Φ_g, we deduce that

$$a(x-lb)=\left(\frac{b}{2\pi}\right)^{\frac{1}{2}}\int_{-\pi/b}^{\pi/b}\Phi_g(x)e^{-ilbg}dg. \tag{5.2}$$

We shall use the notation

$$a^{(l)}(x)\equiv a(x-lb). \tag{5.3}$$

If we multiply (5.2) by $e^{ilbg'}$ and sum over l, using the fact that

$$\sum\nolimits_l e^{ilb(g'-g)}=(2\pi/b)\delta(g'-g), \tag{5.4}$$

we obtain

$$\Phi_g(x)=(b/2\pi)^{\frac{1}{2}}\sum\nolimits_l a^{(l)}(x)e^{ilbg}. \tag{5.5}$$

By means of (5.2) and (5.5), the Wannier functions, $a^{(l)}(x)$, and Bloch waves, $\Phi_g(x)$, can be expressed in terms of one another.

Of course, (5.1) assumes sufficient regularity of $\Phi_g(x)$ for the integral to exist and (5.5) assumes the con-

[8] G. Wannier, Phys. Rev. **52**, 191 (1947).

vergence of the sum over l. To avoid irrelevant mathematical subleties, let us assume from here on that the phases of $\Phi_g(x)$ are so chosen that $\Phi_g(x)$ is a sectionally continuous function of g in the interval $-\pi/b\leqslant g\leqslant\pi/b$. Then the integral (5.1) exists and the sum in (5.5) converges to $\frac{1}{2}(\Phi_{g-0}+\Phi_{g+0})$ for $-\pi/b<g<\pi/b$ and to $\frac{1}{2}(\Phi_{(\pi/b)-0}+\Phi_{(-\pi/b)+0})$ at $g=\pm\pi/b$.

From their definition, the well-known ortho-normality of the $a^{(l)}(x)$ may be verified at once,

$$\int_{-\infty}^{\infty} a^{(l')}(x)^* a^{(l)}(x)dx=\delta_{ll'}. \tag{5.6}$$

6. Reality and Symmetry

In discussing the reality and symmetry of $a(x)$, we must distinguish two cases.

Case A1 or B1:

$$\Phi_0(0)\neq 0.$$

Here we choose the phase of Φ_g such that $\Phi_g(0)$ is real and $\Phi_k(x)$ is analytic for $k=g$. Then $\Phi_g(x)$ is (apart from a possible factor -1) identical to the $\varphi_g(x)$ of Sec. 4. Therefore, on using Eqs. (4.46) and (4.47) in Eq. (5.1), we find

$$a(-x)=\left(\frac{b}{2\pi}\right)^{\frac{1}{2}}\int_{-\pi/b}^{\pi/b}\Phi_{-g}(x)dg=a(x), \tag{6.1}$$

and

$$a^*(x)=\left(\frac{b}{2\pi}\right)^{\frac{1}{2}}\int_{-\pi/b}^{\pi/b}\Phi_{-g}(x)dg=a(x). \tag{6.2}$$

Thus $a(x)$ is both symmetric about $x=0$ and real.

Case A2 or B2:

$$\Phi_0(0)=0.$$

Here we choose the phase of Φ_g such that $\Phi_g(0)$ is purely imaginary and $\Phi_k(x)$ is analytic for $k=g$. This Φ_g is related to the φ_g of Sec. 4 by

$$\Phi_g(x)=\pm i\varphi_g(x). \tag{6.3}$$

Therefore, using Eqs. (4.48) and (4.49), we obtain

$$a(-x)=-\left(\frac{b}{2\pi}\right)^{\frac{1}{2}}\int_{-\pi/b}^{\pi/b}\Phi_{-g}(x)dg=-a(x), \tag{6.4}$$

and

$$a^*(x)=+\left(\frac{b}{2\pi}\right)^{\frac{1}{2}}\int_{-\pi/b}^{\pi/b}\Phi_{-g}(x)dg=a(x). \tag{6.5}$$

This time $a(x)$ is antisymmetric about $x=0$ and real.

Uniqueness

We shall now show that no other choice of phase, which has the property that the corresponding Bloch waves are continuous functions of g, leads to a Wannier function which is both real and either symmetric or antisymmetric about $x=0$.[9]

[9] Apart, of course, from a trivial factor -1; see Eq. (6.15).

Consider then a possible alternative set of Bloch waves,

$$\Phi_g'=e^{i\theta(g)}\Phi_g,\quad -\pi/b\leqslant g\leqslant\pi/b, \tag{6.6}$$

leading to a real and (anti-) symmetric Wannier function. We write

$$e^{i\theta(g)}=\sum_{m=-\infty}^{\infty}A_m e^{-imgb},\quad -\pi/b\leqslant g\leqslant\pi/b. \tag{6.7}$$

The corresponding Wannier function is

$$\begin{aligned} a'(x)&=\left(\frac{b}{2\pi}\right)^{\frac{1}{2}}\int_{-\pi/b}^{\pi/b}\Phi_g'(x)dx \\ &=\sum_{m=-\infty}^{\infty}A_m\left(\frac{b}{2\pi}\right)^{\frac{1}{2}}\int_{-\pi/b}^{\pi/b}\Phi_g(x)e^{-imgb}dg \\ &=\sum_{m=-\infty}^{\infty}A_m a(x-mb). \end{aligned} \tag{6.8}$$

Now using the reality of $a(x)$, we have

$$[a'(x)]^*=\sum_{m=-\infty}^{\infty}A_m^* a(x-mb). \tag{6.9}$$

Since the $a(x-mb)$ form an orthonormal set, we see that $a'(x)$ is real only if

$$A_m=\text{real}. \tag{6.10}$$

Further

$$\begin{aligned} a'(-x)&=\sum_{m=-\infty}^{\infty}A_m a(-x-mb) \\ &=\sum_{m=-\infty}^{\infty}A_{-m}a(-x+mb) \\ &=\pm\sum_{m=-\infty}^{\infty}A_{-m}a(x-mb), \end{aligned} \tag{6.11}$$

depending on whether $a(x)$ is symmetric or antisymmetric. Therefore, if $a(-x)$ is to be symmetric or antisymmetric, we must have

$$A_{-m}=\pm A_m. \tag{6.12}$$

Now using (6.10) and (6.12) in (6.7) gives

$$e^{i\theta(g)}=\begin{cases} A_0+2\sum_{m=1}^{\infty}A_m\cos mgb \\ -2i\sum_{m=1}^{\infty}A_m\sin mgb, \end{cases} \tag{6.13}$$

depending on the sign in (6.12). The second line in (6.13) can be ruled out at once by setting $g=0$. The first line implies that $e^{i\theta(g)}$ is real and hence, because of the assumed continuity of φ_g,

$$e^{i\theta(g)}\equiv\pm 1. \tag{6.14}$$

Therefore,

$$\Phi_g' = \pm\Phi_g. \tag{6.15}$$

7. Asymptotic Behavior of $a(x)$

We begin by discussing the particular Wannier functions of the preceding section, which are real and (anti-) symmetric about $x=0$.

Case A. $\varphi_0(0)$ and $\varphi_{\pi/b}(0)$ Both Nonzero or Both Zero

In this case φ_g is a periodic function of g with period $2\pi/b$. The Φ_g with which we constructed our Wannier functions was related to φ_g of Sec. 4 by a constant ϵ,

$$\Phi_g = \epsilon\varphi_g, \tag{7.1}$$

where $\epsilon=\pm 1$ or $\pm i$ depending on whether $\varphi_0(0)$ did not or did vanish. Since

$$\Phi_g = (b/2\pi)^{\frac{1}{2}} \sum_l a^{(l)}(x) e^{ilbg}, \tag{7.2}$$

we see by comparison with Eq. (4.43) that

$$a^{(l)}(x) \equiv \epsilon\alpha^{(l)}(x). \tag{7.3}$$

Now the functions, $\alpha^{(l)}(x)$, were the coefficients of the Laurent expansion (4.40) which converged for

$$\exp(-b\bar{h}_n) < \lambda < \exp(b\bar{h}_n), \tag{7.4}$$

where $\bar{h}_n$ is defined in Eq. (3.22). Hence

$$\lim_{l\to\infty} \alpha^{(l)}(x)\lambda^l = \begin{cases} 0, & |\lambda| < \exp(b\bar{h}_n) \\ \infty, & |\lambda| > \exp(b\bar{h}_n). \end{cases} \tag{7.5}$$

In view of the connection

$$\alpha^{(l)}(x) = \frac{1}{\epsilon} a(x-lb), \tag{7.6}$$

this is equivalent to

$$\lim_{x\to\infty} a(x)e^{qx} = \begin{cases} 0, & q<\bar{h}_n \\ \infty, & q>\bar{h}_n. \end{cases} \tag{7.7}$$

More loosely speaking, $a(x)$ falls off like $\exp(-\bar{h}_n x)$.

Case B. One and Only One of $\varphi_0(0)$ and $\varphi_{\pi/b}(0)$ is Zero

In this case, φ_g and hence Φ_g is an antiperiodic function of g. The asymptotic behavior of the corresponding Wannier function can be found from (5.2) by performing an integration by parts. For fixed x and large l we find

$$a(x-lb) = \frac{1}{l}\left[\left(\frac{2}{\pi b}\right)^{\frac{1}{2}} (-1)^l i \Phi_{\pi/b}(x)\right] + O\left(\frac{1}{l^2}\right). \tag{7.8}$$

Thus $|a(x)|$ falls off only as x^{-1}.

For many applications such slowly decreasing Wannier functions are most inconvenient. It is therefore fortunate that we find that in Case B, $\varphi_0(b/2)$ and $\varphi_{\pi/b}(b/2)$ are either both nonzero or both zero [see Eqs. (4.54) and (4.55)]. Hence with respect to this new origin we have Case A and can therefore construct Wannier functions $\bar{a}(x)$ with the following properties.

$$\bar{a}(\tfrac{1}{2}b+x) = \pm\bar{a}(\tfrac{1}{2}b-x), \tag{7.9}$$

$$\bar{a}^*(x) = \bar{a}(x), \tag{7.10}$$

$$\lim_{x\to\infty} \bar{a}(x)e^{qx} = \begin{cases} 0, & q<\bar{h}_n \\ \infty, & q>\bar{h}_n. \end{cases} \tag{7.11}$$

8. Existence and Uniqueness of Real, (Anti-) Symmetric, and Short Range Wannier Function

We are now in a position to prove the following theorem: For every band there exists one and only one Wannier function[9] which has all three of the following properties:

1. It is real.
2. It is either symmetric or antisymmetric about either $x=0$ or $x=b/2$.
3. It falls off exponentially.

The quantitative rate of exponential decay of the function which has these properties is given by Eq. (7.7).

To fix our ideas let us take, for example, a band with the following properties: $\varphi_0(0)\neq 0$, $\varphi_{\pi/b}(0)=0$. This is case B1 of Sec. 4 [Eq. (4.51) ff.]. Now in Sec. 6 we have seen that there exists only one choice of $\Phi_g(x)$ which is continuous in g, with the property of leading, via (5.1), to a Wannier function which is real and symmetric or antisymmetric about $x=0$. This Wannier function did not fall off exponentially but like x^{-1}. If we consider also Bloch waves, $\Phi_g(x)$, which are *not* continuous functions of g, the corresponding Wannier function certainly cannot fall off exponentially with x [see Eq. (5.2)]. Thus we conclude that in this case there exists no Wannier function which is real, (anti-) symmetric about $x=0$, and falling off exponentially with x.

However, in the present case we have $\varphi_0(\frac{1}{2}b)\neq 0$ and $\varphi_{\pi/b}(\frac{1}{2}b+b)\neq 0$. Hence relative to $x=\frac{1}{2}b$ this band corresponds to case A1 of Sec. 4. Now in Sec. 6 we have shown that there exists one and only one choice of Φ_g which is continuous in g, with the property of leading via (5.1) to a Wannier function which is real and symmetric or antisymmetric about $x=b/2$. In Sec. 7, we saw that this function fell off exponentially for large x, according to Eq. (7.7). Any other choice of Φ_g, leading to a Wannier function real and symmetric or antisymmetric about $x=b/2$, is necessarily discontinuous in g and hence the corresponding Wannier function does not fall off exponentially.

This completes the proof of our theorem for a band of this type. For the other three types of band exactly similar proofs can be given.

For convenience we list below, for the four possible types of band, the symmetry properties of the corresponding unique Wannier function which has the three properties listed at the beginning of this section.

(A1) $\varphi_0(0)\neq 0,\quad \varphi_{\pi/b}(0)\neq 0:\quad a(-x)=a(x),$ (8.1)

(A2) $\varphi_0(0)=0,\quad \varphi_{\pi/b}(0)=0:\quad a(-x)=-a(x),$ (8.2)

(B1) $\varphi_0(0)\neq 0,\quad \varphi_{\pi/b}(0)=0:$

$$\bar{a}(\tfrac{1}{2}b-x)=\bar{a}(\tfrac{1}{2}b+x), \quad (8.3)$$

(B2) $\varphi_0(0)=0,\quad \varphi_{\pi/b}(0)\neq 0:$

$$\bar{a}(\tfrac{1}{2}b-x)=-\bar{a}(\tfrac{1}{2}b+x). \quad (8.4)$$

The question may be asked if one can possibly obtain a Wannier function falling off more rapidly than Eq. (7.7) by relaxing the requirements of reality and (anti-) symmetry. The answer is clearly "no." For, by Eq. (5.5), the corresponding Bloch waves would be analytic functions of k in a strip wider than Eq. (4.44), and this was shown to be impossible.

9. Nearly Free and Tightly Bound Electrons

If the potential $V(x)$ is weak or if we consider electrons of high energy, the Kramers plot $\mu(E)$ is close to that for free electrons

$$\mu_0(E)=\cos(E^{\frac{1}{2}}b). \quad (9.1)$$

In the free-electron case, the extrema μ_n of $\mu(E)$ occur all at $\mu=\pm 1$. Hence in the "nearly free" case we have

$$|\mu_n|=1+\delta_n, \quad (9.2)$$

where δ_n is a small positive quantity. Therefore, the singularities of E_k and φ_k lie very close to the real axis,

$$h_n=\sqrt{2\delta_n} \quad (9.3)$$

[see Eqs. (3.10) and (3.11)], and hence the exponential decay of the Wannier functions is very slow. This is in accord with the fact that for perfectly free electrons (which are not included in our considerations because their bands intersect), the Wannier function falls off only as x^{-1}.

For very tightly bound electrons of binding energy ϵ_n, it is shown in the Appendix that $\bar{h}_n\rightarrow(-\epsilon_n)^{\frac{1}{2}}$. Therefore, for large x their Wannier function behaves as

$$a_n(x)\sim\exp[-(-\epsilon_n)^{\frac{1}{2}}x], \quad (9.4)$$

that is, it falls off at the same exponential rate as the wave function of an isolated potential well.

ACKNOWLEDGMENTS

The writer takes great pleasure in thanking Mr. S. Michaelson for valuable help with several phases of this paper. The work here reported was begun at the Imperial College of Science and Technology, London, England, while the writer was a National Science Foundation Senior Post-doctoral Fellow. Later it was supported at Carnegie Institute of Technology by the Office of Naval Research. The assistance of these organizations is acknowledged with thanks. The writer also is glad to avail himself of this opportunity to thank the staff of the Department of Mathematics at the Imperial College for the hospitality which he enjoyed during his stay there.

APPENDIX. TIGHTLY BOUND ELECTRONS

Let $V_0(x)$ be a cutoff, symmetric potential,

$$V_0(x)\equiv 0,\quad |x|>c, \quad (A.1)$$

with a number of bound states with eigenvalues

$$\epsilon_n=-q_n^2,\quad n=0,1,\cdots. \quad (A.2)$$

We now consider the solutions of the Schrödinger equation (1.1) corresponding to the periodic potential

$$V(x)\equiv\sum_{l=-\infty}^{\infty}V_0(x-lb) \quad (A.3)$$

in the so-called tight-binding limit,

$$b\rightarrow\infty. \quad (A.4)$$

We wish to determine the magnitude of $\mu(E)$ at those energies E_n where $d\mu/dE=0$. This enables us to locate the singularities of the energy function E_k in the complex k-plane.

For the present purposes it is convenient to use as fundamental system of solutions of the Schrödinger equation (1.1) the functions $u_1(x,E)$ and $u_2(x,E)$ where

$$u_1(-c,E)=1,\quad u_1'(-c,E)=0, \quad (A.5)$$

$$u_2(-c,E)=0,\quad u_2'(-c,E)=1. \quad (A.6)$$

μ is given by the relation [see Eq. (2.9)]

$$\mu(E)=\tfrac{1}{2}[u_1(b-c,E)+u_2'(b-c,E)]. \quad (A.7)$$

Now let us define

$$f_1(E)\equiv u_1(c,E);\quad f_1'(E)\equiv u_1'(c,E), \quad (A.8)$$

$$f_2(E)\equiv u_2(c,E);\quad f_2'(E)\equiv u_2'(c,E). \quad (A.9)$$

Then since in the interval $c<x<b-c$, $V(x)=0$, one finds by matching solutions at $x=c$, that

$$u_1(b-c,E)=\frac{1}{2}\left(f_1+\frac{f_1'}{q}\right)e^{q(b-2c)}+\frac{1}{2}\left(f_1-\frac{f_1'}{q}\right)e^{-q(b-2c)}, \quad (A.10)$$

$$u_2'(b-c,E)=\tfrac{1}{2}(f_2q+f_2')e^{q(b-2c)}-\tfrac{1}{2}(f_2q-f_2')e^{-q(b-2c)}, \quad (A.11)$$

where

$$q=(-E)^{\frac{1}{2}}; \quad (A.12)$$

q is taken positive for negative E. By means of (A.10) and (A.11), we can express μ in terms of f_1, f_1', f_2, and f_2'.

Now let us define the function

$$G(E)\equiv f_1(E)+\frac{f_1'(E)}{q}+f_2(E)q+f_2'(E). \quad \text{(A.13)}$$

Substituting (A.10) and (A.11) into (A.7), we find that as $b\rightarrow\infty$, the condition $d\mu/dE=0$ or $d\mu/dq=0$ gives the following equation for the determination of the E_n, Eq. (2.15):

$$G(E)+\frac{1}{(b-2c)}\frac{d}{dq}G(E)=O(e^{-2qb}). \quad \text{(A.14)}$$

On the other hand, the band edges, where $\mu=\pm1$, are given by

$$G(E)=O(e^{-qb}). \quad \text{(A.15)}$$

Since in the tight-binding limit the band edges coincide with the "atomic" energies, ϵ_n, we must have

$$G(E)=0 \qquad \text{at } E=\epsilon_n; \quad \text{(A.16)}$$

this may also be directly verified.

For large b, the solutions E_n of (A.14) differ only slightly from those of (A.16)

$$E_n=\epsilon_n+\frac{2q_n}{b}+O\left(\frac{1}{b^2}\right). \quad \text{(A.17)}$$

Substituting these values into the expression (A.7) for μ gives, with the aid of (A.10), (A.11), and (A.13)

$$\mu(E_n)=G(E_n)\exp[(-E_n)^{\frac{1}{2}}(b-2c)]+O(e^{-2q_nb})$$
$$=\exp[(-E_n)^{\frac{1}{2}}(b-2c)]$$
$$\times\left[\frac{2q_n}{b}G'(E_n)+O\left(\frac{1}{b^2}\right)\right]. \quad \text{(A.18)}$$

From this we can now determine the distance h_n of the singularities of E_k from the real k-axis [see Eq. (3.12)]:

$$\lim_{b\to\infty}h_n=(-\epsilon_n)^{\frac{1}{2}}. \quad \text{(A.19)}$$

As $\bar{h}_n=\min(h_n,h_{n-1})$ and $\epsilon_n>\epsilon_{n-1}$, we also have

$$\lim_{b\to\infty}\bar{h}_n=(-\epsilon_n)^{\frac{1}{2}}. \quad \text{(A.20)}$$

A Theory of the Electrical Breakdown of Solid Dielectrics.

CLARENCE ZENER, H. H. Wills Physics Laboratory, Bristol.

(Communicated by R. H. Fowler, F.R.S.—Received December 27, 1933, Revised March 1, 1934.)

§ 1. *Introduction.*

In the modern theory of metallic conduction initiated by Bloch, conductors, semi-conductors, and non-conducting crystals may be represented by the same model. In this model each electron is supposed to move freely in the periodic field of the lattice. Owing to this field, not all electronic energy levels are allowed; the allowed levels are grouped into bands, separated by energy intervals which are disallowed. If all the energy levels of a given band are occupied by electrons, then, according to the theory, these electrons can make no contribution to an electric current in the crystal. If all the bands are full, the crystal must be an insulator. Thus in an insulator there exist a number of energy bands which are completely full, and a number of bands of higher energy which, for a perfect crystal at the absolute zero of temperature, are empty.

In a real non-conducting crystal, however, there will be a few electrons in the first unfilled band, owing to thermal excitation, impurities, etc. Their number is, however, too small to give an appreciable current at ordinary field strengths. As the field strength is increased, the current due to these few electrons increases steadily, but it will not show the sudden rise observed in dielectric breakdown. For this sudden rise it is necessary that the number of electrons in an unfilled band should suddenly increase as the field strength passes a critical value. Two distinct mechanisms have been suggested for this sudden increase. Of these, the first† is a process analogous to the electrical breakdown of gases. In the absence of an external field, the few electrons in the upper band are in the lowest energy state of this band; under the action of an electric field, they are raised to higher levels. When one of these electrons reaches a sufficiently high level, it will give up energy to an electron in a lower (full) band, both electrons making a transition to a low level of the upper band. The process will then be repeated; the number of electrons in the

† Hippel, 'Z. Physik,' vol. 67, p. 707, vol. 68, p. 309 (1931); vol. 75, p. 145 (1932).

upper band will thus increase exponentially with time as long as the electric field is maintained.

When a constant electric field is present energy bands have significance only with reference to restricted regions of space, since such an electric field makes each energy band (defined for no external field) of a lattice of infinite extent degenerate with every other energy band, *cf.* fig. 1. Thus in the presence of a constant electric field an electron may pass from one " energy band " into another which, if the external field were absent, would lie above the first band. This second process of excitation is analogous† to the auto-ionization of free atoms by large electric fields. In a gas the effect of auto-ionization is of no practical importance, being masked by the first mechanism of dielectric breakdown. In a solid the analogous process is of more importance than in gases, both because the mean free path of the electrons is shorter than in gases, and because, as we shall show, the electric field can more readily *excite* an electron in a lattice than *ionize* an isolated atom.

The purpose of this paper is to examine this second method of dielectric breakdown. In § 2 the Bloch model of a metal is used to calculate the rate at which electrons escape from the lower into the upper energy bands under the influence of a constant electric field. In § 3 we show that both the magnitude of the breakdown field and the suddenness of the breakdown are obtained by this theory, assuming only the second mechanism to be operative.

§ 2. *Theory.*

The rate at which electrons pass from one band into another under the action of a constant electric field is independent of the thermal motion of the lattice, at least for not too high temperatures. The potential energy of an electron in the lattice will thus be considered a periodic function of the co-ordinates. Only a one-dimensional lattice will be treated, the potential energy of the electron being a function V (x) with period a of the co-ordinate x of the electron.

We shall first determine approximately the motion of an electron in a lattice under the influence of a constant electric field by the method introduced by Bloch.‡ The wave equation with a periodic potential energy has solutions of the form

$$\psi_k(x) = e^{ikx}\, \mathrm{U}_k(x), \tag{1}$$

† Rosenkewitsch and Sinelnikow, 'Z. Physik,' vol. 73, p. 118 (1932); Eisler, 'Z. Physik,' vol. 79, p. 266 (1932).

‡ 'Z. Physik,' vol. 52, p. 555 (1928).

where $U_k(x)$ has the same period as $V(x)$. The condition that ψ_k should remain finite for all values of x imposes the restriction that k be real. But only when E_k lies in the bands of allowed energy referred to in § 1 has the wave equation solutions of the type (1) with a real k. The first energy band lies within the interval $-\pi/a \leqslant k \leqslant \pi/a$ of k.

Bloch further showed that if, when an electric field F is present, the wave function $\psi(x, t)$ is expanded in terms of wave functions of the lowest energy band when $F = 0$, in the form

$$\psi(x, t) = \int_{-\pi/a}^{\pi/a} g(k, t)\, \psi_k(x)\, dk,$$

thus neglecting all energy bands except the first and so excluding from the calculation any possibility of a transition to these bands, then

$$\frac{\partial}{\partial t} |g|^2 = -\frac{2\pi eF}{h} \frac{\partial}{\partial k} |g|^2,$$

or

$$|g|^2 = G\left(k - \frac{2\pi eF}{h} t\right),$$

where G is an arbitrary function. When we remember that $\psi_{-\pi/a} = \psi_{\pi/a}$, we see that $|g|^2$ is a periodic function of t. Since the range of k is $2\pi/a$, and since $|g|^2$ moves in k space with the velocity $2\pi eF/h$, the period of the time variation of $|g|^2$ is $(2\pi/a)/(2\pi eF/h) = h/eFa$; its frequency is thus

$$\nu = eFa/h. \tag{2}$$

We thus conclude that the velocity of the electron, *i.e.*,

$$\frac{\partial}{\partial t} \int \psi x \psi^* \, dx,$$

is also a periodic function of time with the frequency ν, since the identity

$$\frac{\partial}{\partial t} \int \psi\, x\, \psi^*\, dx = \int v_k\, |g|^2\, dk$$

may be readily established when v_k is the velocity of an electron in state k. Thus if we represent the electron by a wave packet confined to the first energy band, the electron moves in the direction of the field until it is reflected by the lattice, then moves in the opposite direction until it is stopped by the field, whereupon the motion is repeated.

We must now calculate the probability per unit time that the electron will make a transition to an excited band. If we take the wave function $\psi(x, t)$ to be a wave packet, then each time the wave packet is reflected by the lattice, a small fraction of the wave packet will pass through into the upper zone, giving the probability p that the electron escapes into the upper zone (*cf.* fig. 1). The probability, γ, per unit time is obtained by multiplying p by the

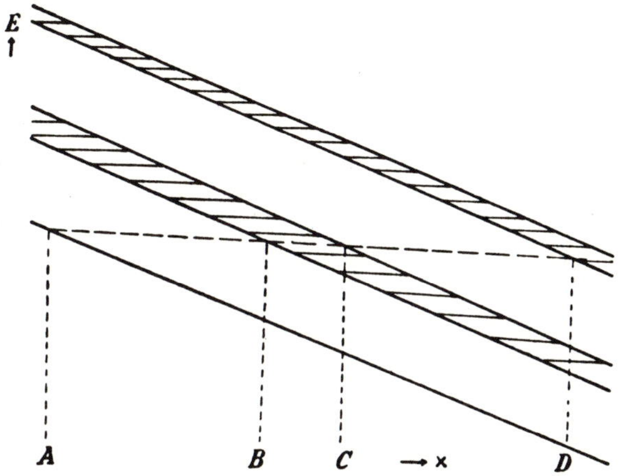

Fig. 1.—" Potential barrier " diagram. The shaded regions represent zones of forbidden energy in the presence of an electric field.

number of times per second that the wave packet is reflected by the lattice, namely, ν, so that

$$\gamma = \nu p.$$

We have now to calculate p. This is most simply done by taking for $\psi(x, t)$ a wave function *periodic* (in time), not a wave packet. The wave equation for an electron with energy E when an external electric field F is present is

$$\left\{\frac{h^2}{8\pi^2 m}\frac{\partial^2}{\partial x^2} - \mathrm{V}(x) + (\mathrm{E} - e\mathrm{F}x)\right\}\psi(\mathrm{E} : x) = 0.$$

An approximate solution of this equation may at once be obtained if we observe that the term $e\mathrm{F}x$ changes only slightly in the distance of a single lattice constant. Such a solution is

$$\psi(\mathrm{E} : x) = e^{i\int \mathrm{K}dx}\,\mathrm{U}_k(x),$$

where K is the same function of $(\mathrm{E} - e\mathrm{F}x)$ as it is of E alone when F is zero, and $\mathrm{U_K}$ is identical to the U_k of (1) with $k = \mathrm{K}$. With a given E our K will now be real *only for certain values* of x, as shown in fig. 1. In the regions where K is complex, ψ (E : x) will decrease, or increase, exponentially. On either side of such a region ψ (E : x) is nearly identical with the wave functions in the absence of a field which belong to two adjacent energy bands. Thus in fig. 1, ψ (E : x) is similar to a wave function belonging to the first energy band in the interval AB, while in the interval CD it is similar to a wave function belonging to the second band. If we take the wave function ψ which *decreases* in the range BC, then we know that the transition probability p is given by†

$$p = |\psi(x_C)/\psi(x_B)|^2.$$

If we write, for values of x in the interval BC,

$$K = \xi(x) + i\eta(x) \tag{4}$$

then we have

$$|\psi(x_C)/\psi(x_B)| \approx e^{-\int_B^C \eta(x)dx} \tag{4a}$$

and hence, by (2) and (3)

$$\gamma = \frac{eFa}{h} e^{-2\int_B^C \eta dx}$$

In order to obtain an explicit expression for K (x), we write our wave equation in the standard form of Hill's equation,

$$\frac{d^2\psi}{ds^2} + (\theta_0 + 2\sum_{n=1}^{\infty} \theta_n \cos 2ns)\,\psi = 0.$$

Here

$$s = \pi x/a,$$

$$\theta_0 = 8ma^2(E - eFz)/h^2,$$

and θ_n is $4ma^2/h^2$ times the coefficient of cos $2\pi nx/a$ in the Fourier expansion of V (x). When θ_1 is small in comparison with unity (with $a = 3 \times 10^{-8}$ cm, this condition means that the first Fourier coefficient must be small in comparison with 4 electron volts), and when the higher coefficients are still smaller, K is determined as a function only of θ_0 and θ_1 by the equation‡

$$\sin^2(aK/2) = \sin^2(\tfrac{1}{2}\pi\,\theta_0^{\frac{1}{2}})\left\{1 + \frac{\pi\theta_1^2}{4\theta_0^{\frac{1}{2}}(1-\theta_0)}\cot(\pi\theta_0^{\frac{1}{2}}/2)\right\} + O(\theta_1^4). \tag{5}$$

In the first energy band in which K is complex, θ_0 is seen from (5) to differ

† Gamow, "Atomic Nuclei and Radioactivity," chap. 2, Oxford (1931); v. Laue, 'Z. Physik,' vol. 52, p. 726 (1928). Houtermans, 'Ergbn. exakt. Naturw.,' vol. 9, p. 133 (1930).

‡ Hill, 'Acta Math.,' vol. 8, p. 27 (1886); M. J. O. Strutt, "Lame'sche-Mathieusche Funktionen," p. 25, Berlin (1932). In the latter reference $2\theta_1$ is erroneously used in place of θ_1. This mistake is easy to make, since θ_1 is defined in Hill's memoir in two ways differing by a factor of two.

only slightly from unity, and K is seen to differ only slightly from $\pi\theta_0^{\frac{1}{2}}/2$, *i.e.*, only slightly from π/a. Hence if we set

$$\left.\begin{aligned} \theta_0 &= 1 + \alpha \\ K &= \pi/a + \beta \end{aligned}\right\} \tag{6}$$

α and β, together with θ_1, may be regarded as small quantities. Substituting (6) in (5), and neglecting powers of α, β, and θ_1 greater than the second, we obtain

$$K = \pi/a \pm i\,(\pi/a)\,(\theta_1^2 - \alpha^2)^{\frac{1}{2}}.$$

If we now choose the origin $x = 0$ to lie midway between B and C, we have, using (6),

$$K = \frac{\pi}{a}\left[1 \pm i\,\frac{8ma^2}{h^2}\left\{V_0^2 - (eFx)^2\right\}^{\frac{1}{2}}\right]. \tag{7}$$

The points B and C are now given by $x = -V_0/eE$ and $x = V_0/eF$, respectively, so that the energy gap ε between the two bands of allowed energy is $2V_0$. Performing the integration of (4*a*), we obtain for the rate γ with which an electron passes from the region AB into CD

$$\gamma = \frac{eFa}{h}\exp -\left\{\frac{\pi^2}{h^2}\frac{ma\varepsilon^2}{|eF|}\right\}. \tag{8}$$

§ 3. *Discussion.*

We have obtained in (8) the rate with which an electron escapes from one energy band into the next energy band. In the derivation of (8) we assumed the system to be one-dimensional, and the first Fourier coefficient of the potential energy of an electron in the lattice to be sufficiently small (considerably less than 4 electron volts when $a = 3 \times 10^{-8}$ cm). It is to be expected that the results obtained for a one-dimensional lattice are true in general for a three-dimensional lattice. Although the Fourier coefficients of most crystals are larger than allowed in the above derivation, it is nevertheless to be expected that the general dependence, given by (8), of the transition rate upon the energy gap ε will still be valid.

When we take the reasonable magnitudes $a = 3 \times 10^{-8}$ cm, $\varepsilon = 2$ electron volts, and let F^* be the electric field in volts cm^{-1}, (8) becomes

$$\gamma = 10^7\,F^*\,10^{-2\times 10^7/F^*}.$$

The dielectric breakdown is thus not to be expected until F^* reaches the order of magnitude of 10^6, as is observed. Further, the breakdown will occur suddenly as F^* is increased, γ being increased by a factor of 100 (in our example) when F^* changes from $1 \cdot 0 \times 10^6$ to $1 \cdot 1 \times 10^6$.

It is to be particularly stressed that for field strengths obtainable in metals, equation (8) shows the transition between different energy bands to be altogether negligible. Thus for $a = 3 \times 10^{-8}$ cm, and $F^* = 1$, an electron will pass an energy gap as small as 2×10^{-3} electron volts only after a period of years.

§ 4. *Summary.*

A theory of the electrical breakdown of solid dielectrics is developed. The breakdown is assumed to be due to the direct excitation of electrons by the electric field. Agreement with experiment is obtained both for the magnitude of the breakdown field and for the suddenness with which the breakdown occurs as the electric field is increased.

Reprinted from THE PHYSICAL REVIEW, Vol. 120, 1175 (1960)
Printed in U. S. A.

Electron Levels in a One-Dimensional Random Lattice

H. L. FRISCH AND S. P. LLOYD
Bell Telephone Laboratories, Murray Hill, New Jersey
(Received July 7, 1960)

Let the potential of a one-dimensional scalar particle be $V(x)=V_0\sum_{-\infty}^{\infty}\delta(x-x_j)$, $-\infty<x<\infty$, where $V_0<0$, and where the sequence (x_j) is random, with a Poisson distribution. The quantity of interest is a certain limiting level distribution, equal numerically to the node density of real solutions $\psi(x)$ of the Schrödinger equation. The random variables $z_j=\psi'(x_j-0)/\psi(x_j)$, $-\infty<j<\infty$, constitute an ergodic stationary Markov process. The stationary density $T(z)$ of the (z_j) satisfies a first-order linear differential-difference equation, and the node density is given (with probability 1) by $\lim_{z\to\infty} z^2T(z)$ (Rice's formula). Numerical results are obtained by integrating the second-order linear differential equation satisfied by the Fourier transform of $T(z)$.

1. INTRODUCTION

WE are concerned with the distribution of energy levels of a one-dimensional electron (scalar particle) moving in a one-dimensional random array of atoms. The atoms, all of one kind, have (randomly) fixed positions, and the electron-atom potentials are assumed to be δ functions. The Schrödinger equation for an electron of mass m and energy E is then

$$-\frac{\hbar^2}{2m}\frac{d^2\psi(x)}{dx^2}+V_0\sum_{j=-\infty}^{\infty}\delta(x-x_j)\psi(x)=E\psi(x), \quad (1)$$

where $V_0<0$ is the strength of each electron-atom interaction (attractive) and $\cdots\leq x_{-1}\leq x_0\leq x_1\leq\cdots$ are the positions of the atoms, randomly distributed on the infinite line. We consider in detail only the case where the x_j's have a Poisson distribution, although our methods are applicable to certain more general distributions (described in Sec. 6). Boundary conditions for (1) are discussed presently.

One model which gives rise to (1) (and the one which led to the present investigation) is the impurity band model of Lax and Phillips.[1] The "atoms" represent impurity atoms in a one-dimensional crystal; the periodic potential of the pure crystal is replaced by a constant (included in E). Another model to which (1) might be applied is a one-dimensional liquid metal. We do not discuss such applications, confining ourselves to a mathematical study of (1). A detailed discussion of the impurity band model, with references to the literature, is given in the Lax and Phillips article.

2. LEVEL DISTRIBUTION

The quantity we seek is the limiting distribution-in-energy of the eigenvalues of (1), defined as follows. Let

[1] M. Lax and J. C. Phillips, Phys. Rev. 110, 41 (1958).

ω denote a sequence $\omega = (\cdots, x_{-1}, x_0, x_1, \cdots)$ of atom positions. We treat each sequence ω as a single point in an infinite-dimensional space Ω. This space Ω of all possible such sequences is a measure space, carrying a probability measure which we denote by $P\{\ \}$; measurable ω functions are "random variables."[2] Expectations [averages over ω with weighting $dP(\omega)$] will be denoted by $\langle \cdots \rangle$.

For each sequence ω let $E_1(L,\omega) \leq E_2(L,\omega) \leq \cdots$ denote the eigenvalues of (1) for a finite interval $0 \leq x \leq L$. That is, for fixed ω, the $E_m(L,\omega)$'s are the values of E for which there exist solutions of (1) on $0 \leq x \leq L$ satisfying, say, $\psi(0) = \psi(L) = 0$.[3] Still with ω fixed, let $\mathfrak{N}_L(E,\omega)$ be defined as a function of E by

$$\mathfrak{N}_L(E,\omega) = (1/L) \times [\text{number of } E_m(L,\omega)\text{'s which satisfy } E_m(L,\omega) \leq E], \quad -\infty < E < \infty;$$

[i.e., a nondecreasing step function which vanishes to the left of $E_1(L,\omega)$ and which jumps by $1/L$ at each $E_m(L,\omega)$]. By the "distribution of levels" we mean the limit

$$\mathfrak{N}(E) = \lim_{L\to\infty} \mathfrak{N}_L(E,\omega) \text{ with probability 1.} \tag{2}$$

As we prove in Appendix 2, the right-hand side in (2) exists and is independent of ω with probability 1, as indicated.[4]

To obtain $\mathfrak{N}(E)$ explicitly we will make use of the fact that $\mathfrak{N}(E)$ is equal to the density of zeros of any real solution of (1).[5,6] For each E and ω let $\psi(x; E,\omega)$ denote the (real) solution of (1) which satisfies boundary conditions $\psi(0; E,\omega) = \xi_0$, $\psi'(0; E,\omega) = \eta_0$, where ξ_0 and η_0, real and not both zero, are arbitrarily given boundary values. If $\nu_L(E,\omega)$ denotes the number of zeros of $\psi(x; E,\omega)$ in the interval $0 \leq x \leq L$, then

$$\lim_{L\to\infty} \mathfrak{N}_L(E,\omega) = \lim_{L\to\infty} [\nu_L(E,\omega)/L] \text{ with probability 1,} \tag{3}$$

independently of ξ_0 and η_0. That both limits exist and are equal whenever either one exists is a theorem in differential equations[5]; probability is not involved. Since the common limit in (3) is constant (as an ω function) with probability 1, it is equal to its expected value with probability 1. Our problem is thus reduced to finding the (ω,x)-average number of zeros per unit length of real solutions of (1).

[2] J. L. Doob, *Stochastic Processes* (John Wiley & Sons, Inc., New York, 1953), p. 599 ff.

[3] The numbers $E_m(L,\omega)$ depend only on those x_j's which happen to satisfy $0 \leq x_j \leq L$, of course. The number of such x_j's is finite with probability 1, implying that the $E_m(L,\omega)$'s are defined with probability 1 (that is, except for a set of ω's of total probability 0).

[4] That is to say, there exists a number $\mathfrak{N}(E)$ which has the following property: The set of all ω for which $\mathfrak{N}_L(E,\omega)$ fails to converge to $\mathfrak{N}(E)$ (as $L \to \infty$, with E and ω fixed) is a subset of Ω which has probability 0.

[5] H. M. James and A. S. Ginzbarg, J. Phys. Chem. **57**, 840 (1953).

[6] H. Schmidt, Phys. Rev. **105**, 425 (1957).

In subsequent sections we work with the dimensionless quantity

$$N(\lambda) = (1/n)\mathfrak{N}(\hbar^2\lambda/2m),$$

where n is the expected density of atoms. There being n atoms per unit length (on the average), $N(\lambda)$ may be regarded as the number of electron levels per atom below energy $E = (\hbar^2/2m)\lambda$. $N(\lambda)$ will depend only on a dimensionless energy ratio λ/κ_0^2 and a dimensionless density parameter n/κ_0, where $\kappa_0 = -mV_0/\hbar^2 (>0)$ is the inverse range of an electron of energy $-(\hbar^2/2m)\kappa_0^2$ bound to an isolated atom. (We follow the Lax and Phillips notation here.)

3. THE PHASE PROCESS

For expository purposes we discuss (1) with reference to still another model, namely, a classical harmonic oscillator disturbed by randomly occurring impulses. We substitute symbol t for symbol x in (1), and regard $\psi(t)$ (real) as the displacement at time t of a unit point mass from an equilibrium position $\psi = 0$ on the ψ axis. With $\xi = \psi$, $\eta = \dot\psi$, (1) may be expressed as

$$\dot\xi(t) = \eta(t),$$
$$\dot\eta(t) = -[\lambda + 2\kappa_0 \sum_{j=-\infty}^{\infty} \delta(t - t_j)]\xi(t), \tag{4}$$

where, again, $\lambda = 2mE/\hbar^2$ and $\kappa_0 = -mV_0/\hbar^2$.

We refer to the times t_j as *hits*. At each hit, the particle coordinate is unchanged (continuity of ψ):

$$\xi(t_j + 0) = \xi(t_j - 0), \tag{5}$$

but the particle momentum receives an increment proportional to the displacement, directed toward the origin:

$$\eta(t_j + 0) - \eta(t_j - 0) = -2\kappa_0 \xi(t_j) \tag{6}$$

[obtained by integrating (4) from $t_j - 0$ to $t_j + 0$]. Between hits the particle moves as a harmonic oscillator with force constant λ:

$$\dot\xi(t) = \eta(t), \quad \dot\eta(t) = -\lambda\xi(t), \quad t \neq t_j. \tag{7}$$

The probability is 0 that two or more hits are ever simultaneous, and we neglect such events.

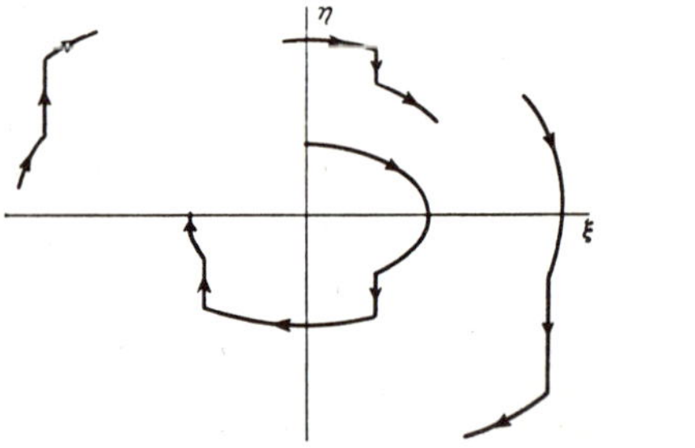

FIG. 1. Phase motion between hits on ellipses $\lambda\xi^2 + \eta^2 = \text{const}$ for given $\lambda > 0$. At hits, motion is instantaneous on vertical segments.

2

Orbits in phase space for $\lambda>0$ and $\lambda<0$ are shown in Figs. 1 and 2, respectively. It is seen that in the case $\lambda>0$ the η-axis crossings (zeros of ψ) occur at a rate at least $\lambda^{1/2}/\pi$, the effect of the hits being to speed the angular motion. When $\lambda<0$, however, the motion (7) is retrograde (counterclockwise) in the sectors $|\eta/\xi|<\sqrt{-\lambda}$, and if there were no hits the phase point would stick at the asymptote $\eta/\xi=+\sqrt{-\lambda}$.

The phase variables $\{(\xi(t,\omega),\eta(t,\omega)),\ -\infty<t<\infty\}$ constitute a two-dimensional stochastic process. The η-axis crossings are determined by the angular part of the motion, however, and since (4) is homogeneous in ξ and η, we may treat the angular part separately.[7] Accordingly, we introduce the variable $z=\eta/\xi=\dot{\psi}/\psi$, and we will show now that the random variables $\{z(t,\omega),\ -\infty<t<\infty\}$ constitute a Markov process. Equations (5) and (6) give

$$z(t_j+0)=z(t_j-0)-2\kappa_0 \tag{8}$$

at hits, while (7) gives

$$\dot{z}=-(z^2+\lambda),\quad t\neq t_j, \tag{9}$$

for the motion between hits. Note that at an η-axis crossing the variable z flies off to $z=-\infty$ and instantly reappears at $z=+\infty$ (compare $z=-c\tan ct$ at its singularities). For any given time τ, Eqs. (8)–(9) determine $z(t,\omega)$ for all times $t>\tau$ as a function of $z(\tau,\omega)$ and the times of the hits occurring after τ. It is a property of the Poisson process, however, that the times of the hits occurring after τ are statistically independent of the times of the hits occurring before τ.[8] From this it can be shown that the times of the hits occurring after τ are independent of the random variables $z(t',\omega)$, $-\infty<t'\leq\tau$. It follows that for any $t>\tau$ the conditional probability distribution of $z(t,\omega)$ given all values $\{z(t',\omega),\ -\infty<t'\leq\tau\}$ is the same as the conditional distribution of $z(t,\omega)$ given only $z(\tau,\omega)$, and this is the Markov property.[9] The z process has

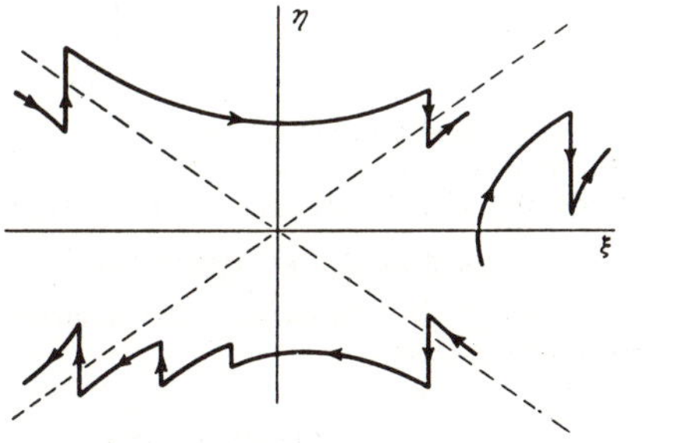

FIG. 2. Phase motion between hits on hyperbolas $\lambda\xi^2+\eta^2=\text{const}$ for given $\lambda<0$. Note that this motion is retrograde (counterclockwise) in the sectors $|\eta/\xi|<\sqrt{-\lambda}$. At hits, motion is instantaneous on vertical segments.

[7] The full (ξ,η) process is discussed in Appendix 1.
[8] See reference 2, pp. 398 ff.
[9] See reference 2, p. 80.

stationary (independent of time origin) transition probabilities; the hit process is stationary, whence (4) admits time shifts.

We wish to find the probability distribution of $z(t,\omega)$, and for this we use the following characterization of the Poisson process.[10] The number of hits in any small time interval $(t,t+dt)$ is statistically independent of the numbers in any other disjoint measurable t sets disjoint from $(t,t+dt)$, and has the probability distribution

$$P\{(t,t+dt)\text{ contains no hit}\}=1-ndt+o(dt),$$

$$P\{(t,t+dt)\text{ contains exactly one hit}\}=ndt+o(dt),$$

$$P\{(t,t+dt)\text{ contains more than one hit}\}=o(dt),$$

where n is the expected number of hits per unit time and $o(dt)$ are negligible higher order quantities.

To simplify the derivation, we assume that the probability distribution of $z(t,\omega)$ has a density:

$$P\{z(t,\omega)\leq\zeta\}=\int_{-\infty}^{\zeta}T(z,t)dz,\quad -\infty<\zeta<\infty,$$

where, of course,

$$T(z,t)\geq 0,\quad \int_{-\infty}^{\infty}T(z,t)dz=1,\quad -\infty<t<\infty. \tag{10}$$

Abusing the terminology somewhat, we refer to the z axis also as "phase space" and a point $z(t,\omega)$ as a "phase point." Let us regard $T(z,t)\delta z$ as the fraction of phase points in an ensemble which lie in an interval $(z,z+\delta z)$ at time t. The phase points which lie in $(z,z+\delta z)$ at time $t+dt$ consist of

(a) those which were in $(z',z'+\delta z')$ at time t and (with probability about $1-ndt$) received no hit during $(t,t+dt)$, where, from (9),

$$z'=z+(z^2+\lambda)dt,$$

$$\delta z'=(dz'/dz)\delta z=(1+2zdt)\delta z;$$

(b) those which were in $(z'',z''+\delta z'')$ at time t and (with probability about ndt) received exactly one hit during $(t,t+dt)$, where, from (8),

$$z''=z+2\kappa_0,$$

$$\delta z''=(dz''/dz)\delta z=\delta z;$$

(c) a negligible proportion which reach $(z,z+\delta z)$ after being hit more than once during $(t,t+dt)$. Conservation of phase points requires

$$T(z,t+dt)\delta z=(1-ndt)T(z',t)\delta z'+ndtT(z'',t)\delta z'',$$

to within higher order differentials. From this we obtain

$$\frac{\partial T(z,t)}{\partial t}=\frac{\partial}{\partial z}[(z^2+\lambda)T(z,t)]+n[T(z+2\kappa_0,t)-T(z,t)]. \tag{11}$$

[10] W. Feller, *Probability Theory and its Applications* (John Wiley & Sons, Inc., New York, 1950), p. 366.

If an initial distribution[11] of phase points is given, say $T(z,0)$, then (11) determines $T(z,t)$ uniquely for all $t>0$. Every solution tends to a limiting density

$$T(z)=\lim_{t\to\infty} T(z,t),$$

the limit being independent of the initial distribution. This stationary density $T(z)$ is the (unique) solution of

$$\frac{d}{dz}[(z^2+\lambda)T(z)]+n[T(z+2\kappa_0)-T(z)]=0 \qquad (12)$$

which satisfies the normalizing condition

$$\int_{-\infty}^{\infty} T(z)dz=1. \qquad (13)$$

Proofs of these statements involve ergodic and other properties of the z process, and are given in Appendix 2. We assume in all that follows that the distribution of phase points is the stationary distribution, i.e., $T(z,t)=T(z)$, $-\infty<t<\infty$.

Differentiation of

$$(z^2+\lambda)T(z)+n\int_z^{z+2\kappa_0} T(\zeta)d\zeta=\text{const} \qquad (14)$$

shows that (14) is an integral-equation version of (12). Equation (14) is simply the assertion that the flux of phase points is constant. For the phase points which pass point z leftward during a small time interval of duration dt due to the motion (9) is $-(1-ndt)T(z)(\dot{z}dt)\approx(z^2+\lambda)T(z)dt$, and the leftward flux at z due to hits is $(ndt)T(\zeta)d\zeta$ from all elements $d\zeta$ in the interval $z\le\zeta\le z+2\kappa_0$. The leftward flux at $z=-\infty$ is the expected η-axis crossing rate $nN(\lambda)$, so that (14) may be put into the form

$$T(z)=\frac{n}{z^2+\lambda}\left[N(\lambda)-\int_z^{z+2\kappa_0} T(\zeta)d\zeta\right]. \qquad (15)$$

Since the integral here vanishes as $z\to\pm\infty$, it follows that

$$N(\lambda)=(1/n)\lim_{z\to\pm\infty} z^2T(z), \qquad (16)$$

which we use later.

Another integrated version of (12) is obtained by integrating with a factor which integrates all terms except the one $T(z+2\kappa_0)$. We define

$$u(z)=e^{(n/k)\arctan(z/k)} \quad \text{if} \quad \lambda=k^2>0,$$

$$u(z)=\left|\frac{z-\kappa}{z+\kappa}\right|^{n/(2\kappa)} \quad \text{if} \quad \lambda=-\kappa^2<0,$$

with $|\arctan x|<\frac{1}{2}\pi$ in the first case. We have $u'(z)/u(z)=n/(z^2+\lambda)$ in either case, and it is easily verified that (12) is equivalent to

$$\frac{d}{dz}\left[\frac{T(z)}{u'(z)}\right]=-\frac{T(z+2\kappa_0)}{u(z)}. \qquad (17)$$

For the case $\lambda=k^2$ we find

$$T(z)=u'(z)\left[e^{-\frac{1}{2}n\pi/k}N(k^2)+\int_z^{\infty}\frac{T(\zeta+2\kappa_0)}{u(\zeta)}d\zeta\right]$$

$$=u'(z)\left[e^{\frac{1}{2}n\pi/k}N(k^2)-\int_{-\infty}^{z}\frac{T(\zeta+2\kappa_0)}{u(\zeta)}d\zeta\right],$$

$$-\infty<z<\infty, \qquad (18)$$

integrating (17) over the indicated intervals and using (16). If we eliminate $N(k^2)$ in (18) we have

$$T(z)=\frac{u'(z)}{1-e^{-n\pi/k}}\left[\int_z^{\infty}\frac{T(\zeta+2\kappa_0)}{u(\zeta)}d\zeta+e^{-n\pi/k}\int_{-\infty}^{z}\frac{T(\zeta+2\kappa_0)}{u(\zeta)}d\zeta\right], \quad -\infty<z<\infty, \qquad (19)$$

while elimination of $T(z)$ gives

$$N(k^2)=\int_{-\infty}^{\infty} T(\zeta+2\kappa_0)\frac{e^{-(n/k)\arctan(\zeta/k)}}{e^{\frac{1}{2}n\pi/k}-e^{-\frac{1}{2}n\pi/k}}d\zeta. \qquad (20)$$

In the case $\lambda=-\kappa^2$ we avoid integrating (17) over $z=\kappa$, since $u(\kappa)=0$. Integrating (17) over $(-\kappa, z)$ for $z<\kappa$ and over (z,∞) for $z>\kappa$, we have

$$T(z)=-u'(z)\int_{-\kappa}^{z}\frac{T(\zeta+2\kappa_0)}{u(\zeta)}d\zeta, \quad -\infty<z<\kappa,$$

$$T(z)=u'(z)\left[N(-\kappa^2)+\int_z^{\infty}\frac{T(\zeta+2\kappa_0)}{u(\zeta)}d\zeta\right], \qquad (21)$$

$$\kappa<z<\infty.$$

The above relations will be used in Sec. 5 to obtain approximations and bounds for $N(\lambda)$. A probabilistic interpretation of (19) and (21) is given in Appendix 2.

4. NUMERICAL METHODS

The methods used to obtain $N(\lambda)$ numerically are based on the system

$$\frac{d^2\varphi(s)}{ds^2}=a(s)\varphi(s), \quad s\neq 0,$$

$$a(s)=\lambda+n\frac{1-e^{-2i\kappa_0 s}}{is}, \qquad (22)$$

$$\varphi(0)=1,$$

$$\lim_{s\to\pm\infty}\varphi(s)=0,$$

[11] We treat ξ_0 and η_0 (and hence $z_0=\eta_0/\xi_0$) as random quantities, statistically independent of the t_j's. The basic random element ω now consists of a specification of ξ_0 and η_0 as well as all of the t_j's.

satisfied by the characteristic function (Fourier transform) $\varphi(s)$ of the probability density $T(z)$:

$$\varphi(s)=\int_{-\infty}^{\infty} e^{isz}T(z)dz.$$

The level distribution $N(\lambda)$ is obtained from

$$\mathrm{Re}\varphi'(0+)=-\mathrm{Re}\varphi'(0-)=-\pi nN(\lambda). \qquad (23)$$

(Proof appears in Appendix 3.)

Three integration schemes used each involve use of an asymptotic formula for $\varphi'(s)/\varphi(s)$, and the accuracy of the numerical results is limited by the accuracy of the asymptotic formula. The values given for $N(\lambda)$ are least accurate when λ is near $-\kappa_0^2$ and μ is small. (Unfortunately, this is a region of considerable interest.)

One method involves numerical integration of (22) from some large s in to $s=0$, the asymptotic formula being used to obtain a starting value for $\varphi'(s)/\varphi(s)$; Eq. (23) then gives $N(\lambda)$. In a closely related method

TABLE I. Level distribution for $n/\kappa_0=10$.

κ/κ_0	$N(-\kappa^2)$
20.00	0.000346
5.00	0.0315
4.672	0.0444
4.472	0.054
4.272	0.063
0.0	0.142
k/κ_0	$N(k^2)$
0.2	0.142
3.162	0.1741
4.472	0.201

the Riccati equation associated with (22),

$$y'(s)+y^2(s)=\lambda+n\frac{1-e^{-2i\kappa_0 s}}{is}, \quad s\neq 0$$

$$\varphi(s)=\exp\left(\int_0^s y(s')ds'\right),$$

is integrated numerically from some large s to $s=0$, the asymptotic formula now giving the starting value of $y(s)$.

A third method is based on the fact that if $\theta(s)$, $0\leq s<\infty$, is an unbounded, zero-free solution of differential equation (22) then

$$\varphi(s)=\theta(s)\int_s^{\infty}\frac{ds'}{\theta^2(s')}\Big/\theta(0)\int_0^{\infty}\frac{ds'}{\theta^2(s')}$$

satisfies the system (22), so that from (23),

$$N(\lambda)=\frac{1}{n\pi}\mathrm{Re}\left[1\Big/\theta^2(0)\int_0^{\infty}\frac{ds'}{\theta^2(s')}-\frac{\theta'(0)}{\theta(0)}\right]. \qquad (24)$$

TABLE II. Level distribution for $n/\kappa_0=1$.

κ/κ_0	$N(-\kappa^2)$
5.00	0.027
2.50	0.043
1.37	0.258
1.18	0.298
1.02	0.335
1.002	0.3349
1.00	0.345
0.98	0.352
0.8	0.392
0.02	0.451
k/κ_0	$N(k^2)$
0.2	0.454
1.0	0.581
1.414	0.629

The integral in (24) can be brought into the form

$$\int_0^{\infty}\frac{ds'}{\theta^2(s')}=\int_0^{s}\frac{ds'}{\theta^2(s')}+\frac{1}{\theta^2(s)[\theta'(s)/\theta(s)-\varphi'(s)/\varphi(s)]}, \qquad (25)$$

and is evaluated numerically by integrating Eq. (22) from $s=0$ $[\theta(0)=1, \theta'(0)=i]$ to some large s and using (57)–(58) to approximate $\varphi'(s)/\varphi(s)$ in the denominator in (25). The numerical solutions $\theta(s)$ were found to be zero-free.

The numerical results are given in Tables I–III, and appear as triangles in Figs. 3–5. For comparison, the open circles and solid circles represent the Monte Carlo and local density model results, respectively, of Lax and Phillips.[1]

5. APPROXIMATIONS

(a) The Optical and Diffusion Approximations

When the hits are fast and weak ($n\to\infty$, $\kappa_0\to 0$, $n\kappa_0\to$ finite) the integral in (15) may be approximated as

$$n\int_z^{z+2\kappa_0}T(\zeta)d\zeta\approx 2n\kappa_0T(z), \qquad (26)$$

TABLE III. Level distribution for $n/\kappa_0=0.1$. Since $N(\lambda)$ is nondecreasing in λ, the value given above for $N(-\kappa^2)$ at $\kappa/\kappa_0=1.0001$ is surely incorrect.

κ/κ_0	$N(-\kappa^2)$
0	0.86
0.67	0.857
0.98	0.729
0.99	0.701
0.999	0.658
0.9999	0.588
1.0001	0.2443
1.001	0.2887
1.01	0.2599
1.02	0.233
1.33	0.0754
2.00	0.00186

5

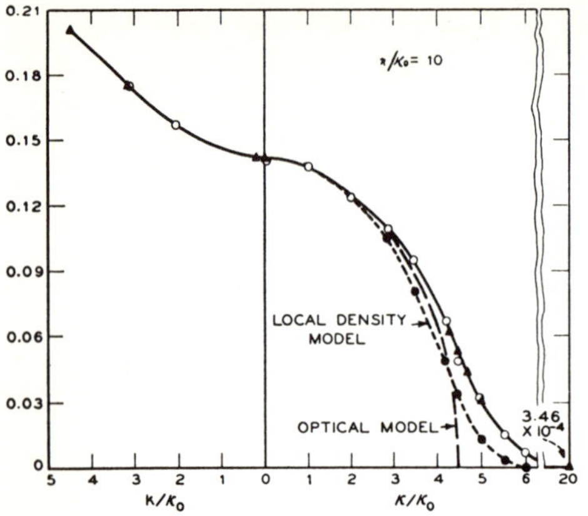

FIG. 3. Level distribution versus $|\lambda|^{\frac{1}{2}}/\kappa_0$ for $n/\kappa_0=10$. ▲=exact value from (22), ○=Monte Carlo value (Lax and Phillips) ●="local density model" (Lax and Phillips).

and the resulting equation is easily solved for

$$T(z)\approx\frac{nN(\lambda)}{z^2+\lambda+2n\kappa_0}. \tag{27}$$

Imposing the normalizing condition (13), we have the optical approximation:

$$N(\lambda)\approx\frac{(\lambda+2n\kappa_0)^{\frac{1}{2}}}{n\pi},\quad \lambda\geq-2n\kappa_0, \tag{28}$$

$$(\approx 0,\quad \lambda\leq -2n\kappa_0).$$

This is also the result given by the "optical model" of Lax and Phillips,[1] whence the terminology.

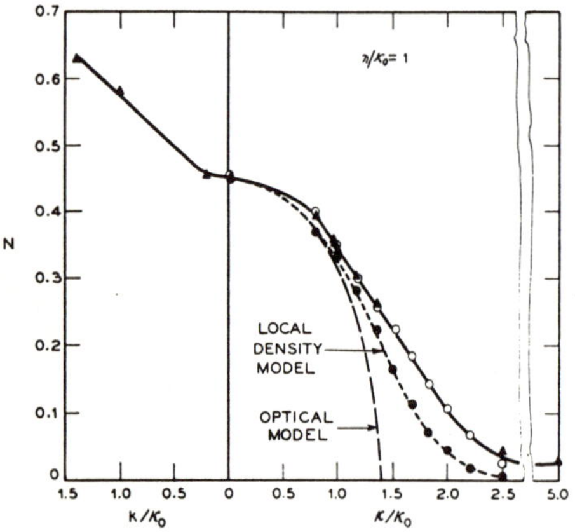

FIG. 4. Level distribution versus $|\lambda|^{\frac{1}{2}}/\kappa_0$ for $n/\kappa_0=1$. ▲=exact value from (22), ○=Monte Carlo value (Lax and Phillips) ●="local density model" (Lax and Phillips).

The approximate solution (27) is not everywhere non-negative when λ is below the optical band edge $-2n\kappa_0$. That this singularity is an artifact of the approximation method is shown by including another term in (26), viz.,

$$n\int_z^{z+2\kappa_0} T(\zeta)d\zeta\approx 2n\kappa_0 T(z)+2n\kappa_0^2 T'(z).$$

We substitute this in (15), solve the resulting differential equation for $T(z)$, and then impose the normalizing condition (13). Omitting the details, the result is the diffusion approximation:

$$N(\lambda)\approx\frac{1}{2\pi^{1/2}3^{1/6}\epsilon^{2/3}\int_0^\infty \exp(-t^6-\mu t^2)dt}, \tag{29}$$

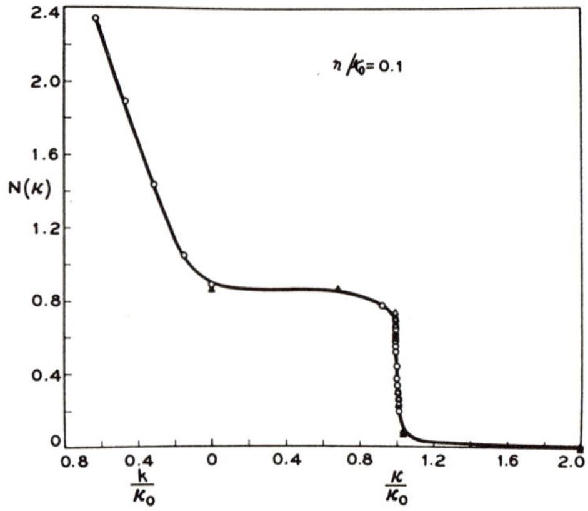

FIG. 5. Level distribution versus $|\lambda|^{\frac{1}{2}}/\kappa_0$ for $n/\kappa_0=0.1$. ▲=exact value from (23), △=exact value from (24), ○=Monte Carlo value (Lax and Phillips).

where

$$\epsilon=\frac{n}{\kappa_0},\quad \mu=\left(\frac{3}{\epsilon^2}\right)^{\frac{1}{3}}\left(\frac{\lambda}{\kappa_0^2}+2\epsilon\right).$$

(We explain the appellation presently.) In Fig. 6 we show the quantity

$$A(\mu)=\left[\int_0^\infty \exp(-t^6-\mu t^2)dt\right]^{-1},$$

together with the approximation

$$B(\mu)=\left[\int_0^\infty \exp(-\mu t^2)dt\right]^{-1}=(4\mu/\pi)^{\frac{1}{2}}$$

[which yields (28)]. We note in passing that according to (29), the number of levels per atom below the

6

optical band edge is

$$N(-2n\kappa_0)\approx\frac{1}{2\pi^{1/2}3^{1/6}\epsilon^{2/3}\Gamma(7/6)}$$
$$\approx 0.2532/\epsilon^{2/3},$$

in agreement with an observation made by Lax and Phillips[1] on the basis of their Monte Carlo results. A comparison of the various computations of $N(\lambda)$ is shown in Fig. 7.

The optical approximation (28) corresponds to replacing the random potential in (1) by its (ω,x)-average value nV_0. Our diffusion approximation (29) corresponds to adding a white Gaussian noise correction term to nV_0, making $(\psi(x),\psi'(x))$ a two-dimensional diffusion process, as follows. The integral of the potential in (1) is

$$Q(x)=\int_0^x V_0\sum_{j=-\infty}^{\infty}\delta(x'-x_j)dx'=V_0P(x),$$

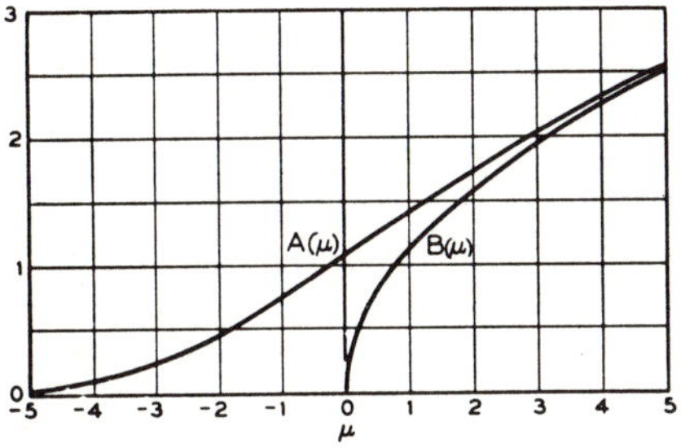

FIG. 6. Ordinates $A(\mu)$ and $B(\mu)$, proportional to the diffusion and optical approximations (29) and (28), respectively. Definition of μ appears with (29).

where $P(x)$ is a Poisson process[12] with density parameter n. If we define a process $\gamma(x)=[P(x)-nx]/\sqrt{n}$ then $\gamma(x)$ is nearly a Brownian motion (Wiener process) when n is large,[13] and the corresponding potential is $nV_0+n^{\frac{1}{2}}V_0\gamma_0'(x)$, where $\gamma_0(x)$ is a Brownian motion ["$\gamma_0'(x)$"="white Gaussian noise"]. This approximation can be used to obtain (29); we omit the details.[14]

(b) The Case $\lambda=k^2>0$

The integral on the right-hand side in (15) lies between 0 and 1, whence

$$\frac{n[N(k^2)-1]}{z^2+k^2}\leq T(z)\leq\frac{nN(k^2)}{z^2+k^2}. \tag{30}$$

We integrate this over $(-\infty,\infty)$ and impose (13),

[12] That is, $P(x)$ increases by unity at each x_j. See reference 2, pp. 98 and 398.
[13] See reference 2, pp. 434–435.
[14] See M. C. Wang and G. E. Uhlenbeck, Revs. Modern Phys. **17**, 332 (1945).

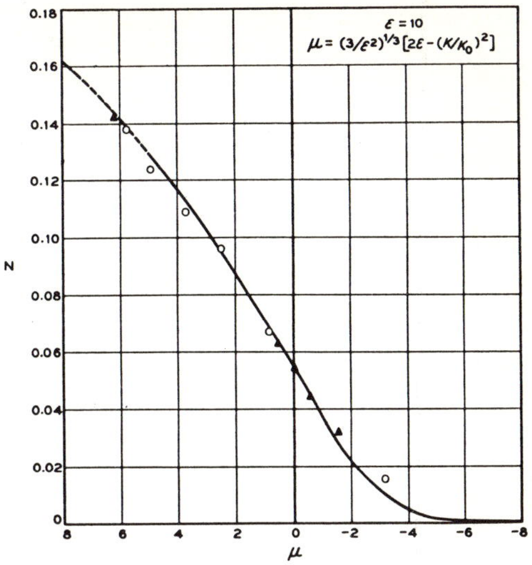

FIG. 7. Comparison of diffusion approximation (29) with exact distribution for $n/\kappa_0=10$. ▲=exact value from (23), ○=Monte Carlo value (Lax and Phillips).

obtaining the bounds $k/(n\pi)\leq N(k^2)\leq 1+k/(n\pi)$. Similarly, from (18),

$$e^{-\frac{1}{2}n/k}N(k^2)u'(z)\leq T(z)\leq e^{\frac{1}{2}n\pi/k}N(k^2)u'(z),$$

giving bounds $(e^{n\pi/k}-1)^{-1}\leq N(k^2)\leq 1+(e^{n\pi/k}-1)^{-1}$. Using $e^x\geq 1+x$, we have the bounds $(\lambda>0)$

$$\frac{k}{n\pi}\leq N(k^2)\leq\frac{1}{1-e^{-n\pi/k}}.$$

For a refined version of the above, we rewrite (20) as

$$N(k^2)=\frac{e^{\frac{1}{2}n\pi/k}}{e^{\frac{1}{2}n\pi/k}-e^{-\frac{1}{2}n\pi/k}}-\int_{-\infty}^{\infty}T(z+2\kappa_0)\frac{e^{\frac{1}{2}n\pi/k}-e^{-(n/k)\arctan(z/k)}}{e^{\frac{1}{2}n\pi/k}-e^{-\frac{1}{2}n\pi/k}}dz.$$

Substituting for $T(z+2\kappa_0)$ the bounds given by (30), we obtain the bounds $(\lambda>0)$

$$\frac{1}{1-e^{-n\pi/k}+\Delta}\leq N(k^2)\leq\frac{1+\Delta}{1-e^{-n\pi/k}+\Delta},$$

$$\Delta=\frac{n}{k}\int_{-\infty}^{\infty}\frac{1-e^{-(n/k)(\frac{1}{2}\pi+\arctan x)}}{[x+(2\kappa_0/k)]^2+1}dx. \tag{31}$$

When $k\gg n$ the numerator in the integral for Δ may be expanded in powers of n/k to give the result

$$N(k^2)\approx\frac{k}{n\pi}+\frac{1}{\pi}\arctan\frac{\kappa_0}{k}+O\left(\frac{n}{k}\right).$$

7

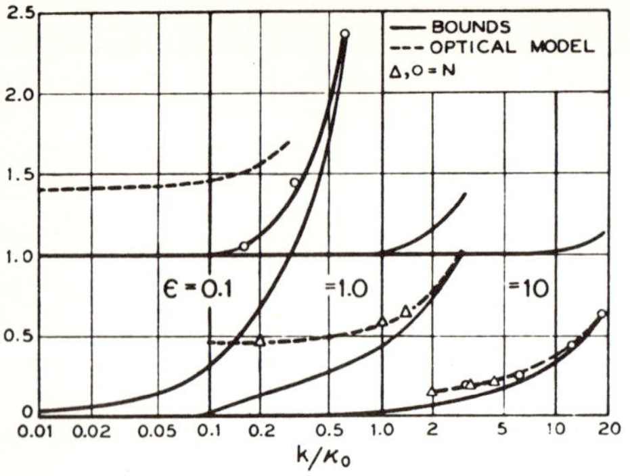

FIG. 8. The bounds (31) for the level distribution. △ = exact value from (22), ○ = Monte Carlo value (Lax and Phillips).

[This can also be obtained by iterating (15).] The bounds (31) are shown in Fig. 8; the integral Δ was evaluated numerically.

(c) The Case $\lambda = -\kappa^2 < 0$

The denominator on the right-hand side in (15) is negative when $|z| < \kappa$, and since $T(z)$ is never negative we must have

$$N(-\kappa^2) \leq \int_z^{z+2\kappa_0} T(\zeta) d\zeta, \quad -\kappa \leq z \leq \kappa. \tag{32}$$

We fit as many disjoint subintervals $z < \zeta < z + 2\kappa_0$ as possible into the interval $(-\kappa, \kappa + 2\kappa_0)$, add integrals (32) for the subintervals, and impose (13), obtaining the bound $(\lambda < 0)$

$$N(-\kappa^2) \leq 1/(q+1),$$
$$q = \text{integer part of } (\kappa/\kappa_0).$$

A bound which is much stronger when q is large is obtained as follows. We rewrite (21) as

$$T(z) = \frac{nu(z)}{\kappa^2 - z^2} \int_{-\kappa}^{z} \frac{T(\zeta + 2\kappa_0)}{u(\zeta)} d\zeta, \quad -\infty < z < \kappa.$$

Now, $u(\zeta)$ is decreasing on $-\kappa < \zeta < \kappa$, whence

$$T(z) \leq \frac{n}{\kappa^2 - z^2} \int_{-\kappa}^{z} T(\zeta + 2\kappa_0) d\zeta, \quad -\kappa < z < \kappa.$$

When $q > 1$ we have

$$\kappa^2 - z^2 \geq 4\kappa_0(\kappa - \kappa_0) \quad \text{if} \quad -\kappa + 2\kappa_0 \leq z \leq \kappa - 2\kappa_0,$$

so that

$$T(z) \leq \alpha \int_{-\kappa}^{z} T(\zeta + 2\kappa_0) d\zeta, \quad -\kappa + 2\kappa_0 \leq z \leq \kappa - 2\kappa_0, \tag{33}$$
$$\alpha = n/4\kappa_0(\kappa - \kappa_0).$$

It is not hard to verify that (33) implies

$$T(z) \leq \frac{\alpha^{r+1}(z+\kappa)(z+\kappa+2r\kappa_0)^{r-1}}{r!},$$
$$-\kappa + 2\kappa_0 \leq z \leq \kappa - 2(r+1)\kappa_0, \tag{34}$$
$$r = 0, 1, \cdots, q-2.$$

[The case $r = 0$ follows from (33) and (13), and the others by induction.] We now apply (32) to the estimate (34), choosing for r the largest value for which (34) holds on an interval of length at least $2\kappa_0$, viz., $r = q - 3$. There results the bound $(\lambda < 0)$

$$N(-\kappa^2) \leq \frac{[2\kappa_0 \alpha (q-1)]^{q-2}}{(q-1)!}$$
$$\leq \frac{(\frac{1}{2}\epsilon)^{q-2}}{(q-1)!}, \tag{35}$$

valid for $q \geq 2$, where, again, $\epsilon = n/\kappa_0$ and $q =$ integer part of (κ/κ_0).

The following intuitive argument gives an approximation even smaller than (35) for large q. Without hits, the velocity of a phase point is $\dot{z} = \kappa^2 - z^2$, and the time required to move from $z = -\kappa + 2\kappa_0$ to $z = \kappa - 2\kappa_0$ is $T = (1/\kappa) \log[(\kappa/\kappa_0) - 1]$, assuming $q > 1$. This time is small when κ is large, and a phase point will spend most of the time in the region $\kappa - 2\kappa_0 \leq z \leq \kappa$, waiting for a favorable succession of hits (see Fig. 2). To escape the interval $-\kappa \leq z \leq \kappa$ the particle must receive about q (or more) hits within a time interval T. (Fewer than q will not carry the point past $z = -\kappa$, and it will slip back to the vicinity of $z = +\kappa$.) Now, the probability that an interval of duration T after any hit contains $q - 1$ or more further hits is[9]

$$p = \frac{e^{-nT}(nT)^{q-1}}{(q-1)!}\left[1 + \frac{nT}{q} + \frac{(nT)^2}{q(q+1)} + \cdots\right]$$
$$\leq \frac{e^{-nT}(nT)^{q-1}}{(q-1)!} \cdot \frac{1}{1-(nT/q)}.$$

If such bursts of hits are infrequent enough their rate of occurrence will be about np. Since each η-axis crossing requires about one burst, we have, finally, the approximation $(\lambda < 0)$

$$N(-\kappa^2) \approx p \approx e^{-nT}(nT)^{q-1}/(q-1)!,$$
$$nT = (n/\kappa)\log(\kappa/\kappa_0 - 1) \approx (\epsilon/q)\log(q-1),$$

valid for $q > 1$, $nT/q \ll 1$.

6. GENERALIZATIONS

There seems to be no way of extending our methods to the corresponding three-dimensional problem; there is no Markov property when the parameter is three-

8

dimensional, and (3) does not generalize (to the best of the authors' knowledge). The one-dimensional case can be generalized somewhat, however.

First, the strength of each interaction can be random. That is, instead of the potential $-(\hbar^2/m)\kappa_0 \sum \delta(x-x_j)$ of (1) we may consider the potential

$$-(\hbar^2/m) \sum \kappa_{0j}\delta(x-x_j),$$

where the κ_{0j}'s are independent random variables with common probability distribution, say $P\{\kappa_{0j}\leq u\}=F(u)$, $-\infty<u<\infty$.[16] The x_j's are to have the Poisson distribution with density n, as before. The analysis follows closely that of the previous sections. In place of (12) we find

$$\frac{d}{dz}[(z^2+\lambda)T(z)]+n\int_{-\infty}^{\infty}[T(z+2\kappa_0)-T(z)]dF(\kappa_0)=0,$$

and the only change in (22)–(23) is

$$a(s)=\lambda+n\int_{-\infty}^{\infty}\frac{1-e^{-2i\kappa_0 s}}{is}dF(\kappa_0).$$

The optical approximation is just (28) again, except that

$$\langle\kappa_0\rangle=\int_{-\infty}^{\infty}\kappa_0 dF(\kappa_0)$$

replaces κ_0. Similarly, the diffusion approximation is given by (29), except that the parameters are

$$\epsilon=\frac{n}{\langle\kappa_0^2\rangle^{\frac{1}{3}}},\quad \mu=\left(\frac{3}{\epsilon^2}\right)^{\frac{1}{3}}\frac{\lambda+2n\langle\kappa_0\rangle}{\langle\kappa_0^2\rangle},$$

with

$$\langle\kappa_0^2\rangle=\int_{-\infty}^{\infty}\kappa_0^2 dF(\kappa_0).$$

With the strengths randomized this way, our Eq. (1) is the Schrödinger equation of the one-dimensional version of a scalar meson pair theory Hamiltonian discussed by Montroll and Potts[17] in their study of interactions of lattice defects.

The distribution of atom positions can also be generalized. We use the temporal description of Sec. 3. Let 1, 2, $\cdots$, ν denote the states of an ergodic Markov chain[18] with transition probabilities p_{ij} and stationary probabilities p_i; that is,

$$p_{ij}\geq 0,\quad \sum_{j=1}^{\nu} p_{ij}=1,\quad i=1,\cdots,\nu$$

$$\sum_{i=1}^{\nu} p_i p_{ij}=p_j,\quad j=1,\cdots,\nu.$$

[16] The case treated in previous sections results when $F(u)=0$ on $-\infty<u<\kappa_0$ and $F(u)=1$ on $\kappa_0\leq u<\infty$.

[17] E. W. Montroll and R. B. Potts, Phys. Rev. **102**, 72 (1956), Eq. (7.1).

[18] See reference 2, Chap. V.

Let n_i denote a hit rate for each state $i=1, 2, \cdots, \nu$. The hit process is determined by the following properties:

(a) Random changes of "state" occur at hits. If the state is i just before a hit, then the probability is p_{ij} that the state is j just after the hit, and the state remains j until the next hit.

(b) If the state is i at time t, the probability that during $(t, t+dt)$ there is

no hit is $1-n_i dt$,
exactly one hit is $n_i dt$,
more than one hit is 0,

neglecting higher order infinitesimals.

(c) The strengths (κ_0) of hits are random variables. At hits where the state changes from i to j, the strength of the hit has (given) probability distribution $P\{\kappa_0\leq u\}=F_{ij}(u)$.

Let $T_i(z)$ denote the stationary joint mixed density of z and i; that is

$$P\{(z\leq\zeta)\ \&\ (\text{state is } i)\}=\int_{-\infty}^{\zeta}T_i(z)dz$$

(at any time). Then, corresponding to (12), we have

$$\frac{d}{dz}[(z^2+\lambda)T_j(z)]-n_jT_j(z)+\sum_{i=1}^{\nu}n_ip_{ij}$$
$$\times\int_{-\infty}^{\infty}T_i(z+2\kappa_0)dF_{ij}(\kappa_0)=0,\quad j=1,2,\cdots,\nu. \tag{36}$$

The normalization is[19]

$$\int_{-\infty}^{\infty}T_i(z)dz=(p_i/n_i)/\sum_{i=1}^{\nu}(p_j/n_j),\quad i=1,\cdots,\nu, \tag{37}$$

and the η-axis crossing rate is

$$\mathfrak{N}(E)=\lim_{z\to\pm\infty}z^2\sum_{i=1}^{\nu}T_i(z). \tag{38}$$

Even though the methods of Sec. 4 are applicable, solution of (36)–(38) appears to be difficult in any cases of interest.

7. ACKNOWLEDGMENTS

The authors wish to acknowledge interesting and helpful discussions and correspondence with Melvin Lax and James C. Phillips, and with Professor Mark Kac and his student Peter Mengert. The numerical analysis described in Sec. 4 was programmed for the IBM 704 by H. T. O'Neil.

[19] The right-hand side is the fraction of time spent in state i (sojourns in state i having mean duration $1/n_i$); p_i is the numerical frequency of i in long random sequences of states.

APPENDIX 1

It should be clear that the two-dimensional phase process $\{(\xi(t,\omega),\eta(t,\omega)), -\infty<t<\infty\}$ is a Markov process. Let us assume for simplicity that the random variables $\xi(t,\omega)$ and $\eta(t,\omega)$ have at time t a joint probability density $R(\xi,\eta,t)$ in the ξ, η-plane, one smooth enough to insure the validity of the following derivation.[20] We regard $R(\xi,\eta,t)\delta A$ as the fraction of phase points in an ensemble which at time t fall in an element of area δA around the point (ξ,η). As in Sec. 3, we obtain easily

$$R(\xi,\eta,t+dt)\delta A=(1-ndt)R(\xi',\eta',t)\delta A' + (ndt)R(\xi'',\eta'',t)\delta A'',$$

where, for the motion (7),

$$\xi'=\xi-\eta dt,$$
$$\eta'=\eta+\lambda\xi dt,$$
$$\delta A'=\frac{\partial(\xi',\eta')}{\partial(\xi,\eta)}\delta A=\delta A,$$

and for the motion (5)–(6),

$$\xi''=\xi,$$
$$\eta''=\eta+2\kappa_0\xi$$
$$\delta A''=\frac{\partial(\xi'',\eta'')}{\partial(\xi,\eta)}\delta A=\delta A,$$

neglecting higher order differentials throughout. There follows

$$\frac{\partial R(\xi,\eta,t)}{\partial t}=\left(\lambda\xi\frac{\partial}{\partial\eta}-\eta\frac{\partial}{\partial\xi}\right)R(\xi,\eta,t) + n[R(\xi,\eta+2\kappa_0\xi,t)-R(\xi,\eta,t)], \quad (39)$$

subject to the normalization conditions

$$R(\xi,\eta,t)\geq 0, \quad \int_{-\infty}^{\infty}\int_{-\infty}^{\infty}R(\xi,\eta,t)d\xi d\eta=1. \quad (40)$$

In contrast to (10)–(11), the system (39)–(40) has no stationary solution.[21] The explanation is that each random orbit spirals out to infinity in the ξ, η-plane, exponentially fast in t, with probability 1. (We prove this assertion in Appendix 2.) In terms of the impurity band model, the amplitude of each random wave function (viz., the quantity $|E[\psi(x;E,\omega)]^2+(\hbar^2/2m)[\psi'(x;E,\omega)]^2|^{\frac{1}{2}}$) increases exponentially fast in x with probability 1, for any E and any initial value and slope.[22]

This is reflected in the behavior of the second moments of ξ and η, as follows. Let $M_{\alpha\beta}(t)$ denote the (α,β) moment of (ξ,η):

$$M_{\alpha\beta}(t)=\int_{-\infty}^{\infty}\int_{-\infty}^{\infty}\xi^\alpha\eta^\beta R(\xi,\eta,t)d\xi d\eta.$$

If the $r+1$ integrals $M_{\alpha\beta}(t)$ of total weight $\alpha+\beta=r$ are absolutely convergent, then they satisfy

$$\dot{M}_{\alpha\beta}=-\beta\lambda M_{\alpha+1,\beta-1}+\alpha M_{\alpha-1,\beta+1} + n\sum_{\gamma=0}^{\beta-1}\frac{(-2\kappa_0)^{\beta-\gamma}\beta!}{\gamma!(\beta-\gamma)!}M_{\alpha+\beta-\gamma,\gamma},$$

$$\beta=r-\alpha,$$
$$\alpha=0,1,\cdots,r,$$
$$r=0,1,\cdots,$$

obtained by integrating (39) with $\xi^\alpha\eta^\beta d\xi d\eta$.

For $r=1$ we have

$$\dot{M}_{10}=M_{01},$$
$$\dot{M}_{01}=-(\lambda+2n\kappa_0)M_{10},$$

with solution

$$M_{10}(t)=M_{10}(0)\cos\Lambda t+M_{01}(0)\Lambda^{-1}\sin\Lambda t$$
$$M_{01}(t)=-M_{10}(0)\Lambda\sin\Lambda t+M_{01}(0)\cos\Lambda t,$$
$$\Lambda=(\lambda+2n\kappa_0)^{\frac{1}{2}}.$$

We note that the approximate $nN(\lambda)$ given by the optical model, (28), is the η-axis crossing rate of the average phase point $(\langle\xi(t,\omega)\rangle,\langle\eta(t,\omega)\rangle)=(M_{10}(t),M_{01}(t))$.

For $r=2$ there obtains

$$\dot{M}_{20}=2M_{11},$$
$$\dot{M}_{11}=-(\lambda+2n\kappa_0)M_{20}+M_{02} \quad (41)$$
$$\dot{M}_{02}=4n\kappa_0^2M_{20}-2(\lambda+2n\kappa_0)M_{11}.$$

Solutions proportional to $e^{\Lambda t}$ exist if Λ satisfies

$$\Lambda^3+4(\lambda+2n\kappa_0)\Lambda-8n\kappa_0^2=0;$$

there is always one positive root, say Λ_1, and roots Λ_2, Λ_3 either both negative or complex conjugates with negative real part. A solution of (41) is

$$M_{20}(t)=\sum k(\Lambda_1,\Lambda_2,\Lambda_3)e^{\Lambda_1 t},$$
$$M_{11}(t)=\sum \tfrac{1}{2}\Lambda_1 k(\Lambda_1,\Lambda_2,\Lambda_3)e^{\Lambda_1 t},$$
$$M_{02}(t)=\sum\left[\frac{\Lambda_1^2}{4}+\frac{2n\kappa_0^2}{\Lambda_1}\right]k(\Lambda_1,\Lambda_2,\Lambda_3)e^{\Lambda_1 t},$$
$$k(\Lambda_1,\Lambda_2,\Lambda_3)=\frac{(8n\kappa_0^2/\Lambda_1)\langle\xi^2\rangle_0+\langle(\Lambda_1\xi+2\eta)^2\rangle_0}{2(\Lambda_1-\Lambda_2)(\Lambda_1-\Lambda_3)},$$

where $\langle\xi^2\rangle_0$ denotes $M_{20}(0)$, etc., and where the sum-

[20] Both here and in Sec. 3, the use of probability measures (set functions) would give only an essentially trivial gain in generality.

[21] If $T(z)$ satisfies (12), then $R(\xi,\eta,t)=(1/\xi^2)T(\eta/\xi)$ satisfies (39), but is not normalizable.

[22] This is different from the case of periodic potentials, where wave functions are bounded in the allowed bands $[N'(\lambda)>0]$ and unbounded in the forbidden bands $[N'(\lambda)=0]$.

mation in each case is over the cyclic permutations of (1,2,3). We observe that the coefficient of $e^{\lambda_1 t}$ is always nonvanishing in each moment, from which it follows that no stationary solution of (39) could have finite second moments. However, as mentioned previously, we prove a stronger result in Appendix 2.

Rice's formula[23] for the expected η-axis crossing rate at time t is

$$\mathfrak{N}_t = \int_{-\infty}^{\infty} R(0,\eta,t)|\eta|d\eta, \tag{42}$$

and we show now that (16) is but a disguised version of the Rice formula. We introduce coordinates

$$\rho = (\xi^2+\eta^2)^{\frac{1}{2}}, \quad z = \eta/\xi,$$

and the marginal density of z, defined as

$$T(z,t) = \int_0^{\infty} \left[R\left(\frac{\rho}{(1+z^2)^{\frac{1}{2}}}, \frac{\rho z}{(1+z^2)^{\frac{1}{2}}}, t\right) + R\left(\frac{-\rho}{(1+z^2)^{\frac{1}{2}}}, \frac{-\rho z}{(1+z^2)^{\frac{1}{2}}}, t\right)\right] \frac{\rho d\rho}{1+z^2}; \tag{43}$$

the factor $\rho/(1+z^2)$ is the Jacobian $\partial(\xi,\eta)/\partial(\rho,z)$. From (43) we have, at least formally,

$$\lim_{z\to\pm\infty} z^2 T(z,t) = \int_0^{\infty} [R(0,\rho,t)+R(0,-\rho,t)]\rho d\rho,$$

and the right-hand side is Rice's integral (42).

APPENDIX 2

The continuous parameter process $\{z(t,\omega), -\infty<t<\infty\}$ is of a type which seems not to have been treated in much detail in the literature. However, there is associated with it a certain discrete parameter process of familiar type.[1] Let the indexing of the hits be such that $\cdots \leq t_0(\omega) \leq 0 < t_1(\omega) \leq \cdots$, and consider the random variables

$$z_0(\omega) = z(0+0,\omega)+2\kappa_0,$$
$$z_j(\omega) = z(t_j(\omega)-0,\omega), \quad j=1,2,\cdots.$$

That is, $z_j(\omega)$ for $j>0$ is the value of $z(t,\omega)$ at the instant before the jth hit. The process $\{z_j(\omega), j=0,1,\cdots\}$ is a Markov process, whose transition probabilities we now ascertain.

We first find the probability $\rho(\zeta,z)dz$ that a phase point starting at ζ and moving according to $\dot{z}=-(z^2+\lambda)$ receives its next hit when it is in the interval $(z, z+dz)$. The duration θ of the time interval to the next hit has probability density $ne^{-n\theta}$ on $0\leq\theta<\infty$,[8] so that if $\theta(\zeta,z)$ denotes the time required to get from ζ to z, then

$$\rho(\zeta,z)dz = ne^{-n\theta(\zeta,z)}\left|\frac{\partial\theta(\zeta,z)}{\partial z}\right|dz = \left|\frac{\partial}{\partial z}e^{-n\theta(\zeta,z)}\right|dz. \tag{44}$$

[23] S. O. Rice, Bell System Tech. J. **23**, 282 (1944).

For the motion $\dot{z}=-(z^2+\lambda)$ we have formally

$$\theta(\zeta,z) = -\int_{\zeta}^{z}\frac{dx}{x^2+\lambda}, \tag{45}$$

but some care is required at this point.

In the case $\lambda=k^2>0$ we obtain from (45)

$$\theta(\zeta,z) = \frac{1}{k}\left[\arctan\frac{\zeta}{k} - \arctan\frac{z}{k}\right],$$

where by arctan we must understand a multiple-valued version of the inverse tangent function. In the ξ, η-plane, a phase point may wind around the origin several times without being hit. This corresponds to traversing the whole z-axis several times without hits, the time required to go from $z=+\infty$ to $z=-\infty$ being π/k. If we add the contributions from the various appropriate branches, there results

$$\rho(\zeta,z) = \frac{u'(z)}{u(\zeta)}\frac{1}{1-e^{-n\pi/k}} \times \begin{Bmatrix} 1 & \text{if } -\infty<z\leq\zeta<\infty \\ e^{-n\pi/k} & \text{if } -\infty<\zeta<z<\infty \end{Bmatrix}, \tag{46}$$

$$u(z) = e^{(n/k)\arctan(z/k)},$$

where now $|\arctan|<\frac{1}{2}\pi$.

When $\lambda=-\kappa^2<0$ we obtain from (45), formally,

$$\theta(\zeta,z) = \frac{1}{2\kappa}\left\{\log\left|\frac{\zeta-\kappa}{\zeta+\kappa}\right| - \log\left|\frac{z-\kappa}{z+\kappa}\right|\right\},$$

but there are complications because the velocity κ^2-z^2 changes sign at $z=\pm\kappa$, and because a point starting at $\zeta<-\kappa$ may reach $z>\kappa$ by passing $-\infty$. Taking these into account, we have

$$\rho(\zeta,z) = \left|\frac{u'(z)}{u(\zeta)}\right| \quad \text{if } -\infty<z<\zeta<-\kappa,$$
$$\text{or if } -\kappa<\zeta<z<\kappa,$$
$$\text{or if } \kappa<z<\zeta<\infty,$$
$$\text{or if } -\infty<\zeta<-\kappa \text{ and } \kappa<z<\infty,$$
$$=0 \quad \text{otherwise},$$

where

$$u(z) = \left|\frac{z-\kappa}{z+\kappa}\right|^{\frac{1}{2}n/\kappa}$$

A point starting at $\zeta=\pm\kappa$ will stay there until the next hit, but this will occur with probability 0 if ζ has a probability density.

Let $T_j(z)$ denote the probability density of the random variable $z_j(\omega)$; we assume that $T_0(z)$ is given. It should be clear that the $T_j(z)$ satisfy

$$T_{j+1}(z)=\int_{-\infty}^{\infty} T_j(\zeta)\rho(\zeta-2\kappa_0, z)d\zeta, \quad j=0, 1, \cdots.$$

If there is a stationary density $T(z)$ of the z_j's then it satisfies

$$T(z)=\int_{-\infty}^{\infty} T(\zeta+2\kappa_0)\rho(\zeta,z)d\zeta. \tag{47}$$

Using the explicit form of $\rho(\zeta,z)$ given above, it is readily verified that (47) is just the Eq. (19) or (21) satisfied by the $T(z)$ of Sec. 3. [To bring (21) to the form (47) one uses

$$N(-\kappa^2)=\int_{-\infty}^{-\kappa} \frac{T(\zeta+2\kappa_0)}{u(\zeta)}d\zeta, \tag{48}$$

obtained by integrating (17) over $(-\infty, -\kappa)$.]

The z_j process satisfies "Condition (Σ)" of Doob,[24] and the z axis constitutes one ergodic set.[25] Furthermore, there exists a finite measure φ of z sets relative to which the higher-order transition probabilities have a uniformly bounded density. [The transition probability densities relative to Lebesgue measure are given by

$$p^{(1)}(\zeta,z)=\rho(\zeta-2\kappa_0, z),$$

$$p^{(j+1)}(\zeta,z)=\int_{-\infty}^{\infty} p^{(j)}(\zeta,\zeta')p^{(1)}(\zeta',z)d\zeta', \quad j=1, 2, \cdots.$$

For the measure φ of Doob[26] we may use

$$\varphi\{A\}=\int_A \varphi'(z)dz,$$

with

$$\varphi'(z)=1/(1+z^2) \quad \text{if} \quad -(n/2)^2\leq\lambda<\infty,$$

$$\varphi'(z)=\frac{1}{|z-\kappa|^{1-\frac{1}{2}(n/\kappa)}[(z^2+1)^{\frac{1}{2}}]^{1+\frac{1}{2}(n/\kappa)}} \quad \text{if} \quad \kappa>\tfrac{1}{2}n,$$

$$(\lambda=-\kappa^2).$$

For $\lambda>0$ it is trivial that

$$\sup_{\substack{-\infty<z<\infty\\ -\infty<\zeta<\infty}} \frac{p^{(1)}(\zeta,z)}{\varphi'(z)}<\infty,$$

and for $\lambda=-\kappa^2<0$, $\kappa\neq\kappa_0$, an extremely tedious calculation shows that

$$\sup_{\substack{-\infty<z<\infty\\ -\infty<\zeta<\infty}} \frac{p^{(2)}(\zeta,z)}{\varphi'(z)}<\frac{const}{|\kappa-\kappa_0|}<\infty;$$

for $\kappa=\kappa_0$ it seems likely that $p^{(3)}(\zeta,z)/\varphi'(z)$ is bounded, but the authors have not attempted to verify this.]

The implication of all this is that there exists a unique normalized solution of (47) [and (12)] given by

$$T(z)=\lim_{j\to\infty} p^{(j)}(\zeta,z).$$

The non-negative limit, independent of ζ, is non-vanishing for every z except possibly $z=\kappa$, and the convergence of $p^{(j)}(\zeta,z)/\varphi'(z)$ to $T(z)/\varphi'(z)$ is exponentially fast in j, uniformly in ζ and z.[26] It is possible to express the solution $T(z,t)$ of (10)–(11) as a functional of $T_0(z)$ and the $p^{(j)}(\zeta,z)$, and to deduce therefrom the limiting behavior of $T(z,t)$. We are not concerned with nonstationary distributions, however, and we omit the details.

Rigorously, the symbol $N(\lambda)$ appearing in various expressions in Secs. 3–5 is to be regarded as the quantity $(1/n)\lim z^2T(z)$. [It is obvious from (14) that the limit exists.] We now give a proof that the limiting t-average η-axis crossing rate exists and is constant with probability 1, the constant being $\lim z^2T(z)$. Let $\gamma_j(\omega)$ denote the number of times that $z(t,\omega)$ reaches $+\infty$ during $t_j(\omega)\leq t<t_{j+1}(\omega)$. [We redefine $t_0(\omega)$ as $t_0(\omega)\equiv0$.] When $\lambda=-\kappa^2$ we have $\gamma_j(\omega)=\gamma(z_j(\omega), z_{j+1}(\omega))$, with

$$\begin{aligned}\gamma(\zeta,z)&=1 \quad \text{if} \quad \zeta<-\kappa+2\kappa_0 \quad \text{and} \quad z>\kappa\\ &=0 \quad \text{otherwise.}\end{aligned}$$

When $\lambda=k^2$, the random variables $\gamma_0(\omega)$, $\gamma_1(\omega)$, $\cdots$ are conditionally independent given all of the $z_k(\omega)$, $k=0, 1, \cdots$, and the conditional distribution of each $\gamma_j(\omega)$ is of geometric type with ratio $e^{-n\pi/k}$ [see (44)–(46)]. The conditional expectation of $\gamma_j(\omega)$ given $z_0(\omega)$, $z_1(\omega)$, $\cdots$ is

$$\langle\gamma_j(\omega)\,|\,z_0(\omega),z_1(\omega),\cdots\rangle=\gamma(z_j(\omega),z_{j+1}(\omega)),$$

where now

$$\begin{aligned}\gamma(\zeta,z)&=\frac{e^{-n\pi/k}}{1-e^{-n\pi/k}} \quad \text{if} \quad z\leq\zeta-2\kappa_0\\ &=\frac{1}{1-e^{-n\pi/k}} \quad \text{if} \quad z>\zeta-2\kappa_0.\end{aligned}$$

The random variable

$$\mathfrak{M}(j,\omega)=\frac{1}{j}\sum_{k=0}^{j-1} \gamma_j(\omega)$$

is the average number of η-axis crossings per hit after j hits, and a slightly modified version of a well-known

---

[24] See reference 2, p. 195.

[25] That is, there exists an integer ν such that for every ζ and every set Z of positive Lebesgue measure, $P\{z_\nu(\omega)\epsilon Z\,|\,z_0(\omega)=\zeta\}>0$. For $\lambda>0$ one may take $\nu=1$; for $\lambda=-\kappa^2$ any $\nu>(\kappa/\kappa_0)+1$ will do [see Sec. 5(c)].

[26] See reference 2, pp. 215–217.

12

ergodic theorem[27] gives

$$\lim_{j\to\infty} \mathfrak{M}(j,\omega)=\langle\gamma(z_k(\omega),z_{k+1}(\omega))\rangle \quad \text{with probability 1,}$$

where the expectation on the right-hand side is to be taken with respect to the stationary distribution of the z_j's. Since the stationary joint density of z_k and z_{k+1} is $T(z_k)\rho(z_k-2\kappa_0, z_{k+1})$, we have

$$\lim_{j\to\infty} \mathfrak{M}(j,\omega)=\int_{-\infty}^{\infty}\int_{-\infty}^{\infty} T(\zeta)\rho(\zeta-2\kappa_0, z)\gamma(\zeta,z)d\zeta dz$$

$$\text{with probability 1.} \quad (49)$$

Let now $N(L,\omega)$ denote the number of hits in a time interval $0\leq t<L$, and let $\nu_L(\omega)$ denote the number of η-axis crossings during $0\leq t<L$. Since

$$N(L,\omega)\mathfrak{M}(N(L,\omega),\omega)\leq\nu_L(\omega)$$
$$\leq[N(L,\omega)+1]\mathfrak{M}(N(L,\omega)+1, \omega),$$

we have for the number of crossings per unit time,

$$\lim_{L\to\infty}\frac{\nu_L(\omega)}{L}=\lim_{L\to\infty}\left[\frac{N(L,\omega)}{L}\cdot\mathfrak{M}(N(L,\omega),\omega)\right]$$

$$=n\lim_{j\to\infty}\mathfrak{M}(j,\omega) \quad \text{with probability 1,}$$

using $\lim N(L,\omega)/L=n$ and $\lim N(L,\omega)=\infty$ with probability 1 as $L\to\infty$. Using the explicit forms given for $\rho(\zeta,z)$ and $\gamma(\zeta,z)$, it is easily verified that the integral on the right-hand side in (49) is just the integral on the right-hand side in (20) or (48). We have thus proved

$$\lim_{L\to\infty}\frac{\nu_L(\omega)}{L}=\lim_{z\to\pm\infty} z^2T(z) \quad \text{with probability 1.}$$

Similar methods show that the (ξ,η) process diverges radially with probability 1. Consider the random variable $\sigma(t,\omega)=|\lambda\xi^2(t,\omega)+\eta^2(t,\omega)|$. For given ω, $\sigma(t,\omega)$ is constant between hits, from (7), while at hits we have (suppressing ω for a moment)

$$\frac{\sigma(t_j+0)}{\sigma(t_j-0)}=\frac{|\lambda\xi^2(t_j)+[\eta(t_j-0)-2\kappa_0\xi(t_j)]^2|}{|\lambda\xi^2(t_j)+\eta^2(t_j-0)|}$$

$$=\frac{|\lambda+(z_j-2\kappa_0)^2|}{|\lambda+z_j^2|}$$

in terms of the random variables z_j previously defined. Assuming that $\sigma(0+0)\neq0$ holds with probability 1, we have

$$\frac{1}{j}\log\frac{\sigma(t_j+0)}{\sigma(0+0)}=\frac{1}{j}\sum_{k=1}^{j}\log\frac{|\lambda+(z_k-2\kappa_0)^2|}{|\lambda+z_k^2|},$$

using $\sigma(t_k-0)=\sigma(t_{k-1}+0)$. The ergodic theorem gives

$$\beta\equiv\lim_{j\to\infty}\frac{1}{j}\log\frac{\sigma(t_j+0)}{\sigma(0+0)}$$

$$=\int_{-\infty}^{\infty}T(z)\log\frac{|\lambda+(z-2\kappa_0)^2|}{|\lambda+z^2|}dz \quad \text{with probability 1}$$

$$=\int_{-\infty}^{\infty}[T(z+2\kappa_0)-T(z)]\log|\lambda+z^2|dz$$

$$=\frac{2}{n}\int_0^{\infty}z[T(z)-T(-z)]dz.$$

[We have used (12) and (14) to obtain the last expression.] It follows that for each $\epsilon>0$ and each ω in a set of probability 1 there exists an integer $J(\epsilon,\omega)$ such that

$$\sigma(t_j(\omega)+0,\omega)>\sigma(0+0,\omega)e^{j(\beta-\epsilon)} \quad \text{for every} \quad j>J(\epsilon,\omega).$$

As we prove in Appendix 3, the number β is always positive; we choose some $\epsilon<\beta$, and our result follows.

APPENDIX 3

We define the function $\varphi_Z(s)$ as

$$\varphi_Z(s)=\int_{-Z}^{Z}e^{isz}T(z)dz,$$

noting that $\varphi_Z(s)$ converges uniformly in s to $\varphi(s)$ as $Z\to\infty$, and that $\varphi_Z(s)$ is an entire function of s for any fixed finite Z. (Here and throughout, s is real.) We multiply (12) by $e^{isz}dz$ and integrate over $(-Z, Z)$; integration by parts and rearrangement gives

$$e^{isZ}(Z^2+\lambda)T(Z)-e^{-isZ}(Z^2+\lambda)T(-Z)$$
$$+n\left(\int_{-Z+2\kappa_0}^{Z+2\kappa_0}e^{-2is\kappa_0}-\int_{-Z}^{Z}\right)e^{isz}T(z)dz$$

$$=is\int_{-Z}^{Z}e^{isz}(z^2+\lambda)T(z)dz$$

$$=is\left[\lambda-\frac{d^2}{ds^2}\right]\varphi_Z(s).$$

For the boundary terms we substitute values given by (14), viz.,

$$(Z^2+\lambda)T(\pm Z)=n\left[N(\lambda)-\int_{\pm Z}^{\pm Z+2\kappa_0}T(z)dz\right],$$

and rearrange to obtain

$$\frac{d^2\varphi_Z(s)}{ds^2}=a(s)\varphi_Z(s)-nI(s)-2nN(\lambda)\frac{\sin sZ}{s}, \quad (50)$$

[27] Reference 2, pp. 218–232, 464–469.

where

$$I(s)=\int_{Z}^{Z+2\kappa_0}\frac{e^{is(z-2\kappa_0)}-e^{isZ}}{is}T(z)dz$$

$$-(\text{expression with } Z\to -Z),$$

and where $a(s)$ is given in (22). We have

$$|I(s)|\leq\int_{Z}^{Z+2\kappa_0}\left|\frac{2\sin\frac{1}{2}s(z-2\kappa_0-Z)}{s}\right|T(z)dz$$

$$+(\text{expression with } Z\to -Z)$$

$$\leq 2\kappa_0\left(\int_{Z}^{Z+2\kappa_0}+\int_{-Z}^{-Z+2\kappa_0}\right)T(z)dz,$$

whence $I(s)$ vanishes uniformly in s as $Z\to\infty$. We now integrate (50) twice from some fixed s_0:

$$\varphi_Z(s)=\varphi_Z(s_0)+(s-s_0)\varphi_Z{}'(s_0)$$

$$+\int_{s_0}^{s}\int_{s_0}^{s'}[a(s'')\varphi_Z(s'')-nI(s'')]ds''ds'$$

$$-2nN(\lambda)\int_{s_0}^{s}\int_{s_0}^{s'}\frac{\sin s''Z}{s''}ds''ds'. \quad (51)$$

As $Z\to\infty$, the first and third terms on the right-hand side of (51) converge to obvious limits, from the uniform convergence properties noted previously. It is easy to show that the fourth term has a limit as $Z\to\infty$, expressible in terms of the Dirichlet integral. Since $\varphi_Z(s)$ itself converges to $\varphi(s)$, it follows that the second term on the right-hand side of (51) also converges to a limit as $Z\to\infty$, of the form $(s-s_0)\times$(function of s_0). {We do not need the fact that the limit is actually $(s-s_0)[\varphi'(s_0+0)+\varphi'(s_0-0)]/2$.} Thus, passing to the limit in (51),

$$\varphi(s)=\varphi(s_0)+(s-s_0)\times(\text{function of } s_0)$$

$$+\int_{s_0}^{s}\int_{s_0}^{s'}a(s'')\varphi(s'')ds''ds'$$

$$-\pi nN(\lambda)s[\operatorname{sgn}s-\operatorname{sgn}s_0], \quad (52)$$

where $\operatorname{sgn}x=x/|x|$ if $x\neq 0$, $=0$ if $x=0$.

One differentiation shows that $\varphi'(s)$ is continuous except at $s=0$, where we have

$$\operatorname{Re}\varphi'(0+0)=-\operatorname{Re}\varphi'(0-0)=-\pi nN(\lambda)$$

and $\operatorname{Im}\varphi'(0+0)=\operatorname{Im}\varphi'(0-0)$. A second differentiation gives (22).

The quantity $\operatorname{Im}\varphi'(0)$ is of some interest, since

$$\operatorname{Im}\varphi'(0)=\int_0^\infty z[T(z)-T(-z)]dz, \quad (53)$$

and this is $\frac{1}{2}n\beta$ of Appendix 2. The following proof of (53) is a slightly modified version of one due to Mengert.[28] We have

$$\operatorname{Im}\varphi(s)=\int_0^\infty[T(z)-T(-z)]\sin sz dz, \quad (54)$$

whence, formally,

$$\operatorname{Im}\varphi'(s)=\int_0^\infty z[T(z)-T(-z)]\cos sz dz. \quad (55)$$

However, it is not hard to show from (15) that $z^4[T(z)-T(-z)]$ is bounded as $z\to\infty$. It follows that the integral on the right-hand side in (55) is uniformly absolutely convergent, and hence that (55) is valid for every s; for $s=0$ we have (53).

To prove that $\operatorname{Im}\varphi'(0)$ is positive, we multiply (22) by $\bar{\varphi}(s)$ and subtract the complex conjugate equation, obtaining

$$\frac{d}{ds}\operatorname{Im}[\varphi'(s)\bar{\varphi}(s)]=-\frac{n(1-\cos 2\kappa_0 s)}{s}|\varphi(s)|^2.$$

Integrating over $(0,\infty)$, we have

$$\operatorname{Im}\varphi'(0)=n\int_0^\infty\frac{1-\cos 2\kappa_0 s}{s}|\varphi(s)|^2ds,$$

using $\bar{\varphi}(0)=1$ and the fact that $\bar{\varphi}(s)\varphi'(s)\to 0$ as $s\to\infty$ (proved in the next paragraph). Since $\varphi(s)$ is continuous and $\varphi(0)=1$, it is clear that $\operatorname{Im}\varphi'(0)>0$.

As the Fourier transform of an integrable function, $\varphi(s)$ vanishes as $s\to\infty$, by the Riemann-Lebesgue theorem. However, the formal derivative of the Fourier integral for $\varphi(s)$ is not convergent, so we proceed as follows. We have, for any fixed $Z>0$, $Z^2+\lambda>0$,

$$\varphi(s)=\int_{-Z}^{Z}e^{isz}T(z)dz+\left(\int_{-\infty}^{-Z}+\int_{Z}^{\infty}\right)e^{isz}T(z)dz$$

$$=\int_{-Z}^{Z}e^{isz}T(z)dz+nN(\lambda)\left(\int_{-\infty}^{-Z}+\int_{Z}^{\infty}\right)\frac{e^{isz}}{z^2+\lambda}dz$$

$$-n\left(\int_{-\infty}^{-Z}+\int_{Z}^{\infty}\right)\frac{e^{isz}}{z^2+\lambda}\int^{z+2\kappa_0}T(x)dxdz,$$

using (15). The first and third terms can now be differentiated under the integral sign, the integrals remaining absolutely convergent after differentiation. According to the Riemann-Lebesgue theorem, these contributions to $\varphi'(s)$ vanish as $s\to\infty$. The second term can be evaluated explicitly in terms of the exponential integral $\int(e^{ix}/x)dx$, and it is not hard to

[28] Peter Mengert (private communication).

14

verify that the derivative of the second term also vanishes as $s \to \infty$, whence, finally, $\varphi'(s) \to 0$ as $s \to \infty$.

To investigate the asymptotic behavior of $\varphi(s)$, we bring (22) to the Riccati form

$$y'(s)+y^2(s)=\lambda+n\frac{1-e^{-2i\kappa_0 s}}{is}, \quad s\neq 0,$$

with the substitution

$$\varphi(s)=\exp\left(\int_0^s y(s')ds'\right);$$

the boundary condition $\varphi(\infty)=0$ becomes

$$\lim_{s\to\infty}\int_0^s \mathrm{Re}\,y(s')ds'=-\infty. \tag{56}$$

Assuming that $y(s)$ can be expanded according to powers of n,

$$y(s)=y_0(s)+ny_1(s)+n^2y_2(s)+\cdots, \tag{57}$$

we find

$$y_0'+y_0^2=\lambda,$$

$$y_1'+2y_0y_1=\frac{1-e^{-2i\kappa_0 s}}{is},$$

$$\vdots \; .$$

These have as solutions

$$y_0(s)=\frac{h'(s)}{h(s)}$$

$$y_1(s)=\frac{1}{h^2(s)}\int^s h^2(s')\frac{1-e^{-2i\kappa_0 s'}}{is'}ds', \tag{58}$$

$$\vdots \; ,$$

where $h(s)=ae^{cs}+be^{-cs}$, $c^2=\lambda$, and a and b are arbitrary constants. The constants and limits of integration will be chosen so that not only is (56) satisfied but also $y_j(\infty)=0$, $j>0$.

For $\lambda=k^2>0$ ($c=k$, $a=0$) there results

$$y_0(s)=-k$$

$$y_1(s)=-\int_s^\infty e^{-2k(s'-s)}\frac{1-e^{-2i\kappa_0 s'}}{is'}ds',$$

$$\vdots \; ,$$

while for $\lambda=-\kappa^2<0$, $\kappa\neq\kappa_0$ ($c=i\kappa$, $b=0$) we have

$$y_0(s)=i\kappa,$$

$$y_1(s)=-\int_s^\infty e^{2i\kappa(s'-s)}\frac{1-e^{-2i\kappa_0 s'}}{is'}ds'$$

$$=-\frac{1}{2\kappa s}+\text{(function integrable at infinity)},$$

$$\vdots \; .$$

In the exceptional case $\kappa=\kappa_0$ the boundary conditions require $2b=a$, giving

$$y_0(s)=i\kappa_0\frac{2e^{i\kappa_0 s}-e^{-i\kappa_0 s}}{2e^{i\kappa_0 s}+e^{-i\kappa_0 s}},$$

$$y_1(s)=\frac{1}{(2e^{i\kappa_0 s}+e^{-i\kappa_0 s})^2}\int_s^\infty \frac{e^{-4i\kappa_0 s'}+3e^{-2i\kappa_0 s'}-4e^{2i\kappa_0 s'}}{is'}ds'$$

$$=-\frac{1}{2\kappa_0 s}+\text{(function integrable at infinity)},$$

$$\vdots \; .$$

In the numerical calculation described in Sec. 4, $y_0(s)+ny_1(s)$ was used as an approximation to $y(s)$. The proof that the expansion $y(s)=y_0(s)+ny_1(s)+\cdots$ is asymptotic in s in the usual sense appears to be difficult.

15

Reprinted from
Proceedings of the Physical Society, Vol. 78, Part 5(ii), No. 504
1961
Printed in Great Britain by J. W. Arrowsmith Ltd., Bristol 3

Existence of Energy Gaps in One-Dimensional Liquids*

BY R. E. BORLAND

Mathematics Division, National Physical Laboratory, Teddington, Middlesex

Communicated by P. Dean; MS. received 31st August 1961

Abstract. It is shown that a definite energy gap exists in a one-dimensional liquid provided that the maximum deviation of the distance between atoms is less than a critical value. Higher gaps successively appear as the disorder is further reduced.

§ 1. INTRODUCTION

IN A RECENT paper Makinson and Roberts (1960) posed the question whether in an infinite chain of atoms lacking long range order there may be energy gaps in which there are no states whatever, or whether the states are always thinly spread with a density approaching zero as the range of order increases to infinity. In this paper we shall prove that, subject to the usual assumptions, energy gaps exist in a chain lacking long range order provided that the short range disorder is sufficiently small. We consider a model in which the minimum distance between atoms is of length a and the maximum distance of length $a+b$. No assumptions are made about the type of disorder so that the results derived are valid whether long range order exists or not.

§ 2. CHAIN OF NEGATIVE DELTA FUNCTIONS

To simplify the proof we consider first the special case of a chain of δ-functions of strength $-k_0$. The extension of the argument to a general atomic potential will be discussed later. Hence we are concerned with the eigenvalues E of the Schrödinger equation for an infinite chain

$$\left[-\frac{d^2}{dx^2}-k_0\sum_i \delta(x-x_i)\right]\psi = E\psi \tag{1}$$

with appropriate boundary conditions, and subject to the condition that

$$a \leqslant x_{i+1}-x_i \leqslant a+b.$$

We shall restrict ourselves to positive electron energies and write $E = k^2$.

It has been shown by James and Ginzbarg (1953) that the number of eigenstates $N(E)$ per atom with energy less than E is equal to the average phase increase per atom divided by π. The phase ϕ is defined by

$$\tan\phi(E, x) = -\frac{c\psi'(E, x)}{\psi(E, x)},$$

and the condition that ϕ changes continuously with x, c being an arbitrary positive constant. In this case c is chosen as $1/k$ in order that the phase increase from the right of the ith δ-function to the left of the $(i+1)$th δ-function is equal to $k(x_{i+1}-x_i)$.

926

The phase ϕ_i' immediately to the right of the ith δ-function can be related to the phase ϕ_i immediately to the left of the δ-function by integrating equation (1) from $x_i - \epsilon$ to $x_i + \epsilon$ and taking the limit as ϵ tends to zero.

We obtain

$$\tan \phi'_i = \tan \phi_i + \alpha \tag{2}$$

where $\alpha = k_0/k$.

We shall now prove that the integrated density of states is equal to unity for the range of k satisfying the inequalities

$$\pi - 2q \leqslant ka \leqslant \pi - kb, \tag{3}$$

where q is the principal value of arctan $\frac{1}{2}\alpha$.

Clearly an integer m exists for which the phase ϕ_i lies either in the interval

$$m\pi - q \leqslant \phi_i < m\pi + \tfrac{1}{2}\pi \tag{4}$$

or in the interval

$$m\pi - \tfrac{1}{2}\pi \leqslant \phi_i < m\pi - q. \tag{5}$$

For reasons which will become obvious we call the interval (4) a stable phase interval and the interval (5) an unstable phase interval.

Consider first the situation in which ϕ_i lies in the stable interval (4). Using equation (2) it follows that ϕ'_i lies in the interval

$$m\pi + q \leqslant \phi'_i < m\pi + \tfrac{1}{2}\pi,$$

since ϕ_i equal to $m\pi - q$ gives ϕ_i' equal to $m\pi + q$. Provided that the inequalities (3) are obeyed ϕ_{i+1} lies in the interval

$$(m+1)\pi - q \leqslant \phi_{i+1} < (m+1)\pi + \tfrac{1}{2}\pi$$

and by repeating the argument ϕ_{i+M} lies in the interval

$$(m+M)\pi - q \leqslant \phi_{i+M} < (m+M)\pi + \tfrac{1}{2}\pi.$$

Therefore the average phase increase is π per δ-function giving an integrated density of states $N(E)$ equal to unity.

Now consider the alternative that ϕ_i lies in the unstable phase interval (5). In this case there is a tendency for the phases to the left of successive δ-functions to drift into a stable phase interval. To see this it is necessary to consider equation (2). This equation shows that the phase increase $\phi' - \phi$ across a δ-function is a monotonically increasing function of ϕ in an unstable phase interval, and a monotonically decreasing function in a stable interval. Hence any tendency for successive phases to slip back, in the sense that $\phi_{i+1} - \pi < \phi_i$, is accelerated when ϕ_{i+1} lies in an unstable phase interval and retarded in a stable phase interval.

This is of course not a rigorous argument and it is possible to construct chains for which ϕ_i lies in an unstable phase interval for all i. However, even for these unlikely chains the average phase increase is still equal to π if the inequalities (3) are obeyed. For this to be otherwise it would be necessary either that successive phases ϕ_i and ϕ_{i+1} lie in the same unstable phase interval or that at least one empty unstable phase interval separates ϕ_i and ϕ_{i+1}. The former case would require that $ka \leqslant \pi/2 - q$ and the latter would require that $ka + kb > \pi$. Both these requirements conflict with the inequalities (3).

The inequalities will in general hold over a range of k provided that the ratio of b to a is sufficiently small. For this range of energy $N(E)$ is equal to unity and consequently no states can exist within this range.

By a trivial modification of the argument it follows that energy gaps will exist for all integers n for which the following inequalities hold over a range of k

$$n\pi - 2q \leqslant ka \leqslant n\pi - kb. \tag{6}$$

The second and higher energy gaps become smaller or non-existent.

§3. GENERALIZATION OF RESULT TO ARBITRARY ATOMIC POTENTIALS

If we take the atoms to be represented by δ-functions of strength $+k_0$ instead of $-k_0$, then similar reasoning applies. Equation (2) becomes instead

$$\tan\phi_i' = \tan\phi_i - \alpha. \tag{7}$$

The phases to the left of succeeding δ-functions drift into and are trapped in the phase intervals

$$m\pi - \tfrac{1}{2}\pi < \phi_i \leqslant m\pi + q \tag{8}$$

for the energy ranges for which the following inequalities hold

$$n\pi \leqslant ka \leqslant n\pi + 2q - kb. \tag{9}$$

Again these inequalities will hold over a range or ranges of energy provided that the ratio of b to a is sufficiently small.

Finally we consider the general problem in which we have a chain of potentials all identical, which are separated by regions of zero potential which have a maximum deviation of length b. Again the phase ϕ' immediately to the right of an atomic potential will be some function $\phi'(\phi, E)$, where ϕ is the phase immediately to the left of the atomic potential, but ϕ' will of course be a different function from that given by equation (2). However, as we prove in the Appendix, it is always possible for a fixed positive electron energy E to replace the atomic potential by a δ-function surrounded by two regions of zero potential which together have the same transformation properties as the atomic potential upon the wave functions to the right and left of it. Thus the phase increase $\phi' - \phi$ for the substitute atomic potential will be the same function of ϕ as given by the original atomic potential apart from an additive integral multiple of π. This multiple is determined by the condition that the phase is a continuous function of x, and the only consequence of this addition is to alter the integrated density of states by an integer. Thus for a fixed energy a problem involving arbitrary atomic potentials can always be transformed into one involving only δ-function potentials. In the transformed problem the distances between adjacent atoms have to be increased (or decreased) by a constant amount.

It is clear from the foregoing work that if the deviation of the original nearest neighbour distances is sufficiently small it is in general possible to find a range or ranges of energy for which the transformed problems give rise to an integrated density of states equal to an integer. These ranges of energy are therefore forbidden bands.

§4. CONCLUSION

The conclusion to be drawn from this work is that forbidden bands in one-dimensional crystals depend essentially upon short range order. Any solution to the wave equation in the forbidden band has a tendency to adjust itself in order to keep in step with the atoms, so that there is only one node between atoms for the first gap, two nodes between atoms for the second gap, and so on.

ACKNOWLEDGMENT

The work described above has been carried out as part of the research programme of the National Physical Laboratory and this paper is published by permission of the Director of the Laboratory.

REFERENCES

James, H. M., and Ginzbarg, A. S., 1953, *J. Phys. Chem.*, **57**, 840.
Makinson, R. E. B., and Roberts, A. P., 1960, *Aust. J. Phys.*, **13**, 437.

APPENDIX

Assume the atomic potential is some function of x between two points x_1 and x_2, and that there is a region of zero potential to the left of x_1 and to the right of x_2. The solution of Schrödinger's equation in the region of zero potential to the left of x_1 can be written as

$$\psi_1 = A_1 \exp[ik(x-x_1)] + B_1 \exp[-ik(x-x_1)]$$

and to the right of x_2 as

$$\psi_2 = A_2 \exp[ik(x-x_2)] + B_2 \exp[-ik(x-x_2)].$$

Since the specification of ψ and its derivative at a point is sufficient to determine the solution of a second-order differential equation, and as

$$\begin{pmatrix} \psi_1(x_1) \\ \psi'_1(x_1) \end{pmatrix} = \begin{pmatrix} 1 & 1 \\ ik & -ik \end{pmatrix} \begin{pmatrix} A_1 \\ B_1 \end{pmatrix} \tag{10}$$

it follows that the vector, whose components are the amplitudes of the waves travelling to the right and left, is sufficient to determine the solution.

We can therefore represent the atomic potential from x_1 to x_2 by a transformation matrix as follows:

$$\begin{pmatrix} A_2 \\ B_2 \end{pmatrix} = \begin{pmatrix} T_{11} & T_{12} \\ T_{21} & T_{22} \end{pmatrix} \begin{pmatrix} A_1 \\ B_1 \end{pmatrix}. \tag{11}$$

We wish to show that we can replace the atomic potential by a δ-function surrounded by two regions of zero potential which together give the same transformation matrix. First we need to know the general form of the transformation matrix for an arbitrary atomic potential.

Regarding the atomic potential as a scattering system we have at the point x_1 an incident wave of amplitude A_1 and a scattered wave of amplitude B_1. Similarly at x_2 we have an incident wave of amplitude B_2 and a scattered wave of amplitude A_2. We can therefore express the vector of scattered wave amplitudes in terms of the vector of incident wave amplitudes by means of a scattering matrix for the potential

$$\begin{pmatrix} B_1 \\ A_2 \end{pmatrix} = \begin{pmatrix} S_{11} & S_{12} \\ S_{21} & S_{22} \end{pmatrix} \begin{pmatrix} A_1 \\ B_2 \end{pmatrix}. \tag{12}$$

It is well known that for such a system described by a real Hamiltonian the scattering matrix must be unitary and symmetric as a consequence of conservation of probabilities and time reversal symmetry. These restrictions give the following relationships between the scattering matrix elements:

$$S_{12} = S_{21} \tag{13a}$$

$$S_{11} = S_{22}^*|S| \tag{13b}$$

$$-S_{12} = S_{12}^*|S| \tag{13c}$$

where the asterisk denotes the complex conjugate and

$$|S| = S_{11}S_{22} - S_{12}^2.$$

On solving equation (12) to obtain equation (11) we find that

$$T_{11} = -\frac{|S|}{S_{12}}, \qquad T_{12} = \frac{S_{22}}{S_{12}}$$

$$T_{21} = -\frac{S_{11}}{S_{12}}, \qquad T_{22} = \frac{1}{S_{12}}.$$

Using equation (13*c*) we find that

$$T_{11} = \frac{1}{S_{12}^*} = T_{22}^*. \tag{14}$$

Equations (13*b*) and (13*c*) give

$$T_{21} = \frac{S_{22}^*}{S_{12}^*} = T_{12}^*. \tag{15}$$

One further restriction on the elements of the transformation matrix is imposed by the condition that the determinant of the matrix must be equal to unity, since

$$T_{11}T_{22} - T_{12}T_{21} = \frac{(1 - S_{22}S_{22}^*)}{S_{12}S_{12}^*}$$

$$= \frac{(1 - S_{22}S_{11}/|S|)}{S_{12}S_{12}^*} = -\frac{S_{12}^2}{|S|S_{12}S_{12}^*} = 1.$$

We see therefore that the transformation matrix for an arbitrary atomic potential is of the form

$$\begin{pmatrix} T_{11} & T_{12} \\ T_{12}^* & T_{11}^* \end{pmatrix},$$

where $T_{11}T_{11}^* - T_{12}T_{12}^* = 1$.

Now let us attempt to represent the transformation properties of this atomic potential by a potential which going from left to right consists of a region of zero potential of length g, a δ-function of strength μ and finally another region of zero potential of length h. The transformation matrix representing a length g of zero potential is

$$\begin{pmatrix} e^{ikg} & 0 \\ 0 & e^{-ikg} \end{pmatrix}.$$

Denote the wave function immediately to the right and left of the δ-function by ψ_+ and ψ_- respectively. We have on integrating the wave equation across the δ-function of strength μ and assuming continuity of ψ

$$\begin{pmatrix} \psi_+ \\ \psi_+' \end{pmatrix} = \begin{pmatrix} 1 & 0 \\ \mu & 1 \end{pmatrix} \begin{pmatrix} \psi_- \\ \psi_-' \end{pmatrix}.$$

Using the equation analogous to (10) in both its direct and inverted form we obtain

$$\begin{pmatrix} A_+ \\ B_+ \end{pmatrix} = \frac{i}{2k} \begin{pmatrix} -ik & -1 \\ -ik & 1 \end{pmatrix} \begin{pmatrix} 1 & 0 \\ \mu & 1 \end{pmatrix} \begin{pmatrix} 1 & 1 \\ ik & -ik \end{pmatrix} \begin{pmatrix} A_- \\ B_- \end{pmatrix}. \tag{16}$$

Hence on multiplying the matrices we see that the transformation matrix for the δ-function of strength μ is

$$\begin{pmatrix} 1 - i\mu/2k & -i\mu/2k \\ +i\mu/2k & 1 + i\mu/2k \end{pmatrix}.$$

Therefore the transformation matrix for the combined potential is

$$\begin{pmatrix} e^{ikh} & 0 \\ 0 & e^{-ikh} \end{pmatrix} \begin{pmatrix} 1 - i\lambda & -i\lambda \\ +i\lambda & 1 + i\lambda \end{pmatrix} \begin{pmatrix} e^{ikg} & 0 \\ 0 & e^{-ikg} \end{pmatrix}$$

where $\lambda = \mu/2k$.

On multiplying these matrices we find that the transformation matrix is

$$\begin{pmatrix} (1 - i\lambda)\, e^{ik(g+h)} & -i\lambda\, e^{-ik(g-h)} \\ i\lambda\, e^{ik(g-h)} & (1 + i\lambda)\, e^{-ik(g+h)} \end{pmatrix}.$$

Now by choosing a value of μ in the range $-\infty$ to 0 or 0 to $+\infty$, the modulus of λ can be made equal to the modulus of T_{12}, and T_{11} and T_{12} can be given their correct phase factors by an independent choice of $g+h$ and $g-h$. Hence we have shown that it is possible to represent the transformation properties of an arbitrary atomic potential by a δ-function surrounded by two regions of zero potential.

The electronic structure of a one-dimensional random alloy*

R. L. AGACY and R. E. BORLAND

Mathematics Division, National Physical Laboratory, Teddington, Middlesex

MS. received 16*th September* 1964

Abstract. The electronic energy spectrum and degree of localization for a one-dimensional random alloy are calculated by numerically solving a modified form of Schmidt's functional equation. It is found that the spectrum is very complex. The results are similar to those obtained by Dean for the closely related vibrational problem.

1. Introduction

In order to shed some light on the behaviour of electrons in random alloys, several authors, for example, James and Ginzbarg (1953), Schmidt (1957) and Faulkner and Korringa (1961), have investigated the density of states in a one-dimensional model. However, none of this work has clearly revealed the very interesting fine structure in the spectrum discussed by Lifshitz in his series of papers (see, for example, Lifshitz 1963). This fine structure has been found to occur in the vibrational spectrum of disordered diatomic chains by Dean (1960).

Schmidt showed how one could solve the stochastic part of the problem by deriving a functional equation which yields the averaged spectrum for the ensemble of infinite one-dimensional disordered chains. This equation has recently been solved numerically for the corresponding vibrational problem (Agacy 1964) and the results obtained confirm the work of Dean. We have now numerically solved this functional equation in a modified form for certain parameters of interest in the electronic case and have used the solution to obtain not only the density of states but also the degree of localization.

The electronic spectrum obtained is very peaky, as is to be expected, because of the mathematical similarity of the corresponding vibrational problem. The results conflict with the smooth spectrum obtained by Faulkner and Korringa. The discrepancy almost certainly arises because the latter authors made the assumption that the averaged spectrum is the spectrum of the arithmetically averaged secular equation.

2. The functional equation

We consider a random alloy containing i different kinds of atoms, a fraction p_j of which are of type j. An atom of type j is represented in our model by a δ-function potential of strength α^j situated at a lattice site $x = na$, where n is an integer. The model is thus described by the ensemble of Schrödinger equations

$$-\frac{d^2\psi}{dx^2} + \sum_{n=1}^{\infty} \alpha_n \delta(x-na)\psi = k^2\psi$$

in which the energy E has been written as k^2, and α_n has the value α^j with probability p_j, and

$$\sum_{j=1}^{i} p_j = 1.$$

As in previous papers on disordered systems by one of the authors (R.E.B.), we follow James and Ginzbarg (1953) by defining the phase ϕ of a real solution of the wave equation by

$$\phi(k, x) = -\arctan\left\{\frac{\psi'(k, x)}{k\psi(k, x)}\right\}$$

together with the auxiliary requirements that

$$-\tfrac{1}{2}\pi \leqslant \phi(k, 0) < \tfrac{1}{2}\pi$$

and that ϕ is a continuous function of x, except where the potential is a δ function. In this latter case ϕ is continuous for each member of the sequence of ordinary approximating functions. James and Ginzbarg have shown that $N(E)$, the integrated density of eigenvalues per atom, is equal to the average phase increase per atom divided by π.

Let ϕ_n be the value of the phase at the point $x = (n+\frac{1}{2})a$ in a particular chain. If the atom at the nth site is of type j, then ϕ_n can be determined as a function of ϕ_{n-1} which we write as

$$\phi_n = T_j(\phi_{n-1}).$$

We write the inverse relation as

$$\phi_{n-1} = T_j^{-1}(\phi_n).$$

It can readily be shown that

$$T_j(\phi) = \tfrac{1}{2}ka+\arctan\left\{\tan(\phi+\tfrac{1}{2}ka)-\frac{\alpha^j}{k}\right\}$$

the branch of the arctan function being determined by the above auxiliary requirements. The functions $T_j(\phi)$ and $T_j^{-1}(\phi)$ are strictly monotonically increasing functions of ϕ with the property that

$$T_j(\phi+\pi) = T_j(\phi)+\pi \tag{1a}$$

$$T_j^{-1}(\phi+\pi) = T_j^{-1}(\phi)+\pi. \tag{1b}$$

The phase increase across the nth cell from $x=(n-\frac{1}{2})a$ to $x=(n+\frac{1}{2})a$ is $T_j(\phi_{n-1})-\phi_{n-1}$ and is a periodic function of ϕ_{n-1} with period π. Hence the average phase increase in the ensemble can be found by determining the distribution of the phases ϕ_n, averaged over all chains and all n, when reduced mod π to the half-open interval $[0, \pi)$. This is essentially the method of Schmidt (1957) who derived a functional equation for the phase distribution function, although he used a slightly different definition of phase from the one used here. Our choice of phase was made because it leads to a simpler form for $T_j(\phi)$ and also enables us to use directly some previously obtained results (Borland 1963, to be referred to as I) for the calculation of the degree of localization of the eigenstates.

We now derive the functional equation appropriate to our choice of phase, using a method similar to that of Schmidt.

Let $H_n(\theta_1, \theta_2)$ be a function which, for

$$0<\theta_2-\theta_1\leqslant\pi$$

is the probability that ϕ_n lies in one of the infinite number of non-overlapping half-open intervals $[m\pi+\theta_1, m\pi+\theta_2)$ where m has all integral values. We introduce a variable

$$0<\psi\leqslant\pi$$

and define $W_n(\psi)$ by the equations

$$W_n(\psi) = H_n(0,\psi) \tag{2a}$$

$$W_n(\psi+r\pi) = W_n(\psi)+r \tag{2b}$$

where r is an integer. As can be seen, $W_n(\psi)$ is the probability that ϕ_n, reduced mod π to the interval $[0, \pi)$, lies in the interval $[0, \psi)$.

Now $T_j^{-1}(\theta)$ is a strictly monotonically increasing function of θ, obeying equation (1*b*), so that

$$0<T_j^{-1}(\psi)-T_j^{-1}(0)\leqslant\pi.$$

It follows that if the nth atom is of type j, then ϕ_n lies in one of the non-overlapping intervals $[m\pi, m\pi+\psi)$ if and only if ϕ_{n-1} lies in one of the non-overlapping intervals $[T_j^{-1}(0)+m\pi, T_j^{-1}(\psi)+m\pi)$. We can therefore relate the distribution of ϕ_n to the distribution of ϕ_{n-1} by the equation

$$H_n(0,\psi) = \sum_j p_j H_{n-1}\{T_j^{-1}(0), T_j^{-1}(\psi)\}. \tag{3}$$

Consider one of the terms $H_{n-1}\{T_j^{-1}(0), T_j^{-1}(\psi)\}$. There exist integers s and t such that

$$0<T_j^{-1}(0)-s\pi\leqslant\pi$$

$$0<T_j^{-1}(\psi)-t\pi\leqslant\pi$$

and t must be equal to s or $s+1$. From the definitions of H_n and W_n we have

$$H_{n-1}\{T_j^{-1}(0), T_j^{-1}(\psi)\} = \begin{cases} W_{n-1}\{T_j^{-1}(\psi)-s\pi\}-W_{n-1}\{T_j^{-1}(0)-s\pi\} & \text{for } t=s \\ W_{n-1}\{T_j^{-1}(\psi)-(s+1)\pi\}-W_{n-1}(0)+W_{n-1}(\pi) & \\ \quad -W_{n-1}\{T_j^{-1}(0)-s\pi\} & \text{for } t=s+1. \end{cases}$$

Using (2*b*) we obtain

$$H_{n-1}\{T_j^{-1}(0), T_j^{-1}(\psi)\} = W_{n-1}\{T_j^{-1}(\psi)\}-W_{n-1}\{T_j^{-1}(0)\}$$

both when t equals s and $s+1$. Hence the functional equation (3) becomes

$$W_n(\psi) = \sum_j p_j W_{n-1}\{T_j^{-1}(\psi)\}+K_n$$

where

$$K_n = -\sum_j p_j W_{n-1}\{T_j^{-1}(0)\}.$$

Now if the distribution functions $W_n(\psi)$ approach a limiting function $W(\psi)$ in the limit of large n, then $W(\psi)$ is a solution of the system

$$W(\psi) = \sum_j p_j W\{T_j^{-1}(\psi)\}+K \tag{4a}$$

$$W(\pi) = 1 \tag{4b}$$

$$W(\psi+r\pi) = W(\psi)+r \tag{4c}$$

$W(\psi)$ is a monotonically increasing function of ψ. (4*d*)

Whether or not $W_n(\psi)$ approaches a limiting form, the average distribution function

$W^n(\psi)$, given by

$$W^n(\psi) = \sum_{j=0}^{n} \frac{W_j(\psi)}{n+1} \tag{5}$$

can be shown to converge to a continuous function $W(\psi)$ satisfying (4), provided that $W_0(\psi)$ is specified. The proof follows very closely that given in the appendix to Schmidt's paper and so we will not repeat it here.

From (5) we see that $W(\psi)$ is the phase distribution function averaged not only over all chains but also over all atoms. In addition it can be shown, using similar arguments to those given in the appendix to Schmidt's paper, that although $W(\psi)$ exists and is continuous, parameters can occur for which the probability density function, which is the derivative of $W(\psi)$, is, if it exists, continuous nowhere. It is for this reason that we derived the functional equation only in terms of the distribution function, rather than via the probability density function as in Schmidt's paper. We shall write the formulae for averaged quantities in terms of the probability density, but in each case we shall integrate by parts so that we use only $W(\psi)$ and thus avoid the difficulties associated with its derivative.

In the following calculation of the average phase increase $\langle\Delta\phi\rangle$ we shall make use of the existence of a phase θ_0 such that $T_j(\theta_0)$ is independent of j. We write

$$\theta_1 = T_j(\theta_0).$$

The values of θ_0 and θ_1 are

$$\theta_0 = \tfrac{1}{2}(\pi - ka) \text{ and } \theta_1 = \tfrac{1}{2}(\pi + ka).$$

Now

$$\begin{aligned}\langle\Delta\phi\rangle &= \sum_j p_j \int_0^{\pi} \frac{dW(\phi)}{d\phi}\{T_j(\phi)-\phi\}d\phi \\ &= \sum_j p_j \int_{\theta_0}^{\theta_0+\pi} \frac{dW(\phi)}{d\phi}\{T_j(\phi)-\phi\}d\phi\end{aligned}$$

as the integrand has period π. Integrating by parts,

$$\begin{aligned}\langle\Delta\phi\rangle = \sum_j p_j W(\phi)\{T_j(\phi)-\phi\}\Big|_{\theta_0}^{\theta_0+\pi} - \sum_j p_j \int_{\theta_0}^{\theta_0+\pi} W(\phi)\frac{dT_j(\phi)}{d\phi}d\phi \\ + \sum_j p_j \int_{\theta_0}^{\theta_0+\pi} W(\phi)d\phi.\end{aligned}$$

But

$$\begin{aligned}\sum_j p_j \int_{\theta_0}^{\theta_0+\pi} W(\phi)\frac{dT_j(\phi)}{d\phi}d\phi &= \sum_j p_j \int_{\theta_0}^{\theta_0+\pi} W\{T_j^{-1}(\theta)\}d\theta \\ &= \int_{\theta_1}^{\theta_1+\pi}\{W(\theta)-K\}d\theta = -\pi K + \int_{\theta_1}^{\theta_1+\pi} W(\theta)d\theta.\end{aligned}$$

Therefore

$$\langle\Delta\phi\rangle = ka\{W(\theta_0+\pi)-W(\theta_0)\}+\pi K - \int_{\theta_1}^{\theta_1+\pi} W(\theta)d\theta + \int_{\theta_0}^{\theta_0+\pi} W(\theta)d\theta.$$

Using (4*c*) it follows that

$$\int_{\theta_0}^{\theta_0+\pi} W(\theta)d\theta - \int_{\theta_1}^{\theta_1+\pi} W(\theta)d\theta = \theta_0 - \theta_1 = -ka.$$

Thus

$$\langle \Delta\phi \rangle = \pi K$$

so that finally

$$K = N(E).$$

3. The degree of localization

Once the phase distribution function has been obtained from the functional equation, it is possible also to calculate the degree of localization using the method of I. The solutions in the region of zero potential to the left and right of the nth δ function are respectively

$$\psi = A_n \cos\{k(x-na+\tfrac{1}{2}a)+\phi_n\}$$

and

$$\psi = A_{n+1} \cos\{k(x-na-\tfrac{1}{2}a)+\phi_{n+1}\}.$$

If the nth δ function is of strength α^j, then it can be shown (equation (9) of I) that

$$\begin{aligned}\frac{A_{n+1}{}^2}{A_n{}^2} &= 1-q_j \sin(2\phi_n+ka)+q_j{}^2\cos^2(\phi_n+\tfrac{1}{2}ka)\\ &= F(\phi_n, q_j)\end{aligned}$$

where $q_j = \alpha^j/k$. Following the arguments given in I, we can determine the degree of localization by calculating the average value over the ensemble of $\ln(A_{n+1}{}^2/A_n{}^2)$, which we shall call $\langle \ln F \rangle$.

Now

$$\begin{aligned}\langle \ln F \rangle &= \sum_j p_j \int_0^\pi \ln\{F(\phi, q_j)\}\frac{dW(\phi)}{d\phi}d\phi\\ &= \sum_j p_j \int_0^\pi \ln\{F(\phi, q_j)\}\frac{d}{d\phi}\left\{W(\phi)-\frac{\phi}{\pi}\right\}d\phi + \frac{1}{\pi}\sum_j p_j \int_0^\pi \ln\{F(\phi, q_j)\}d\phi.\end{aligned}$$

The second term on the right-hand side has the value $\Sigma_j p_j \ln(1+\tfrac{1}{4}q_j{}^2)$, and after integrating the first term by parts we obtain

$$\langle \ln F \rangle = \sum_j p_j \ln(1+\tfrac{1}{4}q_j{}^2) - \sum_j p_j \int_0^\pi \left\{W(\phi)-\frac{\phi}{\pi}\right\}\frac{F'(\phi, q_j)}{F(\phi, q_j)}d\phi. \tag{6}$$

The reason for using the factor $W(\phi)-\phi/\pi$ in the integral rather than simply $W(\phi)$ is that the former is a periodic function, so that one would expect the numerical integration of the integral over its period to have a considerably smaller error than if $W(\phi)$ had been used†.

4. Numerical procedure and results

Our numerical calculations were restricted to a binary alloy containing atoms of type A and B occurring with probabilities 1–c and c respectively. The corresponding

† We are indebted to Dr. R. J. Bell for this suggestion.

strengths of the δ-function potentials are α^A and α^B. The parameters used are

$$\lambda_A = -\tfrac{1}{2}\alpha^A a, \qquad \lambda_B = -\tfrac{1}{2}\alpha^B a, \qquad \nu = 2k/|\alpha^A|.$$

The parameters λ_A and λ_B are measures of the strength of the potentials and ν, which is proportional to the square root of the energy, is a convenient representation of the energy eigenvalue. The integrated spectrum and degree of localization which depend on the energy E will now be denoted by $N(\nu)$ and $\langle \ln F(\nu) \rangle$. For a set of values of ν, $N(\nu)$ was obtained and the density of states $g(\nu)$ determined as a histogram using the formula

$$g(\nu) = \{N(\nu+\tfrac{1}{2}\delta\nu) - N(\nu-\tfrac{1}{2}\delta\nu)\}/\delta\nu.$$

Calculations were performed on the National Physical Laboratory computer ACE for two sets of parameters: (i) $\lambda_A = 1$, $\lambda_B = -0{\cdot}5$ and (ii) $\lambda_A = 2$, $\lambda_B = 0{\cdot}2$, with concentration $c = 0{\cdot}5$ in both cases.

For a fixed value of ν the calculation proceeded as follows. The range of ϕ from 0 to π was divided into M equal intervals. The distribution $W(\phi)$ was represented numerically by function values W_n at the discrete points $\phi^n = n\pi/M$, where n is an integer. Since $W_0 = 0$ and $W_M = 1$, and $W(\phi^n)$ outside the range $(0, \pi)$ is determined from equation (4*c*), there are M–1 unknown values of W_n to be determined in addition to the unknown $K\,(= N(\nu))$, making a total of M unknowns. These M unknowns were determined by the M equations obtained by representing equation (4*a*) numerically at the set of points ϕ^n $(n = 0, 1, \ldots, M-1)$. The value of $W\{T_j^{-1}(\phi^n)\}$ was determined in the following way. The value ϕ^s, say, which was nearest the value of $T_j^{-1}(\phi^n)$ was found. Then $W\{T_j^{-1}(\phi^n)\}$ was taken to be that linear combination of the three values W_{s-1}, W_s and W_{s+1} which makes the interpolation exact if $W(\phi)$ is exactly a quadratic in the interval (ϕ^{s-1}, ϕ^{s+1}).

The integrated spectrum $N(\nu)$ was given directly from the solution of the M simultaneous equations and $\langle \ln F(\nu) \rangle$ was obtained by numerical integration using equation (6).

For the two sets of parameters (i) and (ii) M was taken to be 32. Figures 1 and 2 show $g(\nu)$ plotted as a histogram for ν increasing in steps of $\delta\nu = 0{\cdot}04$ in case (i) and $\delta\nu = 0{\cdot}02$ in case (ii). The degree of localization for these cases is also shown, but for clarity of graphical representation we have plotted $\tfrac{1}{2}\langle \ln F(\nu) \rangle$. The density of states for the ordered diatomic chain is shown (broken curve) on these figures to illustrate the difference between the ordered and disordered alloy.

The accuracy of the calculations was judged by increasing M and examining the convergence of the results. The accuracy varied a great deal in different regions of the spectrum and the convergence with increasing M was not particularly smooth, so that it is difficult to give an overall error estimate. However, the accuracy obtained was sufficient to leave no reasonable doubt that most of the irregularities in the spectrum exist, especially as the fine structure became increasingly resolved as M increased. To indicate the convergence we present some sample results in table 1. We have used the notation g_j for $g(\nu)$ calculated with $M = j$.

There is another way of calculating the integrated spectrum when the concentration c is a rational number. Starting with some chosen boundary conditions the average phase increase is calculated for each chain of the ensemble in which there are exactly cS atoms of type B. The calculation is then repeated with increasing S and the spectra obtained must converge ultimately to the required spectrum. The success of this method depends on how quickly the sequence of spectra converge. We tried this method for case (ii) and found that the spectra converged very slowly above $\nu = \tfrac{1}{2}\pi$. However, they very

Table 1

$\lambda_A = 1,\ \lambda_B = -0{\cdot}5,\ c = 0{\cdot}5$

	g_8	g_{16}	g_{32}	g_{64}	$g_{16}-g_8$	$g_{32}-g_{16}$	$g_{64}-g_{32}$
1·54	0·264	0·250	0·276	0·276	−0·014	+0·026	+0·000
1·58	0·261	0·274	0·276	0·282	+0·013	+0·002	+0·006
1·62	0·347	0·311	0·306	0·309	−0·036	−0·005	+0·003
1·66	0·344	0·327	0·296	0·295	−0·017	−0·031	−0·001
1·70	0·353	0·324	0·360	0·351	−0·029	+0·036	−0·009
1·74	0·318	0·341	0·319	0·326	+0·023	−0·022	+0·007
1·78	0·304	0·320	0·316	0·315	+0·016	−0·004	−0·001
1·82	0·265	0·280	0·295	0·288	+0·015	+0·015	−0·007
1·86	0·244	0·265	0·268	0·271	+0·021	+0·003	+0·003
1·90	0·228	0·256	0·247	0·248	+0·028	−0·009	+0·001
Mean value of $\lvert g_{2j}-g_j \rvert$					0·021	0·015	0·004

quickly settled down to a limiting spectrum in the range of ν from 0 to $\frac{1}{2}\pi$ where the states are well localized and this was the range of greatest interest. Figure 3 shows the spectrum for $S = 14$ which can be compared with figure 2. From a quantitative point of view this method is almost certainly more accurate for the same amount of computing

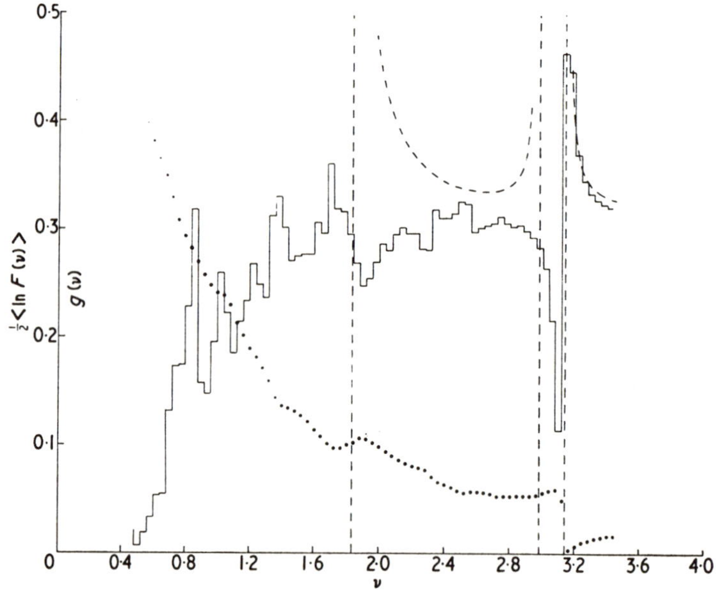

Figure 1. $\lambda_A = 1,\ \lambda_B = -0{\cdot}5,\ c = 0{\cdot}5$. The density of states $g(\nu)$ in histogram form. The dotted curve is $\frac{1}{2}\langle \ln F(\nu)\rangle$ and the broken curve is the density of states for the ordered diatomic chain.

time for the peaky range of the spectrum. However, as a reliable demonstration that the spectrum is irregular, the solution of Schmidt's functional equation has much to recommend it in that it gives directly the averaged spectrum for the infinite chains and the results are subject only to the numerical error in solving the equation. For the smoother part of the spectrum the solution of Schmidt's functional equation is very accurate.

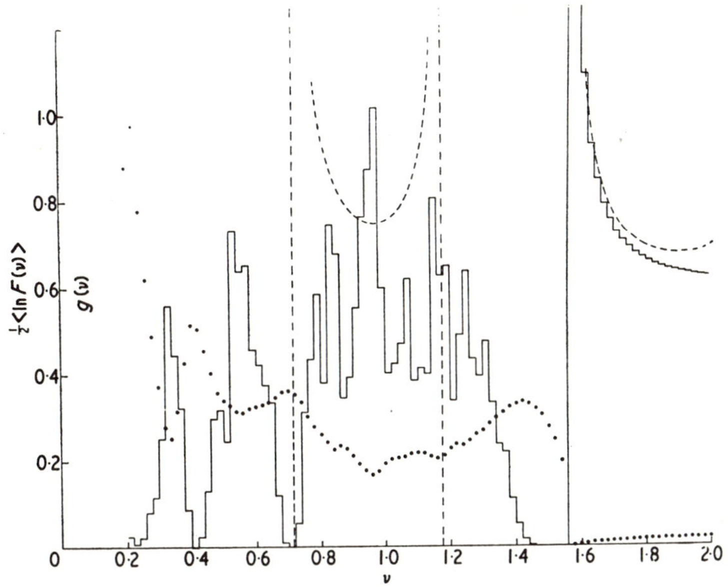

Figure 2. $\lambda_A = 2$, $\lambda_B = 0{\cdot}2$, $c = 0{\cdot}5$. The density of states $g(\nu)$ in histogram form. The dotted curve is $\frac{1}{2}\langle \ln F(\nu) \rangle$ and the broken curve is the density of states for the ordered diatomic chain.

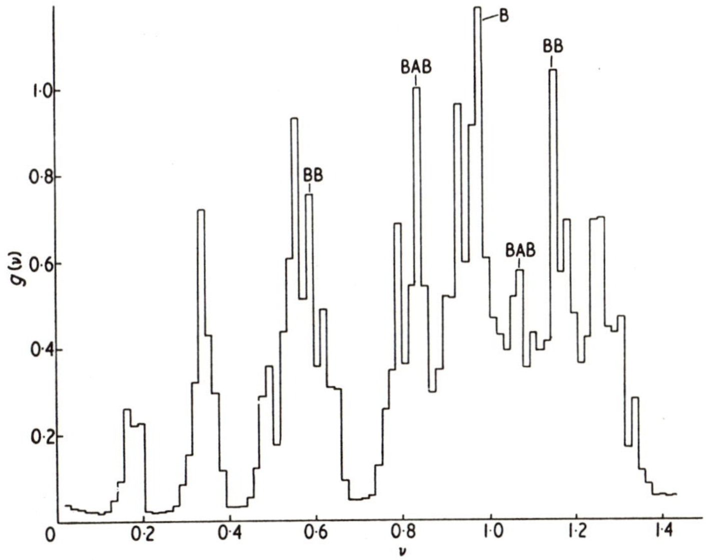

Figure 3. $\lambda_A = 2$, $\lambda_B = 0{\cdot}2$, $c = 0{\cdot}5$. The density of states $g(\nu)$ in histogram form obtained by calculating the average phase advance in the ensemble of all chains of fourteen atoms, which each contain seven A atoms and seven B atoms.

5. Discussion

We see that the electronic spectrum for a one-dimensional random alloy has much similarity to the vibrational spectrum for an isotopically disordered chain, in contrast with the approximate results of Faulkner and Korringa (1961).

By obtaining the degree of localization we have here a rather more complete picture of the behaviour of an electron in this system. What we have specifically calculated is the average logarithmic decay per atom of the square of the envelope of the wave function from its maximum value. As can be seen, the degree of localization has a tendency to decrease in the peaks of the spectrum and increase in the troughs. There is here some similarity to the result found by Roberts and Makinson (1962) that the degree of localization is very strong near the band edges of a one-dimensional liquid.

Dean (1961) found that in the corresponding vibrational problem one could associate peaks in the spectrum with local clusters of atoms. A similar phenomenon occurs in the electronic case. Table 2 shows, for example, the values of ν representing the eigenvalues of states localized on certain clusters placed in a pure chain of A atoms for the parameters of (ii).

Table 2

$\lambda_A = 2,\ \lambda_B = 0{\cdot}2,\ c = 0{\cdot}5$

Cluster	ν
B	0·968
BB	0·595 1·151
BAB	0·845 1·063

As can be seen in figure 3, there are prominent peaks in all these positions. A pure chain of A atoms would have an energy gap in the range $\nu = 0$ to $\nu = \frac{1}{2}\pi$, so that propagating solutions cannot exist in the A lattice. As a result, the A atoms surrounding a cluster have a tendency to isolate the wave functions associated with such a cluster from the rest of the system. That we can clearly identify the peaks with such clusters arises in this case because of the dominating influence of the A atoms in this region of the spectrum. The interpretation of all peaks is, however, not quite so simple. Nevertheless, as a general rule one can say that peaks are most likely to be prominent in a region of the spectrum which is forbidden in a pure lattice of either type of atom. This is not the case if the range is forbidden to both pure lattices simultaneously since then that energy range is also forbidden in the random alloy system (Luttinger 1951). In figure 2 the range of ν from 1·504 to $\frac{1}{2}\pi$ is forbidden by this theorem.

Less prominent peaks can occur in a region of the spectrum which is allowed for both the pure A and pure B lattices. As an example of this latter phenomenon, see figure 1. For these parameters the range of ν from 0·96 to 2·32 is allowed for both pure lattices and yet there is a clear peak in the vicinity of $\nu = 1{\cdot}7$. That this peak is really present and is not due to numerical inaccuracy seems clear from the numerical data presented in table 1.

In some ways the occurrence of peaks of this latter type is one of the most interesting results of this work, since in many real disordered alloys one is chiefly interested in energies which are not forbidden to the pure lattices of either type of atom. Although one cannot extrapolate these results to three dimensions, nevertheless, the existence of this phenomenon in one dimension leads one to suspect that similar behaviour may occur in three dimensions. The occurrence of peaks in the spectrum is linked with the fact that the disorder in a random alloy is of a discrete type, rather than continuous, as in a liquid metal. That is to say that in any small volume of the system the number of different possible configurations of atoms is finite. Provided that the wave functions are also localized, then these local configurations should manifest themselves in discreteness in the spectrum. Wave functions in one-dimensional disordered systems are, so far as we know, always localized (although the variation in localization in different regions of the spectrum can be very large) and there seems no reason why the wave functions should be otherwise in three-dimensional disordered systems. In the vast majority of alloys these peaks are probably so small that their influence is not seen in experimental observations. However, in some alloys where the 'effective scattering potentials' of the two (or more) constituent atoms are dissimilar and at least one of the scattering potentials is large, one should be alert to the possibility that such peaks may contribute to anomalous behaviour.

Acknowledgments

The work described above has been carried out as part of the research programme of the National Physical Laboratory and this paper is published by permission of the Director.

References

AGACY, R. L., 1964, *Proc. Phys. Soc.*, **83**, 591.
BORLAND, R. E., 1963, *Proc. Roy. Soc.* A, **274**, 529.
DEAN, P., 1960, *Proc. Roy. Soc.* A., **254**, 507.
—— 1961, *Proc. Roy. Soc.* A., **260**, 263.
FAULKNER, J. S., and KORRINGA, J., 1961, *Phys. Rev.*, **122**, 390.
JAMES, H. M., and GINZBARG, A. S., 1953, *J. Phys. Chem.*, **57**, 840.
LIFSHITZ, I. M., 1963, *Soviet Physics–JETP*, **17**, 1159.
LUTTINGER, J. M., 1951, *Philips Res. Rep.*, **6**, 303.
ROBERTS, A. P., and MAKINSON, R. E. B., 1962, *Proc. Phys. Soc.*, **79**, 509.
SCHMIDT, H., 1957, *Phys. Rev.*, **105**, 425.

Reprinted from THE PHYSICAL REVIEW, Vol. 138, No. 4A, A1121–A1126, 17 May 1965
Printed in U. S. A.

Existence of Energy Gaps in the Spectrum of a One-Dimensional Atomic Chain

LOWELL DWORIN*
Department of Physics, University of California, San Diego, La Jolla, California
(Received 21 December 1964)

The question of the existence of an energy gap in the spectrum of a one-dimensional atomic chain is shown to be related to the question of the convergence of a certain continued fraction whose elements are functions of the elements of the two-by-two cell matrices $\mathbf{C}^n(E)$ first introduced by Kramers. The implications of a known theorem (Worpitzky's Theorem) in the study of continued fractions are examined for the case of the chain of delta-function potentials and the binary alloy.

I. INTRODUCTION

THE density of states of a one-dimensional chain of atoms has most often been studied in two limiting cases: the chain of equally spaced delta-function potentials[1] (the Kronig-Penney model) or its slightly disordered variants,[2] and the completely random infinite chain of delta-function potentials. It is well known that energy gaps appear in the spectrum of the periodic chain, while the spectrum of the completely random chain, which has been calculated by Frisch and Loyd,[3] does not exhibit energy gaps. In view of these disparate results it is desirable to have some general criterion for the existence of energy gaps in the spectrum of an arbitrary chain. In this spirit Luttinger[4] proved a conjecture of Saxton and Hutner that any energy band that is forbidden to both a pure crystal of A atoms and to a pure crystal of B atoms is forbidden to any substitutional alloy of A and B atoms, provided the atoms are equally spaced and are represented by delta-function potentials differing only in the strength of the potential (the Kronig-Penney substitutional alloy). Another such general criterion for the existence of energy gaps was given by Borland,[5] who showed that gaps appear in the positive energy spectrum of a chain of delta-function potentials of strength $-\kappa_0$ with no long-range order, provided b, the maximum deviation of the interatomic spacing, is below a critical value: If the energy is given by $E_k=\hbar^2k^2/2m$ then there exists an energy gap corresponding to a gap in the allowed values of k given by

$$n\pi-2q\leqslant ka\leqslant n\pi-kb\,,\quad n=1,2,\cdots, \tag{1}$$

where $q=\arctan\kappa_0/2k$ and a is the minimum interatomic distance. In this work, he also shows that for the study of the spectrum any atomic potential may be represented by a delta-function potential of appropriate strength with regions of zero potential of appropriate length on either side, which implies that it should be possible to obtain criteria for the existence of energy gaps in the spectrum of a chain of arbitrary atomic potentials.

In this paper we shall present sufficient, although not necessary, conditions for the existence of energy gaps in the spectrum of an arbitrary chain. In Sec. II, we show that the existence of an energy gap is related to the convergence of a continued fraction whose elements are functions of the elements of the two-by-two unimodular cell matrices first introduced by Kramers.[6] In Sec. III, we apply a known theorem in the study of continued fractions to obtain the desired condition for the existence of an energy gap, and in Secs. IV and V we examine this condition for the case of a chain of delta-function potentials and for the binary alloy in which the atomic potentials are symmetric (this restriction is convenient but not necessary) but otherwise arbitrary. We find that both Borland's condition (1) and Luttinger's theorem may be greatly extended; in particular, there exists no upper limit to the deviation in interatomic spacing in order that an energy gap may exist.

II. THE CELL METHOD

Following James and Ginzbarg,[7] we shall divide our chain of atoms into m cells. To each cell we shall associate a real unimodular matrix $\mathbf{C}^n(E)$, whose coefficients depend upon the energy. To each cell, except the first and last, we shall then associate a function $f(C^n,C^{n-1},C^{n+1})$ of the matrix elements $C(E)$ corresponding both to that cell and to the neighboring cells. An energy gap shall be shown to exist for those energy regions such that the set of $m-2$ inequalities

$$|f(C^n,C^{n-1},C^{n+1})|\leqslant\text{const}\,,\quad n=1,2,\cdots,m-2. \tag{2}$$

is satisfied. If we were to choose a new division of the chain into cells we would obtain a new set of matrices $\mathbf{C}^n(E)$ and hence a new set of conditions to be satisfied, which might indicate the existence of an energy gap not revealed by the first cell division. We shall not attempt to exhaust the possibilities of cell division—rather, we shall choose the most natural assignment of cells in which there is one atom per cell and the cell boundaries are taken to be halfway along the region of zero po-

---

* National Science Postdoctoral Fellow. Supported in part by U. S. Office of Naval Research.

[1] R. de L. Kronig and W. G. Penney, Proc. Roy. Soc. (London) **130**, 49 (1931).

[2] R. D. Mattuck, Phys. Rev. **127**, 738 (1962); J. L. Beeby, *ibid.* **135**, A131 (1964).

[3] H. L. Frisch and S. P. Loyd, Phys. Rev. **120**, 1175 (1960).

[4] J. M. Luttinger, Philips Res. Rept. **6**, 303 (1951).

[5] R. E. Borland, Proc. Phys. Soc. (London) **78**, 926 (1961).

[6] H. A. Kramers, Physica **2**, 483 (1935).

[7] H. M. James and A. S. Ginzbarg, J. Phys. Chem. **57**, 840 (1953).

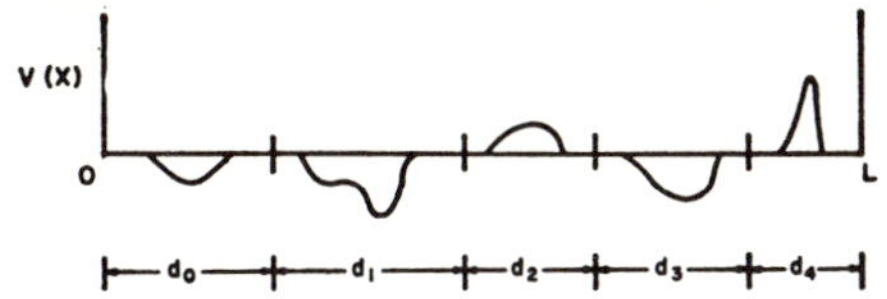

FIG. 1. Division of the chain into cells.

tential (see Fig. 1), leaving the possibility open that some other choice of cell division would lead to the prediction of new energy gaps.

Having divided the chain into m cells, we now consider the two independent real solutions $y_1^n(E,x)$ and $y_2^n(E,x)$ of the wave equation in the nth cell which satisfy the boundary conditions:

$$\begin{aligned} y_1^n(E,0)&=1\,, \quad y_1^{n\prime}(E,0)=0\,,\\ y_2^n(E,0)&=0\,, \quad y_2^{n\prime}(E,0)=1\,, \end{aligned} \tag{3}$$

where we have chosen the left-hand cell boundary as our origin. We now define the real two by two matrix $\mathbf{C}^n(E)$ as follows:

$$\begin{aligned} y_1^n(E,d_n)&=C_{11}^n(E)\,, \quad y_2^n(E,d_n)=C_{12}^n(E)\,,\\ y_1^{n\prime}(E,d_n)&=C_{21}^n(E)\,, \quad y_2^{n\prime}(E,d_n)=C_{22}^n(E)\,, \end{aligned} \tag{4}$$

where d_n is the length of the nth cell. Since the Wronskian of y_2^n and y_1^n is the same at $x=d_n$ as at $x=0$ we have the relation

$$|\mathbf{C}^n(E)|\equiv C_{11}^n(E)C_{22}^n(E)-C_{12}^n(E)C_{21}^n(E)=1\,. \tag{5}$$

Kramers, James,[8] and others have discussed these matrices. The significant feature of their elements, which is responsible for the existence of successive energy gaps followed by regions of allowed energy, is that they are oscillating functions of the energy.

The general solution to the wave equation may be written in terms of the individual cell solutions:

$$\psi(E,x)=a_n y_1^n\Big(E,\, x-\sum_{i=0}^{n-1} d_i\Big)+b_n y_2^n\Big(E,\, x-\sum_{i=0}^{n-1} d_i\Big)\,,$$

$$\text{for} \quad \sum_{i=0}^{n-1} d_i \leqslant x \leqslant \sum_{i=0}^{n} d_i \quad (n=0,1,2,\cdots,m-1)\,. \tag{6}$$

Thus the equations for the continuity of ψ and its derivative at the right-hand side of the nth cell become

$$\begin{aligned} a_{n+1}&=C_{11}^n a_n+C_{12}^n b_n\,,\\ b_{n+1}&=C_{21}^n a_n+C_{22}^n b_n\,, \end{aligned} \tag{7}$$

where

$$\begin{aligned} \psi\Big(E,\sum_{i=0}^{n} d_i\Big)&=a_{n+1}\,,\\ \psi'\Big(E,\sum_{i=0}^{n} d_i\Big)&=b_{n+1}\,. \end{aligned} \tag{8}$$

Still following James and Ginzbarg, we define the phase $\phi(x,E)$ by the equation

$$\tan\phi(E,x)=-\psi'(E,x)/\psi(E,x)\,, \tag{9}$$

where we have set the arbitrary constant K of James and Ginzbarg equal to unity. We require that $-\frac{1}{2}\pi \leqslant \phi(E,0)<\frac{1}{2}\pi$ and that $\phi(E,x)$ change continuously with x (we consider a delta-function potential as the limiting case of an ordinary smooth potential) in order to uniquely specify $\phi(E,x)$. The value of $\phi(E,x)$ is fixed for certain values of x: It is $(m-\frac{1}{2})\pi$ at the mth zero of $\psi(E,x)$ for $x>0$ and $m\pi$ at the succeeding zero of $\psi'(E,x)$. From Eq. (8) and the definition (9) we have

$$\begin{aligned} \tan\phi\Big(E,\sum_{i=0}^{n-1} d_i\Big)&\equiv\tan\phi(n)=-b_n/a_n\,,\\ \tan\phi\Big(E,\sum_{i=0}^{n} d_i\Big)&\equiv\tan\phi(n+1)=-b_{n+1}/a_{n+1}\,. \end{aligned} \tag{10}$$

We may thus write Eq. (7) in the form

$$\tan\phi(n+1)=\frac{\tan\phi(n)-C_{21}^n/C_{22}^n}{C_{11}^n/C_{22}^n-(C_{12}^n/C_{22}^n)\tan\phi(n)}\,. \tag{11}$$

From the condition that $\phi(E,x)$ be a continuous function of x, James and Ginzbarg have shown that:

(i) For a given E, $\phi(n+1)$ is a monotonically increasing function of $\phi(n)$.

(ii) For given $\phi(n)$, $\phi(n+1)$ is a monotonically increasing function of E.

(iii) $$\lim_{E\to-\infty}\phi(n+1)=(m-\tfrac{1}{2})\pi$$

if $(m-\frac{1}{2})\pi\leqslant\phi(n)<(m+\frac{1}{2})\pi$.

We should like to replace Eq. (11) by a linear difference equation. To this end, we define a set of functions $Y(n)$ by the relation

$$\tan\phi(n)=Y(n+1)/Y(n)+C_{11}^n/C_{12}^n\,, \qquad (n=0,1,2,\cdots,m)\,, \tag{12}$$

where we require $Y(0)\neq 0$.

Substituting this expression into Eq. (11), we obtain the equation

$$\begin{aligned} Y(n+2)&+(C_{11}^{n+1}/C_{12}^{n+1}+C_{22}^n/C_{12}^n)Y(n+1)\\ &+1/(C_{12}^n)^2 Y(n)=0\,, \quad (n=0,1,\cdots,m-2)\,. \end{aligned} \tag{13}$$

If we write $\tan\varphi(0)=\alpha$ we find from Eq. (12) that

$$Y(1)/Y(0)+C_{11}^0/C_{12}^0=\alpha\,. \tag{14}$$

We may incorporate this boundary condition into Eq. (13) by defining

$$C_{22}^{(-1)}/C_{12}^{(-1)}=-\alpha \tag{15}$$

and

$$Y(-1)=0\,.$$

[8] H. M. James, Phys. Rev. 76, 1602 (1949).

With these definitions we may let Eq. (13) hold for $n=-1, 0, 1, \cdots, m-2$.

We now write Eq. (13) in the form

$$\frac{Y(n+1)}{Y(n+2)}=\frac{1}{\beta_{n+1}-\alpha_n[Y(n)/Y(n+1)]}, \tag{16}$$

where

$$\beta_{n+1}=-(C_{11}^{n+1}/C_{12}^{n+1})+(C_{22}^{n}/C_{12}^{n}), \quad \alpha_n=1/(C_{12}^{n})^2. \tag{17}$$

By repeated application of Eq. (16) we find that we may write $Y(m-1)/Y(m)$ as the continued fraction

$$\frac{Y(m-1)}{Y(m)}=\cfrac{\beta_{m-1}^{-1}}{1-\cfrac{\alpha_{m-2}\beta_{m-2}^{-1}\beta_{m-1}^{-1}}{1-\cfrac{\alpha_{m-3}\beta_{m-3}^{-1}\beta_{m-2}^{-1}}{1-\cdots\cfrac{}{1-\cfrac{\alpha_2\beta_2^{-1}\beta_3^{-1}}{1-\cfrac{\alpha_1\beta_1^{-1}\beta_2^{-1}}{1-\alpha_0\beta_0^{-1}\beta_1^{-1}}}}}}}. \tag{18}$$

If we choose as one boundary condition $\psi(E,0)=0$ so that $\phi(E,0)=-\frac{1}{2}\pi$ and $\tan\phi(0)=-\infty$, we then find from Eqs. (15) and (17) that $\beta_0^{-1}=0$ so that for this particular boundary condition (18) becomes

$$\frac{Y(m-1)}{Y(m)}=\cfrac{\beta_{m-1}^{-1}}{1-\cfrac{\alpha_{m-2}\beta_{m-2}^{-1}\beta_{m-1}^{-1}}{1-\cdots\cfrac{}{1-\cfrac{\alpha_2\beta_2^{-1}\beta_3^{-1}}{1-\alpha_1\beta_1^{-1}\beta_2^{-1}}}}}. \tag{19}$$

Further, we see from Eq. (12) that if $Y(m-1)/Y(m)=\pm\infty$, then $\tan\phi(m-1)=C_{11}^{m-1}/C_{12}^{m-1}$, and upon substituting this value into Eq. (11) we find that $\tan\phi(m)$ goes to infinity so that $\psi(E,\sum_{i=0}^{m-1}d_i)\equiv\psi(L)=0$.

If we thus choose as our second boundary condition $\psi(L)=0$ we see that the eigenvalue equation for E is, assuming $C_{21}^{m-1}(E)$ to be finite,

$$Y(m-1)/Y(m)=\pm\infty, \tag{20}$$

where $Y(m-1)/Y(m)$ is given by Eq. (19). Conversely, we may say that energy gaps in the spectrum correspond to those regions of E for which the continued fraction (19) is finite. We are thus lead to study the conditions under which a continued fraction of the form of (19) is finite. Since we shall mainly be interested in the case of the infinite chain, we shall require $\lim_{m\to\infty}(Y(m-1)/Y(m))$ to exist and be finite, although the existence of a single limit is not required for the existence of an energy gap. We note that the existence of a limit means that the phase at the end of a sufficiently long chain is essentially independent of the initial phase (boundary condition). Borland[9] has shown that this implies that the amplitude of the wave function whose amplitude and phase are given at the left of the chain increases beyond bound at the right of the chain.

It is of interest to note that the eigenvalue Eq. (20) may be simplified. From the difference Eq. (13) we see that $Y(m-1)$ and $Y(m)$ cannot both be zero if C_{12}^{n} is finite for $n=1, 2, \cdots, m-2$ so that we may write Eq. (20) as

$$Y(m)=0. \tag{21}$$

However, it may be shown[10] that $Y(m)$ is given by the relation

$$Y(m)=\begin{vmatrix} \beta_{m-1} & -\alpha_{m-2}^{1/2} & 0 & \cdots & & & \\ -\alpha_{m-2}^{1/2} & \beta_{m-2} & -\alpha_{m-3}^{1/2} & \cdots & & & \\ 0 & -\alpha_{m-3}^{1/2} & \beta_{m-3} & \cdots & & & \\ \cdot & \cdot & \cdot & \cdots & \cdot & \cdot & \cdot \\ \cdot & \cdot & \cdot & \cdots & \cdot & \cdot & \cdot \\ \cdot & \cdot & \cdot & \cdots & \cdot & \cdot & \cdot \\ & & & \cdots & -\alpha_2^{1/2} & \beta_2 & -\alpha_1^{1/2} \\ & & & \cdots & 0 & -\alpha_1^{1/2} & \beta_1 \end{vmatrix} \tag{22}$$

so that Eq. (20) may be written as a determinental equation.

III. THE WORPITZKY CONDITION

The study of the convergence of continued fractions has occupied mathematicians for over one hundred years. The following definitions are standard. The continued fraction F_m, where

$$F_m=\frac{A_m}{B_m}=\cfrac{a_1}{1+\cfrac{a_2}{1+\cdots\cfrac{}{1+a_m}}}, \tag{23}$$

[9] R. E. Borland, Proc. Roy. Soc. (London) 274, 529 (1963).

[10] H. S. Wall, *Continued Fractions* (D. Van Nostrand Company, Inc. Princeton, New Jersey, 1948), p. 65.

is said to be the mth approximant to the infinite continued fraction

$$F=\cfrac{a_1}{1+\cfrac{a_2}{1+\cfrac{a_3}{1+\cdots}}}. \tag{24}$$

The a_n are called the partial quotients, A_m the mth numerator, and B_m the mth denominator. The continued fraction (22) is said to converge if at most a finite number of its denominators vanish, and if the limit of its sequence of approximants

$$\lim_{m\to\infty} A_m/B_m$$

exists and is finite. Otherwise the continued fraction is said to diverge.

There exist a number of theorems giving conditions under which the continued fraction (22) converges. One of the oldest was given by Worpitzky in 1865[11]:

Let a_2, a_3, $\cdots$ be functions of any variables over a domain D in which

$$|a_{p+1}|\leqslant\tfrac{1}{4},\quad p=1,2,\cdots. \tag{25}$$

Then the following statements hold:

(i) The continued fraction (22) converges uniformly over D.

(ii) The values of the continued fraction and of its approximants are in the circular domain

$$|Z-\tfrac{4}{3}|\leqslant\tfrac{2}{3}. \tag{26}$$

(iii) The constant $\frac{1}{4}$ is the "best" constant and the domain (26) is the "best" domain of values of the fraction and its approximants.

This theorem, which may be proved by induction, can be strengthened somewhat, but as it is the easiest to apply of all its variants, we shall adopt it as our criterion for the existence of energy gaps.

From (19) we see that the conditions (25) become, in our case,

$$|\alpha_n\beta_n^{-1}\beta_{n+1}^{-1}|\leqslant\tfrac{1}{4},\quad n=1,2,\cdots,m-2 \tag{27}$$

or, using (17)

$$|(C_{22}^n+C_{11}^{n+1}(C_{12}^n/C_{12}^{n+1}))\times(C_{11}^n+C_{22}^{n-1}(C_{12}^n/C_{12}^{n-1}))|\geqslant 4,\quad n=1,2,\cdots,m-2. \tag{28}$$

It is not hard to show[12] that if the atomic potential in the nth cell is symmetric then $C_{22}^n=C_{11}^n$, so that in this case (which will be assumed for simplicity to hold in the remainder of the paper) the conditions (28) become

$$|(C_{11}^n+C_{11}^{n+1}(C_{12}^n/C_{12}^{n+1}))\times(C_{11}^n+C_{11}^{n-1}(C_{12}^n/C_{12}^{n-1}))|\geqslant 4,\quad n=1,2,\cdots,m-2. \tag{29}$$

Conditions (28) or (29) are the desired criterion for the existence of an energy gap. In the remainder of the paper we shall discuss these conditions for the case of delta-function potentials located at the center of cells of arbitrary length, and, in greater detail, the application of these conditions to the binary alloy. Before we do so, however, we wish to emphasize the fact that conditions (28) or (29) are only sufficient, not necessary, conditions. Not only might one learn of the existence of new energy gaps by a different method of cell division, or by use of a different convergence theorem, but even the same cell assignments and the use of Worpitzky's theorem may lead to an extension of the known energy gaps. For example, it is possible to show[13] that the convergence of the continued fraction (21) is implied by the convergence of a second continued fraction whose elements are functions of the elements of the first, such that the application of Worpitzky's theorem to the second fraction leads to a second set of convergence conditions:

$$|a_{2n+1}|\leqslant\tfrac{1}{4},\qquad |a_{2n}|\geqslant 25/4,\quad n=1,2,\cdots. \tag{30}$$

IV. THE CHAIN OF SYMMETRIC DELTA-FUNCTION POTENTIALS

It is easy to show that the matrix $\mathbf{C}^n(E)$ for a delta-function potential of strength $(m/\hbar^2)V_0^n=-\kappa_0^n$ located at the center of a cell of length d_n is given by

$$\mathbf{C}^n(E)=\begin{pmatrix} \cos\alpha d_n-\dfrac{\kappa_0^n}{\alpha}\sin\alpha d_n & \dfrac{\sin\alpha d_n}{\alpha}+\dfrac{\kappa_0^n}{\alpha^2}(\cos\alpha d_n-1) \\ -\alpha\sin\alpha d_n-\kappa_0^n(\cos\alpha d_n+1) & \cos\alpha d_n-\dfrac{\kappa_0^n}{\alpha}\sin\alpha d_n \end{pmatrix}, \tag{31}$$

where $\alpha=(2mE)^{1/2}/\hbar$. For positive energies we shall refer to α as k; for negative energies we shall write $\alpha=i\kappa$.

In the case of the periodic chain ($\kappa_0^n=\kappa_0$; $d_n=d$ for all n) condition (25) reduces to the Kramers' condition,

$$|C_{11}^n|\geqslant 1 \tag{32}$$

or, for the delta-function potential,

$$|\cos\alpha d-(\kappa_0/\alpha)\sin\alpha d|\geqslant 1, \tag{33}$$

[11] Reference 10; Theorem 10.1, p. 42. For the periodic case where all $a_p=a$, this condition is also a necessary one.

[12] See, for example, H. M. James, Ref. (8).

[13] Reference 10; Example 3.10, p. 63.

which is just the condition for the existence of an energy gap in the Kronig-Penney model.

In the general case we have the conditions

$$\left|\left\{\left(\cos\alpha d_n-\frac{\kappa_0^n}{\alpha}\sin\alpha d_n\right)+\left(\cos\alpha d_{n+1}-\frac{\kappa_0^{n+1}}{\alpha}\sin\alpha d_{n+1}\right)\frac{\sin\alpha d_n+(\kappa_0^n/\alpha)(\cos\alpha d_n-1)}{\sin\alpha d_{n+1}+(\kappa_0^{n+1}/\alpha)(\cos\alpha d_{n+1}-1)}\right\}\right.$$
$$\times\left\{\left(\cos\alpha d_n-\frac{\kappa_0^n}{\alpha}\sin\alpha d_n\right)+\left(\cos\alpha d_{n-1}-\frac{\kappa_0^{n-1}}{\alpha}\sin\alpha d_{n-1}\right)\right.$$
$$\left.\left.\times\frac{\sin\alpha d_n+(\kappa_0^n/\alpha)(\cos\alpha d_n-1)}{\sin\alpha d_{n-1}+(\kappa_0^{n-1}/\alpha)(\cos\alpha d_{n-1}-1)}\right\}\right|\geqslant 4,\quad n=1,2,\cdots,m-2. \tag{34}$$

It is not hard to see that these conditions may be met, for example, if all $\kappa_0^n=\kappa_0$ and κ_0/k is sufficiently large for arbitrarily large d_n provided $\sin kd_n>0$ or $\sin kd_n<0$ for all n. Even if $\kappa_0/k<1$, it is clear from the periodic nature of the $\mathbf{C}^n(E)$ that if conditions (34) are satisfied for some set of d_n where $k(d_{\max}-d_{\min})\leqslant\pi$, then they are also satisfied for a second set d_n', where $kd_n'=kd_n+2\pi m_n$ (m_n being an integer). In both cases there exists no upper limit to d_n in order that an energy gap exists, and since the maximum distance between the potentials is $d_{\max}$ and the minimum distance is $d_{\min}$, we see that Borland's criterion, Eq. (1), is too restrictive. It is not hard to see that our condition (34) leads to Borland's condition (aside from the possibility of adding $2\pi m_n$ to d_n) in the weak-binding limit $\kappa_0/k\ll\pi$ for $2n\pi-\kappa_0/k\leqslant kd_n\leqslant 2\pi n$ but to a more restrictive condition for $(2n+1)\pi-\kappa_0/k\leqslant kd_n\leqslant(2n+1)\pi$. In this respect Borland's condition is less restrictive than ours.

We note that condition (34) may be satisfied even in the limit $\kappa_0^n\to 0$ for all n, provided $d_n>d_{n+1}$ or $d_n<d_{n+1}$ for all n. However, the magnitude of the energy gap in this case is of order L^{-2}, the natural energy level spacing for free electrons in a region of length L.

V. THE BINARY ALLOY

If there are only two types of symmetric atomic potentials, A and B, with the two corresponding matrices $\mathbf{C}^A$ and $\mathbf{C}^B$, then the most general alloy consists of cell groups of forms:

$$\begin{array}{ll} A\,A\,A, & B\,B\,B,\\ A\,A\,B, & B\,B\,A,\\ A\,B\,A, & B\,A\,B,\\ B\,A\,A, & A\,B\,B; \end{array} \tag{35}$$

and thus the conditions (34) become

$$\begin{array}{ll} |X_A|\geqslant 1; & |X_B|\geqslant 1;\\ |X_A+X_BT|\geqslant 2; & |X_B+X_AT^{-1}|\geqslant 2, \end{array} \tag{36}$$

where

$$X_{(A,B)}=\cos\alpha d_{(A,B)}-(\kappa_0^{(A,B)}/\alpha)\sin\alpha d_{(A,B)} \tag{37}$$

and

$$T=\frac{\sin\alpha d_A+(\kappa_0^A/\alpha)(\cos\alpha d_A-1)}{\sin\alpha d_B+(\kappa_0^B/\alpha)(\cos\alpha d_B-1)}. \tag{38}$$

It is not hard to show, by an extension of an argument of Luttinger, that

$$T=\frac{X_A+1}{X_B+1}f(d_A,d_B), \tag{39}$$

where

$$f(d_A,d_B)=\frac{\sin\alpha(\frac{1}{2}(d_A+d_B))+\sin\alpha(\frac{1}{2}(d_A-d_B))}{\sin\alpha(\frac{1}{2}(d_A+d_B))-\sin\alpha(\frac{1}{2}(d_A-d_B))}, \tag{40}$$

so that the entire $\kappa_0^{(A,B)}$ dependence is in the factor $(X_A+1)/(X_B+1)$.

From Eqs. (32) and (36) we see that our condition for the existence of an energy gap requires the energy to be in a region forbidden to both pure A and pure B. Luttinger considered the case in which $d_A=d_B$ (the Kronig-Penney alloy) and thus, from Eq. (40), $T=(X_A+1)/(X_B+1)$. An examination of the remaining conditions (36) shows that for this case there are no further restrictions; we thus obtain Luttinger's result. Even if $d_A\neq d_B$, Luttinger showed that an energy gap exists if $f(d_A,d_B)$ is positive. Examination of conditions (36) shows that we also obtain an energy gap under that circumstance, provided the condition

$$|X_A|/(2-|X_B|)\geqslant|T|\geqslant(2-|X_A|)/|X_B|\geqslant 0 \tag{41}$$

is also satisfied. In this respect Luttinger's condition is less restrictive than ours. On the other hand, we find that energy gaps also exist for $f(d_A,d_B)<0$ provided one of the following two conditions is satisfied:

$$\begin{array}{ll} \text{(i)} & |X_B|\geqslant 2;\ \ 2\geqslant|X_A|\geqslant 1;\\ & \qquad |T|\geqslant|X_A|/(|X_B|-2),\\ \text{(ii)} & |X_A|\geqslant 2;\ \ 2\geqslant|X_B|\geqslant 1;\\ & \qquad |T|\leqslant(|X_A|-2)/|X_B|. \end{array} \tag{42}$$

If we restrict the configurations of the binary alloy to those in which no more than two A atoms occur in succession, we have the less restrictive conditions:

$$|X_B| \geqslant 1; \quad |X_B + X_A T^{-1}| \geqslant 2; \quad |X_A + X_B T| \geqslant 2; \quad |X_A(X_B + X_A T^{-1})| \geqslant 2, \tag{43}$$

which imply that the energy is in a region forbidden to pure B, with the symmetric conditions if no more than two B atoms occur in succession.

If we further assume that no more than one A atom may occur between neighboring B atoms, we have the still less restrictive conditions

$$|X_B| \geqslant 1; \quad |X_B + X_A T^{-1}| \geqslant 2; \quad |X_A + X_B T| \geqslant 2, \tag{44}$$

with the symmetric conditions if no more than a single B atom lies between two neighboring A atoms.

If we require an A atom to follow a B atom we have the still less restrictive conditions:

$$|X_A + X_B T| \geqslant 2; \quad |X_B + X_A T^{-1}| \geqslant 2, \tag{45}$$

so that an energy gap may be in a region allowed to both pure A and pure B. For the Kronig-Penney alloy, however, an examination of conditions (44) shows that they cannot be satisfied; we are thus unable to prove the existence of an energy gap for values of the energy which are allowed to both pure A and pure B for the Kronig-Penney alloy.

ACKNOWLEDGMENT

The author wishes to thank Professor W. Kohn for suggesting this problem.

ADDENDUM

Conditions (41) and (42), though correct as they stand, may each be extended slightly:

If either $|X_a|$ or $|X_b|$ (or both) >2, the condition (41) becomes the weaker one in which the corresponding lower or upper (or both) limits on $|T|$ are removed.

If both $|X_a|$ and $|X_b| > 2$, but X_a and X_b differ in sign, the conditions (42) are replaced by the stronger ones

$$\text{(i)} \quad |T| \geqslant \frac{|X_a|}{|X_b|-2} \qquad |T| \geqslant \frac{2+|X_a|}{|X_b|}$$

$$\text{(ii)} \quad |T| \leqslant \frac{|X_a|}{2+|X_b|} \qquad |T| \leqslant \frac{|X_a|-2}{|X_b|}\,.$$

The Many-Fermion Problem

Chapter 4

The Many-Fermion Problem

Introduction

Before 1950 little was understood about the nature of solutions to Schrödinger's equation for a large system of interacting fermions. It was suspected that certain properties might best be described by "collective variables" involving the correlated motion of many particles, but no mathematical formulations of this idea existed. Most properties were satisfactorily understood within the "independent particle model" which found its justification in the Hartree-Fock approximation, valid for weakly interacting systems. But it was also known that ordered states of matter such as electronic magnetism, for example, were mediated by relatively strong forces such as the Coulomb repulsion; yet these forces were known to be not significantly weaker in the majority of ordinary, nonordered materials. In the absence of any criteria to indicate when and how the collective behavior might become dominant, the theory of matter could not be considered satisfactory.

Much of the tremendous development of theoretical physics in the past decade has been stimulated by just this problem, and it is by now generally considered to be resolved. Interacting fermions, it is now argued, are essentially schizophrenic: most of the time they are highly individualistic and their motion is well described by the independent-fermion model. But part of the time they engage in collective behavior which depends crucially on the strength and nature of the interactions.

Tomonaga's Model

One of the first important breakthroughs was in connection with a one-dimensional model; it is the theory due to Tomonaga (1950), propounded in the first reprinted article of this chapter, which first showed that interactions can mediate new collective dynamical degrees of freedom. Tomonaga found these modes to be quantized approximately as bosons, and also found that they exhausted most of the potential energy of interaction, so that remaining degrees of freedom (if any) would describe weakly interacting, or noninteracting, fermions—in agreement with the independent particle model. The same conclusions have been reached about electrons in three dimensions, and the interested reader is referred to several excellent texts on the subject[1]; however, even the best of all this work is marred by one or several approximations or hypotheses about convergence of infinite perturbation series, etc. To see this let us explain the mathematical formalism, and give an introduction to Tomonaga's paper.

A two-body interaction among the N fermions is capable of transforming an initial configuration in which the particles are labeled by quantum numbers k_1, $k_2, \ldots, k_N$ into a new configuration in which two (and only two) of the quantum numbers have been altered, e.g., $k_1', k_2', \ldots, k_N$. Successive applications of the interaction can eventually change all the quantum numbers, of course, by changing them a pair at a time; however, the Pauli exclusion principle does place a severe restriction on which new quantum numbers are allowed and in the example above prohibits k_1' and k_2' from coinciding with each other or with any of the primitive quantum numbers $k_3, \ldots, k_N$. On the other hand, the variational principle ensures that in the ground state the most important configurations will be in the neighborhood of the "Fermi sea" which is that set of k's belonging to the lowest set of one-particle energies $e(k)$. Thus each time it operates, the interaction will be capable of placing two fermions outside the Fermi sea, leaving two "holes" behind (in the example above, the fermions are at k_1' and k_2', the holes at k_1 and k_2), or of the opposite (which is, filling in two holes, i.e. returning the two particles at k_1' and k_2' to their initial positions at k_1 and k_2), as well as scattering particles outside the Fermi sea to new positions (e.g. k_1' and k_2' to k_1'' and k_2'') or two holes to new positions (e.g. by scattering two particles within the Fermi sea to two available holes also within the sea). Last but not least, the two-body interaction is capable of simultaneously scattering a particle and a hole to new quantum numbers, or of filling a hole by means of a particle outside the sea while simultaneously scattering another particle or hole. The essential observation by Tomonaga was that the basic building block in the description of all these complicated processes was the simple pair: *particle outside the sea, hole inside the sea*, making up a so-called "elementary excitation." The various processes just described correspond to the creation of two such pairs, or their destruction, or their scattering. To be more precise, let us now introduce the operators c_k and c_k^* which respectively destroy and create a particle with quantum number k, and obey the fermion anticommutation relations:

[1] P. Nozières, "Interacting Fermi Systems," Benjamin, New York, 1963; A. Abrikosov, L. Gorkov, and I. Ddzyaloshinski, "Methods of Quantum Field Theory in Statistical Physics," Prentice-Hall, Englewood Cliffs, New Jersey, 1963; T. D. Schultz, "Quantum Field Theory and the Many Body Problem," Gordon and Breach, New York, 1964.

$$c_k c_{k'} + c_{k'} c_k = 0 = c_k{}^* c_{k'}{}^* + c_{k'}{}^* c_k{}^*$$

These ensure that $c_k{}^* c_k{}^* = 0$, i.e., that two particles cannot be placed in the same state; and finally,

$$c_k c_{k'}{}^* + c_{k'}{}^* c_k = \delta_{k'k'}$$

We shall make no use of these anticommutation relations beyond what is required to establish the properties of the elementary excitations, so the reader who is unfamiliar with the methods of second quantization need not yet despair! The operator which, when operating on the Fermi sea, creates such an elementary excitation is evidently

$$(c_{k'}{}^* c_k)$$

with k inside the Fermi sea and k' outside it, whereas the operator which destroys an elementary excitation is formally the same but with k and k' interchanged. These coupled fermion operators have many of the properties of boson field operators: there is no number conservation law (elementary excitations can be created or destroyed), and two different operators commute (if all their quantum numbers are different). Finally, and most important, if we arbitrarily set commutators of operators with one common subscript such as

$$(c_k{}^* c_{k'})(c_{k'}{}^* c_{k''}) - (c_{k'}{}^* c_{k''})(c_k{}^* c_{k'})$$

equal to zero—i.e., pretend that all elementary excitations commute, except for commutators of operators with two common subscripts:

$$(c_k{}^* c_{k'})(c_{k'}{}^* c_k) - (c_{k'}{}^* c_k)(c_k{}^* c_{k'}) = c_k{}^* c_k - c_{k'}{}^* c_{k'}$$

which are set equal to $+1$ (the value for this operator quantity when operating on the Fermi-sea wave function). Even with these approximations, however, the problem is still not soluble—two more are necessary. One is to restrict ourselves to one dimension, and the other is to replace the kinetic energy p^2 by $|p|$. The second approximation is perhaps the stronger of the two because it results in the fact that no matter how complicated an intermediate state may be, the energy transfer is always proportional to the momentum transfer. It is precisely this point that makes the problem soluble, but at the same time it introduces a tremendous degeneracy into the problem. Whatever unphysical consequences this model may have, they cannot be blamed entirely on one dimension.

Strictly speaking, energy transfer = momentum transfer only if the momentum does not change sign, but Tomonaga argued that for the relevant intermediate states (those near the Fermi surface) this criterion was fulfilled. If we ignore the violations (which must be counted as a fourth approximation) the problem becomes one of normal modes. The reader should turn to the Tomonaga paper for a very beautiful discussion of the physical principles and the mathematical techniques involved.

Exactly Soluble Model

Subsequently, Luttinger (1963) introduced a new, but similar, one-dimensional model in which some of the Tomonaga approximations were introduced into the Hamiltonian *ab initio*. In order to ensure that the energy transfer was always equal to the momentum transfer, Luttinger used the device, which is natural in field theory, of introducing two types of particles—one with $e(p) = p$ and the other with $e(p) = -p$. These interact with each other, but in order to have a finite ground state energy it is necessary to postulate a filled sea of negative energy particles, as in the Dirac theory. This procedure led to some difficulties, however, as indeed it does for the Thirring model of field theory, and the solution which Luttinger gave of his problem did not correspond to the reasonable, but approximate, one which had been studied by Tomonaga. The defect was that Luttinger had not properly filled the negative-energy Dirac sea. The remedy was provided soon thereafter by Mattis and Lieb (1965) who gave the exact solution of Luttinger's model in the second of the reprinted papers in this chapter. We give here some brief comments to clarify this work, although the paper is largely self-contained and can be read without amplification. The calculation proceeds roughly as follows: just as in the Dirac theory of electrons and positrons, one fills the infinite sea of negative energy states and then considers a relabeled set of *particles* in positive energy states, and *holes* in negative energy states. Excitation of these in pairs are the elementary excitations, and a certain linear combination of these elementary excitations describes the collective modes of interest—denoted "plasmons" by analogy with the collective modes of the three-dimensional electron gas (Tomonaga had denoted these modes "phonons"). The Hamiltonian can be formally, and exactly, rewritten as an operator quadratic in the plasmons, and as these plasmons obey rigorous Bose-Einstein commutation relations the problem is thereupon trivially solved, for *general* and *arbitrary interactions*.

Additional insight into this work was afforded by three subsequent investigations. In the first of these, Mattis (1964) extended the Luttinger model to fermions with spin, and showed that aside from the plasmon normal modes there existed a set of "magnon" normal modes, corresponding to oscillations in the spin density at constant charge density. That is, the plasmons corresponded to density oscillations in the spin-up fermions *added* to density oscillations in the spin-down system; whereas the magnons consisted of density oscillations in the spin-up system of *opposite* sign to those in the spin-down system. While electric fields and ordinary potentials affected only the plasmons, the magnons (but not the plasmons) responded to external magnetic fields, and thus determined the Pauli spin susceptibility of the electrons. Effective spin-spin forces (such as the exchange integrals which are introduced into atomic theory to explain Hund's rules concerning the configurations of equivalent electrons) were also found to be mediated by the magnons and were found to affect the Pauli spin susceptibility. When it came to an approximate calculation, relating to the validity of the Hartree-Fock approximation, this paper was in error. This was pointed out by Overhauser (1965) and others (see Mattis, 1965), and it was found that exchange among holes in the negative energy Dirac sea could not be safely neglected. At the same time, Overhauser noted that the plasmon and magnon elementary excitations completely spanned the Hilbert space of excited states. That is, the excitation spectra of the boson Hamiltonians included *all* the excited states of the total fermion Hamiltonian, and consequently the *quasi-particle* part first discussed in Mattis and Lieb (1965) was *void of structure*. This interesting result brings out an important difference between this model and the real world: in real systems the collective modes presumably represent only a fraction of the available degrees of freedom of the interacting fermions, whereas in this soluble one-dimensional model they exhaust all the degrees of freedom. In addition, in three dimensions the plasmons and quasi-particles are not strictly uncoupled; in other words each is only an approximate concept. As we remarked before, however, the difficulty probably does not lie with one dimension *per se*, but with the extreme approximation that $e(p) = p$.

Is Perturbation Theory Valid?

The preceding remarks notwithstanding, the forces of interaction among the fermions have a nonnegligible result on the fermion field operators c_k^* in the Luttinger model, as well as in three dimensions. It is found to be more convenient to analyze the field operator $\Psi^*(x)$ which creates a fermion at the point x; here we summarize a calculation in section 5 of the paper by Mattis and Lieb (1965), *and the equation numbers refer to that paper*. The unitary transformation which diagonalizes the Hamiltonian is

$$e^{i\lambda S} \quad \text{where} \quad \lambda = \text{coupling constant}$$

[Eqs. (4.7)–(4.13)], and therefore in the representation in which the Hamiltonian is diagonal, the field operator becomes

$$\Psi_\lambda^*(x) = e^{i\lambda S}\Psi^*(x)e^{-i\lambda S}$$

After some manipulation of the right-hand side, this can be reduced to the simple form

$$\Psi_\lambda^*(x) = e^{i\lambda\Gamma(x)}\Psi^*(x)$$

[Eqs. (5.4)–(5.9)], with $\Gamma(x)$ calculated to be an Hermitian operator linear in plasmon annihilation and creation operators. This exponential factor may be interpreted as resulting from the backflow, or polarization, of the medium due to the introduction of an interacting fermion at x. (If we generalized to fermions with spin, the backflow would involve magnons as well.) Now if λ, the coupling constant, exceeded a critical value this polarization could conceivably "localize" the fermions sufficiently so as to cause the noninteracting Fermi sea to lose all qualitative significance; recalling that Tomonaga (1950) had to assume that most ground state configurations were not very different from the unperturbed Fermi sea in order to establish his approximate boson commutation relations, we may safely surmise that when this assumption is no longer valid, a breakdown of ordinary perturbation theory occurs. Fortunately the exact solution given by Mattis and Lieb (1965) could be analyzed at all values of λ, and one result of the analysis concerning the breakdown of perturbation theory was reported in section 5 of that paper. There it was demonstrated that above a certain value of λ, the Fermi surface, which is the characteristic property of an ordinary fermion gas (such as in the independent particle model) *was completely obliterated.* A second result follows from formula (A.4) for the ground state energy W as a function of the Fourier transform $v(p)$ of the interaction potential $\lambda v(x_1 - x_2)$:

$$W_{\text{gr.st.}} = \sum_{p>0} p\{[1 + 2\lambda v(p)\pi^{-1}]^{1/2} - 1\} \tag{4.1}$$

Perturbation theory for the ground state energy consists of expanding W in powers of λ, and calculating the coefficients term by term:

$$W_{\text{gr.st.}}^{\text{pert.th.}} = \sum_{n=0}^{\infty} a_n \lambda^n \tag{4.2}$$

where the coefficients a_n are found to be

$$a_n \propto \int dp\, p[v(p)]^n \tag{4.3}$$

(aside from numerical factors) according to the well-known prescriptions of many-body perturbation theory.[1] This may also be found simply by comparison with the exact expression for $W_{\text{gr. st.}}$ given above. In Mattis (1965) it is shown that for all long-ranged potentials which diverge as p^{-r} when $p \to 0$ the coefficients a_n all will become infinite for $n \geq 2/r$, whereas the exact ground state energy is finite for any $r < 4$; and therefore the perturbation series provides no guide to the exact answer in such cases. Another, subtler, point can be made here: even if all a_n are finite, the perturbation series cannot converge if, for any value of p,

$$|\lambda v(p)| > \pi/2 \tag{4.4}$$

whereas the true answer is an analytic in λ provided only that $\lambda v(p) > -\pi/2$ for all p. This includes the previous observation as a special case, and shows most forcefully the danger inherent in the use of weak-coupling schemes beyond their range of validity.

Delta-Function Model

It must be admitted that the model Hamiltonian we have just been discussing is rather different from the Hamiltonian of a real solid. The delta-function model with the conventional quadratic kinetic energy discussed in Chapter 5, The Bose Gas, however, would be much more realistic, but while it can be solved for bosons, its general solution for fermions with spin has thus far not been given. Nevertheless,

McGuire (1965) has given the solution to this interesting model for the special cases of $S = \frac{1}{2}N$ and $S = \frac{1}{2}N - 1$, where S is the total spin. For the latter case he is able to compute exactly the correlation function between the single spin-down particle and one of the remaining $N - 1$ spin-up particles. In part II of his paper, McGuire (1966) is able to solve the one spin-down problem for an *attractive* delta function (the attractive case has not been solved for bosons) and finds, not surprisingly, that bound states exist between the spin-down particle and one of the spin-up particles. There is no doubt that we should be able to learn many lessons from an exact solution of the problem for arbitrary values of S.

Recently, Flicker and Lieb have solved the two-spin-down problem, and it now seems likely that the n-spin-down problem will also yield to analysis. See Flicker (1966).

Lack of Ferromagnetism in One Dimension

In the last of the reprinted papers of this chapter, Lieb and Mattis (1962) proved two theorems concerning certain eigenfunctions and eigenvalues of the Hamiltonian

$$H = -\frac{\hbar^2}{2m} \sum \frac{\partial^2}{\partial x_i^2} + V(x_1, \ldots, x_N) \tag{4.5}$$

Because of the Pauli principle, any energy level of H will depend on the total spin S (which can be $\frac{1}{2}N, \frac{1}{2}N - 1, \ldots, \frac{1}{2}$, or 0). If we denote the ground state energy for any given S by $E(S)$, then the first theorem asserts that $E(S + 1) > E(S)$. The absolute ground state therefore belongs to $S = 0$ or $\frac{1}{2}$. This result was not entirely unexpected as calculations by Mattheiss (1961) and others on special one-dimensional chains with a finite number of fermions had indicated a distinct *anti*ferromagnetic tendency; the theorem in Lieb and Mattis (1962) now served to show that the lack of ferromagnetism was not an accident, and persisted for all signs and strengths of the interaction potential and for any number of fermions, however large. A somewhat weaker form of the one-dimensional theorem could be proved for fermions in three dimensions, provided the interactions were restricted to a special class of nonradial, "separately symmetric," potentials. However, even subject to this restriction, in three dimensions it was found possible for a state of nonvanishing total spin S to be degenerate with the nonmagnetic ground state; the implications of these exact results in the theory of magnetism have been explored in several reviews,[2] and it is generally agreed that these theorems have at least established the need for something better than Hartree-Fock approximation in the prediction of the occurrence of ferromagnetism in real three-dimensional materials.

[2] N. F. Mott, *Adv. Phys.* **13,** 325 (1964); D. C. Mattis, "The Theory of Magnetism: an Introduction to the Study of Cooperative Phenomena," Harper & Row, New York, 1965; C. Herring, "Exchange Interactions among Itinerant Electrons," *in* "Magnetism" (G. T. Rado and H. Suhl, eds.), Volume IV, Academic Press, New York, 1966.

It should be remarked that the one-dimensional theorem does not preclude the possibility of long-range order or a phase transition or other macroscopic attributes of ferromagnetism. It does not deal with these questions, and it would be fruitless to attempt to do so, for we are already reconciled to the fact that cooperative phenomena cannot take place in one dimension. It does imply that the tendency of a system is always antiferromagnetic (contrary to what one would expect on the basis of exchange integrals, for example). Another way of stating this is that if one were convinced that a given system could be represented—as far as its magnetic properties were concerned—by a Heisenberg spin Hamiltonian, then such a Hamiltonian would have to be of the antiferromagnetic type. In short, if one is to explain ferromagnetism on the basis of ordinary Coulomb forces, then the topology of three dimensions must play an essential role.

General

In addition to the reprinted papers, we give in the bibliography all kinds of material related to the many-body problem. In connection with *superconductivity*, references are given to papers by Fröhlich (1954) and Little (1964) proposing models of one-dimensional superconductors. Papers by Wentzel (1951) and Ferrell (1964) argue, for various reasons, that such behavior is impossible in one dimension, and in fact, superconductivity as well as any long-range order is prohibited in the exactly soluble model of Mattis and Lieb (1965). But Engelsgberg and Varga (1964) who have investigated the Tomonaga model in connection with superconductivity note that certain important matrix elements are missing, making it a poor model for superconductivity; the same can be said of the Luttinger-Mattis-Lieb theories, and thus the last word has not been heard on the subject.

In connection with *spin density waves* we include a reference to a paper by Overhauser (1960) proposing that one-dimensional electrons are unstable against an antiferromagnetic spiral spin density configuration, in the Hartree-Fock approximation. Articles by Kohn and Nettel (1960) and Phillips (1960) present arguments against this instability although there is a tendency today to believe that the weak antiferromagnetism of chromium is due to an instability similar to that discussed by Overhauser.

Other references relate to the general topic of *electron gas*, including Luttinger's (1963) model, and various papers on one-dimensional chains of ions plus electrons by Paul (1960) and Mattheiss (1961).

Field Theory

For completeness, we include the topic of relativistic quantum field theory in the bibliography, as in its many-body aspects it is closely related to the present chapter. The papers on Thirring's famous model are arranged in chronological order, but the reader is advised to start his studies with Scarf and Wess (1962), which reviews some of the difficulties encountered in earlier papers. (In particular, the remarks made herein in connection with Luttinger's (1963) solution of his many-body problem apply *mutatis mutandis* to the work of Glaser (1958) on the relativistic problem, with the reasons for the difficulties being given in Scarf and Wess (1962) and also in Sommerfield (1964). Several other field theories are listed chronologically, along with Thirring's.

Bibliography

I. Many-Body Problem

S. Engelsberg and B. B. Varga (1964). "One-Dimensional Electron-Phonon Model," *Phys. Rev.* **136,** A1582.

R. A. Ferrell (1964). "Possibility of One-Dimensional Superconductivity," *Phys. Rev. Letters* **13,** 330.

M. Flicker (1966). Ph.D. Thesis, Yeshiva University (unpublished).

H. Fröhlich (1954). "On the Theory of Superconductivity: the One-Dimensional Case," *Proc. Roy. Soc.* (*London*) **A223,** 296.

W. Kohn and S. J. Nettel (1960). "Giant Fluctuations in a Degenerate Fermi Gas," *Phys. Rev. Letters* **5,** 8.

*E. H. Lieb and D. C. Mattis (1962). "Theory of Ferromagnetism and the Ordering of Electronic Energy Levels," *Phys. Rev.* **125,** 164.

W. A. Little (1964). "Possibility of Synthesizing an Organic Superconductor," *Phys. Rev.* **134,** A1416.

J. M. Luttinger (1963). "An Exactly Soluble Model of a Many-Fermion System," *J. Math. Phys.* **4,** 1154.

J. B. McGuire (1965). "Interacting Fermions in One Dimensions, I, Repulsive Potential," *J. Math. Phys.* **6,** 432.

J. B. McGuire (1966). "Interacting Fermions in One Dimension, II, Attractive Potential," *J. Math. Phys.* **7,** 123.

L. F. Mattheiss (1961). "Antiferromagnetic Linear Chain," *Phys. Rev.* **123,** 1209; "Effective Exchange Integral," *ibid.* 1219.

D. C. Mattis (1964). "Band Theory of Magnetism in Metals in Context of Exactly Soluble Model," *Physics* **1,** 183.

D. C. Mattis (1965). "Reply to A. W. Overhauser," *Physics* **1,** 311.

*D. C. Mattis and E. H. Lieb (1965). "Exact Solution of a Many-Fermion System and Its Associated Boson Field," *J. Math. Phys.* **6,** 304.

A. W. Overhauser (1960). "Giant Spin Density Waves," *Phys. Rev. Letters* **4,** 462.

A. W. Overhauser (1965). "Note on the Band Theory of Magnetism," *Physics* **1,** 307.

D. I. Paul (1960). "Theory of Magnetism and the Ground State Energy of a Linear Chain," *Phys. Rev.* **118,** 92; "Magnetism and the Ground State Energy of a Linear Chain, II, Configuration Interaction Effects," *ibid.* **120,** 463.

J. C. Phillips (1960). "Giant Spin Density Fluctuations," *Phil Mag.* **5,** 1193.

*S. Tomonaga (1950). "Remarks on Bloch's Method of Sound Waves Applied to Many-Fermion Problems," *Progr. Theoret. Phys.* (*Kyoto*) **5,** 544.

G. Wentzel (1951). "Interaction of Lattice Vibrations with Electrons in a Metal," *Phys. Rev.* **83,** 168(L).

II. Thirring's Model of Relativistic Field Theory

W. Thirring (1958). "A Soluble Relativistic Field Theory." "The theory of a self-coupled spinor field with interaction $\lambda\bar{\psi}\psi\bar{\psi}\psi$ is solved in one space and one time dimension. A representation of the eigenstates of the energy is given. Some observable quantities are calculated and found to be finite after renormalization; they can be expanded in powers of λ with a finite radius of convergence." *Ann. Phys.* (*N.Y.*) **3,** 91.

M. Maier and D. Shirkov (1958). *Dokl. Akad. Nauk SSSR* **122,** 45 [English transl. *Soviet Phys.— "Doklady"* **3,** 931 (1959)]; W. Thirring (1958). *Nuovo Cimento* **9,** 1007; V. Glaser (1958) *ibid.* 990; I. Bialynicki-Birula (1958). *ibid.* **10,** 1150; T. Pradhan (1958). *Nucl. Phys.* **9,** 124; F. Scarf (1959). Phys. Rev. **115,** 463; **117,** 868; K. Johnson (1961). *Nuovo Cimento* **20,** 773; F. Berezin (1961). Zhur. *Eksper. Teor. Fiz.* **40,** 885 [English transl. *Soviet Phys.—JETP* **13,** 620 (1961)]; F. Scarf and J. Wess (1962). *Nuovo. Cimento* **26,** 150; C. Sommerfield (1964). *Ann. Phys.* (*N.Y.*) **26,** 1.

III. Other Field Theories

A. Ansel'm (1958). *Zh. Eksperim. i Teor. Fiz.* **36,** 863. "Two fermion fields are considered in a space of one dimension (and in time), with interaction of each field with itself and of the two with each other. The first term in the expansion of the vertex part in an asymptotic series of well-known form is obtained. It is shown that within certain limits the renormalized charge can have arbitrary (nonvanishing) value." [English transl., *Soviet Phys.—JETP* **9,** 608 (1959)].

V. Vaks and A. Larkin (1961). *Zh. Eksperim. i Teor. Fiz.* **40,** 1392. "The methods of the theory of superconductivity are used to find the mass of a particle in renormalizable theories with vanishing bare mass and involving a weak coupling constant. Only the zero solution for the mass is possible in electrodynamics and in the one-dimensional Thirring model; for the one-dimensional model of two interacting fields a nonvanishing solution is obtained. A finite expression for the charge has been obtained for this model." [English transl., *Soviet Phys.—JETP* **13,** 979 (1961).]

P. Federbush (1961). "A field theory is considered . . . in limit as masses of the theory go to zero, it approaches the Thirring model . . . *S*-matrix can be exhibited exactly . . . and leads to no physical scattering . . . renormalized current operator displayed as suitable limit of products of renormalized field operators. The form of the result clearly separates the consistency problem in quantum electrodynamics from that of the 'photon mass'." *Phys. Rev.* **121,** 1247.

P. Federbush (1961). "The field operators in two non-relativistic field theory models satisfy local field equations . . . in this paper we show the same results for two relativistic models." *Progr. Theoret. Phys.* (*Kyoto*) **26,** 148.

J. Schwinger (1962). "It has been remarked that gauge invariance does not necessarily require the existence of a massless physical particle . . . we give a specific model for which an exact solution affirms this possibility . . . it is the situation of electrodynamics in one spatial dimension, where the charge-bearing Dirac field has no associated mass constant. This example is rather unique since it is a simple model for which there is an exact divergence-free solution." *Phys. Rev.* **128,** 2425.

A. Wightman (1964). "Introduction to Some Aspects of the Relativistic Dynamics of Quantized Fields." A review of the subject in the form of notes for lectures at French Summer School of Theoretical Physics, Cargèse, Corsica. Here the reader will find a most complete bibliography of one-dimensional field theory.

Progress of Theoretical Physics Vol. 5, No. 4, July~August, 1950

Remarks on Bloch's Method of Sound Waves applied to Many-Fermion Problems

Sin-itiro TOMONAGA

Institute for Advanced Study
Princeton, N. J.
U.S.A.*

(Received June 29, 1950)

Abstract

The fact implied by Bloch several years ago that in some approximate sense the behavior of an assembly of Fermi particles can be described by a quantized field of sound waves in the Fermi gas, where the sound field obeys Bose statistics, is proved in the one-dimensional case. This fact provides us with a new possibility of treating an assembly of Fermi particles in terms of the equivalent assembly of Bose particles, namely, the assembly of sound quanta. The field equation for the sound wave is found to be linear irrespective of the absence or presence of mutual interaction between particles, so that this method is a very useful means of dealing with many-Fermion problems. It is also applicable to the case where the interparticle force is not weak. In the case of force of too short a range this method fails.

§ 1. Introduction and summary.

The well-known method of Thomas and Fermi provides us with a very practical approximate treatment of many-Fermion problems. Because in this approximation, however, each particle is supposed to move independently, the effect of interaction between particles being simply replaced by an average field of force, one cannot speak of correlations between particles in this rough approximation. This simple method does not apply to problems in which inter-particle correlation plays an important role. A step toward the improvement of the method so as to include the correlation was taken by Euler[1], who calculated the effect of inter-particle interaction, which causes the correlation, by perturbation theory assuming the interaction to be small.

The calculation of such a type is carried out in the following manner: In the zeroth approximation each particle is in some one-particle quantum state, which we shall call a " level ". Let us consider the lowest state for the sake of definitness. In this state all levels up to some highest one, which shall be called the " Fermi maximum ", are each filled by one particle. The perturbation energy, the energy due to the inter-particle force, has non-vanishing matrix

* Now returned to the Tokyo University of Education (Tokyo Bunrika Daigaku), Tokyo.

elements which cause virtual transitions to states in which two particles are excited simultaneously to levels higher than the Fermi maximum.

The state which results from the inter-particle force is thus such one which is a superposition of the zero-order state and various excited states in which holes and particles are present in some levels below and above the Fermi maximum respectively. In the lowest approximation of the perturbation calculation, the numbers of holes and excited particles are two; but, if the calculation is carried out to higher order, which is necessary when the inter-particle force is not small, there appear states in which a considerable number of holes and excited particles are present. The amplitudes of such highly excited states will be appreciable if the inter-particle force is strong.

Such a mixture of excited states gives rise to correlation between particles. We can in this way include the effects of correlations in treating certain problems. In the case of strong inter-particle forces, however, it is necessary to carry through the perturbation calculation up to a very high order, and this is too involved to be practicable. Recently, Nogami[2] proposed a method which applies also to cases of rather strong inter-particle forces. Still his method requires some kind of weakness of interaction because he had to neglect the interactions of holes and of excited particles with each other as well as the interactions between holes and exited particles. This neglect cannot be justified when too many holes and excited particles are present.

In such a situation it is desirable to find some approximate method of dealing with many-Fermion problems different from the perturbation method.

In his famous work on the stopping power of charged particles, Bloch[3] has treated the excited states of the Fermi gas not as states with holes and excited particles, but as states in which the gas oscillates. In this work it was not necessary to treat the oscillation quantum-theoretically. In a later paper[4], he also dealt with a problem in which the quantum aspect of the oscillation was essential. This was a problem in which the density fluctuation played a role. He showed in this work that the density fluctuation of a degenerate Fermi gas (in his theory it was sufficient to treat the gas without inter-particle force) can be calculated in two different ways giving the same result. The one was the orthodox method whereby one calculates directly the expectation value of the operator having expressed in terms of the quantized field variables ψ and ψ^* describing the assembly of the Fermi particles. The other method was to calculate the zero-point amplitudes of sound waves in the gas, the equations of motion for the sound waves being properly chosen. He showed that the correct value of the fluctuation was obtained in this way if the choice of the equations of motion for the sound waves was properly made and if the zero-point amplitudes of the waves had such values as would be expected for a sound field obeying Bose statistics.

If it is proved that the excited states of an assembly of Fermi particles can be in fact described as an excitation of sound waves, where the sound can be described by a Bose field, it will provide us with a new method of treating many-Fermion problems, not dealing directly with the assembly of the Fermi particles, but dealing with the equivalent assembly of Bose particles, i. e. the assembly of sound quanta.

There is a prospect that in this latter method the assumption of the weakness of inter-pariicle forces will not be essential. This prospect lies in the following situation: It is expected that the equations of motion for the sound waves will be linear in the field variables describing the sound field; otherwise we could not speak of waves at all. The field variables for the sound field will be the density ρ of the gas and its properly defined canonical conjugate. Now the essential point is that the linearity of the problem will not be destroyed even when an inter-paticle force is present, because the interaction energy between particles is bilinear in ρ. This fact is in marked contrast to the circumstance that in terms of ψ and ψ^* the interaction energy contains four ψ's, or, more precisely, two ψ's and twe ψ^*'s; thus the field equations for ψ and ψ^* are no longer linear when an inter-particle force is present. This circumstance made it very difficult to treat many-Fermion problems with inter-particle forces. This difficulty will disappear when we deal with the problem in terms of the ρ field but not of the ψ field.

The purpose of this paper is to show that these expectations are really fulfilled. It will be shown that in some approximation, which does not necessarily require the weakness of the interaction, excitated states of Fermi gas are in fact equivalent to corresponding excitations of sound waves, and that the sound is describable by a Bose field whose field equations are linear in the field variables irrespective of the absence or presence of interparticle forces. Thus, Bloch's method of sound waves will be a very useful method for many-Fermion problems.

The possibility of this new method was found independently by Bohm[5], who discusses a very interesting phenomenon of plasma-like oscillations in a degenerate electron gas from a very similar point of view. He, too, gives a proof of the fact that these oscillations are describable by a Bose field in some approximation. He utilizes further the linearity of the field equations to study such a pure correlation phenomenon as plasma oscillations of an electron gas.

The present paper is of a rather more mathematical nature than physical in that it aims mainly at the analysis of the mathematical structure of the method, clarifying the underlying assumptions and the limit of applicability. A mathematically closed and clear-cut presentation of the theory is achieved, however, at the expense of physical usefulness, because, thus far, the author has succeded only in giving a complete formulation for a one-dimensional assembly of particles.

It is rather certain that a similar method applies to the three dimensional case too—this, indeed, has been done by Bohm but the situation is more complicated in this case.

The mathematicl relation between the field of sound quanta on the one hand and the original field of Fermi particles on the other is very similar to the relation between the field of light quanta and the field of neutrinos in the neutrino theory of light[6]. One will find everywhere a marked parallelism between our theory and the neutrino theory of light.

The discussions will be performed in several steps: In § 2 we shall prove that the sound can be described by a Bose field under some assumptions imposed on the states under consideration. Then we shall set up in § 3 the Hamiltonian for the sound field. We shall see that the Hamiltonian is in fact bilinear in the field variables, so that the whole problem is liner. In § 4 we shall show that this Hamiltonian is really equal to the original Hamiltonian for the assembly of Fermi particles, and, therefore, the assembly of sound quanta is equivalent to the original assembly of Fermi particles. In § 5 we will go over to the solution of the eigenvalue problem. This can be done very simply by finding the normal coordinates for the sound field, because the field equations are linear. In § 6 we shall give several general formulae of physical interest which are derived directly by our method. A criterion for the applicability of the method will also be given in this section. In the last section we shall briefly mention the bearing of our results on the plasma oscillations treated by Bohm. Also, the relation between the two kinds of descriptions of the system, one as an assembly of Fermi particles and the other as an assembly of sound quanta, is discussed briefly.

In § 6 we shall see that our method does not work if the inter-particle force is of short range. The range of force must be larger than four times the mean distance between particles. Since sound waves with wave length shorter than the mean distance between particles have no meaning, it is quite conceivable that the method of sound waves fails in describing the event occuring in a small space region comparable with the mean distance between particles. This is the reason why the method fails in the case of short range force. The method is, on the other hand, very suitable for dealing with the case of long-range force in which a considerable number of holes and excited particles are present in the neighborhood of the Fermi maximum. Thus the present method covers a field where the known methods have failed.

§ 2. Approximate commutation relations for the density field.

Let us consider an assembly of Fermi particles in a one-dimensional "box" of length L. Let $\psi(x)$ and $\psi^*(x)$ be the quantized wave functions describing the assembly. The density $\rho(x)$ of the particles is then given by

$$\rho(x)=\psi^*(x)\psi(x). \tag{2.1}$$

We introduce the Fourier transforms ψ_n and $\psi_n{}^*$ of $\psi(x)$ and $\psi^*(x)$ respectively. They are defined by

$$\begin{cases} \psi(x)=\dfrac{1}{\sqrt{L}}\sum\limits_n \psi_n \exp\left(\dfrac{2\pi i}{L}nx\right), \\ \psi^*(x)=\dfrac{1}{\sqrt{L}}\sum\limits_n \psi_n^* \exp\left(-\dfrac{2\pi i}{L}nx\right), \end{cases} \qquad n=0,\ \pm1,\ \pm2,\cdots. \tag{2.2}$$

The Fourier transform of $\rho(x)$, which is defined by

$$\rho(x)=\frac{1}{L}\sum_n \rho_n \exp\left(\frac{2\pi i}{L}nx\right), \tag{2.3}$$

is evidently given by

$$\rho_n=\sum_{n'}\psi_{n'}^*\psi_{n'+n}=\sum_{\bar{n}}\psi^*_{\bar{n}-\frac{n}{2}}\psi_{\bar{n}+\frac{n}{2}}, \tag{2.4}$$

where $\bar{n}$ takes integral values when n is even and half-odd integral values when n is odd.

Our purpose is to describe the system in terms of the density field $\rho(x)$ instead of describing it by the wave field $\psi(x)$. The first task is to find the commutation relations between the field quantity $\rho(x)$ and its properly defined canonical conjugate. For this purpose we separate each ρ into two parts ρ_n^+ and ρ_n^- by means of

$$\begin{cases} \rho_n^+=\sum\limits_{\bar{n}>0}\psi^*_{\bar{n}-\frac{n}{2}}\psi_{\bar{n}+\frac{n}{2}}, \\ \rho_n^-=\sum\limits_{\bar{n}<0}\psi^*_{\bar{n}-\frac{n}{2}}\psi_{\bar{n}+\frac{n}{2}}. \end{cases} \tag{2.5}$$

We have evidently

$$\rho_n=\rho_n^+ + \rho_n^- \tag{2.6}$$

(in case of even n the term with $\bar{n}=0$ is absent in (2.6), but this does not cause any serious error.) We have further

$$\rho_n^+=\rho_{-n}^{+*}, \qquad \rho_n^-=\rho_{-n}^{-*}. \tag{2.7}$$

It will be seen later that the separation of ρ_n by (2.6) corresponds to the separation of the field into parts with positive and negative frequencies, which is the usual procedure in field theory. (See (3.1) and (3.2)).

We now examine the commutation relations of the ρ's. The commutation relations between the ψ's and ψ^*'s are

$$\begin{cases} [\psi_n^*,\ \psi_{n'}]_+=\delta_{n,n'}, \\ [\psi_n^*,\ \psi_{n'}^*]_+=[\psi_n,\ \psi_{n'}]_+=0. \end{cases} \tag{2.8}$$

From these relations we get the commutation relations of the ρ's. For the ρ^+'s they are

$$[\rho_n^+,\ \rho_{n'}^+]=\begin{cases} \sum\limits_{-\frac{n}{2}<\bar{n}\leqq\frac{|n'|}{2}}\psi^*_{\bar{n}-\frac{n}{2}-\frac{n'}{2}}\psi_{\bar{n}+\frac{n}{2}+\frac{n'}{2}} & \text{for } n>0,\ n'\leqq -n, \\ \sum\limits_{\frac{n'}{2}<\bar{n}\leqq\frac{n}{2}}\psi^*_{\bar{n}-\frac{n}{2}-\frac{n'}{2}}\psi_{\bar{n}+\frac{n}{2}+\frac{n'}{2}} & \text{for } n>0,\ -n\leqq n\leqq n, \\ -\sum\limits_{\frac{n}{2}<\bar{n}\leqq\frac{n'}{2}}\psi^*_{\bar{n}-\frac{n}{2}-\frac{n'}{2}}\psi_{\bar{n}+\frac{n}{2}+\frac{n'}{2}} & \text{for } n>0,\ n'\geqq n, \end{cases} \tag{2.9}$$

which, for the special cases of $n'=-n$, reduce to

$$[\rho_n^+, \rho_{-n}^+]= \sum_{-\frac{n}{2}<\bar{n}\leqq\frac{n}{2}} \psi_{\bar{n}}^*\psi_{\bar{n}} \qquad \text{for } n>0. \tag{2.9'}$$

In the same way, we get for the ρ^-'s

$$[\rho_n^-, \rho_{n'}^-]=\begin{cases} -\sum\limits_{\frac{n'}{2}\leqq\bar{n}<\frac{n}{2}} \psi^*_{\bar{n}-\frac{n}{2}-\frac{n'}{2}} \psi_{\bar{n}+\frac{n}{2}+\frac{n'}{2}} & \text{for } n>0,\ n'\geqq -n, \\ -\sum\limits_{-\frac{n}{2}\leqq\bar{n}<-\frac{n'}{2}} \psi^*_{\bar{n}-\frac{n}{2}-\frac{n'}{2}} \psi_{\bar{n}+\frac{n}{2}+\frac{n'}{2}} & \text{for } n>0,\ -n\leqq n'\leqq n, \\ \sum\limits_{-\frac{n'}{2}\leqq\bar{n}<-\frac{n}{2}} \psi^*_{\bar{n}-\frac{n}{2}-\frac{n'}{2}} \psi_{\bar{n}+\frac{n}{2}+\frac{n'}{2}} & \text{for } n>0,\ n'\geqq n, \end{cases} \tag{2.10}$$

and

$$[\rho_n^-, \rho_{-n}^-]= -\sum_{-\frac{n}{2}\leqq\bar{n}<\frac{n}{2}} \psi_{\bar{n}}^*\, \psi_{\bar{n}} \qquad n>0. \tag{2.10'}$$

The commutation relations between the ρ^+'s and ρ^-'s are found to be

$$[\rho_n^-, \rho_{n'}^+]=\begin{cases} \sum\limits_{-\frac{n}{2}<\bar{n}<\frac{n'}{2}} \psi^*_{\bar{n}-\frac{n}{2}-\frac{n'}{2}} \psi_{\bar{n}+\frac{n}{2}+\frac{n'}{2}} & \text{for } -n<n', \\ -\sum\limits_{\frac{n}{2}<\bar{n}<-\frac{n'}{2}} \psi^*_{\bar{n}-\frac{n}{2}-\frac{n'}{2}} \psi_{\bar{n}+\frac{n}{2}+\frac{n'}{2}} & \text{for } n<-n', \end{cases} \tag{2.11'}$$

and

$$[\rho_n^-, \rho_{-n}^+]=0. \tag{2.11'}$$

We now show that the commutation relations (2.9)—(2.11′) can be replaced by simpler ones if the states under consideration satisfy, at least approximately, some conditions which will be specified below. These simplified commutation relations tell us that the density field can be regarded as a Bose-field under these restricting conditions.

We first consider the case of the ideal Fermi gas in which there are no interactions between particles. If the gas is not excited too highly, only particles in the neighborhood of the Fermi maximum are raised to higher levels. There exist holes and excited particles only in the neighborhood of the surface of the Fermi sea. Now, in the case of a non-ideal Fermi gas, the inter-particle forces cause virtual transitions of particles. Thus extra holes and exited particles appear. But, if the range of the inter-particle force is not too short and the force itself is not too strong, these virtual holes and excited particles are still present only in the neighborhood of the Fermi maximum. Such are the states to which we will confine ourselves.

If we confine ourselves to states of such type, we can simplify the commutation relations (2.9)—(2.11′) in the following manner.

Let us consider, for instance, the first commutation relations $[\rho_n^+, \rho_{n'}^+]$ for which $n>0$ and $-n<n'<n$. We notice that the expression on the right-

hand side is a sum of operators each bringing one particle from the level $\bar{n}+\frac{n}{2}+\frac{n'}{2}$ to the level $\bar{n}-\frac{n}{2}-\frac{n'}{2}$. Because the summation over n is extended only between $\frac{n'}{2}$ and $\frac{n}{2}$, the final levels $\bar{n}-\frac{n}{2}-\frac{n'}{2}$ lie in a limited interval between $-\frac{n}{2}$ and $-\frac{n'}{2}$. Now, let $n_{\max}$ denote the value of $|n|$ at the Fermi maximum. Then, if n and $|n'|$ are both sufficiently small compared with $n_{\max}$ (the discussion about how small they should be will be given below), the levels $-\frac{n}{2}$ and $-\frac{n'}{2}$ both lie deep in the bottom of the Fermi sea where there are holes. In such a case the operator $\psi^{+}_{\bar{n}-\frac{n}{2}-\frac{n'}{2}}\psi_{\bar{n}+\frac{n}{2}+\frac{n'}{2}}$ will give a vanishing result because the final level is occupied. Thus, for the states under consideration our commutators $[\rho_n^+, \rho_{n'}^+]$ are équivalent to zero.

We next consider $[\rho_n^+, \rho_{-n}^+]$. In this case, the right-hand side is

$$\sum_{-\frac{n}{2}<\bar{n}\leq\frac{n}{2}} \psi_{\bar{n}}{}^*\psi_{\bar{n}} = \sum_{-\frac{n}{2}<\bar{n}\leq\frac{n}{2}} N_{\bar{n}}\ , \tag{2.9''}$$

where $N_{\bar{n}}$ is the occupation number of the level $\bar{n}$. Since the level $\bar{n}$, which lies between $-\frac{n}{2}$ and $\frac{n}{2}$, lies deep in the Fermi sea if n is small compared with $n_{\max}$, it is occupied by one particle. Then the sum $\sum N_{\bar{n}}$ is simply equal to the number of levels between $-\frac{n}{2}$ and $\frac{n}{2}$, which is just n. So we find that $[\rho_n^+, \rho_{-n}^+]$ is equivalent to n.

A similar consideration applies to the remaining commutators. We can see that the following commutation relations hold in the sense of equivalence:

$$\begin{aligned} &[\rho_n^+, \rho_{n'}^+]=n\delta_{n,-n'} \\ &[\rho_n^-, \rho_{n'}^-]=-n\delta_{n,-n'} \\ &[\rho_n^+, \rho_{n'}^-]=0. \end{aligned} \tag{2.12}$$

In order that these simpler commutation relations can be used instead of the original ones, it was necessary that $|n|$ and $|n'|$ be sufficiently small compared with $n_{\max}$. We now discass this point more quantitatively.

First we notice that the total number N of the particles is related to $n_{\max}$ by

$$N=2n_{\max}+1, \tag{2.13}$$

which we shall make use of in later considerations.

We now assume that in the states under consideration there are no holes under the level specified by $|n|=\alpha n_{\max}$, α being a positive number less than unity. Then, we can easily see that

$$\frac{3}{2}|n|,\ \frac{3}{2}|n'| < \alpha\, n_{max} \tag{2.14}$$

is the required condition for the validity of the simplar commutation relations. (The factor 3/2 is required in order that (2.11) be equivalent to the third relation of (2.12)).

So far as the present consideration is concerned, there is no restriction on α. But we shall see later that α must be 3/4 in order that the whole treatment work consistently. The required conditions will be found to be the following:

(I) In the region $|n| < \frac{3}{4} n_{max}$, there should be no holes.

(II) In the region $|n| > \frac{5}{4} n_{max}$, there should be no excited paiticles.

(III) The absolute values of n in ρ_n^+ and ρ_n^- should not exceed $\frac{1}{2} n_{max}$:

$$|n| < \frac{1}{2} n_{max}. \tag{2.15}$$

Our method works when and only when these conditions are satisfied. The conditions (I) and (II) restrict the states; therefore, it is always necessary in applying our method to verify whether the states obtained as an answer do really satisfy these conditions. The condition (III) requires that no sound waves having shorter wave length than $2L/n_{max}$ play a role in the problem. This condition will be satisfied if the range of inter-particle force is sufficiently long. The reason for the the necessity of the special choice of $\alpha=3/4$ will be given later.

§ 3. Equation of motion and Hamiltonian for the density field.

We first consider the case of non-interacting particles. In this case the change with time of ρ_n^+ and ρ_n^- can be obtained easily because we know the change with time of ψ_n and $\psi_n{}^*$. In the general case we them find a very complicated time dependence of the ρ's. The motions of ρ_n^+ and ρ_n^- are by no means simple harmonic because each term of $\sum\psi^*_{\bar{n}-\frac{n}{2}}\,\psi_{\bar{n}+\frac{n}{2}}$ has a frequency which depends not only on n but also on $\bar{n}$. This fact means that the density does by no means behave like waves.

However, we may here make use of our conditions imposed on the states. According to them holes and excited particles are present in a narrow interval only, from $(3/4)n_{max}$ to $(5/4)n_{max}$, in the neighborhood of n_{max}. This results in non-vanishing matrix elements in $\sum\psi^*_{\bar{n}-\frac{n}{2}}\,\psi_{\bar{n}+\frac{n}{2}}$ being contributed solely by terms with $\bar{n}\pm\frac{n}{2}$ nearly equal to n_{max} or $-n_{max}$ according as we are dealing with ρ_n^+ or ρ_n^-. We may then approximate the frequencies of ψ^* and ψ by their expansion in the neighborhood of $\bar{n}\pm\frac{n}{2}=n_{\max}$ or $\bar{n}\pm\frac{n}{2}=-n_{\max}$ respectively. The frequency of $\psi_{\bar{n}+\frac{n}{2}}$ is, for instance,

$$\frac{1}{2\pi\hbar}\frac{1}{2m}\left(\frac{2\pi\hbar}{L}\right)^2\left(\bar{n}+\frac{n}{2}\right)^2,$$

m being the mass of the particles. If we here expand $\left(\bar{n}+\frac{n}{2}\right)^2$ in the neighborhood of $\bar{n}+\frac{n}{2}=n_{\max}$, we get

$$\left(\bar{n}+\frac{n}{2}\right)^2=\left\{n_{\max}+\left(\bar{n}+\frac{n}{2}-n_{\max}\right)\right\}^2$$
$$=n_{\max}^2+2n_{\max}\left(\bar{n}+\frac{n}{2}-n_{\max}\right)+\cdots.$$

Neglecting the small term $\left(\bar{n}+\frac{n}{2}-n_{\max}\right)^2$, the time dependence of $\psi_{\bar{n}+\frac{n}{2}}$ becomes

$$\psi_{\bar{n}+\frac{n}{2}}\sim\exp i\left[+\frac{1}{\hbar}\left(\frac{2\pi\hbar}{L}\right)^2\frac{n_{\max}^2}{2m}-\frac{1}{\hbar}\left(\frac{2\pi\hbar}{L}\right)^2\frac{n_{\max}}{m}\left(\bar{n}+\frac{n}{2}\right)\right]t.$$

In the same way we have

$$\psi_{\bar{n}-\frac{n}{2}}\sim\exp i\left[-\frac{1}{\hbar}\left(\frac{2\pi\hbar}{L}\right)^2\frac{n_{\max}^2}{2m}+\frac{1}{\hbar}\left(\frac{2\pi\hbar}{L}\right)^2\frac{n_{\max}}{m}\left(\bar{n}-\frac{n}{2}\right)\right]t.$$

Then, combining these two, we get a time dependence for ρ_n^+ of the form

$$\rho_n^+\approx\exp\left[-\frac{i}{\hbar}\left(\frac{2\pi\hbar}{L}\right)^2\frac{n_{\max}}{m}n\right]t. \tag{3.1}$$

By the same consideration using the expansion in the neighborhood of $\bar{n}\pm\frac{n}{2}=-n_{\max}$, we get

$$\rho_n^-\approx\exp\left[+\frac{i}{\hbar}\left(\frac{2\pi\hbar}{L}\right)^2\frac{n_{\max}}{m}n\right]t. \tag{3.2}$$

From (3.1) and (3.2) we see that ρ^+ and ρ^- satisfy the equations of motion

$$\begin{cases}\dot{\rho}_n^+=-\frac{i}{\hbar}\left(\frac{2\pi\hbar}{L}\right)^2\frac{n_{\max}}{m}n\rho_n^+\\ \dot{\rho}_n^-=+\frac{i}{\hbar}\left(\frac{2\pi\hbar}{L}\right)^2\frac{n_{\max}}{m}n\rho_n^-.\end{cases} \tag{3.3}$$

We notice here that the commutation relations (2.12) lead to the fact that the canonically conjugate momentum for

$$\rho_n=\rho_n^++\rho_n^- \tag{3.4}$$

is given by

$$\pi_n = \frac{i}{2}\frac{1}{n}\left(\rho_{-n}^{+} - \rho_{-n}^{-}\right). \tag{3.5}$$

As can be easily verified, we see that

$$[\rho_n, \pi_{n'}] = i\delta_{n,n'}. \tag{3.6}$$

By the equation of motion we then find

$$\pi_n = \frac{\hbar}{2}\left(\frac{L}{2\pi\hbar}\right)^2 \frac{m}{n_{\max}} \frac{1}{n^2}\dot{\rho}_{-n}. \tag{3.5'}$$

The last relation gives a physical meaning to our canonical momenta: As usual, momenta are proportional to the time derivatives of the coordinates. Though the equations of motion of the form of (3.3) hold only for the case of non-interacting particles, we shall see later that (3.5′) holds more generally.

We shall now set up the Hamiltonian for the ρ field. The Hamiltonian is determined by the requirement that

$$\frac{\hbar}{i}\dot{\rho}_n^{\pm} = [\mathfrak{H}, \rho_n^{\pm}] \tag{3.7}$$

yield the equations of motion (3.3) wtih the help of the commutation relations (2.12). We then find that the Hamiltonian has a very simple form:

$$\mathfrak{H} = \left(\frac{2\pi\hbar}{L}\right)^2 \frac{n_{\max}}{m} \sum_{n>0} \left(\rho_{-n}^{+}\rho_n^{+} + \rho_n^{-}\rho_{-n}^{-}\right). \tag{3.8}$$

The order of non-commuting factors is here specified by the condition that $\mathfrak{H}$ has no zero point value.

Though we have set up the Hamiltonian in this way, it is by no means self-evident that this Hamiltonian gives in fact the energy of the system (or eventually energy plus some additive constant). That this is actually the case will be proved in the next section.

So far we have neglected the mutual interaction between particles. It is quite simple to introduce the interaction into the theory, when the interaction force is of the ordinary type, neither of exchange type nor velocity dependent. In this case the interaction energy has the form

$$H_{\text{int}} = \frac{1}{2}\iint \rho(x)\rho(x')J(|x-x'|)\,dx\,dx' - \frac{1}{2}\int \rho(x)J(0)\,dx, \tag{3.9}$$

$J(|x-x'|)$ being the potential of the inter-particle force. The term $\frac{1}{2}\int \rho(x)J(0)dx$ is subtracted in order to remove the interaction of a particle with itself. In terms of the Fourier transform we find

$$H_{\text{int}} = \frac{1}{2}\sum_n{}' \rho_n\rho_{-n}J_n + \frac{1}{2}\rho_0^2 J_0 - \frac{1}{2}\rho_0 J(0), \tag{3.10}$$

where J_n is the matrix element of the potential defined by

$$J_n=\frac{1}{L}\int J(x)\exp\left(\frac{2\pi i}{L}nx\right)dx=J_{-n}. \tag{3.11}$$

The prime on the $\sum'$ symbol means that the term with $n=0$ should be omitted. ρ_0 is the quantity defined by

$$\rho_0=\sum_{\bar{n}}\psi_{\bar{n}}^*\psi_{\bar{n}}=N, \tag{3.12}$$

i.e. the total number of particles.

Since H_{int} commutes with each ρ_n, which is the sum $\rho_n^+ + \rho_n^-$, the equation of motion for ρ_n is not affected by the interaction, while the equations of motion for ρ_n^+ and ρ_n^- separately are affected by the interaction. This fact means that the canonically conjugate momenta π_n are still given by (3.5′) irrespective of the presence or absence of the interaction.

In terms of ρ^+ and ρ^- the interaction Hamiltonian can be expressed as

$$\begin{aligned}H_{\text{int}}=&\sum_{n>0}J_n(\rho_{-n}^+\rho_n^+ + \rho_n^-\rho_{-n}^- + \rho_n^+\rho_{-n}^- + \rho_n^-\rho_{-n}^+)\\ &+\sum_{n>0}nJ_n+\frac{1}{2}N^2J_0-\frac{1}{2}NJ(0).\end{aligned} \tag{3.13}$$

The term $\sum_{n>0} nJ_n$ appeared when we performed the rearrangement of factors in $\rho_n^+\rho_{-n}^+$ and $\rho_{-n}^-\rho_n^-$ into the correct order.

It is the essential point of our method that the energy, the "kinetic part" $\mathfrak{H}$ as well as the "potential part" H_{int}, is bilinear in the $\rho^{\pm}$'s. This fact is in marked contrast to the fact that, in terms of ψ and ψ^*, only the kinetic part is bilinear; the potential part contains four ψ's. This latter fact makes it necessary to solve a complicated non-linear problem if one wishes to deal with a many-particle problem with inter-particle force. Now that we have found the Hamiltonian to be bilinear in the ρ's, irrespective of the presence or absence of the inter-particle force, we have no such difficulty if we deal with the problem in terms of the ρ field not in terms of the ψ field. The problem is simply to perform the principal-axes transformation of the bilinear form, i. e. to find the normal coordinates for the ρ field.

Before we enter into this problem, we shall show that the Hamiltonian $\mathfrak{H}$ is in fact equal to the kinetic energy of the system minus some constant which depends only on the total number of particles.

§ 4. Energy and momentum of the system.

In this section we give the proof that the Hamiltonian $\mathfrak{H}$ really gives the kinetic energy of the system. This can be done by a straightforward calculation in the following manner.

For this purpose we first examine $\sum \rho_{-n}^+\rho_n^+$. We examine the diagonal and the non-diagonal parts of $\sum \rho_{-n}^{+}\rho_{n}^{+}$ separately, where the matrix is supposed to be referred to in the representation in which the occupation numbers of the particles are diagonal.

We have

$$\sum_{\frac{n_{\max}}{2}>n>0} \rho_{-n}^{+}\rho_{n}^{+} = \sum_{\frac{n_{\max}}{2}>n>0} \sum_{\bar{n}>0} \sum_{\bar{n}'>0} \psi^{*}_{\bar{n}+\frac{n}{2}} \psi_{\bar{n}-\frac{n}{2}} \psi^{*}_{\bar{n}'-\frac{n}{2}} \psi_{\bar{n}'+\frac{n}{2}}$$

$$= \sum_{\frac{n_{\max}}{2}>n>0} \Big(\sum_{\bar{n}>0} \sum_{\bar{n}'>0}\Big)' \psi^{*}_{\bar{n}+\frac{n}{2}} \psi_{\bar{n}-\frac{n}{2}} \psi^{*}_{\bar{n}'-\frac{n}{2}} \psi_{\bar{n}'+\frac{n}{2}}$$

$$+ \sum_{\frac{n_{\max}}{2}>n>0} \sum_{\bar{n}>0} (1-N_{\bar{n}-\frac{n}{2}})N_{\bar{n}+\frac{n}{2}}. \tag{4.1}$$

The prime on $(\sum \sum)'$ means that terms with $\bar{n}=\bar{n}'$ are omitted. By this omission the first sum on the right-hand side of (4.1) gives the non-diagonal part and the second sum the diagonal part of $\sum \rho_{-n}^{+}\rho_{n}^{+}$. The summation over n is to be extended only up to $n_{\max}/2$ because of our condition (2.15) (III).

We first observe the diagonal part

$$D^{+} = \sum_{\frac{n_{\max}}{2}>n>0} \sum_{\bar{n}>0} (1-N_{\bar{n}-\frac{n}{2}})N_{\bar{n}+\frac{n}{2}}. \tag{4.2}$$

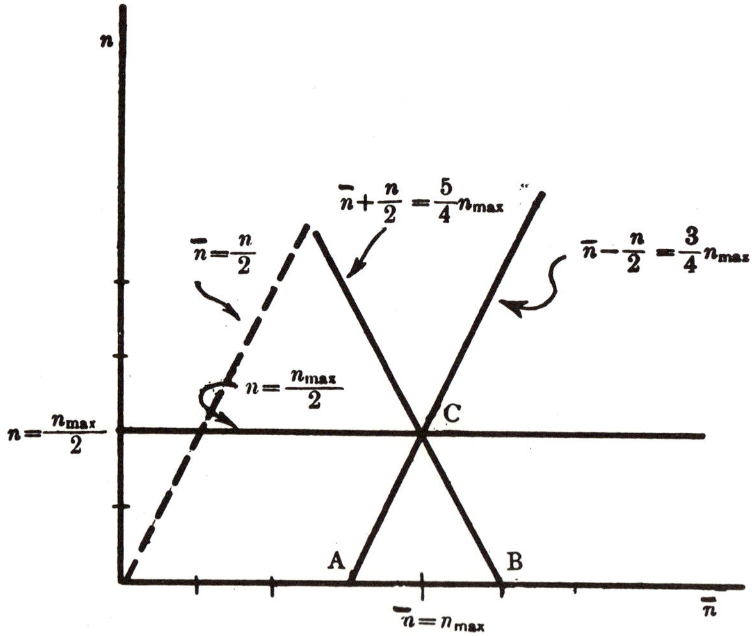

Fig. 1.

We make use of the fact that, because of the first two conditions of (2.15), the summand $(1-N_{\bar{n}-\frac{n}{2}})N_{\bar{n}+\frac{n}{2}}$ has a non-vanishing value only at points $(n, \bar{n})$ lying inside the triangle ABC (see Fig. 1) enclosed by the three straight lines

$$\begin{cases} \bar{n}+\dfrac{n}{2}=\dfrac{5}{4}n_{\max} \\ \bar{n}-\dfrac{n}{2}=\dfrac{3}{4}n_{\max} \\ n=0\ . \end{cases}$$

Since the summand vanishes outside this triangle, two of the boundaries of the domain of summation

$$n=\frac{n_{\max}}{2} \qquad \text{and} \qquad \bar{n}=0 \tag{4.3}$$

can be changed at our will so long as the changed boundaries lie outside the triangle. So we may use the boundaries

$$n=\infty \qquad \text{and} \qquad \bar{n}=\frac{n}{2} \tag{4.3'}$$

instead of the original ones. Then we have

$$D^{+}=\sum_{\infty>n>0}\ \sum_{\bar{n}\geq\frac{n}{2}} (1-N_{\bar{n}-\frac{n}{2}})N_{\bar{n}+\frac{n}{2}}$$

$$=\sum_{n>0}\ \sum_{\bar{n}'\geq n} (1-N_{\bar{n}'-n})N_{\bar{n}'}. \tag{4.4}$$

We now notice that

$$\sum_{n>0}\ \sum_{\bar{n}'\geq n} N_{\bar{n}'}=\sum_{\bar{n}'\geq 0} \bar{n}'\, N_{\bar{n}'} \tag{4.5}$$

and

$$\sum_{n>0}\ \sum_{\bar{n}'\geq n} N_{\bar{n}'-n}\, N_{\bar{n}'}=\frac{1}{2}\Big(\sum_{\bar{n}\geq 0} N_{\bar{n}}\Big)^2-\frac{1}{2}\sum_{\bar{n}\geq 0} N_{\bar{n}}$$

$$=\frac{1}{2}n_{\max}\ (n_{\max}+1). \tag{4.5'}$$

We then get from (4.4)

$$D^{+}=\sum_{n\geq 0} nN_n-\frac{1}{2}n_{\max}\ (n_{\max}+1). \tag{4.6}$$

In the same way we get the corresponding result for $\sum \rho_{\bar{n}}\rho_{-\bar{n}}$

$$D^{-}=\sum_{n\leq 0} |n|\, N_n-\frac{1}{2}n_{\max}\ (n_{\max}+1). \tag{4.7}$$

We now go over to the non-diagonal part:

$$C^{+}=\sum_{\frac{n_{\max}}{2}>n>0} \Big(\sum_{\bar{n}>0}\ \sum_{\bar{n}'>0}\Big)'\ \psi^{*}_{\bar{n}+\frac{n}{2}}\ \psi_{\bar{n}-\frac{n}{2}}\ \psi^{*}_{\bar{n}'-\frac{n}{2}}\ \psi_{\bar{n}'+\frac{n}{2}}\,, \tag{4.8}$$

and show that C^{+} vanishes.

By writing $-n$ instead of n, $\bar{n}'$ instead of $\bar{n}$ and $\bar{n}$ instead of $\bar{n}'$, we see that C^+ can be written also in the following form:

$$C^+ = \sum_{-\frac{n_{\max}}{2}<n<0} \Big(\sum_{\bar{n}>0} \sum_{\bar{n}'>0}\Big)' \psi^*_{\bar{n}+\frac{n}{2}} \psi_{\bar{n}-\frac{n}{2}} \psi^*_{\bar{n}'-\frac{n}{2}} \psi_{\bar{n}'+\frac{n}{2}}, \qquad (4.8')$$

where we have used the fact that $\psi^*_{\bar{n}+\frac{n}{2}} \psi_{\bar{n}-\frac{n}{2}}$ commutes with $\psi^*_{\bar{n}'-\frac{n}{2}} \psi_{\bar{n}'+\frac{n}{2}}$ since $\bar{n} \neq \bar{n}'$. From (4.8) and (4.8′) we get

$$C^+ = \frac{1}{2} \sum'_{-\frac{n_{\max}}{2}<n<\frac{n_{\max}}{2}} \Big(\sum_{\bar{n}>0} \sum_{\bar{n}'>0}\Big)' \psi^*_{\bar{n}+\frac{n}{2}} \psi_{\bar{n}-\frac{n}{2}} \psi^*_{\bar{n}'-\frac{n}{2}} \psi_{\bar{n}'+\frac{n}{2}}. \qquad (4.8'')$$

Now a considerasion similar to the above one shows that, because of the condition (2.15), we can change the boundaries of summation

$$n = \pm\frac{n_{\max}}{2}, \quad \bar{n}=0 \quad \text{and} \quad \bar{n}'=0 \qquad (4.9)$$

into new ones

$$n = \pm\infty, \qquad \bar{n} = -\frac{n}{2}, \qquad \bar{n}' = \frac{n}{2}. \qquad (4.9')$$

We then get

$$C^+ = \frac{1}{2} \sum'_{-\infty<n<\infty} \Big(\sum_{\bar{n}\geqq-\frac{n}{2}} \sum_{\bar{n}'\geqq\frac{n}{2}}\Big)' \psi^*_{\bar{n}+\frac{n}{2}} \psi_{\bar{n}-\frac{n}{2}} \psi^*_{\bar{n}'-\frac{n}{2}} \psi_{\bar{n}'+\frac{n}{2}}; \qquad (4.10)$$

but this C^+ vanishes as is shown in the following manner.

By rearranging factors in (4.10) we get

$$C^+ = -\frac{1}{2} \sum'_{-\infty<n<\infty} \Big(\sum_{\bar{n}\geqq-\frac{n}{2}} \sum_{\bar{n}'\geqq\frac{n}{2}}\Big)' \psi^*_{\bar{n}'-\frac{n}{2}} \psi_{\bar{n}-\frac{n}{2}} \psi^*_{\bar{n}+\frac{n}{2}} \psi_{\bar{n}'+\frac{n}{2}}. \qquad (4.10')$$

If we put here

$$\begin{cases} l = \bar{n}' - \bar{n} \\ \bar{l}' = \frac{1}{2}(\bar{n}' + \bar{n} + n) \\ \bar{l} = \frac{1}{2}(\bar{n}' + \bar{n} - n), \end{cases}$$

we can write (4.10′) in the form:

$$C^+ = -\frac{1}{2} \sum'_{-\infty<l<\infty} \Big(\sum_{\bar{l}\geqq-\frac{l}{2}} \sum_{\bar{l}'\geqq\frac{l}{2}}\Big)' \psi^*_{\bar{l}+\frac{l}{2}} \psi_{\bar{l}-\frac{l}{2}} \psi^*_{\bar{l}'-\frac{l}{2}} \psi_{\bar{l}'+\frac{l}{2}}. \qquad (4.11)$$

That the domains of summation over l, $\bar{l}$ and $\bar{l}'$ are really as specified by the inequalities under the summation symbols can be easily shown. One has only to notice the identities

$$\begin{cases} \bar{n}+\frac{n}{2}=\bar{l}'-\frac{l}{2} \\ \bar{n}'-\frac{n}{2}=\bar{l}+\frac{l}{2}. \end{cases}$$

Comparing (4.11) with (4.10) we find $C^+=-C^+$, which means that

$$C^+=0. \tag{4.12}$$

A similar consideration applies to $\sum \rho_{\bar{n}}^- \rho_{-\bar{n}}^-$. We find that the non-diagonal part of $\sum \rho_{\bar{n}}^- \rho_{-\bar{n}}^-$ vanishes:

$$C^-=0. \tag{4.13}$$

This rather lengthy proof that $C^+=C^-=0$ can be summarized in the following way. Each term in C^+ or C^- corresponds to a simultaneous transition of two particles. If there is a term corresponding to a transition $a \to b$, $c \to d$, there is always a term corresponding to the transition $a \to d$, $c \to b$. Because of the Fermi statistics, the latter term has a sign which is opposite to the former so that they cancel each other when added together.

Substituting (4.6), (4.7), (4.12) and (4.13) into $\mathfrak{H}$, we obtain

$$\mathfrak{H}=\left(\frac{2\pi\hbar}{L}\right)^2 \frac{n_{\max}}{m}\{\sum_n |n|\, N_n - n_{\max}(n_{\max}+1)\}$$

$$=\left(\frac{2\pi\hbar}{L}\right)^2 \frac{n_{\max}}{m}\{\sum_{|n|\leq n_{\max}} |n|\,(N_n-1)+\sum_{|n|>n_{\max}} |n|\, N_n\}. \tag{4.14}$$

On the other hand the kinetic energy of the system is evidently

$$H_{\text{kin}}=\left(\frac{2\pi\hbar}{L}\right)^2 \frac{1}{2m} \sum_n n^2 N_n. \tag{4.15}$$

In the state of perfect degeneracy, i.e., in the state where all levels between $-n_{\max}$ and $n_{\max}$ are occupied and all other levels are empty, this energy becomes

$$H_0=\left(\frac{2\pi\hbar}{L}\right)^2 \frac{1}{2m} \sum_{|n|\leq n_{\max}} n^2. \tag{4.16}$$

We shall now show that, in the same approximation, which is allowed by our condition (2.15), our Hamiltonian $\mathfrak{H}$ is equal to the energy (4.15) minus the constant energy (4.16) : $\mathfrak{H}=H_{\text{kin}}-H_0$.

We calculate $H_{\text{kin}}-H_0$ in the following way: From (4.15) and (4.16) we get

$$H_{\text{kin}}-H_0=\left(\frac{2\pi\hbar}{L}\right)^2 \frac{1}{2m}\{\sum_{|n|\leq n_{\max}} n^2(N_n-1)+\sum_{|n|>n_{\max}} n^2 N_n\}. \tag{4.17}$$

Making use of the assumption that holes and excited particles are present only in the neighborhood of $\pm n_{\max}$, we see that N_n-1 and N_n differ from zero only when n lies in the neighborhood of $\pm n_{max}$. This allows us to use instead of n^2 its expansion:

$$\begin{aligned} n^2 &= \{\pm n_{max} + (n \mp n_{max})\}^2 \\ &\approx n_{max}^2 \pm 2n_{max}(n \mp n_{max}) \\ &= -n_{max}^2 + 2n_{max}|n|. \end{aligned}$$

Substituting this in (4.17), we find

$$H_{kin} - H_0 = -\left(\frac{2\pi\hbar}{L}\right)^2 \frac{n_{max}^2}{m}\{N - (2n_{max}+1)\} + \mathfrak{H}. \tag{4.17'}$$

Then the use of (2.13) gives immediately the required result:

$$H_{kin} - H_0 = \mathfrak{H}. \tag{4.18}$$

This result means that our Hamiltonian $\mathfrak{H}$ is really the kinetic energy of the assembly minus a constant which depends only on the total number of particles. This constant is the energy of the assembly in the case of perfect degeneracy and thus $\mathfrak{H}$ can be interpreted as the deviation of the kinetic energy from this standard value. We express this fact by simply calling $\mathfrak{H}$ " the excitation kinetic energy ".

The total energy of the system is evidently

$$H = \mathfrak{H} + H_{int} + H_0 \tag{4.19}$$

when there are inter-particle forces.

The relations (4.6) and (4.7) can be used to express the momentum of the assembly in terms of ρ. Let the momentum be denoted by G. Then

$$G = \left(\frac{2\pi\hbar}{L}\right) \sum_{n>0} (\rho_{-n}^{+}\rho_n^{+} - \rho_n^{-}\rho_{-n}^{-}). \tag{4.20}$$

The expression for G applics irrespective of the presence or absence of the inter-particle force.

One sees that for the proof of the relation (4.18) the possibility of change of domains of summation, the replacement of (4.3) by (4.3') and the replacement of (4.9) by (4.9'), were essential. These are possible only when the three conditions in (2.15) are simultaneously satisfied. This is the reason why we had to choose the special value of $\alpha = 3/4$. It is easy to see that another choice of α would make it impossible to change the summation domains so that we would not get the relation (4.18).

§ 5. Solution of the eigen-value problem.

We now go over to the solution of the eigen-value problem of our Hamiltonian. As mentioned before, this can be done by performing the principal-axes transformation of the bilinear form representing the Hamiltonian.

Adding (3.8), (3.13) and H_0, the total Hamiltonian is obtained:

$$H=\mathfrak{H}+H_{\text{int}}+H_0$$

$$=\sum_{n>0}\left\{\left(\frac{2\pi\hbar}{L}\right)^2\frac{n_{\max}}{m}+J_n\right\}(\rho_{-n}^+\rho_n^+ + \rho_n^-\rho_{-n}^-)$$

$$+\sum_{n>0}J_n(\rho_n^+\rho_{-n}^- + \rho_n^-\rho_{-n}^+)$$

$$+\sum_{n>0}nJ_n+\frac{1}{2}N^2J_0-\frac{1}{2}NJ(0)+H_0. \tag{5.1}$$

We introduce real coordinates Q_n and real momenta P_n by means of

$$\begin{cases} \rho_n^+ = \sqrt{\dfrac{n}{2}}(Q_n+iP_n) \\ \rho_{-n}^+ = \sqrt{\dfrac{n}{2}}(Q_n-iP_n) \\ \rho_n^- = \sqrt{\dfrac{n}{2}}(Q_{-n}-iP_{-n}) \\ \rho_{-n}^- = \sqrt{\dfrac{n}{2}}(Q_{-n}+iP_{-n}), \end{cases}$$

where the suffix n is considered positive. Then in terms of P's and Q's the Hamiltonian is expressed as

$$H=\sum_n|n|\left\{\left(\frac{2\pi\hbar}{L}\right)^2\frac{n_{\max}}{m}+J_n\right\}\left(\frac{1}{2}P_n^2+\frac{1}{2}Q_n^2-\frac{1}{2}\right)$$

$$+\frac{1}{2}\sum_n|n|J_n(Q_nQ_{-n}-P_nP_{-n})$$

$$+\frac{1}{2}\sum_n|n|J_n+\frac{1}{2}N^2J_0-\frac{1}{2}NJ(0)+H_0. \tag{5.3}$$

In this expression the suffix n takes both positive and negative values and the summation $\sum_n$ is extended over positive as well as negative valus of n.

The commutation relations for the P's and Q's are

$$\begin{cases} [Q_n, P_{n'}]=i\,\delta_{n,n'} \\ [Q_n, Q_{n'}]=[P_n, P_{n'}]=0. \end{cases} \tag{5.4}$$

The transformation of H into principal form can be performed by defining new variables q_n and p_n by means of the canonical transformation

$$P_n=\frac{1}{2}\left\{\left(\frac{T_n+2U_n}{T_n}\right)^{1/4}+\left(\frac{T_n}{T_n+2U_n}\right)^{1/4}\right\}p_n+\frac{1}{2}\left\{\left(\frac{T_n+2U_n}{T_n}\right)^{1/4}-\left(\frac{T_n}{T_n+2U_n}\right)^{1/4}\right\}p_{-n}$$

$$Q_n=\frac{1}{2}\left\{\left(\frac{T_n+2U_n}{T_n}\right)^{1/4}+\left(\frac{T_n}{T_n+2U_n}\right)^{1/4}\right\}q_n-\frac{1}{2}\left\{\left(\frac{T_n+2U_n}{T_n}\right)^{1/4}-\left(\frac{T_n}{T_n+2U_n}\right)^{1/4}\right\}q_{-n} \tag{5.5}$$

with the abbreviations

$$\begin{cases} T_n=\left(\dfrac{2\pi\hbar}{L}\right)^2 \dfrac{n_{\max}}{m}|n|, \\ U_n=|n|J_n. \end{cases} \tag{5.6}$$

It can be easily seen that the p_n and q_n satisfy

$$\begin{cases} [q_n, p_{n'}]=i\delta_{n,n'}, \\ [q_n, q_{n'}]=[p_n, p_{n'}]=0, \end{cases} \tag{5.7}$$

and that H is transformed into

$$H=\sum_n \sqrt{T_n}\left\{\sqrt{T_n+2U_n}\left(\frac{1}{2}p_n^2+\frac{1}{2}q_n^2-\frac{1}{2}\right)+\frac{1}{2}\left(\sqrt{T_n+2U_n}-\sqrt{T_n}\right)\right\}$$

$$+\frac{1}{2}N^2J_0-\frac{1}{2}NJ(0)+H_0. \tag{5.8}$$

The momentum G, expressed in terms of q's and p's, is

$$G=\left(\frac{2\pi\hbar}{L}\right)\sum_n n\left(\frac{1}{2}p_n^2+\frac{1}{2}q_n^2-\frac{1}{2}\right). \tag{5.9}$$

The expressions (5.8) and (5.9) show that our system is in fact equivalent to an assembly of uncoupled harmonic oscillators, or, better, to an assembly of uncoupled sound quanta. Each sound quantum has momentum and energy given by

$$\text{momentum}=\left(\frac{2\pi\hbar}{L}\right)n \qquad n=0,\ \pm 1,\ \pm 2,\cdots \tag{5.10}$$

and

$$\text{energy}=\sqrt{T_n(T_n+2U_n)}$$

$$=\left[\left(\frac{2\pi\hbar}{L}\right)^2\frac{n_{\max}}{m}\left\{\left(\frac{2\pi\hbar}{L}\right)^2\frac{n_{\max}}{m}+2J_n\right\}\right]^{1/2}|n|, \tag{5.10'}$$

and the number of sound quanta with the momentum $\left(\dfrac{2\pi\hbar}{L}\right)n$ is represented by the operator

$$M_n=\frac{1}{2}p_n^2+\frac{1}{2}q_n^2-\frac{1}{2}. \tag{5.11}$$

In terms of M_n the energy and the momentum are

$$H=\sum_n \sqrt{T_n}\left\{\sqrt{T_n+2U_n}\ M_n+\frac{1}{2}(\sqrt{T_n+2U_n}-\sqrt{T_n})\right\} \tag{5.8'}$$

$$+\frac{1}{2}N^2J_0-\frac{1}{2}NJ(0)+H_0,$$

and

$$G=\left(\frac{2\pi\hbar}{L}\right)\sum_n n\,M_n \tag{5.9'}$$

respectively.

In the wave language the relations (5.10) and (5.11′) state that the sound wave with the wave length

$$\lambda_n=\frac{L}{|n|} \tag{5.12}$$

has the frequency

$$\nu_n=\frac{1}{2\pi}\left[\left(\frac{2\pi}{L}\right)^2\frac{n_{\max}}{m}\left\{\left(\frac{2\pi\hbar}{L}\right)^2\frac{n_{\max}}{m}+2J_n\right\}\right]^{1/2}|n|\,. \tag{5.12'}$$

The phase velocity of the sound wave is

$$v_n=\left[\left(\frac{2\pi\hbar}{L}\right)^2\frac{n_{\max}}{m}\left\{\left(\frac{2\pi\hbar}{L}\right)^2\frac{n_{\max}}{m}+2J_n\right\}\right]^{1/2}. \tag{5.13}$$

The phase velocity given by (5.13) is dependent on n and hence on λ_n when J_n does not vanish. This means that, when the inter-particle force is present, our Fermi gas is dispersive. (5.13) shows further that the phase velocity of the sound waves increases or decreases from its value in the ideal gas according as the inter-particle force is repulsive or attractive.

The formula (5.8′) enables us immediately to calculate the energy in various stationaly states. In the lowest state, for instance, where there are no sound quanta, we get

$$E=\frac{1}{2}\sum_n\sqrt{T_n}\,(\sqrt{T_n+2U_n}-\sqrt{T_n})$$
$$+\frac{1}{2}N^2J_0-\frac{1}{2}NJ(0)+H_0\,. \tag{5.14}$$

In the case of an attractive inter-particle force it may happen that E, considered as a function of L, has a minimum for some value of L. Then this will mean that the assembly is capable of forming a stable aggregate.

In the case where J_n is very small, we can expand $\sqrt{T_n+2U_n}$ and thus express E as a power series in J_n. We shall then find that the term linear in J_n together with the term H_0 gives the energy in the usual Tomas-Fermi ap-approximation including the effect of exchange. The energy quadratic in J_n corresponds to the energy obtained by Euler as the second order term of a perturbation calculation.

§ 6. Correlation between particles. Transitions caused by an external perturbing force. Criterion for applicability of the method.

In this section we shall mention some general results of physical interest which can be obtained directly by our theory. We shall give also a criterion for the applicability of our method.

(I) Correlation of position of particles.

As is known there is no correlation between particles in an ideal gas besides that due to the exclusion principle. The interaction between particles causes an extra correlation, which can be calculated very simply.

Mathematically the correlation of position can be exppessed by the quantity

$$C(\xi)=\frac{1}{N^2}\left\langle\int \rho(x)\rho(x+\xi)dx\right\rangle_{Av}, \tag{6.1}$$

$C(\xi)d\xi$ giving the probability of finding another particle at a distance between ξ and $\xi+d\xi$ from a paiticle. This correlation function can be calculated in the following manner.

Expressing (6.1) in terms of the Fourier transform of $\rho(x)$, we have

$$C(\xi)=\frac{1}{L}+\frac{1}{2N^2L}\sum_{n>0}\langle\rho_n\rho_{-n}\rangle_{Av}\cos\left(\frac{2\pi n}{L}\xi\right), \tag{6.2}$$

so that our task is to find $\langle\rho_n\rho_{-n}\rangle_{Av}$. In terms of the q's and p's defined by (5.2) and (5.5) the operator $\rho_n\rho_{-n}$ is

$$\rho_n\rho_{-n}=2n\sqrt{\frac{T_n}{T_n+2U_n}}\left\{\frac{1}{2}(p_n^2+q_n^2)+\frac{1}{2}(p_{-n}^2+q_{-n}^2)+q_nq_{-n}-p_np_{-n}\right\}. \tag{6.3}$$

It then follows that

$$\langle\rho_n\rho_{-n}\rangle_{Av}=2n\sqrt{\frac{T_n}{T_n+2U_n}}(M_n+M_{-n}+1), \tag{6.4}$$

which gives

$$C(\xi)=\frac{1}{L}+\frac{1}{N^2L}\sum_n\left(M_n+\frac{1}{2}\right)\sqrt{\frac{T_n}{T_n+2U_n}}|n|\cos\left(\frac{2\pi n}{L}\xi\right). \tag{6.5}$$

The correlation function $C(\xi)$ consists of two parts. The one is the part expressing the correlation which already exists in the absence of the inter-paiticle force. This part is due solely to the exclusion principle. This part of the correlation can be obtained by putting $U_n=0$ in (6.5). Denoting this part of $C(\xi)$ by $C_0(\xi)$, we have

$$C_0(\xi)=\frac{1}{L}+\frac{1}{N^2L}\sum_n\left(M_n+\frac{1}{2}\right)|n|\cos\left(\frac{2\pi n}{L}\xi\right). \tag{6.6}$$

The second part of $C(\xi)$ is that part of the correlation which is purely due to inter-particle forces. Denoting this second part by $C_1(\xi)$, we get

$$C_1(\xi)=\frac{1}{N^2L}\sum_n\left(M_n+\frac{1}{2}\right)\left(\sqrt{\frac{T_n}{T_n+2U_n}}-1\right)|n|\cos\left(\frac{2\pi n}{L}\xi\right). \tag{6.7}$$

(6.7) can also be written as

$$C_1(\xi)=\frac{\partial}{\partial \xi}\left\{\frac{1}{2\pi N^2}\sum_n\left(M_n+\frac{1}{2}\right)\left(\sqrt{\frac{T_n}{T_n+2U_n}}-1\right)\sin\left(\frac{2\pi|n|}{L}\xi\right)\right\}. \tag{6.7'}$$

Then the quantity

$$D_1(\xi)=\frac{1}{2\pi N^2}\sum_n\left(M_n+\frac{1}{2}\right)\left(\sqrt{\frac{T_n}{T_n+2U_n}}-1\right)\sin\left(\frac{2\pi|n|}{L}\xi\right) \tag{6.7''}$$

can be interpreted as the change of probability due to the inter-particle force of finding another particle within the distance ξ from a particle.

(II) Matrix element of an external perturbing force.

Suppose an external perturbing force is impressed upon the system. Then the perturbation will cause transitions of the system. As is well known, the transition probabilities depend essentially on the matrix elements of the perturbing energy. Now, it is very simple to calculate these matrix elements.

Let the potential of the perturbing force be $V(x)$. Then the perturbation energy is

$$H'=\int\rho(x)V(x)\,dx$$
$$=\sum_{n>0}\{V_n(\rho_n^+ +\rho_n^-)+V_{-n}(\rho_{-n}^+ +\rho_{-n}^-)\}, \tag{6.8}$$

where V_n and V_{-n} are the matrix elements of $V(x)$ defined by

$$\begin{cases} V_n=\dfrac{1}{L}\displaystyle\int V(x)\,\exp\left(\frac{2\pi i}{L}nx\right)dx \\ V_{-n}=V_n^* . \end{cases} \tag{6.9}$$

In terms of the q's and p's it is found that

$$H'=\sum_{n>0}\left(\frac{T_n}{T_n+2U_n}\right)^{1/4}\left[\sqrt{\frac{n}{2}}\,V_n\{(q_n+ip_n)+(q_{-n}-ip_{-n})\}\right.$$
$$\left.+\sqrt{\frac{n}{2}}\,V_{-n}\{(q_{-n}+ip_{-n})+(q_n-ip_n)\}\right]. \tag{6.10}$$

Now, the matrix elements of $q_{\pm n}+ip_{\pm n}$ as well as $q_{\pm n}-ip_{\pm n}$ are all known. They are the matrix elements for harmonic oscillators. Our result (6.10) shows that the matrix elements of H' are given simply by multiplying these universal matrix elements by the factor $(T_n/T_n+2U_n)^{1/4}\sqrt{n/2}\,V_{\pm n}$. When the inter-particle force is absent, the factors are simply $\sqrt{n/2}\,V_{\pm n}$. Our result shows that the effect of the inter-particle force is simply to replace $\sqrt{n/2}\,V_{\pm n}$ by $(T_n+2U_n)^{1/4}\times\sqrt{n/2}\,V_{\pm n}$. The effect of the inter-particle force can thus be dealt with as an apparent change of the perturbing potential by the factor $(T_n/T_n+2U_n)^{1/4}$, the effective potential being expressed as

$$V_{\pm n}^{\text{eff}}=\left(\frac{T_n}{T_n+2U_n}\right)^{1/4}V_{\pm n}. \qquad (6.11)$$

The factor $\left(\frac{T_n}{T_n+2U_n}\right)^{1/4}$ is larger or smaller than unity according as the inter-particle force is attractive or repulsive. This means that in the case of an attractive inter-particle force the effect of an external perturbing force becomes larger, and in the case of a repulsive one the perturbing force becomes less effective as compared with the case of an ideal gas.

(III) Criterion for the applicability of the method.

Because our theory is based on the fundamental conditions of (2.15), it is necessary to see under what circumstances we can be sure that these conditions are satisfied, at least approximately.

It is first clear that the range of the inter-particle force should not be too short in order that the third condition of (2.15) be satisfied. This condition requires that sound waves with wave lengths shorter than $2L/n_{\max}$ play no role. Let us fist consider the lowest state. Then this condition requires that in the sum $\sum_n \sqrt{T_n}\,(\sqrt{T_n+2U_n}-\sqrt{T_n})$ of (5.14) no contribution should arise from terms with $n>n_{\max}/2$. This requires that J_n is negligibly small for $n>n_{\max}/2$. Thus, roughly speaking, the range of the force should be longer than $2L/n_{\max}$ or larger than four times the mean distance of particles.

In the excited states we have to impose a further condition: There should not be any sound quanta with wave length shorter than $2L/n_{\max}$.

In order to see whether or not the first two conditions of (2.15) are satisfied, it is necessary to examine the eigen function which is obtained and see how the holes and excited particles are distributed among the levels. If this eigen function is such that the probability distribution of holes and excited particles is essentially limited between the levels $(3/4)n_{\max}$ and $(5/4)n_{\max}$, the solution obtained will be a good approximation. There is a rough but simple criterion to see under what circumstances such will be the case.

It is to calculate the mean excitation kinetic energy per particle. If this energy is too large, we must expect that there will be a large number of highly excited particles and many holes lying deep in the Fermi sea. If this energy is small, we are sure that the excited particles and holes are present only in the neighborhood of the Fermi maximum.

The excitation kinetic energy per particle is given by

$$\frac{1}{N}\langle\mathfrak{H}\rangle_{Av}=\frac{1}{2n_{\max}+1}\left(\frac{2\pi\hbar}{L}\right)^2\frac{n_{\max}}{m}\sum_{n>0}\langle\rho_{-n}^{+}\rho_n^{+}+\rho_n^{-}\rho_{-n}^{-}\rangle_{Av}, \qquad (6.12)$$

where we have

$$\rho_{-n}^{+}\rho_{n}^{+}+\rho_{n}^{-}\rho_{-n}^{-}=\frac{n}{2}\Big[\Big(\sqrt{\frac{T_n+2U_n}{T_n}}+\sqrt{\frac{T_n}{T_n+2U_n}}\Big)\Big\{\frac{1}{2}(p_n^2+q_n^2)+\frac{1}{2}(p_{-n}^2+q_{-n}^2)\Big\}$$

$$+\Big(\sqrt{\frac{T_n+2U_n}{T_n}}-\sqrt{\frac{T_n}{T_n+2U_n}}\Big)(p_np_{-n}-q_nq_{-n})-2\Big].$$

Taking the expectation value, we then get

$$\frac{1}{N}\langle\mathfrak{H}\rangle_{Av}=\Big(\frac{2\pi\hbar}{L}\Big)^2\frac{1}{2m}\sum_n\Big\{\Big(\sqrt{\frac{T_n+2U_n}{T_n}}+\sqrt{\frac{T_n}{T_n+2U_n}}\Big)M_n$$

$$+\frac{1}{2}\Big(\sqrt{\frac{T_n+2U_n}{T_n}}+\sqrt{\frac{T_n}{T_n+2U_n}}-2\Big)\Big\}\frac{|n|}{2}; \qquad (6.13)$$

in particular, for the lowest state

$$\frac{1}{N}\langle\mathfrak{H}\rangle_{Av}=\Big(\frac{2\pi\hbar}{L}\Big)^2\frac{1}{8m}\sum_n\Big(\sqrt{\frac{T_n+2U_n}{T_n}}+\sqrt{\frac{T_n}{T_n+2U_n}}-2\Big)|n|. \qquad (6.13')$$

We must now determine the value of this mean excitation energy below which our method works. In order to find this critical value, we consider a special state in which all particles in the levels between $(3/4)n_{max}$ and n_{max} are raised to the levels between n_{max} and $(5/4)n_{max}$. In this state the mean excitation energy per particle is $\Big(\frac{2\pi\hbar}{L}\Big)^2\frac{1}{m}\frac{n_{max}^2}{4^2}$. If the mean excitation energy is larger than this amount in some state, it is certain that in this state some particles are excited to levels higher than $(5/4)n_{max}$ or some holes are present in levels below $(3/4)n_{max}$. So our method cannot work consistently in such a case. We find in this way that the necessary condition for the applicability of our method is

$$\frac{1}{N}\langle\mathfrak{H}\rangle_{Av}<\Big(\frac{2\pi\hbar}{L}\Big)^2\frac{1}{m}\frac{n_{max}^2}{16}. \qquad (6.14)$$

In the lowest state (6.14) is

$$\sum_n\Big(\sqrt{\frac{T_n+2U_n}{T_n}}+\sqrt{\frac{T_n}{T_n+2U_n}}-2\Big)|n|<\frac{n_{max}^2}{2}. \qquad (6.15)$$

Of course the condition (6.14) or (6.15) is only necessary but not sufficient. There may be cases in which this condition is satisfied and still our method does not work. But such circumstances will occur rather exceptionally.

The condition (6.15) is certainly satisfied if the inter-particle force is so small that we can expand the square root of the form $\sqrt{T_n+2U_n}$ as a power series in J_n and use only a few terms. This condition is just the one required by the validity of the perturbation theory. But the condition (6.15) is weaker than the condition arising in the perturbation theory, because (6.15) does not require that J_n be uniformly small for all values of n. It requires only the smallness of the sum as a whole, and therefore our method has a wider domain of applicability than the perturbation theory. Because of the factor n outside the bracket, terms with n small do not contribute much to the sum on the left-hand side of (6.15), so that J_n with small n may even be very large, provided that T_n+2U_n does not become negative. This fact means that our method is especially suitable for long-range repulsive interactions. As has been first noticed and worked out by Bohm, the method of sound waves provides a very promising means for the study of an assembly of electrons interacting with each other through a repulsive Coulomb force.

§ 7. Concluding remarks.

In concluding the paper, we shall give a few disconnected remarks.

(I) Plasma-like oscillations.

In order to relate the results of our considerations to the work of Bohm, we shall mention briefly the bearing of our results on plasma-like oscillations of a degenerate electron gas. In this case the inter-particle force is the Coulomb repulsion. By Coulomb force is meant here the fact that J_n is proportional to to $1/n^2$. (Notice that the Poisson equation for a point source is $n^2 J_n=$const. when expressed in terms of the Fourier transform of the potential). According to (5.12′) we find that in this case V_n is independent of n for small n, i. e. for n so small that we can neglect $\left(\frac{2\hbar\pi}{L}\right)^2 \frac{n_{\max}}{m}$ as compared with $2J_n=\text{const}/n^2$. This means that the frequency of the wave is independent of the wave length, which is characteristic of plasma oscillations.

Another remarkable fact which occurs in the case of Coulomb repulsion is the following. According to (6.11) we find that the matrix elements of the external perturbing force become very small for small n. This means that the plasma oscillations are disturbed very little by external perturbing forces. This fact has been anticipated by Bohm in his paper about superconductivity.[7]

(II) Relation between the two kinds of descriptions.

As we have seen, the system can be described either as an assembly of Fermi particles, or as an assembly of sound quanta. Then the question arises: What relation will exist between the states, in one of which the occupation numbers of the Fermi particles have some specified values, and in the other of which the occupation numbers of the sound quanta have some specified values.[8] As the operator $\frac{1}{2}(p_n^2+q_n^2)-\frac{1}{2}$, which represents the number of sound quanta, and the operator $\psi_n^* \psi_n$, which represents the number of Fermi particles, do not commute with each other, it is clear that one cannot assign numerical values simultaneously to both kinds of occupation numbers. In other words, a state in which the occupation numbers of sound quanta have some definite values is a complicated superposition of various states, each of which is specified by a different set of which is specified by a different set of values of occupation numbers of Fermi particles, and *vice versa.* The statistical relation between the two kinds of occupation numbers will be obtained if one can determine the amplitude of each state in this superposition. We shall describe here briefly the general prescription to get this statistical relation.

First consider the case of a system of non-interacting particles. In this case, we can use $\frac{1}{2}(P_n^2+Q_n^2)-\frac{1}{2}$ instead of $\frac{1}{2}(p_n^2+q_n^2)-\frac{1}{2}$. The operator $\mathfrak{H}$, representing the energy of the system, is a function of the number of sound quanta. At the same time, it is a function of the number of the Fermi particles too. This means that a state in which the occupation numbers of sound quantn have some specified values, and a states in which the occupation numbers of Fermi particles have some specified values, are both some eigen-states of the energy operator. This fact results in all states in the superposition mentioned above belonging to the same eigenvalue of the energy. Now, because the number of linearly independent states belonging to one eigenvalue of $\mathfrak{H}$ is finite, one has only to solve a secular equation of finite degree in order to determine the coefficients in the superposition under consideration. So it is always possible, at least in principle, to answer the question about the statistical relation between the two kinds of occupation numbers.

In the case of interacting particles, we must first express the state in which the occupation numbers of the sound quanta have definite values as a superposition of various states, each having some definite set of values of $\frac{1}{2}(P_n^2+Q_n^2)-\frac{1}{2}$. The problem is none other than to expand Hermite functions of q's in terms of other Hermite functions of Q's. We then apply the method mentioned above to each term of this superposition. In this way we find the required statistical relation for the general case.

In a stationary state the occupation numbers of the sound quanta have some definite values, but the occupation numbers of the Fermi particles do not. The stationary state is a very complicated superposition of states with various numbers of holes and excited particles at various levels. This means that the stationary state is very far from a state where only one particle is excited to some higher level. This fact corresponds to Bohr's statement[9] that the one-particle model is a very bad approximation for the stationary states of such an assembly.

(III) A possible application to the case of exchange forces.

Though our method is applicable only to the case of ordinary forces, there seems to be a hope of applying it also to the case of exchange forces. It is to combine it with Ritz's procedure using a trial function whose form is suggested by our method when applied to ordinary forces. This procedure means physically to replace the exchange force by an equivalent ordinary force in such a way as to gives the best approximation.

Unfortunately, the mathematical structure of the method when generalized to the three-dimensional case is not simple and the author has not yet accomplished it. So in this paper we must be content only to present considerations of a rather mathematical nature without entering into real physical problems.

Acknowledgement.

I would like to express my sincere thanks to Professor J. R. Oppenheimer for giving me the opportunity of working at the Institure for Advanced Study and for his kind interest in this work. I am also indebted to Professor D. Bohm for telling me his interesting results before publication and for fruitful conversation.

References

1) H. Euler, ZS. Phys. **105** (1937), 553.
2) M. Nogami, Prog. Theor. Phys. **5** (1950), 65.
3) F. Bloch, ZS. Phys. **81** (1933), 363.
4) F. Bloch, Helv. Phys. Acta **7** (1934), 385.
5) D. Bohm, yet unpublished.
6) P. Jordan, ZS. Phys. **93** (1935), 464; **98** (1936), 759; **99** (1936), 109; R. de L. Kronig, Physica **2** (1935), 491, 854, 968.
7) D. Bohm, Phys. Rev. **75** (1949), 502.
8) R. de L. Kronig, loc. cit.
9) N. Bohr, Nature **137** (1931), 344.

Reprinted from *Printed in U.S.A.*
JOURNAL OF MATHEMATICAL PHYSICS VOLUME 6, NUMBER 2 FEBRUARY 1965

Exact Solution of a Many-Fermion System and Its Associated Boson Field

DANIEL C. MATTIS
International Business Machines Corp., Thomas J. Watson Research Center, Yorktown Heights, New York
AND
ELLIOTT H. LIEB*
Belfer Graduate School of Science, Yeshiva University, New York, New York
(Received 22 September 1964)

Luttinger's exactly soluble model of a one-dimensional many-fermion system is discussed. We show that he did not solve his model properly because of the paradoxical fact that the density operator commutators $[\rho(p), \rho(-p')]$, which always vanish for any finite number of particles, no longer vanish in the field-theoretic limit of a filled Dirac sea. In fact the operators $\rho(p)$ *define* a boson field which is *ipso facto* associated with the Fermi-Dirac field. We then use this observation to solve the model, and obtain the exact (and now nontrivial) spectrum, free energy, and dielectric constant. This we also extend to more realistic interactions in an Appendix. We calculate the Fermi surface parameter $\bar{n}_k$, and find: $\partial\bar{n}_k/\partial k|_{k_F} = \infty$ (i.e., there exists a sharp Fermi surface) only in the case of a sufficiently weak interaction.

I. INTRODUCTION

THE search for a soluble but realistic model in the many-electron problem has been just about as unfruitful as the historic quest for the philosopher's stone, but has equally resulted in valuable byproducts. For example, 15 years ago Tomonaga[1] published a theory of interacting fermions which was soluble only in one dimension with the provision that certain truncations and approximations were introduced into his operators. Nevertheless he had success in showing approximate boson-like behavior of certain collective excitations, which he identified as "phonons." (Today we would denote these as "plasmons," following the work of Bohm and Pines.[2]) Lately, Luttinger[3] has revived interest in the subject by publishing a variant model of spinless and massless one-dimensional interacting fermions, which demonstrated a singularity at the Fermi surface, compatible with the results of the modern many-body perturbation theory.[4]

Unfortunately, in calculating the energies and wavefunctions of his model Hamiltonian, Luttinger fell prey to a subtle paradox inherent in quantum field theory[5] and *therefore did not achieve a correct solution of the problem he himself had posed.* In the present paper we shall give the solution to his interesting problem and calculate the free energy. We shall show the existence of collective plasmon modes, and shall calculate the singularity at the Fermi surface (which may in fact disappear if the interaction is strong enough), the energy of the plasmons, and the (nontrivial) dielectric constant of the system. In an Appendix we shall show how the model may be generalized in such a manner as to remove certain restrictions on the interactions which Luttinger had found necessary to impose.

It is fortunate that solid-state and many-body theorists have so far been spared the plagues of quantum field theory. Second quantization has been often just a convenient bookkeeping arrangement to save us from writing out large determinantal wavefunctions. However there is a difference between very large determinants and *infinitely* large ones; we shall show that one of the important differences *is the failure of certain commutators to vanish* in the field-theoretic limit when common sense and experience based on finite N tells us they *should* vanish! (Here N refers to the number of particles in the field.)

* Research supported by the U. S. Air Force Office of Scientific Research.

[1] S. Tomonaga, Progr. Theoret. Phys. (Kyoto) **5**, 544 (1950).

[2] D. Bohm and D. Pines, Phys. Rev. **92**, 609 (1953).

[3] J. M. Luttinger, J. Math. Phys. **4**, 1154 (1963). Note that we set his $v_0 = 1$, thereby fixing the unit of energy. References to this paper will be frequent, and will be denoted by L (72), for example, signifying his Eq. (72).

[4] J. M. Luttinger and J. C. Ward, Phys. Rev. **118**, 1417 (1960).

[5] Luttinger made a transformation, L (8), which was canonical in appearance only. But in the language of G. Barton [*Introduction to Advanced Field Theory,* (Interscience Publishers, Inc., New York, 1963), pp. 126 *et seq.*] this transformation connected two "unitarily inequivalent" Hilbert spaces, which has as a consequence that commutators, among other operators, must be reworked so as to be well-ordered in fermion field operators. It was first observed by Julian Schwinger [Phys. Rev. Letters **3**, 296 (1959)] that the very fact that one postulates the existence of a ground state (i.e., the filled Fermi sea) *forces* certain commutators to be nonvanishing even though in first quantization they automatically vanish. The "paradoxical contradictions" of which Schwinger speaks seem to anticipate the difficulties in the Luttinger model.

We shall show that these nonvanishing commutators *define* boson fields which must *ipso facto* always be associated with a Fermi–Dirac field, and we shall use the ensuing commutation relations to solve Luttinger's model exactly. Because this model is soluble both in the Hilbert space of finite N and also in the Hilbert space $N = \infty$, with different physical behavior in each, we believe it has applications to the *theory of fields* which go beyond the study of the many-electron problem. The model can be extended to the case of electrons with spin. This has interesting consequences in the band *theory of ferromagnetism*, as will be discussed in some detail in an article under preparation.[5a]

II. MODEL HAMILTONIAN

We recall Luttinger's Hamiltonian[3] and recapitulate some of his results:

$$H = H_0 + H', \tag{2.1}$$

where the "unperturbed" part is

$$H_0 = \int_0^L dx\ \psi^+(x)\sigma_3 p\psi(x) \tag{2.2a}$$

$$= \sum_k (a_{1k}^* a_{1k} - a_{2k}^* a_{2k})k, \tag{2.2b}$$

and the interaction is

$$H' = 2\lambda \iint_0^L dx\, dy\ \psi_1^+(x)\psi_1(x) \times V(x-y)\psi_2^+(y)\psi_2(y) \tag{2.3a}$$

$$= \frac{2\lambda}{L}\sum \delta_{k_1+k_2,k_3+k_4} v(k_3 - k_4) \times a_{1k_1}^* a_{1k_2} a_{2k_3}^* a_{2k_4}. \tag{2.3b}$$

Here ψ is a two-component field and the form (b) of the operator is obtained from (a) by setting

$$\psi = \frac{1}{\sqrt{L}}\sum_k e^{ikx}\begin{pmatrix} a_{1k} \\ a_{2k}\end{pmatrix}$$

and

$$\psi^+ = \frac{1}{\sqrt{L}}\sum_k e^{-ikx}(a_{1k}^*, a_{2k}^*), \tag{2.4}$$

with a_{jk}'s defined to be anticommuting fermion operators which obey the usual relations

$$a_{jk}a_{j'k'} + a_{j'k'}a_{jk} \equiv \{a_{jk}, a_{j'k'}\} = 0$$
$$\{a_{j,k}^*, a_{j'k'}^*\} = 0, \text{ and } \{a_{jk}, a_{j'k'}^*\} = \delta_{jj'}\delta_{kk'}. \tag{2.5}$$

Luttinger noted that for an appropriate operator S_0, the canonical transformation

$$\tilde{H} = e^{i\lambda S_0} H e^{-i\lambda S_0} \tag{2.6}$$

gave the result that

$$\tilde{H} = H_0, \tag{2.7}$$

and consequently that *the spectrum of* $H = H_0 + H'$ *was the same as that of* H_0, *independent of the interaction* $V(x - y)$. This can be explicitly verified for his choice of

$$S_0 = \iint_0^L dx\, dy\ \psi_1^+(x)\psi_1(x)E(x-y)\psi_2^+(y)\psi_2(y), \tag{2.8}$$

where $E(x)$, not to be confused with the energy E, is defined by:

$$\partial E(x-y)/\partial x \equiv V(x-y), \tag{2.9}$$

assuming that

$$\bar{V} \equiv \frac{1}{L}\int_0^L V(x)\, dx = 0. \tag{2.10}$$

In the Appendix we shall show among other things how to generalize to $\bar{V} \neq 0$. It is also simple and instructive to verify Eqs. (2.6) and (2.7) somewhat differently by using the *first* quantization,

$$H_0 = -i\sum_{n=1}^N \frac{\partial}{\partial x_n} + i\sum_{m=1}^M \frac{\partial}{\partial y_m} \tag{2.11}$$

and

$$H' = 2\lambda \sum_{n=1}^N \sum_{m=1}^M V(x_n - y_m), \tag{2.12}$$

where N and M are, respectively, the total number of "1" particles and "2" particles, with coordinates x_n and y_m, respectively. The properly antisymmetrized wavefunctions are given by

$$\Psi = \det|e^{ik_i x_j}|\det|e^{iq_i y_j}| \times \exp\left\{\sum_{n=1}^N\sum_{m=1}^M iE(x_n - y_m)\right\}. \tag{2.13}$$

Using Eqs. (2.9) and (2.10), Ψ is readily seen to obey Schrödinger's equation

$$H\Psi = E\Psi \tag{2.14}$$

with just the unperturbed eigenvalue

$$E = \sum_{n=1}^N k_n - \sum_{m=1}^M q_m. \tag{2.15}$$

The wavenumbers are of the form

$$k_i \text{ or } q_i = 2\pi \text{ integer}/L, \tag{2.16}$$

as required for periodic boundary conditions. This is in exact agreement with the results of Ref. 3, and can also be checked in perturbation theory; first-

[5a] D. Mattis, Physics 1, 184 (1964).

order perturbation theory also gives vanishing results, and indeed, it is easy to verify that to every order in λ the cancellation is complete, in accordance with the exact result given above.

Up to this point, Luttinger's analysis (which we have briefly summarized) is perfectly correct. It is the next step that leads to difficulty. The Hamiltonian discussed so far has no ground-state energy; in order to remove this obstacle, and thereby establish contact with a real electron gas, Luttinger proposed modifying the model by "filling the infinite sea" of negative energy levels (i.e., all states with $k_1 <$ and $q_2 > 0$). Following L(8) we define b's and c's obeying the usual anticommutators, such that

$$a_{1k} = \begin{cases} b_k & k \geq 0 \\ c^*{}_k & k < 0, \end{cases}$$

and

$$a_{2k} = \begin{cases} b_k & k < 0 \\ c^*_k & k \geq 0. \end{cases} \tag{2.17}$$

Using this notation the total particle-number operator becomes

$$\mathfrak{N} = \sum_{\text{all } k} b^+_k b_k - c^+_k c_k \tag{2.17a}$$

(i.e., the number of particles minus the number of holes).

Since the Hamiltonian commutes with $\mathfrak{N}$ we can demand that $\mathfrak{N}$ have eigenvalue N_0. In the noninteracting ground state there are no holes and the b particles are filled from $-k_F$ to k_F where $k_F = \pi(N_0/L) = \pi\rho$. The noninteracting ground-state energy is $N_0\pi\rho$ + energy of the filled sea (W).

The kinetic energy assumes the form

$$H_0 = \sum_{\text{all } k} (b^*_k b_k + c^*_k c_k)\,|k| + W, \tag{2.18}$$

where

$$W = (\sum_{k<0} k - \sum_{k>0} k) \tag{2.18a}$$

is the infinite energy of the filled sea, an uninteresting c number which we drop henceforth in accordance with Luttinger's prescription. The interaction [H', Eq. (2.3) and the operator S_0, Eq. (2.8)] can also be expressed in the new language by means of the substitution (2.17). The reader will no doubt be surprised, as indeed we were, to find that now with the new operators, *Eq.* (2.7), *with* $\tilde{H}$ *defined in* (2.6), *is no longer obeyed.*

Upon further reflection one sees that this must be so, on the basis of very general arguments. In the new Hilbert space defined by the transformation to the particle–hole language (2.17), H is no longer unbounded from below and now has a ground state. A general and inescapable *concavity theorem* states that if $E_0(\lambda)$ is the ground-state energy in the presence of interactions, (2.3), then

$$\partial^2 E_0(\lambda)/\partial\lambda^2 < 0. \tag{2.19}$$

This inequality is incompatible with the previous result, viz. all E = independent of λ, which was possible only in the strange case of a system without a ground state.

The same thing can be seen more trivially using second-order perturbation theory (first-order perturbation theory vanishes). It is easily seen that

$$E_0^{(2)} = -\left(\frac{2\lambda}{L}\right)^2 \sum_k \frac{|v(k)|^2}{2k}\, n_1(k) n_2(-k), \tag{2.20}$$

where $n_1(k)$ and $n_2(k)$ are the number of ways of shifting a particle of type "1" and type "2" respectively by an amount k to an unoccupied state. A simple geometric exercise will convince the reader of the following facts: (1) if we start with a state having a finite number of particles, then n_1 and n_2 are *always* even functions of k (i.e., there are just as many ways to increase the momentum by k as to decrease it by the same amount.) (2) If we start with a filled infinite sea then there is no way to decrease the momentum of the "1" particles nor to increase the momentum of "2" particles. Hence for this second case $n_1(k)n_2(-k)$ is nonzero only for $k > 0$. Thus $E_0^{(2)}$ vanishes for a state with a finite number of particles, but it is negative for a filled sea.

If the reader is unconvinced by perturbation theory, then he can easily prove that E_0 is lowered by doing a variational calculation.

What has gone wrong? We turn to some algebra to resolve this paradox, and following this, present a solution of the field-theoretic problem defined by $H_0 + H'$ in the representation of b's and c's.

III. CASE OF THE FILLED DIRAC SEA

The various relevant operators are given below; the form (a) of each equation will *not* be used in the bulk of the paper, and is just given here for completeness. In the following equations, $p > 0$.

$$\rho_1(+p) \equiv \sum_k a^*_{1\,k+p} a_{1\,k} \tag{3.1a}$$

$$= \sum_{k<-p} c_{k+p} c^*_k + \sum_{-p\leq k<0} b^*_{k+p} c^*_k + \sum_{k\geq 0} b^*_{k+p} b_k, \tag{3.1b}$$

$$\rho_1(-p) \equiv \sum_k a^*_{1\,k} a_{1\,k+p} \tag{3.2a}$$

$$= \sum_{k<-p} c_k c^*_{k+p} + \sum_{-p\leq k<0} c_k b_{k+p} + \sum_{k\geq 0} b^*_k b_{k+p}, \tag{3.2b}$$

$$\rho_2(+p) \equiv \sum_k a^*_{2\,k+p} a_{2\,k} \tag{3.3a}$$

$$= \sum_{k<-p} b^*_{k+p} b_k + \sum_{-p\le k<0} c_{k+p} b_k + \sum_{k\ge 0} c_{k+p} c^*_k, \tag{3.3b}$$

$$\rho_2(-p) \equiv \sum_k a^*_{2\,k} a_{2\,k+p} \tag{3.4a}$$

$$= \sum_{k<-p} b^*_k b_{k+p} + \sum_{-p\le k<0} b^*_k c^*_{k+p} + \sum_{k>0} c_k c^*_{k+p}. \tag{3.4b}$$

Equations (3.1a)–(3.4a) give the density operators in the original representation, so let us calculate in this language a commutator such as (assume $p \ge p' \ge 0$ for definiteness)

$$[\rho_1(-p), \rho_1(p')] = \sum_{k,k'} [a^*_{1\,k} a_{1\,k+p}, a^*_{1\,k'+p'} a_{1\,k'}]$$

$$= \sum_{k=-\infty}^{+\infty} a^*_{1\,k} a_{1\,k+p-p'} - \sum_{k=-\infty}^{+\infty} a^*_{1\,k+p'} a_{1\,k+p} = 0. \tag{3.5}$$

The zero result could have been expected by writing the operators in first quantization:

$$\rho_1(-p) = \sum_n e^{-ipx_n} \quad \text{and} \quad \rho_2(p) = \sum_m e^{ipy_m}, \tag{3.6}$$

whence they evidently commute. Nevertheless, the zero result is achieved in (3.5) only through the almost "accidental" cancellation of two operators, each of which may diverge in the field-theory limit when $N = \infty$. We now show that in that limit the operators in fact no longer cancel, by evaluating the commutator using form (b) for the density operators. It is a matter of only some minor manipulation to obtain the important new result:

$$[\rho_1(-p), \rho_1(p')] = [\rho_2(p), \rho_2(-p')]$$

$$= \delta_{p,p'} \sum_{-p<k<0} 1 = \frac{pL}{2\pi}\delta_{p,p'}, \quad (p' > 0). \tag{3.7a}$$

In addition,

$$[\rho_1(p), \rho_2(p')] = 0. \tag{3.7b}$$

A quick check is provided by evaluating the vacuum expectation value

$$\langle 0|\,[\rho_1(-p), \rho_1(p)]\,|0\rangle$$

$$= \sum_{-p<k,k'<0} \langle 0|\, c_k b_{k+p} b^*_{k'+p} c^*_{k'}\,|0\rangle = pL/2\pi, \tag{3.8}$$

which is exactly what is expected on the basis of the previous equation. Evidently the form (b) of the operators $(2\pi/pL)^{+\frac12}\rho_1(+p)$ and $(2\pi/pL)^{+\frac12}\rho_2(-p)$ have properties of boson raising operators [call them $A^*(p)$ and $B^*(-p)$] and $(2\pi/pL)^{+\frac12}\rho_1(-p)$ and $(2\pi/pL)^{+\frac12}\rho_2(+p)$ have properties of boson lowering operators [$A(p)$ and $B(-p)$], i.e.,

$$[A, B] = [A^*, B] = 0,$$
$$[A(p), A^*(p')] = [B(-p), B^*(-p)] = \delta_{p,p'}. \tag{3.9}$$

The B field is the continuation of the A field to negative p; therefore together they form a *single* boson field defined for all p.

The relationship of the $\rho(p)$'s to Luttinger's $N(x)$'s, L(25), is obtained by using (2.4):

$$N_1(x) = \psi^*_1(x)\psi_1(x) = \frac{1}{L}\sum_p \rho_1(p)e^{-ipx},$$
$$N_2(x) = \psi^*_2(x)\psi_2(x) = \frac{1}{L}\sum_p \rho_2(p)e^{-ipx}. \tag{3.10}$$

IV. SOLUTIONS OF THE MODEL HAMILTONIAN

Before making use of the results of the previous section, we remark that $\rho_1(+p)$ and $\rho_2(-p)$ are exact raising operators of H_0, and $\rho_1(-p)$ and $\rho_2(p)$ are exact lowering operators of H_0 corresponding to excitation energies p. That is,

$$[H_0, \rho_1(\pm p)] = \pm p\rho_1(\pm p),$$
$$[H_0, \rho_2(\pm p)] = \mp p\rho_2(\pm p). \tag{4.1}$$

The identification of the ρ's with boson operators made in the previous section suggested to us the possibility of constructing a new operator T which obeys the same equations (4.1), as H_0. This is indeed possible, if we define T as follows:

$$T \equiv \frac{2\pi}{L}\sum_{p>0} \{\rho_1(p)\rho_1(-p) + \rho_2(-p)\rho_2(p)\} \tag{4.2}$$

[the ρ's being defined here and in the remainder of the paper by Eqs. (3.1b)—(3.4b), i.e., always in the hole-particle representation]. It follows that

$$[T, \rho_1(\pm p)] = \pm p\rho_1(\pm p) \tag{4.3}$$

as required, and similarly for $\rho_2(\mp p)$. Therefore, let us decompose H into two parts

$$H = H_1 + H_2 \tag{4.4}$$

with

$$H_1 = H_0 - T = \Big\{\sum_k |k|\,(b^*_k b_k + c^*_k c_k)$$
$$- \frac{2\pi}{L}\sum_{p>0} \{\rho_1(p)\rho_1(-p) + \rho_2(-p)\rho_2(p)\}\Big\}, \tag{4.5}$$

and

$$H_2 = H' + T$$
$$= \frac{1}{L}\Big[2\lambda \sum_{p>0} \{v(p)\rho_1(-p)\rho_2(p) + v(-p)\rho_1(p)\rho_2(-p)\}$$
$$+ 2\pi\sum_{p>0}\{\rho_1(p)\rho_1(-p) + \rho_2(-p)\rho_2(p)\}\Big] \tag{4.6}$$

with $v(p)$ = real, even function of p. By actual construction, all the ρ operators which appear in H_2

commute with H_1. This will be an important feature in constructing an exact solution of the model. We define an Hermitian operator S,

$$S = \frac{2\pi i}{L} \sum_{\text{all } p} \frac{\varphi(p)}{p} \rho_1(p)\rho_2(-p), \qquad (4.7)$$

where $\varphi(p)$ is also a real, even, function of p to be determined subsequently by imposing a condition that the unitary transformation e^{iS} diagonalize H_2. First we evaluate the effect of such a transformation on various operators. It commutes with H_1,

$$e^{iS}H_1e^{-iS} = H_1 = H_0 - T, \qquad (4.8)$$

because both ρ_1 and ρ_2 appearing in S commute with H_1, as noted above. In the following, p can have either sign:

$$e^{iS}\rho_1(p)e^{-iS} = \rho_1(p)\cosh\varphi(p) + \rho_2(p)\sinh\varphi(p), \qquad (4.9)$$

$$e^{iS}\rho_2(p)e^{-iS} = \rho_2(p)\cosh\varphi(p) + \rho_1(p)\sinh\varphi(p). \qquad (4.10)$$

We have verified that this transformation is a proper unitary transformation and preserves commutation relations (3.7) as well as anticommutation relations (2.5), and the reader may easily check this point. H_2 is brought into canonical form by requiring that in (exp iS) H_2 (exp $-iS$) there be no cross terms such as $\rho_1(p)\rho_2(-p)$. This leads to the equation

$$\tanh 2\varphi = -\lambda v(p)/\pi, \qquad (4.11)$$

which cannot be obeyed unless

$$|\lambda v(p)| < \pi \quad \text{for all} \quad p. \qquad (4.12)$$

Equation (4.12) serves to limit the magnitude of potentials capable of having well-behaved solutions (e.g., a real ground-state energy). For the more realistic potentials discussed in the Appendix, there is also a more realistic bound on $v(p)$: there, $v(p)$ may not be *too* attractive, but it can have any magnitude when it is repulsive, i.e., positive.

With the choice of φ in (4.11), the evaluation of H_2 becomes

$$e^{iS}H_2e^{-iS} = \frac{2\pi}{L}\sum_{p>0} \operatorname{sech} 2\varphi(p)\{\rho_1(p)\rho_1(-p) + \rho_2(-p)\rho_2(p)\} - \sum_{p>0} p(1 - \operatorname{sech} 2\varphi). \qquad (4.13a)$$

The second term is the vacuum renormalization energy

$$W_1 = -\sum_{p>0} p(1 - \operatorname{sech} 2\varphi) = \frac{L}{2\pi}\int_0^\infty dp\, p\left\{\left(1 - \frac{\lambda^2 v^2(p)}{\pi^2}\right)^{\frac{1}{2}} - 1\right\}. \qquad (4.13b)$$

It may be expanded in powers of λ to effect a comparison with Goldstone's many-body perturbation theory[4]; we have checked that they agree to third order.

The problem is now formally solved, for we can find all the eigenfunctions and eigenvalues by studying Eqs. (4.4), (4.8), and (4.13). First notice that the operator T does not depend upon the interaction and that if there is *no interaction* we could write the Hamiltonian either as

$$H = H_0, \qquad (4.14a)$$

or as

$$H = (H_0 - T) + T = H_1 + H_2. \qquad (4.14b)$$

Since H_1 and H_2 commute, every eigenstate, Ψ, of H may be assumed to be an eigenfunction of H_1 and H_2 separately. Moreover, Ψ may also be assumed to be an eigenfunction of each $\alpha_p = A_p^+A_p$ and $\beta_p = B_{-p}^+B_{-p}$ for all $p > 0$, since these operators commute with H and $\mathfrak{N}$.

Evidently (4.14a) and (4.14b) provide two different ways of viewing the noninteracting spectrum. H_0 is quite degenerate: the raising operators of H_0 are the b^+'s and c^+'s. By requiring that Ψ also be an eigenstate of α_p, β_p and H, we are merely attaching quantum numbers to the degenerate levels of H_0. If $\alpha_p\Psi = n_p\Psi$ and $\beta_p\Psi = m_p\Psi$ (where n_p and m_p are of course integers), we say that we have n_p plasmons of momentum p and m_p plasmons of momentum $-p$. With no interaction the energy of a plasmon is

$$\epsilon(p) = |p|. \qquad (4.15)$$

We may speak of H_1 as the quasiparticle part of the Hamiltonian; in H_1 the operator T plays the role of subtracting the plasmon part of the energy from H_0.

When we turn on the interaction, the above description of the energy levels is still valid, except that now we are *forced* to use the form (4.14b) because H_2 is no longer T. The degeneracy of H is partially removed by the interaction, because now the energy of a plasmon is

$$\epsilon'(p) = |p| \operatorname{sech} 2\varphi(p). \qquad (4.16)$$

Notice that the plasmon energy is always *lowered* [and therefore the plasmons cannot propagate faster than the speed of light $c = 1$, i.e., $d\epsilon'/dp \leq 1$. In the more realistic case discussed in the Appendix, the plasmon energy *can* be increased by the interaction although $d\epsilon'/dp \leq 1$ is always obeyed.] by the interaction; if (4.12) is violated the plasmon energy is no longer real and the system becomes unstable. Note, there are no plasmons in the ground state, so that W_1 (4.13), is the shift in the ground-state energy of the system.

There is one important point, however, that requires some elucidation. We would like to be able to say that in view of the fact that H_1, $\alpha(p)$, and $\beta(p)$

conserve particle number, the most general energy level of H (fixed N_0) is the sum of *any* energy of H_1 (same N_0, and no plasmons) plus *any* (plasmon) energy of H_2 (note: the plasmon spectrum is independent of N_0). Were we dealing with a finite-dimensional vector space, such a statement would not be true, for even though H_1 and H_2 commute they could not possibly be independent. Thus, if H_2 had n eigenvalues $e_1, \cdots, e_n$, and if H_1 had an equal number $E_1, \cdots, E_n$ the general total eigenvalue would not be *any* combination of $e_i + E_i$ for this would give too many values (viz. n^2 instead of n.) But we are dealing with an infinite-dimensional Hilbert space and the additivity hypothesis is in fact true for the present model.

To prove this assertion we consider any eigenstate Ψ which is necessarily parameterized by the integers n_p and m_p. Consider the state $\Phi = \{\prod_p (A_p)^{n_p}(B_p)^{m_p}\}\Psi$. The state Φ is nonvanishing and has quantum numbers $n_p = 0 = m_p$. It is also an eigenstate of H_1 with energy $E_1(\Psi)$. In addition (and this is the important point) the state Ψ may be recovered from Φ by the equation

$$\Psi = \text{const} \times \{\prod_p (A_p^+)^{n_p}(B_p^+)^{m_p}\}\Phi.$$

To every state Ψ, therefore, there corresponds a *unique* state Φ from which it may be obtained using raising operators. Conversely, to any eigenstate of H_1 (for fixed N_0) we may apply raising operators as often as we please and obtain a new (nonvanishing) eigenstate. Thus the general energy is an arbitrary sum of quasiparticle and plasmon energies.

It may be wondered where we used the fact that the Hilbert space is infinite-dimensional in the above proof. The answer lies in the boson commutation relations of the A's and B's. It is impossible to have such relations in a finite-dimensional vector space.

The eigenvalues corresponding to these states Φ will be labeled in some order, E_i ($i = 1, 2, \cdots$), so that the total canonical partition function $Z(\lambda)$ and the free energy $F(\lambda)$ are given by

$$\begin{aligned} Z(\lambda) &= e^{-F(\lambda)/kT} \\ &= (\sum_i e^{-E_i/kT})(e^{-W_1/kT}) \prod_{\substack{\text{all } p\\ \neq 0}} \left(\sum_{n=0}^{\infty} e^{-n\epsilon'(p)/kT}\right). \end{aligned} \tag{4.17}$$

The first factor is difficult to evaluate directly. However it can be obtained circuitously by noting that the energies E_i are independent of λ and therefore

$$\begin{aligned} Z(0) &= e^{-F(0)/kT} \\ &= (\sum_i e^{-E_i/kT}) \prod_{\substack{\text{all } p\\ \neq 0}} \left(\sum_{n=0}^{\infty} e^{-n\epsilon(p)/kT}\right). \end{aligned} \tag{4.18}$$

But the second factor can be trivially evaluated, as can $F(0)$ = free energy of noninteracting fermions. Therefore we use (4.18) to eliminate the trace involving the E_i's in (4.17), with the final result:

$$\begin{aligned} F(\lambda) &= F(0) + W_1 \\ &+ 2kT \sum_{p>0} \ln \{(1 - e^{-\epsilon'(p)/kT})/(1 - e^{-\epsilon(p)/kT})\}, \end{aligned} \tag{4.19}$$

where ϵ and ϵ' are given in (4.15) and (4.16). It is noteworthy that the ground state and free energy both diverge in the case of a δ-function potential.

V. EVALUATION OF THE MOMENTUM DISTRIBUTION

In this section we calculate the mean number of particles with momentum k. This quantity is $\bar{n}_k$ and is the expectation value of

$$n_k = b_k^+ b_k \tag{5.1}$$

in the ground state. Since $\bar{n}_k$ is an even function of k we need only consider $k > 0$, and it is further convenient to introduce a Fourier transform so that [using (2.4)]

$$\bar{n}_k = \frac{1}{L} \iint_0^L ds\, dt\, e^{ik(s-t)} I(s, t). \tag{5.2}$$

Here

$$\begin{aligned} I(s, t) &= \langle\Psi|\, \psi_1^+(s)\psi_1(t)\, |\Psi\rangle \\ &= \langle\Psi_0|\, e^{iS}\psi_1^+(s)e^{-iS}e^{iS}\psi_1(t)e^{-iS}\, |\Psi_0\rangle, \end{aligned} \tag{5.3}$$

where S is given by (4.7), Ψ is the new ground state, and Ψ_0 is the noninteracting ground state which is filled with b particles between $-k_F$ and k_F and has no holes (or c particles). This assignment depends on there having been no level crossing, which can be readily verified using (4.7)–(4.13).

In order to calculate the quantity $e^{iS}\psi_1(t)e^{-iS}$ we introduce the auxiliary operator

$$f_\sigma(t) = e^{i\sigma S}\psi_1(t)e^{-i\sigma S}, \tag{5.4}$$

where σ is a c number. We observe that $f_1(t)$ is the desired quantity while

$$f_0(t) = \psi_1(t). \tag{5.5}$$

In addition,

$$\begin{aligned} \partial f/\partial\sigma &= e^{i\sigma S} i[S, \psi_1(t)]e^{-i\sigma S} \\ &= e^{i\sigma S}[2\pi/L \sum_p \rho_2(-p)\varphi(p)p^{-1}e^{ipt}]e^{-i\sigma S} f_\sigma(t), \end{aligned} \tag{5.6}$$

where we have used the commutation relations (3.7) as well as the fact that ψ_1 commutes with ρ_2. Equation (5.6) is a differential equation for $f_\sigma(t)$ and (5.5) is the boundary condition. The solution is

$$f_\sigma(t) = W_\sigma(t)R_\sigma(t)\psi_1(t), \tag{5.7}$$

where

$$W_\sigma(t) = \exp\{2\pi/L \sum_{p>0} [\rho_1(-p)e^{ipt} - \rho_1(p)e^{-ipt}]p^{-1}[\cosh \sigma\varphi(p) - 1]\} \tag{5.8}$$

and

$$R_\sigma(t) = \exp\{2\pi/L \sum_{p>0} [\rho_2(-p)e^{ipt} - \rho_2(p)e^{-ipt}]p^{-1} \sinh \sigma\varphi(p)\} \tag{5.9}$$

The reader may verify that (5.7) satisfies (5.5) and (5.6) by using the commutation relations (3.7). We recall the well-known rule that

$$\exp(A + B) = \exp(A)\exp(B)\exp(-1/2[A, B]) \tag{5.10}$$

when $[A, B]$ commutes with A and B. From here on we shall set $\sigma = 1$ and drop it as a subscript. We note that since $\rho_1(p)^+ = \rho_1(-p)$ and $\rho_2(p)^+ = \rho_2(-p)$,

$$R^+(t) = R^{-1}(t) \quad \text{and} \quad W^+(t) = W^{-1}(t). \tag{5.11}$$

We also note that R and W commute with each other. Thus, (5.3) becomes

$$I(s, t) = \langle\Psi_0|\, \psi_1^+(s)R^{-1}(s)W^{-1}(s)W(t)R(t)\psi_1(t)\, |\Psi_0\rangle = I_1(s, t)I_2(s, t), \tag{5.12}$$

where

$$I_1(s, t) = \langle\Psi_1|\, \psi_1^+(s)W^{-1}(s)W(t)\psi_1(t)\, |\Psi_1\rangle, \quad I_2(s, t) = \langle\Psi_2|\, R^{-1}(s)R(t)\, |\Psi_2\rangle. \tag{5.13}$$

We have used the fact that the ground state is a product state: $\Psi_0 = \Psi_1 * \Psi_2$ where Ψ_1 is a state of the "1" field and Ψ_2 is a state of the "2" field. Ψ_1 is filled with b particles up to $+k_F$ and has no c particles; Ψ_2 is filled with b particles down to $-k_F$ and has no c particles.

Now, using the definition (5.8) and the rule (5.10) we easily find that

$$W^{-1}(s)W(t) = W_-(s, t)W_+(s, t)Z_1(s, t), \tag{5.14}$$

with

$$W_+(s, t) = \exp\{2\pi/L \sum_{p>0} \rho_1(-p)[\cosh\varphi(p) - 1] \times p^{-1}(e^{ipt} - e^{ips})\},$$

$$W_-(s, t) = \exp\{2\pi/L \sum_{p>0} \rho_1(p)[\cosh\varphi(p) - 1] \times p^{-1}(e^{-ips} - e^{-ipt})\},$$

$$Z_1(s, t) = \exp\{2\pi/L \sum_{p>0} [\cosh\varphi(p) - 1]^2 \times p^{-1}(e^{ip(s-t)} - 1)\}. \tag{5.15}$$

Likewise,

$$R^{-1}(s)R(t) = R_-(s, t)R_+(s, t)Z_2(s, t), \tag{5.16}$$

with

$$R_+(s, t) = \exp\{2\pi/L \sum_{p>0} \rho_2(p)[\sinh\varphi(p)] \times p^{-1}(e^{-ips} - e^{-ipt})\},$$

$$R_-(s, t) = \exp\{2\pi/L \sum_{p>0} \rho_2(-p)[\sinh\varphi(p)] \times p^{-1}(e^{ipt} - e^{ips})\},$$

$$Z_2(s, t) = \exp\{2\pi/L \sum_{p>0} [\sinh\varphi(p)]^2 \times p^{-1}(e^{ip(t-s)} - 1)\}. \tag{5.17}$$

We see at once from the definition (3.1b), (3.2b), of $\rho_1(p)$ that, for $p > 0$, $\rho_1(-p)\,|\Psi_1\rangle = 0$. Similarly $\langle\Psi_1|\, \rho(p) = 0$, $\rho_2(p)\,|\Psi_2\rangle = 0$, and $\langle\Psi_2|\, \rho_2(-p) = 0$. Hence,

$$I_2(s, t) = Z_2(s, t)$$

and

$$I_1(s, t) = Z_1(s, t)\langle\Psi_1|\, W_-^{-1}\psi_1^+(s)W_-W_+\psi_1(t)W_+^{-1}\, |\Psi_1\rangle. \tag{5.18}$$

If we now define

$$h_+(y) = 2\pi/L \sum_{p>0} [\cosh\varphi(p) - 1] \times p^{-1}(e^{ipt} - e^{ips})e^{-ipy},$$

$$h_-(y) = 2\pi/L \sum_{p>0} [\cosh\varphi(p) - 1] \times p^{-1}(e^{-ipt} - e^{-ips})e^{ipy}, \tag{5.19}$$

combining (3.10) and (5.15) we have that

$$W_+(s, t) = \exp \int_0^L N_1(y)h_+(y)\, dy, \quad W_-(s, t) = \exp - \int_0^L N_1(y)h_-(y)\, dy. \tag{5.20}$$

Since

$$[\psi_1(x), N_1(y)] = \delta(x - y)\psi_1(x), \quad [\psi_1^+(x), N_1(y)] = -\delta(x - y)\psi_1^+(x), \tag{5.21}$$

it follows that

$$W_+(s, t)\psi_1(t)W_+^{-1}(s, t) = \psi_1(t)\exp[-h_+(t)], \quad W_-^{-1}(s, t)\psi_1^+(s)W_-(s, t) = \psi_1^+(s)\exp[+h_-(s)]. \tag{5.22}$$

Finally,

$$\langle\Psi_1|\ \psi_1^+(s)\psi_1(t)\ |\Psi_1\rangle = 1/L \sum_{p\le k_F} e^{ip(t-s)} \equiv Z_3(s,\ t). \tag{5.23}$$

Combining all these results, we conclude that

$$I(s,\ t) = Z_0(s,\ t)Z_1(s,\ t)Z_2(s,\ t)Z_3(s,\ t), \tag{5.24}$$

where

$$Z_0(s,\ t) = \exp\left(h_-(s) - h_+(t)\right) = \exp\left\{-4\pi/L \sum_{p>0} [\cosh\varphi(p) - 1] \times (1 - e^{ip(s-t)})\right\}. \tag{5.25}$$

In order to make a comparison with Luttinger's calculation of $\bar{n}_k$, we first observe that the functions $Z_i(s,\ t)$ are really functions of $r = s - t$ and that they are periodic in s and t in $(0,\ L)$. We then define the functions $G(r)$ and $Q(r)$ as follows:

$$\exp\left[-Q(r)\right] \equiv G(r) \equiv Z_0(r)Z_1(r)Z_2(r). \tag{5.26}$$

Substituting (5.26), (5.24), and (5.23) into (5.2) we obtain

$$\bar{n}_k = 2\pi/L \sum_{p\le k_F} F(k - p), \tag{5.27}$$

where

$$F(k) = 1/2\pi \int_{-\frac{1}{2}L}^{\frac{1}{2}L} dr\, e^{ikr}e^{-Q(r)} \tag{5.28}$$

$$\cong 1/2\pi \int_{-\infty}^{\infty} dr\, e^{ikr}e^{-Q(r)} \tag{5.29}$$

In (5.29) we have passed to the bulk limit N, $L \to \infty$, not an approximation.

At this point our expression for $\bar{n}_k$ is formally the same as Luttinger's [cf. L (52), L (69)]. The difference is that our Q is different from his. He obtains Q by evaluating an infinite Toeplitz determinant with the result that [L (70)]

$$Q(r) = \lambda^2/2\pi^2 \int_0^\infty dp\, \frac{1 - \cos pr}{p}\, |v(p)|^2. \quad \text{(Luttinger)} \tag{5.30}$$

Our Q, which is the correct one to use, is obtained by combining (5.15), (5.17), and (5.25), replacing sums by integrals in the usual way, and using the definition (4.11) of $\varphi(p)$. The result is

$$Q(r) = \lambda^2/2\pi^2 \int_0^\infty dp\, \frac{1 - \cos pr}{p}\, |u(p)|^2, \tag{5.31}$$

where

$$|u(p)|^2 = (2\pi^2/\lambda^2)\{(1 - (\lambda v(p)/\pi)^2)^{-\frac{1}{2}} - 1\}. \tag{5.32}$$

It is worth noting that (5.30) agrees with (5.31) to leading order in λ^2.

Since we have not yet specified $v(p)$, we may now follow Luttinger's discussion from this point on with the proviso that we use the correct (λ dependent) $u(p)$ instead of $v(p)$. The reader is referred to pages 1159 and 1160 of Luttinger's paper.

There are two main conclusions one can draw. The first is that if we start with a δ-function interaction [so that $v(p)$ and hence $u(p)$] are constants, it can be shown that $\bar{n}_k = \frac{1}{2}$ for all k. Such a result is quite unphysical, but it is not unreasonable because the ground-state energy W (4.13a) diverges when $v(p) =$ constant at large p. Also, the result would be the same if we started with the more physical interaction

$$H' = 1/L \sum_p \{\rho_1(p) + \rho_2(p)\}\{\rho_1(-p) + \rho_2(-p)\}v(p)$$

discussed in the Appendix. This is indeed unfortunate, because relativistic field theories usually begin with local (δ-function) interactions.

The second conclusion is that if one makes a reasonable assumption about $v(p)$, and hence about $u(p)$ and $Q(r)$, one finds that for k in the vicinity of k_F, $\bar{n}_k$ behaves like

$$\bar{n}_k \sim d - e\,|k - k_F|^{2\alpha}\,\sigma(k - k_F), \tag{5.33}$$

where

$$\sigma(k) = 1, \quad k > 0; \qquad = -1, \quad k < 0 \tag{5.34}$$

and d, e, and α are certain positive constants. Now in Luttinger's calculation

$$\alpha = \lambda^2/4\pi^2 v(0)^2, \quad \text{(Luttinger)} \tag{5.35}$$

[cf. L(75)], where $v(0) \equiv \lim_{p\to 0} v(p)$.

If $2\alpha < 1$, then the conclusion to be drawn is that although the interaction removes the discontinuity in $\bar{n}_k$ at the Fermi surface, we are left with a function that has an infinite slope there. There is, so to speak, a residual Fermi surface. In Sec. IV of his paper, Luttinger shows that at least for one example of $v(p)$ perturbation theory gives the same qualitative result as (5.33) with the same value of α, (5.35).

If, on the other hand, $2\alpha > 1$ then there is no infinite derivative at the Fermi surface. $\bar{n}_k$ is perfectly smooth there (although, technically speaking, it is nonanalytic unless $2\alpha =$ odd integer.) In this case virtually all trace of the Fermi surface has been eliminated. But notice that the correct α to use is obtained by replacing $v(0)$ by $u(0) \equiv \lim_{p\to 0} u(p)$ in (5.35), i.e.,

$$2\alpha = \{1 - [\lambda v(0)/\pi]^2\}^{-\frac{1}{2}} - 1. \qquad (5.36)$$

Thus, even subject to the requirement that $|\lambda v(0)|$ be less than π, 2α can become as large as one pleases. Yet perturbation theory predicts (5.35) which yields 2α always less than $\frac{1}{2}$.

We may conclude that a strong enough interaction can eliminate the Fermi surface, while perturbation theory predicts that is always there.

VI. DIELECTRIC CONSTANT

Because the response to external fields of wave vector q only depends on an interaction expression linear in the density operators, we can immediately obtain for the generalized static susceptibility function or *dielectric constant* (response ÷ driving force), for any temperature, T

$$\chi_\lambda(q, T) = \chi_0(q, T)\{\sinh \varphi(q) + \cosh \varphi(q)\}^2 \cosh 2\varphi_q$$
$$= \chi_0(q, T) \frac{1}{1 + \lambda v(q)/\pi} \qquad (6.1)$$

in terms of the "unperturbed" susceptibility $\chi_0(q, T)$. It is also a simple exercise to calculate exactly the time dependent susceptibility in terms of the "unperturbed" quantity.

It is interesting to note that the susceptibility can diverge (which is symptomatic of a phase transformation) only for

$$\lambda v(q) \to -\pi, \qquad (6.2)$$

i.e. only for sufficiently *attractive* interactions and not for repulsive $[v(q) > 0]$ interactions.

Recently Ferrell[6] advanced plausible arguments why a one-dimensional metal cannot become superconducting. We can prove this rigorously in the present model. The electron–phonon interaction is

$$H_{\text{el-ph}} = \sum_p g(p)[\rho_1(p) + \rho_2(p)]\cdot[\xi_p + \xi^+_{-p}], \qquad (6.3)$$

where ξ and ξ^+ are the phonon field operators. In the "filled-sea" limit this coupling is bilinear in harmonic-oscillator operators, and therefore the Hamiltonian continues to be exactly diagonalizable. The new normal modes can be calculated and there is found to be no phase transition at any finite temperature.

APPENDIX

We shall be interested in extending Luttinger's model in two ways. Firstly, we note that the restriction $\bar{V} = 0$ is really not necessary. Turning back to Eqs. (2.13) *et seq.* we impose periodic boundary conditions $\Psi(\cdots, x_i + L, \cdots) = \Psi(\cdots, x_i, \cdots)$, and find that

$$(q + N\lambda\bar{V}) \text{ and } (k + M\lambda\bar{V}) = 2\pi/L \times \text{integer} \quad (A1)$$

replace the usual condition (2.16), where N = number of "1" particles and M = number of "2" particles. However, when $N, M \to \infty$ in the field-theoretic limit the problem evidently becomes ill-defined unless $\bar{V} \equiv 0$.

A less trivial observation concerns the form of the interaction potential. There is no reason to restrict it to the form $\propto \rho_1\rho_2$, and in fact the more realistic two-body interaction

$$H' = \frac{\lambda}{L} \sum_p v(p)\{\rho_1(-p) + \rho_2(-p)\}\{\rho_1(p) + \rho_2(p)\} \qquad (A2)$$

is fully as soluble as the one assumed in the text, for any strength positive $v(p)$, and provided only

$$\lambda v(p) > -\tfrac{1}{2}\pi, \qquad (A3)$$

i.e. provided no Fourier component is *too* attractive. The shift in the ground-state energy is now given by

$$W_2 = \sum_{p>0} p\left\{\left(1 + \frac{2\lambda v(p)}{\pi}\right)^{\frac{1}{2}} - 1\right\}. \qquad (A4)$$

The plasmon energy is now

$$\epsilon''(p) \equiv |p|\,(1 + 2\lambda v(p)/\pi)^{\frac{1}{2}} \qquad (A5)$$

and *for the important case of the Coulomb repulsion*, $v(p) = p^{-2}$, the plasmons describe a relativistic boson field with mass

$$m^* \equiv (2\lambda/\pi)^{\frac{1}{2}} \qquad (A6)$$

and dispersion

$$\epsilon''(p) = (p^2 + m^{*2})^{\frac{1}{2}}. \qquad (A7)$$

Here, too, $d\epsilon''/dp < 1$.

ACKNOWLEDGMENTS

We thank D. Jepsen, L. Landovitz, and T. Schultz for interesting discussions during various stages of this work. We especially thank J. Luttinger for performing a detailed review of his own work and ours, and thus corroborating the results of the present article. While reading our manuscript, Professor Luttinger called our attention to Schwinger's relevant, indeed prophetic remarks (cf. Ref. 5), and we thank him for this reference. We wish to thank Dr. A. S. Wightman, for pointing out to us that P. Jordan first discovered the bosons associated with fermion fields (in his attempts to construct a neutrino theory of light), and for a list of references.[7]

[6] R. A. Ferrell, Phys. Rev. Letters **13**, 330 (1964).

[7] P. Jordan, Z. Physik **93**, 464 (1935); **98**, 759 (1936); **99**, 109 (1936); **102**, 243 (1936); **105**, 114 (1937); **105**, 229 (1937). M. Born and N. Nagendra-Nath, Proc. Ind. Acad. Sci. **3**, 318 (1936). A. Sokolow, Phys. Z. der Sowj. **12**, 148 (1937).

Reprinted from THE PHYSICAL REVIEW, Vol. 125, No. 1, 164–172, January 1, 1962
Printed in U. S. A.

Theory of Ferromagnetism and the Ordering of Electronic Energy Levels

ELLIOTT LIEB AND DANIEL MATTIS
Thomas J. Watson Research Center, International Business Machines Corporation, Yorktown Heights, New York
(Received May 25, 1961; revised manuscript received September 11, 1961)

Consider a system of N electrons in one dimension subject to an arbitrary symmetric potential, $V(x_1,\cdots,x_N)$, and let $E(S)$ be the lowest energy belonging to the total spin value S. We have proved the following theorem: $E(S)<E(S')$ if $S<S'$. Hence, the ground state is unmagnetized. The theorem also holds in two or three dimensions (although it is possible to have degeneracies) provided $V(x_1,y_1,z_1;\cdots;x_N,y_N,z_N)$ is separately symmetric in the x_i, y_i, and z_i. The potential need not be separable, however. Our theorem has strong implications in the theory of ferromagnetism because it is generally assumed that for certain repulsive potentials, the ground state is magnetized. If such be the case, it is a very delicate matter, for a plausible theory must not be so general as to give ferromagnetism in one dimension, nor in three dimensions with a separately symmetric potential.

INTRODUCTION

THIS paper consists mainly in the enunciation and proof of a theorem about the ordering of the energy levels of a system of interacting fermions. As such, our primary concern is with mathematics. Without striving to be pedantic, we have endeavored to construct the proof with care and rigor.

We take advantage of the symmetry properties of electron wave functions belonging to various values of total spin-angular-momentum S. In certain cases we are able to order the ground-state energies belonging to the various spins without any explicit numerical calculations. This circumvents the great difficulties of the N-body problem, such as the applicability of perturbation theory, etc.

Our theorem is not without some theoretical consequences. Notably, whenever it is applicable, there can be no ferromagnetism unless one postulates explicitly spin- or velocity-dependent forces. (The theorem does not apply to electrons in a three-dimensional lattice interacting with Coulomb or central forces; but conversely, correct theories of ferromagnetism should not predict ferromagnetism for interactions which *are* covered by the theorem.) However, as the mathematics stand quite independently of such applications, we shall defer further considerations of the physical consequences of this work to the end of the paper (Sec. IV).

In Sec. I we shall state and prove the theorem for a one-dimensional electron system. In order to pass to higher dimensions, it will be necessary to prove further theorems on one-dimensional systems which have no direct relevance to fermions. These will be discussed in Sec. II. Section III will contain the proof of our theorem for certain specialized problems in two or more dimensions. We have added an Appendix on an analogous theorem for certain one-dimensional chains of three-dimensional atoms; the proof uses a different technique from that in Sec. I, insomuch as we switch to the delta function (or lattice gas) representation and use second quantization.

As a preliminary, we should like to recall a well-known theorem on the two-fermion problem, proved many years ago.[1] Consider the general two-particle Hamiltonian[2] (in any number of dimensions)

$$H=\mathbf{p}_1^2+\mathbf{p}_2^2+V(\mathbf{r}_1,\mathbf{r}_2), \tag{1}$$

where V is any symmetric potential. That is to say, the particles may interact with each other and/or with an external potential, the only proviso being that no spin- or velocity-dependent forces are present. The boundary conditions can be anything so long as they are homogeneous. Since the total spin S is a good quantum number, the ground state is either $S=0$ or $S=1$. The theorem states that the ground state always has $S=0$, a statement borne out by the hydrogen molecule, for example.

Since V is real (hence the necessity for excluding velocity-dependent forces) the eigenfunctions of H are real. An eigenstate ${}_M{}^S\Psi$ with $S=0$ must be of the form[3]

$${}_0{}^0\Psi=\Phi_{Sy}(\mathbf{r}_1,\mathbf{r}_2)[(+-)-(-+)], \tag{2}$$

where Φ_{Sy} is a symmetric real function and where we have used an obvious notation for the spin part of the wave function. Alternatively, a state with $S=1$ (and $M=1$, for example) must be of the form

$${}_1{}^1\Psi=\Phi_A(\mathbf{r}_1,\mathbf{r}_2)[(++)], \tag{3}$$

where Φ_A is antisymmetric. In both cases the symmetry property of Φ is determined by the Pauli principle which states that ${}_M{}^S\Psi$ must be antisymmetric.

Now if the ground-state wave function were $S=1$, consider the trial function obtained by taking the absolute value of Φ_A,

$${}_0{}^0\Psi\equiv|\Phi_A|[(+-)-(-+)], \tag{4}$$

which has $S=0$ and satisfies the Pauli principle. Φ_A is the spatial part of the $S=1$ ground-state function. It is readily verified that the variational energy of ${}_0{}^0\Psi$ is the same as that of the supposed ground state, ${}_1{}^1\Psi$ (we shall return to a proof of this in Sec. I). Thus, by

[1] The authors believe this theorem is due to E. P. Wigner.
[2] $\hbar^2/2m=1$.
[3] We shall use the notation ${}_M{}^S\psi$ for a function with a definite S and M value; ${}^S\psi$ if it has only a definite S value; and ${}_M\psi$ if it has only a definite M value.

reductio ad absurdum, it follows that there is always an $S=0$ eigenfunction having an energy at least as low as the lowest $S=1$ function. In fact, as we shall show later, the trial function given above cannot be an eigenfunction unless V is pathologic (e.g., a repulsive core in one dimension). Thus, if $E(S)$ denotes the lowest energy belonging to a given S value, $E(0)<E(1)$ for two particles.

This paper is an extension of the two-particle theorem to an arbitrary number of particles. We are able to do it completely for one dimension, and in certain cases for other dimensions. The general result is $E(S)\leq E(S+1)$.

I. THE ONE-DIMENSIONAL SYSTEM

We start with the general Hamiltonian[2]

$$H=-\sum_{i=1}^{N}\frac{\partial^2}{\partial x_i^2}+V(x_1,\cdots,x_N), \tag{5}$$

where V is real and symmetric in the N variables $x_1,\cdots,x_N$. Otherwise, it is completely arbitrary. The boundary conditions may be any of the following:

(i) If the particles are in a "box" (i.e., $0\leq x_i\leq L$), then $\Psi=0$ if any $x_i=0$ or L;

(ii) the same as (i) except that $\partial\Psi/\partial x_i=0$ if any $x_i=0$ or L;

(iii) if $-\infty<x_i<\infty$ or $0<x_i<\infty$, then we restrict ourselves to square integrable (bound state) wave functions, assuming these exist. Periodic boundary conditions are excluded, for they require a slight modification of the theorem and a somewhat lengthier analysis.

Although H does not contain the spins explicitly, every eigenfunction belongs to a definite S value, which may take on any of the values $N/2$, $(N/2)-1$, $\cdots$, 0, or $\frac{1}{2}$. If we denote the lowest or ground-state energy belonging to a given S value by $E(S)$, then the theorem to be proved is

Theorem I. If $S>S'$, *then* $E(S)>E(S')$ *unless* V *is pathologic, in which case* $E(S)\geq E(S')$.

The term pathologic potential will be defined in the sequel.

In order to prove this theorem it is first necessary to characterize the spatial part of a wave function of space and spin. To this end, let ${}_M\Psi$ be a wave function satisfying the Pauli principle and having a definite spin azimuthal quantum number M. (That is, $S_z\,{}_M\Psi=M\,{}_M\Psi$, where $S_z=\sum_{i=1}^N S_z^i$.) Then ${}_M\Psi$ may be expanded in the complete set of spin functions having the M value in question. The coefficients of the expansion will be spatial functions. Thus

$${}_M\Psi=\sum_j {}_M\Phi_j(x_1,\cdots,x_N)G_j^M, \tag{6}$$

where G_j^M is a spin function of which a typical one is

$$G_1^M=(\underbrace{--\cdots--}_{p}\underbrace{++\cdots++}_{N-p}),$$

with

$$p=(N/2)-M \tag{8}$$

(i.e., particles $1-p$ have spins down, the rest being up). Since ${}_M\Psi$ is a Pauli function and since the various G_j^M are obtained from each other by a permutation of the spin variables, it follows that the ${}_M\Phi_j$ are related to each other by a permutation of $x_1,\cdots,x_N$. Moreover, the following is easily verified:

(a) If any ${}_M\Phi_j\equiv0$, then all ${}_M\Phi_j\equiv0$.

(b) If Ψ is an eigenfunction of H with energy E, then so is each ${}_M\Phi_j$ separately.

(c) ${}_M\Phi_1$ (henceforth to be denoted simply by ${}_M\Phi$) is nonvanishing and is of the form

$${}_M\Phi=\Phi(x_1,\cdots,x_p|x_{p+1},\cdots,x_N),\quad p=(N/2)-M, \tag{9}$$

where the notation is meant to imply that Φ is antisymmetric in the variables $x_1,\cdots,x_p$ *and* in the variables $x_{p+1},\cdots,x_N$.

(d) Given any spatial function having the symmetry properties of (9) above, it may be used to generate a nonvanishing Pauli function such as ${}_M\Psi\equiv\sum_p(-)^p\times(p\Phi)(pG_1^M)$, where the summation is over all permutations p of the N particles. If Φ is an eigenfunction of H, then so is ${}_M\Psi$.

The next question to consider is what further condition must we impose on ${}_M\Phi$ in order that ${}_M\Psi$ have a definite S value (i.e., $S^2\,{}_M{}^S\Psi=S(S+1)\,{}_M{}^S\Psi$ and $S_z\,{}_M{}^S\Psi=M\,{}_M{}^S\Psi$). For simplicity, let us take $M\geq0$ (i.e., $p\leq N/2$), in which case the S value of Ψ in Eq. (6) could be any of $M, M+1,\cdots,N/2$, or a mixture of these. Suppose we wish $S=M$. Then, a necessary and sufficient condition is $S_+\,{}_M{}^M\Psi=0$ where $S_+=\sum_{i=1}^N S_+^i$. The operator S_+ acts on the G_j^M functions; acting on G_1^M it generates G_1^{M+1} among others. But other G_j^M's also generate G_1^{M+1} and if one sets equal to zero the coefficient of G_1^{M+1} in $S_+\,{}_M{}^M\Psi$, one finds

(e) A necessary and sufficient condition that ${}_M\Psi$ be ${}_M{}^M\Psi$ is that ${}_M{}^M\Phi$ (the coefficient of G_1^M) be of the form (9) and that the bar cannot be moved to the left. By this is meant that ${}_M{}^M\Phi$ cannot be antisymmetrized with respect to the variables $x_p,\cdots,x_N$. In other words,

$$(1-\sum_{j=p+1}^{N}P_{p,j})\,{}_M{}^M\Phi\equiv0, \tag{10}$$

where $P_{p,j}$ is the simple transposition permutation of x_p and x_j.

(f) If the bar can be moved to the left once, but not twice, then Ψ is, in general, a mixture of ${}^{M+1}{}_M\Psi$ and ${}^M{}_M\Psi$, and so forth.

(g) If $M>0$, we can always lower the M value of Ψ by $S_-\,{}_M\Psi=\text{const}\,{}_{M-1}\Psi\neq0$, and hence the bar can always be moved to the right. In other words, if a function is of the form (9), the bar can always be moved to the right if $p<N/2$, but it cannot always be moved to the left. This is a known result of the theory of the

permutation group which we have proved by recourse to the more generally known theory of angular momentum.

(h) Any function satisfying (9) and (10) may be used to generate a nonvanishing ${}_M{}^S\Psi$ as in (d) above. These remarks tell us that the higher the S value of ${}^S\Phi$, the more antisymmetric it must be. For instance, a totally antisymmetric function always belongs to $S=N/2$, but to any $M\leq N/2$.

Now since S_- and S_+ commute with H, $E(S)$ is degenerate with respect to M value; i.e., $H\,{}_M{}^S\Psi=E(S)\,{}_M{}^S\Psi$ implies there exists ${}_j{}^S\Psi$ with the same eigenvalue, where j can take all $2S+1$ values between S and $-S$. We shall therefore prove Theorem I in the following manner: Let $E(M)$ and ${}_M\Psi$ be the lowest eigenvalue and eigenfunction, respectively, of H belonging to a given M value ≥ 0. We shall show that ${}_M\Psi$ is ${}_M{}^M\Psi$ and hence that $E(M)=E(S)$. We shall further show that $E(M)<E(M+1)$ unless V is pathologic. In other words, we shall show that the lowest eigenfunction of H of the form (9) also satisfies Eq. (10).

Let R be the full domain, all $0\leq x_i\leq L$ [assuming we use boundary condition (i) or (ii) for example], and define $R_M\subset R$ by

$$R_M:\ 0\leq x_1\leq\cdots\leq x_p\leq L$$

and

$$0\leq x_{p+1}\leq\cdots\leq x_N\leq L. \tag{11}$$

Consider the Schrödinger equation in R_M;

$$H\varphi=E\varphi, \tag{12}$$

with boundary conditions:

$$\varphi=0 \text{ on the boundary of } R_M. \tag{13}$$

It is well known that the ground-state function φ_0 of Eqs. (12) and (13) satisfies

$$\varphi_0\geq 0 \quad \text{in} \quad R_M. \tag{14}$$

For suppose (14) were not satisfied and consider $\varphi=|\varphi_0|$. Now $\langle\varphi|\varphi\rangle=\langle\varphi_0|\varphi_0\rangle$ and $H\varphi=E\varphi$ everywhere except where φ_0 vanishes, at which points φ is continuous but has discontinuous derivatives. Thus $H\varphi=E\varphi+\delta$ functions, the latter occurring when φ vanishes. Hence

$$\int_{R_M}\varphi(H\varphi)=E\int_{R_M}\varphi\varphi=E\langle\varphi|\varphi\rangle.$$

Therefore φ satisfies the same boundary conditions as φ_0 and has the same variational energy. This implies that among the ground states of (12) and (13) (assuming the possibility of a degeneracy) there is at least one satisfying (14). The following are also true, although the proof is tedious:

(i) If V is bounded, then in fact $\varphi_0\neq 0$ *inside* R_M.

(j) There can be no degeneracy unless $\varphi_0=0$ *inside* R_M.

We can now define the term "pathologic potential." It is a potential with a sufficiently strong infinity to cause φ_0 to vanish *inside* R_M. An infinite repulsive core is an example. Thus there is, in general, no ground-state degeneracy, but even if there is, one of the eigenfunctions satisfies (14).

Let P be a permutation of the variables $x_1,\cdots,x_p$ and Q a permutation of the variables $x_{p+1},\cdots,x_N$, and define $PQ(R_M)$ as the domain defined by the appropriate permutation of the variables in condition (11). All the domains $PQ(R_M)$ are disjoint, except possibly for the boundaries, and together span R. Let φ be any solution to (12) and (13) and define the function Φ everywhere in R by

$$\Phi=(-)^P(-)^Q PQ\varphi \quad \text{in} \quad PQ(R_M). \tag{15}$$

Because of the boundary conditions, (13), it is easily verified that Φ is continuous and has a continuous derivative everywhere in R, and hence satisfies

$$H\Phi=E\Phi \quad \text{in} \quad R. \tag{16}$$

Conversely, any eigenfunction of H in R that satisfies Eq. (9) satisfies (12) and (13) in R_M. Thus if φ_0 is the ground state of (12) and (13), Φ, as defined by (15), is the ground state of H belonging to M.

Now consider

$$ {}_M{}^M\chi=\mathrm{Det}\begin{vmatrix}1x_1\cdots x_1{}^{p-1}\\ \vdots \quad \vdots\\ 1x_p\cdots x_p{}^{p-1}\end{vmatrix}\times\mathrm{Det}\begin{vmatrix}1x_{p+1}\cdots x_{p+1}{}^{N-p-1}\\ \vdots \quad \vdots\\ 1x_N\ \cdots x_N{}^{N-p-1}\end{vmatrix}$$

$$=\prod_{\substack{j,k=1\\ j>k}}^{p}(x_j-x_k)\times\prod_{\substack{j,k=p+1\\ j>k}}^{N}(x_j-x_k). \tag{17}$$

(Except for a totally symmetric Gaussian factor, this is the solution to the problem of noninteracting one-dimensional electrons in a harmonic oscillator potential.) The function ${}_M{}^M\chi$ clearly satisfies (9), (10), and (14). It is easily verified that if ${}_M{}^Sf$ and ${}_M{}^{S'}g$ are any two functions having different S values but the same M value, then

$$\int_R {}_M{}^Sf\,{}_M{}^{S'}g=P!(N-P)!\int_{R_M}{}_M{}^Sf\,{}_M{}^{S'}g, \qquad (p=\tfrac{1}{2}N-M) \tag{18}$$

and further the right-hand side of (18) vanishes if $S\neq S'$. Since ${}_M{}^M\chi$ and the ground state of H belonging to M, ${}_M\varphi$, are both non-negative in R_M, Eq. (18) implies that ${}_M\varphi$ is not orthogonal to ${}_M{}^M\chi$ in R. If the ground state of (12) and (13) is nondegenerate, then ${}_M\varphi$ can belong to only one S value and this S valuə must therefore be $S=M$. If one wants $S=M+1$ or higher, it is necessary to go at least to the first-excited state of (12) and (13). In case of degeneracy, at least one of the ground-state functions belongs to $S=M$.

This completes the proof of Theorem I.

II. GENERALIZATION OF THEOREM I, AND THE "POURING PRINCIPLE"

Thus far we have considered eigenfunctions of H which have the symmetry property (9), the only allowed symmetry class for constructing a Pauli function of space and spin. But there are other symmetry classes with their corresponding eigenfunctions—the totally symmetric function, for example. The latter is a Bose function and plays no role for fermions, unless only two particles are involved.

The most widely known classification of symmetry classes is due to Young; but we shall find it convenient to use a slightly modified version of his scheme.[4] It is well known that every function of 2 variables can be written as the sum of an antisymmetric and a symmetric function which *a fortiori* are orthogonal to each other. Thus,

$$\phi(x_1,x_2)=\tfrac{1}{2}(1+p_{12})\phi+\tfrac{1}{2}(1-p_{12})\phi. \qquad (19)$$

The operator $1-p_{12}$ is said to antisymmetrize ϕ, while the operator $1+p_{12}$ symmetrizes it. If ϕ is a function of $x_1, \cdots, x_N$, the operator $1-p_{12}$ antisymmetrizes it with respect to the variables x_1, x_2; the operator $(1-p_{13}-p_{23})(1-p_{12})$ antisymmetrizes it with respect to x_1, x_2, x_3, and so forth. It is quite possible that ϕ may be antisymmetrized with respect to x_1 and x_2, but not with respect to x_1, x_2, x_3; $(x_1-x_2)x_3$ is an example of such a function.

Now consider a function ϕ of the variables $x_1, \cdots, x_N$ which is of the form

$$\phi(x_1,\cdots|\cdots x_{N-n_1-n_2}|x_{N-n_1-n_2+1},\cdots,x_{N-n_1}|x_{N-n_1+1},\cdots,x_N), \qquad (20)$$

or any of its permutations, by which we mean that it is separately antisymmetric in the n_1 variables $x_{N-n_1+1}, \cdots, x_N$ and in the n_2 variables $x_{N-n_1-n_2+1}, \cdots, x_{N-n_1}$ and so forth, where $n_1 \geq n_2 \geq \cdots$. The bars are to be regarded as unmovable leftwards[5]; e.g., ϕ cannot be antisymmetrized with respect to $x_{N-n_1}, \cdots, x_N$. The function ϕ is said to belong to the symmetry class characterized by the numbers n_1, n_2, etc. For example, a totally symmetric function has N "boxes" with one variable in each, whereas an antisymmetric function has 1 "box" containing all N variables. It is possible to prove the following properties of ϕ:

(a) Any bar may always be moved to the right (i.e., the antisymmetrization procedure which would move the bar to the right gives a nonvanishing result) if the number of variables to the right of the bar is greater than the number to the left.

(b) The largest group of variables with respect to which ϕ may be antisymmetrized is n_1. From among the remainder, the largest group would be n_2, and so forth.

(c) If ϕ and ψ belong to two different symmetry classes, they are orthogonal.

(d) For a function of the form (20), the antisymmetrization operator which moves a bar to the right has as its unique inverse the operator which moves the bar to the left. Therefore, if ψ is obtained from ϕ by moving a bar to the right, it is said to belong to the same class as ϕ.

[4] See, for example, D. E. Rutherford, *Substitutional Analysis* (Edinburgh University Press, Edinburgh, 1948). Young's scheme proceeds by symmetrization, instead of antisymmetrization as used here.

[5] We have departed slightly from the notation in Eq. (9). Formerly the bar was regarded as possibly movable leftwards.

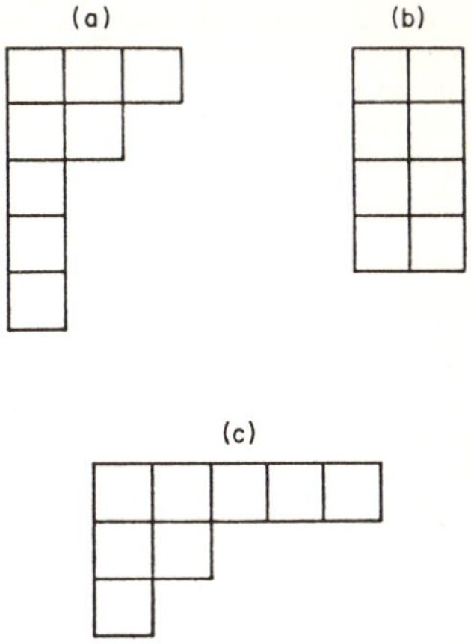

FIG. 1. The tableaux corresponding to several different symmetry classes for eight particles. By the pouring principle $E(a)>E(c)$ and $E(b)>E(c)$, but one cannot say whether (a) or (b) has the lower energy. Note that (c) happens to be the conjugate of (a).

From these remarks we see that by a combination of antisymmetrization and orthogonalization, one can reduce an arbitrary function to a sum of orthogonal functions, each belonging to a different symmetry class. The function in each symmetry class itself is decomposable into a sum of functions of the form (20) and its permutations, although this last decomposition is not unique.

There is a convenient pictorial representation for the symmetry classes called *tableau* illustrated in Fig. 1. One draws a column of n_1 boxes. To the right of it, and starting at the same height, one places a column of n_2 boxes, and so forth. A function of the form (20) is further designated by inserting the appropriate variables in the appropriate column, the order of the variables in any column being immaterial. It will be seen that the lengths of the *rows* decrease from top to bottom; it is therefore possible to define the conjugate to a symmetry class to be the one in which rows are replaced by columns. Thus (a) and (c) in Fig. 1 are conjugate. We shall return to this later, however.

Returning to the problem at hand, it will be appreciated that since H is permutation-invariant, every eigenfunction not only belongs to a definite symmetry class, but is of the form (20). Theorem I, then, tells us about the order of the ground-state energies of functions of the one- and two-column class. Under certain circumstances it is possible to extend this theorem to more than two-column classes. To do this we need one more concept, which we shall call the "pouring principle."

Definition. If α and β are two different symmetry classes, it is possible to *pour α into β* if the bars in the α function can be moved to the right [subject to (*a*) above] so that one gets a function antisymmetric in the same groups of variables as the β function. More formally, if the α tableau has the columns $n_1 \geq n_2 \geq$, and the β tableau has the columns $n_1' \geq n_2' \geq \cdots$, then we require that $n_1 \geq n_1'$; $(n_1 - n_1') + n_2 \geq n_2'$; $(n_1 - n_1' + n_2 - n_2') + n_3 \geq n_3'$; etc., where any missing columns are to be regarded as having $n=0$. If α can be poured into β, we denote this fact by $\alpha \xrightarrow{p} \beta$.

Thus, in Fig. 1, (a) can be poured into (c) and (b) into (c), but neither (a) nor (b) can be poured into the other. Note that if α can be poured into β, then the conjugate of β can be poured into conjugate of α. We can now state the extension of Theorem I in one dimension.

Theorem II. Let α and β be two symmetry classes and let $E(\alpha)$ and $E(\beta)$ be the respective ground-state energies of eigenfunctions of H in the two classes. If α can be poured in β, then $E(\alpha) > E(\beta)$ unless V is pathologic, in which case $E(\alpha) \geq E(\beta)$.

The proof is exactly the same as for Theorem I. One antisymmetrizes the α function until it matches the bars of the β function (a process which does not change the energy of the α function). Next one defines the fundamental region in analogy with (11) above and shows that the ground-state function in this region is positive, and is therefore not orthogonal to a determinantal function which is positive in this region and is known to belong to the β class.

Corollary: Consider[2]

$$H = -\sum_{i=1}^{N} \frac{\partial^2}{\partial x_i^2} - \sum_{i=1}^{M} \frac{\partial^2}{\partial y_i^2} + V(x_1, \cdots, x_N; y_1, \cdots, y_M). \quad (21)$$

If V is symmetric in the x variables, then every eigenfunction belongs to some symmetry class under permutation of these variables. Theorem II obviously is true for this more general problem. If, moreover, V is also separately symmetric in the y variables, although of no particular symmetry under the interchange of an x with a y, then every eigenfunction falls into some symmetry class α in the x variables and some class α' in the y variables. If two functions are characterized by (α,α') and (β,β'), respectively, then $E(\alpha,\alpha') > E(\beta,\beta')$ if α can be poured into β and α' into β'.

III. SEPARATELY SYMMETRIC POTENTIAL IN HIGHER DIMENSIONS

We shall now turn to the proof of the extension of Theorem I to higher dimensions when the potential is separately symmetric. Only two dimensions will be explicitly considered, for the extension to three dimensions is a corollary. The general fermion Hamiltonian is of the form (21) with $M=N$ and V is a symmetric function of the pairs x_1, y_1, etc. A separately symmetric potential is one for which V is, in addition, separately symmetric in the x variables, and in the y variables. The theorem to be proved is:

Theorem III. If V is separately symmetric and if $S > S'$, then $E(S) \geq E(S')$. (Note that Theorem III is not quite so strong as Theorem I, because the equalities can occur without pathologic potentials.)

Before discussing Theorem III, it is necessary to consider the "Kronecker product" of two tableaux. Suppose we have a function of the x's and the y's which is of a definite symmetry class α in the x's and a definite class α' in the y's. Let us consider this function together with all the functions derived from it by permutations, and let us ask if there is some linear combination of these functions which is of a definite symmetry class, β, in the N *pairs* of variables (x_i,y_i). If so, which β classes can be generated and how many independent β functions are there? [By independent β functions we mean functions which cannot be obtained from each other by permutations of the (x_i,y_i) pairs.] This question is analogous to the problem in the theory of angular momentum of combining J_1 and J_2 to give a resultant J. There the answer is given simply by the Clebsch-Gordan theorem: All $J_1 + J_2 \geq J \geq |J_1 - J_2|$ may be produced once and only once. Unfortunately, there is no such simple rule for tableaux except in two special cases: when β is symmetric, or antisymmetric (i.e., one row, or one column).

Lemma I. To generate a totally antisymmetric function in the pairs $x_i y_i$, it is required that the x and y tableaux α and α' be conjugates. (To generate a totally symmetric function in the pairs, the x tableau must be the same as the y tableau.)

This is a well-known result from the theory of permutation groups.[4]

As we saw in Sec. I, the spatial part of a Pauli (antisymmetric) function of definite S must be of the two-column type, where $n_1 = (N/2) + S$ and $n_2 = (N/2) - S$. By Lemma I we are led to believe that spin functions of a given S are of the conjugate two-row type. This is indeed the case. There are $\binom{N}{p}$ different G_M functions, and these will be seen to generate all two-row tableaux in which the first row is $(N/2)+M$ or longer. Because the ${}_M\varphi_j$ in Eq. (6) are all derivable from each other, they will not all be linearly independent. The G_j^M's therefore appear only in certain definite linear combinations; if φ is of the two-column type, these linear combinations can be shown to be of the conjugate two-row type.

Any Pauli eigenfunction of H of definite S value is thus seen to be the triple Kronecker product of a function belonging to the α and α' class in the x and y variables, respectively, and of a two-row function of the spin variables. The resultant must be a one-column function in the triplets (x_i,y_i,s_i), where s_i is the spin variable. The problem can be viewed in two ways: a Kronecker product of α and α' must be of the appropriate two-column type; or a Kronecker product of α and the two-row spin function must be conjugate to α'.

To prove Theorem III, we take the latter view. Let α and α' be the tableaux of the spatial function, giving the lowest energy for a given S value and suppose α' is (c) of Fig. 1, so that the (x_i,s_i) class, which we shall denote by β, is (a) of Fig. 1. A typical representative, χ, would have each of the pairs (x_i,s_i) in some box of the (a) tableau and each of the y_i in some box of the (c) tableau. Suppose x has (x_1,s_1), $\cdots$, (x_5,s_5) in the first column, (x_6,s_6) and (x_7,s_7) in the second, and (x_8,s_8) in the third. It will be seen that χ may be regarded as a five-particle one-dimensional Pauli function in the first five pairs, or as a two-particle Pauli function in the next two pairs.

Now, to prove Theorem III we need the following lemma:

Lemma II. Let χ be an eigenfunction with the following properties: (1) it is of the α and α' symmetry classes in the x and y variables, respectively, and is in fact formed from the lowest eigenfunction of H having these classes; (2) in the (x,s) pairs, it is antisymmetric in the same sets of variables as a function of class β, the conjugate of α', but it is not necessarily itself of the class β; (3) considered as a function of each of the sets of (x,s) pairs in which it is antisymmetric, it has a definite S value (i.e., each column has a definite S value). These S values we shall call $S_1, S_2, \cdots$. Let S be any total S value for all the particles which can be compounded from $S_1, S_2, \cdots$ by the usual Clebsch-Gordan rule. Then $E(S)\leq E(\alpha,\alpha')$.

Proof. By applying the appropriate S_+ and S_- operators of each column the appropriate number of times to χ, we can generate functions having all possible M values in each column and which still have the same energy as χ. These functions are then to be added together with the appropriate coefficients to generate a new function, ${}^S\chi\neq 0$ having the required total S value. ${}^S\chi$ has the same energy as χ. Since the total S^2 operator commutes with all permutations, ${}^S\chi$ may be written as the sum of functions belonging to definite symmetry classes and all having the given S value. One of these tableaux may be β itself, in which case the lemma is proved. If not, then since ${}^S\chi$ is already antisymmetric in the same sets of variables as the β tableau, at least one of the component tableaux, say γ, must be such that $\gamma \xrightarrow{P} \beta$. To make a Pauli function out of this component we should need a function whose y class is γ^+, the conjugate of γ. But γ^+ satisfies $\alpha' \xrightarrow{P} \gamma^+$, and therefore $E(\alpha,\gamma^+)\leq E(\alpha,\alpha')$. Thus, if we carry out the same procedure again with the ground-state function of the (α,γ^+) class, we shall be able to construct a Pauli function having the S value in question and with an energy lower than $E(\alpha,\alpha')$.

Corollary. The ground-state function of a given S value has the properties of χ above, and, moreover, belongs to the symmetry class β in the (x,s) variables.

Proof. If not, then by the procedure of the above proof, we should either be able to lower the energy by changing the y tableau, or else we should end up with the function ${}^S\chi$ having the properties stated in the corollary.

Now we consider the proof of Theorem III. The lowest eigenfunction having a given $S\neq 0$ value is the sum of permutations of a function ${}^S\chi$ having the properties stated in the corollary and therefore has the S values S_1, S_2, etc., in each column of β, the (x,s) tableau. There are three cases to be considered:

(1) If S is not the minimum that can be compounded of S_1, S_2, etc., then we can construct a function of spin $S-1$ and with at least as low an energy.

(2) If S_1, S_2, etc., are already the lowest possible (i.e., either 0 or $\frac{1}{2}$) then (1) above is transparently true.

(3) If S is indeed the minimum of $S_1\oplus S_2\oplus$, etc., then lowering some one of these S_i values, say S_1, would permit us to create a function of total S equal to $S-1$. But there does exist a function having a lower energy than ${}^S\chi$ and having the properties of χ listed in Lemma II, except that the first column has the value S_1-1. To see this, we regard the function having the α tableau in the x's as being the sum of functions each belonging to definite symmetry classes in each of the groups of x variables appearing in each column of β. One of these sets of tableaux must be the respective conjugates of S_1, S_2, etc. Looking at χ from this point of view, it is clear from the results of Secs. I and II that we can lower one or more of the S_i at will and at the same time lower the energy.

This completes the proof of Theorem III for two dimensions; the extension to higher dimensions is obvious: We simply treat the s_i, z_i pairs in the same manner as the spins above. Let us remark, however, on the reason for the lack of strict inequality as we had in Theorem I. Suppose the lowest function of $S=1$ had an (x,s) tableau such as (a) in Fig. 1 with $S_1=S_3=\frac{1}{2}$ and $S_2=0$. Since each column already has its lowest possible S value, the only way in which $S=0$ could have a *lower* energy is by having an (x,s) tableau containing only columns of even length, such as (b) of Fig. 1, with $S=0$ in each. But it is not possible to prove that such a function has, in fact, a lower energy, and therefore it is possible for the ground state as well as for excited states to have a degeneracy in more than one dimension. (This degeneracy can be estimated never to exceed $N^{\frac{1}{3}}$, in three dimensions. It is therefore not an extensive property of the system.)

IV. ON THEORIES OF FERROMAGNETISM

Although we could extend them to other particles obeying various statistics, the results of this paper apply most directly to the problem of interacting electrons, and as such have some bearing on the theory of ferromagnetism.

It is well known that ferromagnetism must be a consequence of the electronic interactions, for a noninteracting electron system always obeys the theorem $E(S) \leq E(S+1)$, regardless of the external potential. This is antiferromagnetism, or at most, paramagnetism. Ferromagnetism is assumed to occur when the ground state belongs to a nonvanishing S whose magnitude is proportional to the size of the system. It is also well known that the direct magnetic spin-spin forces are negligibly weak, so that the spatial forces in conjunction with the Pauli principle are held to be responsible for the phenomenon.

The simplest realistic problem which offers some hope of being soluble is the linear chain of three-dimensional atoms. The atomic states are supposed known when the atoms are infinitely far apart, and the problem is to find the new configurations when overlap becomes important. Peierls[6] considers this very problem in the chapter on ferromagnetism in his book. The approximation which he makes is that there is only one orbital state per atom, and he concludes that the electronic interactions can lead to ferromagnetism. However, recent and more realistic calculations[7–9] on such chains have proved the contrary to be true. Also in Sec. I we showed that under no circumstances can a one-dimensional electron system be ferromagnetic with only space-dependent forces; this includes the special case of a chain of one-dimensional "atoms." In the Appendix we also treat a model applicable to an idealized chain of three-dimensional atoms, with similar results. It therefore seems that a linear chain can be magnetic only if the individual atoms have orbital degeneracy, that is, if the single atom displays a magnetic or truly three-dimensional character; but it is not known whether this is a *sufficient* condition for ferromagnetism to occur in a linear system.

Our theorem has no relevance to atomic magnetism *per se* (Hund's rule) because it does not apply to the central force problem in three dimensions. But if we consider ferromagnetism to be an extensive property of a solid, the theorem does have some relevance. For we shall show that it is not merely sufficient to have (i) a band structure, (ii) strong repulsive interactions, and (iii) three dimensions, to produce ferromagnetism. We shall base ourselves on the results of Sec. III.

Suppose, for example, that highly magnetized states of a noninteracting set of electrons lie rather close in energy to the $S=0$ ground state. If one introduces a repulsive interaction potential into the problem, and treats this by lowest-order perturbation theory, certain terms called the "exchange integral" will favor the magnetized states. To lowest order, one may find that the magnetized states have crossed the $S=0$ state, and the interacting system is supposed to become ferromagnetic. But this conclusion would be fallacious if the effect were cancelled by second-order or higher-order terms in the perturbation series (or if the perturbation expansion did not converge). Indeed, we now give an example based on the previous section, which is a case in point. Consider, for example, the unperturbed Hamiltonian to be

$$H_0 = \sum p_i^2 + \sum_i [V(x_i) + V(y_i) + V(z_i)]. \quad (22)$$

Let $V(x)$ be a periodic potential so that one-electron eigenfunctions are Bloch functions. The potential can be chosen such that the bands display the usual degeneracies and other features of motion in a three-dimensional cubic lattice.

If now we introduce an interaction, say

$$H_I = g^2 \sum_{ij} \left(\frac{1}{|x_i - x_j| + d} + \frac{1}{|y_i - y_j| + d} + \frac{1}{|z_i - z_j| + d} \right), \quad (23)$$

or some other repulsive, separately symmetric interaction, the total Hamiltonian $H_0 + H_I$ is still subject to our theorem and is *not* ferromagnetic. (The theorem does not exclude paramagnetism, however, for the ground state might be degenerate with states of nonvanishing spin angular momentum.) But what are the conclusions we would reach if we were to apply first-order perturbation theory to H_I? This amounts to calculating the expectation value of $H_0 + H_I$ using the Slater determinants appropriate to the unperturbed problem. The unperturbed functions with the most spatial nodes are better, variationally speaking, than those with fewer nodes for *sufficiently large* g^2, and we might be led to conclude that there exist some values of S such that $E(S) < E(0)$, which is erroneous. It is therefore clear that we cannot always trust perturbation theory to properly order the levels, for when it is carried out only to finite order, it might be more accurate for some values of S than for others, depending on the particular features of the problem. The same can be said of variational calculations.

A notorious example of the above is the low-density electron gas with Coulomb interactions which is in a background of compensating positive charge. The expectation value of the Hamiltonian using the unperturbed plane-wave states is lower for the ferromagnetic configuration than for the nonferromagnetic ones, at sufficiently low density. But perturbation theory diverges for this problem,[10] and this ferromagnetism is indeed fictitious. A recent and accurate calculation[11] by

[6] R. E. Peierls, *Quantum Theory of Solids* (Oxford University Press, New York, 1955).
[7] R. K. Nesbet, Phys. Rev. **122**, 1497 (1961).
[8] L. F. Mattheiss, Phys. Rev. **123**, 1209 (1961).
[9] David I. Paul, Phys. Rev. **118**, 92 (1960), and Phys. Rev. **120**, 463 (1960).

[10] M. Gell-Mann and K. A. Brueckner, Phys. Rev. **106**, 364 (1957).
[11] W. J. Carr, Jr., Phys. Rev. **122**, 1437 (1961).

Carr leads that author to conclude that at all but the lowest densities the electronic spins are *antiferromagnetically* aligned.

In concluding, we should recall that our theorem is not valid if there are explicitly spin-dependent forces, or velocity-dependent forces. In the latter case, the eigenfunctions are not real and our method of proof does not apply. Nor does it apply to the Coulomb potential which governs real electrons. But it does serve as a warning that the criterion for ferromagnetism must be rather detailed, and not so broad as to violate the results of this investigation.

We have also found it possible to order many of the energy levels of the Heisenberg Hamiltonian $\sum K_{ij}\mathbf{S}_i\cdot\mathbf{S}_j$ (where the S_i are spins on a lattice in one, two, or three dimensions), by analogy with the calculation in the Appendix. These results will be reported in a subsequent publication.

ACKNOWLEDGMENTS

We should like to thank our colleagues at I.B.M. for many useful discussions, and particularly Dr. D. Jepsen, Dr. T. D. Schultz, and Dr. M. Gutzwiller for their excellent suggestions.

APPENDIX. ONE-DIMENSIONAL LATTICE GAS

We are interested in a theorem analogous to the one in the text, for a chain of *three*-dimensional atoms. This problem cannot be solved in all generality, therefore we are led to consider the following tractable model.

(a) We use a truncated Hamiltonian such that only valence electrons are mobile.

(b) Each atom in the linear chain has only one valence state (capable of double occupancy, however, because of spin degeneracy).

(c) The atoms are at a distance d from their nearest neighbors. This distance is such that only nearest-neighbor overlap is important.

(d) The matrix element for a one-electron hop from site j to $j\pm 1$ is K, a constant. Two-electron hops and exchange effects are neglected. This is equivalent to

(e) assuming that aside from the "hopping" matrix elements the Hamiltonian is diagonal, with an energy calculable by specifying which atoms have empty valence states, which have singly occupied valence state, and which have doubly occupied valence states. These assumptions lead directly to Eq. (A4).

Our model reduces, in a certain limit, to the one-dimensional problem of Sec. I. The Appendix provides therefore an alternate proof for Theorem I.

To see this, let us consider one-dimensional space as consisting of discrete points labeled $i=1, 2, \cdots, N$, separated by a distance d. The length of the chain is therefore Nd. Next, introduce the second-quantized Fermi operators c_{is} and $c_{is}^\dagger$, where $s=$ "up" or "down" according to the spin coordinate, and

$$[c_{is},c_{js'}]_+=0,\quad [c_{is}^\dagger,c_{js'}]_+=\delta_{ij}\delta_{ss'}. \tag{A1}$$

Now consider the Hamiltonian

$$H=-K\sum_{i,s}(c_{i+1,s}^\dagger c_{is}+c_{i-1,s}^\dagger c_{is}-2c_{is}^\dagger c_{.s}) + \sum_{i,j,s,s'} V(|i-j|)c_{is}^\dagger c_{is}c_{js'}^\dagger c_{js'} \tag{A2}$$

If we transform to running waves,

$$H\rightarrow 2K\sum(1-\cos kd)c_{ks}^\dagger c_{ks} +\sum V_g c_{k+gs}^\dagger c_{ks}c_{k'-gs'}^\dagger c_{k's'}, \tag{A.3}$$

where V_g is the Fourier transform of $V(|i-j|)$. In the limit $K^{-\frac{1}{2}}=d=N^{-1}=0$ this reduces to the problem of a one-dimensional electron gas with two-body forces. The Hamiltonian (A2) is a special case of the problem we shall now consider,

$$H=-K\sum_{i,s}(c_{i+1,s}^\dagger c_{is}+\text{H.c.})+V(\cdots n_j\cdots), \tag{A4}$$

where V is an arbitrary symmetric function of the operators $n_j\equiv(c_{j\uparrow}^\dagger c_{j\uparrow}+c_{j\downarrow}^\dagger c_{j\downarrow})$. This Hamiltonian satisfies (a)–(e), is identical with the general Hamiltonian of Sec. I in the limit $K^{-\frac{1}{2}}=d=0$, and can be shown to commute with the spin operators which, in our new representation, are

$$S_z=\tfrac{1}{2}\sum_i(c_{i\uparrow}^\dagger c_{i\uparrow}-c_{i\downarrow}^\dagger c_{i\downarrow}),\quad S_y=\tfrac{1}{2}i\sum(c_{i\uparrow}^\dagger c_{i\downarrow}-c_{i\downarrow}^\dagger c_{i\uparrow}),$$

and

$$S_x=\tfrac{1}{2}\sum(c_{i\uparrow}^\dagger c_{i\downarrow}+\text{H.c.}). \tag{A5}$$

The problem is soluble because there exists a transformation to Pauli (pseudo-spin) variables, in which the Schrödinger equation can be reduced to a series of algebraic equations. We define the Pauli operators as follows:

$$b_{i\downarrow}\equiv c_{i\downarrow}\exp\{\pi i\sum_{j=1}^{i-1} c_{j\downarrow}^\dagger c_{j\downarrow}\},$$
$$b_{i\uparrow}\equiv c_{i\uparrow}\exp\{\pi i[\sum_{j=1}^{N} c_{j\downarrow}^\dagger c_{j\downarrow}+\sum_{j=1}^{i-1} c_{j\uparrow}^\dagger c_{j\uparrow}]\}, \tag{A6}$$

The $b_i^\dagger$'s are given by the Hermitean conjugates of these defining equations. All Pauli operators *commute except* b_{is} and $b_{is}^\dagger$, (for all i and s), which anticommute:

$$b_{is}^\dagger b_{is}+b_{is}b_{is}^\dagger=1. \tag{A7}$$

In terms of these new operators,

$$H=-K\sum(b_{i+1,s}^\dagger b_{is}+\text{H.c.})+V(\cdots n_j\cdots), \tag{A8}$$

where $n_j=b_{j\uparrow}^\dagger b_{j\uparrow}+b_{j\downarrow}^\dagger b_{j\downarrow}$. The Hamiltonian remains simple under this transformation only for this very special case of a linear chain, and nearest-neighbor hops. We now assume $K>0$. Otherwise, K can be made positive by a trivial canonical transformation.

Also, for the moment, let us pretend that the number of electrons is even, and is $2p$. Obviously, p cannot exceed N, the number of sites, as each site can accommodate two electrons at most.

Ground State in $M=0$ Subspace

States of all allowed spin angular momentum can be rotated into the $M=0$ subspace with no change in energy. The ground state here is therefore the ground state of the Hamiltonian. The configurations which form a complete set in this subspace have p electrons with spin up and p with spin down. (A configuration specifies which sites are vacant or occupied); for example, if $p=1$ and $N=2$, the complete set of $M=0$ configurations is

$$\phi_1=b_{1\uparrow}^\dagger b_{1\downarrow}^\dagger|0\rangle, \qquad \phi_2=b_{2\uparrow}^\dagger b_{2\downarrow}^\dagger|0\rangle,$$

$$\phi_3=b_{1\uparrow}^\dagger b_{2\downarrow}^\dagger|0\rangle, \quad \text{and} \quad \phi_4=b_{1\downarrow}^\dagger b_{2\uparrow}^\dagger|0\rangle.$$

The Pauli operators for different particles commute, and therefore, the configurations can all be defined to have the same sign for arbitrary ordering of the operators. The number of distinct configurations is $t\equiv\binom{N}{p}^2$, and we shall label them ϕ_α, where $\alpha=1, 2, \cdots, t$.

Let the ground-state function be

$$\psi_0=\sum f^\alpha\phi_\alpha, \tag{A9}$$

with energy E_0 and real amplitudes f^α. If we denote the eigenvalues of V by V_α,

$$V\phi_\alpha=V_\alpha\phi_\alpha, \quad |V_\alpha|<\infty, \tag{A10}$$

then Schrödinger's equation can be expressed in terms of the amplitudes as

$$-K\sum_{\beta(\alpha)} f^{\beta(\alpha)}=(E_0-V_\alpha)f^\alpha. \tag{A11}$$

The index $\beta(\alpha)$ runs over those configurations for which

$$\langle\beta(\alpha)|H-V|\alpha\rangle\neq 0. \tag{A12}$$

We parenthetically observe that a variational function,

$$\psi=\sum g^\alpha\phi_\alpha, \tag{A13}$$

has variational energy W,

$$W=\frac{-K\sum\sum g^\alpha g^{\beta(\alpha)}+\sum V_\alpha(g^\alpha)^2}{\sum(g^\alpha)^2}\geq E_0. \tag{A14}$$

Clearly, all nonzero amplitudes f^α can be chosen positive in the ground state. For if they oscillate in sign, define a trial function by $g^\alpha\equiv|f^\alpha|$. Then by inspection of (A14), $W\leq E_0$. This is a contradiction unless

$$\psi'\equiv\sum|f^\alpha|\phi_\alpha,$$

is also a ground-state eigenfunction. Therefore,

$$-K\sum_{\beta(\alpha)}|f^{\beta(\alpha)}|=(E_0-V_\alpha)|f^\alpha|. \tag{A15}$$

If any $f^\alpha=0$, then by Eq. (A15) all $f^{\beta(\alpha)}$ also vanish. Since by repeated application of H to any given state, all other states are eventually reached, we can conclude that *all* f^α vanish if some one f^α vanishes. Therefore

$$f^\alpha\neq 0 \text{ for all } \alpha. \tag{A16}$$

(We note that $E_0-V_\alpha<0$ for all α, since the ground-state energy must lie lower than the most favorable potential energy.)

Equations (A11) and (A15) are *incompatible unless* all f^α have the same sign, for if we combine them, we obtain

$$\sum|f^{\beta(\alpha)}|=|\sum f^{\beta(\alpha)}|, \text{ for all } \alpha. \tag{A17}$$

Hence $\psi'=\pm\psi_0$. That the ground state is nondegenerate follows from the observation that all other eigenfunctions of H must be orthogonal to ψ_0 and therefore must have a change of sign, and therefore cannot obey Eq. (A17).

The spin of ψ_0 is found by noting that ψ_0 is *not* orthogonal to the ground state for $V=0$, because they both contain all configurations of Pauli operators with no changes of sign in the amplitudes. The ground state for $V=0$ can be found by inspection of (A3), and belongs to $S=0$. Therefore, the ground state belongs to $S=0$ in general.

Ground State for $M>0$

By a similar procedure, the ground state in any $M>0$ subspace is found to belong to $S=M$. Since each such subspace contains all states of $S\geq M$, this automatically orders the ground states belonging to the various values of S, whether the number of electrons is even or odd. Denoting by $E(S)$ the lowest energy belonging to total spin S, we have therefore proved the following:

Theorem. $E(S+1)>E(S)$.

Note that the restrictions (a)–(e) preclude "double hops," which is related to so-called "exchange." It is therefore reasonable to assume that if ferromagnetism is possible in a linear chain of the sort we have considered, that these neglected exchange matrix elements would be responsible. However, very recent and accurate calculations[7–9] have shown this exchange mechanism to be rather weak. Nesbet[7] finds that the direct "exchange is small compared with the sum of the various antiferromagnetic effects." Mattheiss[8] finds that the true energy levels of such a chain are accurately approximated by the states of the Heisenberg antiferromagnet with nearest-neighbor interactions. Finally, Paul[9] also concludes that linear chains of atoms in s states are nonferromagnetic. It would be interesting to investigate whether an orbital degeneracy on each atom could lead to ferromagnetism for the linear chain in the same way as it appears to be responsible for the magnetic moment of the O_2 molecule.

The Bose Gas

Chapter 5

The Bose Gas

Unlike the situation in classical statistical mechanics, there are very few examples of a Bose gas or liquid, with pairwise interactions among the particles, that can be solved exactly. One wants to consider a Hamiltonian of the form

$$H = (2m)^{-1} \sum_{i=1}^{N} p_i^2 + \sum_{i<j} v(x_i - x_j) \tag{5.1}$$

for N particles constrained to a box of volume V ($= L$ in one dimension), and find the totally symmetric eigenfunctions and their eigenvalues. In addition one would like to be able to calculate the partition function of the system as well as certain correlation functions that are important for neutron and x-ray scattering. It would also be desirable that the system have some kind of calculable phase transition in the hope of elucidating the λ transition of liquid helium. But there exists no explicitly soluble model of the type (5.1) of even a classical gas having a phase transition (except for the two-dimensional Ising model which is a lattice gas) and it seems unlikely that we shall soon have a sensible and exactly soluble model of a quantum-mechanical gas with a phase transition. There does exist, however, the quantum mechanical version of the Kac-Baker long-range force model (see footnote 2 in Chapter 1). The first-order phase transition in this model has been investigated by Burke, Lebowitz, and Lieb (1966).

The sad fact is that there exist only two soluble models of an interacting Bose gas, and of these one is really a special case of the other. Both have purely repulsive potentials. Despite this paucity of models, the situation is better than for a Fermi gas where there are no soluble models of the type (5.1). The special case is the quantum-mechanical version of the Tonks, or hard-core, gas (see Chapter 1 on classical statistical mechanics). In view of the simplicity of the solution, one feels that it must have been known in antiquity. (Recall that the hard-core classical gas was really first analyzed by Zernike and Prins in 1927). The first publication on the subject seems to be a footnote in a paper by Bijl (1937), where the ground state energy of the system is simply stated without derivation. The first systematic treatment, however, was given by Nagamiya (1940), and subsequently by Rubin (1955), Girardeau (1960), and Stachowiak (1960).

The potential in question is

$$\begin{aligned} v(x) &= \infty, \qquad \text{for} \quad |x| < a \\ &= 0, \qquad \text{for} \quad |x| > a \end{aligned} \tag{5.2}$$

and, as usual, there are N particles in a box of length L, with $\rho = N/L$ being the density. We shall also assume units in which $\hbar = 1$, $2m = 1$. First, let us consider the case in which the particle "diameter," a, is zero. Contrary to the classical gas and the three-dimensional hard core Bose gas, letting a be zero does not remove the interaction. Because the wave function must vanish if any two particles touch, the full configuration space is divided into $N!$ regions of the type $R_1 : (0 \leq x_1 \leq x_2 \leq \ldots \leq x_N \leq L)$. Inside R_1, ψ satisfies a free particle Schrödinger equation, but since $\psi = 0$ on the boundary of R_1, ψ must be a determinantal wave function, viz.,

$$\text{in } R_1 : \quad \psi(x_1, x_2, \ldots, x_N) = \det|f_i(x_j)| \tag{5.3}$$

where the f's are a set of N linearly independent solutions of the free particle Schrödinger equation. [Note: We could just as well solve the more general problem having an additional one-body potential, $V = \sum_{i=1}^{N} w(x_i)$, by making the f_i solutions to the single particle Schrödinger equation with a potential $w(x)$.] The f's will be determined by two things: the symmetry requirement of the wave function and the boundary conditions imposed at $x = 0$ and L. If we wished ψ to be totally antisymmetric (one-component fermions) we would let (5.3) represent ψ everywhere in space. If we impose periodic boundary conditions then the f's would be plane waves, viz.,

$$f_i(x) = e^{ik_i x} \tag{5.4}$$

where the k's are of the form $(2\pi/L) \times$ integer. Hard-wall boundary conditions would dictate that $f_i(x) = \sin(k_i x)$ where $k = (\pi/L) \times$ integer. Henceforth we shall consider only periodic boundary conditions (all of the following analysis can in fact be carried through with hard-wall boundary conditions), whence

$$P = \sum_{i=1}^{N} k_i, \qquad E = \sum_{i=1}^{N} k_i^2 \tag{5.5}$$

are the total momentum and energy, respectively, of the state. We note from (5.5) that the energy levels of a one-component fermion system are not altered by introducing point hard cores.

For bosons, on the other hand, we require that ψ *not* change sign when two coordinates are interchanged. Consequently, in any region which is a permutation of R_1:

$$\psi = \pm \det \tag{5.6}$$

where det is as in (5.3) and the plus (+) sign [minus (−) sign] is to be used if the permutation is even (odd). Here, again, the f's will be plane waves as in (5.4) and with periodic boundary conditions the k's will be of the form

$$\begin{aligned} k &= (2\pi/L) \times \text{interger} && \text{for} \quad N \text{ odd} \\ &= (2\pi/L) \times \text{interger} + (\pi/L) && \text{for} \quad N \text{ even} \end{aligned} \tag{5.7}$$

For any N, the ground state eigenfunction for all space is

$$\psi_0 = \prod_{i<j} |\sin(\pi/L)\,(x_j - x_i)| \tag{5.8}$$

It is noteworthy that this function happens to be the probability density function for a sequence of N energy levels in Dyson's orthogonal ensemble of the statistical theory of energy levels (see Chapter 1 on classical statistical mechanics).

The important conclusion to be drawn from (5.5) and (5.7) is that *the boson spectrum for point hard cores is identical to that of noninteracting one-component fermions*. In particular, the boson partition function is the standard Fermi partition function. Strictly speaking, the conclusion is true only for finite N when N is odd, but with hard-wall conditions it is true for any N. Indeed, with hard-wall conditions, every one of the *partitio numerorum* of N possible symmetry classes for N particles has the same spectrum.

The analysis for $a > 0$ is a trivial extension of that given above. The eigenfunctions are nonzero only in a subregion, R_1', of R_1, viz., $0 \leq x_1, \leq x_2 - a \leq x_3 - 2a \leq \ldots \leq x_N - (N-1)a \leq b$, where b is the lesser of $L - (N-1)a$ or $x_1 + L - Na$. If we define the variables $y_j = x_j - (j-1)a$ then

$$\psi = \det|f_i(y_j)| \tag{5.9}$$

[Note: When $a \neq 0$ a system with an additional single particle potential cannot be analyzed as simply as before.] For periodic boundary conditions the f's are plane waves, but the k's satisfy a more complicated condition than (5.7). Let the total momentum of the state be P, given by (5.5). P is always of the form $(2\pi/L) \times$ integer. Then the k's are

$$k_j = [L(1 - \rho a)]^{-1}[-Pa + 2\pi n_j] \tag{5.10}$$

and

$$P = (2\pi/L) \sum_1^N n_j \tag{5.11}$$

where the n_j are, as before, integers for N odd and half-odd integers for N even. Thus, there is a simple one-to-one correspondence between the states for $a \neq 0$ and for $a = 0$ and we see from the energy expression (5.5) that when P is a fixed finite number, in the bulk limit, $L \to \infty$, all energy levels are simply multiplied by the factor $(1 - \rho a)^{-2}$. This conclusion is, indeed, unfortunate. Variation of the core size does not change the character of the spectrum and consequently, in one sense, we are always in the strong coupling limit, even when $a = 0$. In another sense, however, there is something to be learned by varying a. The correlation functions do depend on a in a nontrivial way and they become more and more oscillatory as a is increased. From this point of view we do have a resemblance to the three-dimensional hard-core gas which has been the subject of much recent study.

This simple system is therefore of some importance. Since it behaves as a three-dimensional hard-core gas does at high density it provides a severe test for any approximation scheme that purports to be valid for a system like liquid helium. It is to be regretted that few authors test their methods in this rigorous way.

As for correlation functions (propagators), there are two that are important. One is the structure factor, important for x-ray scattering, defined by

$$S(x_1 - x_2) = L^2 \int \ldots \int \psi^2(x_1, x_2, \ldots, x_N)\, dx_3 \ldots dx_N \tag{5.12}$$

where the integration is over all space and where ψ is the normalized ground state function if $T = 0$, or else ψ^2 is the diagonal part of the density matrix if $T \neq 0$. Since (5.12) is an integral over a Fermi function, it is easy to compute, but we shall calculate instead a generalization of (5.12), namely,

$$G(k, \omega) = (i/N) \sum_n |\langle n|\rho_k|0\rangle|^2(\omega - \omega_n)^{-1} \tag{5.13}$$

Here $|n\rangle$ is any Bose state of the system, $|0\rangle$ being the ground state, ω_n is the energy of $|n\rangle$ measured relative to the ground state energy, and

$$\rho_k = \sum_{j=1}^{N} e^{ikx_j} \tag{5.14}$$

The appropriate generalization of G for temperatures other than zero is well known and can, in fact, be calculated exactly. Plainly, G is analytic in ω except for poles on the real axis. In the bulk limit these poles become dense and form a branch cut. The jump across the cut, viz.,

$$S(k, \omega) = (2\pi)^{-1}[G_+(k, \omega) - G_-(k, \omega)] \tag{5.15}$$

is, aside from some factors, the probability for neutron scattering from the system with momentum transfer k and energy transfer ω. The integral relation

$$1 + \rho\, S(k) = \int_{-\infty}^{\infty} S(k, \omega)\, d\omega \tag{5.16}$$

gives the Fourier transform of the structure factor defined in (5.12).

Despite the physical importance of G, it does not seem to have been computed in the literature. We shall do so here, taking for simplicity the case $a = 0$. To compute the matrix element $\langle n|\rho k|0\rangle$ we observe that both $|n\rangle$ and $|0\rangle$, as given by (5.3) and (5.6), are both $\pm$det, but since only their product enters into the matrix elements we may freely integrate the product of the two determinants over all space without regard to the $\pm$ sign. Because the ω_n's are also the same as for the noninteracting Fermi gas, the function $G(k, \omega)$ is exactly the same as for this latter system. The calculation is thus trivial, and the well-known result (in the bulk limit) is

$$G(k, \omega) = \frac{-i}{4\pi\rho|k|} \ln \frac{\omega - \omega_1}{\omega - \omega_2} \tag{5.17}$$

where

$$\omega_1 = (|k| + \pi\rho)^2 - (\pi\rho)^2, \qquad \omega_2 = |(\pi\rho)^2 - (|k| - \pi\rho)^2| \tag{5.18}$$

The physically important scattering function is then simply

$$\begin{aligned} S(k, \omega) &= \{4\pi\rho|k|\}^{-1}, \qquad && \text{for} \quad \omega_2 < \omega < \omega_1 \\ &= 0, && \text{otherwise} \end{aligned} \tag{5.19}$$

The distribution (5.19) is similar to that for a real gas: for small k its center is at $\pi\rho|k|$ and its width is $2k^2$. In other words, it is sharp, particularly for such a strongly interacting gas. On the other hand, the function $G(k, \omega)$ as given by (5.17) has *no singularities on the second Riemann sheet* (quasi-particle poles). Our result, therefore, is in marked contrast to the standard conjectures about the structure of G. A scattering experiment could be well described in terms of quasi particles, yet there are no quasi-particle poles.

A property related to this scattering function is the spectrum of the Hamiltonian. As we have seen, the energy levels of this Bose gas are the same as for a noninteracting Fermi gas. But the latter spectrum is radically different from the usual Bose spectrum; the principal differences being that each elementary excitation can take place only once, not infinitely often (i.e., there is an "exclusion principle"), and that the elementary excitation spectrum is described by a two-parameter family (the "hole" momentum and the "particle" momentum) and not by a one-parameter $\epsilon(k)$ function. A resolution of this paradox is to describe the spectrum in terms of two types of conventional Bose elementary excitations instead of one. The matter is discussed in detail by Lieb (1963).

The second correlation function of importance is

$$F(x_1 - x_1') = L \int \cdots \int \psi(x_1 x_2 \cdots x_N)\, \psi(x_1' x_2 \cdots x_N)\, dx_2 \cdots dx_N \tag{5.20}$$

F is important for calculating quantities involving kinetic energy. In particular, the average number of particles having a given momentum k is

$$n_k = \rho \int_L F(x) e^{ikx}\, dx \tag{5.21}$$

Now the condensation hypothesis (not to be confused with a condensation in the thermodynamic sense) states that n_0 is a nonzero fraction of N for the ground state. Girardeau (1960) defined a weaker, or generalized, condensation by requiring that

$$\lim_{\epsilon\to 0}\lim_{N\to\infty} N^{-1}\sum_{|k|<\epsilon} n_k$$

be nonzero and, on the basis of an approximate calculation, conclude that the latter existed for the hard-core gas while the former did not (the former, if it exists, always implies the latter). The distinction between the two criteria is the difference between long-long range order and short-long range order. Roughly speaking, for the former we require the $\lim_{N\to\infty} F(\gamma L) \neq 0$ (where γ is any fixed constant <1), while for the latter we require merely that $\lim_{x\to\infty}\lim_{N\to\infty} F(x) \neq 0$. That neither condensation actually exists was shown by Schultz (1963) who proved that $\tilde{F}(x) = \lim_{N\to\infty} F(x)$ was bounded by constant $\times\ |x|^{-4/\pi^2}$. Lenard (1964), using a different analysis, obtained a slightly better bound: $\tilde{F}(x) < |e/\pi\rho x|^{1/2}$ for all x.

We turn now to the generalization of the hard-core gas (Lieb and Liniger, 1963; Lieb, 1963). The pair potential for this model is the penetrable delta function:

$$v(x_1 - x_2) = 2C\,\delta(x_1 - x_2) \tag{5.22}$$

where δ is the usual Dirac delta function and C is a constant. It is well known that a δ-function is equivalent to a discontinuity in the derivative of the wave function whenever any two particles touch, viz.,

$$\left(\frac{\partial}{\partial x_j} - \frac{\partial}{\partial x_k}\right)\psi\Big|_{x_j=x_k+} + \left(\frac{\partial}{\partial x_k} - \frac{\partial}{\partial x_j}\right)\psi\Big|_{x_j=x_k-} = 2C\psi|_{x_j=x_k} \tag{5.23}$$

For a symmetric function the two terms on the left side of (5.23) are the same. The general wave function is analyzed in Lieb and Liniger (1963), but here we will give a somewhat neater expression for the wave function than is given in that paper. In R_1 let φ be a determinantal plane wave function, as in (5.3) and (5.4), which will therefore satisfy the Schrödinger equation *inside* R_1 but not the boundary condition (5.23). But if the eigenfunction ψ be defined in R_1 by

$$\psi = \prod_{j>i} (C + \partial/\partial x_j - \partial/\partial x_i)\varphi \tag{5.24}$$

then, using the fact that φ vanishes whenever any two coordinates are equal, it can be verified immediately that ψ satisfies the boundary conditions (5.23). Once again, the energy and momentum are given by (5.5) and the periodic boundary conditions determine the allowed values of the k's; these conditions are, unfortunately, more complicated than before, viz.,

$$(-1)^{N-1}e^{-ik_jL} = \exp\left[-2i\sum_{s=1}^{N} \tan^{-1}\left(\frac{k_s - k_j}{C}\right)\right] \tag{5.25}$$

for $j = 1$ to N.

At this point it is necessary to impose the important restriction that C be positive (repulsive case). The reason is that only for this case are the solutions to (5.25) real, and their general nature delineated. When C is negative, on the other hand, two things happen. One is that complex solutions become possible, indicating the possibility of two or more particle bound states. Were it possible to understand and catalog these solutions we would have achieved at least one desideratum—namely, a soluble model with attractive potentials—but alas, the attractive case has thus far defied analysis. The second difficulty is that, owing to the absence of a hard core, the system does not saturate, but this need not be an overwhelming drawback. As McGuire (1964) has shown, the ground state wave function in infinite space for this attractive case is

$$\psi = \exp(C/2) \sum_{i<j} |x_i - x_j| \tag{5.26}$$

and the energy is $E_0 = -\frac{1}{12}C^2N^3$. If we keep C fixed and negative, and pass to the bulk limit, the fact that (5.26) does not strictly satisfy the boundary conditions will be immaterial, but the ground state energy will be proportional to N^3. If, on the other hand, we make C proportional to N^{-1} and then let $N \to \infty$, the ground state will be well behaved. We would, of course, have to find a new ground state wave function that satisfied the bound conditions, for (5.26) would be applicable only in the low density limit. Nevertheless, were the problem soluble, the artifice of putting $C \sim N^{-1}$ would presumably yield a model with the desired properties.

Returning to the repulsive, $C > 0$, case, let us briefly summarize the results of Lieb and Liniger:

1. In the bulk limit, there is one nontrivial dimensionless coupling constant, $\gamma = C/\rho$. When $\gamma = \infty$ we have the hard-core case discussed before.

2. It is now possible to test Bogoliubov's perturbation theory[1] for small γ. It turns out to be quite good quantitatively for $\gamma \lesssim 2$, with one important drawback. For any $\gamma \neq 0$ the *true* spectrum of elementary excitations is qualitatively similar to the hard-core spectrum rather than to the free particle spectrum. As was mentioned before, the most natural way to catalog the spectrum is in terms of *two* elementary spectra rather than the single one given by Bogoliubov's theory. For $\gamma \lesssim 2$ his one spectrum agrees quite well with one of the two, but the other is totally unaccounted for in the theory.

3. It might be supposed that this anomaly (if it really is such) is brought about by one dimension. But the fact is that any test one can make on the system does not distinguish one from three dimensions. In other words, it is apparently unimportant whether the particles can "get around" each other, as in three dimensions, or merely "through" each other as for the one-dimensional delta function.

4. As $\gamma \to 0$ (weak coupling) the problem becomes more and more difficult to analyze, even though this is the regime where perturbation theory is applicable.

[1] For a review of the Bogoliubov theory, see E. Lieb, The Bose Fluid *in* "Lectures in Theoretical Physics" (W. Brittin, ed.), Vol. VII. Univ. of Colorado Press, Boulder, Colorado, 1964.

We have met this paradox before in connection with the classical Coulomb gas. The ground state energy is an analytic function of γ, except at $\gamma = 0$, and the nature of the singularity there remains a mystery.

There are also several unanswered questions pertaining to this gas: Although the wave functions are known, they are sufficiently complicated that it has been impossible to compute correlation functions (propagators) as in (5.12), (5.13), and (5.20). One would like to know if condensation, or generalized condensation, exists in the gound state when γ is finite. Also, are there quasi-particle poles for finite γ?

The reader may ask why this Hamiltonian has not been solved for fermions. For totally antisymmetric spatial functions the calculation is trivial—the delta function has no effect whatsoever. But a totally antysymmetric spatial function corresponds to all spins up (total spin is $S = N/2$). For other values of S one has effectively a two-component gas and it is then impossible to restrict one's attention solely to the subregion R_1. McGuire (1965) has in fact solved the problem for one spin down ($S = \frac{1}{2}N - 1$). For the attractive delta function he finds two types of elementary excitations. Flicker and Lieb have solved the two-spin-down problem, and it is likely that the general n-spin-down problem can be solved. See Flicker (1966).

Finally, we refer to McGuire (1964) who has discussed the general solutions to (5.1) for N particles with a delta-function potential (of any sign) in *infinite space*—in other words, the scattering problem. While he treats the case of symmetric interaction (all particles have the same mass and the same interaction)—indeed, this is the *only* case that can be treated analytically, even when N is as small as 3—the scattering problem is far simpler than the problem of a gas in a box. The problem of statistics and especially of boundary conditions are, as we have seen, the principle sources of difficulty—and not the problem of finding solutions to the differential equation

Bibliography

A. Bijl (1937). "Properties of the Condensed Phases of Helium and Hydrogen," *Physica* **4,** 329.

T. Burke, J. Lebowitz, and E. Lieb (1966). "Phase Transition in a Model Quantum System: Quantum Corrections to the Location of the Critical Point," *Phys. Rev.* (to be published).

M. Flicker (1966). Ph.D. Thesis, Yeshiva University (unpublished).

M. Girardeau (1960). "Relationship between Systems of Impenetrable Bosons and Fermions in One Dimension," *J. Math. Phys.* **1,** 516.

*A. Lenard (1964). "Momentum Distribution in the Ground State of the One-Dimensional System of Impenetrable Bosons," *J. Math. Phys.* **5,** 930.

*E. Lieb and W. Liniger (1963). "Exact Analysis of an Interacting Bose Gas, I, The General Solution and the Ground State," *Phys. Rev.* **130,** 1605.

*E. Lieb (1963). "Exact Analysis of an Interacting Bose Gas, II, The Excitation Spectrum," *Phys. Rev.* **130,** 1616.

*J. McGuire (1964). "Study of Exactly Soluble One-Dimensional N-Body Problems," *J. Math. Phys.* **5,** 622.

J. McGuire (1965). "Interacting Fermions in One-Dimension," I, "Repulsive Potential," *J. Math. Phys.* **6,** 432; II, "Attractive Potential," *ibid.* **7,** 123 (1966).

T. Nagamiya (1940). "Statistical Mechanics of One-Dimensional Substances, I," *Proc. Phys.-Math. Soc. Japan* (Nippon Suugaku-Buturigakkwai Kizi, Tokyo) **22,** 705. II, *ibid.* 1034.

R. Rubin (1955). "Quantum-Mechanical Properties of a One-Dimensional System Composed of Hard Spheres," *J. Chem. Phys.* **23,** 1183.

T. Schultz (1963). "Note on the One-Dimensional Gas of Impenetrable Point Particle Bosons," *J. Math. Phys.* **4,** 666.

H. Stachowiak (1960). "Sur un Gaz à Une Dimension de Particules Réciproquement Impénétrables," *Acta Univ. Wratislaviensis (Poland)* **12,** 93.

Reprinted from The Physical Review, Vol. 130, No. 4, 1605–1616, 15 May 1963
Printed in U. S. A.

Exact Analysis of an Interacting Bose Gas. I. The General Solution and the Ground State

Elliott H. Lieb and Werner Liniger
Thomas J. Watson Research Center, International Business Machines Corporation, Yorktown Heights, New York
(Received 7 January 1963)

A gas of one-dimensional Bose particles interacting via a repulsive delta-function potential has been solved exactly. All the eigenfunctions can be found explicitly and the energies are given by the solutions of a transcendental equation. The problem has one nontrivial coupling constant, γ. When γ is small, Bogoliubov's perturbation theory is seen to be valid. In this paper, we explicitly calculate the ground-state energy as a function of γ and show that it is analytic for all γ, except $\gamma=0$. In Part II, we discuss the excitation spectrum and show that it is most convenient to regard it as a double spectrum—not one as is ordinarily supposed.

I. INTRODUCTION

In comparatively recent times, as the history of quantum mechanics goes, a vast body of literature has developed on the quantum-mechanical problem of a gas or liquid of particles interacting via a two-body potential. To attack this problem, perturbation theory has been refined and developed in many elegant forms, too numerous and well known to attempt to summarize here. These results have been mainly of two kinds: attempts, based on summing series to all orders, to make exact statements about the true solution of the problem; and attempts to develop approximation schemes to yield quantitative results. It is fair to say, however, that few of these results may claim to be mathematically rigorous. They are accepted because they agree with our intuitive understanding of the problem.

Faced with such a situation it would seem desirable to find a local, time- and velocity-independent, two-body potential (the sort of potential that actually exists in a gas) such that, starting from the Schrödinger equation, and without introducing any approximations, one could derive the energy levels and wave functions of the system. It would also be desirable that such a model problem be three dimensional.

While there do exist several model many-body Hamiltonians in the literature, which are solvable, with one exception, none of them corresponds to having a simple potential of the character mentioned above. The one exception is the work of Girardeau[1] on a gas of impenetrable bosons in one dimension. He showed that the energy spectrum of such a gas is identical with the spectrum of a noninteracting Fermi gas. It is, indeed, unfortunate that, due to the inability of the particles to get past one another, Girardeau's model can only be thought of as an extreme high-density situation—even though the actual density of his gas may be low. By this is meant that if we attempt to draw a parallel between the one-dimensional hard-core model and the three-dimensional hard-core model, then the former resembles the latter only at very high densities when a cell model becomes applicable. In consequence of this fact, the model cannot be used to check any known form of perturbation theory.

Another drawback of Girardeau's model is that it is essentially a zero-parameter model. It might have been hoped that varying the density, ρ, or the hard-core radius, a, would modify the spectrum in some essential way, but this is not true. If $E_n(p)$ is any energy level of a state having momentum p, then $E_n(p)$ is of the form $[\rho/(1-\rho a)]^2 f_n([(1-\rho a)/\rho]p)$. In other words, the hard cores simply play the role of an excluded volume.

We shall propose here a model that in many respects is similar to Girardeau's, but which overcomes the two difficulties mentioned above. We consider a gas of bosons in one dimension interacting via a repulsive δ-function potential, whose Hamiltonian is given by Eq. (2.1) below. As we show, this problem has one nontrivial parameter, namely, $\gamma=\rho^{-1}c$, where $2c$ is the strength of the δ function. When $\gamma=\infty$ we obviously obtain Girardeau's results since the particles then are impenetrable. When $\gamma=0$ we have the noninteracting Bose gas.

A useful feature of this model is that for small γ Bogoliubov's perturbation theory[2] is valid. This is discussed in Sec. IV. The model agrees with all the predictions of that theory except in one important respect: For all values of γ the most convenient and natural way to view the spectrum of the gas is in terms of a double elementary boson excitation spectrum—not a single one as previous calculations have suggested. Bogoliubov's single spectrum agrees quite well with one of the spectra we obtain, but the other is totally unaccounted for in his theory. The principal value of this model perhaps lies in this startling fact. The whole question of the excitation spectrum, as well as some heuristic reason that this duplicity of the spectrum might have been anticipated, and why it might also exist in three dimensions, are discussed in the following paper.[3]

We may summarize the results of this paper as follows:

[1] M. Girardeau, J. Math. Phys. 1, 516 (1960).

[2] See, for example, *The Many Body Problem*, edited by C. De Witt (John Wiley & Sons, Inc., New York, 1958), pp. 347–355.

[3] E. Lieb, following paper [Phys. Rev. 130, 1616 (1963) (referred to here as II)].

(i) We obtain explicitly the eigenfunctions of the problem and show how they may be derived from the solution of a transcendental algebraic equation. This is the content of Sec. II.

(ii) In Sec. III we find the ground-state energy as a function of γ and show that in the limit of a large system it is analytic in γ except at $\gamma=0$. This shows that perturbation theory can at best hope to give an asymptotic series for the ground-state energy.

Most quantities of interest, such as the ground-state energy, the velocity of sound, the excitation spectra and others, derived both here and in II, have been calculated explicitly and numerical graphs are given. These quantities are derived from solutions to certain integral equations which cannot be found in closed form. The numerical work thus entailed was performed on an IBM 7090 computer. At the end of Appendix B, an outline of the numerical procedure is given. All of the properties of the solutions claimed above, however, such as their analyticity for $\gamma\neq 0$, have been rigorously proved.

II. THE PROBLEM AND THE NATURE OF ITS SOLUTIONS

We begin with the Schrödinger equation for N particles in one dimension interacting via a δ-function potential.[4]

$$\left(-\sum_1^N(\partial^2/\partial x_i^2)+2c\sum_{\langle i,j\rangle}\delta(x_i-x_j)\right)\psi=E\psi, \quad (2.1)$$

where $2c$ is the amplitude of the δ function. The region of space under consideration is

$$R:\ \text{all } 0\leq x_i\leq L. \quad (2.2)$$

The wave function ψ satisfies periodic boundary conditions in each variable. We are interested in the repulsive case so that

$$c\geq 0. \quad (2.3)$$

While the attractive case, $c<0$, has a solution, it is not physically meaningful because there is no saturation. It is easy to demonstrate that in this case, for a fixed density, the ground-state energy is proportional to $-N^2$ instead of to $-N$.

It is well known that a δ-function potential is equivalent to the boundary condition

$$\left(\frac{\partial}{\partial x_j}-\frac{\partial}{\partial x_k}\right)\psi\Big|_{x_j=x_k+}-\left(\frac{\partial}{\partial x_j}-\frac{\partial}{\partial x_k}\right)\psi\Big|_{x_j=x_k-}=2c\psi|_{x_j=x_k}; \quad (2.4)$$

i.e., ψ is continuous whenever two particles touch, but the jump in the derivative of ψ is $2c$. Moreover, since we are interested in symmetric (Bose) wave functions, the two terms on the left side of Eq. (2.4) are by definition equal.

[4] $\hbar=1$, $2m=1$. $\sum_{\langle i,j\rangle}$ means summation over pairs.

We can go one step further than this, however, and observe that if we define the region R_1 by

$$R_1:\ 0\leq x_1\leq x_2\leq\cdots\leq x_N\leq L, \quad (2.5)$$

then knowledge of ψ in R_1 is equivalent to knowledge of ψ in R. Equations (2.1) and (2.4), thus, become

$$-\sum_1^N(\partial^2/\partial x_i^2)\psi=E\psi \text{ inside} R_1, \quad (2.1a)$$

$$(\partial/\partial x_{j+1}-\partial/\partial x_j)\psi|_{x_{j+1}=x_j}=c\psi|_{x_{j+1}=x_j}. \quad (2.4a)$$

The original Schrödinger equation, (2.1) is thus replaced by a Helmholtz equation (2.1a) together with a mixed boundary condition (2.4a) on the boundary of R_1.

The last step is to interpret the periodic boundary conditions on R in terms of ψ defined in R_1. The periodic boundary condition reads in part

$$\psi(0,x_2,\cdots,x_N)=\psi(L,x_2,\cdots,x_N), \quad (2.6)$$

with a similar condition for the derivatives. The argument of the right-hand side of Eq. (2.6) is not in R_1, but by definition

$$\psi(L,x_2,\cdots,x_N)\equiv\psi(x_2,x_3,\cdots,x_N,L). \quad (2.7)$$

Hence, the original periodic condition is equivalent to the following boundary condition on R_1:

$$\psi(0,x_2,\cdots,x_N)=\psi(x_2,\cdots,x_N,L), \quad (2.8a)$$

$$\frac{\partial}{\partial x}\psi(x,x_2,\cdots,x_N)\Big|_{x=0}=\frac{\partial}{\partial x}\psi(x_2,\cdots,x_N,x)\Big|_{x=L}. \quad (2.8b)$$

Equations (2.8), together with Eq. (2.4a), now completely cover the entire boundary of R_1.

We now make the following *ansatz* for ψ: Let $\{k\}=k_1,\cdots,k_N$ be an ordered set of N numbers and define

$$\psi(x_1,\cdots,x_N)=\sum_P a(P)P\exp\left(i\sum_{j=1}^N k_jx_j\right), \quad (2.9)$$

where the summation extends over all permutations of $\{k\}$, and $a(P)$ are certain coefficients depending on P. The wave function ψ is thus a generalization of the one-component Fermi function for which $a(P)$ would be $(-)^P$. It is to be understood that Eq. (2.9) defines ψ *only* in R_1, the extension to the rest of R following from the requirement of total symmetry under all particle permutations.

Can we choose $a(P)$ so that ψ satisfies the Schrödinger equation in R_1? Obviously Eq. (2.1a) is satisfied with

$$E=\sum_1^N k_j^2. \quad (2.10)$$

Let us now examine Eq. (2.4a) when $x_1=x_2$. If all the k's are distinct (as will prove to be the case), then the $N!$ terms in Eq. (2.9) are linearly independent. Let P be the permutation that takes $\{k\}$ into $p, q, k_{\alpha_3},\cdots,k_{\alpha_n}$ and Q be the permutation that takes $\{k\}$ into $q, p, k_{\alpha_3},\cdots,k_{\alpha_n}$, where $q=k_{\alpha_1}$ and $p=k_{\alpha_2}$. If Eq. (2.4a) is to be satisfied, there must be a relationship between $a(P)$ and $a(Q)$ depending on p and q, and not involving the other $a(P)$'s. Denote by y the common value of x_1

and x_2 and substitute the two terms of ψ corresponding to P and Q into Eq. (2.4a). We must be able to satisfy

$$i(q-p)[a(P)-a(Q)]\exp(i(p+q)y+\textstyle\sum_{j=3}^{N}k_{\alpha_j}x_j)$$
$$=c\exp(i(p+q)y+i\textstyle\sum_{j=3}^{N}k_{\alpha_j}x_j)$$
$$\times[a(P)+a(Q)] \quad (2.11)$$

for all values of y, x_3, $\cdots$, x_N. This is, indeed, possible if

$$a(Q)=-a(P)\frac{c-i(q-p)}{c+i(q-p)}=-a(P)\exp(i\theta_{\alpha_1\alpha_2}), \quad (2.12)$$

where

$$\theta_{ij}=\theta(k_i-k_j), \quad (2.13a)$$

and

$$\theta(r)=-2\tan^{-1}(r/c). \quad (2.13b)$$

Assuming r to be real we have

$$\pi\geq\theta(r)\geq-\pi. \quad (2.13c)$$

Thus, by regarding ψ as the sum of $\frac{1}{2}(N!)$ pairs of terms corresponding to permutations such as P and Q above, we can satisfy Eq. (2.4a) for $x_1=x_2$ for each pair separately when the coefficients are given by Eq. (2.12).

The reader will easily verify that we can satisfy Eq. (2.4a) simultaneously for all $N-1$ possibilities $x_j=x_{j+1}$ if we choose the $a(P)$ in the following way: Let

$$a(I)=1. \quad (2.14)$$

Rule. If P takes $\{k\}$ into $\{k\}'=k_{\alpha_1}, \cdots, k_{\alpha_N}$, then rearrange $\{k\}'$ into $\{k\}$ by the process of transposing only adjacent k's. For each transposition, write down the factor $-e^{i\theta_{st}}$ where k_s and k_t are the transposed k's in question, k_s lying to the left of k_t before the transposition. The product of all such factors thus obtained is $a(P)$. There are, in general, many different ways to get $\{k\}'$ into $\{k\}$ but they all give rise to the same set of factors. For example, if

$$P=\begin{pmatrix}123\\321\end{pmatrix},$$

then for both the paths $(321)\rightarrow(231)\rightarrow(213)\rightarrow(123)$ and $(321)\rightarrow(312)\rightarrow(132)\rightarrow(123)$ we obtain the result $a(P)=-\exp i(\theta_{32}+\theta_{31}+\theta_{21})$.

An alternative definition of the above rule which makes it evident that $a(P)$ is indeed independent of the path is the following: Write down k_1, $\cdots$, k_N in a line. Under it write down k_{α_1}, $\cdots$, k_{α_N}. For each k_j draw a straight line between the two points at which it occurs in the two rows. For each crossing point of two lines corresponding to k_α and k_β, $a(P)$ contains the factor $-e^{i\theta_{\alpha\beta}}$ where k_α precedes k_β in the set $\{k\}'$.

We conclude, therefore, that for *any* set $\{k\}$, ψ defined by Eq. (2.9) will satisfy Eqs. (2.1a) and (2.4a) when $a(P)$ is given by the above rule. There is one exception: All the k's in $\{k\}$ must be distinct, otherwise ψ will vanish identically. Condition (2.4a) determines only the form of ψ. The allowed values of the k's will be determined by Eq. (2.8). It is easily verified that these equations are equivalent to the N equations.

$$(-)^{N-1}e^{-ik_jL}=\exp(i\textstyle\sum_{s=1}^{N}\theta_{sj}), \quad \text{(all } j), \quad (2.15)$$

where we have defined

$$\theta_{kk}=0. \quad (2.16)$$

Although Eqs. (2.15) are N equations in N unknowns, there will be many sets of solutions. The problem we have here is quite similar to the one-dimensional Heisenberg model of ferromagnetism with nearest neighbor interactions, first solved by Bethe.[6] Let us agree to order the k's (assuming they are real) so that

$$k_1<k_2<\cdots<k_N. \quad (2.17)$$

By observing that

$$\theta_{ij}=-\theta_{ji}, \quad (2.18)$$

and taking the product of all N equations in (2.15), we find that

$$\textstyle\sum_1^N k_j=2(\pi/L)n, \quad (2.19)$$

where n is an integer. We also find, by virtue of Eq. (2.13a), that if $\{k\}$ is a solution to Eq. (2.15), then the set $\{k'\}$ defined by

$$k_i'=k_i+2\pi n_0/L \quad (2.20)$$

is a solution for any integer n_0. The significance of Eq. (2.20) is the following: Since Eq. (2.9) defines ψ only in R_1, the k's cannot be regarded as true wave vectors. But the sum of the k's *is* a true wave vector and is, in fact, the total momentum of ψ; i.e.,

$$P_{\rm op}\psi=\Big(\sum_1^N(-i\partial/\partial x_j)\Big)\psi=\Big(\sum_1^N k_j\Big)\psi=p\psi \quad (\text{in}R). \quad (2.21a)$$

Equation (2.19) tells us that the total momentum must be an integral multiple of $2\pi/L$—an expected result. Equation (2.20) tells us that for any state with momentum p there is a state with momentum

$$p'=p+2\pi n_0\rho, \quad (2.21b)$$

where

$$\rho=N/L, \quad (2.22)$$

obtained by a simple shift of the k's. Conversely for any state with momentum $|p|\geq\pi\rho$ there is a corresponding state having a momentum in the range

$$-\pi\rho<p\leq\pi\rho. \quad (2.23)$$

We have, thus, to consider only states in the interval (2.23). All other states are obtained by the simple "umklapp," Eq. (2.20), which has the physical meaning of imparting a total momentum to the system while preserving the "internal" state of the system. The integer n_0 in Eq. (2.20) is thus one of the quantum numbers

[6] See R. Orbach, Phys. Rev. **112**, 309 (1958). This contains references to earlier papers.

describing a state. The energy of the new state is

$$E'=\sum_1^N (k_j')^2=E+\frac{4\pi n_0\rho}{N}p+\frac{1}{N}(2\pi n_0\rho)^2, \quad (2.24)$$

a well-known consequence of Galilean relativity.

To elucidate the nature of Eq. (2.15) we must study the two-body problem. To avoid loss of continuity we have relegated this to Appendix A. The principal results are as follows: (a) The *ansatz*, Eq. (2.9), appears to give all eigenstates of the problem; (b) for $\gamma>0$, the case in which we are interested, the k's are always real; (c) for $\gamma=0$ we obtain the noninteracting solution, while for $\gamma=\infty$ we obtain Girardeau's solution[6] with a continuous transition in between.

These same considerations apply to the N-particle case. Since the k's are real, we may order them according to Eq. (2.17). For real $\{k\}$ we can say something about the solution to Eq. (2.15). Dividing two successive equations and equating exponents, we have

$$\delta_j\equiv(k_{j+1}-k_j)L=\sum_{s=1}^N(\theta_{s,j}-\theta_{s,j+1})+2\pi n_j \quad (j=1,2,\cdots,N-1), \quad (2.25)$$

where n_j is an integer depending upon j, defined for $j=1,2,\cdots,N-1$. These n_j's are the generalization of n_1 in Eq. (A3). Since θ_{kj} is a monotonically increasing function of j by Eqs. (2.13), the sum in Eq. (2.25) is negative. Therefore,

$$n_j\geq 1 \quad (\text{all } j), \quad (2.26)$$

if there is to be a solution. Since

$$k_j=k_1+(1/L)\sum_{s=1}^{j-1}\delta_s, \quad (2.27a)$$

and

$$k_\alpha-k_\beta=(1/L)\sum_{s=\beta}^{\alpha-1}\delta_s \quad (\alpha>\beta), \quad (2.27b)$$

the right side of Eq. (2.25) involves only the δ's. Equation (2.25) is thus a self-contained set determining the δ's. If any set $\{n\}$ is chosen satisfying (2.26), there is presumably exactly one solution for the δ's. At least that was the case for two particles and will prove to be true for the cases we shall investigate involving a large number of particles. We also note that

$$\delta_j<2\pi n_j. \quad (2.28)$$

The remaining step is to determine the individual k's so as to satisfy Eqs. (2.15) and (2.23). If the $j=1$ equation of (2.15) is satisfied, the remainder will automatically be by virtue of Eq. (2.25). Thus, if we choose k_1 such that

$$k_1=-\frac{1}{L}\sum_{k=1}^N\theta_{k1}-\frac{2\pi m}{L}+\frac{\epsilon(N)}{L}$$
$$=-\frac{1}{L}\sum_{k=1}^N\theta\left(\frac{1}{L}\sum_{j=1}^{k-1}\delta_j\right)-\frac{2\pi m}{L}+\frac{\epsilon(N)}{L}, \quad (2.29)$$

where

$$\epsilon(N)=\pi \quad \text{for } N \text{ even},$$
$$=0 \quad \text{for } N \text{ odd}, \quad (2.30)$$

and m is some integer, and use Eq. (2.27a) to determine the other k's, Eq. (2.15) will be satisfied. To determine the integer m, which is not arbitrary, we observe that

$$p=\sum_1^N k_j=Nk_1+\frac{1}{L}\sum_{j=1}^{N-1}(N-j)\delta_j$$
$$=\frac{1}{L}\sum_{j=1}^{N-1}(N-j)\delta_j-\rho\sum_{j=1}^N\theta_{j1}-2\pi m\rho+\epsilon(N)\rho. \quad (2.31)$$

Equation (2.23) then fixes m.

As further support for the assertion that Eq. (2.25) has a unique solution for a given $\{n\}$, we shall consider the two limiting cases $c=0$ and $c=\infty$.

$c=\infty$: In this case all the θ's are zero, whence

$$\lim_{c\to\infty}\delta_j=2\pi n_j, \quad (2.32)$$

so that not only is the solution of (2.25) unique, but (2.28) becomes an equality. It is clear that we recover all Girardeau's wave functions and energies; the term $\epsilon(N)$ just gives the extra π mentioned in footnote 6 for the even N case.

$c=0$: By Eq. (2.13)

$$\lim_{c\to 0}\theta(x)=-\pi \quad \text{for } x>0$$
$$=\pi \quad \text{for } x<0, \quad (2.33)$$

and, hence, Eq. (2.25) yields

$$\lim_{c\to 0}\delta_j=2\pi(n_j-1). \quad (2.34)$$

There is, however, a *caveat*: if some $n_j=1$ then $\delta_j\to 0$, but it is then not clear that we can use the limiting form (2.33). We must ask for $\lim_{c\to 0}(\delta/c)$. Only if it is infinity can we use Eq. (2.33) and thereby obtain Eq. (2.34). But if $\delta\nrightarrow 0$ Eq. (2.33) is certainly correct. Hence, by *reductio ad absurdum*, the only consistent conclusion is that Eq. (2.34) is always correct. The solution to Eq. (2.25) is again unique. We have, of course, recovered the well-known Bose functions for free particles. For N even, the $\epsilon(N)$ term in Eq. (2.31), which in this case is unwanted, is compensated by the term $\sum\theta_{j1}$.

We, thus, may conclude that the n_j are a complete set of "internal" quantum numbers for the system. Together with n_0 [Eq. (2.20)] they specify all states. By virtue of Eq. (2.28) moreover, the n_j may be thought of heuristically as the spacing between the k's.

[6] Girardeau obtained the solution to the hard-core problem only for N odd. For N even, however, there is an equally simple solution. One simply uses the Fermi function satisfying *antiperiodic* boundary conditions, i.e., $Lk=\pi+2n\pi$, where n is an integer. This exactly compensates the unwanted change of sign of his A function. For two particles the ground state has $k_2=-k_1=\pi/L$.

The exact spacing depends upon c to be sure, and also upon the relative position of δ_j in the set $\{\delta\}$. The fact that even for $n_j=$const the spacing is not uniform (except for $c=\infty$) gives rise to the principal difficulty in the analysis of the many-particle case. One other peculiarity of the problem should be remarked upon. The limiting case $c=\infty$ proved above to be particularly simple, but for $c\to 0$ the problem became pathological. Indeed, the latter limit will prove to be surprisingly delicate as we shall see.

III. GROUND-STATE ENERGY AND WAVE FUNCTION

We are interested in passing to the limit of a large system. This means N, $L\to\infty$ such that $\rho=$fixed constant. From dimensional considerations the ground-state energy can always be written

$$E_0=N\rho^2 l(N,cL), \tag{3.1}$$

where l is a dimensionless function of its arguments. But if E_0 is to be an extensive variable (as will, indeed, prove to be the case), l can depend only on intensive variables. The only dimensionless intensive variable in the problem is

$$\gamma=c/\rho. \tag{3.2}$$

Therefore,

$$E_0=N\rho^2 e(\gamma). \tag{3.3}$$

It is easy to show directly from Eq. (2.1) that $e(\gamma)$ is a monotonically increasing function of γ. We also expect that $e(0)=0$ (free particles) and $e(\infty)=\pi^2/3$ (Girardeau's solution). We can easily get an upper bound for $e(\gamma)$ by a variational calculation. Using the unperturbed Bose function, $\psi=1$, we obtain

$$e(\gamma)\leq\gamma. \tag{3.4}$$

Using Girardeau's function,[1] we have

$$e(\gamma)\leq\pi^2/3. \tag{3.5}$$

Returning to Eq. (2.25), it is clear that the choice

$$n_j=1 \tag{3.6}$$

gives the ground state, for this choice minimizes the δ's and hence allows the k's to be as close to zero as possible. We also see from the symmetry of the equations that if k is in $\{k\}$, then so is $-k$. Indeed this is always so for $n_j=$const and implies that $p=0$ for such states. Equation (2.31) determines k_1 in this case:

$$-k_1=\frac{1}{NL}\sum_1^{N-1}(N-j)\delta_j=k_N\equiv K(\gamma). \tag{3.7}$$

From Eq. (2.28) *et seq.* we see that

$$K(0)=0,\quad K(\infty)=\pi\rho, \tag{3.8}$$

with presumably a continuous monotonic transition in between.

We are, thus, led to expect that $K(\gamma)$ is an intensive variable and that the process of passing to an infinite system consists in "filling in" more and more points in the set $\{k\}$ between $-K$ and K with the understanding that the spacing between k's may not be uniform. This may be shown as follows: The inequality $k_{j+1}-k_j<2\pi/L$, Eq. (2.28), permits us for a large system to use a Taylor expansion in Eq. (2.25), i.e.,

$$\theta(k_s-k_j)-\theta(k_s-k_{j+1}) =-2c(k_{j+1}-k_j)\frac{1}{c^2+(k_s-k_j)^2}+O(1/L^2), \tag{3.9}$$

and, hence,

$$(k_{j+1}-k_j)=-2c(k_{j+1}-k_j)\frac{1}{L}\sum_{s=1}^{N}[c^2+(k_s-k_j)^2]^{-1} +\frac{2\pi}{L}+O(1/L^2). \tag{3.10}$$

Define

$$k_{j+1}-k_j=1/Lf(k_j), \tag{3.11}$$

by means of which the sum in Eq. (3.10) may be approximated by an integral by Poisson's formula to the required accuracy in L^{-1}. Equation (3.10) becomes

$$2c\int_{-K}^{K}\frac{f(p)}{c^2+(p-k)^2}dp=2\pi f(k)-1. \tag{3.12}$$

The meaning of $f(k)$ is that for a large system

$$Lf(k)dk=\text{number of } k\text{'s in } (k,\,k+dk). \tag{3.13}$$

The subsidiary condition determining the number of particles is

$$\int_{-K}^{K}f(k)dk=\rho, \tag{3.14}$$

while the ground-state energy is given by

$$E_0=\sum_1^N k_j^2=\frac{N}{\rho}\int_{-K}^{K}f(k)k^2dk. \tag{3.15}$$

One final condition implied by Eqs. (3.11) and (3.13) is

$$f(k)\geq 0. \tag{3.16}$$

Let us change variables as follows: Define

$$k\equiv Kx;\quad c\equiv K\lambda;\quad f(Kx)\equiv g(x), \tag{3.17}$$

in terms of which Eqs. (3.12), (3.14), and (3.15) become, respectively,

$$1+2\lambda\int_{-1}^{1}\frac{g(x)dx}{\lambda^2+(x-y)^2}=2\pi g(y), \tag{3.18}$$

$$e(\gamma)=\frac{\gamma^3}{\lambda^3}\int_{-1}^{1}g(x)x^2dx, \tag{3.19}$$

$$\gamma\int_{-1}^{1}g(x)dx=\lambda. \tag{3.20}$$

The program, therefore, consists of the following steps: (i) Solve Eq. (3.18) for a fixed λ; (ii) use Eq. (3.20) to determine λ as a function of γ; (iii) Eq. (3.19) then gives $e(\gamma)$; (iv) Eq. (3.17) gives $K(\gamma)$, i.e.,

$$K=\rho\gamma\lambda^{-1}=\rho\left[\int_{-1}^{1} g(x)dx\right]^{-1} \tag{3.21}$$

Equation (3.18) is an inhomogeneous Fredholm equation of the second kind with an inhomogeneous term that is positive definite (i.e., $+1$). In Appendix B we discuss this equation in considerable detail and prove the following:

(a) For *any* inhomogeneous term there is exactly one solution $g(y)$.

(b) $g(y)$ is an infinitely differentiable (analytic) function of λ for $\lambda>0$.

(c) If the inhomogeneous term is positive definite [as in Eq. (3.18)], $g(y)>0$ for all y.

(d) If the inhomogeneous term is bounded above (below), then $g(y)$ is bounded above (below).

These statements enable us to claim that $g(y,\gamma)$, $e(\gamma)$, and $K(\gamma)$ are analytic functions of γ (except for $\gamma=0$). Since there are then no unusual kinks or points of discontinuous derivatives, it becomes a straightforward matter to evaluate all quantities numerically with confidence. The proof is as follows: (b) and (c) above, together within Eq. (3.20), imply that γ is an analytic function of λ for λ in $(0,\infty)$. Equation (3.19) then implies that e is analytic in λ; Eq. (3.21) implies that K is analytic in λ. The problem is then to prove that (i) λ is an analytic function of γ and that (ii) the range $0<\lambda<\infty$, in fact, covers the required range $0<\gamma<\infty$. By the implicit function theorem, statement (i) will be true if we can show that

$$0\neq\frac{\partial\gamma}{\partial\lambda}=\left[\int_{-1}^{1} g(x)dx\right]^{-2} \times\left\{\int_{-1}^{1} g(x)dx-\lambda\int_{-1}^{1}\frac{\partial}{\partial\lambda}g(x)dx\right\}, \tag{3.22}$$

for all λ. By (b) above, $[\int_{-1}^{1} g(x)dx]^{-2}\neq 0$. As for the two terms in { }, we observe that if we define

$$h(x,\lambda)\equiv\left(\frac{\partial}{\partial\lambda}+\frac{x}{\lambda}\frac{\partial}{\partial x}\right)g(x,\lambda), \tag{3.23}$$

and integrate Eq. (3.18) by parts several times, we easily obtain an equation for h, viz.,

$$2\pi h(y,\lambda)=2\lambda\int_{-1}^{1} dx\frac{h(x,\lambda)}{\lambda^2+(x-y)^2} -2g(1,\lambda)\left(\frac{1}{\lambda^2+(y+1)^2}+\frac{1}{\lambda^2+(y-1)^2}\right) \tag{3.24}$$

By (c) above, the inhomogeneous term in Eq. (3.24) is negative definite and, consequently, $h(x,\lambda)<0$ [by then applying (c) to Eq. (3.24)]. We have, therefore,

$$0>\lambda\int_{-1}^{1} h(x,\lambda)dx =\int_{-1}^{1}\left(\lambda\frac{\partial g(x,\lambda)}{\partial\lambda}-g(x,\lambda)\right)dx+2g(1,\lambda) \tag{3.25}$$

by using the definition (3.23) and integrating by parts. Since $g(1,\lambda)>0$, the integral in Eq. (3.25) is negative definite. But this integral is just the negative of the term in { } in (3.22). Hence, statement (i) above is proved. We see, incidentally, that λ and γ are monotonically increasing functions of each other.

As for (ii) above, we see from Eq. (3.18) that

$$\lim_{\lambda\to\infty} g(x,\lambda)=1/2\pi, \tag{3.26}$$

and, consequently, from Eq. (3.20)

$$\lim_{\lambda\to\infty}\gamma=\infty. \tag{3.27a}$$

As $\lambda\to 0$, on the other hand, the bounds mentioned in (d) above guarantee that $g(x,\lambda)\not\to 0$. [On the contrary $g(x,\lambda)\to\infty$ in this limit.] Equation (3.20) then implies that

$$\lim_{\lambda\to 0}\gamma=0. \tag{3.27b}$$

Equations (3.27) establish (ii) above.

To recapitulate, we have shown that all quantities are analytic functions of γ in $(0,\infty)$. As the above arguments tend to suggest, however, and as we, in fact, show later, there is a very serious singularity at $\gamma=0$. The physical meaning of this singularity is that any interaction, however weak (but volume-independent), leads in the limit of a large system to a system that is in some way basically different from the noninteracting case. If one tries to find a series for $e(\gamma)$ for small γ, such a series can at best be asymptotic—a result already indi-

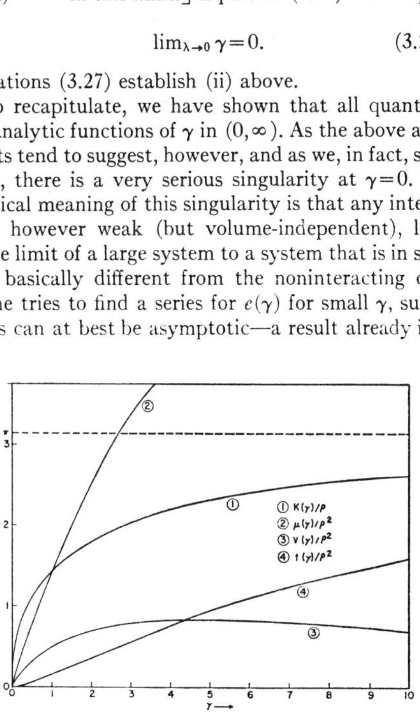

FIG. 1. Various numerically derived properties of the ground state plotted as functions of $\gamma=c/\rho$. K=cutoff momentum, μ=chemical potential, v=potential energy per particle, and t=kinetic energy per particle. As $\gamma\to\infty$: $K\to\pi\rho$; $\mu\to\pi^2\rho^2$, $v\to 0$, and $t\to\frac{1}{3}\pi^2\rho^2$.

cated, but not proved, by the conventional perturbation theoretic treatments.

In Appendix B an outline is given of the numerical solution of Eqs. (3.18–20). While nonpathologic behavior is guaranteed by the above analyticity considerations, the singularity at $\gamma=0$ makes the numerical calculation more and more difficult as this limit is approached.

In plotting results we use units in which $\rho=1$, so that K/ρ varies between 0 and π. We have also eliminated the dependent parameter λ in favor of the independent parameter γ by using Eq. (3.20). [Numerical values of λ as a function of γ can be obtained, if desired, from the relation $\lambda=\gamma(K/\rho)^{-1}$, Eq. (3.21).] Figure 1 shows $K(\gamma)$ while Fig. 2 shows the density function $f(k,\gamma)$ for several values of γ. It will be noticed that as $\gamma\rightarrow 0$ small values of k become relatively more important. Figure 3 shows the end result of the calculation, $e(\gamma)$. The straight dashed line is the zeroth-order perturbation theory result $e=\gamma$, which also coincides with the primitive upper bound obtained before [Eq. (3.4)]. The other dashed curve, $e=\gamma[1-(4/3\pi)\sqrt{\gamma}]$, is the result of Bogoliubov's perturbation theory which we shall discuss presently. It is in fair agreement with the exact answer up to $\gamma=2$. We may, therefore, regard $0\leq\gamma<2$ as the weak coupling region. On the other hand, it is interesting to note that even for $\gamma=10$, $e(\gamma)$ is still quite far from its asymptotic limit, $\frac{1}{3}\pi^2$. In fact, in order to get within 10% of this asymptotic limit γ must be about 37. The intermediate coupling region, by one definition at least, is, therefore, $2<\gamma<37$.

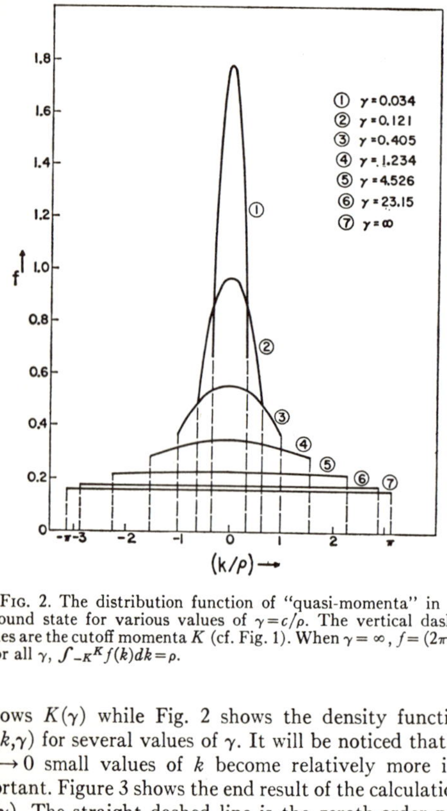

FIG. 2. The distribution function of "quasi-momenta" in the ground state for various values of $\gamma=c/\rho$. The vertical dashed lines are the cutoff momenta K (cf. Fig. 1). When $\gamma=\infty$, $f=(2\pi)^{-1}$. For all γ, $\int_{-K}^{K}f(k)dk=\rho$.

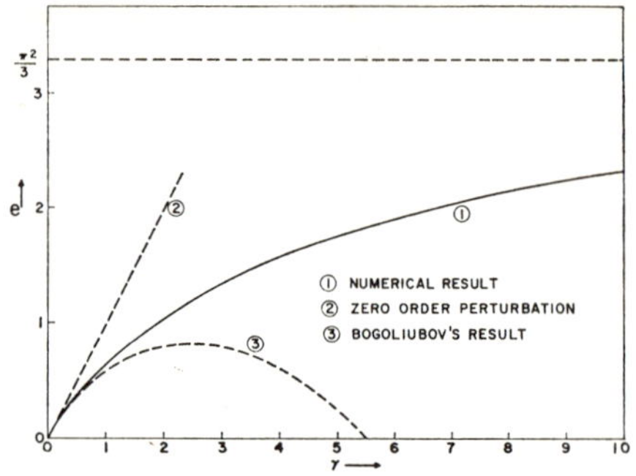

FIG. 3. The ground-state energy $E_0=N\rho^2e(\gamma)$. The full curve gives $e(\gamma)$ obtained numerically. As $\gamma\rightarrow\infty$, $e\rightarrow\frac{1}{3}\pi^2$. Curve 2 is the zero-order perturbation theory result, i.e., the expectation value of the potential in the noninteracting ground state. Curve 3 is the result of Bogoliubov's perturbation theory, which is seen to be good up to $\gamma\sim 2$.

There are other quantities of interest in connection with the ground state; these are also plotted in Fig. 1. There is the chemical potential μ, or energy necessary to add one particle to the system, defined by

$$\mu=\frac{\partial E_0}{\partial N}=\rho^2\left(3e-\gamma\frac{de}{d\gamma}\right). \qquad (3.28)$$

Also interesting is the potential energy per particle v, which by a well-known theorem is

$$v=\frac{c}{N}\frac{\partial}{\partial c}E_0=\rho^2\gamma\frac{de}{d\gamma}, \qquad (3.29)$$

and the kinetic energy per particle

$$t=\frac{1}{N}E_0-v=\rho^2\left(e-\gamma\frac{de}{d\gamma}\right). \qquad (3.30)$$

We conclude this section with a list of the asymptotic forms of the various quantities for large and small γ.

Large γ. This is the easiest case. The denominator in the integral of Eq. (3.18) may be effectively replaced by λ^2. The error thus introduced in the integral is small, but by Eq. (3.26) the integral itself is small compared to 1, so that the resultant error in g will be quite small indeed. This substitution leads to $g(x)=\text{const}$ which may be readily evaluated as

$$g(x)=\lambda(2\pi\lambda-4)^{-1}. \qquad (3.31)$$

Substitution of Eq. (3.31) into Eqs. (3.19) and (3.20)

leads (for large γ) to

$$\lambda=\frac{1}{\pi}(\gamma+2),\quad e=\tfrac{1}{3}\pi^2\left(\frac{\gamma}{\gamma+2}\right)^2,\quad K=\pi\frac{\gamma}{\gamma+2}\rho,$$
$$\mu=\frac{3\gamma+2}{\gamma+2}\rho^2 e,\quad v=\frac{4}{\gamma+2}\rho^2 e,\quad t=\frac{\gamma-2}{\gamma+2}\rho^2 e. \tag{3.32}$$

These results are accurate to 1% for γ as small as 10.

Small γ. As we show in Appendix B, as $\lambda\to 0$ the factor 2π on the right side of Eq. (3.18) becomes an eigenvalue of the integral equation. Consequently, $g(x,\lambda)$ has a singularity at $\lambda=0$. In addition to this difficulty, the kernel of the integral equation also becomes quite pathological. It will be recognized that as $\lambda\to 0$ the kernel becomes a well-known representation for $2\pi\delta(x-y)$ so that in this limit Eq. (3.18) reads: $2\pi g(x)=2\pi g(x)+1$. It is clear, therefore, that in this limit g becomes unbounded—a statement borne out by the limits on g given in Appendix B.

There does not seem to be any simple way to get a systematic, reliable expansion of g as $\lambda\to 0$. We can, however, guess the zeroth order form for g:

$$g(x,\lambda)\sim\frac{1}{2\pi\lambda}(1-x^2)^{1/2},\quad(\lambda\to 0) \tag{3.33}$$

and one can then show, using bounding arguments similar to those in Appendix B, that the correction to (3.33) is of higher order in λ for all x, and is positive definite. Beyond this, we have been unable to obtain an unambiguous correction to (3.33). One of the major difficulties is deciding what happens at the end points, $|x|=1$.

Equation (3.33) allows us to find only the leading term for the quantities mentioned in (3.32). These turn out to be (for small γ)

$$\lambda=\tfrac{1}{2}\sqrt{\gamma},\quad e=\gamma,\quad K=2\rho\sqrt{\gamma},$$
$$\mu=2\rho^2\gamma,\quad v=\gamma,\quad t=0. \tag{3.34}$$

The equation $e=\gamma$ is plotted in Fig. 3. We see that the upper bound for e [Eq. (3.4)] is, in fact, its asymptotic form—a result predicted by elementary perturbation theory.

It will be noticed from Fig. 1, as well as from Eqs. (3.32) and (3.34) ,that for small γ the potential energy dominates the kinetic. The reverse is true for large γ; in fact $v\to 0$ as $\gamma\to\infty$. This behavior is exactly the same as for three dimensions—it is often said that a large potential behaves like a kinetic energy barrier. This fact is supposed by some to be tied up with the ability of particles to "go around" each other in three dimensions. But as we can clearly see, it is also present in one dimension. Thus, the difference between one and three dimensions does not lie here—it is apparently immaterial to the particles whether they can "get around" each other or merely "through" each other.

IV. PERTURBATION THEORY

The well-known perturbation theory of Bogoliubov[2] was at first assumed to be an expansion in the density. That this is incorrect was realized when it was found that the correct parameter for a low-density expansion is the scattering length (in three dimensions) and not $\int v(x)d^3x$ as appears in Bogoliubov's theory. But Bogoliubov's theory *is* asymptotically correct if we regard it as an expansion in the potential. In other words, it may be expected to give the first two terms in the energy correctly for any density if the potential is so weak that it may be treated by the Born approximation. In our case, therefore, the Bogoliubov theory should be correct for small γ. On the other hand, small γ may be thought of as high density—the reverse of the usual (incorrect) assumption mentioned above.

This last observation leads to a second reason for believing in the Bogoliubov result for small γ. One can show (we shall not do so here) that if the Fourier transform of the potential is everywhere nonnegative, then Bogoliubov's theory is correct—at least as far as the ground-state energy is concerned—in the limit of *high* density. The δ-function potential certainly satisfies the above criterion.

As stated in the introduction, one of the uses of an exact model is as a check for approximation schemes. It is indeed fortunate that the present model has a definite range of coupling constant over which the Bogoliubov theory may be unambiguously expected to be correct. Our model, therefore, should serve not only to establish the validity of the Bogoliubov theory but should also serve as an indicator (taking appropriate account of the difference between one and three dimensions) of the range over which the theory is reasonably accurate.

The Bogoliubov prescription easily yields the following results: The spectrum of elementary excitations is given by

$$\epsilon(p)=\rho^2|p/\rho|[(p/\rho)^2+4\gamma]^{1/2}. \tag{4.1}$$

We have purposely written $\epsilon(p)$ in terms of the dimensionless variable p/ρ. The ground-state energy is given by

$$E_0=\tfrac{1}{2}N^2(2c/L)+\tfrac{1}{2}\sum_p[\epsilon(p)-p^2-N(2c/V)]$$
$$=N\rho^2\gamma[1-(4/3\pi)\sqrt{\gamma}]. \tag{4.2}$$

The expression for $e(\gamma)$, (4.2) is plotted in Fig. 3; it is adequate up to approximately $\gamma=2$. Equation (3.34) shows the leading term is exact. While we have not found an analytic expression for the second term, $(4/3\pi)\gamma^{3/2}$, the numerical results of Fig. 3 indicate that it too is correct. It is interesting to note that the exact equation for $e(\gamma)$ is so pathological at $\gamma=0$ that it was an effort to find even the zeroth-order term for $e(\gamma)$, while perturbation theory gives the first two terms by elementary quadrature.

As for the elementary excitations, we shall show in II that Eq. (4.1) is also fairly accurate up to about $\gamma=2$, but that there is another type of elementary excitation unaccounted for by Bogoliubov's theory.

ACKNOWLEDGMENTS

The authors are most grateful to their colleagues at IBM—Dr. D. Mattis, Dr. D. Jepsen, and Dr. T. Schultz, for many stimulating discussions. Above all, they are indebted to Dr. A. Sakakura of this laboratory for his insights into the integral equation of Sec. III.

Approximately half the work of this paper was done while E. Lieb was on leave from IBM as visiting Senior Lecturer at Fourah Bay College, the university college of Sierra Leone, Freetown.

APPENDIX A. THE TWO-BODY PROBLEM

To elucidate the nature of Eqs. (2.15), we shall study the two-body problem. To highlight the basic difference between the repulsive and attractive cases, we shall consider both. For the two-body problem, (2.15) becomes

$$-e^{ik_2L}=e^{i\theta_{21}}=-e^{-ik_1L}. \qquad \text{(A1)}$$

There are several ways to solve this problem and we shall choose one which may be generalized to the N-body case. Define

$$(k_2-k_1)L=\delta, \quad \gamma=\tfrac{1}{2}cL, \qquad \text{(A2)}$$

then, multiplying the two equations (A1), we obtain

$$\begin{aligned}\delta&=2\theta(k_2-k_1)+2\pi n_1 \qquad \text{(A3)}\\ &=-4\tan^{-1}(\delta/2\gamma)+2\pi n_1,\end{aligned}$$

where n_1 is an integer. If we invert Eq. (A3) we obtain

$$\delta/2\gamma=-\tan(\delta/4), \quad n_1 \text{ even} \qquad \text{(A4a)}$$

$$=[\tan(\delta/4)]^{-1}, \quad n_1 \text{ odd}. \qquad \text{(A4b)}$$

It must be remembered that Eqs. (A4) have far more solutions than Eq. (A3) (because $|\theta|<\pi$ by definition). Now if $z\tan z=$real, then z is either real or imaginary, not complex. The same applies if $(1/z)\tan z=$real. Hence, the roots of Eq. (A3) are either real or imaginary. For the real roots we choose δ positive in accordance with (2.17). For $\gamma>0$ (repulsive case), there are, in fact, only real roots as Eqs. (A4) show. For every solution to either Eq. (A4a) or (A4b) there exists an integer n_1 such that Eq. (A3) is satisfied, and conversely for every integer $n_1\geq 1$ there is a unique positive solution to Eq. (A3) given by either (A4a) or (A4b). The single exception is $n_1=0$ for which there can be no solution because the first term on the left side of Eq. (A3) is negative ($\delta=0$ is not allowed). We see that

$$2\pi(n_1-1)<\delta<2\pi n_1. \qquad \text{(A5)}$$

In the limiting cases the roots corresponding to $n_1=1$, 2, 3, etc., are, respectively,

$$\delta=0,\ 2\pi,\ 4\pi,\ \text{etc.} \quad (\gamma=0) \qquad \text{(A6a)}$$

$$\delta=2\pi,\ 4\pi,\ 6\pi,\ \text{etc.} \quad (\gamma=\infty). \qquad \text{(A6b)}$$

We have now to determine k_1 and k_2 separately so as to satisfy Eqs. (A1). Substituting the value of θ_{21} from Eq. (A3) into Eq. (A1) we see that for every root, δ, there is in fact a unique solution satisfying (2.23) given by

$$Lk_2=-Lk_1=\tfrac{1}{2}\delta; \quad p=0, \quad (n_1 \text{ odd}) \qquad \text{(A7a)}$$

$$Lk_2=\tfrac{1}{2}\delta+\pi; \quad Lk_1=-\tfrac{1}{2}\delta+\pi;$$
$$p=2\pi/L=\pi\rho, \quad (n_1 \text{ even}). \qquad \text{(A7b)}$$

Thus, we see that for every integral value of $n_1\geq 1$ there corresponds a unique state having a definite energy and a momentum in the range (2.23). The integer n_1 may, therefore, be considered the second or "internal," quantum number of the problem which, together with n_0, specifies a state. The ground state corresponds to $n_1=1$ and has zero momentum as expected.

For $\gamma<0$ (attractive case), the situation is more complicated and rather surprising. The real roots are much the same as before except that here $n_1=0$ becomes permissible if $\gamma>-2$. Equation (A5) is changed to

$$2\pi n_1<\delta<2\pi(n_1+1), \qquad \text{(A5a)}$$

while Eqs. (A7) remain unaltered. For the imaginary roots, it is wise to return to the original definition of $e^{i\theta}$ [Eq. (2.12)] in order to avoid ambiguity in the definition of $\tan^{-1}$. One finds that for $n_1=1$ there is always a pair of equal and opposite imaginary roots given by Eq. (A4b). Only one of them need be considered, for the other corresponds to interchanging k_1 and k_2. If we write $\delta=i\alpha$ $(\alpha>0)$, then

$$\alpha>-2\gamma. \qquad \text{(A8)}$$

By Eq. (A7a) this state has zero momentum and its energy is

$$E_0=-(1/2L^2)\alpha^2<-\tfrac{1}{2}c^2. \qquad \text{(A9)}$$

This state, which is the ground state of the problem, is the analog of the single bound state for an attractive delta function in infinite space for which $E_0=-\frac{1}{2}c^2$, $\psi=\exp(-\frac{1}{2}c|x_1-x_2|)$. Putting the particles in a box with infinite walls (zero boundary conditions) would reduce the binding energy, however, contrary to Eq. (A9). The periodic boundary conditions are responsible for this anomalous result.

Even more surprising is the appearance of a second bound state. We saw above that for $\gamma>-2$ there is an $n_1=0$ real solution. In this case there is no other bound state. But if $\gamma<-2$ (strong coupling) the real solution for $n_1=0$ disappears in favor of an imaginary solution to Eq. (A4a). For this solution

$$\alpha' < -2\gamma. \tag{A10}$$

By Eq. (A7b) this state has a momentum $p=2\pi/L$ and an energy

$$E_0' = -\frac{1}{2L^2}(\alpha')^2 + \frac{2}{L^2}\pi^2 > -\tfrac{1}{2}c^2. \tag{A11}$$

For $\gamma=-2$ the $n_1=0$ root is $\delta=0$ which must be discarded. On the other hand, the state cannot disappear for a single value of γ. The paradox is resolved by taking the limit $\gamma \to -2$, whereupon we find

$$\psi' = [2+c(x_2-x_1)]\exp\left[i\frac{\pi}{L}(x_1+x_2)\right], \quad (\gamma=-2), \tag{A12}$$

in R_1. Two facts should be noticed about this second bound state. Firstly, it has a nonzero momentum, but it does *not* correspond to the translation of a bound state with $p=0$. Translated states of two particles with zero momentum must have momenta which are integral multiples of $4\pi/L$. Secondly, this state does not disappear in the limit $L\to\infty$ as might have been hoped. In fact, for *any* fixed c we can always choose L large enough such that $\gamma<-2$ and this second state will make its appearance. As $L\to\infty$ both inequalities (A8) and (A10) approach equalities so that E_0 and E_0' becomes asymptotically degenerate and equal to the infinite space binding energy.

Having obtained these wave functions we may ask if the *ansatz* Eq. (2.9) [with the single exception, Eq. (A12)] exhausts all the solutions of the Schrödinger equation. It seems clear that it, in fact, does so, at least for the repulsive case, although we cannot prove that this is true. We are led to this view because in the repulsive case we obtain the well-known wave functions for $\gamma=0$ (free particles) and we obtain Girardeau's solution for $\gamma=\infty$ (hard cores), with a continuous transition from one to the other. For $\gamma<0$, on the other hand, the situation is not quite so clear. As $\gamma\to 0$ we recover the free particle solutions. For $\gamma\neq 0$ we can only observe that as $L\to\infty$ we obtain the well-known bound state and scattering solutions for the attractive δ-function potential.

These same considerations apply to the N-particle problem and we shall therefore suppose that Eqs. (2.9) and (2.15) yield all the wave functions of the problem for the repulsive case.

From consideration of the two-body problem we are led to the following hypothesis: For the repulsive case every solution $\{k\}$ contains only real k's, which we may, therefore, order according to (2.17). For the attractive case, which we shall not consider further, complex k's may appear. One should expect many types of nonreal solutions for $\gamma<0$ corresponding to two-body, three-body, etc., bound states.

APPENDIX B. SOLUTION OF THE BASIC INTEGRAL EQUATION

I. Existence, Uniqueness, and Analyticity of the Solution

(i) The integral equation (3.18) is a special case of

$$r(y) + \int_{-1}^{1} K(y-x)g(x)dx = \sigma g(y), \tag{B1}$$

where

$$K(y) = 2\lambda/(\lambda^2+y^2)$$

and $\lambda>0$. Equation (B1) can be brought into the form

$$u(s) - \mu\int_{-a}^{a} R(s-t)u(t)dt = v(s), \tag{B2}$$

by letting $y=\lambda s$, $x=\lambda t$, $a=1/\lambda<\infty$, $\mu=2/\sigma$, $u(s)=g(y)$, $v(s)=r(y)/\sigma$, and

$$R(s,t) = R(s-t) = 1/[1+(s-t)^2]. \tag{B3}$$

By definition,[7] R is positive definite if

$$J_a = \int_{-a}^{a}\int_{-a}^{a} R(s,t)u(s)u(t)dsdt > 0, \tag{B4}$$

for any square integrable $u(s)\not\equiv 0$. For R to be positive definite with respect to the interval $(-a, a)$ it is sufficient that R be positive definite with respect to $(-\infty, \infty)$ because the square integrable functions with respect to $(-a, a)$ can be thought of as a subset of square integrable functions with respect to $(-\infty, \infty)$ which vanish identically for $|s|>a$. Let

$$\eta(s) = \int_{-\infty}^{\infty} R(s-t)u(t)dt \quad \text{and} \quad u^*(p),\ \eta^*(p)$$

be the Fourier transforms of $u(s)$ and $\eta(s)$, respectively. Then, according to Parseval's equation,[8] and because $\eta(s)$ is real,

$$J_\infty = \int_{-\infty}^{\infty} u(s)\eta(s)ds = \int_{-\infty}^{\infty} u^*(p)\bar{\eta}^*(p)dp.$$

As $\eta(s)$ is defined by a convolution, one has

$$\eta^*(s) = (2\pi)^{1/2}R^*(p)u^*(p),$$

and, thus

$$J_\infty = (2\pi)^{1/2}\int_{-\infty}^{\infty} \bar{R}^*(p)|u^*(p)|^2 dp.$$

---

[7] R. Courant and D. Hilbert, *Methoden der Mathematischen Physik* (Springer-Verlag, Berlin, 1931), Vol. I, p. 105.

[8] W. Schmeidler, *Integralgleichungen mit Anwendungen in Physik und Technik* (Akademische Verlagsgesellschaft, Leipzig, 1955), Vol. 1, pp. 74, 75.

It is well known[9] that

$$R^*(p)=\bar{R}^*(p)=(\pi/2)^{1/2}e^{-|p|}>0,$$

and, thus, $J_\infty>0$ except for $u(s)\equiv 0$, that is (B4) is proved. Since R is positive definite, all eigenvalues of (B1) or (B2) are positive.[10]

(ii) Consider the kernel R and its iterates[11]

$$R^{(1)}(s,t)=R(s,t), \qquad \text{(B5a)}$$

$$R^{(i+1)}(s,t)=\int_{-a}^{a} R^{(i)}(s,\tau)R(\tau,t)d\tau, \quad (i\geqq 1). \qquad \text{(B5b)}$$

From the defintion (B3) of R it is evident that there exist constants $C_j>0$ such that $|\partial^j R^{(1)}/\partial s^j|\leqq C_j$ for any $j\geqq 0$. One has

$$\int_{-a}^{a} R(\tau,t)d\tau=T(a,t)=\arctan(a-t)+\arctan(a+t). \qquad \text{(B6)}$$

It is easy to see that

$$\pi/2<\arctan 2a\leqq T(a,t)\leqq 2\arctan a=\pi-\epsilon, \quad \epsilon>0. \qquad \text{(B7)}$$

Assume now that, for a certain $i\geqq 1$ and any $j\geqq 0$,

$$|\partial^j R^{(i)}(s,t)/\partial s^j|\leqq C_j(\pi-\epsilon)^{i-1}.$$

Then, because $\partial^j R^{(i)}(s,t)/\partial s^j$ is continuous in s for all j, one can differentiate under the integral sign and, using (B7), one gets

$$\left|\frac{\partial^j R^{(i+1)}(s,t)}{\partial s^j}\right|=\left|\int_{-a}^{a}\frac{\partial^j R^{(i)}(s,\tau)}{\partial s^j}\right.$$
$$\left.\times R(\tau,t)d\tau\right|\leqq C_j(\pi-\epsilon)^i. \qquad \text{(B8)}$$

As (B8) holds for $i=0$ it follows by induction that (B8) holds for all $i\geqq 0$. Therefore, the Neumann series and its derivatives with respect to s taken term by term, that is,

$$\frac{\partial^j \rho(s,t)}{\partial s^j}=\sum_{i=0}^{\infty}\mu^i\frac{\partial^j R^{(i+1)}(s,t)}{\partial s^j} \qquad \text{(B9)}$$

for $j\geqq 0$, converge absolutely and uniformly in both variables s and t for all $\mu<1/(\pi-\epsilon)$. The same result holds for the derivatives with respect to t as can be proved in a similar way.

From the uniform convergence of the Neumann series for $j=0$ follows first the existence of the resolvent kernel, $\rho(s,t)$, in particular for $\mu=1/\pi$, and the validity of the representation[12]

$$u(s)=v(s)+\mu\int_{-a}^{a}\rho(s,t)v(t)dt, \qquad \text{(B10)}$$

[9] G. A. Campbell and R. M. Foster, *Fourier Integrals for Practical Applications* (American Telephone and Telegraph Company, New York, 1942), p. 45.
[10] Reference 7, p. 112.
[11] Reference 8, p. 270.
[12] Reference 7, p. 119.

or of the equivalent representation

$$g(y)=\frac{1}{\sigma}r(y)+\frac{1}{\sigma^2}\int_{-1}^{1}\kappa(y,x)r(x)dx, \qquad \text{(B11a)}$$

$$\kappa(y,x)=\sum_{i=0}^{\infty}\sigma^{-i}K^{(i+1)}(y,x). \qquad \text{(B11b)}$$

The eigenvalues of (B2) are, thus, all $\geqq 1/(\pi-\epsilon)$ and those of (B1) are $\leqq 2\pi-2\epsilon$. Property (a) of Sec. III, that is, the existence of a unique solution of Eq. (B1) for $\sigma=2\pi$, is thus proved. From the uniform convergence of the series in (B9) with $j\geqq 1$ follows the analyticity of the resolvent kernel with respect to s and t.

(iii) Let $\lambda=z$ be complex in the expressions for the kernel K and its iterates. Consider a closed circular domain $\bar{D}$ defined by $|z-\lambda_1|\leqq\delta$, with real λ_1, δ and $0<\delta<\lambda_1/\sqrt{2}$. Then, if y is real, it is easy to see that $|K(y;z)|\leqq 2(\lambda_1+\delta)/(\lambda_1-\delta)^2$ for any $z\epsilon\bar{D}$. Let $[\pi-\epsilon_1(\delta)]=2\arctan[2/(\lambda_1-\delta)]$, $\epsilon_1(\delta)>0$. As $\delta\rightarrow 0$, the quantity $q(\delta)=[(\lambda_1+\delta)/(\lambda_1-\delta)]$. $[\pi-\epsilon_1(\delta)]$ decreases monotonically to $(\pi-\epsilon)=2\arctan(2/\lambda_1)$, where $\epsilon_1(\delta)\leqq\epsilon$. Thus, if $0<\epsilon_2<\epsilon$, $q(\delta)\leqq\pi-\epsilon_2$ for sufficiently small δ. Now assume that, for a certain $i\geqq 1$ and real y, x, ξ,

$$|K^{(i)}(y,x;z)|\leqq 2^i\frac{\lambda_1+\delta}{(\lambda_1-\delta)^2}(\pi-\epsilon_2)^{i-1}. \qquad \text{(B12)}$$

Then,

$$|K^{(i+1)}(y,x;z)|=\left|\int_{-1}^{1}K^{(i)}(y,\xi;z)K(\xi,x;z)d\xi\right|$$
$$\leqq 2^i\frac{\lambda_1+\delta}{(\lambda_1-\delta)^2}(\pi-\epsilon_2)^{i-1}\int_{-1}^{1}\frac{2(\lambda_1+\delta)d\xi}{(\lambda_1-\delta)^2+(\xi-x)^2},$$

and

$$\int_{-1}^{1}\frac{2(\lambda_1+\delta)d\xi}{(\lambda_1-\delta)^2+(\xi-x)^2}=\frac{2(\lambda_1+\delta)}{(\lambda_1-\delta)}\int_{-1}^{1}\frac{(\lambda_1-\delta)d\xi}{(\lambda_1-\delta)^2+(\xi-x)^2}$$
$$\leqq 2q(\delta)\leqq 2(\pi-\epsilon_2);$$

thus, (B12) holds for $i+1$. As shown above, (B12) does hold for $i=1$, and by induction it follows that (B12) holds for all $i\geqq 1$. As a consequence, the Neumann series (B11b) with $\lambda=z$ converges uniformly with respect to z in $\bar{D}$. The kernel K and its iterates are analytic functions of z in $\bar{D}$ because $\bar{D}$ does not intersect the imaginary axis, the only place where K and its iterates could have singularities. Thus, for $\sigma\geqq 1/2\pi$, (B11b) represents a uniformly converging series of analytic functions and, as a consequence,[13] $\kappa(y,x;z)$ is analytic in z in the neighborhood of any $\lambda_1>0$. In particular, κ is analytic in λ in the real sense. Finally, from (B11a) follows the analyticity of $g(y)$ as a function of λ. This completes the proof of statement (b) of Sec. III, that is $g(y)$ is analytic in λ for $\sigma=1/2\pi$.

[13] E. T. Whittaker and G. N. Watson, *A Course of Modern Analysis* (Cambridge University Press, New York, 1952), p. 91.

(iv) Using $1/(1+4a^2)\leqq|R(s,t)|\leqq 1$ and inequality (B7) one can prove by induction that

$$[1/(1+4a^2)](\pi/2)^{i-1}\leqq|R^{(i)}(s,t)|\leqq(\pi-\epsilon)^{i-1}.$$

Thus, there exist constants, b_1 and b_2, such that, for $\mu\leqq 1/\pi$,

$$0<b_1=[1/(1+4a^2)]\frac{1}{1-\mu(\pi/2)}\leqq|\rho(s,t)|$$

$$\leqq\frac{1}{1-\mu(\pi-\epsilon)}=b_2<\infty \qquad \text{(B.13)}$$

If there exist constants b_3 and b_4 such that $b_3\leqq v(s)\leqq b_4$, then one obtains from (B10)

$$b_3(1+2ab_1)\leqq u(s)\leqq b_4(1+2a\mu b_2), \qquad \text{(B14)}$$

which proves statement (d) of Sec. III, that is if $r(y)$ is bounded above or below, then so is the solution $g(y)$. The proof of statement (c) of Sec. III [that is, if $r(y)$ is positive definite, then so is the solution $g(y)$] follows immediately from (B14) as b_3 can be chosen positive in this case.

II. Numerical Solution

The basic integral equation (B2) has been solved numerically by applying Simpson's rule to the integral on a grid $\{s_i=-a+(i-1)h;\, h=a/n;\, i=1,\cdots,2n+1\}$. This yields a system of $(2n+1)$ linear algebraic equations for the $(2n+1)$ discrete approximate values $u_i\approx u(s_i)$ which can be solved by a standard method. The quadratures involved in calculating γ and e also have been carried out by Simpson's rule. The functions $K(\gamma)$ and $e(\gamma)$ are obtained in parametric form, that is $(K(\lambda),\gamma(\lambda))$ and $(e(\lambda),\gamma(\lambda))$. To obtain μ, the quantities e and γ are evaluated on a sufficiently fine grid of equidistant λ values. Then,

$$\mu=3e-\gamma\frac{de}{d\gamma}=3e-\gamma\frac{de}{d\lambda}\bigg/\frac{d\gamma}{d\lambda}$$

can be calculated by numerical differentiation.

Reprinted from THE PHYSICAL REVIEW, Vol. 130, No. 4, 1616–1624, 15 May 1963
Printed in U. S. A.

Exact Analysis of an Interacting Bose Gas. II. The Excitation Spectrum

ELLIOTT H. LIEB
Thomas J. Watson Research Center, International Business Machines Corporation, Yorktown, Heights, New York
(Received 7 January 1963)

We continue the analysis of the one-dimensional gas of Bose particles interacting via a repulsive delta function potential by considering the excitation spectrum. Among other things we show that: (i) the elementary excitations are most naturally thought of as a double spectrum, not a single one; (ii) the velocity of sound derived from the macroscopic compressibility is shown to agree with the velocity of sound derived from microscopic considerations, i.e., from the phonon spectrum. We also introduce a distinction between elementary excitations and quasiparticles, on the basis of which we give some heuristic reasons for expecting the double spectrum to be a general feature, even in three dimensions, and not an exception.

I. INTRODUCTION

IN the preceding paper[1] we introduced a soluble model of a Bose gas interacting via a repulsive δ-function potential. We discussed the nature of the eigenfunctions and explicitly calculated the ground-state energy and other properties of the ground state in the limit of a large system.

In this paper we discuss the nature of the excitation spectrum for a large system of N particles. The surprising result, as we stated but did not show in I, is that for all values of the potential strength, the most convenient and natural way to view the spectrum is to regard it as a *double* spectrum of elementary boson excitations. While Bogoliubov's perturbation theory[2,3] gives one of the spectra quite accurately for a weak potential, the second spectrum is entirely unaccounted for (see Figs. 3 and 4). The second spectrum exists only for values of the momentum satisfying $|p|\leq\pi\rho$.

We may summarize the results of this paper as follows: (i) In Sec. II we discuss the nature of the energy spectrum of the problem and show that there are two elementary spectra. These are always well defined and are explicitly calculated. We show that there is no energy gap and that the two spectra have a common slope at $p=0$ which means that they propagate sound at the same velocity. The velocity of sound at absolute zero derived in this way from an atomic picture[4–6] is shown to be identical with the velocity of sound defined by the usual macroscopic considerations [cf. Eqs. (1.1) and (1.4)].

[1] E. Lieb and W. Liniger, Phys. Rev. **130**, 1605 (1963) (referred to here as I).
[2] See I, Sec. IV.
[3] See, for example, *The Many Body Problem*, edited by C. DeWitt (John Wiley & Sons, Inc., New York, 1958), pp. 347–355.
[4] R. P. Feynman, Phys. Rev. **91**, 1291 (1953).
[5] R. P. Feynman, Phys. Rev. **91**, 1301 (1953).
[6] R. P. Feynman, Phys. Rev. **94**, 262 (1954).

The velocity of sound v_s, derived from the excitation spectrum is given by

$$v_s = \lim_{p\to 0} \partial\epsilon(p)/\partial p, \tag{1.1}$$

where $\epsilon(p)$ is the energy of an elementary excitation of momentum p. By a well-known macroscopic argument,[7] on the other hand,

$$v_s = [(-L/m\rho)\partial P/\partial L]^{1/2}, \tag{1.2}$$

where P is the pressure,

$$P = -\partial E_0/\partial L, \tag{1.3}$$

L is the length of the "box," $\rho = N/L$ is the density, m is the mass per particle ($=\frac{1}{2}$ in our units), and E_0 is the ground-state energy.

In I we defined the dimensionless parameter $\gamma = \rho^{-1}c$, where $2c$ is the strength of the δ function. In terms of γ, E_0 may be written $E_0 = N\rho^2 e(\gamma)$. In terms of these quantities, v_s may be written

$$\begin{aligned} v_s &= 2\rho(3e - \gamma\dot{e} + \tfrac{1}{2}\gamma^2\ddot{e})^{1/2} \\ &= 2(\mu - \tfrac{1}{2}\gamma\dot{\mu})^{1/2}, \end{aligned} \tag{1.4}$$

where "dot" denotes differentiation with respect to γ and μ is the chemical potential given by

$$\mu = \frac{\partial E_0}{\partial N} = \rho^2(3e - \gamma\dot{e}). \tag{1.5}$$

Bogoliubov's formula for $\epsilon(p)$ [I(4.1)] gives

$$v_s = 2\rho\gamma^{1/2}, \tag{1.6}$$

while using Eq. (1.4) and Bogoliubov's expression for $e(\gamma)$[I(4.2)] we get

$$v_s = 2\rho[\gamma - (1/2\pi)\gamma^{3/2}]^{1/2}. \tag{1.7}$$

Equations (1.6) and (1.7) are plotted as dashed curves in Fig. 5 along with the correct result for v_s obtained numerically. It will be noted that Eq. (1.7) is far more accurate than Eq. (1.6). In fact, the difference between Eq. (1.7) and the exact answer is so minute up to $\gamma = 10$ that we are unable to distinguish the two graphically. This is indeed remarkable in view of the fact that Bogoliubov's expression for $e(\gamma)$ is quite bad beyond $\gamma = 3$ and becomes negative for $\gamma > (3\pi/4)^2 \simeq 5.5$.

(ii) Based on the results of this model, we are led to the introduction of two distinct quantities—called elementary excitations and quasiparticles—which are defined differently, and are, in fact, different, but which are apparently somehow related. This is discussed in Sec. III.

The quantities that we calculate are given by the solutions to certain integral equations. While these cannot be obtained in closed form we are able to prove directly the statements made above. The quantitative results have been obtained by Werner Liniger on an IBM 7090 computer at the IBM Research Center. The equations considered here are similar to those considered in I. The reader is referred to Appendix B of that paper for a discussion of the numerical methods employed.

[7] See, for example, F. London, *Superfluids* (John Wiley & Sons, Inc., New York, 1954), Vol. II, p. 83.

II. EXCITED STATES AND ELEMENTARY EXCITATIONS

In paper I we discussed the Hamiltonian[8]

$$H = -\sum_1^N \partial^2/\partial x_i^2 + 2c\sum_{\langle i,j\rangle}\delta(x_i - x_j), \tag{2.1}$$

and showed that in the subdomain $0 \le x \le \cdots \le x_N \le L$ all eigenfunctions are of the form

$$\psi(x_1,\cdots,x_N) = \sum_P a(P)P\exp(i\sum_{j=1}^N k_j x_j), \tag{2.2}$$

where the summation is over permutations, P, of $\{k\} = k_1, < k_2, < \cdots, < k_N$, and $a(P)$ is a function of P. We found that for any set $\{k\}$ the $a(P)$'s could be uniquely determined. The condition determining the allowed k's is

$$\delta_j \equiv (k_{j+1} - k_j)L = \sum_{s=1}^N (\theta_{sj} - \theta_{s\,j+1}) + 2\pi n_j \qquad (j = 1, 2, \cdots, N-1) \tag{2.3}$$

$$\theta_{sj} = -2\tan^{-1}[(k_s - k_j)/c], \tag{2.4}$$

and n_j are any integers ≥ 1.

The ground state was obtained from Eq. (2.3) by putting all $n_j = 1$. In this case $-k_1 = k_N = K$ where K is a function of γ. The simplest excitation is obtained by putting some one $n_j = 2$ while the rest remain unity. The effect of this is to increase δ_j by approximately 2π, but at the same time all the other δ's will be shifted slightly. Neglecting the change in spacing, the effect of making $n_j = 2$ can be approximately thought of either as moving k_{j+1} to K (or $-K$) leaving a hole behind, or else as pushing apart the k's, leaving a double space at the jth position.

To orient ourselves, let us consider the $\gamma = \infty$ (hard core) case. As Girardeau[9] has pointed out, the energy spectrum of the Bose gas is then exactly the same as for a noninteracting one-component Fermi gas (the wave functions are quite different of course). The problem is this: How can we describe this Fermi spectrum in conventional boson terms? If we regard the energy levels simply as a collection of numbers, the problem is similar to that of classical spectroscopy. Can we find a small number of elementary energy levels from which all others may be deduced by addition?

The conventional description of the Fermi spectrum is to regard the elementary excitation process as taking a particle from the state q ($|q| < K$) to a state k ($|k| > K$). The energy and momentum of this state are

$$\begin{aligned} \epsilon(k,q) &= k^2 - q^2, \\ p(k,q) &= k - q. \end{aligned} \tag{2.5}$$

[8] $\hbar = 1$, $2m = 1$.

[9] M. Girardeau, J. Math. Phys. **1**, 516 (1960).

The difficulty with this classification scheme is twofold: (a) Every excitation must be described in terms of two parameters instead of merely one as one normally hopes to do for a Bose system. There is no unique $\epsilon(p)$ curve. (b) Too many provisos must be added to the original specification (e.g., there can be at most one excitation of any given type; $|q|$ can be greater than K for high-level compound excitations, etc.).

If we are interested only in low-lying excitations, however, difficulties (a) and (b) above can be remedied by the following scheme. Define two types of one parameter elementary excitations by these rules:

Type I. Take a particle from K to $q>K$ (or alternatively from $-K$ to $q<-K$). This excitation has an energy and momentum given by

$$\epsilon_1=q^2-K^2, \tag{2.6}$$

$$\begin{aligned} p&=q-K, \quad q>K \\ &=q+K, \quad q<-K \end{aligned} \tag{2.7}$$

$$\epsilon_1(p)=p^2+2\pi\rho|p|. \tag{2.8}$$

Type II. Take a particle from $0<q<K$ to $K+2\pi/L$ (or alternatively from $-K<q<0$ to $-K-2\pi/L$). The energy and momentum for this state are given to $O(N^{-1})$ by

$$\epsilon_2=K^2-q^2, \tag{2.9}$$

$$\begin{aligned} p&=K-q, \quad 0<q<K \\ &=-K-q, \quad 0>q>-K \end{aligned} \tag{2.10}$$

$$\epsilon_2(p)=2\pi\rho|p|-p^2. \tag{2.11}$$

For the first type of excitation $-\infty<p<\infty$ while for the second type $-\pi\rho\leq p\leq\pi\rho$ only. These two types of excitations must be supplemented by exactly two other excitations, which we shall call *umklapp* excitations, namely take a particle from $-K$ to $K+2\pi/L$ (or alternatively from K to $-K-2\pi/L$). The energies of the umklapp excitations are $2\pi^2\rho/L$ and, hence, are of order N^{-1}. Their momenta, however, are $2\pi\rho$ and $-2\pi\rho$, respectively.

Any of these excitations may be carried out any number of times and we may simultaneously have as many different types of as many different momenta as we please. For example, one type I excitation of momentum p means taking a particle from K to $K+p$. Two excitations of this same momentum means taking one more particle from $K-2\pi/L$ to $p-2\pi/L$. The momentum of the resulting state is exactly $2p$ while the energy is twice that of one excitation to order N^{-1}. Thus, each of these elementary excitations may be regarded as bosons.

By experimenting with a k-space diagram, the reader may easily convince himself that for any specified combination of multiple excitations of type I (or type II) there corresponds exactly one true state of the system, and conversely. The momentum of the true state is exactly the sum of the momenta of the elementary excitations, while the energy of the true state differs from the sum of the elementary energies only by $O(N^{-1})$ if the total number of excitations is finite (i.e., not of order N).

The difficulty with this classification scheme is that *either* type I *or* type II excitations form a "complete set" in the sense that they are in a one-one correspondence with the true excitation spectrum. Thus, any type II excitation may be thought of as a multiple type I excitation. The correspondence is this: m excitations of type I with momentum $p=2\pi n/L$ corresponds to exactly the same true state as n type II excitations with momentum $p=2\pi m/L$. In particular, one type II excitation of momentum p corresponds to $(1/2\pi)Lp$ type I excitations of momentum $p=2\pi/L$. If p is a fixed momentum independent of L, however, then the corresponding type II excitation may be thought of essentially as an infinite number of type I excitations of vanishing momentum. Since $\epsilon_1(p)$ and $\epsilon_2(p)$ are not proportional to p, the true energy of the type II excitation, $2\pi\rho p-p^2$, is very badly approximated by the $(1/2\pi)Lp$ type I excitations, viz.,

$$(1/2\pi)Lp[2\pi\rho(2\pi/L)-(2\pi/L)^2]\cong 2\pi\rho p.$$

The situation is, indeed, rather complicated and there does not seem to be any very simple way out of the dilemma. On the one hand, one can say that there is only one elementary excitation spectrum (one can choose either type I or type II) which is boson-like and which "generates" all the true states. On the other hand, physical considerations tell us that type I and type II excitations ought to stand on an equal footing, just as holes and particles play a similar role in the theory of the Fermi gas. It is patently absurd to think of a type II excitation as an infinite number of type I excitations of vanishing momentum.

We shall adopt here the "symmetric" point of view which seems to us the more natural way to view the problem. But we must then realize that if we calculate the partition function on this basis we will be double counting. The true free energy is that of the Fermi gas which one can calculate by standard methods. To leading order,

$$F=E_0-(kT)^2A, \tag{2.12}$$

where A is given by a well-known integral. If one calculates the free energy using the double boson spectrum [Eqs. (2.9) and (2.11)], one finds that

$$F'=E_0-2(kT)^2A. \tag{2.13}$$

The factor 2 is a precise reflection of the double counting. Either type I or type II alone would give Eq. (2.12). But if one does not double count, one gets a thoroughly distorted view of the nature of the excitation spectrum.

The situation for $\gamma\neq\infty$ is qualitatively the same as for the $\gamma=\infty$ case we have just discussed. Only the $\epsilon(p)$ functions differ; having discussed the over-all structure of the elementary excitations, we proceed now to calculate $\epsilon_1(P)$ and $\epsilon_2(P)$ for $\gamma\neq\infty$.

Type I Excitations ("Particle" States)

We put $n_j=1(1\leq j\leq N-2)$ and $n_{N-1}\gg 1$. Let $k_N=q$, where $q>K$, and write

$$k_i'=k_i+(1/L)\omega_i,\quad i=1,\cdots,N-1 \qquad (2.14)$$

where k_i' are the new values of the remaining k's. At this point there are two ways to proceed—both leading of the same result. In Eq. (2.14) we could let $k_i=k$'s in the N-particle ground state, or we could let $k_i=k$'s in the $N-1$ particle ground state (ω_i will, of course, be different in the two cases). We shall adopt the latter definition because it leads to a simpler calculation.

Now,

$$\theta(k_j'-k_i')=\theta(k_j-k_i)+\frac{1}{L}(\omega_j-\omega_i)\frac{-2c}{c^2+(k_j-k_i)^2}. \qquad (2.15)$$

Inserting Eq. (2.15) in Eq. (2.3) or I(2.15) we obtain the integral equation

$$\omega(k)+\pi=2c\int_{-K}^{K}\frac{[\omega(r)-\omega(k)]f(r)}{c^2+(r-k)^2}dr-\theta(q-k), \qquad (2.16)$$

where $f(k)$ is the distribution function of the k's in the ground state, [cf. I(3.12)]. We have chosen the term π (instead of 3π, 5π, etc.) in Eq. (2.16) so that as $q\to 0$ the wave function approaches the ground-state wave function for N particles. If we define

$$\omega(k)f(k)\equiv J(k), \qquad (2.17)$$

Eq. (2.16) becomes

$$2\pi J(k)=2c\int_{-K}^{K}\frac{J(r)dr}{c^2+(k-r)^2}-\pi-\theta(q-k). \qquad (2.18)$$

Using Eq. (2.14) the momentum of the state is

$$P=\sum_{j=1}^{N-1}k_j'+q=\sum_{1}^{N-1}k_i+\frac{1}{L}\sum_{1}^{N-1}\omega_i+q$$

$$=q+\int_{-K}^{K}J(k)dk, \qquad (2.19)$$

while the excitation energy is

$$\epsilon_1=\textstyle\sum_{j=1}^{N-1}(k_i')^2-E_0(N)+q^2$$

$$=-\mu+q^2+2\int_{-K}^{K}kJ(k)dk, \qquad (2.20)$$

where $E_0(N)$ is the ground-state energy for N particles. To find $\epsilon_1(p)$ we have to eliminate the parameter q between Eqs. (2.19) and (2.20).

As we stated in I, Sec. III, Eq. (2.18) has a solution which is unique and negative definite (because $\pi+\theta(q-k)$ is positive definite). Since J (and, hence, ω) are negative we conclude, as was to be expected, that adding a particle with momentum $q>K$ to the $N-1$ particle system decreases all the other k's.

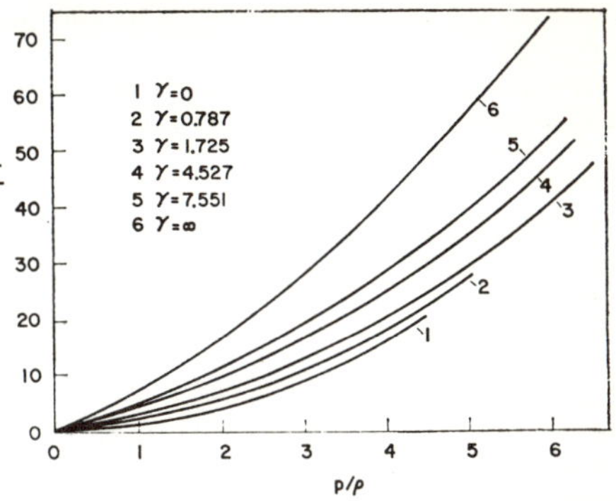

FIG. 1. The type I excitation spectrum, $\epsilon_1(p)$, plotted for small momenta and various values of $\gamma=c/\rho$. When $\gamma=0$, $\epsilon_1=p^2$; when $\gamma=\infty$, $\epsilon_1=p^2+2\pi\rho|p|$.

In Fig. 1 we display the $\epsilon_1(p)$ curve for several values of γ (in dimensionless units). The $\gamma=0$ curve is $\epsilon_1(p)=p^2$ (free particles), while the $\gamma=\infty$ is given by Eq. (2.8). All curves are linear for small p, the region of linearity increasing with γ.

The continuation of these curves for large p is given in the logarithmic plot of Fig. 2. To find an analytic expression for $\epsilon_1(p)$ for large p we expand $\theta(q-k)$ in a Laurent series for large q and retain the leading term. The obvious results are

$$J(k)\simeq-\frac{2\gamma\rho}{q}f(k), \qquad (2.21)$$

$$p\simeq q-2\gamma\rho^2/q, \qquad (2.22)$$

$$\epsilon_1(p)\simeq-\mu+p^2+4\gamma\rho^2. \qquad (2.23)$$

It should be noted that these last equations are true only if γ is finite and $p/\rho\gg\gamma$. If $\gamma=\infty$ on the other

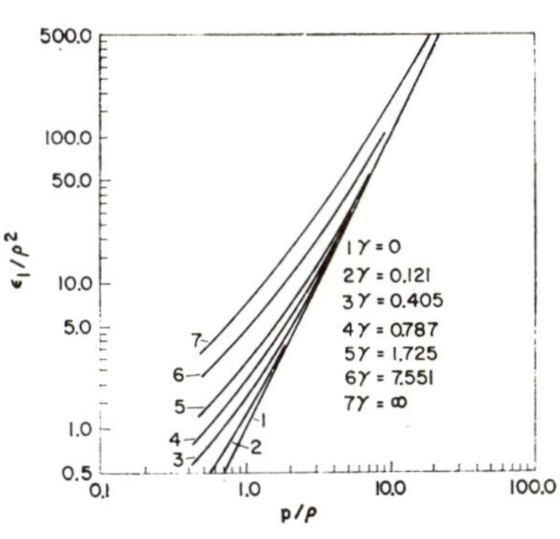

FIG. 2. The type I excitation spectrum for large momentum and various values of $\gamma=c/\rho$. When $\gamma=0$, $\epsilon_1=p^2$; when $\gamma=\infty$, $\epsilon_1=p^2+2\pi\rho|p|$.

hand, $\theta \equiv 0$ and Eqs. (2.6)–(2.8) must be used instead. In this case $J(k) = -\pi f(k) = -\frac{1}{2}$.

Equation (2.21) and (2.23) have a simple physical interpretation. The wave function can be well represented (in an average sense) by

$$\psi_p = \sum_P e^{ipx_N} \psi_0^{N-1}(x_1, \cdots, x_{N-1}), \qquad (2.24)$$

where the summation is on permutations and ψ_0 is the ground state for $N-1$ particles. In second-quantized notation this would be

$$\psi_p = a_p^\dagger | \psi_0^{N-1} \rangle. \qquad (2.25)$$

The energy of this state [Eq. (2.23)] can be interpreted as follows: First, remove a particle from the N-particle ground state. This contributes an energy—μ. Second, put it back in a plane wave state with momentum p—thereby adding p^2 to the energy. But the plane wave also interacts with the remaining $N-1$ particles—this energy is $\rho v(0)$ [$v(p)$ is the Fourier transform (F.T.) of the potential]. In our case this is $2\gamma\rho^2$. Finally, when the wave function is symmetrized, additional terms appear which ordinarily vanish as $p \to \infty$. But one of these terms—the F. T. of the potential times the two-particle correlation function—does not vanish for a δ-function and is in fact just $2\gamma\rho^2$. The total result is Eq. (2.23).

Before turning to excitations of type II ("hole" states) there are several questions to be answered about the behavior of Eqs. (2.18)–(2.20) as $q \to K$.

(a) Do both p and ϵ_1 vanish in this limit? If so, is $\epsilon_1(p)$ linear for small p?

(b) Is Bogoliubov's expression for the velocity of sound correct for small γ?

(c) Do the two expressions, Eqs. (1.1) and (1.4), for the velocity of sound agree for all γ?

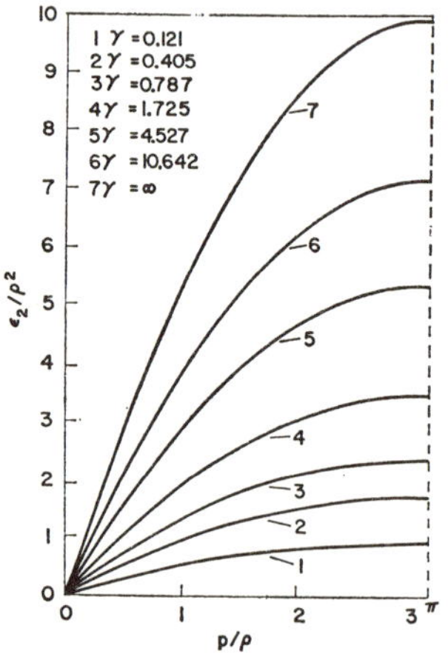

FIG. 3. The type II excitation, ϵ_2, as a function of momentum for various values of $\gamma = c/\rho$. This spectrum exists only up to $|p| = \pi\rho$. When $\gamma = 0$, $\epsilon_2 = 0$; when $\gamma = \infty$, $\epsilon_2 = -p^2 + 2\pi\rho |p|$.

All these questions can be answered in the affirmative, but the proof involves very tedious manipulation of the integral Eqs. I (3.18) and (2.18). These are summarized in the Appendix. Question (b) above is especially significant in view of the great debate as to whether or not there is an energy gap in the excitation spectrum. In our model there is none.

Type II Excitations ("Hole" States)

In this case we set all $n_i = 1$ except for $i = j$ when $n_j = 2$. Let $k_j = q$ (which we may assume to be a continuous parameter) with $0 < q < K$. In analogy with Eq. (2.14) we write

$$\begin{aligned} k_i' &= k_i + (1/L)\omega_i, \quad (i \leq j) \\ k_i' &= k_{i+1} + (1/L)\omega_i, \quad (i > j) \end{aligned} \qquad (2.26)$$

where now $k_i = k$'s in the $(N+1)$ particle ground state.

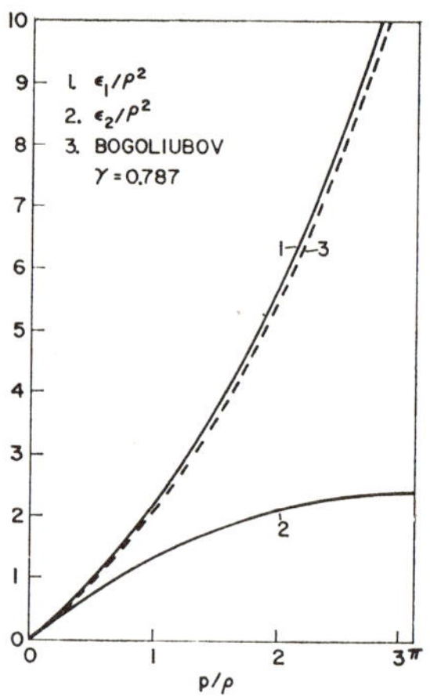

FIG. 4. A comparison plot of the two types of excitations, ϵ_1 and ϵ_2, for $\gamma = 0.787$. The dashed curve is Bogoliubov's spectrum which is quite close to the type I spectrum. The type II spectrum does not exist in Bogoliubov's theory.

As before we derive the equations

$$2\pi J(k) = 2c \int_{-K}^{K} \frac{J(r)dr}{c^2 + (k-r)^2} + \pi + \theta(q-k), \qquad (2.27)$$

$$p = -q + \int_{-K}^{K} J(k)dk, \qquad (2.28)$$

$$\epsilon_2 = \mu - q^2 + 2\int_{-K}^{K} kJ(k)dk. \qquad (2.29)$$

We observe that in this case $J(k)$ is negative definite (to see why this ought to be so, look at the $\gamma = \infty$ limit). When $q = K$, the solution to Eq. (2.27) is the negative of the solution to Eq. (2.18) so that in this limit both p and ϵ_2 vanish. As q decreases both p and ϵ_2 increase until, when $q = 0$, $p = \pi\rho$ [and not $K(\gamma)$ as might have been expected]. The case $q < 0$ need not be considered.

If we think of the Fermi case, what we have done is taken a particle from $q>0$ to K. Taking a particle from $q<0$ to K is, in our picture, to be regarded as the sum of two excitations; taking a particle from $q<0$ to $-K$ followed by an umklapp from $-K$ to K.

In Fig. 3 we show the $\epsilon_2(p)$ curves for various values of γ. The $\gamma=\infty$ case is given by Eq. (2.11). When $\gamma=0$, $\epsilon_2(p)\equiv 0$. Figure 4 compares $\epsilon_1(p)$, $\epsilon_2(p)$ and Bogoliubov's spectrum given in I Eq. (4.1) for $\gamma=0.787$. It will be seen that Bogoliubov's spectrum is a good approximation to $\epsilon_1(p)$ for this value of γ. It will also be noted that ϵ_1 and ϵ_2 have the same slope at $p=0$—a fact which may be easily proved from the integral equations. This common value of the velocity of sound is plotted in Fig. 5 together with Bogoliubov's results, Eqs. (1.6) and (1.7). With regard to the velocity of sound, we have seen that the macroscopic definition, Eq. (1.4), agrees for all γ with the microscopic definition based on the conjecture of Feynman[4–6] that the low-energy excitations are in general longitudinal phonons. But Feynman's additional conjecture that there is only one spectrum is apparently incorrect, at least in the present model. If one excites a sound wave in this system, most probably both types of excitations will be excited, just as in a Fermi gas both holes and particles are excited.

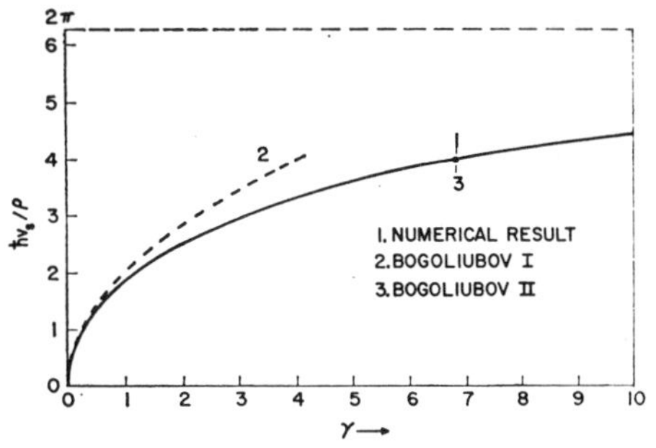

FIG. 5. The velocity of sound, v_s, as a function of $\gamma=c/\rho$. Curve 2 is the result obtained from Bogoliubov's excitation spectrum. Curve 3, which is graphically indistinguishable from the true result up to $\gamma=10$, is the velocity of sound derived from the macroscopic compressibility using Bogoliubov's expression for the ground-state energy.

III. A DISCUSSION OF ELEMENTARY EXCITATIONS AND QUASIPARTICLES

In the literature of the quantum-mechanical many-body problem two phrases are used interchangeably—elementary excitations and quasiparticles. We would like to propose here a distinction between the two.

A gas in a box has certain well-defined energies and wave functions which form a complete set. By definition these do not decay or interact with each other. The energies are real. Furthermore, as we have seen in our model and as is undoubtedly true, in general, the spectrum of low-lying states falls into a pattern. There exists one or more (in our case two) sets of energy vs momentum curves such that: (a) For each point on one of these curves there is an eigenstate; (b) if we add together the energy and momenta corresponding to several points on one or more of the curves, we obtain (to order N^{-1}) a resultant energy and momentum corresponding to an exact wave function of the system. The converse is also true: every state can be thought of as a sum of the elementary states.

These basic energy vs momentum curves we call *elementary excitations*. From this point of view, elementary excitations are a bookkeeping arrangement. There does not exist any simple operator which, acting on the ground state, gives these elementary states, nor can "compound" states be obtained from the elementary ones by simple operators. Nevertheless, when one attempts to diagonalize the many-body Hamiltonian by some method, it is the elementary excitations in the above sense that one is calculating.

Another quantity, called a *quasiparticle* may be defined as a pole in the Fourier transform of a propagator (or Green's function). The one-particle propagator is given in second quantized notation by[10] $G=G_+ + G_-$, where

$$G_+(p,t)=-i\langle\psi_0^N|\,a_p e^{-iHt}a_p^\dagger\,|\psi_0^N\rangle e^{iE_0t},\quad t>0,$$
$$G_-(p,t)=\pm i\langle\psi_0^N|\,a_p^\dagger e^{iHt}a_p\,|\psi_0^N\rangle e^{-iE_0t},\quad t<0,\qquad (3.1)$$

where the $(+)$ sign is for fermions, the $(-)$ sign for bosons. The time Fourier transform of both G_+ and G_- have a branch cut on the real axis and various singularities on the wrong Riemann sheet. The poles nearest the real axis are quasiparticle poles while the other singularities reflect more complex collective modes.

The energy of a quasiparticle (i.e., the position of the pole) is complex, which means that it decays in time. Nevertheless, there is a reputed connection, which has never been made very precise between quasiparticles and elementary excitations in the sense we have used them above. The difference between the two, and the reason one decays and the other does not, lies in the form of the wave functions. The elementary excitations refer to exact eigenfunctions which unfortunately, for an interacting system, do not have the plane wave character of the excitations of a noninteracting gas. The quasiparticle, on the other hand, may be thought of as an attempt to find (inexact) wave functions which do have a plane wave character. The function $a_p^\dagger|\psi_0^N\rangle$ in Eq. (3.1) is such a plane wave type function.

Elementary excitations are more useful for calculating the free energy or any other quantity where only the energy is involved. Quasiparticles are useful for calculating the response of the system to an external influence, for, by the nature of a physical excitation process, plane waves are generally excited at time $t=0$. The propagator G samples the exact states, so to speak, to find a linear combination which looks most like the plane wave state at $t=0$.

[10] V. M. Galitskii and A. B. Migdal, Soviet Phys.—JETP 7, 96 (1958).

For a noninteracting Bose gas, G_- is identically zero. The pole in G_+ lies on the real axis and gives the spectrum $\epsilon(p)=p^2$. When the interaction is turned on, the extent to which this pole moves off the real axis depends upon the extent to which $a_p^\dagger|\psi_0^N\rangle$ is no longer an exact eigenstate of the $N+1$ particle system. It is this pole to which perturbation based calculations direct their attention. In our case we have seen that for large p at least, $a_p^\dagger|\psi_0^N\rangle$ approaches an eigenstate [Eq. (2.25)]. Unfortunately, we are not in a position to calculate G, but if one believes that quasiparticle poles and elementary excitations are similar, than our $\epsilon_1(p)$ curve must be connected with this pole in G_+.

From whence arises our second spectrum, $\epsilon_2(p)$? We believe it is connected with G_-. The negative time part of the propagator must also have singularities, but since it vanishes for zero interaction it would be very difficult indeed for perturbation theory to find them. If $\phi_p=a_p|\psi_0^{N+1}\rangle$ were an exact eigenstate of the N-particle problem, G_- would have a pole on the real axis. For the noninteracting case $\phi_p\equiv 0$, but if we look at our type II wave functions it will be seen that to a rough approximation they are just ϕ_p. Thus, we are led to the view that our type II spectrum corresponds to a quasiparticle pole of the negative time part of the one body propagator.

We do not propose to elucidate here the connection between elementary excitations and quasiparticles—we merely point out that these two types of objects can be defined. Neither will we attempt to evaluate G. But the above discussion is pertinent to the outstanding question of whether our double excitation spectrum has any relevance in three dimensions.

It is to be noted, first of all, that all the quantities we have calculated are smoothly dependent on γ and behave just as one intuitively expects them to behave in any number of dimensions. Although we have frequently called attention to the analogy of this gas with a Fermi system, and although this Fermi gas analogy clearly breaks down in three dimensions, it should not be supposed that the double spectrum result is a consequence of one dimension forcing the Bose system to look like a Fermi system. While it may be argued that the salient difference between one and three dimensions is that in three dimensions particles can "get around each other," the true state of affairs is expressed by the functions $v(\gamma)$ and $t(\gamma)$ plotted in Fig. 1 of I. As we have already mentioned, the fact that the potential is effectively a kinetic energy barrier for large γ, a result that also holds in three dimensions, means that it is really immaterial to the particles whether they can "get around each other" or merely "through each other."

Thus, there appears to be no truly basic physical (as distinguished from mathematical) distinction between our gas and a similar gas in three dimensions. It is quite possible, therefore, that a double (or perhaps more fold) spectrum exists in three dimensions. Another possibility is that in three dimensions the double spectrum does not exist for small potentials, but makes its appearance when the potential becomes sufficiently strong. In the theory of a lattice of coupled harmonic oscillators,[11] for example, it is known that one imperfect light mass may cause an isolated energy level to appear above the perfect lattice band. In one dimension this always happens; in three dimensions it occurs only if the imperfect mass is sufficiently small. If this second possibility—the onset of a double spectrum for large potentials—were to occur, then clearly perturbation theory would be hard pressed to predict it.

We turn finally to an argument based on the previous considerations of this section. For the very same reasons that it is believed that the positive time part of the propagator has a pole near the real axis, the negative time part too may be presumed to have a similar pole. For what reason is this pole any less "elementary" than its counterpart in the positive time part of the propagator?

ACKNOWLEDGMENTS

The author is most grateful to his colleagues at IBM, Dr. D. Mattis, Dr. D. Jepsen, and Dr. T. Schultz, for many stimulating discussions. Above all, he is indebted to Dr. A. Sakakura for helping him to understand elementary excitations and quasiparticles. He is also indebted to W. Liniger for carrying out the numerical analysis.

APPENDIX: THE EXCITATION SPECTRUM AT LOW MOMENTUM AND THE VELOCITY OF SOUND

We justify here the claims made in Sec. II about the excitation spectrum of type I (a similar analysis applies to type II). We want to show the absence of a gap and the equivalence of the two definitions of v_s, Eqs. (1.1) and (1.4).

We first transform Eq. (2.18) to dimensionless form by defining

$$k=Kx;\quad J(Kx)=j(x);\quad q=Ks;\quad c=K\lambda;\quad \text{(A1)}$$

where $s\geq 1$. Expanding the inhomogeneous term in Eq. (2.18) for $s\sim 1$, we obtain

$$2\pi j(x)=2\lambda\int_{-1}^{1}\frac{j(y)dy}{\lambda^2+(x-y)^2}-\pi+2\tan^{-1}\left(\frac{1-x}{\lambda}\right)+(s-1)\frac{2\lambda}{\lambda^2+(1-x)^2}. \quad \text{(A2)}$$

We write

$$j(x,s)=j_0(x)+(s-1)j_1(x), \quad \text{(A3)}$$

and define

$$2\phi_1(x)=j_0(x)+j_0(-x), \quad \text{(A4)}$$

$$2\phi_2(x)=j_0'(x)+j_0'(-x), \quad \text{(A5)}$$

$$2\phi_3(x)=j_1(x)+j_1(-x), \quad \text{(A6)}$$

$$2\phi_4(x)=j_1(x)-j_1(-x), \quad \text{(A7)}$$

[11] A. Maradudin, P. Mazur, E. Montroll, and G. Weiss, Rev. Mod. Phys. 30, 175 (1958).

where "prime" means differentiation with respect to x. We easily obtain the equations

$$2\pi\phi_1(x,\lambda)=\int \mathrm{Ker}\phi_1-\pi+\tan^{-1}\left(\frac{1-x}{\lambda}\right)+\tan^{-1}\left(\frac{1+x}{\lambda}\right), \quad \text{(A8)}$$

$$2\pi\phi_2(x,\lambda)=\int \mathrm{Ker}\phi_2-\lambda\left[\frac{1}{\lambda^2+(1-x)^2}+\frac{1}{\lambda^2+(1+x)^2}\right]\times[1+j_0(1)-j_0(-1)], \quad \text{(A9)}$$

$$2\pi\phi_3(x,\lambda)=\int \mathrm{Ker}\phi_3+\lambda\left[\frac{1}{\lambda^2+(1-x)^2}+\frac{1}{\lambda^2+(1+x)^2}\right], \quad \text{(A10)}$$

$$2\pi\phi_4(x,\lambda)=\int \mathrm{Ker}\phi_4+\lambda\left[\frac{1}{\lambda^2+(1-x)^2}-\frac{1}{\lambda^2+(1+x)^2}\right], \quad \text{(A11)}$$

where $\int$ Ker means the integral operator appearing in Eq. (A2).

The function $g(x,\lambda)$ introduced in I satisfies

$$2\pi g(x)=\int \mathrm{Ker}g+1, \quad \text{(A12)}$$

while the function

$$h(x,\lambda)=\left(\frac{\partial}{\partial\lambda}+\frac{x}{\lambda}\frac{\partial}{\partial x}\right)g \quad \text{(A13)}$$

satisfies

$$2\pi h(x)=\int \mathrm{Ker}h-2g(1,\lambda)\left[\frac{1}{\lambda^2+(1-x)^2}+\frac{1}{\lambda^2+(1+x)^2}\right]. \quad \text{(A14)}$$

A simple manipulation of Eq. (A12) shows that the function $g'=\partial g/\partial x$ satisfies

$$2\pi g'=\int \mathrm{Ker}g'-2\lambda g(1,\lambda)\left[\frac{1}{\lambda^2+(1-x)^2}-\frac{1}{\lambda^2+(1+x)^2}\right]. \quad \text{(A15)}$$

Comparing Eqs. (A10) and (A11) with Eqs. (A14) and (A15) we find that

$$\phi_3(x,\lambda)=-\frac{\lambda}{2g(1,\lambda)}h(x,\lambda), \quad \text{(A16)}$$

and

$$\phi_4(x,\lambda)=-\frac{1}{2g(1,\lambda)}g'(x,\lambda). \quad \text{(A17)}$$

Equation (A8) may be solved by inspection, i.e.,

$$\phi_1(x,\lambda)=-\tfrac{1}{2}. \quad \text{(A18)}$$

Comparison of Eqs. (A9) and (A10), together with the observation that

$$\int_{-1}^{1} dx\,\phi_2(x,\lambda)=j_0(1)-j_0(-1), \quad \text{(A19)}$$

yields the result

$$\phi_2(x,\lambda)=-\phi_3(x,\lambda)\left[1+\int_{-1}^{1} dx\,\phi_3(x,\lambda)dx\right]. \quad \text{(A20)}$$

Integrating Eq. (A14) with respect to x and using Eq. (A16) we find that

$$\int_{-1}^{1} dx\frac{\partial g}{\partial\lambda}=-\frac{2}{\lambda}g(1,\lambda)\left[1+\int_{-1}^{1} dx\,\phi_3(x,\lambda)\right]+\frac{1}{\gamma}, \quad \text{(A21a)}$$

$$\int_{-1}^{1} dx\,x^2\frac{\partial g}{\partial\lambda}=-\frac{2}{\lambda}g(1,\lambda)\left[1+\int_{-1}^{1} dx\,x^2\phi_3(x,\lambda)\right]+\frac{3}{\lambda}\int_{-1}^{1} dx\,x^2 g(x,\lambda), \quad \text{(A21b)}$$

where we have used the definition of λ/γ given in I (3.20):

$$\frac{\lambda}{\gamma}=\int_{-1}^{1} dx\,g(x,\lambda)=\frac{\rho}{K}. \quad \text{(A22)}$$

Now, since

$$e(\gamma)=\frac{\gamma^3}{\lambda^3}\int_{-1}^{1} g(x,\lambda)x^2dx, \quad \text{(A23)}$$

[cf. I Eq. (3.19)], the definition of μ, Eq. (1.5) together with Eqs. (A21), is equivalent to

$$\mu=K^2\left[1+\int_{-1}^{1} dx\,\phi_3(x)x^2\right]\Bigg/\left[1+\int_{-1}^{1} dx\,\phi_3(x)\right]. \quad \text{(A24)}$$

In terms of the functions ϕ_1 to ϕ_4, Eq. (2.19) for the momentum takes the form (for $s\sim 1$)

$$p=K\left[1+\int_{-1}^{1} dx\,\phi_1(x)\right]+K(s-1)\left[1+\int_{-1}^{1} dx\,\phi_3(x)\right]. \quad \text{(A25)}$$

The expression for ϵ_1, Eq. (2.20), takes the form (for $s\sim 1$)

$$\epsilon_1=-\mu+K^2+K^2\int_{-1}^{1} dx\,\phi_2(x)(1-x^2) +2K^2(s-1)\left[1+\int_{-1}^{1} dx\,x\phi_4(x)\right] \quad \text{(A26)}$$

Inserting Eqs. (A18), (A20), and (A24) into Eqs. (A25) and (A26) we find that

$$\lim_{s\to 1} p=0;\quad \lim_{s\to 1}\epsilon_1=0; \quad \text{(A27)}$$

thereby proving the assertion that there is no energy gap and that the lower end of the spectrum is in fact $p=0$.

Inserting Eqs. (A17) and (A22) into Eq. (A26), we find that for small s (i.e., small p)

$$\epsilon_1(p)=p\left\{g(1,\lambda)\left[1+\int_{-1}^{1} dx\,\phi_3(x)\right]\right\}^{-1}\rho. \quad \text{(A28)}$$

By Eq. (1.1), the velocity of sound, v_s, is, therefore, the factor $\{\ \}^{-1}\rho$ in Eq. (A28). It is always positive since ϕ_3 is positive. An alternative expression for v_s can be obtained from Eq. (A21a), viz.,

$$v_s=\frac{2}{\lambda}\left[\frac{1}{\gamma}-\int_{-1}^{1} dx\frac{\partial}{\partial\lambda}g(x,\lambda)\right]\rho. \quad \text{(A29)}$$

In I we showed that for λ small, $g(x,\lambda)\sim(1/2\pi\lambda)(1-x^2)^{1/2}$ and $4\lambda^2=\gamma$. Substituting this into Eq. (A29) we obtain

$$v_s\simeq 2\gamma^{1/2}\rho,\quad (\text{small }\gamma), \quad \text{(A30)}$$

thereby confirming Bogoliubov's result, Eq. (1.6).

The last step is to confirm that Eqs. (1.4) and (A29) agree. We must calculate $d\mu/d\gamma=(d\lambda/d\gamma)(d\mu/d\lambda)$. The factor $d\lambda/d\gamma$ can be obtained from Eq. (A22) and (A21a), viz.,

$$\frac{d\lambda}{d\gamma}=\frac{1}{2}\frac{\lambda^2}{\gamma^2}\left\{g(1,\lambda)\left[1+\int_{-1}^{1} dx\,\phi_3(x)\right]\right\}^{-1}. \quad \text{(A31)}$$

To obtain $d\mu/d\lambda$ from Eq. (A24) the essential point is to evaluate $(\partial/\partial\lambda)\phi_3(x,\lambda)$. To this end we define the auxiliary function

$$m(x,\lambda)=\left(\frac{\partial}{\partial\lambda}+\frac{x}{\lambda}\frac{\partial}{\partial x}\right)\phi_3(x,\lambda). \quad \text{(A32)}$$

In analogy with Eqs. (A13) and (A14), this function satisfies

$$2\pi m(x,\lambda) =\int \mathrm{Ker}\,m-[1+2\phi_3(1,\lambda)]\left[\frac{1}{\lambda^2+(1-x)^2}+\frac{1}{\lambda^2+(1+x)^2}\right] +2\left[\frac{1-x}{[\lambda^2+(1-x)^2]^2}+\frac{1+x}{[\lambda^2+(1+x)^2]^2}\right]. \quad \text{(A33)}$$

On the other hand, as we see from Eq. (A15), the function $g''=\partial^2 g/\partial x^2$ satisfies

$$2\pi g''=\int \mathrm{Ker}\,g-2\lambda g'(1,\lambda)\left[\frac{1}{\lambda^2+(1-x)^2}+\frac{1}{\lambda^2+(1+x)^2}\right] -4\lambda g(1,\lambda)\left[\frac{1-x}{[\lambda^2+(1-x)^2]^2}+\frac{1+x}{[\lambda^2+(1+x)^2]^2}\right]. \quad \text{(A34)}$$

Comparing Eqs. (A10), (A33), and (A34) we find that

$$m(x,\lambda)=-\frac{1}{2\lambda g(1,\lambda)}g''(x,\lambda) -\frac{1}{\lambda}\left[1+2\phi_3(1,\lambda)+\frac{g'(1,\lambda)}{g(1,\lambda)}\right]\phi_3(x,\lambda). \quad \text{(A35)}$$

In analogy with Eqs. (A21), $d\mu/d\lambda$ may be written in terms of $m(x,\lambda)$ which in turn may be written in terms of ϕ_3, g, g', and g'' by using Eq. (A35). We shall not carry out the tedious manipulation here, but one eventually finds that Eq. (1.4) reduces to Eq. (A29).

An alternative expression for v_s may be obtained by comparing Eqs. (A22), (A28), and (A31). One easily finds that

$$v_s=2\rho\frac{\gamma^3}{\lambda^3}\frac{d\lambda}{d\gamma}=2K-2\gamma\frac{\partial K}{\partial\gamma}. \quad \text{(A36)}$$

Reprinted from *Printed in U.S.A.*
JOURNAL OF MATHEMATICAL PHYSICS VOLUME 5, NUMBER 7 JULY 1964

Momentum Distribution in the Ground State of the One-Dimensional System of Impenetrable Bosons*

A. LENARD
Plasma Physics Laboratory, Princeton University, Princeton, New Jersey
(Received 4 February 1964)

Girardeau has shown that an exact analytical formula may be given for the ground-state wavefunction of a system of one-dimensional impenetrable bosons. Starting with this formula, we give a mathematically rigorous analysis leading to the determination of major features of the momentum distribution in the limit of an infinitely large system.

1. INTRODUCTION

THE study of quantum mechanical many-body systems has great contemporary interest, and in recent years much knowledge has accumulated in this field. However, the results of most investigations, while plausible, are not established with standards of mathematical rigor. This state of affairs, unsatisfactory as it may be, is probably inherent in the subject. Therefore, any not completely trivial system whose reasonably detailed properties are subject to exact analysis has an intrinsic interest, even if it is not representative of any real physical situation.

The purpose of this paper is the rigorous discussion of a specific problem of this kind. We shall study the nature of the momentum distribution in the ground state of a system consisting of a great number of one-dimensional impenetrable particles satisfying Bose–Einstein statistics. In a recent paper Girardeau[1] has shown that for this system explicit analytic formulas may be given for all stationary-state wavefunctions in coordinate-space representation. The momentum distribution in any of these stationary states is defined by the well-known rules of quantum mechanical transformation theory. However, even the simplest momentum-space properties are given in terms of many-dimensional integrals of the type encountered in classical statistical mechanics, so that no effective information is immediately available. Of interest is the usual "thermodynamic limit," namely the limit in which both the number of particles and the size of the container tend to infinity, proportionally to each other. We shall show that for our system it is possible to establish a number of interesting specific facts in this limit.

The problem of the momentum distribution was already attacked by Girardeau[1] who applied an approximation method suggested by the work of Bogoliubov.[2] He found a singular behavior at the origin in momentum space reminiscent of, though not identical with, the behavior of an ideal Bose system (Bose–Einstein condensation). This claim was disproved by Schultz[3] who observed a mathematical correspondence between the Girardeau model and a certain asymptotic limit of an exactly soluble "spin system." Using this correspondence, Schultz was able to show that in the ground state of the Girardeau model no Bose–Einstein condensation exists, not even in the generalized sense proposed by Girardeau. Our work confirms this conclusion. We do not use the spin analogy of Schultz, but proceed in a straightforward manner from the ground-state wavefunction given by Girardeau and, by a step-by-step process, transform the quantities of interest into forms for which powerful techniques of analysis become applicable. It will be seen that an especially important role is played by certain results of Szegö in the theory of Toeplitz matrices.[4]

An interesting question is the relationship of the present paper to studies of more general and realistic models of interacting Bose particles, and in particular, to the three-dimensional model with hard-core repulsive interparticle forces. It has been suggested in the past that the singular behavior of Bose systems observed at low temperatures is due to Bose–Einstein condensation. Such a theory is tenable only if a strong short-range interparticle repulsion is insufficient to change the qualitative features of the momentum distribution from those of an ideal gas, though, of course, it must change them quantitatively. Whether this is or is not the case is not known with certainty, although approximate calculations suggest that it is. Simple qualitative comparison of the one- and the three-dimensional problems

* This work was supported by the U. S. Atomic Energy Commission.
[1] M. Girardeau, J. Math. Phys. 1, 516 (1960).
[2] N. N. Bogoliubov, J. Phys. USSR 11, 23 (1947).
[3] T. D. Schultz, J. Math. Phys. 4, 666 (1963).
[4] U. Grenander and G. Szegö, *Toeplitz Forms and their Applications* (University of California Press, Berkeley, California, 1958).

suggests that in the latter case the momentum distribution is "sharper" at the origin. Thus a discovery of Bose–Einstein condensation in the one-dimensional model studied here would have been taken as very strong evidence of such condensation in the three-dimensional model. Unfortunately the facts are otherwise and so our investigation throws no light on the three-dimensional problem. It only shows that very strong interparticle repulsion can (at least in one dimension) sufficiently alter the momentum distribution to remove its most characteristic feature, the singularity at the origin.

In the following section we give a precise description of the problem and define the quantities which form the subject of the investigation. In Sec. 3 we derive the representation of the one-particle density matrix as a determinant of a large Toeplitz matrix. In Sec. 4 we present the proof of an important inequality due to Szegö. In Sec. 5 we transform the determinant representing the density matrix into a form where—from the point of view of the thermodynamic limit—it appears as the well-known "discrete approximation" to the Fredholm determinant of a certain kernel given in simple explicit form. These results are used in Sec. 6 to prove theorems about the thermodynamic limit. It is shown that the limiting momentum distribution exists in a mathematically precise sense, and certain bounds for its behavior for large and small momenta are derived. In Sec. 7 we discuss the relevance of these theorems from the point of view of the criteria of Penrose and Onsager[5] as well as Girardeau[1] for the presence or absence of Bose–Einstein condensation. Finally, in Sec. 8 we compare our results with the previous work of Girardeau[1] and Schultz.[3] Some mathematical material used in the text is collected in three appendices.

2. PRELIMINARIES

The quantum mechanical problem we shall study is defined by the following conditions:

(i) The wavefunction satisfies the free-particle Schrödinger equation for the motion of N particles in one dimension. We assume $N \geq 2$.

(ii) The wavefunction is symmetrical with respect to interchange of particle coordinates (Bose–Einstein statistics).

(iii) The wavefunction satisfies periodic boundary conditions with period L.

(iv) The wavefunction vanishes whenever two particle coordinates coincide.

[5] O. Penrose and L. Onsager, Phys. Rev. **104**, 576 (1956).

Condition (iii) may be interpreted to mean that the particles are constrained to move on a circle of circumference L. Condition (iv) is the expression of the mutual impenetrability of the particles. Recently, Lieb[6] studied a more general problem where this condition was replaced by the milder one that the logarithmic derivative of the wavefunction with respect to any of the coordinates increase discontinuously by a specified amount γ when the chosen coordinate crosses another one. Our model corresponds to the limit $\gamma \to \infty$, the other limit $\gamma \to 0$ being the case of free particles.

The problem defined by the conditions (i)–(iv) has a close relationship to the problem of *free* particles satisfying the exclusion principle.[1] The stationary-state wavefunctions of the two problems differ only by a multiplicative function which assumes only the two values ± 1, and the energy eigenvalues are identical.[1] In the following we shall be concerned exclusively with the ground state. Its wavefunction is

$$\psi_{N,L}(x_1, x_2, \cdots, x_N) = (N!\,L^N)^{-\frac{1}{2}} \,\Big|\det_{1\leq n,m\leq N} e^{2\pi i n x_m/L}\Big|. \quad (1)$$

This may also be written in the factored form

$$\psi_{N,L}(x_1, x_2, \cdots, x_N) = (N!\,L^N)^{-\frac{1}{2}} \prod_{1\leq n<m\leq N} |e^{2\pi i x_n/L} - e^{2\pi i x_m/L}|. \quad (2)$$

It is real, positive, and translation-invariant, properties which depend only on the fact that the particles satisfy Bose–Einstein statistics, periodic boundary conditions, and that it is the wavefunction of the ground state.[7] Although Girardeau gave (1) only for odd N, it is equally correct for even N.[8]

We shall be interested in the one-particle reduced density matrix (referred to in the following simply as the density matrix) given by

$$\rho_{N,L}(x - x') = N \int_0^L dx_1 \int_0^L dx_2 \cdots \int_0^L dx_{N-1} \times \psi_{N,L}(x_1, \cdots, x_{N-1}, x)\psi^*_{N,L}(x_1, \cdots, x_{N-1}, x'). \quad (3)$$

It is so normalized that $\rho_{N,L}(0) = N/L$ is the particle number density. It is real and positive and also of positive type which means that its Fourier coefficients with respect to the variable $\xi = x - x'$ are nonnegative. We write

[6] E. H. Lieb and W. Liniger, Phys. Rev. **130**, 1605 (1963).
[7] E. H. Lieb, Phys. Rev. **130**, 2518 (1963), footnote 15.
[8] Reference 6, footnote 6.

$$\rho_{N,L}^{(n)} = \int_0^L \rho_{N,L}(\xi)e^{-2\pi in\xi/L}\,d\xi = \frac{N}{L}\int_0^L dx_1 \cdots$$
$$\times \int_0^L dx_{N-1}\left|\int_0^L dx\psi_{N,L}(x_1, \cdots, x_{N-1}, x)e^{-2\pi inx/L}\right|^2$$
$$(n = 0, \pm 1, \pm 2, \cdots). \quad (4)$$

The Fourier coefficients have an important physical significance. $\rho_{N,L}^{(n)}$ is the quantum mechanical mean (expectation value) of the number of particles having momentum[9]

$$k_n = 2\pi n/L. \quad (5)$$

We note that it is actually independent of L. For the sake of keeping its physical meaning clear we shall, however, not drop the subscript L.

We now lay down the following precise definition. Let

$$F_{N,L}(k) = \frac{1}{N}\sum_{-\infty<n<(kL/2\pi)} \rho_{N,L}^{(n)}. \quad (6)$$

This function will be referred to in the following as the *momentum distribution function*. It is the mean fraction of particles whose momenta are less than k, regarded as a function of k; the subscripts N and L calling attention to the total number of particles and the box size on which it depends parametrically. In terms of the momentum distribution function, we now have

$$\frac{L}{N}\rho_{N,L}(\xi) = \int_{-\infty}^{\infty} e^{ik\xi}\,dF_{N,L}(k). \quad (7)$$

The reason for introducing the definition (6) and for the somewhat fanciful way of writing what is basically the Fourier series of the density matrix is the following. The momentum distribution function belongs to a class known as *probability distribution functions*. Another class of functions is formed by the Fourier–Stieltjes transforms, in the manner of (7), of probability distribution functions, known as *characteristic functions*. There is a one-to-one correspondence between functions belonging to the two classes, and important properties of functions belonging to one class mirror themselves in corresponding properties of the other class. Of particular significance is the fact that limiting operations preserve this correspondence. We are here interested in the "thermodynamic limit"[10]

$$N \to \infty, \quad L = lN \to \infty \quad (l > 0 \text{ constant}). \quad (8)$$

[9] We put $\hbar = 1$ throughout.

[10] Since we are dealing with a single quantum state for the whole system, namely the ground state, this terminology is actually a misnomer. But there is no point in avoiding it as long as its meaning is clearly understood.

While our interest lies in the momentum distribution, it turns out that it is simpler to investigate the density matrix and from its limiting properties draw the appropriate conclusions by means of general theorems. For the mathematical details on the subject of distribution functions and their characteristic functions, the reader must be referred to the literature,[11] but we have collected in Appendix 1 the statements of all needed facts in a form sufficiently complete for the task on hand.

We mention here one simple property of the momentum distribution which follows from the fact that the ground-state energy is identical to the ground-state energy of free Fermions with the same values of N and L. Since the ground-state energy is also expressible as

$$N\int_{-\infty}^{\infty} k^2\,dF_{N,L}(k) = \sum_{n=-\infty}^{\infty} \rho_{N,L}^{(n)}\left(\frac{2\pi n}{L}\right)^2, \quad (9)$$

we have for this quantity[12]

$$N\int_{-\infty}^{\infty} k^2\,dF_{N,L}(k)$$
$$= \left(\frac{2\pi}{L}\right)^2 \sum_{q=1}^{N}\left(q - \frac{N+1}{2}\right)^2 = N\frac{\pi^2}{3}\frac{N^2-1}{L^2}. \quad (10)$$

Thus, in the thermodynamic limit,[13]

$$\lim_{N,L\to\infty} \int_{-\infty}^{\infty} k^2\,dF_{N,L}(k) = \frac{\pi^2}{3l^2}. \quad (11)$$

3. DETERMINANTAL REPRESENTATION OF DENSITY MATRIX

We shall now derive an algebraic representation for the density matrix, obtained by carrying out the integrations in (3).

Since the box size L enters throughout only as a scale factor, we write

$$\rho_{N,L}(\xi) = (1/L)R_N(2\pi\xi/L). \quad (12)$$

The function $R_N(\alpha)$ will be shown to be equal the determinant of a matrix with $N - 1$ rows and columns whose elements are functions of α alone.

Our starting point is the identity

$$\psi_{N,L}(x_1, \cdots, x_{N-1}, x_N)$$
$$= (NL)^{-\frac{1}{2}}\psi_{N-1,L}(x_1, \cdots, x_{N-1})$$
$$\times \prod_{1\le n\le N-1} |e^{2\pi ix_n/L} - e^{2\pi ix_N/L}|, \quad (13)$$

[11] M. Loeve, *Probability Theory* (D. Van Nostrand Company, New York, 1955), Chap. IV.

[12] For even N the wave vectors in the fermion problem must be half-odd integral multiples of the basic unit $2\pi/L$, cf. Ref. 8.

[13] This notation will be employed for the limit (8).

obtained by comparing $\psi_{N,L}$ and $\psi_{N-1,L}$ in the form given in (2). If we then substitute (13) into (3) and make use of the determinantal form (1) for $\psi_{N-1,L}$, we obtain

$$R_N(\alpha) = [(N-1)!\,(2\pi)^{N-1}]^{-1} \times \int_0^{2\pi} d\theta_1 \cdots \int_0^{2\pi} d\theta_{N-1} \left| \det_{1\leq n,m\leq N-1} e^{in\theta_m}\right|^2 \times \prod_{1\leq r\leq N-1} |e^{i\theta_r} - e^{i\frac{1}{2}\alpha}|\,|e^{i\theta_r} - e^{-i\frac{1}{2}\alpha}|. \tag{14}$$

Now we expand the determinant in the integrand in the usual way into its $(N-1)!$ terms

$$\det_{1\leq n,m\leq N-1} e^{in\theta_m} = \sum_P \delta \prod_{1\leq n\leq N-1} e^{i\nu_n\theta_n}. \tag{15}$$

Here P is a permutation on the first $N-1$ integers, ν_n is the integer into which P sends n, δ is the parity of P, and the sum goes over all P. This results in the formula

$$R_N(\alpha) = [(N-1)!]^{-1} \sum_P \sum_{P'} \delta\delta' \times \prod_{1\leq n\leq N-1} \left\{\frac{1}{2\pi}\int_0^{2\pi} d\theta\, e^{i(\nu_n-\nu_n')\theta} \times |e^{i\theta} - e^{i\frac{1}{2}\alpha}|\,|e^{i\theta} - e^{-i\frac{1}{2}\alpha}|\right\}. \tag{16}$$

Let

$$f(\theta,\alpha) = |e^{i\theta} - e^{i\frac{1}{2}\alpha}|\,|e^{i\theta} - e^{-i\frac{1}{2}\alpha}| = 2\,|\cos\theta - \cos\tfrac{1}{2}\alpha|, \tag{17}$$

and

$$c_n(\alpha) = \frac{1}{2\pi}\int_0^{2\pi} e^{-in\theta} f(\theta,\alpha). \tag{18}$$

In (16) we have then

$$R_N(\alpha) = [(N-1)!]^{-1} \times \sum_P \sum_{P'} \delta\delta' \prod_{1\leq n\leq N-1} c_{\nu_n-\nu_n'}(\alpha). \tag{19}$$

It is easy to see that the summation over one of the running permutation variables is trivial and merely serves to cancel the factorial in front. But the remaining sum is just the expansion of a determinant. Thus we have

Theorem 1.[14]

$$R_N(\alpha) = \det_{1\leq n,m\leq N-1} c_{n-m}(\alpha). \tag{20}$$

The calculations of this paper are based on this purely algebraic representation of the density matrix.

[14] This has been found independently by Professor F. J. Dyson of the Institute for Advanced Study, Princeton, New Jersey.

We evaluate the elements of the determinant explicitly. The integrals (18) are elementary, but some care has to be exercised on account of the occurrence of the absolute-value sign and the fact that the cases $n = 0$ and $n = \pm 1$ have to be treated separately. The result may be written in the following simple manner:

$$c_n(\alpha) = 2\delta_{n,0}\cos\tfrac{1}{2}\alpha - \delta_{n,1} - \delta_{n,-1} + \frac{2}{\pi}\left[\frac{\sin\frac{1}{2}(n+1)\,|\alpha|}{n+1} + \frac{\sin\frac{1}{2}(n-1)\,|\alpha|}{n-1} - \frac{2\cos\frac{1}{2}\alpha\sin\frac{1}{2}n\,|\alpha|}{n}\right]. \tag{21}$$

This is valid for

$$-\pi \leq \alpha \leq \pi \tag{22}$$

and all n, with the proviso that terms which appear with vanishing denominators are to be interpreted formally by continuity. For instance,

$$c_0(\alpha) = 2\cos\tfrac{1}{2}\alpha + (2/\pi)[2\sin\tfrac{1}{2}|\alpha| - |\alpha|\cos\tfrac{1}{2}\alpha], \tag{23}$$

and similarly for $n = \pm 1$.

For the special case $\alpha = 0$, we have

$$c_n(0) = \begin{cases} 2 & (n=0) \\ -1 & (n=\pm 1) \\ 0 & (\text{otherwise}). \end{cases} \tag{24}$$

The determinant (20) is easily evaluated by expanding according to the first two rows. This results in the recursion formula

$$R_N(0) = 2R_{N-1}(0) - R_{N-2}(0), \tag{25}$$

whose solution is

$$R_N(0) = N. \tag{26}$$

The significance of Theorem 1 lies in the following. We have succeeded in carrying out the integrations involved in the definition of the density matrix and reduced it to the purely algebraic representation of a determinant with elements which are simple elementary functions. That this is possible is actually not too surprising since multiple integrals of related type have been evaluated before.[15] However, our problem is as yet far from solved because we are interested in limiting behavior for large N and this

[15] See, for instance, the articles by F. J. Dyson, J. Math. Phys. **3**, 140, 157, and 166 (1962). The perceptive reader will notice that our ground-state wavefunction has an interpretation in terms of Dyson's one-dimensional "Coulomb gas on a circle" which, in turn, is related to the eigenvalue distribution of his random matrices.

cannot be read off the determinantal representation any more than it can in the original representation as a multiple integral.[16] Fortunately, the actual limit of interest [cf. (8) and (12)] demands that $\alpha \to 0$ *simultaneously* with $N \to \infty$, and we shall see below that the determinant (20) is a good starting point for its calculation.

The representation given in Theorem 1 has another fortunate feature. The determinant is of a very special type. Its elements are constant along any of the lines parallel to the principal diagonal, and they are the Fourier coefficients of a real, non-negative function. Matrices having this property are called Toeplitz matrices and have a very detailed analytic theory.[4] We make use of this theory to derive an important inequality for $R_N(\alpha)$. This is done in the next section; then the thermodynamic limit is investigated.

4. AN IMPORTANT INEQUALITY

The purpose of this section is to establish the following result.

Theorem 2.

$$R_N(\alpha) < \operatorname*{Min}_{1<R<\infty} \{R^{2N}(R^2 - 1)^{-\frac{1}{2}} \times (R^4 - 2R^2 \cos\alpha + 1)^{-\frac{1}{4}}\} \tag{27}$$

This theorem is due to Szegö.[17] Before presenting its proof we discuss the reasons for its importance in the present context.

Let us first inquire what qualitative information may be obtained from it. For $\alpha = 0$, the minimizing value of R is

$$R = [1 + 1/(N-1)]^{\frac{1}{2}}, \tag{28}$$

and the inequality reads

$$R_N(0) < N[1 + 1/(N-1)]^{N-1} \tag{29}$$

In this case we know $R_N(0) = N$, so that the inequality yields no new information, but we see that it does not "deteriorate" as $N \to \infty$, since the bound overshoots the actual value by a factor no more than $e = 2.7182 \cdots$.

For $\alpha \neq 0$ the direct minimization leads to a cubic equation for R^2 which, while it can be solved, yields a result too complicated to be useful. Instead, we may proceed by a slight weakening of the inequality, replacing the factor $(R^4 - 2R^2 \cos\alpha + 1)^{-\frac{1}{4}}$ by the larger quantity $|2 \sin \frac{1}{2}\alpha|^{-\frac{1}{2}}$. Then the minimizing value of R is

$$R = [1 + 1/(2N-1)]^{\frac{1}{2}}, \tag{30}$$

and the inequality reads

$$R_N(\alpha) < \left|\frac{N}{\sin\frac{1}{2}\alpha}\right|^{\frac{1}{2}} \left(1 + \frac{1}{2N-1}\right)^{N-\frac{1}{2}} < \left|\frac{eN}{\sin\frac{1}{2}\alpha}\right|^{\frac{1}{2}} \tag{31}$$

Observe the square-root dependence on N. Together with (26) this shows that, in the limit $N \to \infty$, $R_N(\alpha)$ develops a large peak in the neighborhood of the origin $\alpha = 0$. The width of this peak can, of course, not be determined from the results so far obtained, since (31) only gives an upper bound. Nevertheless, the form of this inequality strongly suggests that in the thermodynamic limit when αN remains fixed as $N \to \infty$ one remains, in effect, on top of the peak and $R_N(\alpha)$ remains of order N. To the extent that we anticipate the existence of the thermodynamic limit of $\rho_{N,L}(\xi)$ with fixed ξ, this is to be expected, but the preceding consideration does show that the inequality of Theorem 2 is in some sense "uniformly good" for all α in spite of the fast variation of $R_N(\alpha)$ near $\alpha = 0$ in the case of large N.

Again, anticipating the existence of this limit, we now write[18]

$$\lim_{N,L\to\infty} \rho_{N,L}(\xi) = \lim_{N\to\infty} \frac{1}{N} R_N\left(\frac{2\pi\xi}{N}\right) = \rho(\xi) \tag{32}$$

Applying the inequality (31) and passing to the limit we then obtain

$$\rho(\xi) \leq |e/\pi\xi|^{\frac{1}{2}} \tag{33}$$

This inequality is useful for the following reason. In the course of proving the existence of the limit (32), an infinite series representation for the function $\rho(\xi)$ emerges [cf. (58) below]. While not a power series in ξ, nevertheless the nature of this series is such that, from its first few terms, only the behavior of $\rho(\xi)$ for small $|\xi|$ may be deduced. The inequality (33) is the only information we have on its behavior for *large* $|\xi|$.

The proof[17] of Theorem 2 is based on Szegö's investigations of Toeplitz matrices.[4] In Appendix 2 we have collected the information from this theory needed for our purpose.

---

[16] In fact, for some purposes the multiple integral seems to offer advantages. Dyson (in an unpublished lecture at the Eastern Theoretical Physics Conference of October 1963) has shown how a not rigorous but very suggestive argument may be based on it, indicating an asymptotic property for large N.

[17] Communicated privately to the author by Professor G. Szegö.

[18] Here and in the following we set $L = N$ which corresponds merely to a choice for the unit of length, but for conceptual clarity we occasionally refer to L in the notation.

In the notation of Toeplitz theory,

$$R_N(\alpha) = D_{N-2}(f), \qquad (34)$$

where[19] the function $f(\theta)$ is given by (17). Let R be an arbitrary positive number larger than one, and let

$$f_R(\theta) = |Re^{i\theta} - e^{i\frac{1}{2}\alpha}| \cdot |Re^{i\theta} - e^{-i\frac{1}{2}\alpha}|. \qquad (35)$$

Then $f(\theta) < f_R(\theta)$ and, therefore, by Corollary 1 of Szegö's Theorem A [cf. Appendix 2],

$$D_{N-2}(f) < D_{N-2}(f_R). \qquad (36)$$

Corollary 2 shows that for $n = N-2, N-1, N, \cdots$

$$D_{N-2}(f_R) \le G(f_R)^{N-1}[D_n(f_R)/G(f_R)^{n+1}], \qquad (37)$$

and the right-hand side of this inequality does not decrease with n. Now, the function $f_R(\theta)$ satisfies the conditions of Theorem B and so the right-hand side of (37) has a finite limit as $n \to \infty$, and this limit provides an upper bound for $R_N(\alpha)$. It turns out, luckily, that the limit, as specified by Theorem B can be given in elementary terms. The analytic function assigned to $f_R(\theta)$ by Theorem B is simply

$$g_R(z) = (z - Re^{i\frac{1}{2}\alpha})^{\frac{1}{2}}(z - Re^{-i\frac{1}{2}\alpha})^{\frac{1}{2}}, \qquad (38)$$

as one verifies by checking the relevant properties which define it uniquely. Thus

$$h_n = -(1/nR^n)\cos\tfrac{1}{2}(\alpha n), \qquad (39)$$

and

$$\lim_{n\to\infty} \frac{D_n(f_R)}{G(f_R)^{n+1}} = \exp \sum_{n=1}^{\infty} n\,|h_n|^2$$
$$= \left(1 - \frac{1}{R^2}\right)^{-\frac{1}{2}}\left(1 - \frac{2}{R^2}\cos\alpha + \frac{1}{R^4}\right)^{-\frac{1}{2}}. \qquad (40)$$

This, together with

$$G(f_R) = |g_R(0)|^2 = R^2, \qquad (41)$$

shows that

$$D_{N-2}(f) < R^{2N}(R^2 - 1)^{-\frac{1}{2}}$$
$$\times (R^4 - 2R^2\cos\alpha + 1)^{-\frac{1}{2}}. \qquad (42)$$

But $R > 1$ was arbitrary and the right-hand side of (42) is a continuous function of R which tends to ∞ both as $R \to 1$ and as $R \to \infty$, so that there is a minimum on the interval $1 < R < \infty$. This completes the proof of Theorem 2.

5. THE THERMODYNAMIC LIMIT

We now transform the determinant (20) into a form where the limit (32) can be carried out.

[19] In the remainder of this section we suppress the dependence on α which is regarded as a fixed parameter. For the notation used here the reader is advised to consult Appendix 2 or Ref. 4.

We begin by writing

$$c_n(\alpha) = c_n(0) + b_n(\alpha), \qquad (43)$$

where $c_n(0)$ is given by (24), and $b_n(\alpha)$ is obtained from (21),

$$b_n(\alpha) = -4\delta_{n,0}\sin^2\frac{\alpha}{4} + \frac{8}{n\pi}\sin^2\frac{\alpha}{4}\sin\frac{n}{2}|\alpha|$$
$$+ \frac{2}{\pi}\left[\frac{\sin\frac{1}{2}(n+1)|\alpha|}{n+1} + \frac{\sin\frac{1}{2}(n-1)|\alpha|}{n-1} - 2\frac{\sin\frac{1}{2}n|\alpha|}{n}\right]. \qquad (44)$$

We now write[20]

$$R_N(\alpha) = \det c_{n-m}(\alpha)$$
$$= [\det c_{n-m}(0)]$$
$$\times [\det(\delta_{n,m} + \sum_r M^{(N)}_{n,r} b_{r-m}(\alpha))], \qquad (45)$$

where $M^{(N)}_{n,r}$ is the $N-1$ by $N-1$ matrix inverse to $c_{n-m}(0)$. The first factor is just $R_N(0) = N$, and therefore we have[18]

$$\rho_{N,L}(\xi) = \det\left(\delta_{n,m} + \sum_r M^{(N)}_{n,r} b_{r-m}\left(\frac{2\pi\xi}{N}\right)\right). \qquad (46)$$

One may verify explicitly that

$$M^{(N)}_{n,m} = \begin{cases} \dfrac{m(N-n)}{N} & (1 \le m \le n \le N-1), \\[2ex] \dfrac{n(N-m)}{N} & (1 \le n \le m \le N-1). \end{cases} \qquad (47)$$

It will become evident in a moment that it is convenient to rewrite this as follows:

$$M^{(N)}_{n,m} = N\omega\left(\frac{n}{N}, \frac{m}{N}\right), \qquad (48)$$

where

$$\omega(s,t) = \begin{cases} s(1-t) & (0 \le s \le t \le 1), \\ t(1-s) & (0 \le t \le s \le 1) \end{cases} \qquad (49)$$

is a continuous function on the unit square $0 \le s$, $t \le 1$. Thus

$$\rho_{N,L}(\xi)$$
$$= \det\left(\delta_{n,m} + N\sum_r \omega\left(\frac{n}{N}, \frac{r}{N}\right) b_{r-m}\left(\frac{2\pi\xi}{N}\right)\right). \qquad (50)$$

We now inspect closely the three terms of the matrix product $N\sum_r$ in (50) obtained from the three terms of $b_n(\alpha)$ as given in (44).

[20] To keep the notation uncluttered we suppress the range of indices in the sums and determinants. It is the first $N-1$ positive integers.

The first term gives

$$-4N \sin^2 \frac{\pi\xi}{2N}\, \omega\left(\frac{n}{N}, \frac{m}{N}\right) = -\frac{1}{N}\pi^2\xi^2\omega\left(\frac{n}{N}, \frac{m}{N}\right) + O\left(\frac{1}{N^3}\right). \quad (51)$$

Here and in the following we mean by the O-symbol a quantity which is bounded by a constant times the indicated argument, where the "constant" may depend on ξ but not on n and m. This ensures that we can investigate limiting behavior as $N \to \infty$ and fixed ξ *uniformly* in n and m.

Next we consider the second term in (44). Its contribution to the matrix product in (50) is

$$\frac{8}{\pi}\sin^2 \frac{\pi\xi}{2N} \sum_r \omega\left(\frac{n}{N}, \frac{r}{N}\right) \frac{\sin \pi\, |\xi|\, (r/N - m/N)}{r/N - m/N}. \quad (52)$$

In the limit of large N, the sum over r when divided by N appears as a Riemann approximating sum to an integral and the error is easily estimated by the customary methods. Thus (52) becomes

$$\frac{1}{N}\, 2\pi\xi^2 \int_0^1 d\sigma\, \omega\left(\frac{n}{N}, \sigma\right) \frac{\sin \pi\, |\xi|\, (\sigma - m/N)}{\sigma - m/N} + O\left(\frac{1}{N^2}\right). \quad (53)$$

Finally, we come to the third term in (44) and its contribution to the matrix product. The computation is greatly facilitated by the observation that the square bracket in (44) contains a *second difference* with respect to the subscript n. The sum over r in (50) may then be treated by the "summation by parts" procedure, the discrete analog of integration by parts. Great simplification arises because—apart from boundary terms which, of course, have to be duly taken into account—we have to deal with the second difference of the other matrix factor, but $M^{(N)}_{n,r+1} + M^{(N)}_{n,r-1} - 2M^{(N)}_{n,r} = \delta_{n,r}$ by definition of the matrix M [cf. (24)]. A careful manipulation reduces this contribution to

$$\frac{2}{\pi N}\left[\frac{(n/N) \sin (\pi\, |\xi|\, (1 - m/N))}{1 - m/N} + \frac{(1 - n/N) \sin (\pi\, |\xi|\, m/N)}{m/N} - \frac{\sin (\pi\, |\xi|\, (n/N - m/N))}{n/N - m/N}\right]. \quad (54)$$

Summing up (51), (53), and (54), we may write then

$$\rho_{N,L}(\xi) = \det\left[\delta_{n,m} + \frac{1}{N} K\left(\frac{n}{N}, \frac{m}{N}; \xi\right) + O\left(\frac{1}{N^2}\right)\right]. \quad (55)$$

The function K is defined as follows:

$$K(s, t; \xi) = -\pi^2\xi^2\omega(s, t) + 2\pi\xi^2 \int_0^1 d\sigma\, \frac{\omega(s, \sigma) \sin \pi\, |\xi|\, (\sigma - t)}{\sigma - t} + \frac{2}{\pi}\left[\frac{s \sin (\pi\, |\xi|\, (1 - t))}{1 - t} + \frac{(1 - s) \sin (\pi\, |\xi|\, t)}{t} - \frac{\sin (\pi\, |\xi|\, (s - t))}{s - t}\right]. \quad (56)$$

It is a continuous function of s and t on the unit square $0 \leq s, t \leq 1$, and depends on ξ parametrically. Since the left-hand side of Eq. (55) is a periodic function of ξ we must, strictly speaking, note that the equation as written is valid only in the interval $-\frac{1}{2}L \leq \xi \leq \frac{1}{2}L$; but this limitation becomes irrelevant as $N = L \to \infty$ and so we may think of ξ as arbitrary.

Now we observe that the determinant (55) is—apart from the $O(N^{-2})$ error terms in the elements—precisely the discrete approximation to the Fredholm determinant belonging to the integral kernel $K(s, t; \xi)$.[21] Thus we have shown the following.

Theorem 3. *For all fixed ξ, the thermodynamic limit of the density matrix exists and is equal to*

$$\rho(\xi) = 1 + \sum_{n=1}^{\infty} \frac{1}{n!} \int_0^1 ds_1 \cdots \times \int_0^1 ds_n \det_{1\leq i,j\leq n} K(s_i, s_j; \xi) \quad (57)$$

where the kernel $K(s, t; \xi)$ is given by (56).

In the next section we shall derive a number of specific conclusions from this fact.

6. THE LIMITING MOMENTUM DISTRIBUTION

We would now like to conclude that a limiting momentum distribution $F(k)$ exists satisfying

$$\rho(\xi) = \int_{-\infty}^{\infty} e^{ik\xi}\, dF(k), \quad (58)$$

which is the analog of (7) for a finite system, and that—in some sense—

$$\lim_{N,L\to\infty} F_{N,L}(k) = F(k). \quad (59)$$

Actually there are two separate questions involved. First, is there a probability distribution function $F(k)$ such that (58) is true? Second, if so, does (59)

[21] G. Kowalewski, *Einführung in die Determinantentheorie* (Chelsea Publishing Company, New York, 1948), 3rd ed., Paragraph 115. We prove in Appendix 3 that the error terms may be ignored.

hold for this function and in what precise sense? The answer to these questions is contained in the continuity theorem for distribution functions and characteristic functions (cf. Appendix 1). It asserts that the answer to the first question is affirmative and that (59) holds in the sense of a pointwise limit *at all continuity points of* $F(k)$ under the single additional assumption that $\rho(\xi)$ is continuous at the origin.

It is not difficult to verify that this is so, but we actually prove a much stronger property of which the continuity at the origin is a trivial consequence and which gives important additional information.

Theorem 4. *$\rho(\xi)$ is an entire function of the variable $|\xi|$.*

To be quite precise, this means that there is an entire function of the complex variable $f(z)$ such that $\rho(\xi) = f(|\xi|)$. To prove this, we consider the kernel (56) as a function of z, that is to say a function $Q(s, t; z)$ which arises when $|\xi|$ is everywhere formally replaced by z. This function is manifestly an entire function. Its power-series expansion is

$$Q(s, t; z) = -\pi^2 z^2 \omega(s, t) + \sum_{n=1}^{\infty} A_n(s, t) \frac{z^{2n+1}}{(2n+1)!}, \tag{60}$$

where the coefficients $A_n(s, t)$ are certain uniformly bounded functions of s and t on the unit square $0 \leq s, t \leq 1$. It follows that each term

$$\frac{1}{n!} \int_0^1 ds_1 \cdots \int_0^1 ds_n \det_{1 \leq i, j \leq n} Q(s_i, s_j; z) \tag{61}$$

in the expansion (57) of $f(z)$ is an entire function $f_n(z)$, say, of the complex variable z. Let r be an arbitrary positive number and let $M = M(r)$ denote the maximum absolute value of $Q(s, t; z)$ for $|z| = r$ and s, t on the unit square. According to the determinant inequality of Hadamard[22] we have then

$$|f_n(z)| < n^{\frac{1}{2}n} M^n / n!. \tag{62}$$

From this and the Weierstrass "M-test"[23] it follows that the series

$$f(z) = 1 + \sum_{n=1}^{\infty} f_n(z) \tag{63}$$

converges uniformly for $|z| = r$, and this in turn implies[24] that its sum represents an analytic function of z for $|z| \leq r$. But r was arbitrary, hence $f(z)$ is an entire function. This proves Theorem 4.

The first few terms of the convergent power-series expansion guaranteed by this theorem are easily calculated. We have[25]

$$K(s, t; \xi) = -\pi^2 \xi^2 \omega(s, t) + \tfrac{2}{3}\pi^2 |\xi|^3 s(1 - s) + O(|\xi|^5), \tag{64}$$

and therefore

$$\rho(\xi) = 1 - \frac{\pi^2}{6} \xi^2 + \frac{\pi^2}{9} |\xi|^3 + \frac{\pi^4}{120} \xi^4 + O(|\xi|^5). \tag{65}$$

The calculation of higher terms is straightforward in principle but becomes prohibitive in practice on account of the rapidly increasing algebraic complexity.

Naturally, (65) implies the continuity of $\rho(\xi)$ at the origin. Therefore, $\rho(\xi)$ is a characteristic function, and the corresponding distribution function $F(k)$ is given by the inversion formula of (58), namely,

$$F(k_1) - F(k_2) = \lim_{\lambda \to \infty} \int_{-\lambda}^{\lambda} \frac{e^{-ik_2\xi} - e^{-ik_1\xi}}{2\pi i \xi} \rho(\xi)\, d\xi. \tag{66}$$

This formula holds for all continuity points k_1 and k_2 of $F(k)$ and defines it uniquely. Moreover, (59) holds in the sense of a pointwise limit, again at all continuity points of $F(k)$. This is the most that can be said to follow from the uniqueness and continuity theorems (cf. Appendix 1) and our Theorems 3 and 4.

We shall now show that more can be said upon making use of the inequality (33). Substitute this inequality on the right-hand side of (66) and make use of

$$|\sin \tfrac{1}{2}(k_1 - k_2)\xi| \leq \mathrm{Min}\,\{1, |\tfrac{1}{2}(k_1 - k_2)\xi|\}. \tag{67}$$

Thus

$$\begin{aligned} |F(k_1) - F(k_2)| &< e^{\frac{1}{2}} \pi^{-\frac{3}{2}} \int_{-\infty}^{\infty} |\xi|^{-\frac{3}{2}} \mathrm{Min}\,\{1, |\tfrac{1}{2}(k_1 - k_2)\xi|\}\, d\xi \\ &= 4 e^{\frac{1}{2}} \pi^{-\frac{3}{2}} |k_1 - k_2|^{\frac{1}{2}} \end{aligned} \tag{68}$$

This inequality is a Hölder condition with index $\frac{1}{2}$. It implies, in particular, the continuity of $F(k)$ as seen by letting $k_1 \to k_2$. Now, if a sequence of probability distribution functions tends to a *con-*

---

[22] E. T. Whittaker and C. N. Watson, *A Course of Modern Analysis* (Cambridge University Press, Cambridge, England, 1958), 4th ed., Paragraph 11.11. The statement and proof must be slightly modified to make it valid for determinants with complex elements.

[23] Reference 22, Paragraph 3.4.

[24] Reference 21, Paragraph 5.3.

[25] It will be observed that $K(s, t; \xi)$ is not a symmetrical kernel. The author has tried in vain to modify the argument of Sec. 5 so that a representation in terms of a symmetrical kernel should emerge. Such a representation is, however, implicit in the work of Schultz, Ref. 3, and is discussed in Sec. 8 of this paper.

tinuous probability distribution function, then the convergence is uniform over the whole real axis.[26] We summarize our results as follows.

Theorem 5. *In the thermodynamic limit, the momentum distribution function $F_{N,L}(k)$ converges uniformly to a continuous limiting distribution $F(k)$ which satisfies the Hölder condition (68).*

The physically interesting information is the continuity of the limiting distribution. Its proof depends crucially on the Szegö inequality, Theorem 2, of which it forms the principal application.

The discontinuities of $F_{N,L}(k)$ which have the physical interpretation of the mean fractional occupation numbers of the various momentum states shrink to zero in the thermodynamic limit. How fast? Our next theorem gives an estimate.

Theorem 6. *The discontinuities of $F_{N,L}(k)$ are less than*

$$N^{-\frac{1}{2}}\left(1+\frac{1}{2N-1}\right)^{N-\frac{1}{2}}\frac{1}{\pi}B(\tfrac{1}{4},\tfrac{1}{2}). \tag{69}$$

Proof: Because of the positiveness of $\rho_{N,L}(\xi)$ the largest discontinuity of $F_{N,L}(k)$ occurs at the origin [cf. (6) and (4)], and that is

$$\frac{\rho_{N,L}^{(0)}}{N}=\frac{1}{2\pi N}\int_{-\pi}^{\pi}R_N(\alpha)\,d\alpha. \tag{70}$$

Substitution of the inequality (31) and evaluation of the integral proves the theorem. The numerical coefficient $\pi^{-1}B(\frac{1}{4},\frac{1}{2})$ is less than 2.

Finally, we give a result on the behavior of the limiting momentum distribution for $|k|\to\infty$. In effect, it gives a bound to the rate at which $F(k)$ approaches its asymptotic values.

Theorem 7. *No absolute moments of $F(k)$ of order $\delta\geq 3$ exist. The second moment is finite and equal to $\frac{1}{3}\pi^2$.*

The proof depends on the differentiability theorem for characteristic functions (cf. Appendix 1). For, if any absolute moment of order $\delta\geq 3$ would exist (i.e., would be finite), then $\rho(\xi)$ would be three times continuously differentiable; but it is not, on account of the appearance of the $|\xi|^3$ term in (65). It is, however, twice differentiable, and

$$\int_{-\infty}^{\infty}k^2\,dF(k)=-\rho''(0)=\tfrac{1}{3}\pi^2. \tag{71}$$

This quantity is just the mean kinetic energy per particle in the thermodynamic limit [cf. also (11)].

[26] A proof of this fact is given in Appendix 1.

7. BOSE-EINSTEIN CONDENSATION

The phenomenon of Bose–Einstein condensation (or condensation in momentum space) for an *ideal* gas of Bose particles was discovered long ago,[27] but it was only relatively recently that proposals were put forward for mathematically precise criteria to distinguish the occurrence from the nonoccurrence of such condensation in the physically more interesting but much more difficult problem of a Bose gas with interparticle forces. We shall consider the application of two such criteria to the present one-dimensional model.

Penrose and Onsager[28] formulate the criterion in terms of the largest eigenvalue of the density matrix. According to this criterion, there is no condensation when the largest eigenvalue divided by the total number of particles tends to zero in the thermodynamic limit. Since

$$\int_0^L\rho_{N,L}(x-x')e^{2\pi inx'/L}\,dx'=\rho_{N,L}^{(n)}e^{2\pi inx/L}, \tag{72}$$

the eigenvalues in question are just the quantities $\rho_{N,L}^{(n)}$ $(n=0,\pm1,\pm2,\cdots)$ and the largest of them is $\rho_{N,L}^{(0)}$. This divided by N is just the largest discontinuity of the momentum distribution function. Thus we see that *Theorem 6 signifies the fulfillment of the Penrose–Onsager criterion for the absence of Bose–Einstein condensation.* It actually does more: it provides the specific bound $O(N^{-\frac{1}{2}})$ for the mean fraction of zero-momentum particles in the limit $N=L\to\infty$.

In his paper on the one-dimensional gas of Bose particles, Girardeau[1] suggested a more stringent criterion. This may be formulated in our notation as follows[29]:

$$\lim_{k\to+0}\limsup_{N,L\to\infty}\,[F_{N,L}(k)-F_{N,L}(-k)]=0. \tag{73}$$

It is fulfilled because of the existence and continuity of the limiting momentum distribution (Theorem 5). The essence of Girardeau's criterion lies in the order of the two limiting processes. If the order is reversed one obtains the Penrose–Onsager criterion. The latter is always fulfilled when the Girardeau criterion is, because

$$\lim_{k\to+0}\limsup_{N,L\to\infty}\,[F_{N,L}(k)-F_{N,L}(-k)]$$

$$\geq\lim_{k\to+0}\limsup_{N,L\to\infty}\,[\rho_{N,L}^{(0)}/N]=\limsup_{N,L\to\infty}\,[\rho_{N,L}^{(0)}/N]. \tag{74}$$

[27] A. Einstein, Sitzber. Preuss. Akad. Wiss. Physik-Math. Kl., 261 (1924); 3, 18 (1925).

[28] O. Penrose and L. Onsager, Phys. Rev. 104, 576 (1956).

[29] Where we write "lim sup" Girardeau has "lim". This change was made to render the criterion logically independent of the existence of a limiting momentum distribution.

The implication does not hold the other way, however, as can be demonstrated by counter examples.

The $O(N^{-\frac{1}{2}})$ bound of Theorem 6 and the $O(|k_1 - k_2|^{\frac{1}{2}})$ bound of Theorem 5 are not unconnected. Both are rooted in the Szegö inequality (31) or ultimately (27). It is possible to write a single inequality which displays this connection. It is derived by writing the difference $F_{N,L}(k_1) - F_{N,L}(k_2)$ in the form of a finite sum from (6) and replacing the $\rho_{N,L}^{(n)}$ by their definition (4) as Fourier coefficients. This gives

$$F_{N,L}(k_1) - F_{N,L}(k_2) = \frac{1}{2\pi N}\int_0^{2\pi} R_N(\alpha) \times e^{-i(\alpha/2)(n_1+n_2+1)} \frac{\sin \frac{1}{2}\alpha(n_1 - n_2)}{\sin \frac{1}{2}\alpha}\, d\alpha, \tag{75}$$

where n_1 is the largest and n_2 the smallest integer satisfying

$$k_2L/2\pi \leq n < k_1L/2\pi. \tag{76}$$

Making use of the inequalities (31) and (67), one obtains then

$$F_{N,L}(k_1) - F_{N,L}(k_2) < \text{const}\,[(L/N)(k_1 - k_2)]^{\frac{1}{2}}. \tag{77}$$

This inequality implies both the Hölder condition (68) of Theorem 5, and Theorem 6; the former by letting $N = L \to \infty$, the latter by putting $L(k_1 - k_2)$ equal to some positive number less than 2π.

It should be stressed that our conclusions must not be interpreted to mean that the momentum distribution shows no distinguished features at the origin in the thermodynamic limit. It is characteristic of our results that they provide bounds; bounds which are strong enough to satisfy physically motivated criteria for the absence of Bose–Einstein condensation, but not precise enough to reveal the qualitative properties of the momentum distribution. It seems quite likely that $F(k)$ is actually a quite smooth function of k (perhaps even analytic) outside the origin, and that near the origin its behavior is

$$F(k) = \tfrac{1}{2} + \text{const}\,\frac{k}{|k|^{\frac{1}{2}}} + o(|k|^{\frac{1}{2}}) \quad (k \to 0). \tag{78}$$

Whether there really is a "singularity" of this, or perhaps some even milder, type at the origin in momentum space remains an unsolved problem.[30]

8. COMPARISON WITH PREVIOUS WORK

The problem of the momentum distribution has been already considered in the past.[1,3] Here we shall consider some aspects of this previous work in the light of our own results.

Girardeau treats the problem of the momentum distribution with an approximation method due to Bogoliubov.[2] This method is not well adapted to the investigation of the question of Bose–Einstein condensation because in a sense it assumes an affirmative answer. This is so because the zero momentum state is distinguished at the outset by the assumption that it has a very much larger average occupation than the other momentum states. In addition to this difficulty of principle, Girardeau was forced to make certain approximations in his calculation whose validity is open to question. His conclusion was that the Penrose–Onsager criterion for the absence of Bose–Einstein condensation is satisfied, but not the more strict criterion (73), indeed this was his reason for proposing the latter in the first place. The suggested dependence of the average zero-momentum occupation number on the total number of particles[31]

$$\rho_{N,L}^{(0)} \sim N/\ln N \quad (N \to \infty) \tag{79}$$

is clearly contradicted by our Theorem 6. The incorrectness of Girardeau's result was already proved by Schultz.[3]

Schultz started by posing the quantum mechanical problem in the formalism of second quantized fields. Then he replaced the continuum of space points in the box $0 \leq x \leq L$ by a large but finite number of equidistant points in this interval so that the commutation rules became a finite algebraic system. He then made the crucial observation that this "discretized" problem is from the mathematical point of view completely identical to a certain problem of interacting spins, a problem whose solution had already been worked out.[32] In particular, the density matrix, or what is the same, the function $R_N(\alpha)$ was given in the following form[33]:

$$\frac{|\alpha|}{\pi} R_N(\alpha) = \lim_{\nu\to\infty} \nu \det_{1\leq q,p\leq \nu} \left[-\delta_{q,p+1} + \frac{1}{\nu} G_N\!\left(\frac{p+1-q}{N};\alpha\right)\right], \tag{80}$$

where

$$G_N(s;\alpha) = \frac{|\alpha|}{\pi}\frac{\sin \frac{1}{2}(N\alpha s)}{\sin \frac{1}{2}(\alpha s)}. \tag{81}$$

The essence of the method is the extra limiting

[30] The difficulty is that we have not been able to do better than the inequality (33) as regards the $|\xi| \to \infty$ behavior of $\rho(\xi)$. A sufficient sharpening of it would allow the derivation of (78).

[31] Reference 1, Eq. (A50).

[32] E. H. Lieb, T. Schultz, and D. Mathis, Ann. Phys. (N. Y.) 16, 407 (1961).

[33] Reference 3, Eqs. (2), (11), (14), and (15).

process, visible in (80), which corresponds to going from the discrete to the continuum. It must be carried out independently of, and prior to, the thermodynamic limit. Leaving ν fixed and finite in (80) one gets an approximation to $R_N(\alpha)$ equal to replacing the integrals in (14) by their Riemann approximating sums with a subdivision of the interval $(0, 2\pi)$ into ν equal parts.[34]

By analyzing the nature of the ν by ν matrix which occurs in (80), Schultz was able to derive an inequality which corresponds to our (31) but is somewhat weaker. With its help he showed that (73) is fulfilled, thus controverting the claim of Girardeau.

Regarding the determinant in (80) we observe the following. It is precisely equal to the algebraic complement of the element in the lower left-hand corner ($q = 0$, $p = \nu$) of the larger determinant with $\nu + 1$ rows and colums,

$$\det_{0\le q,p\le\nu}\left[\delta_{q,p} - \frac{1}{\nu}G_N\left(\frac{q-p}{\nu};\alpha\right)\right]. \tag{82}$$

From the point of view of the limiting process $\nu \to \infty$, this is of the Fredholm type; and not only its limiting behavior but also that of its *minors* can be found exactly.[35] One obtains[36]

$$\frac{|\alpha|}{\pi}R_N(\alpha) = G_N\begin{pmatrix}0\\1\end{pmatrix};\alpha\Big) + \sum_{n=1}^{\infty}\frac{(-1)^n}{n!}\int_0^1 ds_1\cdots \times\int_0^1 ds_n G_N\begin{pmatrix}0 & s_1 & \cdots & s_n\\ 1 & s_1 & \cdots & s_n\end{pmatrix};\alpha\Big), \tag{83}$$

where

$$G_N\begin{pmatrix}s_0 s_1 & \cdots & s_n\\ t_0 t_1 & \cdots & t_n\end{pmatrix};\alpha\Big) = \det_{0\le i,j\le n} G_N(s_i - t_j;\alpha) \tag{84}$$

is the Fredholm symbol belonging to the kernel G_N defined by (81). We shall now show that the series in (83) is actually finite because

$$G_N\begin{pmatrix}s_0 s_1 & \cdots & s_n\\ t_0 t_1 & \cdots & t_n\end{pmatrix};\alpha\Big) = 0 \tag{85}$$

identically in the variables whenever $n \ge N$. This fact has its origin in the identity

$$\frac{\sin\frac{1}{2}(N\alpha s)}{\sin\frac{1}{2}(\alpha s)} = \sum_r e^{ir\alpha s}, \tag{86}$$

the summation ranging over $r = -\frac{1}{2}(N-1),\cdots, \frac{1}{2}(N-3), \frac{1}{2}(N-1)$. Thus the matrix of $n+1$ rows and columns whose determinant is (84) may be regarded as a matrix product of two rectangular but in general not square matrices, one having $n+1$ rows and N columns, and the second one N rows and $n+1$ columns. It is a general fact of the algebra of determinants[37] that the determinant of such a product vanishes when $N < n+1$.

It is rather remarkable that the expression given by (83) for $R_N(\alpha)$ represents the same function as the determinant (20) derived in the present paper, different though—to all appearances—it may seem. An explanation lies, perhaps, in the circumstance that the *elements* of a determinant may be altered in manifold ways without changing its *value*. Nevertheless, it would be interesting to see a short proof of the identity of these two expressions. We may remark in this connection that their general character *is* the same, namely a finite combination of trigonometric functions of $|\alpha|$ and powers of $|\alpha|$. This is obvious[38] in the case of (20), and in the case of (83) it follows from the identity (86), together with the fact that integration of functions belonging to the class indicated always produces functions again belonging to that class. A simple proof of the identity would seem to require an ingenious exercise in formal manipulation.

Our own representation of $R_N(\alpha)$ seems simpler because no integrations occur in it, but it is conceivable that (83) is more useful for certain purposes. For one, it is easier to carry out for it the thermodynamic limit (32). Instead of the elaborate calculations of Sec. 5, we now only have to notice that

$$\lim_{N\to\infty} G_N\left(s;\frac{2\pi\xi}{N}\right) = \frac{2\sin\pi s|\xi|}{\pi s} = g(s;\xi), \tag{87}$$

and in (83) we can go to the limit term by term.[39] This gives

$$2|\xi|\rho(\xi) = g\begin{pmatrix}0\\1\end{pmatrix};\xi\Big) + \sum_{n=1}^{\infty}\frac{(-1)^n}{n!} \times\int_0^1 ds_1\cdots\int_0^1 ds_n\, g\begin{pmatrix}0 & s_1 & \cdots & s_n\\ 1 & s_1 & \cdots & s_n\end{pmatrix};\xi\Big), \tag{88}$$

with

$$g\begin{pmatrix}s_0 s_1 & \cdots & s_n\\ t_0 t_1 & \cdots & t_n\end{pmatrix};\xi\Big) = \det_{0\le i,j\le n} g(s_i - t_j;\xi). \tag{89}$$

Again, (88) looks very different from the representation given in Theorem 3, but it is easily verified, for instance, that the behavior for small $|\xi|$ is given

[34] This was pointed out to the author in a correspondence by Dr. Schultz.

[35] Reference 21, Paragraph 117.

[36] The possibility of the transformation from (80) to (83) seems not to have been noticed by Schultz.

[37] Reference 21, Sec. 34.

[38] Powers of $|\alpha|$ arise from $c_n(\alpha)$ with $n = 0, \pm 1$.

[39] The rigorous justification of the legitimacy of this step should not be difficult.

correctly by (65). Whether any properties of $\rho(\xi)$ can be better revealed in the representation (88) is not known.

ACKNOWLEDGMENT

The author wishes to express his indebtedness to Professor C. N. Yang whose remarks on the problem of Bose–Einstein condensation at the spring meeting (1963) on statistical mechanics at Yeshiva University induced him to undertake this investigation. He wishes to thank Professor F. J. Dyson for several fruitful conversations and much encouragement. And, above all, he is grateful to Professor G. Szegö for the correspondence which resulted in Theorem 2, and for his consent to let his contribution become an essential part of this paper.

APPENDICES

A1. Distribution Functions and Characteristic Functions

A (cumulative) probability distribution function $F(k)$ is a real nondecreasing function such that its total variation is from 0 to 1. Such a function has at most a denumerable number of ordinary jump discontinuities and is continuous at all other points. Its value at the discontinuity points may be conventionally fixed by continuity from the left, for example.

The characteristic function $\phi(\xi)$ belonging to the probability distribution function $F(k)$ is defined by the Fourier–Stieltjes transformation

$$\phi(\xi) = \int_{-\infty}^{\infty} e^{ik\xi}\, dF(k). \tag{90}$$

This relationship is unique, i.e., two probability distribution functions to which belongs the same characteristic function can differ only (inessentially) at discontinuity points. There is an explicit inversion formula

$$F(k_1) - F(k_2) = \lim_{\lambda\to\infty} \int_{-\lambda}^{\lambda} \frac{e^{-i\xi k_2} - e^{-i\xi k_1}}{2\pi i\xi}\, \phi(\xi)\, d\xi, \tag{91}$$

the implication being that the limit exists and is equal to the difference on the left, provided k_1 and k_2 are continuity points. (Uniqueness Theorem.)

One reason for the importance of this one-to-one relationship between these two classes of functions is that limiting operations preserve this correspondence. More specifically, let $\phi_n(\xi)$ be a sequence of characteristic functions belonging to the probability distribution functions $F_n(k)$, such that

$$\lim_{n\to\infty} \phi_n(\xi) = \phi(\xi) \tag{92}$$

exits (pointwise) and is continuous at $\xi = 0$. Then there is a probability distribution function $F(k)$ such that (i) its characteristic function is $\phi(\xi)$, and (ii) the limiting relation

$$\lim_{n\to\infty} F_n(k) = F(k) \tag{93}$$

holds at every continuity point of it. (Continuity Theorem.)

The mode of convergence in (93) is considerably sharper when $F(k)$ is continuous. In this case the convergence is *uniform for all* k. To prove this, let $\epsilon > 0$ be arbitrary and let k_i ($i = 1, 2, \cdots$) be a finite number of points at which $F(k)$ assumes values which are integral multiples of $\frac{1}{2}\epsilon$. Let then $N = N(\epsilon)$ be so large that all differences $F_n(k_i) - F(k_i)$ are less in magnitude than $\frac{1}{2}\epsilon$ for all $n \geq N$; this can be done because i runs through a finite number of values only. If now k is arbitrary, it is easy to see that from the nondecreasing nature of the distribution functions it follows that $F_n(k)$ and $F(k)$ differ by less than ϵ for all $n \geq N$.

There is a simple relationship between the moments of the probability distribution function $F(k)$ and the derivatives of $\phi(x)$ at the origin. If the absolute moment

$$\int_{-\infty}^{\infty} |k|^{\delta}\, dF(k) \tag{94}$$

exists (i.e., is finite) then the characteristic function corresponding to $F(k)$ has everywhere continuous derivatives of all orders not exceeding δ. If δ is an even integer the value of the integral (94) is obtained from (90) by formal differentiation. (Differentiability Theorem.)

We have quoted these facts only to the extent needed in the text (Sec. 6). For more complete statements and proofs the reader is referred to the literature.[11]

A2. Szegö's Theory of Toeplitz Determinants[4]

A Toeplitz matrix is a finite matrix whose elements are given by

$$c_{n-m} = \frac{1}{2\pi} \int_{-\pi}^{\pi} f(\theta) e^{-i(n-m)\theta}\, d\theta \qquad (0 \leq n,\, m \leq N), \tag{95}$$

where $f(\theta)$ is a real, nonnegative function. Its determinant is denoted by $D_N(f)$. An important role is played by the geometric mean

$$\exp \frac{1}{2\pi} \int_{-\pi}^{\pi} \ln f(\theta)\, d\theta, \tag{96}$$

which is denoted by $G(f)$.

Szegö has shown that the ratio of two succeeding Toeplitz determinants may be characterized as the solution of a certain minimum problem.

Theorem A. *The ratio $D_N(f)/D_{N-1}(f)$ is equal to the minimum with respect to the N complex numbers $u_1, u_2, \cdots, u_N$ of the integral*

$$\frac{1}{2\pi} \int_{-\pi}^{\pi} f(\theta) \times |1 + u_1 e^{i\theta} + u_2 e^{2i\theta} + \cdots + u_N e^{Ni\theta}|^2\, d\theta. \tag{97}$$

This theorem is proved[40] by consideration of certain orthogonal polynomials associated with the function $f(\theta)$. It has two important consequences.

Corollary 1. *If $\bar{f}(\theta) > f(\theta)$ then $D_N(\bar{f}) > D_N(f)$.*

Corollary 2. *The sequence $D_N(f)/G(f)^{N+1}$ ($N = 0, 1, 2, \cdots$) is nondecreasing.*

The first follows from Theorem A by the general principle that if one real function is always larger than another then the minimum of the first one is larger than that of the second. Corollary 2 is shown as follows. Write the minimizing trigonometric polynomial of (97) in the factored form

$$1 + u_1 e^{i\theta} + \cdots + u_N e^{Ni\theta} = (1 - A_1 e^{i\theta}) \cdots (1 - A_N e^{i\theta}), \tag{98}$$

where the A's are certain complex constants depending on f and N. Now, the integral (97) is an arithmetical mean of a function of θ and therefore is never less[41] than the geometrical mean of the same function. On the other hand, the geometrical mean of a product is the product of the geometrical means. Now, for each of the factors in (98) one shows by an elementary integration that

$$G(|1 - Ae^{i\theta}|^2) = \text{Max}\,\{1, |A|^2\} \geq 1, \tag{99}$$

and hence

$$D_N(f)/D_{N-1}(f) \geq G(f), \tag{100}$$

from which the assertion follows.

In view of Corollary 2 there are two posibilities. As $N \to \infty$, either the sequence tends to infinity or else it has a finite limit. Both possibilities actually occur,[42] but for a wide class of functions $f(\theta)$ the limit is finite. This is the content of the second of Szegö's theorems which we make use of.

Theorem B. *Let $f(\theta)$ be strictly positive and let its derivative $f'(\theta)$ satisfy a Hölder condition with some index less than 1. Then the sequence $D_N(f)/G(f)^{N+1}$ tends to the finite limit*

$$\exp \sum_{n=1}^{\infty} n\, |h_n|^2 \tag{101}$$

The numbers h_n are coefficients in a power-series expansion

$$\ln g(z) = \sum_{n=0}^{\infty} h_n z^n, \tag{102}$$

$g(z)$ being defined uniquely (up to an irrelevant constant phase factor) by the properties (i) $g(z)$ is analytic, regular and not zero in the circle $|z| < 1$, and (ii) the squared absolute value of $g(re^{i\theta})$ tends to $f(\theta)$ as $r \to 1$.

Szegö's proof[43] is a subtle construction of analysis based on approximation of functions by trigonometric polynomials. An alternative proof based on different ideas has been given by Kac.[44]

We note that in the application contemplated (Sec. 4) it was impossible to apply Theorem B directly to the Toeplitz determinant $R_N(\alpha)$ because the $f(\theta)$ belonging to it is $2\,|\cos\theta - \cos\frac{1}{2}\alpha|$, and this does not satisfy the requisite conditions.

A3. A Theorem on Determinants

In order to complete the proof of Theorem 3 it is necessary to show that in the limit $N \to \infty$ the "error terms" of (55) may be ignored. We do this by proving a proposition which shows that a sufficient condition for this is that they be $o(N^{-\frac{1}{2}})$.

Let A_N and B_N be given N by N matrices for $N = 1, 2, 3, \cdots$ such that the elements of A_N are uniformly $O(N^{-1})$ and the elements of B_N are uniformly $o(N^{-\frac{1}{2}})$ as $N \to \infty$. Let I_N be the N by N unit matrix. Then the difference between the determinant of $I_N + A_N + B_N$ and the determinant of $I_N + A_N$ tends to zero as $N \to \infty$. We prove this by making use of the celebrated inequality of Hadamard,[22] according to which the magnitude of a determinant is not more than the product of the Euclidean norms of its row vectors. The difference of our two determinants may be written with the help of the mean-value theorem

[40] Reference 4, p. 38.

[41] G. H. Hardy, J. E. Littlewood, and G. Polya, *Inequalities* (Cambridge University Press, Cambridge, England, 1934), Theorem 184.

[42] An example of the first possibility is furnished by our $R_N(0) = D_{N-2}(f) = N$ with $f(\theta) = 2(1 - \cos\theta)$.

[43] Reference 4, p. 76.

[44] M. Kac, Duke Math. J. **21**, 501 (1954).

of differential calculus as follows:

$$\Delta_N = [(d/d\theta)\det(I_N + A_N + \theta B_N)]_{\theta=\theta_N}, \tag{103}$$

where $0 \leq \theta_N \leq 1$. According to the rule of differentiating determinants, this is

$$\Delta_N = \sum_{1\leq n,m\leq N}\sum (B_N)_{n,m}(-1)^{n+m}\det Q_N^{n;m}, \tag{104}$$

where $Q_N^{n;m}$ is that $N-1$ by $N-1$ matrix which arises from the matrix $I_N + A_N + \theta_N B_N$ by striking the nth row and the mth column. We consider first the N terms $n = m$ in the sum (104). All rows of $Q_N^{n;n}$ consist of one "large" element $1 + O(N^{-1})$ and $N-2$ "small" elements $O(N^{-1})$. The Euclidean norms of these rows are therefore

$$\left\{\left[1 + O\left(\frac{1}{N}\right)\right]^2 + (N-2)\left[O\left(\frac{1}{N}\right)\right]^2\right\}^{\frac{1}{2}} = 1 + O\left(\frac{1}{N}\right). \tag{105}$$

Hence

$$|\det Q_N^{n;n}| = \left[1 + O\left(\frac{1}{N}\right)\right]^{N-1} = O(1), \tag{106}$$

and the contribution of N such terms to Δ_N is $No(N^{-\frac{3}{2}})O(1) = o(N^{-\frac{1}{2}})$. On the other hand, consider the terms in the sum $n \neq m$. The matrix $Q_N^{n;m}$ now has one row, namely the mth, with no large element, so that the norm of this row is

$$\left\{(N-1)\left[O\left(\frac{1}{N}\right)\right]^2\right\}^{\frac{1}{2}} = O\left(\frac{1}{N^{\frac{1}{2}}}\right), \tag{107}$$

while its other rows have norms (105). Thus, for $n \neq m$,

$$|\det Q_N^{n;m}| = O\left(\frac{1}{N^{\frac{1}{2}}}\right)\left[1 + O\left(\frac{1}{N}\right)\right]^{N-2} = O\left(\frac{1}{N^{\frac{1}{2}}}\right). \tag{108}$$

There are $N^2 - N$ such terms in the sum (104), so their total contribution is $(N^2 - N)o(N^{-\frac{3}{2}})O(N^{-\frac{1}{2}}) = o(1)$. Thus $\lim \Delta_N = 0$, as was to be shown.

Reprinted from *Printed in U.S.A.*
JOURNAL OF MATHEMATICAL PHYSICS VOLUME 5, NUMBER 5 MAY 1964

Study of Exactly Soluble One-Dimensional N-Body Problems

J. B. McGuire
University of California, Los Angeles, California
(Received 9 August 1963)

In this paper it is shown that several cases of one-dimensional N-body problems are exactly soluble. The first case describes the motion of three one-dimensional particles of arbitrary mass which interact with one another via infinite-strength, repulsive delta-function potentials. It is found in this case that the stationary-state solution of the scattering of the three particles is analogous to an electromagnetic diffraction problem which has already been solved. The solution to this analogous electromagnetic problem is interpreted in terms of particles. Next it is shown that the problem of three particles of equal mass interacting with each other via finite- but equal-strength delta-function potentials is exactly soluble. This example exhibits rearrangement and bound-state effects, but no inelastic processes occur. Finally it is shown that the problem of N particles of equal mass all interacting with one another via finite- but equal-strength delta functions is exactly soluble. Again no inelastic processes occur, but various types of rearrangements and an N-particle bound state do occur. These rearrangements and the N-particle bound state are illustrated by means of a series of sample calculations.

I. INTRODUCTION

SINCE the advent of quantum theory, physicists have relied on exactly soluble problems to describe some of the strange effects which have quantum mechanical origin. The way in which the potential enters the Schrödinger wave equation makes this equation soluble only for a very limited class of potentials, and with the exception of the Coulomb potential and the harmonic-oscillator potential, the exactly soluble problems are not particularly good imitations of the interactions which exist in the physical world. On the other hand, these exactly soluble problems illustrate a broad range of effects which are present in the physical world, and therefore at least allow us a qualitative description of the processes which can occur, and perhaps an insight into perturbation and approximation methods which can be used in more physical problems.

One would hope that exact solutions of N-body problems would be of help in producing similar insights into qualitative effects and possible approximation methods for problems of this type. There are, however, additional mathematical difficulties introduced by the presence of more than two particles which have made the exact solution of an N-body problem a more elusive goal.[1,2] These mathematical difficulties are related to the broad range of physical effects which are possible due to the presence of more than two particles.

Let us discuss the kinds of effects which may occur in N-body problems. We know that the many-particle wavefunction will contain all of the information about two-particle interactions because we may isolate two particles by putting the other particles so far away that their influence on the remaining two is negligible. Under these circumstances we will recover the two-particle wavefunction. More complicated effects arise when the N particles are close together in space and time.

Our task is to discuss those effects which arise from the proximity of the N particles, so let us focus our attention on the simplest problem which contains these effects, the three-body problem. Even here we expect a large number of physical effects. We expect finite probabilities for any two particles with an attractive potential between them to be bound in the final state, even though all of the particles are free in the initial state. Also, there will be finite probabilities for the particles to be scattered from a free state to another free state with a different distribution of energy among the particles. If a free particle is incident on a bound state we would expect that this free particle could ionize, excite, or perhaps replace a bound particle. In general we would expect finite matrix elements between any initial and final state which have the same energy.

In view of the many effects which exist in problems of this type it is not surprising that exact solutions or even reliable approximation methods are difficult to find. In order to construct exactly soluble problems we are going to be forced to make many simplifying assumptions. We are going to deal with a three-body problem where all of the particles move in only one dimension and interact with one

[1] E. Lieb and H. Koppe, Phys. Rev. **116**, 367 (1959).
[2] R. Jost, Commun. Math. Helv. **28**, 173 (1954).

another through delta-function potentials. We will later argue that this does not *a priori* restrict the number of physical effects which can occur, except for the fact that two particles which interact through an attractive delta-function potential have only a single bound state, and therefore an incident third particle cannot excite to another bound state, but only to the continuum.

II. FORMULATION OF THE PROBLEM

We consider the Hamiltonian

$$H = -\frac{\hbar^2}{2}\left(\frac{1}{M_1}\frac{d^2}{dx_1^2} + \frac{1}{M_2}\frac{d^2}{dx_2^2} + \frac{1}{M_3}\frac{d^2}{dx_3^2}\right)$$
$$+ A\,\delta(x_1 - x_2) + B\,\delta(x_2 - x_3) + C\,\delta(x_1 - x_3),$$

which arises when three particles of mass M_1, M_2, M_3 at positions x_1, x_2, x_3 interact with one another via delta-function potentials with strengths A, B, C which depend on coordinate differences between particles.

If we make the change of variables

$$z = (M_1 + M_2 + M_3)^{-1/2}(M_1x_1 + M_2x_2 + M_3x_3),$$
$$y = \frac{[M_3(M_1 + M_2)]^{1/2}}{(M_1 + M_2 + M_3)^{1/2}}\left(\frac{M_1x_1 + M_2x_2}{M_1 + M_2} - x_3\right),$$
$$x = (M_1M_2)^{1/2}/(M_1 + M_2)^{1/2}(x_1 - x_2),$$

the resulting Hamiltonian will be

$$H = -\frac{\hbar^2}{2}\left(\frac{d^2}{dz^2} + \frac{d^2}{dy^2} + \frac{d^2}{dx^2}\right) + A\,\delta\left(\frac{1}{\mu_{12}}\,x\right)$$
$$+ B\,\delta\left(\frac{1}{\mu_{23}}\,[x\cos\alpha + y\sin\alpha]\right)$$
$$+ C\,\delta\left(\frac{1}{\mu_{13}}\,[x\cos\beta - y\sin\beta]\right),$$
$$\frac{1}{\mu_{ij}} = \left[\frac{1}{M_i} + \frac{1}{M_j}\right]^{1/2},$$
$$\tan\alpha = [(M_1 + M_2 + M_3)M_2/M_1M_3]^{1/2},$$
$$\tan\beta = [(M_1 + M_2 + M_3)M_1/M_2M_3]^{1/2}.$$

Transformations of this type are discussed in the Appendix. Formulas are derived which are valid for N particles and not restricted to one dimension.

If we remove the center-of-mass motion of all three particles and eliminate the time from Schrödinger's equation, the stationary-state equation for the internal motion of the three particles will be

$$\Bigg[-\frac{\hbar^2}{2}\left(\frac{d^2}{dx^2} + \frac{d^2}{dy^2}\right)$$
$$+ A\,\delta\left(\frac{1}{\mu_{12}}\,x\right) + B\,\delta\left(\frac{1}{\mu_{23}}\,[x\cos\alpha + y\sin\alpha]\right)$$
$$+ C\,\delta\left(\frac{1}{\mu_{13}}\,[x\cos\beta - y\sin\beta]\right) - E\Bigg]\psi = 0.$$

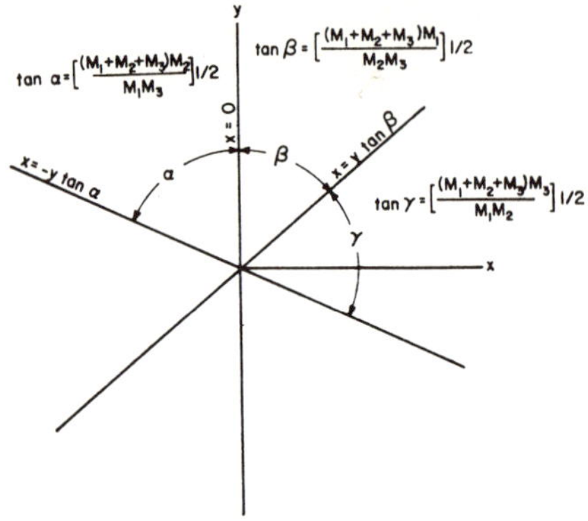

FIG. 1. Potential diagram for three particles interacting in one dimension.

This differential equation may be interpreted as describing the motion of a single particle in a two-dimensional space. Interpreting the differential equation this way may seem awkward since it tends to obscure the true nature of the physical problem, but we will find that the interpretation in terms of particles is not difficult once we have the solution to this mathematically equivalent problem.

The potential in which the single two-dimensional particle moves is zero everywhere except on the lines $x = 0$, $x = y\tan\beta$, $x = -y\tan\alpha$, as shown in Fig. 1. We see that we must solve $(\nabla^2+k^2)\psi=0$ everywhere except on the boundaries provided by the "line" delta functions. It is well known that the boundary condition on such a delta-function line is that there is a discontinuity in the normal derivative of the wavefunction which is equal to the strength of the delta function times the value of the wavefunction on the boundary.

Even this problem is an extremely difficult one mathematically and only limited progress has been made toward its solution. In Sec. IV we will solve a special case of this problem where the masses of the particles and the strengths of the delta functions are chosen in a particular way.

Another special case of interest has already been solved for us. If we take the strength of the delta-function interactions to be infinite, then the wavefunction must approach zero on the delta-function boundaries. This problem is analogous to the diffraction of electromagnetic waves from wedges and corners made of conducting materials, and is soluble for arbitrary angles between the delta-function lines and hence for arbitrary masses of the interacting

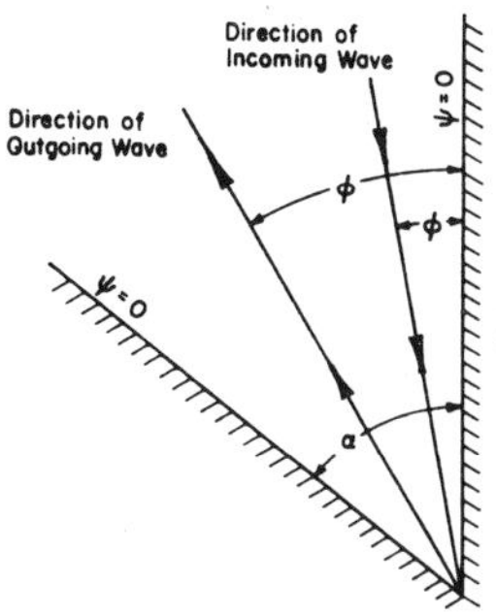

FIG. 2. Coordinate system for the electromagnetic diffraction problem.

particles. We propose to make a quantum mechanical interpretation of the solution to this analogous electromagnetic problem.

III. INFINITE-STRENGTH DELTA FUNCTIONS

A. Discussion of the Solution

The analysis of the infinite-delta-function problem is simplified by the fact that the wavefunction is confined to one of the wedges of Fig. 1. We interpret this as meaning that the particles stay in a particular order in the one dimension. The particles cannot transmit through one another because they cannot penetrate the infinite-strength delta-function wall.

Most readers will recall that certain wedge problems may easily be solved by the method of images, which is equivalent to tracing rays through a wedge until the wave vector for the ray is pointed in such a direction that it will not hit one of the sides of the wedge again. For a wedge of arbitrary angular opening, there will be two such rays emerging from the wedge corresponding to the bifurcation of the incident wave by the two sides of the wedge. If, however, the angular opening of the wedge is π/n, the reflected rays will emerge in parallel with one another and fill all of the space within the wedge. In this case the entire solution to the problem requires only the sum of the incident plus the reflected waves.

For a wedge of arbitrary angular opening, the outgoing waves will emerge in different directions and either overlap or not fill all of the space within the wedge, and thus something must be added to the solution to fit the continuity conditions along the so-called "boundaries of geometric optics" which are the terminators of the regions filled by the outgoing waves.

Diffraction problems of this type have received extensive treatment in the literature beginning with Sommerfeld's paper in 1896.[3] The interested reader can trace the literature from Oberhettinger[4] (our principal source), who has given a particularly convenient treatment for quantum mechanical interpretation.

The solution, as we have argued that it should, consists of all of the reflections in the various regions of space plus a diffraction term which fits the continuity conditions along the boundaries of geometric optics. For the scattering solution we are interested only in the far-field part of this diffraction term. For a discussion of the general boundary conditions and form of the solution of problems of this type, see Ref. [5]. We write the far-field solution as

$$\psi = \psi_{\text{incident}} + \text{two-body reflections} + f(\varphi, \varphi', \alpha)(e^{ikr}/r^{1/2}) + O(r^{-3/2}),$$

where φ and φ' are, respectively, the angles of the incoming and outgoing k vectors, and α is the angular opening of the wedge. The coordinate system is shown in Fig. 2.

Oberhettinger has provided an expansion from which we may calculate f,

$$f(\varphi, \varphi'; \alpha) = \frac{\lambda^{1/2}}{4} \frac{\sin(\pi^2/\alpha)}{\alpha} \times \left[\frac{\sin\frac{\pi\varphi}{\alpha} \sin\frac{\pi\varphi'}{\alpha}}{\left(\sin\frac{\pi\varphi}{\alpha}\sin\frac{\pi\varphi'}{\alpha}\right)^2 - \left(\cos\frac{\pi\varphi}{\alpha}\cos\frac{\pi\varphi'}{\alpha} - \cos\frac{\pi^2}{\alpha}\right)^2} \right].$$

Notice that $f(\varphi, \varphi'; \pi/n) = 0$, so that, as asserted earlier, there is no diffraction when the wedge is of angular opening π/n. The singularity of f along the boundaries of geometric optics, where $\pi \pm (\varphi \pm \varphi') = 2n\alpha$, is required to fit continuity conditions.

Before going further let us analyze a particular example to practice interpreting this solution in terms of particles.

B. Analysis of a Particular Example

As an example we will take the interaction between two particles of equal mass and a third particle of infinite mass. The two light particles interact with each other, but we will assume that only one of the light particles interacts with the massive particle. All interactions are infinite-strength, repulsive delta functions.

The potential diagram for this problem is shown in Fig. 3. We must solve $(\nabla^2 + k^2)\psi = 0$ with $\psi = 0$ on the lines $x = 0$ and $x = y$. The interpretation of the solution of this problem is simplified

[3] A. Sommerfeld, Math. Ann. **47**, 317 (1896).

[4] F. Oberhettinger, J. Res. Natl. Bur. Std. **61**, 343 (1958).

[5] E. Gerjuoy, Phys. Rev. **109**, 1806 (1958).

by the fact that the x coordinate of the potential diagram is the position of the x particle relative to the massive particle (the x particle is the one which interacts with the massive particle), and the y coordinate is the coordinate of the y particle relative to the massive particle.

If the incoming wave is in Region I no three-body reactions occur since Region I is a wedge of π/n. The total solution in this region is a succession of the two-body problems. The experimental arrangement corresponding to Region I would have the y particle starting to the left of the x particle as shown in Fig. 3.

In Region II the situation is more interesting because the angular opening is $\frac{3}{4}\pi$. The experimental arrangement which corresponds to this region has the x particle starting to the left of the y particle, again as shown in Fig. 3.

We will assume that the incoming beam is "collimated" in the sense that the x and y particles in the incoming beam are adjusted so as to be at the origin at about the same time. This introduces a correlated distribution in the incoming state, that is the probability of finding an x particle per unit length depends upon where the y particle is. The collimation is necessary because otherwise the probability per unit volume to find the x particle at x_0 and the y particle at y_0 would depend on interference terms between the incident wave, the two-particle reflected waves and the true three-body waves. By introducing a collimation we have allowed the possibility of positioning the detector outside the beam and the two-particle reflections where the true three-particle effects are directly observable without interference.

As is usual in problems of this type, the probability per second to be scattered into some "solid" angle is proportional to the incident flux. This flux is neither the x-particle flux nor the y-particle flux, but the magnitude of the vector flux in the two-dimensional space. We will assume that the incident beam is normalized such that

$$\psi_{\text{incident}} = (\rho_x \rho_y)^{\frac{1}{2}} e^{ik_x x} e^{ik_y y},$$

where ρ_x and ρ_y are, respectively, the number of x particles per unit length and the number of y particles per unit length. Under these circumstances the magnitude of the flux[6] is $\rho_x \rho_y (v_x^2 + v_y^2)^{\frac{1}{2}}$ and

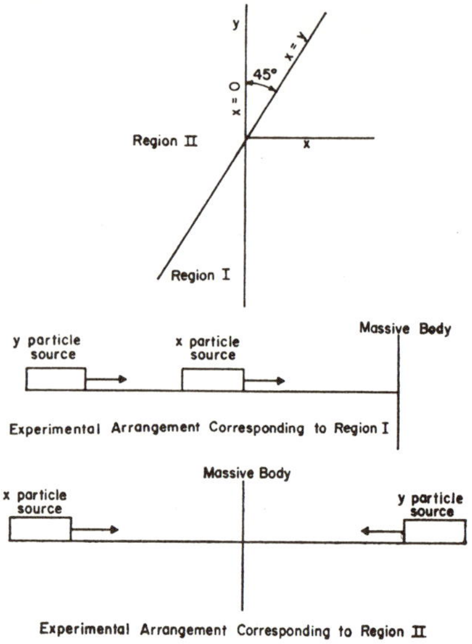

FIG. 3. Potential diagram and corresponding experimental arrangements for the particular example.

the reaction rate is

$$w = \rho_x \rho_y (v_x^2 + v_y^2)^{\frac{1}{2}} \, |f(\varphi)|^2$$

per second per unit solid angle.

A possible experiment would be to place an x-particle momentum detector to the left of the massive particle and measure the outgoing momentum distribution of x particles. This distribution would have two high peaks corresponding to the geometrical reflections of the incoming beam, but it would also have x particles of every possible momentum from zero up to the maximum possible consistent with the conservation of energy. The height of the peaks of the distribution would be proportional to the total number of the incoming particles, whereas the distributed portion of the spectrum would be proportional to the beam flux.

A second type of experiment would be to count coincidences of particle x situated between x_0 and $x_0 + \Delta x_0$ and particle y situated between y_0 and $y_0 + \Delta y_0$. The coincidence rate for this experiment is computed from the reaction rate given above where $\tan \varphi' = x_0/y_0$. For fixed $\Delta x_0 \, \Delta y_0$, the coincidence rate would decrease as

$$(x_0^2 + y_0^2)^{-\frac{1}{2}}.$$

C. Summary of the Infinite-Delta-Function Results

The goals of analyzing the infinite-strength delta-function problem were limited. Of the three-body

[6] It can be verified that this quantity is the scale factor for the reaction rate by calculating the reaction rate from the "golden rule" or by analyzing what happens to each Fourier component of a situation where a packet of x particles is incident from the left and a packet of y particles is incident from the right.

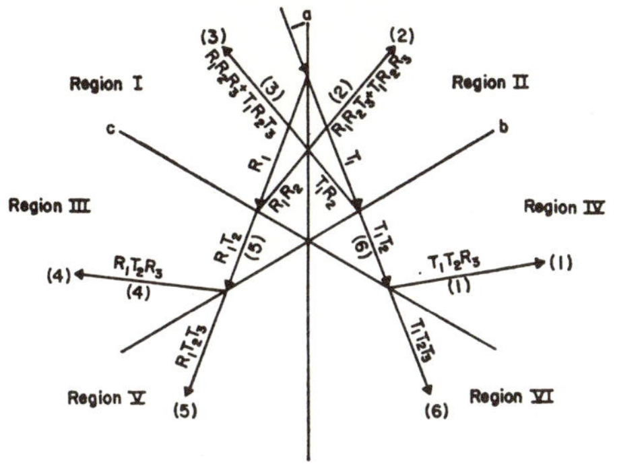

FIG. 4. Ray diagram which applies when the incoming (6) hits potential (a) first.

physical effects outlined in the Indrotuction the only one we expected to see was the redistribution of energy among the three particles, and this we have seen in our particular example. We could not have expected to see any of the other effects, because by using infinite-repulsive delta-functions for all of the interactions we have guaranteed that there will be no bound states to be ionized or rearranged.

We have, however, learned a good deal about the structure of solutions of problems of this type, and we could at least make a guess as to the form the solutions might take if the delta-function walls were transmissive. We would guess that the wavefunction would consist of all of the transmitted and reflected waves in the various regions of space, plus diffracted waves which fit the continuity conditions along the boundaries of geometric optics. In spite of these insights no one has yet been able to construct a general solution to the problem where the interpotential angles are arbitrary.

Suppose we could construct a problem which bears the same relations to the finite-strength delta-function case as does the π/n wedge to the infinite-delta-function case, that is a problem in which there is no diffraction. If such a problem exists, all of the angles between the potential walls must be π/n because the transmission from wedge to wedge would assure that there would be some probability to get into a diffracting (non π/n) wedge. Since the three interpotential angles of Fig. 1 must add up to 180° there are only three possibilities for mass ratios where all angles are π/n. These possibilities are:

(1) The masses of two like particles are infinitesimal compared with that of a third particle. The interpotential angles in this case are 45°, 45°, and 90°.

(2) Particle 1 is of infinite mass compared to particle 2 which in turn is three times the mass of particle 3 (90°, 60°, 30°).

(3) All three particles have equal mass (all angles are 60°).

We have examined all three cases and it turns out that the first two possibilities will always diffract if the strengths of all of the potentials involved are finite, but as we will now proceed to show, the third possibility will not diffract if the strengths of all the potentials are the same.

IV. EXPLICIT SOLUTION TO A PROBLEM WITH FINITE-STRENGTH DELTA-FUNCTION POTENTIALS

A. Free-Particle Solution

The Hamiltonian for the internal motion of three-one dimensional particles of equal mass interacting with one another by equal-strength delta-function potentials is

$$H = -\left(\frac{d^2}{dx^2} + \frac{d^2}{dy^2}\right) - g\delta(x) - g\delta\left(\frac{1}{2}x + \frac{\sqrt{3}}{2}y\right) - g\delta\left(\frac{1}{2}x - \frac{\sqrt{3}}{2}y\right).$$

We have chosen units so that $\hbar = M = 1$. If c is the "true" strength of the delta-function potentials, the equivalent strength is

$$-g = \sqrt{2}\,c.$$

The potential diagram for this problem is three-line delta functions intersecting at 60° angles. The method of solution will be to trace rays through this complex of delta functions and verify that there are no boundaries of geometric optics and hence no diffraction.

We again wish to take literally the mathematical equivalence of this Hamiltonian to a single particle in a two-dimensional space, and return to the interpretation of one-dimensional particles after we have solved the problem.

The potential diagram and the rays which result are shown in Figs. 4 and 5. Any sequence of reflections of the incident ray result in one of six rays as shown in Fig. 6. As is indicated in Fig. 6, there are three possible angles of incidence for these rays to strike a potential. These angles are φ, $60° + \varphi$, and $60° - \varphi$.

The rays transmit or reflect with an amplitude which is dependent only on the component of momentum perpendicular to the potential surface, that is, the sine of the angle of incidence. As is usual in problems of this type we do not have to consider path-length effects because two nearby pieces of the phase front travel the same distance between incoming and outgoing wave.

For a delta-function potential of strength g the transmission coefficient may easily be shown to be[7]

$$T = \frac{2ik/g}{2ik/g + 1} = \frac{S}{S + 1}, \qquad S = \frac{2ik}{g}.$$

Similarly, the reflection coefficient is

$$R = \frac{-1}{2ik/g + 1} \frac{-1}{S + 1},$$

where k is the component of the wave vector perpendicular to the delta-function surface.

We denote the six possible plane waves ψ_1 through ψ_6. Their momentum vectors are shown in Fig. 6. For convenience we have labeled the potentials a, b, and c and numbered the Regions I through VI.

In Fig. 4 we consider the incoming wave to be of Type 6 in Region I. The incoming wave may strike potential a first. If it does so it has an amplitude T_1 to be transmitted into Region II and an amplitude R_1 to be reflected into a Type-5 wave and stay in Region I. If the plane wave is transmitted through potential a, it will then hit potential b and again be transmitted or reflected, and so on. Notice that each ray interacts only three times, once at each angle of incidence, before becoming an outgoing ray. This is a consequence of equal-massed particles. Figure 5 illustrates the sequence of reflections we would obtain if the incoming Type-6 plane wave in Region I struck potential b first.

By following rays through the potential complex

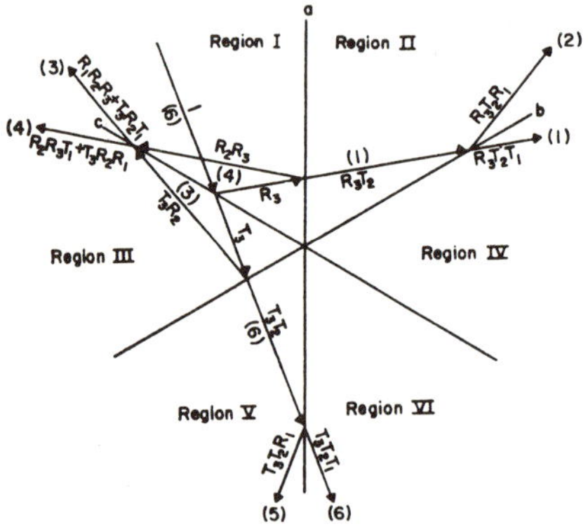

FIG. 5. Ray diagram which applies when the incoming (6) hits potential (c) first.

[7] P. Morse and H. Feshbach, *Methods of Theoretical Physics* (McGraw-Hill Book Company, Inc., New York, 1953), Vol. II, p. 1644.

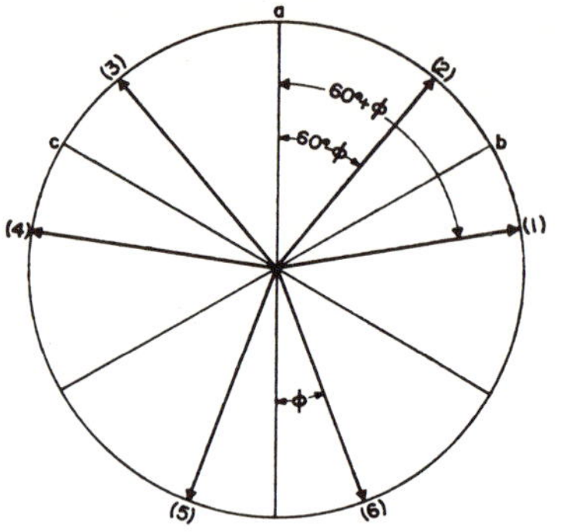

FIG. 6. Representation of the six plane waves which may be generated by reflections in the potential complex.

it is possible to evaluate the amplitude for each type of wave (i.e., Types 1 through 6) to be present in each region. As we have seen in the case of the infinite-strength delta functions, the incoming wave is bifurcated depending on which potential wall it hits first. As we saw in the infinite-delta-function case, the two halves of the plane wave must reunite to form a complete plane wave or diffraction will result. In this problem the two halves of the plane wave must be parallel and fill all of space and be equal in magnitude and phase.

It would seem at first that this problem would contain diffraction because the outgoing 2 in Region II is made up of the sum of two amplitudes from the potential a side and only one amplitude from the b side. From the a side we have

$$T_3R_2R_1 + R_3R_2T_1 = (s_3 + s_1)/(s_1 + 1)(s_2 + 1)(s_3 + 1),$$

but

$$s_2 - s_3 = (2ik/g)[\sin(60° + \varphi) - \sin(60° - \varphi)] = (2ik/g)\sin\varphi = s_1.$$

Thus

$$s_1 + s_3 = s_2$$

$$T_3R_2R_1 + R_3R_2T_1 = s_2/(s_3 + 1)(s_2 + 1)(s_1 + 1) = R_1T_2R_3,$$

which is exactly equal to the contribution from the b side; thus there is no diffraction. A similar situation occurs in Region III and the same relationship between s's again shows that there is no diffraction. In all of the other wedges it is clear that the two halves have the same magnitude and phase because the amplitudes of the two halves are made up of

TABLE I. Amplitude of plane waves in the various regions for the free wave solution.

Wave type	Region I	Region II	Region III
6	1	T_1	T_3
5	R_1		R_1T_2
4	R_2R_3		$T_3R_2R_1 + R_3R_2T_1 = R_1T_2R_3$
3	$R_1R_2R_3 + T_3R_2T_1$	T_1R_2	T_3R_2
2	R_1R_2	$T_1R_2R_3 + R_1R_2T_3 = R_3T_2R_1$	
1	R_3	R_3T_2	
Wave type	**Region IV**	**Region V**	**Region VI**
6	T_1T_2	T_3T_2	$T_1T_2T_3$
5		$T_3T_2R_1$	
4			
3			
2			
1	$T_1T_2R_3$		

the product of three complex numbers which are the same for both halves.

Since there is no diffraction, the solution to the problem may be specified by giving the amplitude for each of the six plane waves in each of the six regions. These amplitudes are given in Table I.

B. Interpretation of the Solution in Terms of Particles

Let us now interpret our solution in terms of particles. If we have the three particles oriented along a line, say in the order 1 2 3 from left to right, and we are interested in the scattering of these three particles off one another, we see first that particle 1 must be traveling to the right faster than particle 2, which in turn must be traveling to the right faster than particle 3. This is because we are interested in a scattering problem and the initial state of a scattering problem must be such that if the state is projected backwards in time there are no collisions. If particle 1 were traveling slower than particle 2 and we projected this state backwards in time, there would be a collision between 1 and 2 at some time in the past. This "no collision in the past" condition is the condition that the incoming plane wave be aimed into a wedge in such a way that the tail of the k vector not intersect any of the delta-function walls.

In the potential diagram we recall that the incoming plane wave was bifurcated by the two walls bounding the Region-I wedge. In terms of particles this means that there are two possible first interactions among the three particles, viz., particle 1 may hit particle 2 or particle 2 may hit particle 3.

When two particles of equal mass collide in one dimension, the amplitude to reflect is the amplitude that the particles retain their original order along the one-dimensional line and the amplitude to transmit is the amplitude that they exchange positions along the line. Each of the six wedges of the potential diagram represent a given order of the particles along the line. If Region I is the order 1 2 3 from left to right, then Region II must be the order 2 1 3 because to get from Region I to Region II, particle 1 must transmit through particle 2 since a is the potential between particles 1 and 2.

In two-body collisions between particles of equal mass no new velocities are generated. That is to say that the particles in the incoming state have the same velocities as the particles in the outgoing state, although the particles may switch velocities during the collision.

What we have demonstrated in our problem is that there are no new velocities generated even though there are three particles present. If we make any small change in our problem, such as letting the strength of one of the delta functions change or one of the masses be slightly different from the other two, the character of the solution will change radically due to the presence of diffraction, and there will be an infinity of new velocities brought into the problem.

We now see that we have calculated the scattering, or S matrix for this problem, which is simply a 6×6 matrix, the elements of which tell how each of the six possible initial permutations of the particles on a line couple to each of the final six permutations of the particles. To specify this 6×6 matrix entirely, it is sufficient to write down how one permutation, say 1 2 3, propagates into the six possible outgoing permutations. This is shown in Table II. The remaining elements of the 6×6 matrix may be derived by a relabeling of the particles in the initial state.

TABLE II. Elements of the S matrix.

Wavefunction	Amplitude	Region
$\exp i(k_1x_1 + k_2x_2 + k_3x_3)$	1 (Incoming Wave)	Region I: $x_1 < x_2 < x_3$
$\exp i(k_3x_1 + k_2x_2 + k_1x_3)$	$\frac{-1 - s_1s_3}{(s_1 + 1)(s_2 + 1)(s_3 + 1)}$	Region I: $x_1 < x_2 < x_3$
$\exp i(k_2x_1 + k_3x_2 + k_1x_3)$	$\frac{s_2}{(s_1 + 1)(s_2 + 1)(s_3 + 1)}$	Region II: $x_2 < x_1 < x_3$
$\exp i(k_3x_1 + k_1x_2 + k_2x_3)$	$\frac{s_3}{(s_1 + 1)(s_2 + 1)(s_3 + 1)}$	Region III: $x_1 < x_3 < x_2$
$\exp i(k_1x_1 + k_3x_2 + k_2x_3)$	$\frac{-s_1s_2}{(s_1 + 1)(s_2 + 1)(s_3 + 1)}$	Region IV: $x_2 < x_3 < x_1$
$\exp i(k_2x_1 + k_1x_2 + k_3x_3)$	$\frac{-s_2s_3}{(s_1 + 1)(s_2 + 1)(s_3 + 1)}$	Region V: $x_3 < x_1 < x_2$
$\exp i(k_1x_1 + k_2x_2 + k_3x_3)$	$\frac{s_1s_2s_3}{(s_1 + 1)(s_2 + 1)(s_3 + 1)}$	Region VI: $x_3 < x_2 < x_1$

C. Rearranged Solutions

Although we have seen that there is no diffraction in the free wave solution there is still the possibility that a particle which is bound in the initial state may be free in the final state.

Suppose we choose particles 1 and 2 to be bound in the initial state. The boundary conditions at infinity requires that there be no incoming waves in particles 1 and 2. The only way this can happen is to make both T_1 and R_1 infinite by choosing $s_1 = -1$ so that the ratio of the amplitudes of the incoming to the outgoing waves in particles 1 and 2 is zero. In the limit as s_1 approaches -1, the ratio of T_1 to R_1 is unity.

Let us evaluate the entries in Table I in this limit, that is, we set $T_1 = R_1 = 1$ and any product of amplitudes which does not contain T_1 or R_1 is set equal to zero. This result is given in Table III. The amplitudes given in Table III may be verified to constitute a solution.

In order to interpret the entries in Table III we return to Fig. 4. The incoming 6 in Region I now is at an imaginary angle of incidence with respect to potential a. Its amplitudes to either transmit or reflect are infinite and equal. This transmission and reflection together represent an incoming wave, bound in potential a. This incoming wave is a decaying exponential in both the positive and negative x direction and a propagating exponential in the y direction.

The outgoing 2 in Region II makes the same imaginary angle to potential b as does the 5 in Region I to potential a. This outgoing 2 together with the outgoing 1 in Region IV form a bound-state wavefunction in the direction perpendicular to b which is propagating in the direction parallel to b. Notice that all of the intermediate states, such as the two in Region I tend to zero exponentially at infinity in the region in which they exist.

There are three outgoing waves, a bound state in potential b propagating up to the right parallel to potential b, a bound state in potential c propagating up to the left parallel to c, and a bound state in potential a propagating down parallel to potential a.

The interpretation in terms of particles may be made without difficulty since we know that potential b, for example, is the potential between particles 1 and 3 and if the outgoing wave is bound in potential b, particles 1 and 3 must be bound together

TABLE III. Amplitudes of waves in various regions for the rearranged solution.

Wave Type	Region I	Region II	Region III
6		1	
5	1		T_2
4			$R_2T_3 + R_3R_2$
3	$R_2R_3 + R_2T_3$	R_2	
2	R_2	$R_2R_3 + R_2T_3$	
1			

Wave Type	Region IV	Region V	Region VI
6	T_2		T_3T_2
5		T_3T_2	
4			
3			
2			
1	T_2R_3		

and particle 2 free. This would represent a rearrangement of our initial state which had particles 1 and 2 bound and particle 3 free.

We follow this line of reasoning and conclude that the amplitude to leave the vertex bound in potential a is the amplitude for no rearrangement to occur. This amplitude is given from Table III,

$$T_3T_2 = \left(\frac{s_3}{s_3+1}\right)\left(\frac{s_2}{s_2+1}\right) = \frac{s_3-1}{s_3+1}.$$

The amplitude to go up to the right along potential b and the amplitude to go up to the left along potential c are interpreted as the amplitudes for the particles 1 and 2, respectively, to have been replaced by the incoming particle 3. They are

$$T_2R_3 = R_2R_3 + R_2T_3 = -s_2/(s_3+1)(s_2+1),$$

since $s_1 = -1$;

$$\tfrac{1}{4}[2i(k_\nu^2 - \tfrac{1}{4}g^2)^{\frac{1}{2}}]\sin\varphi = -1,$$

$$\sin\varphi = \frac{+ig}{2(k_\nu^2 - \frac{1}{4}g^2)^{\frac{1}{2}}},$$

$$s_2 = \frac{2ik\sin(60° + \varphi)}{g} = \frac{\sqrt{3}\,ik_\nu}{g} - \frac{1}{2},$$

$$s_3 = \frac{2ik\sin(60° - \varphi)}{g} = \frac{\sqrt{3}\,ik_\nu}{g} + \frac{1}{2}.$$

The corresponding probabilities are:

Probability that 3 replaces 1 = Probability that 3 replaces 2,

$$\left(\frac{3ky^2}{g^2} + \frac{9}{4}\right)^{-1}$$

$$\text{Probability of no rearrangement} = \frac{(3k_\nu^2/g^2 + \frac{1}{4})}{(3k_\nu^2/g^2 + \frac{9}{4})}.$$

Note the following results of this rearrangement solution:

(1) There is no ionization, that is there is no amplitude for the final state of the system to be three free particles. This is intimately connected with the lack of diffraction in the free wave solutions.

(2) There is no reflection. If some particle is incident from the left in the initial state, some particle will be moving to the right in the final state with the same velocity as the initially incident particle.

(3) Even if the incident particle is moving toward the bound pair with an infinitesimal velocity, it has a probability of $\frac{1}{9}$ to transmit through the bound pair.

D. The Bound State of Three Particles

In addition to the free wave and rearranged solutions there is one totally bound state of the three particles. The condition for this state is that there be no incoming waves in any of the particles. This is achieved by imposing the further condition that $s_3 = -1$ on the rearranged solution. All of the outgoing waves then have equal amplitude. Their k vectors are pure imaginary and are pointed along the bisectors of the angles of the six wedges. Apart from the normalization factor, this wavefunction may be written as

$$\psi = n\exp\{-\sqrt{2}\,g[|x_1 - x_2| + |x_2 - x_3| + |x_1 - x_3|]\},$$

where x_1, x_2, x_3 are the positions of the three particles along the one-dimensional line. This wavefunction is totally symmetric to the interchange of any pair of particles and its energy is $E = -\frac{1}{2}g^2$.

All of the properties of the outgoing wavefunction for the rearranged state and the bound state may be deduced directly from the S matrix by simply considering the behavior of the S matrix at the values of k for which it has a pole when it is analytically continued to complex or imaginary k. The discussion in this section was carried out in terms of the wavefunction for purposes of clarity, henceforth we shall discuss rearrangements and bound states from the analytically continued S matrix.[8]

V. N-PARTICLE SOLUTION

In this section we will show that the corresponding N-particle problem is exactly soluble, that is, the problem of an arbitrary number of particles of equal mass all interacting with one another via equal-strength delta-function potentials.

The Hamiltonian is

$$H = -\frac{\hbar^2}{2M}\sum_{i=1}^{N}\frac{d^2}{dx_i^2} + C\sum_{i>j=1}\sum^{N}\delta(x_i - x_j).$$

We will continue to use the units

$$\hbar = M = 1, \qquad \sqrt{2}\,C = -g.$$

[8] We should note here that there is a universal peculiarity bred into this problem which is retained in all of the problems we shall discuss subsequently. This peculiarity is that there are no bound-state solutions where any two particles are moving with zero relative velocity. From an inspection of the S matrix it would appear that the condition $s_1 = -1$, $s_3 = 0$, $s_2 = -1$ is also a state which has no incoming waves. If one applies this condition and looks at the wavefunction, one finds that it does not satisfy the boundary conditions on the delta-function surfaces. One can construct a wavefunction which does satisfy the boundary conditions on the delta-function surfaces by a careful limiting process, but this wavefunction increases exponentially at infinity in certain domains.

It is apparently quite impossible to continue to think of the N-particle problem as an equivalent one-particle problem in a multidimensional space, for the dimensionality of this equivalent space is $N - 1$. An alternative point of view exists, however, in which the difficulty of increasing dimensionality may be avoided.

Suppose we consider the space–time plot of a two-particle problem. The particles enter at some momentum which dictates the slope of the line in space–time. When the two particles collide they either transmit or reflect, but since no new velocities are generated, the space–time plot looks as shown in Fig. 7. If particle 1 started on the left and particle 2 on the right, then the transmission coefficient is the amplitude for particle 1 to come out on the right and particle 2 to come out on the left. The reflection coefficient is the amplitude for particle 1 to come out on the left and particle 2 to come out on the right.

We should remark here that there is no intention of changing our formulation of the many-particle problem from the stationary-state type to that of space–time. We intend only to argue that by interpreting the space–time plots we may derive all of the information which would be available in a ray tracing argument such as we used in Sec. IV.

Now let us consider the three-particle problem. There are two possible topologically different three-particle space–time diagrams which are again shown in Fig. 7. These two diagrams correspond exactly to the bifurcation of the incoming plane wave with which we dealt in the previous section. If the particles are ordered 1 2 3 from left to right, then the diagram on the left is the diagram which occurs when particle 1 strikes particle 2 first and the diagram on the right is the diagram which occurs when particle 2 strikes particle 3 first. It is now obvious that there is one collision at each of the three possible relative velocities and that there are exactly three collisions between incoming and outgoing waves. As a matter of fact even the "miraculous" property that $s_1 + s_3 = s_2$ is now evident because

$$s_1 = \sqrt{2}\, i(k_1 - k_2)/g,$$
$$s_3 = \sqrt{2}\, i(k_2 - k_3)/g;$$

thus

$$s_1 + s_3 = \sqrt{2}\, i(k_1 - k_3)/g = s_2.$$

What we have shown in the previous section is that, as far as the outgoing waves are concerned, it does not matter which of the two possible diagrams is used, for both give exactly the same result.

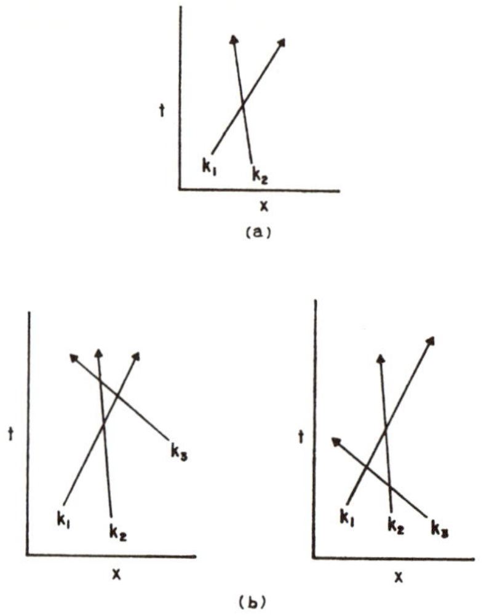

FIG. 7. Space–time plots for (a) two- and (b) three-particle problems.

If we were to change one of the particle masses or one of the delta-function strengths, the two diagrams would not give the same result and diffraction would occur.

In order to calculate the amplitude of the outgoing waves, let us invent two operators T_{ij}^{ll+1} and R_{ij}^{ll+1} which are to operate on some permutation of particles along the line. The indices $ll + 1$ label the position of an adjacent pair of particles which are interacting, and i and j label the k vectors with which the particles are interacting. The operator T interchanges the particle in the lth slot and the particle in the $l + 1$ slot with the amplitude t_{ij} where

$$t_{ij} = \frac{\sqrt{2}\, i(k_i - k_j)/g}{\sqrt{2}\, i(k_i - k_j)/g + 1} = \frac{s_{ij}}{s_{ij} + 1}.$$

The operator R leaves the same particles in the l and $l + 1$ slot, with the amplitude r_{ij} where

$$r_{ij} = -1/(s_{ij} + 1).$$

We denote the order of the particles by (1 3 2) meaning that particle 1 is in the first slot (that is, it is to the left of all of the other particles), particle 3 is in the second slot, and particle 2 is in the last slot. Thus, for example

$$T_{12}^{12}(132) = [s_{12}/(s_{12} + 1)](312),$$
$$R_{12}^{12}(132) = [-1/(s_{12} + 1)](132).$$

We use the three-particle diagrams of Fig. 7

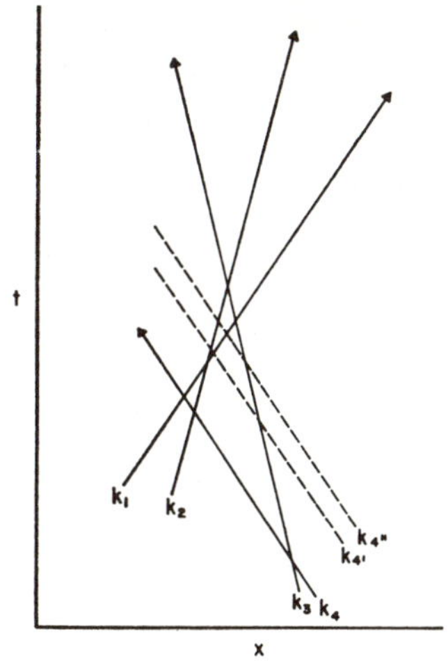

FIG. 8. Four-particle space–time plot.

to tell us in what order these operators work. For example the three-particle diagram on the left in Fig. 7 implies the order

$$(T_{23}^{12} + R_{23}^{12})(T_{13}^{23} + R_{13}^{23})(T_{12}^{12} + R_{12}^{12}),$$

which operates on some linear combination of the initial permutations of the three particles.

The three-particle diagram on the right in Fig. 7 implies the order

$$(T_{12}^{23} + R_{12}^{23})(T_{13}^{12} + R_{13}^{12})(T_{23}^{23} + R_{23}^{23}).$$

It is easily verified that these operators on any permutation of the particles give exactly the same result.

If we go on to four particles it will require fourteen diagrams to fill all space, and there will be six operator products going from incoming to outgoing states. We will now show that the outgoing waves are the same from each of the possible diagrams.

In order to do this let us start with a typical four-particle diagram such as the one shown in Fig. 8. This diagram implies the sequence of six operators

$$O_{24}^{23}\, O_{13}^{34}\, O_{12}^{23}\, O_{14}^{12}\, O_{24}^{23}\, O_{34}^{34},$$

where

$$O_{ij}^{l\,l+1} = T_{ij}^{l\,l+1} + R_{ij}^{l\,l+1}.$$

Suppose we now imagine moving the bottom line, that is, the k_4 line, up the page. We generate a new sequence when this line passes the collision between k_1 and k_2 indicated by the dotted line in Fig. 8. This second sequence of collisions gives exactly the same result as the first because all we have done is change the order of operators and the diagram in exactly the way we changed them in the three-body problem. If we continue to move the k_4 line up the page, another new sequence will be generated when the k_4 line reaches the position indicated by the second dotted line. This sequence involves an interchange of the operators O_{12}^{12} and O_{34}^{34} which commute because they have no slot in common. Thus this diagram gives exactly the same result as do the first two.

A continuation of this argument will show that every possible diagram contributes exactly the same outgoing waves. The argument does not depend on the number of particles, for all that is ever required is to move lines across intersections or to move intersections past commuting intersections. In order to show there is no diffraction, one must show that the amplitudes for every intermediate state which may be reached by more than one route are equal. The argument for these states proceeds in exactly the same way, and requires no more than the operators discussed above.

VI. N-PARTICLE CALCULATIONS

A. The Three-Particle Problem

We are now in a position to calculate amplitudes for various N-particle processes with relative ease. We could now draw some convenient sequence diagram with N k lines, write the appropriate operator sequence, and generate the S matrix by the operator rules of the preceding section. We know that we may pick any sequence whatever to generate the S matrix, for all sequences yield the same result. If it is desired to study scatterings in which particles are bound to one another we simply take ratios of elements of the S matrix where the elements are evaluated at the poles which correspond to the desired bound state.

In order to make the method more clear, let us redo the three-particle problem by the methods of Sec. V.

Let us evaluate the three-particle S matrix by using the space–time sequence diagram on the left in Fig. 7. We will assume that the incoming wave has the particles in the order (123) from left to right. Now

$$\begin{aligned} S(123) &= (T_{23}^{12} + R_{23}^{12})(T_{13}^{23} + R_{13}^{23})(T_{12}^{12} + R_{12}^{12})(123) \\ &= (T_{23}^{12} + R_{23}^{12})(T_{13}^{23} + R_{13}^{23})[t_{12}(213) + r_{12}(123)] \\ &= (T_{23}^{12} + R_{23}^{12})[t_{13}t_{12}(231) + r_{13}t_{12}(213) \end{aligned}$$

$$+ t_{13}r_{12}(132) + r_{13}r_{12}(123)]$$

$$= (t_{23}r_{13}t_{12} + r_{12}r_{13}r_{12})(123)$$

$$+ (r_{23}r_{13}t_{12} + t_{23}r_{13}r_{12})(213) + (r_{23}t_{13}r_{12})(132)$$

$$+ r_{23}t_{13}t_{12}(213) + t_{23}t_{13}r_{12}(312) + t_{23}t_{13}t_{12}(321),$$

if we let

$$s_{12} = s_1, \qquad s_{23} = s_3, \qquad s_{13} = s_2.$$

The elements above are exactly the same as the elements of the S matrix as given in Table II.

Of course, the other sequence diagram of Fig. 7 gives the same result. As we have seen previously, all of the scattering amplitudes for all possible processes as well as the bound-state energies may be calculated from analytic continuation of the S matrix.

The evaluation of S matrices for more particles is a straightforward but tedious process. We will consider here some processes whose amplitudes may be calculated without calculating the entire S matrix.

B. Four-Particle Processes

Suppose we consider the scattering of a pair of bound particles incident on a second bound pair of particles. We will denote the incoming state as

$$|(12)(34)\rangle$$

where the (12) indicates that particles one and two are bound and the order inside the "ket" indicates the order along the one dimension from left to right.

Let us first calculate the amplitude for the bound aggregates to pass through one another, that is, for the outgoing state to be

$$\langle(34)(12)|.$$

We know that the wavefunction is symmetric to the interchange of particles 1 and 2, for the single bound-state wavefunction of two particles bound by a delta-function potential is symmetric. If one of the particles has momentum k_1 and the other has momentum k_2 we know that

$$\sqrt{2}\, i(k_1 - k_2)/g = -1,$$

for this is the condition that the two particles be bound. Similarly the condition

$$\sqrt{2}\, i(k_3 - k_4)/g = -1$$

is the condition that particles 3 and 4 be bound. Let

$$\sqrt{2}\, i(k_2 - k_3)/g = s.$$

If the bound aggregate (12) is to pass through (34), clearly both particles 1 and 2 must transmit through particles 3 and 4. The amplitude for particle 2 to transmit through particle 3 is

$$\langle 32 \mid 23\rangle = \frac{\sqrt{2}\, i(k_2 - k_3)/g}{\sqrt{2}\, i(k_2 - k_3)/g + 1} = \frac{s}{s+1}.$$

The amplitude for 2 to transmit through 4 is

$$\langle 42 \mid 24\rangle = (s-1)/s.$$

Similarly,

$$\langle 31 \mid 13\rangle = (s-1)/s,$$

$$\langle 41 \mid 14\rangle = (s-2)/(s-1).$$

The amplitude for all four of these events is the product of their respective amplitudes and is the amplitude for the aggregate (12) to pass through the aggregate (34). Thus,

$$\langle(34)(12) \mid (12)(34)\rangle = (s-1)(s-2)/s(s+1).$$

Let us write this amplitude in terms of the energy in the center-of-mass system. By use of the formula in the Appendix which relates the energy to the s variables we can write

$$-4E_{\text{tot}}/g^2 = [s-1]^2 + 1.$$

The factor $+1$ on the right-hand side is just the binding energy of the two pairs of particles, and in this problem it is more convenient to represent the solution in terms of the kinetic energy of incidence of the bound aggregates in their center of mass; thus if we remove the binding energy term

$$-4E/g^2 = [s-1]^2,$$

$$s = 1 + 2iE^{\frac{1}{2}}/g.$$

The amplitude for transmission is then

$$\langle(34)(12) \mid (12)(34)\rangle = \frac{(2iE^{\frac{1}{2}}/g)(2iE^{\frac{1}{2}}/g - 1)}{(2iE^{\frac{1}{2}}/g + 1)(2 + 2iE^{\frac{1}{2}}/g)},$$

and the probability of transmission is

$$|\langle(34)(12) \mid (12)(34)\rangle|^2 = \frac{E/g^2}{E/g^2 + 1}.$$

So the probability for the (12) aggregate to pass through the (34) aggregate is zero at zero energy and monotonically increasing to unity at infinite energy.

We have several other possibilities for outgoing states. It is possible for one of the particles of the bound aggregate (12) to switch places with one of the bound particles of the aggregate (34). One way this could happen is for the first interaction to be a reflection and for the last three interactions to be transmissions. Thus we have the amplitude

$$\langle(13)(24) \mid (12)(34)\rangle = (s-1)(s-2)/s^2(s+1),$$

which yields the probability

$$|\langle(13)(24) \mid (12)(34)\rangle|^2 = \frac{4E/g^2}{(4E/g^2+1)(4E/g^2+4)}.$$

This probability is zero both at zero and at infinite energy. It is a "resonance" probability having a maximum at $E = \frac{1}{2}g^2$ where the probability of the production of the $\langle(13)(24) \mid$ state is $\frac{1}{9}$.

There is no easy way to calculate the amplitude for the reflection of the (12) aggragate off the (34) aggregate; one must return to the sequence diagrams and analyze the S matrix at the appropriate poles. We will simply state the result

$$\langle(12)(34) \mid (12)(34)\rangle = 2(s-2)/s^2(s+1),$$

$$|\langle(12)(34) \mid (12)(34)\rangle|^2 = \frac{4}{(4E/g^2+1)(4E/g^2+4)}.$$

C. The Many-Particle Bound State

From the four-particle amplitudes worked out above, we can see that there is a four-particle bound state. Every element of the S matrix is proportional to $1/[s(s+1)]$; thus if we let $s = -1$ there are only outgoing waves, and we will have a bound state. It would also appear that $s = 0$ would give a bound state, but as we have seen in the three-particle case there are no solutions where two particles have zero relative velocity.

The condition for the four-particle bound state is

$$\frac{\sqrt{2}\, i(k_1-k_2)}{g} = \frac{\sqrt{2}\, i(k_2-k_3)}{g} = \frac{\sqrt{2}\, i(k_3-k_4)}{g} = -1.$$

One may show using the many-particle S matrix that the condition for an N-particle bound state is

$$\sqrt{2}\, i(k_j - k_{j+1})/g = -1 = s_j$$

for all j.

Using this condition we may evaluate the energy of the N-particle bound state using the formula derived in the Appendix which relates $k_i - k_{i+1}$ to the internal energy,

$$E = \frac{g^2}{4}\sum_{n=1}^{N-1}\frac{1}{n(n+1)}\left[\sum_{l=1}^{n} l s_l\right]^2$$
$$= -\tfrac{1}{48}(g^2)N(N^2-1).$$

There is no saturation; the energy decreases as N^3. The wavefunction is symmetric to the interchange of any pair of particles and the average density of particles in the vicinity of the center of mass is of the order of Ng.

D. Scattering of One-Particle by $N-1$ Bound Particles

As a final example of an N-body calculation let us consider the scattering of one free particle by $N-1$ bound particles. By the usual method the relation between s and kinetic energy in the center of mass is found to be

$$s = \frac{2iN^{\frac{1}{2}}E^{\frac{1}{2}}}{g(N-1)^{\frac{1}{2}}} + \frac{N-2}{2} = \frac{\sqrt{2}\, i(k_1-k_2)}{g}.$$

The amplitude for the incident particle to pass through the bound aggregate of $N-1$ particles is the product of its amplitude to pass through each particle individually, and is

$$\langle(2\cdots N)(1) \mid (1)(2\cdots N)\rangle$$
$$= \left(\frac{s}{s+1}\right)\left(\frac{s-1}{s}\right)\left(\frac{s-2}{s-1}\right)\cdots\frac{[s-(N-2)]}{s-(N-3)}$$
$$= \frac{s-(N-2)}{s+1} = \frac{2i(NE)^{\frac{1}{2}}/g(N-1)^{\frac{1}{2}} - \frac{1}{2}(N-2)}{2i(NE)^{\frac{1}{2}}/g(N-1)^{\frac{1}{2}} + \frac{1}{2}N}.$$

The amplitude for the incident particle to replace one of the bound particles is the same for all bound particles, and is

$$\langle(13\cdots N)(2) \mid (1)(2\cdots N)\rangle$$
$$= \langle(124\cdots N)(3) \mid (1)(2\cdots N)\rangle$$
$$= \left(\frac{-1}{s+1}\right)\left(\frac{s-1}{s}\right)\cdots\frac{s-(N-2)}{s-(N-3)}$$
$$= \left[\frac{-1}{\dfrac{2i(NE)^{\frac{1}{2}}}{g(N-1)^{\frac{1}{2}}} + \frac{1}{2}N}\right]\left[\frac{\dfrac{2i(NE)^{\frac{1}{2}}}{g(N-1)^{\frac{1}{2}}} - \frac{1}{2}(N-2)}{\dfrac{2i(NE)^{\frac{1}{2}}}{g(N-1)} + \frac{1}{2}(N-2)}\right].$$

We noted previously that one of the peculiar things about the three-particle solution was that the amplitude for the incident particle to transmit through the bound aggregate was nonzero even at zero energy. Here we see that this transmission amplitude is always nonzero for one particle incident on $N-1$ particles, and in fact the amplitude to transmit approaches 1 for infinite N. Thus in the limit of large N nothing happens to the incident particle; it simply passes through this extremely dense bound aggregate as though it were not there.

VII. SUMMARY

In the Introduction we stated that one of our objectives in studying exactly soluble N-body problems was the illustration of physical effects. We have succeeded in this objective to some degree, for we have seen a number of the possible effects outlined in the Introduction. In the infinite-strength delta-function case we have seen particles redistribute their energy among the particles in a way which cannot be understood by a sequence of the two-body interactions. For infinite-strength delta functions we have illustrated the possibility of rearrangement of particles and the existence of N-particle bound states.

On the other hand we have not been able to illustrate any inelastic processes which involve dis-association or recombination of particles out of or into bound states. We have, however, learned to associate these processes with some generalization of diffraction processes in a multidimensional space. This multidimensional diffraction must be dealt with in any successful approximation method so at least in this sense we have provided some insight into the approximation methods which might be used in more physical problems. Moreover, we now have an exactly soluble problem involving a rearrangement of particles which can be used to check the existing approximation methods, and perhaps lead to a better understanding of the lack of convergence which seems to be implicit in problems of this type.[9]

Finally one must wonder about the statistical mechanics of a one-dimensional system of particles of equal mass which interact through equal-strength delta-function potentials. If the particles were bosons or distinguishable particles and the potentials attractive, the problem would make no sense, for the system would collapse into the N-particle bound state independent of temperature. The case of repulsive bosons has recently been worked out by Lieb[10] who, independent of this work, constructed the totally symmetric wavefunction for an arbitrary number of particles of equal mass interacting via finite-strength, repulsive delta-function potentials. The situation with attractive or repulsive fermions remains open and should prove to be an interesting area of further research.

ACKNOWLEDGMENTS

The author wishes to thank Professor David S. Saxon and Dr. Edward Gerjuoy for thir encouragement, interest, and assistance throughout the course of this work. The financial support of the National Science Foundation and the Air Force Office of Scientific Research, Physical Sciences Directorate, is also gratefully acknowledged.

APPENDIX. DISCUSSION OF COORDINATE SYSTEMS FOR N-BODY PROBLEMS

Consider an N-body Hamiltonian of the form

$$H = -\frac{\hbar^2}{2}\sum_{i=1}^{N}\frac{1}{M_i}\frac{\partial^2}{\partial x_i^2} + \sum_{j>i}^{N}\sum v_{ij}(x_i - x_j).$$

We will take the x_i to be one-dimensional variables, but the results we derive will be independent of dimensionality and may be extended to more dimensions by simply substituting the vector quantities ∇_i^2 for $\partial^2/\partial x_i^2$ and $\mathbf{x}_i$ for x_i.

We wish to make a change of variables which will allow us to separate out the center of mass of the entire system. In addition we will find that a more unified view of N-body problems is attained if we pick a "rationalized" coordinate system so that the second derivative terms in the new variables all have the same coefficient. The following has been shown to be such a coordinate system[11,12]:

$$z_1 = \frac{(M_1M_2)^{\frac{1}{2}}}{(M_1 + M_2)^{\frac{1}{2}}}(x_1 - x_2),$$

$$z_2 = \frac{M_3^{\frac{1}{2}}(M_1 + M_2)^{\frac{1}{2}}}{(M_1 + M_2 + M_3)^{\frac{1}{2}}}\left(\frac{M_1x_1 + M_2x_2}{M_1 + M_2} - x_3\right),$$

$$z_n = \frac{M_{n+1}^{\frac{1}{2}}\left(\sum_{i=1}^{n} M_i\right)^{\frac{1}{2}}}{\left(\sum_{i=1}^{n+1} M_i\right)^{\frac{1}{2}}}\left[\frac{\sum_{i=1}^{n} M_ix_i}{\sum_{i=1}^{n} M_i} - x_{n+1}\right], \quad n < N,$$

$$z_N = \sum_{i=1}^{N} M_ix_i \Big/ \left(\sum_{i=1}^{N} M_i\right)^{\frac{1}{2}}.$$

Pick the first new coordinate to be the relative coordinate between any two particles multiplied by the square root of the reduced mass of those two particles. Pick the second new coordinate to be the coordinate of a third particle relative to the center of mass of the first two multiplied by the square root of the reduced mass of particle three and the sum of the masses of the first two, etc. The last coordinate is the center of mass of the whole system multiplied by the square root of the sum of the masses.

We can verify that this transformation has the property that the coefficients of the new second-derivative terms are equal by observing that the transformation between z and x can be written as an orthogonal matrix times a diagonal matrix, where

[9] R. Aaron, R. D. Amado, and B. W. Lee, Phys. Rev. **121**, 319 (1961).

[10] E. Lieb and W. Liniger, Phys. Rev. **130**, 1605 (1963).

[11] D. W. Jepsen and J. O. Hirschfelder, Proc. Natl. Acad. Sci. U. S. **45**, 249 (1959).

[12] J. O. Hirschfelder and J. S. Dahler, Proc. Natl. Acad. Sci. U. S. **42**, 363 (1956).

the elements of the diagonal matrix are the square roots of the masses of the particles. That is,

$$z = UM^{\frac{1}{2}}x,$$

where

$$M^{\frac{1}{2}} = M_i^{\frac{1}{2}}\delta_{ij},$$

and

$$U_{ni} = (M_{n+1}^{\frac{1}{2}})(M_i^{\frac{1}{2}}) \Big/ \left(\sum_{j=1}^{n+1} M_j\right)^{\frac{1}{2}} \left(\sum_{j=1}^{n} M_j\right)^{\frac{1}{2}}, \quad N > n \geq i,$$

$$U_{nn+1} = -\left[\left(\sum_{j=1}^{n} M_j\right)^{\frac{1}{2}} \Big/ \left(\sum_{j=1}^{n+1} M_j\right)^{\frac{1}{2}}\right], \qquad N > n,$$

$$U_{Ni} = M_i^{\frac{1}{2}} \Big/ \left(\sum_{j=1}^{N} M_j\right)^{\frac{1}{2}},$$

$$U_{ni} = 0, \qquad i > n + 1.$$

The operator $\partial/\partial x$ transforms by the rule

$$\partial/\partial x = (M^{\frac{1}{2}}U')\,\partial/\partial z.$$

Thus the quadratic form

$$\left(\frac{\partial}{\partial x}\right)'(M)^{-1}\frac{\partial}{\partial x} = \sum_{i=1}^{N}\frac{1}{M_i}\frac{\partial^2}{\partial x_i^2}$$
$$= \left(\frac{\partial}{\partial z}\right)' UM^{\frac{1}{2}}(M)^{-1}M^{\frac{1}{2}}U'\frac{\partial}{\partial z} = \sum_{i=1}^{N}\frac{\partial^2}{\partial z_i^2}.$$

To find the arguments of the potentials we invert the transformation between x and z for form $x_i - x_j$. The result is

$$x_i - x_j = \left(\frac{M_i + M_j}{M_i M_j}\right)^{\frac{1}{2}} \times \left[\sum_{n=i}^{j-2}\frac{(M_i M_j)^{\frac{1}{2}}}{(M_i + M_j)^{\frac{1}{2}}}\frac{M_{n+1}^{\frac{1}{2}}}{\left(\sum_{l=1}^{n} M_l\right)^{\frac{1}{2}}\left(\sum_{l=1}^{n+1} M_l\right)^{\frac{1}{2}}} z_n + \frac{\left(\sum_{l=1}^{j} M_l\right)^{\frac{1}{2}} M_i^{\frac{1}{2}}}{(M_i + M_j)^{\frac{1}{2}}\left(\sum_{l=1}^{j-1} M_l\right)^{\frac{1}{2}}} z_{j-1} - \frac{M_j^{\frac{1}{2}}\left(\sum_{l=1}^{i-1} M_l\right)^{\frac{1}{2}}}{(M_i + M_j)^{\frac{1}{2}}\left(\sum_{l=1}^{i} M_l\right)^{\frac{1}{2}}} z_{i-1}\right].$$

The factor of the square root of the reduced mass is introduced so that the sum of the squares of the coefficients of the z's add up to unity. These coefficients may be looked at as the "direction cosines" of the potential in the multidimensional space.

Another result which we shall find useful is the transformation law between momenta in the two systems. This transformation is

$$\partial/\partial z = [U(M^{\frac{1}{2}})^{-1}]\,\partial/\partial x,$$
$$P_z = U(M^{\frac{1}{2}})^{-1}P_x,$$

which leads to

$$P_{z_n} = \frac{M_{n+1}^{\frac{1}{2}}}{\left(\sum_{i=1}^{n+1} M_i\right)^{\frac{1}{2}}\left(\sum_{n=1}^{n} M_i\right)^{\frac{1}{2}}}\sum_{i=1}^{n}(P_{x_i} - P_{x_{n+1}}), \qquad n < N.$$

The total energy of internal motion is

$$E = \frac{\hbar^2}{2}\sum_{n=1}^{N-1} P_{z_n}^2.$$

We shall need to calculate this energy for the case when all of the masses are equal. For convenience we set $h = 1$ and $M = 1$. This leads to

$$E = \frac{1}{2}\sum_{n=1}^{N-1}\left[\frac{1}{n^{\frac{1}{2}}(n+1)^{\frac{1}{2}}}\sum_{i=1}^{n}(P_{x_i} - P_{x_{n+1}})\right]^2,$$

which can be written

$$E = \frac{1}{2}\sum_{n=1}^{N-1}\frac{1}{n(n+1)}\left[\sum_{i=1}^{n} ik_i\right]^2,$$

where

$$k_i = P_{x_i} - P_{x_{i+1}}.$$

Magnetism

Chapter 6

Magnetism

Origins of the Magnetic Force

The theory of magnetism is concerned with the properties of atoms and electrons. It is their collective behavior which allows microscopic magnetic moments to be set up and to be stable at low temperatures, and it is also their collective behavior which orients these microscopic moments parallel to each other in ferromagnets, and in some oscillatory manner in antiferromagnets. It is with these internal properties that we are concerned in the present chapter, and not with the magnitude or distribution of the resultant magnetic field.

Chain molecules of real three-dimensional atoms provide us with good examples of one-dimensional magnetic behavior. If the valence electrons occupy nondegenerate s-orbitals, the motion of these electrons from one atom to the next is essentially that discussed in the Appendix of Lieb and Mattis (1962b) (discussed and reprinted in Chapter 4) and *a nonferromagnetic coupling must result;* an example is the hydrogen molecule. The ground state of such chains has been studied by Mattheiss (1961) and Paul (1960), and it is generally found that both the ground state and the elementary excited states are well described by an effective "Heisenberg Hamiltonian" which couples nearest-neighbor spins antiferromagnetically and neglects all the electronic degrees of freedom.

If, on the other hand, two or more valence states are allowed to the conduction electrons on each atom, the situation is analogous to several interpenetrating and interacting chains, and the resultant behavior is at once more metallic and more *ferromagnetic;* an example is the oxygen molecule. We might expect that an effective "Heisenberg Hamiltonian" with ferromagnetic coupling among nearest-neighbor spins will satisfactorily describe the magnetic degrees of freedom of the chain, although not the electronic ones of course.

By *ferromagnetism* in one dimension is meant the tendency of nearest-neighbor spins to be parallel, and by *antiferromagnetism* is meant their tendency to be antiparallel. There is no question of long-range order although such long-range order might exist at finite temperatures if the magnetic forces were of infinite range instead of being restricted to nearest neighbors, as shown by Baker (1963). But in marked contrast to their two- or three-dimensional analogs, the one-dimensional magnetic chains with nearest-neighbor forces will be ordered, if at all, only in their ground state; at any finite temperature, they will always be in the high temperature disordered phase and therefore the free energy and partition function will be analytic functions of the temperature, except possibly at $T = 0$.

In this chapter we shall treat the quantum mechanics and statistical mechanics of some models selected either for their simplicity or, as in the case of Bethe's solution of the antiferromagnetic chain, for the challenge which their solution presented to students of interacting systems. Further background may be found in the papers cited, and in a text on the subject.[1]

Algebra of Spin Operators

The spin operators obey the following commutation relations:

$$S_x S_y - S_y S_x = i\hbar S_z$$
$$\text{(and cyclic permutations of } x, y, z) \tag{6.1}$$

S_x, S_y, and S_z may be thought of as the components of the atomic magnetic moment, in appropriate units (if we neglect some subtleties concerning the Landé g-factor), and Eq. (6.1) follows from the quantization of atomic and intraatomic motions. In addition to the Cartesian components, we should label each spin operator according to the atom in the chain to which it refers; thus the operator $S_{x,i}$ obeys Eq. (6.1) with respect to $S_{y,i}$ and $S_{z,i}$ but *commutes* with all operators referring to other atoms, e.g., $S_{x,i+1}$. Associated with each spin there is a scalar magnitude, s_i,

$$S_{x,i}^2 + S_{y,i}^2 + S_{z,i}^2 = \hbar^2 s_i(s_i + 1) \tag{6.2}$$

Henceforth we shall assume all atoms have the same spin, $s_i = s$. The allowed, values of s are quantized by virtue of Eqs. (6.1) and (6.2), and may be shown to be restricted to the half-integer values $s = \frac{1}{2}, 1, \frac{3}{2}, 2, \ldots$. *The extreme quantum limit* is $s = \frac{1}{2}$, for which the operators have the representation of 2×2 *Pauli spin matrices:*

$$S_{x,i} = \frac{1}{2}\begin{bmatrix} 0 & 1 \\ 1 & 0 \end{bmatrix}, \qquad S_{y,i} = \frac{i}{2}\begin{bmatrix} 0 & -1 \\ 1 & 0 \end{bmatrix}, \qquad S_{z,i} = \frac{1}{2}\begin{bmatrix} 1 & 0 \\ 0 & -1 \end{bmatrix} \tag{6.3}$$

operating on a basis set of "up" and "down" *spinors:*

$$\text{"up"} = \begin{bmatrix} 1 \\ 0 \end{bmatrix}, \qquad \text{"down"} = \begin{bmatrix} 0 \\ 1 \end{bmatrix} \tag{6.4}$$

[1] D. C. Mattis, "The Theory of Magnetism." Harper and Row, New York, 1965.

which describe the two possible states of a spin-$\frac{1}{2}$ atom. The operators (6.3) can affect only the state of the ith atom; therefore a direct product with the unit operators for all other spins $j \neq i$ is implied in (6.3).

The extreme *classical* limit, on the other hand, is the correspondence limit of $\hbar \to 0$ and $s \to \infty$, in which limit *all* operators commute and behave as the components of a classical vector. See Fisher (1964). Many calculations are trivial in the classical limit, and, moreover, the large spin of certain magnetic atoms (Mn with $s = 5/2$ and Gd with $s = 7/2$) sometimes makes the classical limit a valid approximation, whereas the small spins of some other atoms (e.g., Ni with $s = 1/2$) places them in the extreme quantum limit. But the behavior of magnetic systems seems to be qualitatively similar for all values of s, and therefore either (or both) limiting cases are fairly representative of all.

Ferromagnetic Heisenberg Chain

The properties of the Heisenberg ferromagnetic chain are known in several important limiting cases. Specifically, one studies the Hamiltonian,

$$\begin{aligned} \mathcal{H} &= -J \sum_{i=1}^{N} (S_{x,i}S_{x,i+1} + S_{y,i}S_{y,i+1} + S_{z,i}S_{z,i+1}) + \mathfrak{h} \sum_i S_{z,i} \\ &= -J \sum_i \mathbf{S}_i \cdot \mathbf{S}_{i+1} + \mathbf{h} \cdot \sum_i \mathbf{S}_i \end{aligned} \tag{6.5}$$

where $\mathfrak{h}$ is the applied magnetic field, in appropriate units, and $J > 0$.

The ground state consists of all spins parallel and is $2Ns + 1$-fold degenerate when $\mathfrak{h} = 0$; but for nonvanishing field the state of all spins down is preferred, and has energy

$$E_0 = -JNs^2 - \mathfrak{h}Ns \tag{6.6}$$

We may also easily guess the eigenstates which differ from the ground state by one spin deviation. Denoting the ferromagnetic ground state by $|0)$, we construct the basis set for one spin deviation (unnormalized):

$$\Phi_n = (S_{x,n} + iS_{y,n})|0) \qquad \text{for} \quad n = 1, \ldots, N \tag{6.7}$$

By translational invariance the eigenstates of $\mathcal{H}$ must be plane waves:

$$\Psi_k \equiv \sum_n e^{ikn}\Phi_n \qquad \text{(unnormalized)} \tag{6.8}$$

The energy of these plane waves is readily found to be

$$E_k = E_0 + \mathfrak{h} + 2J(1 - \cos k), \qquad k = (2\pi/N) \times \text{integer} \tag{6.9}$$

The energy of two-spin-deviate states can be found by eliminating the center of mass motion,[2] but the three- or more-spin-deviate states involve us in a generally

[2] This problem was solved by M. Wortis, *Phys. Rev.* **132,** 85 (1963); N. Fukuda and M. Wortis, *Phys. Chem. Solids* **24,** 1675 (1963), and is fully discussed on p. 137 ff of footnote 1.

insoluble many-body problem. One conclusion which may be drawn, however, is that not all highly excited states can be accurately approximated by superpositions of "elementary excitations" of the type (6.8)–(6.9). In this respect the one-dimensional chain differs from the three-dimensional magnetic solid in which the very same elementary excitations are known as *magnons* and accurately describe the properties of the magnetic system which are important at low temperatures.

It has been found that the properties of the infinite magnetic chain are well approximated by those of finite chains of length $N \sim 10$. These have been studied in detail by Bonner and Fisher (1964) whose calculations provide us with quantitative results on the thermodynamic properties of the present model at finite temperature. Earlier, Fisher (1964) had investigated the classical limit, and by application of the methods given in Chapter 1 of this book, obtained all the thermodynamic properties analytically. Most other investigations of ferromagnetism have been carried for out three-dimensional systems and are thus outside the scope of this chapter.

Antiferromagnetic Heisenberg Chain

One of the first many-body problems to have been given an exact solution was the ground state of the Heisenberg antiferromagnetic chain,

$$\mathcal{H} = +J \sum_{i=1}^{N-1} \mathbf{S}_i \cdot \mathbf{S}_{i+1} + J\mathbf{S}_N \cdot \mathbf{S}_1 \tag{6.10}$$

in the special case $s = \frac{1}{2}$. It is the principal subject of study in this chapter. The ground state eigenfunction was discovered by Bethe (1931), whereupon the ground state energy could be calculated (Hulthén, 1938) and generalized to various anisotropic models by Orbach (1958) and Walker (1959). The derivation is reproduced in sections II and III of the reprinted paper by Griffiths (1964a). [The reader will note how similar this solution is to the ground state of the exactly soluble model of interacting bosons in the reprinted paper of Lieb and Liniger (1963).]

For the convenience of the student, we offer here a qualitative introduction to the reasoning behind Bethe's celebrated solution. Other, more detailed, summaries are also available.[1] Consider the two-spin-deviate problem,

$$\Psi = \sum_{n,m} a_{nm}(S_{x,n} + iS_{y,n})(S_{x,m} + iS_{y,m})|0) \tag{6.11}$$

in which it is desired to determine the coefficients a_{nm} such that Ψ is an eigenfunction of (6.10). The eigenvalue equation is quickly found to be

$$\begin{aligned}(E - E_0 + 2J)a_{nm} &- (J/2)(a_{nm+1} + a_{nm-1} + a_{n+1,m} + a_{n-1,m}) \\ &= (-J/2)(\delta_{n,m-1} + \delta_{n,m+1})(a_{nn} + a_{mm} - a_{nm} - a_{mn})\end{aligned} \tag{6.12}$$

We recall that we are dealing with spins $\frac{1}{2}$, and by (6.3) we can readily establish that the configurations with $n = m$ do not exist and are therefore excluded from (6.11). As a consequence the amplitude a_{nn} or a_{mm} can be given any value we please without affecting Ψ or E. The convenient choice made by Bethe was

$$a_{nn} + a_{n\pm1,n\pm1} = a_{nn\pm1} + a_{n\pm1,n} \tag{6.13}$$

This allows the right-hand side of the equation of motion (6.12) to vanish for all values of n and m; consequently the solutions of this equation are simple plane waves:

$$a_{nm} \sim \alpha e^{ikn+iqm} + \beta e^{iqn+ikm} \tag{6.14}$$

The energy is given straightway by Eq. (6.12),

$$(E - E_0 + 2J) - J(\cos k + \cos q) = 0 \tag{6.15}$$

similar to (6.9) [but with k, q *not* of the form $2\pi \times$ integer$/N$, being, instead, chosen so as to satisfy Eq. (6.13) in addition to the periodic boundary conditions; and $J \to -J$].

This approach can be generalized to an arbitrary number of spin-deviate states. Along with Bethe, let us assume the solution has the following form:

$$a(n_1, \ldots, n_r) = \exp\left(i \sum k_j n_j + \tfrac{1}{2} \sum_{j<j'}\sum \phi_{jj'}\right) + \text{all permutations} \tag{6.16}$$

and then determine the wave vectors k_j and the phase shifts $\phi_{jj'}$ in such a manner as to satisfy Eq. (6.13) for every pair of nearest-neighboring spin deviates n_i and n_j (previously labeled n and m), as well as periodic boundary conditions.

The precise form of the equations to which this procedure leads (in the limit $N \to \infty$) and the solution of these equations is adequately discussed in Hulthén (1938), Orbach (1958), and Walker (1959), and in the two reprinted papers: des Cloizeaux and Pearson (1962) and Griffiths (1964a). The proof that the ground state belongs to total spin zero, i.e., $r = \frac{1}{2}N$ in Eq. (6.16) is given in Lieb and Mattis (1962a).

The theory of the elementary excitations of the Heisenberg antiferromagnet is given in des Cloizeaux and Pearson (1962) who combine their theory with numerical calculations on chains of up to 48 spins. Just as the ground state belongs to total spin zero, it also belongs to total momentum $\Sigma k_j = 0$. On the other hand, the elementary excitations studied by des Cloizeaux and Pearson belong to total spin 1, and carry momentum $k_j = k$, where

$$T\psi = e^{ik}\psi \tag{6.17}$$

with ψ the excited state eigenfunction and T the operator which translates the entire chain by one lattice spacing. For the dispersion law which they discover, and for comparison with the linearized theory of Anderson,[3] we refer the interested reader to their reprinted paper.

The magnetization as a function of the applied field is the subject of Griffiths' (1964a) investigation, which is also reprinted. Griffiths calculates the ground state energy in each subspace characterized by exactly r spin reversals. Lieb and Mattis

[3] P. W. Anderson, *Phys. Rev.* **86**, 694 (1952).

(1962a) have proved that the ground state eigenfunction in each such subspace automatically belongs to total spin $S = |\frac{1}{2}N - r|$, so that Griffiths is obtaining $E_0(S)$, the curve of the lowest internal energy as a function of given total spin. As the total spin is proportional to the total magnetization, the susceptibility χ at zero temperature is obtained from the formula

$$\chi \equiv \frac{S}{\mathfrak{h}} = \frac{1}{2}\left(\frac{\partial E_0(S)}{\partial(S^2)}\right)^{-1} \tag{6.18}$$

in which S is the solution of the minimization condition,

$$\frac{d}{dS}[E_0(S) - \mathfrak{h}S] = 0 \tag{6.19}$$

and $\mathfrak{h}$ is, as before, the applied magnetic field in appropriate units. The study of finite-length chains permits this work to be extended to finite temperatures, as in Bonner and Fisher (1964). Griffiths (1964b) has found rigorous upper and lower bounds on the free energy at $T \geq 0$. Other approximate theories, based in the main on finite-temperature perturbation theory, have been given by Inawashiro and Katsura (1964), and Bulaevskii (1963); these works are all in general agreement, and it may be safely concluded that the problem is now quantitatively well understood from many different points of view.

Some Exactly Soluble Models

The susceptibility $\chi(T)$, the free energy $F(T)$, the long-range order, and many other quantities in which one might conceivably be interested cannot be obtained rigorously in the Heisenberg model for spins $\frac{1}{2}$, except at $T = 0$. Because this model itself is the result of many physical and mathematical approximations,[1] there is no reason not to further modify it if in so doing one obtains an exactly soluble model which has approximately the same features as the original but which lends itself to more extensive analysis.

The Ising model is such a modification, which consists of eliminating all the terms in $S_{x,i}$ and $S_{y,i}$ in the Hamiltonian (6.5) or (6.10), so that the Hamiltonians involve only the operators $S_{z,i}$ which all commute with one another. Historically, the Ising model came first, fully a decade before the Heisenberg model, and was proposed by W. Lenz on the basis of Bohr's old quantum theory. Several years afterward, the properties of such a one-dimensional magnetic chain were computed and published by Lenz's student, E. Ising (1925). [This paper contains the wrong inference that because at finite temperature spontaneous long-range order and magnetism were not possible in the linear chain, the model was incapable of showing a phase transition in three dimensions. Later work by Bethe and Peierls, and Onsager showed that this model could show a phase transition not only in three dimensions, but also in two. A complete bibliography is now available,[4] to which the reader is referred.]

[4] S. G. Brush, "History of the Lenz-Ising Model," UCRL-7940, Livermore, California (1964).

Two entirely different modifications are considered in the paper by Lieb, Schultz, and Mattis (1961), part of which is reprinted here. They are of such a nature as to lead to simple and exact solutions for all the eigenstates and eigenvalues, and to permit the direct computation of such subtle quantities as long-range order. Yet both models are genuinely quantum-mechanical in the sense that all operators in the Hamiltonian do not commute with one another. In both these soluble models the nearest-neighbor forces, the spins $\frac{1}{2}$, and the linear topology are necessary ingredients for obtaining an exact solution. The first of these is denoted the XY model, previously considered by Nambu (1950) in connection with the two-dimensional Ising model.[5] As it is adequately explained in the reprinted paper, we present herewith an alternative explanation which shows the connection with Bethe's solution and with the theory of interacting bosons discussed in Chapter 5.

The Hamiltonian considered in the isotropic XY model is

$$\mathcal{H} = J \sum_i (S_{xi}S_{xi+1} + S_{yi}S_{yi+1}) \tag{6.20}$$

Again, we may consider the subspace with exactly r spin deviates; first let $r = 2$, set up a wave function such as (6.11), and calculate the analog of (6.12). It is readily found to be

$$Ea_{nm} - (J/2)(a_{nm+1} + a_{nm-1} + a_{n+1,m} + a_{n-1,m}) = 0 \tag{6.21a}$$

subject to the "boundary condition" on the unphysical amplitudes,

$$a_{nn} = a_{mm} = 0 \tag{6.21b}$$

and the usual cyclic boundary conditions at the end of the chain. Once more, the solution is a linear combination of products of plane waves, e.g.,

$$a_{nm} \sim (e^{ikn}e^{iqm} - e^{iqn}e^{ikm}) \tag{6.22}$$

The minus sign $(-)$ guarantees that the wave function vanishes when $n = m$. The wave vectors k and q must now be chosen such as to satisfy cyclic boundary conditions. The generalization to $r \neq 2$ is straightforward. We recognize that a determinantal wave function automatically satisfies the condition (6.21b), and therefore the solution

$$a(n_1, \ . \ . \ . \ , n_r) \sim \pm\det\|e^{ik_1n_1} \ \cdot \ \cdot \ \cdot \ e^{ik_rn_r}\| \tag{6.23}$$

is of the same general form as the Bethe solution (6.16), with phase shifts appropriate to the present problem. The signs of the wave function must be chosen such that $a(\cdot \cdot \cdot n_i \cdot \cdot \cdot n_j \cdot \cdot \cdot) = +a(\cdot \cdot \cdot n_j \cdot \cdot \cdot n_i \cdot \cdot \cdot)$, and the wave vectors must be chosen appropriate to cyclic boundary conditions; see Chapter 5 on bosons for a more extensive discussion.

[5] The relationship of the two-dimensional Ising model to a quantum-mechanical, one-dimensional linear chain is reviewed in T. Schultz, D. Mattis, and E. Lieb, *Rev. Modern Phys.* **36,** 856 (1964), and also in footnote 1.

The extension of this model to an anisotropic interaction, the study of the effects of end boundary conditions, the introduction of a second-quantized fermion field solution to these problems, and the computation of long-range order are some of the topics covered in Lieb, Schultz, and Mattis (1961). In this paper is also discussed a second soluble model, the *Heisenberg-Ising chain*, which is not found elsewhere in the literature. The discussion of the long-range order is omitted in the extract we have reprinted here, but two appendices: (A) "To diagonalize a general quadratic form in Fermi operators" and (B) "Nondegeneracy of the ground state and absence of an energy gap in the Heisenberg model," have been retained in full. The magnetic susceptibility for the XY model has been computed by Katsura (1962), but the corresponding properties in the authors' *Heisenberg-Ising* model have never been exploited.

Applications of One-Dimensional Theories

The experimental evidence on magnetic chain molecules is referenced in footnote 3 of Bonner and Fisher (1964), and we might also note the recent experiments of Wagner and Friedberg[6] on $Mn(HCOO)_2 \cdot 2\,H_2O$.

But the one-dimensional models have a more important role than the explanation of some few one-dimensional physical systems. Rather, they serve as tests of approximate theories devised for the study of insoluble three-dimensional systems. The quantum-mechanical chains have an additional importance: they have proved invaluable in the solution of the two-dimensional Ising model,[5] which could otherwise be understood only on the basis of some very fancy algebraic analysis.

Finally, one-dimensional theories can teach us something about the prerequisites for ferromagnetism in three dimensions. The reader is referred to the subsection "Lack of Ferromagnetism in One Dimension" in Chapter 4.

Bibliography

I. Ferromagnetic Spin Waves

F. Bloch (1930). *Z. Physik* **61,** 206.
J. C. Slater (1930). *Phys. Rev.* **35,** 509.
M. Wortis (1963). *Phys. Rev.* **132,** 85.

II. Antiferromagnetic Heisenberg Chain, Spins $\frac{1}{2}$

H. Bethe (1931). *Z. Physik* **71,** 205.
*J. Bonner and M. Fisher (1964). *Phys. Rev.* **135,** A640.
L. Bulaevskii (1963). *Soviet Phys.—JETP* **16,** 685.
*J. des Cloizeaux and J. Pearson (1962). *Phys. Rev.* **128,** 2131.
H. Falk and T. Ruijgrok (1961). *Physica* **27,** 710.
H. Falk and T. Ruijgrok (1965). *Phys. Rev.* **139,** A1203.
*R. Griffiths (1964a). "Magnetization Curve at Zero Temperature for the Antiferromagnetic Heisenberg Linear Chain," *Phys. Rev.* **133,** A768.

[6] C. Wagner and S. Friedberg, *Phys. Letters* **9,** 11 (1964).

R. Griffiths (1964b). *Phys. Rev.* **136,** A751.
L. Hulthén (1938). *Arkiv Mat. Astron. Fys.* **26A,** No. 11.
S. Inawashiro and S. Katsura (1964). *J. Math. Phys.* **5,** 1091.
E. Lieb and D. Mattis (1962a). *J. Math. Phys.* **3,** 749.
L. Mattheiss (1961). *Phys. Rev.* **123,** 1209.
R. Orbach (1958). *Phys. Rev.* **112,** 309.
T. Ruijgrok and S. Rodriguez (1960). *Phys. Rev.* **119,** 596.
L. R. Walker (1959). *Phys. Rev.* **116,** 1089.

III. Some Exactly Soluble Models

G. Baker, Jr. (1963). "Ising Model with a Long-Ranged Interaction in the Presence of Residual Short-Range Interactions," *Phys. Rev.* **130,** 1406.
M. Fisher (1964). "Magnetism in One-Dimensional Systems: the Heisenberg Model for Infinite Spin," *Am. J. Phys.* **32,** 343.
S. Katsura (1962). "Statistical Mechanics of Anisotropic Linear Heisenberg Model (i.e., *XY model*)," *Phys. Rev.* **127,** 1508.
*E. Lieb, T. Schultz, and D. Mattis (1961). "Two Soluble Models of an Antiferromagnetic Chain," *Ann. Phys.* (*N.Y.*) **16,** 407.
Y. Nambu (1950). "A Note on the Eigenvalue Problem in Crystal Statistics," *Progr. Theoret. Phys.* (*Kyoto*) **5,** 1.

IV. Physical Origins of Magnetic Coupling in One Dimension

*E. Lieb and D. Mattis (1962b). "Theory of Ferromagnetism and the Ordering of Electronic Energy Levels," *Phys. Rev.* **125,** 164.
L. Mattheiss (1961). "Antiferromagnetic Linear Chain," *Phys. Rev.* **123,** 1209; "Effective Exchange Integral," *ibid.* 1219.
D. Paul (1960). "Theory of Magnetism and the Ground State Energy of a Linear Chain," I, *Phys. Rev.* **118,** 92; II, *ibid.* **120,** 463.

Reprinted from ANNALS OF PHYSICS, Volume 16, No. 3, December 1961

ANNALS OF PHYSICS: **16**, 407–466 (1961)

Two Soluble Models of an Antiferromagnetic Chain

ELLIOTT LIEB, THEODORE SCHULTZ, AND DANIEL MATTIS

Thomas J. Watson Research Center, Yorktown, New York

Two genuinely quantum mechanical models for an antiferromagnetic linear chain with nearest neighbor interactions are constructed and solved exactly, in the sense that the ground state, all the elementary excitations and the free energy are found. A general formalism for calculating the instantaneous correlation between any two spins is developed and applied to the investigation of short- and long-range order. Both models show nonvanishing long-range order in the ground state for a range of values of a certain parameter λ which is analogous to an anisotropy parameter in the Heisenberg model. A detailed comparison with the Heisenberg model suggests that the latter has no long-range order in the isotropic case but finite long-range order for any finite amount of anisotropy. The unreliability of variational methods for determining long-range order is emphasized. It is also shown that for spin ½ systems having rather general isotropic Heisenberg interactions favoring an antiferromagnetic ordering, the ground state is nondegenerate and there is no energy gap above the ground state in the energy spectrum of the total system.

I. INTRODUCTION

For an infinite chain of spins interacting with nearest neighbors via a Heisenberg interaction, the exact energy eigenstates were found, in principle, many years ago by Bethe (*1*) and the ground-state energy was found somewhat later by Hulthén (*2*). The problem has nevertheless occasioned a persistent theoretical interest. This is because the exact method of Bethe does not seem capable of generalization to the more interesting cases of two and three dimensions. The aim has generally been to construct approximate methods that give accurate results in one dimension, as determined by a comparison with the known exact results, but which can be generalized to two or three dimensions with some degree of confidence and simplicity. Unfortunately, the crucial test for any approximate method is how well it describes the long-range order; but it is precisely this test which has been impossible, because the long-range order has never been calculated exactly by the method of Bethe.

There is, in fact, still considerable doubt about the nature of the long-range order in one or more dimensions. On the one hand, spin wave methods used by Anderson (*3*) and Kubo (*4*) have predicted long-range order in two or three dimensions, although Anderson (but not Kubo) predicts no long-range order in one dimension. Variational methods of Kasteleijn (*5*) generalized to two and three dimensions by Taketa and Nakamura (*6*), and of Marshall (*7*), all related to a variational method of Hulthén (*2*), indicate no long-range order in one, two, and three dimensions for the completely isotropic interaction. On the other hand, they all predict the onset of long-range order for a certain critical amount of anisotropy—the same amount at which a kink is predicted for the short-range order. Exact calculations by Orbach (*8*) on the anisotropic case, generalizing the method of Bethe, show the kink to be spurious for the one-dimensional case, but throw no further light on the long-range order. Walker (*9*) has observed that the ground-state energy calculated by Orbach has a power series expansion around the limiting case of the completely anisotropic interaction (Ising limit) which seems to give a good representation of the ground-state energy even for the isotropic case, although he observes that a similar expansion for the long-range order suggests the long-range order might vanish when the anisotropy falls below a certain critical value. Ruijgrok and Rodriguez (*10*) have developed a variational method for the one-dimensional case which gives a good value for the ground-state energy, significantly better than previous variational methods, and which also predicts finite long-range order. Davis (*11*) has performed a perturbation theoretic calculation which indicates, to the order to which it has been carried, a long-range order in any number of dimensions even for the isotropic case.

There are two questions raised by all these investigations. First, can a purely isotropic Heisenberg interaction between nearest neighbors produce long-range order in any number of dimensions? Second, if such an ordering tendency exists in, say, two or three dimensions, will it also exist in only one dimension, or would the absence of order in one dimension prove nothing at all about order in two or three dimensions?

The purpose of the present paper is to gain further insight into the effects of anisotropy in one dimension. Two models will be constructed which can be solved exactly in considerable detail and which bear a reasonably close resemblance to the Heisenberg model. The investigation of the first of these models strongly suggests that the isotropic Heisenberg model has no long-range order but that such order exists for any finite amount of anisotropy. Both models emphasize the subtle nature of long-range order and the insufficiency of variational methods as a reliable approach to this question. The question of the relation of these results to two- and three-dimensional cases is left completely open.

In Section II we consider the first of these models, the "*XY* model." It is shown that the Hamiltonian can be expressed as a quadratic form in creation and annihilation operators for fermions, and this quadratic form can be diagonalized, thereby giving the complete set of states, excitation energies, and partition sum. In terms of these operators, general expressions for the order between any two spins are derived* involving a kind of Green function, G_{ij}, which can be explicitly

*These calculations are omitted here.

evaluated. The general relation between the order and G_{ij} is equally valid for the Heisenberg model, but G_{ij} itself is then not explicitly calculable. The short-, intermediate-, and long-range order are calculated for various situations, and it is shown that only for the isotropic case does the long-range order vanish. The model is compared with the Heisenberg model for one special case from which reasonable surmises may be made about the latter.

In Section III we consider the "Heisenberg-Ising model" in a similar way. Although the excited states and statistical mechanics are considerably more complicated, the conclusions about long-range order are much the same. An application of the method of Ruijgrok and Rodriguez to this model* emphasizes the care with which one must interpret the results of variational calculations with regard to the ground-state energy.

II. THE *XY* MODEL

A. Formulation

The first model consists of N spin ½'s (N even) arranged in a row and having only nearest neighbor interactions. It is

$$H_\gamma = \sum_i [(1+\gamma)S^x{}_i S^x{}_{i+1} + (1-\gamma)S^y{}_i S^y{}_{i+1}], \tag{2.1}$$

where γ is a parameter characterizing the degree of anisotropy in the xy-plane. $S^x{}_i$, $S^y{}_i$ and $S^z{}_i$ may be represented by the usual Pauli spin matrices ($\hbar = 1$):

$$S^x{}_i = \frac{1}{2}\begin{pmatrix}0 & 1\\ 1 & 0\end{pmatrix}; \quad S^y{}_i = \frac{1}{2}\begin{pmatrix}0 & -i\\ i & 0\end{pmatrix}; \quad S^z{}_i = \frac{1}{2}\begin{pmatrix}1 & 0\\ 0 & -1\end{pmatrix}.$$

Because the Hamiltonian only involves the x- and y- components of the spin operators, we call this model the *XY* model.

The ends of the chain may be treated in at least two different but physically reasonable ways:

(i) as free ends, in which case the range of the summation index is $1 \leqq i \leqq N-1$, a situation that is convenient for discussing the long-range order;

(ii) as a cyclic chain, in which case $1 \leqq i \leqq N$ and $S^x{}_{N+1} \equiv S^x{}_1$, $S^y{}_{N+1} \equiv S^y{}_1$. This problem, or a slight variation of it (see below), is most convenient for for calculating interesting physical quantities other than the long-range order.

The *XY* model is exactly soluble for all values of γ, although we shall consider only the range $-1 \leqq \gamma \leqq 1$. Furthermore, it is strikingly similar to the generalized Heisenberg model described by the Hamiltonian

$$H_\gamma = \sum_i [(1+\gamma)S^x{}_i S^x{}_{i+1} + (1-\gamma)(S^y{}_i S^y{}_{i+1} + S^z{}_i S^z{}_{i+1})]. \tag{2.1'}$$

As $\gamma \to 1$ both models tend to the Ising model in which the x-components of spin are completely order and the y- and z-components are completely disordered. For $|\gamma| \neq 1$ both models are genuinely quantum mechanical because

*These calculations are omitted here.

different components of $\mathbf{S}_i$ appearing in H do not commute. The effect of the "transverse terms" (those multiplied by $1 - \gamma$ in either Hamiltonian) in both models is to oppose the ordering of the x-components but to favor the ordering of the y-components (and, in the Heisenberg model, of the z-components, too). Hitherto, it has not been clear for the Heisenberg model just how strong these two effects are: for any particular positive value of γ, do the transverse terms either establish any long-range order among the y- and z-components or destroy the long-range order of the x-components (which would imply absence of long-range order for the y- and z-components as well)? For the XY model, as we shall see, the transverse terms do neither until the limiting case $\gamma = 0$, when they succeed in destroying the order of the x-components. This result is highly suggestive for the Heisenberg model, a subject we discuss in detail in Section II F.

To solve the XY model, we first introduce the raising and lowering operators

$$a_i^\dagger = S^x_i + iS^y_i \quad \text{and} \quad a_i = S^x_i - iS^y_i \tag{2.2a}$$

in terms of which the Pauli spin operators are

$$S^x_i = (a_i^\dagger + a_i)/2; \qquad S^y_i = (a_i^\dagger - a_i)/2i; \qquad S^z_i = a_i^\dagger a_i - \tfrac{1}{2} \tag{2.2b}$$

and the Hamiltonian is

$$H_\gamma = \tfrac{1}{2}\sum_i [(a_i^\dagger a_{i+1} + \gamma a_i^\dagger a_{i+1}^\dagger) + \text{h.c.}]. \tag{2.3}$$

These operators partly resemble Fermi operators in that

$$\{a_i, a_i^\dagger\} = 1; \qquad a_i^2 = (a_i^\dagger)^2 = 0; \tag{2.4a}$$

and they partly resemble Bose operators in that

$$[a_i^\dagger, a_j] = [a_i^\dagger, a_j^\dagger] = [a_i, a_j] = 0, \qquad i \neq j. \tag{2.4b}$$

It is therefore not possible to diagonalize the quadratic form appearing in (2.3) directly with a canonical transformation; principal axis transformations of the a's and $a^\dagger$'s do not preserve this mixed set of canonical rules. However, it is possible to transform to a new set of variables that are strictly Fermi operators and in terms of which the Hamiltonian is just as simple.[1] Let

[1] This transformation from a set of Pauli spin operators to a set of fermion creation and annihilation operators dates back at least as far as the classical paper on second quantization of fermion fields by Jordan and Wigner (*12*). It is described in that context, for example, by Kramers (*12*). It was used as the basis for approximate calculations of the ground-state energy of the isotropic Heisenberg model independently by Meyer (*12*), and by Rodriguez (*12*). The more sophisticated variational calculation of an approximate ground-state energy and long-range order by Ruijgrok and Rodriguez (*10*) is also based on this transformation. A somewhat different procedure, based on the same idea of converting nearest neighbor pairs of "paulions" to nearest neighbor pairs of fermions, was applied to precisely $H_{\gamma=0}$ by Nambu (*12*). The spirit of our paper is, however, entirely different from the work of the previous

$$c_i \equiv \exp\left[\pi i \sum_{1}^{i-1} a_j^\dagger a_j\right] a_i \tag{2.5a}$$

and

$$c_i^\dagger \equiv a_i^\dagger \exp\left[-\pi i \sum_{1}^{i-1} a_j^\dagger a_j\right]. \tag{2.5b}$$

Then

$$c_i^\dagger c_i = a_i^\dagger a_i, \tag{2.6}$$

so that the inverse transformation is simply

$$a_i^\dagger = \exp\left[-\pi i \sum_{1}^{i-1} c_j^\dagger c_j\right] c_i, \tag{2.7a}$$

$$a_i^\dagger = c_i^\dagger \exp\left[\pi i \sum_{1}^{i-1} c_j^\dagger c_j\right]. \tag{2.7b}$$

The c's and $c^\dagger$'s are Fermi operators:

$$\{c_i, c_j^\dagger\} = \delta_{ij}, \tag{2.8a}$$

$$\{c_i, c_j\} = \{c_i^\dagger, c_j^\dagger\} = 0. \tag{2.8b}$$

Because $c_j^\dagger c_j$ is an occupation number having values 0 or 1,

$$\exp(\pi i c_j^\dagger c_j) = \exp(-\pi i c_j^\dagger c_j). \tag{2.9}$$

Furthermore, for $i = 1, 2, \ldots, N-1$,

$$a_i^\dagger a_{i+1} = c_i^\dagger c_{i+1} \tag{2.10a}$$

and

$$a_i^\dagger a_{i+1}^\dagger = c_i^\dagger c_{i+1}^\dagger, \tag{2.10b}$$

so that, for the case of free ends, the Hamiltonian is

$$H_\gamma = \tfrac{1}{2} \sum_{1}^{N-1} [(c_i^\dagger c_{i+1} + \gamma c_i^\dagger c_{i+1}^\dagger) + \text{h.c.}]. \tag{2.11}$$

For the cyclic chain, we need also

$$a_N^\dagger a_1 = -c_N^\dagger c_1 \exp(i\pi\mathfrak{N}) \neq c_N^\dagger c_1 \tag{2.12a}$$

and

$$a_N^\dagger a_1^\dagger = -c_N^\dagger c_1^\dagger \exp(i\pi\mathfrak{N}) \neq c_N^\dagger c_1^\dagger, \tag{2.12b}$$

authors, and the results are rather more extensive. Because the cases of particular interest ($\gamma \neq 0$, free ends) contain complications not previously encountered, we present here a detailed exposition of the mathematical tricks needed for the ultimate solution.

where

$$\mathfrak{N} \equiv \sum_1^N c_j^\dagger c_j = \sum_1^N (S^\gamma{}_j + \tfrac{1}{2}). \tag{2.13}$$

The Hamiltonian is

$$H_\gamma = \tfrac{1}{2} \sum_1^N [(c_i^\dagger c_{i+1} + \gamma c_i^\dagger c_{i+1}^\dagger) + \text{h.c.}] - \tfrac{1}{2}[(c_N^\dagger c_1 + \gamma c_N^\dagger c_1^\dagger) + \text{h.c.}] (\exp (i\pi\mathfrak{N}) + 1). \tag{2.11'}$$

That is, in terms of the Fermi operators c_i and $c_i^\dagger$, H_γ no longer has a simple cyclic structure. For large systems we may neglect the correction term proportional to $\exp (i\pi\mathfrak{N}) + 1$ in which case we call it the "c-cyclic" problem (the original problem being the "a-cyclic" problem). Actually it is not difficult to solve the a-cyclic problem exactly, but we shall first consider the simpler c-cyclic one.

In all cases, the Hamiltonian is a simple quadratic form in Fermi operators and can be exactly diagonalized. The particular simplicity of H_γ depends on the fact that the spins can be arranged in a definite order, that interactions occur only between neighboring spins in this ordering, and that the z-components of spin do not enter. If the interactions were to extend to nth nearest neighbors, H_γ would involve a polynomial of order $2n$ in the c's. In two-dimensional models, it can be readily seen that any ordering and nontrivial scheme of interactions must lead to a Hamiltonian involving a polynomial roughly of order $2N$ for a system of N^2 spins. Thus we are making maximum use of the nearest neighbor character of the interactions and the one-dimensionality of the system.

B. Ground-State Energy, Elementary Excitations, and Free Energy

The diagonalization of quadratic forms such as occur in (2.11) or (2.11′) is discussed in Appendix A. The Hamiltonian is reduced to the diagonal form

$$H_\gamma = \sum_k \Lambda_k \eta_k^\dagger \eta_k + \text{constant} \tag{2.14}$$

by the linear transformation

$$\eta_k = \sum_i \frac{\phi_{ki} + \psi_{ki}}{2} c_i + \frac{\phi_{ki} - \psi_{ki}}{2} c_i^\dagger, \tag{2.15a}$$

$$\eta_k^\dagger = \sum_i \frac{\phi_{ki} + \psi_{ki}}{2} c_i^\dagger + \frac{\phi_{ki} - \psi_{ki}}{2} c_i, \tag{2.15b}$$

where ϕ_k and ψ_k, considered as N-component vectors, are real solutions to certain matrix equations. For the c-cyclic problem, the relevant matrices

$$\mathbf{A} = \frac{1}{2}\begin{pmatrix} 0 & 1 & & & & & 1 \\ 1 & 0 & 1 & & & \mathbf{0} & \\ & \cdot & \cdot & \cdot & & & \\ & & \cdot & \cdot & \cdot & & \\ & & & \cdot & \cdot & \cdot & \\ & \mathbf{0} & & & 1 & 0 & 1 \\ 1 & & & & & 1 & 0 \end{pmatrix}, \tag{2.16a}$$

$$\mathbf{B} = \frac{1}{2}\gamma\begin{pmatrix} 0 & 1 & & & & & -1 \\ -1 & 0 & 1 & & & \mathbf{0} & \\ & \cdot & \cdot & \cdot & & & \\ & & \cdot & \cdot & \cdot & & \\ & & & \cdot & \cdot & \cdot & \\ & \mathbf{0} & & & -1 & 0 & 1 \\ 1 & & & & & -1 & 0 \end{pmatrix}, \tag{2.16b}$$

$(\mathbf{A} - \mathbf{B})(\mathbf{A} + \mathbf{B}) =$

$$\frac{1}{4}\begin{pmatrix} 2(1+\gamma^2) & 0 & 1-\gamma^2 & \cdot & \cdot & \cdot & 1-\gamma^2 & 0 \\ 0 & 2(1+\gamma^2) & 0 & 1-\gamma^2 & & & & 1-\gamma^2 \\ 1-\gamma^2 & 0 & 2(1+\gamma^2) & 0 & 1-\gamma^2 & & \mathbf{0} & \\ \cdot & \cdot & \cdot & \cdot & \cdot & \cdot & & \\ \cdot & & \cdot & \cdot & \cdot & \cdot & \cdot & \\ \cdot & \mathbf{0} & & 1-\gamma^2 & 0 & 2(1+\gamma^2) & 0 & 1-\gamma^2 \\ 1-\gamma^2 & & & & 1-\gamma^2 & 0 & 2(1+\gamma^2) & 0 \\ 0 & 1-\gamma^2 & & & & 1-\gamma^2 & 0 & 2(1+\gamma^2) \end{pmatrix}. \tag{2.16c}$$

The vectors are the real solutions of the eigenvalue equation

$$\boldsymbol{\phi}_k(\mathbf{A} - \mathbf{B})(\mathbf{A} + \mathbf{B}) = \Lambda_k^2 \boldsymbol{\phi}_k\,. \tag{2.17}$$

A complete set of solutions is

$$\phi_{kj} = \begin{matrix} (2/N)^{\frac{1}{2}} \sin kj \\ (2/N)^{\frac{1}{2}} \cos kj \end{matrix} \tag{2.18a}$$

belonging to the set of eigenvalues

$$\Lambda_k^2 = 1 - (1 - \gamma^2) \sin^2 k, \tag{2.18b}$$

where

$$k = 2\pi m/N, \qquad m = -\tfrac{1}{2}N, \cdots, 0, 1, \cdots, \tfrac{1}{2}N - 1. \tag{2.18c}$$

In (2.18a) we take the upper solution for ϕ_{kj} if $k > 0$, the lower solution if $k \leqq 0$. For $\Lambda_k \neq 0$, $(A - 7a)$ gives

$$\psi_{kj} = \Lambda_k^{-1} (\cos k\, \phi_{kj} + \gamma \sin k\, \phi_{-kj}), \tag{2.18d}$$

while if $\Lambda_k = 0$

$$\psi_{kj} = \pm\, \phi_{kj}\,. \tag{2.18d$'$}$$

$\Lambda_k = 0$ is an eigenvalue only if $\gamma = 0$ and $N/4$ is an integer. To simplify the discussion, let us assume that $N/4$ is not an integer and therefore that $\Lambda_k \neq 0$.

The sign of Λ_k being arbitrary, we shall take it always to be positive. This corresponds to a particle-hole picture for the η-particles, where the ground state has no elementary fermions and the elementary fermion excitations both above and below the Fermi surface have positive energies.[2] Λ_k is shown in Fig. 1 for the isotropic and extreme anisotropic (Ising) cases and for one intermediate case. Remark that only for the isotropic case is there no energy gap.

The ground state Ψ_0 is the state with no elementary excitations:

$$\eta_k \Psi_0 = 0, \quad \text{all } k. \tag{2.19}$$

The ground-state energy, according to (A-12) is

$$E_0 = -\tfrac{1}{2} \sum_k \Lambda_k\,. \tag{2.20}$$

In the limit $N \to \infty$, the sum can be replaced by an integral giving

$$\begin{aligned} E_0/N &= -(\tfrac{1}{4}\pi) \int_{-\pi}^{\pi} dk\, [1 - (1 - \gamma^2) \sin^2 k]^{\frac{1}{2}} \\ &= -(1/\pi)\mathcal{E}(1 - \gamma^2), \end{aligned} \tag{2.21}$$

where $\mathcal{E}(k^2)$ is one of the complete elliptic integrals (*13*). E_0/N goes smoothly between the limiting cases

$$E_0/N = -1/\pi, \quad \text{isotropic case}, \quad \gamma = 0, \tag{2.22a}$$

and

$$E_0/N = -\tfrac{1}{2}, \quad \text{Ising case}, \quad \gamma = 1. \tag{2.22b}$$

[2] The "Fermi surface" consists of the points $k = \pm\pi/2$. The alternate picture in which all excitations are "particles" (in the isotropic case, a "particle" is an up spin) would be obtained by letting Λ_k have the sign of $\cos k$. The Fermi sea would then be defined by $|k| > \pi/2$. The particle-hole picture is preferred because it somewhat simplifies the algebra.

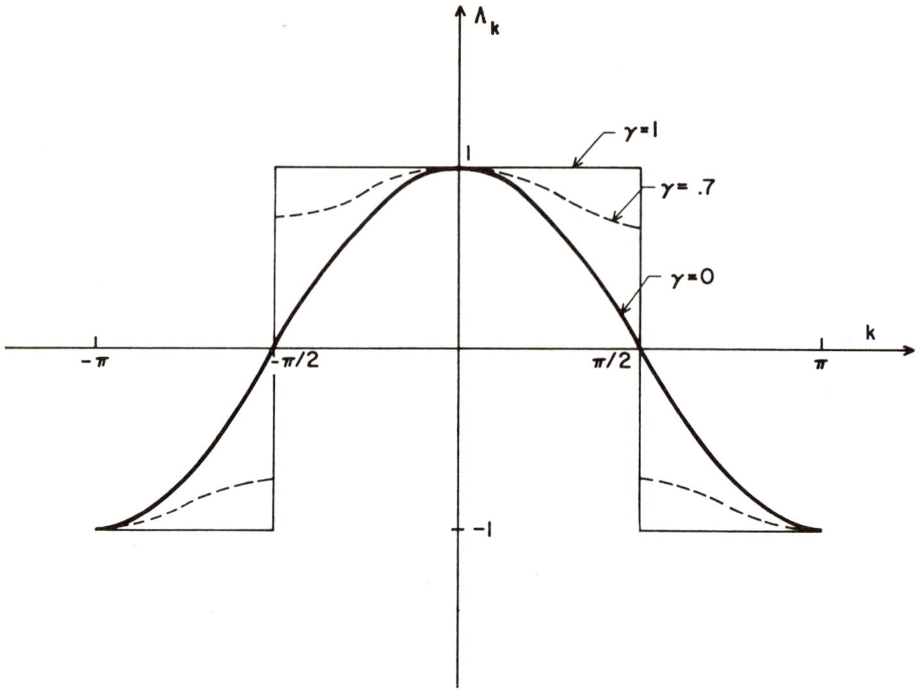

FIG. 1. Energy of elementary excitations in XY model as function of wave vector for three different degrees of anisotropy.

At this point let us remark on the simplification resulting from the consideration of the c-cyclic rather than the a-cyclic problem. According to (2.11′), the Hamiltonian for the a-cyclic problem is complicated by the presence of the term

$$- \tfrac{1}{2}[(c_N{}^\dagger c_1 + \gamma c_N{}^\dagger c_1{}^\dagger) + \text{h.c.}] (\exp (i\pi\mathfrak{N}) + 1).$$

Although $\mathfrak{N}$ is not invariant under the transformation (2.15), its evenness or oddness is invariant, so that exp $(i\pi\mathfrak{N})$ is invariant. Now in the ground state of the c-cyclic problem, and in all states with an even number of excitations, the number of c-particles is odd (assuming $N/4$ is not an integer, the k's are occupied symmetrically around $k = 0$, except that $k = \pi$ but not $k = -\pi$ is occupied). Therefore, the additional term gives zero acting on such states and they remain eigenstates of the a-cyclic problem. States with an odd number of excitations, on the other hand, have $\mathfrak{N}$ even, giving the additional term

$$- \tfrac{1}{2}(c_N{}^\dagger c_1 + \gamma c_N{}^\dagger c_1 + \text{h.c.})$$

in the Hamiltonian. This has the effect of making changes of order $1/N$ in the

k's, ϕ_k's, and ψ_k's, all of which can be exactly calculated and are negligible in the calculation of real physical quantities. Strictly speaking though, the elementary excitations are not independent for the a-cyclic problem because of this dependence on the evenness or oddness of the total number of excitations present, and this is why we have preferred to consider the c-cyclic problem.

The free energy is the grand potential of such a system of noninteracting fermions with zero chemical potential ($\beta = 1/kT$):

$$F_\gamma/N = -kT\left[\ln 2 + \frac{2}{\pi}\int_0^{\pi/2} dk \ln \cosh\left(\tfrac{1}{2}\beta\Lambda_k\right)\right]. \tag{2.23}$$

In the isotropic limit we obtain

$$F_{\text{isotrop.}}/N = -kT\left[\ln 2 + \frac{2}{\pi}\int_0^{\pi/2} dk \ln \cosh\left(\tfrac{1}{2}\beta \cos k\right)\right]; \tag{2.24a}$$

while in the Ising limit, we obtain the classical result,

$$F_{\text{Ising}}/N = -kT(\ln 2 + \ln \cosh \tfrac{1}{2}\beta). \tag{2.24b}$$

Neither case exhibits any singular behavior as a function of temperature, a result to be expected in view of the one-dimensional nature of the model.

III. THE HEISENBERG-ISING MODEL

A. Formulation of Ground-State Problem

The second model consists of $2N$ spin $\tfrac{1}{2}$'s also arranged in a row and having only nearest neighbor interactions. The interactions are alternately Ising and isotropic Heisenberg interactions, so that the Hamiltonian for a chain with free ends is

$$\begin{aligned} H_\lambda &= \sum_1^N \mathbf{S}_{2i-1}\cdot\mathbf{S}_{2i} + 2\lambda \sum_1^{N-1} S^z_{2i}S^z_{2i+1} \\ &= H_0 + H_1 . \end{aligned} \tag{3.1}$$

The parameter λ is to be considered variable but positive and characterizes the relative strength of the two types of interaction.[8]

The particular simplicity of this Hamiltonian is noticed if the representation diagonalizing H_0 is introduced. For the ith pair of spins we introduce the four eigenfunctions of $\mathbf{S}_{2i-1}\cdot\mathbf{S}_{2i}$:

[8] The symbol λ is chosen because this parameter appears in many expressions in exactly the same way as does λ for the XY model. In both models, λ ranges from 0 to ∞.

$$\Phi_{11}^{i} = \uparrow\uparrow, \qquad \Phi_{1-1}^{i} = \downarrow\downarrow,$$
$$\Phi_{10}^{i} = \frac{1}{\sqrt{2}}(\downarrow\uparrow + \uparrow\downarrow), \quad \text{and} \quad \Phi_{00}^{i} = \frac{1}{\sqrt{2}}(\downarrow\uparrow - \uparrow\downarrow); \tag{3.2}$$

where the first and second arrows refer to the $(2i - 1)$st and $2i$th spins, respectively, and represent states of a single spin in the positive and negative z-direction. The first subscript of Φ^i refers to the quantum number J_i and the second to M_i for the ith pair. Application of either $S^z_{2i-2}\, S^z_{2i-1}$ or $S^z_{2i}\, S^z_{2i+1}$ to any of these four states leaves the values of M_i unchanged. Thus the assignment of one of the three possible values $(\pm 1, 0)$ to each of the N M_i's defines a subspace of the 2^{2N} dimensional space of all states. The Hamiltonian, having no matrix components between states in different subspaces, can be diagonalized separately in each. It will be shown in Section III E that the ground state is in the subspace for which $M_i = 0$ for all i, a subspace we now consider.

We are faced with a situation that is formally similar to that encountered in the XY model: a set of dynamical systems (in this case a pair of spins) each having two possible states (in this case Φ_{10} and Φ_{00}) and interacting only with the nearest neighbor systems. If we call the states Φ_{10}^i the "up" states and Φ_{00}^i the "down" states, we can formally introduce raising and lowering operators for the ith pair having the usual properties:

$$a_i^{\dagger}\Phi_{00}^{i} = \Phi_{10}^{i}, \qquad a_i^{\dagger}\Phi_{10}^{i} = 0, \tag{3.3a}$$

and

$$a_i\Phi_{00}^{i} = 0, \qquad a_i\Phi_{10}^{i} = \Phi_{00}^{i}, \tag{3.3b}$$

so that

$$\{a_i, a_i^{\dagger}\} = 1 \quad \text{and} \quad a_i^2 = a_i^{\dagger 2} = 0. \tag{3.4}$$

It is to be emphasized that a_i and $a_i^{\dagger}$ do not lower and raise a particular spin as they did for the XY model. It is only a *formal* analogy between the states Φ_{10}^i and Φ_{00}^i on the one hand, and the up and down orientations of a single spin ½ that we are exploiting.

Because of the fundamental relations

$$S^z_{2i-1}\Phi_{10}^{i} = -\Phi_{00}^{i}, \qquad S^z_{2i-1}\Phi_{00}^{i} = -\Phi_{10}^{i} \tag{3.5a}$$

and

$$S^z_{2i}\Phi_{10}^{i} = \Phi_{00}^{i}, \qquad S^z_{2i}\Phi_{00}^{i} = \Phi_{10}^{i}, \tag{3.5b}$$

we may represent S^z_{2i-1} and S^z_{2i} in the $M_i \equiv 0$ subspace by means of a_i and $a_i^{\dagger}$:

$$S^z_{2i-1} \to -\tfrac{1}{2}(a_i^{\dagger} + a_i); \qquad S^z_{2i} \to \tfrac{1}{2}(a_i^{\dagger} + a_i). \tag{3.6}$$

Because the diagonal energies of the states Φ_{10}^i and Φ_{00}^i are, respectively, $\frac{1}{4}$ and $-\frac{3}{4}$, the ith term in H_0 can also be expressed in terms of a_i and $a_i^\dagger$:

$$\mathbf{S}_{2i-1}\cdot\mathbf{S}_{2i} \to a_i^\dagger a_i - \tfrac{3}{4}. \tag{3.7}$$

In the subspace defined by $M_i \equiv 0$, the total Hamiltonian is then

$$H_\lambda = -\tfrac{3}{4}N + \sum_1^N a_i^\dagger a_i - \tfrac{1}{2}\lambda \sum_1^{N-1} (a_i^\dagger + a_i)(a_{i+1}^\dagger + a_{i+1}), \tag{3.8}$$

where, in addition to (3. 4), we have

$$[a_i,a_j] = [a_i^\dagger,a_j^\dagger] = [a_i,a_j^\dagger] = 0. \tag{3.9}$$

The eigenstates and associated energies of H_λ can be found exactly as for the XY model by introducing a complete set of Fermi operators, through (2.5), in terms of which H_λ is given by

$$H_\lambda = -\tfrac{3}{4}N + \sum_1^N c_i^\dagger c_i - \tfrac{1}{2}\lambda \sum_1^{N-1} (c_i^\dagger - c_i)(c_{i+1}^\dagger + c_{i+1}). \tag{3.10}$$

B. Ground State of the Cyclic Chain

It is convenient in this section to consider the "c-cyclic" case obtained by letting $\sum_1^{N-1} \to \sum_1^N$ in H_1 and defining

$$c_{N+1} \equiv c_1 \quad\text{and}\quad c_{N+1}^\dagger \equiv c_1 . \tag{3.11}$$

The matrices relevant to the diagonalization of H_λ are then

$$\mathbf{A} - \mathbf{B} = \begin{pmatrix} 1 & & & -\lambda \\ -\lambda & 1 & \mathbf{0} & \\ & \cdot & \cdot & \\ & & \cdot \quad \cdot & \\ & & & \cdot \quad \cdot \\ \mathbf{0} & & -\lambda & 1 \end{pmatrix} = (\mathbf{A} + \mathbf{B})^T \tag{3.12}$$

and

$$(\mathbf{A} - \mathbf{B})(\mathbf{A} + \mathbf{B}) = \begin{pmatrix} 1+\lambda^2 & -\lambda & & & -\lambda \\ -\lambda & 1+\lambda^2 & -\lambda & \mathbf{0} & \\ & \cdot & \cdot & \cdot & \\ & & \cdot & \cdot & \cdot \\ & \mathbf{0} & \cdot & \cdot & \cdot \\ -\lambda & & & -\lambda & 1+\lambda^2 \end{pmatrix}. \tag{3.13}$$

For $\lambda \neq 1$ the normal modes are characterized by the functions

$$\phi_{kj} = \begin{cases} (2/N)^{\frac{1}{2}} \sin kj \\ (2/N)^{\frac{1}{2}} \cos kj \end{cases} \tag{3.14a}$$

and

$$\psi_{kj} = \Lambda_k^{-1} [(1 - \lambda \cos k)\phi_{kj} - \lambda \sin k \, \phi_{-kj}], \tag{3.14b}$$

belong to the eigenvalues

$$\Lambda_k = [(1 + \lambda)^2 - 4\lambda \cos^2 \tfrac{1}{2}k]^{\frac{1}{2}}, \tag{3.14c}$$

where

$$k = 2\pi m/N \tag{3.14d}$$

and

$$\begin{aligned} m &= -\tfrac{1}{2}N, \cdots, 0, \cdots, \tfrac{1}{2}N - 1 \quad \text{for } N \text{ even}, \\ m &= -\tfrac{1}{2}(N-1), \cdots, 0, \cdots \tfrac{1}{2}(N-1) \quad \text{for } N \text{ odd}. \end{aligned} \tag{3.14e}$$

We take the upper solution for ϕ_{kj} if $k > 0$, the lower solution if $k \leqq 0$. For $\lambda = 1$ and $m = 0$, we have $\Lambda_k = 0$. For this particular k,

$$\phi_{kj} = N^{-\frac{1}{2}}, \qquad \psi_{kj} = \pm N^{-\frac{1}{2}}. \tag{3.14'}$$

The sign, which is arbitrary, will be taken positive.

It is convenient to introduce the parameter γ, ranging from $+1$ to -1, by

$$\gamma = (1 - \lambda)/(1 + \lambda); \tag{3.15}$$

then

$$\Lambda_k = (1 + \lambda)\,[1 - (1 - \gamma^2) \cos^2 \tfrac{1}{2}k]^{\frac{1}{2}}, \tag{3.16}$$

which resembles the spectrum of the XY model.

We see that except for $\lambda = 1$ or $\gamma = 0$ the spectrum of elementary excitations has an energy gap. It should be emphasized, however, that the ground state together with states of all possible combinations of these elementary excitations are but a small subset of the complete set of stationary states, all the rest having one or more $M_i = \pm 1$. Thus, the behavior of the system for finite temperatures is not immediately apparent from the knowledge of the excitation spectrum (3.14c).

The ground-state energy, according to (A-12) is

$$E_0 = -\tfrac{3}{4}N + \tfrac{1}{2}\left(\sum_1^N 1 - \sum_k \Lambda_k\right). \tag{3.17}$$

As $N \to \infty$,

$$E_0/N = -\tfrac{1}{4} - (2/\pi)[1/(1+\gamma)]\,\mathcal{E}(1 - \gamma^2). \tag{3.18}$$

We notice that E_0/N is not analytic at $\lambda = 1$ ($\gamma = 0$) although it has power series expansions in λ and $1/\lambda$.

APPENDIX A. TO DIAGONALIZE A GENERAL QUADRATIC FORM IN FERMI OPERATORS

We wish to diagonalize the quadratic form

$$H = \sum_{i,j} [c_i^\dagger A_{ij} c_j + \tfrac{1}{2}(c_i^\dagger B_{ij} c_j^\dagger + \text{h.c.})], \tag{A-1}$$

where the c_i's and $c_i^\dagger$'s are Fermi annihilation and creation operators and H is Hermitian. The Hermiticity of H requires that $\mathbf{A}$ be a Hermitian matrix, while the anticommutation rules among the c_i's require that $\mathbf{B}$ be an antisymmetric matrix. In the situations of interest *here*, one can always arrange that $\mathbf{A}$ and $\mathbf{B}$ are real.

We try to find a linear transformation of the form

$$\eta_k = \sum_i (g_{ki} c_i + h_{ki} c_i^\dagger), \tag{A-2a}$$

$$\eta_k^\dagger = \sum_i (g_{ki} c_i^\dagger + h_{ki} c_i), \tag{A-2b}$$

with the g_{ki} and h_{ki} real, which is canonical (i.e., the η_k's and $\eta_k^\dagger$'s should also be Fermi operators) and which gives for H the form

$$H = \sum_k \Lambda_k \eta_k^\dagger \eta_k + \text{constant}. \tag{A-3}$$

If this is possible, then

$$[\eta_k , H] - \Lambda_k \eta_k = 0. \tag{A-4}$$

Substituting (A-2) in (A-4) and setting the coefficients of each operator equal to zero, we obtain a set of equations for the g_{ki} and h_{ki} :

$$\Lambda_k g_{ki} = \sum_j (g_{kj}A_{ji} - h_{kj}B_{ji}), \tag{A-5a}$$

$$\Lambda_k h_{ki} = \sum_j (g_{kj}B_{ji} - h_{kj}A_{ji}). \tag{A-5b}$$

These are simplified by introducing the linear combinations

$$\phi_{ki} = g_{ki} + h_{ki} \tag{A-6a}$$

and

$$\psi_{ki} = g_{ki} - h_{ki} \tag{A-6b}$$

in terms of which the coupled equations are

$$\phi_k(\mathbf{A} - \mathbf{B}) = \Lambda_k \psi_k \tag{A-7a}$$

and

$$\psi_k(\mathbf{A} + \mathbf{B}) = \Lambda_k \phi_k \tag{A-7b}$$

in an obvious matrix notation. Either ψ_k or ϕ_k can be eliminated from (A-7) giving either

$$\phi_k(\mathbf{A} - \mathbf{B})(\mathbf{A} + \mathbf{B}) = \Lambda_k^2 \phi_k \tag{A-8a}$$

or

$$\psi_k(\mathbf{A} + \mathbf{B})(\mathbf{A} - \mathbf{B}) = \Lambda_k^2 \psi_k . \tag{A-8b}$$

For $\Lambda_k \neq 0$, either (A-8a) or (A-8b) is solved for ϕ_k or ψ_k and the other vector is then obtained from (A-7a) or (A-7b).

For $\Lambda_k = 0$, both ϕ_k and ψ_k are determined by (A-8), or more simply by (A-7), their relative sign being arbitrary. Changing the sign of ψ_k , but not of ϕ_k , interchanges g_{kj} and h_{kj} , hence η_k and $\eta_k^\dagger$, and thus interchanges the definitions of occupied and unoccupied for this zero-energy mode. That the choice of definition is arbitrary is not surprising, because it has no effect on the energy.

Because $\mathbf{A}$ is symmetric and $\mathbf{B}$ is antisymmetric, $(\mathbf{A} + \mathbf{B})^T = \mathbf{A} - \mathbf{B}$, so that both $(\mathbf{A} - \mathbf{B})(\mathbf{A} + \mathbf{B})$ and $(\mathbf{A} + \mathbf{B})(\mathbf{A} - \mathbf{B})$ are symmetric and at least positive semi-definite. Thus all the Λ_k's are real and it is possible to choose all the ϕ_k's and ψ_k's to be real as well as orthogonal. If the ϕ_k's are normalized vectors $(\sum_i \phi_{ki}^2 = 1)$, then the ψ_k's are also automatically normalized when $\Lambda_k \neq 0$ or can be so chosen when $\Lambda_k = 0$. This ensures that

$$\sum_i (g_{ki}g_{k'i} + h_{ki}h_{k'i}) = \delta_{kk'} \tag{A-9a}$$

and

$$\sum_i (g_{ki}h_{k'i} - g_{k'i}h_{ki}) = 0, \tag{A-9b}$$

the necessary and sufficient conditions that the η_k's and $\eta_k^\dagger$'s be canonical Fermi operators.

The constant in H can be determined by substituting (A-3) in (A-1) or, less tediously, from the invariance of tr H under the canonical transformation (A-2). From (A-1)

$$\text{tr}\, H = 2^{N-1} \sum_i A_{ii} \tag{A-10}$$

while from (A-3)

$$\text{tr}\, H = 2^{N-1} \sum_k \Lambda_k + 2^N \times \text{constant}. \tag{A-11}$$

The constant is thus $\frac{1}{2}(\sum_i A_{ii} - \sum_k \Lambda_k)$ and

$$H = \sum_k \Lambda_k \eta_k^\dagger \eta_k + \frac{1}{2}\left(\sum_i A_{ii} - \sum_k \Lambda_k\right). \tag{A-12}$$

APPENDIX B. NONDEGENERACY OF THE GROUND STATE AND ABSENCE OF AN ENERGY GAP IN THE HEISENBERG MODEL

We prove two exact theorems about the ground state and excitation spectrum for a Heisenberg model with nearest neighbor interactions in one dimension. The generalization to longer range interactions and higher-dimensional lattices is indicated. A further generalization to particles of spin $\neq \frac{1}{2}$ and a discussion of the ordering of excited state energy levels has been submitted for publication in the *Journal of Mathematical Physics* by Lieb and Mattis.

THEOREM 1. For a linear chain of spin $\frac{1}{2}$'s with nearest neighbor antiferromagnetic Heisenberg interactions, the ground state is nondegenerate (hence $S = 0$).

Proof. We first remark that this is a stronger theorem than that due to Marshall (7), who proved that there is a singlet ground state, but who did not exclude the possibility of there being several degenerate ground states, some of which may not be singlets. We consider the Hamiltonian

$$H = \sum S^z{}_i S^z{}_j + \tfrac{1}{2} \sum S^+{}_i S^-{}_j + S^-{}_i S^+{}_j, \tag{B-1}$$

where i and j are on the A and B sublattices respectively and the sums run over all interacting pairs. It is convenient to make the canonical transformation rotating all spins on the B sublattice ($S^x{}_j \to -S^x{}_j$, $S^y{}_j \to -S^y{}_j$, $S^z{}_j \to S^z{}_j$), so that instead of (B-1) we consider

$$H' = \sum S^z{}_i S^z{}_j - \tfrac{1}{2} \sum S^+{}_i S^-{}_j + S^-{}_i S^+{}_j . \tag{B-2}$$

Because $[S_z , H] = 0$, let us consider only states having $S_z = 0$ and show that only one such state has the ground-state energy. A complete set of states in the $S_z = 0$ subspace is the set of configurations in which $N/2$ spins are up and $N/2$ spins are down. We denote these states by Φ_μ where $1 \leqq \mu \leqq \binom{N}{\frac{1}{2}N}$. Any eigenfunction Ψ of H' can be expanded as

$$\Psi = \sum C_\mu \Phi_\mu , \tag{B-3}$$

and Schrödinger's equation in this representation reduces to a set of coupled linear equations

$$(E - \epsilon_\mu) C_\mu = \tfrac{1}{2} \sum_{\mu'(\mu)} C_{\mu'} , \tag{B-4}$$

where

$$\epsilon_\mu \Phi_\mu = (\sum S^z{}_i S^z{}_j) \Phi_\mu ,$$

and the $\Phi_{\mu'(\mu)}$ are the set of configurations which connect to Φ_μ via the interaction. Because the Hamiltonian is real, we may assume without loss of generality that all C's are real. Essential to the proof is the following lemma:

LEMMA 1. For any ground state with $S_z = 0$, all $C_\mu \neq 0$.

Proof. Suppose the contrary, i.e., for some ground state Ψ_0 having the ground-state energy E_0,

$$C_\mu = 0 \quad \text{for} \quad \mu = \mu_1 , \cdots , \mu_r . \tag{B-5a}$$

For these C's (B-2) reduces to

$$0 = \tfrac{1}{2} \sum_{\mu'(\mu)} C_{\mu'}, \qquad \mu = \mu_1 , \cdots , \mu_r . \tag{B-5b}$$

Now in at least one of these equations, say the μ_pth, some of the $C_{\mu'}$'s $\neq 0$ (otherwise H would break into blocks with no matrix elements connecting $\Phi_{\mu_1} , \cdots , \Phi_{\mu_r}$ with the other configurations, which is readily seen to be impossible); therefore (B-5b) implies that there are nonzero C's of *both* signs. Consider then the trial function Ψ_0':

$$\Psi_0' = \sum | C_\mu | \Phi_\mu . \tag{B-6}$$

On the one hand, Ψ_0' is not an eigenstate because

$$|C_{\mu_p}| = 0 \quad \text{but} \quad \sum_{\mu\ (\mu_p)} |C_{\mu'}| \neq 0, \tag{B-7}$$

so that, from the variational principle, we have for its energy

$$E_0' > E_0 . \tag{B-8}$$

On the other hand, explicit evaluation gives

$$E_0' = \sum \epsilon_\mu C_\mu^2 - \tfrac{1}{2}\sum_\mu \sum_{\mu'(\mu)} |C_\mu|\,|C_{\mu'}| \tag{B-9a}$$

and

$$E_0 = \sum \epsilon_\mu C_\mu^2 - \tfrac{1}{2}\sum_\mu \sum_{\mu'(\mu)} C_\mu C_{\mu'} , \tag{B-9b}$$

from which it follows that

$$E_0' \leqq E_0 . \tag{B-10}$$

The contradiction between (B-8) and (B-10) proves the lemma.

We now prove a lemma which is a stronger version of the lemma due to Peierls and used by Marshall.

LEMMA 2. For every ground state with $S_z = 0$, all C_μ's have the same sign.

Proof. For Ψ_0 to be a ground state, the equality must hold in (B-10). This occurs if, and only if, all the terms $C_\mu C_{\mu'}$ occuring in (B-9b) are positive, (they are all nonzero by lemma 1), i.e., the coefficients of all configurations connected through the interaction with each other should have the same sign. But as we have remarked, each configuration is ultimately connected with every other through repeated applications of the interaction, proving the lemma.

It is now obvious that there can be only one ground state with $S_z = 0$; otherwise, the several states would all have all positive coefficients and so could not be orthogonal to one another. Now Marshall has shown that *at least* one ground state has $S = 0$. The existence of another ground state, whatever its multiplicity, would imply that there is a second ground state with $S_z = 0$, which we have shown to be impossible. The entire proof is immediately generalizable to any number of dimensions and any lattice which is decomposable into two equivalent sublattices with antiferromagnetic Heisenberg interactions between spins on different sublattices and ferromagnetic Heisenberg interactions between spins on the same sublattice. Also, although Marshall's proof, used above, requires periodic boundary conditions to ensure translational invariance, this restriction may also be dropped if the lattice has reflection symmetry about some plane (so that there is a transformation mapping the A and B sublattices into each other but leaving the Hamiltonian unchanged).

Next we investigate the nature of the excitation spectrum and prove

THEOREM 2. There is an excited state for the cyclic linear chain with nearest neighbor Heisenberg interactions having vanishingly small excitation energy in the limit that the length of the chain becomes infinite.

Proof. Consider the state

$$\Psi_k = \exp\,(ik \sum nS^z{}_n)\Psi_0 \equiv \mathcal{O}^k\Psi_0\,. \tag{B-11}$$

We first show that if $k = (2\pi/N) \times$ odd integer, Ψ_k is orthogonal to the ground state. Consider the unitary operator U_x that displaces all the spins by one site cyclically:

$$U_x\mathbf{S}_iU_x^{-1} = \mathbf{S}_{i+1}\,, \qquad \mathbf{S}_{N+1} = \mathbf{S}_1\,. \tag{B-12}$$

Because

$$[H,U_x] = 0,$$

if Ψ_0 is an eigenstate of H, so is $U_x\Psi_0$. By the nondegeneracy of Ψ_0

$$U_x\Psi_0 = e^{i\alpha}\Psi_0\,. \tag{B-13}$$

Thus

$$\begin{aligned}\langle\Psi_0 \mid \Psi_k\rangle &= \langle\Psi_0 \mid \mathcal{O}^k \mid \Psi_0\rangle \\ &= \langle\Psi_0 \mid U_x\mathcal{O}^kU_x^{-1} \mid \Psi_0\rangle.\end{aligned} \tag{B-14}$$

But

$$U_x\mathcal{O}^kU_x^{-1} = \mathcal{O}^k \exp\,(ikNS^z{}_1) \exp\,(-ik\sum_1^N S^z{}_n). \tag{B-15}$$

Because Ψ_0 is a singlet

$$\exp\,(-ik\sum_1^N S^z{}_n)\Psi_0 = 0. \tag{B-16}$$

Furthermore, in the most convenient representation

$$S^z{}_1 = \begin{pmatrix} \frac{1}{2} & 0 \\ 0 & -\frac{1}{2} \end{pmatrix} \tag{B-17a}$$

so that

$$\exp(ikNS^z{}_1) = \begin{pmatrix} -1 & 0 \\ 0 & -1 \end{pmatrix} \tag{B-17b}$$

in this or any other representation, providing

$$k = 2\pi m/N, \quad m \text{ an odd integer.} \tag{B-18}$$

Thus

$$\langle\Psi_0 \mid \Psi_k\rangle = -\langle\Psi_0 \mid \Psi_k\rangle = 0. \tag{B-19}$$

Note that we have not proved the Ψ_k's to be orthogonal among themselves. The energy of Ψ_k is also readily calculated.

$$\langle\Psi_k \mid H \mid \Psi_k\rangle = \langle\Psi_0 \mid \mathcal{O}_x^{k-1} H \mathcal{O}^k \mid \Psi_0\rangle. \tag{B-20}$$

Using,

$$\mathcal{O}^{k-1} S^x{}_n \mathcal{O}^k = S^x{}_n \cos kn + S^y{}_n \sin kn, \tag{B-21a}$$

$$\mathcal{O}^{k-1} S^y{}_n \mathcal{O}^k = -S^x{}_n \sin kn + S^y{}_n \cos kn, \tag{B-21b}$$

and

$$\mathcal{O}^{k-1} S^z{}_n \mathcal{O}^k = S^z{}_n\,, \tag{B-21c}$$

we find

$$\begin{aligned}\langle\Psi_0 \mid \mathcal{O}^{k-1} H \mathcal{O}^k \mid \Psi_0\rangle = \Big\langle \Psi_0 \Big| H &+ (\cos k - 1) \sum_1^N (S^x{}_n S^x{}_{n+1} + S^y{}_n S^y{}_{n+1}) \\ &+ \sin k \sum_1^N (S^x{}_n S^y{}_{n+1} - S^y{}_n S^x{}_{n+1}) \Big| \Psi_0 \Big\rangle.\end{aligned} \tag{B-22}$$

Use has been made of (B-18) in the term arising from $\mathbf{S}_N \cdot \mathbf{S}_1$. Consider the terms on the right in (B-22), one by one, for $k = 2\pi/N$.

(i)
$$\langle\Psi_0 \mid H \mid \Psi_0\rangle = E_0\,. \tag{B-23a}$$

(ii)
$$\begin{aligned}(\cos k - 1) &\Big\langle \Psi_0 \Big| \sum_1^N (S^x{}_n S^x{}_{n+1} + S^y{}_n S^y{}_{n+1}) \Big| \Psi_0 \Big\rangle \\ &= \left[-\tfrac{1}{2}\left(\frac{2\pi}{N}\right)^2 - O(N^{-4})\right] \sum_1^N \langle\Psi_0 | S^x{}_n S^x{}_{n+1} + S^y{}_n S^y{}_{n+1} | \Psi_0\rangle \\ &\leq \left(\frac{2\pi}{N}\right)^2 \frac{N}{2} + O(N^{-3}).\end{aligned} \tag{B-23b}$$

(iii)
$$\begin{aligned}\sin k &\Big\langle \Psi_0 \Big| \sum_1^N (S^x{}_n S^y{}_{n+1} - S^y{}_n S^x{}_{n+1}) \Big| \Psi_0 \Big\rangle \\ &= -i \sin k\, \langle\Psi_0 | [\textstyle\sum nS^z{}_n, H] | \Psi_0\rangle = 0.\end{aligned} \tag{B-23c}$$

Thus for $k = 2\pi/N$,

$$\langle\Psi_k \mid H \mid \Psi_k\rangle \leqq E_0 + (2\pi^2/N), \tag{B-24}$$

and there is no energy gap.

REFERENCES

1. H. BETHE, *Z. Physik* **71,** 205 (1931).
2. L. HULTHÉN, *Arkiv. Mat. Astron. Fysik* **26A,** No. 11(1938).
3. P. W. ANDERSON, *Phys. Rev.* **86,** 694 (1952).
4. R. KUBO, *Revs. Modern Phys.* **25,** 344 (1953).
5. P. W. KASTELEIJN, *Physica* **18,** 104 (1952).
6. H. TAKETA AND T. NAKAMURA, *J. Phys. Soc. (Jap.)* **11,** 919 (1956).
7. W. MARSHALL, *Proc. Roy. Soc.* **A232,** 48 (1955).
8. R. ORBACH, *Phys. Rev.* **112,** 309 (1958).
9. L. R. WALKER, *Phys. Rev.* **116,** 1089 (1959).
10. T. W. RUIJGROK AND S. RODRIGUEZ, *Phys. Rev.* **119,** 596 (1959).
11. H. L. DAVIS, *Phys. Rev.* **120,** 789 (1960).
12. P. JORDAN AND E. WIGNER, *Z. Physik* **47,** 631 (1928);
    H. A. KRAMERS, "Quantum Mechanics" (translated by D. TER HAAR), p. 328 ff. Interscience, New York, 1957;
    K. MEYER, *Z. Naturforsch.* **11A,** 865 (1956);
    S. RODRIGUEZ, *Phys. Rev. 116*, 1474 (1960);
    Y. NAMBU, *Progr. Theoret. Phys. Japan* **5**, 1 (1950).
13. E. JAHNKE AND F. EMDE, "Tables of Functions," 4th ed., p. 73. Dover, New York, 1945.
14. G. C. WICK, *Phys. Rev.* **80,** 268 (1950).
15. E. WHITTAKER AND G. WATSON, "Modern Analysis," Am. ed., p. 212. Cambridge Univ. Press, London and New York, 1948.
16. T. MATSUBARA, *Prog. Theoret. Phys. Japan* **14,** 351 (1955);
    D. J. THOULESS, *Phys. Rev.* **107,** 1162 (1957).

Reprinted from THE PHYSICAL REVIEW, Vol. 135, No. 3A, A640–A658, 3 August 1964
Printed in U. S. A.

Linear Magnetic Chains with Anisotropic Coupling

JILL C. BONNER
Department of Physics, Royal Holloway College, Englefield Green, Surrey, England

AND

MICHAEL E. FISHER
Wheatstone Physics Laboratory, King's College, London W. C. 2, England
(Received 11 February 1964; revised manuscript received 8 April 1964)

Linear chains (and rings) of $S=\frac{1}{2}$ spins with the anisotropic (Ising-Heisenberg) Hamiltonian

$$\mathcal{H}=-2J\sum_{i=1}^{N}\{S_i^zS_{i+1}^z+\gamma(S_i^xS_{i+1}^x+S_i^yS_{i+1}^y)\}-g\beta\sum_{i=1}^{N}\mathbf{H}\cdot\mathbf{S}_i$$

have been studied by exact machine calculations for $N=2$ to 11, $\gamma=0$ to 1 and for ferro- and antiferromagnetic coupling. The results reveal the dependence on finite size and anisotropy of the spectrum and dispersion laws, of the energy, entropy, and specific heat, of the magnetization and susceptibilities, and of the pair correlations. The limiting $N\rightarrow\infty$ behavior is accurately indicated, for all γ, in the region $kT/|J|\geqslant 0.5$ which includes the maxima in the specific heat and susceptibility. The behavior of thermal and magnetic properties of infinite chains at lower temperatures is estimated by extrapolation. For infinite antiferromagnetic chains the ground-state degeneracy, the anisotropy gap, and the magnetization, perpendicular susceptibility, and pair correlations at $T=0$ are similarly studied. Estimates of the long-range order suggest that it vanishes only at the Heisenberg limit $\gamma=1$ and confirm the accuracy of Walker's perturbation series in γ.

1. INTRODUCTION

EXPERIMENTAL and theoretical studies have shown that in many magnetic materials, the magnetic part of the Hamiltonian may be quite accurately represented as a set of localized spins $\mathbf{S}_i$ with bilinear interactions. For certain highly anisotropic systems,[1] the coupling energy can be approximated by the pure Ising form $-2JS_i^zS_j^z$, but for most systems the anisotropy, although important, is not very large and the pure Heisenberg coupling $-2J\mathbf{S}_i\cdot\mathbf{S}_j$ is more realistic. Between these extremes is the anisotropic "intermediate" coupling $-2J\{\gamma\mathbf{S}_i\cdot\mathbf{S}_j+(1-\gamma)S_i^zS_j^z\}$ which reduces to the Ising case when $\gamma=0$ and the Heisenberg case when $\gamma=1$.

With pure Ising interaction the partition function of a finite or infinite chain of spins in a parallel magnetic field, and thence the thermal and magnetic properties, may be calculated exactly.[2] For simple two-dimensional Ising lattices exact solutions may be obtained in zero field.[2] For the Heisenberg, or intermediate γ spin Hamiltonians, however, exact closed formulas for the finite-temperature behavior have not been found even for the linear chain despite much theoretical effort (see below). Although, naturally, greatest interest attaches to three-dimensional lattices, the properties of linear chains with non-Ising spin coupling are of both experimental and theoretical significance.

Experimentally a number of crystals are known,[3] for example copper tetramine sulfate monohydrate,[3a,b,4] in which the magnetic ions are arranged in chains with strong interactions within each chain but rather weak interactions between chains. Except at the lowest temperatures, the chains should be almost independent and theoretical values based on a one-dimensional model may be confronted directly with experimental measurements. Furthermore, the experimental evidence on various cupric quinone complex salts[3c,d] indicates the existence of independent *finite* chains of 10 to 20 magnetic ions. A one-dimensional model is similarly valid for magnetically active polymeric molecular chains of finite or indefinitely great length.[5]

Theoretically the linear chain Heisenberg-Ising model is interesting as one of the simplest many-body systems in which quantum effects play a vital part. This is especially so for the antiferromagnetic chain where the calculation of even the ground-state energy is not easy, although it has been performed exactly.[6–9] Accurate

[1] A. H. Cooke, D. T. Edmonds, C. B. P. Flinn, and W. P. Wolf, Proc. Phys. Soc. (London) **74**, 791 (1959); M. Ball, M. T. Hutchings, M. J. M. Leask, and W. P. Wolf, Proceedings of the Eighth International Congress on Low-Temperature Physics (to be published); M. Ball, M. J. M. Leask, W. P. Wolf, and A. F. G. Wyatt, J. Appl. Phys. **34**, 1104 (1963).

[2] C. Domb, Advan. Phys. **9**, 149 (1960). The perpendicular susceptibility (in zero field) may also be calculated exactly with Ising coupling in one and two dimensions. See Refs. 43 and 44.

[3] (a) T. Haseda and A. R. Miedema, Physica **27**, 1102 (1961); (b) T. Watanabe and T. Haseda, J. Chem. Phys. **28**, 323 (1958); (c) T. Haseda, A. R. Miedema, H. Kobayashi, and E. Kanda, J. Phys. Soc. Japan **17**, Suppl. B-I, 518 (1962); (d) H. Kobayashi, T. Haseda, E. Kanda, and S. Kanda, J. Phys. Soc. Japan **18**, 349 (1963); (e) L. Berger, S. A. Friedberg, and J. T. Schriempf, Phys. Rev. **132**, 1057 (1963).

[4] R. B. Griffiths, Phys. Rev. following paper **135**, A659 (1964).

[5] V. L. Ginsburg and V. M. Fain, Dokl. Acad. Nauk SSSR **131**, 785 (1960); L. A. Blyumenfel'd, A. E. Kalmanson, and S. P'ei-Ken, *ibid.* **124**, 1144 (1959); L. A. Blyumenfel'd and V. A. Benderskii, *ibid.* **133**, 1451 (1960) [English transls.: Soviet Phys.—Doklady **5**, 328 (1960); **4**, 260 (1959); **5**, 919 (1961)]; L. A. Blyumenfel'd, Biofizika **4**, 515 (1959) [English transl.: Biophysics USSR **4**, 3 (1959)].

[6] H. A. Bethe, Z. Physik **71**, 205 (1931).

[7] L. Hulthén, Arkiv Mat. Astron. Fysik **26A**, No. 11 (1938).

[8] R. L. Orbach, Phys. Rev. **112**, 309 (1958).

[9] L. R. Walker, Phys. Rev. **116**, 1089 (1959).

values for other properties would be valuable as an aid to judging various approximate theoretical treatments which are also applicable to more complex systems. From the viewpoint of statistical mechanics, it is also of interest to study how the limiting behavior for infinite systems is approached by finite systems. Such insight is useful since, with the advent of fast digital computers, it is feasible to perform exact or Monte Carlo calculations on finite models of many types of physical system.

With this motivation we have undertaken a numerical study of finite chains and rings of N spins with the Hamiltonian,

$$\mathcal{H}=-2J\sum_{i=1}^{N}\{S_i^zS_{i+1}^z+\gamma(S_i^xS_{i+1}^x+S_i^yS_{i+1}^y)\} -g\beta\sum_{i=1}^{N}\mathbf{H}\cdot\mathbf{S}_i \quad (1.1)$$

for spin $S=\frac{1}{2}$. We have computed the energy levels and eigenvectors and thence the thermal and magnetic properties for varying anisotropy γ and varying magnetic field H for rings of size $N=2$–11 (and for some open chains) for both ferro- and antiferromagnetic coupling ($J=+|J|$ and $J=-|J|$, respectively). (Our results for the antiferromagnetic entropy in a magnetic field for varying γ have already been reported and discussed in the light of the third law of thermodynamics.[10])

Previously, Orbach[11] had computed the energy levels and eigenvectors for $N=2$, 4, 6, 8 and (unpublished) $N=10$ for the case $\gamma=1$, and for some intermediate γ values in the case of $N=8$ (also unpublished), but he did not compute the thermal or magnetic properties. Independently of the present authors Griffiths[12] has calculated the energy levels for $N=3$, 5, 7, 9, and 10 for $\gamma=1$. He evaluated the energies, entropies, specific heats, and susceptibilities in zero field for $N=2$–10, and compared them with the series expansions. We are deeply indebted to both Dr. R. L. Orbach and Dr. R. B. Griffiths for sending us their own unpublished calculations and for helpful discussions and correspondence.

Theoretical work on the Hamiltonian (1.1) with $\gamma=1$ dates back to 1930 when Bloch[13] introduced the concept of a spin wave and gave the exact eigenstates for one overturned spin ($\sum S_i^z=NS-1$) for ferromagnetic interactions. Using a similar approach, Bethe[6] obtained the exact eigenstates corresponding to interacting spin waves for an arbitrary number of overturned spins on a linear chain ($\gamma=1$). He showed how the states may be classified as "unbound" or "bound." Hulthén, in a comprehensive paper,[7] used Bethe's solutions to obtain an integral equation from which he obtained the exact value for the antiferromagnetic ground-state energy of the infinite linear chain. He also obtained the exact ground states for rings of $N=4$, 6, 8, and 10 spins and estimated the susceptibility at $T=0$.

The arguments of Bethe and Hulthén were extended by Orbach[8] to the full anisotropic Hamiltonian (1.1), which had also been considered by Kasteleyn.[14] Kasteleyn used a variational method which indicated a critical anisotropy constant $\gamma_c=0.483$ beyond which the zero-temperature long-range order (present when $\gamma=0$) vanished identically. The variational short-range order and energy showed singularities in γ at the same point. With the aid of an exact integral equation Orbach showed that these latter singularities were spurious but he was unable to calculate the long-range order exactly. Walker,[9] in an important paper, obtained an analytic solution of Orbach's integral equation which revealed that the antiferromagnetic ground state $E_0(\gamma)$ as a function of γ was nonanalytic at $\gamma=1$. This implies that the Heisenberg limit ($\gamma=1$) is a special point perhaps analogous to a critical point.[15] Walker also gave perturbation series in powers of γ^2 for the energy and for the short-range and long-range order. His series for the long-range order seems to indicate that it does not vanish for γ less than about 0.9.

Perturbation calculations equivalent to those of Walker have been presented by Boon,[16] who also considered two- and three-dimensional lattices. Alternative perturbation procedures have been proposed by Davis, Rodriguez, Frank, Mannari, and Mills, Kenan, and Korringa.[17–19] Ruijgrok and Rodriguez[20] developed a variation method which was rather accurate for the ground state but which yielded a finite long-range order even for $\gamma=1$.

Earlier calculations based on the spin-wave approach have been reviewed by Van Kranendonk and Van Vleck.[21] One of the most striking predictions is that for a pure Heisenberg antiferromagnet the lowest energy states should obey a dispersion law of the form

---

[10] J. C. Bonner and M. E. Fisher, Proc. Phys. Soc. (London) **80**, 508 (1962).

[11] R. L. Orbach, Ph.D. dissertation, University of California, 1959 (unpublished); Phys. Rev. **115**, 1181 (1959).

[12] R. B. Griffiths, mimeographed reports, Stanford University, 1961, and La Jolla, California, 1963 (unpublished).

[13] F. Bloch, Z. Physik **61**, 206 (1930).

[14] P. W. Kasteleyn, Physica **18**, 104 (1952).

[15] The fact that $\gamma=1$ is a mathematical singularity (nonanalytic point) of $E_0(\gamma)$, follows immediately from Walker's observation that his formula exhibits a pole in any interval, however small, of the open segments $|\gamma|>1$. It should be noted, however, that the singularity at $\gamma=1$ is not visible in graphs of $E_0(\gamma)$ for $\gamma\leqslant 1$ contrary to what might be expected. One may regard Kasteleyn's variational transition point $\gamma_c=0.48$ (where the energy was nonanalytic as in any phase transition) as an approximation to the "true transition point" at $\gamma=1$.

[16] M. H. Boon, Nuovo Cimento **21**, 885 (1961).

[17] H. L. Davis, Phys. Rev. **120**, 789 (1960). These results are presented as a power series in $\gamma(=1-\alpha$ in Davis's notation), but the higher coefficients are inexact since the perturbation Hamiltonian is not just the transverse part of (1.1).

[18] S. Rodriguez, Phys. Rev. **116**, 1474 (1959); R. L. Mills, R. P. Kenan, and J. Korringa, Physica **26**, S204 (1960).

[19] D. Frank, Z. Physik **146**, 615 (1956); I. Mannari, Progr. Theoret. Phys. (Kyoto) **19**, 201 (1958).

[20] T. W. Ruijgrok and S. Rodriguez, Phys. Rev. **119**, 596 (1960).

[21] J. Van Kranendonk and J. H. Van Vleck, Rev. Mod. Phys. **30**, 1 (1958).

$\epsilon(k)=E-E_0\sim|\sin k|$.[21–24] This conclusion was tested against the numerical spectra for finite rings of $N=6$ and 8 spins by Mattheiss,[25] who found an approximately $|\sin k|$ dependence, but with a modified amplitude. More recently, des Cloizeaux and Pearson[26] extended the computation to the lowest antiferromagnetic states for rings of 16 and 48 spins. They were also able to derive the limiting dispersion law $\epsilon(k)=\pi|J||\sin k|$ for the corresponding class of states of an infinite chain.

The theoretical knowledge of the ground and lower excited states of Heisenberg-Ising chains is not matched by a corresponding knowledge of the thermodynamic behavior. For example, the positions of the maxima in the specific heat and susceptibility do not previously seem to have been estimated with any accuracy. (Although recently, Bulaevskii[27] has reported a finite temperature Hartree-Fock calculation which reproduces the general features quite well.)

It is interesting, however, that the spin-$\frac{1}{2}$ model obtained by deleting the parallel or Ising terms $S_i^zS_{i+1}^z$ from Eq. (1.1) to leave only the transverse Hamiltonian is completely soluble. Lieb, Schultz, and Mattis[28] investiagated the spectrum in detail while Katsura[29] studied the thermal and magnetic properties including the behavior in a field. Although the transverse Hamiltonian is somewhat artificial, the behavior of this model resembles in many respects the results we have found with the full Hamiltonian (1.1).[30]

It is also possible to calculate by elementary methods the zero-field free energy, correlations and susceptibility for isotropic Heisenberg chains in the limit of infinite spin.[31] The low-temperature behavior of the thermal properties in this case is somewhat unrealistic since in the classical limit $S=\infty$ the specific heat necessarily goes to a nonzero value as $T\rightarrow 0$. The susceptibility, however, correlates quite closely with the results we find for the $S=\frac{1}{2}$ Heisenberg chains (except that a nonphysical contradiction of the third law is shown again by a nonzero slope for small T).

The plan of this paper is as follows: In Sec. 2, the machine computations are outlined and the possibilities of extrapolation from finite N to the limit $N\rightarrow\infty$ are illustrated by examining the antiferromagnetic ground state as a function of γ. The thermal properties, in particular the specific heat, are discussed in Sec. 3 for antiferromagnetic and ferromagnetic rings. The magnetic properties are described in Sec. 4. Attention is given especially to the magnetization curve at $T=0$ and the corresponding parallel and perpendicular susceptibilities. In Sec. 5 details of the energy spectrum are discussed and the antiferromagnetic anisotropy gap and the asymptotic degeneracy of the ground state are estimated. The spin-wave dispersion laws for ferro- and antiferromagnetic chains are illustrated. The short-range antiferromagnetic order and further pair correlation functions $\langle S_0^zS_l^z\rangle$ are described in Sec. 6. The long-range order is estimated by extrapolation and compared with previous approximations. Our results are for the most part presented graphically but tables of the thermal and magnetic properties for $\gamma=1$ and 0.5 and $N=9$, 10, and 11 have been prepared and will be made available on request to the authors. (For reasons of economy they are not reproduced in this article.)

2. FINITE CHAIN CALCULATIONS AND THE ANTIFERROMAGNETIC GROUND STATE

For a finite system of N ($S=\frac{1}{2}$) spins the problem of calculating the energy levels reduces to the diagonalization of the $2^N\times 2^N$ matrix representing the Hamiltonian. (As basic states it is covenient to use direct products of "UP" and "DOWN" single-spin states: These are, of course, eigenstates of the Ising Hamiltonian $\gamma=0$.) By classifying the states by the value of S^z, the z component of the total spin, the matrix splits into $N+1$ blocks of order 1, N, $\binom{N}{2}$, $\cdots\binom{N}{\frac{1}{2}N}$, $\cdots N$, 1 (N even). A further reduction can be obtained for closed chains (rings) of spins by using the translational invariance. Each level is then also classified by a (total) wave number $k=2\pi r/N$ with $r=0, 1, 2, \cdots, N-1$ (k is only determined up to a multiple of 2π).

The largest value of N which can be handled numerically is limited by the size of the largest block matrix in relation to the speed and capacity of the electronic computer available. Our calculations were performed on the now relatively slow University of London Ferranti "Mercury" which restricted us to $N=11$ or less. To economize on computing time many of the calculations for intermediate values of γ and magnetic field, etc., were, in fact, performed with $N=8$, or 10, once the trend with increasing N was clear.[10] The diagonalization, yielding both energy levels and eigenstates, was performed by standard subroutines and the results checked against the other available computations.[11,12,32] The calculation of the thermodynamic properties was performed by direct evaluation of the appropriately weighted partition sums.

The possibilities of estimating the properties of

[22] P. W. Anderson, Phys. Rev. **86**, 694 (1952).
[23] R. Kubo, Phys. Rev. **87**, 568 (1952).
[24] T. Oguchi, Phys. Rev. **117**, 117 (1960).
[25] L. F. Mattheiss, Phys. Rev. **123**, 1209 (1961).
[26] J. des Cloizeaux and J. J. Pearson, Phys. Rev. **128**, 2131 (1962).
[27] L. N. Bulaevskii, Zh. Eksperim. i Teor. Fiz. **43**, 968 (1962); see also V. L. Ginsburg and V. M. Fain, *ibid.* **39**, 1323 (1960); **42**, 180 (1962) [English transls.: Soviet Phys.—JETP **16**, 685 (1963); **12**, 923 (1961); **15**, 131 (1962)].
[28] E. Lieb, T. Schultz and D. Mattis, Ann. Phys. (N. Y.) **16**, 407 (1961).
[29] S. Katsura, Phys. Rev. **127**, 1508 (1962).
[30] S. Katsura (preprint) has discussed the full Hamiltonian by treating the parallel terms as a perturbation.
[31] M. E. Fisher, Am. J. Phys. **32**, 343 (1964).

[32] For the Heisenberg limit $\gamma=1$ one has a further check since the total spin S is also conserved so that many levels from different blocks coincide. It was not found possible to use this added invariance to reduce the calculations.

TABLE I. Energy of the antiferromagnetic ground state $\mathcal{E}_0(N)$.

N	Even	N	Odd
4	−1.00000	3	−0.50000
6	−0.93425	5	−0.74721
8	−0.91277	7	−0.81577
10	−0.90309	9	−0.84384
12	−0.8979	11	−0.85799
⋮		⋮	
∞	−0.88629	∞	−0.88629

infinite chains from those of finite chains are revealed by a study of the pure Heisenberg antiferromagnetic ground state energies. The antiferromagnetic ground state E_0 is equal to $-2N|J|\ln 2+\frac{1}{2}N|J|$ for the Hamiltonian (1.1) with $H=0$ and $\gamma=1$ in the limit $N\rightarrow\infty$. It is convenient to define normalized ground-state energies for finite N

$$\mathcal{E}_0(N)=E_0(N)/N|J|$$

and these energies are tabulated numerically in Table I. (The values for $N=4$, 6, 8, and 10 were given exactly by Hulthén,[7] that for $N=12$ by Ledinegg and Urban.[33]) In Fig. 1 the energies are plotted versus $1/N$ (circles) and versus $(1/N)^2$ (squares). The values form two sequences (N odd or even) and it is evident that the limiting energy is approached linearly with $(1/N)^2$. A relation of the form

$$\mathcal{E}_0(N)=\mathcal{E}_0(\infty)+a/N^2 \qquad (2.1)$$

holds quite accurately even down to $N=4$ or 5 with $a_{\text{odd}}\simeq -2a_{\text{even}}$.

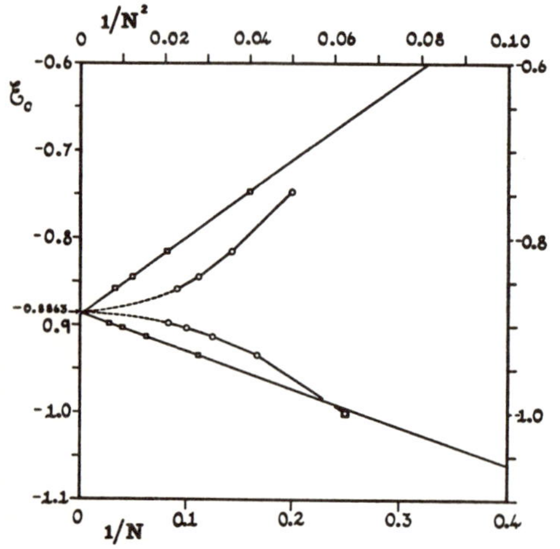

FIG. 1. Antiferromagnetic ground-state energies versus $1/N$ (circles) and versus $(1/N)^2$ (squares) for pure Heisenberg rings ($\gamma=1$).

[33] E. Ledinegg and P. Urban, Acta Phys. Austriaca **6**, 257 (1953).

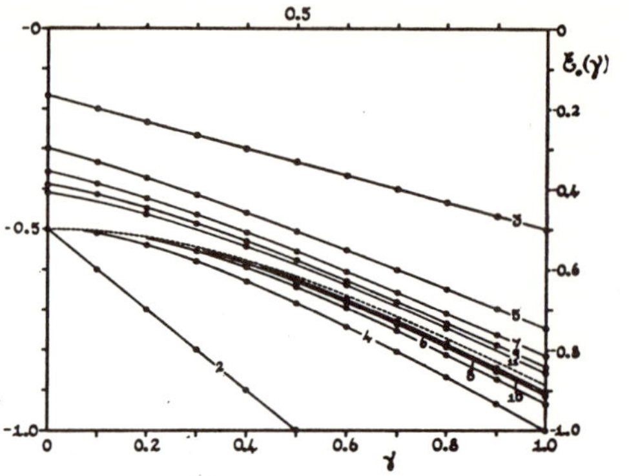

FIG. 2. Antiferromagnetic ground state as a function of γ for finite and infinite rings. The exact limiting curve due to Orbach (Ref. 8) is shown by the dashed curve.

If the limiting value had not been known, it could have been estimated to about 0.1% accuracy by linear extrapolation with $1/N^2$. Heisenberg chains with open ends show a slower convergence, apparently linear with $1/N$. For pure Ising ($\gamma=0$) rings the ground state is exact for N even but approaches the limit as $1/N$ for N odd on account of the "misfit seam." In Fig. 2 the approach of the antiferromagnetic ground state to the limit is shown as a function of γ for $N=2$–11. The limiting curve (due to Orbach[8]) is rather well defined by these results.

Encouraged by the relatively simple and regular behavior of the finite N results found here, we may go on to examine properties for which the exact $N=\infty$ limits are unknown with the reasonable expectation that careful extrapolation to large N will not be misleading.

3. THERMAL PROPERTIES

In this section we consider the thermal properties firstly of antiferromagnetic and then of ferromagnetic chains.

Antiferromagnetic Coupling

In Fig. 3 is shown the energy per spin in zero field as a function of temperature for antiferromagnetic pure Ising and Heisenberg rings of $N=2$–11 spins. As for the ground states, odd and even rings form two distinct sequences. For the Ising case ($\gamma=0$), where the exact limit is known, the two sequences approach the limiting curve monotonically from above and below. There seems no reason to doubt that the same situation prevails at $\gamma=1$. The energy of an infinite Heisenberg chain is thus defined to an accuracy of better than ±0.5% down to temperatures of $kT/|J|=0.5$ by the mean of the curves for $N=10$ and 11. (These two curves and the estimated limiting curve have been tabulated and are available on request.) Below this temperature the true limiting curve is less certain, but since the value for

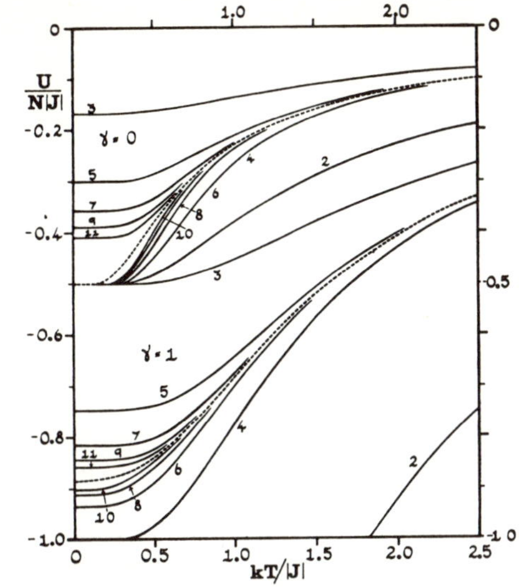

FIG. 3. Zero-field energy as a function of temperature for finite antiferromagnetic Ising and Heisenberg chains. The limiting ($N=\infty$) curve is shown dashed in each case.

$T=0$ is known exactly, even a roughly estimated limiting curve would be accurate to $\pm1\%$. We return to this point shortly in considering the entropy.

The entropy for the pure antiferromagnetic Heisenberg chains is shown in Fig. 4. The convergence appears to be similar to that for the energy and the limit seems well determined down to $kT/|J|=0.5$ by interpolation between the curves for odd and even N. For odd N at $T=0$ the ground state is fourfold degenerate and the entropy per spin thus goes to zero only as $2k\ln 2/N$.

To estimate the nature of the low-temperature behavior for $N\to\infty$, we may reasonably postulate a power law for the limiting free energy per spin (or, more directly, for the limiting internal energy). Thus, suppose[34]

$$U(T)-U(0)=AN|J|(kT/|J|)^{\alpha}, \tag{3.1}$$

where A and α are fixed (as $T\to 0$). Then for the specific heat we have

$$C/Nk=A\alpha(kT/|J|)^{\alpha-1} \tag{3.2}$$

and for the entropy

$$S(T)/Nk=A\alpha(\alpha-1)^{-1}(kT/|J|)^{\alpha-1}. \tag{3.3}$$

By combining (3.1) and (3.3) we obtain

$$U(T)-U(0)=(1-\alpha^{-1})TS(T) \tag{3.4}$$

so that a plot of energy $U/N|J|$ versus $TS/N|J|$ for $N\to\infty$ should be linear with slope determined by the index α. Such a plot is shown in Fig. 5 for $N=10$ and $N=11$ and supports the view that the limiting curve would indeed be linear. The solid line which is drawn from the exact ($N\to\infty$) ground state with slope 0.515 should be close to any reasonable "best" fit and yields the estimate $\alpha\simeq 2.06\pm0.03$. This value is largely determined by the data for temperatures in the range $kT/|J|=0.30$–0.60. The dashed line, on the other hand, is of slope $\frac{1}{2}$ and corresponds to $\alpha=2$ exactly. It seems likely that this is the true limiting value which would be obtained by fitting data for larger N at lower temperatures.[35]

If we assume $\alpha=2$, the amplitude A may be estimated by various methods; for example, from the temperature variation of $U/N|J|$ and $TS/N|J|$, from plots of $U/N|J|$ versus $(S/Nk)^2$ using the weighted means,[35] and from the slope of the weighted entropy means at low temperatures. We thus obtain the approximate results

$$U(T)\approx U(0)+0.175N|J|(kT/|J|)^2, \tag{3.5}$$

$$S(T)\approx 0.35Nk(kT/|J|), \quad (\gamma=1) \tag{3.6}$$

valid up to $kT/|J|=0.4$ to 0.5.[36] These are shown as

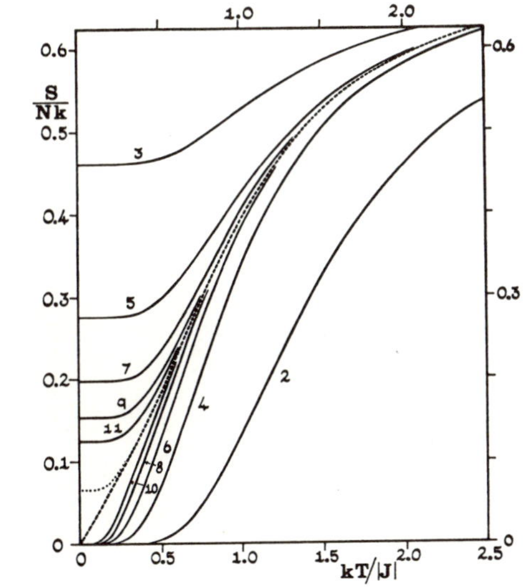

FIG. 4. Entropy versus temperature for antiferromagnetic Heisenberg chains ($\gamma=1$). The dotted curve is the mean of $N=10$ and $N=11$ weighted as in (a) (Ref. 35), and the dashed curve is the estimated limit.

[34] Our method is a slight modification of a procedure devised by Griffiths (Ref. 12) who plotted $U(T)$ versus $S(T)$ and estimated the index α by fitting B and β in the power law $U-U(0)=BS^{\beta}$.

[35] Griffiths (Ref. 12) estimated $\alpha=2.1$, but also suggested that $\alpha=2$ was probably the exact value. This conclusion is supported more closely by analyzing the trends of the weighted means (a) $[NP_N+(N-1)P_{N-1}]/(2N-1)$, (b) $\frac{1}{2}P_N+\frac{1}{2}P_{N-1}$, (c) $[(N-1)P_N+NP_{N-1}]/(2N-1)$ (where P is the thermodynamic property in question), which converge more rapidly.

[36] Our conclusions regarding the amplitude A agree closely with those of Griffiths. The spin-wave theory of the low-temperature behavior is given by R. Kubo, Phys. Rev. **87**, 568 (1952); J. Van Kranendonk and J. H. Van Vleck, Rev. Mod. Phys. **30**, 1 (1958).

dashed curves in Figs. 3 and 4 and seem to be the most reasonable estimates of the low-temperature behavior.

It is interesting that the functional forms of (3.5) and (3.6) agree with the predictions of simple antiferromagnetic spin-wave theory but the corresponding amplitude[36] $A_{s.w.} = \pi/6 = 0.52$ is too great by a factor of about three. A T^2 law for the energy is exact for the pure transverse Hamiltonian discussed by Katsura.[29]

The corresponding specific heats for pure antiferromagnetic Heisenberg chains are shown in Fig. 6. For temperatures above $kT/|J| = 0.5$ the convergence again appears to be monotonic from above and below and the limiting curve (shown dashed) can be estimated quite accurately through the maximum of height

$$C_{max}/Nk \simeq 0.350 \quad (\gamma = 1) \tag{3.7}$$

at

$$kT_{max}/|J| \simeq 0.962 \quad (\gamma = 1). \tag{3.8}$$

At lower temperatures the convergence must be more complex since, for finite N, the specific heat always vanishes exponentially fast as $T \to 0$, owing to the finite energy gap between the ground and first excited states. As $N \to \infty$, however, the states close up (as $1/N$) and merge into a continuum, which runs from the ground state (see below). The probable limiting low-temperature behavior follows from (3.5) and (3.6), which yield

$$C(T) \simeq 0.35Nk(kT/|J|) \quad (\gamma = 1), \tag{3.9}$$

and this has been used in deriving the estimated curve for the limit $N = \infty$ shown by the dashed line in Fig. 6.

The variation of the antiferromagnetic specific heat with anisotropy may be seen from Fig. 7 which shows the specific heats for rings of $N = 8$ spins. The exact limiting curve for $\gamma = 0$ and the estimated limit for $\gamma = 1$ are plotted as dashed lines to indicate the degree to which the results for $N = 8$ approach the limit. The effect of increasing γ from the Ising value $\gamma = 0$ is to shift the specific heat maximum to higher temperatures, to reduce its height and to broaden the peak. At low temperatures ($kT/|J| \simeq 0.1$) the curves for finite N (even) and γ in the range 0.1–0.6 display anomalous small peaks and points of inflection. These are "small number effects" due to the finite splitting of the degenerate Ising ground state by the transverse terms in the Hamiltonian. For fixed γ and increasing N, this splitting diminishes and eventually goes to zero as $N \to \infty$ (see Sec. 5). Correspondingly, the anomalies move to lower temperatures and are reduced in magnitude, finally disappearing in the limit $N = \infty$.

For $\gamma \neq 1$ and large but finite N, the specific heat curves at low temperature (but above the anomalies) vanish exponentially fast, roughly as $\exp[-\Delta E_A(\gamma)/kT]$, where $\Delta E_A(\gamma)$ is the limiting anisotropy gap between the ground state and the first excited states. The value of this gap is estimated in Sec. 5; it approaches zero as $\gamma \to 1$ and the Ising value, $2|J|$, as $\gamma \to 0$.

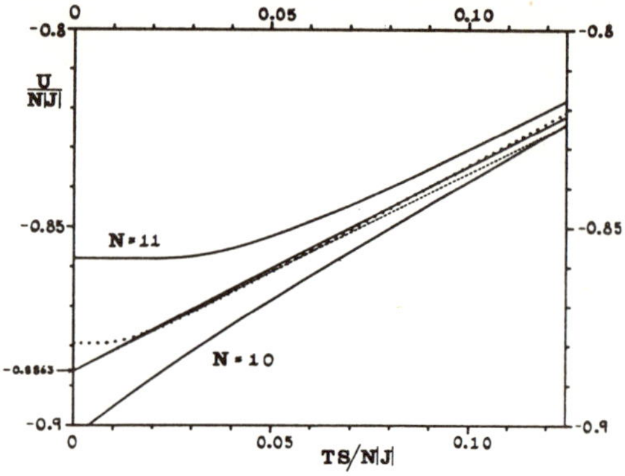

FIG. 5. Energy versus the product of temperature and entropy for $N = 10$ and $N = 11$ Heisenberg antiferromagnetic chains. The line of crosses is the 11/10 mean weighted as in (a) (Ref. 35).

[Owing to the existence of "bound states" of indefinitely large groupings of adjacent "overturned spins," the number of states just above the gap might be of order N^2 rather than N. Consequently, it is possible that the true limiting specific heats might rise faster than $\exp[-\Delta E_A/kT]$. If this were the case, $\Delta E_A(\gamma)$ should be replaced by an "effective gap" $\Delta E_A^*(\gamma) < \Delta E_A(\gamma)$ as happens at the Ising limit $\gamma = 0$ where, in fact, one finds that $\Delta E_A^*(0) = \frac{1}{2}\Delta E_A(0)$.]

The variation of specific heat with magnetic field may be studied just as for zero field. The convergence at temperatures above $kT/|J| \simeq 0.3$ is found to be

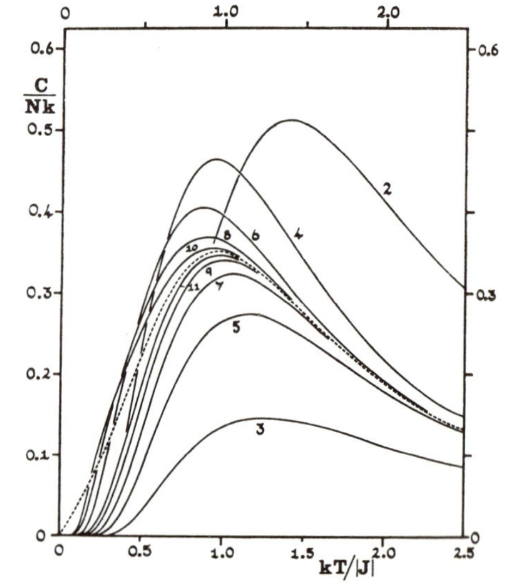

FIG. 6. Variation of specific heat with temperature for antiferromagnetic Heisenberg chains: finite N solid lines; estimated limit $N = \infty$, dashed line.

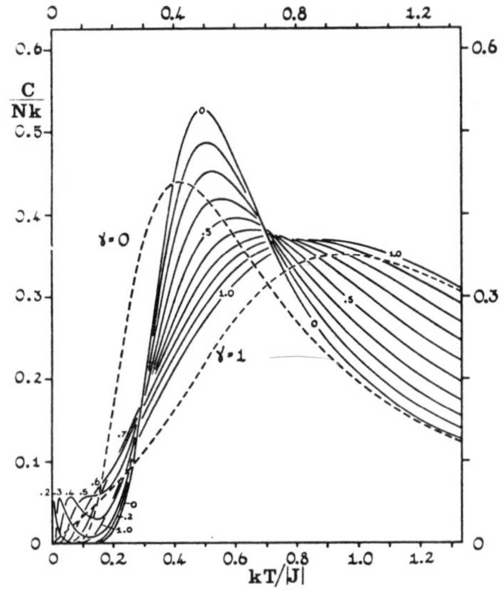

FIG. 7. Antiferromagnetic specific heat for rings of $N=8$ spins for various anisotropies γ. The dashed curves are the exact and estimated limiting curves ($N=\infty$) for $\gamma=0$ and $\gamma=1$, respectively.

quite rapid especially if the consecutive means for $N=2n$ and $N=2n-1$ are considered. Accordingly, in Fig. 8, we have plotted for the pure Heisenberg case only the mean specific heats for $N=10$ and 9 at various fields (except at $H=0$ where the estimated limit is shown). These means should represent the true limiting curves to within 1 or 2% down to temperatures of $kT/|J|\simeq 0.5$. With increasing field the maximum shifts to lower temperatures and falls in height. However, at the field $H_s(g\beta/|J|)=4$, which represents a "critical field" above which the antiferromagnet becomes fully magnetized at zero temperature (i.e., saturation occurs, see Sec. 4), there remains only a broad low maximum at a higher temperature. The behavior is similar for other values of γ as can be seen from Fig. 9, where the specific heat for $\gamma=0.5$ is plotted. The critical field in this case is $H_s=3(|J|/g\beta)$. Above the critical field the initial rise is governed by the energy gap $\Delta E_H(\gamma)=Hg\beta-2(1+\gamma)|J|$ and it may be observed that the maximum starts to increase in height again.

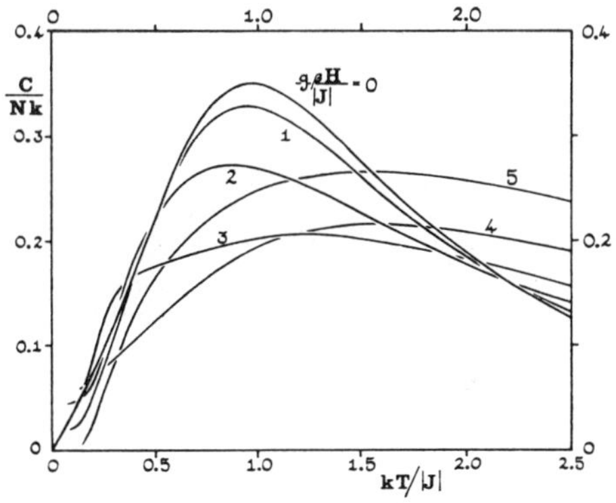

FIG. 8. Variation of the antiferromagnetic specific heat with magnetic field for $\gamma=1$. Except for $H=0$, the curves are the means of the values for $N=9$ and 10.

Ferromagnetic Coupling

We turn now to ferromagnetic chains: Values of the energy, entropy, and specific heat for the longest chains and $\gamma=1$ and 0.5 have been tabulated and are available on request. The energy (per spin) is plotted versus the temperature for Heisenberg chains of $N=3, 4, \cdots, 10, 11$ spins in Fig. 10. In contrast to the antiferromagnetic case, the curves for finite N do not appear to bracket the limiting curve, but rather increase monotonically towards it. This makes extrapolation at low temperatures more difficult although the limit seems quite accurately defined down to $kT/J=0.5$. (All curves, of course, approach the exact limiting ground state $\mathcal{E}=-\frac{1}{2}$).

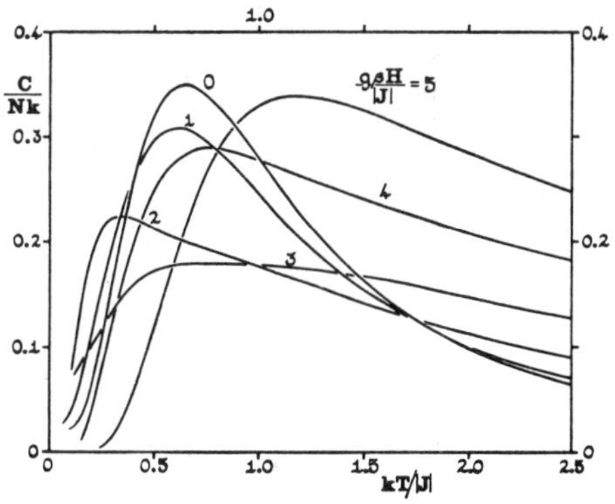

FIG. 9. Variation of the antiferromagnetic specific heat with field as in Fig. 8, but for $\gamma=0.5$.

The curves for the entropy display a more complicated convergence, successive curves crossing at low temperatures. The ground state of the ferromagnet when $\gamma=1$ has spin $NS=\frac{1}{2}N$, and hence degeneracy $N+1$. Consequently, the zero-point entropy is $(k/N)\ln(N+1)$, which approaches zero rather slowly. We may, nonetheless, attempt to estimate the low-temperature behavior of the energy and entropy by the power law analysis presented in Eq. (3.1) to (3.4). The corresponding $U(T)$ versus $TS(T)$ plots are not very straight and their slopes increase monotonically with N.[37] The maximum slopes for $N=9$ and 10 correspond to an index in (3.1)

[3] The curves are concave downwards except for very small values of TS/NJ (below 0.04), where an exponential decay sets in.

of $\alpha = 1.42$ and 1.43, respectively, while the slopes of the tangents through the origin correspond to $\alpha = 1.312$ and 1.324. Rough extrapolation linearly in $1/N$ suggests that the true index might lie in the neighborhood of $\alpha = 1.45$–1.50, but the corresponding pure power laws would only be followed closely below temperatures of $kT/J \simeq 0.25$. Simple spin-wave theory predicts $\alpha = \frac{3}{2}$, and one might expect this to be accurate since the ground and first excited states are described exactly in the ferromagnetic case. However, as is known, spin-wave theory does not give any account of the lower lying "bound" states which arise for further overturned spins. Nevertheless, our results do indicate that $\alpha = \frac{3}{2}$ might be exact. Accepting this, one may estimate the amplitudes

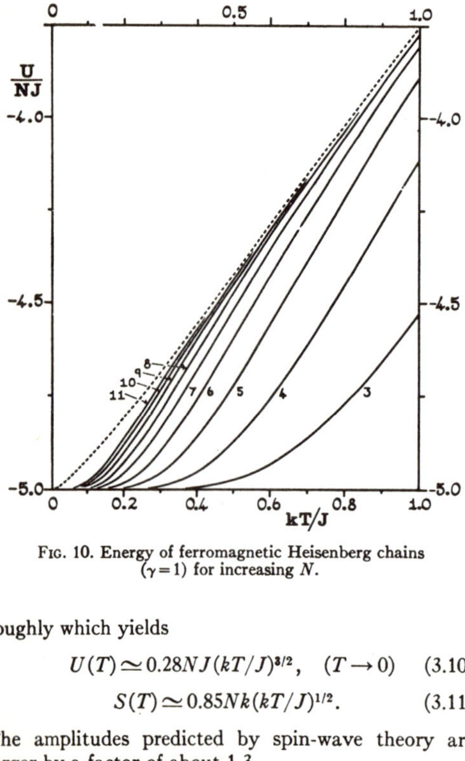

FIG. 10. Energy of ferromagnetic Heisenberg chains ($\gamma = 1$) for increasing N.

roughly which yields

$$U(T) \simeq 0.28NJ(kT/J)^{3/2}, \quad (T \to 0) \qquad (3.10)$$

$$S(T) \simeq 0.85Nk(kT/J)^{1/2}. \qquad (3.11)$$

The amplitudes predicted by spin-wave theory are larger by a factor of about 1.3.

The specific heats for the Heisenberg rings are shown in Fig. 11 as solid curves. The convergence is monotonic decreasing at temperatures above $kT/J = 0.3$ for $N \geqslant 8$.

It seems clear that a *double* maximum will develop when $N = 13$ or 14 and that the maximum of the limiting curve will not derive from the maxima in the curves for $N \leqslant 11$. The broken curves in Fig. 11 correspond to *open* chains of 3 and 4 spins. Although they do not seem to converge very rapidly, they do approach the limit from below and seem to give a truer representa-

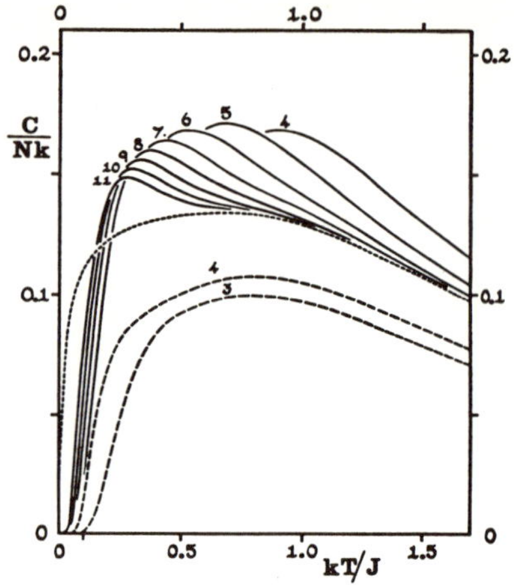

FIG. 11. Specific heats for ferromagnetic Heisenberg rings (solid curves) and open chains (broken curves). The dashed curve is the estimated limiting curve.

tion of its shape. (Calculations for longer open chains were not performed since, as translational symmetry is absent, they would have required more machine time than available.) Guided by these results and the low-temperature estimates following from (3.10) [which indicate that $C(T) \sim T^{1/2}$] and remembering that the integrated area under the specific heat curve must check with the energy one may estimate with moderate accuracy the limiting ferromagnetic specific heat curve.

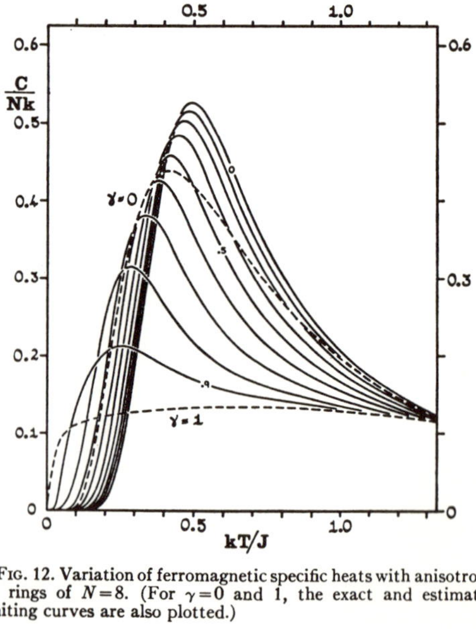

FIG. 12. Variation of ferromagnetic specific heats with anisotropy for rings of $N = 8$. (For $\gamma = 0$ and 1, the exact and estimated limiting curves are also plotted.)

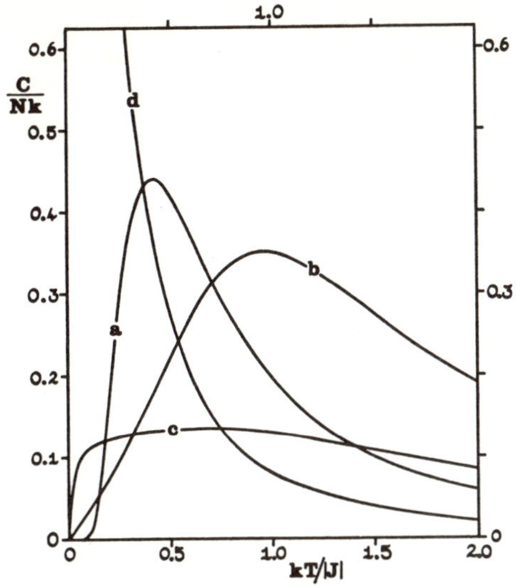

FIG. 13. Comparison of specific heats for infinite chains for (a) $S=\frac{1}{2}$ Ising coupling, (b) $S=\frac{1}{2}$ Heisenberg antiferromagnetic coupling, (c) $S=\frac{1}{2}$ Heisenberg ferromagnetic coupling, and (d) $S=\infty$ Heisenberg coupling (classical spins).

This is shown dashed in Fig. 11; the rather broad peak has a maximum height

$$C_{\max}/Nk\approx 0.134, \tag{3.12}$$

at

$$kT_{\max}/J\approx 0.70. \tag{3.13}$$

The nature of the variation of specific heat with anisotropy can be gauged from Fig. 12, which (except for $\gamma=1$ where the estimated limit is plotted) shows the results for rings of $N=8$ spins at intervals of $\Delta\gamma=0.1$. The effects of a magnetic field (for $\gamma=1$) are rather similar to those due to anisotropy so that we do not present a figure. With a field $H(g\beta/J)=0.5$, the specific heat maximum increases sharply to a height of about $C/Nk=0.37$ at $(kT/J)=0.61$. Further increase of the field to $H(g\beta/J)=1.0$ and 2.0 increases the maximum to $C/Nk=0.42$ and 0.45 and raises the corresponding temperatures to $(kT/J)=0.94$ and 1.45, respectively. These figures are derived for $N=8$ but should not differ significantly from the limiting results since convergence is faster when $H\neq 0$ or $\gamma\neq 1$, since the ferromagnetic ground state is then nondegenerate, and is separated by a finite energy gap, even in the limit $N=\infty$.

Finally, in Fig. 13 are compared on the same scale the exact and estimated specific heats for infinite chains with (a) $S=\frac{1}{2}$ Ising coupling, ferro- or antiferromagnetic, (b) $S=\frac{1}{2}$ Heisenberg antiferromagnetic coupling, (c) $S=\frac{1}{2}$ Heisenberg ferromagnetic coupling, and (d) $S=\infty$ (classical) Heisenberg ferro- or antiferromagnetic coupling.[31] The large difference between the relatively sharp antiferromagnetic $S=\frac{1}{2}$ Heisenberg specific heat and the low broad ferromagnetic specific heat is striking testimony to the lower stability of the isotropic ferromagnetic coupling. Curve (d) for $S=\infty$ continues to rise monotonically to a maximum at $T=0$ of height $C/Nk=1$.

4. MAGNETIC PROPERTIES

Antiferromagnetic Coupling

In Fig. 14 are plotted the antiferromagnetic susceptibilities for isotropic Heisenberg coupling ($\gamma=1$) in zero field for finite rings of $N=3, 4, 5, \cdots, 11$ spins. (The tabulated values for $N=10$ and 11 are available.) The limiting curve is apparently bracketed by the curves for odd N, which approach monotonically from above, and those for even N, which approach monotonically from below. The convergence is rather rapid above $kT/|J|=0.6$, at which temperature the values for $N=10$ and 11 differ by only 5%, and their mean probably differs from the true limiting curve by less than 1%. The susceptibility displays a rounded maximum of height (for $N\to\infty$)

$$\chi_{\max}/(g^2\beta^2/|J|)\simeq 0.07346, \quad (\gamma=1) \tag{4.1}$$

at

$$kT_{\max}/|J|\simeq 1.282. \quad (\gamma=1). \tag{4.2}$$

The situation at low temperatures as zero is approached is more complicated. For finite chains with N even, the antiferromagnetic ground state for all values

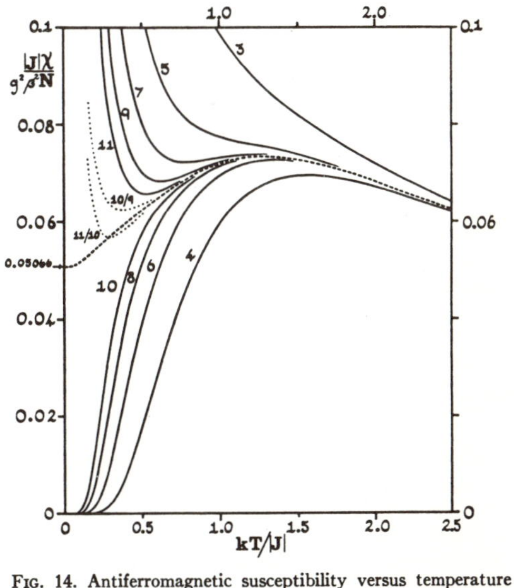

FIG. 14. Antiferromagnetic susceptibility versus temperature for finite Heisenberg rings (solid curves) and the estimated limit for infinite rings (dashed curve). The dotted curves are means of $N=9$ and 10, and $N=10$ and 11 weighted as in (a) and (c) (Ref. 35), respectively.

of γ is characterized by total spin component $S^z=0$, and there is a finite gap to the first excited state of nonzero S^z. Consequently, for all chains of even N (finite), the susceptibility parallel to the z axis approaches zero exponentially fast as $T\rightarrow 0$, at a rate governed by the energy gap. On the other hand, for finite odd N, the ground state, or rather degenerate ground states, have $S^z=\pm\frac{1}{2}$. (Clearly, $S^z=0$ is impossible.) The susceptibility for odd N thus diverges as $1/T$ as T approaches zero. However, as N increases the (relative) amplitude of the divergence falls in magnitude, and it sets in at a lower temperature. These remarks may be verified explicitly in the case $\gamma=0$ (Ising limit), where the exact

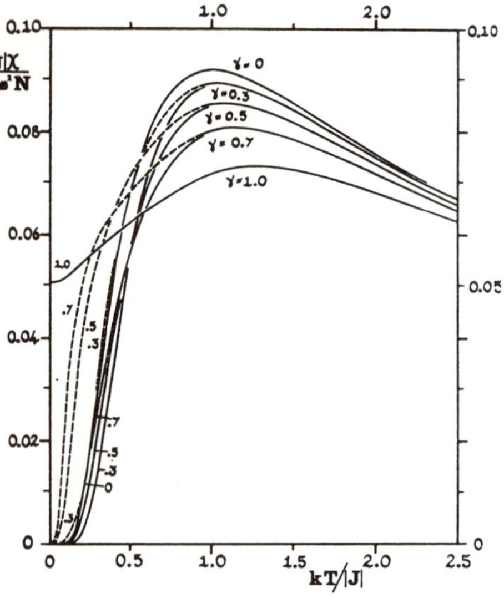

FIG. 15. Variation of the parallel antiferromagnetic susceptibility with anisotropy for rings of $N=8$ spins (solid curves). The estimated limiting values are shown by broken curves. Note that for $\gamma=0$ and $\gamma=1$ only the exact and estimated limiting curves are plotted.

result for the parallel susceptibility of a finite chain is

$$\chi_{\parallel}(\gamma=0)=\frac{g^2\beta^2N}{4kT}e^{-|J|/kT}\left\{\frac{1-(-\tanh K)^N}{1+(-\tanh K)^N}\right\} \qquad (4.3)$$

with

$$K=|J|/2kT. \qquad (4.4)$$

In this case, as is well known, the limiting susceptibility, which is approached both for even N and odd N $(T>0)$, goes to zero exponentially fast (see curve for $\gamma=0$ in Fig. 15).

Now for γ nonzero, but less than unity, the anisotropy gap between the ground state(s) and first excited states persists even in the limit $N\rightarrow\infty$ (see next section). Consequently, the limiting behavior should be similar to the Ising case with $\chi_{\parallel}(T)\rightarrow 0$ as $T\rightarrow 0$. Figure 15

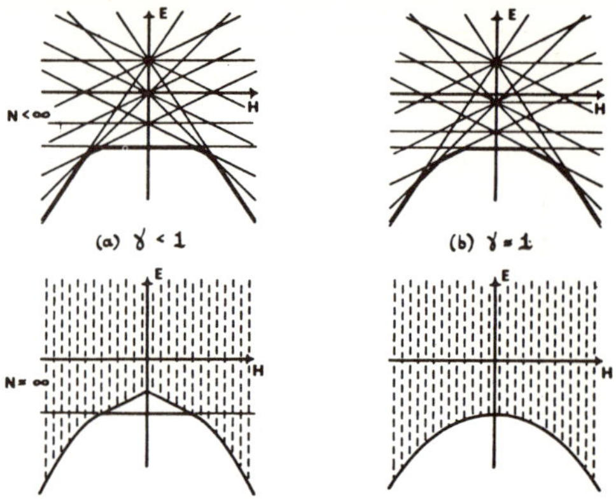

FIG. 16. Sketch of the energy levels versus magnetic field for finite and infinite isotropic $(\gamma=1)$ and anisotropic $(\gamma<1)$ systems (Ref. 39). The shading indicates a continuum of levels.

shows the parallel susceptibilities for rings of $N=8$ spins for $\gamma=0.3$, 0.5, and 0.7 (solid lines) and the approximate limiting curves (broken lines) roughly estimated on the basis of the limiting anisotropy gap. These curves should be accurate to within 5 or 10% down to $kT/|J|=0.3$. Evidently the effect of increasing γ is to shift the susceptibility peak to a higher temperature, to reduce its height and to increase its width. The positions of the maxima are for $\gamma=0$, $\theta_{\max}=kT_{\max}/|J|=1$; for $\gamma=0.3$, $\theta_{\max}\simeq 1.03$, for $\gamma=0.5$, $\theta_{\max}\simeq 1.07$ and for $\gamma=0.7$, $\theta_{\max}\simeq 1.14$. (For small γ the variation is quadratic in γ.)

For the isotropic pure Heisenberg case $\gamma=1$, however, the behavior as $T\rightarrow 0$ is different. As N increases, the gap between the ground state and the first excited states with $S^z\neq 0$ shrinks and approaches zero as $1/N$.[38] In the limit $N=\infty$, there is no anisotropy gap. This

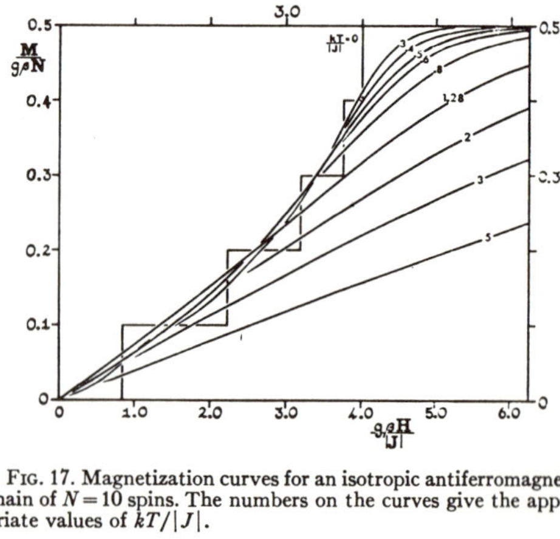

FIG. 17. Magnetization curves for an isotropic antiferromagnetic chain of $N=10$ spins. The numbers on the curves give the appropriate values of $kT/|J|$.

[38] For simplicity we describe only the case for N even.

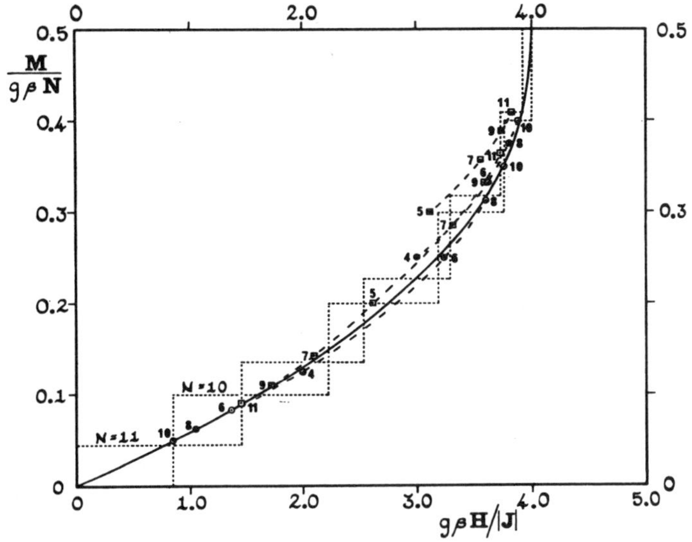

FIG. 18. Estimation of the limiting zero-temperature magnetization curve from the midpoints of the steps of the finite N curves [circles N even and squares N odd]. The zero-point step functions corresponding to $N=10$ and 11 are shown dotted.

implies that the limiting susceptibility for $\gamma=1$ can approach a finite *nonzero* value χ_0 at $T=0$.

To illustrate the detailed mechanism by which this occurs, we have sketched in Fig. 16 the relevant energy levels for (a) $\gamma<1$ and (b) $\gamma=1$ for finite systems[39] and for the corresponding infinite systems as a function of magnetic field. The susceptibility at $T=0$ is essentially the curvature of the ground state $E_0(H)$ since $\chi(H) = (\partial^2/\partial H^2)E_0(H)$. [Similarly the magnetization is just the slope of $E_0(H)$.] For finite N, the ground state (marked by a bold curve) consists of a series of straight lines and consequently (for N even) the curvature for small H is always zero, so that $\chi_N(0)\equiv 0$ for all γ. In the limit $N\rightarrow\infty$, however, the ground-state curve for $\gamma=1$ is smooth with continuous slope and definite (in general nonzero) curvature. When $\gamma<1$, on the other hand, the limiting curve consists of a horizontal straight line out to a finite field strength H_a at which the anisotropy gap vanishes. The curvature, and hence the susceptibility, near $H=0$ remains zero.

It is worth remarking that the existence of a nonzero χ_0 for isotropic coupling can be demonstrated rigorously for chains in the limiting case of infinite spin.[31]

To estimate roughly the value of the limiting zero-point susceptibility χ_0 for $S=\frac{1}{2}$, one may examine the trend of the means of $\chi_N(T)$ for $N=9$ and 10, and $N=10$ and 11 in the range $kT/|J|=0.4$ to 0.8 (see Fig. 14). These curves suggest that χ_0 lies between 0.045 and 0.06 (in units of $g^2\beta^2N/|J|$). Indeed, some time ago, Hulthén, by means of an approximate calculation, estimated that $\chi_0\simeq 0.0591g^2\beta^2N/|J|$.[40] In an attempt to improve this estimate, we examined the magnetization curve for the finite chains, which is also of interest in its own right.

Figure 17 shows the magnetization for a chain of $N=10$ spins as a function of field for different temperatures. For temperatures above $kT/|J|=0.3$ the curves are smooth and investigation of the convergence with N suggests that the limiting curves are well approximated. For lower temperatures the magnetization displays oscillations and approaches a step function at $T=0$, the discontinuities being $\Delta M/M_{\max}=2/N$. This behavior is, of course, just what follows from our previous discussion of the ground state when N is finite.

Despite the discontinuities, one notices that the midpoints of the vertical and horizontal parts of the steps lie near a smooth curve which presumably approximates the limiting ($N=\infty$) zero-temperature magnetization curve. This is confirmed by Fig. 18 where the midpoints for a number of the longer chains are plotted. The solid line shows the estimated limiting curve which should be accurate to within about 1% of the saturation value. This is supported by the analytic calculations of Griffiths.[41]

As mentioned previously, the magnetization at $T=0$ attains its saturation value at a finite critical field $H_s=4|J|/g\beta$. Below this critical field the magnetization appears to follow a square root law

$$M/M_{\max}=1-A[1-(H/H_s)]^{1/2} \qquad (4.5)$$

as $H\rightarrow H_s$ with $A\approx 1.2$–$1.3\approx 4/\pi=1.2732$. (For further discussion, see below.) This behavior is also a

[39] Accurate graphs of all the energy levels as a function of field are given in Ref. 10 for $N=8$.
[40] Reference 7, p. 78.

[41] R. B. Griffiths (private communication); Phys. Rev. 133 A768 (1964).

feature of the transverse ($J^{\parallel}=0$) model solved rigorously by Katsura.[29] In that case $A=2\sqrt{2}/\pi=0.9003$.

The slope of the limiting magnetization curve at $H=0$ is the zero-point susceptibility χ_0. It seems likely that the midpoints of the finite N magnetization steps approach the limiting curve as fast as $1/N$. Consequently, if the limiting magnetization $M(H)$ has a Taylor series expansion in H about $H=0$, one would expect the gradients $g_{1,N}=\frac{1}{2}\Delta M_{1,N}/\Delta H_{1,N}$ of the lines from the origin to the midpoints of the first steps (for N even), and similar gradients to other steps, to approach the limiting slope χ_0 as fast as $1/N$. Examination of the gradients $g_{j,N}$ for small j and $N=4$ to 11 [42] does yield roughly linear plots versus $1/N$ which, if extrapolated, suggest that $\chi_0/(g^2\beta^2N/|J|)$ lies in the range 0.0555 ± 0.0020. This is significantly lower than Hulthén's estimate of 0.0591.

This estimate rests on the assumption that the derivative of $M(H)$ does not vary too rapidly near $H=0$. Since this work was performed, however, Griffiths[41] has produced strong (though not quite rigorous) theoretical arguments based on an analysis of the "unbound" antiferromagnetic spin-wave states which show (a) that $M(H)$ is nonanalytic at $H=0$ ($\partial^2M/\partial H^2$ diverging sharply to $+\infty$ as $H\to0$) and, (b) that

$$\chi_0/(g^2\beta^2N/|J|)=0.050661\cdots\approx1/2\pi^2. \qquad (4.6)$$

In view of Walker's results[9] mentioned in the Introduction, which show that the point $\gamma=1$, $T=0$, $H=0$ is in some ways analogous to a phase transition point,[15] one should not really be surprised that $M(H)$ is nonanalytic at $H=0$. The divergence of $\partial^2M/\partial H^2$ shows why extrapolation of the gradients $g_{j,N}$ linearly with $1/N$ leads to an overestimate of χ_0. [Essentially $\chi(H)$ for small H, of order $1/N$, is well approximated by the $g_{j,N}$ but does not itself approximate χ_0 well unless H is exponentially small.] It seems very probable that $1/2\pi^2$ is the exact constant in (4.6) so that the estimate from the gradients is 10% high.

Accepting (4.6) as correct we may complete the estimation of the antiferromagnetic susceptibility for $\gamma=1$ down to zero temperature. The result is shown in Fig. 14 (as a dashed curve) and in Fig. 15. Above $kT/|J|=0.60$ it should be accurate to within 1% but in the region $kT/|J|=0.05$ to 0.50 the error might perhaps rise to 5%.

The zero-temperature magnetization curve for other values of γ may be studied as for $\gamma=1$. The critical field H_s is determined by the intersection of the energy level for total $S^z=\frac{1}{2}N$ (a component of the zero-field ferromagnetic ground state) with the lowest level for $S^z=\frac{1}{2}N-1$ (single ferromagnetic spin wave)[39] both of which are known exactly. Thus,

$$H_s(\gamma)=2(1+\gamma)|J|/g\beta. \qquad (4.7)$$

[42] The gradients from the origin to the midpoints of the vertical parts of the steps are given by

$$g_{j,N}=(j-\tfrac{1}{2})\Delta M_{j,N}/\Delta H_{j,N} \quad \text{for } N \text{ even and}$$
$$g_{j,N}=j\Delta M_{j,N}/\Delta H_{j,N} \quad \text{for } N \text{ odd.}$$

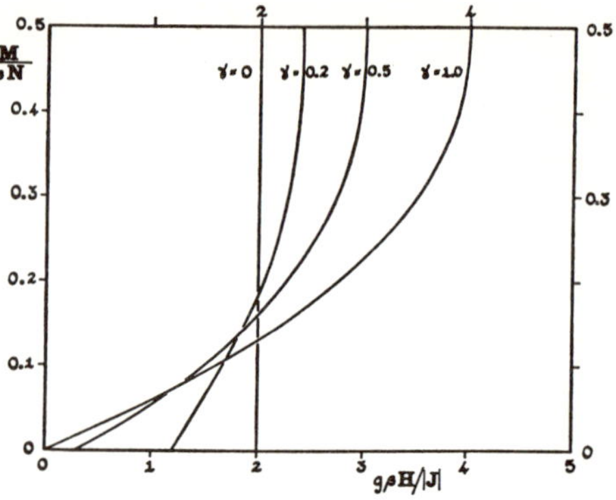

FIG. 19. Dependence of zero-temperature magnetization curves on anisotropy.

In order to study the form of the magnetization curve near saturation, it is necessary to consider the width of the first magnetization step away from saturation (see Fig. 18). The position $H=H_1$ of this step is given by the intersection of the lowest level for $S^z=\frac{1}{2}N-1$ with the lowest level for $S^z=\frac{1}{2}N-2$ which in the limiting case, $N\to\infty$, is given by $-4|J|(1+\gamma)$ [Eq. (5.4), Sec. 5, with $k=2\pi$]. It is seen, however, that use of the limiting curve again yields H_s; it appears, therefore, that the width of this step is directly determined by the energy discrepancy ΔE_1 between the lowest lying level for $S^z=\frac{1}{2}N-2$ (finite N) and its limiting value. An estimate of the lowest, finite N, energy level as a function of γ may be obtained, for example, by applying second-order perturbation theory to the appropriate submatrices of the Hamiltonian (1.1). One finds that for large N, $\Delta E_1\simeq2\gamma\pi^2|J|/N^2$. Corresponding to a magnetization step of $\Delta M=g\beta$ or $\Delta M/M_{\max}=2/N$, we thus find a magnetic field step $\Delta H_1=\frac{1}{2}(H_s-H_1)=\Delta E_1/2g\beta\simeq\gamma\pi^2|J|/g\beta N^2$. On eliminating N we obtain

$$\frac{\Delta M}{M_{\max}}\approx\left[\frac{8(1+\gamma)}{\pi^2\gamma}\right]^{1/2}\left(\frac{\Delta H_1}{H_s}\right)^{1/2},$$

which suggests that when $N\to\infty$ we may write more generally, as $H\to H_s$,

$$M/M_{\max}=1-A(\gamma)[1-(H/H_s)]^{1/2} \qquad (4.8)$$

with

$$A(\gamma)=(4/\pi)[(1+\gamma)/2\gamma]^{1/2}.$$

This argument is, of course, not rigorous but it should be at least qualitatively correct and its accuracy is supported by the results for finite N and by agreement with Griffiths' analysis at $\gamma=1$.[42]

Figure 19 shows the complete estimated limiting magnetization curve for the Ising limit $\gamma=0$, where it is a simple step function, and for intermediate values of γ.

When $\gamma<1$ we must, as mentioned, distinguish between a *perpendicular* and a *parallel* susceptibility.

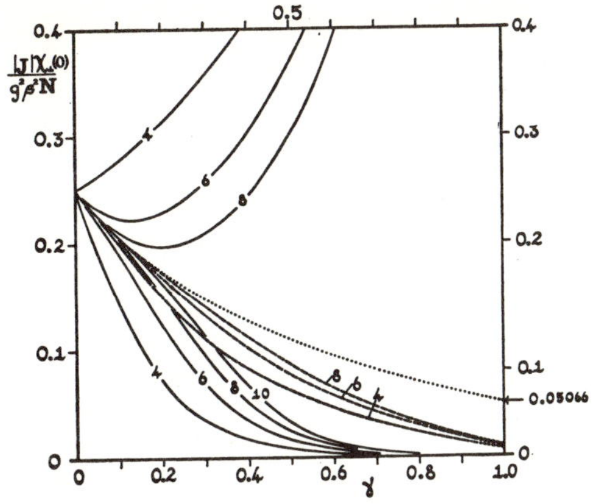

FIG. 20. Perpendicular susceptibilities for the ground and first $S^z=0$ excited states for finite rings (solid curves) and the geometric means (broken curves). The dotted curve represents the probable limit for $N\to\infty$.

For the Ising limit we have the exact result[43,44]

$$\chi_\perp(T)=\frac{g^2\beta^2N}{4|J|}[\tanh K+K\,\mathrm{sech}^2K]\quad(\gamma=0),\quad(4.9)$$

where K is defined in (4.4). From this we see that $\chi_\perp(0)$ for $\gamma=0$, is nonzero. In the limit $N\to\infty$ this will remain true for all γ and in fact as $\gamma\to1$, $\chi_\perp(0,\gamma)\to\chi_0$. For this reason the susceptibility in the isotropic case is perhaps best regarded as a ***perpendicular*** susceptibility.

For finite N the situation is again more complicated. The lowest set of solid curves in Fig. 20 show $\chi_{\perp N}(0,\gamma)$ for rings of $N=4$, 6, 8, and 10 spins. (The derivation requires the numerical calculation of matrix elements with the ground-state eigenvector.) As $\gamma\to1$, for N even and finite, $\chi_{\perp N}(0,\gamma)$ approaches zero as the previous arguments have shown it must. (For odd N it diverges at $\gamma=1$.) Furthermore, for $\gamma>0.2$ the convergence for N increasing is evidently very slow. The upper solid curves in Fig. 20 represent the perpendicular susceptibilities calculated for the lowest excited states. In the limit $N\to\infty$ these are expected to approach the ground-state results. The broken curves are the geometric means of these two sets of values and they seem to be converging somewhat more rapidly, at least for $\gamma<0.5$. Accepting the value (4.6) for χ_0, the limiting zero-point perpendicular susceptibility $\chi_\perp(0,\gamma)$ must resemble in general form the dotted curve in Fig. 20. The slope of the limiting curve at $\gamma=0$ is known exactly,[43] but for $0.2<\gamma<0.5$ the dotted curve is probably accurate only to within 10%.

Ferromagnetic Coupling

The susceptibility for ferromagnetic chains is conveniently discussed in terms of the deviations from Curie's law. Quite generally we can write for the parallel susceptibility

$$\chi_N(T)=[g^2\beta^2N/4kT]\xi_N(T),\quad(4.10)$$

where

$$\xi_N(T)=(4/N)\langle[\sum_{i=1}^{N}S_i^z]^2\rangle,\quad(4.11)$$

the angular brackets denoting the canonical average. At high temperatures $\xi_N(T)$ approaches unity as $1/T$. As the temperature falls $\xi_N(T)$ rises monotonically (for $J>0$) and, for finite N, levels off at a value determined by the properties of the ground state. For anisotropic chains ($\gamma<1$) the ferromagnetic ground state is twofold degenerate with $\sum S_i^z=\pm\frac{1}{2}N$ so that $\xi_N(0)=N$. In the isotropic ($\gamma=1$) case, on the other hand, the ground state has total spin $S=\frac{1}{2}N$ and hence is $(N+1)$-fold degenerate, $\sum S_i^z$ taking the values $\frac{1}{2}N$, $\frac{1}{2}N-1$, $\cdots-\frac{1}{2}N+1$, $-\frac{1}{2}N$. For finite pure Heisenberg chains, therefore, $\xi_N(T)$ rises to a maximum value $\xi_N(0)=\frac{1}{3}(N+2)$.

In the limit $N\to\infty$ we see that for all γ, $\xi(T)$ diverges as $T\to0$. For Ising chains ($\gamma=0$) this divergence is exponentially fast since we have rigorously

$$\xi(T)=\exp[J/kT],\quad(\gamma=0).\quad(4.12)$$

This rapid divergence finds its origin in the anisotropy gap which ensures that all the pair correlation functions $\langle S_i^zS_j^z\rangle$ approach their zero-point values exponentially fast. For this reason we expect the limiting divergence for other values of $\gamma<1$ will also be of the form $\exp[\alpha J/kT]$ with α depending on the limiting anisotropy gap.

For pure Heisenberg coupling, however, a power law might be expected. Indeed, in the limit $S=\infty$ ($\gamma=1$), the susceptibility diverges as $1/T^2$ so that $\xi(T)\sim1/T$ as $T\to0$.[31] Fig. 21 is a log-log plot of $[\xi_N(T)-1]$ versus the temperature for finite isotropic chains of $N=3$, 4, $\cdots$, 10, 11 spins. The curves evidently approach the

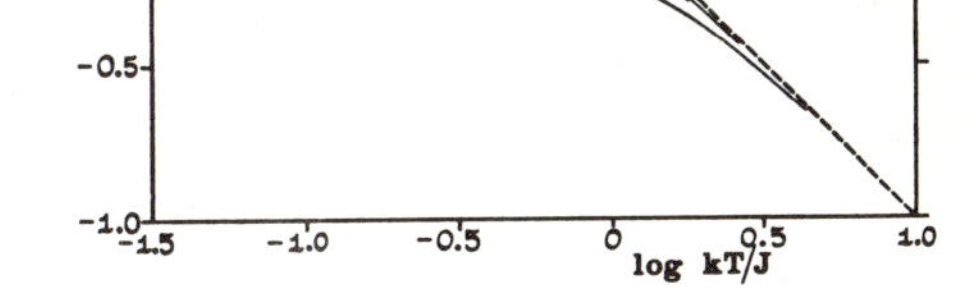

FIG. 21. Log-log plot of the reduced ferromagnetic susceptibility $\xi(T)-1$ for finite isotropic chains.

[43] M. E. Fisher, Physica 26, 618 (1960).
[44] M. E. Fisher, J. Math. Phys. 4, 124 (1963), and Ref. 29.

limit monotonically from below. The convergence is quite rapid and for $kT/J>0.3$ the limiting curve is indicated quite accurately. At high temperatures the log-log plot becomes linear with slope unity as shown by the broken line in Fig. 21. This simply confirms the $1/T$ deviations from Curie's law. At lower temperatures in the range $kT/J=0.25$–1.0, the limiting curve is again almost linear but with slope close to $\frac{4}{5}$ (see dotted line in Fig. 21). If the curves for larger N continued this trend it would imply a divergence of $\xi(T)$ like $1/T^{4/5}$ and of $\chi(T)$ like $1/T^{9/5}$ as $T\rightarrow 0$. It is quite possible, however, that the true asymptotic behavior sets in only below $kT/J=0.2$. Nevertheless, the qualitative behavior is clearly rather similar to that for $S=\infty$.

The nature of the variation of the susceptibility with magnetic field can be seen in Fig. 22 which shows the ferromagnetic susceptibility of isotropic chains of $N=8$ spins in various fields. The convergence with N is appreciably more rapid in a field than for $H=0$.

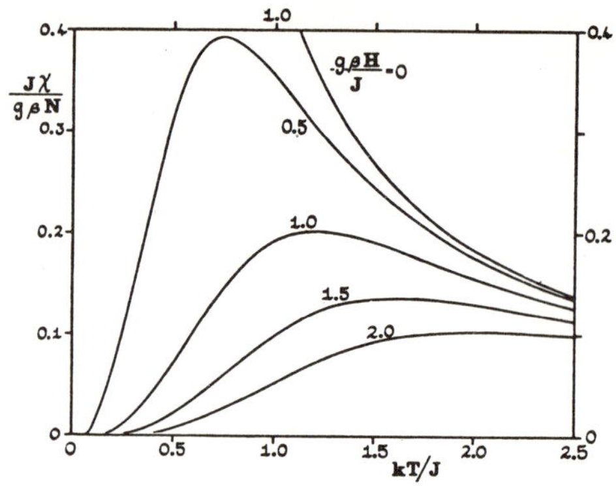

FIG. 22. Ferromagnetic susceptibility for rings of $N=8$ spins with isotropic coupling for different magnetic fields.

5. SPECTRUM

In the following account we confine our attention to a discussion of the spectrum of energy levels in the absence of an applied field, i.e., the eigenvalues of the Hamiltonian (1.1) with $H=0$.

Ferromagnetic Coupling

The ferromagnetic ground state for all γ is, of course, exactly $E_0=-\frac{1}{2}NJ$ corresponding to $\sum S_i^z=\frac{1}{2}N$ (all spins aligned). With one "overturned spin" we have a single "spin wave" of energy

$$E-E_0=\epsilon(k)=2J[1-\gamma\cos k], \tag{5.1}$$

where

$$k=2\pi r/N, \quad r=0, \pm 1, \pm 2, \cdots.$$

With two overturned spins, the simple spin-wave approximation predicts

$$E-E_0=\epsilon(k_1)+\epsilon(k_2), \tag{5.2}$$

but this is not exact owing to the spin-wave "interactions." For anisotropic linear chains the interaction of

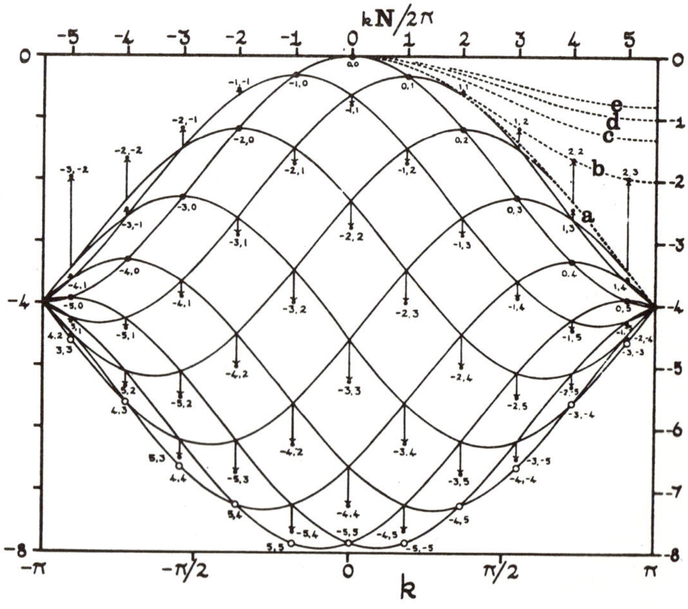

FIG. 23. Energy levels for two overturned spins (interacting spin waves) for a ring of $N=11$ spins and $\gamma=1$. The arrows indicate the deviations from the levels for two independent spin waves. Note that we are taking an "antiferromagnetic view" of the energy levels ($J=-|J|$), and each level has been normalized by subtraction of an energy $-\frac{1}{2}NJ$.

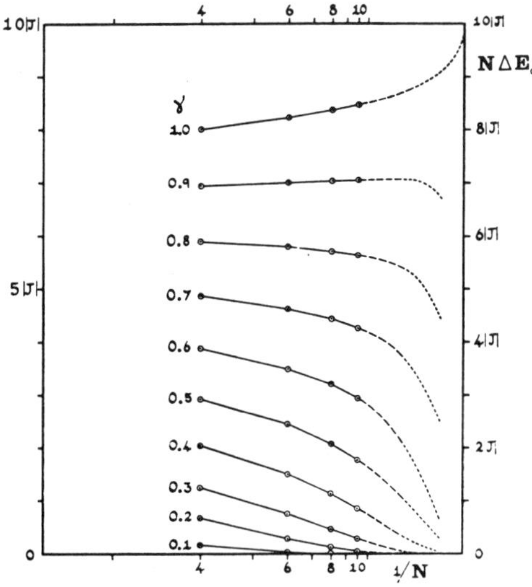

FIG. 24. Dependence of the splitting ΔE_0 of the antiferromagnetic ground state on γ and N.

two spin waves has been analyzed theoretically by Orbach,[8] who followed Bethe's analysis for $\gamma=1$.[6]

A graphic understanding of these results can be obtained from Fig. 23, which shows energy levels for a ring of $N=11$ spins for $\gamma=1$. The solid curve through the dark circles is the single spin-wave dispersion law (5.1). The other solid curves represent the superimpositions (5.2), and their intersections are thus the spin-wave approximations to the levels for two overturned spins. The arrows indicate the energy shifts due to the interactions, the exact energy levels lying at the points of the arrowheads.

The open circles represent "missing states," i.e., levels predicted by (5.2) which do not occur for the true Hamiltonian (essentially because two spins cannot be overturned at the same site). For most combinations of k_1 and k_2 the interactions are evidently repulsive and the spin waves "scatter" and remain "unbound" (category I states[8]). However, for $k_1 \simeq k_2 \simeq \frac{1}{2}k$, the interactions are strongly attractive and the spin waves condense into a bound state (category II) given by,[8]

$$\epsilon_{II}(k)=2J[1-\tfrac{1}{2}\gamma^2(1+\cos k)]. \tag{5.3}$$

This is indicated by the dotted line (b) in Fig. 23. As the form of this dispersion relation suggests, the corresponding eigenfunction relates to two closely associated overturned spins traveling around the ring together. The dotted line (a) in Fig. 23 is the $N=\infty$ limit of the energies of the unbound states namely,[8]

$$\epsilon_I^{\min}(k)=4J(1-\gamma\cos\tfrac{1}{2}k), \tag{5.4}$$

which is just the envelope of the curves (5.2).

One may in a similar way study numerically the bound states for r overturned spins. The lowest states seem to follow the relation

$$\epsilon^{(r)}(k)\simeq(2J/r)[1-\cos k] \tag{5.5}$$

when $\gamma=1$, as illustrated in curves (c), (d), and (e) of Fig. 23, corresponding to 3, 4, and 5 overturned spins, respectively. For $\gamma\neq 1$ the amplitudes of the $\cos k$ term appear to decrease as γ^r or more rapidly.

Antiferromagnetic Coupling

The approach of the antiferromagnetic ground-state energies for finite chains to the limiting value was discussed in Sec. 2. The question of the degeneracy of the ground state has not, however, been considered. When N is even[45] the antiferromagnetic ground state at the Ising limit ($\gamma=0$) is twofold degenerate. This degeneracy, however, is split by the transverse terms in the Hamiltonian, although, as is well known, the splitting only arises in Nth-order perturbation theory. This suggests that in the limit $N\to\infty$, the ground state should again become degenerate, at least in some sense.

To investigate this point we may examine the ground-state splitting $\Delta E_0(N)$ as a function of N. If ΔE_0 varies as $1/N$ for N large, the levels close up, but only at a rate characteristic of a continuum of levels in the limit of $N=\infty$. Conversely, if ΔE_0 vanishes more rapidly than $1/N$, e.g., as $1/N^2$, then we may consider the levels as asymptotically degenerate even if they lie within a continuum. In Fig. 24, the product $N\Delta E_0(N)$ has been plotted versus $1/N$ for $N=4$, 6, 8, and 10, and for values of γ in the range 0 to 1. For $\gamma\leqslant 0.5$, $N\Delta E_0$ is rapidly decreasing, and there seems little doubt that the limit is zero. For $\gamma=0.6$, 0.7, and 0.8, the decrease is slower but the rate increases for larger N, and it seems probable that the limit is again zero, as suggested by the broken lines. (These lines are purely suggestive and are not to be taken as numerical extrapolations.) At $\gamma=0.9$ the values of $N\Delta E_0$ at first increase slightly with N, but for $N=8$ and 10 they are almost equal, and we believe that larger values of N would again yield a product decreasing at first slowly, but eventually rapidly. At the limit $\gamma=1$, on the other hand, $N\Delta E_0$ seems to be rising steadily and approaching a definite limit at $1/N=0$, consistent with a state lying in a continuum bounded by the ground state. (Consideration of the des Cloizeaux-Griffiths analysis[26,41] suggests that this limit should be $\pi^2|J|$).

In summary, we feel that the evidence of Fig. 24 definitely suggests that for all $\gamma<1$ (i.e., anisotropic coupling) ΔE_0 decreases more rapidly than $1/N$ ($N\Delta E_0\to 0$), so that the limiting ground state may be said to be (twofold) degenerate. In the isotropic limit, however, ΔE_0 decreases only as $1/N$ and the anti-

[45] When N is odd, the Ising ground state is $2N$-fold degenerate, but for $\gamma>0$, only a fourfold degeneracy corresponding to $\Sigma S_i^z=\pm\frac{1}{2}$ remains. We will not consider this case further.

ferromagnetic ground state should be regarded as nondegenerate.

For an Ising chain the first excited states lie at an energy $\Delta E_A(0)=2|J|$ above the ground states, but at the Heisenberg limit the ground state is the limit of a continuum, and there is no anisotropy gap. (This may be seen explicitly from the theoretical calculation of the antiferromagnetic spin-wave states for $\gamma=1$ by des Cloizeaux and Pearson.[26]) To estimate the limiting anisotropy gap $\Delta E_A(\gamma)$ for intermediate values of γ, we have studied (a) the difference $\Delta E_A(\gamma,N)$ between the "lowest excited state" $E_s(\gamma,N)$ [actually two degenerate ($S^z=\pm1$) states which must be distinguished from the higher component of the split ground state] and the finite N ground state $E_0(\gamma,N)$, and (b) the gap $\Delta E_A'(\gamma,N)$ between $E_s(\gamma,N)$ and the *limiting* ground state $N|J|\mathcal{E}_0(\gamma,\infty)$. At the Heisenberg limit $E_s(\gamma,N)$ becomes a continuum state and the fact that $\Delta E_A(1,N)$ and $\Delta E_A'(1,N)$ fall to zero like $1/N$ may be checked from the values for small finite N. For other values of γ also, linear extrapolation with $1/N$ seems to be appropriate in estimating the limit. The two sets of values form two distinct sequences but, as one would expect, they tend to the same limiting value. In the case $\gamma=1$, the lowest excited state $E_s(1,N)$ becomes degenerate with the continuum state ($S^z=0$) split off the degenerate ($\gamma=0$) ground state, and hence $N\Delta E_A(1,N)=N\Delta E_0$ tends to a limit [which, as mentioned, is probably $\pi^2|J|$ although $N\Delta E_A'(1,N)$ appears to tend to a slightly different value]. Figure 25 shows the variation with γ of the gap for finite N and the estimated limit $\Delta E_A(\gamma)$, which is probably accurate to within $\pm0.05J$. The main feature of interest is the very slow increase in the gap as $(1-\gamma)$ increases from zero to 0.3 or 0.4.

The problem of the antiferromagnetic spin-wave spectrum has recently been studied in detail by des Cloizeaux and Pearson,[26] who computed numerically the lowest category I states for $\gamma=1$ and rings of $N=6$, 8, 16, and 48 spins. They showed analytically that the limiting dispersion law is

$$\epsilon(k)=E^{(1)}(k)-E_0=|J|\pi|\sin k|. \tag{5.6}$$

Extrapolation of the appropriate energy differences versus $1/N$ for N up to 11 yields the amplitude $|J|\pi$ to within 5 to 10%, which confirms the validity of extrapolating the spectral properties.

Mattheiss[25] has given a plot of all the energy levels versus k for $N=6$ and 8. The results for higher values of N are more complex since there are more levels but they do not differ much qualitatively and so we do not present a figure. Mattheiss remarked that for even $N=2m$, the antiferromagnetic ground state corresponded to $k=0$ when m was even, but to $k=\pi$ when m was odd. For odd $N=2m+1$, we have found that the two degenerate ground states are at $k\simeq\pm\frac{1}{2}\pi$, and the state at $k=0$ or π is the higher according as m is odd or even.[46] As Mattheiss says, there seems to be no direct physical interpretation of these facts.

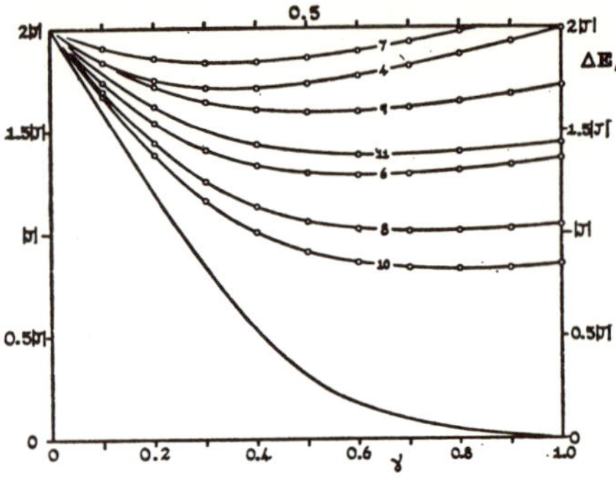

FIG. 25. Antiferromagnetic anisotropy gap $\Delta E_A(\gamma,N)$ and the estimated limit $\Delta E_A(\gamma)$.

It would be of interest to decide how well the higher antiferromagnetic states can be represented as superpositions of spin waves obeying (5.6). The situation seems appreciably more complex than for the ferromagnetic chains, and although the exact spectra do display levels that lie roughly on curves $|J|\pi(|\sin k_1|+|\sin k_2|)$ with $k=k_1+k_2$, we have not been able to discover any simple numerical correlations, and some low-lying states (presumably "bound states") are definitely not representable in this way.

6. ANTIFERROMAGNETIC CORRELATIONS AND ORDER

For an antiferromagnetic chain at zero temperature the pair-correlation functions

$$\omega_l(N)=4\langle S_0^zS_l^z\rangle=(4/N)\sum_{j=1}^{N}\langle S_j^zS_{j+l}^z\rangle \tag{6.1}$$

can be calculated from the detailed expansion of the ground-state wave function in the basis of functions of definite "up" and "down" spins. Numerical results for $N=4$, 6, 8 and 10, and $\gamma=0.1, 0.3, \cdots, 1.0$ are given in Table II. For fixed l the convergence in the case of N even is not regular: for $\gamma<0.3$ it is monotone increasing and for $\gamma>0.8$ monotone decreasing, the curves for different N crossing in the intermediate region. However, the convergence is quite rapid for $\gamma<0.3$ for all l, and for $l<\frac{1}{4}N$ in the case of $\gamma>0.8$. In the case of N odd the convergence for $l<\frac{1}{2}N$ is monotone increasing for all γ but rather slow. (These values are not tabulated.) The case of $l=\frac{1}{2}N$ is of special interest and will be discussed later.

Despite the relatively erratic convergence, extrapolation procedures for the limit $N\to\infty$ may be attempted and their accuracy checked against Orbach's exact calculation[8] of the limiting zero-temperature short-

[46] More precisely, the ground states are at $k=\pm\pi m/N$ for m even and $k=\pm\pi(m+1)/N$ for m odd.

TABLE II. Antiferromagnetic ground state correlation functions $\omega_l(N,\gamma)$. [Note that $\omega_0 \equiv 1$.]

N	l	$\gamma=0.1$	$\gamma=0.3$	$\gamma=0.5$	$\gamma=0.7$	$\gamma=0.9$	$\gamma=1.0$
10	1	−0.99001	−0.90791	−0.77658	−0.67994	−0.62221	−0.60206
	2	0.98005	0.81790	0.56583	0.38839	0.28782	0.25407
	3	−0.97998	−0.81203	−0.54351	−0.35953	−0.26199	−0.23117
	4	0.97988	0.80441	0.51181	0.31041	0.20558	0.17307
	5	−0.97989	−0.80475	−0.51510	−0.31865	−0.21839	−0.18781
8	1	−0.98972	−0.90139	−0.77381	−0.68355	−0.62819	−0.60852
	2	0.97948	0.80498	0.55947	0.39250	0.29464	0.26104
	3	−0.97936	−0.79910	−0.54364	−0.37551	−0.28238	−0.25194
	4	0.97921	0.79102	0.51598	0.33311	0.23186	0.19883
6	1	−0.98821	−0.89102	−0.77359	−0.69291	−0.64155	−0.62284
	2	0.97645	0.78410	0.55640	0.40440	0.31068	0.27735
	3	−0.97648	−0.78616	−0.56561	−0.42298	−0.33826	−0.30902
4	1	−0.98113	−0.88125	−0.78868	−0.72542	−0.68282	−0.66667
	2	0.96225	0.76249	0.57735	0.45084	0.36564	0.33333

range order $\omega_1(\gamma)$. In Fig. 26 this exact result (a) is compared with an extrapolation estimate (b) based on sequences of means, 10/9, 9/8, 8/7, etc., which show monotone increasing behavior for all γ. As may be seen, the agreement is very good (to within 1%) and a similar technique has therefore been applied to estimate the limiting curve for the cases $|\omega_2(\gamma)|$, $|\omega_3(\gamma)|$, and $|\omega_4(\gamma)|$ (see Fig. 27). However, as the value of l successively increases, the accuracy is expected to fall off somewhat, since there are fewer points to extrapolate.

Also shown in Fig. 26 are sums of Walker's perturbation series[9]

$$\omega_1 = 1-\gamma^2+\tfrac{3}{4}\gamma^4-(7/2^6)\gamma^8-(9/2^7)\gamma^{10} -(11/2^9)\gamma^{12}+(13/2^{11})\gamma^{14}+\cdots \quad (6.2)$$

truncated at γ^{14} [curve (c)] and at γ^6 [curve (d)]. In the former case the error is detectable only for $\gamma>0.85$ and reaches a maximum of only 6% at $\gamma=1$. In the latter case, however, the deviations are significant for $\gamma\gtrsim 0.7$. Curve (e) is derived from Davis's perturbation expansion[17] and is evidently less accurate than the extrapolations once $\gamma>0.40$ and seriously in error for $\gamma>0.85$.

An appreciation of the decay of the correlations with distance l (as well as of the convergence in N) can be gained from Fig. 28 which shows the finite ring correlations for $\gamma=0.3$, 0.5, and 1.0. An alternating effect is evident, the values of $|\omega_l|$ for even l being lower relative to those for odd l than might be expected. For $l>\tfrac{1}{2}N$ the correlations, of course, start increasing as the points 0 and l approach one another around the closed ring. Nevertheless, it is clear for $\gamma=0.3$, and reasonably so for $\gamma=0.5$, that the correlations for $l<\tfrac{1}{2}N$ are decaying to a constant level of about 0.8 and 0.5, respectively. These values may be identified with the long-range order $\omega_\infty(\gamma)$ defined by

$$\omega_\infty(\gamma)=\lim_{l\to\infty}\left|\lim_{N\to\infty}\omega_l(\gamma,N)\right|. \quad (6.3)$$

To estimate $\omega_\infty(\gamma)$ we have formed the minimum

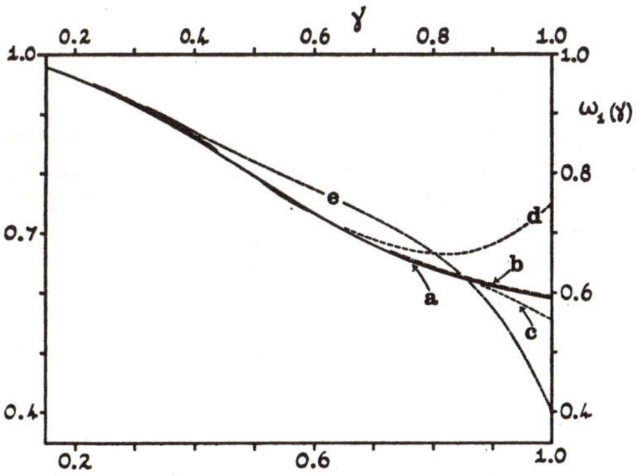

FIG. 26. Comparison of (a) the exact value of $\omega_1(\gamma)$ with (b) present extrapolations, (c) Walker's perturbation series to γ^{14}, (d) to γ^6 and (e) Davis's perturbation series.

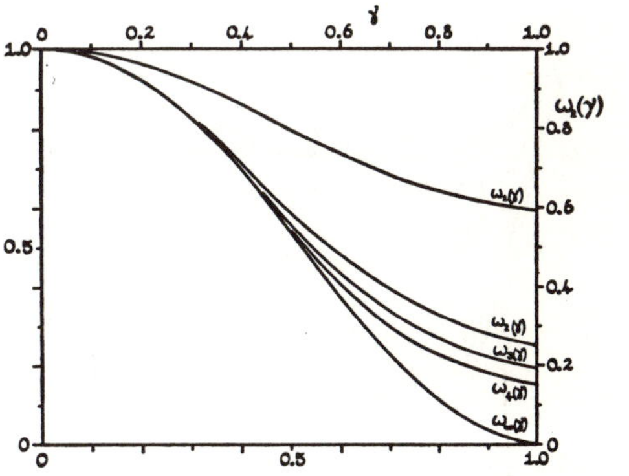

FIG. 27. Variation of the pair correlations $|\omega_l|$ with γ for $l=1$ (exact) and $l=2$, 3, 4 (estimated) and $l=\infty$, the final estimate of the long-range order.

means for N even

$$\omega_{\min}(\gamma,N)=\tfrac{1}{2}|[\omega_{\frac{1}{2}N}(\gamma,N)-\omega_{\frac{1}{2}N-1}(\gamma,N)]| \quad (6.4)$$

and attempted to extrapolate to the limit $N=\infty$, which should agree with (6.3). These means are taken to reduce the alternation effects.

In this case we felt it was not advisable to use the results for odd N in forming the estimate because, with the above definition of long-range order, interference effects around the chain occur near the Ising limit. (Note that for N odd and $\gamma=0$, $\omega_l(N)=1-2l/N$, $0\leqslant l<\frac{1}{2}N$.) For this reason we do not expect the limiting result for ω_∞ to have accuracy comparable with our estimate for ω_1. (An estimate for ω_1 using results for even N only gives a result accurate to about 2.5%.) Figure 29 shows a plot of $\omega_{\min}(N)$ versus $1/N$. The points for $N>4$ appear reasonably collinear but care must be taken in extrapolation since curvature is to be expected for $\gamma<0.7$. (A similar, though less marked, effect occurs in the case of the short-range order.) For $\gamma<0.3$, the over-all range of variation is slight and an estimate can be made with some confidence: The limiting values here must, in fact, lie very close to the values for $N=10$. For $\gamma>0.7$ a linear extrapolation would seem to be reliable. For intermediate γ some attempt has been made to allow for curvature effects but we would not claim very high accuracy for our results in this range. In fact, for $0.4<\gamma<0.6$, we have probably underestimated by several percent. We feel justified in concluding, however, that the long-range order vanishes only at the Heisenberg limit $\gamma=1$. In view of our conclusions regarding the asymptotic degeneracy of the ground state (and the vanishing of the anisotropy gap at $\gamma=1$) this is perhaps not unexpected.

To compare our estimate of the long-range order with other approximations we must recognize that most authors have used as an order parameter the so-called

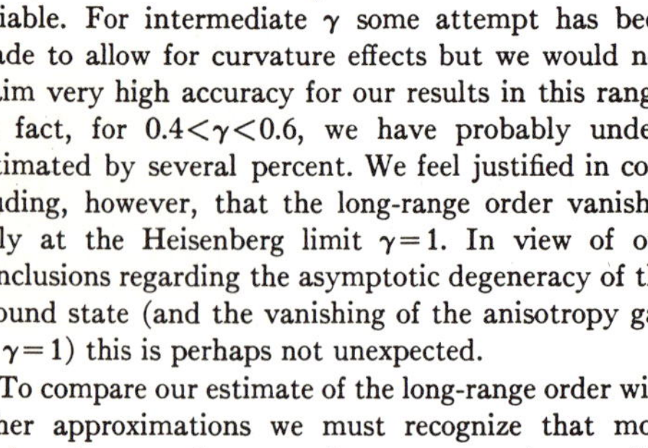

FIG. 28. Variation of $|\omega_l(N)|$ with l for $\gamma=0.3$, 0.5, and 1.0. and rings of $N=4$, 6, 8, and 10 spins.

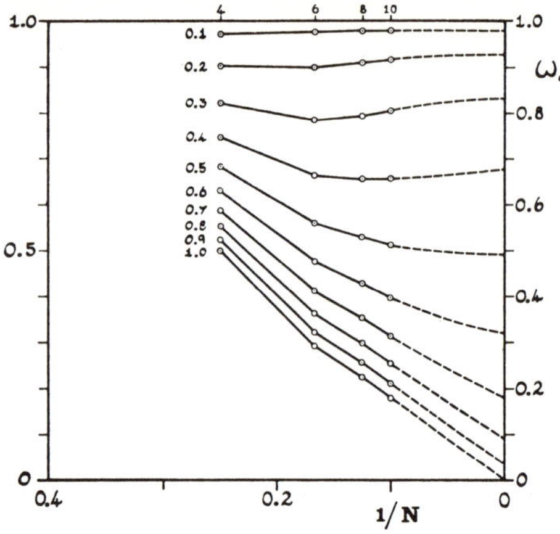

FIG. 29. Plot of $\omega_{\min}(\gamma,N)$ versus $1/N$. Extrapolation yields an estimate of the long-range order.

"sublattice magnetization"[9,14,17,47]

$$\sigma=(2/\tfrac{1}{2}N)\langle\sum_{j\text{ even}} S_j^z\rangle=2\langle S_0^z\rangle, \quad (6.5)$$

where the angular brackets denote the canonical average. If this formula is interpreted literally, it is easily shown (on the grounds of spin reversal symmetry) that σ always vanishes identically. Although this seems to have given rise to some confusion in the literature,[20,48,49] the situation really parallels that in the ferromagnetic case which is quite well understood.[50] If the (reduced) spontaneous magnetization of a ferromagnet is defined simply by $M_0=(2/N)\langle\sum_i S_i^z\rangle$, it also vanishes identically. The correct definition is made with the aid of a nonzero magnetic field, namely,

$$M_0(T)=\lim_{H\to0+}\lim_{N\to\infty}(1/N)M_N(T,H), \quad (6.6)$$

where

$$M_N(T,H)=2\sum_{j=1}^{N}\langle S_j^z\rangle_N=2N\langle S_0^z\rangle_N. \quad (6.7)$$

By considering the finiteness of the susceptibility per spin in the limit $H\to0$ [which implies the convergence of the sum $\sum_l(\langle S_0^zS_l^z\rangle-\langle S_0^z\rangle^2)$] one can then see that

$$M_0^2(T)=\lim_{N\to\infty}(1/N)^2\sum_{i,j=1}^{N}\langle S_i^zS_j^z\rangle,$$
$$=\lim_{l\to\infty}\lim_{N\to\infty}\langle S_0^zS_l^z\rangle=\omega_\infty(T), \quad (6.8)$$

[47] W. Marshall, Proc. Roy. Soc. (London) A232, 48 (1955).
[48] G. W. Pratt, Jr., Phys. Rev. 122, 489 (1961).
[49] N. Karayianis, C. A. Morrison, and D. E. Wortman, Phys. Rev. 126, 1443 (1962).
[50] See, for example, the discussion of the Ising model by G. F. Newell and E. W. Montroll, Rev. Mod. Phys. 25, 353 (1953).

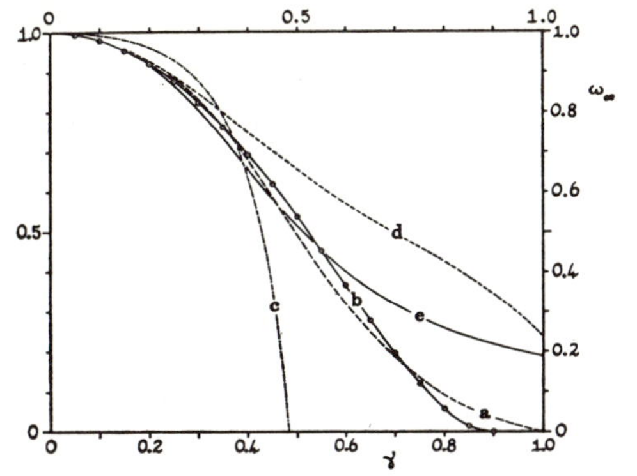

FIG. 30. Comparison of different approximations for the long-range order $\omega_\infty(\gamma)$: (a) extrapolation estimates, (b) Walker's perturbation series to γ^6, (c) Kasteleyn's variational formula, (d) Davis's perturbation series, (e) $\omega_{\min}$ for a finite chain of 10 spins.

i.e., the long-range order is just equal to the square of the spontaneous magnetization.

For an antiferromagnetic one must introduce a "staggered" magnetic field H^* by adding to the Hamiltonian a term

$$\mathcal{H}^* = H^* \sum_{j=0}^{N-1} (-)^j S_j^z. \tag{6.9}$$

With the aid of the corresponding "staggered magnetization" the sublattice magnetization may be defined properly by

$$\begin{aligned} \sigma &= \lim_{H^*\to 0+} \lim_{N\to\infty} (1/N) M_N^*(H^*), \\ &= \lim_{H^*\to 0+} [\langle S_0^z\rangle_\infty^* - \langle S_1^z\rangle_\infty^*]. \end{aligned} \tag{6.10}$$

In a similar way one can then conclude that the long-range order defined by (6.3) is related to σ by[51]

$$\omega_\infty = \sigma^2. \tag{6.11}$$

We thus see that the two-order parameters are equivalent. One may, of course, avoid the somewhat artificial introduction of a staggered field by sticking to the definition (6.3) in terms of the correlation functions which, as argued by Ruijgrok and Rodriguez,[20] also has other features to recommend it. In practice, however, it is often easier to compute σ (or M_0) than to compute ω_∞ directly.

With these preliminaries we may compare our estimate of $\omega_\infty(\gamma)$ [curve (a) in Fig. 30] with Walker's exact perturbation expansion

$$\sigma(\gamma) = 1 - \gamma^2 - \tfrac{1}{4}\gamma^4 - \tfrac{1}{16}\gamma^6 + \cdots \tag{6.12}$$

[curve (b)],[9] with Kasteleyn's variational formula [curve (c)],[14] and with Davis's perturbation formula [curve (d)].[17] Also shown in Fig. 30 are the values $\omega_{\min}(10)$ [curve (e)]. Up to $\gamma = 0.3$, all approximations agree well with the exact series but at this point Davis's approximation (d) starts deviating seriously and predicts a relatively large nonzero value for $\omega_\infty(1)$. Kasteleyn's approximation (c) falls away sharply to zero at $\gamma_c = 0.483$, which is surely incorrect. Our estimated curve falls some 10% below the series value in the region $\gamma = 0.45$ to 0.65, although the trend is very similar. In this region the series is probably still converging rapidly, as suggested both by the numerical magnitude of the terms, and by the agreement of the exact short-range order with the corresponding series up to γ^6 (Fig. 26). Above $\gamma = 0.7$ we must expect the series to deviate from the true value and indeed the series for $\omega(\gamma)$ (rather than for $[\sigma(\gamma)]^2$) then yields lower values and has its zero at $\gamma = 0.817$ rather than at $\gamma = 0.897$. As already observed our extrapolated $\omega_\infty(\gamma)$ does not vanish until $\gamma = 1$. Although we suspect this is the true situation it would be of great interest to have more precise information—ideally the rigorous answer—in the range above $\gamma = 0.8$! At present it seems fair to conclude that the exact series expansion provides the best approximation up to $\gamma = 0.75$ and that the true long-range order is unlikely to vanish for $\gamma < 0.85$ and probably vanishes at $\gamma = 1$.

In Fig. 27, therefore, we have plotted together with $\omega_1(\gamma)$ [exact] and our results for $|\omega_2|$, $|\omega_3|$, and $|\omega_4|$, what we think is the best estimate so far for $\omega_\infty(\gamma)$. This curve follows the series expansion up to $\gamma = 0.6$, and thereafter has the same form as our extrapolation curve (a) in Fig. 30, approaching the latter from above and in close agreement for $\gamma > 0.8$.

The excellent convergence of Walker's perturbation series in γ suggests that this approach should be developed further for two- and three-dimensional lattices where convergence seems to be even better. We remark in passing that it is possible to calculate exactly the γ^2 correction to the energy and specific heat for all temperatures in the case of two-dimensional lattices for which the Ising problem has been solved, by expressing the perturbation in terms of the multiple correlation functions for $\gamma = 0$.[52]

ACKNOWLEDGMENTS

Weare grateful for the interest of Professor C. Domb and Dr. M. F. Sykes and for the assistance of members of the University of London Computer Unit. One of us (J.C.B.) is indebted to the Department of Scientific and Industrial Research for the award of a maintenance grant. As mentioned in the Introduction, we have greatly benefited from correspondence and discussion with Dr. R. L. Orbach and Dr. R. B. Griffiths, who also made their own computations freely available to us. We wish to apologize to Dr. Griffiths for inadvertently delaying publication of his own work.

[51] Strictly, it may be necessary to define the long-range order as an average over the two sublattices [e.g., if the limit in (6.3) differs as $l \to \infty$ through even or through odd integers].

[52] The technique is similar to that expounded in Ref. 43.

Reprinted from THE PHYSICAL REVIEW, Vol. 128, No. 5, 2131–2135, December 1, 1962
Printed in U. S. A.

Spin-Wave Spectrum of the Antiferromagnetic Linear Chain

JACQUES DES CLOIZEAUX* AND J. J. PEARSON†
University of California, San Diego, La Jolla, California

(Received July 30, 1962)

The methods of Bethe and Hulthén are used to build spin-wave states for the antiferromagnetic linear chain. These states, of spin 1 and translational quantum number k, are eigenstates of the Hamiltonian $H=\Sigma_j \mathbf{S}_j\cdot\mathbf{S}_{j+1}$ with periodic boundary conditions. For an infinite chain, their spectrum is $\epsilon_k=(\pi/2)|\sin k|$, whereas Anderson's spin-wave theory gives $\epsilon_k=|\sin k|$. For finite chains it has been verified by numerical computation that these states are the lowest states of given k, but no rigorous proof has been given for an infinite chain.

I. INTRODUCTION

THE problem of determining the ground state and the lowest excited states of the Heisenberg Hamiltonian for an antiferromagnet is of great physical interest but is very difficult to solve exactly in three dimensions. Consequently, it is useful to consider the more tractable case of the antiferromagnetic linear chain, which can serve as a check on approximations to the general case and may also, in certain crystals, approach the actual physical situation. The eigenstates of the nearest-neighbor Heisenberg Hamiltonian,

$$H=\sum_j(\mathbf{S}_j\cdot\mathbf{S}_{j+1}-\tfrac{1}{4}), \quad (1)$$

for a chain of N atoms with periodic boundary conditions have been studied by Bethe[1] and Hulthén.[2] They have proposed a solution for the antiferromagnetic ground state and calculated its energy for an infinite chain.

In this paper we consider the spin-wave excitations from this ground state, defining them as the lowest energy eigenstates of H corresponding to a given translational quantum number k. We specify the form of these states and calculate their energy spectrum as a function of k for both finite and infinite chains using the methods of Bethe and Hulthén. In order to make our results as convincing as possible, we proceed systematically from the case of short chains, where all the eigenstates can be calculated explicitly, to longer chains where this would be impractical. Each assumption, as it is introduced, is verified for shorter chains, and the final results are seen to be extremely reasonable.

II. SPIN-WAVE SPECTRUM FOR SHORT CHAINS (APPLICATION $N=6$, $N=8$)

In order to find the rigorous spin-wave spectrum for short chains, it is necessary to calculate the energy of the eigenstates and also their quantum numbers S (total spin) and k. The value of k corresponding to an eigenfunction Ψ is defined by the relation

$$T\Psi=e^{ik}\Psi, \quad (2)$$

where T is the operator translating the entire chain through one lattice spacing. The spin-wave state of wave number k is taken to be the lowest energy state of quantum number k. We expect these spin-wave states to have spin 1. It is convenient to plot their excitation energy ϵ above the ground state as a function of $q=k-k_0$, where k_0 is the value of k associated with the ground state ($k_0=0$ if $N/2$ is even; $k_0=\pi$ if $N/2$ is odd). For chains of 6 and 8 atoms Mattheiss[3] has given the quantum numbers of all the states, and, in agreement with his results, we plot in Fig. 1 and 2 the spin-wave spectrum for these chains.[4] All these states do indeed have spin 1.

III. SPIN-WAVE STATES AND BETHE-HULTHÉN STATES (APPLICATION $N=16$)

It is rather lengthy to compute all the states for chains with more than 10 atoms, and, therefore, it would be useful to find a method of selecting from among all the states the particular spin-wave states we wish to consider. A classification of the eigenstates

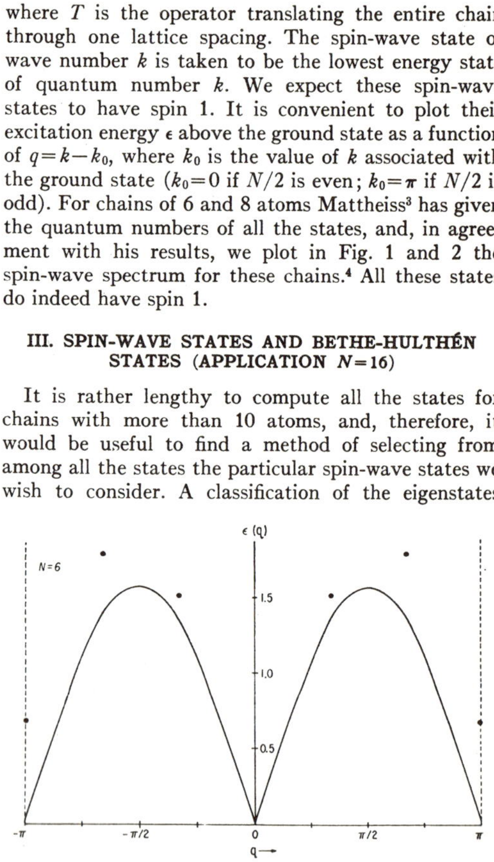

FIG. 1. Spin-wave spectrum for a chain of six atoms. The solid curve shows the spectrum for the infinite chain.

* On leave from the Centre d'Etudes Nucléaires de Saclay, Gif-sur-Yvette, France.

† Present address: Centre d'Etudes Nucléaires de Saclay, Gif-sur-Yvette, France.

[1] H. Bethe, Z. Physik **71**, 205 (1931) and in *Handbuch der Physik*, edited by S. Flügge (Springer Verlag, Berlin, 1933), Vol. XXIV/2, p. 604.

[2] L. Hulthén, Arkiv Mat. Astron. Fysik **26A**, 1 (1938).

[3] L. F. Mattheiss, Phys. Rev. **123**, 1209 (1961).

[4] For chains of 6, 8, and 10 atoms, R. Orbach computed all the energy levels [Phys. Rev. **115**, 1181 (1959)], but, unfortunately, not the corresponding k values, and consequently failed to obtain the correct spectrum.

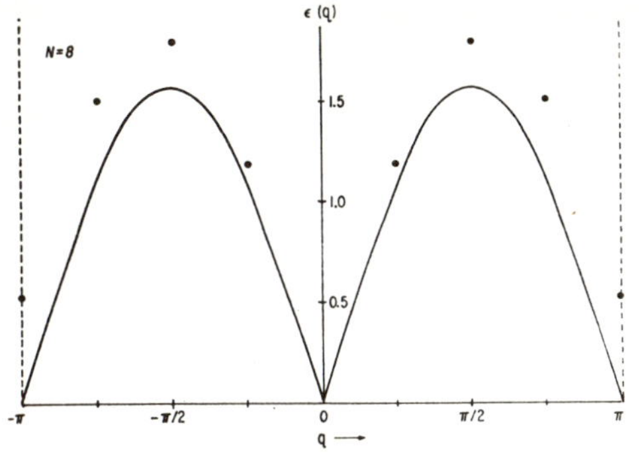

FIG. 2. Spin-wave spectrum for a chain of eight atoms. The solid curve shows the spectrum for the infinite chain.

introduced by Bethe is helpful for this purpose. It is possible to write all the eigenstates of the Heisenberg Hamiltonian in the following way:

$$\Psi=\sum a(n_1\cdots n_r)\Phi(n_1\cdots n_r), \quad (3)$$

where $\Phi(n_1\cdots n_r)$ is obtained from the ferromagnetic ground state by reversing the spins at sites $n_1\cdots n_r$ and Bethe assumed that the $a(n_1\cdots n_r)$ for a state of spin $S=\frac{1}{2}N-j$ have the form

$$a(n_1\cdots n_r)=\sum_{P_\mu}\sum_{P}\exp i\left(\sum_{j=1}^{s}k_{P_j}n_{\mu_j}+\tfrac{1}{2}\sum_{j<l}\varphi_{Pj,\,Pl}\right). \quad (4)$$

$\sum_P$ indicates a summation over all the permutations of the numbers $k_1\cdots k_s$ and $\sum_{P_\mu}$ a summation over all possible choices of s values n_μ from the set $n_1\cdots n_r$. The k's and the φ's are solutions of the coupled equations

$$2\cot(\varphi_{jl}/2)=\cot(k_j/2)-\cot(k_l/2), \quad -\pi\leq\varphi_{jl}\leq\pi. \quad (5)$$

$$Nk_j=2\pi\lambda_j+\sum_l\varphi_{jl}, \quad (6)$$

where each λ_j is an integer. The energy of such a state is

$$E=-\sum_{j=1}^{s}(1-\cos k_j), \quad (7)$$

and the corresponding wave number k is

$$k=\sum_j k_j=(2\pi/N)\sum_j\lambda_j. \quad (8)$$

The k_j's and the φ_{jl}'s can be either real or complex. We shall restrict our considerations to the class C for which all the k_j's and all the φ_{jl}'s are real, since a discussion by Bethe of the ordering of the energy levels suggests that the lowest energy state for a given k should belong to C. Each state of C is specified by a set of $s=\frac{1}{2}N-S$ integers λ_j satisfying the following conditions[5]:

$$0<\lambda_j<N, \quad \lambda_j-\lambda_{j-1}>1. \quad (9)$$

[5] These conditions apply, in fact, only to the states with maximum projection of S on the z axis. The other degenerate states of the same spin are obtained by including additional λ's all of which are zero.

In particular, the ground state, being a spin 0 state, corresponds to $\frac{1}{2}N$ integers λ_j which, according to the above rules, must be

$$1, 3, 5, \cdots(N-1). \quad (10)$$

The spin-wave states are to be sought among the spin 1 states with real k_j. Each of these states is defined by $(\frac{1}{2}N-1)$ integers λ_j satisfying the above conditions. In this way $N(N+2)/8$ states can be obtained. We verified that for chains of 6 and 8 atoms these states do indeed include the spin-wave states which we found directly (Figs. 1 and 2). A calculation of this class of states for a chain of 16 atoms was performed on a Control Data 1604 computer. The energies ϵ of the lowest states are plotted as a function of q in Fig. 3.

IV. GENERAL PRESCRIPTION FOR SPIN-WAVE STATES (APPLICATION $N=48$)

The previous numerical calculations show that, for chains of 6, 8, and 16 atoms, the states of lowest energy and wave number q are given by the following choice of the λ_j:

$$\text{For } q>0\text{: } 1, 3, 5, \cdots(N-2n-1), \quad (N-2n+2), \cdots(N-2), \quad (11)$$

$$\text{For } q<0\text{: } 2, 4, 6, \cdots(2n-2), (2n+1), \cdots(N-1),$$

with $n=N|q|/2\pi$. For reasons of symmetry, the energy ϵ must be an even function of q. Assuming the above prescription for the choice of the λ_j to be valid in general, we computed numerically the spin-wave spectrum for a chain of 48 atoms. The results are plotted in Fig. 4. The results indicate fairly well the shape of the spectrum for a long chain. The spectrum has a periodicity of 2π due to the translational symmetry of the chain as expected, but also a nearly exact double periodicity of π. This behavior is surprisingly similar to that found in Anderson's two-sublattice spin-wave theory.[6] Let us remark, however, that in Anderson's

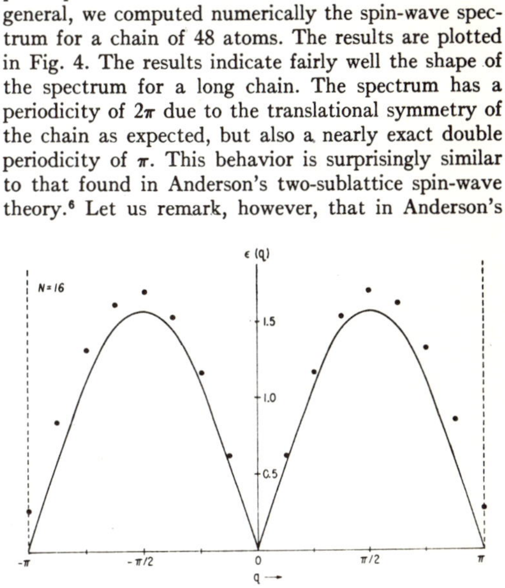

FIG. 3. Spin-wave spectrum for a chain of sixteen atoms. The solid curve shows the spectrum for the infinite chain.

[6] P. W. Anderson, Phys. Rev. 86, 694 (1952).

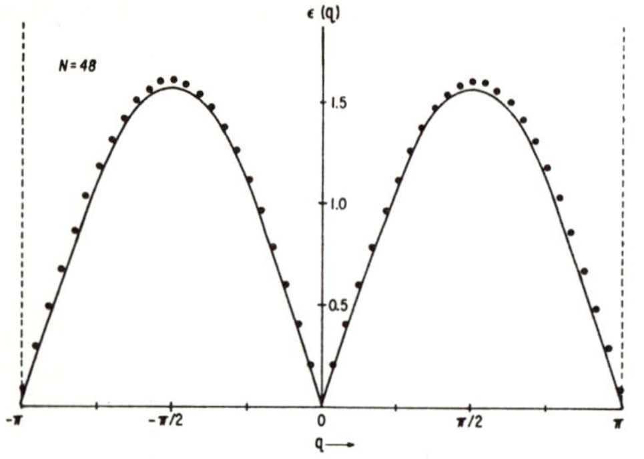

FIG. 4. Spin-wave spectrum for a chain of 48 atoms. The solid curve shows the spectrum for the infinite chain.

theory this double periodicity is artificially produced by doubling the basic translation and consequently halving the size of the Brillouin zone.

V. THE INFINITE CHAIN

In a manner analogous to the Hulthén calculation of the antiferromagnetic ground state it is possible also to calculate the spin-wave spectrum for an infinite chain. This is done by using the prescription of Sec. IV and by replacing discrete indices by continuous variables in Eqs. (4), (5), and (6). This process must be done carefully without leaving out terms of the order $1/N$ because, whereas the energy E_0 of the ground state is roughly proportional to N, the excitation energy which we want to calculate remains of the order of 1. For reasons of convenience, we compute the energy of a spin-wave state having $S_z=0$ and corresponding to a negative value of q. For this state, the sequence of the λ_j's is

$$0,\ 2,\ 4,\ \cdots(2n-2),\ (2n+1),\ \cdots(N-1), \tag{12}$$

again with $n=N|q|/2\pi$.

Let the continuous variable x be defined by

$$x=(2j-1)/N,\quad 0<x<1 \quad \text{with}\quad j=1,2,\cdots N/2. \tag{13}$$

If we define functions $k(x)=k_j$, $\lambda(x)=\lambda_j/N$, and $\varphi(x,y)=\varphi_{jl}$, then Eqs. (4), (5), and (6) become

$$2\cot\frac{\varphi(x,y)}{2}=\cot\frac{k(x)}{2}-\cot\frac{k(y)}{2}, \tag{14}$$

$$k(x)=2\pi\lambda(x)+\tfrac{1}{2}\int_0^1\varphi(x,y)dy, \tag{15}$$

$$E=-\frac{N}{2}\int_0^1[1-\cos k(x)]dx. \tag{16}$$

According to (12), $\lambda(x)$ is given by

$$\lambda(x)=(2j-2)/N=x-1/N \text{ for } j\leq n \text{ or } x<|q|/\pi, \tag{17}$$

$$\lambda(x)=(2j-1)/N=x \quad \text{for}\quad j>n \quad\text{or}\quad x>|q|/\pi. \tag{18}$$

Therefore, we may write

$$\lambda(x)=x-(1/N)\Theta(|q|/\pi-x), \tag{19}$$

where $\Theta(x)$ is the step function. Let us remark that, according to (10), the function $\lambda_0(x)$ corresponding to the antiferromagnetic ground state would be

$$\lambda_0(x)=x. \tag{20}$$

Eliminating $\varphi(x,y)$ between Eqs. (14) and (15), we obtain

$$k(x)=2\pi[x-(1/N)\Theta(|q|/\pi-x)]$$
$$+\int_0^1 \operatorname{arc\,cot}\{\tfrac{1}{2}\cot[k(x)/2]-\tfrac{1}{2}\cot[k(y)/2]\}dy, \tag{21}$$

with $-(\pi/2)\leq \operatorname{arc\,cot} k<(\pi/2)$. The differentiation of this equation with respect to k yields

$$1=\pi\frac{dx}{dk}[1+(2/N)\delta(x-|q|/\pi)]+\{\sin[k(x)/2]\}^{-2}$$
$$\times\int_0^1\{4+[\cot(k(x)/2)-\cot(k(y)/2)]^2\}^{-1}dy. \tag{22}$$

Following Hulthén, we make the change of variable

$$\xi\equiv\cot[k(x)/2] \tag{23}$$

$$-dx/d\xi=f(\xi). \tag{24}$$

Letting ξ_0 be the value of ξ corresponding to $x=|q|/\pi$, we obtain

$$\delta(x-|q|/\pi)=\delta(\xi-\xi_0)|d\xi/dx|=\delta(\xi-\xi_0)[1/f(\xi)]. \tag{25}$$

After this transformation, the equation becomes

$$f(\xi)+(2/\pi)\int_{-\infty}^{+\infty}f(\eta)[4+(\xi-\eta)^2]^{-1}d\eta$$
$$=(2/\pi)(I+\xi^2)^{-1}-(2/N)\delta(\xi-\xi_0). \tag{26}$$

On the other hand, the energy is given by

$$E=-N\int_{-\infty}^{+\infty}f(\xi)[1+\xi^2]^{-1}d\xi. \tag{27}$$

By using $\lambda_0(x)$ instead of $\lambda(x)$, we would obtain the Hulthén integral equation for the function $f(\xi)$ corresponding to the ground state. This equation can be derived from (26) by omitting the term in $(1/N)$. Equation (26) can now be solved easily. Putting

$$f(\xi)=f_0(\xi)+\Delta f(\xi) \tag{28}$$

we may write

$$\Delta f(\xi)=-(2/N)\int_{-\infty}^{+\infty}R(\xi,\eta)\delta(\eta-\xi_0)d\eta$$
$$=-(2/N)R(\xi,\xi_0), \tag{29}$$

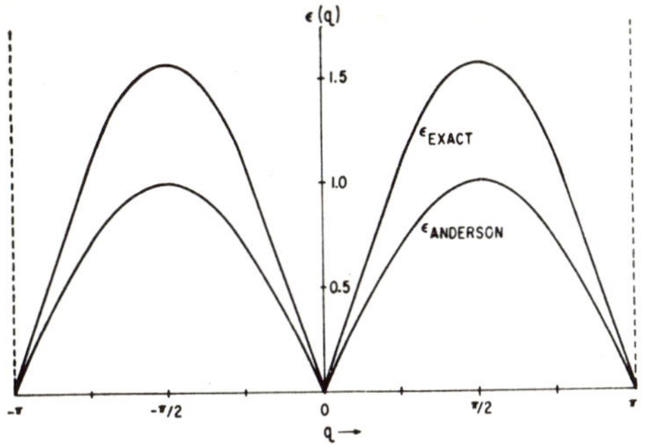

FIG. 5. The spin-wave spectrum in the exact case and in Anderson's approximation.

where $R(\xi,\eta)$ is the resolvent of Eq. (26) and has been calculated by Hulthén

$$R(\xi,\eta)=\delta(\xi-\eta)+\sum_{n=1}^{\infty}\frac{(-1)^n}{\pi}\frac{2n}{(2n^2)+(\xi-\eta)^2}. \tag{30}$$

Now putting

$$E=E_0+\epsilon, \tag{31}$$

we deduce from (27), (29), and (30)

$$\begin{aligned}\epsilon&=-N\int_{-\infty}^{\infty}\Delta f(\xi)[1+\xi^2]^{-1}d\xi\\&=2\sum_{n=0}^{\infty}\frac{(-1)^n(2n+1)}{(2n+1)^2+\xi_0^2}\\&=\pi/2\cosh(\tfrac{1}{2}\pi\xi_0).\end{aligned} \tag{32}$$

Now we must express ξ_0 as a function of q. To do this we recall the relation between x and ξ given by Eq. (23) and the following boundary conditions:

$$\begin{aligned}x&=0 \quad \text{for} \quad \xi=-\infty,\\x&=1 \quad \text{for} \quad \xi=+\infty.\end{aligned} \tag{33}$$

This equation can be simplified by replacing $f(\xi)$ by $f_0(\xi)$, since the terms omitted are of order $1/N$ and may safely be neglected. As was shown by Hulthén,

$$f_0(\xi)=1/2\cosh(\tfrac{1}{2}\pi\xi), \tag{34}$$

and consequently Eq. (23) becomes

$$-dx/d\xi=1/2\cosh(\tfrac{1}{2}\pi\xi). \tag{35}$$

Integrating this equation and taking into account the boundary conditions we get

$$\sinh(\tfrac{1}{2}\pi\xi)=\cot(\pi x). \tag{36}$$

Remembering that ξ_0 is the value of ξ for which $x=|q|/\pi$, we obtain

$$\sinh(\tfrac{1}{2}\pi\xi_0)=\cot(|q|). \tag{37}$$

From this result and from Eq. (32) it follows that the spin-wave spectrum is given by

$$\begin{aligned}\epsilon&=\pi/2\cosh(\tfrac{1}{2}\pi\xi_0)\\&=\tfrac{1}{2}\pi|\sin q|.\end{aligned} \tag{38}$$

This result agrees very well with the spectra found directly for finite chains, and has been plotted on each figure. It is interesting to note that the double periodicity mentioned above is now exact. This fact facilitates comparison with the approximate expression obtained by Anderson, which automatically exhibits this double periodicity.

$$\epsilon_{\text{Anderson}}=|\sin q|. \tag{39}$$

Also, our spin-wave states, being spin 1 states, are threefold degenerate, whereas Anderson's spin-wave states have only two components. Anderson's approximation is really not valid for a linear chain and spin $\frac{1}{2}$, but it gives a surprisingly good value for the energy of the antiferromagnetic ground state ($E_{0\,\text{Anderson}}=-0.6817$; $E_{0\,\text{Exact}}=-0.6931$). Therefore, it is interesting to note the discrepancy between his spin-wave spectrum and the exact one. Both spectra are plotted in Fig. 5.

VI. SUMMARY AND CONCLUSION

For the sake of clarity it may be useful now to summarize the work of the preceding paragraphs. In Sec. II we have considered the case of short chains of 6 and 8 atoms. The energies and the quantum numbers (total spin S and wave number k) have been calculated. The ground state has spin 0 and wave number k_0 ($k_0=0$ if $N/2$ is even, $k_0=\pi$ if $N/2$ is odd). For a given value of $k\neq k_0$, the states of lowest energy have spin 1 and therefore can be identified with spin-wave states. In Sec. III we introduce a class C of eigenstates originally presented by Bethe and Hulthén, and we observe that the states previously identified as spin-wave states for 6 and 8 atom chains are included in this class. On this basis we have assumed that for any value of N the spin-wave states belong to C. In accordance with this assumption we have calculated the states C of spin 1 for a 16 atom chain and from them selected as spin waves the states of lowest energy. In Sec. IV we have introduced a prescription for determining in advance which states in the class C are the spin-wave states. We have verified this prescription by comparison with the results obtained for the 6, 8, and 16 atom chains and have used it to calculate the spin-wave spectrum of a 48 atom chain. In Sec. V the same prescription has been used to obtain the spin-wave spectrum for an infinite chain and has been found to have the form $\epsilon=\frac{1}{2}\pi|\sin q|$ where $q=k-k_0$. This result differs from the approximate result of Anderson by the factor $\pi/2$.

Aside from questions of rigor, two points of interpretation remain unresolved. The first one concerns the real origin of the double periodicity mentioned above. As we have shown, this double periodicity is a feature of Anderson's two-sublattice theory. It appears, however, that it has nothing to do with long-range order, first because it is possible to find models without long-range order which display this double periodicity, and second because the ground state itself is probably not ordered. The second point concerns the number of spin-wave states. In the rigorous case, the spin-wave states, being eigenstates of spin 1, are triply degenerate and correspond to values of k ranging from $-\pi$ to π. On the other hand, in a two-sublattice model, for each value of k, ranging from $-\pi/2$ to $\pi/2$, there are only two spin waves. It is not obvious how these two points of view can be reconciled.

ACKNOWLEDGMENTS

The authors are grateful to Professor Walter Kohn for numerous helpful discussions and for the hospitality extended to them by the Physics Department of the University of California at San Diego.

Reprinted from The Physical Review, Vol. 133, No. 3A, A768–A775, 3 February 1964
Printed in U. S. A.

Magnetization Curve at Zero Temperature for the Antiferromagnetic Heisenberg Linear Chain*

Robert B. Griffiths†
University of California, San Diego, La Jolla, California
(Received 9 September 1963)

The highest and lowest energies as a function of the total spin are computed for the class of "unbound" states in the Bethe formalism for the linear chain of spin-$\frac{1}{2}$ atoms with a Heisenberg exchange interaction between nearest neighbors. The lowest energies are used to compute the magnetization curve for the infinite antiferromagnetic chain in the limit of zero temperature. At zero temperature and in zero field, the magnetic susceptibility has the value 0.050661 $g^2\mu^2/J$, where μ is the Bohr magneton, g the electron g factor, and the interaction between neighboring spins is of the form $2J\mathbf{S}_1\cdot\mathbf{S}_2$.

I. INTRODUCTION

The properties of antiferromagnetic insulators are often discussed on the basis of the Heisenberg model of exchange between neighboring atoms. Calculations for two and three dimensions invariably proceed by means of approximations whose validity is difficult to judge. Hence, there is some interest in examining the one-dimensional case, for which a certain amount of progress has been made toward an exact solution.

Bethe[1] showed that the eigenvalue problem for a chain of N spin-$\frac{1}{2}$ atoms with Hamiltonian

$$\mathcal{H}=2J\sum_{i=1}^{N}\mathbf{S}_i\cdot\mathbf{S}_{i+1}, \qquad (1)$$

$$\mathbf{S}_{N+1}\equiv\mathbf{S}_1,$$

where $\mathbf{S}_i$ is the spin operator for the ith atom and J the "exchange integral," could be reduced to that of solving a set of coupled transcendental algebraic equations. Using this procedure, Hulthén[2] calculated the exact ground-state energy for an infinite antiferromagnetic ($J>0$) chain; and des Cloizeaux and Pearson[3] have recently obtained the energies of the lowest lying excitations or "spin waves."

The class C[4] of "unbound" states in the Bethe formalism is of particular interest for the antiferromagnetic chain, since it contains the ground state and the des Cloizeaux and Pearson spin waves. In this paper we compute (in Sec. III) the minimum and maximum energies of states in class C with a given total spin S, as a function of S. There is good reason to believe that the states of minimum energy in class C are also the lowest of all levels (for a given S) in the antiferromagnetic chain. Assuming this to be the case, we calculate (in Sec. IV) the magnetization as a function of magnetic field at zero temperature.

Section II contains a summary of the Bethe formalism for the eigenvalue problem. A proof of the existence of real solutions to the Bethe equations for states in class C is found in Appendix A. The derivation of Hulthén's integral equation, upon which the work in Sec. III B, C, and D depends, is summarized in Sec. III A.

The results of our computations do not agree with those of Ledinegg and Urban.[5] The reason for this discrepancy is discussed in Appendix B.

II. THE BETHE EQUATIONS

Let E_F and E_{AF} be the largest and smallest eigenvalues of the Hamiltonian (1) with $J>0$, corresponding to the ferro- and antiferromagnetic ground states, respectively:

$$E_F=\tfrac{1}{2}NJ,\qquad E_{AF}=NJ(\tfrac{1}{2}-2\ln 2). \qquad (2)$$

The second equation[2] holds in the limit $N\to\infty$. For a state with energy E, define the normalized energies[6]

$$\epsilon=\tfrac{1}{2}(E_F-E)/JN,\qquad \eta=\tfrac{1}{2}(E-E_{AF})/JN=\ln 2-\epsilon. \qquad (3)$$

With "up" and "down" defined with respect to the positive z axis, let $\Psi(n_1,n_2,\ldots n_r)$ be a state for which the spins $n_1, n_2, \ldots, n_r$ are down and all other spins are up. Any eigenstate of (1) with z component of spin equal to $\frac{1}{2}N$-r may be written as a linear combination of such states:

$$\Psi=\sum a(n_1,n_2,\ldots,n_r)\Psi(n_1,n_2,\ldots,n_r), \qquad (4)$$

with summation over all sets of r (distinct) indices n_j.

The eigenfunctions discussed by Bethe[1] are of the form

$$a(n_1,\cdots,n_r)=\sum_{P=1}^{r!}\exp i\left(\sum_{j=1}^{r}k_{Pj}n_j+\tfrac{1}{2}\sum_{j<l}\phi_{PjPl}\right), \qquad (5)$$

where the summation extends over all permutations of the integers 1, 2, . . ., r, and Pj is the image of j under

---

* Supported in part by the U. S. Office of Naval Research.
† National Science Foundation Postdoctoral Fellow.
[1] H. Bethe, Z. Physik **71**, 205 (1931).
[2] L. Hulthén, Arkiv Mat. Astron. Fysik **26A**, No. 11, (1938).
[3] J. des Cloizeaux and J. J. Pearson, Phys. Rev. **128**, 2131 (1962).
[4] This notation comes from Ref. 3 and is explained in detail in Sec. II.
[5] E. Ledinegg and P. Urban, Acta Phys. Austriaca **6**, 257 (1953).
[6] Our definition of ϵ is smaller than that given in Refs. 1, 2, and 3 by a factor of N^{-1}.

the Pth permutation. The "wave vectors" k_j (real or complex) satisfy the equations

$$Nk_j=2\pi\lambda_j+\sum_{l(\neq j)}\phi_{jl}\qquad j=1,2,\cdots,r, \tag{6}$$

where the λ's are integers between 0 and N-1, and the ϕ's are defined by

$$\begin{aligned}\cot\tfrac{1}{2}\phi_{jl}&=\tfrac{1}{2}(\cot\tfrac{1}{2}k_j-\cot\tfrac{1}{2}k_l),\\ -\pi&\leq\phi_{jl}\leq\pi.\end{aligned}\tag{7}$$

The energy of the state (5) is equal to

$$\epsilon=N^{-1}\sum_{j=1}^{r}(1-\cos k_j). \tag{8}$$

The order of the λ's associated with a state (5) through Eq. (6) is unimportant. Furthermore, a state with some λ's equal to zero has the same energy and total spin as the corresponding state in which the zero λ's have been eliminated.[2] Hence, without loss of generality we may assume that

$$0<\lambda_1\leq\lambda_2\leq\cdots\leq\lambda_r<N. \tag{9}$$

By "class C" we denote either the sets $\{\lambda_j\}$ satisfying, in addition to (9), the restriction

$$\lambda_{j+1}\geq\lambda_j+2, \tag{10}$$

or the eigenstates (5) corresponding to these sets. The importance of class C comes from the following properties.

1. For each set $\{\lambda_j\}$ in class C there is a solution to (6) for which all the k_j are real and no two are equal. [When two k's are equal, the state (5) vanishes identically.]
2. For a given total spin S, the state with the lowest energy ($J>0$) belongs to class C.

Appendix A contains a rigorous proof of the first property, together with a general discussion of the behavior of the solutions $\{k_j\}$ of (6) as a function of the sets $\{\lambda_j\}$. The second property has not yet been proved, but it is very plausible on the basis of arguments by Bethe[1] and Orbach,[7] and calculations on finite chains by des Cloizeaux and Pearson,[3] and the author.[8]

For an antiferromagnetic chain, the energy of a state is lowest when ϵ [see (3)] is largest, and vice versa. Now by Eq. (8), ϵ is largest, for a given r, when the k_j cluster near the center of the interval $[0,2\pi]$. This may be expected, according to plausible arguments in Appendix A, when the λ_j cluster near the center of $[0,N]$, subject to the constraint (10). Hence, for a given total spin $S=\frac{1}{2}N$-r, the state of minimum energy in class C should correspond to

$$\lambda_1=\tfrac{1}{2}N-r+1,\quad \lambda_2=\lambda_1+2,\quad \lambda_3=\lambda_2+2,\ \ldots,\quad \lambda_r=\tfrac{1}{2}N+r-1. \tag{11}$$

[7] R. Orbach, Phys. Rev. **112**, 309 (1958).
[8] R. B. Griffiths (unpublished).

By a similar argument, the state of maximum energy in class C should correspond to

$$\lambda_1=1,\quad \lambda_2=3,\ \ldots,\quad \lambda_{r/2}=r-1,\quad \lambda_{r/2+1}=N-(r-1),\quad \lambda_{r/2+2}=N-(r-3),\ \ldots,\quad \lambda_r=N-1. \tag{12}$$

For $S=0$ there is only one state in class C, the antiferromagnetic ground state,

$$\lambda_1=1,\quad \lambda_2=3,\quad \lambda_3=5,\ \ldots,\quad \lambda_{N/2}=N-1. \tag{13}$$

We have tacitly assumed in writing (11), (12), and (13) that N and r are even; minor modifications are required if one or the other is odd.

There also exist solutions of (6) for which two or more of the k_j are not real but complex. For the case $r=2$, Bethe[1] showed that complex k values result in a state (5) for which $|a(n_1,n_2)|^2$ decreases exponentially as $|n_2-n_1|$ increases, and thus has the form of two spin waves interacting to form a bound state. We shall, for convenience, refer to any state for which some of the k_j are complex as a "bound" state, and to states for which all k_j are real as "unbound" states. Some of the unbound states are not contained in class C, though Bethe's discussion indicates that these are relatively few in number as N becomes infinite.

III. THE HULTHÉN INTEGRAL EQUATION

A. The Antiferromagnetic Ground State

In the limit of large N it is reasonable to replace (6), (7), (8) for the set λ_j given by (13) by the equations

$$k(x)=2\pi x+\tfrac{1}{2}\int_0^1\phi(x,y)dy, \tag{14}$$

$$\cot\tfrac{1}{2}\phi(x,y)=\tfrac{1}{2}[\cot\tfrac{1}{2}k(x)-\cot\tfrac{1}{2}k(y)], \tag{15}$$

$$-\pi\leq\phi\leq\pi,$$

$$\epsilon_0=\tfrac{1}{2}\int_0^1[1-\cos k(x)]dx, \tag{16}$$

where λ_j has been replaced by Nx and k_j by $k(x)$.

By differentiating both sides of (14) with respect to x, Hulthén[2] obtained an integral equation linear in the function dx/dk:

$$1=\pi\frac{dx}{dk}+\csc^2(\tfrac{1}{2}k)\int_0^{2\pi}\frac{(dx/dk')dk'}{4+(\cot\frac{1}{2}k-\cot\frac{1}{2}k')^2}, \tag{17}$$

which, by means of the substitution

$$\begin{aligned}\xi&=\cot\tfrac{1}{2}k,\\ f_0(\xi)&=-dx/d\xi,\end{aligned}\tag{18}$$

may be rewritten in the form

$$f_0(\xi)=g_0(\xi)-\int_{-\infty}^{\infty}K(\xi-\eta)f_0(\eta)d\eta\,, \tag{19}$$

where

$$g_0(\xi)=(2/\pi)(1+\xi^2)^{-1}, \tag{20}$$

$$K(\xi-\eta)=(2/\pi)[4+(\xi-\eta)^2]^{-1}, \tag{21}$$

and further,

$$\epsilon_0=\tfrac{1}{2}\pi\int_{-\infty}^{\infty}f_0(\xi)g_0(\xi)d\xi\,. \tag{22}$$

For our purposes it is convenient to write Hulthén's solution of Eq. (19) in the form

$$f_0(\xi)=g_0(\xi)-\int_{-\infty}^{\infty}R(\xi-\eta)g_0(\eta)d\eta \tag{23a}$$

$$=\tfrac{1}{2}\operatorname{sech}\tfrac{1}{2}\pi\xi\,, \tag{23b}$$

$$\epsilon_0=\ln 2\,, \tag{24}$$

where the resolvent kernel $R(x)$ may be written as an infinite sum,

$$R(x)=\pi^{-1}\sum_{n=1}^{\infty}2n(-1)^{n+1}(4n^2+x^2)^{-1}, \tag{25}$$

or as an integral,

$$R(x)=(4\pi)^{-1}\int_{-\infty}^{\infty}[1+(y+x)^2]^{-1}\operatorname{sech}(\tfrac{1}{2}\pi y)dy\,, \tag{26}$$

and satisfies the equation

$$\int_{-\infty}^{\infty}K(\xi-\eta)R(\eta-\zeta)d\eta=K(\xi-\zeta)-R(\xi-\zeta)\,. \tag{27}$$

B. States of Minimum Energy

For the sets of integers given by (11), the Hulthén equations (14) and (16) must be replaced by

$$k(x)=2\pi x+\tfrac{1}{2}\int_{(1/2)-\rho}^{(1/2)+\rho}\phi(x,y)dy\,, \tag{28}$$

$$\epsilon=\tfrac{1}{2}\int_{(1/2)-\rho}^{(1/2)+\rho}[1-\cos k(x)]dx\,, \tag{29}$$

where ρ is determined by the condition

$$\tfrac{1}{2}N-S=r=N\rho\,. \tag{30}$$

Also, define σ by

$$\sigma=S/N=\tfrac{1}{2}-\rho\,. \tag{31}$$

Equation (28) may be transformed by the same procedure previously applied to (14); we shall denote $-dx/d\xi$ by $f(\xi)$[9]:

$$f(\xi)=g_0(\xi)-\int_{-\alpha}^{\alpha}K(\xi-\eta)f(\eta)d\eta\,, \tag{32}$$

$$\epsilon=\tfrac{1}{2}\pi\int_{-\alpha}^{\alpha}f(\xi)g_0(\xi)d\xi\,, \tag{33}$$

with limits of integration determined by the requirement

$$\rho=\tfrac{1}{2}\int_{(1/2)-\rho}^{(1/2)+\rho}dx=\tfrac{1}{2}\int_{-\alpha}^{\alpha}f(\xi)d\xi\,. \tag{34}$$

In Eqs. (32)–(34), only the values of $f(\xi)$ for $|\xi|\leq\alpha$ are employed, although (32) also *defines* $f(\xi)$ for $|\xi|>\alpha$. Multiply the right- and left-hand sides of (32) by $R(\zeta-\xi)$ and integrate with respect to ξ from $-\infty$ to $+\infty$. By interchanging the order of integration on the right-hand side, and using (27), (23a), (32), and the fact that $f(\xi)=f(-\xi)$ and $R(x)=R(-x)$, one obtains the equation

$$f(\xi)=f_0(\xi)+\int_{\alpha}^{\infty}[R(\xi-\eta)+R(\xi+\eta)]f(\eta)d\eta\,. \tag{32a}$$

Similarly, in place of (33) and (34) one has

$$\eta=\epsilon_0-\epsilon=\pi\int_{\alpha}^{\infty}f_0(\xi)f(\xi)d\xi\,, \tag{33a}$$

$$\sigma=\tfrac{1}{2}-\rho=\tfrac{1}{2}\int_{\alpha}^{\infty}f(\xi)d\xi\,. \tag{34a}$$

[Multiply both sides of (32a) by $g_0(\xi)$ and integrate from $-\infty$ to $+\infty$. Upon interchanging the order of integration, and using (22), (23a), and (33), one obtains (33a). To obtain (34a), integrate both sides of (32) from $-\infty$ to $+\infty$.]

The asymptotic behavior of η and σ for large values of α determines the magnetic susceptibility in small magnetic fields at zero temperature, and hence we shall examine it in some detail using Eqs. (32a)–(34a). For large α, it is clearly an excellent approximation to replace $f_0(\xi)$ [see (23b)] by $\exp(-\tfrac{1}{2}\pi\xi)$ in (32a) and (33a), since only values of f_0 for $\xi\geq\alpha$ are required. Even for $\alpha=3$, the fractional error is less than 10^{-4}. By use of this approximation and the function

$$p(\xi)=e^{\pi\alpha/2}f(\xi+\alpha)\,, \tag{35}$$

Eqs. (32a)–(34a) may be rewritten in the form

$$\mathrm{p}(x)=e^{-\pi x/2}+\int_0^{\infty}[R(x-y)+R(x+y+2\alpha)]\mathrm{p}(y)dy\,, \tag{36}$$

[9] L. Hulthén (Ref. 2) obtained Eq. (32), but did not attempt to solve it.

$$\eta = b(\alpha)e^{-\pi\alpha}, \qquad (37)$$

$$\sigma = a(\alpha)e^{-\pi\alpha/2}, \qquad (38)$$

where

$$a(\alpha) = \tfrac{1}{2}\int_0^\infty p(x)dx, \qquad (39)$$

$$b(\alpha) = \pi\int_0^\infty e^{(-1/2)\pi x}\, p(x)dx. \qquad (40)$$

Since the kernel $R(x)$ decreases as x^{-2} for large x, it is reasonable to suppose that the function $p(x)$ is determined primarily by the first kernel in the integrand in (36) when α is large. One can, in fact, show (we shall not give the proof here) that as α becomes infinite,[10]

$$a(\alpha) = a_0[1 + (2\pi\alpha)^{-1}] + O(\ln\alpha/\alpha^2), \quad b(\alpha) = b_0 + O(\alpha^{-2}), \qquad (41)$$

$$da(\alpha)/d\alpha = O(\alpha^{-2}), \quad db(\alpha)/d\alpha = O(\alpha^{-2}), \qquad (42)$$

where a_0 and b_0 are obtained from (39) and (40) when $p(x)$ is replaced by $q(x)$, the solution to the equation

$$q(x) = e^{(-1/2)\pi x} + \int_0^\infty R(x-y)q(y)dy. \qquad (43)$$

Equation (43) has been solved numerically[11] using a computer program with several internal checks. We believe the results,

$$a_0 = 0.48394, \quad b_0 = 1.15573 \qquad (44)$$

are correct to five decimal places. Combining (37), (38), (41), and (44), we have, for large α or for small σ,

$$\eta = \{4.9348[1 - (\pi\alpha)^{-1}] + O(\ln\alpha/\alpha^2)\}\sigma^2, \qquad (45a)$$

$$= 4.9348\left(1 + \frac{1}{2\ln\sigma}\right)\sigma^2 + O\left[\frac{\sigma^2 \ln(|\ln\sigma|)}{(\ln\sigma)^2}\right]. \qquad (45b)$$

It is clear that η is not an analytic function of σ at $\sigma = 0$. The constant 4.9348 agrees with $\pi^2/2$ to five significant figures, and there are certain plausible (though far from rigorous) arguments based on the spin-wave spectrum of des Cloizeaux and Pearson[3] which suggest that the value should be $\pi^2/2$ exactly.

The asymptotic behavior given by (37), (38), and (41) is in agreement with numerical calculations of η and σ for finite α (discussed below in part D), but in disagreement with results published by Ledinegg and Urban,[5] which we believe to be in error. For further discussion, see Appendix B.

C. States of Maximum Energy

For a given total spin S, the states of maximum energy in class C are associated with sets of integers of the form (12). The integral equations in the limit $N \to \infty$ are obtained in close analogy to the results in part B, so we shall omit details. Corresponding to Eqs. (32)–(34) one has

$$f(\xi) = g_0(\xi) - \int_\alpha^\infty [K(\xi-\eta) + K(\xi+\eta)]f(\eta)d\eta, \qquad (46)$$

$$\epsilon = \pi\int_\alpha^\infty f(\xi)g_0(\xi)d\xi, \qquad (47)$$

$$\rho = \int_\alpha^\infty f(\xi)d\xi, \qquad (48)$$

and, corresponding to Eqs. (32a)–(34a),

$$f(\xi) = f_0(\xi) + \int_{-\alpha}^{\alpha} R(\xi-\eta)f(\eta)d\eta, \qquad (46a)$$

$$\eta = \pi\int_0^\alpha f_0(\xi)f(\xi)d\xi, \qquad (47a)$$

$$\sigma = \tfrac{1}{2}\int_0^\alpha f(\xi)d\xi. \qquad (48a)$$

When α or σ is small we have

$$\eta \sim \pi\sigma, \qquad (49)$$

whereas when α is large, and therefore ρ is small

$$\epsilon \sim \tfrac{2}{3}\pi^2\rho^3. \qquad (50)$$

The results (49) and (50) have been confirmed by numerical computations.

It must be emphasized that the energy is a maximum *only* for states in class C. In general, for a given value of S, there will be states not belonging to class C with energies both less than and greater than the maximum for class C.

D. Numerical Solutions of the Integral Equations

The integral equations (32) and (46a) were approximated by a set of 41 coupled linear algebraic equations which were solved using a Control Data Corporation 1604 digital computer. The results are shown in Fig. 1 where the normalized energies ϵ and η [see (3)] are plotted as a function of $\sigma = S/N$, where S is the total spin of the state, and $\rho = \frac{1}{2} - \sigma$.

[10] Capital O stands for "of the order of" in the sense used by E. T. Whittaker and G. N. Watson, *A Course of Modern Analysis* (Cambridge University Press, Cambridge, England, 1927), 4th ed., p. 11.

[11] The actual computations were carried out, purely as a matter of convenience, on an integral equation equivalent to (43), but which used the kernel K instead of R.

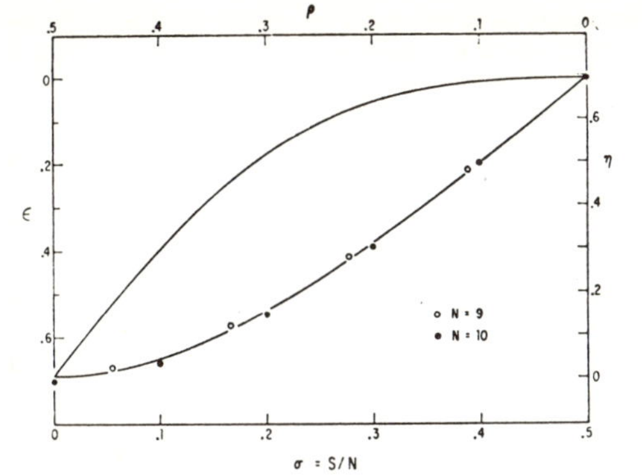

FIG. 1. Highest and lowest energies for states in class C as a function of the total spin S. The circles represent the lowest energies for finite chains containing 9 and 10 atoms.

Table I contains values of σ and η for the lower curve in Fig. 1. If the arguments alluded to in Sec. II are correct, these energies, for a given σ, should not only be the lowest in class C, but also the lowest for all states in the antiferromagnetic chain. Fortunately, there are some independent checks on this property.

TABLE I. The minimum energy η of levels in class C for various values of $\sigma = S/N$. At zero temperature, σ is proportional to the magnetization and $2d\eta/d\sigma$ is proportional to the applied magnetic field.

σ	η	$2\frac{d\eta}{d\sigma}$	σ	η	$2\frac{d\eta}{d\sigma}$
0.000000	0.00000000	0.00000	0.17822	0.12262	2.529
0.001280	0.00000759	0.02362	0.19293	0.14177	2.677
0.002407	0.00002668	0.04412	0.2088	0.16357	2.828
0.004531	0.00009387	0.08235	0.2258	0.18829	2.977
0.006218	0.00017608	0.11245	0.2441	0.2162	3.125
0.008538	0.0003303	0.15349	0.2636	0.2475	3.268
0.011728	0.0006198	0.2094	0.2845	0.2823	3.405
0.016117	0.0011630	0.2855	0.2955	0.3012	3.470
0.02216	0.002182	0.3887	0.3068	0.3209	3.533
0.02600	0.002989	0.4534	0.3184	0.3417	3.593
0.03050	0.004095	0.5285	0.3304	0.3634	3.650
0.03579	0.005607	0.6156	0.3428	0.3861	3.703
0.04200	0.007678	0.7166	0.3554	0.4097	3.753
0.04931	0.010509	0.8333	0.3685	0.4343	3.798
0.05789	0.014377	0.9677	0.3818	0.4598	3.840
0.06799	0.019654	1.1220	0.3955	0.4862	3.877
0.07986	0.02684	1.2980	0.4095	0.5135	3.909
0.09382	0.03659	1.4975	0.4239	0.5416	3.937
0.11020	0.04979	1.7214	0.4385	0.5706	3.960
0.11943	0.05802	1.8425	0.4535	0.6002	3.977
0.12943	0.06754	1.9695	0.4687	0.6306	3.990
0.14025	0.07855	2.102	0.4842	0.6616	3.997
0.15194	0.09125	2.240	0.5000	0.6931	4.000
0.16458	0.10586	2.383			

Energies for finite chains containing $N=2, 3, 4, \ldots$, 10 atoms have been computed directly from the Hamiltonian (1).[8,12,13] The lowest energies as a function of S lie above the lower curve in Fig. 1 for $N=3$, 5, 7, and 9, and below the curve for $N=2$, 4, 6, 8, and 10. Points for $N=9$, 10 are shown in Fig. 1 (open and solid circles, respectively) and lie quite close to the computed curve, which should represent the limit as N becomes infinite.

[12] R. Orbach, Phys. Rev. **115**, 1181 (1959).
[13] J. C. Bonner and M. E. Fisher (to be published).

A strict upper bound on the minimum energy curve is provided by the work of Bulaevskii.[14] At zero temperature he, in effect, finds, in a particular representation, the lowest diagonal element of the Hamiltonian (1) in a subspace containing all states with a given z component of total spin. Of course, the lowest diagonal element must be larger than the lowest eigenvalue. The energy so obtained is, in our notation,

$$\eta_B = \sigma^2 - \pi^{-1}\cos\pi\sigma - (\pi^{-1}\cos\pi\sigma)^2 + \ln 2 - \tfrac{1}{4}. \quad (51)$$

A strict lower bound on the minimum energy curve may be obtained if we write the Hamiltonian (1) as

$$\mathcal{H} = \mathcal{H}_0 + \mathcal{H}_1, \quad (52)$$

where

$$\mathcal{H}_0 = 2J\sum_{i=1}^{N}(S_i^xS_{i+1}^x + S_i^yS_{i+1}^y), \quad (53)$$

$$\mathcal{H}_1 = 2J\sum_{i=1}^{N} S_i^zS_{i+1}^z, \quad (54)$$

and the superscripts denote the x, y, and z components of the spin operators. The Hamiltonian (53) has been solved exactly[15] and the minimum energy obtained as a function of the z component of total spin. Of course (54) is just the Ising Hamiltonian. The sum of the minimum energies for these two Hamiltonians,

$$\eta_C = |\sigma| - \pi^{-1}\cos\pi\sigma + \ln 2 - \tfrac{1}{2}, \quad (55)$$

is a lower bound on the minimum energy for (1), since the lowest eigenvalue of the sum of two Hermitian matrices is larger than the sum of the lowest eigenvalues of the summands.[16]

We therefore expect that

$$\eta_C(\sigma) \leq \eta(\sigma) \leq \eta_B(\sigma), \quad (56)$$

and, indeed, the values of η in Table I lie within the specified bounds. For instance, at $\sigma = 0.3554$ we have

$$\eta_B = 0.4103,$$
$$\eta = 0.4097,$$
$$\eta_C = 0.4089.$$

For larger values of σ the bounds are even closer together, since $\eta_B - \eta_C$ goes to zero as $(\frac{1}{2}-\sigma)^4$ as σ

[14] L. N. Bulaevskii, Zh. Eksperim. i Teor. Fiz. **43**, 968 (1962) [translation: Soviet Phys.—JETP **16**, 685 (1963)].
[15] E. Lieb, T. Schultz, and D. Mattis, Ann. Phys. (N. Y.) **16**, 407 (1961); S. Katsura, Phys. Rev. **127**, 1508 (1962); see also Ref. 14.
[16] Pointed out, for example, by P. W. Anderson, Phys. Rev. **83**, 1260 (1951).

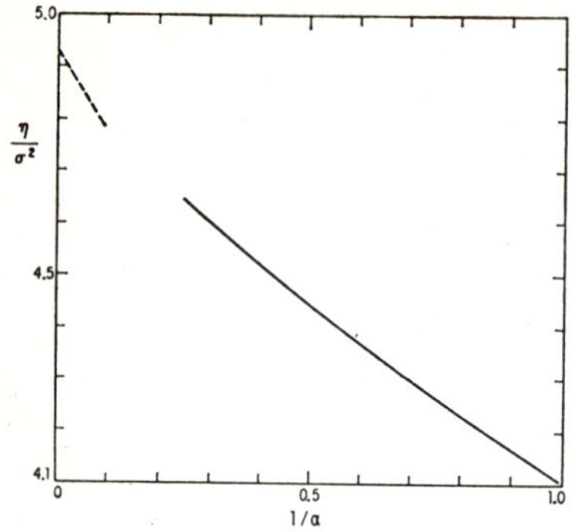

FIG. 2. The ratio η/σ^2 for the states of minimum energy in class C as a function of $1/\alpha$. The dashed line at the left side gives the asymptotic behavior as α^{-1} goes to zero.

approaches $\frac{1}{2}$. These bounds are not of much value when σ is small.

The numerical calculations also confirm the asymptotic behavior of η and σ for large α given by Eqs. (37)–(42). In Fig. 2 the ratio

$$b(\alpha)/a(\alpha)^2=\eta/\sigma^2$$

is plotted as a function of $1/\alpha$. Results for $\alpha>4$ were not very accurate and have not been included in the graph. But there is no reason to doubt that the curve approaches the asymptotic limit, shown by a dotted line, smoothly as α^{-1} approaches zero.

IV. MAGNETIC MOMENT AND SUSCEPTIBILITY AT ZERO TEMPERATURE

Let there be a magnetic field H along the positive z axis. The Zeeman energy

$$\mathcal{H}_z=g\mu H\sum_{i=1}^{N} S_i^z \tag{57}$$

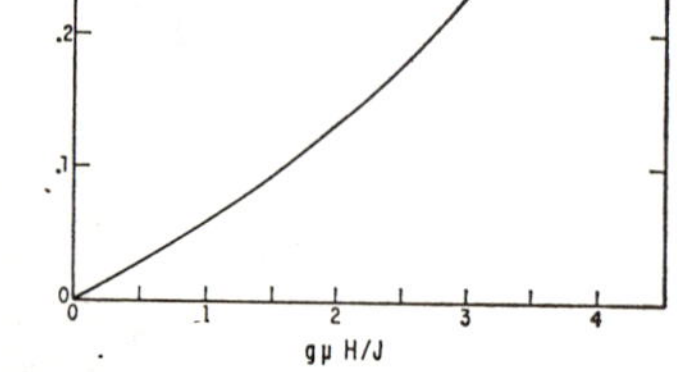

FIG. 3. The magnetization as a function of magnetic field for the antiferromagnetic chain at zero temperature.

commutes with the exchange energy (1). Here μ is the Bohr magneton, g the electron g factor, and S_i^z the z component of the ith spin. The lowest level of the chain with a given total spin S will have an energy

$$E_M(S)=E(S)-g\mu HS, \tag{58}$$

where $E(S)$ is the lowest energy in the absence of a magnetic field,

$$E(S)=2NJ\eta(S/N)+E_{AF}. \tag{59}$$

By $\eta(S/N)=\eta(\sigma)$ we mean the function corresponding to the lower curve in Fig. 1 and the values in Table I.

Let $S_0=N\sigma_0$ be the value of S for which $E_M(S)$ is a minimum. It is determined by setting the derivative with respect to S of the right-hand side of (58) equal to zero, with the result

$$g\mu H/J=2\eta'(\sigma_0). \tag{60}$$

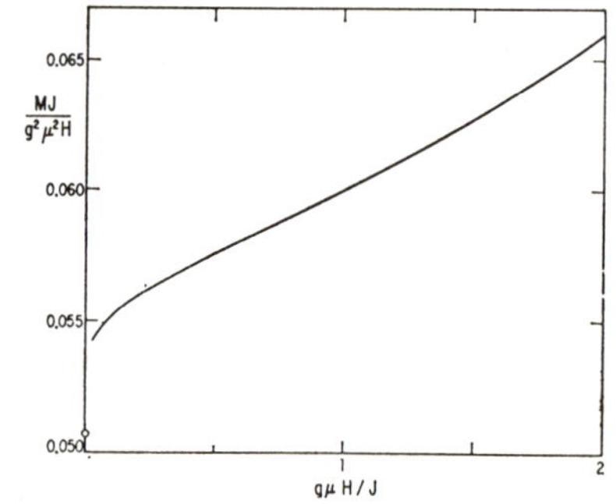

FIG. 4. The ratio of magnetization to magnetic field as a function of the field, at zero temperature. The circle at the left side [obtained from Eq. (62)] shows the ratio in the limit as H goes to zero.

Here $\eta'(\sigma)$ stands for $d\eta/d\sigma$; this quantity is tabulated as a function of σ in Table I. At zero temperature, the free energy of the chain F is equal to $E_M(S_0)$, and the average magnetization per spin, M, is given by

$$M=-N^{-1}dF/dH=g\mu\sigma_0. \tag{61}$$

Figure 3 shows M as a function of H. As H goes to zero, the ratio M/H approaches a limiting value,[17] the zero field susceptibility:

$$\chi=0.050661g^2\mu^2/J. \tag{62}$$

The numerical constant in (62), $a_0^2/4b_0$, is equal to $(2\pi^2)^{-1}$ to five significant figures. [See the remark following (45b).]

The ratio M/H as a function of H is shown in Fig. 4. A peculiar feature of this curve is that it approaches the limiting value at $H=0$ with infinite slope [as may be

[17] The reader may verify this using (37), (38), (41), and (42) to evaluate $d\eta/d\sigma$ for large values of α.

verified by means of (41) and (42)], a result of the fact that $\eta(\sigma)$ is not an analytic function near $\sigma=0$.

V. CONCLUSION

The principal results of the present paper are found in Fig. 1 (and Table I) which shows the upper and lower limits of the energies of states in class C as a function of the total spin of the state, in the limit $N \to \infty$. From the lower limit we have computed the magnetization as a function of field in the limit of zero temperature.

The magnetization curve is exact in the sense that we have made no approximations in treating the Hamiltonian (1) other than the neglect of terms which vanish (or are expected to vanish) as N becomes infinite relative to the terms retained. Nevertheless, certain hypotheses which enter into the Bethe formalism and our use thereof are, at present, supported by plausible arguments rather than rigorous proof. Perhaps we may say that the energies and magnetization curve computed above are "exact" in the same sense that Hulthén's value for the ground-state energy or the des Cloizeaux and Pearson spin-wave spectrum are "exact."

The problem of determining the magnetization and other thermodynamic quantities at low (nonzero) temperatures for the antiferromagnetic chain remains unsolved. (At high temperatures, the numerical results obtained for finite chains are probably adequate for most purposes.[8,13]) Unfortunately, if this problem is to be attacked within the framework of the Bethe formalism, it appears necessary to calculate energies for states outside class C, a rather difficult task since the k_j are in general complex.

ACKNOWLEDGMENTS

The author is indebted to Professor W. Kohn and Professor S. Ulam for helpful discussions, and to Dr. M. E. Fisher for stimulating and informative correspondence. He wishes to express gratitude for the hospitality extended to him by the Physics Department of the University of California at San Diego, and for the financial support of a National Science Foundation Postdoctoral Fellowship.

APPENDIX A. EXISTENCE OF SOLUTIONS OF EQUATION (6) FOR STATES IN CLASS C

It is convenient to regard (6) as a nonlinear transformation of the vector $(k_1, k_2, \ldots, k_r)$ into another vector with components

$$k_j' = 2\pi N^{-1}\lambda_j + N^{-1} \sum_{l(\neq j)} \phi(k_j,k_l)\,, \tag{A1}$$

where for ϕ_{jl} we have written $\phi(k_j,k_l)$. A solution to (6) is a fixed point of the transformation (A1). If we attempted to solve (A1) by iteration, a sensible starting value for the k_j would be $2\pi N^{-1}\lambda_j$, which lies within the interval $(0,2\pi)$. The function $\phi(k_j,k_l)$ is positive for $k_l > k_j$ and negative for $k_j > k_l$ if k_j and k_l are real and fall in the interval $(0,2\pi)$. Hence the second term on the right side of (A1) represents an "attractive force" between pairs of wave vectors, whereas the term $2\pi N^{-1}\lambda_j$ tends to "anchor" k_j near its starting position.

When the starting values of k_j and k_{j+1} are too close together, successive iteration may result in a "collision": $k_{j+1}=k_j$. But a solution to (6) for which two k's coincide is trivial: The corresponding wave function (5) vanishes identically. A possible remedy in this situation is to make k_j and k_{j+1} complex. However, condition (10) for states in class C ensures that the initial values of the k_j are far enough apart to make "collisions" impossible, and hence one may expect a solution to (6) for which all the k's are real. We shall now make these intuitive notions more precise.

Let V be the real r-dimensional space of vectors of the form $(k_1, k_2, \ldots, k_r)$ and let K be the subset of those vectors whose components satisfy the inequalities

$$\begin{gathered} k_1 \geq 2\pi N^{-1}, \quad k_r \leq 2\pi(1-N^{-1}), \\ k_{j+1}-k_j \geq 2\pi N^{-1}, \quad j=1, 2, \ldots, r-1. \end{gathered} \tag{A2}$$

The function $\phi(k_j,k_l)$ [see (7)] is, in absolute value, less than or equal to π. It is continuous for all k_j,k_l in the interval $(0,2\pi)$ except at the points $k_j=k_l$. Thus, the transformation (A1) is continuous on the set K. The inequality

$$k_{j+1}'-k_j' \geq 2\pi N^{-1}(\lambda_{j+1}-\lambda_j) - 2N^{-1}\phi(k_j, k_{j+1}) \tag{A3}$$

is a consequence of the fact that the function $\phi(k_j,k_l)$ in either the region $k_j<k_l$ or the region $k_j>k_l$ is monotone increasing in k_j and decreasing in k_l.

The λ_j for states in class C satisfy the inequalities (9) and (10). This, together with the inequality (A3) and the properties of ϕ mentioned above, implies that the k_j' satisfy the inequalities (A2). In other words, the transformation (A1) carries the set K into itself. The set K is closed, convex, and bounded; and, since the transformation (A1) is continuous on K, there must exist at least one fixed point by Brouwer's theorem.[18] We have not been able to show that the solution is unique.

APPENDIX B. THE LEDINEGG-URBAN CALCULATION

For convenience, we enclose equation numbers from the paper by Ledinegg and Urban[5] in square brackets, and transcribe the equations in our notation. They obtain the following asymptotic behavior for σ and η as a function of α for large α using Eqs. (32)–(34):

$$\sigma \sim 0.518 e^{(-1/2)\pi\alpha}, \tag*{[38d]}$$

$$\eta \sim 0.398 e^{(-1/2)\pi\alpha}/\alpha^2. \tag*{[37]}$$

The result [38d] agrees with our calculations—see (38), (41), and (44)—except for the numerical constant.

---

[18] G. T. Whyburn, *Analytic Topology*, American Mathematical Society Colloquium Publications, Vol. XXVIII (American Mathematical Society, New York, 1942), p. 243.

However, [37] is in disagreement with (37) and (41); in particular, the exponent in the former is half that in the latter. The numerical solutions to Eqs. (32)–(34) support (37) and (41) rather than [37].

Ledinegg and Urban obtain their asymptotic estimate as follows. The function $f(\xi)$ in (32) (their Eq. [23]) is set equal to

$$f(\xi)=f_0(\xi)+\phi(\xi), \qquad [24a]$$

where $f_0(\xi)$ is the solution of (19) given by (23). They assume that

$$|\phi(\xi)|/f_0(\xi)\ll 1, \qquad [24c]$$

stating that this inequality will be justified through an explicit calculation of the function ϕ. The justification for [24c] is contained in their Eq. [33a] which indicates that [24c] is correct, but *only* for $|\xi|<\alpha$. But they have already used [24c] for $|\xi|$ *greater than* α in order to deduce an approximate integral equation for ϕ, [26b]. Thus their treatment is not internally consistent.

On the basis of our results it may be shown that $f(\xi)$ decreases as ξ^{-2} for large ξ; on the other hand, $f_0(\xi)$ decreases exponentially. Hence, in fact, [24c] does not hold for $\xi>\alpha$, and the integral equation [26b] is not correct.

Time-Dependent Phenomena

Chapter 7

Time-Dependent Phenomena

It is our main purpose in including this chapter to provide a diversified bibliography in the field of many-body one-dimensional dynamics and nonequilibrium mechanics. Topics I and II in the bibliography following are of greatest general interest, as they deal with the approach of a complex system to thermodynamic equilibrium.

In the first of the reprinted papers, Ford (1961) discusses when and how energy-sharing among the normal modes of a linear chain can occur, explaining why Fermi, Pasta, and Ulam (1955) failed to obtain equipartition in their famous and pioneering computer study of a nonlinear system. In a subsequent study, Ford and Waters (1963) discuss the conditions for rapid approach to equilibrium of weakly anharmonic coupled oscillators.

In the third of the reprinted papers, Jepsen (1965) gives general and rigorous formulas on the motion in a Tonk's gas (hard rods), and obtains various results concerning Poincaré cycles, delocalization of a particle with time, and electrical conductivity (when one particle is charged). Of the many papers which have sought to derive the properties of Brownian motion from microscopic considerations, we reprint only the most recent, by Ford, Kac, and Mazur (1965). The reader should also note earlier and parallel papers by Rubin (1960), Turner (1960), and others listed in the bibliography. We have not attempted to do full justice to this field due to lack of space, although it has a rich history over the past 60 years going back to the original investigations of Gibbs and of Einstein. Other related subjects are included in the bibliography.

Bibliography

I. General Problems of Irreversibility in Mechanical Systems, Poincaré Cycles, and Approach to Equilibrium

E. Fermi, J. Pasta, and S. Ulam (1955). "Studies of Nonlinear Problems," *Los Alamos Sci. Lab. Rept.* LA-1940.

J. Ford (1958). "Approach of One-Dimensional Systems to Equilibrium," *Phys. Rev.* **112,** 1445.

*J. Ford (1961). "Equipartition of Energy for Nonlinear Systems," *J. Math. Phys.* **2,** 387.

*J. Ford and J. Waters (1963). "Computer Studies of Energy Sharing and Ergodicity for Nonlinear Oscillator Systems," *J. Math. Phys.* **4,** 1293.

H. L. Frisch (1956). "Poincaré Recurrences," *Phys. Rev.* **104,** 1.

P. C. Hemmer (1959). "Dynamical and Stochastic Types of Motion in the Linear Chain," Thesis, Trondheim, Norway.

P. C. Hemmer, L. C. Maximon, and H. Wergeland (1958). "Recurrence Time of a Dynamical System," *Phys. Rev.* **111,** 689.

*D. W. Jepsen (1965). "Dynamics of a Simple Many-Body System of Hard Rods," *J. Math. Phys.* **6,** 405.

G. Klein and I. Prigogine (1958). "Sur la Mécanique Statistique des Phénomènes Irréversibles, I and II," *Physica* **19,** 74 and 89.

P. Mazur and E. Montroll (1960). "Poincaré Cycles, Ergodicity, and Irreversibility in Assemblies of Coupled Harmonic Oscillators," *J. Math. Phys.* **1,** 70.

E. Montroll and K. Shuler (1957). "Studies in Nonequilibrium Rate Processes. I, The Relaxation of a System of Harmonic Oscillators," *J. Chem. Phys.* **26,** 454.

H. Mori (1958). "Time-Correlation Functions in the Statistical Mechanics of Transport Processes." *Phys. Rev.* **111,** 694.

R. Northcote and R. B. Potts (1964). "Energy Sharing and Equilibrium for Nonlinear Systems," **5,** 383.

E. Teramoto and C. Suzuki (1955). "The Statistical Mechanical Aspect of *H*-Theorem," *Progr. Theoret. Phys. (Kyoto)* **14,** 411.

E. Teramoto (1962). "Heat Flow in the Linear Chain of Harmonically Coupled Particles," *Progr. Theoret. Phys. (Kyoto)* **28,** 1059.

N. Zabusky (1962). "Exact Solutions for the Vibrations of a Nonlinear Continuous Model String," *J. Math. Phys.* **3,** 1028.

II. Comparison of Studies of Mechanical Motion with Predictions of Statistical Models, Including Brownian Motion

*G. W. Ford, M. Kac, and P. Mazur (1965). "Statistical Mechanics of Assemblies of Coupled Oscillators," *J. Math. Phys.* **6,** 504.

R. J. Rubin (1956). "Random Motion of a Heavy Particle Substituted in a One-Dimensional Crystal Lattice," *in Proc. Intern. Symp. on Transport Processes in Statistical Mechanics (Brussels) Aug. 27–31, 1956*, pp. 155–160. Wiley (Interscience), New York.

R. J. Rubin (1960). "Statistical Dynamics of Simple Cubic Lattices. Model for the Study of Brownian Motion," *J. Math. Phys.* **1,** 309.

M. Toda and Y. Kogure (1962). "Statistical Dynamics of Systems of Interacting Oscillators," *Progr. Theoret. Phys. (Kyoto) Suppl.* **23,** 157.

R. E. Turner (1960). "Motion of a Heavy Particle in a One-Dimensional Chain," *Physica* **26,** 269; "Temporal Fluctuations in a Classical Linear System," *ibid.* 274.

III. Lifetime Effects in the Anharmonic Lattice

C. Manohar (1965). "Anharmonic Lattice in One Dimension: Energy Shift and Lifetime of an Excitation," *Phys. Rev.* **138,** A731.

A. A. Maradudin (1962). "On the Lifetime of Phonons in One-Dimensional Crystals," *Phys. Letters* **2,** 298.

IV. Temporal Properties of the Disordered Lattice

J. Hori (1962). "Long-Time Behavior of a Random Lattice," *J. Math. Phys.* **3,** 382.

J. Hori (1962). "Remarks on Time-Dependent Behavior of Disordered Lattice," *Progr. Theoret. Phys.* (*Kyoto*) *Suppl.* **23,** 223.

S. Kashiwamura and E. Teramoto (1962). "Effect of an Isotopic Impurity on the Heat Flow in One-Dimensional Coupled Harmonic Oscillators," *Progr. Theoret. Phys.* (*Kyoto*) *Suppl.* **23,** 207.

S. Kashiwamura (1962). "Statistical Dynamical Behaviors of a One-Dimensional Lattice with an Isotopic Impurity," *Progr. Theoret. Phys.* (*Kyoto*) **27,** 571.

A. A. Maradudin, G. H. Weiss, and D. W. Jepsen (1961). "Nonequilibrium Processes in Isotopically Disordered Crystals," *J. Math. Phys.* **2,** 349.

E. Teramoto and S. Takeno (1960). "Time Dependent Problems of the Localized Lattice Vibration," *Progr. Theoret. Phys.* (*Kyoto*) **24,** 1349.

V. Motion of Dislocations

W. Atkinson and N. Cabrera (1965). "Motion of a Frenkel-Kontorowa Dislocation in a One-Dimensional Crystal," *Phys. Rev.* **138,** A763.

J. H. Weiner and W. T. Sanders (1964). "Peierls Stress and Creep of a Linear Chain," *Phys. Rev.* **134,** A1007.

J. H. Weiner (1964). "Dislocation Velocities in a Linear Chain," *Phys. Rev.* **136,** A863.

JOURNAL OF MATHEMATICAL PHYSICS VOLUME 6, NUMBER 4 APRIL 1965

Statistical Mechanics of Assemblies of Coupled Oscillators*

G. W. FORD†
Department of Physics, University of Michigan, Ann Arbor, Michigan
M. KAC
The Rockefeller Institute, New York, New York
AND
P. MAZUR
Lorentz Institute for Theoretical Physics, Leiden, The Netherlands
(Received 25 September 1964)

It is shown that a system of coupled harmonic oscillators can be made a model of a heat bath. Thus a particle coupled harmonically to the bath and by an arbitrary force to a fixed center will (in an appropriate limit) exhibit Brownian motion. Both classical and quantum mechanical treatments are given.

1. INTRODUCTION

OUR aim here is to study a simple mechanical model, a chain of coupled harmonic oscillators, in order to come to a deeper understanding of some of the phenomena associated with Brownian motion.[1,2] With this model we are able to carry through the program one would like to achieve with more realistic interactions. This program, which really goes back to Gibbs, goes as follows[3]:

(i) Solve the equations of motion of the mechanical system consisting of a Brownian particle coupled to heat bath. The solution consists of expressions for the coordinates and momenta at time t in terms of the initial coordinates and momenta.

(ii) Assume the initial coordinates and momenta of the heat bath to be distributed according to some statistical distribution, e.g., that of the canonical ensemble.

(iii) Show that the coordinate and momentum of the Brownian particle, as functions of time, will then represent stochastic processes (whose statistical properties arise from the initial distribution of the heat bath) of the kind familiar from standard theories.

This is a very ambitious program, and it is no wonder that it can be carried out only for the simplest models.

* Part of the work reported in this paper was done while one of us (M. K.) was H. A. Lorentz Visiting Professor at the University of Leiden. When this work was reported at Yeshiva University in the Fall of 1963 Dr. N. L. Balazs informed us that he had independently obtained some of the results concerning the classical case.

† Supported in part by a grant from the National Science Foundation.

[1] There is an extensive literature on the motion of coupled oscillators, mostly concerned with motion in a lattice with nearest-neighbor interactions. Some of the more recent articles which have a bearing on our work are: P. Mazur and E. Montroll, J. Math. Phys. **1**, 70 (1960); P. C. Hemmer, "Dynamic and Stochastic Types of Motion in the Linear Chain," thesis, Norges Tekniske Høgskole, Trondheim, Norway (1959); R. J. Rubin, J. Math. Phys. **1**, 309 (1960); **2**, 373 (1961); M. Toda and Y. Koguri, Suppl. Progr. Theoret. Phys. (Kyoto) **23**, 157 (1962); R. E. Turner, Physica **26**, 274 (1960).

[2] The classic papers on the phenomenological theory of Brownian motion are: G. E. Uhlenbeck and L. S. Ornstein, Phys. Rev. **36**, 823 (1930); M. C. Wang and G. E. Uhlenbeck, Rev. Mod. Phys. **17**, 323 (1945).

[3] See the article by H. Wergeland in *Fundamental Problems in Statistical Mechanics*, edited by E. G. D. Cohen (North-Holland Publishing Company, Amsterdam, 1962).

We have a good idea of what the results of this program should be, since, after all, Brownian motion is a throughly studied experimental phenomenon with a satisfactory phenomenological theory. In general we expect to show:

(i) The approach to equilibrium. In particular, the distribution of momentum of the Brownian particle should approach the Maxwellian distribution.

(ii) The description of this approach to equilibrium should be *contracted*, i.e., should involve only a small number of the possible variables describing the system. Another way of saying this is that there should be a reduced description in terms of which the stochastic process is Markoffian.

(iii) What changes occur when one adopts a quantum description of the system. Here we have less of an idea of what we should expect, but somehow the basic features of the stochastic process should be preserved in the quantum description.

We can be still more explicit about what we mean by a contracted description of Brownian motion; we mean the Langevin equation of motion. For a Brownian particle of mass m acted upon by an outside force $F(x)$ this equation is

$$\dot{p} = -fp/m + E(t) + F(x), \tag{1}$$

where $p = m\dot{x}$ is the momentum of the Brownian particle, f the friction constant, and $E(t)$ is the random force due to the heat bath. This random force is a purely random Gaussian process characterized by

$$\langle E(t)\rangle = 0, \quad \langle E(t)E(t')\rangle = 2fkT\,\delta(t - t'),$$

where T is the temperature of the heat bath and k is Boltzmann's constant. Note that the Langevin equation is a contracted description in the sense that the heat bath is described by only two parameters, the friction constant and the temperature, and that only the first two time derivatives of the position x of the Brownian particle appear.[4]

In Sec. 2 we discuss the dynamics of a system of coupled oscillators. There we *formally* carry through the program for the case of an arbitrary coupling of the oscillators. In Sec. 3 the arbitrary coupling of a linear chain is considered, and we show that there is a coupling for which, in the limit of an infinite chain, the resulting stochastic process is Markoffian. Taking the chain of oscillators with this coupling as the heat bath, we derive in Sec. 4 the Langevin equation for a Brownian particle. In Sec. 5 we discuss the quantum description of the system, and in Sec. 6 we discuss the quantum Langevin equation. Finally, in Sec. 7 we consider the Brownian motion of a quantum oscillator.

[4] That is, there are no memory effects.

2. DYNAMICS OF A SYSTEM OF COUPLED OSCILLATORS

Consider a system of $(2N + 1)$ coupled oscillators with Hamiltonian:

$$H = \frac{1}{2}\sum_{j=-N}^{N} p_j^2 + \frac{1}{2}\sum_{j,k=-N}^{N} q_j A_{jk} q_k. \tag{2}$$

Here q_j and p_j are, respectively, the canonical coordinate and momentum of the jth oscillator. The mass of each oscillator has been taken to be unity. The interactions of the oscillators are characterized by the $(2N + 1) \times (2N + 1)$ symmetric matrix $\mathbf{A}$, whose elements are the A_{jk}. At present, we make no special assumptions about this matrix except that it has no negative eigenvalues. The canonical equations of motion may be conviently written in matrix notations as follows:

$$\dot{\mathbf{q}} = \mathbf{p}, \qquad \dot{\mathbf{p}} = -\mathbf{A}\mathbf{q}. \tag{3}$$

Here $\mathbf{p}$ and $\mathbf{q}$ are $(2N + 1)$-rowed column matrices whose elements are the p_j and q_j, respectively. The formal solution of the equations of motion is

$$\begin{aligned} \mathbf{q}(t) &= \cos(\mathbf{A}^{\frac{1}{2}}t)\mathbf{q}(0) + \mathbf{A}^{-\frac{1}{2}}\sin(\mathbf{A}^{\frac{1}{2}}t)\mathbf{p}(0), \\ \mathbf{p}(t) &= -\mathbf{A}^{\frac{1}{2}}\sin(\mathbf{A}^{\frac{1}{2}}t)\mathbf{q}(0) + \cos(\mathbf{A}^{\frac{1}{2}}t)\mathbf{p}(0), \end{aligned} \tag{4}$$

where, e.g.,

$$\cos \mathbf{A}^{\frac{1}{2}}t = \sum_{n=0}^{\infty} \frac{(-)^n}{(2n)!}\mathbf{A}^n t^{2n}. \tag{5}$$

We now assume that at $t = 0$ the system is in equilibrium at temperature T. That is, we assume that the $q_j(0)$ and $p_j(0)$ are distributed according to the canonical distribution

$$D(\mathbf{q}(0), \mathbf{p}(0)) = (2\pi/\beta)^{2N+1}(\det \mathbf{A})^{-\frac{1}{2}}e^{-\beta H(\mathbf{q}(0),\mathbf{p}(0))}, \tag{6}$$

where $\beta = (kT)^{-1}$ and $\det \mathbf{A}$ is the determinant of $\mathbf{A}$. Note that there is a difficulty here, since $\det \mathbf{A} = 0$ if $\mathbf{A}$ has zero eigenvalues. We therefore assume for the time being that $\mathbf{A}$ has no zero eigenvalues. The expectation of any function $F(\mathbf{q}(0), \mathbf{p}(0))$ is given by

$$\langle F\rangle = \int \cdots \int dq_{-N}(0)\cdots dq_N(0)\,dp_{-N}(0)\cdots dp_N(0) \times F(\mathbf{q}(0), \mathbf{p}(0))D(\mathbf{q}(0), \mathbf{p}(0)). \tag{7}$$

Now we ask, what are the properties of the stochastic variables $q_j(t)$ and $p_j(t)$ which result from (4) under the distribution (6)? First of all, it is clear that the process is Gaussian. This follows from the fact that the distribution (6) is Gaussian and that the relation (4) is linear. That the process is stationary follows from the Liouville theorem of mechanics, which states that

$$D(\mathbf{q}(t), \mathbf{p}(t)) = D(\mathbf{q}(0), \mathbf{p}(0)). \tag{8}$$

It is well known that the statistical properties of such a stationary Gaussian process are completely described by the pair correlation functions. In our case these are obtained in Appendix 1; the results are[5]

$$\langle p_j(t)p_k(t+\tau)\rangle = kT\,||\cos \mathbf{A}^{\frac{1}{2}}\tau||_{jk}, \tag{9a}$$

$$\langle q_j(t)p_k(t+\tau)\rangle = -kT\,||\mathbf{A}^{-\frac{1}{2}}\sin \mathbf{A}^{\frac{1}{2}}\tau||_{jk}, \tag{9b}$$

$$\langle q_j(t)q_k(t+\tau)\rangle = kT\,||\mathbf{A}^{-1}\cos \mathbf{A}^{\frac{1}{2}}\tau||_{jk}. \tag{9c}$$

Note that the position correlation (9c) involves the inverse of **A**, which does not exist if **A** has zero eigenvalues.

If we fix our attention on a single oscillator, say the one with index 0, the momentum autocorrelation is

$$\langle p_0(t)p_0(t+\tau)\rangle = kT\,||\cos \mathbf{A}^{\frac{1}{2}}\tau||_{00}. \tag{10}$$

This is the autocorrelation of a stationary Gaussian process in one variable. It is well known that such a process is Markoffian if and only if the autocorrelation is an exponential, i.e.,

$$\langle p_0(t)p_0(t+\tau)\rangle = kTe^{-f|\tau|}, \tag{11}$$

where f is a positive constant. The question we turn to in the next section is that of finding an interaction matrix **A** for which (10) assumes the form (11).

3. THE INTERACTION MATRIX

In our model we assume the $(2N+1)$ oscillators are identical and that they are arranged in a chain with cyclic boundary conditions. This means that the interaction matrix A is a symmetric cyclic matrix.[6] The elements of such a matrix can be written in the form

$$A_{mn} = \frac{1}{2N+1}\sum_{k=-N}^{N}\omega_k^2 \times \exp\left\{i\frac{2\pi}{2N+1}k(m-n)\right\}, \tag{12}$$

where the symmetry of A requires

$$\omega_k^2 = \omega_{-k}^2. \tag{13}$$

The eigenvalues of this matrix are the quantities ω_s^2, $s = -N, -N+1, \cdots, N$. That is

$$\mathbf{A}\boldsymbol{\xi}^{(s)} = \omega_s^2\boldsymbol{\xi}^{(s)}, \tag{14}$$

where the eigenvector $\boldsymbol{\xi}^{(s)}$ is a $(2N+1)$-rowed column matrix whose elements are

$$\xi_n^{(s)} = (2N+1)^{-\frac{1}{2}}\exp\{i[2\pi/(2N+1)]sn\}. \tag{15}$$

These properties follow from the elementary formula

$$\sum_{k=-N}^{N}\exp\left(ik(m-n)\frac{2\pi}{2N+1}\right) = \delta_{m,n}, \qquad -N \le m, n \le N. \tag{16}$$

With this formula we can also readily demonstrate that if $F(\mathbf{A})$ is a function of the matrix **A**, then

$$||F(\mathbf{A})||_{m,n} = \frac{1}{2N+1}\sum_{k=-N}^{N}F(\omega_k^2) \times \exp\left\{i\frac{2\pi}{2N+1}k(m-n)\right\}. \tag{17}$$

Note, incidentally, that the special case of nearest-neighbor interactions is that for which

$$\omega_s^2 = \bar{\omega}^2\sin^2[\pi s/(2N+1)]. \tag{18}$$

Consider next the limit $N \to \infty$, the infinite chain. If we make the additional assumption that ω_s^2 is slowly varying function of s, then (12) becomes

$$\begin{aligned} A_{mn} &= \frac{1}{2\pi}\int_{-\pi}^{\pi}d\theta\, f(\theta)e^{i(m-n)\theta} \\ &= \frac{1}{2\pi}\int_{-\pi}^{\pi}d\theta\, f(\theta)\cos(m-n)\theta, \end{aligned} \tag{19}$$

where

$$f(\theta) = \{\omega_s^2\}_{s=(2N+1)\theta/2\pi}. \tag{20}$$

The relation (17) becomes in this limit

$$||F(\mathbf{A})||_{m,n} = \frac{1}{2\pi}\int_{-\pi}^{\pi}d\theta\, F(f(\theta))\cos(m-n)\theta. \tag{21}$$

We are now ready to turn to the problem posed at the end of the last section; that of finding an interaction matrix A for which

$$||\cos \mathbf{A}^{\frac{1}{2}}t||_{00} = e^{-f|t|} \tag{22}$$

Using the result (17) we see that for a finite matrix

$$||\cos \mathbf{A}^{\frac{1}{2}}t||_{00} = \frac{1}{2N+1}\sum_{k=-N}^{N}\cos\omega_k t. \tag{23}$$

For any choice of the ω_k this is a quasiperiodic function and cannot be of the form (22). However, in the limit of large N, we can use (21) which gives

$$||\cos \mathbf{A}^{\frac{1}{2}}t||_{00} = \frac{1}{2\pi}\int_{-\pi}^{\pi}d\theta\cos\{[f(\theta)]^{\frac{1}{2}}t\}. \tag{24}$$

Taken with our requirement (22), this becomes an integral equation for $f(\theta)$. The answer is essentially

[5] We use the notation $||\mathbf{M}||_{jk}$ for the element in the jth row and kth column of a matrix **M**.

[6] See, e.g., G. Kowalewski, *Determinantentheorie* (Chelsea Publishing Company, New York, 1948), 3rd ed., p. 105.

unique and is[7]

$$f(\theta) = f^2 \tan^2 \tfrac{1}{2}\theta. \tag{25}$$

There is a difficulty here, however; when (25) is inserted in the expression (19) for the matrix elements, the expression diverges! What we must do is employ a second limiting process (after the limit $N \to \infty$), defining

$$f_{\omega_L}(\theta) = \begin{cases} f^2 \tan^2 \dfrac{\theta}{2}, & |\theta| < \theta_L, \\ 0, & \theta_L \le |\theta| \le \pi. \end{cases} \tag{26}$$

Here

$$\omega_L \equiv f \tan \tfrac{1}{2}\theta_L \tag{27}$$

is a high-frequency cutoff in the spectrum of eigenfrequencies which ensures that the matrix elements (19) are finite. This frequency cutoff corresponds to a "microscopic interaction time" ω_L^{-1} which we assume is very small compared with the "macroscopic relaxation time" f^{-1}. The result (22) holds strictly only in the limit $\omega_L \to \infty$. Alternatively, we can say that, for $\omega_L \gg f$ the result (22) holds for times long compared with ω_L^{-1}.

Our model, then, is that the interaction matrix elements are given by (19) with $f(\theta)$ given by (26) with $\omega_L \gg f$. If in (24) we make the change of variable $\omega = f \tan \frac{1}{2}\theta$, we find

$$\|\cos \mathbf{A}^{\frac{1}{2}} t\|_{00} = \frac{1}{\pi} \int_{-\omega_L}^{\omega_L} d\omega \frac{f}{\omega^2 + f^2} \cos \omega t. \tag{28}$$

In the limit $\omega_L \to \infty$ this becomes $e^{-f|t|}$ and, therefore, the Gaussian process $p_0(t)$ becomes also Markoffian.

4. THE LANGEVIN EQUATION

Having seen that our model leads to a Gaussian Markoffian stochastic process for the collection of coupled oscillators, we are led to ask whether it also leads to the Langevin equation for the motion of a single particle coupled to a heat bath consisting of such oscillators. In this section we see that this is indeed the case.

We select from the chain of $(2N + 1)$ oscillators, the particle with index 0 to be the Brownian particle; the remaining $2N$ oscillators represent the heat bath. The outside force on this particle we denote by

$$F(t) \equiv F(q_0(t)). \tag{29}$$

If we define $\mathbf{F}(t)$ to be a $(2N + 1)$-rowed column matrix whose elements are all zero except for the zeroth element which is $F(t)$, then in the notation of Sec. 2 the equations of motion for coupled "particle and heat path" are

$$\dot{\mathbf{q}} = \mathbf{p}, \qquad \dot{\mathbf{p}} = -\mathbf{A}\mathbf{q} + \mathbf{F}(t). \tag{30}$$

The formal solution of these equations is

$$\mathbf{q}(t) = \cos \mathbf{A}^{\frac{1}{2}} t\, \mathbf{q}(0) + \mathbf{A}^{-\frac{1}{2}} \sin \mathbf{A}^{\frac{1}{2}} t\, \mathbf{p}(0) + \int_0^t dt' \frac{\sin \mathbf{A}^{\frac{1}{2}}(t - t')}{\mathbf{A}^{\frac{1}{2}}} \mathbf{F}(t'), \tag{31a}$$

$$\mathbf{p}(t) = -\mathbf{A}^{\frac{1}{2}} \sin \mathbf{A}^{\frac{1}{2}} t\, \mathbf{q}(0) + \cos \mathbf{A}^{\frac{1}{2}} t\, \mathbf{p}(0) + \int_0^t dt' \cos \mathbf{A}^{\frac{1}{2}}(t - t') \mathbf{F}(t'). \tag{31b}$$

If, now, we take the zeroth element of Eq. (30) for $\dot{\mathbf{p}}$ and substitute (31a) we get

$$\dot{p}_0 = -\sum \|\mathbf{A} \cos \mathbf{A}^{\frac{1}{2}} t\|_{0i}\, q_i(0) - \sum_i \|\mathbf{A}^{\frac{1}{2}} \sin \mathbf{A}^{\frac{1}{2}} t\|_{0i}\, p_i(0) - \int_0^t dt' \|\mathbf{A}^{\frac{1}{2}} \sin \mathbf{A}^{\frac{1}{2}}(t - t')\|_{00} F(t') + F(t).$$

Next, we eliminate $p_0(0)$ between this equation and the zeroth element of Eq. (31b). The result can be written in the form

$$\dot{p}_0 - F(t) = -\gamma(t) p_0 + E(t) + \int_0^t dt' [\gamma(t) - \gamma(t - t')] \times \|\cos \mathbf{A}^{\frac{1}{2}}(t - t')\|_{00} F(t'), \tag{32}$$

where

$$\gamma(t) = \frac{\|\mathbf{A}^{\frac{1}{2}} \sin \mathbf{A}^{\frac{1}{2}} t\|_{00}}{\|\cos \mathbf{A}^{\frac{1}{2}} t\|_{00}} = -\frac{d}{dt} \log \|\cos \mathbf{A}^{\frac{1}{2}} t\|_{00}, \tag{33}$$

and

$$E(t) = -\sum_i \{\gamma(t) \|\mathbf{A}^{\frac{1}{2}} \sin \mathbf{A}^{\frac{1}{2}} t\|_{0i} + \|\mathbf{A} \cos \mathbf{A}^{\frac{1}{2}} t\|_{0i}\} q_i(0) + \sum_i \{\gamma(t) \|\cos \mathbf{A}^{\frac{1}{2}} t\|_{0i} - \|\mathbf{A}^{\frac{1}{2}} \sin \mathbf{A}^{\frac{1}{2}} t\|_{0i}\} p_i(0). \tag{34}$$

Note that the coefficient of $p_0(0)$ vanishes in this expression. Equation (32) is the equation of motion for the Brownian particle. The right-hand side is the net force exerted on the Brownian particle by the other particles, i.e., by the heat bath. The first term represents a frictional force with time-dependent "friction coefficient" $\gamma(t)$, the second term represents a fluctuating force $E(t)$ depending upon the initial state of the heat bath, and the third

[7] What is unique is the *spectrum* of eigenfrequencies: $g(\omega) = 2\omega/(\pi f'(\theta))$, where θ is the function of ω obtained by inverting the equation $\omega^2 = f(\theta)$. For (25), $g(\omega) = (2f/\pi) \cdot (\omega^2 + f^2)^{-1}$.

term represents a memory effect depending upon the past history of the motion of the Brownian particle.

If we assume that the interaction between the Brownian particle and the heat bath is invariant under translations, we have

$$\sum_j A_{ij} = 0, \tag{35}$$

and this in turn implies that (34) may be written

$$E(t) = -\sum_j{}' \{\gamma(t)\ ||\mathbf{A}^{\frac{1}{2}} \sin \mathbf{A}^{\frac{1}{2}}t||_{0j} + ||\mathbf{A} \cos \mathbf{A}^{\frac{1}{2}}t||\}[q_j(0) - q_0(0)] + \sum_j{}' \{\gamma(t)\ ||\cos \mathbf{A}^{\frac{1}{2}}t||_{0j} - ||\mathbf{A}^{\frac{1}{2}} \sin \mathbf{A}^{\frac{1}{2}}t||_{0j}\}p_j(0), \tag{36}$$

where the prime on the sum denotes the omission of the term $j = 0$. Thus the fluctuating force depends only upon the initial coordinates of the particles of the heat bath *relative* to the initial coordinate of the Brownian particle, and is independent of the initial coordinate and momentum of the Brownian particle.

Consider now what happens when the matrix of interactions is that of the model discussed in Sec. 3, in which the matrix elements are given by (19) with $f(\theta)$ given by (26) in the limit $\omega_L \gg f$. For this model

$$||\cos \mathbf{A}^{\frac{1}{2}}t||_{00} = e^{-f|t|} \tag{37}$$

and therefore, from (33) we find that

$$\lim \gamma(t) = f, \tag{38}$$

which is a constant.

This in turn implies that the last term on the right-hand side of (32) (the memory-effect term) becomes, in the limit, identically 0!

With these results, (32) takes the form

$$\dot{p}_0 - F(t) = -fp_0 + E(t), \tag{39}$$

with

$$E(t) = -\sum_j ||f\mathbf{A}^{\frac{1}{2}} \sin \mathbf{A}^{\frac{1}{2}}t + \mathbf{A} \cos \mathbf{A}^{\frac{1}{2}}t||_{0j}q_j(0) + \sum_j ||f \cos \mathbf{A}^{\frac{1}{2}}t - \mathbf{A}^{\frac{1}{2}} \sin \mathbf{A}^{\frac{1}{2}}t||_{0j}p_j(0). \tag{40}$$

Equation (39) is the Langevin equation.

It remains to prove that the statistical properties of $E(t)$ become (again in the limit $N \to \infty$, $\omega_L \gg f$) those of a purely random Gaussian process. This depends, of course on the statistical assumptions concerning initial positions and momenta.

We would like to require that at $t = 0$ the heat bath is in equilibrium at temperature T and the simplest way of doing this is to assume that the initial distribution is the canonical distribution (6). This however is, strictly speaking, impossible since (35) implies that $\omega_0 = 0$ so that det $\mathbf{A} = 0$ and the canonical distribution becomes improper. The difficulty is not serious and can be remedied, e.g., by slightly modifying the matrix $\mathbf{A}$; i.e., replacing it by ($\mathbf{I}$ is the unit matrix)

$$\mathbf{A} + \varepsilon_N \mathbf{I}, \tag{41}$$

where ε_N, though positive for every finite N, approaches 0 as $N \to \infty$ (as fast as one pleases). Now (35) is only approximately true and the canonical distribution (6) is proper. We hope that the reader does not become unduly confused by our use of the symbol $\mathbf{A}$ to denote three different matrices. We use it to denote the finite cyclic matrix (12), the modification (41), and last but not least for the infinite cyclic matrix (defined only with the cutoff ω_L). Clearly, since the distribution (6) of the $q_j(0)$ and $p_j(0)$ is Gaussian, $E(t)$ is a Gaussian process. We can form its covariance with the help of the results of Appendix 1, and, in the limit considered throughout this paper, we find that

$$\langle E(t)E(t')\rangle = kT\ ||(f^2 + \mathbf{A}) \cos \mathbf{A}^{\frac{1}{2}}(t - t')||_{00}. \tag{42}$$

But the matrix $\mathbf{A}$ is given by the model of Sec. 3; hence, using (21), we find

$$\langle E(t)E(t')\rangle = \frac{kT}{2\pi} \int_{-\pi}^{\pi} d\theta \left(f^2 + f^2 \tan^2 \frac{\theta}{2}\right) \cos f \tan \frac{\theta}{2} (t - t') = \frac{kTf}{\pi} \int_{-\infty}^{\infty} d\omega \cos \omega(t - t').$$

This last integral is the well known expression for the Dirac delta function, so we have

$$\langle E(t)E(t')\rangle = 2fkT\,\delta(t - t'). \tag{43}$$

Thus $E(t)$ is a purely random, Gaussian, stochastic process and Eq. (39) is the Langevin equation for Brownian motion.

In order that the equation of motion (32) become the Langevin equation it is necessary that (i) the friction constant $\gamma(t)$ be independent of time; (ii) the stochastic process $E(t)$ be a purely random Gaussian process; (iii) the memory effects disappear.

We feel that it is striking that, for our model, these three properties are intimately related. Undoubtedly this fact is of much more general significance.

5. THE MOTION OF COUPLED QUANTUM OSCILLATORS

We turn now to the question of the changes required for a quantum mechanical description of the motion of coupled oscillators. The answer is that much of our previous discussion is, formally at any rate, entirely unchanged. Thus in our discussion of Sec. 2, the Hamiltonian (2) is unchanged, but the coordinates q_i and momenta p_i are now operators whose commutation rules are

$$[q_i, q_k] = [p_i, p_k] = 0, \qquad (44)$$
$$[q_i, p_k] = i\hbar\delta_{ik}.$$

The equations of motion (3) are the equations of motion in the Heisenberg picture, and their solution (4) relates the Heisenberg operators at time t to the operators at the initial time.

At $t = 0$, we assume the system is in equilibrium at temperature T. In the quantum description this means that the initial state of the system is described by the density matrix corresponding to the canonical ensemble:

$$\rho(\mathbf{q}(0), \mathbf{p}(0)) = \exp\{-\beta H(\mathbf{q}(0), \mathbf{p}(0))\}. \qquad (45)$$

The expectation of any function $F(\mathbf{q}(0), \mathbf{p}(0))$ of the operators $\mathbf{q}(0)$, $\mathbf{p}(0)$ given by

$$\langle F\rangle = \frac{\mathrm{Tr}\,\{F(\mathbf{q}(0), \mathbf{p}(0))\rho(\mathbf{q}(0), \mathbf{p}(0))\}}{\mathrm{Tr}\,\{\rho(\mathbf{q}(0), \mathbf{p}(0))\}}. \qquad (46)$$

Just as in the classical case, we now consider the properties of the *stochastic operators* $q_i(t)$ and $p_i(t)$ which result from the equation of motion and the initial density matrix. These properties we describe in terms of the correlation functions, the simplest being the pair correlation functions. These are obtained in Appendix 2; the results are

$$\langle p_i(t)p_k(t+\tau)\rangle = \left\|\frac{\hbar\mathbf{A}^{\frac{1}{2}}}{2}\left[\coth\frac{\hbar\mathbf{A}^{\frac{1}{2}}}{2kT}\cos\mathbf{A}^{\frac{1}{2}}\tau + i\sin\mathbf{A}^{\frac{1}{2}}\tau\right]\right\|_{ik} \qquad (47a)$$

$$\langle q_i(t)q_k(t+\tau)\rangle = \left\|\frac{\hbar}{2\mathbf{A}^{\frac{1}{2}}}\left[\coth\frac{\hbar\mathbf{A}^{\frac{1}{2}}}{2kT}\cos\mathbf{A}^{\frac{1}{2}}\tau + i\sin\mathbf{A}^{\frac{1}{2}}\tau\right]\right\|_{ik} \qquad (47b)$$

$$\langle q_i(t)p_k(t+\tau)\rangle = \left\|\frac{\hbar}{2}\left[-\coth\frac{\hbar\mathbf{A}^{\frac{1}{2}}}{2kT}\sin\mathbf{A}^{\frac{1}{2}}\tau + i\cos\mathbf{A}^{\frac{1}{2}}\tau\right]\right\|_{ik}. \qquad (47c)$$

Note that in the limit $\hbar = 0$, these expressions become identical with the classical expressions (9). As indicated in Appendix 2, the higher correlations are given by the rule:

Correlations of an odd number of q's and p's vanish. The correlation of an even number of q's and p's is equal to the sum of products of pair correlations, the sum being over all possible pairings of the operators, *with order preserved*.

Except for the italicized proviso that the order be preserved, this rule is identical with the classical rule for a Gaussian random process.[8] The stochastic operators also have the classical *stationarity property*; the correlations depend only upon the time differences. There are, however, obvious differences between the properties of the stochastic operators and the properties of the corresponding classical stochastic process. The principal differences arise from the fact that the quantum operators do not commute. Thus there are many (indeed, an infinity!) of correlations of the quantum operators which can be associated with a given classical correlation, corresponding to different orderings of the operators. As a simple example,

$$\langle \epsilon qp + (1-\epsilon)pq\rangle_{\text{quantum}} \to \langle qp\rangle_{\text{classical}},$$

where ϵ is an arbitrary complex number. Another difficulty comes from the fact that the product of two noncommuting Hermitian operators is not Hermitian, and, therefore, does not correspond to a physical observable. We see this difficulty explicitly in the correlations (47), which are complex functions, whereas the expectation value of a physical observable should be real.

The difficulties we mention would be largely resolved if we had a conventional definition of the product of operators with the following properties:

(a) The product is independent of the order of the operators.

(b) The product of a number of Hermitian operators is itself Hermitian.

(c) The classical pair decomposition rule for expressing higher correlations in terms of pair correlations holds.

For our system of coupled oscillators there is such a conventional product, namely the ordered product or *normal product* introduced in quantum field theory[9]; it is defined as follows. In Appendix 2 we show how the operators $\mathbf{q}(0)$ and $\mathbf{p}(0)$ can be expanded in terms of creation and annihilation operators for the normal modes;

$$\mathbf{q}(0) = i\sum_s \boldsymbol{\xi}^{(s)}\left(\frac{\hbar}{2\omega_s}\right)^{\frac{1}{2}}(a_s - a_s^*),$$
$$\mathbf{p}(0) = \sum_s \boldsymbol{\xi}^{(s)}\left(\frac{\hbar\omega_s}{2}\right)^{\frac{1}{2}}(a_s + a_s^*). \qquad (48)$$

[8] See, e.g., Wang and Uhlenbeck, Ref. 2.
[9] See, e.g., G.-C. Wick, Phys. Rev. 80, 268 (1950).

Here $\xi^{(s)}$ is the eigenvector of the interaction matrix $\mathbf{A}$, and ω_s^2 is the associated eigenvalue

$$\mathbf{A}\xi^{(s)} = \omega_s^2\xi^{(s)}. \qquad (49)$$

The operator a_s is the annihilation operator for the sth normal mode and a_s^* is the corresponding creation operator. Their commutation rules are

$$[a_s, a_r^*] = \delta_{sr}, \quad [a_s, a_r] = [a_s^*, a_r^*] = 0. \qquad (50)$$

The time-dependent operators are expressed in terms of the a_s and a_s^* by inserting (48) in (4). We find

$$\begin{aligned} \mathbf{q}(t) &= i \sum_s \xi^{(s)}\left(\frac{\hbar}{2\omega_s}\right)^{\frac{1}{2}}(a_s e^{-i\omega_s t} - a_s^* e^{i\omega_s t}), \\ \mathbf{p}(t) &= \sum_s \xi^{(s)}\left(\frac{\hbar\omega_s}{2}\right)^{\frac{1}{2}}(a_s e^{-i\omega_s t} + a_s^* e^{i\omega_s t}) \end{aligned} \qquad (51)$$

The normal product of a number of the operators a_s and a_s^* is defined to be that product in which all the a_s^* are written to the left of all the a_s. Because of the commutation rules (50), this defines a unique order. The normal product of a number of operators $q_i(t)$ and $p_i(t)$ is the product in which the expansions (51) are used and each product of the a_s and a_s^* is written in normal form. We denote the normal product by a colon placed before and after the product of factors. As an example,

$$\begin{aligned} :q_i(t_1)p_k(t_2): \equiv i\frac{\hbar}{2}\sum_s\sum_r \left(\frac{\omega_r}{\omega_s}\right)^{\frac{1}{2}}\xi_i^{(s)}\xi_k^{(r)}[a_s a_r e^{-i(\omega_s t_1+\omega_r t_2)} \\ - a_s^* a_r e^{i(\omega_s t_1-\omega_r t_2)} + a_r^* a_s e^{-i(\omega_s t_1-\omega_r t_2)} \\ - a_s^* a_r^* e^{i(\omega_s t_1+\omega_r t_2)}]. \end{aligned}$$

The normal product fulfills our requirements. It us clearly independent of the order of the factors, and, since a_s^* is the Hermitian conjugate of a_s, the normal product of a number of Hermitian operators is Hermitian. Using the results given in Appendix 2, we can show that the pair correlations of normal products of the $q_i(t)$ and $p_k(t)$ are

$$\begin{aligned} \langle :p_i(t)p_k(t+\tau): \rangle &= ||P(\hbar\mathbf{A}^{\frac{1}{2}}/kT) \cos \mathbf{A}^{\frac{1}{2}}\tau||_{jk}, \\ \langle :q_i(t)q_k(t+\tau): \rangle &= ||P(\hbar\mathbf{A}^{\frac{1}{2}}/kT)\mathbf{A}^{-1} \cos \mathbf{A}^{\frac{1}{2}}\tau||_{jk}, \\ \langle :q_i(t)p_k(t+\tau): \rangle &= -||P(\hbar\mathbf{A}^{\frac{1}{2}}/kT)\mathbf{A}^{-\frac{1}{2}} \sin \mathbf{A}^{\frac{1}{2}}\tau||_{jk}. \end{aligned} \qquad (52)$$

Here we have introduced the Planck function

$$P(x) \equiv kTx/(e^x - 1). \qquad (53)$$

When $x = \hbar\omega/kT$ the Planck function is the average energy, relative to the ground state, of a quantum oscillator of frequency ω. As $x \to 0$, $P(x) \to kT$, the classical equipartition energy.

The higher correlations of normal products are given by the well-known rule for a Gaussian random process:

The correlation of the normal product of an odd number of q's and p's vanishes. The correlation of the normal product of an even number of q's and p's is equal to the sum of products of pair correlation of normal products, the sum being over all possible pairings.

What we can say, then, is that the correlations of normal products of our stochastic operators are identical with those of a *stationary Gaussian process* whose pair correlations are given by (52).[10]

Finally, if we fix our attention on a single oscillator, the one with index 0, the momentum autocorrelation is

$$\langle :p_0(t)p_0(t+\tau): \rangle = ||P(\hbar\mathbf{A}^{\frac{1}{2}}/kT) \cos \mathbf{A}^{\frac{1}{2}}\tau||_{00}. \qquad (54)$$

Just as in the corresponding classical process, we can ask whether there is an interaction matrix $\mathbf{A}$ for which (54) is an exponential and, therefore, the corresponding Gaussian process is Markoffian. The answer is that we can, but that it is temperature-dependent, and, therefore, not of physical interest. If we use the model discussed in Sec. 3, in which the matrix elements are given by (19) with $f(\theta)$ given by (26) in the limit $\omega_L \gg f$, we find (54) becomes

$$\langle :p_0(t)p_0(t+\tau): \rangle = \frac{2}{\pi}\int_0^\infty d\omega\, P\left(\frac{\hbar\omega}{kT}\right)\frac{f}{\omega^2 + f^2}\cos\omega\tau. \qquad (55)$$

In the limit $\hbar \to 0$, this becomes identical with the classical result obtained earlier, and the process becomes Markoffian.

6. THE QUANTUM LANGEVIN EQUATION

The formal manipulations we used in Sec. 4 to derive the Langevin equation are unchanged when we interpret the q's and p's as quantum operators. In particular, when the matrix of interactions is that given by the model of Sec. 3, in which the matrix elements are given by (19) with $f(\theta)$ given by (26) in the limit $\omega_L \gg f$, we obtain an operator equation of motion which is formally identical with Langevin equation (39). That is

$$\dot{p}_0 - F(t) = -fp_0 + E(t), \qquad (39')$$

where

$$\begin{aligned} E(t) = &-\sum_j ||f\mathbf{A}^{\frac{1}{2}} \sin \mathbf{A}^{\frac{1}{2}}t + \mathbf{A} \cos \mathbf{A}^{\frac{1}{2}}t||_{0j}\, q_j(0) \\ &+ \sum_j ||f \cos \mathbf{A}^{\frac{1}{2}}t - \mathbf{A}^{\frac{1}{2}} \sin \mathbf{A}^{\frac{1}{2}}t||_{0j}\, p_j(0). \end{aligned} \qquad (40')$$

[10] Essentially the same point is made in connection with the quantum description of statistical light beams by E. C. G. Sudarshan, Phys. Rev. Letters 10, 277 (1963). See also R. J. Glauber, Phys. Rev. Letters 10, 84 (1963).

The operator Langevin equation is an equation of motion for the time-dependent Heisenberg operators $p_0(t)$ and $q_0(t)$. The operator $F(t)$ is the external force operator,

$$F(t) \equiv (i\hbar)^{-1}[p_0(t), V(q_0(t), t)], \tag{56}$$

with $V(q_0, t)$ the (time-dependent) potential of the external force. The random-force operator $E(t)$ is in fact independent of the operators $p_0(0)$ and $q_0(0)$ since their coefficients in the expression (40′) vanish for our model of the interaction matrix. Because of the commutation rules (44), this means that

$$[q_0(0), E(t)] = [p_0(0), E(t)] = 0. \tag{57}$$

We assume that the statistical state of the initial coordinates and momenta of the heat bath is described by the density matrix (45). Just as in the classical case, this means that at $t = 0$ the heat bath (i.e., the oscillators other than the Brownian particle) is in equilibrium with a fictitious force-free Brownian particle. Since the random-force operator $E(t)$ is independent of the initial coordinate and momentum of the Brownian particle, its stochastic properties are unaffected by the dependence of the density matrix upon these operators. The covariance of the normal product is readily obtained using the results (52), we find

$$\langle :E(t)E(t+\tau):\rangle = \|(f^2 + \mathbf{A})P(\hbar\mathbf{A}^{\frac{1}{2}}/kT)\cos\mathbf{A}^{\frac{1}{2}}\tau\|_{00}. \tag{58}$$

But for our model, we may use the general expression (21), with $f(\theta)$ given by (25). Hence

$$\langle :E(t)E(t+\tau):\rangle = \frac{1}{2\pi}\int_{-\pi}^{\pi} d\theta\, f^2(1 + \tan^2 \tfrac{1}{2}\theta) \times P\left(\frac{\hbar f}{kT}|\tan\tfrac{1}{2}\theta|\right)\cos(f\tau\tan\tfrac{1}{2}\theta)$$

$$= \frac{2f}{\pi}\int_0^\infty d\omega\, P\left(\frac{\hbar\omega}{kT}\right)\cos\omega\tau. \tag{59}$$

Correlations of higher normal products of $E(t)$ are again given by the rule for a Gaussian random process. Hence, the stochastic properties of the random operator $E(t)$, as expressed by the correlations of normal products, are identical with those of a stationary Gaussian process whose covariance is given by (59). In the limit $\hbar \to 0$, this covariance approaches the corresponding classical covariance (43), which is the covariance of a purely random Gaussian process. However, for finite $\hbar$, (59) is the covariance of a Gaussian process which is not even Markoffian. This is the chief difference between the quantum and classical Langevin equations for our model.

As an elementary application, consider the motion in a constant field of force,

$$F(t) = \mathcal{E}. \tag{60}$$

The solution of the operator Langevin equation is

$$p_0(t) = e^{-ft}p_0(0) + f^{-1}\mathcal{E}(1 - e^{-ft}) + \int_0^t dt'\, e^{-f(t-t')}E(t'). \tag{61}$$

If we average this expression over the initial state of the heat bath, we find[11]

$$\langle p_0(t)\rangle = e^{-ft}p_0(0) + f^{-1}\mathcal{E}(1 - e^{-ft}), \tag{62}$$

since $\langle E(t)\rangle = 0$. After a long time ($t \gg f^{-1}$) we find

$$\langle p_0(t)\rangle \sim f^{-1}\mathcal{E}. \tag{63}$$

This is the analog for our model of Ohm's law. The left-hand side is the "current," which in the steady state is proportional to the applied field, and inversely proportional to the "resistance," i.e., the friction constant.

The mean-square "fluctuation current" may also be obtained from (60)

$$\langle :[p_0(t) - \langle p_0(t)\rangle]^2:\rangle = \int_0^t dt'\, e^{-f(t-t')}\int_0^t dt''\, e^{-f(t-t'')}\langle :E(t')E(t''):\rangle. \tag{64}$$

Note that we have used the normal product in the definition of the squared fluctuation. Using the expression (58) for the covariance of $E(t)$, we find

$$\langle :[p_0(t) - \langle p_0(t)\rangle]^2:\rangle = \frac{2f}{\pi}\int_0^\infty d\omega\, P\left(\frac{\hbar\omega}{kT}\right)\left|\frac{1 - e^{-(f-i\omega)t}}{f - i\omega}\right|^2 \tag{65}$$

where we have used the elementary result

$$\int_0^t dt'\int_0^t dt''\, e^{-f(t-t')}e^{-f(t-t'')}\cos\omega(t' - t'') = |(1 - e^{-(f-i\omega)t})/(f - i\omega)|^2. \tag{66}$$

In the steady state, i.e., when $t \gg f^{-1}$, (65) becomes

$$\langle :[p_0(t) - \langle p_0(t)\rangle]^2:\rangle = \frac{2f}{\pi}\int_0^\infty d\omega\, P\left(\frac{\hbar\omega}{kT}\right)\frac{1}{\omega^2 + f^2}. \tag{67}$$

This is the analog for our model of the Nyquist formula, which relates the noise power spectrum

[11] The result of the average over the states of the heat bath is still in general an operator function of $q_0(0)$ and $p_0(0)$. We trust that our definition of this average, which involves an average over the initial coordinate and momentum of a fictitious Brownian particle, is not confusing.

to the resistance and the absolute temperature.[12] Note that if we had used the ordinary product instead of the normal product, we would find

$$\begin{aligned}&\langle[p_0(t) - \langle p_0(t)\rangle]^2\rangle \\ &\equiv \langle[p_0(t) - \langle p_0(t)\rangle]\cdot[p_0(t) - \langle p_0(t)\rangle]\rangle \\ &= \langle:[p_0(t) - \langle p_0(t)\rangle]^2:\rangle + \frac{2f}{\pi}\int_0^{\omega_L} d\omega \frac{\hbar\omega}{2}\frac{1}{\omega^2 + f^2}. \end{aligned} \quad (68)$$

The added integral is clearly related to the zero-point fluctuations of the heat bath. It is also divergent in the limit $\omega_L \to \infty$, so we have retained this cutoff explicitly. Which of the possible definitions of the product corresponds to an experimental measurement of the fluctuation spectrum? This, in the last analysis, must be determined by the experiment itself. We tend to the opinion that the normal product is the physically appropriate definition, since it leads to a noise spectrum which vanishes at absolute zero.

7. BROWNIAN MOTION OF A QUANTUM OSCILLATOR

As a second application of the quantum Langevin equation, we consider the case of the harmonic oscillator, for which the external force is[13]

$$F(t) = -\kappa^2 q_0(t), \quad (69)$$

where κ is the natural frequency of the oscillator. The quantum Langevin equation (39′) becomes

$$\dot{p}_0 + \kappa^2 q_0 = -fp_0 + E(t), \quad (70)$$

to which we must append the equation

$$\dot{q}_0 = p_0. \quad (71)$$

The solution of this pair of coupled equations is

$$\begin{aligned} q_0(t) &= e^{-\frac{1}{2}ft}\{[\cos \nu t + (f/2\nu) \sin \nu t]q_0(0) \\ &\quad + \nu^{-1} \sin \nu t p_0(0)\} \\ &\quad + \int_0^t dt'\, e^{-\frac{1}{2}f(t-t')}\frac{1}{\nu}\sin \nu(t - t')E(t'), \\ p_0(t) &= e^{-\frac{1}{2}ft}\{-(\kappa^2/\nu) \sin \nu t q_0(0) \\ &\quad + [\cos \nu t - (f/2\nu) \sin \nu t]p_0(0)\} \\ &\quad + \int_0^t dt'\, e^{-\frac{1}{2}(t-t')}[\cos \nu(t - t') \\ &\quad - (f/2\nu) \sin \nu(t - t')]E(t'). \end{aligned} \quad (72)$$

[12] For elementary discussion of the Nyquist formula and its quantum generalization see C. Kittel, *Elementary Statistical Physics* (John Wiley & Sons, Inc., New York, 1958), pp. 141–153. See also J. Lawson and G. E. Uhlenbeck, *Threshold Noise Signals* (McGraw-Hill Book Company, Inc., New York, 1950), especially pp. 64–79.

[13] The Brownian motion of a quantum oscillator is considered in a paper by J. Schwinger, J. Math. Phys. **2**, 407 (1961).

Here

$$\nu = (\kappa^2 - \tfrac{1}{4}f^2)^{\frac{1}{2}}, \quad (73)$$

and we restrict our discussion to the underdamped case, where $\kappa > \frac{1}{2}f$, so ν is real and positive.

The mean motion of the oscillator is described by the operators obtained by averaging (72) and (73) over the initial states of the heat bath. Since $\langle E(t)\rangle = 0$, we have

$$\begin{aligned} \langle q_0(t)\rangle &= e^{-\frac{1}{2}ft}\{[\cos \nu t + (f/2\nu) \sin \nu t]q_0(0) \\ &\quad + \nu^{-1} \sin \nu t p_0(0)\}, \\ \langle p_0(t)\rangle &= e^{-\frac{1}{2}ft}\{-\kappa^2\nu^{-1} \sin \nu t q_0(0) \\ &\quad + [\cos \nu t - (f/2\nu) \sin \nu t]p_0(0)\}. \end{aligned} \quad (74)$$

These are just the operator solutions of the average of the equations of motion (70) and (71). We see from (74) that these operators vanish for $t \gg f^{-1}$; the mean motion of the oscillator vanishes for times long compared with the "macroscopic relaxation time" f^{-1}.

To see more precisely what we mean by the operators (74), consider the coordinate representation. The Heisenberg state of the oscillator is then specified by a time-independent wavefunction $\psi[q_0(0)]$, and the initial momentum operator is represented by $-i\hbar\, \partial/\partial q_0(0)$. The simplest kind of question we can ask about the Heisenberg operators is their *expectation value*, denoted by a subscript "ex" to the operator. For example,

$$\begin{aligned} \langle q_0(t)\rangle_{\text{ex}} &\equiv \int_{-\infty}^{\infty} dq_0(0)\ \psi^*[q_0(0)]\langle q_0(t)\rangle\psi[q_0(0)] \\ &= e^{-\frac{1}{2}ft}\{(\cos \nu t + (f/2\nu) \sin \nu t)[q_0(0)]_{\text{ex}} \\ &\quad + \nu^{-1} \sin \nu t[p_0(0)]_{\text{ex}}\}. \end{aligned} \quad (75)$$

As a simple illustration, if the wavefunction is

$$\psi[q_0(0)] = (\alpha/\pi)^{\frac{1}{4}}e^{-\frac{1}{2}\alpha[q_0(0)-x_0]^2}, \quad (76)$$

then

$$[q_0(0)]_{\text{ex}} = x_0, \qquad [p_0(0)]_{\text{ex}} = 0, \quad (77)$$

and

$$\begin{aligned} \langle q_0(t)\rangle_{\text{ex}} &= e^{-\frac{1}{2}ft}[\cos \nu t + (f/2\nu) \sin \nu t]x_0, \\ \langle p_0(t)\rangle_{\text{ex}} &= -e^{-\frac{1}{2}ft}\kappa^2\nu^{-1} \sin \nu t x_0. \end{aligned} \quad (78)$$

The operator describing the fluctuation of the displacement of the oscillator about the mean displacement is

$$q_0(t) - \langle q_0(t)\rangle = \int_0^t dt'\, e^{-\frac{1}{2}f(t-t')}\frac{\sin \nu(t - t')}{\nu}E(t'). \quad (79)$$

The mean-square fluctuation in displacement we express as the mean of the square of (79), written as a normal product. Using (59), we find

$$\langle :[q_0(t) - \langle q_0(t)\rangle]^2 : \rangle = \frac{2f}{\pi}\int_0^\infty d\omega\, P\left(\frac{\hbar\omega}{kT}\right)$$
$$\times \left|\frac{1 - e^{-(\frac{1}{2}f - i\omega)t}[\cos\nu t + (\frac{1}{2}f - i\omega)\nu^{-1}\sin\nu t]}{\omega^2 - \kappa^2 + if\omega}\right|^2. \quad (80)$$

Here we have used the following result of an elementary but tedious integration

$$\int_0^t dt' \int_0^t dt''\, e^{-\frac{1}{2}f(t-t')}\,\frac{\sin\nu(t-t')}{\nu}$$
$$\times\, e^{-\frac{1}{2}f(t-t'')}\,\frac{\sin\nu(t-t'')}{\nu}\cos\omega(t'-t'')$$
$$= \left|\frac{1 - e^{-(\frac{1}{2}f - i\omega)t}[\cos\nu t + (\frac{1}{2}f - i\omega)\nu^{-1}\sin\nu t]}{\omega^2 - \kappa^2 + if\omega}\right|^2. \quad (81)$$

The time dependence of the mean fluctuation in displacement, as expressed by (80), is rather complicated. It does, however, have the simple feature that, for times long compared with the relaxation time f^{-1}, it approaches an equilibrium value

$$\langle :[q_0(t) - \langle q_0(t)\rangle]^2 : \rangle_{\text{eq}}$$
$$= \frac{2f}{\pi}\int_0^\infty d\omega\, \frac{P(\hbar\omega/kT)}{(\omega^2 - \kappa^2)^2 + \omega^2 f^2}. \quad (82)$$

In the classical limit we use the property of the Planck function (53)

$$P(\hbar\omega/kT) \to kT, \quad \text{as} \quad \hbar \to 0. \quad (83)$$

The remaining integral in (82) is elementary:

$$\frac{2f}{\pi}\int_0^\infty d\omega\, \frac{1}{(\omega^2 - \kappa^2)^2 + \omega^2 f^2} = \kappa^{-2}, \quad (84)$$

and we find

$$\langle :[q_0(t) - \langle q_0(t)\rangle]^2 : \rangle_{\text{eq}} \to kT/\kappa^2, \quad \text{as} \quad \hbar \to 0, \quad (85)$$

which is the classical equipartition result.

Another simple limit of (82) is the weak-coupling limit, in which the coupling of the Brownian particle to the heat bath is weak compared with the oscillator coupling. That is, $\kappa \gg f$. In this limit the resonance denominator in (82) becomes sharply peaked at $\omega = \kappa$, with a width $\approx f$. Hence, we can evaluate the Planck function at $\omega = \kappa$ and perform the remaining integral using (84). We find

$$\langle :[q_0(t) - \langle q_0(t)\rangle]^2 : \rangle_{\text{eq}} \to \frac{1}{\kappa^2} P\left(\frac{\hbar\kappa}{kT}\right), \quad \text{for} \quad \kappa \gg f. \quad (86)$$

This is the well-known Planck result for the mean-square displacement of an independent oscillator of frequency κ at temperature T.

APPENDIX 1: CORRELATIONS IN A SYSTEM OF COUPLED CLASSICAL OSCILLATORS

Consider first the correlations of the initial values of the coordinates and momenta, whose distribution is [c.f. Eq. (6)]

$$D(\mathbf{q}(0), \mathbf{p}(0)) = (2\pi/\beta)^{2N+1}(\det \mathbf{A})^{-\frac{1}{2}}$$
$$\times \exp\left\{-\frac{\beta}{2}\left[\sum_i p_i^2(0) + \sum_{i,k} q_i(0)A_{ik}q_k(0)\right]\right\}. \quad (1)$$

Since this is a Gaussian distribution, all higher correlations can be expressed in terms of the pair correlations. These are ($\beta = 1/kT$):

$$\langle p_i(0)p_k(0)\rangle = kT\delta_{ik},$$
$$\langle p_i(0)q_k(0)\rangle = 0, \quad (2)$$
$$\langle q_i(0)q_k(0)\rangle = kT\, ||\mathbf{A}^{-1}||_{ik}.$$

These expressions are consequences of well-known integral formulas for Gaussian distributions.[14]

The pair correlations for the time-dependent coordinates and momenta are found from (4), using (2). We have for the momentum correlation:

$$\langle p_i(t)p_k(t+\tau)\rangle = \sum_{m,n}\{||\mathbf{A}^{\frac{1}{2}}\sin \mathbf{A}^{\frac{1}{2}}t||_{im}$$
$$\times\, ||\mathbf{A}^{\frac{1}{2}}\sin \mathbf{A}^{\frac{1}{2}}(t+\tau)||_{kn}\langle q_m(0)q_n(0)\rangle$$
$$+\, ||\cos \mathbf{A}^{\frac{1}{2}}t||_{im}\, ||\cos \mathbf{A}^{\frac{1}{2}}(t+\tau)||_{kn}\langle p_m(0)p_n(0)\rangle\}$$
$$= kT\{||\sin \mathbf{A}^{\frac{1}{2}}t \sin \mathbf{A}^{\frac{1}{2}}(t+\tau)$$
$$+ \cos \mathbf{A}^{\frac{1}{2}}t \cos \mathbf{A}^{\frac{1}{2}}(t+\tau)||_{ik},$$

where we have used the fact that the matrix $\mathbf{A}$ is symmetric. Using the formula for the cosine of the difference of two angles, we find

$$\langle p_i(t)p_k(t+\tau)\rangle = kT\, ||\cos \mathbf{A}^{\frac{1}{2}}\tau||_{ik}. \quad (3)$$

In a similar way, we can also show

$$\langle q_i(t)p_k(t+\tau)\rangle = -kT\, ||\mathbf{A}^{-\frac{1}{2}}\sin \mathbf{A}^{\frac{1}{2}}\tau||_{ik}, \quad (4)$$
$$\langle q_i(t)q_k(t+\tau)\rangle = kT\, ||\mathbf{A}^{-1}\cos \mathbf{A}^{\frac{1}{2}}\tau||_{ik}. \quad (5)$$

Since the process is Gaussian, the higher correlations are given by the following rule[8]:

The correlation of an odd number of coordinates and momenta vanish. The correlation of an even number of coordinates and momenta is equal to the

[14] See, e.g., H. Cramer, *Mathematical Methods of Statistics* (Princeton University Press, Princeton, New Jersey, 1945), p. 118.

sum of products of pair correlations, the sum being over all pairings.

For example,

$$\langle q_i(t_1)q_k(t_2)q_l(t_3)p_m(t_4)\rangle = \langle q_i(t_1)q_k(t_2)\rangle\langle q_l(t_3)p_m(t_4)\rangle + \langle q_i(t_1)q_l(t_3)\rangle\langle q_k(t_2)p_m(t_4)\rangle + \langle q_i(t_1)p_m(t_4)\rangle\langle q_k(t_2)q_l(t_3)\rangle.$$

APPENDIX 2: CORRELATIONS IN A SYSTEM OF COUPLED QUANTUM OSCILLATORS

The system of oscillators is described by the Hamiltonian

$$H = \frac{1}{2}\sum_i p_i^2 + \frac{1}{2}\sum_{i,k} A_{ik}q_iq_k, \tag{1}$$

where the operators q_i and p_i satisfy the commutation relations

$$[q_i, p_k] = i\hbar\delta_{ik}, \qquad [q_i, q_k] = [p_i, p_k] = 0. \tag{2}$$

The expectation value of any operator F with respect to the canonical ensemble at temperature T is defined by

$$\langle F\rangle = \mathrm{Tr}\,\{Fe^{-\beta H}\}/\mathrm{Tr}\,\{e^{-\beta H}\}, \tag{3}$$

where the trace operation is in the space of the eigenfunctions of the Hamiltonian operator. We are interested mainly in the case where the operator F is a product of q's and p's.

As a first step in the evaluation of such traces, consider the eigenvalues and eigenvectors of the matrix $\mathbf{A}$:

$$\sum_k A_{ik}\xi_k^{(s)} = \omega_s^2\xi_i^{(s)}. \tag{4}$$

The eigenvectors are assumed normalized so that

$$\sum_i \xi_i^{(s)}\xi_i^{(r)} = \delta_{sr}, \qquad \sum_r \xi_i^{(r)}\xi_k^{(r)} = \delta_{ik}. \tag{5}$$

We now introduce the operators

$$a_s = (2\hbar\omega_s)^{-\frac{1}{2}} \sum_i \xi_i^{(s)}(p_i - i\omega_s q_i),$$
$$a_s^* = (2\hbar\omega_s)^{-\frac{1}{2}} \sum_i \xi_i^{(s)}(p_i + i\omega_s q_i). \tag{6}$$

The commutation relations for these operators follow from (2) and (5). We find

$$[a_s, a_r^*] = \delta_{sr}, \qquad [a_s, a_r] = [a_s^*, a_r^*] = 0. \tag{7}$$

The inversion of (6) is readily accomplished using the relations (5). We find

$$q_i = i\sum_s \xi_i^{(s)}\left(\frac{\hbar}{2\omega_s}\right)^{\frac{1}{2}}(a_s - a_s^*),$$
$$p_i = \sum_s \xi_i^{(s)}\left(\frac{\hbar\omega_s}{2}\right)^{\frac{1}{2}}(a_s + a_s^*). \tag{8}$$

Inserting these expressions in the Hamiltonian (1) and using the relations (4) and (5) we find

$$H = \sum_s \hbar\omega_s(a_s^*a_s + \tfrac{1}{2}). \tag{9}$$

The operator $a_s^*a_s$ is the number operator for the sth normal mode; its eigenvalues are the nonnegative integers. The operator a_s is the step-down (annihilation) operator, and the operator a_s^* is the step-up (creation) operator; they have matrix elements only between eigenstates of the number operator which differ by unity.[15]

The evaluation of the expectation value (3) when F is a product of the a's and a^*'s is straightforward. Clearly, the only nonvanishing expectation values are for products containing an equal number of a^*'s and a's. The simplest of these are the pair expectation values. Thus

$$\langle a_s^*a_r\rangle = \delta_{sr}\,\mathrm{Tr}\,\{a_s^*a_se^{-\beta H}\}/\mathrm{Tr}\,\{e^{-\beta H}\}$$
$$= \delta_{sr}\frac{\sum_{n=0}^{\infty} n\exp\left[-\frac{\hbar\omega_s}{kT}\left(n+\frac{1}{2}\right)\right]}{\sum_{n=0}^{\infty}\exp\left[-\frac{\hbar\omega_s}{kT}\left(n+\frac{1}{2}\right)\right]}$$
$$= \delta_{sr}\left[\exp\left(\frac{\hbar\omega_s}{kT}\right) - 1\right]^{-1}$$

or

$$\langle a_s^*a_r\rangle = \tfrac{1}{2}\delta_{sr}[\coth \hbar\omega_s/2kT) - 1]. \tag{10}$$

Using the commutation relation (7), we have

$$\langle a_sa_r^*\rangle = \tfrac{1}{2}\delta_{sr}[\coth(\hbar\omega_s/2kT) + 1]. \tag{11}$$

The results for higher products are summarized by the following rule: The expectation value of a product of a's and a^*'s is equal to the sum of products of pair expectation values, the sum being over all possible pairings with the order of each pair preserved.

For example:

$$\langle a_s^*a_ra_ua_v^*\rangle = \langle a_s^*a_r\rangle\langle a_ua_v^*\rangle + \langle a_s^*a_u\rangle\langle a_ra_v^*\rangle.$$

We do not prove this rule here, since the most convincing demonstration is by example.

We turn now to the consideration of expectation

[15] These properties of the operators are discussed in many textbooks on quantum mechanics. See, e.g., A. Messiah, *Quantum Mechanics* (North-Holland Publishing Company, Amsterdam, 1961), Chap. 12.

values of products of q's and p's. Because of the linear relations (8) we have the same rule for these expectations:

The expectation value of an odd number of q's and p's vanishes. The expectation value of an even number of q's and p's is equal to the sum of products of pair expectation values, the sum being over all pairings which preserve the order of the pair.

For example:

$$\langle q_i q_j p_k q_l \rangle = \langle q_i q_j \rangle \langle p_k p_l \rangle + \langle q_i p_k \rangle \langle q_j q_l \rangle + \langle q_i q_l \rangle \langle q_j q_k \rangle.$$

The pair correlations are readily obtained from (10) and (11), using (8). Thus

$$\langle q_i q_k \rangle = \sum_s \frac{\hbar}{2\omega_s} \coth \frac{\hbar\omega_s}{2kT} \xi_i^{(s)} \xi_k^{(s)}.$$

Using (4) we see this can be written

$$\langle q_i q_k \rangle = \left\| \frac{\hbar}{2\mathbf{A}^{\frac{1}{2}}} \coth \frac{\hbar \mathbf{A}^{\frac{1}{2}}}{2kT} \right\|_{ik}. \tag{12}$$

Similarly, we find

$$\langle p_i p_k \rangle = \| \tfrac{1}{2}\hbar \mathbf{A}^{\frac{1}{2}} \coth (\hbar \mathbf{A}^{\frac{1}{2}}/2kT) \|_{ik}, \tag{13}$$

$$\langle q_i p_k \rangle = -\langle p_i q_k \rangle = \tfrac{1}{2} i\hbar \delta_{ik}. \tag{14}$$

Consider now the time-dependent correlation functions in which the operators at time t are expressed in terms of the initial operators through the relations (4), considered here as formal solutions of the Heisenberg equations of motion. Again we have the rule that correlations of an odd number of operators vanish while a correlation of an even number of operators is equal to a sum of products of pair correlations, the sum being over all pairings which preserve the order of the pair. For example,

$$\begin{aligned}\langle q_i(t_1) q_k(t_2) p_l(t_3) q_m(t_4) \rangle &= \langle q_i(t_1) q_k(t_2) \rangle \langle p_l(t_3) q_m(t_4) \rangle \\ &+ \langle q_i(t_1) p_l(t_3) \rangle \langle q_k(t_2) q_m(t_4) \rangle \\ &+ \langle q_i(t_1) q_m(t_4) \rangle \langle q_k(t_2) p_l(t_3) \rangle.\end{aligned}$$

The pair correlations are

$$\langle p_i(t) p_k(t+\tau) \rangle$$
$$= \| \tfrac{1}{2}\hbar \mathbf{A}^{\frac{1}{2}} [\coth (\hbar \mathbf{A}^{\frac{1}{2}}/2kT) \cos \mathbf{A}^{\frac{1}{2}}\tau + i \sin \mathbf{A}^{\frac{1}{2}}\tau] \|_{ik}, \tag{15}$$

$$\langle q_i(t) q_k(t+\tau) \rangle$$
$$= \| \tfrac{1}{2}\hbar \mathbf{A}^{-\frac{1}{2}} [\coth (\hbar \mathbf{A}^{\frac{1}{2}}/2kT) \cos \mathbf{A}^{\frac{1}{2}}\tau + i \sin \mathbf{A}^{\frac{1}{2}}\tau] \|_{ik}, \tag{16}$$

$$\langle q_i(t) p_k(t+\tau) \rangle$$
$$= \| \tfrac{1}{2}\hbar [-\coth (\hbar \mathbf{A}^{\frac{1}{2}}/2kT) \sin \mathbf{A}^{\frac{1}{2}}\tau + i \cos \mathbf{A}^{\frac{1}{2}}\tau] \|_{ik}. \tag{17}$$

The derivation of these expressions goes exactly as the derivation of the corresponding classical correlations, obtained in Appendix 1, but using the expressions (12), (13), and (14) for the initial expectation values.

Reprinted from
JOURNAL OF MATHEMATICAL PHYSICS VOLUME 6, NUMBER 3 MARCH 1965
Printed in U.S.A.

Dynamics of a Simple Many-Body System of Hard Rods

D. W. JEPSEN
IBM Watson Research Center, Yorktown Heights, New York
(Received 17 July 1964)

General formulas are given for the exact calculation of the nonequilibrium properties of the one-dimensional system of equal-mass hard rods both for a finite but large system and in the limit of infinite size. Only properties which depend upon labeling one or more of the particles are nontrivial in this system. Various results are obtained on Poincaré cycles, delocalization of a particle with time and electrical conductivity when one particle is charged.

INTRODUCTION

IN order to understand the dynamical behavior of systems containing a large number of particles, it is important to have a number of sample systems in which the dynamical equations can be handled exactly. We give here a study of Poincaré cycles, pseudostochastic behavior, and nonequilibrium properties of the one-dimensional hard-rod system, which is perhaps the simplest model that can be considered. Some of the simpler properties have been considered previously by Frisch and others.[1] The other studies of this type in the literature that we have seen are restricted to various harmonic oscillator assemblies, and we feel that a study of a system whose mechanics are not based on the theory of normal modes should be important for contrast, in spite of the fact that it is less physical than these other models.

The discussion of many ergodic properties in this system is quite convenient. We also obtain some results on Poincaré cycles not really restricted to this system and compare them with other known results. It appears to be possible to calculate almost any desired property of this system, although, of course, some calculations are messy. Unfortunately, most of the most interesting properties of real many-body systems are both degenerate and untypical in this system. However, the Brownian-motion-like path of a given particle appears to be one property that does bear attention.

DESCRIPTION OF THE DYNAMICS OF THE SYSTEM

We consider N point particles, all of the same mass, constrained to move along a line like beads on a string. The particles do not penetrate each other, so that they retain their ordering along the line. When a pair of particles collide, their energy and momentum are conserved so that they merely exchange velocities. We assume that there are no three body collisions or that they are so improbable as to have negligible effect during the time in which we are observing the system. This can be achieved by assuming hard-sphere or other very sharp repulsive forces between the particles and negligible attraction. We assume, as stated above, that the particles are mass points; the same procedure can also be used for particles of finite size simply by subtracting off the distances of closest approach from the interparticle distances and relating these new distances to the mass-point model. If the system has periodic boundary conditions, as we assume in this work, this equivalence is not quite exact because after these new distances are introduced, the center of mass of the system (to the extent that it can be defined) should still move in a box of the original size.

In order to follow the dynamics of the system in complete detail we make use of the following trick. We plot the positions of the particles versus time as shown in Fig. 1. If no collision occurs, the motion of one of the particles is represented by a straight line, or "trajectory", starting on the X axis at the initial position of the particle, and with a slope equal to the reciprocal of its velocity. When a collision occurs, two of these lines cross, and since collisions are elastic, the two particles merely exchange trajectories. Neither trajectory is changed in direction. This makes the dynamics of the system simple to follow; we need only keep track of which particle is on each line in the diagram at any given time.

To this end, we number the particles and the trajectories. We pick one of the particles, number it zero, and put our origin of coordinates at the position of this particle at the initial time. We number the rest of the particles 1, 2, $\cdots$, $N - 1$, starting with the neighbor to the right of particle zero, and continuing from left to right. Since one particle never passes through another, this ordering

[1] H. L. Frisch, Phys. Rev. **104**, 1 (1956); E. Teramoto and C. Suzuki, Progr. Theoret. Phys. (Kyoto) **14**, 411 (1955).

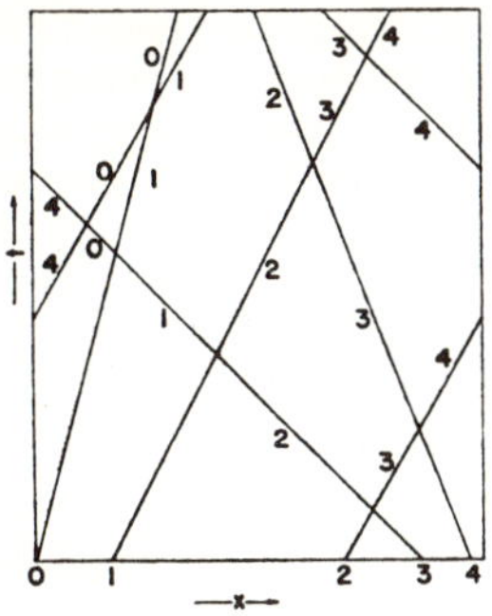

FIG. 1. Diagram showing the nature of the dynamics of the system.

is maintained for all time. We give each trajectory a number equal to the number of the particle occupying it at the initial time.

We now define $A_{jk}(t)$ to be equal to one if particle j is on trajectory k at time t and to be zero otherwise. A knowledge of $A_{jk}(t)$ for each j and k constitutes a complete solution of the dynamics of the system in a very convenient form. If, instead of studying a single system, we form an ensemble over the initial conditions, the average $\langle A_{jk}(t)\rangle$ of $A_{jk}(t)$ becomes the probability that particle j is on trajectory k at time t.

Since the particles remain in the same order around the ring (we assume periodic boundary conditions), it is possible to determine which particle is on a given trajectory simply by counting how many trajectories have crossed this trajectory from each side up to the given time. Each time a trajectory crosses it from the right, the particle on the given trajectory at this time is replaced by its neighbor to the right, which is numbered one higher. Similarly, after a collision from the left, the number of the particle on the trajectory is decreased by 1. Notice that this is true independent of the history of each of the trajectories involved. We shall refer to a collision from the left as a "negative" collision. As a result of these considerations we find that after trajectory k has been crossed n times by other trajectories (in the special sense where the number of negative crossings is subtracted), it will be occupied by particle number $k + n$. In this discussion particle zero is the particle with number one larger than $N - 1$ and $N - 1$ is one smaller than zero, i.e., we must count the particles modulo N.

We now show how these considerations can be used to calculate the $A_{jk}(t)$'s. We define $r_n(h, k, t)$ to be equal to 1 if trajectory h crosses trajectory k exactly n times between the initial time and time t, and to be zero otherwise. The number n can be positive, negative, or zero. If we average r_n over an ensemble of initial conditions, $\bar{r}_n$ will give the probability that trajectory h crosses trajectory k a total of n times. Next we define the characteristic functions for the probability distributions,

$$s(u; h, k, t) \equiv \sum_{n=-\infty}^{\infty} r_n(h, k, t)e^{inu}. \tag{1}$$

In terms of these quantities, $A_{jk}(t)$ is given by

$$A_{jk}(t) = \frac{1}{N}\sum_{l=0}^{N-1} \exp\left[-\frac{2\pi i}{N}(j-k)l\right] \times s\left(\frac{2\pi l}{N}; 0, k, t\right)s\left(\frac{2\pi l}{N}; 1, k, t\right) \times s\left(\frac{2\pi l}{N}; 3, k, t\right)\cdots s\left(\frac{2\pi l}{N}; N-1, k, t\right)$$

with

$$s(u, k, k, t) \equiv 1. \tag{2}$$

To show this, we replace the s's by the sums containing the r_n's and use the fact that

$$\frac{1}{N}\sum_{l=0}^{N-1} \exp\left[-\frac{2\pi i}{N}nl\right] = \begin{cases} 1 & n = 0 \pmod N, \\ 0 & \text{otherwise;} \end{cases} \tag{3}$$

$$A_{jk} = \sum_{n_0, n_1, \cdots = -\infty}^{\infty} \left[\frac{1}{N}\sum_{l=0}^{N-1} \exp\left\{\frac{2\pi i}{N} l(n_0 + n_1 + \cdots n_{N-1} - j + k)\right\}\right] r_{n_0}r_{n_1}\cdots r_{n_{N-1}}. \tag{4}$$

We state the argument in terms of probabilities, since a single system is actually a special case in which the probabilities are either zero or one. If the expression in the brackets is left out for the moment, the resulting sum adds together the probabilities of all the possible combinations of crossings which trajectory k could undergo between the initial time and time t. Putting in the expression in brackets restricts the summation to those combinations which put particle j on trajectory k. The sum of these probabilities is just $A_{jk}(t)$.

The expression for $s(u; h, k, t)$ to be used in this equation takes two forms depending upon whether h is greater than or less than k. If particle h starts to the right of k, s can be written in the form

$$s(u; h, k, t) = S[u, w_{kh}], \quad k < h, \tag{5}$$

with

$$S[u, w] = e^{inu} \quad \text{when} \quad (n-1)L < w \leq nL \quad \text{for each } n \tag{6}$$

and

$$w_{kh} = x_k - x_h + (v_k - v_h)t, \tag{7}$$

where x_i and v_i are the initial position and the velocity of the ith trajectory. If h starts to the left of k we have

$$s(u; h, k, t) = e^{-iu}S[u, w_{kh}], \quad h < k. \qquad (8)$$

In terms of S, Eq. (4) becomes

$$A_{jk} = \frac{1}{N} \sum_u \exp[-iju] \prod_{h=0}^{N-1} S[u, w_{kh}]$$

with

$$u = \frac{2\pi l}{N}, \quad \text{and} \quad \sum_u = \sum_{l=0}^{N-1}. \qquad (9)$$

We note that S satisfies the equation

$$S[u, w + L] = e^{iu}S[u, w]. \qquad (10)$$

Since exp $[iu]$ is always taken equal to one of the Nth roots of unity, $S[u, w]$ is periodic in w with period NL.

Unfortunately the fact that the dynamics of the system are so simple makes most of the characteristics important in the usual many-body systems trivial and degenerate in this one. For example, the velocity distribution does not change with time since there is always one particle with each of the initial velocities. More generally it can be stated that the only properties of this system that will be interesting are those which depend upon attaching labels to one or more of the particles. If the particles are not distinguishable we may treat the system more simply by allowing them to exchange identities during collision, which in turn is equivalent to allowing them to pass through each other without interaction. Hence in this case we may expect the same results as for a noninteracting gas.

CALCULATIONS WITH A FIXED SET OF INITIAL CONDITIONS

Although we cannot study the approach of the velocity distribution toward equilibrium we may assume that the velocities at the initial time have already been randomized by some otherwise unconsidered process such as three-body collision and study randomization in configuration space. This is of some interest even for noninteracting gases. The procedures illustrated first on the noninteracting gas will later also be applied on properties of the hard-particle system.

It is convenient to make the assumption that the velocities of the N particles satisfy no relation of the form

$$\sum_{i=0}^{N-1} n_i v_i L = 0, \qquad (11)$$

where the n_i's are positive or negative integers with perhaps some, but not all, of them zero. When this assumption is fulfilled we may apply a theorem due to Kronecker and Weyl to calculate time averages of various properties of this system following the procedure used by Montroll and Mazur[2] for a harmonic oscillator system. By the time average expectation we mean the expectation

$$\langle f[x_0(t) \cdots x_{N-1}(t)]\rangle = \lim_{T\to\infty} \frac{1}{T} \int_0^T f[x_0(t) \cdots x_{N-1}(t)]\, dt \qquad (12)$$

with $x_i(t) = x_i + v_i t$. The Kronecker–Weyl theorem states that if a bounded Riemann integrable function of N variables, $f(x_0 \cdots x_{N-1})$, is periodic in each argument with period 2π and if the previously stated assumption holds then

$$\lim_{T\to\infty} \frac{1}{T} \int_0^T f\, dt = \frac{1}{(2\pi)^N} \int^{2\pi} dx_0 \cdots \int_0^{2\pi} dx_{N-1}\, f. \qquad (13)$$

This theorem justifies the use of the usual uniform average in the configuration space of this system.

We use the theorem first to describe the Poincaré cycle behavior of the trajectories. We say that a set M of trajectories has returned to an ϵ neighborhood of their original positions, or alternatively has suffered a recurrence, at time t if the inequality

$$f(t) = \sum_{i\in M} \dashv x_i(t) - x_i(0) \vdash < \epsilon \qquad (14)$$

is satisfied. By $\dashv y \vdash$ we mean the function which is equal to the absolute value of y for $-L/2 < y \leq L/2$ and is defined outside this interval by the requirement that it be periodic with period L.

We may calculate the number of recurrences in a given period of time by counting the number of zeros of the function $f(t) - \epsilon$ and dividing by two. For this reason following Montroll and Mazur we define the recurrence time T_r by the equation

$$\frac{1}{T_r} = \frac{1}{2} \lim_{T\to\infty} \frac{1}{T} \int_0^T \sum_\alpha \delta(t - t_\alpha)\, dt, \qquad (15)$$

where the t_α are the zeros of $f(t) - \epsilon$. This can be transformed to

$$\frac{1}{T_r} = \frac{1}{2} \lim_{T\to\infty} \frac{1}{T} \int_0^T \delta[f(t) - \epsilon]\, |f'(t)|\, dt. \qquad (16)$$

The theorem can be applied to this expression if we replace the delta function with any well-behaved

[2] P. Mazur and E. Montroll, J. Math. Phys. 1, 70 (1960).

approximation. This gives

$$\frac{1}{T_r} = \frac{1}{2}\frac{1}{L^M}\int_{-L/2}^{L/2} dx_1 \cdots \int dx_M\, \delta\Big(\sum_{i\in M} |x_i| - \epsilon\Big) \times \Big|\sum_{i\in M} v_i \,\mathrm{sgnp}\, x_i\Big| \qquad (17)$$

where

$$\mathrm{sgnp}\, y = 1 \qquad 0 < x < L/2$$
$$-1 \qquad -L/2 < x < 0. \qquad (18)$$

Sgnp y is defined outside this interval by the requirement that it be periodic with period L. This formula is easily evaluated to give

$$\frac{1}{T_r} = \frac{1}{2(M-1)!\,L}\left(\frac{\epsilon}{L}\right)^{M-1} \sum_{\pm} \Big|\sum_i \pm v_i\Big|, \qquad (19)$$

provided $\epsilon \leq L/2$. In this formula we take the sum over all combinations of + and − signs in the sum of velocities inside the absolute value sign. It can be shown using the central-limit theorem of probability theory that if the number of trajectories is large and a small number of v's are not much larger than the rest that

$$\sum_{\pm} \Big|\sum_i \pm v_i\Big| \sim 2^M\left(\frac{2\sigma^2}{\pi}\right)^{\frac{1}{2}}, \qquad \sigma^2 = \sum_i v_i^2, \qquad (20)$$

and hence the recurrence time approaches

$$T_r \sim (2\epsilon\pi/M^{\frac{1}{2}}\sigma)(ML/2\epsilon e)^M. \qquad (21)$$

Hemmer, Maximon, and Wergeland[3] have also given a method for finding the recurrence time in a problem equivalent to this one. Their definition of the recurrence time is the same as ours except that a recurrence is defined to be a return to the volume in M space defined by

$$\dashv x_i(t) - x_i(0) \vdash\, < a_i \quad i = 1, 2, 3, \cdots, M \qquad (22)$$

for any set of positive a_i's rather than the volume defined by (14). The recurrence time that they obtain for this volume is

$$T_r = \prod_{i=1}^{M}\left(\frac{L}{2a_i}\right) \Big/ \sum_{i=1}^{M} \frac{|v_i|}{2a_i}. \qquad (23)$$

If we choose the a_i's all equal and adjust the common value so that the volume of this hypercube is equal to that of the "2^M-a hedron" of (14), this formula gives

$$T_r = (L/2\epsilon)^M[2\epsilon M!/(M!)^{1/M} \sum_i |v_i|]. \qquad (24)$$

If we further assume a Maxwell distribution of velocities so that $\bar{v} = (8/3\pi M)^{\frac{1}{2}}\sigma$, we find that this formula differs from (21) by an additional factor

$$(3M)^{\frac{1}{2}}/2(M!)^{1/M} \cong \tfrac{1}{2}e(3/M)^{\frac{1}{2}} \quad \text{for large } M. \qquad (25)$$

This difference is not large considering that the formula contains factors as large as $(LM/2\epsilon)^M$. The method of Hemmer, Maximon, and Wergeland can also be adapted to the region of Eq. (14) to yield precisely Eq. (19). A third method of defining the recurrence time due to Smoluchowski has been shown by Kac[4] to give the same result as Hemmer, Maximon, and Wergeland.

These same methods can be used to determine properties of the system which depend upon knowing which particle is on which trajectory except that the expressions to be evaluated are periodic in the usual variables with a period NL reflecting this property of $S(u, x)$. Single-particle properties are easily calculated. We find that the time average single-particle distribution is uniform throughout the system independent of the initial spatial distribution. A particle spends $1/N$th of its time on each trajectory, consequently the probability that it will have a velocity in a given range is the fraction of the trajectories with a velocity in this range. On the average, particle i requires a time of NL/v_i to return to its original position on its original trajectory. The time required for a particle to return to a set of velocities θ is given by

$$\tau \cong \left[\rho \int_{\bar\theta} P(v_1)\, dv_1 \int_{\theta} P(v_2)\, dv_2\, |v_1 - v_2|\right]^{-1}, \qquad (26)$$

where $P(v)$ is the probability density for finding trajectories in a range about v, $\rho = N/L$ is the density of the system, and where $\bar\theta$ is the compliment of the set θ.

Properties of the system which depend upon two-particle correlations are only slightly more difficult to calculate. The recurrence time for two particles to come back to their original trajectories and close to their original positions is given approximately by

$$\frac{1}{T_r^{(2)}} = \lim_{T\to\infty} \frac{1}{2}\frac{1}{T}\int_0^T \sum_{\alpha} \delta(t - t_\alpha) A_{jj}(t) A_{kk}(t)\, dt, \qquad (27)$$

where the t_α's are the roots of the function

$$f(t) - \epsilon = \dashv x_j(t) - x_j(0)\vdash + \dashv x_k(t) - x_k(0)\vdash - \epsilon. \qquad (28)$$

This formula counts only those zeros of $f(t) - \epsilon$ which occur when A_{jj} and A_{kk} are both equal to 1. The formula is approximate because it is not clear

[3] P. C. Hemmer, L. C. Maximon, and H. Wergeland, Phys. Rev. 111, 689 (1958).

[4] M. Kac, Phys. Rev. 115, 1 (1959).

whether we should assume that there are one or two zeros of the function for each recurrence. Equation (27) can be transformed to

$$\frac{1}{T_r^{(2)}} = \frac{1}{2(NL)^N} \int_0^{NL} \cdots \int_0^{NL} dx_0 \cdots dx_{N-1} \times \delta(|x_j| + |x_k| - \epsilon) \times |v_j \operatorname{sgnp} x_j + v_k \operatorname{sgnp} x_k| A_{jj} A_{kk} \qquad (29)$$

where in each of the A's w_{kh} is set equal to $x_k - x_h$. This expression contains a special case of the quantity $\mathfrak{A}_{j'jk'k}$

$$\mathfrak{A}_{j'jk'k} = \frac{1}{(NL)^{N-2}} \int_0^{NL} \cdots \int_0^{NL} A_{j,j} A_{k,k} \prod_{\substack{h=0 \\ h \neq j,k}}^{h=N-1} dx_h \qquad (30)$$

to be evaluated, which is of some interest in itself. It gives the time average probability of finding particle j' on trajectory j which is at a given position and particle k' on trajectory k at another given position. Written out more fully, the expression for $\mathfrak{A}_{j'jk'k}$ is

$$\mathfrak{A}_{j'jk'k} = \frac{1}{(NL)^{N-2}} \times \int_0^{NL} \cdots \int_0^{NL} \frac{1}{N^2} \sum_{u_1 u_2} \exp\left[-i(j'u_1 + k'u_2)\right] \times \prod_{h=0}^{N-1} S[u_1, x_j - x_h] S[u_2, x_k - x_h] \prod_{h \neq j,k} dx^h. \qquad (31)$$

The integral over a typical x_h in this expression has the form

$$I = \frac{1}{NL} \int_0^{NL} S[u_1, x_j - x_h] S[u_2, x_k - x_h] \, dx_h \qquad (32)$$

and using Eq. (10),

$$I = e^{i(u_1+u_2)} \frac{1}{NL} \int_0^{NL} S[u_1, x_j - x_h + L] \times S[u_2, x_k - x_h + L] \, dx_h.$$

In this expression we may introduce $x_h' = x_h - L$ as a new variable of integration and maintain the same limits because the integral is over a full period to obtain

$$I = e^{i(u_1+u_2)} I$$

or

$$I = 0 \quad \text{unless} \quad u_2 = -u_1 \bmod (2\pi). \qquad (33)$$

We use this formula to eliminate u_2 from the expression and find that the result is now clearly periodic in each x_h with period L so that each spacial average can now be carries out over this shorter range. Following this the evaluation can be carried out straightforwardly to obtain

$$\mathfrak{A}_{j'jk'k} = \frac{1}{N} \frac{(N-2)!}{(N-k'+j'-1)!\,(k'-j'-1)!} \times \left(\frac{x_k - x_j}{L}\right)^{k'-j'-1} \left(1 - \frac{x_k - x_j}{L}\right)^{N+j'-k'-1}$$

$$\text{for } k' > j' \text{ and } x_k > x_j \qquad (34)$$

and

$$\mathfrak{A}_{j'jk'k} = \frac{(N-2)!}{N(N-k'+j'-1)!\,(k'-j'-1)!} \times \left(\frac{x_j - x_k}{L}\right)^{N-k'+j'-1} \left(1 - \frac{x_j - x_k}{L}\right)^{k'-j'-1}$$

$$\text{for } k' > j' \text{ and } x_k < x_j. \qquad (35)$$

The results for $j' < k'$ are obtained by interchanging labels. In terms of $\mathfrak{A}$ the two particle recurrence time is

$$\frac{1}{T_r^{(2)}} = \frac{1}{2L^2} \int_{-L/2}^{L/2} dx_j \int_{-L/2}^{L/2} dx_k \, \delta(|x_j| + |x_k| - \epsilon) \times |v_j \operatorname{sgnp} x_j + v_k \operatorname{sgnp} x_k| \, \mathfrak{A}_{jjkk}. \qquad (36)$$

To simplify the work, we consider only the limit of an infinite system so that N and L approach infinity in such a way that $\rho = N/L$, the density of the system remains finite. Then $\mathfrak{A}$ is given for $k' > j'$ by

$$\mathfrak{A}_{j'jk'k} \sim \frac{1}{N} \frac{[\rho(x_k - x_j)]^{k'-j'-1}}{(k'-j'-1)!} e^{-(x_k - x_j)\rho}, \quad x_k > x_j,$$

$$\mathfrak{A}_{j'jk'k} \sim 0 \qquad (x_k < x_j). \qquad (37)$$

This second expression becomes zero because the particles cannot change their order along the line in an infinite system. Substituting this in Eq. (36) we obtain the result

$$\frac{1}{T_r^{(2)}} = \frac{\rho}{2N^3} |v_k - v_j| \frac{(\rho\epsilon)^{\nu+1}}{\nu!} e^{-\rho\epsilon} + \frac{\rho^{\nu+2}}{2N^3 \nu!} |v_k + v_j| \int_0^{\epsilon} y^{\nu} e^{-\rho y} \, dy$$

where $\nu = k - j - 1$ and

$$\frac{1}{\nu!} \int_0^{\epsilon} y^{\nu} e^{-\rho y} \, dy = \frac{1}{\rho^{\nu+1}} \left[1 - \sum_{i=0}^{\nu} \frac{(\rho\epsilon)^i}{i!} e^{-\rho\epsilon}\right]. \qquad (38)$$

The result which we have given for $\mathfrak{A}_{j'jk'k}$ can actually be obtained more simply. We note that $\mathfrak{A}_{j'jk'k}$ is the probability that particle j' is on trajectory j at a given position x_j and that there are exactly $\nu = k' - j' - 1$ trajectories between this point and the given point x_k where trajectory k is located, so that particle k' falls on this trajectory. From (13) the probability that there are ν trajectories between these points is given by

$$\frac{(N-2)!}{\nu!\,(N-\nu-2)!}\left(\frac{x_k - x_i}{L}\right)^{\nu}\left(1 - \frac{x_k - x_i}{L}\right)^{N-\nu-2}, \quad (39)$$

and the probability that particle j' is on trajectory j is just $1/N$ no matter how the trajectories are arranged by a study of the single-particle average we have discussed. Combining these results gives (34) and (35). The extension from two particles to M particles is obvious.

ENSEMBLE CALCULATIONS

We now consider the calculation of the properties of an ensemble of systems obtained by assuming suitable spatial and velocity distributions for the particles at the initial time. It might appear that any results found in this way should be obtainable from the formalism we have already given by virtue of the equivalence of time and ensemble averages. This is not the case, however, because we are able to study the dynamics of an ensemble in the limit of a large volume and large number of particles. For this infinite system the Kronecker–Weyl theorem is not available. Perhaps this is one of the more interesting points to be illustrated by a model. Unfortunately the mathematics becomes somewhat messy so we limit the discussion to properties which can be written in the form

$$F = \langle f[v_0(0); x_j(t), v_j(t)]\rangle. \quad (40)$$

We assume, as indicated, the f is a function of the coordinate and momentum of the jth particle at time t and also of the velocity of particle zero at the initial time. Note that we have assumed that particle zero starts at the origin at zero time. The function f may also depend on t explicitly without causing trouble. The quantity F can be written quite simply in terms of the A_{jk}'s and the initial values of the variables of the system,

$$F = \sum_k \langle f(v_0; x_k + v_k t, v_k) A_{jk}\rangle. \quad (41)$$

We assume that at the initial time the particles are uniformly distributed in space except that the zeroth particle is at the origin and that each has the velocity distribution $g(v)$. The thermal distribution g_T,

$$g_T(v) = (m\beta/2\pi)^{\frac{1}{2}} \exp[-m\beta v^2/2], \quad (42)$$

is of particular interest. Since the velocities of the particles are independent, it is possible to simplify formulas by expressing them in terms of the quantity

$$R[u, x_k + v_{kt} - x_h] \equiv \int_{-\infty}^{\infty} g(v_h) S[u, x_k + v_k t - x_h - v_h t]\, dv_h. \quad (43)$$

Thus we have

$$F = \frac{1}{N}\sum_u \frac{(N-1)!}{L^{N-1}} \int_0^L dx_{N-1} \times \int_0^{x_{N-1}} dx_{N-2} \cdots \int_0^{x_2} dx_1 \Bigg\{ \sum_{k\neq 0} \int_{-\infty}^{\infty} g(v_k)\, dv_k \times \prod_{\substack{h=1\\ h\neq k}}^{N-1} (R[u, x_k + v_k t - x_h]) \int_{-\infty}^{\infty} g(v_0)\, dv_0 \times S[u, x_k + v_k t - v_0 t] f(v_0; x_k + v_k t, v_k) + \int_{-\infty}^{\infty} g(v_0)\, dv_0 \int_0^L dx_{N-1} \cdots \int_0^{x_2} dx_1 \times \prod_{h=1}^{N-1} (R[u, +v_0 t - x_h]) f(v_0; v_0 t, v_0) \Bigg\}. \quad (44)$$

This can be simplified by using the fact that

$$\int_0^L dx_n \int_0^{x_n} dx_{n-1} \cdots \int_0^{x_2} dx_1 = \frac{1}{n!}\int_0^L dx_n \int_0^L dx_{n-1} \cdots \int_0^L dx_1 \quad (45)$$

if the integrand is symmetric in these arguments. By interchanging integrations, using this relation and carrying out the sum over k explicitly the expression can be simplified considerably:

$$F = \frac{1}{N}\sum_u \frac{N-1}{L} \exp[-iju] \times \int_0^L dx_k \int_{-\infty}^{\infty} g(v_k)\, dv_k Q(u, x_k + v_k t, v_k) \times \left[\frac{1}{L}\int_0^L R[u, x_k + v_k t - x_h]\, dx_k\right]^{N-2} + \frac{1}{N}\sum_u \exp[-iju] \int_{-\infty}^{\infty} g(v_0)\, dv_0 f(v_0, v_0 t, v_0) \times \left[\frac{1}{L}\int_0^L R[u, x_k + v_k t - x_h]\, dx_h\right]^{N-1}$$

$$Q(u, x_k + v_k t, v_k) \equiv \int_{-\infty}^{\infty} g(v_0) f(v_0, x_k + v_k t, v_k) \times S[u, x_k + v_k t - v_0 t]\, dv_0. \quad (46)$$

Next it is convenient to consider the origin to be at the center of the interval. This is accomplished by the transformation,

$$\int_0^L dx_k = \int_0^{L/2} dx_k + \int_{L/2}^L dx_k,$$

$x_k' = x_k - L$ in second integral,

$$S(u, w' + L) = e^{+iu}S[u, w'],$$

$$\int_0^L dx_k = \int_0^{L/2} dx_k + e^{-iu}\int_{-L/2}^0 dx_k', \tag{47}$$

where the integrand is now the original function in both integrals. Now

$$\frac{1}{L}\int_0^L R[u, w - x_k]\, dx_k$$

$$= \int_{-\infty}^{\infty} g(v_k)\, dv_k \frac{1}{L}\int_0^L S(u, w - v_k t - x_k)\, dx_k \tag{48}$$

and

$$\frac{1}{L}\int_0^L S[u, w' - x_k]\, dx_k = 1 + (w'/L)(1 - e^{-iu}) \times S[u, w'] \quad \text{for} \quad -L < w' < L, \tag{49}$$

hence if contributions from outside this range are negligible, as is true if $g(v_k) \cong 0$ when $|w - v_k t| > L$, we have

$$\frac{1}{L}\int_0^L R[u, w - x_k]\, dx_k \cong 1 + \frac{1}{L}(1 - e^{-iu})T[u, w], \tag{50}$$

with

$$T[u, w] = \int_{-\infty}^{\infty} (w - v_k t)S[u, w - v_k t]g(v_k)\, dv_k. \tag{51}$$

We expect these conditions to be fulfilled for large L and N.

Now consider the limit as N and L approach infinity with $\rho = N/L$ held finite. This gives

$$F = \rho\frac{1}{2\pi}\int_0^{2\pi} du\, e^{-iju} \times \left[\int_0^{\infty} dx_k + e^{-iu}\int_{-\infty}^0 dx_k\right]\int_{-\infty}^{\infty} g(v_k)\, dv_k$$
$$\times Q(u, x_k + v_k t, v_k) \exp\{\rho(1 - e^{-iu})T[u, x_k + v_k t]\}$$
$$+ \frac{1}{2\pi}\int_0^{2\pi} du\, e^{-iju}\int_{-\infty}^{\infty} g(v_0)\, dv_0\, f(v_0, v_0 t, v_0)$$
$$\times \exp\{\rho(1 - e^{-iu})T[u, v_0 t]\}, \tag{52}$$

where we have used the fact that

$$\frac{1}{N}\sum_u \to \frac{1}{2\pi}\int_0^{2\pi} du, \tag{53}$$

and have assumed that $Q \to 0$ for x_k large in order that we have the limit

$$[1 + (1/L)(1 - e^{-iu})T]^N \to \exp\{\rho(1 - e^{-iu})T\} \tag{54}$$

for all values of x_k which contribute.

The integration over u can be carried out if the expressions for T and Q are simplified.

$$T(u, w) = -A(w) + (w + A)e^{iu},$$
$$Q(u, w, v_k) = B + Ce^{iu},$$

with

$$A(w) = \int_{w/t}^{\infty} (-w + v_k t)g(v_k)\, dv_k,$$

$$B = \int_{w/t}^{\infty} g(v_0)f(v_0, w, v_k)\, dv_0, \tag{55}$$

$$C = \int_{-\infty}^{w/t} g(v_0)f(v_0, w, v_k)\, dv_0,$$

where we have used the assumption

$$\int_{-\infty}^{\infty} vg(v)\, dv = 0. \tag{56}$$

We make use of the formula

$$J^{(m)} \equiv \frac{1}{2\pi}\int_0^{2\pi} du\, e^{-imu} \times \exp\{\rho(1 - e^{-iu})(-A + Ge^{+iu})\}$$
$$= (G/A)^{m/2}I_m[2\rho(AG)^{\frac{1}{2}}]e^{-(A+G)\rho}, \tag{57}$$

where I_m is a Bessel function of imaginary argument. This formula is easily obtained by integrating the usual generating function for Bessel functions,

$$\exp\left[\frac{x(t - t^{-1})}{2}\right] = \sum_{n=-\infty}^{\infty} t^n J_n(x), \tag{58}$$

around the origin with a t^{-m-1} factor and changing variables.

The results obtained by using $\delta[y - x_j(t)]$ for f are particularly interesting. This gives the probability $P(y)$ of finding particle j at position y at time t. The result is

$$P(y) = \{2\rho(1 - B_0)B_0H^jI_j + \rho(1 - B_0)^2H^{j-1}I_{j-1}$$
$$+ \rho B_0^2H^{j+1}I_{j+1} + (1/t)H^j g(y/t)I_j\}\exp[-(2A(y)+y)\rho]$$

where

$$B_0 = \int_{y/t}^{\infty} g(v_0)\, dv_0 \quad H = \left[\frac{y + A(y)}{A(y)}\right]^{\frac{1}{2}}$$
$$I_n = I_n(2\rho\{A(y)[y + A(y)]\}^{\frac{1}{2}}) \tag{59}$$

We examine the behavior described by this complicated formula only asymptotically in the limit of large time so that we may set $y/t \simeq 0$ throughout the expression. However we shall assume that y is also large of order $t^{\frac{1}{2}}$ to keep in certain contributions which would otherwise be neglected. In this limit

$$A \simeq t \int_0^\infty vg(v)\, dv,$$
$$B_0 \simeq \int_0^\infty g(v)\, dv. \tag{60}$$

A standard asymptotic formula for the Bessel function of large argument gives

$$J^{(m)} \cong \left(\frac{G}{A}\right)^{m/2} \frac{1}{[4\pi\rho(AG)^{\frac{1}{2}}]^{\frac{1}{2}}} e^{2(AG)^{\frac{1}{2}}\rho - (A+G)\rho}$$
$$= \left(\frac{G}{A}\right)^{m/2} \frac{1}{[4\pi\rho(AG^{\frac{1}{2}})]^{\frac{1}{2}}} e^{-(A^{\frac{1}{2}} - G^{\frac{1}{2}})^2\rho} \tag{61}$$

and since $G = y + A \simeq A$,

$$J^{(m)} \cong (4\pi\rho A)^{-\frac{1}{2}} e^{-y^2\rho/4A}$$

and if

$$D = \frac{1}{\rho} \int_0^\infty vg(v)\, dv,$$

we have

$$P(y) \simeq (4\pi Dt)^{-\frac{1}{2}} e^{-y^2/4Dt}. \tag{62}$$

Thus the long-time effect of the motion is to cause the particles about the origin to appear to diffuse away. This, of course, is not true diffusion since there is no way for a particle to get past its nearest neighbors. This delocalization phenomenon is not unexpected here since it also occurs in one-dimensional harmonic-oscillator assemblies.[5]

We can obtain precisely this value of D from an extremely crude statistical model of this process. We assume that the trajectories are "on the average" side by side with a distance $1/\rho$ between them. The particle we are observing performs a random walk on these trajectories, stepping right or left with equal probability at a rate equal to the rate at which the trajectories move this distance to collide with each other. This rate is

$$\rho \int_{-\infty}^{\infty} |v|\, g(v)\, dv = 2\rho \int_0^\infty vg(v)\, dv. \tag{63}$$

The usual theory described, for example, by Chandrasekhar[6] gives for the "diffusion constant" of this motion the expression

$$D = \tfrac{1}{2}nl^2;$$
$$n = \text{rate of stepping}, \quad l = \text{length of step}. \tag{64}$$

Now if we assume that the process is such a random walk over a line of points a distance $1/\rho$ apart with a jumping rate of $\rho\langle|v|\rangle$ as above we get the same result as in the foregoing analysis.

Another quantity of interest is the correlation function $\langle v_0(0)v_i(t)\rangle$. The behavior of correlations of this type is of general interest because of the role they play in the theory of stochastic processes. A trick is required to calculate this quantity using the general formalism we have given because there are contributions to the integral over x_k for large values of the argument, and we are required to make a transformation on the expression in Eq. (46) before taking the limit of an infinite system. We restrict the $g(v)$ under consideration to the thermal distribution $g_T(v)$ given in Eq. (42). Then the expression of Eq. (46) with f taken as $v_0(0)v_i(t)$ has the general form

$$F = \left(\frac{m\beta}{2\pi}\right)^{\frac{1}{2}} \int_{-\infty}^{\infty} v_k\, dv_k\, e^{-(m\beta/2)v_k^2} \int_0^L dx_k\, \phi(x_k + v_k t) \tag{65}$$

$+$ second term.

Integrating by parts in v_k we obtain

$$F = \frac{1}{m\beta}\left(\frac{m\beta}{2\pi}\right)^{\frac{1}{2}} \int_{-\infty}^{\infty} e^{-(m\beta/2)v_k^2}\, dv_k$$
$$\times \int_0^L dx_k \frac{\partial}{\partial v_k} \phi(x_k + v_k t) + \text{second term} \tag{66}$$

and, since in this expression $\partial/\partial v_k = t\partial/\partial x_k$, we have

$$F = \frac{t}{m\beta} \int_{-\infty}^{\infty} dv_k\, g_T(v_k)[\phi(L + v_k t) - \phi(v_k t)]$$
$$+ \text{second term}. \tag{67}$$

Here we may make use of Eq. (10) and then take the limit as N approaches infinity with ρ finite as before. The final result can be written as

$$\langle v_0(0)v_i(t)\rangle$$
$$= \frac{\rho t}{2\pi m\beta} \int_{-\infty}^{\infty} dv[H^{i+1}I_{i+1} + H^{i-1}I_{i-1} - 2H^iI_i]$$
$$\times \exp\,[-m\beta v^2 - (2A(vt) + vt)] + \left(\frac{m\beta}{2\pi}\right)^{\frac{1}{2}}$$
$$\times \int_{-\infty}^{\infty} v^2\, dv\, H^iI_i \exp\{-(m\beta/2)v^2 - [2A(vt) + vt]\rho\},$$

with

$$H = [A(vt) + vt/A(vt)]^{\frac{1}{2}},$$
$$I_n = I_n(2\rho\{A(vt)[vt + A(vt)]\}^{\frac{1}{2}}). \tag{68}$$

We note from the definition of A in Eq. (60) that H is actually independent of t and that the argument of the Bessel function is linear in t.

---
[5] R. J. Rubin, J. Math. Phys. 1, 309 (1960); 2, 373 (1961).
[6] S. Chandrasekhar, Rev. Mod. Phys. 15, 5 (1943).

A rather messy calculation shows that for large t,

$$\langle v_0(0)v_0(t)\rangle \cong \left(\frac{m\beta}{2\pi}\right)^{\frac{1}{2}}\left(-1+\frac{5}{2\pi}\right)\frac{1}{\rho^3 t^3}. \qquad (69)$$

Unlike harmonic oscillators this correlation approaches an asymptotic value even for a finite system and without oscillation. The initial thermal average appears to be responsible for the loss of Poincaré cycles. Since this correlation goes to zero for large times $v_0(t)$ viewed as a stochastic process must be ergodic by a theorem of Montroll and Mazur.[2] In essence the vanishing of this correlation for large times indicates a type of irreversibility in this system. If a single particle of the system is given an electrical charge the motion of this particle in an external electric field will give rise to an electrical conductivity. This conductivity can be expressed in terms of the motion of the system in the absence of the external field by the Kubo formula[7]

$$\sigma(\omega) = e^2\beta \int_0^\infty e^{i\omega t}\langle v_0(0)v_0(t)\rangle\, dt. \qquad (70)$$

This formula is sometimes written with an additional factor to give convergence of the integral for large t but from what we have just seen such a factor is clearly unnecessary for this system. The time integration indicated in the formula can easily be carried out on the expression for the correlation we have given, yielding a result which still contains a messy integration over v. For low frequencies we find that

$$\sigma \cong e^2\beta\left[D + \text{linear term in } \omega + (m\beta/2\pi)^{\frac{1}{2}} \times\left(-1+\frac{5}{2\pi}\right)\frac{1}{2\rho^3}\omega^2 \log\omega + O(\omega^2)\right]. \qquad (71)$$

At zero frequency the conductivity is related to the diffusion coefficient as is required by the Einstein relation[7]; the $\omega^2 \log \omega$ term is easily shown to be implied by the asymptotic dependence given in Eq. (69). It would be interesting to know whether such terms exist for actual physical systems.

[7] R. Kubo, Proc. Phys. Soc. Japan **12**, 570 (1957).

Other initial conditions can also be treated. Let us assume that at the initial time the first M particles labeled from zero to $M - 1$ are restricted to lie in the interval $[0, L/2]$ and that they have a Maxwellian velocity distribution with temperature T_1. The rest of the particles labeled from M to $N - 1$ are assumed to be in the interval $[L/2, L]$ at time $t = 0$ and have a Maxwellian velocity distribution with temperature T_2. These initial conditions model the usual experimental arrangement used for studying gaseous or liquid diffusion. The mathematics which we have described for uniform initial conditions can also be carried through in this case with slight additional complication. The most striking feature of the analysis is that the condition that the pressures in the two regions be equal to each other, which is necessary in order to make observations of diffusion in the usual experiments, is replaced for this model by the somewhat different condition that

$$\rho_1(T_1)^{\frac{1}{2}} = \rho_2(T_2)^{\frac{1}{2}}, \qquad (72)$$

where ρ_1 and ρ_2 are the initial densities in the two regions. This condition actually states that the number of trajectories crossing the boundary between the two regions per unit time be the same from the left as from the right. Given in this way, the condition is not surprising. Carrying through the analysis we find that a particle starting on the boundary of the two regions again has a Gaussian probability distribution asymptotically with time, but with the diffusion constant D given by

$$D = [4\rho_1/(\rho_1+\rho_2)^2](2\pi m\beta_2)^{-\frac{1}{2}} = [4\rho_2/(\rho_1+\rho_2)^2](2\pi m\beta_1)^{-\frac{1}{2}}. \qquad (73)$$

It appears that a variety of other properties of this system could be calculated if required.

ACKNOWLEDGMENT

I should like to thank Dr. Elliott Montroll for suggesting this work and making encouraging comments.

Reprinted from the JOURNAL OF MATHEMATICAL PHYSICS, Vol. 2, No. 3, 387–393, May–June, 1961
Printed in U. S. A.

Equipartition of Energy for Nonlinear Systems

JOSEPH FORD*
Physics Department, University of Miami, Coral Gables, Florida
(Received January 11, 1961)

A system of harmonic oscillators weakly coupled by nonlinear forces will not achieve equipartition of energy as long as the uncoupled frequencies ω_k are linearly independent on the integers, i.e., as long as there is no collection of integers $\{n_k\}$ for which $\Sigma n_k\omega_k=0$ other than all $n_k=0$. This result is shown to follow from the general form of the Kryloff and Bogoliuboff series solution to the equations of motion. Physically, the linear independence of the uncoupled frequencies means that none of the interacting oscillators drives another at its resonant frequency, and this lack of internal resonance precludes appreciable energy sharing in the limit as the coupling tends to zero. It is shown that the lack of equipartition of energy observed by Ulam, Fermi, and Pasta for certain nonlinear systems may be explained in terms of the preceding remarks. Moreover, a Kryloff and Bogoliuboff series solution to the appropriate equations of motion is shown to yield qualitative agreement with the Ulam, Fermi, and Pasta computer solution. Finally, a particular system of linear differential equations is solved which illustrates a mechanism whereby oscillator systems may achieve equipartition of energy.

I. INTRODUCTION

IT is generally believed that when a large number of independent linear harmonic oscillators are suitably coupled by nonlinear forces, an approach to equilibrium will ensue.[1] For a finite number of oscillators, deviations from equilibrium are expected[2] and the system will periodically return to its initial conditions because of Poincaré recurrences. However, when the number of oscillators is large, the Poincaré period is presumed large[3] and deviation from equilibrium rare.[2] Ulam, Fermi, and Pasta[4] tried to illustrate the expected approach to equilibrium by observing the equipartition of energy among normal modes for a system of one-dimensional oscillators obeying equations of the type

$$\ddot{y}_i=(y_{i+1}-2y_i+y_{i-1})+\alpha[(y_{i+1}-y_i)^2-(y_i-y_{i-1})^2], \quad (1)$$

where i runs over the positive integers from unity to as high as 64. Here α is to be chosen sufficiently small that the nonlinear terms can be treated as a small perturbation. Using MANIAC I at Los Alamos, they numerically solved Eq. (1) and then calculated the energy in each normal mode as a function of time. These calculations revealed no tendency toward equipartition of energy among the normal modes. Various initial conditions were used and cubic and broken linear as well as quadratic couplings were considered without materially changing the outcome. Energy was shared by only a few modes in a periodic fashion reminiscent of the energy sharing between two vertical pendulums hanging from the same horizontal string. A plot of the normal mode energies as a function of time for a typical case appears in Fig. 2 .

The results of the computer calculations of UFP thus appear to be in sharp contrast, if not in actual contradiction, to the widely held notions concerning the approach to equilibrium. Strictly speaking, there is no contradiction here because the theories on this subject usually attempt such generality that they make no effort to categorize the specific types of nonlinear forces needed to cause an approach to equilibrium. Thus, one could argue that the simple algebraic nonlinearity of Eq. (1) is not of the appropriate type. However, since the nonlinear force that does give rise to equilibrium might be expected to possess a power series expansion valid for small displacements, the nonlinear coupling of Eq. (1) should be adequate to illustrate the approach to equilibrium. Consequently, aside from any element of contradiction, a number of physicists have been puzzled by the failure of Eq. (1) to lead to equilibrium behavior.

This paper provides an explanation for the non-equilibrium behavior of Eq. (1) using a perturbation technique due to Kryloff and Bogoliuboff.[5,6] Section II states the basic reason for the failure of the UFP solution and then illustrates the mathematical techniques to be used by applying them to an interesting linear problem which can be solved exactly. Section III then solves the nonlinear equations obtaining qualitative agreement with the computer solution. The final section summarizes the main conclusion of the paper by illustrating how they follow from the general form of the Kryloff and Bogoliuboff series solution.

II. HARMONIC CHAIN

The reason why the UFP oscillator system fails to approach equilibrium can be explained by such a simple physical argument that it is worth exposing immediately unincrusted by mathematical detail. This is particularly cogent in view of the fact that the Kryloff-Bogoliuboff technique is subject to question. Equation (1) can be

* Present address: Physics Department, Johns Hopkins University, Baltimore 18, Maryland.

[1] For a discussion of recent work on this problem and for references to past work, see I. Prigogene, *Proceedings of the International Symposium on Transport Processes in Statistical Mechanics* (Interscience Publishers, Inc., New York, 1958).

[2] P. Mazur and E. Montroll, J. Math. Phys. 1, 70 (1960).

[3] See, for example, the discussion by J. A. McLennan, Jr., Phys. Fluids 2, 92 (1959), and his accompanying bibliography.

[4] S. Ulam, E. Fermi, and J. Pasta, "Studies of nonlinear problems I," Los Alamos Sci. Lab. Rept. LA-1940 (1955), hereafter referred to as UFP.

[5] N. Kryloff and N. Bogoliuboff, *Introduction to Nonlinear Mechanics* (Princeton University Press, Princeton, New Jersey, 1947).

[6] N. Minorsky, *Introduction to Nonlinear Mechanics* (Edwards Brothers, Inc., Ann Arbor, Michigan, 1947).

transformed to normal mode coordinates x_k via the transformation

$$y_l=(2/N)^{\frac{1}{2}}\sum_{k=1}^{(N-1)} x_k \sin(kl\pi/N) \tag{2}$$

to read

$$\ddot{x}_k=-\omega_k^2x_k+\alpha\sum_{r,s=1}^{(N-1)} A_{krs}x_rx_s, \tag{3}$$

where $k=1, \cdots, (N-1)$, where the A_{krs} are determinable constants, and where

$$\omega_k=2\sin(k\pi/2N), \tag{4}$$

provided $y_0\equiv y_N\equiv 0$.

For purposes of discussion, we may suppose that the normal coordinates x_k describe the small amplitude motion of a system of pendulums coupled by nonlinear springs. We would like to determine whether or not the pendulums described by Eq. (3) can exchange energy. In order to do this, first consider just two pendulums coupled by a linear spring. Here appreciable energy exchange will occur only when the lengths of the pendulums are very nearly equal, i.e., when $(\omega_1-\omega_2)\approx 0$. Stated another way, appreciable energy sharing occurs only when one pendulum can drive the other pendulum at its resonant frequency. Kryloff and Bogoliuboff refer to this type of resonance as internal resonance. When the coupling between the two pendulums is nonlinear, one must expect a Fourier analysis of the motion to reveal the presence of sum and difference frequencies $(n_1\omega_1\pm n_2\omega_2)$. Consequently, in the presence of nonlinear forces, we must expect internal resonance and energy sharing when $(n_1\omega_1+n_2\omega_2)\approx 0$, where n_1 and n_2 are integers which may be positive or negative. Thus, the only effect of nonlinearity in regard to energy sharing is to generalize the condition for internal resonance from $(\omega_1-\omega_2)\approx 0$ to $(n_1\omega_1+n_2\omega_2)\approx 0$. The question of how closely $(n_1\omega_1+n_2\omega_2)$ must approximate zero depends on the strength of the coupling α between pendulums. As $\alpha\rightarrow 0$, the frequency sums must also tend to zero. In what follows, we generally assume that α is very small and, therefore, we write the resonance condition as an equality.

On returning to Eq. (3), we now expect appreciable energy sharing only if the $\{\omega_k\}$ are such as to cause internal resonance, i.e., only if

$$\sum_{k=1}^{(N-1)} n_k\omega_k=0 \tag{5}$$

for some nonzero collection of integers $\{n_k\}$. Numbers ω_k satisfying Eq. (5) for nontrivial $\{n_k\}$ are said to be linearly dependent on the integers. The question of energy sharing for Eq. (3) then turns on whether or not the frequencies ω_k of Eq. (4) are linearly dependent on the integers. Hemmer[7] has shown that the frequencies ω_k of Eq. (4) are linearly dependent except for N equal to a prime or a power of 2. The only cases considered by UFP were for N equal a power of 2. Thus, there are no Eq. (5) dependency relations for the UFP chains and hence no appreciable energy sharing. The austerity of this conclusion must be altered slightly since the actual values of α used by UFP did not approximate $\alpha\approx 0$; therefore, the equal sign of Eq. (5) must be replaced by the "approximately equal to" sign and some energy sharing occurs. From Eq. (4) we have that $\omega_1\approx\omega_2/2\approx\omega_3/3\approx\omega_4/4\ldots$, where the approximation becomes poorer as one reads to the right. Thus, as the value of α is increased from zero, one would expect appreciable energy sharing between higher and higher modes. UFP happened to choose a value of α such that appreciable energy sharing occured among the first few modes.

These heuristic conclusions are supported in detail by the Kryloff-Bogoliuboff calculations of the following two sections. From the standpoint of mathematical rigor, however, it is worthwhile to note that these conclusions also find support in the work of Balescu[8] who shows that the nonlinear Eqs. (3) do not possess any analytic constants of the motion other than the total energy provided, among other things, that the ω_k are linearly dependent. Consequently, when the ω_k are linearly independent, as is true for the UFP chain, we must anticipate the possibility of finding analytic constants of the motion other than the total energy; and in particular, one must anticipate the possibility that each normal mode energy E_k generates an analytic constant of the motion ϕ_k of Eq. (3) given by

$$\phi_k=E_k(t)+\sum_{n=1}^{\infty}\alpha^n\phi_k^{(n)}(t), \tag{6}$$

such that, for $\alpha\ll 1$, $E_k(t)\approx\phi_k$. In what follows, we confine our attention to the Kryloff and Bogoliuboff approach. We hope to discuss the full application of Balescu's work to these oscillator systems in another place.

Since we have now made it amply clear that there is, in general, no energy sharing without internal resonance, we might be expected in some later section to solve a nonlinear system of equations possessing internal resonance and show that equipartition of energy obtains. Unfortunately, except for the case of linear forces, the Kryloff and Bogoliuboff method becomes unmanageable in the presence of internal resonance; consequently, this task must be left for subsequent work on a computer. In view of these analytical difficulties, it is worthwhile to ask whether or not a system of pendulums weakly coupled by linear springs can provide any insight into the question of equipartition of energy. We now investigate such a linear system. In order to provide maximum internal resonance, we require that all the ω_k be equal. If we further assume

[7] P. Chr. Hemmer, Dynamic and stochastic types of motion in the linear chain, thesis, Trondheim, 1959.

[8] R. Balescu, Bull. acad. roy. Belg. Cl. Sci. 42, 622 (1956).

nearest neighbor coupling, the differential equations governing the system may be written

$$\ddot{x}_k = -\omega^2 x_k + \alpha(x_{k+1} + x_{k-1}), \tag{7}$$

where again $k=1, \cdots, (N-1)$ and where $x_0 \equiv x_N \equiv 0$. Here the length of each pendulum has been adjusted so that the coefficient of each y_k is $(-\omega^2)$. Equation (7), being linear, cannot exhibit ergodic behavior; nonetheless, for selected initial conditions, we might observe equipartition of energy among the x_k degrees of freedom because of internal resonance.

We now compute the solution to Eq. (7) by using the Kryloff and Bogoliuboff technique. That the method leads to a correct solution for this case can be verified by exact integration of Eq. (7). We begin by observing that for $\alpha=0$, Eq. (7) has the solution

$$x_k = A_k \cos(\omega t + \varphi_k), \tag{8}$$

where A_k and φ_k are constants. If we now seek a power series solution to Eq. (7) of the form

$$x_k = A_k \cos(\omega_k t + \varphi_k) + \sum_{l=1}^{\infty} \alpha^l x_{kl}, \tag{9}$$

we encounter secular terms of the form

$$x_k \sim t \sin(\omega t + \varphi_k).$$

Since we are interested in using a truncated form of series (9), such terms are inadmissible. x_k is known to be a bounded, multiple-periodic function, while these secular terms are unbounded and aperiodic. Kryloff and Bogoliuboff observe that these secular terms arise because of a need to shift the frequency and vary the amplitude of the zeroth order terms in solution (9). In order to anticipate this amplitude and frequency modulation from the onset, Kryloff and Bogoliuboff suggest that one seek a power series solution of the form

$$x_k = A_k \cos\tau_k + \sum_{l=1}^{\infty} \alpha^l x_{kl}(A_1, \cdots, A_N, \tau_1, \cdots, \tau_N), \tag{10a}$$

where

$$\dot{A}_k = \sum_{l=1}^{\infty} \alpha^l A_{kl}(A_1, \cdots \tau_N) \tag{10b}$$

and

$$\dot{\tau}_k = \omega + \sum_{l=1}^{\infty} \alpha^l \omega_{kl}(A_1, \cdots, \tau_N). \tag{10c}$$

Here the x_{kl} are required to be periodic functions of each τ_j of period 2π. The series (10a) is to be put into Eq. (7). Equating the coefficient of each power of α to zero leads to a system of differential equations which formally can be solved sequentially for the x_{kl} as functions of the A_l and τ_l. The A_{kl} and ω_{kl} of Eqs. (10b) and (10c) are determined as functions of the A_l and τ_l at each step such that they eliminate secular terms. The resulting coupled differential Eqs. (10b) and (10c) can then be solved, at least for the linear case, yielding A_k and τ_k as functions of time. Carrying out this calculation for Eq. (7) requires some straightforward, but laborious, algebraic and trigonometric manipulations which we omit; to first order in α, the solution is found to be

$$x_k = \left[\frac{2}{N}\right] \sum_{r,s=1}^{(N-1)} \sin\left(\frac{kr\pi}{N}\right) \sin\left(\frac{rs\pi}{N}\right) \times [x_s(0) \cos\Omega_r t + \omega^{-1} \dot{x}_s(0) \sin\Omega_r t], \tag{11a}$$

where

$$\Omega_r = \omega - \omega^{-1}\alpha \cos(r\pi/N), \tag{11b}$$

and where $x_s(0)$ and $\dot{x}_s(0)$ are the initial position and speed of the sth pendulum.

Having found the solution to Eq. (7), we may now determine if there is some initial state which leads to equipartition. At $t=0$, let all $\dot{x}_k(0)$ be zero and let all $x_k(0)=0$ except for $x_1(0)=1$; thus, all the energy is initially possessed by the end pendulum. Equations (11a) and (11b) may then be written

$$x_k = (2/N) \sum_{r=1}^{(N-1)} \sin(r\pi/N) \sin(kr\pi/N) \times \cos(\omega - \beta_r)t, \tag{12a}$$

where

$$\beta_r = \omega^{-1}\alpha \cos(r\pi/N). \tag{12b}$$

It is obvious that Eq. (12a) satisfies the initial conditions if one verifies that

$$\sum_{l=1}^{(N-1)} \sin(lr\pi/N) \sin(ls\pi/N) = (N/2)\delta_{rs}, \tag{13}$$

where δ_{rs} is the Kronecker δ. For convenience, we assume that $(N-1)$ is even. Equation (12a) then becomes for k odd

$$x_k = 4N^{-1} \cos\omega t \sum_{r=1}^{(N-1)/2} \sin(r\pi/N) \sin(kr\pi/N) \cos\beta_r t, \tag{14a}$$

and for k even

$$x_k = 4N^{-1} \sin\omega t \sum_{r=1}^{(N-1)/2} \sin(r\pi/N) \sin(r\pi k/N) \sin\beta_r t. \tag{14b}$$

Each pendulum oscillates with a frequency equal to its uncoupled frequency and with an amplitude which consists of a sum of amplitude modulations, each modulating at a distinct frequency. The total energy E_T of the pendulum system divided by $\omega^2/2$ equals unity. Let E_i be the energy of the ith pendulum given by

$$E_k = 2^{-1}[\dot{x}_k^2 + \omega^2 x_k^2] \tag{15}$$

to zeroth order in α. Putting Eqs. (14a) and (14b) into Eq. (15) yields

$$(2E_k/\omega^2)=N^{-1}+\{\text{Multiple-periodic time varying terms}\} \quad (16a)$$

for $i=2, 3, \cdots, N-2$. For either end particle, we find

$$(2E_{\text{end}}/\omega^2)=N^{-1}(\tfrac{3}{2})+\{\text{Multiple-periodic time varying terms}\}. \quad (16b)$$

By averaging Eqs. (16a) and (16b) over a near period, we see that equipartition certainly occurs on the time average. By itself, however, this is a rather weak result since it does not preclude, for instance, each pendulum in turn, having all the energy for $1/N$th of the time. However, when the number $(N-1)$ of oscillators is large, we evidently have the much stronger result that each oscillator almost always possesses $1/N$th of the total energy. Moreover, this result is valid for a single system starting from a definite state and there is no need to average the motion over some ensemble of initial states as has been required in the past.[2] These points can be understood heuristically in the following way. The multiple-periodic terms in Equation (16a) consist of several sums S which can be brought to the general form

$$S=\sum_{l=1}^{\sim N} B_l \cos[t\cos(l\pi/N)], \quad (17)$$

where each B_l is a real number. Since most of the numbers $\cos(l\pi/N)$ are linearly independent on the integers, the sum S may be thought of as a sum of random variables similar to that appearing in the problem of a random walk. Consequently, as $N\to\infty$, we expect the probability distribution density for values of S to be a sharply peaked Gaussian with values of S at or near zero being overwhelmingly probable.

A detailed proof of these points, modeled on the arguments presented in Appendix III of the paper by Mazur and Montroll,[2] will be given elsewhere along with a more detailed consideration of this particular linear system. Our main purpose here is to try to present some insight into how internal resonance can bring about equipartition of energy, and the multiple-periodic amplitude modulation of Eqs. (14a) and (14b) certainly illustrates a possible mechanism for providing equipartition.

III. ANHARMONIC CHAIN

We now apply the Kryloff and Bogoliuboff method to obtain a solution to Eq. (3) which may be compared with the UFP results. Equation (3) contains constants A_{krs} not explicitly defined. Consequently, we now write out the normal mode differential equations in complete detail.

$$\ddot{x}_k=-\omega_k^2x_k+\frac{2^{\frac{3}{2}}\alpha}{N^{\frac{1}{2}}}\Bigg\{\sum_{r=N-(k-1)}^{N-1} x_rx_{2N-r-k}\sin\left[\frac{\pi(r+k)}{N}\right]\times\left\{-1+\cos\left[\frac{\pi r}{N}\right]\right\}$$
$$+\sum_{r=1}^{N-k-1} x_rx_{r+k}\left\{\sin\left[\frac{\pi(r+k)}{N}\right]-\sin\left[\frac{r\pi}{N}\right]-\sin\left[\frac{k\pi}{N}\right]\right\}$$
$$+\sum_{r=1}^{k-1} x_rx_{k-r}\sin\left[\frac{\pi(k-r)}{N}\right]\left\{-1+\cos\left[\frac{r\pi}{N}\right]\right\}\Bigg\}, \quad (18)$$

where $k=1, 2, \cdots, N-1$, and where

$$\omega_k=2\sin(k\pi/2N). \quad (19)$$

Equations (18) need be solved only for values of $N=2^k$, where k is a positive integer, for which the ω_k are linearly independent on the integers. We thus seek a Kryloff and Bogoliuboff solution of the form

$$x_k=A_k\sin\tau_k+\sum_{l=1}^{\infty}\alpha^l x_{kl}(\tau_1,\cdots,\tau_{N-1}), \quad (20a)$$

where

$$\tau_k=\Omega_k t+\phi_k \quad (20b)$$

and

$$\Omega_k=\omega_k+\sum_{l=1}^{\infty}\alpha^l\Omega_{kl}. \quad (20c)$$

The x_{kl} are required to be periodic functions of period 2π in each variable τ_j. The Ω_{kl} are constants independent of time determined so as to eliminate secular terms. The A_k and ϕ_k are constants determined by the initial conditions. On comparing Eqs. (20) with Eqs. (10), one sees that Eq. (20) retain the capacity for frequency modulation while discarding the mechanism for amplitude modulation. The lack of amplitude modulation, of course, means that one does not expect any appreciable energy sharing between normal modes. For large N and small k, however, we see from Eq. (19) that

$$\omega_1\approx\omega_2/2\approx\omega_3/3\approx\text{etc.};$$

hence, for these values of N and k, the ω_k are "almost" linearly dependent and, strictly speaking, we should at least allow for the possibility of some internal resonance, i.e., amplitude modulation. To allow for amplitude modulation, however, the Kryloff and Bogoliuboff method requires that one solve coupled nonlinear equations at least as difficult as Eqs. (18) themselves. We are, therefore, reduced to finding solution (20) while bearing in mind its limitations.

On putting Eqs. (20) into Eqs. (18), we find after some laborious algebraic and trigonometric manipulations that to first order in α

$$x_{2l+1}=A_{2l+1}\sin\tau_{2l+1}+\frac{\alpha}{(2N)^{\frac{1}{2}}}\Bigg\{\sum_{r=N-l}^{N-1}A_rA_{(2N-r-2l-1)}\times\Bigg\{\cot\left[\frac{(2l+1)\pi}{4N}\right]\cos[\tau_r-\tau_{(2N-r-2l-1)}]+\tan\left[\frac{(2l+1)}{4N}\right]\cos[\tau_r+\tau_{(2N-r-2l-1)}]\Bigg\}+\sum_{r=1}^{N-2l-2}A_rA_{r+2l+1}\Bigg\{-\cot\left[\frac{(2l+2)\pi}{4N}\right]\times\cos[\tau_r-\tau_{r+2l+1}]-\tan\left[\frac{(2l+1)}{4N}\right]\times\cos[\tau_r+\tau_{r+2l+1}]\Bigg\}+\sum_{r=1}^{l}A_rA_{2l+1-r}\times\Bigg\{-\tan\left[\frac{(2l+1)\pi}{4N}\right]\cos[\tau_r-\tau_{2l+1-r}]-\cot\left[\frac{(2l+1)\pi}{4N}\right]\cos[\tau_r+\tau_{2l+1-r}]\Bigg\}\Bigg\},\quad(21a)$$

and

$$x_{2l}=A_{2l}\sin\tau_{2l}+\frac{\alpha}{(2N)^{\frac{1}{2}}}\Bigg\{A_{N-l}{}^2\Bigg\{\tfrac{1}{2}\cot\left[\frac{l\pi}{2N}\right]+\tfrac{1}{2}\tan\left[\frac{l\pi}{2N}\right]\cos[2\tau_{N-l}]\Bigg\}+\sum_{r=N-l+1}^{N-1}A_rA_{(2N-r-2l)}\times\Bigg\{\cot\left[\frac{l\pi}{2N}\right]\cos[\tau_r-\tau_{(2N-r-2l)}]+\tan\left[\frac{l\pi}{2N}\right]\cos[\tau_r+\tau_{(2N-r-2l)}]\Bigg\}+\sum_{r=1}^{N-2l-1}A_rA_{r+2l}\Bigg\{-\cot\left[\frac{l\pi}{2N}\right]\cos[\tau_r-\tau_{r+2l}]-\tan\left[\frac{l\pi}{2N}\right]\cos[\tau_r+\tau_{2l+r}]\Bigg\}+\sum_{r=1}^{l}A_rA_{2l-r}\times\Bigg\{-\tan\left[\frac{l\pi}{2N}\right]\cos[\tau_r-\tau_{2l-r}]-\cot\left[\frac{l\pi}{2N}\right]\times\cos[\tau_r+\tau_{2l-r}]\Bigg\}+A_l{}^2\Bigg\{-\tfrac{1}{2}\tan\left[\frac{l\pi}{2N}\right]-\tfrac{1}{2}\cot\left[\frac{l\pi}{2N}\right]\cos[2\tau_l]\Bigg\}\Bigg\}.\quad(21b)$$

Here $\tau_k=\omega_kt+\varphi_k$ since $\Omega_k=\omega_k$ to first order in α.

Let us now specialize solutions (21a) and (21b) to the case $N=32$, $\alpha=\frac{1}{4}$, and all $x_k(0)$ and $\dot{x}_k(0)$ equal to zero except for $x_1(0)$. These, of course, are the conditions for one of the solutions obtained by UFP. Since solutions (21a) and (21b) are expected to be valid only when relatively little energy finds its way into the higher modes, we neglect all x_k and A_k for $k\geqq 6$. Under these conditions, Eqs. (21a) and (21b) take the form

$$x_{2l+1}=A_{2l+1}\cos[\omega_{2l+1}t]+\tfrac{1}{32}\cot\left[\frac{(2l+1)\pi}{128}\right]\times\Bigg\{-\sum_{r=1}^{5}A_rA_{r+2l+1}\cos[(\omega_r-\omega_{r+2l+1})t]+\sum_{r=1}^{l}A_rA_{2l+1-r}\cos[(\omega_r+\omega_{2l+1-r})t]\Bigg\},\quad(22a)$$

$$x_{2l}=A_{2l}\cos[\omega_{2l}t]+\tfrac{1}{32}\cot\left[\frac{l\pi}{64}\right]\Bigg\{-\sum_{r=1}^{4}A_rA_{r+2l}\times\cos[(\omega_r-\omega_{r+2l})t]+\sum_{r=1}^{l-1}A_rA_{2l-r}\times\cos[(\omega_r+\omega_{2l-r})t]+\frac{A_l{}^2}{2}\cos[2\omega_lt]\Bigg\},\quad(22b)$$

where $l\leqq 2$. Here we have neglected the tan terms of Eqs. (21a) and (21b) since they are small relative to the cot terms, and we have deferred until later the problem of determining the A_k in terms of $x_l(0)$. We see that Eqs. (22a) and (22b) already satisfy $\dot{x}_l(0)=0$.

Before evaluating the A_k, we show that solutions (22a) and (22b) exhibit a relatively short-term periodic behavior independent of the value of the A_k. To illustrate this behavior, we calculate the energy in the first normal mode as given following UFP by

$$E_1=\dot{x}_1^2/2+(\omega_1^2x_1^2)/2.\quad(23)$$

(This expression neglects the energy in the coupling which is proportional to α and higher powers of α; this neglect is no doubt justified if the resulting normal mode energies always sum to the correct total energy.) The calculation for the energy in the higher modes reveals the same short-term periodicity, and, therefore, we perform the calculation only for E_1. Putting Eq. (22a) into Eq. (23) yields, to first order in α,

$$\begin{aligned}(2E_1/\omega_1^2)=A_1^2-\tfrac{5}{2}\{&A_1^2A_2\cos[(2\omega_1-\omega_2)t]\\&+A_1A_2A_3\cos[(\omega_1+\omega_2-\omega_3)t]\\&+A_1A_2A_4\cos[(\omega_1+\omega_3-\omega_4)t]\\&+A_1A_4A_5\cos[(\omega_1+\omega_4-\omega_5)t]\\&+A_1A_5A_6\cos[(\omega_1+\omega_5-\omega_6)t],\end{aligned}\quad(24)$$

where we set $(\omega_2-\omega_1)=\omega_1$, $(\omega_4-\omega_5)=\omega_1$, etc., when such terms appear in an amplitude factor. Now from Eq.

(19), we have

$$(2\omega_1-\omega_2)=4\sin(\pi/64)-2\sin(\pi/32). \qquad (25)$$

Expanding each sin term in Eq. (25) to third order in its argument, we have

$$(2\omega_1-\omega_2)=2(\pi/64)^3. \qquad (26)$$

Making the same calculation for each frequency factor shows that each, to third order, is an integral multiple of $2(\pi/64)^3$. Equation (24) thus may be written

$$\begin{aligned}(2E_1/\omega_1^2)=A_1^2-\tfrac{5}{2}\{&A_1^2A_2\cos\Omega t+A_1A_2A_3\cos 3\Omega t\\ &+A_1A_3A_4\cos 6\Omega t+A_1A_4A_5\cos 10\Omega t\\ &+A_1A_5A_6\cos 15\Omega t\}, \qquad (27)\end{aligned}$$

where $\Omega=2(\pi/64)^3$. Moreover, it is easy to show that this periodicity holds also for the E_1 terms second order in α where frequency factors like $[2\omega_2-\omega_1-\omega_3]=2\Omega$ arise.

We now must evaluate the A_k of Eqs. (22a) and (22b) for the case in which all $x_k(0)$ equal zero except $x_1(0)$. Setting $t=0$ in Eqs. (22a) and (22b), one finds

$$x_{2l+1}(0)=A_{2l+1}+\tfrac{1}{32}\cot[(2l+1)\pi/128] \times\left\{-\sum_{r=1}^{5}A_rA_{r+2l+1}+\sum_{r=1}^{l}A_rA_{2l+1-r}\right\}, \qquad (28a)$$

and

$$x_{2l}(0)=A_{2l}+\tfrac{1}{32}\cot[l\pi/64] \times\left\{-\sum_{r=1}^{5}A_rA_{r+2l}+\sum_{r=1}^{l-1}A_rA_{2l-r}+A_l^2/2\right\}, \qquad (28b)$$

where $l\leqq 2$. Now $A_k\approx\alpha^{-1}$ is as large a value of A_k as it is prudent to use in Eqs. (28a) and (28b) without closely investigating higher order terms in the α-series expansion. The coefficient of α^n in a Kryloff and Bogoliuboff series in general contains powers of the A_k^m with $m>n$. Thus, setting $A_k\approx\alpha^{-1}$ introduces a diverging factor in the coefficient of α^n which the smallness of α^n alone does not overcome. In view of this fact, we chose to solve Eqs. (28a) and (28b) by setting $A_1=1$ and then computing the remaining A_k subject to the conditions $x_k(0)=0$, $k=2, \cdots, 5$. On carrying out this procedure, one calculates from Eqs. (28a) and (28b) that $A_2=-0.249$, $A_3=0.088$, $A_4=-0.035$, and $A_5=0.013$, while $x_1(0)=1.35$. On putting these values of the A_k into Eqs. (22a) and (22b) and calculating the energy in the first three normal modes via the equations,

$$E_k=\dot{x}_k^2/2+(\omega_k^2x_k^2)/2 \qquad (29)$$

provides the data for the curves plotted in Fig. 1. The UFP computer curves to be compared with Fig. 1 are presented in Fig. 2. All data are the same except for the magnitude of $x_1(0)$ which equals 1.35 for the series solution and 250 for the UFP solution. Quantitatively, then, the two solutions are not comparable; qualitatively, however, they are similar. Both solutions appear to be periodic with periods the same order of magnitude, and for both solutions the amount of energy sharing decreases monotonically with increasing mode number. The greater amount of energy sharing present in the UFP solution indicates the presence of internal resonance for these first few modes. The Kryloff and Bogoliuboff solution indicates its own need to include internal resonance, i.e., amplitude modulation, for this comparison by its evident divergence for $x_1(0)=250$.

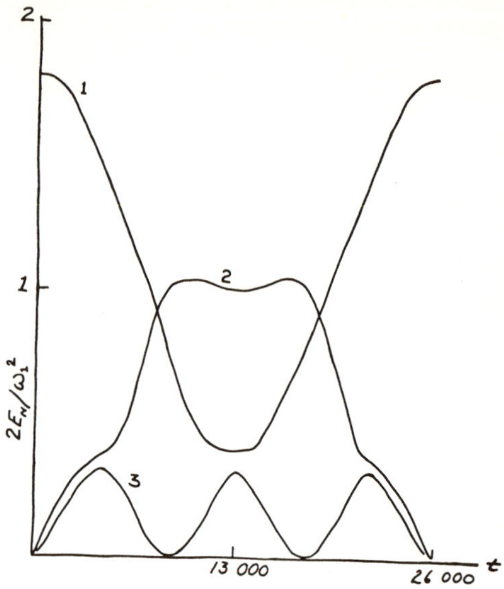

FIG. 1. A plot of the energy in the first three normal modes for $N=32$, $\alpha=\frac{1}{4}$, and $x_1(0)=1.35$ as predicted by the Kryloff and Bogoliuboff solution. Initially, mode 1 possesses the total energy. The energy in modes higher than mode 3 is negligible on this scale.

IV. CONCLUSIONS

The central purpose of this paper has been to expose a reason for the failure of the UFP computer calculation to reveal equipartition of energy. This purpose has certainly been achieved. We now add one additional note which also summarizes the reasons for the failure. We have remarked that equipartition for very weakly coupled oscillator systems can never occur as long as the uncoupled frequencies are linearly independent, i.e., as long as

$$\sum_k n_k\omega_k\neq 0$$

for any nonzero collection of integers $\{n_k\}$. Moreover, we have heuristically connected this notion with the concept of resonance remarking that $\sum_k n_k\omega_k\neq 0$ means that the interacting oscillators cannot drive each other at resonance. It is interesting to note that all these ideas are revealed in the form of the Kryloff and Bogoliuboff solution. The solution to Eqs. (18) of the last section may be written in the form

$$x_{2l+1}=A_{2l+1}\sin\tau_{2l+1}+\frac{\alpha}{(2N)^{\frac{1}{2}}}\sum_{r=1}^{N-2l-2}A_rA_{r+2l+1}$$
$$\times\frac{\cos(\tau_r-\tau_{r+2l+1})}{[\omega_{2l+1}{}^2-(\omega_r-\omega_{r+2l+1})^2]}\left\{\sin\left[\frac{\pi(r+k)}{N}\right]\right.$$
$$\left.-\sin\left[\frac{r\pi}{N}\right]-\sin\left[\frac{k\pi}{N}\right]+\cdots,\right. \quad (30)$$

where the resonance denominator

$$[\omega_{2l+1}{}^2-(\omega_r-\omega_{r+2l+1})^2]$$

has been explicitly written out. Equations (21a) and (21b) of the previous section hid these resonance denominators in the cot and tan factors. Solution (30) clearly indicates that as $\alpha\to 0$, there will be no energy sharing unless some resonance denominator is zero, i.e., unless the uncoupled frequencies are linearly dependent. Moreover, the form of the resonance denominators almost demands an interpretation in terms of internal resonance. Seen in this light, the lack of equipartition of energy for the "linearly independent" UFP oscillator system is a property only of the form of the Kryloff and Bogoliuboff solution and hence might be expected to be independent of the convergence properties of these solutions.

That the UFP oscillator systems cannot exhibit equipartition is clear both from our heuristic arguments and the basic work of Balescu.[8] It is not expected that this result will be affected one way or another by the convergence questions regarding the Kryloff-Bogoliuboff series. The validity of Sec. III, however, does rest on such questions. Unfortunately, no really satisfactory answer has been given to the question of convergence. The author chooses to believe that the similarity of the series solution to the accurate computer solution is an indication of its validity, but this is not offered as a compelling argument. The validity of the solution found in Sec. III could most easily be verified using a computer, and it is hoped that this may be accomplished in the near future.

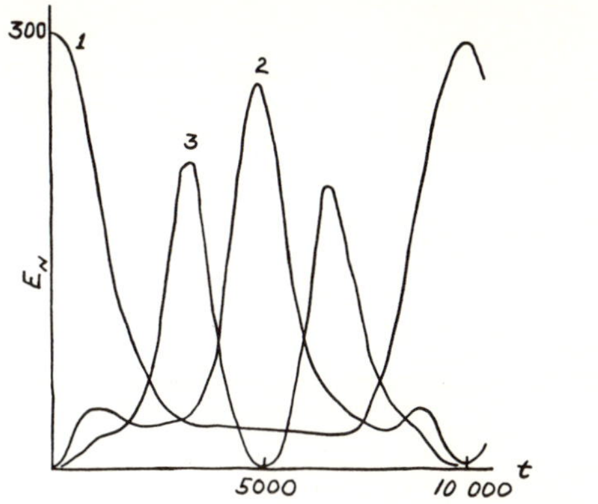

FIG. 2. A plot of the energy in the first three normal modes for $N=32$, $\alpha=\frac{1}{4}$, and $x_1(0)\approx 250$ as computed by UFP. Initially, mode 1 possesses the total energy. The energy in modes higher than mode 5 is negligible on this scale.

As to the future, by using as a guide the linear problem of Sec. II, the form of the Kryloff and Bogoliuboff series, and the criteria given in the work of Balescu, it should be possible to find a system of differential equations which can be solved on a computer to demonstrate equipartition of energy. Indeed such calculations might shed light on the whole question of the approach to equilibrium. Perhaps even more interesting, however, is the possibility of generalizing or modifying the Kryloff and Bogoliuboff technique to cover the case of internal resonance. Even a formal solution to the problem, independent of the covergence question, would be useful in view of the fact that a computer might be used to establish the range of usefulness of such a solution.

ACKNOWLEDGMENTS

It is a pleasure to thank Professor T. H. Berlin for suggesting this problem. The author also gratefully acknowledges the financial support given for this work by the National Science Foundation.

Reprinted from
JOURNAL OF MATHEMATICAL PHYSICS VOLUME 4, NUMBER 10 OCTOBER 1963
Printed in U.S.A.

Computer Studies of Energy Sharing and Ergodicity for Nonlinear Oscillator Systems*

JOSEPH FORD AND JOHN WATERS
School of Physics, Georgia Institute of Technology, Atlanta, Georgia
(Received 7 January 1963)

Weakly coupled systems of N oscillators are investigated using Hamiltonians of the form

$$H = \tfrac{1}{2}\sum_{k=1}^{N}(p_k^2 + \omega_k^2 q_k^2) + \alpha \sum_{i,k,l=1}^{N} A_{ikl}q_iq_kq_l,$$

where the A_{ikl} are constants and where α is chosen to be sufficiently small that the coupling energy never exceeds some small fraction of the total energy. Starting from selected initial conditions, a computer is used to provide exact solutions to the equations of motion for systems of 2, 3, 5, and 15 oscillators. Various perturbation schemes are used to predict, interpret, and extend these computer results. In particular, it is demonstrated that these systems can share energy only if the uncoupled frequencies ω_k satisfy resonance conditions of the form

$$\sum n_k\omega_k \lesssim \alpha$$

for certain integers n_k determined by the particular coupling. It is shown that these systems have N normal modes, where a normal mode is defined as motion for which each oscillator moves with essentially constant amplitude and at a given frequency or some harmonic of this frequency. These systems are shown to have, at least, one constant of the motion, analytic in q, p, and α, other than the total energy. Finally, it is demonstrated that the single-oscillator energy distribution density for a 5-oscillator linear and nonlinear system has the Boltzmann form predicted by statistical mechanics. Thus, these nonlinear systems are shown to have many features in common with linear systems. In particular, it is unlikely that they are ergodic. From the standpoint of statistical mechanics, it is argued that this lack of ergodicity may be a welcome feature.

1. INTRODUCTION

THERE is a long standing desire to solemnize the union of particle mechanics and statistical mechanics by exhibiting a mechanical system whose exact solution can be shown to yield an approach to equilibrium. Due to the work of Poincare,[1] Fermi,[2] and others,[3] physicists tend to believe that weakly coupled nonlinear systems will exhibit that ergodic behavior deemed necessary[3] for an approach to equilibrium. In accepting ergodicity, however, one gives up almost all hope of analytic solution. The solutions to the equations of motion are expected to be incredibly complex, and the constants of the motion are assumed to be pathological monstrosities.

As a first step in circumventing this difficulty, Fermi, Pasta, and Ulam[4] used a computer to investigate the question of equipartition of energy for weakly coupled oscillator systems using Hamiltonians of the form

$$H = \frac{1}{2}\sum_{k=1}^{N}(p_k^2 + \omega_k^2 q_k^2) + \alpha \sum_{i,k,l=1}^{N} A_{ikl}q_iq_kq_l, \tag{1}$$

where the A_{ikl} are constants defined later and where α is chosen sufficiently small that the coupling energy never exceeds some small fraction of the total energy. To their surprise, FPU found that these nonlinear oscillator systems yielded very little energy sharing at all. Certainly, the systems considered did not exhibit ergodic behavior. Since nonergodic systems might be susceptible to analysis, one of us[5] was led to investigate these nonlinear oscillator systems using perturbation theory. This heuristic investigation indicated that the FPU systems failed to show energy sharing because of an unfortuitous choice of uncoupled frequencies ω_k. Moreover, the available perturbation theories indicated that even energy sharing oscillator systems would not be ergodic.

If one believes that ergodicity is necessary for an approach to equilibrium, then apparently one must abandon these oscillator systems. But the work of Peierls,[6] for example, on lattice thermal conductivity makes one reluctant to believe that these systems do not exhibit an approach to equilibrium. The immediate suggestion then is

* Work supported by the National Science Foundation.
[1] H. Poincare, *Methodes Nouvelles de la Mechanique Celeste*, (Dover Publications, Inc., New York, 1957), Vol. 1, p. 233.
[2] E. Fermi, Z. Physik **24**, 261 (1923).
[3] For references see D. ter Haar, *Elements of Statistical Mechanics* (Rinehart and Company, New York, 1954), Appendix I.
[4] E. Fermi, J. Pasta, and S. Ulam, Los Alamos Scientific Laboratory Report. LA-1940 (1955), hereafter referred to as FPU.
[5] J. Ford, J. Math. Phys. **2**, 387 (1961).
[6] R. E. Peierls, Ann. Physik **3**, 1055 (1929).

that one should give up ergodicity. In addition to the data we shall present in the following sections, there is a growing amount of evidence that many physically interesting systems are not ergodic. Resibois and Prigogine[7] have discovered constants of the motion for gas systems. Kolmogorov[8] claims to have shown that nonlinear systems, more general than the systems considered here, are not ergodic. Statistical mechanics itself yields indirect evidence on this question. Numerous authors[9] have pointed out that only the reduced probability densities f_n, $n \ll N$, are needed for most physical calculations. Thus, the N-particle probability density f_N is not required to approach equilibrium and there is no need for ergodicity.

Perhaps the most important consequence of abandoning ergodicity, however, is that it opens the door for an analytic solution of the approach to equilibrium. It might then be expected that the improvement wrought by nonlinear systems over linear ones in the study of irreversible behavior is more modest than popularly supposed. Thus, for example, the only effect of a weak nonlinearity superimposed on a linear system might be the bringing to equilibrium of certain low-order correlation functions known[10] not to approach equilibrium for the linear system. In what follows, we attempt to show that the weakly coupled oscillator systems of Hamiltonian (1) indeed have that simple behavior expected of nonergodic systems while retaining the capacity for an approach to the equilibrium. It is our hope that these investigations will lay the groundwork for a more general study of irreversibility. We now outline the material to be covered in the sections which follow.

The question of energy sharing for the nonlinear oscillator systems of Hamiltonian (1) is investigated in Sec. II. The criteria for energy sharing are derived using perturbation theory. Their validity is demonstrated using the exact solutions to the equations of motion obtained on a computer for systems of 2, 3, 5, and 15 oscillators. Equipartition of energy is then demonstrated to occur for the 5-oscillator system.

In Sec. III, the energy-sharing oscillator systems of Sec. II are shown to have normal modes, where a normal mode is defined as motion in which each oscillator moves at essentially constant amplitude (energy) and at a given frequency or some harmonic of this frequency. The amplitudes of these normal modes are predicted using perturbation theory; the validity of the calculations is then demonstrated using a computer for systems of 3 and 5 oscillators. The computer solutions indicate that these normal modes are stable in the sense that a system started near one of its normal modes remains near it. Attention is then turned to constants of the motion. Perturbation theory yields an expression for a constant of the motion, analytic in q, p, and α, distinct from the total energy. The validity of the perturbation calculation for the case of the 5-oscillator system is verified. Thus, these nonlinear systems are shown to have many properties in common with linear systems. In particular, they are nonergodic and are susceptible to analysis.

Section IV compares the computer solution of Hamiltonian (1) for $N = 5$ with the exact solution of the linear Hamiltonian

$$H = \frac{1}{2} \sum_{k=1}^{5} (p_k^2 + q_k^2) - \alpha \sum_{k=1}^{4} q_{k+1} q_k. \qquad (2)$$

For both cases, all energy was initially given to one oscillator. The single-oscillator energy distribution density is calculated from these curves, i.e., the probability density that any single oscillator will have energy between E and $E + dE$, and a Boltzmann distribution is obtained for both linear and nonlinear systems. Thus for both systems, four particles evidently form an adequate heat bath for the fifth particle.

Section V reviews our conclusions and offers suggestions for future investigations.

We conclude this introduction on a note of caution. We have used a computer to provide exact, but limited, information on just a few of the systems described by Hamiltonian (1), and the perturbation theories we use are presented without a hint of rigor. Nonetheless, it is our belief that the necessary rigor can be supplied and these results have a rather wide range of validity. Our optimism, however, is certainly open to question; and we offer our results in the hope of spurring further study of these systems.

II. ENERGY SHARING

In this section we investigate the question of energy sharing for systems of 2, 3, 5, and 15 oscilla-

[7] P. Resibois and I. Prigogine, Bull. Acad. Roy. Belg. Class. Sci. **46**, 53 (1960).

[8] A. N. Kolmogorov, *Proceedings of the International Congress of Mathematicians, Amsterdam,* (North-Holland Publishing Co., Amsterdam, 1957), Vol. 1, pp. 315–333. Kolmogorov's article is in Russian. There is a detailed English review by J. Moser, Math. Rev. **20**, 675 (1959).

[9] See, for example, P. Mazur and E. Montroll, J. Math. Phys. **1**, 70 (1960).

[10] G. Klein and I. Prigogine, Physica **19**, 1053 (1953).

tors. Using perturbation theory, we obtain the criteria for energy sharing. The validity of these criteria is then tested by numerically solving the equations of motion on a computer. In order that the main result not be lost in the following welter of mathematical detail, let us now state that the oscillators of Hamiltonian (1) share energy only if the uncoupled frequencies ω_k satisfy resonance conditions of the form

$$\sum n_k\omega_k \lesssim \alpha, \tag{3}$$

where the n_k are integers determined by the particular coupling used. This sharpens a result previously stated by one of the authors.[5]

We are interested in continuing the investigations begun by FPU and thus we examine systems which have the same cubic coupling that they used. Now the Hamiltonian for the systems they considered may be written in the form

$$H = \frac{1}{2}\sum_{k=1}^{N} P_k^2 + \frac{1}{2}\sum_{k=0}^{N}(X_{k+1} - X_k)^2 + \frac{\alpha}{3}\sum_{k=0}^{N}(X_{k+1} - X_k)^3, \tag{4}$$

where $X_0 \equiv X_{N+1} \equiv 0$. But is is well-known that a linear transformation will diagonalize the quadratic terms of Hamiltonian (4), thus bringing it to the form of Hamiltonian (1). Hence, we refer to Eq. (4) as the FPU particle Hamiltonian and to Eq. (1), obtained by substituting the necessary linear transformation into Eq. (4), as the FPU oscillator Hamiltonian.

Since all the essential features of energy sharing appear in a 2-oscillator system, we develop the theory for this simplest case. The FPU oscillator Hamiltonian for two oscillators has the form

$$H = \tfrac{1}{2}(p_1^2 + p_2^2 + \omega_1^2 q_1^2 + \omega_2^2 q_2^2) + 2^{-\frac{1}{2}}\alpha(q_1^2 q_2 - q_2^3). \tag{5}$$

The linear transformation which converts Hamiltonian (4) for $N = 2$ into Hamiltonian (5) determines the ω_k; i.e., $\omega_1 = 1.0$ and $\omega_2 = \sqrt{3}$. However, we wish to determine if there are any ω_1, ω_2 pairs which yield energy sharing using the same cubic coupling; thus, we treat ω_1, ω_2 as free parameters. The equations of motion which follow from Hamiltonian (5) are

$$\ddot{q}_1 = -\omega_1^2 q_1 - \sqrt{2}\,\alpha q_1 q_2,$$

$$\ddot{q}_2 = -\omega_2^2 q_2 - 2^{-\frac{1}{2}}\alpha(q_1^2 - 3q_2^2). \tag{6}$$

In Appendix I, we use Wigner–Brillouin perturbation theory[11] and find that, to order α^2, Eq. (6) has the formal solution

$$\begin{aligned} q_1 = {} & A_1 \cos\tau_1 - \alpha\left\{\frac{A_1A_2\cos(\tau_1 + \tau_2)}{\sqrt{2}\,[\omega_1^2 - (\Omega_1 + \Omega_2)^2]} + \frac{A_1A_2\cos(\tau_1 - \tau_2)}{\sqrt{2}\,[\omega_1^2 - (\Omega_1 - \Omega_2)^2]}\right\} \\ & + \alpha^2\left\{\frac{A_1^3\cos 3\tau_1}{4[\omega_2^2 - 4\Omega_1^2][\omega_1^2 - 9\Omega_1^2]} - \frac{3A_1A_2^2\cos(\tau_1 + 2\tau_2)}{4[\omega_2^2 - 4\Omega_2^2][\omega_1^2 - (\Omega_1 + 2\Omega_2)^2]}\right. \\ & - \frac{3A_1A_2^2\cos(\tau_1 - 2\tau_2)}{4[\omega_2^2 - 4\Omega_2^2][\omega_1^2 - (\Omega_1 - 2\Omega_2)^2]} + \frac{A_1A_2^2\cos(\tau_1 + 2\tau_2)}{2[\omega_1^2 - (\Omega_1 + \Omega_2)^2][\omega_1^2 - (\omega_1 + 2\Omega_2)^2]} \\ & \left. + \frac{A_1A_2^2\cos(\tau_1 - 2\tau_2)}{2[\omega_1^2 - (\Omega_1 - \Omega_2)^2][\omega_1^2 - (\Omega_1 - 2\Omega_2)^2]}\right\}, \end{aligned} \tag{7a}$$

$$\begin{aligned} q_2 = {} & A_2\cos\tau_2 - \alpha\left\{\frac{(A_1^2 - 3A_2^2)}{2\sqrt{2}\,\omega_2^2} + \frac{A_1^2\cos 2\tau_1}{2\sqrt{2}\,[\omega_2^2 - 4\Omega_1^2]} - \frac{3A_2^2\cos 2\tau_2}{2\sqrt{2}\,[\omega_2^2 - 4\Omega_2^2]}\right\} \\ & + \alpha^2\left\{\frac{A_1^2A_2\cos(2\tau_1 + \tau_2)}{2[\omega_1^2 - (\Omega_1 + \Omega_2)^2][\omega_2^2 - (2\Omega_1 + \Omega_2)^2]} + \frac{A_1^2A_2\cos(2\tau_1 - \tau_2)}{2[\omega_1^2 - (\Omega_1 - \Omega_2)^2][\omega_2^2 - (2\Omega_1 - \Omega_2)^2]}\right. \\ & - \frac{3A_1^2\cos(2\tau_1 - \tau_2)}{4[\omega_2^2 - 4\Omega_1^2][\omega_2^2 - (2\Omega_1 - \Omega_2)^2]} - \frac{3A_1^2A_2\cos(2\tau_1 + \tau_2)}{4[\omega_2^2 - 4\Omega_1^2][\omega_2^2 - (2\Omega_1 + \Omega_2)^2]} \\ & \left. + \frac{9A_2^3\cos 3\tau_2}{4[\omega_2^2 - 4\Omega_2^2][\omega_2^2 - 9\Omega_2^2]}\right\}, \end{aligned} \tag{7b}$$

$$\begin{aligned} \Omega_1^2 = {} & \omega_1^2 - \alpha^2\left\{\frac{A_1^2 - 3A_2^2}{2\omega_2^2} + \frac{A_1^2}{4[\omega_2^2 - 4\Omega_1^2]} \right. \\ & \left. + \frac{A_2^2}{2[\omega_1^2 - (\Omega_1 + \Omega_2)^2]} + \frac{A_2^2}{2[\omega_1^2 - (\Omega_1 - \Omega_2)^2]}\right\}, \end{aligned} \tag{7c}$$

---

[11] The application of Wigner–Brillouin perturbation methods to oscillator systems is discussed in a paper by E. Atlee Jackson, J. Math. Phys. 4, 551 (1963). We first heard of the W–B method through him, and we are indebted to him for allowing us to read a preprint of this paper.

$$\Omega_2^2 = \omega_2^2 - \alpha^2\left\{\frac{A_1^2}{2[\omega_1^2 - (\Omega_1 + \Omega_2)^2]} + \frac{A_1^2}{2[\omega_1^2 - (\Omega_1 - \Omega_2)^2]} - \frac{3(A_1^2 - 3A_2^2)}{2\omega_2^2} + \frac{9A_2^2}{4[\omega_2^2 - 4\Omega_2^2]}\right\}. \quad (7d)$$

Here $\tau_1 = \Omega_1 t + \theta_1$, and $\tau_2 = \Omega_2 t + \theta_2$. The constants A_1, A_2, θ_1, and θ_2 are determined by initial conditions. Thus, the Wigner–Brillouin perturbation scheme converts the problem of solving differential Eq. (6) to the problem of solving the algebraic Eqs. (7c) and (7d) for the perturbed frequencies Ω_1 and Ω_2. Fortunately, our task is even simpler since we may use Eq. (7) to discuss energy sharing without obtaining an exact solution of Eqs. (7c) and (7d).

An approximate solution to Eqs. (7) is obtained by setting $\Omega_1 = \omega_1$ and $\Omega_2 = \omega_2$ in all the denominators. Provided that none of these denominators is zero or of the order of α, we see from Eqs. (7) that, for small α, $q_1 = A_1 \cos \tau_1$ and $q_2 = A_2 \cos \tau_2$. In this case the oscillators move almost as if uncoupled and there is no appreciable energy sharing. In terms of the uncoupled frequencies, no energy sharing occurs provided

$$n_1\omega_1 + n_2\omega_2 \gg \alpha, \quad (8)$$

where n_1 and n_2 range over the small integral values which appear in Eqs. (7). Examining Eq. (7c) in this approximation,

$$\Omega_1^2 = \omega_1^2 - \alpha^2\left\{\frac{A_1^2 - 3A_2^2}{2\omega_2^2} + \frac{A_1^2}{4(\omega_2^2 - 4\omega_1^2)} + \frac{A_2^2}{2[\omega_1^2 - (\omega_1 + \omega_2)^2]} + \frac{A_2^2}{2[\omega_1^2 - (\omega_1 - \omega_2)^2]}\right\}, \quad (9)$$

we see that a small denominator will occur only if $(2\omega_1 - \omega_2) \approx \alpha$. The same conclusion is reached examining Eq. (7d). Thus, when $(2\omega_1 - \omega_2) \approx \alpha$, we would anticipate that the exact solution of Eqs. (7c) and (7d) would have the form

$$\Omega_1 = \omega_1 + K_1\alpha + \cdots,$$
$$\Omega_2 = \omega_2 + K_2\alpha + \cdots, \quad (10)$$

where K_1 and K_2 are constants. As a consequence, for example, one of the terms of order α in Eq. (7a) may be written

$$-\frac{\alpha A_1 A_2 \cos(\tau_1 - \tau_2)}{\sqrt{2}\,[\omega_1^2 - (\Omega_1 - \Omega_2)^2]} \approx -KA_1A_2 \cos(\tau_1 - \tau_2), \quad (11)$$

where K is some constant; i.e., what had been a first-order term becomes a zeroth-order term. But this modification of our zeroth-order term means that, even for α arbitarily small, energy sharing will occur.

When $2\omega_1$ is not on the order of ω_2, we see from Eq. (9) that $\Omega_1 = \omega_1 + K_3\alpha^2$ and $\Omega_2 = \omega_2 + K_4\alpha^2$. In this case our only hope for energy sharing lies with terms of order α^2 in Eq (7a). Examining the denominators of these second-order terms, we see that none of these denominators will be of order α^2 unless $(\omega_1 - \omega_2) \approx \alpha^2$. If $\omega_1 \approx \omega_2$, then some of the second-order terms become zeroth order, and energy sharing occurs. As might be expected, the amount of energy sharing for the case $\omega_1 = \omega_2$ is much less than that for the case $2\omega_1 = \omega_2$. This point is verified by the computer solutions discussed below and by the theory presented in Section IV.

We have now exhausted the possibilities of energy sharing for Eq. (6). Unless $2\omega_1 \approx \omega_2$ or $\omega_1 \approx \omega_2$, we have seen that there is no energy sharing through second order. In nth order, $n > 2$, a frequency relation such as $3\omega_1 - 5\omega_2 = 0$ might, at worst, make some denominator become of the order of α^2; but the α^n in the numerator would prevent the term from contributing to zeroth order.

The generalization of these arguments for a larger number of oscillators or for a different type coupling is straightforward. For the cubic coupling of Eq. (1), the conditions for energy sharing are summarized by Eq. (3). We might mention in passing that while the Wigner–Brillouin method is useful for discussing energy sharing, its usefulness for a quantitative solution at resonance, e.g., $2\omega_1 = \omega_2$, is probably limited due to the fact that corrections to the zeroth-order term arise in most of the higher-order terms. Nonetheless, this point needs investigation.

We now turn to a discussion of the verification of the preceding theory by exact solution of the differential equations on a computer. These exact solutions were used to compute the energy E_k of each oscillator as given by

$$E_k = \tfrac{1}{2}(p_k^2 + \omega_k^2 q_k^2). \quad (12)$$

The numerical data were then plotted to give E_k as a function of time. In all cases $\alpha = 0.1$ and the coupling energy in the cubic terms seldom exceeds 10% of the total energy; thus it is legitimate to speak of individual oscillator energies.

In Figs. 1 and 2 we plot E_1 and E_2 as determined from the computer solution of Eq. (6) for various ω_k near the $2\omega_1 = \omega_2$ resonance. Our object here is

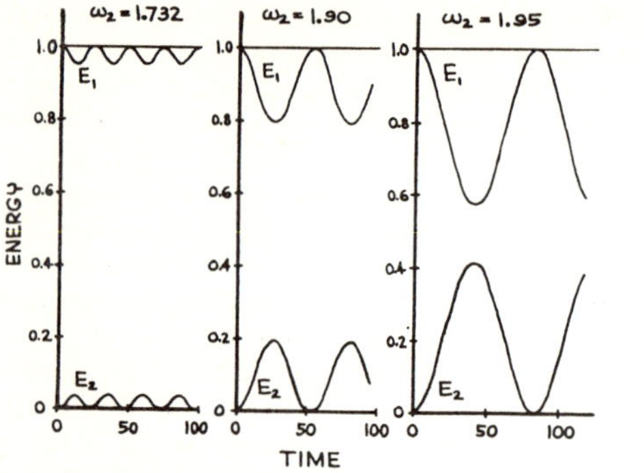

FIG. 1. A plot of the single-oscillator energies for the 2-oscillator system showing the dependence of energy sharing on frequency near the $\omega_2 = 2\omega_1$ resonance. Here $\omega_1 = 1.0$, and ω_2 is varied, starting from its FPU value of $\sqrt{3}$. For all curves, initially one unit of kinetic energy was given to oscillator 1. Here $T_1 = (2\pi/\omega_1) = 6.28$ time units.

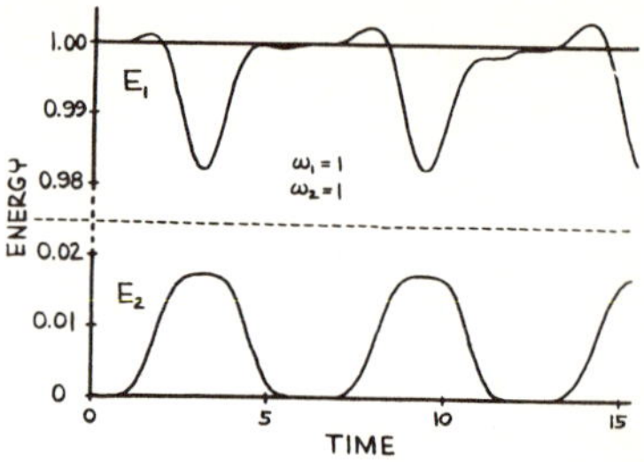

FIG. 3. A plot of the E_k for the weak resonance at $\omega_1 = \omega_2$ for the same system as in Figs. 1 and 2. Initially, one unit of kinetic energy was given to oscillator 1. Oscillator 2 never receives more than 0.02 units of energy.

not only to show that energy sharing occurs, but also to demonstrate the sharpness of the resonance. For computational definiteness, we set $\omega_1 = 1.0$ and varied ω_2. First, we set $\omega_2 = \sqrt{3}$, the FPU frequency. Then ω_2 is set equal to 1.90, 1.95, 1.99, and 2.00. The initial conditions for all these curves were the same; all positions and momenta were zero except $p_1 = \sqrt{2}$, making the total energy $E_T = 1.0$. A measure of the time scale may be realized by noting that the uncoupled period of oscillator one is $T_1 = (2\pi/\omega_1) = 6.28$.

Figures 3 and 4 plot E_1 and E_2 for Eq. (6) at the resonance $\omega_1 = \omega_2 = 1.0$. In Fig. 3, initially only $p_1 = \sqrt{2}$ was nonzero, making $E_T = 1.0$. For Fig. 4, initially only $p_1 = 1.0$ and $p_2 = 1.0$ were nonzero, again making $E_T = 1.0$. As previously noted, not much energy sharing occurs for this resonance; but it should be noted that the energy sharing which does occur will continue to occur no matter how small the value of α, excluding $\alpha = 0$ of course.

Figures 5 and 6 plot E_1, E_2, and E_3 for the three oscillator system whose oscillator Hamiltonian is

$$H = \tfrac{1}{2}\sum_{k=1}^{3}(p_k^2 + \omega_k^2 q_k^2) + \alpha\{(1 - 2^{-\frac{1}{2}})q_1^2 q_2 - (1 + 2^{-\frac{1}{2}})q_2 q_3^2 + \sqrt{2}\, q_1 q_2 q_3\}. \qquad (13)$$

Hamiltonian (13) was obtained, of course, by substituting the necessary linear transformation into Hamiltonian (4). Figure 5 plots the E_k found using the FPU frequencies $\omega_1 = (2 - \sqrt{2})^{\frac{1}{2}} = 0.765$, $\omega_2 = \sqrt{2} = 1.41$, and $\omega_3 = (2 + \sqrt{2})^{\frac{1}{2}} = 1.85$. Here only $p_1 = \sqrt{6}$ is nonzero, making $E_T = 3.0$. The shortest uncoupled period is $T_3 = (2\pi/\omega_3) = 3.4$; thus, this system has a recurrence time of about 14 T_3. Note that E_3 gets almost no energy. Figure 6 plots the E_k using the frequencies $\omega_1 = 0.6$, $\omega_2 = 1.2$, $\omega_3 = 1.8$. This is the resonance $\omega_1 = \frac{1}{2}\omega_2 = \frac{1}{3}\omega_3$

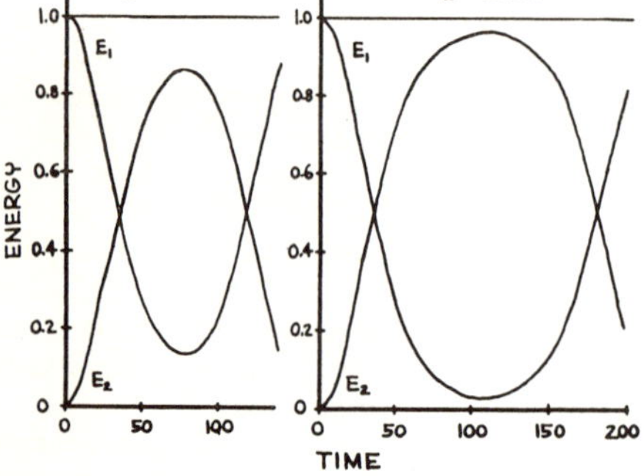

FIG. 2. This figure is a continuation of Fig. 1 and plots the E_k for $\omega_2 = 1.99$ and 2.00.

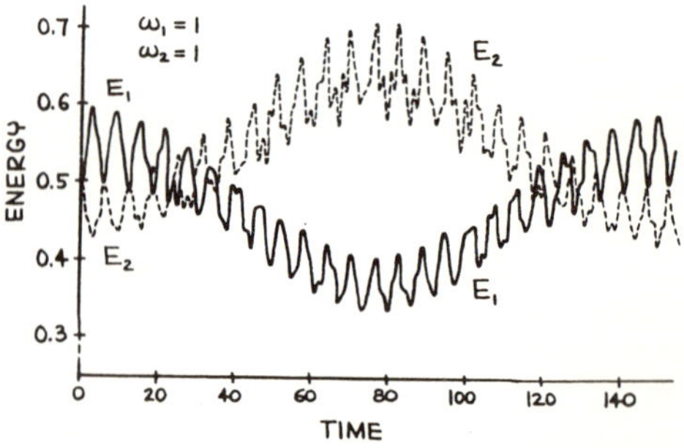

FIG. 4. Plots the E_k for different initial conditions for the same system as in Fig. 3. Here initially one unit of kinetic energy was equally divided between the oscillators. Again, there is very little energy sharing.

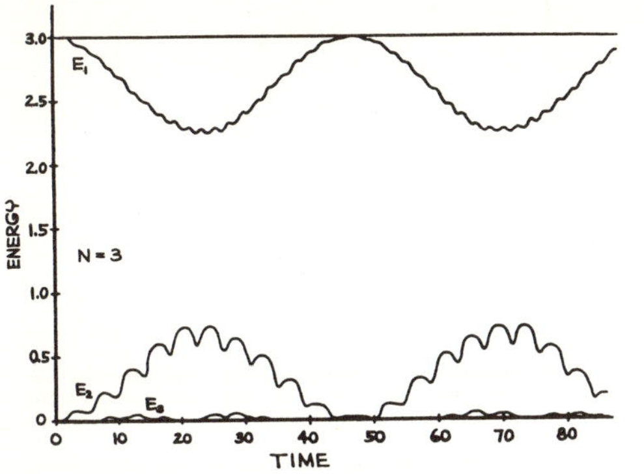

FIG. 5. The E_k for the 3-oscillator system using the FPU frequencies $\omega_1 = 0.765$, $\omega_2 = 1.41$, and $\omega_3 = 1.85$. Initially, three units of kinetic energy were given to oscillator 1. The recurrence time for this system is about $14T_3$, where $T_3 = (2\pi/\omega_3)$.

determined from the Wigner–Brillouin series. Note that these resonant frequencies represent only a slight shift from the FPU frequencies, again indicating the sharpness of the resonance. The initial conditions for Fig. 6 are the same as those for Fig. 5.

Figures 7 and 8 plot the E_k for the oscillator Hamiltonian

$$H = \tfrac{1}{2} \sum_{k=1}^{5} (p_k^2 + \omega_k^2 q_k^2) + \alpha\{0.078 q_1^2 q_2 + 0.5 q_2^2 q_4 - 1.078 q_2 q_5^2 - 0.5 q_4^3 + 0.422 q_1 q_2 q_3 + 0.732 q_1 q_3 q_4 + q_1 q_4 q_5 + 1.578 q_2 q_3 q_5 - 2.732 q_3 q_4 q_5\}. \quad (14)$$

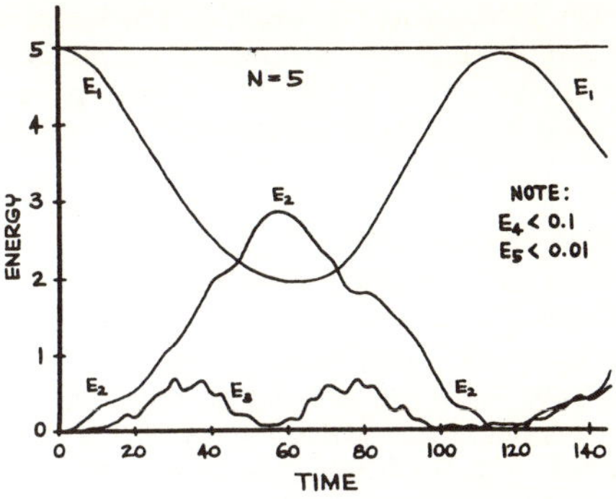

FIG. 7. The E_k for the 5-oscillator system using the FPU frequencies $\omega_k = 2\sin(k\pi/12)$. Initially, five units of kinetic energy were given to oscillator 1. The system has a recurrence time of about $35T_5$, and oscillators 4 and 5 receive little energy.

For both figures, only $p_1 = \sqrt{10}$ is initially nonzero, making $E_T = 5.0$. Figure 7 plots the E_k using the FPU frequencies $\omega_1 = (2 - \sqrt{3})^{1/2}$, $\omega_2 = 1.0$, $\omega_3 = \sqrt{2}$, $\omega_4 = \sqrt{3}$, $\omega_5 = (2 + \sqrt{3})^{1/2}$. Note that E_4 and E_5 are negligible on the scale shown and that the recurrence time is about $35T_5$, where $T_5 = (2\pi/\omega_5) = 3.25$. The Wigner–Brillouin series indicate that there should be a resonance at $\omega_1 = \frac{1}{2}\omega_2 = \frac{1}{3}\omega_3 = \frac{1}{4}\omega_4 = \frac{1}{5}\omega_5$. The frequencies used for Fig. 8 were $\omega_1 = 0.4$, $\omega_2 = 0.8$, $\omega_3 = 1.2$, $\omega_4 = 1.6$, and $\omega_5 = 2.0$. As before, these frequencies are shifted only slightly from the FPU frequencies. The increase in energy sharing, however, is dramatic. The total energy given to this resonant system is 5 units; thus if equipartition of energy occurs, each oscillator should have unit energy on the time

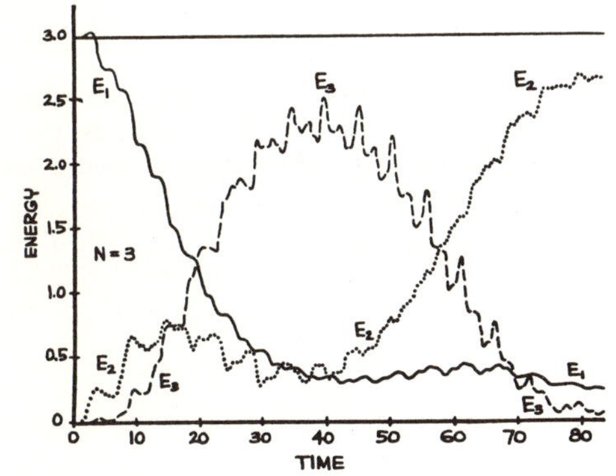

FIG. 6. The E_k for the same system as in Fig. 5 but using the resonant frequencies $\omega_1 = 0.6$, $\omega_2 = 1.2$, $\omega_3 = 1.8$. Note the increase in energy sharing and the disappearance of the short-term recurrence.

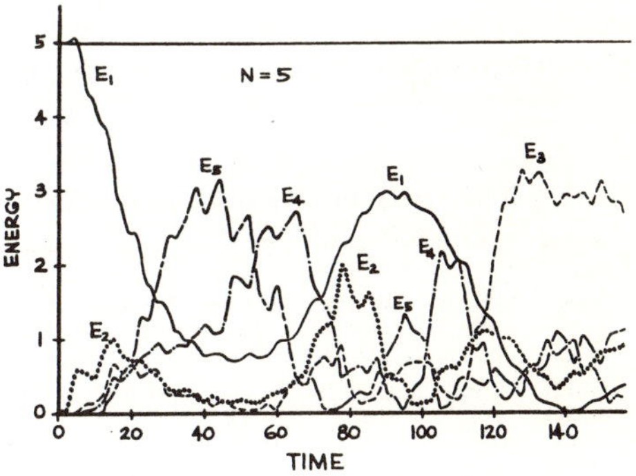

FIG. 8. The E_k using the resonant frequencies $\omega_k = (2k/5)$ for the same system as in Fig. 7. The frequencies are shifted only slightly from their FPU values; however, the increase in energy sharing is dramatic.

average. Table I shows the extent to which equipartition occurs, computed by finding the time average E_k over various time intervals. It perhaps should be noted that the solution for this 5-oscillator system was obtained from $t = 0$, to $t = 1000$. Thus, Fig. 8 and Table I consider only part of the total solution available.

Finally, Fig. 9 plots three of the E_k for the oscillator Hamiltonian for 15 oscillators. The nonresonant FPU case is not plotted since it would look quite similar to the curves appearing in the FPU paper. The resonant curves of Fig. 9 were obtained using the frequencies $\omega_k = (2k/15)$. Only $p_1 = \sqrt{10}$ was nonzero, making $E_T = 5.0$. We have plotted only E_2, E_7, and E_{15} in order to show, without the clutter of 15 interlaced curves, that energy spreads throughout the system.

From the foregoing, we may conclude that the Wigner–Brillouin series is a reasonable method of predicting energy sharing for weakly coupled oscillator systems. It may even be capable of giving quantitative data, although we have not fully investigated this point.

III. NORMAL MODES

In the preceding section we have seen that resonant nonlinear systems share energy, starting from certain initial conditions. Now it is well-known that linear resonant systems share energy, starting from some initial conditions; but if a system of oscillators weakly coupled by linear forces is initially excited in one of its normal modes, the energy of each oscillator remains constant. Thus, not all initial conditions lead to energy sharing. We thus are led to ask whether the same situation holds for weak nonlinear coupling.

Recall that we find linear normal modes by assuming a solution $q_k = A_k \cos \Omega t$, i.e., by assuming that each oscillator can move at the same frequency. The constant amplitudes A_k, as well as the allowed Ω, are then determined from the equations of motion. We now apply the same technique to the systems discussed in Sec. II. Because these systems are nonlinear, we slightly generalize our assumed solution and write $q_k = A_k \cos n\Omega t$; i.e., we allow for the possibility that the kth oscillator moves at some harmonic of Ω. We now illustrate the method for a resonant 3-oscillator system; we then turn to the computer verification of the theory.

TABLE I. Time-average single-oscillator energies for the resonant nonlinear 5-oscillator system.

Time interval ΔT	Average $\langle E_1\rangle$	Average $\langle E_2\rangle$	Average $\langle E_3\rangle$	Average $\langle E_4\rangle$	Average $\langle E_5\rangle$
0 → 100	2.05	0.59	0.34	0.98	1.14
100 → 200	1.12	0.91	1.48	0.91	0.69
200 → 300	0.70	1.41	0.85	0.70	1.43
300 → 400	1.09	0.76	1.46	0.81	0.99
0 → 400	1.24	0.92	1.03	0.85	1.06
100 → 400	0.97	1.03	1.26	0.81	1.04
200 → 400	0.90	1.09	1.16	0.75	1.21

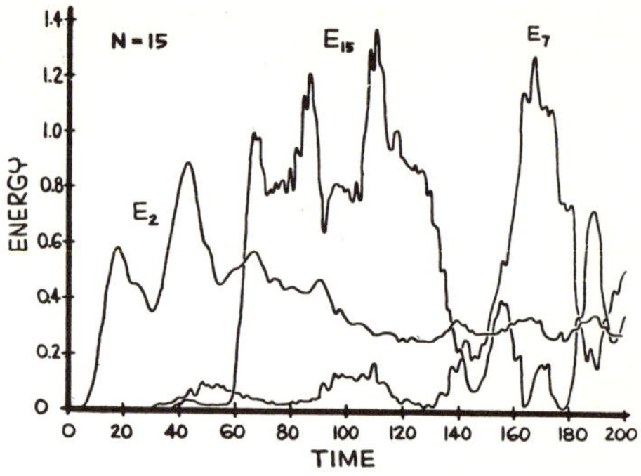

FIG. 9. Three of the E_k for the resonant 15-oscillator system using the frequencies $\omega_k = (2k/15)$. The curves plotted are typical of all 15 curves. Initially, five units of kinetic energy were given to oscillator 1. The comparison FPU curves are not plotted since they would strongly resemble the curves of Figs. 5 and 7.

The equations of motion for the 3-oscillator system which follow from Hamiltonian (13) are

$$\ddot{q}_1 + \omega_1^2 q_1 = \alpha\{(\sqrt{2} - 2)q_1 q_2 - \sqrt{2}\, q_2 q_3\},$$
$$\ddot{q}_2 + \omega_2^2 q_2 = \alpha\{(1 + 2^{-\frac{1}{2}})q_3^2 - \sqrt{2}\, q_1 q_3 + (-1 + 2^{-\frac{1}{2}})q_1^2\},$$
$$\ddot{q}_3 + \omega_3^2 q_3 = \alpha\{(\sqrt{2} + 2)q_2 q_3 - \sqrt{2}\, q_1 q_3\}. \quad (15)$$

If the ω_k are such that Eqs. (15) describe nonresonant motion, then, for small α, the oscillators move as if independent. The three normal modes are then easily seen to be

$$(1)\quad q_1 = A_1 \cos \omega_1 t, \quad q_2 = 0, \quad q_3 = 0;$$
$$(2)\quad q_1 = 0, \quad q_2 = A_2 \cos \omega_2 t, \quad q_3 = 0; \quad (16)$$
$$(3)\quad q_1 = 0, \quad q_2 = 0, \quad q_3 = A_3 \cos \omega_3 t.$$

However, the existence of normal modes for the nonresonant motion is not particularly interesting; the real interest lies in determining if normal modes continue to exist in the presence of resonance.

Thus, we now consider the case $\omega_1 = \frac{1}{2}\omega_2 = \frac{1}{3}\omega_3$,

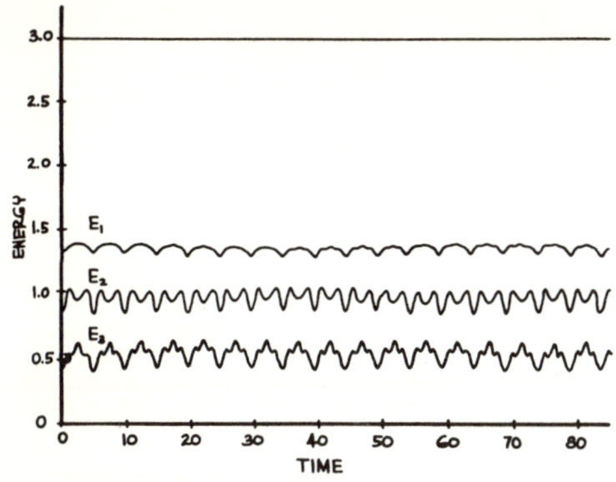

FIG. 10. A plot of the E_k for the first normal mode for the resonant 3-oscillator system. A total energy of three units was chosen for comparison with the energy-sharing curves of Fig. 6.

and we look for a solution of the form

$$\begin{aligned} q_1 &= A_1 \cos \Omega t + \alpha q_{11} + \cdots, \\ q_2 &= A_2 \cos 2\Omega t + \alpha q_{21} + \cdots, \qquad (17) \\ q_3 &= A_3 \cos 3\Omega t + \alpha q_{31} + \cdots, \end{aligned}$$

where we shall require that the first-order terms be negligible as $\alpha \to 0$. Putting Eq. (17) into Eq. (15), we find, to order α,

$$\begin{aligned} &(-\Omega^2 + \omega_1^2)A_1 \cos \Omega t + \alpha\{\ddot{q}_{11} + \omega_1^2 q_{11}\} \\ &\quad = \alpha\{-0.293A_1A_2 \cos \Omega t - 0.293A_1A_2 \cos 3\Omega t \\ &\qquad - 0.707A_2A_3 \cos \Omega t - 0.707A_2A_3 \cos 5\Omega t\}, \qquad (18a) \\ &(-\Omega^2 + \omega_1^2)4A_2 \cos 2\Omega t + \alpha\{\ddot{q}_{21} + 4\omega_1^2 q_{21}\} \\ &\quad = \alpha\{1.71A_3^2 + 1.71A_3^2 \cos 6\Omega t \\ &\qquad - 0.707A_1A_3 \cos 2\Omega t - 0.707A_1A_3 \cos 4\Omega t \\ &\qquad - 0.147A_1^2 - 0.147A_1^2 \cos 2\Omega t\}, \qquad (18b) \\ &(-\Omega^2 + \omega_1^2)9A_3 \cos 3\Omega t + \alpha\{\ddot{q}_{31} + 9\omega_1^2 q_{31}\} \\ &\quad = \alpha\{1.71A_2A_3 \cos \Omega t + 1.71A_2A_3 \cos 5\Omega t \\ &\qquad - 0.707 \cos \Omega t - 0.707A_1A_2 \cos 3\Omega t\}. \qquad (18c) \end{aligned}$$

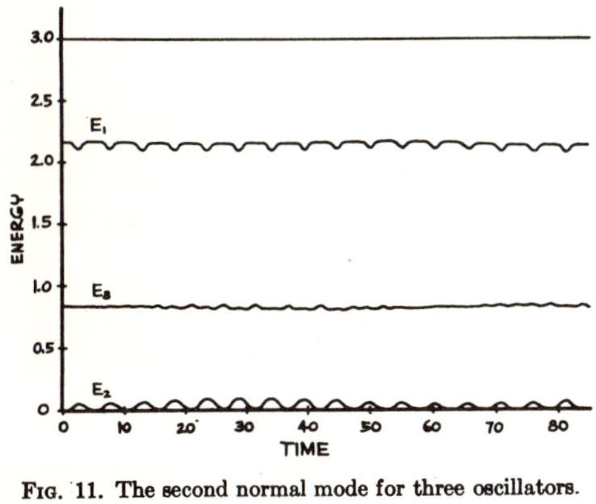

FIG. 11. The second normal mode for three oscillators.

In order to be able to solve for q_{11}, q_{21}, and q_{31} without obtaining small denominators, we must get rid of the cos Ωt terms in Eq. (18a), the cos $2\Omega t$ terms in Eq. (18b), and the cos $3\Omega t$ terms in Eq. (18c). These terms will be eliminated, allowing us to obtain a solution of the desired form in which higher-order terms are negligible as $\alpha \to 0$, provided

$$\begin{aligned} &(\Omega^2 - \omega_1^2)A_1 \\ &\qquad - 0.293\alpha A_1A_2 - 0.707\alpha A_2A_3 = 0, \\ &-0.147\alpha A_1^2 \\ &\qquad + 4(\Omega^2 - \omega_1^2)A_2 - 0.707\alpha A_1A_3 = 0, \\ &\qquad -0.707\alpha A_1A_2 + 9(\Omega^2 - \omega_1^2)A_3 = 0. \qquad (19) \end{aligned}$$

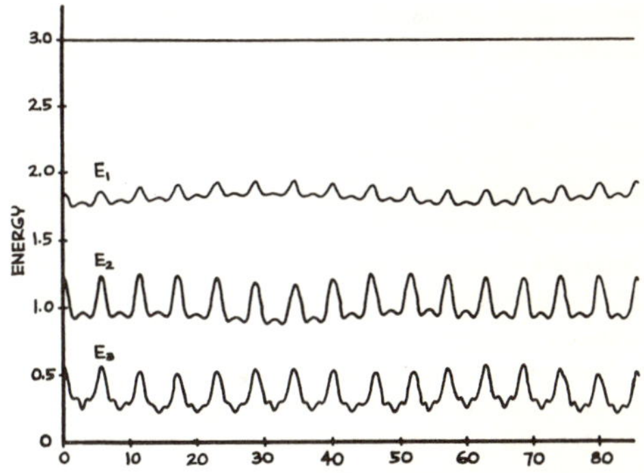

FIG. 12. The third normal mode for three oscillators. Here the initial conditions were not so well chosen and there is a long-term oscillation about the accurate values.

These, of course, are just nonlinear versions of the linear secular equations which arise in finding linear normal modes. Solving Eq. (19), we find the three normal modes:

$$\begin{aligned} &(1) \quad q_1 = 5.4A_3 \cos \Omega t, \quad q_2 = 2.2A_3 \cos 2\Omega t, \\ &\qquad q_3 = A_3 \cos 3\Omega t, \text{ and } \Omega^2 = \omega_1^2 + 0.933\alpha A_3; \\ &(2) \quad q_1 = -4.8A_3 \cos \Omega t, \quad q_2 = 0, \\ &\qquad q_3 = A_3 \cos 3\Omega t, \text{ and } \Omega^2 = \omega_1^2; \\ &(3) \quad q_1 = 5.4A_3 \cos \Omega t, \quad q_2 = -2.2A_3 \cos 2\Omega t, \\ &\qquad q_3 = A_3 \cos 3\Omega t, \text{ and } \Omega^2 = \omega_1^2 - 0.933\alpha A_3. \qquad (20) \end{aligned}$$

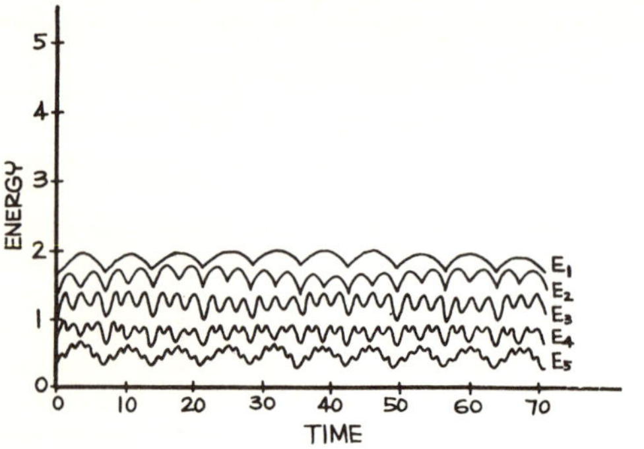

FIG. 13. First normal mode of the 5-oscillator resonant system.

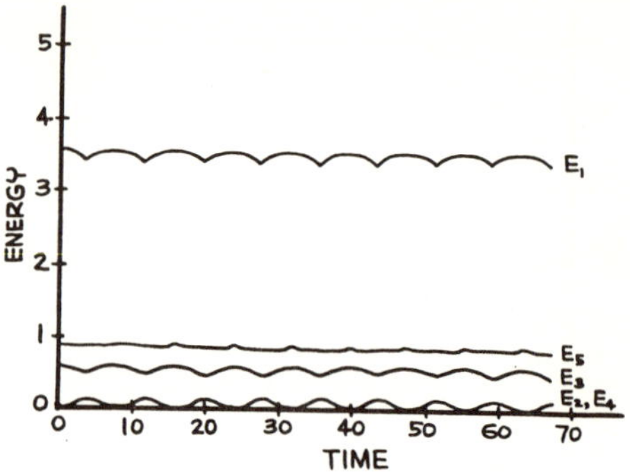

FIG. 15. Third normal mode for five oscillators.

Note that these nonlinear normal modes have two things in common with linear normal modes. First, mode 1 has no nodes, mode 2 has one node, mode 3 has two nodes; second, the frequency splitting is symmetric about ω_1. These same similarities were also observed in the calculations for the resonant 5-oscillator system. Now, because of nonlinearity, the complete solution of Eq. (15) is not just a sum of these three normal modes. However, the computer solutions, to be discussed shortly, indicate that these modes are stable in the sense that a system, started near one of these modes, remains near it. This, at least, suggests that the breakdown of superposition may not be complete. At the present time, we are unable to shed further light on this point.

We now turn to the computer check on the previous theory. Figures 10–12 plot the $E_k = \frac{1}{2}\omega_k^2 A_k^2$ for normal modes 1, 2, and 3, respectively, of Eq. (20). In Eq. (20), A_3 is arbitrary; A_3 is chosen such that the total energy $E_T = 3.0$. A close look at these curves reveals that there is a rapid and a slow oscillation of each E_k. The rapid oscillation is due to higher order terms neglected in Eq. (20). The slow variation is due to the error in calculating the A_k from Eq. (20); i.e., the initial conditions used were slightly off their precise normal values. When the initial E_k deviate from their precise normal mode values, the E_k then oscillate about these precise values. This effect is most noticeable in Fig. 12. It is this oscillation about normal mode values, observed for these and other curves, which leads us to believe that these systems have stable normal modes.

Using the same mathematical technique as for three oscillators, we calculated the normal modes for the resonant 5-oscillator system obeying Hamiltonian (14). Figures 13–17 plot the no-node, one-node, two-node, three-node, and four-node normal modes, respectively. Due to algebraic difficulties, the initial conditions used for Figs. 14 and 16 are rather crude approximations to the accurate normal mode values. No attempt was made to improve

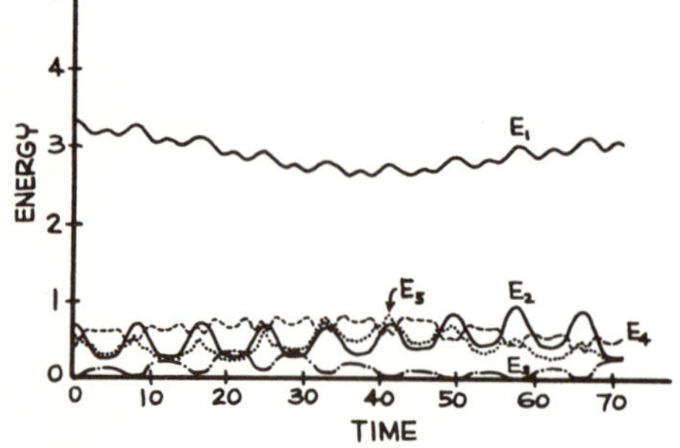

FIG. 14. Second normal mode for five oscillators. Note the long-term oscillation.

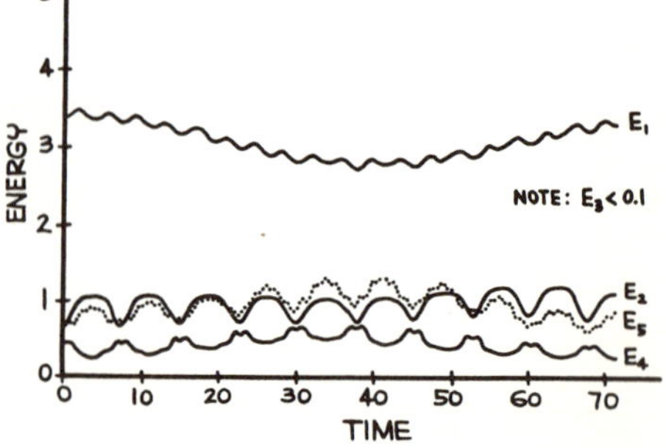

FIG. 16. Fourth normal mode for five oscillators. Again, note the long-term oscillation.

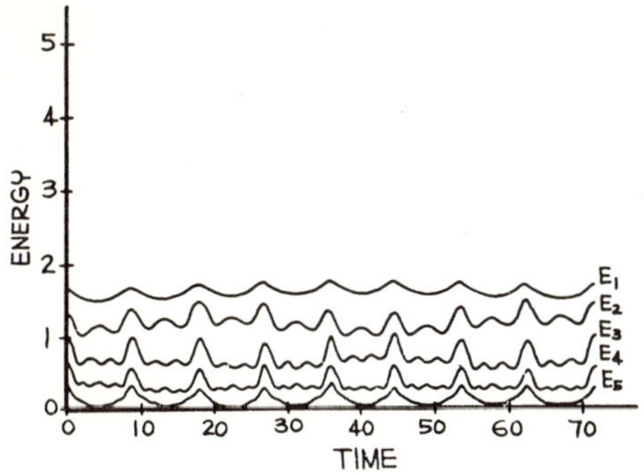

FIG. 17. Fifth normal mode for five oscillators. Note the almost mirror symmetry of Fig. 17 and Fig. 13. Mode one is the completely "in-phase" mode and mode five is the completely "out-of-phase" mode. There are a number of such similarities between linear and nonlinear normal modes.

the approximation, since we wanted to demonstrate the stability of these modes.

Thus, we have shown that the same resonant systems discussed in Sec. II which shared energy for certain initial conditions do not share energy for other initial conditions. We have also established that one may have a small amount of quantitative confidence in these perturbation schemes.

If one accepts the stability of these normal modes, then he concludes that these systems are not ergodic. A system started in a given state may cover a large section of the energy surface, but regions of nonzero measure will be left uncovered. In all the current attempts to derive a master equation for the approach to equilibrium, certain assumptions about initial conditions have been made. We suggest that these initial-condition assumptions are not just mathematical conveniences. They are needed in order to minimize the importance of "normal mode" solutions which form a set of nonzero measure in the totality of solutions. The physical justification for eliminating these normal modes rests with the fact that it is easy to excite one end of a system, but it is hard to excite just one normal mode.

Secs. II and III have now shown that the nonlinear systems obeying Hamiltonian (1) have certain striking similarities to linear systems. We conclude this section by asking whether or not Hamiltonian (1) has, in common with linear systems, N constants of the motion analytic in q, p, and α. Since the weakly coupled oscillators for a nonresonant system do not share very much energy, we see that for a nonresonant system the E_k of Eq. (12) are N constants of the motion; i.e., they are the zeroth-order term in a power series expansion of the constants of the motion.[12] In view of the fact that normal modes exist even in the presence of resonance, one would be almost surprised if N analytic constants of the motion did not also continue to exist in the presence of resonance. The question of analytic constants for 2-oscillator systems has been considered by Whittaker,[12] and he succeeds in finding a formal, analytic integral distinct from the total energy. Cherry[13] considers the case of N oscillators. He proves that there is always at least one analytic integral distinct from the energy. Thus far, we have not been able to improve on these results other than to remark that their proof goes most easily using the canonical formalism of Birkhoff.[14] However, using a computer we can observe whether or not this second, formal constant of the motion is really constant. Before discussing the computer result, we illustrate a simple method for finding this second constant of the motion.

Consider the case of the resonant 2-oscillator Hamiltonian

$$H = \tfrac{1}{2}(p_1^2 + p_2^2 + q_1^2 + 4q_2^2) + 2^{-\frac{1}{2}}\alpha(q_1^2 q_2 - q_2^3). \quad (21)$$

Introduce the canonical transformation

$$q_1 = (2J_1)^{\frac{1}{2}} \cos \theta_1, \qquad q_2 = (J_2)^{\frac{1}{2}} \cos \theta_2,$$
$$p_1 = -(2J_1)^{\frac{1}{2}} \sin \theta_1, \quad p_2 = -2(J_2)^{\frac{1}{2}} \sin \theta_2. \quad (22)$$

Hamiltonian (21) then becomes

$$H = J_1 + 2J_2 + 2^{-\frac{1}{2}}\alpha[J_1 J_2^{\frac{1}{2}} \cos \theta_2 + \tfrac{1}{2}J_1 J_2^{\frac{1}{2}}$$
$$\times \cos(\theta_2 + 2\theta_1) + \tfrac{1}{2}J_1 J_2^{\frac{1}{2}} \cos(\theta_2 - 2\theta_1)$$
$$-(\tfrac{3}{4}J_2^{\frac{3}{2}}) \cos \theta_2 - (\tfrac{1}{4}J_2^{\frac{3}{2}}) \cos 3\theta_2]. \quad (23)$$

In zeroth order we expect $\theta_1 = t$, $\theta_2 = 2t$; thus among the first-order terms in Eq. (23), all will be rapidly varying except the term involving $\cos(\theta_2 - 2\theta_1)$. A quick check reveals that this slowly varying term has zero Poisson bracket with the term $(J_1 + 2J_2)$. Thus, we anticipate that one can find a constant of the motion Φ of the form

$$\Phi = J_1 J_2^{\frac{1}{2}} \cos(\theta_2 - 2\theta_1) + \alpha\Phi_1 + \cdots . \quad (24)$$

---

[12] E. T. Whittaker, *Analytic Dynamics* (Dover Publications, Inc., New York, 1944), Chap. 16.
[13] T. M. Cherry, Proc. Cambridge Phil. Soc. **22**, 510 (1925).
[14] G. D. Birkhoff, *Dynamical Systems* (American Mathematical Society Colloquium Publications, New York, 1927), p. 82. The generalization of Birkhoff's theorem needed for this purpose is stated as Lemma 4 in a paper by J. Moser, Commun. Pure Appl. Math. **11**, 81 (1958).

Transforming back to the q, p variables, we find

$$\Phi = \tfrac{1}{2}(q_1^2 q_2 - q_2 p_1^2 + q_1 p_1 p_2) + \alpha\Phi_1 + \cdots, \quad (25)$$

the desired analytic constant of the motion. In passing, we note that the two constants of the motion which can be determined for a 2-oscillator system allow one to determine[12] the complete solution. Using this method one can obtain quantitative agreement with the curves plotted in Figs. 1 through 4. In particular, one can show that the $\omega_1 = \omega_2$ resonance yields little energy sharing.

Using the method outlined above, we found the slowly varying first-order terms in Hamiltonian (14) for the $\omega_1 = \frac{1}{2}\omega_2 = \frac{1}{3}\omega_3 = \frac{1}{4}\omega_4 = \frac{1}{5}\omega_5$ resonance. The sum of these terms, expressed in the q, p coordinates, is just the desired constant of the motion Φ. Using the same initial conditions that led to the complete energy sharing of Fig. 8, the computer calculated this Φ. The results are plotted in Fig. 18; the plot of Φ is labeled "resonant coupling energy" since it is these slowly varying terms in the coupling which give rise to the small denominators in the Wigner–Brillouin series. For comparison, the rapidly varying total coupling energy is plotted. Note that the total coupling energy seldom exceeds 10% of the total energy $E_T = 5.0$. Figure 18 plots these two curves for 80 time units, where $T_5 = (2\pi/\omega_5) = 3.14$. We have plotted these curves out to $t = 1000$, and the segment presented is typical.

IV. SINGLE-OSCILLATOR ENERGY DISTRIBUTION FUNCTION

In this section we compare the behavior of a resonant 5-oscillator linear system with that of the resonant 5-oscillator nonlinear system obeying Hamiltonian (14) with $\omega_1 = \frac{1}{2}\omega_2 = \frac{1}{3}\omega_3 = \frac{1}{4}\omega_4 = \frac{1}{5}\omega_5 = 0.4$. The linear Hamiltonian used was

$$H = \frac{1}{2}\sum_{k=1}^{5}(p_k^2 + q_k^2) - \alpha(q_1q_2 + q_2q_3 + q_3q_4 + q_4q_5). \quad (26)$$

For both systems, five units of kinetic energy were initially given to particle number one. A plot of E_1 for both systems appears in Fig. 19, and the plot of E_5 for both systems appears in Fig. 20. For the

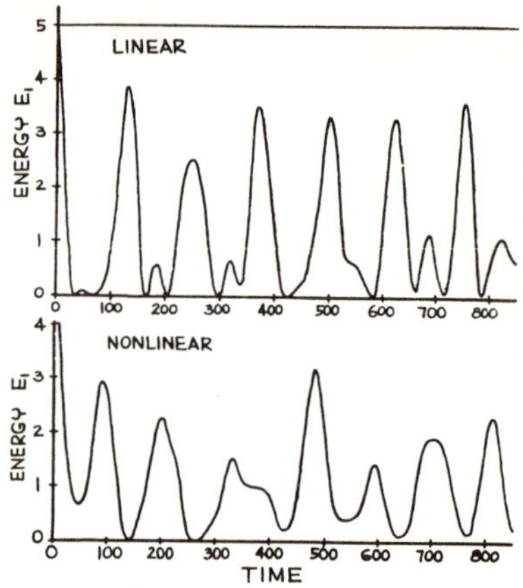

FIG. 19. A comparison of the E_1 curves for a resonant 5-oscillator linear and nonlinear system. For both systems initially, five units of kinetic energy were given to oscillator 1. The nonlinear curve is a continuation of the curve in Fig. 8.

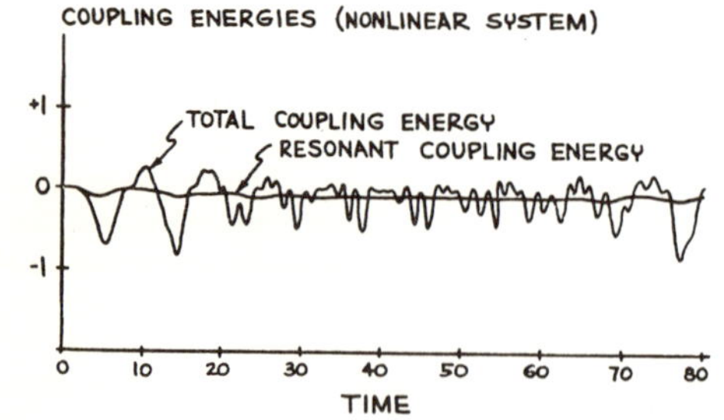

FIG. 18. The curve labeled "resonant coupling energy" is a plot of a constant of the motion distinct from the total energy for the same 5-oscillator system as in Fig. 8. This constant was predicted by perturbation theory and is essentially Whittaker's Adelphic integral. For comparison the total coupling energy is plotted. Note that the total coupling energy seldom exceeds 10% of the total energy of five units.

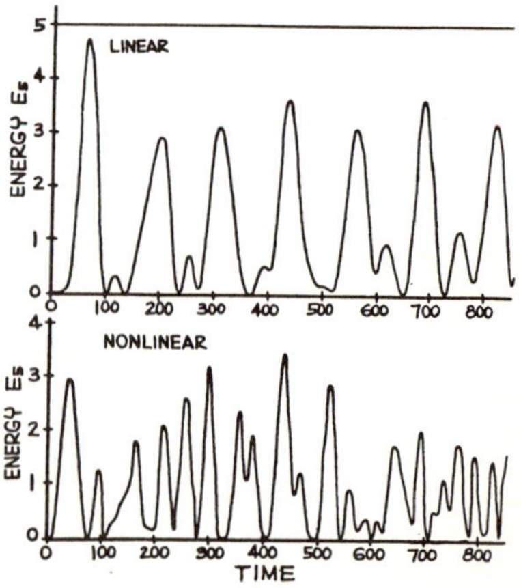

FIG. 20. A comparison of the E_5 curves for the same system as in Fig. 19.

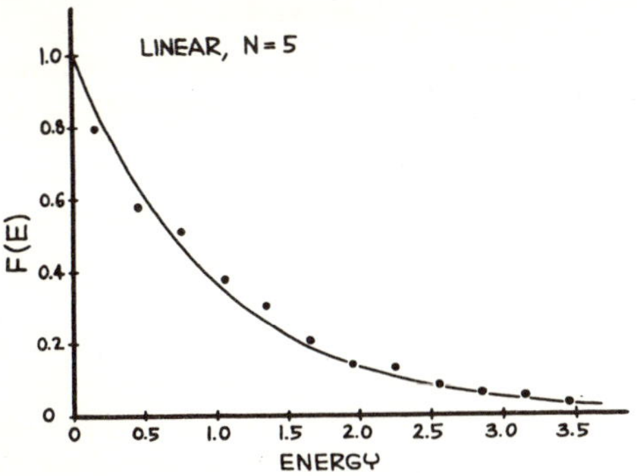

FIG. 21. The dots represent the $f(E)$ data points for the linear system computed from the E_k curves. Figures 19 and 20 present two of these curves. The solid curve is a plot of $f(E) = e^{-E}$ predicted by statistical mechanics.

linear system all the uncoupled periods are equal to 6.28. For the nonlinear system, $T_1 = (2\pi/\omega_1) = 15.7$, and $T_5 = (2\pi/\omega_5) = 3.14$. These curves present at least two interesting features. First, both systems appear to reach a steady state after the elapse of only a small number of uncoupled periods; and second, these E_k curves show a rather striking similarity. Certainly there are differences, but bearing in mind the totally different coupling involved, the similarities are more impressive than the differences.

In view of this similarity, it is interesting to compare the single-oscillator energy distribution densities $f(E)$ for these two systems. Here we define $f(E)$ to be the fractional amount of time a single oscillator spends in the energy interval $(E, E + dE)$. First, we computed $f(E)$ data points using $\Delta E = 0.3$ for each oscillator. Since we found that all the $f(E)$ plots were more or less the same, for presentation we took the average of the linear $f(E)$ data and plotted the data points as the dots shown in Fig. 21. The composite $f(E)$ data points for the nonlinear system are shown as dots in Fig. 22.

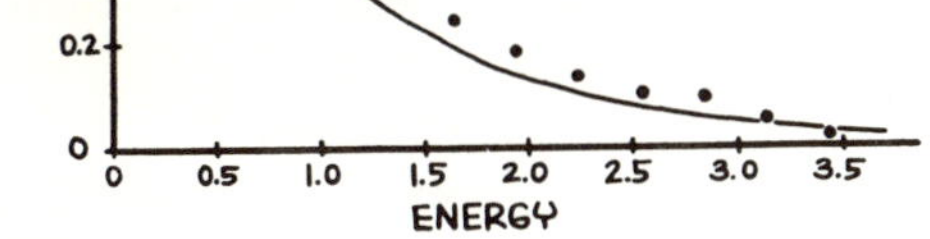

FIG. 22. Plot of the same information as Fig. 21 for the resonant nonlinear oscillator system.

According to equilibrium statistical mechanics, the single oscillator energy distribution density for a system at temperature T is

$$f(E) = (kT)^{-1}e^{-E/kT} \tag{27}$$

where $kT = \langle E \rangle$ and $\langle E \rangle$ equals the average energy of a single oscillator. For these 5-oscillator systems, $kT = \langle E \rangle = E_T/5 = 1.0$. Thus according to statistical mechanics, $f(E) = e^{-E}$. The solid curve in Figs. 21 and 22 is a plot of e^{-E}.

Thus, the data points represent the time average $f(E)$ while the solid curve represents the phase average $f(E)$. It is interesting to note the agreement of these two averages for both these systems. It thus appears that for these systems, four oscillators form an adequate heat bath for the fifth oscillator.

V. CONCLUSIONS

It is believed that a thorough understanding of oscillator systems obeying Hamiltonians of the form

$$H = \tfrac{1}{2} \sum_{k=1}^{N} (p_k^2 + \omega_k^2 q_k^2) + \alpha \sum_{j,k,l=1}^{N} A_{jkl} q_j q_k q_l, \tag{28}$$

might shed light on some of the unresolved questions of statistical mechanics. This paper has investigated those properties of these systems which, it is hoped, are basic to a complete understanding. Our work relies heavily on numerical solutions to the equations of motion. Clearly, a computer can provide these solutions only for selected initial conditions and for relatively short time intervals. The analytical methods we have used to imply the long-time behavior are open to serious convergence questions, and it remains to be shown whether or not they provide an adequate treatment of the small-denominator problem. Nonetheless, whenever we have been provided with a check, we have found agreement between analytical and numerical solution. Consequently, it is with guarded optimism that we offer our conclusions.

Oscillator systems weakly coupled by nonlinear forces can share energy only if there is resonance. For weak coupling, resonance will occur provided

$$\sum n_k \omega_k \lesssim \alpha \tag{29}$$

for certain values of the n_k determined by the particular coupling. Starting from certain initial conditions, some of the resonances predicted using

Eq. (29) will lead to equipartition of energy. It is interesting to note that the time required to achieve a more or less uniform spread of energy is surprisingly short. In this paper, we have primarily investigated resonances leading to complete energy sharing. Further study of the weak resonances is needed.

Resonant oscillator systems with N oscillators have N normal modes. The computer solutions indicate that these normal modes are stable. The normal-mode frequency spectrum and the normal-mode node count is quite similar to that for linear systems. We have verified that these oscillator systems have at least one constant of the motion analytic in q, p, and α other than the total energy. We suspect that there are N analytic constants, but we have not been able to find all of them. However, one could easily explain our failure since it is possible that these constants are analytic in q and p but nonanalytic in α.[15] In any event, it is unlikely that these oscillator systems are ergodic. Thus, weakly coupled oscillator systems are seen to share many of the properties of linear systems. This observation has been reinforced by showing that the single-oscillator energy distribution densities for a particular choice of linear and nonlinear system are quite similar. Finally, we have shown that the single-oscillator energy distribution for both systems has the Boltzmann form predicted by statistical mechanics.

Our efforts in this paper have been directed toward laying the groundwork for further study of irreversible processes. We have been able to demonstrate that certain simple phase functions approach equilibrium. It remains to be shown that a greater variety of phase functions approach equilibrium. The possibility of providing an understanding of lattice thermal conductivity using these models is a particularly interesting prospect. There are a host of unanswered mathematical questions concerning these systems. The need for a more adequate perturbation scheme is especially pressing. The partial success of the perturbation methods we have used indicates that these systems may eventually succumb completely to analysis; although a computer is likely to remain a useful adjunct to analysis for some time to come. In conclusion, it is our belief that knowledge of these simple nonlinear systems will measurably increase our understanding of irreversibility and that the problems these systems present, though difficult, are on the verge of solution.

[15] T. M. Cherry, Proc. Cambridge Phil. Soc. 22, 287 (1925).

APPENDIX I

In this appendix we derive the Wigner–Brillouin solution through order α^2 of the equations

$$\begin{aligned} \ddot{q}_1 + \omega_1^2 q_1 &= -\sqrt{2}\,\alpha q_1 q_2, \\ \ddot{q}_2 + \omega_2^2 q_2 &= -2^{-\frac{1}{2}}\alpha(q_1^2 - 3q_2^2). \end{aligned} \tag{30}$$

We assume a solution of the form

$$\begin{aligned} q_1 &= A_1 \cos \tau_1 + \alpha q_{11} + \alpha^2 q_{12} + \cdots, \\ q_2 &= A_2 \cos \tau_2 + \alpha q_{21} + \alpha^2 q_{22} + \cdots, \end{aligned} \tag{31}$$

where $\tau_1 = \Omega_1 t + \theta_1$ and $\tau_2 = \Omega_2 t + \theta_2$ and where Ω_1 and Ω_2 are frequencies to be determined. Substituting Eq. (31) into Eq. (30), we obtain

$$(-\Omega_1^2 + \omega_1^2)A_1 \cos \tau_1 + \alpha\{\ddot{q}_{11} + \omega_1^2 q_{11} + \sqrt{2}\, q_{10}q_{20}\} + \alpha^2\{\ddot{q}_{12} + \omega_1^2 q_{12} + \sqrt{2}(q_{10}q_{21} + q_{11}q_{20})\} = 0, \tag{32a}$$

$$\begin{aligned} (-\Omega_2^2 + \omega_2^2)A_2 \cos \tau_2 &+ \alpha\{\ddot{q}_{21} + \omega_2^2 q_{21} + 2^{-\frac{1}{2}}(q_{10}^2 - 3q_{20}^2)\} \\ &+ \alpha^2\{\ddot{q}_{22} + \omega_2^2 q_{21} \\ &+ \sqrt{2}\,(q_{10}q_{11} - 3q_{20}q_{21})\} = 0, \end{aligned} \tag{32b}$$

where $q_{10} = A_1 \cos \tau_1$ and $q_{20} = A_2 \cos \tau_2$.

Setting the coefficient of α in Eq. (32) to zero yields

$$\begin{aligned} \ddot{q}_{11} + \omega_1^2 q_{11} &= -2^{-\frac{1}{2}}[A_1A_2 \cos(\tau_1 + \tau_2) + A_1A_2 \cos(\tau_1 - \tau_2)], \\ \ddot{q}_{21} + \omega_2^2 q_{21} &= -2^{-\frac{1}{2}}[(A_1^2 - 3A_2^2) + A_1^2 \cos 2\tau_1 - 3A_2^2 \cos 2\tau_2]. \end{aligned} \tag{33}$$

The solution of Eq. (33) may be written

$$\begin{aligned} q_{11} &= -\frac{A_1A_2 \cos(\tau_1 + \tau_2)}{\sqrt{2}\,[\omega_1^2 - (\Omega_1 + \Omega_2)^2]} - \frac{A_1A_2 \cos(\tau_1 - \tau_2)}{\sqrt{2}\,[\omega_1^2 - (\Omega_1 - \Omega_2)^2]}, \\ q_{21} &= -\frac{(A_1^2 - 3A_2^2)}{2\sqrt{2}\,\omega_2^2} - \frac{A_1^2 \cos 2\tau_1}{2\sqrt{2}\,[\omega_2^2 - 4\Omega_1^2]} + \frac{3A_2^2 \cos 2\tau_2}{2\sqrt{2}\,[\omega_2^2 - 4\Omega_2^2]}. \end{aligned} \tag{34}$$

Setting the coefficient of α^2 in Eq. (32) equal to zero yields

$$\ddot{q}_{12} + \omega_1^2 q_{12} = \Bigg\{\frac{(A_1^2 - 3A_2^2)A_1 \cos \tau_1}{2\omega_2^2} + \frac{A_1^3 \cos \tau_1}{4[\omega_2^2 - 4\Omega_1^2]} + \frac{A_1^3 \cos 3\tau_1}{4[\omega_2^2 - 4\Omega_1^2]}$$

$$- \frac{3A_1A_2^2 \cos(\tau_1 + 2\tau_2)}{4[\omega_2^2 - 4\Omega_2^2]} - \frac{3A_1A_2^2 \cos(\tau_1 - 2\tau_2)}{4[\omega_2^2 - 4\Omega_2^2]}$$

$$+ \frac{A_1A_2^2 \cos \tau_1}{2[\omega_1^2 - (\Omega_1 + \Omega_2)^2]} + \frac{A_1A_2^2 \cos(\tau_1 + 2\tau_2)}{2[\omega_1^2 - (\Omega_1 + \Omega_2)^2]}$$

$$+ \frac{A_1A_2^2 \cos \tau_1}{2[\omega_1^2 - (\Omega_1 - \Omega_2)^2]} + \frac{A_1A_2^2 \cos(\tau_1 - 2\tau_2)}{2[\omega_1^2 - (\Omega_1 - \Omega_2)^2]}\Bigg\}, \tag{35a}$$

$$\ddot{q}_{22} + \omega_2^2 q_{22} = \Bigg\{+\frac{A_1^2A_2 \cos \tau_2}{2[\omega_1^2 - (\Omega_1 + \Omega_2)^2]}$$

$$+ \frac{A_1^2A_2 \cos(2\tau_1 + \tau_2)}{2[\omega_1^2 - (\Omega_1 + \Omega_2)^2]} + \frac{A_1^2A_2 \cos \tau_2}{2[\omega_1^2 - (\Omega_1 - \Omega_2)^2]}$$

$$+ \frac{A_1^2A_2 \cos(2\tau_1 - \tau_2)}{2[\omega_1^2 - (\Omega_1 - \Omega_2)^2]} - \frac{3(A_1^2 - 3A_2^2)A_2 \cos \tau_2}{2\omega_2^2}$$

$$- \frac{3A_1^2A_2 \cos(2\tau_1 - \tau_2)}{4[\omega_2^2 - 4\Omega_1^2]} - \frac{3A_1^2A_2 \cos(2\tau_1 + \tau_2)}{4[\omega_2^2 - 4\Omega_1^2]}$$

$$+ \frac{9A_2^3 \cos \tau_2}{4[\omega_2^2 - 4\Omega_2^2]} + \frac{9A_2^3 \cos 3\tau_2}{4[\omega_2^2 - 4\Omega_2^2]}\Bigg\}. \tag{35b}$$

The terms involving cos τ_1 in Eq. (35a) and the terms involving cos τ_2 in Eq. (35b) must be eliminated since they will lead to small denominators of the type $(\omega_1^2 - \Omega_1^2)$ and $(\omega_2^2 - \Omega_2^2)$. The zeroth-order terms in Eq. (32) are set up for the specific purpose of eliminating these terms. The cos τ_1 in Eq (35a) are grouped with the zeroth-order term in Eq. (32a); the cos τ_2 terms of Eq. (35b) are grouped with the zeroth-order terms of Eq. (32b). Ω_1 and Ω_2 are then chosen such that they eliminate these terms. Carrying out this procedure leads to Eqs. (7c) and (7d) of Sec. II. Having eliminated the dangerous terms from Eqs. (35a) and (35b), we may solve for q_{12} and q_{22} just as we solved Eq. (33). Carrying out these steps and putting the results into Eq. (31) leads to Eqs. (7a) and (7b) of Sec. II. Clearly, this procedure can be continued to arbitrary order in α. We eliminate dangerous cos τ_1 and cos τ_2 terms in any order by grouping them with the zeroth-order terms, and then proceed to solve for the q_{1n} and q_{2n} in the usual fashion.